ENVIRONMENTAL SCIENCE

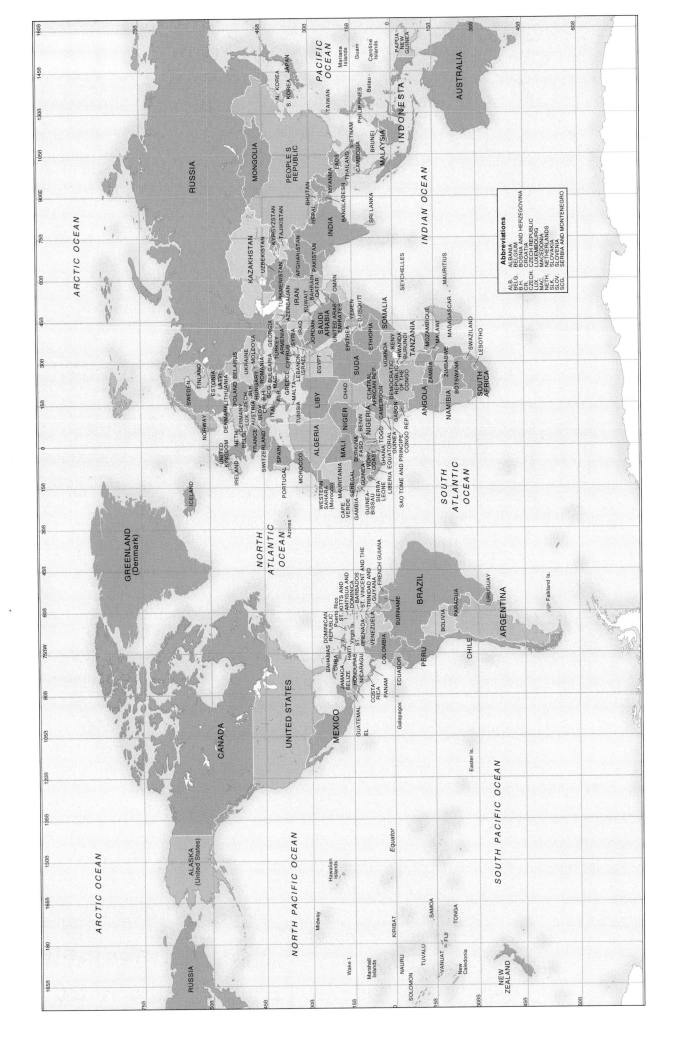

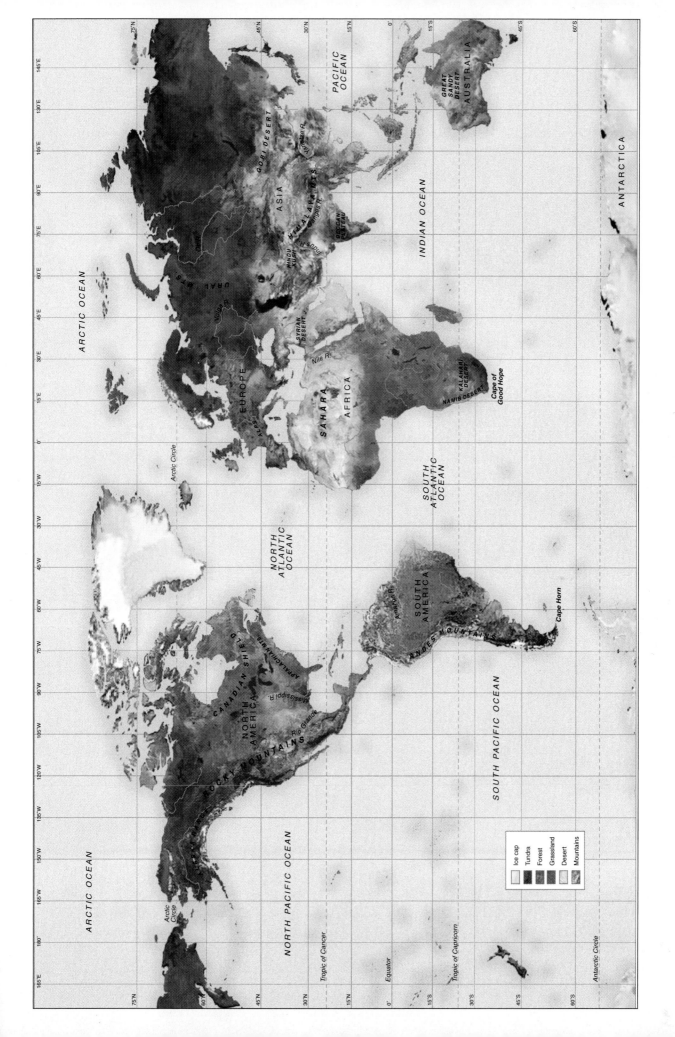

SEVENTH EDITION

ENVIRONMENTAL SCIENCE

DANIEL D. CHIRAS
Colorado College

JONES AND BARTLETT PUBLISHERS
Sudbury, Massachusetts
BOSTON   TORONTO   LONDON   SINGAPORE

World Headquarters

Jones and Bartlett Publishers
40 Tall Pine Drive
Sudbury, MA 01776
978-443-5000
www.jbpub.com

Jones and Bartlett Publishers Canada
6339 Ormindale Way
Mississauga, Ontario L5V 1J2
Canada

Jones and Bartlett Publishers
International
Barb House, Barb Mews
London W6 7PA
United Kingdom

Jones and Bartlett's books and products are available through most bookstores and online booksellers. To contact Jones and Bartlett Publishers directly, call 800-832-0034, fax 978-443-8000, or visit our website www.jbpub.com.

Substantial discounts on bulk quantities of Jones and Bartlett's publications are available to corporations, professional associations, and other qualified organizations. For details and specific discount information, contact the special sales department at Jones and Bartlett Publishers via the above contact information or send an email to specialsales@jbpub.com.

ISBN-13: 978-0-7637-0860-3
ISBN-10: 0-7637-0860-7

Production Credits

Chief Executive Officer: Clayton E. Jones
Chief Operating Officer: Don W. Jones, Jr.
President, Higher Education and Professional Publishing: Robert W. Holland, Jr.
V.P., Design and Production: Anne Spencer
V.P., Sales and Marketing: William Kane
V.P., Manufacturing and Inventory Control: Therese Connell
Acquisitions Editor, Science: Cathleen Sether
Managing Editor, Science: Dean W. DeChambeau
Editorial Assistant, Science: Molly Steinbach
Senior Production Editor: Louis C. Bruno, Jr.
Senior Photo Researcher: Kimberly Potvin
Marketing Manager: Andrea DeFronzo
Text and Cover Design: Anne Spencer
Illustrations: Precision Graphics, Magellan Geographix, Graphic World Publishing Services
Composition: Graphic World, Inc.
Printing and Binding: Courier Kendallville
Cover Printing: Courier Kendallville
Cover Photo: © Comstock Images/Creatas

Library of Congress Cataloging-in-Publication Data
Chiras, Daniel D.
Environmental science / Daniel D. Chiras.—7th ed.
    p. cm.
Includes bibliographical references and index.
ISBN 0-7637-0860-7 (alk. paper)
1. Environmental sciences. 2. Sustainable development. I. Title.
GE140.C48 2006
363.7—dc22
                    2005034987
6048

Printed in the United States of America
10 09 08 07 06 10 9 8 7 6 5 4 3 2    Printed on Recycled Paper

To my dad for starting me on the conservation path.

BRIEF CONTENTS

Frontmatter Photo Credits

p. iv-v © Comstock Images/Creatas; p. x Courtesy of Roger Taylor/National Renewable Energy Laboratory; p. vii © Ana Vasileva/ShutterStock, Inc.; p. xi © Kim Pin Tan/ShutterStock, Inc.; p. xii Courtesy of GSFC/NASA; p. xiii Courtesy of Dan Chiras; p. xv Courtesy of Mark Wolfe/FEMA

CONTENTS

PREFACE

ENVIRONMENTAL SCIENCE courses vary widely. Some professors prefer to teach more of the basic science. That is, they prefer to discuss basic science before they delve into environmental issues such as air pollution, water pollution, global climate change, species, and extinction. Other professors delve more into ethics, economics, and other aspects of various topics. In between these two approaches is a wide range of options.

This book is meant to serve the diverse approaches of professors. Like other textbooks on the subject, this one offers an in-depth look at the environmental problems facing the world. It also offers a variety of solutions—not only some of the more traditional ones but also a new generation of policies and actions that could help us achieve lasting results. They're solutions that involve all stakeholders: businesses, governments, and individuals.

This book has evolved dramatically over the past twenty-five years in response to profound changes in my thinking. Perhaps the most important change has been the shift toward a more comprehensive and more systematic way of viewing environmental problems. One of the useful discoveries I, and many others, have made over many years of study is that most environmental problems spring from a common set of root causes. This book discusses the common root causes and ways to address them.

Over the past decade and a half, many of us have discovered that we've been addressing environmental problems a bit superficially. That is to say, most efforts have been stopgap measures that treat the symptoms while tending to neglect the underlying root causes. In recent years, it has become clear to environmental scientists and others that to create lasting solutions we must address root causes. One of the root causes is that the systems we depend on for food, energy, materials, water, waste treatment, and other services are unsustainable. To solve environmental problems and create an enduring human presence (a sustainable future), we must rethink and restructure basic human systems, such as transportation, energy, waste management, water supply, industry, and housing. This book shows the reader how this can be done.

Another important realization I and many others have made is that to be sustainable solutions to environmental problems must make sense from social and economic perspectives, not just an environmental one. One-sided solutions don't work. This shift to solutions that make sense from all three perspectives will take enormous creative energy on our part in the years to come.

This edition reflects a pressing need for sustainable solutions—measures that confront the root causes of the environmental crisis and create a society that meets its needs without bankrupting the Earth. This book includes discussions meant to help students understand why systems need to be revised and how we go about this challenging task.

In this new edition, I have added new material, updated statistics, and polished the writing. As in the first six editions, I worked diligently to ensure that this text remains user-friendly. My goal was to continue to present the most important facts and concepts in a clear and exciting fashion, not bog readers down with endless statistics and pointless detail. As always, I wanted to present both sides of the issues by offering Point/Counterpoints on controversial issues. Even though this text presents a strong case for sustainability and systems reform, it is left to you, the reader, to make up your own mind as to the need and desirability of such an approach. My efforts to make this book as unbiased as possible support my objective of letting students make up their own minds about our predicament and ways to extricate ourselves from it. Critical thinking skills presented in the book also help students learn to analyze issues and make up their own minds.

Themes

All textbooks have a central theme or, in some cases, a set of themes that shape the presentation. This book is no different. It is molded by six central ideas: focus on key principles, sustainability, addressing root causes, systems reform, critical thinking, and action.

FOCUS ON KEY PRINCIPLES. Environmental science is a vast field, requiring many years to master. Most students, however, can become proficient in the subject much more quickly when they are taught key concepts. Although I've always tried to focus the text on key principles, in this edition I've highlighted the principles in special Key Concept boxes at the end of each section. They encapsulate the material covered in each section and provide an excellent tool for studying.

SUSTAINABILITY. The main theme of this book is that the long-term well-being of this planet and its inhabitants is in jeopardy and that to create an enduring human presence we must make a massive course change. We must transition to a sustainable society. A sustainable society seeks balance between human and ecological needs. Its economic systems serve people and the planet. Creating a sustainable society may be our only realistic hope for surviving on a finite planet, but it will not evolve without foresight, planning, and action.

ADDRESSING ROOT CAUSES AND SYSTEMS REFORM. Creating a sustainable future will require serious efforts to understand and confront the root causes of the environmental crisis. This book outlines those causes and shows ways to address them. Part of this struggle will involve efforts to revamp human systems—make sustainable systems to provide energy, transportation, and waste management, for example.

CRITICAL THINKING. This text also stresses critical thinking skills—learning to think critically about issues, a task that is essential to creating a sustainable society.

ACTION. Finally, this text emphasizes an often-overlooked point, that building a sustainable future requires actions by all of us. Air pollution is not just caused by inadequate laws or corporate neglect, it is the result of our own often wasteful lifestyles. Because we are all part of the problem, we must all be part of the solution. Individual action is as essential as responsible corporate and government policies and practices.

Organization

This book is divided into six parts.

PART I introduces students to four of the principal themes of this book: sustainability, addressing root causes, restructuring human systems, and critical thinking. It lays an important foundation for the rest of your study.

PART II introduces the student to basic principles of ecology necessary for understanding environmental issues. These chapters elaborate on six operating principles of sustainability—ideas that will help us revamp modern society one system at a time.

PART III begins the discussion of environmental issues, dealing with one of the most pressing of all, the population crisis. This part examines the impact of rapid population growth—one of the root causes of modern environmental problems—and explores culturally acceptable ways of slowing it down.

PART IV deals with a variety of resource issues, such as wildlife extinction and energy, and outlines strategies for solving them sustainably. In this material, I attempt to show how we can revamp some of the vital human systems such as energy and waste management.

PART V discusses pollution and legal, technical, and personal solutions for it, including both traditional and sustainable strategies.

PART VI, the capstone of the book, attempts to place the population, resource, and pollution crises in a social context. It reexamines ethics and explores economics and government in more detail.

Special Features (What's New, Too!)

Over the years, this book has undergone some dramatic changes to help make the study of environmental science more interesting, more meaningful, and more memorable. Below is a list of features students should find helpful:

Study Skills

Immediately following the preface is a brief section on study skills. The study skills section includes numerous simple but effective tips that help students improve their memory, note-taking skills, reading abilities, test-taking abilities, and much more. Study skills can help all students, even A students, become more efficient learners. Skills learned here will carry over to virtually every other course students take and will be helpful throughout life.

Foundation Tools

This book offers several Foundation Tools, conceptual models that help students understand how the world works. These models are designed to encourage systems thinking and help students organize facts into a solid conceptual framework. Below is a brief description of each model.

POPULATION, RESOURCES, AND POLLUTION This model shows how populations of organisms, like ourselves, affect their environment and how their actions, in turn, affect the populations themselves.

IMPACT ANALYSIS This model shows the various impacts that humans have on various components of the environment.

RISK ANALYSIS This model presents an overview of a process called risk assessment.

ROOT CAUSES This model shows the many factors that contribute to the crisis of unsustainability and helps students identify key leverage points for change.

Examining Both Sides of the Issue

In keeping with my long-standing belief that students must examine all sides of an issue, I have continued to present information from opposing viewpoints. You will find this approach exercised in the text itself, where I examine contradictory positions, and in the Point/Counterpoints.

POINT/COUNTERPOINTS AND VIEWPOINTS Complex environmental issues often result in hotly contested debates:

- Should wolves be reintroduced into the wild?
- Is population growth good or bad for us?
- Are current laboratory tests for carcinogens valid?
- Is the spotted owl worth saving?
- Should private landowners be reimbursed by the government for restrictions on land use required to save an endangered species?
- Is extinction of wild species something we should worry about?
- Are cancer risks overstated?

These and many other timely issues are debated in Point/Counterpoint essays or discussed in Viewpoints written by such luminaries as Amory and Hunter Loves, Norman Myers, the late Garrett Hardin, the late Julian Simon, Lewis Regenstein, David Eaton, Dave Armstrong, and others. These editorials can stimulate individual thinking as well as classroom discussion on complex problems. They're also a perfect avenue for developing one's critical thinking skills.

Spotlights on Sustainable Development

To give students further insight into timely issues of our day, I've included numerous case studies throughout the text called "Spotlights on Sustainable Development" and have included some exciting new ones in this edition. These sections highlight examples of what individuals, businesses, and governments are doing to create a sustainable future. They provide guidance and hope.

Key Concepts

In this edition, key concepts are highlighted in boxes at the end of each section. These brief encapsulations are designed to help students understand the crucial concepts that form the foundation of environmental science. They serve as a great tool for reviewing chapters as well as studying for tests. Read them before you read the text, after completing a chapter, and while you're studying for tests. They're the glue that holds this book together and forms a lifetime of understanding.

Key Facts

Throughout this edition, I highlight key facts. Key facts represent some of the most important statistics in the book. They're the ones most worth memorizing.

Critical Thinking

As pointed out earlier, critical thinking is one of the central themes of this book. Critical thinking enables students to analyze complex issues and make rational decisions.

A number of important critical thinking rules are discussed in Chapter 1. Each chapter also begins with a brief critical thinking exercise, which asks students to critically analyze an issue, a research finding, or an assertion. Students are also asked to exercise critical thinking skills after Point/Counterpoints, and many of the discussion questions at the end of each chapter call on students to put their knowledge and their critical thinking skills to use.

Focus on Sustainability and Systems Reform

While the main focus of the book is environment, not development, I attempt to show how development must be revamped to become sustainable. I describe basic human systems and why they are unsustainable. Moreover, I outline a variety of ways that we can make them environmentally, socially, and economically sustainable.

Expanded Basic Science

In recent editions, I have expanded the coverage of basic science, especially information to help students understand global climate. You'll find detailed discussion, for instance, of weather, ocean currents, aquatic ecology, and other important facts that help you better understand the world we live in.

Updated Coverage

The seventh edition has been thoroughly updated with new discoveries, new concepts, new environmental laws, and the most recent statistics on resources, population, and pollution. I have added information on changes in environmental laws and regulations. I have also included important new information on peak oil and natural gas, renewable energy options, environmentally friendly transportation, and environmentally friendly building.

In this edition, I have added several spectacular new spotlights on sustainable development that highlight exciting efforts to address pressing environmental concerns sustainably. In addition, I have included an important section on energy and energy laws to the chapter on nonrenewable energy. Many new photographs and tables have been added as well.

Reorganized Coverage

In this edition, I have made some significant changes in the organization of the book. I have combined the early information on environmental protection and sustainable development previously covered in three chapters into one chapter (Chapter 2) and moved the information on science, scientific method, and critical thinking to Chapter 1. This reorganization allows students an opportunity to learn important critical thinking skills before examining the larger issues like environmental protections and sustainable development.

Personal Actions

In each chapter, you'll find a table of actions, titled "Individual Actions Count" that list a wide assortment of things you can do to help build a sustainable future.

Global Focus and Expanded Coverage of International Issues and Solutions

Environmental problems affect us locally, in our own homes and towns and cities. They affect larger regions, too, often covering many states or crossing international borders. And, of course, they often span the globe. As in previous editions, I present a wealth of information on local, regional, and global problems and solutions. Special efforts were made to include examples from Canada and other countries in Europe, Asia, and Africa.

Ancillaries

Jones and Bartlett Publishers offers an impressive variety of traditional print and interactive multimedia supplements to assist instructors and aid students in mastering environmental science. Additional information and review copies of any of the following items are available through your Jones and Bartlett Sales Representative or by going to **www.bioscience.jbpub.com**.

Environmental Science, 7th Edition Web Site

The seventh edition is linked to an extensive web site developed exclusively by the author and Jones and Bartlett Publishers. Students will find a variety of study aids and resources at **http://environment.jbpub.com**, all designed to explore the concepts and controversies of environmental science in more depth. The central on-line learning component, *eLearning* is an interactive study guide with a variety of activities to help students review class material. Students will find study aids and resources designed to explore the concepts and controversies of environmental science in more depth. The on-line study guide includes chapter reviews, study quizzes, virtual flash cards to help students learn important terms and concepts, and summaries of Point/Counterpoints. It also includes additional questions for the Point/Counterpoints in the book and links to web sites that present views on each side of the issues so students can evaluate the arguments more fully and clarify their own opinions. Also available on the web site:

- **Individual Actions Count!** builds upon the same feature in the text by providing links to environmental/government organizations, career information, freeware, and activities.
- **Learn More** offers links to web sites that offer additional coverage of material covered in the various chapters.
- **Links to Your Region** provides links to regional web sites so that students can learn more about environmental issues specific to their own area of the United States or Canada.

Instructor's ToolKit CD-ROM

The *Instructor's ToolKit CD-ROM* contains a suite of files to help professors teach their courses. The materials are cross-platform for Windows and Macintosh systems. All the files on the CD are ready for online courses using the **WebCT** or **Blackboard** formats.

- The **PowerPoint™ Image Bank** provides the illustrations, photographs, and tables (to which Jones and Bartlett Publishers holds the copyright or has permission to reprint digitally) inserted into PowerPoint™ slides. With the Microsoft PowerPoint™ program, you can quickly and easily copy individual image slides into your existing lecture slides.
- The **PowerPoint™ Lecture Outline Slides** presentation package provides lecture notes and images for each chapter of *Environmental Science , 7th Edition.* A PowerPoint™ viewer is provided on the CD. Instructors with the Microsoft PowerPoint™ software can customize the outlines, art, and order of presentation.

- **The Instructor's Manual** is provided as a text file. Mark Aronson from Scott Community College has updated and expanded the Instructor's Manual for this edition. The Instructor's Manual contains chapter outlines, objectives and key terms as well as lecture outlines and suggestions for presenting each chapter. We have also included sample syllabi to demonstrate a few of the many approaches that can be taken to an environmental science course.
- The **Test Bank**, also prepared by Mark Aronson, is provided as a text file with over two thousand questions in a variety of formats.

Transparencies

Qualified adopters may request a set of high-quality color transparencies. The standard set contains over one hundred of the most frequently used illustrations and tables from the text.

ACKNOWLEDGMENTS

Although I have spent thousands of hours researching and writing it, this book is really the product of thousands of researchers and scholars in anthropology, biology, chemistry, demography, natural resources, political science, economics, ecology, and dozens of other disciplines. Their findings and their thoughts form the foundation on which this book rests. To them, a world of thanks!

A genuine thanks also to the staff at Jones and Bartlett. A very special thanks to Dean DeChambeau, my editor, who has offered moral support and encouragement throughout the projects. Thanks also to Lou Bruno, Anne Spencer, and Kimberly Potvin for their work on the manuscript, design, and photo research. Their creativity, bright ideas, and hard work have been vital to the success of this book. Thanks also to Shellie Newell for her assistance with copyediting. As always, it is has been a pleasure to work with the Jones and Bartlett staff in this complex and sometimes tedious production. Also a special thanks to my research assistant, Linda Stuart, who helped update the book. Her persistence and attention to detail were much appreciated. Updating statistics can be tiring, frustrating, and at times overwhelming and I'm forever thankful for her assistance. Also, a word of thanks to the many people in government agencies and nonprofit groups who shared articles, reports, and data with us. Although they are too numerous to mention here, their part in this project is much appreciated.

Finally, many manuscript reviewers and survey respondents provided helpful and constructive criticism on all editions of this book. I am very thankful for their helpful comments.

Daniel D. Chiras
Evergreen, Colorado

David Armstrong, University of Colorado, Boulder
Robert Auckerman, Colorado State University
Terry Audesirk, University of Colorado, Denver
Nancy Bain, Ohio University
Michael Bass, Mary Washington College
Richard Beckwitt, Framingham State College
Mark C. Belk, Brigham Young University
Charles Blalack, Kilgore College
Bayard H. Brattstrom, California State University, Fullerton
Lester Brown, Worldwatch Institute
Anya Butt, Central College
Ann S. Causey, Auburn University
Jay Clymer, Marywood University
Sheree E. Cohn, Saint Cloud State University
Donald Collins, Montana State University
Peter Colverson, Mohawk Valley Community College
Anne Cummings Pikes Peak Community College
Craig B. Davis, Ohio State University
Lola Deets, Penn State, Erie, The Behrend College
JodyLee Estrada Duek, Pima Community College
Sally DeGroot, St. Petersburg Junior College
Joseph Farynairz, Mattatuck Community College
David Fluharty, University of Washington
Gary J. Galbreath, Northwestern University
Ted Georgian, St. Bonaventure University
Sigurdur Greipsson, Troy State University
James H. Grosklags, Northern Illinois University
Richard Haas, California State University, Fresno
Zac Hansom, San Diego State University
William S. Hardenbergh, Southern Illinois University
John P. Harley, Eastern Kentucky University
John N. Hoefer, University of Wisconsin, La Crosse
James Hornig, Dartmouth College
William Hoyt, Saint Joseph's College of Maine
Gary James, Orange Coast College
Pat Hilliard Johnson, Palm Beach Community College
John A. Jones, Miami Dade Community College
Alan R. P. Journet, Southeast Missouri State University
Thomas L. Keefe, Eastern Kentucky University
Suzanne Kelly, Scottsdale Community College
Thomas G. Kinsey, State University College at Buffalo
Ronald Kolbash, Ohio University
Mary Hurn Korte, Concordia University of Wisconsin
Kip Kruse, Eastern Illinois University
Steve LaDochy, California State University, Los Angeles
David Lovejoy, Westfield State College

Timothy F. Lyon, Ball State University
Ted L. Maguder, St. Petersburg Junior College, Clearwater
Ravindra Malik, Albany State University
Heidi Marcum, Baylor University
Richard Meyer, University of Arkansas
Myles Miller, Philadelphia University
Sheila Miracle, Southeast Kentucky Community and Technical College
Glenn P. Moffat, Foothill College
Charles Mohler, Cornell University
Joseph Moore, California State University, Northridge
Bryan C. Myres, Cypress College
Paul E. Nowack, University of Michigan
Natalie Osterhoudt, Broward Community College
Nancy Ostiguy, California State University
John H. Peck, St. Cloud State University
Charles R. Peebles, Michigan State University
Michael Picker, Western Director for the National Toxics Campaign
David Pimental, Cornell University
Michael J. Plewa, University of Illinois, Urbana-Champaign
Michael Priano, Westchester Community College
Joseph Priest, Miami University
Martha W. Rees, Baylor University
Robert J. Robel, Kansas State University
Jack Schelein, York College, City University of New York
Michael P. Shields, Southern Illinois University
Cynthia Simon, University of New England
Doris Shoemaker, Dalton College
Rocky Smith, Colorado Environmental Coalition
Steven Soldheim, University of Wisconsin, Madison
Stephen Sousa, Andover High School
Laura Tamber, Nassau Community College
Roger E. Thibault, Bowling Green State University
J. Rafael Toledo, University of North Texas
Leland Van Fossen, DeAnza College
Bruce Webb, Animal Protection Institute
Ross W. Westover, Canada College
Jeffrey R. White, Indiana University
Ray E. Williams, Rio Hondo College
Larry Wilson, Miami Dade Community College
Stephen W. Wilson, Central Missouri State University
Susan Wilson, Miami Dade Community College
Timothy Winter, Louisiana State University, Shreveport
Robert Wiseman, Lakewood Community College
Richard J. Wright, Valencia Community College
Paul A. Yambert, Southern Illinois University
Astatkie Zikarge, Texas Southern University

DAN CHIRAS received a Ph.D. in reproductive physiology from the University of Kansas Medical School. In 1976, he accepted a teaching and research position at the University of Colorado, Denver, where he taught a variety of courses, including general biology, cell biology, histology, reproductive biology, and endocrinology. Over the years, Dr. Chiras developed a number of courses on the environment, including a graduate course on global environmental issues and the environmental and health effects of pollution. He has also taught undergraduate science modules on air pollution, nuclear power, noise pollution, impacts from coal development, and strategies for sustainability.

While at the university, Dr. Chiras worked on several EPA projects, including a study of the health impacts of chlorinated organics from wastewater treatment and the impacts of coal mining in the West. He also prepared an assessment of the impacts of shale oil development in Colorado for the Department of Energy.

In 1981, Dr. Chiras resigned his full-time position at the university to pursue a writing career. Since that time, he has published nearly 250 articles in journals, magazines, encyclopedias, and other publications, including *Environment, American Biology Teacher, The Amicus Journal, Solare Today, Natural Home, Sustainable Futures, Colorado Outdoors, Home Power* and *Environmental Carcinogenesis and Ecotoxicology Reviews.*

Dr. Chiras has written the environment section for World Book Encyclopedia's Science Year since 1993. He wrote an extensive article titled "The Population Explosion" in *Science Year* 1998. Dr. Chiras has written articles on ecology, the environment, environmental issues, air pollution, and population articles for *Encyclopedia Americana.*

Dr. Chiras has published twenty-two books on ecological design, environmental science, sustainable systems design, and renewable energy. Included in this list are a college-level textbook titled *Natural Resource Conservation* (with John P. Reganold) and a high school textbook titled *Environmental Science: Framework for Decision Making.* Dr. Chiras has also published two college-level biology textbooks: *Human Biology* and *Biology: The Web of Life.*

Dr. Chiras has also written several books for a more general audience, *Beyond the Fray: Reshaping America's Environmental Response,* which offers advice on building a truly effective response to the environmental crisis, and *Lessons from Nature: Learning to Live Sustainably on the Earth,* which outlines ways to build a sustainable future. He has also assembled an anthology titled *Voices for the Earth: Vital Ideas from America's Best Environmental Books.* This book includes 14 essays summarizing the key ideas of books nominated by a national committee as America's best environmental books.

Dr. Chiras's more recent book include *The Natural House: A Complete Guide to Healthy, Energy-Efficient, Environmental Homes, The Solar House, The New Ecological Home, Superbia!, 31 Ways to Create Sustainable Neighborhoods, The Homeowner's Guide to Renewable Energy, The Natural Plasters Book,* and *EcoKids: Raising Children Who Care for the Earth.*

Dr. Chiras is currently a visiting professor in the environmental program at Colorado College, where he teaches courses. Dr. Chiras has also served as a visiting professor at the University of Washington in Seattle, where he taught the introductory environmental science course.

Besides teaching and writing, Dr. Chiras has played an active role in the environmental movement. He has served on the Board of Directors of the Colorado Environmental Coalition for five years and was president of this coalition of 40 environmental groups for two years. In 1988, he cofounded Friends of Curbside Recycling, which was instrumental in convincing the city of Denver to begin a curbside recycling program. In 1989, he cofounded Speakers for a Sustainable Future, which offered slide programs on recycling, water conservation, and sustainability. In 1993, Dr. Chiras cofounded another nonprofit organization, the Sustainable Futures Society. He served on the board of directors of the Advanced Technology Environmental Education Library, which has produced a Web site that serves as a repository for environmental information.

Dr. Chiras has spoken to a wide range of audiences, including the National Association of County Agricultural Agents; the American Society of Interior Designers; the American Home Economics Association; Architects, Designers, Planners for Social Responsibility; the Colorado Renewable Energy Society; and others.

In addition to his active scientific pursuits, Dr. Chiras is a gardener, musician, songwriter, river runner, and bicyclist. He lives with his two children, Skyler and Forrest, in a nearly 100% self-sufficient solar home in the Colorado Rockies, where he practices what he preaches.

STUDY SKILLS

College is a demanding time. For many students, term papers, tests, reading assignments, and classes require a new level of commitment to their education. At times, the workload can become overwhelming.

Fortunately, there are many ways to lighten the load and make time spent in college more profitable. This section offers some helpful tips on ways to enhance your study skills. It teaches you how to improve your memory, how to become a better note taker, and how to get the most out of what you read. It also helps you prepare better for tests and become a better test taker.

Mastering these study skills will require some work, mostly to break old, inefficient habits. In the long run, though, the time you spend now learning to become a better learner will pay huge dividends. Over the long haul, improved study skills will save you lots of time and help you improve your knowledge of facts and concepts. That will no doubt lead to better grades and very likely a more fruitful life.

General Study Skills

- Study in a quiet, well-lighted space. Avoid noisy, distracting environments.
- Turn off televisions and radios.
- Work at a desk or table. Don't lie on a couch or bed.
- Establish a specific time each day to study, and stick to your schedule.
- Study when you are most alert. Many students find that they retain more if they study in the evening a few hours before bedtime.
- Take frequent breaks-one every hour or so. Exercise or move around during your study breaks to help you stay alert.
- Reward yourself after a study session with a mental pat on the back or a snack.
- Study each subject every day to avoid cramming for tests. Some courses may require more hours than others, so adjust your schedule accordingly.
- Look up new terms or words whose meaning is unclear to you in the glossaries in your textbooks or in a dictionary.

Improving Your Memory

You can improve your memory by following the PMC method. The PMC method involves three simple learning steps: (1) paying attention, (2) making information memorable, (3) correlating new information with facts you already know.

STEP 1 Paying attention means taking an active role in your education-taking your mind out of neutral. Eliminate distractions when you study. Review what you already know and formulate questions about what you are going to learn before a lecture or before you read a chapter in the text. Reviewing and questioning help prime the mind.

STEP 2 Making information memorable means finding ways to help you retain information in your memory. Repetition, mnemonics, and rhymes are three helpful tools.

- Repetition can help you remember things. The more you hear or read something, the more likely you are to remember it, especially if you're paying attention. Jot down important ideas and facts while you read or study to help involve all of the senses.
- Mnemonics are useful learning tools to help remember lists of things. I use the mnemonic CARRRP to remember the biological principles of sustainability: conservation, adaptability, recycling, renewable resources, restoration, and population control.
- Rhymes and sayings can also be helpful when trying to remember lists of facts.
- If you're having trouble remembering key terms, look up their roots in the dictionary. This often helps you remember their meaning.
- You can also draw pictures and diagrams of processes to help remember them.

STEP 3 Correlating new information with the facts and concepts you already know helps tie facts together, making sense out of the bits and pieces you are learning.

- Instead of filling your mind with disjointed facts and figures, try to see how they relate with what you already know. When studying new concepts, spend some time tying information together to get a view of the big picture.
- After studying your notes or reading your textbook, go back and review the main points. Ask yourself how this new information affects your view of life or critical issues and how you may be able to use it.

Becoming a Better Note Taker

- Spend 5 to 10 minutes before each lecture reviewing the material you learned in the previous lecture. This is extremely important!
- Know the topic of each lecture before you enter the class and spend a few minutes reflecting on facts you already know about the subject that's going to be discussed.
- If possible, read the text before each lecture. If not, at least look over the main headings in the chapter, read the topic sentence of each paragraph, and study the figures. If your chapter has a summary, read it too.
- Develop a shorthand system of your own to facilitate note taking. Symbols such as = (equals), > (greater than), < (less than), w/ (with), and w/o (without) can save lots of time so you don't miss the main points or key facts.
- Develop special abbreviations to cut down on writing time. E might stand for energy, AP might be used for air pollution, and AR could be used to signify acid rain.
- Omit vowels and abbreviate words to decrease writing time (for example: omt vwls & abbrvte wrds to dcrs wrtng tme). This will take some practice.
- Don't take down every word your professor says, but be sure your notes contain the main points, supporting information, and important terms.
- Watch for signals from your professor indicating impor-

tant material that might show up on the next test (for example, "This is an extremely important point …").

- If possible, sit near the front of the class to avoid distractions.
- Review your notes soon after the lecture is over while they're still fresh in your mind. Be sure to leave room in your notes written during class so you can add material you missed. If you have time, recopy your notes after each lecture.
- Compare your notes with those of your classmates to be sure you understood everything and did not miss any important information.
- Attend all lectures.
- Use a tape recorder if you have trouble catching important points.
- If your professor talks too quickly, politely ask him or her to slow down.
- If you are unclear about a point, ask during class. Chances are other students are confused as well. If you are too shy, go up after the lecture and ask, or visit your professor during his or her office hours.

How to Get the Most Out of What You Read

- Before you read a chapter or other assigned readings, preview the material by reading the main headings or chapter outline to see how the material is organized.
- Pause over each heading and ask a question about it.
- Next, read the first sentence of each paragraph. When you have finished, turn back to the beginning of the chapter and read it thoroughly.
- Take notes in the margin or on a separate sheet of paper. Underline or highlight key points.
- Don't skip terms that are confusing to you. Look them up in the glossary or in a dictionary. Make sure you understand each term before you move on.
- Use the study aids in your textbook, including summaries and end-of-chapter questions. Don't just look over the questions and say, "Yeah, I know that." Write out the answer to each question as if you were turning it in for a grade, and save your answers for later study. Look up answers to questions that confuse you. This book has questions that test your understanding of facts and concepts. Critical thinking questions are also included to sharpen your skills.

Preparing for Tests

- Don't fall behind on your reading assignments, and review lecture notes as often as possible.
- If you have the time, you may want to outline your notes and assigned readings. Try to prepare the outline with your book and notes closed. Determine weak areas, then go back to your text or class notes to study these areas.
- Space your study to avoid cramming. One week before your exam, go over all of your notes. Study for two nights, then take a day off that subject. Study again for a couple of days. Take another day off from that subject. Then make one final push before the exam, being sure to study not only the facts and concepts, but also how the facts are related. Unlike cramming, which puts a lot of information into your brain for a short time, spacing will help you retain information for the test and for the rest of your life.
- Be certain you can define all terms and give examples of how they are used.
- You may find it useful to write flash cards to review terms and concepts.
- After you have studied your notes and learned the material, look at the big picture-the importance of the knowledge and how the various parts fit together.
- You may want to form a study group to discuss what you are learning and to test one another.
- Attend review sessions offered by your instructor or by your teaching assistant, but study before the session and go to the session with questions.
- See your professor or class teaching assistant with questions as they arise.
- Take advantage of free or low-cost tutoring offered by your school or, if necessary, hire a private tutor to help you through difficult material. Get help quickly, though. Don't wait until you are hopelessly lost. Remember that learning is a two-way street. A tutor won't help unless you are putting in the time.
- If you are stuck on a concept, it may be that you have missed an important point in earlier material. Look back over your notes or ask your tutor or professor what facts might be missing and causing you to be confused.
- If you have time, write and take your own tests. Include all types of questions.
- Study tests from previous years, if they are available legally.
- Determine how much of a test will come from lecture notes and how much will come from the textbook.
- Purchase a study guide, if one is available, and use it to review material and test your knowledge.
- Check out your instructor's Web site or the author's Web site. They often have valuable study aids, such as review questions and practice quizzes.

Taking Tests

- Eat well and get plenty of exercise and sleep before tests.
- Remain calm during the test by breathing deeply.
- Arrive at the exam on time or early.
- If you have questions about the wording of a question, ask your professor.
- Skip questions you can't answer right away, and come back to them at the end of the session if you have time.
- Read each question carefully and be sure you understand its full meaning before answering it.
- For essay questions and definitions, organize your thoughts first on the back of the test before you start writing.

Now take a few moments to go back over the list. Check off those things you already do. Then, mark the new ideas you want to incorporate into your study habits. Make a separate list, if necessary, and post it by your desk or on the wall and keep track of your progress.

Mt. Rainier, Washington

Environmental Science and Critical Thinking

A problem well stated is a problem half solved.

—*Charles Kettering*

If you are like most people, you've probably heard a lot of conflicting information about the environment and environmental issues. Teachers, friends, news reporters, writers, and scientists may have warned you of serious problems. Others may have offered an opposite view, saying things are not nearly as bad as some people would have you believe. What is the truth? In this book, you will find ample scientific evidence to suggest that the problems are real. In fact, a survey of U.S. environmental trends that I conducted covering the 30 years showed that the vast majority (nearly 70%) of environmental indicators are pointing in the wrong direction. They indicate movement away from a stable, healthful relationship with the planet we call home. Although there are some very encouraging signs of improvement—about 20% of the trends showed improving conditions—there is still cause for serious concern and action. I invite you to explore the issues in this book with an open mind, looking carefully at all aspects of each one. We'll examine the scientific evidence behind the issues and hear what the experts from all sides of the issues have to say. We'll explore traditional solutions and a new brand of responses that could help us find lasting solutions to pressing problems—solutions that are good for people, the environment, and the economy. Let's begin with some of the encouraging trends.

1.1 Encouraging Signs/Continuing Challenges

All across the world, change is underway. It's not ordinary change, either. It is profound change in the ways societies conduct their everyday affairs—ways that are leading to a new wave of environmental protection with lasting impacts. In Holland, for example, industries have banded together to produce a nationwide Green Plan. Drafted in conjunction with several key government agencies, the Green Plan calls on industries to reduce toxic emissions drastically and pollutants voluntarily (see Spotlight on Sustainable Development 1-1). This bold new plan will, if its goals are reached, practically eliminate hazardous wastes and a host of additional pollutants. What is encouraging to many is that the plan does not require regulations and fines, which governments typically use to get companies to improve their environmental performance. Rather, companies are left largely to their own devices to find economical ways to meet national goals. The government is there to lend a hand, providing financial resources and technical assistance if needed. If companies do not meet goals, fines may be imposed. Because of Holland's success, many other nations, including Bhutan, New Zealand, Nepal, Mexico, and Singapore, have launched similar programs.

In the United States, impressive change is also in the works. For example, a number of programs have been established by the **U.S. Environmental Protection Agency (EPA)**, a federal agency charged with writing many environmental regulations and enforcing environmental laws passed by the Congress. One program was designed to promote voluntary reductions in hazardous waste production by major manufacturers; hundreds of major corporations signed up and slashed hazardous waste production by millions of tons per year. The second program was designed to reduce electrical consumption by businesses and government agencies. So far, thousands of the nation's largest corporations, including Boeing, and many governmental agencies have signed up to participate in this program (FIGURE 1-1). Participating companies, such as those in Holland, are working with the EPA to cut electrical demand through conservation. All efforts are voluntary. The EPA provides technical assistance.

FIGURE 1-1 **EPA Green Lights program.** Boeing Corporation, based in Seattle, Washington, joined the EPA Green Lights program (now part of the EPA's Energy Star Program) and saves over $12 million per year by reducing electrical energy consumption. Every eight years, this would buy one Boeing 757 jet.

1-1 The Netherlands' Green Plan Revolutionizes the Way Industries Function

The Netherlands is a tiny country, about one-fourth the size of the state of Illinois (FIGURE 1). It is, however, a giant among nations. Why this glowing praise for such a small country?

In 1989, the country adopted a National Environmental Policy Plan commonly called its Green Plan, after widespread public alarm over many environmental trends. An even tougher plan came into existence several years later. Known as the NEPP 2, it gave the Dutch government more power to write and enforce environmental laws and to earmark impressive sums of money to research and redesign industrial processes. The most exciting aspect of the Green Plan, though, is that it allows for a new approach—a set of voluntary agreements with industry and other sectors of society responsible for producing the bulk of the nation's pollution. The National Environmental Policy Plan has undergone two additional revisions to ensure success. Why was all of this necessary?

Although the Netherlands has some of the most stringent environmental laws in the world, experience had shown that they weren't working well. Because of this, the government took a bold step: It admitted that there are limitations to regulating complex and interdependent environmental problems on an issue-by-issue basis. In place of this unworkable system of rules and regulations, the government decided to identify all major environmental problems and meet with the key players (industries and citizen groups, for example) to reach agreement on establishing bold new targets and timetables for drastically reducing pollution. The government let the companies select ways to reduce pollution; all that it asked of industry was a commitment to meet ambitious government targets and timetables. Industry groups happily signed the agreements, recognizing that the responsibility of meeting the targets was theirs and that failure to meet the goals would result in stiff penalties. The Dutch, therefore, combined the carrot (incentive) and stick (penalty) approaches in hopes of restructuring industry for sustainability.

This approach, while not perfect, was widely hailed by many major Dutch industrialists. They liked the agreement because it gave them free rein to find solutions that made the most sense to them and their businesses. The net effect

FIGURE 1 The Netherlands is leading the world in sustainable development.

What motivates companies to reduce hazardous wastes and use energy more efficiently? Many adopted these programs to become better environmental neighbors, but they also did it to save money, sometimes hundreds of thousands of dollars a year in energy bills.

In other encouraging news, cities, states, and nations throughout the globe are developing programs to monitor social, economic, and environmental conditions over time to see if things are improving or deteriorating. One leader in this effort is the state of Oregon, which has established an extensive set of benchmarks, 90 indicators that cover a wide assortment of conditions, from teen pregnancy to air pollution to commuting time on highways. It uses the yearly measurements to tell whether its policies and practices are working. Key trends are examined, and decisions regarding policies and financial expenditures are made on the basis of the data. Such measures help create a more effective and more economically efficient form of government while improving social, economic, and environmental conditions.

KEY CONCEPTS

Although there are some signs of improvement, the vast majority of the environmental trends show signs of movement away from sustainability.

What has brought about all of the changes discussed previously? Quite simply, these and a great many other changes underway have been brought about by a growing recognition that the current paths many nations are following are not sustainable. That is to say, many individuals have realized we can't go on as we have been. In my analysis of 25 environmental trends spanning a period of 30 years, mentioned earlier, I found that nearly 70%, including energy consumption, population growth, waste production, and loss of wildlife habitat, showed decline. About 20% of the trends showed some improvement. The rest of the conditions were neither improving nor deteriorating (FIGURE 1-2).

Worldwide, trends have caused concern among many political leaders, activists, citizens, scientists, and business lead-is a kind of customized implementation that avoids the one-size-fits-all solutions found in industrial nations' environmental control policies, which meet with considerable resistance in the United States because they often impose prohibitive and sometimes unnecessary costs on businesses.

The voluntary approach was also popular because it helped companies become more efficient and more profitable. In addition, the Netherlands now stands in a good position to become an exporter of home-grown technologies created by the country's industries to meet the goals of the nation's ambitious Green Plan. As other nations see the need for this shift, the Dutch pollution prevention and sustainable technology industry could profit enormously.

The Netherlands' Green Plan relies on four basic principles. First is life cycle management, which makes companies responsible for products throughout their life cycle—from production to consumption to disposal. Under this principle, producers are responsible for disposing of products after their useful life span has expired. This has spawned interest in manufacturing products that are easily recycled.

Another essential component of the strategy is energy efficiency. By using energy more efficiently, companies can save huge sums of money—and gain a competitive edge in the market by producing goods and services more cheaply than international competitors.

Sustainable technologies are also an important component of this plan. The government has launched a major program to develop sustainable technologies and make existing technologies more readily available to people and industries.

Finally, success also hinges on improving public awareness. The government has launched a media blitz that echoes the important theme: "A better environment begins with you."

Today, over 250,000 Dutch businesses participate in the Green Plan, focusing on ways to conserve energy, produce more environmentally friendly products, and reduce pollution emissions. Although the initial reports on the success of the project are encouraging, some problems have arisen. This is to be expected. The government called for 70 to 90% reductions in many forms of pollution while doubling the nation's economic output by the year 2010. Such a task is not easy; one wouldn't expect this to be a painless or trouble-free route.

The successes of this ambitious program have been impressive: ozone-depleting substances have been phased out. Industry has reduced its waste disposal by 60%. Recycling has increased to 70%, and sulfur dioxide emissions have been reduced by 70%.

Not all environmental problems facing this small country were the result of industry. Realizing this, the Dutch government has revised its plan to address the citizens and enlist their aid in reducing consumption. They're even working on land use and transportation issues with a greater emphasis on consumer responsibility. All in all, this tiny, progressive country stands as a model for the rest of the world.

Go to http://environment.jbpub.com to learn more.

ers. Population is growing. Forests are being cut down. Species are disappearing, and global pollution is increasing. Although global statistics can become overwhelming, consider what will happen in the next 24 hours. During this period, 230,000 people (nearly a quarter of a million!) will be added to the world population. In the next 24 hours, a huge amount of tropical rain forest will be leveled to make room for farms, roads, dams, mines, and towns. The daily loss comes to about 440 square kilometers (170 square miles)–that's equivalent to a swath 85 miles long and 2 miles wide. Unfortunately, very few of these trees will ever be replanted—only about 1 for every 10 trees felled. In this same period, 180 square kilometers (70 square miles) of land will turn to desert, in large part because of poor farming practices, overgrazing, and a warming trend that may be caused by pollutants—largely from industrial societies—that trap heat in the Earth's atmosphere. The list goes on.

These and a host of other environmental trends presented throughout this book suggest to many observers that the current course of virtually all nations of the world is un-

sustainable. This is not to say that we are doomed, however. Stating we're on an unsustainable course is simply another way of asserting that things can't go on as they have been. We must change our ways. To avoid problems—potentially serious problems—and to create a better future, we will very likely need to chart a new course. Thankfully, many of the world's people are realizing this. You should not give up hope. There are many very favorable trends underway and new and promising developments. New technologies are also helping solve problems, as are individual actions (**FIGURE 1-3**). Another encouraging sign is that more and more people have come to realize that there are a number of fundamental driving forces—or root causes—that have created

The world population expands by about 230,000 people per day. Each day, we lose a section of tropical rain forest measuring 70 miles long by 2 miles wide.

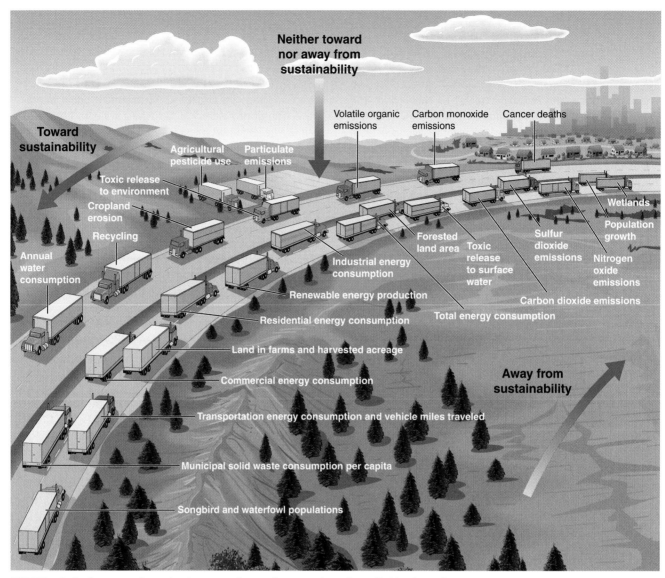

FIGURE 1-2 **Environmental trends.** Summary of 25 environmental trends studied by the author.

FIGURE 1-3 **Solar electricity.** Photovoltaic (PV) modules produce electricity from sunlight. On these homes in Brazil, PVs provide electricity for two indoor compact fluorescent lights.

the present conditions. Two of the most important are continuing growth in both economic output and population.

Encouragingly, many nations have responded swiftly to the population challenge. They have developed educational programs and other voluntary measures that are helping halt the growing tide of human population. In 1995, the United Nations announced that world population growth was slowing, thanks to these efforts. The growth rate, although still high, continues to decline.

Many countries have begun to take steps to create economies that are better for the environment. Important agreements have been reached to curb global pollution, protect forests, and protect wild species. Although there's a lot more to do, the world seems to be moving in the right direction.

KEY CONCEPTS

There are many positive changes in perception and policy that have led to real progress in solving local, regional, and global environmental problems. However, the problems still outweigh the solutions.

1.2 What Is Environmental Science?

This book focuses on the environmental problems that we face—issues in three principal areas: pollution, natural resources, and population. However, it also focuses on solutions, that is, what individuals, businesses, and governments can do about our many environmental problems. The topics covered in this book fall within the domain of a relatively new science, known as *environmental science*. What is environmental science?

Environmental science is, in broadest terms, a branch of science that seeks to understand the many ways that we affect our environment and the many ways that we can address these issues. Environmental science draws on the research and expertise of specialists from numerous traditional sciences, including biology, ecology, chemistry, geology, engineering, and physics. The study of environmental issues and solutions also draws heavily on the humanities: economics, political science, anthropology, history, law, sociology, and even psychology. Thus, environmental science is a multidisciplinary field of endeavor.

To many observers, myself included, a full understanding of issues and solutions requires this broad, multidisciplinary approach. We won't solve the complex environmental issues that we face today through science and technology. We must understand and address the human element.

This book takes a multidisciplinary approach. It discusses the science behind the issues, technologies that will help us address problems, but also draws on a wide range of other disciplines to help you develop a deeper understanding of issues and solutions.

Environmental science, like engineering, is an applied science. Environmental scientists seek knowledge to expand our body of knowledge, but also to enable us to develop solutions to our problems. Those solutions may involve new technologies, but they may also involve new ways of thinking, new behaviors, and new laws and regulations. Today, numerous environmental scientists realize that many of solutions proposed since the 1960s were stop-gap measures—that is, they solved the problems but only temporarily. This realization has led many to re-think previous ideas and propose more lasting, preventive measures—solutions that are good for people, the economy, and the environment. We call these sustainable solutions. In this book, we focus on traditional, stop-gap–type solutions and the newer set of sustainable solutions. As you shall see, these solutions are part of a new movement called **sustainable development**. You'll see the connection in the next chapter.

1.3 Science and Scientific Method: Keys to Understanding Environmental Issues and Creating Lasting Solutions

The term *science* comes from the Latin word *scientia*, which means "to know or to discern." Technically, the term **science** refers to a body of knowledge derived from observation, measurement, study, and experimentation—and to the process of accumulating such knowledge.

Throughout this book you will learn both fundamental principles of science and many of the scientific concepts and facts behind environmental issues. Some of the most important scientific principles come from the science of **ecology**, the study of ecosystems (Chapters 4–6). An **ecosystem** is a system composed of living things (animals, plants, microorganisms) and the interrelated physical and chemical environment. Before we can delve into the realm of science, however, it is helpful to understand how scientific information is acquired.

Scientific Method: The Basis of Good Science

Scientific study of the world around us generally occurs in an orderly fashion. As a rule, scientific discovery begins with observations and measurements of the subject under study (**FIGURE 1-4**). From such observations, scientists often formu-

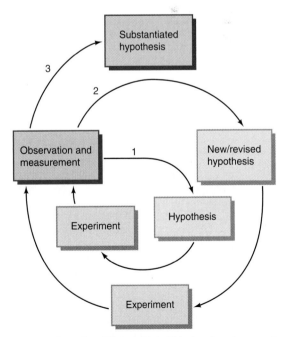

FIGURE 1-4 The scientific method. This drawing shows a simplified view of the scientific method, beginning with observation and measurement. In some cases, the original hypothesis may have to be revised many times.

late tentative explanations, or **hypotheses**. Consider an example: Scientists hypothesize that rattlesnakes shake their rattles to warn large animals to stay away. It is not an aggressive act, but a protective measure that reduces the chances of a snake's being stepped on and killed. This hypothesis is based on observations of wild rattlesnakes. Hypotheses (generalizations) such as this one that are based on observation and measurement are derived by inductive reasoning. **Inductive reasoning** occurs any time a person uses facts and observations to arrive at general rules or hypotheses. This is very common in scientific study. Charles Darwin based his theory of natural selection on observations. He was engaged in inductive reasoning.

Inductive reasoning is the opposite of **deductive reasoning**, in which one arrives at a specific conclusion drawn from general principles. Deductive reasoning is common in mathematics, philosophy, politics, and ethics. General principles are used to examine and evaluate decisions. Consider an example: Scientists have found that brightly colored frogs of the tropics are generally poisonous, a fact established by scientific research (inductive reasoning). Brightly colored snakes are also often poisonous. Studies suggest that this is a kind of warning coloration. Wary predators know to avoid such colorful meals. If you were to encounter a brightly colored snake or frog, you might assume that it is poisonous (**FIGURE 1-5**). You would have used deductive reasoning to do this. You would have used a general principle to derive a conclusion about a specific organism.

> KEY CONCEPTS
>
> Scientific knowledge is often gained through a deliberate process of observation and measurement, which leads scientists to formulate hypotheses that can then be tested through experiments. This is a form of inductive reasoning.

Experimentation: Testing Hypotheses

After a hypothesis is made, the scientist must determine if it is valid. This is usually done by experimentation. Properly performed, experiments provide information that either supports or refutes the hypothesis under question. As shown in Figure 1-4, if a hypothesis is refuted, a modified or new one is generally substituted. The new hypothesis must then be tested.

Scientific method, then, consists of observations and measurements that result in hypotheses. These are tested through experimentation and refined as needed. Surprisingly, most people use scientific method nearly every day of their lives. Suppose, for example, that every time you went to a friend's house you became ill. To determine what made you sick, you'd probably start by making a mental list of all of the activities you engaged in at your friend's house during your visits. After making the list and weeding out activities that seem irrelevant, you might conclude that your problems began when you petted your friend's cat. That's your hypothesis.

To test this hypothesis, you could perform an experiment. For instance, you might ask your friend to keep the cat outdoors when you were there. If you come home feeling sick again, you might alter your hypothesis. Maybe it's the new carpet in her apartment. Of course, scientific study is much more involved than this example, but the process is much the same.

Many scientific experiments are carried out on animals, plants, and other living organisms. Studies such as these are often designed to test the effects of a chemical substance—for example, the effect of a pollutant on an animal's lungs. In experiments of this nature, it is necessary to establish two distinct groups of organisms, an experimental group and a control group. The **experimental group** is the one that you treat or "experiment on." In a study of the effects of air pollution on laboratory mice, for example, the experimental group would be exposed to certain amounts of air pollution. The **control group** is treated identically, except that it is not exposed to the pollution. By setting up experiments in this manner, scientists can test the effect of one and only one variable, which is known as the **independent variable**. Any observed differences in the two groups should result from the treatment.

> KEY CONCEPTS
>
> Experiments enable scientists to test hypotheses and gain new knowledge, but experiments must be carried out very carefully. Good experiments often require control and experimental groups.

Studies on Human Effects: A Special Kind of Experimentation   In experiments outside the laboratory, it is not always easy to isolate one variable. Scientists studying areas in which acidic air pollutants are being deposited in rain, for example, have found many biological effects. However, other pollutants such as ozone may be responsible for some of the effects. Sorting out the specific pollutants involved can be quite complicated.

Studying the effects of pollution on humans is much more difficult than analyzing the effects of chemicals on laboratory animals described previously. Why? First, ethics generally prohibits experimentation on humans that would

FIGURE 1-5 **Predator beware.** Many brightly colored reptiles and amphibians are poisonous, like this arrow poison frog from the tropical rain forest. Predators learn to avoid them.

cause pain and suffering. Second, unlike laboratory mice, which are genetically identical, humans represent a genetically diverse group of organisms. Setting up control and experimental groups containing large numbers of genetically similar humans who have been exposed to the same conditions throughout life is impossible. Consequently, scientists often rely on **epidemiological** (EP-eh-DEEM-ee-ah-LO-jeh-cul) studies to study the effects of chemical pollutants on people.

Literally translated, **epidemiology** (EP-eh-DEEM-ee-OL-oh-gee) is the study of epidemics caused by infectious disease organisms. Today, however, epidemiology refers to a branch of science that also studies the health effects of environmental pollutants, food additives, radiation, and a variety of other agents. Epidemiological studies generally rely on statistical analysis of data on populations that have been exposed to various potentially harmful substances. Consider an example: Suppose a scientist wanted to know the cause of a kind of cancer in humans. If she were an epidemiologist, she would generally begin by searching through death certificates from hospitals to identify patients who had succumbed to the cancer in question. She would then interview their next of kin to learn about the deceased's personal habits, lifestyle, and possible exposures. She would then put all of the information into a computer and sort through it to see if she could find any commonalities. For example, a large percentage of the people might have been employed in a chemical factory. The job is not finished yet, though. Before any conclusions could be drawn, the researcher would assemble a control group as identical to the first group as possible. The control group would consist of people of the same age, sex, and perhaps occupation, except that these people would have died from causes other than the cancer under study. By analyzing their life histories and comparing the results with those from the cancer group, she might be able to pinpoint the cause of the cancer in the experimental group.

Epidemiological studies may also involve comparisons of living subjects. For example, a scientist might compare the health records of workers exposed to a certain pollutant in a factory to workers at the same company who are not exposed to the pollutant (for example, office workers). This may turn up differences that could be attributed to workplace exposure.

Science requires a lot of experimentation to validate hypotheses. No one study can be taken as the final word. However, by repeating experiments and achieving the same results each time, scientists can gain confidence in their hypotheses. (Such repetition may seem like overkill to some people, but it is essential to the scientific process.)

KEY CONCEPTS

Studies on the effects of chemicals on humans are often performed by comparing carefully selected groups of individuals exposed at work or at home to individuals of the same sex and age who were not exposed. These are known as epidemiological studies.

Scientific Theories and Paradigms

Scientific knowledge grows little by little. As facts accumulate from observation and experimentation, researchers gain an understanding of how things work. They often formulate **theories**, explanations that account for many different facts, observations, and hypotheses. Theories are scientifically acceptable principles, broad generalizations regarding natural, physical, and chemical phenomena. Unlike hypotheses, theories generally cannot be tested by single experiments because they encompass many different pieces of information. Atomic theory, for instance, explains the structure of the atom and fits numerous observations.

Science is not static, however. It constantly grows and evolves. Thus, scientists sometimes find that new interpretations emerge, replacing entrenched ideas. This sometimes forces scientists to alter or abandon well-established theories. Consequently, scientists must be open-minded about theories; they must be willing to replace their most cherished theories with new ones, as more information accumulates.

Perhaps the best known example of a changing theory is the **Copernican revolution** (coe-PURR-nick-in) in astronomy. The Greek astronomer Ptolemy (TOLL-eh-me) hypothesized in A.D. 140 that the Earth was at the center of the solar system. This is called the *geocentric view*. The moon, the stars, and the sun, he argued, all revolve around the Earth. In 1580, however, Nicolaus Copernicus (coe-PURR-nick-us) showed that the many observations of astronomers of the time were better explained by assuming that the sun was the center of the solar system. Copernicus was not the first to suggest this *heliocentric* view. Early Greek astronomers had proposed the idea, but it gained little attention until his time.

Over time, scientists have created philosophical and theoretical frameworks within which theories, laws, and generalizations are formulated. These frameworks that underlie all scientific disciplines are called *paradigms*, a term popularized by the philosopher and science historian Thomas Kuhn. A **paradigm** is a basic model of reality in science. For example, evolution is a paradigm of the biological sciences. Atomic theory is the dominant paradigm of modern chemistry.

Paradigms govern the way scientists think, form theories, and interpret the results of experiments. They also govern the way nonscientists think. Once a paradigm is accepted, it is rarely questioned. New observations are generally interpreted according to the paradigm; those that are inconsistent with it are often ignored or disputed. In some cases, though, observations that fail to fit the paradigm may amass to a point at which they can no longer be ignored, causing scientists to rethink their most cherished beliefs and, sometimes, discard them. This is called a **paradigm shift**.

The term *paradigm* is commonly used today in a more general way, referring simply to the way we see the world. Many observers, for instance, speak about the **dominant social paradigm** of modern society when they refer to the set of beliefs that govern our lives. As you will see in Chapter 3, our attitudes and behaviors often flow from our beliefs. Unsustainable systems, for instance, are largely a result of our beliefs.

FIGURE 1-6 **Science is Fun for the Mind!** Robert Bakker, one of America's most creative scientists is a man of great energy, enthusiasm, and creativity.

Paradigms also shape our language, thought, and perceptions. Slogans and common sayings reflect our paradigms and are repeated and reinforced day after day. Perhaps the most obvious expression of the dominant social paradigm is the growth-is-essential philosophy. You hear it in the business world every day when businesspeople proclaim, "A company must grow or die." Politicians echo a similar sentiment. Systems analyst Donella Meadows wrote in her book *The Global Citizen*, "Your paradigm is so intrinsic to your mental process that you are hardly aware of its existence, until you try to communicate with someone with a different paradigm." Perhaps the most difficult problem is that people often get attached to their paradigms and fiercely resist any effort to change them. Because of the importance of our paradigms, success in building a sustainable future may require concerted efforts to overcome this resistance.

Although this description of science and scientific method shows that they require analytical skills and reasoning, they are also creative and fascinating endeavors. Scientific research may sound dull and boring to many people, but the truth is that, in many cases, it is extremely exciting. Far from being the left-brained, analytical types depicted in many movies and books, scientists are often highly creative and—believe it or not—intuitive people (FIGURE 1-6). They see hidden connections in the chaos of experimental data and make leaps of logic that, upon testing, often turn out to be true.

KEY CONCEPTS

As scientific knowledge accumulates, scientists are able to formulate theories that explain their observations. Overarching theories or paradigms emerge and profoundly shape scientific thought and our perceptions of reality.

Science and Values

Science is all about facts and figures that explain our world. Scientists endeavor to remain objective, letting the data determine their hypotheses and theories. Ultimately, scientists hope to find objective truths—assertions about the world around us that can be backed up by data they've acquired through careful experimentation and observation. Because of this, most scientists argue that science is amoral—outside the sphere in which moral judgments are made; that is, it is not about right and wrong.

Scientists bring human qualities to their work, including their own values and preconceived notions that can taint their observations. Thus, scientists are not always as objective as one might hope. Today, scientists can be found to argue both sides of controversial issues. They may look at the same data and come away with markedly different conclusions because their biases often get in the way of objective analysis. Numerous Point/Counterpoints in this book illustrate this situation.

Just as human values affect scientific interpretation of data, science can also affect public values. A good understanding of ecological science, for instance, can help reshape human values concerning nature and our part in the natural world.

Societal values can be based on accurate scientific knowledge. This book will help expand your knowledge of natural and human systems, helping you see things more clearly. This may help you to rethink your values.

One value-shaping role of science rests in its ability to help us see hidden connections—for example, the role of forests in maintaining atmospheric carbon dioxide levels and global climate. This book shows the many ways that natural systems support human life. Knowing how important the planet is to our future cannot help but change our attitudes and, possibly, our actions.

KEY CONCEPTS

Scientists seek to be objective in their work so that science is fact based and value free. However, biases do cloud objectivity at times. Nonetheless, scientific knowledge can also affect human values.

1.4 Critical Thinking Skills

One of the chief tools a good scientist (and many a great thinker) has is a skill called *critical thinking*. **Critical thinking** is the capacity to distinguish between beliefs (what we think is true) and knowledge (facts that are backed by accurate observation and valid experimentation). Thus, critical thinking helps us separate judgment from facts. It is the most ordered kind of thinking of which people are capable. Critical thinking involves subjecting facts and conclusions to careful analysis, looking for weaknesses in logic and other errors of reasoning. Critical thinking skills are essential in analyzing a wide range of environmental problems, issues, and information.

Table 1-1

Summary of Critical Thinking Rules

Gather all information.
- Dig deeper.
- Learn all you can before you decide.
- Don't mistake ignorance for perspective.

Understand all terms.
- Define all terms you use.
- Be sure you understand terms and concepts others use.

Question how information/facts were derived.
- Were they derived from scientific studies?
- Were the studies well conceived and carried out?
- Were there an adequate number of subjects?
- Was there a control group and an experimental group?
- Has the study been repeated successfully?
- Is the information anecdotal?

Question the source of the information.
- Is the source invested in the outcome of the issue?
- Is the source biased?
- Do underlying assumptions affect the viewpoint of the source?

Question the conclusions.
- Do the facts support the conclusion?
- Correlation does not necessarily mean causation.

Expect and tolerate uncertainty.
- Hard-and-fast answers aren't always possible.
- Learn to be comfortable with not knowing.

Examine the big picture.
- Study the whole system.
- Look for hidden causes and effects.
- Avoid simplistic thinking.
- Avoid dualistic thinking.

So how does one think critically? Clearly, there is no single formula. However, most critical thinkers would agree that several steps contribute to this process. These steps or principles of critical thinking, summarized in **Table 1-1**, will help you analyze arguments, research findings, issues, and statements you read.

> KEY CONCEPTS
>
> Critical thinking is an acquired skill that helps us analyze issues and discern the validity of experimental results and assertions.

Gather All Information

The first requirement of critical thinking is to gather all of the information you can before you make a decision. In order to understand most issues, it is often necessary to seek out additional information. When you do, you often find a slightly different picture of reality. This is especially true in environmental debates, in which advocates tend to simplify issues or to present information that supports their case while ignoring contradictory findings. Although you'll rarely have all the information you need to make a decision, you can always get more. So get in the habit of learning all you can before you make a decision. Don't mistake ignorance for perspective.

> KEY CONCEPTS
>
> Critical thinking requires one to know as much information about an issue as possible before rendering an opinion or making a decision.

Understand All Terms

The second rule of critical thinking is that when analyzing any issue, solving any problem, or judging the accuracy of someone's statements, you must understand all terms. In some instances, you will find that the people presenting a case have inaccurate or incomplete understanding. A good example is the term *sustainability*, which is currently being used by people who don't really understand what it means. They even use the word to construct oxymorons, such as *sustainable economic growth*. In a finite system, as you will soon see, growth cannot be sustained.

> KEY CONCEPTS
>
> To think critically about an issue, one must understand the terms and concepts related to it.

Question the Methods

The third principle of critical thinking is to question the methods by which information has been acquired. Were the facts derived from experimentation? Was the experiment performed correctly? Did the experimenters use an adequate number of subjects? Did the experiment have a control group? Were the control and experimental groups treated identically except for the experimental variable?

Beware of anecdotal information—stories of isolated incidents that are often made to appear as if they represent the truth. A newscast showing an angry mob, for example, may give the impression that the entire country is in turmoil. In reality, only a few dozen people may be upset.

> KEY CONCEPTS
>
> Critical thinking requires that we know how information has been acquired and that we question the methods by which it was derived.

Question the Source

A fourth requirement of critical thinking is to question the source of the facts. Ask yourself who is giving the information. When business owners say that pollution from their factories isn't causing any harm, beware. Environmentalists are not immune to bias and are also fair game for your critical thinking skills.

Be on the lookout for bias and underlying assumptions, even in this text. Some of the most well-entrenched assumptions and myths are listed in Table 1-2. Study them and be aware of them. Also, look at the Individual Actions Count! table. It will help you examine your own values and assumptions for bias.

KEY CONCEPTS

Critical thinking requires one to search for hidden biases and assumptions that may influence one's understanding of an issue or interpretation of data.

Question the Conclusions

The fifth requirement of critical thinking is to question the conclusions derived from facts. Do the facts support the conclusions? Are other interpretations possible? One of the earliest studies on lung cancer, for instance, showed a correlation between lung cancer and sugar consumption. The researchers who performed the study found that people with the highest rate of lung cancer ate the most sugar. However, upon reexamination it turned out that cigarette smokers consumed more sugar than nonsmokers. The real cause of cancer was cigarette smoking.

This example illustrates a key scientific principle worth remembering: Correlation doesn't necessarily mean causation. A correlation is an apparent connection between two variables. For example, just because people living near a chemical plant have a high rate of leukemia (cancer of the blood) doesn't mean that the factory is causing it.

KEY CONCEPTS

Critical thinking requires us to question the conclusions drawn from facts to see if other interpretations might be possible.

Tolerate Uncertainty

The sixth rule of critical thinking is to tolerate uncertainty. Although this may seem contradictory at first, it really isn't. As noted earlier, science is a dynamic process. Theories come and go. Scientific knowledge is constantly being refined. What we know, however, is dwarfed by our ignorance. Although knowledge of poisons and hazardous materials is immense, in actuality very little is known about the toxic effects of chemicals on humans. How do they interact? Are impacts affected by diet?

Table 1-2

Eighteen Myths and Assumptions of Modern Society

1. People do not shape their future; it happens to them.
2. Individual actions don't count.
3. People care only about themselves and money; they can't be counted on to take action on the part of a good cause unless they'll gain.
4. Conservation is sacrifice.
5. For every problem there is only one solution; find it, correct the problem, and all will be well.
6. For every cause, there is one effect.
7. Technology can solve all problems.
8. Environmental protection is bad for the economy.
9. People are apart from nature.
10. The key to success is through the control of nature.
11. The natural world is here to serve our needs.
12. All growth is unqualifiably good.
13. We have no obligation to future generations.
14. Favorable economics justifies all actions; if it's economical it's all right.
15. The systems in place today were always here and will always remain.
16. Happiness stems from material possession.
17. Results can be measured by the amount of money spent on a problem.
18. Slowing the rate of environmental destruction and pollution solves the problems.

Source: Adapted from Donella Meadows (1991). *The Global Citizen.* Washington, D.C.: Island Press; and Daniel D. Chiras (1990). *Beyond the Fray: Reshaping America's Environmental Response.* Boulder, CO: Johnson Books.

Some current debates on environmental issues involve a fair amount of uncertainty. Hard-and-fast answers just aren't available yet. Sometimes we have to make decisions with limited data.

KEY CONCEPTS

Our knowledge of the world around us is evolving, and, thus, it is necessary to accept uncertainty as an inevitable fact of life and make decisions with the best information possible.

Individual Actions Count

- **Values.** Make a list of your environmental values. What do you think is right or wrong, environmentally speaking? (Example: It is wrong to waste resources.) Enter additional environmental values in a notebook over the next week.
- **Sources of Your Values.** After you have made a list of values, spend some time thinking about their origin. Where did your values come from? Did they come from an influential teacher, a parent, a minister, a friend, your readings, or some other source?
- **Applying Values.** Make a list of actions (e.g., cycling or walking to school rather than driving) that express your values. Make a list of those that don't. How can the latter be changed?

Examine the Big Picture

The seventh rule of critical thinking is always to examine the big picture. Don't get trapped in simplistic thinking. Dualistic thinking is one of the most common forms of simplistic thinking. Dualistic thinking is reasoning that is black-and-white, right-or-wrong oriented. Dualistic thinking sees two options for every issue and nothing in between.

This book presents several conceptual models that will help you understand the big picture. These learning tools organize a great deal of information into a simple format.

To understand the big picture, we must often look at whole systems, both human and natural. For example, it is important to understand how value systems influence the systems (infrastructure) that serve our needs, such as our economic system. The economy is largely based on a belief that the Earth is an unlimited supply of resources, all for human use. Economic expansion is perceived as a positive thing in a world of infinite resources. Agricultural systems in many countries are a reflection of the view that the key to success is through domination and control of nature. We level grasslands, plant acre after acre of the same crop, and then control the insects that arise to take advantage of this bonanza with an arsenal of potentially harmful pesticides (FIGURE 1-7).

Human systems obviously have profound effects on natural systems. One of the goals of this book is to help you become a better systems thinker. This is one of the most valuable skills architects of a sustainable future must possess.

FIGURE 1-7 **The control of nature.** A belief that humans are superior to other living things and that domination and control of natural systems are the key to success profoundly influences how we design human systems such as agriculture. Sustainability calls for systems to be ecologically intelligent.

KEY CONCEPTS

To become a critical thinker it is necessary to examine the big picture—relationships and entire systems.

To be ignorant of one's ignorance is the malady of the ignorant.

—*Bronson Alcott*

CRITICAL THINKING

Exercise Analysis

Critical thinking requires students to think carefully and objectively and to assimilate facts and concepts. At the heart of critical thinking, however, is something much more important. Your dictionary probably offered several definitions of the word critical. For example, it may have defined it as "finding fault," which is clearly not appropriate in this instance. The dictionary may have also defined critical as "being characterized by careful analysis."

Careful analysis requires that we examine issues very thoughtfully, that we study the facts that back up our opinions. It also means that we look for loopholes in people's logic. Critical thinking helps us separate beliefs that may have little factual basis from facts. Chapter 2 covers the subject in more detail.

One of the rules of critical thinking illustrated by this exercise is that to be a critical thinker you must understand terms you encounter. Be sure a speaker explains the terms he or she is using. Don't be timid about asking for an explanation. Also, be sure you understand new words you're reading. Don't launch into an argument until you really understand what the other person is saying. One thing you might try when you're talking with someone is a process called active listening—repeating what you think someone is trying to say. This technique is very simple and is one way you can be sure you truly understand another person's point before you propose your own ideas.

CRITICAL THINKING AND CONCEPT REVIEW

1. Companies throughout the world are recognizing that good environmental practices are good for business. Make a list of reasons why this may be true.

2. In what ways can science help us solve problems sustainably? What are the limitations of science in problem solving?

3. Describe the scientific method. What are the steps, and why is each one necessary?

4. Describe the components of a good scientific experiment.

5. A graduate student interested in studying the effects of air pollutants on lichens (a fungus that grows on rocks) designed an experiment as follows: He proposed to bring rocks with lichens into the laboratory, where he would expose them to urban air pollution. (His university was located downtown in a major metropolitan area.) He then proposed to circulate outside air into the laboratory and measure the effects. What is wrong with this experimental design? How would you design a better experiment?

6. What is the difference between a theory and a hypothesis?

7. Numerous studies on the natural toxin known as aflatoxin, which is found in peanuts, show that it is mutagenic and carcinogenic; that is, it causes mutations in the genetic material and cancer in some organisms. Drawing on these studies, scientists have labeled aflatoxin as a potentially dangerous substance. What kind of reasoning is this, inductive or deductive?

8. "From the facts of the case," said Sherlock Holmes, famous investigator of fictional repute, "I can deduce that the butler did it." Is Holmes using the word *deduce* correctly in this context?

9. In what ways do our values affect science? How can science affect our values?

10. Make a list of the critical thinking skills outlined in this chapter. Describe and give an example of each one. Over the next week, make a list of all the times you use these rules.

11. Study the list of myths and assumptions in Table 1-2. Which ones do you agree with? Which ones do you disagree with?

KEY TERMS

control group
Copernican revolution
critical thinking
deductive reasoning
dominant social paradigm
ecology
ecosystem
epidemiological
epidemiology
experimental group
hypotheses
independent variable
inductive reasoning
paradigm
paradigm shift
sustainable development
theories
U.S. Environmental Protection Agency (EPA)

REFERENCES AND FURTHER READING

The References and Further Reading section at the end of this book contains a list of sources for the information discussed in this chapter and recommendations for further reading.

Connect to this book's website:
http://environment.jbpub.com/
The site features eLearning, an online review area that provides quizzes, chapter outlines, and other tools to help you study for your class. You can also follow useful links for in-depth information, research the differing views in the Point/Counterpoints, or keep up on the latest environmental news.

Rice terraces in Bali.

CHAPTER 2 | Environmental Protection and Sustainable Development

There are no passengers on spaceship earth. We are all crew.
—*Marshall McLuhan*

Human civilization has experienced two major revolutions—changes in the ways people lived on the Earth. The first was the **Agricultural Revolution**, a long, progressive shift from hunting and gathering to subsistence-level farming to a system of highly mechanized farming. The Agricultural Revolution started around 10,000 B.C. The second major revolution was the **Industrial Revolution**, the shift from manufacturing goods by hand to machine manufacturing powered by coal—and later by other fossil fuels such as oil and natural gas. This period started in the late 1700s in England and in the 1800s in the United States.

The successes of the Agricultural and Industrial Revolutions are many: broad prosperity, material goods, and a better way of life, to name a few. The gains made during these periods have not been without environmental costs, however, and include pollution, habitat destruction, and species extinction. Today, people are recognizing that changes are needed to create a sustainable society. We may, in fact, be entering a new cultural revolution, a **Sustainable Revolution**: a long period in which human systems that evolved during the previous 12,000 years are redesigned to make them much more benign and sustainable. These strategies are vital if we are going to address the many environmental problems now facing human society. This chapter describes the history of environmental protection and shows why environmental protection is crucial to our future. It shows the vital link between environmental and sustainable development. It also outlines the rationale for systems reform, gives an overview of the key systems that support our lives, and ends with a discussion of ways to revamp systems for sustainability.

2.1 Environmental Protection and Sustainable Development

In this book, we'll examine dozens of important environmental issues and look at many ways we can protect the environment. We'll study numerous laws and pollution control devices, for example, and will look at innovative ways to protect forests and farmlands. We'll examine ways to help protect species from becoming extinct and ways to lower the impact of our fossil fuel-based economy. But unlike most other textbooks, this one takes a broader look. Rather than view environmental protection as an isolated set of actions, I view it as an integral act, a part of every decision we make about our society and how it develops. In short, we will look at how we can meet human needs while protecting the environment. We call this process **sustainable development**.

> **KEY CONCEPTS**
>
> Increasingly, environmental protection is being incorporated more broadly into all human actions and into the process of development. Meeting our needs while protecting the environment is called *sustainable development*.

To understand the need for this broader approach and the importance of learning about sustainable development in a course on the environment, let's take a look at the evolution of environmental protection efforts in the United States and other developed nations.

The Evolution of Environmental Protection

FIGURE 2-1 shows the stages through which environmental protection has progressed in the United States and other industrial nations. As illustrated, environmental protection in the United States began in a rather piecemeal fashion, taking root as individual cities and towns began to address local environmental problems. In the early stages, starting

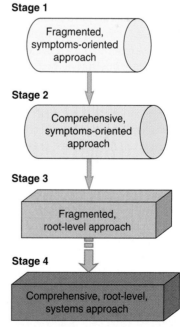

FIGURE 2-1 The evolution of environmental protection. Environmental protection has evolved considerably over the past 4 decades. Now efforts are underway to incorporate it into sustainable development—that is, make it part of the design of new technologies, new homes, new vehicles, and so on. This effort could dramatically reshape basic human systems.

in the 1950s, most of the problems revolved around lakes and rivers that were polluted by human sewage or industrial waste. The response to these problems was largely technological. Most of the solutions involved the installation of **pollution control devices**, equipment that removed pollutants from the waste stream. However, local officials soon discovered that pollution control devices were almost always inadequate. Why? Even though a single factory might be required to clean up, pollution from sewage treatment plants or factories upstream rendered its efforts meaningless. It soon became clear that neighboring cities and towns had to take action as well, but they weren't always willing to pass the necessary regulations. Consequently, environmental protection efforts soon graduated to the state level. States began to enact laws requiring statewide compliance.

The first stage of environmental protection in the United States involved cities, towns, and later state governments. Important as these efforts were, they focused on a limited number of problems such as water and air pollution. Efforts were uncoordinated. Each municipality and state had its own regulations and ways of solving problems. The effort was also *symptoms oriented*; that is, solutions were primarily designed to treat symptoms of the problems (not underlying root causes). Pollution control devices, commonly called **end-of-pipe controls**, were placed on factories and power plants to remove air pollutants from smokestacks. But they, in turn, produced vast amounts of waste that had to be disposed of. In time, the states found that their efforts were not enough. They failed to halt pollution entering from neighboring states either in the air or in rivers that crossed state boundaries. The federal government was then drawn into the fray, and Congress enacted numerous laws starting in the 1960s to ensure uniform environmental protection in all states. The federal government created programs to monitor pollution emissions and levels of pollutants in the air and water and to to encourage compliance among businesses (including penalties). In the 1960s and 1970s, a fairly comprehensive nationwide program of environmental protection emerged. Although it was not complete, it covered many problems.

In recent years, environmental protection has escalated to an international level, for the same reason that it moved from local to state to national governments: Pollution does not honor arbitrary human boundaries. International protection efforts became essential as people began to realize that what happens to the environment in one part of the world may have serious impacts elsewhere. In the 1980s and 1990s, numerous multilateral and international agreements addressed global problems.

The involvement of the federal government, followed by international accords involving many nations, constitutes the second stage of environmental protection. Clearly more comprehensive than earlier efforts, it was during this phase that the United States established uniform national standards and federal inducements to encourage states to undertake environmental protection. Over time, additional environmental issues were addressed.

Stage two represents a major improvement in environmental protection. It was much more comprehensive. More problems were covered, and more players entered the scene. Although these changes represent a major step forward in the evolution of environmental protection, the main focus remained on treating pollution after the fact, with end-of-pipe controls. In other words, despite the fact that the scope of environmental protection had broadened considerably, the second stage remained largely a symptoms-level approach.

KEY CONCEPTS

Environmental protection has evolved from piecemeal local efforts to a much more comprehensive global strategy involving high levels of cooperation among states and nations covering a wide assortment of environmental problems.

The Next Generation of Environmental Protection Efforts

The second phase of environmental protection resulted in many gains. In recent years, however, many people have come to realize that many environmental protection measures, although effective in the short term, are inadequate over the long haul. End-of-pipe controls have slowed the rate of environmental destruction, but they have not stopped it. We're even finding that reductions in pollution from automobiles, power plants, and factories are being offset by increases in population size and economic activity.

Recognizing the shortcomings of the environmental response, many nations have entered into the third stage of environmental protection. They are experimenting with solutions that address what many observers believe are the root causes of the environmental crisis: population growth, inefficiency in resource use, heavy dependence on fossil fuels, and our "throw-away" mentality (Chapter 3). As was mentioned in the first section of Chapter 1, in the United States, the Environmental Protection Agency launched bold experiments in environmental protection. The EPA's 33/50 program, for instance, enlisted major corporations in hazardous waste reduction. Participants slashed their output of hazardous wastes by 50% or more through fairly simple and cost-effective measures, which is detailed in Chapter 23. The EPA's Green Lights program promotes corporate energy efficiency in lighting, and its Energy Star program promotes energy efficiency in computers, printers, monitors, and televisions. Similar programs have been instituted to promote more energy-efficient housing and appliances (FIGURE 2-2). The hazardous waste programs address the problem of our throw-away mentality. Other programs address some of our inefficient technologies.

Other nations have launched similar programs, some much bolder than those under way in the United States. Clearly, then, a handful of nations are on the threshold of the third stage of environmental protection shown in Figure 2-1. Nevertheless, root-level efforts are in their infancy. We have a long way to go to be meaningfully engaged in the third stage of environmental protection.

FIGURE 2-2 EPA's Energy Star program. The U.S. EPA enrolled many manufacturers of computers, monitors, printers, televisions, and appliances in a voluntary program to design and manufacture energy-efficient products. The most efficient products come with an Energy Star label (lower left on control panel).

KEY CONCEPTS

Although environmental policy and environmental protection efforts have evolved dramatically in the past three decades, most solutions dealt with symptoms. Efforts are now under way to address the root causes of the problems.

Creating Sustainable Solutions

Important as it is, entry into the third state of environmental protection probably won't ensure an enduring human future. To do so, we must move into a fourth stage, and quickly (Figure 2-1). The fourth stage would be comprehensive in nature, involving actions at all levels: local, state, national, and international. Moreover, it would entail activities by all players—not just government, but businesses and citizens as well. Although it would address root causes, as in stage three, its emphasis would be on restructuring basic human systems—among them energy, industry, agriculture, transportation, housing, and waste management. This is called sustainable development. Why is this necessary?

As you will see in Chapters 2 and 3, the environmental problems facing the world are the result of population pressure and unsustainable systems. We will examine the population dilemma in Chapter 8. In Chapters 2 and 3, you will see that environmental problems are a symptom of "flaws" in the systems we have designed to meet basic human needs. Systems of agriculture, industry, housing, transportation, energy production, waste management, and water supply were designed

when numbers were small, resource supplies were vast, and our understanding of environmental issues was still growing. But things have changed. The human population has exceeded six billion, and the systems that have served us well are spawning problems of global impact. In this book, I discuss ways that we can revamp these systems to reduce or eliminate environmental problems—that is, to make them sustainable.

Although ample evidence suggests that we are not presently on a sustainable course and that we are undermining our future, we are not doomed. We have the brainpower and knowhow to invent machines and systems that are sustainable—that produce what we need without destroying the planet. The challenge will be immense, however. The transition to a sustainable society will require fundamental redesign, deep changes that may take a century or more to put into effect. We're just on the threshold of that change now.

The fourth stage involves integrating environmental protection into every decision we make. It insists that all policy is environmental. Environmental considerations become a key part of the development path we choose—hence the name *sustainable development*. End-of-pipe controls are still important; they're a kind of first aid. Ultimately, pollution control devices must be replaced by strategies of meeting needs that produce little or no pollution at all. Gas-powered automobiles, for example, may be replaced by mass transit and electric cars powered by photovoltaic panels mounted on rooftops. This book focuses on all solutions—traditional and new—showing the role each has to play in building an enduring human presence.

Despite substantial progress in sustainable development, in recent years the United States has begun to back pedal on important environmental issues, refusing to support international efforts to curb global climate change, loosening air pollution standards, and weakening a host of other hard-won environmental protections.

KEY CONCEPTS

All environmental problems result from the fact that human systems such as energy production and agriculture are unsustainable. They are inefficient in their use of resources, and most of them rely heavily on finite supplies of fossil fuels whose combustion creates many problems.

2.2 Meeting Human Needs While Protecting the Environment

In this section, we take a closer look at sustainable development and the principles that underlie this exciting idea. Our focus will be on ways that environmental protection can be integrated into human activity or development.

What Is Sustainable Development?

Sustainable development is a term rich with meaning. Although dozens of definitions have been proposed over the past few years, one of the most widely used was published in *Our Common Future*, a book written by the World Commission on Environment and Development (WCED). The WCED was established by the United Nations in the 1980s to examine the relationship between the environment and development and to suggest ways to make the two compatible. The WCED defined *sustainable development* as "development that meets the needs of the present without compromising the ability of future generations to meet their needs." Because future generations depend on the environment, meeting our needs while ensuring future generations the ability to meet their needs means we must safeguard the environment.

> The WCED defined *sustainable development* as "development that meets the needs of the present without compromising the ability of future generations to meet their needs."

To understand sustainable development, we begin by dissecting the term into its parts, beginning with the word *sustainable*. According to *Webster's New World Dictionary*, sustainable means "able to be sustained." To sustain means "to keep in existence, to maintain, and endure." According to Webster's, *development* is "a step or a stage in advancement or improvement." *In short, sustainable development is a way of improving or advancing our culture in a way that can be maintained over the long haul.*

Then Vice President Al Gore noted at a national conference in 1993 that the term *sustainable development* implies that "there's something called *unsustainable development*." He and many others would agree that our current development path fits that bill. You will see evidence of this throughout this book.

> Sustainable development is a way of improving or advancing our culture in a way that can be maintained over the long haul.

Sustainable development, although something of a newcomer on the political scene, is not a new idea by any stretch of the imagination. In fact, 90 years ago, U.S. President Theodore Roosevelt alluded to the concept in his annual message to Congress, a speech that has since become known as the State of the Union address. Roosevelt noted, "To waste our natural resources, to skin and exhaust the land instead of using it (so) as to increase its usefulness, will [undermine the prospects] of our children." He went on to talk about "the very prosperity which we ought by right to hand down to [future generations] amplified and developed." In other words, we ought to meet our needs without foreclosing on future generations. Roosevelt was not the first to hold such a view, either. Native American cultures and indigenous peoples the world over have espoused a similar view and lived accordingly for thousands of years before Roosevelt's time.

Sustainable development, then, refers to improvements or advancements in human well-being that are enduring. Advocates of sustainable development concern themselves with strategies designed to meet all human needs—not just the need for a clean, healthy environment, but also needs for respectable work, good pay, recreation, peace, freedom from harm, and a host of other factors. Individuals clearly differ in the goals they set for human development, but several goals seem to be almost universal. These include a long and healthy life, good education, and a decent standard of living. The list also includes political freedom and a guarantee of human rights. Freedom from violence and meeting survival needs for food, shelter, water, and clothing are also basic goals of human development. The challenge today is to meet these and other needs while protecting the environment.

Sustainable development strategies seek ways to forge lasting relationships between humans and the environment, ones that protect and restore ecosystems rather than destroy and deplete them. Environmental protection is therefore vital to the success of any state, local, or national development plan. Sustainable development also seeks ways to create enduring relationships among people (for example, cooperative rather than adversarial) and a new economic system that is kind to the Earth and to all individuals, all genders, and all races. These measures are as important as sound environmental policies.

FIGURE 2-3 is a concept map that encapsulates much of the thinking on the social, economic, and environmental conditions essential to creating a humane, sustainable society. Take a few moments to study it. We won't deal with the social aspects of sustainable development in this book, but concentrate primarily on economics and the environment. It is important to remember, though, that creating a sustainable future requires much more than efforts to protect the environment.

KEY CONCEPTS

Sustainable development is a means of meeting present needs in ways that do not impair future generations—and other species—from meeting their needs. Because the environment is essential to satisfying the needs of present and future generations, environmental protection is a key to its success.

Satisfying the Triple Bottom Line

For many years, social, economic, and environmental policies have been devised in isolation. Economic decisions were made with little regard for the environmental impact. Environmental policies were made with little regard for economics. Sustainable development requires an integration of social, economic, and environmental goals. As I like to tell my students, all policy is environmental; that is, to be successful, all social and economic policy must be crafted in a way that ensures environmental protection. By the same token, environmental policy must make sense from social and economic perspectives. FIGURE 2-4 illustrates this important concept. The three circles on the left side of the figure show three realms—social, economic, and environmental. In traditional government policymaking, the three are often considered separately. Solutions for each one are devised without regard to the others. Environmental policy, for instance, may

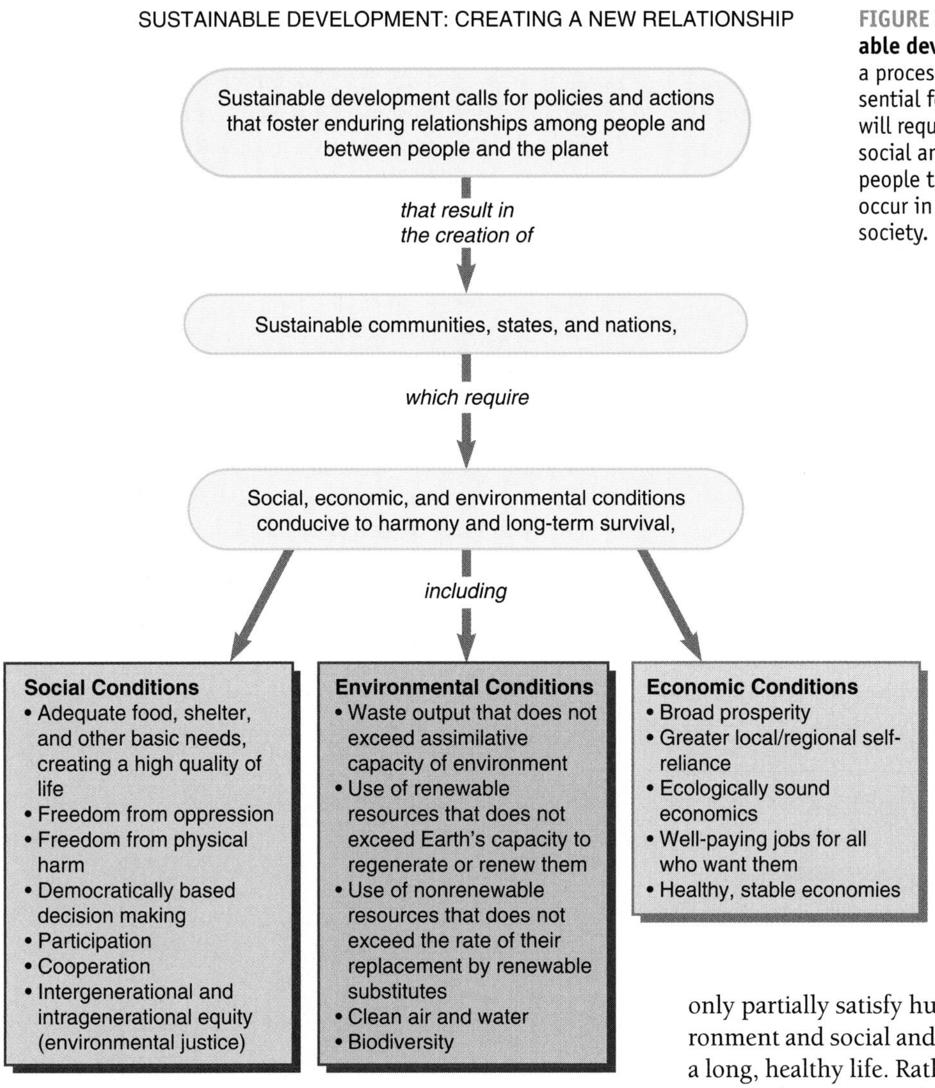
SUSTAINABLE DEVELOPMENT: CREATING A NEW RELATIONSHIP
Sustainable development calls for policies and actions that foster enduring relationships among people and between people and the planet
that result in the creation of
Sustainable communities, states, and nations,
which require
Social, economic, and environmental conditions conducive to harmony and long-term survival,
including
Social Conditions
• Adequate food, shelter, and other basic needs, creating a high quality of life
• Freedom from oppression
• Freedom from physical harm
• Democratically based decision making
• Participation
• Cooperation
• Intergenerational and intragenerational equity (environmental justice)
Environmental Conditions
• Waste output that does not exceed assimilative capacity of environment
• Use of renewable resources that does not exceed Earth's capacity to regenerate or renew them
• Use of nonrenewable resources that does not exceed the rate of their replacement by renewable substitutes
• Clean air and water
• Biodiversity
Economic Conditions
• Broad prosperity
• Greater local/regional self-reliance
• Ecologically sound economics
• Well-paying jobs for all who want them
• Healthy, stable economies

FIGURE 2-3 Conditions created by sustainable development. Sustainable development is a process that seeks to promote conditions essential for an enduring human presence. This will require changes in environmental as well as social and economic aspects of our lives. Some people think that profound changes will have to occur in all three realms to create a sustainable society.

call for changes that adversely affect the economy. In contrast, sustainable development calls for solutions that make sense from all three perspectives simultaneously, as shown on the right side of Figure 2-4. The area of intersection in this drawing represents a new way of thinking and making policy. However, sustainable solutions are not compromises that only partially satisfy human needs for a safe, healthy environment and social and economic conditions conducive to a long, healthy life. Rather, sustainable solutions strike out in new directions, taking advantage of opportunities to achieve the social, economic, and environmental conditions vital to a sustainable future. By satisfying the triple bottom line (that is, economic, environmental, and social goals), sustainable development could conceivably reshape human civilization. Numerous examples of sustainable solutions will be presented in this book.

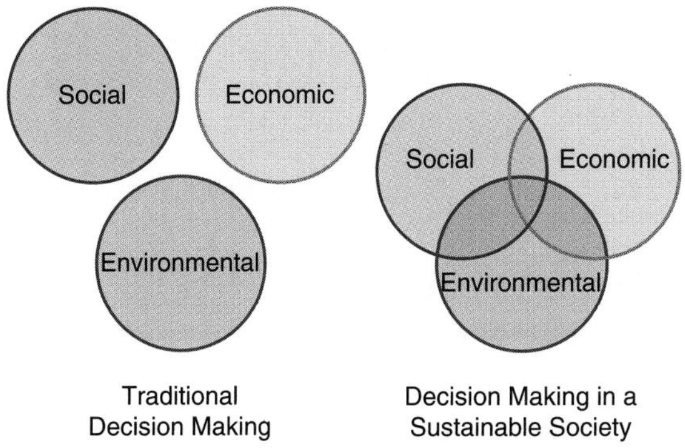
Social
Economic
Environmental
Social
Economic
Environmental
Traditional Decision Making
Decision Making in a Sustainable Society

FIGURE 2-4 Sustainable solutions. For many years, social, economic, and environmental issues were addressed independently. Solutions made in one sector often had adverse impacts in the others. Sustainable development calls for solutions that make sense from all three perspectives simultaneously.

Sustainable development is having a profound effect on people throughout the world. It is a common theme of conferences and workshops. It has also been adopted by numerous political entities, agencies, and organizations as a central focus. Included in this rather impressive list are the United Nations General Assembly, the U.S. Agency for International Development, the United Nations Development Program, and the World Bank. Environmentalists—and a growing number of business leaders and economists—have embraced the notion and have begun to explore and implement policies and actions needed to create a much more enduring human way of life. How do we create a sustainable society?

KEY CONCEPTS

Sustainable development requires strategies that satisfy social, economic, and environmental goals simultaneously.

Principles of Sustainable Development

Sustainable development is based on 10 key principles, which derive from three areas: ecology, social thought and ethics, and economics. In this section, we examine each one very briefly and reflect on ways they will help ensure a clean, healthy environment, a sustainable society, and ultimately a better future.

Ecological Principles of Sustainable Development We begin our exploration of the principles of sustainable development with those derived from the realm of ecology. Four ecological principles are central to the concept of sustainable development: dependence, biophysical limits, living within the carrying capacity of the environment, and interdependence.

Principle 1: Dependence Many people think of the environment as something only birdwatchers and hikers care about, something that is of little worldly interest to the average person. Nothing could be further from the truth. Humans depend on the Earth and its ecosystems for a wide array of goods and services. These goods and services are vital to our personal well-being and also contribute significantly to the global econ-omy. In the words of the economist Robert Goodland, the environment is the source of all material inputs, feeding the economic system, and it is the sink for all its wastes. In fact, so important is the environment that many economists refer to the Earth's natural systems (ecosystems) as the **biological infrastructure**—or **infra-infrastructure**—of modern society because the Earth and the natural systems it contains make possible all of the **physical infrastructure** of our society—the buildings, dams, ports, highways, and so on (**FIGURE 2-5**). The Earth is the source of all wealth. It provides us with these goods and services free of charge. However, even though these services are free, they are infinitely costly to replace.

If one recognizes human dependence on the Earth and on ecosystems, it becomes evident that environmental protection and environmentally sustainable development are forms of self-protection—an insurance policy for long-term human prosperity. In short, planet care is the ultimate form of self-care.

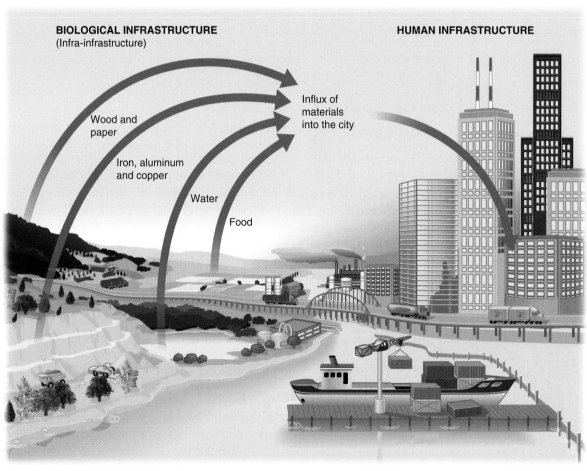

FIGURE 2-5 Infrastructure and infra-infrastructure. The Earth and the environment provide fuel for the economy and provide the materials needed to create our infrastructure—our highways, buildings, water systems, and so on. The Earth and the environment are the infra-infrastructure of our society. Damaging and depleting the environment detract from our possibilities in the long run. The main goal of sustainable development is to create a society and an infrastructure, a set of systems, that protects and even enhances the infra-infrastructure.

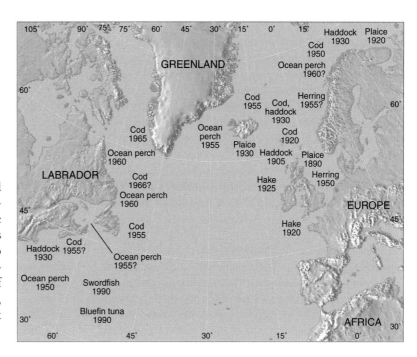

FIGURE 2-6 Ocean in peril. This map of North Atlantic fisheries shows the dates when major fisheries were depleted and thus no longer commercially fishable.

Principle 2: Biophysical Limits The Earth and its ecosystems provide a wide assortment of resources. Some resources such as soil and forests are **renewable**—capable of being regenerated. Others such as iron ore are **nonrenewable**—not able to be regenerated. Both types of resources have limits. For example, there are limits to the amount of oil, a nonrenewable resource, that we can extract, and there are limits to how many fish can be caught without depleting their population.

A growing body of evidence shows that humankind has transcended certain biological and physical limits and stands perilously close to others. An example of our transgression of limits: Overfishing in the Atlantic Ocean has already eliminated more than 25 commercially important fisheries (**FIGURE 2-6**). As evidence that we stand on the brink of transcending other limits:

> Overfishing has already eliminated more than 25 commercial fisheries in the Northern Atlantic since the early 1900s.

According to a 2005 report from the United Nation's Food and Agriculture Organization, 52% of the world's marine fish stocks are fully exploited—that is, they are at or very close to their maximum sustainable production limit. Sixteen percent are overexploited, and 7% are depleted. One percent is recovering from depletion.

In addition to these examples of biological limits, a growing body of scientific evidence indicates that chemical limits are being exceeded by pollution generated by human society. In both the industrial and nonindustrial nations of the world, polluted rivers no longer support fish or support a fraction of their original populations. The buildup of carbon dioxide (a pollutant produced by the combustion of oil, coal, and natural gas) in the atmosphere and the changes in the climate of planet Earth are other examples. The partial destruction of the ozone layer by chlorofluorocarbons is another example of limits that modern society has transcended. Much of the writing on sustainable development recognizes the presence as well as the proximity of limits.

KEY CONCEPTS

The renewable and nonrenewable resources that support our lives have very real limits. Many signs that we have exceeded limits of the planet's resources are evident.

Principle 3: Living Within Limits Although there are limits to resources and the waste assimilation capacity of the Earth, we can live and prosper. To do so, we must address rapid population growth and design systems that operate within limits. Both efforts must go hand in hand. As you will see in Part II, one of the keys to sustainable living is learning to live within the carrying capacity of the environment. **Carrying capacity** can be defined as the number of organisms an environment can support indefinitely. Carrying capacity is determined by two elements: resource supplies and the capacity of the environment to absorb, assimilate, and/or detoxify wastes. Proponents of sustainable development call for policies and practices that temper human activities to honor the very real limits in the Earth's carrying capacity—limits to the amount of arable land, fish in the sea, and oil in the Earth's crust, for example. Systems must be designed with limits in mind. You'll learn more about this daunting task in upcoming chapters.

KEY CONCEPTS

Living sustainably means finding ways of prospering within limits.

Principle 4: Interdependence The fourth ecological principle is that of interdependence. **Humans and the Earth exist in a precarious interdependence.** Clearly, just as we humans are dependent on the environment, the fate of the global environment is dependent on us. Our influence has grown so great—and continues to grow with incredible speed—that we humans have become custodians of an entire planet. What befalls the Earth befalls humanity.

KEY CONCEPTS

The future of the biosphere, upon which we humans depend, is in our hands.

Social/Ethical Principles of Sustainable Development Living sustainably on the Earth is a biological challenge faced by all species. As noted in the previous section, living sustainably in the environment requires species to thrive within the limits of the resource supplies and waste assimilation capacity of the Earth—the source and sink functions. Although the challenge of living sustainably on the planet is basically the same for all species, for humans, the task is complicated by a variety of factors, among them ethics, economics, and politics. This section describes three key social or ethical principles: intergenerational equity, intragenerational equity, and ecological justice. Keep them in mind as you consider various solutions to environmental problems.

Principle 5: Intergenerational Equity One of the most important principles of sustainable development is **intergenerational equity**—fairness to future generations. The doctrine of intergenerational equity calls on us to live in a way that does not rob future generations of the opportunity to prosper. By managing resources well, for example, we ensure that there are ample supplies for future generations. The notion of intergenerational equity could have profound influences on the technologies we develop and the ways human systems are designed and operated. As a simple example: Honoring intergenerational equity might dictate houses powered by sunlight and wind and made from many recycled materials. Why? It helps to ensure future generations a fair share of the Earth's resources.

> **KEY CONCEPTS**
> Intergenerational equity calls on us to live in ways that honor the needs of future generations.

Principle 6: Intragenerational Equity In the international arena, debates over global environmental protection have drawn attention to another important ethical principle, the notion of *intragenerational equity*. The doctrine of intragenerational equity calls on us to act in ways that satisfy our needs while safeguarding the welfare of all who are alive today. What we do is tempered by a concern for potential impacts on other people—our neighbors, our fellow citizens, and global citizens as well.

Intragenerational equity is based on a realization that all members of the world community share the same air and water—and that abuse of these resources is often felt far from the site of impact. Soil erosion on a Kansas farm, for instance, may lead to siltation in a downstream reservoir in Arkansas and flooding along the banks of the Mississippi River. Impacts of human actions can even extend globally. Ozone depletion caused by the release of CFCs (chlorofluorocarbons) from refrigerators in Canada, for instance, may lead to skin cancer in Andean peasants who have never even seen a refrigerator.

The lines of impact are reciprocal. That is to say, the developed nations can be profoundly distressed by resource abuse in the developing nations. Deforestation of tropical rain forests in Africa to accommodate population growth, for example, could have a profound impact on the climate of European nations by altering the movement of moist air masses that affect precipitation patterns hundreds of miles northward.

> **KEY CONCEPTS**
> Intragenerational equity calls on us to act in ways that honor the rights and needs of all people alive today.

Principle 7: Ecological Justice A small but growing number of proponents of sustainable development support the idea of **ecological justice**, the notion that the Earth is the rightful property of all species—not just humans. According to this view, humans have an obligation not just to other people, but to all living things, present and future. We examine this controversial idea again in Chapter 11.

> **KEY CONCEPTS**
> The notion of ecological justice says that all species have a right to a clean environment and adequate resources.

Political Principles of Sustainable Development Numerous principles of a political nature underpin the current thinking on sustainable development; three of the most important are

Individual Actions Count

Individual Actions Count!—General Guidelines

- **Be frugal.** Buy only what you need. Be a conscientious consumer. Don't buy unnecessary things.
- **Be efficient.** Support legislation and nonprofit organizations that promote energy efficiency. Use all resources in your day-to-day life efficiently (for example, ride the bus or walk to school or work).
- **Be a recycler.** Support legislation and nonprofit organizations that promote recycling. Recycle all wastes that you can, and buy products made from recycled materials, too.
- **Support renewable resource use.** Support legislation and nonprofit organizations that promote renewable energy; wherever possible, use renewable energy and other renewable resources yourself.
- **Help restore the environment.** Support nonprofit groups, and take an active part in restoring damaged ecosystems yourself.
- **Help control population growth.** Support organizations that promote population stabilization and growth management strategies.

the need for widespread participation, the need for global co-operation, and the need to address the root causes of problems.

Principle 8: Participation   A sustainable future will require much more than new laws and regulations. It will also require the efforts of responsible businesses. However, the goals cannot be met without broad citizen support and actions. Citizens are an integral part of the economic system as consumers of goods and services. By becoming conscientious consumers—buying green (environmentally responsible) products, but also buying less—individuals can help stimulate the transition to sustainability. Citizens can take hundreds of additional actions to promote a sustainable future. Driving fuel-efficient vehicles, carpooling, riding buses, bicycling, or walking—all make significant contributions. Recycling is very important, too (see the Individual Actions Count! table for more ideas). In sum, the transition to a sustainable society requires the participation of all of us. As the opening quote of this chapter reminds us, "There are no passengers on spaceship earth. We are all crew."

Building a sustainable society will require participation by governments, businesses, and individuals.

Principle 9: Cooperation   Ensuring an enduring human presence will require a level of cooperation unwitnessed in human history. Regional agreements among nations, such as the multilateral treaties designed to clean up the Baltic Sea, are examples of the kind of cooperation needed to create a sustainable global human presence. International accords signed in the past decade, such as the global warming agreement and the ozone treaty, are additional examples of cooperation (Chapter 20). Improvements in these and other treaties, as well as new agreements, are needed for further progress. Cooperation is also needed on a smaller scale—among nonprofit organizations, governments, citizens, businesses, and even opposing political parties.

Environmental protection and sustainable development will require cooperation of all participants.

Principle 10: Addressing the Root Causes   One emerging realization in business and government is that many environmental solutions enacted over the past three decades have addressed the symptoms of environmental problems while overlooking fundamental root causes. Treating symptoms may help in the short term, but it won't solve the problems. A patient who is under enormous stress might be relieved by medicine, but if the patient doesn't learn to reduce stress, he or she may never get better. Treating symptoms is a trait or tendency found in virtually all areas of human endeavor from medicine to crime fighting to educational reform. In the arena of environmental protection, the examples of our short-sighted approach are many: smokestack scrubbers, catalytic converters on automobiles, hazardous waste incinerators, sewage treatment plants, and so on. Over the years, it has become clear to many observers that these and other measures are in-

adequate for several reasons. Most importantly, they address such problems as pollution and hazardous waste too late—after they are produced, at the end of the pipe. A pollutant attacked at the point of origin can be eliminated, but after it is produced, it is too late: It becomes a social, economic, and environmental liability. Unfortunately, most nations have based their environmental protection strategy on policies that attack pollutants after they've been generated.

This conventional approach to environmental protection—treating the symptoms—has several additional shortcomings. One is that it results in short-term gains that may be offset by other forces—among them continuing population growth, increasing economic activity, and increasing per capita consumption. This is especially evident in the United States and other Western nations. Take for example, the catalytic converter, a device used on automobiles to reduce the emissions of certain pollutants (notably, hydrocarbons and carbon monoxide) spewing from tail pipes. Catalytic converters work well and are responsible for a dramatic decline in the pollution our cars produce. They're even responsible for a general cleansing of the air in many urban centers. However, continuing expansion of the U.S. population and the ever-increasing number of vehicle miles traveled per year (increasing by more than 50 billion miles per year) could overwhelm the gains resulting from the use of catalytic converters (FIGURE 2-7).

FIGURE 2-7 **Rush hour?** Although substantial gains have been made in reducing pollution from individual automobiles, the growth in the automobile fleet and the dramatic annual increase in the number of miles we travel threaten to overwhelm progress resulting from increased vehicle efficiency and pollution control devices.

Another problem is that the symptoms approach has resulted in a shifting of pollutants from one medium to another. **Smokestack scrubbers**, a type of pollution control device that captures sulfur oxide gases from coal-fired power plants, convert air pollutants into a toxic sludge. This waste is typically dumped into landfills, where toxic substances contained in it may leak into groundwater.

Other solutions have been much too narrow—that is, they address only a small portion of the problem. For example, although catalytic converters reduce hydrocarbon and carbon monoxide emissions, they do nothing to reduce emissions of carbon dioxide—a pollutant that may be causing the Earth's atmosphere to heat up (Chapter 20).

Fortunately, a growing number of policymakers are beginning to realize that lasting solutions require policies and practices that confront the root causes of the problems we face. Even prominent business leaders recognize the importance of root-level solutions such as energy efficiency as means to comply with or even exceed environmental performance standards, and many recognize that such an approach is not detrimental to business but is highly profitable. For a quick review of the principles discussed in this section, see Table 2-1.

Table 2-1

Ecological Principles

Dependence—Humans are dependent on a clean, healthy environment for many goods and services vital to our personal and economic well-being.

Biophysical Limits—Very real biophysical limits exist in the planet's ability to supply resources and absorb wastes from human civilization.

Living within the Carrying Capacity of the Environment—Living sustainably on the Earth will require steps to live within the Earth's biophysical limits.

Interdependence—The fate of the environment is in our hands. What we do or don't do will profoundly affect the natural environment in ways that could have serious consequences for virtually all living things, including ourselves.

Social/Ethical Principles

Intergenerational Equity—Present generations have an obligation to meet their needs in ways that do not foreclose on future generations.

Intergenerational Equity—Present generations also have an obligation to act in ways that do not prevent or impair others who are alive today from meeting their needs.

Ecological Justice—Human actions should not endanger other species, which also have an inherent right to the resources they need to survive.

Political Principles

Participation—Building a sustainable future will require unprecedented participation from all sectors of society.

Cooperation—Creating a sustainable society will require cooperation among many different participants.

Addressing the Root Causes—Successful solutions will require efforts that address the root causes of the problems.

To create a sustainable society, we must focus on strategies that address the root causes of environmental problems.

Growth and Development: Understanding the Differences

One of the key issues that must be addressed to protect the environment in the coming years is growth—economic growth and population growth. Can we reach a sustainable state if population and the economy continue to grow? There are two schools of thought with widely differing views on the issue of economic growth: progrowth and antigrowth.

We will address economic growth here. (Population growth is discussed in Chapters 8 and 9.) The antigrowth school questions policies that promote continued, unlimited economic growth. Their "growth principle" might read something like this: Continued economic growth within a finite system is unsustainable. To create a sustainable human presence, economies must eventually reach a stable state. Growth—expansion of the economy—must come to an end. Rather than continued growth, we should focus on ways to improve the quality of our lives to reduce continued resource extraction. They refer to this as *development*.

In contrast, the proponents of growth believe that continued economic growth is absolutely essential to building a sustainable future. Continued economic growth will, they say, promote an improvement in the standard of living needed to eliminate poverty and slow the growth of the human population, especially in the developing nations. How else can we improve the economic well-being of the world's people if we cannot continue to grow economically?

Before we explore these diametrically opposed views, a few definitions are in order. First, **economic growth** is defined as an expansion of human economic activity. One barometer of economic growth is the consumption of resources such as energy. Growth strategies typically result in an expansion of infrastructure—additional buildings, power plants, roads, water projects, and so on. This expansion, in turn, requires more mineral extraction and energy consumption. It results in the additional harvest of natural resources (forests, for example) and additional pollution, all of which can contribute to environmental degradation.

Development, as used in this book, implies a qualitative improvement. It is measured by a betterment in both the quality of our lives and the condition of the environment. Development involves measures that ensure adequate housing and work space. It is aimed at providing more efficient transportation, sufficient energy, and education. It often includes steps to improve cultural opportunities, health care, and freedom from physical danger. Sustainable development seeks to achieve these goals while safeguarding the environment. Development is consequently not a simple numerical definition, but a complex quality-of-life issue.

As noted earlier, advocates of continued economic growth view it as an essential means of raising the standard of living of the world's vast and growing population of the poor. They see continued economic growth as a desirable path for the wealthy nations of the world, as well as for the less fortunate developing nations. The rationale is that economic growth in countries such as the United States will benefit all the world's people. Economic growth in the developing nations would benefit many people within those nations.

Antigrowth or, more appropriately, development advocates recognize the importance of reducing poverty but argue that continued economic growth in nations such as the United States will cause further unsustainable deterioration of the environment and won't help the people of the developing nations very much, if at all. Economic expansion in the developing nations is essential, as long as the benefits are truly distributed among the population. Chapter 26 outlines strategies for achieving sustainable economic development in the developing nations of the world.

The belief in the benefits of economic growth enjoys such widespread acceptance that it is difficult for many to consider alternative economic strategies. This book presents some of these alternatives to stimulate your thinking. Nevertheless, the controversy over growth is bound to endure for decades. In the meantime, a compromise position seems to be emerging: Most people agree that economic growth, which is inevitable, should maximize the quality of people's lives but minimize—or, better yet, avoid—undesirable environmental and social impacts. This strategy calls on us to develop and deploy the least environmentally disruptive technologies and systems designs possible. You will have to develop your own opinions about growth. It is not an issue that is easy to resolve. I encourage you to gather facts and give this thoughtful consideration as you read the rest of this book and ponder environmental problems and solutions.

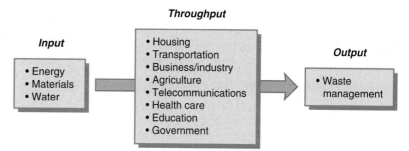

FIGURE 2-8 **Human systems.** Our lives depend on a variety of systems, such as waste management, transportation, energy, and agriculture. Some systems provide inputs needed to make others run. All systems produce waste, handled by our waste management system.

KEY CONCEPTS

Growth and development are fundamentally different goals. Growth results in an increase in material production and consumption that may be unsustainable in the long run; development is a strategy that calls for improvements in culture that do not necessarily require further increases in resource consumption, pollution, and environmental destruction.

2.3 Human Settlements: Networks of (Unsustainable) Systems

Your life and mine are supported by a handful of systems that operate behind the scenes to provide us with a mind-boggling range of goods and services. A few moments of reflection on these systems may prove helpful in understanding just how dependent we are on them. Let's begin with your breakfast. If you ate at home, you probably cooked your breakfast on a stove or in a microwave oven supplied with electricity from a power plant. The power plant, in turn, is probably fueled by coal from a coal mine. The transmission lines that supply your energy, the power plants, and

the mines are all part of an extensive **energy supply system**. Where did the food come from? Most likely from a grocery store. Before that, though, the food was processed and packed in a food processing plant, possibly from corn or wheat grown in an Iowa or Kansas farm field. All of these components—the farms, food processing facilities, and grocery stores—are part of an elaborate **food production system**.

FIGURE 2-8 shows the many interdependent systems that work for us. In addition to the two systems noted in the previous paragraph, there is an extensive waste management system, consisting of landfills, recycling facilities, incinerators, compost facilities, and trucks that move waste from its source to its many destinations. All countries have extensive systems for providing water and housing. Of course, there are also **manufacturing systems** for producing goods and a **transportation system** for conveying people and goods.

Gluing all of this together are an **economic system** and a **system of government**. Increasingly, an informational system has emerged to convey information on the way the other systems are working. An **educational system** provides the necessary training (at least, we hope) for people to function in this complex web of systems.

Supporting these human systems is the Earth and environment, the **ecological system**. As noted in Chapter 1, the Earth and environment are really the infra-infrastructure of society, for they provide the goods and services that make all of human society possible. The ecological system also serves as a sink for our wastes.

Human settlements are made up of many interdependent systems. They usually function so well that we don't have to think about them. In fact, most of us take the systems for granted. This is to be expected, for they have been designed to make our lives easier so that we can go about whatever we do unburdened by such concerns as growing our own food or repairing the roads so that we can get back and forth to work.

The invisibility of the systems is a benefit to most of us, but it also has a downside. Because the systems are invisible, it is difficult for many people to think in terms of systems. It is also difficult for some people to understand the central assertion of this book—that the fundamental problem facing the world's people is that the systems that support our lives are unsustainable. You may be having some trouble with this idea. After all, every time you flick on a light switch, there's electricity. Every time you go to the grocery store, there's an abundance of food on the shelves. Every time you turn on a faucet, there's clean, safe water to drink or wash with. How can I assert that these systems are unsustainable?

KEY CONCEPTS

Human settlements are made up of many interdependent systems, such as transportation, energy, and waste management. Growing evidence shows that these systems are not sustainable in the long run.

2.4 What's Wrong with Human Systems: Why Are They Unsustainable?

Just because a system seems to be reliably supplying goods and services does not mean it is sustainable. On the surface, many systems seem to be performing well because they pro-vide us with what we need and at a relatively decent price. It is what's taking place behind the scenes—what's happening to the environment, for instance—that makes them unsustainable. Consider the U.S. transportation system.

People and goods are transported via an extensive network of flight paths, highways, railways, and navigable waterways covering many millions of miles (Figure 2-8). They are transported in all sorts of vehicles: jets, planes, cars, buses, trucks, motorcycles, bicycles, trains, and boats. Together, the components of the transportation system represent an investment of trillions of dollars. A brand new Boeing 757 costs $100 million. The transportation industry employs millions of people.

The transportation system has made our lives infinitely easier. Those who can afford it can jet around the globe in the time it used to take a person on horseback to ride a few hundred miles. Unfortunately, this system destroys open space and wildlife habitats. It produces millions of tons of harmful pollutants that are imperiling human health and the health of the entire planet (Chapters 19 and 20). Each year, for example,

individual Actions Count

Creating a More Sustainable Home

The principles of sustainable development can be applied to a home and your lifestyle. Here are some examples:

Conservation (efficiency)

Energy

- Install additional insulation in ceilings and other places.
- Install insulated curtains or window shades.
- Caulk and weather-strip to eliminate drafts around windows, doors, and other locations.
- Install ceiling fans for summer and winter use.
- Wrap your water heater with an insulated blanket.
- Wrap exposed hot-water pipes with insulation.
- Buy energy-efficient appliances when old ones wear out.
- Keep temperature setting at 78°F in the summer and 68°F in winter.
- Turn off lights when you're out of the room for any length of time.
- Install compact fluorescent lightbulbs in the most commonly used light fixtures.
- Live close to work.
- Consider alternative transportation such as bicycling, walking, carpooling, or bus riding.
- Buy an energy-efficient vehicle.

Water

- Install water-efficient showerheads.
- Take shorter showers.
- Take showers rather than baths.
- Don't brush teeth or shave with the water running.
- Water lawns early or late in the day.

Recycling

- Set up a recycling center in your house or garage.
- Buy recycled products.
- Compost yard waste.

Renewable Resources

- In sunny climates, add a greenhouse on the south side for wintertime heating.
- In sunny climates, add a photovoltaic array to your roof to supplement electrical energy.
- In windy areas, add a small wind generator to supplement electrical energy.
- Install an energy-efficient woodstove.

Restoration

- In rural areas, return part of your lawn to wildlife habitat for birds and butterflies.
- In arid locations, plant native vegetation that can withstand drought.
- Replant areas that have not regrown.
- Buy wood from sustainably harvested forestry operations.

Population Stabilization

- Limit your family size to one or two children.
- Support family planning efforts at home and abroad.

the combustion of gasoline and other liquid fuels in the global transportation system produces billions of tons of the gaseous pollutant known as carbon dioxide (CO_2). Much of this enormous quantity of carbon dioxide cannot be absorbed by the oceans and plant life. Because carbon dioxide captures heat like a pane of glass in a greenhouse, its accumulation in the atmosphere may be slowly causing the Earth to heat up. This heating could accelerate in years to come, as many scientists predict, causing a serious threat to all life on the planet. This warming of the Earth's atmosphere, referred to as **global warming**, could have devastating effects on human populations (Chapter 20).

Supplying energy to the transportation system has also resulted in significant pollution of the world's oceans with oil (Chapter 21). Oil drilling in pristine areas is having many adverse effects on the local environment and local species, further endangering the Earth's other inhabitants and adding to the system's overall unsustainability. Supplies are also limited. In the long run, the Earth's oil reserves are bound to run out, again making this resource and the system that depends on oil unsustainable (Chapter 14). Many examples of the unsustainable impacts of our systems are provided in this book.

A study of human systems suggests that they are unsustainable for three basic reasons: (1) They produce levels of pollution that exceed the local, regional, and even global capacity to absorb and render them harmless; (2) they deplete finite nonrenewable resources; and (3) they use renewable resources such as forests faster than they can naturally regenerate. In short, human systems exceed the carrying capacity of the planet to supply resources and deal with our wastes. Ultimately, they erode the planet's ability to support life. Signs of these three problems can be seen in cities and towns and nations across the planet.

The Challenges Ahead

This book's main theme is that current systems are unsustainable and that to live sustainably on the planet we must revamp the systems that support our lives. The challenge of making human systems sustainable is enormous. It can be distilled into two basic challenges, however. The first part of the challenge is to modify or retrofit what already exists—for example, making existing homes and buildings more efficient. The second challenge occurs when building anew. As opportunities present themselves to design new components of the systems we've discussed, we must create new, sustainable components that incorporate the best materials, technologies, and designs to minimize resource demand and impact. New houses provide an opportunity to create struc-

tures with a fraction of the impact on the environment of existing homes. They could, for example, be made from waste materials such as straw bales. (**FIGURE 2-9a**) or recycled materials such as automobile tires (**FIGURE 2-9b**). Some builders are even using recycled paper, which they mix with sand and cement to form blocks of papercrete, which then can be laid up like bricks in walls. Instead of building new highways or expanding existing highways, we can lay down rail lines on highway medians to move much larger numbers of people to and from their work with minimum environmental impact. Although this requires the use of resources and results in the production of pollution, it is far more desirable than the alternative. The goal here is to achieve ways to change or grow that produce sustainable impacts and ensure a strong, vibrant economy and a better future. See Spotlight on Sustainable Development 2-1 for an example of sustainable housing.

(a)

(b)

FIGURE 2-9 Alternative housing. (a) Straw bale house and **(b)** packed tire home made from discarded automobile tires and dirt from the site, described in Spotlight on Sustainable Development 2-1.

2-1 Treading Lightly on the Earth: Creating Sustainable Homes for the 21st Century and Beyond

The home featured in FIGURE 1 is my home. It uses only a fraction of the natural resources of a typical home, yet it offers the amenities most of us have grown accustomed to. It's part of a new wave of *sustainable housing*. Why?

FIGURE 1 **The author's sustainable home.**

In designing and building the house, I tried to incorporate four of the operational principles of sustainable development: conservation (efficiency), recycling, renewable resource use, and restoration. As you read through this description, try to pinpoint the specific principle featured in each description.

First of all, this home generates all of the electricity needed by me, my two sons, and an in-home office—via a small wind generator and solar cells, or photovoltaics, mounted on the roof. This system and superefficient appliances and lighting made it unnecessary for me to hook up to the electrical grid, saving a thousand dollars or so in hookup costs and a thousand dollars or more a year in electrical bills.

The home is also an extremely efficient user of electricity. Every appliance and electronic device in the house operates on a fraction of the electricity required in most conventional appliances and consumer electronics. The lighting system, for instance, requires only about a quarter of the electrical energy of a typical house. The television uses half of the energy of most TVs. The refrigerator uses about one-tenth of the energy of a typical fridge.

This house is dependent on the sun for wintertime heating. South-facing windows permit ample sunlight to enter the house in the fall, winter, and early spring (FIGURE 2). The tile floors and thick interior mass walls of the house store the solar heat and radiate it back evenly, keeping the house at a fairly constant temperature during the winter. A su-

perinsulated ceiling also helps hold the heat in during cold winter nights. Most of the north side of the house is sheltered by an earth berm, which also helps maintain a constant temperature. The roof over the bedrooms and my office is completely underground. It's planted in native vegetation and is called a living roof. Native vegetation on the roof helps it blend in with the surrounding meadow and helped us return more of the site to vegetation.

During the summer, the house is naturally cooled by the Earth. Earth sheltering helps keep the house naturally cool,

FIGURE 2 **Solar heating.** Sunlight streams through the windows, warming the house even on the coldest winter days.

like most basements. But it isn't a dark, dank basement. The house is very well lighted by south-facing windows (Figure 2).

Unlike most homes in the region, this one has no well. Rainwater and snowmelt are captured by the roof, stored in a 5000-gallon underground tank, and pumped into the house as needed. Drinking water is purified by a filtration system. Gray water—water from the washing machine—is isolated from black water (water from the toilets) and used to supply an extensive set of planters in the house and an outside garden.

Further adding to its sustainability, the home is built largely of recycled products, including discarded automobile tires. Yes, 800 discarded tires. As shown in FIGURE 3, the back walls and many of the interior walls of this house are made of automobile tires, packed with dirt and piled on top of one another like bricks. After the tire walls were erected, we covered them with stucco, creating attractive thermal mass walls that absorb heat from sunlight and radiate it back into the rooms at night. By using tires, I've greatly reduced the amount of timber needed to build this house. In fact, by our estimates we used about 50 to 60% less lumber than a typical house of this size.

(a)

(b)

FIGURE 3 **Tire walls.** Interior walls are made of tires packed with dirt. Stucco covers the tires, creating an aesthetic look. **(a)** Close-up and **(b)** a view of the entire home.

FIGURE 4 **Barn wood cabinets.** These cabinets are made of wood salvaged from a 150-year-old barn. The wood is remilled, stained, and treated with a water-based (nontoxic) finish.

The house is insulated with 100% recycled cellulose insulation made from old newspaper. The roof is made of steel that can be recycled when its useful life span is over—approximately 50 years. The floor tile is made from recycled feldspar mine waste, and the carpeting is made from recycled plastic pop bottles (PET plastic). The carpet pad is made from cloth scraps from a clothing factory. The cabinets and vanities are made from old barn wood from a 150-year-old barn that was slated for demolition (FIGURE 4). The wood was refinished and stained. The paint on the walls is 50% reclaimed paint—paint that was discarded by homeowners during the county's monthly hazardous-waste roundup.

So, you ask, how much did it cost? Surprisingly, this house cost just a little less than most other homes being built in the area— about $95 per square foot. New homes were going for $100 to $110 per square foot in this area at the time my house was built. Interestingly, virtually all of the recycled building materials were competitively priced. Tires came to us free of charge.

Living sustainably also requires personal actions that all of us can engage in. My boys and I, for example, recy-

cle virtually all of our household waste. We also compost all organic material in an outdoor compost facility. Because of these actions, we produce only about one garbage bag of trash every month. We use cloth bags when we go shopping and purchase recycled products (toilet paper, for example). We're careful in how we use water and energy. The superefficient showerhead uses a half gallon per minute, compared to a normal shower head which might use 5 to 10 gallons per minute.

I write nearly full time now and work out of my home, so I don't have to commute. When I must travel to meetings or other events and to access skiing or hiking, I drive an energy-efficient Toyota hybrid Prius Gen II, which gets around 50 miles per gallon year round. We eat low on the food chain. We're not vegetarians but have a lot of meatless meals.

At this writing, I've lived in the house for ten years and find it to be working very well. It's a bit colder than I had anticipated, and I'm working on ways to change that. I installed insulated window shades, for example. We also have run out of water periodically, so we've installed a more efficient washing machine, which has helped us cut down on demand. Most people aren't able to build a sustainable home from scratch. Many things can be done to create a more sustainable living space, though. The Individual Actions Count! table lists some things you can do when you acquire your first home. You may want to help your parents implement these ideas, too. These suggestions are organized around the operating principles of sustainable development.

2.5 Applying the Principles of Sustainable Development

The principles of sustainable development outlined in this chapter and summarized in Table 2-1 provide a philosophical foundation for sustainable development, a strategy vital to solve the many environmental problems facing the world today. Think of them as directive or guiding principles. Although they provide some general direction, they do not suggest specific actions needed to create a sustainable future. It is one thing to say that we should recognize that the world has limits or promote intergenerational equity, for instance, but it is quite another to figure out how we actualize these goals. The question that arises then is, how do we put ideals into effect? In other words, how do we *operationalize* the directive principles?

Operating Principles

This section offers some suggestions for living well while protecting, even enhancing, the environment—that is, making sustainable development happen. Students and instructors are encouraged to add their own ideas to this list of seven operating principles. To be sustainable:

1. We must **stabilize our population**. Our population cannot continue to grow indefinitely.

2. We must **better manage how we grow**. In other words, we cannot spread willy-nilly across the landscape, usurping prime agricultural land needed to feed people and provide other valuable services such as recreation.

3. We must **use resources much more efficiently**. All resources, from fossil fuels to building materials to drinking water, must be used more efficiently. Resource efficiency provides a great economic benefit to businesses and clearly helps societies ensure adequate supplies for future generations.

4. We must turn to **clean, renewable energy supplies**, such as solar and wind energy. They provide the fuel we need at a fraction of the environmental impact.

5. We must **recycle** everything we can. Waste is intolerable in a world of limited resources, and recycling can make good economic sense. It also helps to ensure a supply of resources for future generations. Not only must we recycle, we must **manufacture a large portion of our goods with recycled materials**.

6. We must **restore natural systems** that have been damaged in years past. Moreover, we cannot continue to exploit the land, deplete natural systems, and then move on, as we have through much of history.. To support people and the other living things that share this planet with us, we must replant forests, rebuild wetlands, and restore farmlands, grasslands, and pastures.

7. We must **manage resources sustainably**. That is to say, we must do a much better job of managing forests, grasslands, and other natural systems. Short-term profit at the expense of long-term productivity only leads us further from the path of sustainability.

KEY CONCEPTS

Creating a society that is sustainable requires a number of steps: (1) population stabilization, (2) growth management, (3) efficiency, (4) renewable energy use, (5) recycling, (6) restoration, and (7) sustainable resource management.

Applying the Operating Principles to Human Systems

FIGURE 2-10 shows a simple yet effective tool that allows us to apply these operating principles of sustainable development to the arduous task of restructuring basic human systems such as transportation and waste management to create a clean, healthy environment and to help humankind prosper in the long term. This simple chart can help citizens, business owners, and government officials concentrate their efforts and optimize limited budgets by allowing them to determine the most effective ways to achieve environmental sustainability.

Figure 2-10 lists the systems (for example, energy) in the first column. The top row lists the operating principles of sustainability (such as conservation). Each box in this table can be filled with ideas to put that principle into action within a given system. Dozens of ideas could be included. The ideas could include policies and practices that operationalize the principles of sustainability outlined earlier.

The initial work with the table should be considered brainstorming and thus should not be subject to critical thinking, criticism, or prioritization. After a full set of ideas has been generated, the next step is to think critically about them. Discard those ideas that don't make sense and choose the highest priority items—those that are most effective and affordable. Remember that they should make sense from social, economic, and environmental perspectives—a fundamental prerequisite of sustainable development (Chapter 1).

Next, you can devise an action plan. The plan may include a timetable, a list of key players (for example, business leaders and activists), and resources. It might include education programs through schools, businesses, and government. It might also contain some indicators to plot progress in achieving goals. The action plan might list anticipated obstacles and address ways to overcome them.

As an example, consider the box in Figure 2-10 that relates the conservation principle to housing/buildings. Remember, however, that sustainability is a twofold challenge: One goal is to revamp existing buildings; the second is to make sure that all new buildings use resources efficiently. The necessary policies and actions that a town might include to make this happen could involve changes in local building codes to improve insulation in new homes and remodeling projects, energy-efficient lighting designs, and educational programs for builders.

The systems approach outlined here is a way to streamline the sustainable development process. It gives people a framework to organize and address complicated issues in a

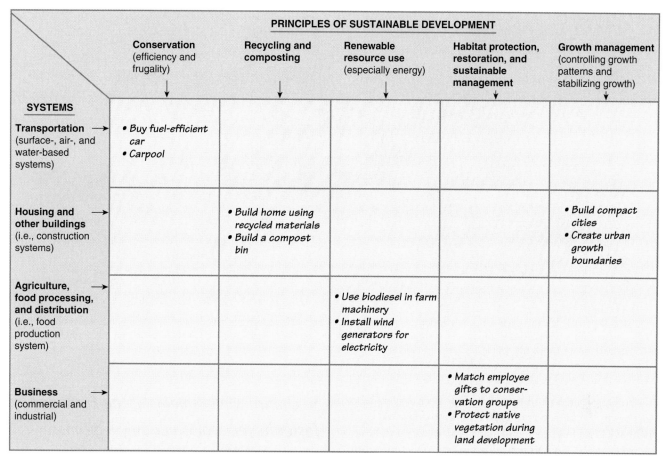

	PRINCIPLES OF SUSTAINABLE DEVELOPMENT				
SYSTEMS	**Conservation** (efficiency and frugality) ↓	**Recycling and composting** ↓	**Renewable resource use** (especially energy) ↓	**Habitat protection, restoration, and sustainable management** ↓	**Growth management** (controlling growth patterns and stabilizing growth) ↓
Transportation (surface-, air-, and water-based systems) →	• Buy fuel-efficient car • Carpool				
Housing and other buildings (i.e., construction systems) →		• Build home using recycled materials • Build a compost bin			• Build compact cities • Create urban growth boundaries
Agriculture, food processing, and distribution (i.e., food production system) →			• Use biodiesel in farm machinery • Install wind generators for electricity		
Business (commercial and industrial) →				• Match employee gifts to conservation groups • Protect native vegetation during land development	

FIGURE 2-10 Matrix showing the systems and principles of sustainability.

systematic fashion. It can be used by cities, towns, state governments, and even large national governments to meet needs while protecting the environment and saving money. One important piece that's missing from the discussion is how to address connections among systems, which are often interdependent. Energy, for instance, is a component of all other systems. It is as vital to the transportation sector as it is to industry, housing, and food production. Thus, decisions in one system often affect others, so certain systems might be best addressed in conjunction with others. For example, creating a sustainable transportation system cannot be done without discussing housing (at least the placement of housing subdivisions) and business development (notably the location of new businesses). To streamline the process, people involved in drafting sustainable development plans are encouraged to group interrelated systems and tackle them together.

You may have noted that this approach omits some familiar environmental issues—for example, loss of open space and air quality. The reason for this is quite simple, although sometimes difficult to grasp: As noted in Chapter 1, air pollution and loss of open space are *symptoms* of unsustainable systems (**FIGURE 2-11**). Air pollution is best addressed by creating sustainable systems of transportation, housing, and industry—for the same reason that symptoms of a disease are best addressed by finding out what is fundamentally wrong

and treating the person for that. As an example, severe neckaches and eyestrain in a computer operator are best addressed by changing the position he or she sits in while at the computer, not by giving the person a larger dose of pain reliever.

Air pollution might be sustainably addressed in a number of ways, for example, by encouraging alternative forms of transportation such as buses and carpools, creating energy-efficient houses and factories, using renewable energy resources, and emphasizing more compact forms of development so that people have shorter commutes.

Sustainability won't occur until cities and towns, states, nations, and businesses get serious about restructuring human systems. It won't occur until we recognize that the problems many communities face are symptoms of deeper problems—unsustainable systems that erode the life-support systems of the planet. But keep reading this book and decide for yourself.

Sustainable development will require root-level solutions. This concept can be applied to all problems—social, economic, and environmental. Unfortunately, most political leaders and citizens don't give much thought to the systems that undergird their communities and the ways these systems could be restructured to create a lasting future. Also, many people may find the assertion that human systems are unsustainable difficult to grasp, much less believe.

FIGURE 2-11 **It's the systems!** Air pollution, waste, and species extinction are symptoms of overpopulation and unsustainable human systems. Redesigning these systems could help reduce many of the symptoms of the environmental crisis and put us back on a sustainable course.

To help overcome these barriers, we need leaders who can convince others that a sustainable approach is required and that it is not just a matter of doing more of the same. Leaders need not be government officials—citizens, nonprofit organizations, and progressive-minded business owners all can play a role.

To cherish what remains of the Earth and foster its renewal is our only legitimate hope of survival.

—Wendell Berry

CRITICAL THINKING

Exercise Analysis

Labels are valuable to human beings. They help us communicate about the world around us. They also help us create an identity. ("I'm a scientific writer, a father, and a musician.") Labels can be a form of thought stopper, though. A thought-stopping label hinders expression and thought in one of several ways. Labels can put people into categories that may or may not be accurate. To be introduced as an "avid environmentalist" at a dinner held by a nonprofit environmental organization, for instance, might have a very different effect than, say, being introduced similarly at a Chamber of Commerce luncheon. A label can be a barrier to communication. You might have a perfectly reasonable viewpoint, but many people in the Chamber audience might not listen.

Labels also simplify things and may mischaracterize the person who is being labeled. To say that a person is an "environmentalist" is sometimes taken to mean that he or she cares only about the environment. To say that someone is a "corporate type" may wrongly imply that the person cares only about business. Few of us have such a narrow realm of interest.

The lesson in all this is to be careful about the labels you use and the way you interpret the labels others use. Assume that people are more than their labels may imply. Labels may create inaccurate pictures. It is better to ask someone what he or she believes.

CRITICAL THINKING AND CONCEPT REVIEW

1. Define the term *sustainable development*. Why is it important to environmental protection? Make a list of practices that you think would contribute positively to sustainable development efforts.

2. Describe the evolution of environmental protection in the United States. What forces have caused environmental protection efforts to change over the past 30 to 40 years? What changes are needed to make environmental protection truly effective? Why is sustainable development essential to environmental protection?

3. List the 10 principles of sustainable development outlined in this chapter and briefly describe each one. Do you agree with all of these principles? Why or why not?

4. What is meant by an end-of-pipe control? Give reasons why you think such an approach may have been favored in the past. Why are such controls proving to be inadequate?

5. Summarize the views of the progrowth and antigrowth advocates. Why is growth so important to the issue of environmental protection?

6. Do you believe that economies and populations can continue to grow indefinitely? Why or why not?

7. Make a list of five of the major activities you were involved in today—for example, riding the bus to school, eating meals, and so on. Then make a list of the major systems that made these activities possible—for example, transportation system, food production system, and communication system. Describe how human systems are interdependent.

8. This chapter argues that human systems are unsustainable. Why? Cite evidence. Can you think of other reasons?

9. What is meant by the statement that the invisibility of systems makes it difficult for people to understand that they are unsustainable? How does this invisibility contribute to our natural inclination to focus on symptoms of problems rather than on the systems themselves? Why is this a problem?

10. Restructuring human systems (physical infrastructure and social and economic systems) will require two broad types of actions outlined in this chapter. What are they? Give some examples of each.

11. This chapter presents a technique to help people apply the principles of sustainable development to systems reform. Draw a diagram of the matrix and explain how it is used. What are the advantages of this approach? What are the disadvantages?

12. Why is it necessary to understand and address the root causes of unsustainability to build a sustainable society?

13. List the principles of sustainable development and describe each one. Then give an example of a policy or practice that would operationalize each principle in a system of your choice—for example, housing or waste management.

14. Using the matrix presented in this chapter (Figure 2-10), take a system such as energy production and apply the various principles to it. Come up with as many ideas as you can. Pick out your best ideas. What criteria helped you determine which ideas were better than others?

KEY TERMS

agricultural revolution
biological infrastructure
biophysical limits
carrying capacity
dependence
development
ecological justice
ecological system
economic growth
economic system
educational system

end-of-pipe controls
energy supply system
food production system
generational equity
global warming
industrial revolution
infra-infrastructure
interdependence
intergenerational equity
intragenerational equity
living within limits

manufacturing systems
nonrenewable
physical infrastructure
pollution control devices
renewable
smokestack scrubbers
sustainable development
sustainable revolution
system of government
transportation system

REFERENCES AND FURTHER READING

The References and Further Reading section at the end of this book contains a list of sources for the information discussed in this chapter and recommendations for further reading.

Connect to this book's website:
http://environment.jbpub.com/
The site features eLearning, an online review area that provides quizzes, chapter outlines, and other tools to help you study for your class. You can also follow useful links for in-depth information, research the differing views in the Point/Counterpoints, or keep up on the latest environmental news.

This housing development in a Southern California desert could have been made more sustainable if the developer had installed solar panels to take advantage of the year-round sun, allowed for stores within walking distance, and encouraged the planting of water-conserving plants instead of thirsty grass.

CHAPTER 3

Understanding the Root Causes of the Environmental Crisis

There are a thousand hacking at the branches of evil to one who is striking at the root.

—Henry David Thoreau

It seems that just about everyone has an opinion on the causes of environmental issues, the main subject of this book. To some people, the current environmental dilemma is largely the result of overpopulation—too many people. To others, it is the fault of those corporations whose concern for the environment ranks well below their interest in making money. Others may attribute it to inept or corrupt government officials. Many think that the environmental problems of the world result primarily from overconsumption and waste. Modern, environmentally destructive technology is seen as the culprit by some. The list goes on. I'm sure you have your own thoughts on the subject.

If the truth be known, the current environmental crisis or, more appropriately, the **crisis of unsustainability** is the result of a great many factors. Population is a big player. So is consumption. Corporate policies play a part, too. There are many other factors as well. This chapter describes the major reasons for environmental unsustainability and presents a diagram of the complex web of cause and effect to help you understand how we arrived at our current situation.

The chapter's exploration of the roots of the environmental crisis is intended to help you not only comprehend things better but also understand key leverage points where change can be effected. The main theme of this chapter is that if we know the root causes and address them, we have a far better chance of arriving at lasting solutions and creating a better future.

3.1 Roots of the Environmental Crisis

Books and articles on the root causes of the environmental crisis reveal many interesting ideas. This chapter combines the insights of historians, social scientists, economists, psychologists, anthropologists, theologians, and others into a coherent picture. It also discusses my own biological/evolutionary hypothesis. We study the ideas of these thinkers one at a time, and then combine them to show a map of cause and effect. Remember as you read this material that books could be written on each subject. I present the main ideas and opposing ideas as well, to give you as balanced a view as possible. As you read the various hypotheses, imagine them as pieces of a larger puzzle, all leading to the problem we call the crisis of unsustainability.

> **KEY CONCEPTS**
>
> Many different hypotheses have been presented to explain the root causes of the environmental crisis. When combined, these hypotheses do a good job of explaining environmental troubles.

Religious Roots

An early, influential hypothesis on the roots of the environmental crisis was presented by UCLA historian Lynn White in the 1960s. He argued that the roots of our environmental crisis are "largely religious." In his classic paper entitled "The Historical Roots of Our Ecological Crisis," White argued that the emergence of science and technology four generations ago in western Europe and North America spawned an era of massive environmental manipulation that was paralleled by enormous environmental problems. However, White argued that Western-style science and technology "got their start, acquired their character, and achieved world dominance in the Middle Ages." The most important influence on science and technology at this time was Christianity's view of human dominance over nature.

White noted the familiar passage in Genesis 1:28 that directs humankind to "fill the earth and subdue it, and have dominion over the fish of the sea and over the birds of the air and over every living thing." In Christian teaching, White contended, no item in physical creation had any purpose except to serve man's purposes.

White also noted that during the Middle Ages Christianity promoted a sense of dualism, a view of humanity as apart from nature. In a paganistic society, humans viewed themselves as a part of the whole.

White notes that the disastrous environmental problems facing the world cannot be solved by simply applying more Western-style science and technology to our problems, which, in his view, are rooted in Christian domination and duality. Professor White writes that "we are not, in our hearts, part of the natural process. We are superior to nature, contemptuous of it, willing to use it for our slightest whim."

Widely circulated in academic circles in the 1960s and 1970s, White's ideas were embraced by many scholars, lay people, and even some theologians. Not all theologians and scholars, such as René Dubos and Gagriel Fackre, however, agreed with his hypothesis, citing biblical verses that foster stewardship and protection. Another critic of White's premise was Lewis Moncrief, a professor at Michigan State University. Moncrief finds no fault in White's contention that human ecology is deeply conditioned by religious belief, "but to argue that it is the primary conditioner of human behavior toward the environment is much more than the data that [White] cites to support this proposition will bear." Moncrief points out that White himself notes in his classic paper that humans have drastically altered the environment since antiquity and cites several examples of pre-Christian environmental destruction that undermine his argument.

René Dubos pointed out that in both ancient Eastern and Western civilizations, environmental destruction was, and still is, rampant (**FIGURE 3-1**). The deforestation, soil erosion, and overgrazing prevalent in some cultures suggest that Christian beliefs have had no monopoly on ecological damage. In other words, exploitation of the natural world may be a universal human trait. If environmental destruction has occurred since antiquity, White's thesis must be questioned. White answers his critics by pointing out that although the Bible proclaimed it humans' duty to protect the Earth, little heed was paid to such instructions.

What can we make of the debate? Obviously, opinions vary. While some are skeptical of White's ideas there are those who contend that religion may have played a role—and may still be playing a role—in shaping our thoughts in ways that are detrimental to the environment and our long-term future. A Gallup poll found that one third of Americans believe the Bible is literally true. Why is this significant? Because roughly 50 million of these Americans are expecting the Apocalypse soon. There are those who believe any individual effort, political activism, or research to address serious long-term environmental issues is pointless. Some oppose actions on the grounds that environmental destruction will help hasten the Apocalypse.

FIGURE 3-1 Destruction in the East. Environmental damage is not a Western phenomenon. All cultures have had a profound effect on the environment, even those whose philosophy embraces the protection of nature as in the case of the Three Rivers Gorge Dam in China. **(a)** The river before dam construction began. **(b)** The dam under construction.

(a)

(b)

> **KEY CONCEPTS**
>
> Some scholars think that early Christian teaching shaped many people's attitudes toward nature, which in turn fostered the creation of exploitive systems of science and technology that are largely responsible for the destruction of the environment. Others argue that humans have a long history of environmental destruction dating back long before the advent of Christianity.

Some Cultural Roots: Democracy, Industrialization, and Frontierism

Lewis Moncrief (mentioned previously) attributes the environmental crisis to two dominant forces: the spread of democracy and industrialism. Democracy began to take hold after the French Revolution. Prior to that time, land and resources in the Western world were owned primarily by royalty or the Roman Catholic Church. They bestowed control of the land resources on the nobility and others who pledged their allegiance. Much of the land was protected as royal property. However, the French Revolution marked the unraveling of these feudal systems of land tenure. As a result, land became vested in the hands of more and more individuals.

At about the same time, but over a more extended period, another kind of revolution was taking place that would dramatically change the world. Known as the **Industrial Revolution**, this shift from manual labor to energy-intensive machine production resulted from the application of science and technology to the service of humankind. With this change, the productive capacity of each worker greatly increased. It became possible to mass-produce goods (**FIGURE 3-2**).

Moncrief contends that the spread of democracy and industrialization resulted in a more equitable distribution of wealth among the human population and, more important, rising affluence and consumption. Pollution and environmental decay followed (**FIGURE 3-3**).

Moncrief asserts that the colonists of North America were quick to adopt the "technical and social innovations" taking place in Europe. In America, national policy was designed to transmit both ownership of the land and many of the nation's natural resources from the federal government to private individuals. Huge parcels of land were given to settlers. In the 1800s, mining laws were designed to transfer land to those willing to make the effort to exploit it. (Incidentally, these mining laws are still in effect today!) In short, much of the land and the natural resources of the nation came to be controlled by citizens and businesses. Decisions that affect the environment were made by millions of private owners. Industrialization and the spread of wealth in the United States further added to the destruction of the environment, as it did in Europe.

Moncrief also touches on an important dimension of the environmental crisis, frontierism. **Frontierism** is a way of life that is based on an assumption of plenty. To most frontierspeople—and particularly to early farmers—many of the natural resources we now fight to protect were originally perceived as obstacles rather than as assets. Forests needed to be cleared to permit farming. Marshes needed to be drained.

Part of the reason for this obstacle mentality was a false perception of the inexhaustibility of the vast resources of the new nation, a frontier ethic. "After all," observes Moncrief, "if a section of timber was put to the torch to clear it for farming, it made little difference because there was still plenty to be had." Frontier thinking continues today as a major factor in the ever-worsening environmental crisis (**FIGURE 3-4**).

In the frontier society, the relationship of humans to nature was indifferent at best, and often antagonistic. Antagonism and indifference prevailed in large part because of a lack of knowledge on the part of settlers about the long-term implications of the type of changes being made to the environment. Lacking an understanding of ecology and the consequences of abusing the life-support systems of the planet, frontier societies plundered one ecosystem after another. Ignorance, another dynamic in the complex web of causation, is alive and well today. Although our understanding of ecology and the impacts of humans on the environment has reached an all-time high, many people remain ignorant of the impacts, and this ignorance drives many environmentally unsound practices—building energy-inefficient homes, for instance. Ignorance, therefore, occupies a prominent place in the network of causation.

FIGURE 3-2 **Industrial production.** Industrialization made it cheaper to produce goods and made them more accessible to a large number of people, thus increasing our demand for resources and our impact on the environment.

FIGURE 3-3 **Industrial pollution from factories.** Scenes like this are common in many less developed nations and were very common in the developed nations until the 1960s and 1970s.

FIGURE 3-4 **Frontierism continues.** Tropical rain forests are cut down in record number each year to supply wood and make way for farms. The belief that there are plenty more makes this resource highly vulnerable to humans.

KEY CONCEPTS

Some scholars believe that the spread of democracy, which put land ownership and wealth in the hands of many, and the Industrial Revolution, which made mass production of goods possible and spread wealth throughout society, are at the root of the environmental crisis.

Frontierism, a belief in the inexhaustibility of resources, may have also been a root cause of the environmental crisis.

Biological and Evolutionary Roots

Philosopher Bertrand Russell once wrote that "every living thing is a sort of imperialist, seeking to transform as much as possible of the environment into itself and its seed." Although the vast majority of the world's species do not intentionally seek to convert the resources of the environment into their kind, populations do naturally expand in proportion to available resources—water, soil, sunlight, and so on. Organisms are biological imperialists. In this chapter, I use the term **biological imperialism** to describe an expansionist tendency.

Countless examples of biological imperialism exist in the natural world. This characteristic of living things is amply illustrated by the spread of a vine known as *kudzu* (pronounced CUD-zoo), which grows rampant in the southeastern United States, smothering farm fields, forests, and abandoned buildings (FIGURE 3-5a).

The water hyacinth (HIGH-ah-sinth) is another biological imperialist extraordinaire (FIGURE 3-5b). Introduced to the United States in 1884 from South America as an aquatic ornamental by a private collector, the water hyacinth escaped from the pond where it grew. After entering the warm, nutrient-rich waterways of Florida, it spread throughout the extensive network of canals and rivers that lace the state. Aided by a remarkable reproductive capacity—10 plants can multiply to 600,000 in 8 months—the water hyacinth today chokes many of Florida's waterways. The plant has crowded out native species and made navigation virtually impossible in some water bodies. From Florida, the plant has spread throughout much of the southern United States and today covers nearly two million acres of rivers and lakes along the southern tier of states from Florida to California.

Biological imperialism refers to a fundamental tendency for populations of organisms to proliferate in response to available resources. Bacteria in a petri dish reproduce and colonize. Organisms don't choose to be such imperialists. It is simply a biological fact of life: When conditions are favorable, organisms prosper and reproduce successfully—and populations expand. Biological imperialism is aided by the availability of nutrients, sunlight, moisture, and a host of other factors.

In nature, checks and balances tend to prevent many populations from reaching their full potential. However, some species fare better than others, especially when placed in foreign environments where the natural checks and balances that evolved over millions of years are not present. *High reproductive rates and a host of other special characteristics make some organisms superb imperialists.* The success of any one species depends on the structural, functional, and behavioral characteristics (called *adaptations*) it arrives with or acquires through evolution (Chapter 6).

Humans are superior colonizers. They have special features such as brainpower and technology to remake the planet—to create conditions and expand food supplies in ways that enable them to prosper and expand.

My hypothesis is that the major environmental transgressions of the past and the present owe their origin in large part to a sort of human biological imperialism. From its meager origins in Africa, humankind has expanded widely. Today, human civilization in one form or another inhabits virtually every biome on the Earth and makes use of many aquatic life zones. Hardly a square foot of the planet has been spared the human imprint.

Early environmental transgressions in North America—for example, the extinction of many species of mammals—may have occurred as the first settlers spread across the continent about 10,000 years ago (FIGURE 3-6). According to some scientists, the extinction of these species can be attributed in large part to our successful planetary colonization (human biological imperialism) aided by the use of very primitive technology, such as fire and early weapons. Human populations expanded as a result, migrating throughout the world, often wreaking havoc along the way.

As technology became more advanced, the pace and severity of environmental destruction quickened. The past 60 years have seen a further acceleration of the Earth's destruction as technology and per capita consumption, particularly in the Western world, have increased dramatically.

Over the years, many technologies have helped us to overcome natural forces that might otherwise have controlled human population expansion. *The bottom line is that technologies have permitted us to expand almost uncontrollably.* Agricultural technologies have dramatically increased food production. Medical technology has greatly reduced outbreaks of many infectious diseases, such as the plague, that once took a massive toll on human populations. Together, these and other advances led to an explosion in the human population. As our numbers increase, so does our impact.

(a)

(b)

FIGURE 3-5 **Biological imperialism.** Two supercolonizers, **(a)** the kudzu and **(b)** the water hyacinth.

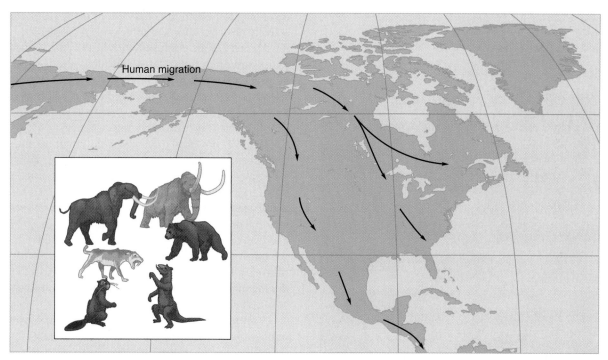

FIGURE 3-6 Early inhabitants of North America, who entered by a land connection with Siberia about 10,000 to 12,000 years ago, may have wiped out many species of megafauna.

Technological liberation from biological limits has unleashed our imperialistic tendencies, creating a global environmental crisis. *In short, technological liberation has translated a natural biological urge into a global environmental disaster.*

Technological development was made possible by two evolutionary developments unique to humans: the evolution of the brain and the evolution of manual dexterity. From an environmental perspective, one of the most important developments in human nervous system evolution was the emergence of the frontal lobe. The frontal lobe, which is part of the forebrain, houses several functions, among them (1) ideation (the ability to create ideas) and (2) the ability to plan and think about the future. Both functions are vital to technological development. Of equal importance to the development of technology were refinements of the brain that gave humans the ability to communicate and build tools and weapons.

The ability to create ideas, plan for the future, invent tools, communicate ideas, and manipulate objects were all essential to the development of technology. These uniquely human abilities today give us a power to do many great things—to till the Earth to grow crops, for instance.

Appreciated by many for expanding our capacity to improve on nature, these abilities are also responsible, in large part, for much destruction. In some cases, the destruction wrought by technology is intentional, as manifest in tropical deforestation. In others, it is unintentional, as manifest in ozone depletion and acid deposition. From an environmental standpoint, the distinction is irrelevant. Either way, this damage could ultimately reduce the Earth's carrying capacity for humans and the millions of species that share this planet with us.

KEY CONCEPTS

Human populations, like those of other organisms, expand if there are adequate supplies of resources and no other controls. The expansion of some organism populations is facilitated by special characteristics. For humans, technology has greatly facilitated population growth and greatly increased our environmental impact.

Psychological and Economic Roots

Many psychological factors also contribute to the environmental crisis by aiding and abetting our biological imperialist tendencies. Although psychology may seem far afield from environmental science, it is important to understand if we are to solve problems.

In their book, *New World, New Mind*, ecologist Paul Ehrlich and psychologist Robert Ornstein note that the human nervous system is "designed" by evolution to respond to immediate physical danger. The survival of early cave dwellers, they assert, depended largely on quick reactions to imminent threats. This survival tool served our ancestors well for over 3 million years—99% of our evolutionary history. Ehrlich and Ornstein argue, though, that the human nervous system today is rather ill equipped to respond to trends that threaten our long-term future.

The human brain may be something of a double-edged sword. The frontal lobe, discussed earlier, helps us plan for the future, engineer, and design technologies. However, it may not equip us with a very adequate response to dangers unless they are immediate and life-threatening. Although some environmental problems create an imminent threat to us, most pose long-term, less obvious threats to the health and

FIGURE 3-7 Tropical deforestation. (a) The loss of tropical rain forests creates local environmental problems of enormous magnitude, but for most of the world's people the impacts are long term, such as climate change. **(b)** The rain forests help determine climate hundreds of miles away.

on to take responsibility or action. In many cases, they will actively obstruct efforts to solve crucial problems.

Denial can be confused with healthy skepticism. Scientists are often naturally skeptical about ideas. They require adequate data to support new views. Just because some scientists don't believe in the latest theory of planetary disaster doesn't mean that they are necessarily in denial. They may simply be waiting for convincing evidence.

Another powerful influence in the environmental crisis is the seemingly near obsession with economic growth that is prevalent in many of the more developed countries. Most Western industrial nations, which produce the bulk of the world's goods and services, are predicated on ever-increasing material production and consumption. With the growth-is-good (indeed essential) logic so deeply imbedded in many societies, modern Western culture has evolved what many think is a lopsided view of human progress based on producing and consuming more and more, regardless of the cost to the planet or to future generations.

Continued human expansion and economic growth are impossible in a finite system. In his book *The Poverty of Affluence*, Paul Wachtel observes that growth is "an omnipresent symbol of the good." With this mindset, it is "very difficult for us to accept any idea of a limit to growth as implying anything other than stagnation. Our emphasis on growth leads us to equate contentment with complacency." Why is this penchant for growth so prevalent?

Numerous authors have studied the psychological roots of the economic growth philosophy and overconsumption. Andrew Bard Schmookler, a leading thinker in this area, contends that the modern culture of mass consumption is developed around a core of unfulfilled longing. People buy to make themselves feel better—to satisfy internal longings. Modern advertisers have capitalized on our sense of inadequacy. According to Schmookler, advertisers promise that the goods we purchase will make us happy or successful.

In addition, the many ads that bombard people in consumer societies suggest that success is measured in the kinds of cars we drive, the size of our bank accounts, and the restaurants in which we dine. Most of us, however, can never fully match the beautiful or handsome models portrayed in ads. There's always some area that we can improve, but our efforts to placate an unsatisfiable longing for adequacy are like trying to fill a sieve with sand. From a marketer's standpoint, the effect is remarkable: We never stop consuming in our vain attempts to fill the emptiness. The economy expands as a result. Unfortunately, the more we acquire to offset the sense of personal inadequacy, the more environmental damage we cause.

Psychologists tell us that the roots of the sense of inadequacy can be traced to other factors. The lack of self-esteem that advertisers feed off in our culture results, in large part, from inadequate family systems and deficiencies in child rearing. We may feel inadequate because of the messages we received as children. Inadequate child rearing results in part from the lack of parental education on the subject.

welfare of humans—among them global warming, acid deposition, and tropical deforestation (**FIGURE 3-7**). Although our response to immediate danger has been vital to our survival, we must also learn to respond to long-term dangers.

Denial is another psychological factor that contributes to our inability to respond to the environmental crisis. Many environmental controversies have been marked by long-standing denial that has hindered progress toward lasting solutions.

Denial is often prevalent among those who have a vested interest in maintaining the status quo, as well as those of limited perspective and knowledge. Such individuals often find it difficult to believe that human activities could endanger the biosphere and humankind. Individuals who have difficulty accepting the reality of a problem cannot be counted

Advertisers also play off a powerful human desire to conform. At the root of conformity is the obvious and pervasive need to belong. In order to belong, a person must have the right car, the right clothes, the right cologne or perfume. In other words, to belong, one must consume. When styles change and one's sense of conformity and belonging are threatened, people buy more. In such a culture, it is better to be in debt to the hilt than to be with*out*.

Underlying the need for conformity and belonging may be the erosion of family and community. *Homo sapiens* is by nature highly social. Our early evolutionary success can be traced in part to social groups that permitted cooperative hunting and food-gathering ventures and joint efforts in warding off predators. As our communities dissolve into frenetic megalopolises, belonging at a deep and genuine level is replaced by superficial membership in society facilitated by conformity with transient fashions orchestrated by advertising firms.

Consumption also stems from the joy of novelty and the usefulness and convenience of many new products. In addition, some consumption results from the relative costs of mass production versus labor for repair. How many of us have had to throw away a $50 portable CD player because it would cost $65 to replace a faulty part?

The growth-oriented economic system is also shaped by an intense competitive drive in humans. In modern society, competitiveness is culturally propagated through advertisements, education, and the public utterances of our political and business leaders. The telecommunications industry, for instance, sells its wares on the value of the competitive edge gained by such innovations as overnight delivery, faxes, e-mail, cell phones and other forms of "instantaneous" communication—features that give one company a leg up on the competition. Improvements in the American educational system have been deemed important primarily for the competitive edge they could give the U.S. economy in the global marketplace.

Acquisitiveness and greed also drive economic systems that exploit people and the planet with detrimental consequences. Garrett Hardin discussed the consequences of greed in a classic paper entitled "The Tragedy of the Commons." In England, Hardin noted that cattle growers grazed their livestock freely on the commons, land open to all. The commons fell into ruin as the users became caught in a blind cycle of greed. To increase their personal wealth, individuals expanded their herds. Each additional cow meant more income at only a small cost. Hardin notes that the cattle growers were rewarded for doing wrong. Although they may have realized that increasing their herd might lead to overgrazing and deterioration of the pasture, they also knew that the negative effects of overgrazing would be shared by all members of the community. Each herdsman had more to gain than to lose by expanding his herd. This shortsighted thinking (another root cause) resulted in a spiraling decay of the commons. Hardin summarized the situation as follows: "Each man is locked into a system that compels him to increase his herd without limit in a world that is limited." As

each pursues what is best for himself, the whole system is pushed toward ruin.

Tragedies of the commons date back to the days of ancient Greece, when Aristotle recognized that property shared freely by many people often received the least care. Early civilizations toppled forests and carelessly overgrazed their cattle on rangelands, just as the tribespeople of the Sahel (the region bordering the southern Sahara) do today. History, however, shows that early civilizations paid dearly for their disregard. The skeletons of buildings from ancient cities still stand in deserts that were the once-rich forests and grasslands of the Fertile Crescent (FIGURE 3-8). Much of Iran and Iraq, now barren desert, once supported cattle, farms, and forests. Greece and Rome fell, historians believe, partly because of the misuse of their lands. The tragedy of the commons is the cumulative result of the self-serving tendencies of humans (greed), as well as of ignorance, competitiveness, and lack of regulation.

Hardin's analysis of the tragedy of the commons, although important, gives the impression that common resources are alone in being mismanaged. In many places

FIGURE 3-8 **Remnant of past action.** This abandoned city in the desert was once a thriving region, but environmental abuses caught up with residents and destroyed their homeland.

private lands are no better cared for because of short-term profit or poor management (in the case of former communist nations).

All of the psychological factors discussed so far influence the way we conduct business, government, and our personal lives throughout the world. They influence the very laws that govern economic activity. They affect the way we have settled the land and made our place on the Earth. Beliefs and values are, in essence, the philosophical underpinnings of all human systems—from law to the economy to government. These psychological factors have either contributed to our biological imperialist tendencies or failed to create mechanisms to temper them.

KEY CONCEPTS

Human attitudes and beliefs are also responsible for many unsustainable practices. Denial, apathy, inability to respond to subtle threats, greed, acquisitiveness, and other factors influence our economic systems, laws, and way of life in profound ways. In short, they worsen our biological imperialist tendencies.

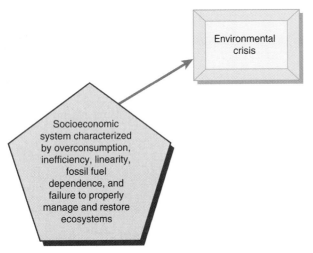

FIGURE 3-9 It's the systems. Unsustainable systems are largely to blame for environmental ills. But the "shape" of these systems is influenced by many other factors.

Putting It All Together

It should be clear from the forgoing discussion that many factors contribute to the environmental problems facing the world's people today. People are complex, as are the systems that support our lives. So are the issues. An analysis of thinking on root causes shows that there are many. We can't point to a single root cause, or even a couple of root causes. To understand how the factors fit together, let's build a model, one piece at a time, drawing on the information you've just read.

FIGURE 3-9 shows the environmental crisis and its link to human systems designed to meet our many needs, a subject discussed in Chapter 2. Ultimately, many believe it is these systems that are responsible for the problems. As you will learn in the next chapters, these systems have many characteristics that contribute to environmental problems. They are dependent on fossil fuels, for example, and many are highly inefficient.

Human systems unleash population growth and bring us power unimagined even 100 years ago. They unleash our biological imperialism. **FIGURE 3-10** draws the connections. Democratization and the spread of wealth through industrialization, shown in Figure 3-10, are also responsible for the makeup of the modern systems.

FIGURE 3-11 shows how our inability to respond to long-term trends, based on economic self-interest, disbelief, denial, and ignorance, influence the socioeconomic system. **FIGURE 3-12** illustrates how the obsession with economic growth contributes to the socioeconomic system that is responsible for the crisis. Many factors contribute to this obsession, as described earlier. Among them are our sense of inadequacy, competitiveness, acquisitiveness, frontier ethics, and laws and regulations. These factors, in turn, are influenced by a host of other factors.

If all of the components are placed together, the model shown in **FIGURE 3-13** emerges. It helps illustrate how complex the problems really are. But if you understand each one, then it is not that difficult to see how they all fit together.

KEY CONCEPTS

People, the systems that support them, and environmental issues are complex. So are the cause-and-effect relationships.

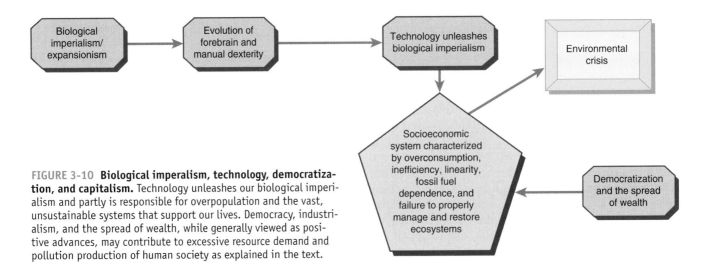

FIGURE 3-10 Biological imperalism, technology, democratization, and capitalism. Technology unleashes our biological imperialism and partly is responsible for overpopulation and the vast, unsustainable systems that support our lives. Democracy, industrialism, and the spread of wealth, while generally viewed as positive advances, may contribute to excessive resource demand and pollution production of human society as explained in the text.

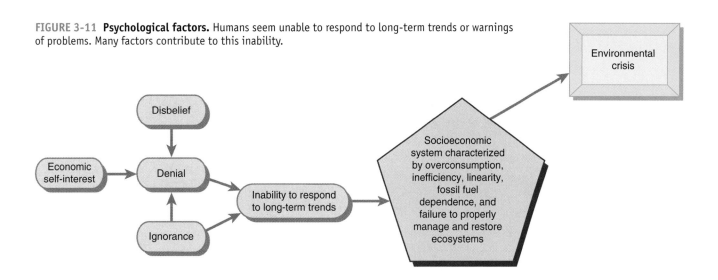

FIGURE 3-11 Psychological factors. Humans seem unable to respond to long-term trends or warnings of problems. Many factors contribute to this inability.

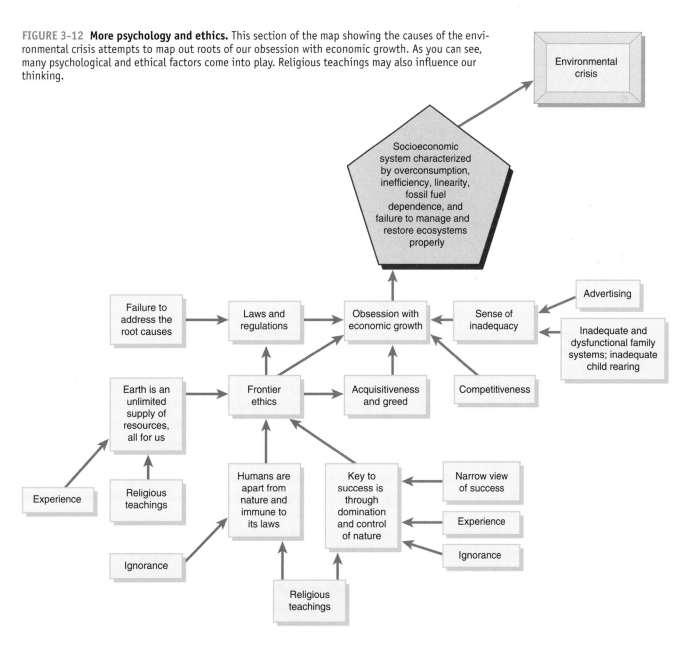

FIGURE 3-12 More psychology and ethics. This section of the map showing the causes of the environmental crisis attempts to map out roots of our obsession with economic growth. As you can see, many psychological and ethical factors come into play. Religious teachings may also influence our thinking.

FOUNDATION TOOL

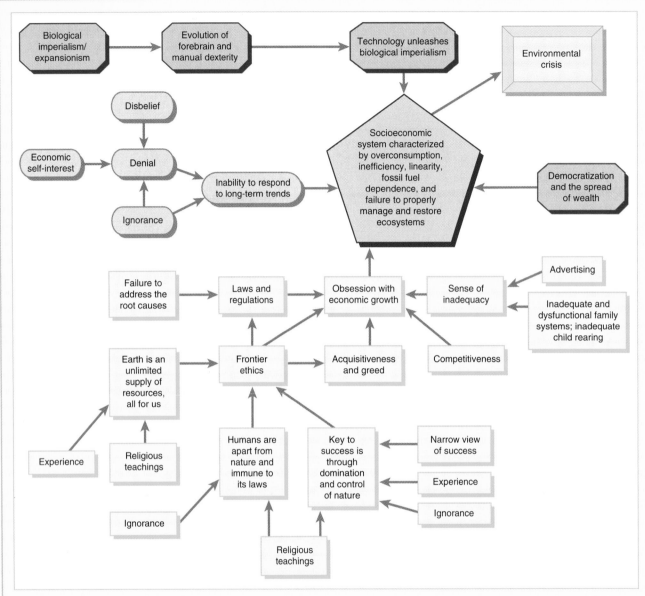

FIGURE 3-13 Many factors contribute to the environmental crisis. They interact in many ways to produce a socioeconomic system that threatens its own existence.

3.2 Leverage Points

Charles Kettering once wrote that a "problem well understood is a problem half solved." One of the key benefits of analyzing the root causes of a problem is that it helps us recognize key leverage points—places where actions would prove most effective. Take a moment to study Figure 3-13; while you do this, think about areas in which action would be most effective.

One area that would become a major point of focus in our efforts to transition to a more sustainable human way of life is technology.

KEY CONCEPTS

Understanding the root causes of the environmental crisis helps us understand key leverage points that can be addressed to help put society back onto a sustainable course.

Making Human Systems and Technologies Sustainable

If the environmental crisis could be distilled to a single sentence, it would be this: *The world's environmental woes result from the fact that human technologies and human systems are fundamentally unsustainable.* Put another way, the environmental problems we face result from the fact that the systems that supply our many needs were designed in ways that do not respect the limits of the Earth's carrying capacity. Human systems may have been sufficient in a world of two billion people, but today, with the world's population exceeding the six billion mark and growing by approximately 80 million per year, evidence strongly suggests that the ways we produce goods and services, house our people, handle our waste, and so on are fundamentally unsustainable. Just as hazardous waste and pollution are a sign of failed technology, the world's many environmental problems are symptoms of systems that are inadequately designed for a world of limits. Chapter 2 explores reasons why human systems are unsustainable and outlines many practical ways to make them sustainable. In the next section, I present an alternative worldview, a sustainable ethic to ponder as you study this subject. This ethic may be one of the most fundamental leverage points for change.

KEY CONCEPTS

Unsustainable human systems are a result of many factors. Building a sustainable society will require a restructuring of basic human systems. Changes in technology will be essential to this transition.

A New Worldview: Changing Our Perceptions, Values, and Beliefs

The material in the previous sections indicates the importance of our perceptions, values, and beliefs. As you saw in Chapter 2, science can shape our perceptions—how we see the world and the nature of the problems facing us—and our understanding of many issues. This book could change your values and beliefs. Creating the broad cultural support for sustainability, however, will very likely require much more in the way of education. One recommendation for changing these important factors is to institute environmental courses at all levels of the educational system—from kindergarten through college—throughout the world. These could help play a role in educating all people. They could help develop a more accurate picture of the state of affairs—how serious the environmental crisis is—and could help people understand the many options we have to forge sustainable lifestyles and a sustainable economy. Further ideas on changing our values are outlined throughout this text and are summarized in Chapter 25. Many people believe that fundamental changes in our ethics are required.

KEY CONCEPTS

Many observers believe that creating a sustainable society will require profound changes in our understanding of issues, through education. Changes in values are also essential.

Unsustainable Ethics

Many people agree that building a sustainable future will very likely require a new system of values or **ethics**. The system of values of a person, religion, group, or even a nation determines how it acts. The prevalent system of Earth values in Western culture, as well as others, is the **frontier ethic**. It embraces a rather narrow view of humans in the environment and an even narrower view of the purpose of nature. The frontier ethic is characterized by three tenets: (1) The Earth is an unlimited supply of resources for exclusive human use—there's always more, and it's all for us; (2) humans are apart from nature and immune to its laws; and (3) human prosperity and well-being result from our efforts to control the environment. In other words, nature is something to overcome or conquer. We get ahead by reshaping natural systems to our liking. Let's consider each of these tenets in more detail.

KEY CONCEPTS

In most Western nations, human values express a sense of frontierism that causes us to pursue our own interests at the expense of the environment and the long-term future of human society.

The First Tenet of Unsustainable Ethics The view of the Earth as an unlimited supply of resources for exclusive human use no doubt evolved in prehistoric time when human numbers were small and the Earth's resources did indeed appear inexhaustible. The massive increase in economic activity and the upsurge in population growth in the last 200 years, however, have brought us face to face with the limits of the planet. Since 1920, dozens of ocean fisheries have been depleted. Dozens more are now threatened. In the past 100 years, half of the world's rain forests have been destroyed, along with countless species of plants and animals. Within most of our lifetimes, tropical rain forests could be wiped out. Oil and natural gas supplies are also likely to run out.

Despite growing evidence that our frontiers have vanished, many people still persist in viewing the Earth as an unlimited supply of resources.

KEY CONCEPTS

One of the main tenets of the frontier ethic is that the Earth is an unlimited supply of resources for exclusive human use—that is, there's always more, and it's all for us.

The Second Tenet of Unsustainable Ethics Humankind positioned itself outside the realm of nature for thousands of years. Today, many people continue to view humans as separate from nature, and they persist in thinking that we can do what we please without harm. The French philosopher and writer Albert Camus summed our philosophy of separation best when he wrote, "Man is the only creature that refuses to be what he is." In reality, says ecologist Raymond Dasmann, "A human apart from environment is an abstraction—in reality no such thing exists." Our lives are intricately linked to the natural world. How?

A huge portion of the oxygen we breathe each year is replenished by plants and algae. All of the food we eat comes from plants, soil, water, and air. Even plastic comes from organisms (phytoplankton) that lived on the Earth several hundred million years ago. To think of ourselves outside of the realm of nature is not only foolish, it is suicidal.

KEY CONCEPTS

The second tenet of the frontier ethic is that humans are apart from nature, not a part of it. As a result, we often assume that we are immune to natural laws that govern other species.

The Third Tenet of Unsustainable Ethics Industrial nations view nature as a force that must be conquered and subjugated. Matthew Arnold, a 19th-century British poet, summed up the modern view best when he wrote: "Nature and man can never be fast friends. Fool, if thou canst not pass her, rest her slave!" That is, if we can't subdue nature, we will become a slave to natural forces.

Believing in the need to reign supreme, humankind has conquered rivers and subdued the wilderness. In the past few decades, it has become clear to many that our acts of conquest have not come without a cost. In fact, many of the environmental problems we face today stem from our efforts to control and dominate nature. Such impacts are often called *ecological backlashes*.

A good example of ecological backlash can be seen on today's chemical-intensive farms. To increase crop production, many farmers use an assortment of pesticides, chemicals used to kill organisms that farmers view as pests. However, research has shown that pesticide use may be a Pyrrhic victory—one that comes at a high cost. Why? Since World War II, chemical pesticide application has increased 10-fold, while pest damage has doubled. Today, 540 species of insects are resistant to at least one chemical pesticide, and 20 insect pests are resistant to every known pesticide.

KEY CONCEPTS

The third tenet of sustainable ethics is that success comes through domination and control of nature.

Some Impacts of Frontier Thinking   Frontier ethics profoundly affects how people act and, as the previous material has shown, has led to numerous environmental problems. It also affects our attitudes about the seriousness of environmental problems. Projections of declining fossil fuel supplies, for instance, generally spark little interest among people convinced that supplies will never run out—or that if they do, scientists will find substitutes.

The frontier ethic also influences how people solve environmental problems. The solution to impending oil shortages, for example, is to increase our search for new supplies, even if it takes us into pristine wildernesses like Alaska's Arctic National Wildlife Refuge. Our faith in unlimited resources prevents many among us from finding ways to use resources more efficiently. Why be efficient if resources are unlimited?

The frontier mentality also permeates our own lives, influencing personal goals and expectations. Most of us tend to make our buying decisions on the basis of what we can afford. Very rarely do we ask whether a decision to purchase a product helps or hinders the environment. Rarely do we think about the product's effects on the long-term habitability of the planet. If the world is an unlimited supply of resources, why should we?

Over the years, the unquestioning acceptance of frontier ethics has spawned a way of life that, while successful in many ways, threatens our own future. Truly, frontier ethics has been a fundamental driving force in the creation of human systems that are out of sync with the natural systems that support them.

Societies that are out of sync with natural systems are unsustainable. They grow crops at the expense of their soils.

They may also destroy wetlands, which are a vital habitat for many species of fish—which in turn are an important source of human food. Many other examples of this phenomenon are given in this book.

In contrast, sustainable societies satisfy their needs for food, shelter, and other resources without destroying the natural systems upon which they depend. For example, they grow crops without damaging their soils. They harvest trees while protecting forests and without clogging nearby streams and lakes with sediment eroded from the land. Such communities exist in harmony with nature (see Spotlight on Sustainable Development 3-1).

A sustainable society does not merely prevent destruction; it also seeks ways to enhance natural systems. Sustainable farming, for example, increases the organic content of the soil, which makes land more productive. Ultimately, creating a sustainable society means fitting human societies within the economy of nature. It is about creating a better future, a stronger nation, and broad prosperity.

KEY CONCEPTS

Frontier thinking influences how we design and operate all human systems, from the economy to government to education to waste management.

Toward a Sustainable Ethic   Creating a sustainable society that lives within the Earth's means may require the world's people to adopt a new, sustainable ethic. **Sustainable ethics** contrast sharply with the frontier ethics described earlier. They assert that (1) the Earth has a limited supply of resources, and they're not all for us; (2) humans are a part of nature, subject to its laws; and (3) success stems from efforts to cooperate with the forces of nature. I like to add a fourth element of sustainable ethics to remind us of the importance of the natural world. It is that our future depends on creating and maintaining a healthy, well-functioning global ecosystem. An **ecosystem**, described in detail in Chapter 4, consists of a community of organisms and all of the interactions between them and their physical environment.

As you can see, the principles of sustainable ethics are the opposite of the principles of frontier ethics. Just as frontier ethics leads us to exploitive behavior, sustainable ethics could lead to a less exploitive human presence, one that could endure for thousands of years and greatly improve our chances of long-term success. Sustainability is not about limiting our future but ensuring it—creating a better world.

KEY CONCEPTS

Creating a sustainable society that protects the environment will require a new value system—one that respects limits, sees humans as a part of the natural world, and recognizes the need to cooperate with nature, not dominate it.

Never does nature say one thing and wisdom another.

—Juvenal

3-1   Ashland, Oregon, Shows That Water Conservation Works

Ashland is a small city in Oregon with 20,600 inhabitants. Like a growing number of cities and towns, Ashland is showing the value of conservation as a means of saving money and providing for the future needs of people, with little or no environmental impact. Ashland's story is an inspiring example of critical thinking—more specifically, of questioning conventional ideas and conventional means of supplying resources.

This story begins in 1989, when city officials started discussions about a potentially severe problem: future water shortages that would be caused when the city lost badly needed water rights in just 8 years. This would reduce its supply of drinking water considerably. Accordingly, the officials contacted a consulting firm that, after reviewing the problem, proposed the construction of an $11 million dam on nearby Ashland Creek (for which the city held water rights). According to the consultants, the dam and reservoir would supply 2 million liters (500,000 gallons) of water a day for the city and prevent potential water shortages.

Having adopted progressive energy conservation standards 8 years earlier (after reading the writings of Amory and Hunter Lovins of the Rocky Mountain Institute), the city officials thought that conservation might be a better way of making up for the potential loss. City officials also realized that their town of 20,600 would be financially hard-pressed to pay for an $11 million dam. Consequently, they contacted another consulting firm and asked them to look into the feasibility of making up for the lost water by using conservation—a program to install water-efficient showerheads and faucets and low-flush toilets. After some study, the company found that at one-twelfth the cost of the dam—or $825,000—the city could phase in a water conservation program that could save the required 2 million liters (500,000 gallons) of water a day relatively painlessly. The net effect: They wouldn't need to dam the creek. They'd also save more than $10 million! Their solution made sense from economic, social, and environmental perspectives—and is a tribute to the value of critical thinking: questioning traditional ways.

For years, water shortages have been addressed by building new dams. But supplies of water and many other resources can be stretched considerably by using what we already have more efficiently. Such efforts often cost a fraction of traditional approaches and are an important part of the strategy for sustainable development.

CRITICAL THINKING

Exercise Analysis

This exercise is very personal, and I won't attempt to cover what you should or should not think. That's up to you, and I guarantee it will probably change as you learn more about the environment. I can assure you that the more you think about what you believe, the better off you'll be. The more open you are to new ideas and alternative viewpoints, the more apt you are to make good judgments. Be patient, though. It may take a long time to arrive at the truth. (This is another rule of critical thinking!)

I will, however, spend a few moments talking about where your ideas and values come from. Most of the ideas expressed in this exercise are expressions of values. These come from many sources. Our parents, for example, often help forge our values. Sometimes they teach us through example—positive and negative. Sometimes they lecture us in an attempt to instill values. We also acquire values from teachers, ministers, friends, relatives, government leaders, speakers, books, documentaries, movies, and advertisements. Almost everything we read or see or hear has some value statement behind it. Values often shape our ideas and actions.

As noted in Chapter 2, science can also shape values. You will be learning a great deal about the science of the environment and sustainability in this book, facts that may change both your thinking and your answers to the questions posed during this exercise.

CRITICAL THINKING AND CONCEPT REVIEW

1. Everyone seems to have a personal opinion on the root causes of the environmental crisis—or even if there is a crisis in the first place. Why do you think this is so?

2. Describe the views of Professor Lynn White on the religious roots of the environmental crisis. Do Christian teachings about dominance and about humans being apart from nature (duality) influence the way you think? How might science and technology be influenced by these Christian teachings?

3. Describe some of the criticisms of Lynn White's hypothesis. Do you see any flaws in his arguments? Does this mean that his hypothesis is invalid or may be just part of the picture?

4. Describe the way in which some scholars think that the spread of democracy and industrialization, especially mass production, may have contributed to the environmental crisis.

5. I use the term *biological imperialism* in this chapter to describe a natural biological tendency of species. What is this tendency? How has this tendency in humans contributed to the environmental crisis? What factors make humans superior biological imperialists? How do values and beliefs affect our actions? *(Hint: You can think of both positive and negative influences on human beliefs.)*

6. If technology is so central to the environmental crisis, what can be done to make technology work for us, instead of undermining our future?

7. Make a list of the psychological factors discussed in this chapter and explain each one. Then list ways these psychological factors affect our laws, government, businesses, and personal lives.

8. Make a list of the systems that underlie your community and describe why you think they might be unsustainable in the long run. To help you think about this, you might consider their impacts on the local, regional, and global economy, society, and environment.

KEY TERMS

biological imperialism
crisis of unstainability
ecosystem

ethics
frontier ethic
Frontierism

Industrial Revolution
sustainable ethics

REFERENCES AND FURTHER READING

The References and Further Reading section at the end of this book contains a list of sources for the information discussed in this chapter and recommendations for further reading.

Connect to this book's website:
http://environment.jbpub.com/
The site features eLearning, an online review area that provides quizzes, chapter outlines, and other tools to help you study for your class. You can also follow useful links for in-depth information, research the differing views in the Point/Counterpoints, or keep up on the latest environmental news.

A northern coniferous forest, the taiga biome

CHAPTER 4 — Principles of Ecology: How Ecosystems Work

Never does nature say one thing, and wisdom another.

—*Juvenal*

Most of us live our lives seemingly apart from nature. We make our homes in cities and towns, surround ourselves with concrete and steel, and drown out the sound of birds with our noise. The closest many urbanites get to nature is a romp with the family dog on the grass in the backyard. A lucky few come in much closer contact with the great outdoors through hiking, camping, canoeing, and kayaking. For many of these people, though, nature is still viewed as something apart from humans—a thing to protect to preserve a few pristine places for people to enjoy.

4.1 Humans and Nature: The Vital Connections

Hard as it may be for many people to accept, human beings are part of the fabric of life. We are a part of nature. We are dependent on the Earth and natural systems in thousands of ways and are an integral part of the cycles of nature. Consider our de-

Exercise

The information gained from various fields of science such as ecology is often loosely translated in the public arena. Terms are sometimes misinterpreted. Facts are taken out of context. New findings are given more credence than they deserve, and old, disproved ideas remain in the popular thinking for a long time. As you read this chapter, make a list of terms, ideas, concepts, and facts you encounter that contradict what you thought was true.

pendence first by taking a look around the room in which you are sitting. Everything in that room comes from the Earth or a natural system. The clothes you wear, your morning tea or coffee, and even the cornflakes you ate for breakfast are products of the Earth—the soil, water, air, and plants.

Like all other species, humans depend on the soil, air, water, sun, and a host of living organisms to survive. Each year, in fact, human beings (and other animals) consume enormous quantities of oxygen, which is used in the cells of our bodies to break down food molecules to generate energy. Oxygen is produced by plants and algae. Without these organisms, humans and other animals could not survive. Trees, grasses, and other plants also provide a host of additional free services. For example, plants protect the watersheds near our homes, preventing flooding and erosion. Swamps purify the water in streams and lakes—water many of us drink. Birds help to control insect populations.

Clearly, nature serves us well. Although many of us have isolated ourselves from nature, we still depend on nature in many ways. We have not emancipated ourselves from it at all. We also influence natural cycles, and therefore, as Chapter 1 explained, not only do we depend on nature, the fate of natural systems depends on us.

KEY CONCEPTS

Humans are a part of nature, dependent on natural systems for a variety of economically important resources and ecological services essential to our survival and long-term prosperity.

4.2 Ecology: The Study of Natural Systems

This chapter explores *ecology*, the study of living organisms and the web of relationships that binds all of us together in nature. Professor Garrett Hardin, a world-renowned ecologist, wrote that ecology takes as its domain the entire living world. Ecologists study how organisms interact with one another and how they interact with the *abiotic*, or nonliv-

ing, components of the environment (sunlight, for example). Throughout this chapter, we explore our connections to the living world. We examine the ways in which human systems depend on natural systems and the ways in which humans affect them. One of the goals of this chapter is to help you understand how nature works and how we can work better with nature to create a sustainable future. You will find that a great many of the lessons learned from the study of ecology can be applied to human society. Before we begin our journey, however, let me say a few words about the term *ecology*.

Ecology is probably one of the most misused words in the English language. Banners proclaim, "Save Our Ecology." Speakers argue that "our ecology is in danger," and others talk about the "ecological movement." These common uses of the word *ecology* are incorrect. Why?

Ecology is a branch of science. It describes and quantifies the web of interactions in the environment. But ecology is not synonymous with the word *environment*. Thus, we can save our ecology department and ecology textbooks, but we cannot save our ecology. Our ecology is not in danger, but our environment is. You cannot join the ecology movement, but you would be a welcome addition to the environmental movement.

KEY CONCEPTS

Ecology is a field of science that seeks to describe relationships between organisms and their chemical and physical environment.

4.3 The Structure of Natural Systems

In order to understand ecology, you must study the structure of natural systems. We begin with the biosphere.

The Biosphere

The science of ecology focuses much of its attention on biological systems, examining their components and interactions. The largest biological system is the **biosphere** (BI-oh-sfere), the skin of life on planet Earth. As shown in FIGURE 4-1, the biosphere forms at the intersection of air, water, and land. In fact, all living organisms consist of components derived from these three realms. The carbon atoms in body proteins, for example, come from carbon dioxide in the atmosphere. Carbon dioxide is captured by plants and made into food molecules by a process known as **photosynthesis**. Animals eat plants; the food molecules then become the building blocks of proteins and other important molecules in animals. The minerals in the bones of animals come from the soil, again through plants. Water comes directly from streams and lakes and plant matter.

The biosphere extends from the bottom of the ocean, approximately 11,000 meters (36,000 feet) below the surface, to the tops of the highest mountains, about 9000 meters (30,000 feet) above sea level. Although that may seem like a long way, it's really not. In fact, if the Earth were the size of an apple, the biosphere would be about the thickness of

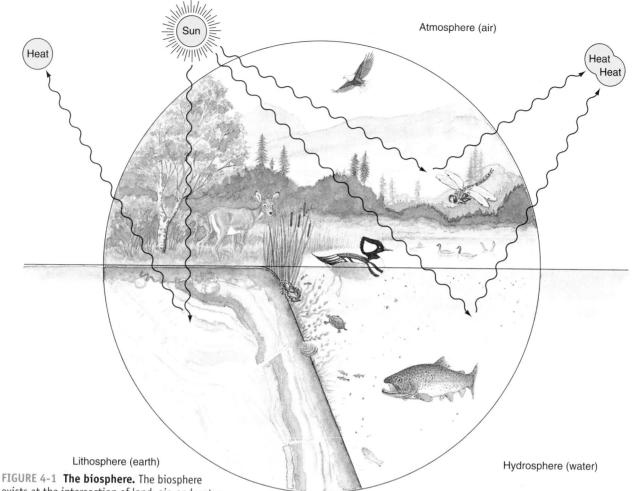

Heat

Sun

Atmosphere (air)

Heat
Heat

Lithosphere (earth)

Hydrosphere (water)

FIGURE 4-1 **The biosphere.** The biosphere exists at the intersection of land, air, and water. Organisms derive essential minerals and other substances from these three spheres.

its skin. Although life exists throughout the biosphere, it is rare at the extremes, where conditions for survival are less than optimum. Most living things are concentrated in a narrow band extending from less than 200 meters (600 feet) below the surface of the ocean to about 6000 meters (20,000 feet) above sea level.

The biosphere is a **closed system**, much like a sealed terrarium. By definition, a closed system receives no materials from the outside. The only outside contribution is sunlight, which is vital to the health and well-being of virtually all life.

Sunlight powers almost all life on the planet. Even the energy released by the combustion of coal, oil, and natural gas (which we use to power our homes and factories) comes from sunlight that fell on the Earth several hundred million years ago.

Because the Earth is a closed system, all materials necessary for life must be recycled. The carbon dioxide you exhale, for instance, may be used by a rice plant during photosynthesis next month in Indonesia. Those carbon dioxide molecules will be incorporated into carbohydrate produced by the plant and stored in the seed. Consumed by an Indonesian boy, the carbohydrate will be broken back down during cellular energy production. The carbon dioxide molecules are released into the atmosphere. Without this and dozens of other

recycling processes, all life on the planet would grind to a halt. Protecting the environment, then, helps to protect global recycling systems on which we all depend.

KEY CONCEPTS

The biosphere is an enormous biological system, spanning the entire planet. The materials within this closed system are recycled over and over in order for life to be sustained. The only outside contribution to the biosphere is sunlight, which provides energy for all living things.

Biomes and Aquatic Life Zones

The biosphere consists of terrestrial and aquatic systems. Viewed from outer space, the Earth—the terrestrial portion of the biosphere—resembles a giant jigsaw puzzle, consisting of large landmasses among vast expanses of ocean. The landmasses, or continents, can be divided

> Although the biosphere extends from the bottom of the ocean to the tops of the tallest mountains, most living things are concentrated in a narrow band extending from less than 200 meters (600 feet) below the surface of the ocean to about 6000 meters (20,000 feet) above sea level.

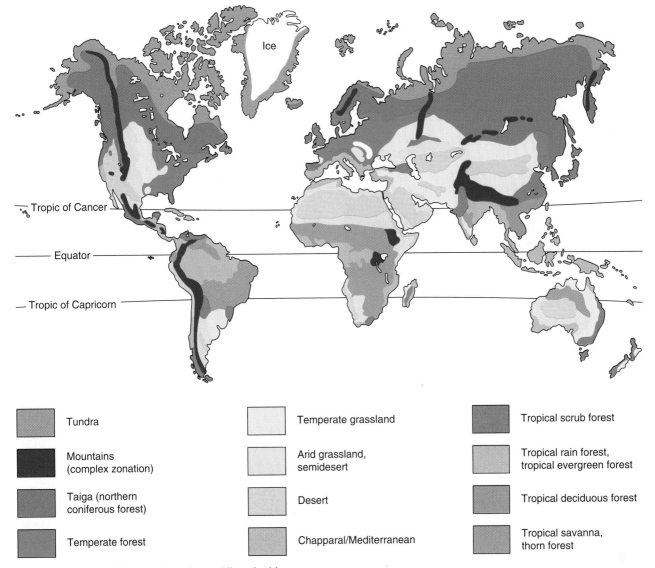

Ice

Tropic of Cancer

Equator

Tropic of Capricorn

▢ Tundra	▢ Temperate grassland	▢ Tropical scrub forest
▢ Mountains (complex zonation)	▢ Arid grassland, semidesert	▢ Tropical rain forest, tropical evergreen forest
▢ Taiga (northern coniferous forest)	▢ Desert	▢ Tropical deciduous forest
▢ Temperate forest	▢ Chapparal/Mediterranean	▢ Tropical savanna, thorn forest

FIGURE 4-2 Biomes. This map shows the world's major biomes.

into fairly large regions called *biomes* (**FIGURE 4-2**). A **biome** is a terrestrial portion of the biosphere characterized by a distinct climate and a particular assemblage of plants and animals adapted to it.[1] Chapter 5 describes the biomes in detail. This section provides an overview.

In a biome, abiotic conditions—such as soil type, temperature, and rainfall—determine the plant communities that can survive. These, in turn, determine which animals can subsist. As illustrated in Figure 4-2, the North American continent contains seven major biomes, five of which are discussed here. Starting in the north is the **tundra** (TON-dra), a region of long, cold winters and rather short growing seasons (**FIGURE 4-3a**). The rolling terrain of the tundra supports grasses, mosses, lichens, wolves, musk oxen, and other animals adapted to the bitter winter cold. Trees cannot grow on the tundra because of the short growing season and be-

cause the subsoil (called *permafrost*) remains frozen year round, preventing the deep root growth necessary for trees.

Immediately south of the tundra lies the **taiga** (TIE-ga), also known as the **northern coniferous** or **boreal, forest**. The taiga's milder climate and longer growing season result in a greater diversity and abundance of plant and animal life than exists on the tundra. Evergreen trees, bears, wolverines, and moose are characteristic species (**FIGURE 4-3b**).

East of the Mississippi River lies the **temperate deciduous forest biome**, characterized by an even warmer climate and more abundant rainfall (**FIGURE 4-3c**). Broad-leaved trees make their home in this biome. Opossums, black bears, squirrels, and foxes are characteristic animal species.

West of the Mississippi lies the **grassland biome** (**FIGURE 4-3d**). Inadequate rainfall and periodic drought prevent trees from growing on the grasslands, except near rivers, streams, and human habitation. Over the years, deep-rooted grasses have evolved on the plains. These grasses can withstand fire, drought, and grazing even in arid grasslands. Coyotes, hawks, and voles are characteristic animal species.

---

[1]Although each biome has its specific climate, there can be considerable climatic variation within a given biome.

(a)

(b)

(c)

(d)

(e)

In the Southwest, where even less rain falls, is the **desert** biome (FIGURE 4-3e). In contrast to what many people think, the desert often contains a rich assortment of plants and animals uniquely adapted to aridity and heat. Cacti, mesquite trees, rattlesnakes, and a variety of lizards all make their home in this seemingly inhospitable environment. The scorching hot deserts

FIGURE 4-3 **North America's five major biomes. (a)** Tundra, **(b)** taiga, **(c)** temperate deciduous forest, **(d)** grassland, and **(e)** desert.

of Saudi Arabia receive less moisture than the desert around Tucson and therefore contain far fewer plants and animals.

The oceans can also be divided into distinct zones, known as **aquatic life zones**. The aquatic equivalent of biomes, each of these regions has a distinct environment and characteristic plant and animal life adapted to conditions of the zone. Both freshwater and saltwater (marine) systems exist. Freshwater systems include lakes, rivers, ponds, and marshes. The four major marine aquatic life zones are coral reefs, estuaries (the mouths of rivers, where fresh and salt water mix), the deep ocean, and the continental shelf, all of which are discussed in the next chapter.

Humans inhabit all biomes on Earth but are concentrated in those in which conditions are mildest and most conducive to growing food. Within many biomes is a wide assortment of natural resources such as minerals and fuel. Although they are vital to our economies and well-being, these resources are but a fraction of the benefit we gain from the biome. Plants and animals, for example, provide food and great enjoyment to many. Microorganisms in the soils of the biome detoxify wastes and recycle nutrients, keeping the soil rich and productive. The soil itself serves as a growing medium for all plants—crops as well as forests. Besides providing timber, trees provide oxygen, remove air pollutants, and protect the soil from erosion, thus reducing sediment pollution in surface waters. Vegetation also reduces flooding.

KEY CONCEPTS

The biosphere consists of distinct regions called *biomes* and aquatic life zones, each with its own chemical and physical conditions and unique assemblage of organisms. Humans inhabit all biomes, but are most prevalent in those with the mildest climates.

What Is an Ecosystem?

The biosphere is a chemical, physical, and biological system that encompasses the entire surface of the planet. Therefore, the biosphere is often referred to as a global **ecological system**, or **ecosystem**. If the biosphere is a global ecosystem, biomes are regional ecosystems. For the sake of convenience, ecologists often limit their studies to smaller portions of a biome—individual forests, ponds, or meadows. All of these ecosystems, no matter how big or small, consist of two components: the living or **biotic** components and the nonliving or **abiotic** components. Numerous interactions exist among these various components.

KEY CONCEPTS

Ecosystems are biological systems consisting of organisms and their environment.

Abiotic Components of Ecosystems and the Range of Tolerance The abiotic components of an ecosystem are the physical and chemical factors necessary for life—sunlight, precipitation, temperature, and nutrients. In most ecosystems, the abiotic conditions vary during the day and often shift from one season to the next. To live in most ecosystems, then, organisms must be able to survive a range of conditions. The range of conditions to which an organism is adapted is called its **range of tolerance**. As **FIGURE 4-4** shows, organisms do best in the **optimum range**. Outside of that are the **zones of physiological stress**, where survival and reproduction are possible but not optimal. Outside of these zones are the **zones of intolerance**, where life for that organism is implausible. As a rule, organisms that have a wide range of tolerance are

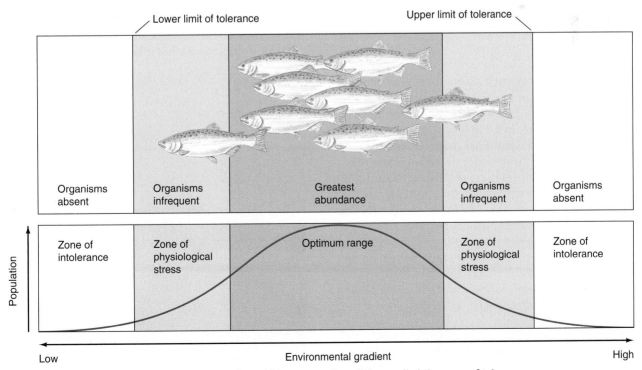

FIGURE 4-4 **Range of tolerance.** Organisms can live within a range of conditions, called the range of tolerance, but they thrive in the optimum range.

(a)

(b)

FIGURE 4-5 **Upsetting the ecological balance. (a)** Dams like the mighty Glen Canyon Dam in Arizona not only inundate upstream areas, they often change the water temperature in the downstream section. **(b)** The cool water flowing out of the bottom of the reservoir created by Glen Canyon Dam has endangered the razorback sucker and several other native fish species.

more widely distributed than organisms with a narrow range of tolerance.

Many examples of the range of tolerance can be given. Think of your own tolerance for temperature. Most people are comfortable around 23°C (70°F). When the temperature becomes much warmer, you enter the zone of physiological stress. If it increases even more, death occurs. This is the zone of intolerance. The same occurs at the lower end of the temperature scale.

It is important to note that the tolerance of any given species for an environmental factor may vary with the age of the organism. Newly hatched salmon, for instance, may be much more vulnerable to pollutants in the water than adult salmon are. The tolerance of individuals in any population also varies. Thus, some humans are more tolerant than others to heat or cold.

Humans often alter the abiotic components of their environment. Dams, for example, create lakes in streambeds, altering water temperature and water flow, with devastating effects on native fish populations. Dams also seriously alter downstream conditions. In the Colorado River, for instance, native razorback suckers and other species that once thrived in the warm waters of the river are now **endangered species** (in danger of extinction) because of the large dams that release extremely cold water from the bottom of the reservoir (FIGURE 4-5a). This water is too cold for the sucker and other natives (FIGURE 4-5b). (The Point/Counterpoint in this chapter gives two opposing views on human-caused extinction.)

Natural events can also alter the biotic and abiotic conditions of the environment. Floods, tropical storms, and volcanic eruptions all change conditions for varying periods, with sometimes dramatic effects on native species.

KEY CONCEPTS

Organisms thrive within a range of abiotic conditions; altering those conditions can have severe consequences and can even cause extinction.

Limiting Factors Although species are sensitive to all of the abiotic factors in their environment, the one factor that is in short supply, called a **limiting factor**, tends to regulate population size. In freshwater lakes and rivers, dissolved phosphate is a limiting factor. Phosphate is needed by plants and algae for growth, but phosphate concentrations are naturally low. As a result, plant and algal growth is held in check. When phosphate is added to a body of water—say, by detergents released in the effluent of a sewage treatment plant—plants and algae proliferate. Algae often form dense surface mats, blocking sunlight (FIGURE 4-6).

Plants that are rooted on the bottom of the water body may perish for lack of sunlight. Because these plants produce oxygen, their demise often causes oxygen levels in the deeper waters to decline, killing fish and other aquatic organisms. On land, precipitation tends to be the limiting factor. At any

FIGURE 4-6 **Algal bloom.** This pond is choked with algae due to the abundance of plant nutrients from human sources.

given temperature, the more moisture that falls, the richer the plant and animal life.

KEY CONCEPTS

Organisms require many different abiotic factors to survive, but one factor—*the limiting factor*—tends to be critical to survival and growth of a population. Altering concentrations of limiting factors can result in dramatic fluctuations in populations.

Biotic Components Ecosystems consist of numerous organisms, including bacteria, plants, fungi, and animals. These are the biotic components of the ecosystem. Organisms of the same species within a biome or aquatic life zone usually occupy a specific region. This group of organisms is called a **population**. In any given ecosystem, populations of different organisms exist together in an interdependent **biological community**.

All organisms in a community, including humans, are part of the web of life. Organisms interact in many ways. Some organisms prey on others. Others are preyed upon. Still others compete with fellow members of the community of life.

KEY CONCEPTS

Organisms are the biotic components of ecosystems; they form an interdependent community of life.

Niche and Competition If asked to give a brief description of yourself, you would probably begin by describing the place where you live. You might discuss the work you do, the friends you have, and other important relationships that describe your place in human society. Now that you're discovering your part in nature, you may even describe your place in the community of life.

A biologist would do much the same when describing an organism. He or she would start with a description of the place where an organism lives—that is, its **habitat**. Next, he or she would describe how the organism fits into the ecosystem, its **ecological niche**, or simply **niche**. An organism's niche consists of its relationships with its environment—both the abiotic and biotic components. This niche includes what an organism eats, what eats it, its range of tolerance for various environmental factors, and other important elements. The niche of an organism is its functional role compared with the habitat, which is its "address."

Organisms in a community occupy the same habitat, but most of them have quite different niches, a phenomenon that minimizes competition. The fact that organisms occupy separate niches provides for a wider use of an ecosystem's resources, especially food.

Niches do overlap somewhat. For example, two species may feed on some of the same foods. Coyotes and foxes, for example, both feed on rabbits and mice. Coyotes also feed on larger prey, however—even deer. Foxes tend to feed on smaller prey, including reptiles and amphibians.

The more two species' niches overlap, the more they compete. When niches overlap considerably, competition becomes intense, and one species usually suffers. If two species occupy identical niches, competition will eliminate one of them. As a result, two species cannot occupy the same niche for long. This rule is called the **competitive exclusion principle**.

The concept of the niche is very important to the sustainable management of natural resources. For example, successful control of an insect pest on crops is best achieved through an understanding of the species' niche. An analysis of an insect's niche might show that certain birds or insects feed on the pest. By encouraging these beneficial species—say, by providing trees for the birds—farmers can reduce pest populations naturally. They could also save enormous sums of money on chemical pesticides and help protect the environment.

KEY CONCEPTS

Competition occurs between species occupying the same habitat if their niches overlap; although competition is a naturally occurring process, natural systems have evolved to minimize it.

Humans: Competitors Extraordinaire Humans, like other organisms, compete with one another for a variety of resources. People also compete with the many other species that share this planet with us. For example, we compete with sea otters and seals over salmon. When we graze a cow on a pasture, our livestock compete with woodchucks and rabbits for food.

The competition between people and other species is fairly lopsided. As noted in Chapter 3, humans possess a marked advantage over most other species. This edge on the competition stems primarily from our technological prowess—high-powered rifles, bulldozers, and chain saws being three examples. Fishing nets and sonar give commercial fishers a marked advantage over their two major competitors, seals and sea otters.

As the human population grows, as our demand for food and resources climbs, and as our technological prowess increases, our competitive advantage will only increase, but the effect may not be advantageous to human civilization in the long run. Already, commercial overfishing has depleted dozens of the world's fisheries (Figure 2-6). Many other fisheries are now in danger, as noted in Chapter 2. Overfishing has had a ripple effect on other species, reducing populations of seals and other fish-eating animals.

On every continent and in every nation, humans are outcompeting other species. According to various estimates, 40 to 100 species become extinct every day, largely because of tropical deforestation. Unless we do something, hundreds of thousands of species will become extinct in the next decade. Many scientists warn that none of us will be immune to the impacts of such widespread biological impoverishment. Cutting down the rain forests, for example, could alter global climate. Why? Rain forests absorb massive amounts of carbon dioxide and thus help control atmospheric levels of this important greenhouse gas. As forests are destroyed, carbon dioxide levels increase, and the Earth's atmosphere could become warmer (Chapter 20). We could feel the effect in higher utility bills to cool our homes because

Humans Are Accelerating Extinction

David M. Armstrong

Dr. David M. Armstrong is a mammologist who teaches environmental biology at the University of Colorado at Boulder and has written several books on mammals and ecology of the Rocky Mountain region.

Evolution is the process of change in gene pools through time. When one gene pool becomes reproductively independent, a new species has formed. Such speciation generates species; extinction takes them away. Simply put, extinction is a failure to adapt to change, the termination of a gene pool, the end of an evolutionary line.

Extinction is a natural process. Most species that ever lived are now extinct. The 3 to 30 million species on Earth today are no more than 1 to 10% of the species that have evolved since life began about 3.5 billion years ago. So why are thoughtful people concerned about endangered species? History makes it clear that, given enough time, each species will become extinct.

The concern is that today the natural process of extinction proceeds at an unnatural rate. Let us estimate by how much human activity has accelerated rates of extinction. The average life span of species is 1 to 10 million years. Assume (to be conservative) that the average longevity of species of higher vertebrates is 1 million years. In round numbers, there are 13,000 species of birds and mammals. So, on average, one species ought to go extinct each century. However, between 1600 and 2000, at least 36 species of mammals and 94 species of birds became extinct—32 species per year, or 32 times the natural rate.

What does it mean to increase a rate by 32 times? Exceed the 55 mph speed limit by 32-fold, and you are moving 1760 miles per hour, over twice the speed of sound. The difference between natural rates of extinction and present, human-influenced rates is analogous to the difference between a casual drive and Mach 2! Is that a problem? You decide: concern is a moral construct, not a scientific one.

Several human activities have contributed—mostly inadvertently—to accelerating rates of extinction. The dodo and the passenger pigeon were extinguished by overhunting. Wolves and grizzly bears were exterminated over much of their ranges as threats to livestock. The black-footed ferret was driven to the verge of extinction because prairie dogs, its staple food, were poisoned as an agricultural pest. The smallpox virus was exterminated in the "wild" (but sur-

vives in a half-dozen laboratories). This is the closest that humans have come to deliberate elimination of an organism, and note that thoughtful scientists with the power to destroy smallpox chose not to do so, electing instead to manage it with care.

Habitat change is the most important cause of endangerment and extinction. Clearing forests for agriculture has decimated the lemurs of Madagascar. Pesticides led to the decline of the peregrine falcon. Introducing exotic species (like goats on the Galapagos and mongooses in Hawaii) displaces native animals and plants. Developing the Amazon Basin is a habitat alteration, and a cause of extinction, on an unprecedented scale.

Many urge saving species for their aesthetic value. Whooping cranes are beautiful, and part of their beauty is that they are products of a marvelous evolutionary process. Most concern about accelerated extinction, however, stresses economic value. A tiny fraction of seed plants are used commercially. An obscure plant like jojoba eventually could be a source of oil more reliable than the Middle East. Wild grasses have yielded genes that have improved disease resistance in wheat. Some animals like musk ox, kudu, and whales could contribute protein to our diet. Species may have medical value; penicillin, after all, was once an obscure mold on citrus fruit. Some sensitive species are monitors of environmental quality, and the presence of healthy populations of many species may promote greater stability or resilience of ecosystems. Naturalist Aldo Leopold noted that humans have a way of "tinkering" with the ecosphere. But he suggested that the first rule of tinkering ought to be that one never throws away any of the parts.

"Extinction is forever" and impoverishes both ecosystems and the potential richness of human life. Borrowing again from Aldo Leopold, I believe our concern about unnaturally rapid extinction is part of a "right relationship" between people and the landscapes that nurture and inspire them.

Biologist Sir Julian Huxley noted that "we humans find ourselves, for better or worse, business agents for the cosmic process of evolution." We hold power over the future of the biosphere, the power to destroy or preserve. German philosopher George Hegel noted that freedom implies responsibility. I agree. The power to destroy species implies a responsibility to preserve them. The question of human-accelerated extinction boils down to a simple ethical question, "Does posterity matter?" Some of us have ethics that are human centered. We ask simply, "Do my children deserve a life as rich, with as much opportunity, as mine?" If they do, then we have the responsibility to choose restraint. (Perhaps my life has been rich enough without the dodo, but I am reluctant to make that judgment for future generations.)

Evolution is the formation of new species from preexisting ones by a process of adaptation to the environment. Evolution began long ago and is still going on. During evolution those species better adapted to the environment replaced the less well adapted species. It is this process, repeated year after year for millennia, that has produced the present mixture of wild species. Perhaps 95% of the species that once existed no longer exist.

Human activities have eliminated many wild species. The dodo is gone, and so is the passenger pigeon. The whooping crane, the California condor, and many other species are on the way out. The bison is still with us because it is protected, and small herds are raised in semicaptivity. The Pacific salmon remains because we provide fish ladders around our dams so it can reach its breeding places. The mountain goat survives because it lives in inaccessible places. But some thousands of other animal species, to say nothing of plants, are extinct, or soon will be.

Some nature lovers weep at this passing and collect money to save species. They make lists of animals and plants that are in danger of extinction and sponsor legislation to save them.

I don't. What the species preservers are trying to do is stop the clock. It cannot and should not be done. Extinction is an inevitable fact of evolution, and it is needed for progress. New species continually arise, and they are better adapted to their environment than those that have died out.

Extinction comes from failure to adapt to a changing environment. The passenger pigeon did not disappear because of hunting alone, but because its food trees were destroyed by land clearing and farming. The prairie chicken cannot find enough of the proper food and nesting places in the cultivated fields that once were prairies.

And you cannot necessarily introduce a new species, even by breeding it in tremendous numbers and putting it out into the wild. Thousands of pheasants were bred and set out year after year in southern Illinois, but in the spring of each year there were none left. Another bird, the capercaillie, is a fine, large game bird in Scandinavia, but every attempt to introduce it into the United States has failed. An introduced species cannot survive unless it is preadapted to its new environment.

A few introduced species are preadapted, and some make spectacular gains. The United States has received the English sparrow, the starling, and the house mouse from Europe, and also the gypsy moth, the European corn borer, the Mediterranean fruit fly, and the Japanese beetle. The United States gave Europe the gray squirrel and the muskrat, among others. The rabbit took over in Australia, at least for a time.

The rabbit and the squirrel were successful on new continents because their requirements are not as narrow as those of other species that failed. Today, adjustment to human-made environments may be just as difficult as adjustment to new continents. The rabbit and the squirrel have succeeded in adjusting to the backyard habitat, but most wild animals have disappeared.

Extinction Is the Course of Nature

Norman D. Levine
Norman D. Levine is a professor emeritus at the College of Veterinary Medicine and Agricultural Experiment Station, University of Illinois at Urbana. His research interests include parasitology, protozology, and human ecology. This article is provided by the American Institute of Biological Sciences (© 1989).

Human-made environments are artificial. People replace mixed grasses, shrubs, and trees with rows of clean-cultivated corn, soybeans, wheat, oats, or alfalfa. Variety has turned into uniform monotony, and the number of species of small vertebrates and invertebrates that can find the proper food to survive has become markedly reduced. But some species have multiplied in these environments and have assumed economic importance; the European corn borer in this country is an example.

Would it improve Earth if even half of the species that have died out were to return? A few starving, shipwrecked sailors might be better off if the dodo were to return, but I would not be. The smallpox virus has been eliminated, except for a few strains in medical laboratories. Should it be brought back? Should we bring panthers back into the eastern states? Think of all the horses that the automobile and tractors have replaced, and of all the streets and roads that have been paved and the wild animals and plants killed as a consequence. Before people arrived in America about 10,000 years ago, the animal–plant situation was quite different. What should we do? Should we all commit suicide?

Evolution exists, and it goes on continually. People are here because of it, but people may be replaced someday. It is neither possible nor desirable to stop it, and that is what we are trying to do when we try to preserve species on their way out. It can be done, I think, but should we do it to them all? Or to just a few, as we are doing now?

Critical Thinking Questions

1. Summarize and critically analyze the key points of both authors. Do you see any flaws in their reasoning?
2. Which viewpoint do you adhere to? Why?

You can link to web sites that represent both sides through Point/Counterpoint: Furthering the Debate at this book's internet site, http://environment.jbpub.com/. Evaluate each side's argument more fully and clarify your own opinion.

of noticeably hotter summers. If predictions come true, we could face much higher food costs as hotter summers reduce crop production. If conditions became bad enough, massive food shortages could occur. Although all this human intrusion is rather depressing, there is hope. We can find

> Scientists estimate that 40 to 100 species become extinct every day, largely as a result of tropical deforestation.

ways of coexisting with nature. Understanding how natural systems operate and patterning human systems after sustainable natural systems could help us redesign our infrastructure. See Spotlight on Sustainable Development 4-1 for an example.

KEY CONCEPTS

Humans are a major competitive force in nature. Our advanced technologies and massive population size permit us to outcompete many species. Destroying other species through competition, however, can be disadvantageous in the long run.

>> SPOTLIGHT ON SUSTAINABLE DEVELOPMENT

4-1 Sustainable Sewage Treatment: Mimicking Nature

Residents of eastern Mexico City produce more than 300 tons of feces every day, much of which is deposited on city streets, vacant lots, and alleys. The feces dry and are often pulverized by cars and trucks. They soon become entrained into the city's dust, creating a monumental health hazard all too common in many poor countries.

Since the advent of cities and towns, dealing with human waste has proved to be a huge challenge. Even in the rich, industrialized countries of the world, modern waste treatment practices leave something to be desired. In most cases, much of the sludge that remains after treatment is trucked off to landfills. The liquid waste remaining in sewage treatment plants is chemically treated and dumped into streams and lakes.

Traditional waste treatment methods not only pollute our environment, they also waste valuable nutrients, which come from the foods we eat and ultimately the soil on which crops are grown. In nature, plant and animal waste is returned to the soil, where it nourishes new plant life.

Over the years, scientists and others have sought more environmentally compatible—and sustainable—ways of dealing with human waste. Biologist John Todd has led this effort through the invention of waste disposal systems that mimic nature's ways. Todd designed and built his first solar-powered sewage treatment plant at Sugarbush ski resort in Vermont. In this system, raw sewage enters a solar-heated greenhouse and then flows into cylinders, where naturally occurring bacteria convert the ammonia in the sewage into nitrate, a plant nutrient. The effluent then enters special channels where algae consume the nitrates. The algae are part of an artificial ecosystem containing freshwater shrimp (that feed on algae) and fish (that feed on the shrimp). Snails in the system consume the sludge (organic wastes that fall to the bottom) and also serve as food for fish.

At the far end of the greenhouse is a small marsh containing organisms that remove additional impurities before the water is released into a nearby stream (FIGURE 1). In this artificial marsh are many plants that absorb toxic substances.

In Providence, Rhode Island, Todd installed a much larger system that handles up to 16,000 gallons of raw sewage per day flowing through a greenhouse containing 1200 aquariums, which house a variety of organisms to purify the water. This system, which costs one third as much as an equivalent sewage treatment plant, has negligible environmental impact.

Another pioneer in ecologically sound waste disposal is environmental engineer Bill Wolverton. In the mid-1970s, while working at NASA, Wolverton began experiments to purify wastewater from NASA facilities using lagoons filled with prolific, nutrient-hungry water hyacinths, which have become a "pest" in lakes, streams, and rivers in the southern United States. One system he designed for a 4000-person NASA facility in Mississippi has saved the agency millions

FIGURE 1 Proponents of this new technology point out that malfunctions at conventional plants can be serious and sometimes require the evacuation of nearby residents to avoid poisonous chlorine gas emitted from them. A malfunction in a solar-aquatic plant may kill off some organisms, but it poses no threat to people.

4.4 Ecosystem Function

Life on land and in the Earth's waters is possible principally because of the existence of the **producers**, organisms such as the algae and plants. These organisms absorb sunlight and use its energy to synthesize organic foodstuffs from atmospheric carbon dioxide and water via photosynthesis. These organic molecules are used by the producers themselves, but they also provide nourishment for all the other organisms. Producers, therefore, form the foundation of the living world.

Another large group of organisms is the **consumers**. Ecologists place consumers into four general categories, depending on the type of food they eat. Some, such as deer, elk, and cattle, feed directly on plants and are called **herbivores** (ERB-ah-voors). Others, such as wolves, feed on herbivores and other animals and are known as **carnivores** (CAR-neh-voors). Humans and a great many other animal species subsist on a mixed diet of plants and animals and are known as **omnivores** (OM- neh-voors). Another group feeds on animal waste or the remains of plants and animals and are called **detritivores** (dee-TREH-teh-

of dollars in sewage fees. Since he began, Wolverton has designed more than a hundred systems in the southern United States.

Researchers are also experimenting with smaller constructed wetlands for biological treatment that we can use in our own backyards. Many are designed so waste water percolates through a bed of gravel under a layer of soil and plants, and thus, there is no way sewage could reach the surface.

Another pioneer in biological waste treatment is Tom Watson, who lives in New Mexico. Watson specializes in household waste treatment centers using inexpensive plastic infiltrators in a pumice bed (FIGURE 2). Plants growing in the overlying soil send their roots into pumice, where they feed on water and nutrients from household waste, including toilet water. Bacteria in the pumice decompose organic matter, releasing nutrients for plants.

Biological treatment systems have pros and cons. Those being built by John Todd can operate in virtually any climate. These systems can also be scaled up to any size simply by adding more greenhouses. Watson's systems and constructed wetlands can also function well in a variety of climates, but because they're not protected from the weather, they may have problems in extremely cold climates.

Biological treatment systems, both large and small, are well suited to developing countries because they offer a low-cost option for treating waste. In fact, some developing nations are currently considering installing biological systems to upgrade existing facilities. They're finding that the necessary upgrades can be made at moderate cost. Moreover, the facilities do not require as much management or use as much energy as conventional systems. They even produce cleaner treated water than their high-tech counterparts—and at a lower cost.

One concern is that in large biological treatment facilities plants may absorb heavy metals such as mercury and lead, which would need to be disposed of in a way that was safe. Todd's colleague, Alan Liss, notes, "It's better to have a small amount of highly toxic plants than a huge

amount of moderately toxic sludge." Eliminating such wastes from the waste stream through preventive measures would easily solve this problem.

Larger biological systems such as those being built by John Todd and Bill Wolverton require a level of biological knowledge that few sewage treatment engineers have. They represent a radically different approach for society, even though natural systems have been purifying water for millennia. This appearance of being different makes it difficult to convince city officials to give them a try when they know that existing technology will do the job. Some facilities produce enormous amounts of plant waste, sometimes containing hazardous substances including heavy metals that must be safely disposed of.

Despite their drawbacks, biological systems that mimic nature can be adapted for most or all situations, and their use will undoubtedly rise in the years to come as cities and towns look for more sustainable ways of treating waste.

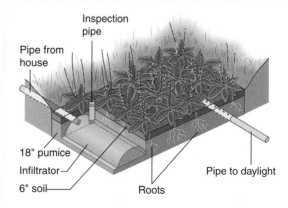

FIGURE 2 The Watson Wick Filter. In this simple, effective set up waste water enters the pumice bed, where bacteria go to work on waste. Plant roots absorb nutrients and water. Water coming out of the system is relatively free of nutrients. Tests so far have shown them to be more effective than conventional sewage treatment methods.

voors) or **decomposers**. This group includes many bacteria, fungi, and insects.

Food Chains and Food Webs

Biological communities consist of numerous food chains. A **food chain** is a series of organisms, each one feeding on the organism preceding it (**FIGURE 4-7**). All organisms in the community are members of one or more food chains.

Biologists recognize two general types of food chains: grazer and decomposer. **Grazer food chains** begin with plants and algae. These organisms are consumed by herbivores, or grazers. Herbivores, in turn, may be eaten by carnivores.

Decomposer food chains begin with dead material—either animal wastes (feces) or the remains of plants and animals. These are consumed by insects, worms, and a host of microorganisms such as bacteria. These organisms are re-sponsible for the decomposition of the waste and the return of its nutrients to the environment for reuse (recycling).

In ecosystems, decomposer and grazer food chains are tightly linked (**FIGURE 4-8**). Thus, waste from the grazer food chain enters the decomposer food chain. Nutrients liberated by the decomposer food chain enter the soil and water and are reincorporated into plants at the base of the grazer food chain.

Food chains exist only on the pages of textbooks; in a community of living organisms, food chains are part of a much more complex network of feeding interactions, **food webs** (**FIGURE 4-9**). Food webs present a complete picture of the feeding relationships in any given ecosystem.

As with so many other topics in ecology, an understanding of food chains and food webs is essential to living sustainably on the planet. For instance, efforts to protect the ozone layer, which shields the Earth from harmful ultraviolet radiation, are important to protect people from cancer, but they are also important because they protect phytoplankton in the world's oceans. Phytoplankton, small microscopic photosynthesizers, form the base of aquatic food chains. Ultraviolet radiation can kill phytoplankton and thus cause a collapse of aquatic food chains.

FIGURE 4-7 Simplified grazer food chains. These drawings show terrestrial (land-based) and aquatic grazer food chains.

Grazer food chain

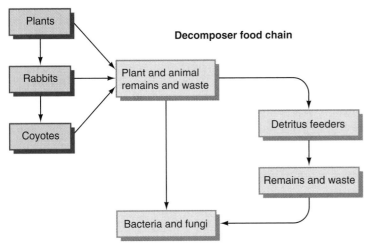

Decomposer food chain

FIGURE 4-8 A grazer food chain and a decomposer food chain, showing the connection between the two.

KEY CONCEPTS

Food and energy flow through food chains that are part of much larger food webs in ecosystems.

The Flow of Energy and Nutrients Through Food Webs

Energy and nutrients both flow through food webs, but in very different ways. Let us begin with energy. As you just learned, solar energy is the main driving force in nature. It is first captured by plants and then used to produce organic food molecules. Energy from the sun is then stored in these molecules. In the food chain, organic molecules pass from plants to animals. In both plants and animals, the molecules are broken down. This process releases stored solar energy, which is used to power numerous cellular activities.

During cellular energy release, a good portion of the energy stored in organic food molecules is lost as heat. Heat escaping from plants and animals is radiated into the atmosphere and then into outer space. It cannot be recaptured and reused by plants or animals. Because all solar energy is eventually converted to heat, energy is said to flow unidirectionally through food chains and food webs. Put another way, energy cannot be recycled.

In contrast, nutrients flow cyclically; that is, they are recycled. Nutrients in the soil, air, and water are first incor-

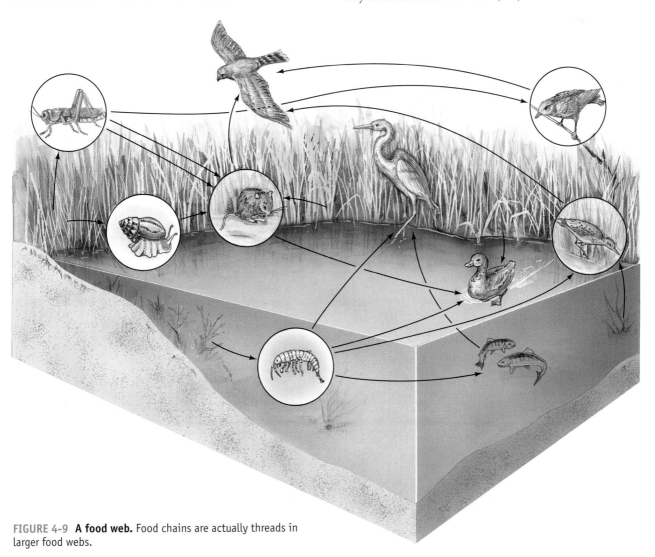

FIGURE 4-9 **A food web.** Food chains are actually threads in larger food webs.

porated into plants and algae and are then passed from plants to animals in various food chains. Nutrients in the food chain eventually reenter the environment through waste or the decomposition of dead organisms.

Every time you exhale, for example, you release carbon dioxide, a waste product of cellular energy production. Carbon dioxide reenters the atmosphere for reuse. Thus, through the act of breathing you play an important role in the global recycling system that makes life possible. Other wastes must be decomposed before releasing their nutrients. The feces of a rhinoceros, for example, are broken down by bacteria, which liberate carbon dioxide, nitrogen, and minerals.

Nutrients also reenter the environment through the decomposition of dead organisms. When a plant or animal dies, bacteria and fungi devour its organic remains. This process is called **decomposition.** Although these microorganisms absorb many nutrients released during this process, some nutrients escape and enter the soil and water for reuse. (Of course, when a bacterium dies, it also breaks down, releasing nutrients into its environment.)

One way or another, nutrients eventually make their way back to the environment for reuse. As a result, each new generation of organisms relies on the recycling of material in the biosphere. Every atom in your body has been recycled since the beginning of life on Earth. Perhaps some of those atoms were in the very first cells.

Food chains are biological avenues for the flow of energy and the cycling of nutrients in the environment. Energy flows in one direction through food chains, but nutrients are recycled.

Trophic Levels

Ecologists classify the organisms in a food chain according to their position, or **trophic level** (literally, "feeding" level). The producers are the base of the grazer food chain and belong to the first trophic level. The grazers are part of the second trophic level. Carnivores that feed on grazers are in the third trophic level, and so on.

Most terrestrial food chains are limited to three or four trophic levels. Longer terrestrial food chains are rare because food chains generally do not have a large enough producer base to support many levels of consumers. Why not?

Plants absorb only a small portion of the sunlight that strikes the Earth (only 1 to 2%), which they use to produce organic matter, or biomass. Technically, **biomass** is the dry weight of living material in an ecosystem. The biomass at the first trophic level is the raw material for the second trophic level. The biomass at the second trophic level is the raw material for the third trophic level, and so on.

As shown at the top of **FIGURE 4-10**, not all of the biomass produced by plants is converted to grazer biomass. At least

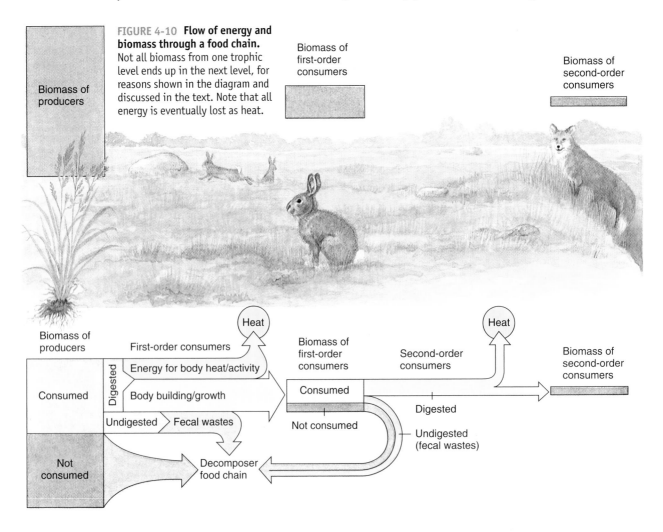

FIGURE 4-10 Flow of energy and biomass through a food chain. Not all biomass from one trophic level ends up in the next level, for reasons shown in the diagram and discussed in the text. Note that all energy is eventually lost as heat.

three reasons account for the incomplete transfer of biomass from one trophic level to the next. First, some of the plant material, such as the roots, is not eaten. Second, not all of the material that the grazers eat is digested. Third, some of the digested material is broken down to produce energy and heat and therefore cannot be used to build biomass in the grazers (Figure 4-10). As a rule, only 5 to 20% of the biomass at any trophic level is passed to the next level. (The amount varies depending on the organisms involved in the food chain.)

When plotted on graph paper, the biomass at the various trophic levels forms a pyramid, the **biomass pyramid** (**FIGURE 4-11**). Because biomass contains energy (stored in the bonds between atoms), the biomass pyramid can be converted into a graph of the chemical energy in the various trophic levels. This graph is called an **energy pyramid**.

In most food chains, the number of organisms also decreases with each trophic level, forming a **pyramid of numbers**. Knowledge of ecological pyramids helps us understand why people in many developing countries subsist on a diet of grains (corn, rice, or wheat) rather than meat. As shown in **FIGURE 4-12**, in the grain → human food chain on the right, 20,000 kilocalories of grain can feed 10 people for a day. (A kilocalorie is the same unit as a Calorie.) If that amount of grain is fed to a steer and the beef is then fed to humans, however, only 1 person can subsist on the original 20,000 kilocalories. Why? In the grain → steer → human food chain, the 20,000 kilocalories fed to the cow that day produce only 2000 kilocalories of food, barely enough to feed

1 person for a day (assuming a 10% transfer of biomass). Although people don't eat meat-only diets, this oversimplified example does illustrate a key point: the shorter the food chain, the more food is available to top-level consumers. (This is not a criticism of the cattle industry. Cattle often feed on grasses that grow on land too poor to support crops, although they are often fattened on grain before being slaughtered.)

> Twenty thousand kilocalories of grain can feed 10 people a day, but when fed to cattle that are then slaughtered to provide meat, the grain supports only one person per day.

This simple rule has profound implications for the human race. The human population increases by about 80 million people a year. Feeding these people poses an enormous challenge. How can new residents be fed most efficiently?

The most efficient food source will be crops such as corn, rice, and wheat that are fed directly to people. It is far less efficient to feed corn and other grains to cattle and other livestock that are slaughtered for human consumption. Vegetarianism, say proponents, is not only good for your health but good for the environment because it requires less grain production than a nonvegetarian diet. It should be pointed out, however, that vegetarianism is not environmentally benign. Plowing former grasslands or forests to grow food for people has a profound impact on the environment. Fertilizer

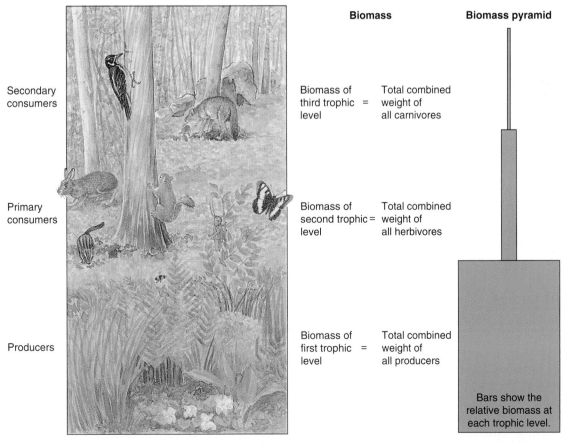

FIGURE 4-11 Biomass pyramid. In most food chains, biomass decreases from one trophic level to the next higher one.

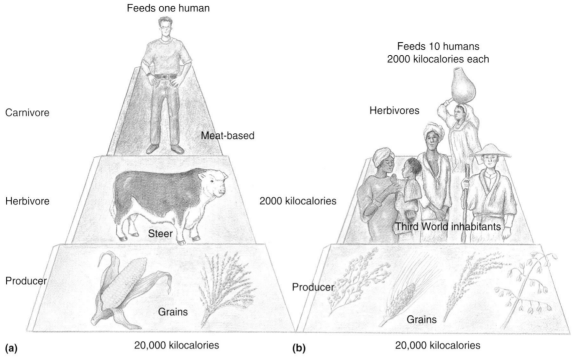

FIGURE 4-12 **Energy pyramids in two food chains. (a)** The typical meat-based diet. The 20,000 kilocalories of corn fed to cows produces only 2000 kilocalories of meat. An adult needs only about 2000 calories per day. **(b)** In a shorter food chain, 20,000 kilocalories can feed 10 people directly. This is the reason many people in developing nations subsist primarily on a vegetarian diet. Although few, if any, people eat a meat-only diet, this example does illustrate an important point: more food is available to those societies that eat lower on the food chain.

and pesticide use can have an enormous negative effect as well. Many wild animals are killed to grow grains.

Eating lower on the food chain is not always possible or advisable. In the case of grain-fed cattle, it may be both; but range-fed beef or elk are quite different. They feed on plants that are, of course, not edible for humans, and they live on land that would be poor farmland. Consuming meat from these sources may be a better option from an environmental perspective.

KEY CONCEPTS

The position of an organism in a food chain is called its *trophic level*. Producers are on the first trophic level. Herbivores are on the second level. Carnivores are on the third level. The length of a food chain is limited by the loss of energy from one trophic level to another. The largest number of organisms is generally supported by the base of the food chain, the producers.

Nutrient Cycles

The sustainability of natural systems results primarily from the dependence on the sun and the reliance on the recycling of nutrients to ensure an adequate supply. The term **nutrients** is used here to refer to all ions (charged atoms) and molecules used by living organisms.

In ecosystems, nutrients flow from the environment through food webs and are then released back into the environment. This circular flow constitutes a **nutrient cycle**, also known as a **biogeochemical cycle**.

Nutrient cycles can be divided broadly into environmental and organismic phases (**FIGURE 4-13**). In the **environmental phase**, a nutrient exists in the air, water, or soil, or sometimes in two or more of them simultaneously. In the **organismic phase**, nutrients are found in the biota—the plants, animals, and microorganisms.

Dozens of global nutrient cycles operate continuously to ensure the availability of chemicals vital to all living things, present and future. Unfortunately, however, a great many human activities disrupt nutrient cycles. These activities can profoundly influence the survival of species, including our own.

This section examines two of the most important nutrient cycles, the carbon and nitrogen cycles, and the ways they are being altered.

KEY CONCEPTS

Nutrients are recycled in global nutrient cycles. In these cycles, nutrients alternate between organisms and the environment. Humans can disrupt nutrient cycles in many ways, with profound impacts on ecosystems and our own future.

The Carbon Cycle   The carbon cycle is shown in **FIGURE 4-14** in a slightly simplified form. To understand how it operates, we begin with carbon dioxide. In the environmental phase of the cycle, carbon dioxide resides in two reservoirs, or sinks: the atmosphere and the surface water (oceans, lakes, and rivers). As illustrated, atmospheric carbon dioxide is absorbed by plants and other photo-

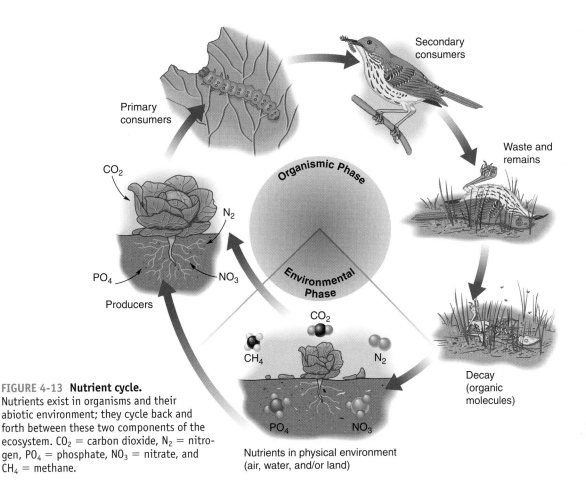

FIGURE 4-13 Nutrient cycle.
Nutrients exist in organisms and their abiotic environment; they cycle back and forth between these two components of the ecosystem. CO_2 = carbon dioxide, N_2 = nitrogen, PO_4 = phosphate, NO_3 = nitrate, and CH_4 = methane.

synthetic organisms in terrestrial ecosystems, thus entering the organismic phase of the cycle. These organisms convert carbon dioxide into organic food materials, which are passed along the food chain. Carbon dioxide reenters the environmental phase via cellular energy production (cellular respiration) of the organisms in the grazer and decomposer food chains.

For tens of thousands of years, our ancestors lived in relative harmony with nature. Because their numbers were small and their technology fairly primitive, they had little impact on the environment. With the advent of the Industrial Revolution, however, human beings began to interfere with natural processes on a large scale. One of the victims has been the global carbon cycle. The widespread combustion of fossil fuels (which releases carbon dioxide) and rampant deforestation (which reduces carbon dioxide uptake) have overloaded the cycle with carbon dioxide.

For many years before the Industrial Revolution, global carbon dioxide production equaled carbon dioxide absorption by plants and algae. Today, seven billion tons of carbon dioxide are added to the atmosphere each year. Three

> Human activities contribute about seven billion tons of carbon dioxide into the atmosphere each year. Three fourths of it comes from the combustion of fossil fuels; the remaining portion comes from deforestation.
>
> In the past 100 years, global atmospheric carbon dioxide levels have increased by about 25%.

quarters of the increase results from the combustion of fossil fuels such as the gasoline in our cars; the remaining quarter stems from deforestation, which reduces the amount of carbon dioxide absorbed by the planet's plants. Making matters worse, many forests are burned after cutting, further adding to the carbon dioxide levels in the atmosphere.

In the past 100 years, global atmospheric carbon dioxide levels have increased by around 30%. In the atmosphere, carbon dioxide traps heat escaping from Earth and reradiates it to the Earth's surface. As carbon dioxide levels increase, global temperatures rise. Such a rise could shift rainfall patterns, destroy agricultural production in many regions, and wipe out thousands of species. A rising global temperature might cause glaciers and the polar ice caps to melt, raising the sea level and flooding many low-lying coastal regions. Fortunately, there are many cost-effective strategies for reducing our dependence on fossil fuel. The most notable are energy efficiency and the use of renewable fuel—solar energy and wind, for example. Transitioning to these clean, reliable, and cost-effective fuels could help us ensure a healthy economy, continued prosperity, and a better future.

KEY CONCEPTS

The carbon cycle is vital to the survival of the Earth's many species. It is the basis of food and energy production in the living world. It is also vital to maintaining global temperature. The carbon cycle is currently being flooded with excess carbon dioxide as a result of the combustion of fossil fuels and deforestation, which could have devastating effects on climate and ecosystems.

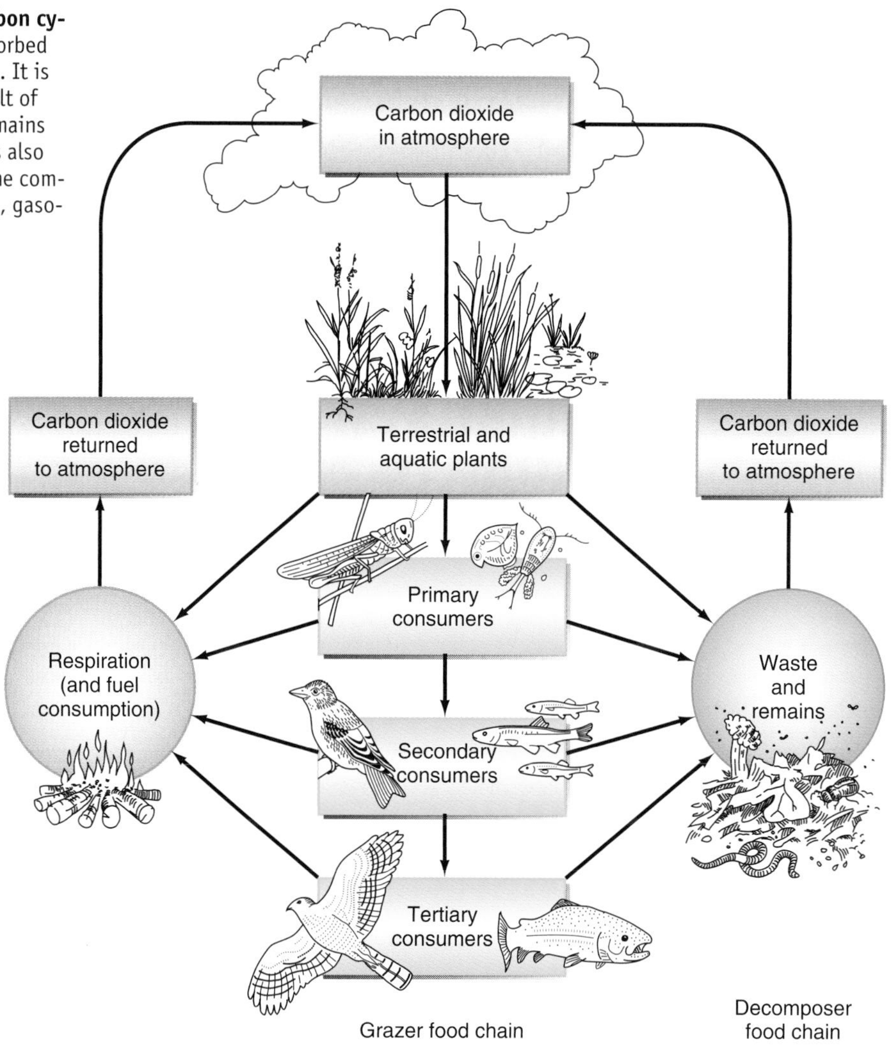

FIGURE 4-14 A simplified view of the carbon cycle. Carbon dioxide in the atmosphere is absorbed by plants and passed through the food chain. It is released back into the environment as a result of the decomposition of the waste and dead remains of plants, animals, and other organisms. It is also released by cellular energy production and the combustion of organic materials such as coal, oil, gasoline, and wood.

The Nitrogen Cycle   Nitrogen is an element that is essential to many important biological molecules, including amino acids, DNA, and RNA. The Earth's atmosphere contains enormous amounts of it, but atmospheric nitrogen is in the form of nitrogen gas (N_2), which is unusable to all but a few organisms. As a result, atmospheric nitrogen must first be converted to a usable form, either nitrate or ammonia.

The conversion of nitrogen to ammonia is known as **nitrogen fixation**. It occurs in terrestrial and aquatic environments. As **FIGURE 4-15** shows, the roots of leguminous plants (peas, beans, clover, alfalfa, vetch, and others) contain small swellings called *root nodules*. Inside the nodules are bacteria that convert atmospheric nitrogen to ammonia (ammonium ions). Ammonia is also produced by bacteria called *cyanobacteria* that live in the soil. Once ammonia is produced, other soil bacteria convert it to nitrite and then to nitrate. Nitrates are incorporated by plants and used to make amino acids and nucleic acids. All consumers ultimately receive the nitrogen they require from plants.

Nitrate in soil also comes indirectly from the decay of animal waste and the remains of plants and animals. As shown on the right side of Figure 4-15, decomposition returns ammonia to the soil for reuse. Ammonia is converted to nitrite, then to nitrate and reused. Some nitrate, however, may be converted to nitrite and then to nitrous oxide (N_2O) by denitrifying bacteria, as illustrated in Figure 4-15. Nitrous oxide is converted to nitrogen and released into the atmosphere.

Humans alter the nitrogen cycle in at least four ways: (1) by applying excess nitrogen-containing fertilizer on farmland, much of which ends up in waterways; (2) by disposing of nitrogen-rich municipal sewage in waterways; (3) by raising cattle in feedlots adjacent to waterways; and (4) by burning fossil fuels, which release a class of chemicals known as nitrogen oxides into the atmosphere. The first three activities increase the concentration of nitrogen in the soil or water, upsetting the ecological balance. Nitrogen oxides released into the atmosphere by power plants, automobiles, and other sources are converted to nitric acid, which falls with rain or snow. Besides changing the pH (acidity) of soil and aquatic ecosystems, nitric acid also adds nitrogen to surface waters and may be responsible for 25% of the nitrogen pollution in some coastal waters in the United States.

Nitrogen, like phosphorus, is a plant nutrient. It stimulates the growth of aquatic plants and causes rivers and lakes to become congested with dense mats of vegetation, making them unnavigable. Sunlight penetration to deeper lev-

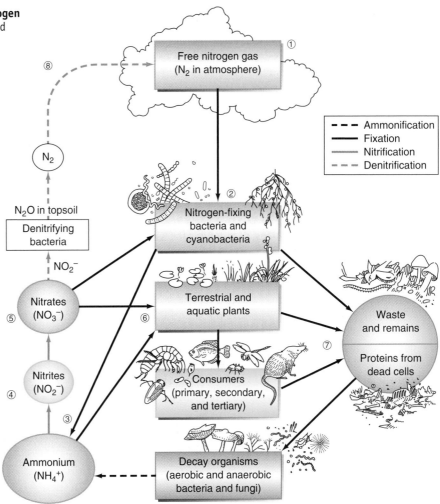

FIGURE 4-15 A simplified view of the nitrogen cycle. Nitrogen in the atmosphere is converted into ammonium ions by bacteria in the soil. Ammonia is converted to nitrates and taken up by plants.

els is also impaired by the growth of plants, causing oxygen levels in deeper waters to decline. In the autumn, when aquatic plants die and decay, oxygen levels may fall further, killing aquatic life.

Nothing can survive on the planet unless it is a cooperative part of a larger global life.
—Barry Commoner

CRITICAL THINKING

Exercise Analysis

It is difficult for me to know the list of misconceptions you came up with, but here are a few that might have arisen along the way. The exact meaning of the term *ecology* may have been one. Many people use it very loosely. You may not have understood the term *niche*, either. Like many people, you may have thought of it as a physical place an organism occupied, its *habitat*.

I suspect that your sense of the word *environment* has shifted. Now you can see that it includes biotic and abiotic components. The environment of an organism is quite complex.

I suspect that the discussion of the range of tolerance may have helped you to understand how our activities and natural events alter an organism's chances of survival.

What about energy? Has your understanding of energy changed? Do you see food chains and food webs now in a different light? Rather than being a simple matter of one organism feeding on another, food chains and food webs are elaborate pathways for the flow of nutrients and energy.

You may have gained new perspective on nutrient cycles. You may already have known that they existed, but did you know that humans are altering them in major ways?

CRITICAL THINKING AND CONCEPTS REVIEW

1. "Humans are a part of nature." Do you agree or disagree with this statement? Support your answer.
2. Define the term *ecology* and give examples of its proper and improper use.
3. The Earth is a closed system. What does this mean, and what are the implications of this fact?
4. Define the following terms: *biosphere, biome, aquatic life zone*, and *ecosystem*.
5. Define the term *range of tolerance*. Using your knowledge of ecology, give some examples of ways in which humans alter the abiotic and biotic conditions of certain organisms. Describe the potential consequences of such actions.
6. Describe ways in which humans alter conditions within their own range of tolerance.
7. What is a limiting factor? Give some examples.
8. Define the following terms: *habitat, niche, producer, consumer, trophic level, food chain*, and *food web*.
9. A hunting advocate in your state is proposing the introduction of a foreign species, one very similar to deer, that he encountered in Russia on a hunting expedition. He thinks the introduced species will provide additional hunting opportunities and additional tax revenue for the state, which will be good for the economy. The governor is in favor of the proposal. Write a letter to the governor explaining what needs to be known about this species before it should be considered for introduction.
10. Explain why the biomass at one trophic level is less than the biomass at the next lower trophic level.
11. Outline the flow of carbon dioxide through the carbon cycle, and describe ways in which humans adversely influence the carbon cycle.
12. Using what you have learned about ecology, describe why it is important to protect natural ecosystems and other species.
13. With the knowledge you have gained, explain why it is beneficial to set aside habitat to protect an endangered species.
14. Looking back over the principles you have learned in this chapter, write a set of guidelines for human society that would help us live sustainably on the Earth.
15. Reread the Point/Counterpoint in this chapter. Which view do you agree with? Why? Is your support of one view based on values or science, or both?.

KEY TERMS

abiotic	ecological system	nutrients
aquatic life zones	ecology	omnivores
biogeochemical cycle	ecosystem	optimum range
biological community	endangered species	organismic phase
biomass	energy pyramid	photosynthesis
biomass pyramid	environmental phase	population
biome	food chain	producers
biosphere	food webs	pyramid of numbers
biotic	grassland biome	range of tolerance
carnivores	grazer food chains	taiga
closed system	habitat	temperate deciduous forest biome
competitive exclusion principle	herbivores	trophic level
consumers	limiting factor	tundra
decomposer food chains	niche	zones of intolerance
decomposers	nitrogen fixation	zones of physiological stress
desert biome	northern boreal forest	zones of tolerance
detritivores	northern coniferous forest	
ecological niche	nutrient cycle	

REFERENCES AND FURTHER READING

The References and Further Reading section at the end of this book contains a list of sources for the information discussed in this chapter and recommendations for further reading.

Connect to this book's website:
http://environment.jbpub.com/
The site features eLearning, an online review area that provides quizzes, chapter outlines, and other tools to help you study for your class. You can also follow useful links for in-depth information, research the differing views in the Point/Counterpoints, or keep up on the latest environmental news.

A grizzly bear catching fish in a river.

Principles of Ecology: Biomes and Aquatic Life Zones

Let us permit nature to have her way: she understands her business better than we do.

—*Montaigne*

The Earth formed approximately 4.5 billion years ago. At first, the planet was a barren mass of rock and ice. Over billions of years, however, life evolved. Today, the Earth's surface is carpeted with a rich and diverse array of life forms existing in a complex web of life. As anyone who has traveled across any major continent knows, however, the biosphere varies from one region to the next. Put another way, the biosphere is divided into distinct regions, which ecologists call **biomes**. Biomes differ from one another in climate—that is, in rainfall, sunlight, temperature, and other abiotic factors. These differences lead to marked variations in the types and abundance of species that live in the Earth's many biomes.

Exercise

An article in *Science News* reports on studies by a scientist who examined sonar measurements of ice thickness in the Arctic Sea near the North Pole. He studied assessments made by two British submarines that took similar routes across the Arctic in 1976 and 1987. During the first trip, the average ice thickness was 5.3 meters (17.5 feet), and on the second trip, it was 4.5 meters (14.85 feet).

To some scientists, these data suggest that the Arctic ice is melting, a sign that global warming may be occurring. What questions would you ask to determine the validity of this conclusion? Write them down. What additional information might be helpful in analyzing the possibility that the polar ice caps are shrinking?

The Earth's surface is also covered by massive bodies of water. These, too, have distinct regions, known as **aquatic life zones**, the characteristics of which are determined by abiotic factors.

This chapter describes the major biomes and aquatic life zones and outlines some threats to them. It also offers suggestions for protecting the world's rich biological legacy and points out how important protection is for the survival of people and the many other species that share this planet with us. Before we discuss these topics, however, let's review some facts about weather and climate, which determine the biotic community of a region.

5.1 Weather and Climate: An Introduction

Most of us have heard the terms *weather* and *climate* many times in our lives, but have you ever stopped to consider what they mean and what factors determine what the weather and climate are like?

Weather refers to the daily conditions in our surroundings, including temperature and rainfall. Weather changes constantly from week to week, day to day, and even hour to hour. **Climate** is the average weather over a long period—approximately 30 years.

The climate of a region determines what plants can live in an area. Dry climates such as those found in the desert are home to cacti and other plants that are adapted to survive with little rainfall. Wetter climates such as those of the tropical rain forest are home to a wide variety of plants for which water conservation is not a major concern.

Plant life forms the base of grazer food chains and is the source of nutrients and energy for all other species. It also helps to determine the characteristic animal life.

KEY CONCEPTS
Weather refers to daily conditions such as rainfall and temperature. Climate is the average weather over a long period. Climate determines the plant and animal life of a region.

Major Factors that Determine Weather and Climate

Several factors determine a region's weather patterns and, ultimately, its climate. First is the amount of light and heat that strike different parts of the Earth. These come from the sun. As shown in FIGURE 5-1, the Earth can be divided into three climatic zones: tropics, the temperate zones, and the polar regions. The **tropics** lie on either side of the equator between 30° north and 30° south latitudes. They receive the most light and heat energy from the sun and are the warmest. The **temperate zones** lie between 30° and 60°, both north and south latitudes. They receive less heat and sunlight and are therefore cooler. The **poles** receive the least amount of solar energy and are the coolest.

The unequal heating of the Earth—and the three major climatic zones that occur as a result—are the product of the Earth's tilt (Figure 5-1). Tropical regions receive the most direct sunlight and are, therefore, the hottest. Temperate and polar regions receive sunlight that filters through more atmosphere, which reduces overall heating. The unequal heating of the Earth's surface creates air and water currents that profoundly influence the climate of the Earth. Let's begin with the wind currents.

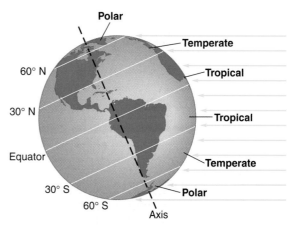

FIGURE 5-1 **Climate zones.** The Earth's surface can be roughly divided into the three major climate zones shown here.

When air is heated, it expands and becomes less dense. As shown in **FIGURE 5-2a**, hot air in the tropics rises and moves toward the poles, transporting heat and moisture from the equator to the poles. Cold air from the poles tends to move southward in the northern hemisphere toward the equator. These phenomena create a general air circulation pattern between the equator and the poles that tends to distribute excess heat from the equator throughout the globe.

The Coriolis Effect and Topography

Air circulation is actually a bit more complicated than the previous discussion might lead you to believe. In the northern hemisphere, for example, warm air from the equator rises and moves northward. As it moves northward, much of its moisture is lost as rainfall over the tropics. As the air moves northward it also cools, and some of it sinks back to the Earth's surface (**FIGURE 5-2b**). This air flows back toward the equator, creating winds that blow toward the equator called the **trade winds** (**FIGURE 5-2c**).

The rest of the equatorial air continues northward over the temperate zone. This air is relatively dry, having lost much of its moisture in the tropics. However, as it travels northward it picks up moisture from the temperate zone and deposits it as rainfall and snow. As Figure 5-2b shows, the air circulating toward the north pole splits into higher and lower level winds—both flowing generally northward. When the equatorial air reaches the poles, it is cold and very dry. It then begins its way back to the equator (Figure 5-2b).

This general circulation distributes heat and moisture throughout the planet. It makes the poles warmer than they might otherwise be and makes the tropics a bit cooler. The distribution of heat and moisture caused by air flow also helps determine weather. However, this distribution is com-plicated a bit by the fact that Earth spins on its axis. This, in turn, shifts the wind direction and helps to create distinct climatic zones within the major regions.

Figure 5-2c shows three distinct wind patterns, one in each zone of the northern hemisphere. The trade winds blow toward the equator, but they don't blow directly south. In the Northern hemisphere, they blow from northeast to southwest. The low-level winds in the temperate zone, called the **westerlies**, generally blow from south to north (as shown in Figure 5-2b) and are deflected a bit, blowing southwest to northeast. The **polar easterlies**, cold Arctic air that sinks back toward the equator, blow from northeast to southwest.

Scientists call the deflection of wind currents by the spin of the Earth the **Coriolis** (CORE-ee-ole-iss) **effect**. To understand this phenomenon, suppose that a punter kicks a football 50 yards south; the ball is in the air for five seconds, and it flies in a perfectly straight course. If this were so, you would expect it to land in a perfectly straight line from the kicker. It does not. In actuality, it lands a little less than half an inch off target. The reason for this is the Earth's spin. While the ball is in the air in a straight path, the Earth is spinning underneath it. The path of the ball thus appears to be deflected. Air flowing along the Earth's surface is also deflected because of the Earth's spin. Thus, the trade winds, the westerlies, and the easterlies are produced by the Earth's rotation. Wind currents, instead of flowing straight north or south (as they would if the Earth did not spin), actually flow across continents. Why is this worth noting?

As noted previously, wind currents help to determine weather patterns. In the United States, moist air tends to flow from the Pacific Ocean (southwesterly) across the continent (toward the northeast). Topography then comes into play. Mountain ranges along the West Coast and in the interior, for example, rob some of the moisture carried in this air, creating deserts or semiarid lands on their downwind side. This occurs because mountains thrust warm, moist air upward. As it rises it cools, and the moisture condenses and falls as precipitation. When the air comes down the other side of the mountain range, it is drier. Little precipitation falls, and

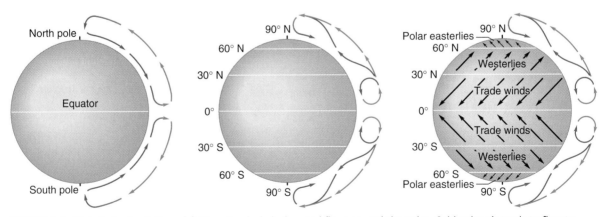

FIGURE 5-2 Global air circulation. (a) Warm tropical air rises and flows toward the poles. Cold polar air tends to flow toward the equator. **(b)** Air circulation actually is more complicated than shown in part (a). Warm tropical air loses some of its heat and sinks, creating a circulation pattern that brings air back toward the equator. This creates the trade winds. **(c)** Wind patterns caused by the spin of the earth.

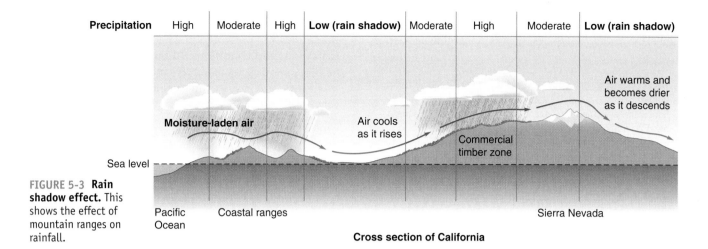

Precipitation | High | Moderate | High | **Low (rain shadow)** | Moderate | High | Moderate | **Low (rain shadow)**

Air warms and becomes drier as it descends

Moisture-laden air

Air cools as it rises

Commercial timber zone

Sea level

Pacific Ocean

Coastal ranges

Sierra Nevada

Cross section of California

FIGURE 5-3 Rain shadow effect. This shows the effect of mountain ranges on rainfall.

the area is either desert or semidesert (**FIGURE 5-3**). This phenomenon is called the *rain shadow effect*. As the air flows across the Great Plains, it picks up and deposits moisture downwind. Rainfall increases as one travels eastward. This increase in precipitation, in turn, explains why the prairie due east of the Rocky Mountains starts off as short grass, which requires little rain, but gradually becomes tallgrass prairie and then deciduous forest from eastern Kansas eastward.

> ### KEY CONCEPTS
> Weather within the major climatic zones is altered by wind flow patterns, which are profoundly influenced by the spin of the Earth. Weather is also affected by topography, especially mountain ranges.

Ocean Currents

Climate is also influenced by ocean currents. At the equator, water warmed by the sun rises and tends to drift toward the poles, creating huge currents. The Gulf Stream, shown in **FIGURE 5-4**, is a good example. The warm water of the Gulf Stream flows northward, affecting the climate of landmasses that it passes near. The Gulf Stream, for instance, warms England and the rest of Europe and produces much milder winters than would be expected otherwise. Northern Japan and Alaska are also much warmer than one would predict based on their location because of warm-water currents traveling up from the equator.

As the warm-water currents move northward, however, they begin to cool. As the waters cool, they start to sink and then move back toward the equator, in much the same way that cold Arctic air tends to flow toward the equator. Cold water returning to the equator creates broad, deep currents in the opposite direction of the warm-water currents. As shown in Figure 5-4, the cold water Humboldt current flows up from the south pole, bringing cold, deep waters toward the equator. This water eventually emerges at the surface in regions called **upwellings**. Because it is rich in nutrients, it supports a diverse aquatic community of great ecological and economic value. Every 10 years or so, the Humboldt current turns warm, a phenomenon called **El Niño**. This creates heavy rain and rough seas, which can have a devastating effect on coastal communities and often affects weather worldwide. (Today, El Niños are occurring more frequently, quite possibly as a result of global climate change.)

Weather is much more complicated than discussed here, but this introduction gives you an overview that may help you understand why particular biomes are located where

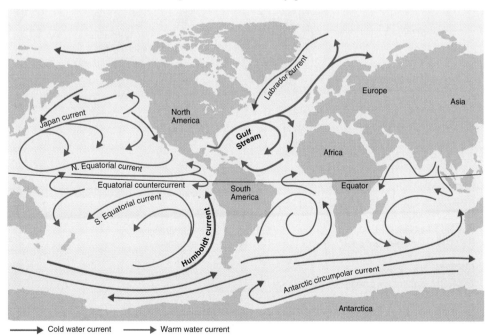

FIGURE 5-4 Major ocean currents.

Cold water current   Warm water current

they are. It will also help you to understand some major global environmental issues, such as global climate change.

KEY CONCEPTS

Warm water from the equator flows toward the poles, warming landmasses near which it passes. As it flows northward, it cools. Cool water eventually sinks, and flows back toward the equator, creating a huge global circulation pattern.

5.2 The Biomes

We begin our journey through the Earth's diverse biomes (FIGURE 5-5) in the north and work our way south. On this journey, you will learn about the climate of several key biomes and the unique biological characteristics (or adaptations) that permit plants and animals to thrive within them. As you will see, each biome is characterized by a dominant form of vegetation. However, within any given biome, regional differences in climate (and other factors such as soil) can alter the composition and abun-

dance of species. The study of biomes underscores an important lesson, presented in Chapter 6: Life can adapt to a wide range of conditions. In the frozen tundra, for example, some organisms survive in air temperatures as low as −70°C (−90°F). In contrast, some desert species can tolerate 50°C (120°F).

KEY CONCEPTS

The Earth's surface can be divided into biologically distinct zones called biomes, each with a distinct climate and unique assemblage of plants and animals. Nevertheless, regional variations occur within each biome.

The Tundra

Stretching across the northernmost portions of North America, Europe, and Asia is the Arctic **tundra**. The Arctic tundra is the northern limit to plant growth. Covering about 10% of the land mass of the Earth, the tundra lies between a region of perpetual ice and snow to the north and a band of coniferous forests to the south. The tundra is a treeless tract

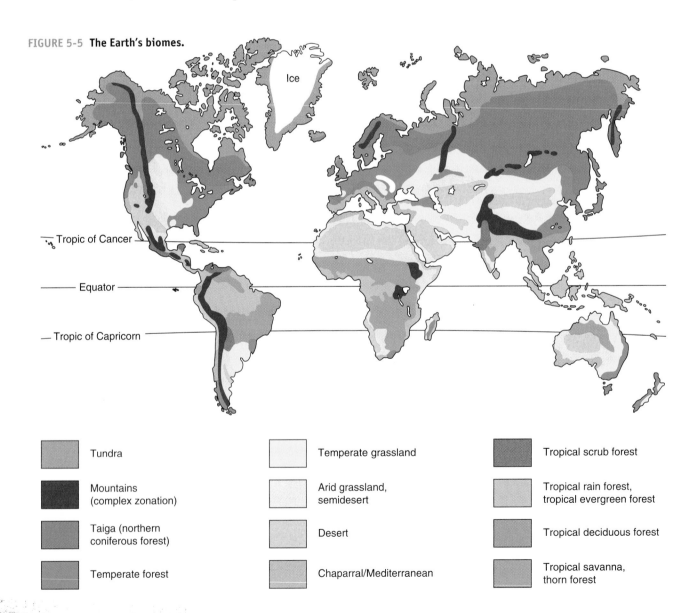

FIGURE 5-5 **The Earth's biomes.**

Tundra

Mountains (complex zonation)

Taiga (northern coniferous forest)

Temperate forest

Temperate grassland

Arid grassland, semidesert

Desert

Chaparral/Mediterranean

Tropical scrub forest

Tropical rain forest, tropical evergreen forest

Tropical deciduous forest

Tropical savanna, thorn forest

characterized by grasses, shrubs, and matlike vegetation (mosses and lichens) adapted to the harsh climate.

The Arctic tundra receives very little precipitation (less than 25 centimeters, or 10 inches, a year), and most precipitation occurs in the summer. During the long, cold winters, the mean average temperature remains well below zero for months on end. Because Arctic summers are short and winters are so cold, the deeper layers of soil remain frozen throughout the year and are called **permafrost**.

The permafrost and the harsh winters of the Arctic tundra prevent deep-rooted plants such as trees from growing there. In the brief Arctic summer, however, the long days and relatively warm temperatures permit the superficial layers of soil to melt. Because evaporation is low and because water from melted snow and ice cannot percolate downward into the soil in the summer (due to the permafrost), the tundra becomes dotted with shallow ponds, lakes, and bogs (**FIGURE 5-6**).

During the summer months, the tundra comes alive with insects (mosquitoes and black flies) and birds that migrate north to nest on the rolling plains. The birds feed on the swarms of insects, raise their young, and then migrate south with their offspring, often in great flocks.

Despite the harsh conditions, a variety of animals live year round on the tundra. Ptarmigan, musk oxen, and arctic hares, for example, are adapted to the extreme cold (**FIGURE 5-7**). Animals survive by living in burrows or by having thick layers of insulation or a large body size (which retains heat well). Some animals, like caribou, are migratory, moving southward when winter comes.

(a)

(b)

FIGURE 5-7 **Tundra species. (a)** The ptarmigan and **(b)** musk ox are both well adapted for life in the cold Arctic tundra.

The tundra is a fragile environment, easily damaged by human actions. The short growing season provides little time for vegetation to recover from damage caused by mining, oil and gas development, and other human activities. Tire tracks that destroy vegetation on the tundra take many decades to heal. Damage caused by spills of oil or hazardous waste may take far longer.

Nowhere is the threat to the tundra more worrisome than in northern Alaska. Alaska's northern coastline is approximately 1760 kilometers (1100 miles) long, and until recently, all of it except a 184-kilometer (115-mile) stretch was open to oil and gas exploration and development. In 2005, however, the U.S. Congress opened a small portion of this area to development, even though it is part of the Arctic National Wildlife Refuge (ANWR), one of the last great wilderness areas on the planet.

Many conservationists and citizens fear that the region will be turned into a network of roads, airports, power plants, oil platforms, waste ponds, buildings, and gravel pits—all to

FIGURE 5-6 **The tundra.** In the summer, water collects on the surface because evaporation is low and because water percolation into the ground is prevented by the permafrost.

FIGURE 5-8 **Arctic oil.** Oil development on Alaska's North Slope causes significant alteration of the delicate tundra, as shown here.

supply oil to make fuel for cars and other motorized vehicles (**FIGURE 5-8**). Nonetheless, proponents of oil development in the ANWR believe that wildlife and oil development can peacefully coexist despite estimates by biologists suggesting that full-scale oil development will result in a 20 to 40% decline in the large caribou herds that spend their summers in the region. Biologists also predict that full-scale oil development will wipe out half of the existing musk ox population. Populations of grizzly bears, polar bears, wolverines, and other animals are likely to suffer enormously. Oil development in the region would also result in air pollution, water pollution, and noise. Oil spills and hazardous waste disposal on the tundra are common in nearby Prudhoe Bay. In fact, the U.S. Department of Interior has recorded over 17,000 oil spills in the Arctic since 1973. Proponents are currently planning on developing only a small portion of ANWR, but critics are worried that after development commences, proponents will call for further development that could be devastating to this pristine area. For more on this controversial issue, see Chapter 14.

> **KEY CONCEPTS**
>
> The Arctic tundra, the northernmost biome, is characterized by the harshest climate. Because the growing season is so short, life on the Arctic tundra is extremely vulnerable to human actions.

The Taiga

Just south of the tundra is a wide band of cone-bearing, or coniferous, trees (pine, fir, and spruce). It extends across Canada, part of Europe, and Asia and constitutes the **taiga** (TIE-ga), or **northern coniferous forest biome** (Figure 5-5). The taiga, which a student of mine once defined as a "carnivorous forest" on his test, has a longer growing season than its northern neighbor, the tundra. It also receives far more precipitation.

In the summer, the subsoil of the taiga thaws, permitting deep-rooted plants such as trees to live. But the summer growing season is still short in comparison with that in the southern biomes (grassland and temperate deciduous forest). The rather cold, snowy winters and limited growing season of the taiga have resulted in numerous adaptations. Few organisms illustrate these adaptations as well as **conifers**, trees that remain green throughout the year. Conifers have narrow, pointed leaves called *needles*. Needles contain photosynthetic cells and are retained throughout the year, allowing conifers to continue to photosynthesize (albeit slowly) throughout the winter. The presence of needles throughout the year also permits the plant to make the most of the growing season. Unlike the deciduous hardwoods of the south, which lose their leaves in the winter and must develop new ones each growing season, conifers can take immediate advantage of the sunlight, moisture, and warmer weather of spring.

Retaining needles may help conifers survive in the taiga's short growing season, but it also creates a problem: The needles tend to capture snow. Instead of breaking under the weight of snow, though, the branches of conifers are adapted to bend and shed snow. Another adaptation of great importance to conifers is the waxy coating found on their needles. This layer greatly reduces evaporation. Because water transport up the tree from the ground is restricted during the winter, the waxy coating prevents the needles from drying out and helps the tree survive.

As anyone who has flown east to west over Canada will attest, the dense coniferous forests of the taiga extend for miles on end. These forests are interspersed with meadows, lakes, ponds, and rivers. Together, they support a variety of species such as bears, moose, and deer. Many smaller mammals also make their home there, including foxes, wolverines, and snowshoe hares. Because of the relatively cold winter and short growing season, however, species diversity in the taiga is fairly low compared with that in biomes to the south.

Along the west coast of North America, the forests of the taiga are bathed in moisture from the Pacific Ocean. Because the ocean moderates temperature, this region enjoys a long growing season. Consequently, forests in this area contain some of the world's largest trees. Heavy pressure from timber companies and consumers, however, is now destroying many of these magnificent trees and the species that live in them (**FIGURE 5-9**).

Efforts are under way to reduce the overcutting, but in many areas of Canada and the United States, powerful logging companies are resisting. The U.S. Congress, however, reduced cutting in the Tongass National Forest of Alaska and set aside a large tract of the forest for permanent protection.

FIGURE 5-9 **Clear-cutting.** This forest has been clear-cut. It once contained trees 500 to 700 years old. Old-growth forests such as these contain many species that cannot survive in younger forests.

The Temperate Deciduous Forest Biome

The temperate deciduous forest biome is located in the eastern United States, Europe, and northeast China. In the United States, this biome is home to about half of the human population. Characterized by abundant precipitation and a long growing season (5 to 6 months), this biome supports a wide variety of plants and animals.

The dominant plants of the temperate deciduous forest biome are **deciduous trees**—the broad-leaved trees that shed their leaves each year in the fall. Maple, oak, black cherry, and beech trees are examples (**FIGURE 5-10**). The loss of leaves in the fall is believed to be an adaptation that greatly reduces evaporation at a time when the supply of liquid water is limited.

In the spring, new leaves develop from buds. In the relatively brief period after the ground has thawed and before the leaves have fully developed, numerous species of wildflowers sprout on the sunny forest floor. The shade provided by deciduous trees, however, greatly limits plant growth on the forest floor throughout most of the rest of the growing season.

The temperate deciduous forest biome is characterized by a deep, rich soil. The richness of the soil results from mineral and organic nutrients. The nutrients are drawn up through the roots and incorporated in the leaves. During the fall, the leaves detach and fall to the ground. There, they decay and release their nutrients into the soil.

The fertile soil and abundant plant life of the forest support a rich and varied population of insects, microorganisms,

FIGURE 5-10 **Temperate deciduous forest.** Broad-leaved trees in this biome lose their leaves each fall. Years of leaf litter have produced rich soils that humans have long used for farming.

birds, reptiles, amphibians, and mammals (**FIGURE 5-11**). Common mammals include the raccoon, white-tailed deer, red fox, and black bear.

Ernest Hemingway once wrote that "a continent ages quickly once we come." In fact, few biomes have been so heavily altered by human activities as the temperate deciduous forest. Large-scale destruction of the forests began when colonists first settled the continent. Early settlers cleared the forests to grow crops on the rich soil. They also cut trees to make room for homes, towns, orchards, and roads. In the United States, only about 10% of the land east of the Mississippi is still forested, and only about 0.1% of the original forests remains. Much of the existing forest consists of young trees that have regrown after successive harvests.

Unfortunately, many early settlers failed to practice soil conservation on their farms. Heavy rains often washed the soil away. Moreover, farmers usually did little to replenish soil nutrients. Many farmers depleted the land, destroyed the soil that had taken centuries to form, and then moved westward.

(a)

Table 5-1	
Precipitation in Major Biomes	
Biome	**Annual Precipitation Centimeters (Inches)**
Tundra	Under 25 (10)
Taiga	38 to 100 (15 to 40)
Temperate deciduous	75 to 150 (30 to 60)
Grassland	25 to 75 (10 to 30)
Desert	Under 25 (10)
Tropical rain forest	150 to 400 (60 to 160)

(b)

FIGURE 5-11 **Animals of the temperate deciduous forest.**
(a) Black bear, **(b)** white-tailed deer.

FIGURE 5-12 **Grassland biome.** This biome does not receive enough moisture to support trees. The only trees found in the grasslands, therefore, were planted by humans or live along streams.

The Grassland Biome

Grasslands exist in temperate and tropical regions in areas that receive intermediate levels of precipitation—that is, less precipitation than forested regions but more than deserts (**Table 5-1**). Grasslands are found in North America, South America, Africa, Europe, Asia, and Australia.

In North America, grasslands form a continuous, wedge-shaped zone extending from the Gulf of Mexico northward through Canada to the taiga (Figure 5-5). A small area of grassland is found in the Great Basin, between the Cascade Mountains on the west and the Rockies on the east.

All grasslands bear a remarkable similarity (**FIGURE 5-12**). Most are on flat or slightly rolling terrain. Carpeted in thick grasses, the soils are probably the richest in the world as a result of thousands of years of plant growth and decay.

In the United States, the grasslands begin just east of the Rocky Mountains. The grasslands of eastern Colorado, however, are not as rich as those further east because of lower rainfall, as explained earlier. The grasslands of the eastern plains of Colorado, Wyoming, and Montana are known as the **short-grass prairie**. Further east, as rain and snowfall increase, the short-grass prairie gives way to the tallgrass prairie of the Dakotas, Nebraska, and eastern Kansas. Today, most of the tall-grass prairie has been destroyed by farming so that only small patches remain.

Grasslands are usually devoid of trees because they lack sufficient annual precipitation and experience periodic drought. Thus, the only trees found in the Great Plains of the United States are those that have been planted around homes and farms or those that live along streams and rivers.

Grasses, however, require less water than trees, and many species of grass send their roots deep into the Earth, drawing on reliable moisture supplies far below the surface. Grasses are also well adapted to periodic fires ignited by lightning. Fires burn the plant above the ground but do not harm the roots, from which new life can spring.

Because the soils of the grassland biome are so rich, they have been heavily exploited by people, primarily for agriculture, a practice that profoundly alters the plant and animal life. The grassland biome of North America, for example, once supported an enormous population of bison—perhaps as many as 60 million animals. Roaming the prairies, bison herds were so large that it would take men traveling on horseback 3 to 4 days to pass through a single herd. These animals were killed for food and sport and squeezed out of their habitat by the spread of farms.

The grassland biome of North America also supported grizzly bears and elk. As humans settled the grasslands and began to farm and raise cattle, grizzlies were exterminated. Elk herds fled to the protection of the Rocky Mountains, where they remain today. Many species—including deer, pronghorn antelope, badgers, and coyotes—still live on the grasslands of North America, sharing habitat with cattle, horses, sheep, and people.

In the grasslands, as in the temperate deciduous biome, farming has probably had the most significant impact of all human actions. Today, many of the world's tallgrass prairies have been plowed under to plant corn, wheat, soybeans, and other crops. The short-grass prairie, being drier and less productive, has fared much better. Except where irrigation water has allowed farmers to cultivate the land, much of the short-grass prairie still exists. This land, however, is often overgrazed by cattle. Continual overgrazing eliminates many hardy grasses, and in the semiarid grasslands of North America (in Colorado, for example), it creates dry soil conditions suitable for the growth of weedy species such as sagebrush. Because cattle will not eat it, sagebrush thrives in overgrazed fields, eventually taking over. Throughout the West, once productive pastures have become overgrown with sagebrush (FIGURE 5-13). As you travel in eastern Wyoming and Colorado, look for fields of sagebrush; chances are that they are the result of overgrazing. With a little care, such an ecological tragedy could have been avoided.

Poor agricultural practices on farms and ranches in grasslands throughout the world have resulted in widespread soil erosion and desertification in the grassland biome (Chapter 12). The most dramatic evidence of abuse came in the 1930s in the western and midwestern United States. An extended drought, combined with fencerow-to-fencerow planting, spawned one of the most significant environmental

FIGURE 5-13 **Sagebrush.** Overgrazed western rangeland is often overrun with this hardy species, the sagebrush, which is edible only to a limited number of species such as pronghorn antelopes.

disasters of human history: the **dust bowl.** Millions of tons of topsoil were lost from U.S. farms in huge dust storms.

The dust bowl stimulated a rash of conservation efforts to help protect farmland. Perhaps the most significant step was the planting of long rows of trees alongside fields to reduce wind erosion (FIGURE 5-14). Today, farmers sometimes leave wheat stubble and residues from crops on the ground over the winter to protect soils. Unfortunately, many of the improvements in soil management made in the immediate post–dust bowl era are being lost. Many farmers are once again planting from fencerow to fencerow in an effort to increase their output. Not surprisingly, devastating dust storms are becoming more and more common in some states, such as Texas and California.

KEY CONCEPTS

The grassland biome occurs in regions of intermediate precipitation—enough to support grasses but not enough to support trees. On most continents, the rich soil of the biome has been heavily exploited by humans for agriculture.

FIGURE 5-14 **Shelterbelts.** Rows of trees planted along the margins of farm fields help reduce wind erosion and drying.

The Desert Biome

Deserts exist throughout the world. Some cover vast regions. The Sahara, for example, stretches across northern Africa and is about the size of the United States. In North America, deserts exist primarily on the downwind side of mountain ranges.

Deserts, although dry, are not devoid of precipitation (Table 5-1). Rain often comes in violent downpours, however, causing flash flooding and severe erosion that have sculpted the magnificent canyons of the desert Southwest. Many species of desert wildflowers are well adapted to the infrequent but intense spring rains. These species have an accelerated life cycle, in which they grow and produce flowers within a few days of a thunderstorm, turning the desert into a colorful garden almost overnight (FIGURE 5-15).

Plants that live in the desert must also be adapted to tolerate a wide range of temperatures. In the desert, temperatures may reach 50°C (120°F) during the day and then drop to near freezing at night. The temperature in the desert fluctuates so much in large part because of the absence of moisture in the atmosphere, which reduces heat retention by the nighttime sky. Thus, even though the sun warms up the desert floor during the day, the heat escapes quickly at night.

Plants in the desert are adapted to low soil moisture. Many desert plants, such as cacti, have shallow root systems that extend laterally from the plant. Running just below the surface, these roots absorb rain and melted snow and then transport the water to the main body of the plant, where it is stored. Other desert plants, such as the mesquite tree, have deep taproots that extend downward, sometimes over 30 to 60 meters (100 to 200 feet), to moist soil. Taproots anchor the plant and also provide a relatively reliable source of water.

Water absorbed by the roots of many desert plants is stored in succulent, water-retaining tissues, giving the plant an ample supply on which to draw during the rainless months. In most desert plants, water supplies are protected by the presence of thick outer layers and waxy coats that reduce moisture loss.

Desert plants are often widely spaced on the desert floor, which reduces competition for water and ensures an adequate supply. How do plants space themselves? Some plants release growth-inhibiting chemicals into the soil that deter competitors from taking root in a region around each plant.

The thorns of cacti are yet another adaptation that reduces water loss. Thorns protect cacti from being eaten, but they also give some degree of protection from the sun by providing shade and reflecting some sunlight from the plant (FIGURE 5-16a). White, fluffy hairs found on some species of cacti also reflect sunlight and provide shade (FIGURE 5-16b).

Many insects and other animals also make the desert biome their home. Like the plants, the animals are well adapted to desert conditions. The thick scales of snakes and lizards, for example, minimize water loss, permitting these

(a)

(b)

FIGURE 5-16 **The cactus. (a)** Spines protect cacti from hungry animals and also provide shade. **(b)** Some cacti have white hairs that block the sun.

FIGURE 5-15 **Desert garden.** Sudden spring rains cause desert flowers to bloom.

creatures to thrive in the dry, hot conditions (FIGURE 5-17a). Lizards and other species, such as mice, also survive by avoiding daytime heat. They rest in caves or burrows and venture forth only at night. The ringtail, for example, sleeps all day and comes out at night to find its food (FIGURE 5-17b).

Because water is a rare commodity in the desert, many species acquire the moisture they need from cellular energy production. Water released during this process is called *metabolic water*. In some species, metabolic water provides nearly all of the water needed to survive. The kangaroo rat receives *all* of the moisture it needs from this process and from the plants and insects it eats. Although cellular energy production does not produce large amounts of water, it can be enough if species possess additional adaptations that help them conserve body water. The kangaroo rat, for example, excretes a highly concentrated urine, a physiologic adaptation to the hot, dry conditions of the desert. Snakes and

(a)

(b)

FIGURE 5-17 Animals of the desert. (a) Gila monster and **(b)** ringtail.

lizards also excrete a highly concentrated urine that reduces water loss.

Deserts can turn cold. Winter in the desert often brings freezing temperatures and snow for several days and sometimes weeks on end. Thus desert plants must also be able to tolerate cold.

Large cities have sprung up in many of the world's deserts. To supply inhabitants, food is trucked in from farms hundreds of miles away. Water is often pumped from deep aquifers or is transported in extensive pipelines. Phoenix, Arizona, for example, receives much of its water from the Colorado River, several hundred miles to the northwest, through a gigantic (and costly) canal and pipeline.

The continuing expansion of cities in the deserts of the world and the growing water demand have created serious problems, however. Southeast of Phoenix, for instance, over 300 square kilometers (120 square miles) of land has subsided (sunk) more than 2 meters (6 feet) because of intensive groundwater withdrawal. Cracks in the Earth's surface have developed in subsided areas. Some of the largest cracks are 3 meters (10 feet) wide and 3 meters deep and run for 300 meters, three times the length of a football field. Subsidence and cracks in the Earth's surface can break pipelines and can destroy homes and highways. Depletion of groundwater can eliminate the supply of water for native plant species.

Each year, millions of acres of new desert form on semiarid grasslands. Research suggests that deserts are expanding principally because of human actions. Livestock overgrazing destroys grasses in semiarid regions. The loss of vegetation may reduce rainfall, creating desertlike conditions. Climate change, both natural and human induced, may also be contributing to the spread of deserts. Stopping desertification will require dramatic improvements in land management, especially grazing, and reductions in greenhouse gases. See the Individual Actions Count! table for suggestions on ways you can help.

KEY CONCEPTS

The desert biome is characterized by dry, hot conditions, but often abounds with plants and animals adapted to the heat and lack of moisture. Unfortunately, deserts of the world are expanding as a result of human activities such as overgrazing livestock and the production of greenhouse gases.

The Tropical Rain Forest Biome

Heading south to the equator, we find one of the most endangered biomes on Earth, the tropical rain forest (FIGURE 5-18). Tropical rain forests exist near the equator in South and Central America, Africa, and Asia (Figure 5-5). By far the most complex and diverse of all the Earth's biomes, tropical rain forests support a wealth of plants, animals, and microorganisms. A small tropical island, for example, may have as many butterfly species (500 to 600) as the entire United States. Two hundred fifty different tree species exist in a single hectare (2.5 acres) of rain forest; in the temperate deciduous forest biome of the United States, a similar tract may have 20 to 30 species.

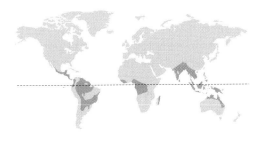

FIGURE 5-18 **Tropical rain forest.** This endangered biome covers vast areas and is rich in species.

The tropics actually contain a variety of different ecosystems, of which the rain forest is the largest and best known. With 200 to 400 centimeters (60 to 160 inches) of rain falling per year, trees of the tropical rain forest grow to heights of 60 meters (200 feet) or more. The tops of the tallest trees in a tropical rain forest form a dense canopy that blocks much of the incoming sunlight. Shorter trees form a lower canopy that intercepts most of the remaining light. As a result, only about 1% of the incoming sunlight reaches the ground. Because so little light strikes the forest floor, ground-level vegetation is sparse. Only those species adapted to low light (for example, African violets, philodendrons, and ferns) can survive in such conditions. The lack of ground-level vegetation relegates most life to the treetops. Dwelling in this zone are monkeys, birds, and countless species of insects.

The tropical rain forest is a paradox to the untrained observer. Although it is the richest and most diverse biome on Earth, most of its soils are thin and nutrient poor. If a section of forest growing on such poor soil is cleared to make room for crops or cattle ranches, it will generally produce for only 4 or 5 years before its nutrients are depleted. Large-scale farming and ranching on many rain forest soils are generally doomed to fail.

> Approximately half of the world's tropical rain forests have already been cut down. Each year, an additional 17 million hectares or 42 million acres of rain forest are leveled.

An understanding of basic ecology explains the paradox of the tropical rain forest, in which the world's richest biome grows on some of the world's poorest soils. To begin, we ask: Why is the soil so poor? In tropical rain forests, dead trees, leaves, and other forms of biomass (dead animals) that fall to

individual Actions Count

- Learn more about your biome—including threats and actions people are taking to protect and restore it.
- Support environmental groups that protect habitats.
- Join groups that help restore ecosystems through tree planting, streambed reclamation, and other activities.
- Support growth management activities that help concentrate growth rather than permitting urban sprawl.
- Reduce resource consumption in any way you can to protect ecosystems.

the ground are consumed by countless insects. The material that the insects do not eat is rapidly decomposed by bacteria, fungi, and other organisms. Nutrients released into the soil by bacterial decay are quickly absorbed by the roots of trees located just beneath the ground's surface. Thus, virtually all of the nutrients are locked up in plants and animals. In essence, then, tropical soils are so poor because life is so rich.

The lush vegetation of the tropics protects the soil from erosion. Thus, clearing the land for agriculture can be a chancy proposition. Heavy rainfall in the wet season erodes soils no longer protected by trees and other vegetation, rendering the new farmland useless and choking nearby streams and rivers with sediment. The loss of the thin topsoil also makes recovery from damage more difficult and protracted. Some forestry scientists believe that it could take large forest clearings 500 years or more.

Many rain forest soils are also useless for farming and ranching because they contain large amounts of iron. These soils are called **lateritic** (LA-ter-it-ick) **soils** (*later* is Latin for "brick"). When cleared and exposed to sunlight, lateritic soils bake as hard as bricks. The impenetrable crust that develops 1 to 2 years after clearing makes the soils virtually impossible to cultivate.

Tropical forests are the lungs of the planet. They "breathe in" vast amounts of carbon dioxide and release oxygen vital to animals. By absorbing carbon dioxide, tropical rain forests help reduce global warming, a problem discussed in Chapter 20. Tropical rain forests also help regulate planetary climate. One computer simulation showed that widespread destruction of rain forests in Africa could dramatically reduce rainfall as far north as Europe. The impact on agriculture would be substantial.

Once covering a region approximately the size of the United States, only about half of the rain forest remains. Despite worldwide concern for the fate of tropical rain forests, timber cutting continues. A 1980 study by the United Na-

tions suggested that about 11 million hectares (27 million acres) of tropical rain forest were being cleared each year. A more recent UN study suggests that tropical deforestation is far greater—about 17 million hectares (42 million acres) per year, an area about the size of the state of Washington. At this rate, the remaining tropical forests will be gone in 100 years—and with them, tens of thousands, perhaps millions, of species. Computer studies suggest that the widespread destruction of tropical forests could upset global rainfall patterns and could affect global climate by reducing the amount of carbon dioxide removed from the atmosphere each year.

KEY CONCEPTS

The tropical rain forest is the richest and most diverse biome on Earth because of its abundant rainfall and warm climate. About half of the world's rain forest has been destroyed; huge tracts could be eliminated in the near future, with devastating effects on climate and plants and wildlife, if current trends continue.

Altitudinal Biomes

Temperature and precipitation (rain and snow) profoundly influence life in biomes, affecting species diversity and the abundance of life. The relationship between climate and life is dramatically illustrated in mountainous terrain. If you walked from the top of a mountain to its base, you would very likely progress through several distinct life zones called **altitudinal biomes** (FIGURE 5-19).

Altitudinal biomes mirror the latitudinal biomes described in this chapter and are a reflection of differences in temperature and precipitation. Starting at the top of the mountain, for example, one encounters a cold, treeless region similar to the Arctic tundra. Known as the **alpine tundra**, this area is characterized by a short growing season and

FIGURE 5-19 **Altitudinal biomes.**

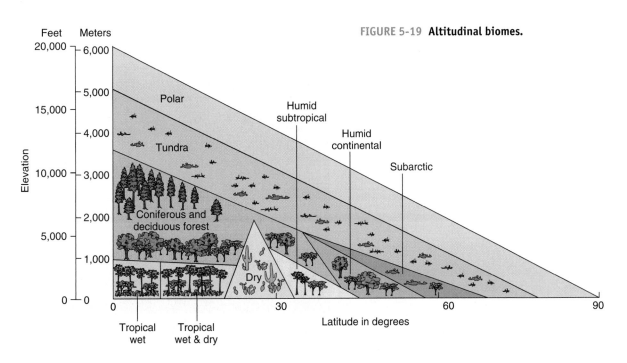

FIGURE 5-20 **Alpine tundra.** This biome is, in many ways, like the Arctic tundra, except that it receives far more moisture. Because the growing season is short, plants remain small.

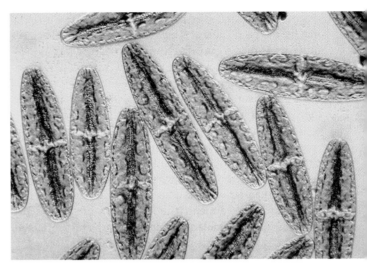

long, cold winters (**FIGURE 5-20**). But unlike the Arctic tundra, the alpine tundra receives lots of moisture. Most of the precipitation comes in the form of snow, and most blows away, ending up at the next lower level.

Climbing downward, one encounters a band of conifers akin to the taiga. In the Rocky Mountains, this zone borders on grassland. On the west slope of the Cascade Mountains, however, the taigalike region borders deciduous forests.

KEY CONCEPTS

Because climate varies with altitude, the distribution and abundance of life also change.

5.3 Aquatic Life Zones

On land, precipitation and temperature are the chief determinants of the distribution and abundance of life. In aquatic life zones, however, water is abundant and temperature relatively constant. The abundance and diversity of life forms are determined principally by energy and nutrients.

Aquatic life zones may be either freshwater or saltwater. In both freshwater and saltwater aquatic life zones, many food chains begin with a group of producer organisms, the **phytoplankton** (FIE-toe-plank-ton). Phytoplankton are microscopic, free-floating, photosynthetic organisms, mostly algae and diatoms (**FIGURE 5-21**). These organisms capture solar energy, using it to produce carbohydrates from carbon dioxide dissolved in the water.

Phytoplankton are consumed by microscopic **zooplankton** (ZOE-oh-plank-ton), single-celled protozoans, and multicellular crustaceans (**FIGURE 5-22**). Zooplankton form the second trophic level of many aquatic food chains. Zooplankton, in turn, are consumed by small fish, which are a food source for larger fish and a variety of other organisms.

KEY CONCEPTS

Aquatic systems are divided into distinct regions, known as *aquatic life zones,* which may be freshwater or saltwater. The abundance of life is determined by energy and nutrient levels. Phytoplankton form the base of aquatic food chains.

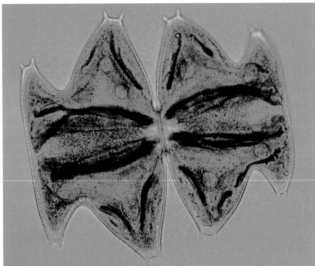

FIGURE 5-21 **Phytoplankton.** Unicellular algae like these are called phytoplankton and form the base of many aquatic food chains.

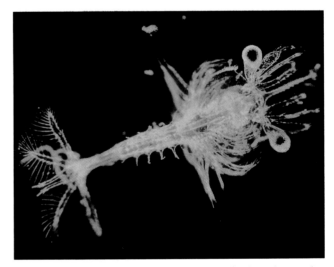

FIGURE 5-22 **Zooplankton.** These organisms feed on algae and are consumed by small fishes.

Freshwater Lakes

Freshwater aquatic life zones include lakes, ponds, rivers, and streams. We'll begin by looking at ponds and lakes.

Ponds are relatively small, shallow bodies of water. Because they are shallow, sunlight often penetrates to the bottom, providing plenty of energy for all aquatic life. Lakes are deeper than ponds and generally contain four distinct zones. The first is the **littoral** (LIT-tore-el) **zone**, from the Latin word *litoralis*, meaning "seashore" or "coast" (**FIGURE 5-23**). The littoral zone consists of the shallow waters at the margin of a lake, where rooted vegetation often grows. It contains abundant phytoplankton and rooted vegetation. Some rooted vegetation never breaks the surface; it is submerged throughout its life. Others—for example, cattails—extend above the water's surface. Still other plants—such as water lilies—have leaves and flowers that float on the water's surface.

As Figure 5-23 shows, the **limnetic** (limb-NET-tick) **zone** is a region commonly called "open water." Extending downward to the point at which light no longer penetrates, the limnetic zone is the main photosynthetic body of a lake. In many lakes, the limnetic zone supports abundant phytoplankton—so many, in fact, that phytoplankton biomass may exceed the biomass of rooted vegetation along the shoreline. The limnetic zone is therefore something of a biological factory, producing the food that supports most other aquatic life forms. It is also a major source of oxygen, required by zooplankton, many bacteria, and animals such as fishes.

The **profundal zone** lies beneath the limnetic zone in deeper lakes and extends to the bottom (Figure 5-23). Because no sunlight penetrates the profundal zone, conditions are not favorable for plant and algal growth. Without sunlight and plants, oxygen levels remain fairly low. Fishes can survive in the region, but they rely on food produced in the limnetic and littoral zones.

The bottom of a lake is the **benthic zone** (Figure 5-23). It is home to organisms that tolerate cool temperatures and low oxygen levels—such as snails, clams, crayfish, various aquatic worms, and insect larvae (including those of the mayfly, dragonfly, and damselfly). The larvae emerge during the spring and become free-flying insects. Mayflies, for example, generally breed within a day of emergence and then die. Many of them are eaten by fish and are therefore an important source of food in freshwater aquatic ecosystems.

KEY CONCEPTS

Lakes are divided into four regions—the littoral zone, the limnetic zone, the profundal zone, and the benthic zone—with very different conditions and, consequently, very different life forms.

Lake Turnover   During the summer in temperate climates (such as North America), the water of most lakes forms three distinct layers characterized by different temperatures (**FIG-**

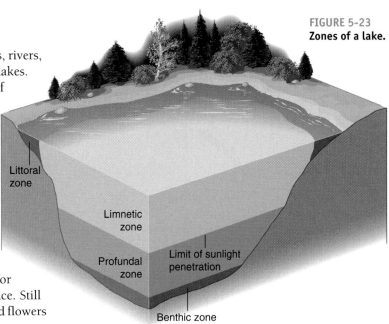

FIGURE 5-23
Zones of a lake.

Littoral zone

Limnetic zone

Profundal zone

Limit of sunlight penetration

Benthic zone

URE 5-24). The warm surface water of a lake is called the **epilimnion** (ep-eh-LIM-knee-on) (**FIGURE 5-24a**). Just beneath the epilimnion is a region of abrupt temperature change, the **thermocline** (thur-moe-CLINE). Swimmers can experience the thermocline by diving toward the bottom of a lake. Below the thermocline is a region of fairly uniform temperature, the **hypolimnion** (high-poe-LIM-knee-on). In the summer, the hypolimnion contains dense, cold water.

As Figure 5-24a shows, the three summertime layers also differ in oxygen levels. The oxygen concentration is highest in the epilimnion, where light penetration and photosynthesis are greatest, and falls rather rapidly from the surface to the bottom as photosynthesis declines.

The thermal layering of lakes is not static. During the fall, for example, the surface waters gradually cool, and the water temperature of a lake eventually becomes fairly uniform from top to bottom (**FIGURE 5-24b**). In other words, thermal stratification disappears. In the fall, winds often churn the water, causing a thorough mixing of surface and deep water. As a result, oxygen levels also become fairly uniform from top to bottom. This mixing of surface and bottom waters is known as the **fall overturn**.

In late fall and early winter, the surface waters of a lake cool even more. Ice may form on the lake, often covering the entire surface. Because ice is less dense than liquid water, it floats; this leaves an ice-free zone in deeper lakes, where fish live. At this time, water oxygen levels are fairly uniform from top to bottom, although levels are lowest in the deepest waters and at the bottom (**FIGURE 5-24c**).

During the winter, many species such as turtles burrow in the mud, where they hibernate. Fishes, however, remain active and continue to feed throughout the winter, living on oxygen produced during the summer months. Fortunately, cold water reduces their metabolism and thus their demand for oxygen. Algae, which are present in reduced number, replenish the oxygen so long as sunlight can penetrate the ice.

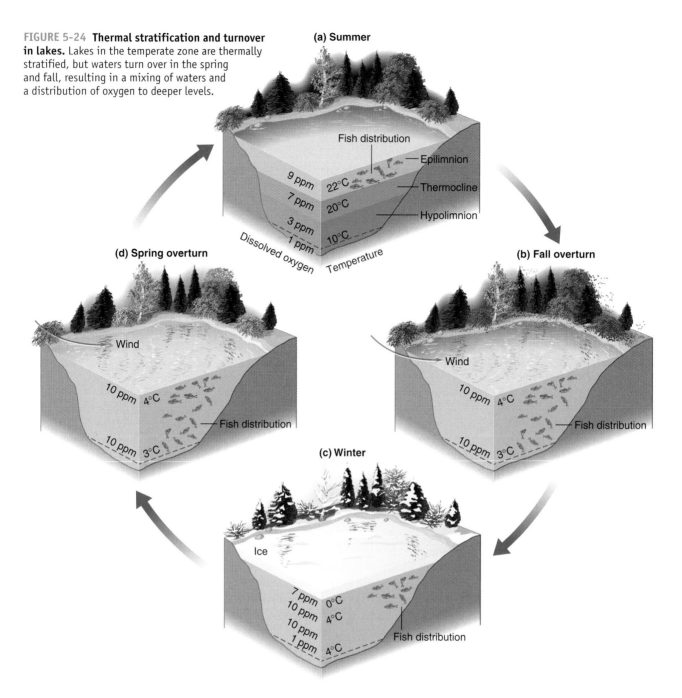

FIGURE 5-24 Thermal stratification and turnover in lakes. Lakes in the temperate zone are thermally stratified, but waters turn over in the spring and fall, resulting in a mixing of waters and a distribution of oxygen to deeper levels.

(a) Summer

Fish distribution
Epilimnion
9 ppm 22°C
7 ppm 20°C Thermocline
3 ppm
1 ppm 10°C Hypolimnion
Dissolved oxygen Temperature

(d) Spring overturn

Wind
10 ppm 4°C
Fish distribution
10 ppm 3°C

(b) Fall overturn

Wind
10 ppm 4°C
Fish distribution
10 ppm 3°C

(c) Winter

Ice
7 ppm 0°C
10 ppm 4°C
10 ppm
1 ppm 4°C
Fish distribution

In the spring the ice melts, and the water temperature becomes uniform once again. Winds agitate the water, causing a mixing known as the **spring overturn** (FIGURE 5-24d). As the days get longer and warmer, however, the surface waters begin to warm, and the lake again becomes thermally stratified.

Rivers and Streams

Rivers and streams are complex ecosystems. As with lakes, no two streams are alike. In many areas, streams begin in mountains or hilly terrain, collecting water that falls to the Earth as rain or snow. The region drained by a stream is called a **watershed**. Small streams join to form rivers, and rivers flow downhill to the sea.

Streams and rivers are generally well oxygenated because they have a relatively large surface area (relative to their water volume) to absorb oxygen from the air. Current also facilitates oxygenation. Current velocity is determined by the gradient—that is, the steepness of the terrain. Fast-moving currents of mountain streams produce waves and rapids as the water collides with rocks or drops over ledges. This agitation greatly increases oxygenation. For these and other reasons, photosynthesis is a less important source of oxygen in rivers and streams than it is in ponds and lakes. Certain fish are adapted to different currents. Trout, for example, generally inhabit cold, oxygen-rich mountain streams where the current is quite rapid. Black bass live in rivers, where the waters are warmer, less oxygenated, and slower.

Unlike many lakes and ponds and terrestrial biomes, streams are rather open ecosystems; that is, they receive a great many nutrients from bordering ecosystems. In some streams, much of the biologically available energy actually comes from nearby terrestrial vegetation, such as leaves that fall into the stream. Animal feces, insects, stems, nuts, and other biomass may also be washed into streams during rainstorms. All of this material feeds the aquatic food web. Many primary consumers in streams, therefore, are detritivores—organisms that feed on waste or remains of plants and animals (detritus). Streams do have their own producers, mostly algae and rooted vegetation, but in some cases, these organisms play only a minor role in providing food.

> **KEY CONCEPTS**
>
> Rivers and streams are complex ecosystems that rely more on agitation for oxygenation of their waters than lakes do. Many nutrients in streams that support aquatic life come from neighboring terrestrial ecosystems. The quality of water in a stream is profoundly influenced by activities in the watershed.

Protecting Freshwater Ecosystems

Lakes are repositories for pollutants and are highly vulnerable to them, especially if their waters are replaced slowly. Streams generally fare better than lakes because of their flow, which tends to whisk pollutants away. Despite this natural purging, streams and rivers are also quite vulnerable to pollutants from human sources if those sources are numerous. The Rhine River of Europe, for example, once supported 150 species of fishes. Today, only about 15 species remain, in large part because of pollutants from the hundreds of factories, towns, farms, and sewage treatment plants dotting its banks in the seven countries through which it flows.

Protecting rivers and streams requires measures to control pollution, that is, to reduce the release of pollutants into water bodies. One way they are achieving this goal is by waste minimization. **Waste minimization**, or **waste reduction**, involves measures that greatly reduce or eliminate waste. Waste minimization is a preventive measure with great benefit to people and the planet. Many companies are finding that simple, often inexpensive changes in their manufacturing processes will eliminate the need for toxic chemicals—or greatly reduce the output of them. Where there is no waste, there is no need for waste treatment. Waste minimization, therefore, can save companies enormous amounts of money in waste treatment and disposal costs, and it eliminates future liability for harm done by pollution. Spotlight on Sustainable Development 5-1 shows ways Christmas tree farmers can reduce pollutants.

> **KEY CONCEPTS**
>
> Like lakes and ponds, streams are self-purging, but are extremely vulnerable to pollution if sources exceed the capacity to self-cleanse.

Saltwater Life Zones

The oceans cover over 70% of the Earth's surface. Like freshwater systems, the ocean can be divided into ecologically distinct life zones. The sections that follow discuss several of the most important ones. In the ocean, as in all other bod-ies of water, the distribution and abundance of life are dependent on many factors, but the most important are energy and nutrients.

The Coastal Life Zones

The ocean can be crudely divided into two groups of life zones, those lying near the coasts (the coastal life zones) and those of the deeper ocean. We begin our exploration of the seas by examining three coastal life zones: estuaries, seashores, and coral reefs.

> **KEY CONCEPTS**
>
> The coastlines are highly productive waters characterized by abundant sunlight and a rich supply of nutrients, both of which contribute to an abundance of life forms.

Estuaries and Coastal Wetlands: The Estuarine Zone Estuaries (ES-stu-air-ees) are the mouths of rivers, places where freshwater mixes with saltwater. Estuaries are very rich life zones because streams and rivers transport many nutrients from the land and incoming tides carry nutrients into them from the ocean. These nutrients support abundant plant and algal growth and sizable populations of fish and molluscs (clams, oysters, and scallops).

Most estuaries are located near **coastal wetlands**—salt marshes, mangrove swamps, and mud flats (**FIGURE 5-25**). Together, estuaries and coastal wetlands form the **estuarine zone**. For most people, coastal wetlands are seen as muddy, smelly places of little value. However, studies show that approximately two thirds of the world's commercially valuable fishes and molluscs (shellfish) depend on the estuarine zones at some point in their life cycle. For example, many marine fishes spawn in the estuarine zone. Their eggs are laid on the bottom and are extremely susceptible to the deposition of sediment. The larvae that hatch from the eggs feed off

FIGURE 5-25 Coastal wetland. Aerial view of an estuary and mangrove swamp.

5-1 Dreaming of a Green Christmas

Each year, Americans buy millions of Christmas trees, which are grown on tree farms throughout the country. In Watauga County, North Carolina, alone, half a million trees worth $12.2 million are sold in a single season. Trees are grown in fields cleared from native forest (FIGURE 1). Although the industry helps to boost many local economies, tree farms displace native wildlife. Because of the heavy use of chemical pesticides, they can also threaten waterways and species, including humans, living nearby.

In 1992, residents of Boone, North Carolina, in the Appalachian Mountains, found that the heavy use of chemicals to control both insects and weeds on local Christmas tree farms might be seeping into local water supplies and might also be linked to an elevated rate of childhood leukemia in the region. Although health officials could not say with certainty that there was a correlation between pesticide use and leukemia, hundreds of farmers in Watauga County decided to take action—drastically slashing the amount of pesticide they used. In some instances, farmers were able to eliminate certain pesticides.

Weeds and pests still persist, but they're being controlled by natural biological control methods. The results have been encouraging. Groundwater supplies are improving. Wildlife are rebounding in fields and nearby woodlands and streams. One local tree farmer noted that he's seen more wildlife in the last 5 years than he saw 30 years ago.

This effort to help improve both human and environmental health was successful partly because the county agricultural extension office developed a series of workshops for local farmers who wanted to learn alternative ways of controlling pests. Some solutions have been quite simple. In years past, farmers sprayed fields with diazinon or chlordane to kill grubs that dine on the roots of fir trees. Wooly aphids, another pest, were controlled with yet another potent pesticide.

To control weeds, farmers sprayed fields in the late winter with herbicides. This wipes out weeds before they sprout. They then mowed the grass around the trees. It turns out, however, that grubs prefer short grass for laying eggs. Today, many farmers are forgoing the early herbicide treatment and letting the grass grow longer. Grubs have been effectively controlled. Farmers save on mowing costs and herbicide use.

FIGURE 1 Tree farm in Watauga County, North Carolina.

phytoplankton and zooplankton that flourish in the nutrient-rich waters of the estuary. The estuarine zone also provides food, shelter, and breeding grounds for millions of waterfowl and fur-bearing animals such as muskrats.

Unfortunately, the estuarine zone is one of the most altered ecosystems on Earth. Dams built on rivers—to control flooding, supply drinking or irrigation water, or provide recreational activities—reduce the flow of freshwater into estuaries. Reservoirs behind dams also capture sediment that would otherwise travel to the estuary. Sediment contains many nutrients vital to estuarine phytoplankton. Dams, therefore, literally starve the organisms at the base of the estuarine food chain.

The impact of dams is aptly illustrated by the Aswan High Dam, built in the early 1960s on the Nile River in Egypt to provide irrigation water for farms. Although the dam provides many benefits, it has nearly halted the flow of nutrients into the river's estuary. This in turn caused a collapse in the sardine fishery in the Mediterranean Sea, as phytoplankton perished. The commercial sardine catch in the Mediterranean dropped from 18,000 tons per year before the dam was built to about 500 tons per year today.

Coastal wetlands have been drained and filled with dirt to build homes, highways, recreational facilities, and factories. These activities destroy habitat for fish and other species. In the United States, over 40% of the coastal wetlands have been destroyed. In California, an estimated 90% have been destroyed.

> In the United States, approximately 40% of all coastal wetlands have been destroyed.

KEY CONCEPTS

Estuaries are nutrient-rich zones at the mouths of rivers, often associated with coastal wetlands, together forming the estuarine zone. The estuarine zone is highly productive and of great value to humans and other species. Human activities severely threaten this important biological asset.

The Shoreline The Earth's coastlines are generally rocky or sandy regions that support a variety of specially adapted organisms. Abundant sunlight and nutrients account for much of the biological diversity in this zone. Rocky coastlines are home to seaweed (algae), sea urchins, barnacles, and sea stars (**FIGURE 5-26**). Many of these organisms anchor themselves to the rocks and are thus able to withstand turbulence generated by waves. Organisms that do not attach themselves, such as sea urchins, live in rocky crevices protected from water currents.

> " The commercial sardine catch in the Mediterranean Sea has fallen from 18,000 tons per year to around 500 tons as a result of the construction of the Aswan Dam, which has nearly halted the flow of nutrients into the sea. "

The coastline is also home to a variety of shorebirds that dine on insects, crustaceans, and other organisms. In addition, the coastlines serve as nesting sites for turtles and are home to breeding colonies of seals and walruses. Popular recreation and building sites, many of the world's coastlines have been severely altered. Turtle nesting sites have been destroyed in many places. Protecting the coastlines from offshore oil spills and human encroachment is extremely important.

KEY CONCEPTS

The shorelines of the world are rocky or sandy regions that are home to a surprising number of organisms adapted to the tides and the turbulence created by wave action.

Coral Reefs Coral reefs are found in relatively warm and shallow waters in the tropics or nearby regions (subtropics). A **coral reef** consists of calcium carbonate or limestone produced by various species of algae and by colonies of organisms, the stony corals (**FIGURE 5-27**). These organisms have stiff calcium carbonate outer skeletons that persist long after their death, creating a base on which other corals grow. Coral reefs take thousands of years to form.

Coral reefs are the aquatic equivalent of the tropical rain forest. They are home to a dazzling variety of organisms, many of which are colorful beyond imagination (**FIGURE 5-28**).

Like estuaries and coastal wetlands, coral reefs are also highly vulnerable and experiencing considerable damage. Ships running aground on the coral reefs of Florida, for example, are damaging these fragile structures. Each year, thou-

FIGURE 5-26 Tidal pool inhabitants. Numerous species live along the rocky shorelines of the world's oceans.

FIGURE 5-27 Coral reef. The coral reef is bathed in sunlight and found in warm water. These features and ample nutrients make it the richest of all marine aquatic zones.

(b)

FIGURE 5-28 Coral reef inhabitants. (a) Coral reefs house some of the most spectacular fishes known to science. **(b)** Brightly colored anemones filter food from the water.

(a)

sands of divers visit the coral reefs of Florida. Careless divers also break delicate coral with their swim fins. Over the years, many of the most popular reefs have been severely damaged.

Sediment is a particularly troublesome pollutant in coral reefs. It clouds seawater, which reduces photosynthesis in a group of microscopic organisms (called *dinoflagellates*) that live in a mutualistic relationship with many species of corals. Decreasing photosynthesis decreases the amount of food available to coral, and reefs grow more slowly. Particularly heavy sedimentation can bury a reef, choking the life out of it. Several large reefs near Honolulu, in fact, have been destroyed by sediment washed from construction sites, highways, and farms. Housing development on the Florida Keys is also increasing sedimentation, threatening nearby coral reefs. Many of the world's coral reefs are dead or dying as a result of warming ocean waters thought to be associated with global climate change brought on by human activities.

KEY CONCEPTS

Coral reefs are the aquatic equivalent of the tropical rain forests and are being rapidly destroyed.

The Marine Ecosystem

FIGURE 5-29 shows a cross-section of the ocean floor. As illustrated, the ocean floor slopes downward away from landmasses and then drops more steeply. The gradually sloping region is called the **continental shelf**. The more steeply falling region is the **continental slope**, and the bottom of the deep ocean is called the **abyssal plain**.

As Figure 5-29 shows, the ocean is divided into four ecologically distinct life zones: the neritic, euphotic, bathyal, and abyssal. The **neritic** (neh-RIT-ick) **zone** is equivalent to the littoral zone of lakes and lies above the continental shelf, which varies in width from 10 to 200 miles away from dry land. The neritic zone contains relatively shallow water and consequently receives abundant sunshine. Its waters are relatively warm and well oxygenated.

Nutrients in the neritic zone come from streams and rivers that flow into the ocean. They are also supplied by upwelling, the transport of nutrient-rich water from the floor of the ocean up the continental slope. These nutrients, which had been deposited from shallower water above, support abundant phytoplankton, zooplankton, and fishes.

FIGURE 5-29 **Ocean zones.**

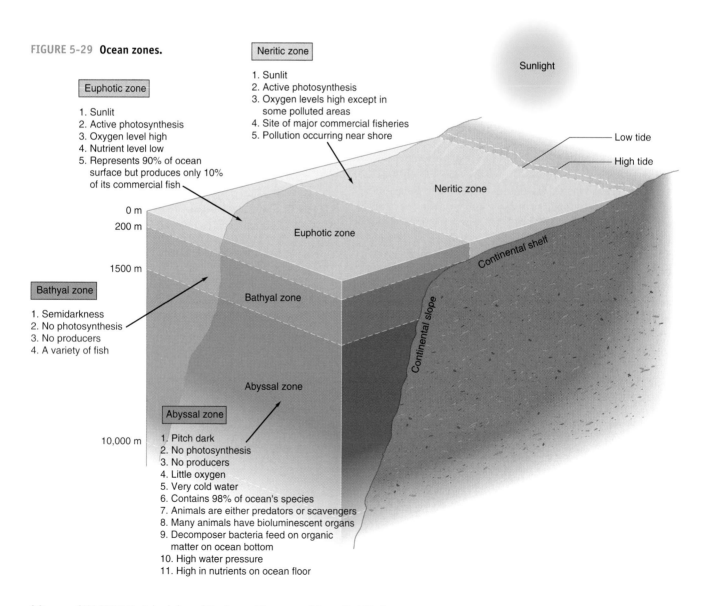

Neritic zone

1. Sunlit
2. Active photosynthesis
3. Oxygen levels high except in some polluted areas
4. Site of major commercial fisheries
5. Pollution occurring near shore

Euphotic zone

1. Sunlit
2. Active photosynthesis
3. Oxygen level high
4. Nutrient level low
5. Represents 90% of ocean surface but produces only 10% of its commercial fish

Bathyal zone

1. Semidarkness
2. No photosynthesis
3. No producers
4. A variety of fish

Abyssal zone

1. Pitch dark
2. No photosynthesis
3. No producers
4. Little oxygen
5. Very cold water
6. Contains 98% of ocean's species
7. Animals are either predators or scavengers
8. Many animals have bioluminescent organs
9. Decomposer bacteria feed on organic matter on ocean bottom
10. High water pressure
11. High in nutrients on ocean floor

Sunlight

Low tide

High tide

Neritic zone

Continental shelf

Continental slope

Euphotic zone

Bathyal zone

Abyssal zone

0 m
200 m
1500 m
10,000 m

In the neritic zone, sunlight normally penetrates to the ocean bottom. Ample sunlight supports large populations of algae and rooted plants, which, in turn, support a great many other species. In fact, most commercial fishing operations in the ocean concentrate their operations in the neritic zone.

The **euphotic** or **photic** (you-FO-tic or FO-tic) **zone** is the oceanic equivalent of the limnetic zone of lakes. It is the open-water region that extends to the lower limits of sunlight penetration, or about 200 meters (650 feet) below the ocean's surface. Abundant sunlight supports numerous species of phytoplankton, including diatoms, dinoflagellates, and others. Phytoplankton support a variety of zooplankton, mostly minute crustaceans.

Phytoplankton also produce plenty of oxygen. Given the high levels of oxygen and ample supply of sunlight, one might think that the euphotic zone would be extremely productive. Unfortunately, the waters are not very rich in nutrients, and the zone is therefore not very productive. Thus, even though the euphotic zone covers 90% of the ocean's surface, it produces only about 10% of the commercial fish taken each year.

Beneath the euphotic zone is a region of semidarkness, the **bathyal** (BATH-ee-el) **zone**. The bathyal zone is too dark to support photosynthesis. It is therefore characterized by a lack of photosynthetic organisms and low oxygen levels. Despite these factors, the bathyal zone is home to a variety of fish and other organisms, such as shrimp and squid, which feed on organisms "raining down" from above.

Beneath the bathyal zone is the abyssal zone. The **abyssal** (ah-BISS-el) **zone** is a region of complete darkness. It contains no photosynthetic organisms and is characterized by low oxygen levels. Most animals that live there are either predators or scavengers. To live in this zone, an animal must be adapted to extremely cold water, high water pressure, low oxygen, and complete darkness.

Sediment in the abyssal zone is often rich in nutrients. The deep-ocean floor is populated by a variety of bizarre-looking creatures that make use of these nutrients. Some species of fishes have evolved luminescent organs that shine in the dark, presumably helping them attract food and mates.

KEY CONCEPTS

The marine ecosystem consists of four ecologically distinct life zones, similar to those found in lakes.

Nature's laws affirm instead of prohibit. If you violate her laws you are your own prosecuting attorney, judge, jury and hangman.

—*Luther Burbank*

CRITICAL THINKING

Exercise Analysis

Scientists have asked many questions about these studies. First, could the differences in ice thickness have been the result of normal variation? (Possibly.) Were the measurements of the two submarines taken during the same time of year? (No.) Did the submarines follow the same path? (Actually, they followed slightly different routes.) Could winds have altered ice buildup between the two periods? (Possibly. Winds do alter ice buildup.) Have other studies shown thinning? (Yes. A study of ice measurements by two submarines on a virtually identical route across the Canada Basin during 1958 and 1970 showed that the mean ice cover was thinner in 1970 by 0.69 meter, about 2 feet.)

What other information/data might be useful in analyzing the hypothesis that the world's polar ice caps are shrinking? One bit of information that would be very helpful would be the extent of the ice sheets—that is, the square kilometers of ice sheet, plotted over time. Recent studies of this phenomenon show a marked decrease in the extent of the Arctic ice sheet, valuable habitat of the polar bear.

CRITICAL THINKING AND CONCEPTS REVIEW

1. What is a biome? Why is one biome so different from another? Give some examples.
2. Using what you have learned in the book, and especially the last two chapters, critically analyze the following comment. "Environmentalists speak about ecosystems as being fragile; yet look at the tundra. The conditions there are as harsh as any on Earth."
3. Design an experiment to determine how long it would take for the Arctic tundra to recover from human impact caused by truck traffic on the unprotected vegetation.
4. Animals are generally well adapted to the biome in which they live. Give some examples of plants and animals and their adaptations.
5. Explain why there are regional differences in vegetation and animal life within a biome.
6. Which biome do you live in? Describe it. How has it been altered?

7. List the pros and cons of oil development in the ANWR. Which side do you take on the issue? Why?

8. Critically analyze the following comment. "The United States has created a rich agricultural industry by plowing grasslands and clearing and plowing temperate forests. There is no reason to believe that countries in the tropics can't follow the same example. The rain forests are, after all, the richest terrestrial biome known to science."

9. Why do many deserts form downwind from mountain ranges?

10. Explain why the desert biome is spreading.

11. What are the primary determinants of species composition and abundance in terrestrial and aquatic habitats?

12. Describe each of the four zones of a lake (littoral, limnetic, profundal, and benthic) in terms of light avail-

ability, nutrient levels, oxygen concentration, and life forms.

13. Describe thermal stratification in deep lakes. Why does it occur during the summer but disappear throughout most of the rest of the year?

14. Why is the current so important in determining the abundance and type of life in a river?

15. Explain how streams and rivers recover from pollution and why natural recovery methods are so often overwhelmed.

16. What is an upwelling? How is it similar to an estuary? How is it different?

17. Describe ways in which human activities affect coral reefs and the estuarine zone.

KEY TERMS

abyssal plain
abyssal zone
alpine tundra
altitudinal biomes
aquatic life zones
bathyal zone
benthic zone
climate
coastal wetlands
conifers
continental shelf
continental slope
coral reef
Coriolis effect
deciduous trees
dustbowl
El Niño

epilimnion
euphotic (or photic) zone
estuarine zone
estuaries
fall overturn
grasslands
hypolimnion
lateritic
limnetic zone
littoral
neritic zone
northern coniferous biome
permafrost
phytoplankton
polar easterlies
poles
profundal zone

short-grass prairie
spring overturn
taiga
tallgrass prairie
temperate zones
thermocline
trade winds
tropics
tundra
upwellings
waste minimization
waste reduction
watershed
weather
westerlies
zooplankton

REFERENCES AND FURTHER READING

The References and Further Reading section at the end of this book contains a list of sources for the information discussed in this chapter and recommendations for further reading.

Connect to this book's website:
http://environment.jbpub.com/
The site features eLearning, an online review area that provides quizzes, chapter outlines, and other tools to help you study for your class. You can also follow useful links for in-depth information, research the differing views in the Point/Counterpoints, or keep up on the latest environmental news.

The South American toucan is part of the network of all living organisms.

CHAPTER 6   Principles of Ecology: Self-Sustaining Mechanisms in Ecosystems

And this our life, exempt from public haunt, finds tongues in trees, books in running brooks, sermons in stones, and good in everything.

—*Shakespeare*

In 1884, the water hyacinth was introduced into Florida from South America as an ornamental plant. Housed in a private pond, this lovely flowering plant, which grows on the water's surface, was accidentally released into a nearby river. From there, it spread throughout the state's extensive network of canals and rivers. Aided by a remarkable ability to reproduce—10 plants can multiply to 600,000 in 8 months—the hyacinth now chokes waterways throughout much of the state. This hearty colonizer crowds out native plants and grows so thick that it has made navigation impossible in some rivers (FIGURE 6-1).

FIGURE 6-1 Beautiful but obnoxious. The water hyacinth was intentionally introduced into southern Florida but was accidentally released into its waterways. Since its release, the plant has proliferated wildly and today chokes canals and streams throughout the state and much of the southern United States.

From Florida, the water hyacinth continued to spread, and it now inhabits the rivers and lakes of much of the southern United States. Today, it occupies nearly 800,000 hectares (2 million acres) of rivers and lakes from Florida to California. In the three states where the plant infestation is the worst (Florida, Louisiana, and Texas), officials spend $11 million a year to keep waterways navigable.

The story of the water hyacinth illustrates a common phenomenon: the disruption of a natural ecosystem by a seemingly harmless human action—the importation of an alien species. It also serves as a warning to humans that seemingly harmless actions can have severe ecological repercussions. The economic cost of such acts can also be staggering. See the Point/Counterpoint in this chapter for a debate on related issues.

This chapter continues the discussion of ecosystems by exploring mechanisms in natural systems that have permitted them to persist for millions of years. An understanding of these mechanisms can help us manage natural systems better—and perhaps manage ourselves better as well. This material also permits us to explore the principles of sustainability: conservation, recycling, restoration, population control, and adaptability, which were introduced in Chapter 2.

6.1 Homeostasis: Maintaining the Balance

Life is a bit of a balancing act. Take the human body as an example. Inside each of us are hundreds of mechanisms that evolved to help maintain fairly constant internal conditions—that is, to keep our bodies in balance. These mechanisms help us maintain body temperature, blood sugar levels, water concentration, and many other things.

The body's state of internal constancy is called **homeostasis** (HOE-me-oh-STAY-siss). The word *homeostasis* comes from two Greek words, *homeo*, which means the "same," and *stasis*, meaning "standing." Literally translated, *homeostasis* means "staying the same." Homeostasis, however, is not to be misconstrued as unwavering constancy. Internal conditions do not stay the same from one minute to the next. Rather, homeostasis is best described as a fairly constant state. Scientists refer to homeostasis as a state of **dynamic equilibrium**. This means that conditions are dynamic—ever

Environmentalists Are Subversive to Progress

William Tucker
The author, a critic of the environmental movement, has written numerous articles on environmental issues. His book *Progress and Privilege* is a thought-provoking and controversial discussion of environmentalism.

One of the key realizations of ecology is that the Earth is a kind of living system governed by many self-regulating (homeostatic) mechanisms. The Earth is in a state of equilibrium. If pushed too far in any one direction, the self-regulating mechanisms can become overloaded and break down, resulting in radical changes.

In its scientific aspects, ecology seems to offer an extraordinary broadening of our understanding of life on the planet. Yet with its transfer into the public domain, it has become little more than a sophisticated way of saying, "We don't want any more progress." Somehow this exciting discipline has been translated into a very conservative social doctrine. People have often waved the flag of "ecology" as a new way of saying that nature must be preserved and human activity minimized. Ecology is sometimes viewed as "subversive" to technological progress. It supposedly tells us that our ignorance of natural systems is too great for us to proceed any further with human enterprise. Just as nationalistic conservatives always try to throw a veil of reverence around such concepts as "patriotism" and "national tradition," so environmentalists try to maintain the same indefinable quality around ecosystems.

The lesson environmentalists drive home is that because we do not understand ecosystems in their entirety, and never will, we dare not touch them. Our knowledge is too limited, and nothing should be done until we understand more fully the implications of our actions.

To say that ecology is the science that does not yet grasp the complex interrelationships of organisms is like trying to define medicine as the science that does not yet know how to cure cancer. Environmentalists emphasize the negative parts of the discipline because it fits their concept that we have already had enough technological progress.

The lessons of ecology tell us many things. They tell us that organisms cannot go on reproducing uncontrollably. But they also tell us that many organisms have developed behavioral systems that keep their populations from exploding. The laws of ecology tell us that we cannot throw things away into the environment without having them come back to haunt us. But they also tell us that nature evolved intricate ways of recycling wastes long before human beings appeared, and ecosystems are not as fragile as they seem.

In fact, the whole notion of "fragile ecosystem" is somewhat contradictory. If these systems are so fragile, how could they have survived this long? If ecology teaches us anything, it enhances our appreciation of how resilient nature is and how tenaciously creatures cling to life in the most severe circumstances. This, of course, should not serve as an invitation for us to see how efficiently we can wipe them out. But it does suggest that the rumors of our powers for destruction may be exaggerated.

The environmentalist interpretation of ecology has been that ecosystems have somehow perfectly evolved and that human intervention always leads to degradation. It should be clear that even if a particular ecosystem did represent biological perfection, that is not reason in and of itself to preserve it at the expense of human utility. Our ethical position cannot be one of completely detached aesthetic appreciation. We must first be human beings in making our ethical judgments. We cannot be completely on the side of nature.

We are not a group of imbeciles aimlessly poking into the backs of watches or tossing rocks into the gears of Creation. There is purpose to what we do, and it is essentially the same as nature's. We are trying to rearrange the elements of nature for our own survival, comfort, and welfare. We can certainly act stupidly, but we can also act out of wisdom. It is foolish to argue that everything is already perfect and must be left alone. To portray humans as meddling outsiders in an already perfected world is nonsense. In going to this extreme to reaffirm nature, we only deny that we are a part of it.

Environmental writers suggest that we practice an "ecological ethic," extending our moral concerns to other animals, plants, ecosystems, and the entire biosphere. I would accept this proposal, with one important qualification: that is, that our ethical concerns still retain a hierarchy of interest. We should extend our moral concerns to plants, trees, and animals, but not at the expense of human beings. Our first obligation is to humanity. We should avoid actions that are destructive to the biosphere, but we must recognize that at some point our interests are going to impinge upon other living things.

You can link to websites that represent both sides through Point/Counterpoint: Furthering the Debate at this book's internet site, http://environment.jbpub.com/. Evaluate each side's argument more fully and clarify your own opinion.

Environmentalists Support Sustainable Progress

Daniel D. Chiras

Contrary to what some might have you think, the environmental movement is not composed of a homogeneous group of people who uniformly object to all human progress. The environmental movement attracts a following linked by a common interest in protecting the environment.

My experience within the movement, however, shows that environmental protection means many things to many people. People sympathetic with the plight of the natural world range from those who call for "complete preservation," which tolerates little human intervention, to the "preserve-it-in-parks" folks, who seek to protect small pieces of the environment for future generations to enjoy, while developing most of the rest for human use.

William Tucker's essay in this Point/Counterpoint casts environmentalists as narrow-minded obstructionists who condemn all human progress and technological development. No doubt, many businesspeople share this view—and for good reason. Environmentalists have threatened many a proposed dam, highway, and mine. In this narrow vein, few could deny that environmentalism is subversive to progress as some see it.

The truth, however, is not so simple. Progress is not one thing to all people. In fact, much of the environmental debate hinges, quite precariously, on the difference in how people view progress.

To most environmentalists, progress has a meaning much broader than economic growth, faster jets, and new gadgets to make life a little easier. Progress means prospering within the limits of nature, living on Earth without destroying the air, water, and soil on which our lives depend. Progress means finding a way of living sustainably on the planet.

Most environmentalists embrace progress, but no human endeavor can be counted a success if it destroys the Earth's life-support systems.

Tucker's assertion that environmentalists don't want any more progress misses the point. Environmentalists want lots of progress. We want progress in recycling and tapping into renewable energy in a big way. We want progress in using energy and other resources more efficiently. We want progress in restoring ecosystems that we depend on and in controlling human numbers.

As for Tucker's assertion that environmentalists think that nature must be preserved and human activity minimized, my experience in environmental protection shows that most environmentalists are working to keep society from destroying the Earth's life-support systems. The knowledge and skills necessary to fit into the web of life and succeed are readily available. Restraint and proper design are necessary, but stopping all human activity is not.

Ecology has taught environmentalists to look to the future with their eyes wide open, rather than narrowly focused on material welfare, economic wealth, and convenience. In attempting to rearrange the elements of nature to satisfy our needs, we must tread carefully, finding ways to meet our needs without impairing the ability of future generations—and other species—to meet theirs.

As for the claim that environmentalists see society as a stupid meddler whose every activity leads to degradation, let me say this. History teaches us that human activities have had profound, even devastating, effects on the natural world and human society. The dust bowl days, the great deserts spreading through Africa, and destruction of the once rich Fertile Crescent are blatant reminders of the severity of our intrusions. More recent studies show that our impact continues on a grand scale, as witnessed by the threats of deforestation, species extinction, global warming, and stratospheric ozone depletion.

Human intervention doesn't always lead to degradation, but it is also not a force to be trivialized. Ecosystems are not as fragile as some would have you believe, but they can be ruined on a massive scale. Witness the widespread loss of wetlands and rain forest throughout the world. Nature may be resilient, but against widespread human intervention, it is often powerless.

The crux of the disagreement between environmentalists and "developmentalists" lies in each group's view of progress. In my view, many environmentalists have a more sustainable view of progress. Environmentalism is not subversive. It advocates an alternative form of progress that seeks to foster a sustainable relationship with the planet. Our interests are likely to impinge on the interests of other species, but our challenge is to find ways to minimize these conflicts. By adopting the principles of sustainability that I've outlined in Chapter 2 of this book we can create more efficient and effective government, a better future, broader prosperity, and a stronger country. Through mutual responsibility, we can live well without ripping apart the planet. Let's stop fighting over the environment and find a new way of life and worldview that serve humans and the environment simultaneously.

Critical Thinking Questions

1. Summarize the main points in each essay. Were arguments adequately supported by facts?
2. Use your critical thinking skills to analyze and comment on these views.
3. Whose views most closely represent yours? Why?

changing—but they stay more or less the same over long periods. Consider an example.

Blood sugar (glucose) levels oscillate a bit around a set point of approximately 100 milligrams per milliliter. Blood sugar levels increase after a meal as sugars enter the bloodstream, but within an hour or so, they return to the set point. As we increase activity, which burns sugar, levels drop. Then internal storage depots release sugar so that levels return to stay fairly constant. Blood sugar is said to be in a *dynamic equilibrium*.

Shifts in external conditions—for example, temperature—can also cause internal changes. However, as in the case of blood sugar levels, internal processes will generally restore conditions to their normal state. In sum, then, homeostasis isn't about maintaining absolutely constant conditions. It is about maintaining conditions within a range that is required for survival.

KEY CONCEPTS

Homeostasis is a state of relative constancy vital to the survival of organisms.

Homeostasis in Natural Systems

Ecosystems also possess a number of homeostatic mechanisms that help to maintain relatively constant conditions. Predation is one such mechanism. Predators are animals that consume other organisms, their prey. Wolves, for example, are predators of deer. Deer are predators of grass. Predators help to hold prey populations in check. Diseases also help to ensure homeostasis in ecosystems. Predators and diseases are therefore beneficial to prey populations, for they help to maintain their number within the carrying capacity of the environment.

Ecosystem homeostasis is readily observable to those who study tropical rain forests and other ecosystems, such as coral reefs, where conditions remain more or less constant year round. It is less obvious in temperate climates such as the deciduous forest biome, where seasonal climate changes result in dramatic shifts in conditions. It is important to note that even though climatic conditions shift, ecosystems in the temperate climate remain fairly constant from one year to the next. Thus, if you studied a forest ecosystem near your home each spring for an extended period, you would find that (1) the same species were present each year and (2) the population size of each species was approximately the same from one year to the next.

This is not to say that conditions are static (unchanging). In fact, things do change. Just like the human body, ecosystems exist in a state of dynamic equilibrium. Over the course of a year, some organisms die and are replaced by newborns. Some organisms may migrate out; others move in. Population sizes may increase a bit if the weather is favorable, or they may decline if the weather turns bad for a long time. Despite these changes, the system remains more or less the same over the long haul.

KEY CONCEPTS

Like organisms, ecosystems possess many mechanisms that help to keep natural systems in a state of relative constancy.

Factors That Contribute to Ecosystem Homeostasis

Numerous factors affect the organisms in any ecosystem. These factors fall into two groups: those that tend to increase population size, which I call **growth factors**, and those that tend to decrease it, **reduction factors** (FIGURE 6-2). Factors that increase or decrease population size of any given organism can be biological or biotic (as in the case of predators) or nonbiological or abiotic (as in the case of rainfall or temperature). At any given moment, the population size of a given organism is determined by the interplay of numerous biotic and abiotic growth and reduction factors. Because

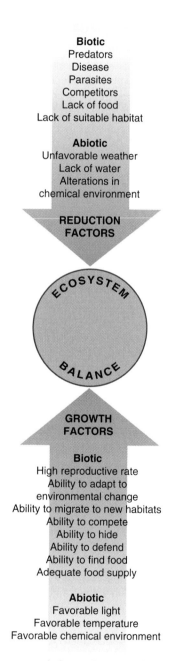

FIGURE 6-2 **Ecosystem balance.** Ecosystem homeostasis is affected by both forces that tend to increase population size and forces that tend to decrease it. Growth and reduction factors consist of both abiotic and biotic components.

ecosystems contain many species, the balance within an ecosystem can be crudely related to the sum of the individual population balances.

To understand this concept, let's examine an elk population in the foothills of the Rocky Mountains. Numerous biotic factors contribute to the growth of elk populations—for example, an abundance of food. Favorable abiotic conditions also stimulate increases in the elk population size. Ample sunshine, mild temperatures, and an abundance of spring rain, for example, promote the growth of the elk's primary food source: the rich grasses that grow in the meadows dotting the foothills. This, in turn, enhances survival and reproduction.

In nature, growth is generally opposed by a number of abiotic and biotic factors. Ecologists describe these opposing factors (reduction factors) collectively as **environmental resistance.** In the elk populations of Colorado, predators, disease, parasites, and competition tend to reduce the elk population size, as do unfavorable weather and lack of food and water. In general, when environmental resistance is low, organisms such as the elk proliferate.

In most relatively undisturbed ecosystems, the interplay of growth and reduction factors creates a crude balance or state of homeostasis. Consider another example.

In the grasslands of Kansas lives a mouse-like rodent known as the prairie vole (**FIGURE 6-3**). Its population size depends on many factors, among them light, rainfall, available food supply, predation, temperature, and disease. As an undergraduate student, I conducted laboratory experiments on prairie voles, and found that the optimum conditions for reproduction are low temperature (slightly above freezing), long days (14 hours of light a day), and plenty of water and food. Under these conditions a female gives birth to a litter of seven pups every 3 weeks for several years. For the captive-raised vole, motherhood is no picnic!

Fortunately, in the wild, optimum conditions rarely occur simultaneously. In the summer, for instance, when food, water, and day length are optimal, the temperature is much too warm for optimum reproduction. In the winter, when the temperature is optimal for breeding, the days are short and the food supply is low. Reproduction is less than optimum, and the population is held in check.

The Resilience of Ecosystems

In ecosystems, normal fluctuations in environmental conditions such as rainfall and temperature influence all of the species, from the tiniest bacterium to the largest carnivore. Thus, abiotic changes result in biotic shifts. But are these changes permanent? Consider the prairie vole again.

Among prairie voles, an unusually cool summer results in an increase in reproduction, resulting in an increase in the population size. As the population of prairie voles increases, though, populations of hawks and coyotes, which prey on voles, expand. If conditions favorable to vole reproduction are sustained, a new balance might be achieved—with substantially larger populations of voles *and* their predators.

If, as is often the case, normal weather returns in subsequent years, reproduction among the voles would decrease, causing the population to decline. The hawk and coyote populations, which have increased in previous years, would continue to feed on voles, resulting in further declines. But as the vole population declined, the predators themselves would decline, bringing the system back into equilibrium.

The ability of an ecosystem to recover from temporary changes in conditions is referred to as **resilience.** It is essential to the maintenance of life on Earth.

Resisting Changes from Human Activities

Human impact on the environment may vary from relatively minor to severe. As a rule, minor damage can be repaired by homeostatic mechanisms. Sewage accidentally dumped into a stream, for example, adds organic chemicals to the water (**FIGURE 6-4**). These substances are food for naturally occurring bacteria. In a healthy stream, the population of these bacteria is normally small. If organic food supplies increase, however, the bacterial population expands. Because the bacteria consume oxygen as they devour organic waste, the level of dissolved oxygen in the stream plummets. This, in turn, may kill aquatic organisms and fish (or causes them to migrate to better waters).

The impact of a spill can be substantial because of the chain reaction of effects. However, if the spill is a one-time event,

FIGURE 6-3 The prairie vole. The prairie vole is a small, mouse-like rodent that lives on the grasslands of much of the midwestern United States. Like all other species, its population is regulated by a myriad of biotic and abiotic factors.

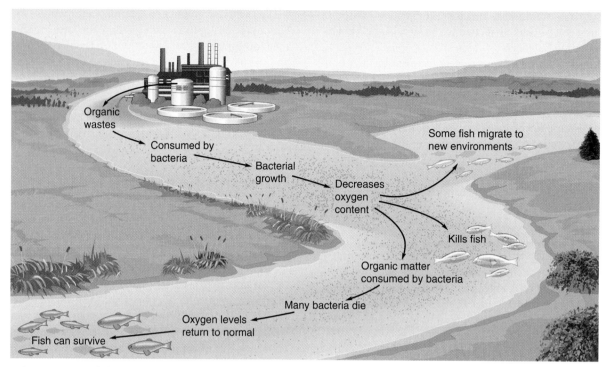

FIGURE 6-4 **Regaining balance.** The release of organic wastes from a sewage plant causes the bacterial populations to flourish. As they increase, oxygen levels decline, causing troubles for fish and other aquatic organisms. Oxygen levels and fish populations return to normal in the stream, thanks both to naturally occurring bacteria that decompose the organic matter in the sewage and to the water flow. Continued dumping or dumping by a downstream sewage treatment plant, however, could seriously hinder the stream's recovery.

the system will cleanse itself. The flow of water in the stream will purge it of pollutants. Bacteria will destroy the organic matter. These two factors will cause the organic nutrient levels to decline. As food supplies decline, the bacteria will die off. As the bacteria decline, the levels of dissolved oxygen will begin to increase (because of oxygen uptake at the surface). Populations of fish and other aquatic organisms will recover.

Problems arise when human activities push ecosystems too far. Continued pollution of the stream, for example, may strain the limits of resilience. In some cases, damage may be irreversible. Deforestation in the tropics, for instance, can also result in long-term damage. Although the system may recover if left alone for long periods, the recovery takes hundreds of years.

KEY CONCEPTS

The ability of ecosystems to recover from small changes minimizes and sometimes negates the impacts of human actions. In many instances, though, human actions can overwhelm the recuperative capacity of natural systems.

Population Control and Sustainability

As you may recall from Chapter 2, population control is essential to the sustainability of ecosystems. It is not surprising then to find numerous population control mechanisms at work in ecosystems. These mechanisms, which are part of environmental resistance, help to maintain populations within an ecosystem's carrying capacity.

In most cases, population control mechanisms are extrinsic—that is, they are factors that influence an organism from the outside. Predation, disease, and adverse weather, for example, are extrinsic factors that help to regulate population size. In many cases, these factors are a consequence of changes in population itself. Population growth, for instance, causes an increase in density, which may result in food shortages or more rapid spread of disease.

Intrinsic factors are rare. They arise from behavioral mechanisms *within* a species. Consider an example. Wolves live in packs that occupy a specific habitat—their territory. For reasons not well understood, only the dominant male and female of the pack breed and produce young. This intrinsic mechanism helps to control the size of the wolf pack within its habitat.

Humans are one of nature's many species and are also subject to natural extrinsic controls. Epidemics (outbreaks) of disease, for instance, can reduce populations. Adverse weather can take its toll as well, a phenomenon witnessed in the sweltering summer of 2003 when 20,000 people died in France as a result of intense heat. Humans also have intrinsic methods to control population such as abstinence and contraceptives, which we use to reduce our numbers voluntarily.

In nature, organisms are held within the carrying capacity of their environment. As explained in Chapter 3, even though humans are subject to the laws of nature, especially population control, we have a remarkable ability to break laws. We have, for instance, found ways to conquer disease and expand the food production capacity of the Earth. But our ability to expand the Earth's carrying capacity in some areas often has serious consequences—for example, species extinction, pollution, and climatic disturbance. Most experts agree that to create a sustainable human society, we humans must learn to live within the Earth's carrying capacity. We must respect limits and live within them. (This is one of the principles of sustainable development discussed in Chapter 2.)

First and foremost on the list of things to do is reduce population growth. Robert Gilman, a leading proponent of sustainability, argues that the challenge is to "set our own limits or have limits disastrously imposed on us." (Chapters 8 and 9 discuss population in more detail.)

KEY CONCEPTS

Natural systems are sustained in large part by intrinsic and extrinsic mechanisms that help to maintain populations within the carrying capacity of the environment. To create a sustainable society, many experts believe that humans must also control population size.

Species Diversity and Stability

Ecosystem stability may also be enhanced by species diversity. Roughly speaking, **species diversity** is a measure of the number of species living in a community. The more species, the more diversity. As a general rule, ecologists have found that the higher the species diversity, the greater an ecosystem's stability. This conclusion is based on observations that extremely complex ecosystems, such as rain forests, remain unchanged almost indefinitely if undisturbed. A stable system has fairly constant numbers of organisms and species. In stable systems, small changes such as the harvest of fruits and nuts by native peoples or the killing of an animal for food have little impact on the forest.

In contrast, ecosystems with fewer species such as the Arctic tundra tend to be less stable. They experience sudden, drastic shifts in population size. Other greatly simplified ecosystems such as fields of wheat and corn are also extremely vulnerable to change, and they deteriorate rapidly if abiotic or biotic factors shift very much.

FIGURE 6-5 illustrates the reason some ecologists think there is a connection between stability and diversity. This diagram shows two food webs, one in a simple (less diverse) ecosystem and the other in a more complex or diverse ecosystem. Because the number of species in a complex ecosystem is greater, the food webs are more complex. In a complex ecosystem, the elimination of one species would have little effect on the rest of the food web. In contrast, because the number of species in the food web of a simple ecosystem is small, the elimination of one species would very likely have major repercussions.

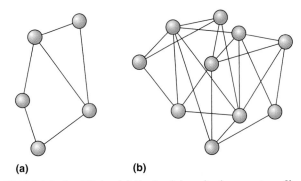

(a)   **(b)**

FIGURE 6-5 Food Webs. (a) Food web in a simple ecosystem. Circles represent organisms. Note the dearth of links in the simplified web. **(b)** Food web in a complex ecosystem. Many ecologists contend that complex ecosystems are more stable because the increased number of links reduces the importance of any one species.

This principle actually provides an opportunity to exercise our critical thinking skills. Some ecologists point out that the stability of the tropical rain forest is not the result of species diversity but of constant climate. On the relatively simple and unstable tundra, the climate shifts dramatically from season to season. These shifts may be responsible for the tundra's relative instability.

FIGURE 6-6 supports this alternative interpretation. It illustrates that species diversity among mammals varies with latitude in North and Central America. In the frozen northern regions of Canada and Alaska, species diversity is low. Heading south, diversity increases until one reaches the

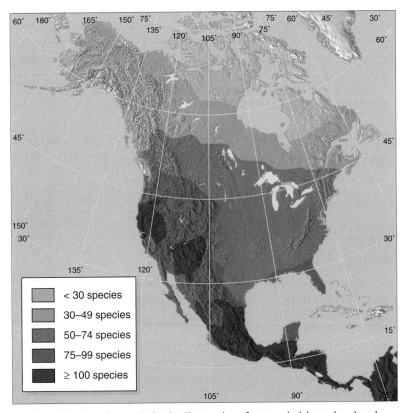

	< 30 species
	30–49 species
	50–74 species
	75–99 species
	≥ 100 species

FIGURE 6-6 Diversity vs. latitude. The number of mammals (shown here) and most other species varies considerably with latitude. The highest diversity is found in the tropics; the lowest is found in the Arctic tundra. (After Simpson, 1964)

tropics of Central America, where the highest diversity is found. The relationship between species diversity and latitude is also found in plants and virtually all other kingdoms (major groups of organisms including plants, animals, and fungi). Latitude, therefore, is an important factor affecting species diversity. The connection between latitude and species diversity is climate. Quite clearly, the milder the climate, the more species that live there. The more stable the climate, the more stable the ecosystem. Digging a little deeper helps us get a more accurate picture.

Truthfully, no one knows for sure whether species diversity fosters stability. However, we do know that simplifying ecosystems by reducing species diversity makes systems less stable and more vulnerable to outside influences. This is discussed later in the chapter.

KEY CONCEPTS

Although ecologists still debate the reasons why some ecosystems are stable and some are not, we do know that reductions in species diversity can destabilize ecosystems.

6.2 Natural Succession: Establishing Life on Lifeless Ground

As noted earlier in the chapter, small shifts in the biotic and abiotic factors in an ecosystem, whether caused by natural forces or humans, are very common and readily correctable. Drastic shifts also occur. Although such changes can cause serious ecological disruption, they may be repairable. The process of repair is called *secondary succession*; it is one type of natural succession. The other type is called *primary succession*.

Natural succession is a process in which a biotic community forms on a lifeless or relatively lifeless piece of ground. (It also occurs in aquatic systems.) To understand it, let's begin with a look at primary succession.

KEY CONCEPTS

Ecosystems can form on barren or relatively lifeless ground by a process called *natural succession*.

Primary Succession: Starting from Scratch

Primary succession occurs when ecosystems form on previously barren terrain, for example, after the retreat of glaciers. As a glacier begins to melt, it exposes a barren, lifeless terrain composed largely of rocks. Over time, biological communities form on this barren surface, transforming the landscape into a verdant garden. The process that leads to the formation of a mature ecosystem on lifeless land is called **primary succession.** Consider an example.

In North America when the great glaciers began to retreat 15,000 years ago, large areas of barren land and rock were exposed. The exposed rock first became populated with lichens. Lichens are primitive organisms, actually a type of fungi, that house photosynthetic bacteria in their bodies, which permit them to survive, grow, and reproduce on barren rock (FIGURE 6-7a). Clinging to a rock's surface, lichens live off moisture from the rain and carbon dioxide from the air, which they use to produce organic nutrients via photosynthesis. Lichens also secrete (release) carbonic acid, a weak acid that dissolves rock. Carbonic acid liberates nutrients from the rock, which lichens use. It also forms small mineral particles that will, over time, become part of the soil. Tiny insects join the lichens. Together, the insects and lichens that invade newly exposed rock form a **pioneer community,** the first community to become established on the barren land (FIGURE 6-7b). Lichens and insects of the pioneer community thrived for a long time. Over time, they changed their once-barren environment. Lichens captured wind-blown dirt particles. Dead lichens crumbled. Insects died, too. Together, the particles of dirt and the decaying bodies of lichens began to form soil. In time, enough soil developed so that mosses took root. Eventually, the mosses overran the lichens' habitat, displacing them. Together, the mosses, new insects, and bacteria combined to form a new **intermediate community.**

These communities persisted for a long period, building soil very slowly. Eventually, however, the mosses and their cohorts were replaced by herbs and small shrubs, and yet another intermediate community was formed. As the soil depth increased, larger shrubs were able to take hold. Eventually, a forest began to grow. The forest is believed to be the final stage in this process and is therefore referred to as a *climax community.* **Climax communities** are relatively stable ecosystems believed to be the end point of natural succession.

Primary succession also occurs on newly formed volcanic islands. The Hawaiian Islands, for example, arose from lava flows from deep within the Earth. After the lava cooled, rock formed. This rocky landscape was colonized by lichens, and the process of succession began. As soil began to form, plants took root. The first plants to colonize the islands most likely came from seeds deposited in the feces of seabirds that happened upon the islands. Other plants are thought to have come from neighboring continents and islands. How did they get there? Biologists hypothesize that uprooted vegetation may have drifted to the newly formed islands over many thousands of years. Some of them took root there, turning the once-barren mass of rock into a rich tropical garden.

KEY CONCEPTS

Primary succession is a process in which mature ecosystems form on barren ground where none previously existed.

Secondary Succession: Natural Ecosystem Restoration

Natural succession also occurs in the wake of ecological disturbance, ranging from mild to severe. This process is called **secondary succession.** Secondary succession is the sequential development of biotic communities after the complete or partial destruction of an existing community. In other words, secondary succession is nature's way of restoring damaged ecosystems.

Biological communities may be destroyed by natural events—for example, volcanic eruptions, floods, droughts, and fires (FIGURE 6-8). The eruption of Mt. Saint Helens in 1980 and the devastating fires in Yellowstone in 1988 are examples. Biological communities are also commonly destroyed by human activities such as agriculture, intentional flooding, fire, or mining.

Secondary succession generally takes place more rapidly than primary succession. Why? The development of soil that takes place in primary succession is unnecessary be- cause the soil was not destroyed. Thus, an important component of life on Earth is present. The soil is generally able to accept seed—or already contains enough seeds—so that the process of succession can occur much more rapidly.

Abandoned farm fields provide an excellent opportunity to observe secondary succession (FIGURE 6-9). Abandoned farmland is first invaded by hardy species such as crabgrass or broom sedge, depending on the area. These plants are well adapted to survive in bare, sun-baked soil. In the eastern United States, crabgrass, insects, and mice invade abandoned fields, forming the pioneer community. Crabgrass is soon joined by tall grasses and other plants. The newcomers, however, eventually produce so much shade that they eliminate the sun-loving crabgrass. Tall grasses and other plants dominate the ecosystem for a few years and become home to mice, woodchucks, rabbits, insects, and birds.

In time, pine seeds settle in the area, and seedlings begin to spring up in the open field. Like crabgrass, the pine trees flourish in the sunny fields. Unlike crabgrass, they have the ability to rise above their competitors, the grasses. Over the next three decades, pines begin to shade out the grasses and other plants. As a forest forms, animals such as woodchucks that feed on grasses move on to more hospitable environments. Squirrels and chipmunks, which prefer a wooded habitat, colonize the new ecosystem.

(a)

FIGURE 6-7 **Primary succession.** (a) Lichens are fungi that grow on rocks and slowly erode their surface. In this photo taken in Missouri, the dark blotches on the rocks are lichens. The green plants are mosses that have established a foothold on the rock. (b) A representation of primary succession on Isle Royale, located in Lake Superior. Rock exposed by the retreat of glaciers is colonized by lichens and then mosses, followed by other plant communities and leading to a climax community. One biotic community replaces another until a mature community is formed. During succession, the plants of each community alter their habitat so drastically that conditions become more suitable for other species.

Balsam fir, paper birch, and white spruce

Jack pine, black spruce, and aspen

Heath mat

Small herbs/shrubs

Lichens/ mosses invade

Exposed rocks

(b) Pioneer community → Intermediate communities → Climax community

FIGURE 6-8 **(a)** Mount St. Helens. This volcano in southwest Washington was surrounded by a scenic northern coniferous forest prior to its eruption in 1980. **(b)** The aftermath of the eruption. One hundred and fifty thousand acres of forest were flattened in just a few minutes. The ridge in the lower left was once densely covered; the side facing the mountain was blasted away down to bedrock. **(c)** More than nine years later, vegetation was just starting to return in areas less than six miles from the mountain. **(d)** The Mount St. Helens area today (photo taken August 19, 2005).

Shade from the pines, however, gradually creates an inhospitable environment for pine seedlings and a favorable environment for shade-tolerant hardwood trees such as maple and oak. As a result, hardwoods begin to take root. Over time, they begin to tower over the pines, and their shade gradually kills many of their predecessors. The climax stage consists primarily of hardwoods.

KEY CONCEPTS

Secondary succession is a long-term repair process that takes place after an ecosystem is destroyed by natural or human causes. It occurs more rapidly than primary succession because there's generally no need to form soil.

Changes During Succession: An Overview

Succession is a biological race to make optimal use of available resources—especially sunlight, soil nutrients, and water. As noted previously, the success of pioneer communities and many intermediate communities is transient. In both primary and secondary succession, organisms in the pioneer communities create conditions conducive to the establishment of other species. The species of the intermediate community, in turn, create conditions conducive to others. This occurs until a climax community forms.

During succession, animal populations shift with the changing plant communities. The early stages of succession

Abandoned farmland				10–30 years
0–1 years				Pine forest is established
Crabgrass colonizes first				
1–3 years				30–70 years
Tall grass/ herbaceous plants establish				Hardwoods invade
3–10 years				70+ years
Pines invade				Hardwood climax forest is formed
				Succession complete

FIGURE 6-9 Secondary succession. Here, abandoned eastern U.S. farmland is gradually replaced by crabgrass, which in turn gives way to other herbaceous plants. Pine trees move in, and over time a mature hardwood ecosystem is formed. Note that many non-plant species (not shown here) also appear in these communities. In the early stages, for example, insects, mice, woodchucks, and seed-eating birds would be found. In the later stages, as trees come to dominate, squirrels and chipmunks (which prefer a wooded habitat) invade the new ecosystem.

are characterized by a rather low species diversity (Table 6-1). Because there are fewer species early on, food webs tend to be simple, and populations tend to be volatile. In contrast, mature biotic communities have a high species diversity and relatively stable populations.

As ecosystems become increasingly complex, the food chains become woven into more complex food webs. In immature and intermediate stages, grazer food chains account for the bulk of the biomass flow. In climax ecosystems, however, most of the biomass flows through the detritus food

Table 6-1

Characteristics of Mature and Immature Ecosystems

Characteristic	Immature Ecosystem	Mature Ecosystem
Food chains	Linear, predominantly grazer	Weblike, predominantly detritus
Net productivity	High	Low
Species diversity	Low	High
Niche specialization	Broad	Narrow
Nutrient cycles	Open	Closed
Nutrient conservation	Poor	Good
Stability	Low	Higher

Source: Modified from E. Odum (1969). "The Strategy of Ecosystem Development," *Science* 164: 262–270. Copyright 1969 by the American Association for the Advancement of Science.

chain. In fact, in a mature forest less than 10% of the net primary productivity is consumed by grazers.

Interestingly, in mature ecosystems nutrients are recycled more efficiently than in immature or intermediate ecosystems, both of which tend to lose a considerable amount of their nutrients because of erosion and other factors. This fact is extremely important to those who manage forests. Leaving a mature stand of trees on a hillside on the banks of a stream, for example, will minimize soil erosion and could protect the stream from filling with sediment.

Environmental resistance also changes over time during natural succession in each community. In pioneer communities, for example, there is little environmental resistance to early colonizers such as crabgrass. Over time, environmental resistance increases. This increase is caused in large part by increasing competition from other species that become established (**FIGURE 6-10**). In an abandoned farm field, for example, crabgrass and the grasses that invade later compete for sunlight. The grasses, which can grow taller, eventually win the struggle, shading out the crabgrass.

Climax communities that emerge from secondary succession contain organisms that exist in a complex web of life controlled by a variety of growth and reduction factors. These communities tend to stay in relative balance over long periods if undisturbed. These relatively stable assemblages of organisms are sustained in large part because of their efficient use of resources, recycling of nutrients, and dependence on a reliable, renewable resource base. They also sustain themselves because they are capable of restoring damage and maintaining population balance through a variety of population control mechanisms. These principles, as noted earlier in the book, can be used as guidelines for managing natural resources and restructuring our own society for sustainability.

Those of you who will go on to study ecology will learn that the view of secondary succession just presented is the classical interpretation. New research, however, has thrown into question a few of the key tenets of this theory. Research, for example, shows that succession varies considerably in different regions and with different types of vegetation. In the Rocky Mountains in New Mexico, Colorado, Wyoming, and Montana, for example, avalanches sometimes destroy stands of conifers. The opening in the forest created by an avalanche first fills with grasses, wildflowers, and aspen trees. Over the years, dense stands of aspen grow. Next, shade-tolerant fir and spruce grow up in the forests, eventually re-establishing the climax community. All of this complies with the theory. However, in this example, species diversity in the intermediate community—the aspen forest—actually exceeds that of the mature system. Aspen forests support grasses, herbs, wildflowers, and a great many birds and mammals, such as deer and elk. As the climax community becomes established, species diversity falls because the dense forests inhibit growth of grasses and other sun-loving plants that provide food for animals. The needle-covered floors of a fir and spruce forest support relatively few species.

New research also shows that the sequential replacement of one community by another is not always as clear-cut as was once thought. In the succession of an eastern farm field to forest, for example, pine trees are often present early in the vegetational development. Because they grow slowly and are often browsed (grazed), however, their presence is not noticeable until later stages.

KEY CONCEPTS

During succession, biotic and abiotic conditions change. Pioneer and intermediate communities alter conditions so much so that they promote the growth of new communities that eventually replace them. During succession, two of the most notable changes are an increase in species diversity and ecosystem stability.

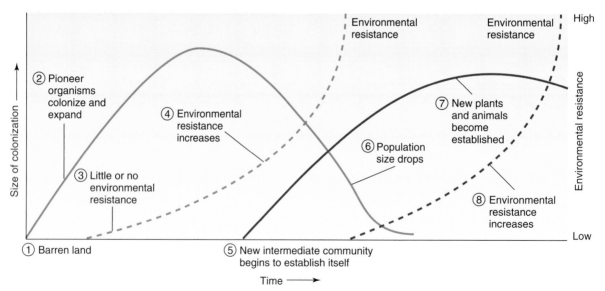

FIGURE 6-10 Changing conditions during succession. A graphic representation of the population growth and decline of a pioneer community (solid black line) and the rise of an intermediate community (red line) during succession. Environmental resistance (dashed black line) is created as the pioneer community becomes established. This, in turn, makes conditions conducive to growth of the intermediate community. Follow numbers to understand the progression of events. Environmental resistance will increase as the intermediate community becomes established, too (dashed red line).

6.3 Evolution: Responding to Change

Most people think of evolution as a process in which life formed and new life-forms emerged, which is, of course, an accurate view. They overlook an important point, though: Evolution is a life-sustaining process. Put another way, life on Earth has sustained itself, in part, because populations of plants, animals, and other organisms have changed or evolved as conditions changed, often dramatically. Evolutionary change provided organisms with new means of adapting to their environment—new ways of eking out an existence or adjusting to changes in biotic and abiotic conditions.

Evolution has two widely recognized outcomes. In some cases, it results in modifications in existing species that make them better suited to their environment—that is, better able to survive and reproduce and thus sustain their kind. Changes in the coloration of a species of butterfly, for example, might make it better able to avoid being eaten by predatory birds. Evolution also results in the formation of new species. Birds, for example, evolved from reptiles (**FIGURE 6-11**). Fish gave rise to the first amphibians.

Most of what scientists know about evolution comes from fossils—preserved bones, imprints of organisms captured in rock, or footprints of early animals. By examining fossils embedded in rock strata whose age can be determined through various means, scientists have been able to construct a biological history of the Earth (Figure 6-11).

> **KEY CONCEPTS**
>
> The ability of species to evolve in response to changes in the biotic and abiotic conditions has contributed to the sustainability of life on Earth.

Evolution by Natural Selection: An Overview

The theory of evolution is often attributed to Charles Darwin, a 19th-century British naturalist. Interestingly, Darwin did not originate the idea; it dates back to ancient Greece. In fact, the theory was already widely discussed in Darwin's time. What Darwin and another scientist Alfred Russell Wallace did (working independently) was to propose a mechanism by which evolutionary changes occurred—the **theory of evolution by natural selection.** This theory states that species evolve as a result of natural selection.

Natural selection is a kind of weeding out that takes place in populations of organisms. Within any given population, some members are better adapted to their environment than others. That is, they possess structural, functional, and

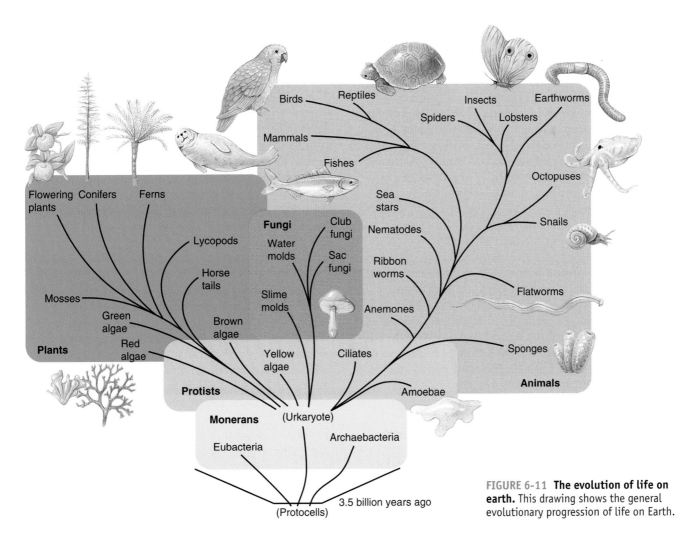

FIGURE 6-11 The evolution of life on earth. This drawing shows the general evolutionary progression of life on Earth.

behavioral characteristics that give them an advantage over other members of the population. These characteristics are called **adaptations**. Why are some members of a population better adapted than others?

The reason for the structural, functional, and behavioral differences is genetic. In most instances, adaptations increase an individual's chances of surviving and reproducing. For instance, predators that are faster than others in their population are more likely to acquire the food they need to survive. They are likely to be stronger and more successful in reproduction. Others that are less fortunate in their genetic endowment produce fewer offspring. Consequently, the predators that out hunt their cohorts will leave more surviving offspring in the population. Future generations will have a higher percentage of individuals possessing the favorable adaptation.

Like many other great ideas, natural selection took many years to be understood and appreciated. Not until the 1940s, nearly a hundred years after they developed the concept, did Darwin and Wallace's ideas on natural selection become widely accepted.

KEY CONCEPTS

Natural selection is the driving force behind evolution. It consists of natural forces that select for those members of a population that are superior in one or more features—advantages that increase their chances of surviving and reproducing.

Genetic Variation: The Raw Material of Evolution

As the previous section showed, evolution occurs as a result of two factors: the presence of genetically determined traits (adaptations) that give organisms an advantage over others and natural selection. Adaptations result from **genetic variation**, naturally occurring differences in the genetic composition of organisms in a population that may give some members of a population an advantage over others. Genetic variation comes from several different sources, three of which are described here.

In sexually reproducing species, genetic variation results from mutations in the genetic material, or DNA, of the germ cells (ova in females and sperm in males). DNA is a storehouse of information that controls the structure and function of the cells of the body. DNA consists of many segments called **genes**, each of which plays a specific role in regulating cell structure and function. Mutations in the genes may be caused by ultraviolet radiation, chemicals in the environment, and cosmic rays from the sun. Some may occur randomly. Some germ cell mutations are neutral—that is, they have no effect on an organism's offspring. Others are harmful. If not repaired, they can be passed on to an organism's offspring and may result in birth defects, cancer, or, in some instances, death. Mutations that result in beneficial adaptations in the offspring increase their survivability— for example, by improving their ability to escape predators, tolerate cold temperatures, or find food. Increased survivability, in turn, increases their chances of breeding and passing on this favorable trait.

Genetic variation may also arise during cell divisions that take place in the production of germ cells, or **gametes**. During this process, genetic material may be transferred from one chromosome (long strand of DNA with associated protein) to another. This process, called **crossing over**, results in new and sometimes favorable genetic combinations.

Another source of new genetic combinations is sexual reproduction. **Sexual reproduction** occurs when offspring are produced by a union of sperm from males and ova from females. Each offspring contains half of each parent's genetic information. Offspring, therefore, represent a new genetic combination that may provide benefits.

Genetic variation within a population results in structural, functional, and behavioral variation among the individuals of that population. As noted earlier, this results in some members being better adapted to environmental conditions than others. For these reasons, biologists refer to variation as the raw material of evolution. Organisms with favorable adaptations are said to possess a **selective advantage**. They are more likely to survive and reproduce.

KEY CONCEPTS

Genetic differences in organisms of a population, called *genetic variation,* are the raw material of evolution. Genetic variation results in naturally occurring differences in the genetic composition of organisms in a population that may give some members of a population an advantage over others. Genetic variation comes from mutations, sexual reproduction, and a process known as crossing over, which occurs during the formation of sperm and ova. Genetic variation produces variation in populations in structure, function, and behavior.

Natural Selection: Nature's Editor

Sociologist Andrew Schmookler once wrote that evolution employs no author but only an extremely patient editor. By that he meant that although evolution may appear to move in certain directions, it is not consciously directed. Variation occurs naturally. Beneficial genetically based adaptations tend to persist because they are selected for by the abiotic and biotic conditions of the environment. In other words, the environment preserves (*selects*) those organisms in a population with traits that confer some advantage over the rest. Ultimately, natural selection results in organisms better adapted to their environment.

Natural selection is really about the survival and reproduction of the fittest. **Fitness** is commonly thought of as a measure of strength. To a biologist, though, fitness is actually a measure of reproductive success—and thus the genetic influence an individual has on future generations. By definition, the fittest individuals leave the largest number of descendants in subsequent generations. Because they make up a higher proportion of future generations, their influence on the genetic makeup of those generations is greater than that of less fit individuals. The fittest individuals are those best adapted to environmental conditions.

It is important to remember that fitness is not necessarily dependent on superior strength. Greater fitness may be conferred by species' ability to hide or digest food better than others. One important measure of fitness is the efficiency with which organisms use resources. Adaptations that make an organism's offspring more efficient in their use of water in a desert environment, for instance, confer a tremendous advantage on its offspring. Numerous adaptations also contribute to energy conservation in many species, especially those that live in harsh environments such as the Arctic tundra. In a world of growing human population and dwindling resources, human survival will also very likely require the evolution of similar efficiency measures. But evolution in human society will be a conscious or deliberate process, a kind of cultural evolution, whereas evolution in the nonhuman world is not a thought-driven or directed process.

> **KEY CONCEPTS**
> Natural selection weeds out the less fit organisms of a population, leaving behind the fittest.

Speciation: How New Species Form

The foregoing discussion shows why the evolutionary process helps to change—and ultimately sustain—species in an ever-changing environment. This section discusses another important process, **speciation**: how new species arise.

Speciation occurs most commonly when members of a species are separated from one another. In scientific language, this is referred to as **geographic isolation**. Geographic isolation may occur when a mountain range or a river forms in an organism's habitat. If isolated by impenetrable physical barriers and exposed to different environmental conditions, two new populations may evolve. If the populations are separated long enough, their members may lose their ability to interbreed, a phenomenon called **reproductive isolation.**

When geographic isolation leads to reproductive isolation, new species have formed. Scientists call the formation of new species in different regions **allopatric speciation** (*allopatric* is derived from the Greek for "other" and "fatherland"). New species may also form without geographical isolation, though. This process is common in plants and is termed **sympatric speciation** ("same fatherland"), a subject beyond the scope of this book.

> **KEY CONCEPTS**
> Evolution may result in the formation of new species. When members of a population are separated by a physical barrier, they may evolve along different lines, forming separate species.

Coevolution and Ecosystem Balance

The discussion of natural selection to this point has focused primarily on abiotic factors as the chief agents of evolutionary change. Numerous studies, however, show that organisms themselves can also be important agents of natural selection. Predators, for example, can put pressure on their prey, which results in changes in the prey. Consider a hypothetical example.

Barn owls hunt for mice at night. If a population of barn owls evolves better mechanisms to detect prey—for example, improved night vision—it would put additional pressure on the mice. This change would first be felt by the slower mice, who would fall victim to the owls. The fastest mice would survive, and they and their offspring would form a larger and larger portion of the mouse population. The second result of the owl's improved performance would therefore be a new, genetically superior population of mice that was less apt to be snatched by the owls with improved night vision. Faced with a faster population of mice, the owls might evolve even greater hunting skills. So not only can one organism affect the evolutionary course of another, but the relationship can be reciprocal.

Biologists refer to this process as **coevolution**. During coevolution, each species serves as an agent of natural selection that affects the evolution of the other. Coevolution can be likened to an arms race between predator and prey because improvements in a predator's ability are matched by an improvement in its prey—much as countries try to outmatch each other during arms races.

Coevolution may also occur between plants and the animals that feed on them. This process prevents one species from destroying another. It is, therefore, an important element of sustainability in natural systems. Without it, species might destroy their own food supply.

Understanding evolution is important to understanding environmental issues, especially species extinction. Reductions in populations threaten the survivability of a species in part because they can reduce genetic variation. They may, for example, eliminate organisms that could survive other changes wrought by human society. (You'll see why in Chapter 11.) Perhaps the most important lesson in the study of evolution is that to sustain itself, a species must change in response to changing realities. Organisms generally have no choice; either they evolve, or they die out. For humans, all of the signs suggest that change is needed. But our change must be cultural in nature and purposefully directed. This book is largely about the kinds of cultural changes we need to make to create an enduring human presence on planet Earth.

> **KEY CONCEPTS**
> Organisms can evolve in concert with one another. Changes in one organism result in changes in the other. This process is called *coevolution*.

6.4 Human Impact on Ecosystems

The next chapter takes you on a journey through time to learn how human society has changed and how these changes have affected our relationship with the environment. This section helps to set the stage for that discussion by introducing you to the fundamental ways humans affect the environment. FIGURE 6-12 shows that humans alter the environment via two basic routes: changes in abiotic and biotic factors of the natural world.

> **KEY CONCEPTS**
> Human activities alter the environment by changing its biotic and abiotic components--directly or indirectly.

FIGURE 6-12 **Concept map.** Impact Analysis model, showing the range of impacts caused by human activities. Note that human activities impact two broad areas, abiotic and biotic conditions. Also note that alterations to the abiotic conditions (air, water, land) can also impact biotic conditions.

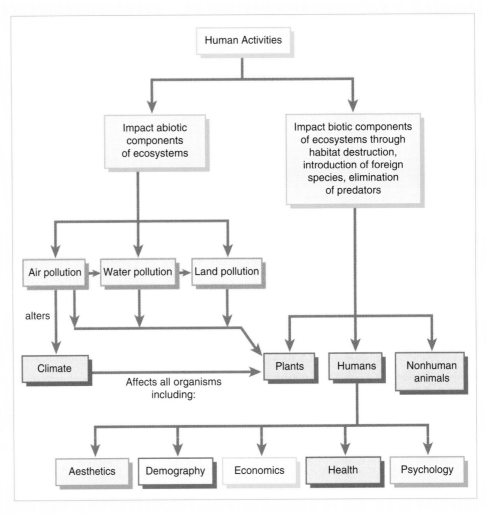

Altering Abiotic Factors

Let's begin with the abiotic factors. As illustrated in Figure 6-12, some human activities produce pollutants that contaminate the air, water, and land, changing the chemical composition of the environment. Some air pollutants alter the climate. Changes to the chemical environment and the climate have profound impacts on other species as indicated by the lines connecting air pollution, water pollution, land pollution, and climate to plants, humans, and nonhuman animals. These may stress organisms, making survival problematic. They may also impair reproduction, which lessens a population's chances of survival, or they may kill organisms outright. Spotlight on Sustainable Development 6-1 shows some positive influences humans can have on natural systems.

Some chemical pollutants may be identical to naturally occurring substances found in the environment. If their concentrations exceed an organism's range of tolerance it may suffer (Chapter 4). Nitrates and phosphates released from sewage treatment plants are good examples. Other chemical pollutants from human sources are entirely foreign to natural systems. These are synthetics such as pesticides. Because they are not naturally occurring, organisms generally have no natural defenses against them. That is, they have no way of detoxifying them. These substances can cause tumors, reduce reproduction, or make other changes. One substance found in detergents and pesticides, called nonylphenols, when released into the waterways in the Pacific Northwest may be partly responsible for the decline in salmon populations. This substance, research suggests, alters the hormonal system of young salmon, making it impossible to adapt to salt water as they migrate from rivers to the ocean.

Pollution from human activities can also affect the physical conditions of ecosystems. Chlorofluorocarbons, for instance, deplete the ozone layer and increase the amount of ultraviolet radiation striking the Earth. Power plants release hot water that changes a stream's temperature, with adverse effects on aquatic species. Carbon dioxide may alter global climate, resulting in an increase in the Earth's temperature with potentially serious effects on human health, as outlined in Chapter 20.

KEY CONCEPTS

Human activities often alter the chemical and physical nature of the environment; that is, the abiotic conditions, with profound effects both on us and on the species that share this planet with us.

6-1   Restoration Ecology: Sustaining the Biosphere

One of the most often overlooked solutions to living sustainably is restoring the damage humans have created over the past several hundred years. There's much talk about solar energy, energy efficiency, ways of controlling growth, and preventing pollution, but very little attention has been given to the need to restore the lands and waters that have been so badly damaged over the past 200 to 300 years.

Restoring natural systems, farmland, waterways, and lakes is important, not only to protect wild species but also to ensure an adequate food supply for the world's people. Such efforts are good for the economy and help create a better future. These systems also provide many ecological benefits, such as oxygen production and flood control with many economic and noneconomic benefits.

Fortunately, the scientific community has been at work on this subject for a long time. In 1985, scientist Ed Garbisch began the long and costly process of restoring an ancient freshwater wetland in New Jersey. In a $4 million project, Garbisch and his workers seeded marsh grass, cut channels to restore water flow, and built knolls for nesting sites for waterfowl. Despite initial skepticism, Garbisch and his coworkers restored the swamp to its previous condition.

Garbisch and others like him have given rise to a new field of ecology, called **restoration ecology**, the scientific study of how ecosystem recovery occurs and how it can be facilitated. Restoration ecology is one branch of science known as **conservation biology**, which seeks to understand natural ecosystems and ways to protect and restore them.

In many respects, restoration ecology is akin to rehabilitative medicine. Some of the earliest restoration projects were designed to reclaim land that had been surface mined for coal and other minerals. Today, however, ecologists and others are working to restore marshes, tropical rain forests, streams, and prairies throughout the world to help create a sustainable future (FIGURE 1).

Ecologist John Berger has spearheaded the movement in the United States to restore battered ecosystems. Berger founded a nonprofit group called Restoring the Earth, dedicated to repairing damaged ecosystems.

One of the most ambitious projects currently under way is that of Professor Daniel Janzen, a University of Pennsylvania biologist. With others, Janzen hopes to reforest 3900 hectares (9600 acres) of dry tropical forest in northwest Costa Rica. Janzen's project will take a hundred years or so to complete, one fifth of the time it would take for nature to reestablish a forest that has been cut and burned over the years.

Ecological restoration offers economic, environmental, and aesthetic benefits. For example, protecting an eroding shoreline with concrete or rocks can cost $500 per meter. In contrast, planting a 7-meter (21-foot) wide strip of salt marsh to protect the shoreline may cost only $50–$80 per meter. This can result in a savings of $100,000 or more to owners of shoreline property. Plants can also turn a barren beachfront into a dense, lush marsh that attracts a variety of colorful birds.

Creating a marsh costs $5000 to $25,000 per hectare ($2000 to $10,000 per acre). Although a marsh's value cannot easily be calculated, biologists in Louisiana estimate that a hectare of salt marsh is worth about $210,000 solely for its ability to reduce water pollutants such as sediment. Moreover, marshes are one of the world's most productive ecosystems. Today, approximately one half of the U.S. salt marshes have been lost. When a salt marsh disappears, shorelines erode rapidly, fish populations collapse, birds vanish, wildlife retreats, and some of nature's remarkable plant communities are destroyed.

Some environmentalists worry that restoration legitimizes further environmental destruction. The Sierra Club, for instance, points out that developers who want to build on existing wetlands frequently offer to replace them with wetlands "created" elsewhere.

A closer look at the issue—so essential to critical thinking—suggests several reasons that support the conservation group's position. First, natural wetlands are complex ecosystems. A flooded field planted with some swamp vegetation is a far cry from the natural system it replaced. Although it may come to resemble a true swamp in a decade, it will very likely have much lower species diversity. Second, rebuilding wetlands or other ecosystems is a complex and costly task that requires expert attention and follow-up. All in all, many think that it is best to protect what we have and not rely on strategies designed to compensate for losses by creating new wetlands.

FIGURE 1 Restored wetland in the southeastern United States.

Altering Biotic Factors

As illustrated in Figure 6-12, humans also alter biotic components of the natural world. As noted in this figure, we destroy habitat or severely alter it when we build homes and highways. We also introduce foreign species such as the water hyacinth to areas where there are no natural controls to hold them in check. These new species may outcompete native species, eliminating them entirely. In some instances, we eliminate predators such as mountain lions and wolves, which may cause prey populations to skyrocket, or we introduce predators for which prey species have no natural defense. We also accidentally introduce disease organisms. These changes affect plants and animals and humans as well, as indicated in the diagram.

A few examples illustrate the severity of these seemingly innocuous alterations that have environmental and economic impacts. First, consider the Africanized honeybee. Africanized honeybees, commonly known as *killer bees*, were introduced to South America in 1956 by a well-meaning geneticist, Warwick Kerr. He imported the bees to Brazil to develop a successful stock of honey producers to replace the docile European honeybees, which had fared poorly in the tropical climate. Kerr hoped that interbreeding the two might yield a more successful tropical strain.

Knowing their aggressiveness, Kerr isolated the bees in screened-in hives. However, in 1957 a visitor unwittingly lifted the screen, allowing 26 queens and their entourages to escape. Trouble soon began. The killer bee population quickly spread, moving a remarkable 350 to 500 kilometers (200 to 300 miles) a year. As they spread, they began interbreeding with local honeybees, eventually destroying the honey industry in Brazil and other countries. The bees also viciously attacked people, horses, and livestock that crossed their paths, stinging them severely and sometimes killing their victims.

The venom of killer bees is not any more toxic than that of native honey bees. It is just that the bees attack in great numbers and pursue enemies for greater distances than native honey bees. In addition, after a hive has been disturbed, the bees remain agitated for 24 hours, attacking people and animals within a quarter of a mile from their hive.

Killer bees arrived in Texas in 1990. By 1999, killer bees had been reported in more than 100 counties in Texas, 6 in New Mexico, 14 in Arizona, 1 in Nevada, and 3 in California. The bees have colonized a region of Southern California spanning 88,000 square kilometers (34,000 square miles), including Orange, Imperial, Riverside, and San Bernardino counties and parts of San Diego County. Officials destroyed hundreds of commercial beehives in an attempt to wipe out the killer bees that may have mixed with the colonies. At least five people were killed by the bees in the United States from 1990 to 1999. Scientists fear that if the bee spreads into northern climates and breeds with its tamer cousin, thousands of hives would have to be destroyed. This would cripple the $140-million-a-year honey industry, but it would be even more devastating to farmers, for each year America's honeybees pollinate 90 major crops worth an estimated $19 billion a year.

In 2004, killer bees had continued to expand northward, invading an additional 26 counties in Texas alone. They can now also be found throughout Arizona. The bees have also continued to spread northward in California but at a slow rate. Attacks are fairly common news in some parts of the country.

Many biologists hope that the northward spread of the killer bee would be halted by the colder climate. A study published in 1988, however, showed that killer bees could survive at 0°C (32°F) for 6 months. Some scientists now fear that the bees could migrate as far north as Canada, causing widespread damage to the honey industry and to bee-pollinated crops.

> American honeybees are an economic asset of enormous value. They pollinate 90 major crops such as apples that are worth an estimated $19 billion a year.

Another example is the zebra mussel, accidentally introduced into North America by tankers arriving from Europe (FIGURE 6-13). Already widespread in Russia, the zebra mussel has invaded many streams from New York to Minnesota, where it sets up home in water pipes coming from power plants and factories. There the mussel proliferates and can greatly reduce the flow of water. To keep their facilities running, plant managers must treat their pipes more frequently with chemicals that kill the mussels, adding to their product cost and polluting waterways with potentially toxic chemicals. The mussels also feed on microscopic phytoplankton, which form the base of the food chain that supports fish populations in lakes.

Russian scientists have tried many different tactics to eliminate the mussel, but none has proved very successful. A group of scientists from Virginia recently proposed introducing blue crabs (which feed on zebra mussels) from Chesapeake Bay, in hopes that they would help control zebra mussels without becoming a pest themselves.

The sagas of the Africanized honeybee and the zebra mussel are two of a number of severe biological disruptions created by the introduction of a foreign species into a new region. Foreign species often proliferate in their new homes because there are few if any competitors or predators to con-

FIGURE 6-13 **Zebra mussel.** The zebra mussel is an alien species that has wreaked havoc on the waterways of the northeastern United States.

trol them. Fortunately, not all introductions have adverse effects. The ring-necked pheasant and chukar partridge, both aliens in the United States, have done well in some areas. In other cases, alien species have perished without a trace. Hardy species such as the killer bee, however, are the ones that demand our attention and remind us to be vigilant.

Another example of human alteration of biotic components of the environment is the elimination of predators. Predators have never fared well in human societies. Early hunters and gatherers killed them for food and because they viewed them as competition for prey. Modern societies have carried on the tradition, killing bears, eagles, hawks, wolves, coyotes, and mountain lions with a vengeance, often with serious ecological consequences.

Problems also arise when predators are introduced into new habitat. The mosquito fish, a native of the southeastern United States, for example, has been introduced into many subtropical regions throughout the world because it eats the larvae of mosquitoes and thus helps to control malaria, a mosquito-borne disease. Unfortunately, the mosquito fish also feeds heavily on zooplankton. When zooplankton populations decline, algae proliferates and forms thick mats that reduce light penetration and plant growth in waterways.

The recent outbreaks of the deadly ebola virus in Africa, Europe, and the United States illustrate the dangers of spreading disease organisms, or **pathogens**. Pathogens are harmful microorganisms, mostly viruses and bacteria, that are a natural part of ecosystems. As a rule, they are held in check by natural forces. When introduced into new environments where there are no natural controls, though, they can spread quickly, killing people and other species.

One of the classic tragedies of pathogen introduction began in the late 1800s, when a fungus that infects Chinese chestnuts was introduced accidentally into the United States. The fungus arrived in a shipment of Chinese chestnut trees purchased by the New York Zoological Park. Chinese chestnuts are immune to the fungus, but the American chestnuts is not. Once a valuable commercial tree found in much of the eastern United States, this species was virtually eliminated between 1910 and 1940 (FIGURE 6-14).

KEY CONCEPTS

Many human activities have a direct effect on the biotic components of ecosystems. Introduction of foreign species is particularly troublesome because these species may proliferate without control, causing major economic and environmental damage.

Simplifying Ecosystems

When changes occur in the biotic and abiotic conditions of an ecosystem, species diversity is often a victim. Ecologists refer to the loss of species diversity as **ecosystem simplification**. One example of this phenomenon is the conversion of forests and grasslands to farmland. Grassland contains many species of plants and animals. When plowed under and planted in one crop, called a **monoculture**, the field becomes highly simplified. It contains few species—the crop and a few varieties of weed. Monocultures are vital to food production but are highly vulnerable to insects, diseases, drought, and wind.

The reasons for this susceptibility are many. Consider the vulnerability of monocultures to insects and plant pathogens. Monocultures generally stretch for miles, providing a virtually unlimited food source for insects and plant pathogens, especially viruses and fungi. As crops grow, food supplies for the pests increase dramatically, favoring rapid growth and spread of their populations. Viruses, fungi, and insects can become major pests. A more diverse field, one containing a number of different crops, reduces the food supply for insects and tends to hold populations in check (Chapters 10 and 22).

Monocultures also tend to be genetically homogeneous. They contain one crop species with the same genetic makeup. If it is not resistant to pests or disease, the crop is highly vulnerable. Given an unlimited food supply with little resistance, insects and diseases can spread rapidly.

Protecting monocultures from harm leads to many environmentally harmful practices—for example, the application of chemical pesticides to control insects, a practice that may have severe environmental impacts (Chapter 22).

KEY CONCEPTS

Tampering with abiotic and biotic factors tends to reduce species diversity and thus simplify ecosystems, which makes them considerably more vulnerable to natural forces.

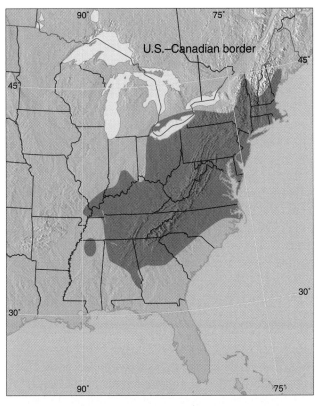

FIGURE 6-14 **Decline of the American chestnut.** Former range of the American chestnut, a species nearly wiped out by the accidental introduction of a harmful fungus.

Why Study Impacts?

Figure 6-12, described in the preceding paragraphs, helps us understand and perhaps even predict the impacts of human activities on entire systems. It also alerts us to the potential dangers of different options. In our frontier society, impact analysis is generally used to identify existing or potential impacts so that we can find ways to **mitigate** (offset) damage. For example, hydropower projects flood recreational streams. The loss of recreation opportunities such as kayaking, rafting, and fishing resulting from the construction of a dam might be compensated for by opening a previously inaccessible stream to recreational users.

In the United States, all major construction projects on federal land (for example, U.S. Forest Service land) or supported by federal dollars require analyses of potential impacts in the form of an **Environmental Impact Statement (EIS)**. The EIS investigates the environmental consequences of proposed actions, as well as the potential social, cultural, and archeological impacts. Many states require similar analyses for projects funded by state money or projects on state-owned land.

As it stands, the EIS requires a review of other options. In most cases the options are ignored, however, and efforts are concentrated on ways to mitigate damage created by the project. After completion, the EIS is subject to public review. The federal agencies responsible for the project then either approve or deny it. Permits that are granted for the work often include provisions to mitigate damage.

Although this system may sound good, it is not a sustainable approach because, as mentioned, most EISs focus on ways to mitigate, not prevent, the impact of projects. To promote sustainable development, some critics argue that the EIS should require proponents to thoroughly analyze several options and select the option that provides the services at the very least economic *and* environmental cost. Instead of a new dam, for instance, a massive water-conservation program might be the recommended choice. Water conservation costs less and can yield more water than a new dam. The environmental impact of water conservation would be minimal compared to the huge impact of building a massive dam to flood a stream or river. This alternative system is called **least impact analysis** or **sustainability analysis**. It is preventive in nature and may help us shift to much more sustainable development alternatives that are good for people, the economy, *and* the environment.

KEY CONCEPTS

Being able to predict impacts permits us to select the least harmful and most sustainable development options.

For the first time in the whole history of evolution, responsibility for the continued unfolding of evolution has been placed upon the evolutionary material itself. . . . Whether we like it or not, we are now the custodians of the evolutionary process on earth. Within our own hands—or rather, within our own minds—lies the evolutionary future of the planet.

—Peter Russell

CRITICAL THINKING

Exercise Analysis

The first step in evaluating this proposal would require you to gather additional information. You might find it helpful to examine potential ecosystem effects of introducing a foreign species. Has this species been introduced into similar habitats before? What were the effects? Is the deer adapted to the region into which it will be introduced? Is its native habitat similar to its new habitat? Would it have enough food?

Second, assuming conditions are appropriate, you may want to estimate how the transplanted deer population would grow. Would numbers increase so much that the deer would destroy its new home? Could hunters keep the population under control to prevent it from destroying the range? Would predators need to be introduced to control numbers? What would their effect be?

Third, you would have to study native grazers to determine if the introduction of deer would affect them. In addition, it would be wise to determine if the introduced deer carries diseases that could affect native or domestic animals.

Fourth, you may want to determine if the deer would remain in the region or if they would migrate into other areas, affecting other large grazers such as white-tailed deer and elk. Can you think of other concerns?

As this example shows, there are many factors to take into consideration besides the possible increase in income from hunting licenses and other incidental expenses. Critical thinking requires us to gather as much information as possible and to take a careful look at all the possibilities—that is, to examine the big picture. In Montana and Wyoming, for example, red deer from central Europe were introduced by game ranchers. Some have escaped and formed wild populations that are now interbreeding with native elk populations. Although the idea of introducing the deer seemed good at the outset, many proponents are having second thoughts.

CRITICAL THINKING AND CONCEPTS REVIEW

1. Define the term *ecosystem homeostasis*.
2. If you were to examine a mature, undisturbed ecosystem over the course of 30 years at the same time each year, would you expect the number of species and the population size of each species to be the same from year to year? Why or why not?
3. What is environmental resistance? What role does it play in maintaining ecosystem balance?
4. Give evidence that species diversity affects ecosystem stability. Is there any evidence to contradict this idea? What is it?
5. What is a mature ecosystem? What are its major features? How does it differ from an immature ecosystem?
6. Do you think that human communities resemble mature or immature ecosystems? If immature, are there any dangers one should be aware of?
7. Describe temporary imbalances caused in ecosystems you are familiar with. Indicate how the ecosystems respond to them.
8. The process of secondary succession is nature's way of restoring itself. In some cases, restoration may be protracted or impossible. Why?
9. What is the difference between primary and secondary succession? Why does secondary succession generally occur more rapidly than primary succession?
10. Explain why organisms in the pioneer and intermediate communities are replaced by others.
11. Describe how introducing alien species into an ecosystem can affect ecosystem stability. Give some examples.
12. Give some examples of ways humans tamper with abiotic components of ecosystems.
13. Discuss why simplified ecosystems (monocultures) are highly susceptible to pests.
14. Define the following terms: *evolution, natural selection, genetic variation*.
15. Describe how geographical isolation results in speciation.
16. In what ways has the study of ecology broadened your view of life? Has it made you reconsider any of your views?
17. Make a list of the major principles you have learned in this chapter. Then note how each one could be useful in managing natural resources. Describe what applications they might have for designing sustainable human systems.

KEY TERMS

adaptations
allopatric speciation
climax communities
coevolution
conservation biology
crossing over
dynamic equilibrium
ecosystem simplification
Environmental Impact Statement
environmental resistance
fitness
gametes
genes

genetic variation
geographic isolation
growth factors
homeostasis
intermediate community
least impact analysis
mitigate
monoculture
natural selection
natural succession
pathogens
pioneer community
primary succession

reduction factors
reproductive isolation
resilience
restoration ecology
secondary succession
selective advantage
sexual reproduction
speciation
species diversity
sustainability analysis
sympatric speciation
theory of evolution

REFERENCES AND FURTHER READING

The References and Further Reading section at the end of this book contains a list of sources for the information discussed in this chapter and recommendations for further reading.

Connect to this book's website:
http://environment.jbpub.com/
The site features eLearning, an online review area that provides quizzes, chapter outlines, and other tools to help you study for your class. You can also follow useful links for in-depth information, research the differing views in the Point/Counterpoints, or keep up on the latest environmental news.

Earth supplies a wide assortment of goods, services, and human enjoyment.

CHAPTER 7

Human Ecology: Our Changing Relationship with the Environment

For 200 years we've been conquering nature. Now, we're beating it to death

—Tim McMillan

In 1948, the noted British astronomer Sir Frederick Hoyle predicted that "once a photograph of the Earth, taken from the outside, is available . . . a new idea as powerful as any in history will let loose." It was not too many years later that the first photograph of Earth from outer space came to us, and Hoyle's prediction bore fruit. Standing in sharp contrast to the darkness of space, the brilliant sphere with its gossamer veil of clouds was breathtaking and yet disturbing—for it showed our home, which we had always seen as inexhaustible, as a tiny body alone in a vast universe. Sparkling in the sun's rays, the Earth seemed exquisite, fragile, and vulnerable.

We know now the Earth is indeed vulnerable. Its climate is in danger of being severely altered. Its life-support systems—which provide food, clean water, and oxygen—are being destroyed at a record pace, dooming tens of thousands of species to extinction. How did we end up in such a precarious situation? To answer this question we journey back in time to trace human cultural evolution. This brief survey of several million years of our life on Earth provides important insights into the modern environmental crisis. It will conclude with two conceptual models designed to improve your understanding of human impacts and your systems thinking abilities, skills vital to building a sustainable future.[1] As you read this material, take time to think about the changes in human systems—how they have moved away from sustainable designs—and how they can be revamped to create a sustainable future.

7.1 Human Biological Evolution: Some Highlights

The humorist Will Cuppy once remarked that "all modern men are descended from a wormlike creature, but it shows more in some people." The truth be known, fossil evidence supports the view that humans evolved not from worms, but from ape-like creatures known as **dryopithecines** (dry-oh-PITH-eh-seens), shown in FIGURE 7-1. Dryopithecines probably gave rise to the modern apes (the chimpanzees, gibbons, and gorillas) and the first humanlike creatures, **australopithecines** (oss-TRAL-oh-PITH-eh-seens).

Australopithecines emerged about 3.5 million years ago. Roaming the grasslands of southern Africa, they met their needs for food by gathering fruits, nuts, roots, and other plant materials. These first human-like organisms also hunted and killed other animals.

As Figure 7-1 shows, our early ancestors evolved through several stages. *Homo sapiens*, the species to which we belong, emerged about 400,000 years ago.

Homo sapiens, our species, emerged an estimated 400,000 years ago.

The evolutionary history of humans is marked by a number of significant developments vital to our understanding of the modern human predicament. One very important development was the evolution of **bipedal** (two-legged) locomotion, which requires less energy than quadrupedal (four-legged) locomotion and may have given our early ancestors a competitive advantage over other animals. Many scientists believe that because the front appendages were no longer needed for walking, the hands improved over time through natural selection. This improvement in manual dexterity clearly made it easier for our predecessors to make and use tools and weapons and to manipulate objects.

Bipedalism and changes in the physical structure of the hand were not solely responsible for improvements in manual dexterity. During the long evolutionary history of humans, brain size also increased. This in turn improved hand–eye coordination and enhanced our ability to make tools and shape our environment. These developments obviously set the stage for human technological development, which has contributed so mightily to human progress. As noted in Chapter 3, the development of technology has been a boon to humans that has helped to overcome environmental resistance. Irrigation practiced by early residents of the Tigris–Euphrates valleys (the Fertile Crescent), for example, reduced the threat of food shortages. Improvements in sanitation and medical science have controlled many infectious diseases, such as the plague, that once killed millions of people. Expanded food production and other benefits of technology fueled a massive increase in the human population in the past 200 years.

Humans also developed the ability to create and communicate ideas. This attribute—when combined with the ability to design technologies, plan for the future, and manipulate the environment—gave human society a power unequaled in the biological world, a power that has grown ever stronger in recent decades. Seen by many as a means of expanding the Earth's carrying capacity and enhancing our survival and well-being, these evolutionary developments

[1]Systems thinking refers to one's capacity to grasp interactions and complexity of systems—for example, human systems.

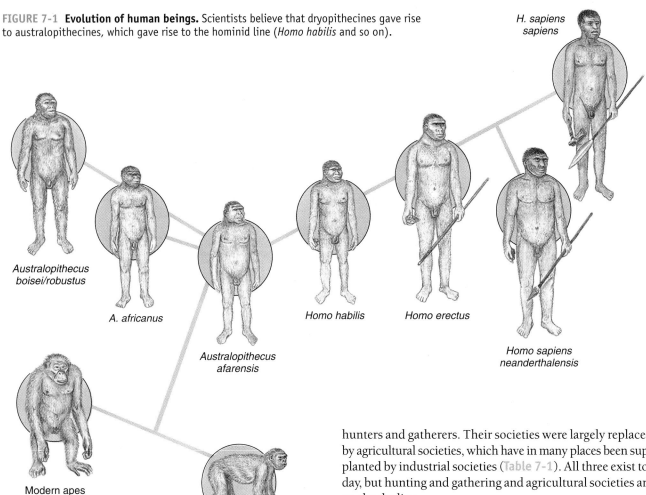

FIGURE 7-1 **Evolution of human beings.** Scientists believe that dryopithecines gave rise to australopithecines, which gave rise to the hominid line (*Homo habilis* and so on).

H. sapiens sapiens

Australopithecus boisei/robustus

A. africanus

Australopithecus afarensis

Homo habilis

Homo erectus

Homo sapiens neanderthalensis

Modern apes

Dryopithecus

have also contributed to the current problems facing humankind. The importance of these changes in our relationship to the environment was discussed in Chapter 3, on root causes. Our goal today is to learn to redirect our thinking to create a way of life and new technologies that promote a sustainable way of life.

KEY CONCEPTS

Humans evolved from primates and, over time, have evolved remarkable abilities that give us a distinct advantage over many other species. These advantages have permitted us to colonize much of the world and reshape the environment, sometimes to our own detriment.

7.2 Human Cultural Evolution: Our Changing Relationship with Nature

Humans have evolved biologically over a period of 3.5 million years. During this period, our predecessors' social systems also changed. The earliest humanlike creatures were

hunters and gatherers. Their societies were largely replaced by agricultural societies, which have in many places been supplanted by industrial societies (Table 7-1). All three exist today, but hunting and gathering and agricultural societies are on the decline.

As the following material will show, each distinct social structure used various forms of technology to manipulate the environment, but the impacts of each stage in cultural development vary considerably. The most primitive forms of technology, and the least damaging, existed in hunting and gathering societies; the most advanced and Earth-threatening are found in industrial societies. The following sections highlight important changes in the human–environment relationship over our long cultural evolution—especially changes in how we view and subsequently treat the natural world.

KEY CONCEPTS

Humans have evolved culturally through three distinct phases: hunting and gathering, agricultural, and industrial. During this time, our interaction with the environment and our impact on it have shifted dramatically away from sustainability.

Hunting and Gathering Societies

Hunting and gathering societies were the dominant form of social organization throughout most of our evolutionary history. In fact, for 99% of the time we have spent on Earth, humans have gathered fruits, seeds, and berries and have hunted animals. Growing evidence suggests that hunters and gatherers probably gained a substantial amount of their meat by scavenging—picking the bones of animals that had been

Table 7-1

Classification of Human Social Systems

Social System	Features
Hunting and gathering	1. The people were nomadic or semipermanent.
	2. They benefited from their intelligence and ability to manipulate tools and weapons.
	3. They were knowledgeable about the environment and skilled at finding food and water.
	4. On the whole, they were generally exploitive of their resources.
	5. The environmental impact was generally small because of low population density and lack of advanced technology.
	6. They lived healthy lives, were well fed, and experienced low disease rates.
	7. Their widespread use of fire may have caused significant environmental damage in some locations.
Agriculture	1. Farmers were generally either subsistence level or urban based.
	2. They benefited from new technologies to enhance crops and resource acquisition needed for their survival.
	3. They were knowledgeable about domestic crops and animals.
	4. They were highly exploitive of their resources.
	5. The impact of subsistence-level farming was significant, but because population size was small, damage was minimized. The impact of urban-based agriculture was much larger because of new technologies, trade in food products, increasing population, and lack of good land-management practices.
	6. Disease was more common among city dwellers because of increased population density.
	7. Poor agriculture, overgrazing, and excessive timber cutting caused widespread environmental damage.
Industry	1. Industry includes early and advanced forms.
	2. It relies on new technologies, energy, energy-intensive forms of transportation, tremendous input of materials, reduced number of workers, and, recently, biotechnology.
	3. Mass production and modern technology are transferred to the farm.
	4. Industry is highly exploitive, more so than earlier societies, and devoted to maximum material output and consumption.
	5. Impact is enormous and includes pollution, species extinction, waste production, and dehumanization.
	6. Humans become subject to infectious disease and new industrial-age diseases including ulcers, heart disease, and mental illness.
	7. Widespread environmental damage results from industry, agriculture, and population growth.

killed by others or that had died from natural causes. Only in the last few fleeting moments of geological time have we turned to agriculture, then to industry.

> Humans were hunters and gatherers for 99% of human history.

Studies of present-day hunters and gatherers suggest that members of these societies had a profound knowledge of the environment. They knew where to find water, edible plants, and animals; how to predict the weather; and what plants had medicinal properties. Studies of present-day hunter-gatherers also suggest that, contrary to popular conception, in many regions of the world, especially the warmer climates, our early ancestors did not live with the constant threat of starvation, nor did they spend the greater part of their lives in search of food. Studies also suggest that they were healthy, well nourished, and suffered from few diseases.

Many hunter–gatherer scavengers were nomads who foraged for plants and captured a variety of animals by using primitive traps and weapons. Hunter-gatherers fashioned tools from sticks, stones, and bones. Because their technology was primitive, they did not enjoy a great advantage over other species, and their populations never grew very large.

Judging from existing hunting and gathering societies, anthropologists believe that hunter-gatherers had a deep reverence for the environment and the plants and animals on which they depended. These people understood that they were a part of the Earth and dependent on it.

Some hunter-gatherers developed semipermanent lifestyles, setting up homes near hunting or fishing grounds that could provide a year-round supply of food. These groups were more likely to cause noticeable damage. New research on hunting and gathering societies also suggests that many groups also grew their own food and raised animals to feed themselves. Food may have been traded with other groups, a step that marked the advent of the first system of commerce. Cave dwellers in Europe 28,000 to 10,000 years ago, for example, probably participated in extensive networks to trade food and other valuable commodities.

On balance, the hunter-gatherers had little impact on the environment, and their way of life was generally sustainable, as witnessed by their long history. Their existence was sustainable because it did not violate the principles of sustainability (Chapter 2). Their numbers were held in check by natural forces. Their demands were small. Because they generally lived nomadic lifestyles, the damage they created was easily repaired. They lived off the land's renewable resource base and recycled their waste, so they fit well within the economy of nature. Moreover, they seem to have exhibited a reverence for the Earth. In short, they had the lifestyle and the ethics necessary for sustainability.

It is important to point out that although hunter-gatherers were relatively benign, they did cause considerable damage in some areas. In North America, hunter-gatherers may have been responsible for the extinction of many species of large animals after the last ice age. Species lost during this period include the cave bear, giant sloth, mammoth, giant bison, mastodon, saber-toothed tiger, and giant beaver (FIGURE 7-2). These animals may have been killed directly, driven out of their preferred habitats, or wiped out as their prey were de-

FIGURE 7-2 **Extinct mammals of North America.** These species became extinct during the Pleistocene, possibly as a result of overhunting by humans.

>> SPOTLIGHT ON SUSTAINABLE DEVELOPMENT

7-1   Agroforestry: Tapping Ancient Wisdom

The destruction of tropical forests to make way for highways, towns, and farms is an issue of worldwide concern. Not only does it wipe out thousands of wild species, it also destroys vital economic assets, increases soil erosion, and contributes to global warming.

Population growth is a key factor in the loss of rain forest, as is economic development. About 25% of the rain forest cut down each year can be attributed to the search for new agricultural land to accommodate population growth. Migration to deforested regions is often viewed as a means of relieving overcrowding in urban settings and landlessness in agricultural regions.

As Chapter 22 explains, government policies in the industrial and nonindustrial nations of the world also contribute significantly to the rapid deforestation, despite official endorsements of conservation goals. In the developing countries, for example, governments seeking to raise foreign exchange earnings or to finance economic development programs often turn to their forests to grow cash crops such as tea and coffee, which they can export to the industrialized nations.

As a result, rain forests are often used for brief periods, exhausted, and then abandoned. The inhabitants move on, starting a whole new cycle of exploitation. To curtail deforestation, many experts believe that we need to find ways to use tropical rain forests sustainably—so that they can continue to be productive for many years. This could stop the cycle of exploitation and abandonment. One candidate is agroforestry, a sustainable management system that combines agriculture with tree crops or other forest plants (FIGURE 1). These activities occur on the same par-

cel of land, either simultaneously or sequentially. Agroforestry maintains or improves the environmental quality of the land, provides income and food, and reduces the need to exploit new land. It can be practiced on forestland that is already degraded and cleared. It therefore combines key elements of sustainability. It is good for society, the environment, and the economy.

Traditional agroforestry systems have evolved over many years. In fact, forests have been prudently managed by indigenous peoples for many generations without any apparent loss of species diversity or deterioration of soil quality.

FIGURE 1 **Agroforestry.** By planting trees and crops on the same plot, farmers can create a sustainable system of food and fiber production. It provides for human needs and eliminates further cutting of tropical rain forests.

---

stroyed. In the Great Plains region of North America, some Indian tribes ignited grass fires to drive buffalo over cliffs for slaughter, killing many more than were needed. Such a wasteful action does not fit the image of a wise steward of the land. Compared with today's societies, though, hunter-gatherers were relatively benign.

KEY CONCEPTS

Throughout most of human evolutionary history, our ancestors were hunter-gatherers who lived in relative harmony with the Earth. Primitive technology, small population size, and a nomadic lifestyle were the three main features of these societies that held human impact to sustainable levels.

Agricultural Societies

Anthropologists believe that agricultural societies emerged between 10,000 and 6000 B.C., although evidence shows that some hunter-gatherers were tending to crops long before this time. The roots of agriculture are generally traced to the rain forests of Southeast Asia (**FIGURE 7-3**), where early agriculturalists practiced slash-and-burn or *swidden* agriculture. In the **slash-and-burn agriculture** of this region, farmers cleared small sections of jungle to plant crops. Because rain forest soils are generally nutrient poor, crops failed after several years, and plots were abandoned. Farmers moved on to clear neighboring forests. In the abandoned fields, native species invaded and returned the land to its original state. (For a discussion of this practice in modern times, see Spotlight on Sustainable Development 7-1.) The early agricultural societies of Southeast Asia also domesticated many animals, such as pigs and fowl. These became vital food sources, greatly supplementing crops.

Agriculture emerged 10,000 to 6000 B.C.

One form of agroforestry is swidden agriculture, briefly described in this chapter. In swidden agriculture, people clear small plots of rain forest in which they plant food and tree crops. When the soil loses its fertility, the plots are abandoned and allowed to regrow. This restores the soil in time, making it useful once again.

The Bora, an Amazonian tribe native to northeastern Peru, practice swidden agriculture. Some scientists believe that large areas of the Amazon forest may have been managed in this manner by considerably larger populations in the past.

Swidden farming is not restricted to native peoples in the Amazon. In fact, the Kenyah people, who live in a remote section of Indonesia, have practiced it for at least 200 years. The Kenyah have converted much of the virgin forest into a mosaic of secondary forest of different ages, but they have retained uncut reserves in which they harvest products that are rare or nonexistent in secondary forest.

Many of the agroforestry systems studied to date have been practiced by relatively remote tribes, suggesting that these systems are suitable only for subsistence production. Recent studies, however, prove that they can also be important cash producers near urban centers. In Tanshiqacu, Peru, for example, a nontribal group of 2000 people of mixed Amazonian and European ancestry produce food, fiber, handicraft materials, and charcoal from the forests. Their products appear in local markets as well as in markets in the Peruvian capital, Lima. Most households maintain several agroforestry fields at the same time and sell a variety of products, a tactic that ensures a year-round income.

In Java, farmers clear forests and plant teak trees, whose wood is exported to many developed nations. Among the teak trees, they plant rice and corn. This system of agroforestry has been modified to include horticultural species, animal food, and fuelwood crops, which provide additional income.

In the Philippines, the rural poor venture into steep, mountainous terrain to clear forests to make room for crops. To prevent widespread destruction of these regions, governmental agencies constructed numerous demonstration plots during the 1980s to introduce agroforestry practices to the rural people. Because deforestation of steep hillsides leads to soil erosion, the first step in these projects was to stabilize the soil by laying bamboo poles along the contour of the land. The bamboo barriers capture soil, preventing it from washing down the slope. Fast-growing trees were then planted along the contours. The trees provide wood for fuel and leaves for animal feed and soil enrichment. The seeds of these trees are used to produce traditional medicines. Fruit trees such as mango and food crops such as pineapple and cassava were planted in the fields. Today, the agroforestry plots provide a year-round food and fuel supply for the people. By stabilizing the soil, they reduce the need to cut more forest.

As these examples illustrate, agroforestry provides many benefits. It helps protect virgin forest and provides a sustainable supply of food, fuel, and fiber. It also reduces soil erosion and land destruction, and it benefits the rural poor, providing (in addition to food) income and employment opportunities. Agroforestry is also being practiced in developed nations such as the United States. Agroforestry is not a panacea; it's just one of many strategies for living sustainably on the planet.

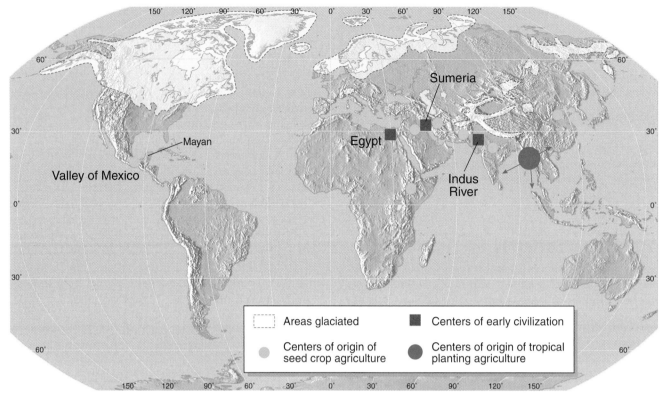

FIGURE 7-3 Roots of agriculture. This map shows the areas where tropical planting and seed crop agriculture originated. The Mayans of Mexico were probably the first people in the Western hemisphere to grow seed crops.

Seed crops originated in a wide region extending from China to eastern Africa (Figure 7-3). The first farmers cleared rich woodlands, but agricultural efforts were fairly limited until the development of the plow. Coming into use in the Middle East around 3000 B.C., the first plows were nothing more than tree limbs with branches that cut through the topsoil. In subsequent years, more elaborate plows were developed. Pulled by oxen, these plows enabled farmers to cultivate grassland soils whose thick sod was difficult to work with previous implements. This dramatically increased crop production. As in Southeast Asia, a variety of domesticated animals such as goats, sheep, and cows supplemented the human diet.

The plow gave agricultural societies the means to increase dramatically the productivity of the land. As a result, farmers achieved a greater degree of control over their destiny and could expand beyond the limits previously set by the natural food supply.

Before the development of the plow, most farmers fed themselves and their families. Farming was a subsistence activity. As food became mass-produced, two major changes occurred. First, the human population began to swell. Since more food was available, survival became less problematic. Second, because fewer people were needed to work to provide food, many migrated from farms to villages and cities, where they took up crafts and small-scale manufacturing. As cities and towns formed, they became centers of trade, commerce, government, and religion.

The growth of population, cities and towns, and small-scale industry placed greater demands on the environment for resources—wood, metals, and stone. Heightened exploitation accompanied by poor land management resulted in widespread destruction of the natural environment. Many fertile areas were destroyed by overgrazing, excessive timber cutting, and poor agricultural practices.

The shift to mass-produced food had a potentially more harmful effect: It severed, in large part, the link to nature. Unlike hunter-gatherers and early subsistence farmers—who depended on a wide variety of plants and animals for food and who had an impressive working knowledge of the environment—the new generation of farmers relied on a relatively small number of plants and animals. For many people, especially city dwellers, a profound knowledge of the environment was no longer necessary. The emergence of technological agriculture also ushered in dramatic changes in the way people treated the environment. Most important, domination of nature replaced the harmonious ways of earlier peoples.

The environmental impact of agricultural societies and the large urban centers they supported were enormous. Archeological and historical records show that overgrazing, widespread destruction of forests, and poor farming practices changed many once-productive regions into barren landscapes. Crop failures and the invasion of displaced people caused the demise of numerous ancient civilizations. These occurrences were especially evident in the Middle East, North Africa, and the Mediterranean from 5000 B.C. to A.D. 200. The Babylonian Empire, for example, once occupied most of what is now Iran and Iraq. At the outset, this land was covered with productive forests and grasslands. Huge herds of cattle, goats, and sheep overgrazed the grasslands and eventually destroyed the natural vegetation. Forests were cut to provide timber and create pasture. The loss of grassland and forests decreased rainfall and eventually parched the land. Sediment washed from the barren soils, robbing them of nutrients and filling irrigation canals. These changes and a succession of invading armies eventually destroyed this once great empire.

In summary, agricultural societies were relatively benign at first, but as technologies improved, agriculture became a major source of environmental damage. The success of agriculture resulted in an increase in population size and ushered in the age of commerce. As commerce began to flourish, material demands increased, and waste from shops and small factories began to accumulate outside cities. Local resources began to decline. Finally, as commerce and technologically based agriculture took hold, humans began to see the natural world in a much different way—more as a source of wealth and a force to dominate and control. Agriculturalists lost the physical and spiritual connection with nature that characterized their hunting and gathering predecessors. The human population began to increase, and the systems that supported humans, agriculture, and the economy became less sustainable.

KEY CONCEPTS

Agriculture started as a subsistence activity, but the plow and other forms of technology gave our ancestors the ability to produce excess food. This, in turn, displaced farm workers, who took up crafts and trades in cities and towns, and it fostered a dramatic increase in human population. The growth of human population and the emergence of cities and towns as centers of commerce had a significant negative impact on the environment.

The Industrial Society

The **industrial society** resulted from a major shift in manufacturing, from small-scale production by hand to large-scale production by machine—a change commonly referred to as the **Industrial Revolution** (FIGURE 7-4). Industrial societies emerged in England in the 1700s and in the United States in the 1800s.

The Industrial Revolution was made possible by new technologies but also by an abundant supply of fuel, notably coal. The introduction of coal-powered machines made manufacturing much more energy intensive and much less labor intensive. Industrialization also led to the mass production of products. This meant that more goods were available to more people. As the demand for goods rose and industries grew, the influx of fuel, food, minerals, and timber into the city rose sharply.

Manufacturing technologies that made the Industrial Revolution possible were the outcome of many scientific and engineering advances. Although the technologies improved the economic output of cities and towns, they were complex and often made work meaningless and boring. They also produced large quantities of smoke, ash, and other wastes. The shift to machine production, therefore, changed the workplace, the city, and the surrounding countryside that supplied the resources.

Mechanization also swept through farms during the Industrial Revolution (FIGURE 7-5). Technological advances such as Jethro Wood's cast-iron plow with interchangeable parts and Cyrus McCormick's reaper brought on a rapid increase in agricultural production. One of the most significant advances was the invention of the internal combustion engine, which

(a)

(b)

FIGURE 7-4 **Changing ways. (a)** Carpets being made by hand in Nepal. **(b)** This factory turns out thousands of carpets per year using the latest technology. The Industrial Revolution made goods more readily available, which increased the demand on resources and resulted in widespread pollution and habitat destruction.

FIGURE 7-5 Industrialization on the farm. Modern agriculture depends heavily on machinery, energy, and additional resources. Large fields are worked to achieve maximum output.

made horse-drawn implements obsolete. The motor-powered tractor could plow as much land in a week as one of our forebears could work in a lifetime using hand tools. Another significant advance was the development of fertilizers, which allowed an increase in agricultural productivity (output per hectare). Plant breeding produced higher yield crops that also contributed to a rise in productivity. Because of more efficient farming methods, fewer farm workers were needed. Unemployed workers migrated to cities, swelling their populations.

New medicines and better control of infectious disease through improved sanitation also occurred during the Industrial Revolution. These important developments enhanced human survival. Because the death rate declined and people began to live longer, the human population began another upward turn.

Like the Agricultural Revolution, the shift to an industrial society further changed our relationship with nature. Fewer and fewer people were in contact with nature as city life became more advantageous. In addition, control over nature became increasingly important. Industrial people came to view themselves as more and more apart from nature. The prevailing attitude is best summarized by the 17th-century English philosopher John Locke, who wrote, "The negation of nature is the way toward happiness." Locke argued that people must become "emancipated from the bonds of nature."

These notions have been passed from generation to generation throughout the industrial age. Locke also preached unlimited economic growth and expansion, with the belief that individual wealth was socially important for a harmonious society, an idea that persists today.

> The Industrial Revolution began in England in the 1700s and in the United States in the 1800s.

It must be pointed out that the Industrial Revolution provided many benefits. On the whole, people ate better, lived better, and began to live longer. Many new products were available, and there was more leisure time, but there was a price to be paid in environmental deterioration for the unsustainable way of life that emerged. Populations increased, and systems for providing human needs became less and less sustainable.

KEY CONCEPTS

The Industrial Revolution increased population size, resource demand, pollution, and environmental destruction and dramatically altered the human–environment interaction. Attitudes toward nature shifted even farther away from stewardship.

The Advanced Industrial Age

Industrialization, well established by World War II, expanded rapidly after the end of the war, creating an **Advanced Industrial Society.** Several features distinguish this new phase. First and foremost is a marked rise in production and consumption, which have resulted in huge increases in energy and resource demand by virtually all sectors of our society. Consider energy. In hunting and gathering societies, individuals require only about 2000 to 5000 kilocalories of energy per day. A kilocalorie (KILL-oh-CAL-er-ee) is a measure of energy. One kilocalorie is equal to 1000 calories, which is the amount of energy needed to raise one gram of water one degree Celsius. Kilocalorie and calorie, a measure used by dieters, are equivalent. The 2000 to 5000 kilocalories used by hunter-gatherers provide metabolic energy and additional heat to cook food. Early agricultural societies, however, required twice that amount per person to grow crops—about 10,000 kilocalories. Advanced agricultural societies required twice as much again, or about 20,000 kilocalories per person per day. Increased energy use resulted from improvements in agriculture and the growth of crafts, commerce, and cities. In industrial societies per capita energy use climbed again, increasing to 60,000 kilocalories per day per person. Modern industrial society, however—with its mechanized farming, energy-intensive industries, and affluent lifestyles—has doubled that amount, raising energy consumption to about 120,000 kilocalories per person per day. Some nations such as the United States and Canada consume, on average, even more: 250,000 kilocalories per person per day.

The problem with this is that fossil fuel combustion is the source of most air pollution. The production of fossil fuels can also result in significant amounts of water pollution, as witnessed by the 1989 Exxon Valdez oil spill off the coast of Alaska. Another problem with the energy we use is that it produces carbon dioxide. As we discuss in Chapter 20, carbon dioxide gas released in large quantities now threatens to change the global climate in ways that could disrupt agriculture.

A second characteristic of the advanced industrial society is a shift toward synthetics, such as plastics, and nonrenewable resources, such as oil and metal. Synthetics create problems in nature because bacteria in soil and water, which decompose naturally occurring substances, are frequently unable to break them down. Consequently, synthetics may persist in the environment for decades. The persistent insecticide DDT, for example, can accumulate in the fatty tissues of birds and disrupt reproduction (Chapter 22). Plastic pollution in water has proven to be a problem for some aquatic organisms (Chapter 21).

The advanced industrial society seems to be caught up in an ever-escalating production-consumption cycle that is, say critics, highly unsustainable. Domination of nature continues as the central theme of modern industrial societies, and economic growth retains its commanding allure—despite evidence that both threaten the long-term future of our planet and our long-term well-being.

In this intense period of economic and population growth, the unsustainability of virtually all human systems—from agriculture to energy production to water supply to waste management—is becoming clear. The challenge today, outlined in Chapter 2, is to restructure those systems to ensure a better future. Environmental protection efforts must be integrated with development efforts to create lasting change and an enduring human presence.

KEY CONCEPTS

In recent times, industrialization has grown rapidly. Resource demand and environmental destruction have reached unsustainable levels.

7.3 The Population, Resources, and Pollution Model

The last chapter pointed out that human activities affect both the abiotic and biotic conditions of the environment. As our culture has shifted away from hunting and gathering to agriculture and then to industrial societies, the impact on these conditions has grown to a level that many scientists believe is unsustainable.

This section provides an ecological model of human activity that shows *how* impacts arise. It is called the **Population, Resources, and Pollution (PRP) model** (FIGURE 7-6). The PRP model illustrates several important relationships between humans and their environment and offers a big-picture view of the human–environment interaction. Not only is this vital to becoming a critical thinker, but also, it helps promote systems thinking, which is vital to sustainable development.

The PRP model (Figure 7-6) shows that humans, like all other organisms, acquire resources from the environment. The acquisition of resources—for example, through coal mining—alters both the abiotic and biotic conditions of ecosystems. Surface coal mines, for example, disrupt wildlife habitat. They can also lead to soil erosion that pollutes nearby

FOUNDATION TOOL

FIGURE 7-6 **Population, Resources, and Pollution.** This model applies to all organisms. It outlines how organisms interact with their environment. The + signs by the arrows indicate a positive feedback loop, in which one activity enhances another. The – signs indicate a negative feedback loop, in which one activity adversely affects another.

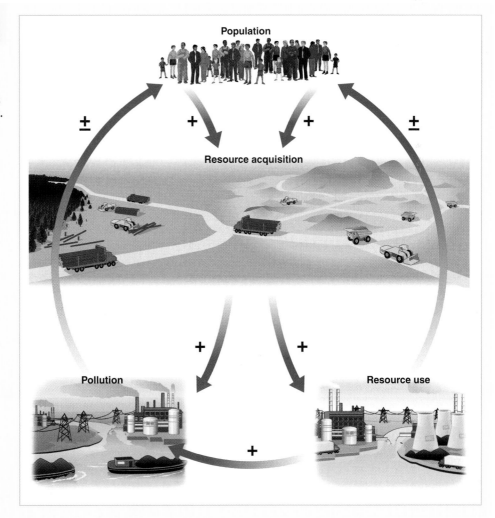

streams—hence, the arrow between resource acquisition and pollution in the model.

The resources we extract from the Earth are put to use. After being mined, for example, coal is burned in power plants. Minerals extracted from the Earth are fashioned into finished products such as automobiles. Such activities can produce substantial amounts of pollution, indicated in the model by an arrow connecting resource use to pollution.

As we ponder society's contribution to building a sustainable society, it is important to remember our personal role, too. Although we rarely think of it, everything we purchase comes from a mine, forest, field, stream, or the ocean. The acquisition of the raw materials needed to make the products we purchase creates some physical damage. The conversion of those raw materials into finished goods results in pollution of air, water, and soil.

Some pollutants are **biodegradable**—that is, broken down by bacteria and other organisms in air, water, and soil. Others are **nonbiodegradable**—not broken down by organisms. Both forms can significantly alter abiotic and biotic conditions of ecosystems.

Scientists also classify pollutants by the medium they contaminate—for example, air, water, and land. Thus we have air pollution and water pollution. In recent years, scientists have found that pollution readily crosses the boundaries between these media, a phenomenon called **cross-media contamination**. Air pollution, for example, washes from the sky and is deposited in lakes and forests. Some water pollutants evaporate from lakes and streams and enter the atmosphere.

The negative impacts of development on the human population are indicated on the PRP model by minus signs

---

>> SPOTLIGHT ON SUSTAINABLE DEVELOPMENT

7-2   Monfort Boys' School: Mining Human Wastes to Feed People

Fish are a popular source of protein in many countries. Today, in fact, the world's fishing fleets catch over 100 million metric tons of fish from the world's oceans each year. This is made possible because of advanced technologies such as sonar, which help fishing fleets find and capture fish.

Unfortunately, commercial overfishing has depleted several dozen of the world's richest fisheries—just in the Atlantic Ocean alone—areas of the ocean once served up huge fish catches. Many other fisheries are now in danger. In fact, of the world's 15 major oceanic fisheries, 11 are in decline. So heavily harvested are many of these fisheries that if we stopped fishing immediately, it would take them from 5 to 20 years to recover.

Overfishing has had a ripple effect on other species, reducing populations of seals and other fish-eating animals. Thus, not only are we threatening our own food supply, we're eliminating the supply of the many species that share this planet with us. Unfortunately, the story does not end here. Each year, millions of tons of fish netted by commercial fishing operations are thrown overboard because they are the wrong species or sex or are too small. International fishing regulations require this to protect various species and populations.

We can also turn to fish farming—growing fish and shellfish—to increase food production. Fish and shellfish can be raised in artificial environments such as ponds or in natural bodies of water (for example, in bays, where they are contained by nets) then harvested for food when they've reached market size. Worldwide, fish farms produce about 22 million metric tons of food a year.

Important as they are, fish farming is not without its problems. Fish need to be fed, for example, and fish food needs to be grown. That takes resources and a considerable amount of energy. Moreover, fish in ponds produce lots of waste that can pollute the waters into which their ponds are emptied, waters that we may eventually drink or swim in or waters upon which species depend.

There is environmentally sustainable way to raise fish for human consumption, however. Consider the work of a group called ZERI, an acronym that stands for Zero Emissions Research and Initiatives, founded by Gunter Pauli, a European businessman with a firm commitment to the environment. Pauli's organization is pioneering an ecosystems approach to raising fish—and many other foods as well—for human consumption in ways that do not harm the environment or, preferably, enhance it. His group helps people in less developing nations use waste from one activity, for example, brewing beer, to produce fish, fruit, vegetables, and animal food.

At the Monfort Boys' School, a Catholic Technical school in the island nation of Fiji, located in the Pacific north of New Zealand, officials use waste grain from a local commercial brewery to feed an entire human-made ecosystem, reaping a huge profit in the process. The waste grain begins its travel through this human-made food chain as a substrate for a commercial mushroom venture. It could be fed directly to livestock such as pigs, but the waste grain is not well digested as is. Enzymes released by the mushrooms, however, render the waste more digestible. Thus, after the school has harvested the mushrooms, the biologically processed waste grain is mixed with pig food and fed to the hogs. Waste grain is also mixed with duck and chicken feed, cutting down the need to purchase expensive feeds for them as well.

on the arrows that lead from pollution and resource use to population. This response is an example of a **negative feedback loop**, in which one factor leads to a decrease in a second factor. Negative feedback loops are the chief means of controlling biological systems and regulating homeostasis (Chapter 6). Many human systems are controlled by negative feedback. A familiar example is the furnace in your home, which is regulated by a thermostat. When the temperature drops below the setting on the thermostat, the furnace is turned on. Heat from the furnace then warms the room air until the air temperature reaches the desired level. When it does, the furnace shuts off.

The negative feedback loops shown on the PRP model are of great concern to scientists. They could in fact limit future growth catastrophically. Continued soil erosion from the world's farmland resulting from poor agricultural practices, for example, could cause a marked decline in food production that could have a devastating effect on the global human population.

As the previous discussion of cultural evolution pointed out, resource acquisition and use (especially that permitted by technology) also have a positive effect. Put another way, they enhance survival and promote human population growth. The efficient harvest of food and fiber, for example, has made it possible for large numbers of people to inhabit the Earth. This relationship is indicated in the model by the plus signs on the arrows leading from resource use to population. It is an example of a positive feedback loop, in which one factor leads to the growth of a second factor, which in turn stimulates the first one in a repetitive cycle.

Positive feedback mechanisms are rare in nature; in the human–environment system, they may cause serious prob-

Hogs are occasionally slaughtered, and the meat can be sold or eaten by the school's 100 students. However, that's not the end of the story. In fact, it is only the beginning. Waste from the pigpens is washed out twice daily and drains into a huge settling tank. Here, the waste is broken down by anaerobic bacteria, microorganisms that break down organic material in the absence of oxygen. One by-product of this process is methane, a combustible gas that can be burned to produce light and to cook food. The farm produces enough methane for a family of six.

The sludge, a liquid waste is drawn off by bucket, is used to fertilize crops—bananas and various vegetables—that are then sold for food or consumed by the students. Liquid waste from the settling tank is then fed into a series of settling ponds where naturally occurring algae consume much of the nitrogen and phosphorus in the waste. The algae-rich water is then delivered to a fish pond containing six different species of fish. The fish consume the algae, living entirely off this free food source. At the end of each year, the fish are harvested and sold for food. Interestingly, ducks and chickens, which are fed waste grain, are raised in coops suspended over the fish ponds. Their waste freely drops into the fish ponds, adding additional nutrients to the waters. In this ecological production system, all waste is food for something else.

At one time, waste from the hog farm polluted nearby mangrove swamps, killing aquatic life. Today, when the water leaves this "facility," it is so clean that the nearby mangrove swamps are showing promising signs of recovery. Species of fish and crabs wiped out by pollution are now returning.

Besides producing food from waste, this farm also generates a fair amount of income. This system, in fact, can produce an annual income of as much as $60,000 for a family of six . . . get this . . . working a half a day a week each! Compare that to the typical income of a worker in the less developed world—averaging about $2000 a year. Needless to say, this system is being seriously considered by a great many other people in other countries.

What can you and your family do to help save endangered fish populations?

One thing that will help immediately is to buy fish whose populations are not in danger. See the Monterey Aquarium's website (at www.mbayaq.org) for a list of fish (Seafood Watch) that you can eat without guilt.

Another important step you can take is to cut down on shrimp consumption or avoid it altogether. Shrimp harvesting is one of the most environmentally damaging activities in which humans engage. Trawlers drag huge nets along the bottom of the ocean, scooping up shrimp and a host of other sea creatures. As noted earlier, for every pound of shrimp that is harvested, there are 5 pounds of other sea life that are swept up. Most are dead before they are thrown back overboard. Shrimp nets also scrape up the bottom of the ocean, removing plants and leaving the ocean floor in a state of ecological disarray.

Another thing you can do is to purchase fish raised on the farm, fish farm, that is.

Yet another option is for us to eat lower on the food chain. Most Americans eat way more meat protein than they need. By reducing fish and other meats from the diet, or eliminating it, we can reduce our demand on fish farms and natural fisheries. A good vegetarian diet can provide all of the protein one needs.

Adapted with permission from Dan Chiras, *EcoKids: Raising Children Who Care for the Earth*. Gabriola Island, B.C.: New Society Publishers, 2004.

lems, for they can create devastating cycles of depletion and environmental destruction. For example, although fossil fuel energy has increased our capacity to produce food, it could be creating a dangerous rise in global temperature and a massive shift in rainfall that might devastate crop production on much of the existing farmland (Chapter 20).

The PRP model represents a fundamental ecological relationship true to all living organisms. It's as relevant to humans as it is to black bears. What is more, it provides some insight into the human–environment interaction. It allows us to predict the impacts of human actions. If you change one variable in the model, the rest change. If you add more people, for example, resource use is bound to rise. Acquire more resources, and pollution is likely to climb. The model also provides insight into sustainable solutions. To solve rising pollution levels, resource demand might be cut. To control resource depletion, population growth control measures might be desirable.

KEY CONCEPTS

Human populations acquire and use resources from the environment to produce goods and services that provide us with many benefits. These activities, however, degrade the environment by altering abiotic and biotic conditions.

7.4 The Sustainable Society: The Next Step

For most of our evolutionary past, humankind has lived in a sustainable relationship with nature. In the past few centuries, though, human civilization has undergone many changes that have shifted it off its previous sustainable course. These changes in both attitude and way of life arose during the immensely successful Agricultural and Industrial Revolutions.

One thing that this book makes clear is that major changes are needed to create a sustainable future. New systems of transportation, housing, waste management, industrial production, and the like are needed to protect Earth's systems, which provide us with the goods and services on which we depend.

Fortunately, many of the necessary changes are already under way. Many examples are given in this book. The Spotlights on Sustainable Development present uplifting examples (see Spotlight on Sustainable Development 7-2). Although change is in its infancy, the signs are encouraging, leading some observers to believe that human society may indeed be in the very early stages of a **Sustainable Revolution**, the next step in human cultural evolution. These changes designed to restore and protect natural systems and to undo the mistakes of the past may take decades, perhaps centuries to complete. Change is occurring all around you, though. Local governments are drafting sustainable development programs. Businesses are finding ways to eliminate hazardous waste and slash energy consumption. Individuals are taking steps, as well, to reduce consumption, recycle waste, and reuse products. Books could be written on the changes under way. Some individuals fear that these efforts to build a sustainable society will move society backward, undoing decades of progress. Others argue that they will propel us forward, using the knowledge we have of Earth's limits—and of ecology and sustainability—to create sustainable systems for meeting our needs. The result of such efforts are many: stronger, more secure nations; reduced political tension; stronger economies, a cleaner environment, and healthier people. All in all, sustainability could create a better future for all.

KEY CONCEPTS

Many steps are under way to create a more enduring way of life. These changes may be the early signs of a new cultural shift—a Sustainable Revolution—designed to restructure human systems to honor the limits of the natural systems on which all living things depend.

The danger is not what nature will do with man, but what man will do with nature.

—Evan Esar

CRITICAL THINKING

If you are like me, you probably had to do a little research to answer this question well. Gathering more information is one of the most important rules of critical thinking. Without adequate information, it is difficult to answer a question accurately or to form reliable opinions. Unfortunately, as you may have discovered, all too many people volunteer their opinions or make crucial decisions without sufficient information. Remember the critical thinking rule: Don't mistake ignorance for perspective.

Your research into the Spaceship Earth analogy probably revealed many similarities and differences between a spaceship and the Earth. Like the Earth, the spaceship is a fairly self-contained unit. It floats around in outer space, and everything the occupants need is aboard. I found, though, that the similarities pretty much ended there.

I knew that a spaceship carries its food from another source—the Earth. Food is not generated on board; it is contained in packages. The Earth, however, produces its own food, from plants that take sunlight, carbon dioxide, and water to make food molecules (via photosynthesis).

Another difference is that a spaceship contains a special filter that captures carbon dioxide from the respiration of the astronauts. This carbon dioxide comes from the breakdown of food molecules in the cells of their bodies. On Earth, inhabitants release carbon dioxide, but it is absorbed by plants, which use it to produce food. The food molecules are consumed by animals, and the carbon dioxide is released once again. So on Earth, carbon dioxide is recycled. In a spaceship, it is not.

Oxygen is consumed by animal life both on spaceships and on Earth. The oxygen on the spaceship, however, is released from huge storage tanks filled before the spaceship blasts off. On Earth, oxygen is supplied by plants, which generate oxygen during photosynthesis. Thus, oxygen is recycled on Earth but not on the shuttle.

Water is consumed by animal life, both on Earth and in the spaceship. It is also released as we exhale. On a spaceship, water is collected and discarded. Drinking water is supplied in foods and drinks. On Earth, water is recycled. It enters the atmosphere and rains down upon the land, where it is absorbed by plants.

Finally, on the spaceship, waste matter (feces and urine) are collected. Urine is jettisoned into space, but solid matter is stored in huge tanks that are emptied when the astronauts return to Earth. On Earth, human waste (in modern technological societies) is typically collected and treated. Some of the material is returned to the land—that is, recycled by applying it to pasture and soils.

Thus, it turns out that the Earth as a spaceship has some validity, but it is really a rather poor analogy. It makes for a great image but falls short of scientific accuracy.

CRITICAL THINKING AND CONCEPTS REVIEW

1. What developments occurring during human biological evolution have contributed to the modern environmental crisis? In your opinion, do these developments mean that we are doomed to continue living unsustainably? How can they be used to our advantage?

2. Describe each of the major steps in human cultural evolution. How are they similar? How are they different? How did the human–environment interaction shift over time? How did human systems shift over time away from sustainability?

3. In your view, what general changes are needed in modern society to restructure systems that evolved during the Agricultural and Industrial Revolutions to create a sustainable future?

4. Describe the Population, Resources, and Pollution model. Illustrate through example some of the positive and negative feedback loops.

5. How have the elements of the PRP model shifted as human culture evolved?

6. Has the PRP model presented in this chapter clarified your thinking on environmental problems and solutions? How? Can you see how using it would help expedite efforts to build a sustainable society?

7. One reviewer commented that the material in this chapter gives the impression that any way of life "above" that of the hunter-gatherer was unsustainable. Do you agree with this assessment? Can human society be made sustainable? What will it take to create a sustainable way of life?

8. Before one can devise strategies to create a sustainable society, he or she must ask, "What are we trying to sustain?" How would you answer this question, given the material you have encountered so far in this book and in your course?

9. Using your critical thinking skills, debate the statement, "Efforts under way to create a sustainable society are bound to negate the progress of the past 200 years and should be stopped."

KEY TERMS

Advanced Industrial Society
australopithecines
biodegradable
bipedal
cross-media contamination

dryopithecines
Industrial Revolution
industrial society
negative feedback loop
nonbiodegradable

Population, Resources, and Pollution
model
slash-and-burn agriculture
Sustainable Revolution

REFERENCES AND FURTHER READING

The References and Further Reading section at the end of this book contains a list of sources for the information discussed in this chapter and recommendations for further reading.

Connect to this book's website:
http://environment.jbpub.com/
The site features eLearning, an online review area that provides quizzes, chapter outlines, and other tools to help you study for your class. You can also follow useful links for in-depth information, research the differing views in the Point/Counterpoints, or keep up on the latest environmental news.

Hong Kong is one of the most populated regions of the world.

CHAPTER 8

Population: Measuring Growth and Its Impact

One generation passeth away, and another generation cometh, but the earth abideth forever.

—*Ecclesiastes 1:4*

In the 60 seconds it took you to find this book on your bookshelf and turn to this page, the human population of planet Earth increased by approximately 160 people. That's the equivalent of about 2.70 new world residents every second! At this remarkable rate of growth, nearly 1.6 million people join the human population every week and 83 million people join every year. In 3 years, the human population increases by an amount nearly equivalent to the entire U.S. population.

In 2005, world population reached an all-time high of nearly 6.5 billion. By the year 2025 it could reach nearly eight billion and could be expanding by about 100 million a year. Ninety percent of this growth takes place in the poorer nations, where over 80% of the world's people currently live.

A U.S. senator speaking on the population issue noted that 90% of the current growth in the world population is taking place in less developed countries such as India and China. In order to protect the environment, he said, population growth in these countries needs to be stabilized—and soon. When asked about population growth in the United States and other more developed nations, he replied, "The U.S. population is growing at a slow rate, only about 1% per year. The developed nations, on the whole, are growing at an even slower pace, about 0.1% per year—much, much slower than the growth rate in the developing nations. Therefore, population growth is not a problem in the United States or other developed countries."

Do you think his emphasis on stabilizing population growth in the less developed countries and ignoring growth in the United States is justified? Why or why not?

After you have completed the exercise, answer these questions: What data did you have to call on to make your argument? What critical thinking rules did you use to judge this assertion?

8.1 The Growing Human Population

Throughout most of our history the human population has been small (FIGURE 8-1). Two thousand years ago, in fact, there were about 200 to 300 million people in the world. By 1850, though, world population had climbed to one billion (Table 8-1). But then things began to change, and rapidly. In the next 80 years the population doubled. Forty-five years later it had doubled again, reaching four billion. What caused the rapid upsurge in population growth?

KEY CONCEPTS

The human population has grown slowly for most of human history; the rapid increase in population size is a relatively recent phenomenon.

Why Has the Human Population Grown So Large?

As noted in Chapter 7, hunting and gathering and early agricultural practices could not support large populations. Disease, famine, and war also caused numbers to remain low. In the course of human history, however, several important changes occurred that altered our prospects. Among them were the development of technology (tools and weapons), the Agricultural Revolution, the Industrial Revolution, and the opening of the New World for settlement. Each advance helped to unleash human population from the constraints of environmental resistance. Improvements in agriculture, such as the plow and irrigation during the Agricultural Revolution, for example, increased food supplies, which spurred human population growth. The most dramatic influence of technology on food supplies, however, occurred during the Industrial Revolution as new forms of machinery were introduced. Many new technologies improved food production, boosted food supplies, and permitted the

> In 2005, the world population reached 6.5 billion, and the U.S. population was 295 million. Canada's population was nearly 32 million.

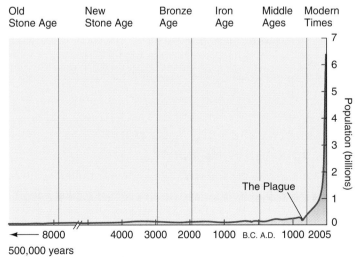

FIGURE 8-1 **World population growth.** This graph dramatically illustrates the growth curve of the world population. Human numbers were held in check by famine, disease, war, and primitive technological development for most of human history. Advances in technology and health care have caused a massive increase in human numbers in the last 200 years.

Table 8-1		
Global Population Growth		
Population Size	**Year**	**Time Required to Double**
1 billion	1850	All of human history
2 billion	1930	80 years
4 billion	1975	45 years
8 billion (projected)	2025	50 years

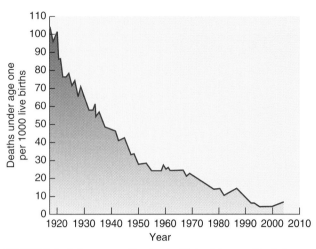

FIGURE 8-2 **Infant mortality in the United States.** The infant mortality rate has declined from 1915 to 1997. The consistent decline has contributed to a marked increase in average life expectancy.

population to increase. Modern medicine also enhanced the human condition. New drugs such as penicillin dramatically lowered the death rate, especially in the early years of life (FIGURE 8-2). Although we later learned that it caused serious environmental problems, the pesticide DDT was used in the tropics to combat malaria-carrying mosquitoes, saving millions of lives each year. Finally, improvements in sanitation and water purification also boosted population growth by decreasing the incidence of infectious diseases. As a result of these and other developments, death rates in many countries dropped. This, in turn, resulted in a substantial increase in population growth.

The benefits of technological advances mentioned in the previous paragraph were first felt in the more developed nations in Europe and later North America. But the decrease in death rates in these countries was not matched by a decrease in birth rates. It took over 150 years for most countries to bring the two in line. The result was a rapid increase in population in the developed countries during this period.

The benefits of the Agricultural and Industrial Revolutions came later to the less developed countries (LDCs) of Asia, Latin America, and Africa—in fact, not until after World War II. Prior to this time, high growth rates and high death rates were commonplace in these nations. The introduction of modern medicine, advanced agricultural techniques, and sanitation caused a sudden decrease in death rates in LDCs after World War II. Because birth rates remained high, growth really took off and the populations of the LDCs began to expand.

> ### KEY CONCEPTS
> The human population has skyrocketed in the last 200 years primarily because of a worldwide lowering of death rates without a corresponding decrease in birth rates. Death rates plummeted primarily as a result of increases in food supply and better medicine and sanitation.

Expanding the Earth's Carrying Capacity: An Ecological Perspective

Each of the advances cited previously here—technology, agriculture, medicine, and others—has helped alter the delicate balance between population growth and population reduction factors. The net effect has been an increase in those factors that stimulate growth and a decrease in those that hold growth in check. This, in turn, has caused a dramatic increase in the Earth's carrying capacity for human beings.

As noted in Chapters 2 and 6, the **carrying capacity** of the Earth is the number of organisms (in this case, humans) the Earth can support over the long haul. The carrying capacity of the Earth is determined by two major factors: resource availability, including food supplies, and the environment's capacity to absorb and detoxify wastes.

All ecosystems have a specific carrying capacity for each population. The carrying capacity is not static, however. In other words, it fluctuates from year to year, depending on climate and other factors noted in Chapter 6. More important, carrying capacity can be altered by organisms. As a rule, most species have little effect on the carrying capacity of their surroundings. Humans, however, have a remarkable ability to expand the Earth's human carrying capacity. Although such changes enhance the survival and reproduction of our kind, the gains we make often occur at the expense of other organisms. The more fish we harvest from the sea through advanced fishing techniques, the fewer there are for sea mammals.

Expansion of the carrying capacity affects humans as well. Importing water via pipelines to desert communities, for example, increases the regional carrying capacity. However, dramatic increases in the human population may result in sizable increases in pollution. At a certain point, the amount of waste produced cannot be assimilated by natural mechanisms. People begin to suffer.

> ### KEY CONCEPTS
> Technological advances lower environmental resistance and promote population growth, increasing the carrying capacity for humans. This, in turn, generally decreases the prospects of other species and may cause adverse effects in human populations as well.

What Is the Earth's Carrying Capacity for Humans?

Much of the debate over the environment often revolves around arguments concerning the Earth's carrying capacity for human beings. Current estimates range from 500,000 to 1000 billion people—170 times the present population. The reason for this vast discrepancy lies in the assumptions one makes about the standard of living and the level of technological development. If everyone lived frugally, recycling resources, driving solar-powered cars, living in small solar homes, using energy efficiently, and so on, the Earth could support a fairly large population. But in the absence of environmental lifestyles and substantial changes in industrial production, the Earth's carrying capacity would be much smaller.

Where are we today? Many scientists believe that the 2005 human population of 6.5 billion already exceeds the carrying capacity. Polluted air, depleted ocean fisheries, species extinction, denuded landscapes, deforestation, widespread starvation, and poverty are signs that humanity has pushed beyond the carrying capacity of the Earth. While recognizing these problems, critics say that new laws and improvements in technology could reduce these problems and perhaps even allow the Earth to support many more people. By using energy more efficiently or converting to solar energy, for example, they say that many more people could live on the Earth. Is this realistic?

As you study the material presented in this book and in your course, you will see that the problems faced by the world's people are massive. Many will be difficult to solve. Many are growing worse. Adding more people could worsen the problems. A larger human population, even one using resources more efficiently, will require massive amounts of food, wood, water, fuel, and building materials. According to one scientific estimate, the human population currently consumes about 40% of the Earth's primary terrestrial productivity—that is, 40% of the plant matter produced on land. A doubling of human population *without any increase in the standard of living* would virtually turn the entire planet into a farm to support humans, an outcome that would cause widespread extinction of plants and animals and further loss of the ecological services so essential to human survival. A doubling of human population would also generate huge amounts of pollution and waste. (For a debate on the pros and cons of further population growth, see the Point/Counterpoint in this chapter.)

> "The human population currently consumes about 40% of the Earth's plant matter—trees and other vegetation—to provide food, shelter, clothing, and other needs."

KEY CONCEPTS

Determining the Earth's carrying capacity for humans is a task fraught with difficulty. Some people think that we have not reached the carrying capacity. Others believe that the human population already exceeds the Earth's long-term carrying capacity and that further growth will only further undermine the ecological health of the planet and our own prosperity and well-being.

Too Many People, Reproducing Too Quickly?

The impacts of the human population can be felt everywhere—in both cities and rural lands. In fact, every environmental problem discussed in this book is rooted in population, although not everyone agrees with this assertion. See what you think after you study the chapter and look at both sides of the issue.

Scientists who study populations use the term **overpopulation** to refer to a condition in which a population of organisms—any organism—exceeds the carrying capacity of its environment. In short, the population exceeds the capacity of the environment to supply resources to all and to assimilate waste. Based on this definition, they say that humans suffer from overpopulation. Is this true? Let's take a look at the issue.

KEY CONCEPTS

Population is at the root of virtually all environmental problems, including pollution and resource depletion, as well as many social and economic problems. The size of the population and the rate of growth both have significant impacts on environmental problems and solutions.

Overpopulation: Urban and Rural Despair In 1819, the British poet Shelley wrote, "Hell is a city much like London." If he were alive today, the poet might have put it differently: "Hell is a city much like Cairo, Calcutta, Shanghai, Bangkok, London, Los Angeles, Mexico City, and dozens of others." In both the rich and poor nations of the world, many people live in deplorable conditions without adequate water and sanitation. Social problems run rampant. Crowding in urban centers has been implicated in a variety of social, mental, and physical diseases. Many social psychologists assert that divorce, mental illness, drug and alcohol abuse, and social unrest result, in part, from stress caused by crowding. Prenatal death and rising crime rates may also be attributed partly to overcrowding. They label this malaise the **inner city syndrome.** Research on animals supports the contention that crowding is not a healthy condition.

In addition to the social problems, urban centers frequently exceed the capacity of the environment to assimilate wastes. Cities throughout the world are centers of intense pollution from automobiles, factories, power plants, and sewage treatment plants. Even pollution from homes contributes to urban air pollution. Pollution affects people's health and has significant effects on crops and the natural environment. In China, for example, studies suggest that air pollution from urban and rural sources reduces crop production by 5 to 30%, depending on the level of pollution. By cleaning the air, the nation might eliminate its need for imported grain.

Large urban populations also place considerable demands on the outlying countryside for resources such as fuel, water, and food. Also, as cities spread, they expand onto farmland and open space, literally devouring the resources they need to survive and prosper.

A surplus of people strains many rural areas, too. Rural Bangladesh, Kenya, Ethiopia, Mexico, and other less developed nations, for example, reel under the burden of massive populations. Overpopulation in these and other countries results in unsanitary living conditions, water shortages, food shortages, and disease. In Africa, trees around rural villages are cut down for firewood, creating a barren landscape that extends many miles from the villages. Grasses

Is More Always Better?

Garrett Hardin
The author is a renowned environmentalist, writer, and lecturer. He taught at several universities and served as chairman of the board and CEO of the Environmental Fund. He wrote numerous books and articles on environmental ethics and is best known for his article "The Tragedy of the Commons."

To get at the heart of the question "Is more better?" study the daily flow of water over Niagara Falls. You will find that twice as much water flows over the falls during each daylight hour as during each nighttime hour. There in a nutshell you have the population problem.

Puzzled? You should be. The connection between Niagara Falls and population is not obvious. Before we can understand it we need to review a little biology.

For every nonhuman species there is an upper limit to the size of a population. Near the maximum, individuals are not so well off as they are at lower densities. Starvation appears. Crowded animals often fight among themselves and kill their offspring. Wildlife managers and advocates agree that the maximum is not the optimum.

What about humans? Will we be happiest if our population is the absolute maximum the Earth can support? Few people say so explicitly, but some argue that "more is better!"

Admittedly, we need quite a few people to maintain our complex civilization. A population the size of Monaco's, with about 30,000 people, could never have enough workers for an automobile assembly line. But Sweden, with nearly 9 million people, turns out two excellent automobiles. Nine million is a long way from 6.2 billion, the population of the world today.

Some say, "More people—more geniuses." But is the number of practicing geniuses directly proportional to population size? England today is 12 times as populous as it was in Shakespeare's day, but does it now boast 12 Shakespeares? For that matter does it have even one?

Consider Athens in classical times. A city of only 40,000 free inhabitants produced what many regard as the most brilliant roster of intellectuals ever: Solon, Socrates, Plato, Aristotle, Euripides, Sophocles—the list goes on and on. What city of 40,000 in our time produces even a tenth as much brilliance?

Of course, the free populace of Athens was served by ten times as many slaves and other nonfree classes. This left the 40,000 free people to apply themselves to intellectual and artistic matters. Modern peoples are given creative freedom by labor-saving machines, certainly a more desirable form of slavery. But where are our geniuses?

Business economists are keenly aware of "economies of scale," which reduce costs per unit as the number of units manufactured goes up. Communication and transportation, however, suffer from diseconomies of scale. The larger the city, the higher the monthly phone bill. Crimes per capita increase with city size. So do the costs of crime control. All these suggest that more may not be better.

Democracy requires effective communication between citizens and legislators. In 1790, each U.S. senator represented 120,000 people; in 2004, the figure was 4.0 million. At which time was representation closer to the ideal of democracy? To communicate with his or her constituents each senator now has an average of 60 paid assistants. President Franklin D. Roosevelt had a staff of 37 in 1933, when the population was 125 million. Ronald Reagan had a staff of 1700 in 1981, when the population was 230 million. We have to ask whether democracy can survive unrestrained population growth.

Let's look at another aspect of the more-is-better argument. An animal population is limited by the resources available to it. With humans, a complication arises. Though the quantities of minerals on Earth are fixed, improved technology periodically increases the quantity of resources available to us. In the beginning copper ores we mined contained 20% copper; now we are using ores with less than 1%. Available copper has increased but not the total amount of copper on Earth.

Let's return to Niagara Falls. Less water flows over the falls at night because more water is diverted to generate electricity when people aren't looking at the falls. It would be possible to use all the water to generate electricity, but then there would be no falls for us to look at. As a compromise, the volume of water "wasted" falling over the falls is reduced only at night. Therefore, the turbines and generators are not fully used 24 hours of the day, which means that local electricity costs are just a bit higher.

If the population continues to grow, the day may come when electricity is so scarce and expensive that the public will demand that Niagara Falls be shut down so that all the water can be used to generate electricity. Similar dangers face every aesthetic resource. Wild rivers can be dammed to produce more electricity, and estuaries can be filled in to make more building sites for homes and factories. The maximum is never the optimum. With human populations, quantity (of people) and quality (of life) are trade-offs. Which should we choose—the maximum or the optimum?

Classical economic theory holds that population growth reduces the standard of living: the more people, the lower the per capita income, all else being equal. However, many statistical studies conclude that population growth does not have a negative effect on economic growth. The most plausible explanation is the positive effect additional people have on productivity by creating and applying new knowledge.

Because technological improvements come from people, it seems reasonable to assume that the amount of improvement depends in large measure on the number of people available. Data for developed countries show clearly that the larger the population, the greater the number of scientists and the larger the amount of scientific knowledge produced.

There is other evidence of the relationship between population increase and long-term economic growth: an industry, or the economy of an entire country, can grow because population is growing, because per capita income is growing, or both. Some industries in some countries grow faster than the same industries in other countries or than other industries in the same country. Comparisons show that in the faster growing industries the rate of increase of technological practice is higher. This suggests that faster population growth, which causes faster growing industries, leads to faster growth of productivity.

The phenomenon economists call "economy of scale"—greater efficiency of larger-scale production where the market is larger—is inextricably intertwined with the creation of knowledge and technological change, along with the ability to use larger and more efficient machinery and greater division of labor. A large population implies a bigger market. A bigger market is likely to bring bigger manufacturing plants, which may be more efficient than smaller ones and may produce less expensive goods.

A bigger population also makes profitable many major social investments that would not otherwise be profitable—railroads, irrigation systems, and ports. For instance, if an Australian farmer were to clear a piece of land for farming a distance away from neighboring farms, he might have no way to ship his produce to market. He might also have trouble finding workers and supplies. When more farms are established nearby, however, roads will be built that link him with markets in which to buy and sell.

We often hear that if additional people have a positive effect on per capita income and output, it is offset by negative impacts such as pollution, resource shortages, and other problems. These trends are myths. The only meaningful measure of scarcity is the economic cost of goods. In almost every case the cost of natural resources has declined throughout human history relative to our income.

The Case for More People

Julian L. Simon

The author was professor of business administration at the University of Mayland at College Park. He wrote several important books on population and economics, including *The Ultimate Resource* and *Population Matters*.

Conventional wisdom has it that resources are finite. But there is no support for this view. There is little doubt in my mind that we will continue to find new ore deposits, invent better production methods, and discover new substitutes, bounded only by our imagination and the exercise of educated skills. The only constraint on our capacity to enjoy unlimited raw materials at acceptable prices is knowledge. People generate that knowledge. The more people there are, the better off the world will be.

Critical Thinking Questions

1. Is our population now below or above the optimum? To answer this question, make two lists. On one, list all of the things that you would expect to be better if the population doubled; on the other, list all of the things that would be worse.

2. On which list would you put the availability of wilderness? What about the noise level? Amount of democracy? Amount of pollution? Per capita cost of pollution control? Availability of parking spaces? Personal freedom?

3. When you are through, compare your list with those of your friends. What value judgments account for the differences? Can these differences be reconciled? How?

4. Outline the main points and supporting evidence given by both authors. Which views do you agree with? Why? Are your reasons based on feelings or facts?

You can link to websites that represent both sides through Point/Counterpoint: Furthering the Debate at this book's internet site, http://environment.jbpub.com/. Evaluate each side's argument more fully and clarify your own opinion.

are overgrazed by livestock, and once-fertile land is turned into desert (**FIGURE 8-3**).

The problems caused by overpopulation are especially evident in Bangladesh, which lies just east of India along the Bay of Bengal. No larger than Wisconsin, Bangladesh houses 25 times as many people, 144 million in 2005. Eighty-five percent of the population lives outside of cities.

Rural and poor, Bangladesh is the most densely populated nation in the world. Over the years, cattle have severely overgrazed the land. Desperate peasants have stripped away the vegetation in search of food, fuel, and shelter. When the rains come, floods follow because denuded hillsides cannot hold the rain back. Soil from these badly abused lands washes into streams. Sediment-choked streams flow to the sea and deposit their silt in river deltas. In search of farmland and a place to live, people flock to the deltas by the tens of thousands. As almost any ecologist will tell you, the deltas belong to rivers and the sea, not to people. When the heavy rains and hurricanes come, as they do most years, high waters flood the deltas, driving people away and destroying their farms and homes. Thousands of lives are lost each year from storms. In 1985, for example, a hurricane brought a 5-meter (15-foot) storm surge that crashed inland, devastating makeshift homes and farms on the deltas. The storm left a quarter of a million people homeless and killed an estimated 4000 to 15,000 people and countless livestock. In 1988, a similar disaster struck. Heavy rains flooded two thirds of the land, leaving 25 million people homeless. In 1991, the disaster was repeated again, with an estimated 250,000 people killed. Such disasters continue to this day.

On the surface, the disasters in Bangladesh look like severe natural disasters. On closer examination, however, it is clear that much of the blame can be pinned on an underlying human problem—overpopulation, which forces people onto marginal lands unsuitable for human habitation.

Rural overpopulation feeds the urban crisis. Throughout the world, dismayed peasants and their families who are unable to survive on their farms migrate in large numbers to the cities in search of jobs and security. In Mexico City, thousands of rural farmers and their families arrive each year. Unfortunately, what awaits them and their counterparts in other countries is intense crowding, skyrocketing unemployment, poverty, crime, inadequate food, and pollution.

KEY CONCEPTS

The massive size of the human population causes environmental problems evident in urban and rural areas. These include shortages of resources, environmental deterioration, and a host of social problems.

Reproducing Too Quickly

Hunger, starvation, disease, poverty, illiteracy, pollution, unemployment, and barren landscapes are, to many observers, signs that the human population is already exceeding the Earth's carrying capacity. Although legislative and technological efforts are being mounted to solve these problems, continued growth makes it difficult to keep even with them, let

FIGURE 8-3 **Rural despair.** Overpopulation, desertification, and drought have diminished the resources available to these nomadic people in northern Africa. Many other rural poor living at or beyond the carrying capacity of their environment face similar problems.

alone improve conditions. Nowhere is this problem more evident than in Africa.

The population of Africa is currently growing at a rate of 2.4% per year—or doubling every 29 years. In order to keep up with this growth rate, the amount of food must be doubled every 29 years—so must the number of doctors, nurses, and teachers, as well as water supplies and transportation facilities. Even then, say some scientists, conditions would remain the same. To improve the already deplorable conditions, the growth in facilities and services would have to increase faster than the population.

Population growth makes social, economic, and environmental problems more difficult to solve and often worsens the impact we have on the planet and our society. Population growth may, for example, cause hunger, starvation, poverty, and illiteracy to spread. It causes local, regional, and global pollution and species extinction to increase and it may cause further shortages in resources.

Although global population growth has slowed and certain countries have shown dramatic declines in growth—some are even shrinking—growth continues at a rapid pace. As noted in the introduction, the human population is expanding by about 83 million

> The human population is expanding by around 83 million people per year. Ninety percent of this growth is occurring in the less developed nations, but the more developed nations still use the majority of the Earth's resources and produce the bulk of its pollution.

> The growth rate of the human population is about 1.2%. Growth rates have begun a steady decline. In 1965, it was about 2%. Although the population is still expanding rapidly, the decline is seen as an encouraging trend.

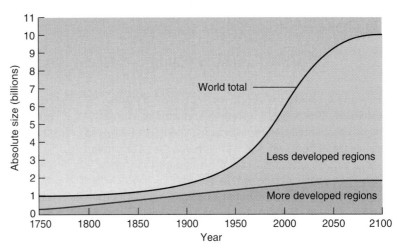

FIGURE 8-4 Graph of global population growth. Note that the vast majority of growth is occurring in the less developed nations, a trend which began after World War II in large part because of the introduction of modern medicines and other measures that lowered death rates. Birth rates continued at high levels, resulting in massive growth. Measures are now being taken to reduce birth rates and slow the rate of growth.

people per year. While the decline in growth may be comforting to some critics, it is still a potentially dangerous force.

Population growth is occurring at different rates in different parts of the world. The world's fastest-growing areas are, in decreasing order, Africa, Latin America, and Asia. These regions account for nearly 90% of the annual growth in the human population (FIGURE 8-4). This leads many observers to focus their attention on measures to control growth in these regions. However, as you will see in Chapter 9, growth in the more developed countries, while less robust, has major impacts on the condition of the planet.

Robert McNamara, president of the World Bank for 14 years, maintains that "short of nuclear war itself, population growth is the gravest issue the world faces. . . . If we do not act, the problem will be solved by famine, riot, insurrection, and war." Solving the population, pollution, and resource "trilemma" say many experts will require a substantial and immediate decrease in the population growth by means discussed in Chapter 9. To see what one country is doing, see Spotlight on Sustainable Development 8-1 in this chapter.

> **KEY CONCEPTS**
>
> All of the social, economic, and environmental problems of cities and rural areas are aggravated by rapid population growth. The large size of many populations makes it difficult for governments to keep up with current demands and continued rapid growth makes it nearly impossible to improve conditions and create a sustainable human presence.

8.2 Understanding Populations and Population Growth

Newspapers and magazines sometimes present a confusing array of statistics on population, making it difficult for the public to understand this global issue. With a little effort you can learn enough demography, or population science, to weed out the fallacies from the facts of the population debate.

Measuring Population Growth— Growth Rates and Death Rates

One of the most important population measurements is the growth rate. What does it mean when demographers report that a population is growing at a rate of 3% per year? Is this something to be concerned about? To answer this question, let's first see how a population's growth rate is determined.

The growth rate of the world population, also known as the **natural increase**, is calculated by subtracting the num-

ber of deaths in a population from the number of births in any given year, as shown below:

Growth Rate =
Crude Birth Rate − Crude Death Rate

Crude birth rate is the number of births per 1000 people in a population. Crude death rate is the number of deaths per 1000 people. Worldwide, the crude birth rate is currently 21/1000, and the crude death rate is 9/1000. The difference between the birth and death rates is the population growth rate, in this case 12/1000. This means that every year 12 people join the world population for every 1000 people in the population.

To convert this to a percentage, you simply multiply by 100, as shown below:

Growth Rate = 12/1000 × 100 = 1.2%

The relationship between birth rates and death rates determines whether the world's population is growing, shrinking, or staying the same. Some important growth rates are listed in Table 8-2. This table also shows the doubling times, which are discussed shortly.

Table 8-2		
Growth Rate and Doubling Time		
Region	**Growth Rate (%)**	**Doubling Time (years)***
World	1.2	58
More Developed Countries	0.1	700
Less Developed Countries	1.5	47
Africa	2.4	29
Asia	1.3	54
North America	0.5	140
Latin America	1.6	44
Europe	−0.2	—
Oceania	1.0	70
*Discrepancies in doubling time are the result of rounding off growth rates.		

8-1   Thailand's Family Planning Success Story

Paul and Anne Ehrlich noted in their book *The Population Explosion,* "The population/resource/environment predicament was created by human actions, and it can be solved by human actions. All that is required is the political and social will." Thailand is a stellar example of what can happen when people decide to take action.

In 1971, Thailand adopted a national population policy. In the next 15 years, the country's population growth rate plummeted from 3.2 to 1.6%. During that period, the use of contraceptives by married couples rose from 15 to 70%. Thailand's success continued for another decade. By 1997, its growth rate had plummeted to 1.1%. By 2005, it had fallen to 0.7%.

Thailand's success stems from many factors. The first is religious in nature. Ninety-five percent of the Thai people are Buddhists, and Buddhist scripture warns that "many children make you poor." In addition, the Buddhist religion embraces family planning. In an effort to support population control, Buddhist monks distribute contraceptives with cards that remind the Thai people that "many births cause suffering."

Another factor is cultural openness. The Thai people are open to new ideas, and relationships between men and women are egalitarian. Consequently, women have equal say in decisions about family matters.

A third reason for the success in family planning is political. The government of Thailand encourages nationwide family planning and offers considerable financial support. Over the years, Thailand's government has made a wide range of contraceptives available to the public and has worked cooperatively with an influential nonprofit agency, the Population and Community Development Association of Thailand or PDA, to elevate family planning to a national goal.

PDA was founded in 1974 by Mechai Viravaidya (FIGURE 1). Mechai's creative and high-profile ways of promoting family planning in Thailand are considered by some as the single most important factor in the country's remarkable success in population control. Mechai promotes condom use by handing them out in public at any opportunity he can find. He even sponsors a "cops and rubbers" program, during which police officers hand out condoms on New Year's Eve. Today, condoms are affectionately known in Thailand as "mechais."

Because of Mechai's efforts, condoms, birth control pills, IUDs, and spermicidal foams are all available in bus terminals. In addition, his organization has opened vasectomy clinics across the country and celebrates the King's birthday each year by offering free vasectomies. Sterilization is the most widely used form of contraception in Thailand.

Mechai has also launched an economic development program. As a former government economist, he realized that income-generating alternatives could help change his people's attitude toward large families. With German financing, the PDA introduced a revolving loan scheme to support efforts to develop clean drinking water supplies. However, loans were only available to those individuals who participated in family planning programs. Money provided by the fund helped Thai people install thousands of toilets and containers to catch rainwater for drinking. The PDA also ini-

---

The balance between birth rates and death rates in various regions of the world determines global population growth. Each of these rates is influenced by a variety of factors. The birth rate in a given population, for instance, depends on (1) the age at which men and women get married, (2) their educational level, (3) whether a woman works after marriage, (4) whether a couple uses reliable contraceptives, (5) the number of children a woman and her husband want, (6) cultural values, and (7) religious beliefs.

These factors can also be viewed as leverage points for controlling human numbers. For example, efforts to reduce population growth often rely on improving educational and employment opportunities for men and women, as well as making contraceptives available.

Death rates are also influenced by a variety of factors and are equally important in determining population growth. As you learned earlier, food supplies, proper sanitation, water purification, and modern medicine have a profound effect on death rates.

KEY CONCEPTS

Global population growth is determined by subtracting the crude death rate from the crude birth rate.

Doubling Time

Another important measurement of population dynamics is the **doubling time**, the time it takes for a population to double. The following formula is used to determine doubling time:

$$\text{Doubling Time} = 70/\text{Growth Rate (\%)}$$

In this equation, 70 is a demographic constant. Using this equation, you can quickly convert growth rates into doubling times. A 2% growth rate would yield a doubling time of 35 years.

KEY CONCEPTS

Doubling time is the time it takes a population to double in size. It is determined by dividing 70 by the growth rate. Surprisingly, even relatively small growth rates result in rapid doubling.

tiated a loan program to support farmers who are participating in family planning programs. Loans were offered at rates far better than those traditionally available to farmers.

FIGURE 1 Mechai Viravaidya is one of the key proponents of family planning in Thailand. His dedication has helped make Thailand a family planning success story.

The government of Thailand also sponsors a loan program to individuals in various communities. Loan funds, in fact, are apportioned on the basis of a village's use of contraceptives. The total amount of the loan fund increases as the level of contraceptive use goes up.

Today, widespread commitment of the Thai people to family planning is believed by many to be associated with both the PDA's and the government's economic development programs. The PDA's programs have been so successful in Thailand that the nation now offers a 3-week training course to representatives from other less developed countries. Two of the many participants are Bangladesh, where family planning is considered inappropriate, and the Philippines, which is predominantly Catholic. The training course is designed to allow participants to develop ways to adapt the PDA's strategies for use in their countries.

Thailand's success in family planning has been paralleled by economic success, with a doubling of per capita income in the country since 1971. Economic progress has allowed the government to expand its family planning services and medical facilities throughout the country. Economic improvements have meant that families live better and have a wider variety of choices.

Thailand is still a long way from a stable population. Currently, its 65 million population is slated to double in fewer than 90 years. But efforts to date have been remarkable, proving that when people recognize the problems of overpopulation and decide to take action, they can make a big difference.

Growth Rates and Doubling Times The world is crudely divided into the more developed countries and the less developed (or developing) countries (Table 8-3). The **more developed countries**, such as the United States, Canada, Australia, and Japan are industrial nations with a strong economic base. Many have strong agricultural economies as well, which are highly mechanized. The population growth rates of the more developed countries have decreased dramatically over the years so that most developed nations are growing at fairly slow rates—on the average around 0.1% per year in 2005. The doubling time of the more developed nations is around 700 years. Averages can be deceiving, however. Some more developed

> The growth rate of the more developed nations is 0.1%, yielding a doubling time of around 700 years. The growth rate of the less developed nations is 1.5%, which yields a doubling time of 47 years.

countries such as Canada and the United States have relatively fast growth rates of about 0.3 to 0.6% per year, respectively. Others, such as Denmark and Belgium, have extremely slow growth rates, 0.1% per year. Some, such as Greece, have stopped growing altogether. A number of others, such as Germany, Sweden, and Hungary, are shrinking, or, in the language of demographers, are experiencing "negative growth."

In contrast, the **less developed nations** are growing much more rapidly—at an average of 1.5% per year in 2005 (doubling time = 47 years). This average hides some dangerous trends, though—countries that are growing at over 3% per year. Nigeria, for instance, currently contains about 138 million people and is growing at a rate of 2.9% per year, giving it a doubling time of 24 years.

KEY CONCEPTS

Growth rates in the more developed countries are relatively low. In the less developed nations, growth rates are generally much higher.

Table 8-3

Comparison of More Developed and Less Developed Countries (1997)

Feature	More Developed	Less Developed
Standard of living	High	Low
Per capita food intake	High (3100–3500 cal/day)	Low (1500–2700 cal/day)
Crude birth rate	Low (11/1000 population)	High (25/1000 population)
Crude death rate	Low (10/1000 population)	Low (9/1000 population)
Growth rate	Low (less than 0.1%)	High (1.7%)
Doubling time	High (809 years)	Low (40 years)
Infant mortality	Low (8/1000 births)	High (63/1000 births)
Total fertility rate	At or below replacement level (1.5)	High (3.2)
Life expectancy at birth	High (75 years)	Lower (64 years)
Urban population	High (75%)	Low (38%)
Wealth (per capita GNP) (1995 U.S. dollars)	High ($19,500)	Low ($1,260)
Industrialization	High	Low
Energy use per capita	High	Low
Illiteracy rate	Low (1–4%)	High (25–75%)

Source: Selected data from *US in the World.* Washington, DC: Population Reference Bureau, 1997.

The Total Fertility Rate and Replacement-Level Fertility

The growth rate and doubling time of a population are vital to understanding its future. Another important measure of human population, which helps us understand trends in growth, is the total fertility rate. The **total fertility rate** (TFR) is the number of children women in a population are expected to have in their lifetimes based on current trends in childbearing. In the United States, for example, the total fertility rate was 2.0 in 2005; this means that, on average, each woman in the reproductive age group of 15 to 45 is expected to have 2 children over her lifetime. Canada's TFR is 1.5. In India, a much poorer nation, women have a TFR of 3.0.

Women in many countries have achieved **replacement-level fertility**, a total fertility rate at which couples produce exactly the number of children needed to replace themselves. In the United States and other more developed countries, replacement-level fertility is a TFR of 2.1 children. That means that 10 women must have 21 children to replace themselves and their partners. The reason for the additional child is that in these countries 1 of every 21 children dies before reaching reproductive age. Replacement-level fertility is higher in less developed nations because death rates for children are higher.

Today, nearly 80 nations (out of 200) have reached or fallen below replacement-level fertility. Most of these nations are in Europe. However, even though a nation's population has reached replacement-level fertility, that does not mean that it has stopped growing. A population stops growing only when the death rate equals the birth rate and when the net migration is zero—that is, when the number of people entering the country equals the number leaving it. Demographers refer to this state as **zero population growth**, or ZPG.

In the United States, the TFR fell below replacement level in 1972, thanks in large part to modern contraception, rising affluence, and widespread public awareness of the population dilemma. Even though the TFR has remained below replacement-level fertility since that time, demographers project that the U.S. population will continue to grow well into the future.

The U.S. population continues to grow for two reasons: First, each year numerous people move to the United States from other countries. Some of these people come legally—about 2200 per day; others arrive illegally—about 1000 per day. Legal and illegal immigration add about 1.2 million people each year to the U.S. population and is responsible for about 30% of the annual growth.

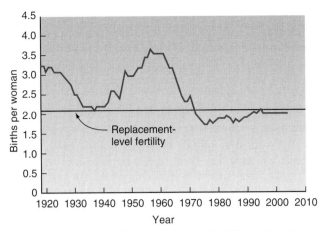

FIGURE 8-5 Total fertility rate in the United States. Total fertility rate (TFR) is the number of children women are expected to have, based on the current age-specific fertility rates (fertility rates in each age group). The TFR in the United States has fluctuated widely with changing economic conditions. In the decade of the Great Depression, women had a low TFR. In the several decades following World War II, the TFR shot up; in 1972 it fell below the replacement level, where it has remained, in recent years it has begun to edge upward.

The rest of the growth rate comes from within. To understand why, we begin with **FIGURE 8-5**, which shows the total fertility rate for the United States for the last 85 years. After World War II, the total fertility rate climbed dramatically and remained high for several decades. This increase resulted in a **postwar baby boom**, a rapid increase in childbirth that in turn resulted in a long-term increase in the U.S. population. The children of the baby-boom era form something of a tidal wave in the American population. Today, many of these individuals are having their own children. Even though they are having smaller families than their parents—on average, about two children each—the number of women giving birth is still on the rise. In other words, more women are having children. This delayed effect is called the **lag effect**.

Because of this lag effect and because of migration (discussed shortly), populations can grow long after they have reached replacement-level fertility. Stabilization is possible *only* if a number of conditions are met: (1) the number of women entering the reproductive age group levels off or declines, (2) the total fertility rate remains below replacement level, and (3) immigration is held in check. Small changes in any of these factors can have drastic effects on a population. As testimony to the effect of relatively small changes, in the late 1980s, the U.S. Census Bureau projected that the U.S. population would stabilize at about 300 million by 2050. As a result of changes in immigration policy that increased the number of legal immigrants and because of higher than expected fertility rates among U.S. women, the U.S. population is now expected to reach 420 million by 2050, up from 296.5 million in 2005, with stabilization nowhere in sight.

Migration

The previous sections have dealt primarily with global population growth, which is determined by birth rates and death rates. As alluded to in the previous section, to calculate the growth of specific regions—cities, towns, states, and countries—we must take into account the number of people moving out of and into the region to set up residence—that is, **migration**.

Migration has two forms—immigration and emigration. The term **immigration** refers to the movement of people into a country; **emigration** refers to movement out. **Net migration** is the difference between the two. Population growth in a country will stabilize if the growth rate and net migration are zero.

KEY CONCEPTS

The growth of a town, city, state, or region is determined by two factors: growth rate (natural increase) and migration—the movement of people into and out of the population.

The Pros and Cons of Immigration Immigration is one of the hottest topics today in the United States, in part because legal and illegal immigration into the United States accounts for 30% of the annual growth. Proponents of growth controls argue that efforts to stabilize the U.S. population are doomed unless the country reduces all forms of immigration.

> Immigration accounts for about 30% of the annual growth in the U.S. population.

Public opinion polls in the United States in the late 1980s showed that a majority favors reducing immigration of all kinds. Accordingly, Congress passed a law in 1986 levying penalties on those who knowingly employ illegal immigrants and requiring job seekers to provide proof of citizenship or legal immigration status. Four years later, though, a new Congress switched its position and passed the 1990 Immigration Act, which increased the number of legal immigrants by 35%—increasing from 530,000 in 1991 to nearly 700,000 a year from 1992 to 1994. Currently, about 800,000 legal immigrants enter the United States each year. Most of them are from Asia (primarily China) and Mexico.

Opponents of liberal immigration policies argue that such policies have many negative effects. They say that it increases the tax burden on already struggling cities such as New York, Miami, and Los Angeles, where the bulk of the immigrants end up. Social programs in these and other cities are designed to support low-income families, and immigrants often require assistance. Opponents argue that immigration also worsens crowding and internal strife in racially torn neighborhoods. It increases resource demand and puts additional stress on school districts that are already coping with severe budget problems. It also increases competition for jobs.

Proponents of more liberal immigration policies argue that immigration offers many benefits to a country. It increases cultural diversity and provides workers for low-income jobs, thus keeping the cost of running businesses down, which means cheaper products for consumers. It also helps immigrants escape political persecution, squalid living conditions, and poverty and enables families to improve their economic standing and educational opportunities.

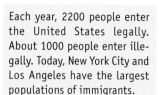

Each year, 2200 people enter the United States legally. About 1000 people enter illegally. Today, New York City and Los Angeles have the largest populations of immigrants.

Internal migration, the movement of people within a country from one region to another, is also of concern to scientists and environmentalists. Such movements may have enormous impacts on local economies, cultures, and environments. In the United States, for example, in the past 3 decades people have migrated en masse from the northeastern, north central, and midwestern regions to the South (mostly Florida) and West (Arizona, Wyoming, Utah, Alaska, Idaho, Colorado, New Mexico, Texas, and California). This movement, referred to by many as the "sunning of America," is caused by (1) the expansion of industries in the West and Southwest; (2) a decline in industries of the northeastern and north central states, especially auto and steel manufacturing; (3) a desire for a warmer climate; (4) a desire for a lower cost of living; (5) a preference for abundant recreation; (6) an aspiration for a less hectic, less crowded lifestyle; and (7) the growth of retirement communities.

The massive migration to the South and West had many important economic benefits for these regions. Builders, restaurant owners, bankers, and other members of the service sector of the economy all benefited economically from the growth. Ironically, the influx of people into the Sunbelt destroyed many of the values the migrants sought. With growth rates in the range of 3 to 5% per year, many western cities and states are being swamped by new people. Air pollution in Sunbelt cities has often worsened with increasing population; the clean air that many came for has disappeared. Traffic has become congested. The rapid rise in demand for housing sent the cost of homes skyward. The West now has the most expensive new homes in the nation. The breakneck speed of growth made it difficult for local governments to provide water, schools, sewage treatment plants, and transportation facilities. Spreading cities engulfed smaller outlying communities and changed the tempo of life. The exodus of retirement-age individuals to warmer climates also puts strains on existing health care facilities and results in a significant shift in the types of medical care that are required.

Internal migrants also affect the places they come from. Since many of the outmigrants are young, educated, and skilled workers, they often create a significant drain on human resources in the areas they left.

Population Histograms

Demographers use growth rates, fertility rates, doubling times, and net migration to explain the dynamics of the human population—where it is and where it is going. However, few techniques shed as much light on a subject as a nicely drawn graph.

One particularly helpful graphic tool is the population **histogram,** a bar graph that offers a profile of a population. The population histogram displays the age and sex composition of a population. The area of each horizontal bar on the histogram represents the size of a certain age and gender group.

As illustrated in FIGURE 8-6, three general profiles exist: expansive, constrictive, and stationary. Kenya and Mexico are two countries that are expanding. They both have a large number of young people. If they produce more offspring than their parents did, the population will continue to expand at the base. If, on the other hand, family size decreases, the base of the histogram will begin to constrict. This is what is happening to the Austrian population; its population histogram is constrictive. Sweden presents an entirely different picture. For many years Swedish couples have been having the same number of children as their parents had; as a result, Sweden's population is nearly stationary.

Population histograms can be used to predict population size and age structure—information that's useful for planning schools, hospitals, retirement homes, and so on. For example, government officials looking at an expansive population such as that in Kenya would be wise to consider special efforts to expand educational facilities to accommodate new children. Officials in Austria, with its constrictive population profile, would be well advised to plan for an increasing number of retirees.

A population histogram of the world is expansive. Surprisingly, 30% of the world's people today are under the age of 15 years. If we're going to create a sustainable future, many experts agree that something must be done to curb childbearing in this group.

Exponential Growth

So far you've learned that many populations are large and growing and that the size and growth rate have many implications to people, the economy, and the environment. You've learned how demographers measure growth and predict the future. To comprehend how serious population growth is, you must understand a phenomenon called *exponential growth.* **Exponential growth** occurs when something increases by a fixed percentage every year. For example, a sav-

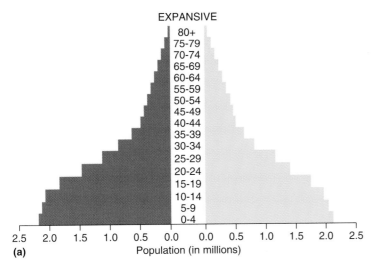

EXPANSIVE

(a) Population (in millions)

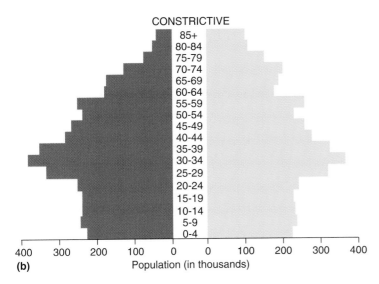

CONSTRICTIVE

(b) Population (in thousands)

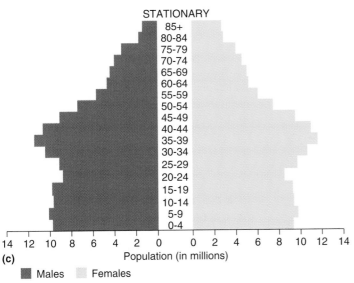

STATIONARY

(c) Population (in millions)

■ Males   Females

FIGURE 8-6 Three population profiles. (a) Expansive populations have a large percentage of young people and are expanding at the base. **(b)** Constrictive populations have a tapering base, resulting from lower total fertility. **(c)** Stationary or nearly stationary populations show no expansion or constriction. This results from couples having only the number of children that will replace them. The world population histogram is expansive.

ings account with 5% compounded interest grows exponentially. The remarkable characteristic of all exponential growth is that early growth in absolute numbers is quite slow. Once growth "rounds the bend," however, the item being measured—whether money in a bank account or number of people in a population—begins to increase more and more rapidly.

To appreciate the true nature of exponential growth, consider an example. Suppose that your parents had opened a $1000 savings account in your name on the day you were born. Also suppose that this account earned 10% interest and that all of the earned interest was applied to the balance, where it also earned interest. At this rate of growth, your account will double every 7 years. Consequently, when you were 7 years old, your account would have been worth $2000 (**FIGURE 8-7**). By the age of 14 years, it would have increased to $4000. By the age of 42 years, you would have $64,000. If you left the money in a little longer, you'd find it growing faster and faster. At the age of 49 years the account would be worth $128,000; at the age of 56 years you'd have a quarter of a million dollars. In 7 more years, you'd have half a million dollars. But if you waited until you were 70 years old, your account would be worth over $1 million.

Looking back over your account records, you'd find that during the first 49 years it grew from $1000 to $128,000, but that during the last 21 years it increased by nearly $900,000! The rate of growth was constant over the entire period, and the account doubled every 7 years; yet it was not until the balance in your account "rounded the bend" of the growth curve that the growth seemed to take off.

> About 30% of the world's population is under the age of 15 years.

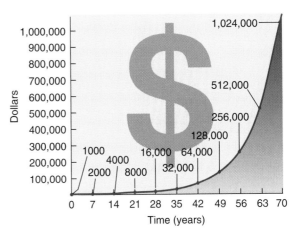

FIGURE 8-7 Exponential growth in savings. This graph plots the growth of a savings account with an initial deposit of $1000 and an interest rate of 10% per year. Notice how slowly the bank account grows until it "rounds the bend" of the exponential growth curve.

How did this rapid upturn occur, even though the rate of growth was the same? The reason for this is simple: Once the base amount becomes extremely large, even small percentage increases result in substantial absolute increases. For instance, the 1.2% increase in a world of 6.5 billion people results in a net increase of about nearly 83 million people a year! At what appears to be a seemingly slow rate of growth (1.2%), world population could double in just 58 years.

> "The problem isn't just that the population is increasing exponentially. It's that population, resource use, pollution, and environmental damage are all increasing at exponential rates."

The problem isn't just that the population is increasing exponentially. It's that population, resource use, pollution, and environmental damage are all increasing at exponential rates. If you accept the argument that the human population has exceeded the Earth's carrying capacity, further rapid growth only pushes us past the point of sustainability.

Systems analysts refer to this phenomenon as an **overshoot**. As you will see later, an overshoot could result in a collapse of the system. Although population growth is slowing, many scientists expect social, economic, and environmental conditions to deteriorate as the population continues to expand.

KEY CONCEPTS

Human populations are growing exponentially. Globally, the human population has "rounded the bend" of the exponential growth curve so that even small percentage increases result in huge numbers of new world residents. It is this exponential growth that is cause for great concern for as our population increases, so does our demand for resources, our waste production, and our environmental damage.

8.3 The Future of World Population: Some Projections and Concerns

Perhaps the questions most often asked of demographers are "What is the future of the world population?" and "How big will it get?" As the previous discussion has shown, many factors determine whether a population will grow or shrink and how quickly. The factors that determine future population growth—such as birth rates, fertility, and death rates—can change dramatically. Several less developed countries such as Thailand, for example, have mounted massive family planning programs that caused population growth to fall dramatically. Consequently, demographers can only guess about the future of the human population. Population projections, they like to remind us, are mathematical calculations based on assumptions about current levels and future trends. Although demographers may not know for sure how large the population will be in the year 2050, they do know that it will very likely be a lot larger than it is today. FIGURE 8-8 presents several possible scenarios for the world and U.S. populations. Let's start with global population. As illustrated in FIGURE 8-8a, estimates for global population growth vary from a low of about 7.3 billion in 2050 to a high of about 10.7 billion. In the low scenario, the population is only 0.9 billion more than today and has begun to decline. In the high scenario, the population is 4.3 billion more than today and is continuing to grow. Which trend is the most likely? Most experts think the middle one—an increase of 2.5 billion is likely. That's 38% increase in population.

FIGURE 8-8b shows three scenarios for U.S. population. The lowest line shows the U.S. population stabilizing at 290 million in 2050 if the TFR is 1.8 and net migration is held to 500,000 per year—a highly unlikely scenario. The middle curve, based on the present immigration policy and current TFRs, shows a continued rise to

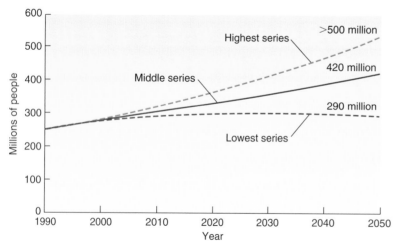

(a) Projected world population growth (billions)

(b) Projected U.S. population growth (millions)

FIGURE 8-8 Projected population growth. (a) Future world population, based on a variety of assumptions. **(b)** The future U.S. population depends on future total fertility rate and annual net migration.

nearly 420 million. The third curve shows a more dramatic rise, to over 500 million, resulting from increased immigration and higher fertility rates among Hispanics and Asians.

At the present time, the middle scenario seems the most likely possibility. This projection calls for an increase of nearly 126 million Americans by 2050—well within the expected lifetime of most readers of this book. One can only imagine how 126 million more people will affect traffic, air pollution, water pollution, wildlife habitat, energy demand, and other areas. One can also see the need to find new ways of meeting human needs—ways that are far less damaging to the environment. The mandate for redesigning basic human systems, described in Chapters 2 and 7, seems all the more pressing given this possibility. Changes in systems alone, though, may not be enough to build a sustainable society. Most experts agree that population growth must be slowed, if not stopped. Some believe that the human population must actually decrease over the long run, a subject addressed in Chapter 9. Without a halt in growth, gains made in efficiency and other areas such as recycling and restoration will very likely be offset by rising population size.

> **KEY CONCEPTS**
>
> Predicting the future size of the world's population is fraught with difficulty. It is likely, though, that the population of the world will increase dramatically before it stabilizes. This increase could bring about massive changes in the environment.

Population Growth in the Less Developed World: Why Should We Worry?

Many people in the more developed nations consider population growth in the less developed nations an issue of little importance to them. What difference does it make to the average North American how fast the less developed countries are growing?

A survey of the impacts of growth in the less developed countries clearly shows that residents of the industrial North do indeed have something to worry about. For example, rapid population growth in Latin America has resulted in a sharp increase in the size of its workforce, but many workers cannot find jobs at home and thus enter the United States illegally in search of work. Although this cultural infusion is seen as a positive impact by some, these individuals compete with U.S. workers for low-level jobs. New residents also strain U.S. schools and hospitals, which are not well equipped to accommodate Spanish-speaking immigrants.

Illegal immigrants also strain the natural environment. An estimated 3,650,000 illegal workers and their families annually enter the United States, each needing water, food, shelter, and other goods and services. This places additional burdens on already stressed land, air, and water.

Population growth abroad creates other impacts on our lives. For instance, 54 million people lived in the Middle East 30 years ago. Today, the population has climbed to over 190 million. The rapid growth of population and high pop-

ulation density fan religious, ethnic, and political turmoil, which often leads to bloodshed (FIGURE 8-9). Regional turmoil threatens access to one of the world's largest oil reserves, on which the United States, Japan, and Europe rely. The 1991 Persian Gulf War is a case in point.

Growing population abroad increases the production of food for export. Although that may be good for farmers, it also places additional stresses on agricultural soils (Chapter 10). It also means additional pesticide use, energy consumption, and demand for irrigation water—all of which impact our environment.

Expanding population abroad also results in the destruction of forests, fields, and wetlands, as well as the many species that live in them. The loss of species has numerous economic implications. Many new drugs, for instance, could come from plants in the rain forests that are now being cut down to make way for more people.

Population growth in the less developed nations also threatens global climate. The loss of rain forests, to accomodate growing populations in tropical nations, for example, is partly responsible for rising carbon dioxide levels that contribute to global warming (Chapter 20). Deforestation of rain forests in central Africa may have a profound effect on rainfall patterns in Europe. And as these countries strive for economic prosperity, they will produce more and more pollution that contributes to global problems. China, home to more than 1.3 billion people, is the second leading producer of carbon dioxide, a major contributor to global climate change (Chapter 20).

> **KEY CONCEPTS**
>
> Population growth in the less developed nations of the world has many social, economic, and environmental impacts—serious impacts that affect them as well as us.

Rapid population growth in the less developed nations contributes to the vicious cycle of poverty gripping many nations, making it difficult for them to pay back loans to more

FIGURE 8-9 Strife in the Middle East. Rapid population growth and high population density in the Middle East contribute to the region's turmoil.

FIGURE 8-10 **Patterns of population growth.**
(a) Smooth transition to a stable population size.
(b) Gradual drop-off caused by population exceeding the carrying capacity. **(c)** Population crash caused by irreparable change to the ecosystem. No doubt all three patterns will be seen in individual countries.

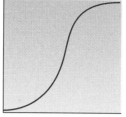

(a) S-shaped or sigmoidal

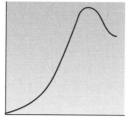

(b) Domed

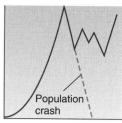

Population crash

(c) Irruptive

developed countries. Today, billions of dollars in international debt remain unpaid, in part because countries cannot cope with their population growth. When countries fail to pay back loans to U.S. banks, consumers suffer. Banking services invariably become more costly. Savings and loan failures in the United States in the early 1990s resulted in part from loans made to less developed nations that were unable to make payments.

Why worry? In our highly interconnected world, population growth in the less developed nations creates problems everywhere. There are few boundaries that this problem does not cross.

A World of Possibilities

The human population will likely grow larger in the near term. Over the long haul, most experts believe that growth cannot continue. Population size will rise to a certain level and then stabilize or, very likely, fall. The curves in FIGURE 8-10 show some of the potential patterns.

In FIGURE 8-10a, the population transitions smoothly and rather gradually to a stable size. Such curves are said to be sigmoidal or S-shaped. This is the most optimistic of all possibilities. World population may follow a domed pattern as shown in FIGURE 8-10b, climbing for a while and then falling to a more sustainable level. A third and less desirable pattern, known as an *irruptive pattern*, is shown in FIGURE 8-10c. In this scenario, population grows until it goes so far beyond the carrying capacity that it crashes. It could either fall precipitously or fall to a lower level and then rise and fall in a dangerous cycle.

No one knows what will happen to world population. The only opinion on which the experts approach agreement is that world population cannot grow indefinitely. Some countries will very likely make a smooth transition to a stable state; others may experience periodic crashes caused by epidemics in crowded urban populations. Still others may fall to low population levels because of starvation, disease, and widespread environment damage.

KEY CONCEPTS

Most experts agree that the human population cannot grow indefinitely. Some countries will very likely make a smooth transition to a stable population size; others may experience periodic crashes that will eliminate large numbers of people. Still others may overshoot the carrying capacity and destroy their ability to support people so drastically that their populations may fall to much lower levels.

H. G. Wells once wrote that "human history becomes more and more a race between education and catastrophe." The current population/resource/pollution bind clearly illustrates this fact. What we decide today will have far-reaching effects that determine the kind of lives our children and theirs will have. What we do and don't do will literally shape the future of the world and determine the future of many species that live on this planet with us.

You cannot escape the responsibility of tomorrow by evading it today.
—Abraham Lincoln

CRITICAL THINKING

Exercise Analysis

Stabilizing population growth in the less developed nations is extremely important. Local conditions in many countries are deplorable and environmental damage is significant. However, critical thinking skills you learned in Chapter 2 suggest the importance of examining the big picture—that is, looking at issues in their entirety. This is one of the facts you will find: Although population growth in less developed nations accounts for 90% of the annual increase, an increase of about 76 million people, residents of the industrial nations actually have a far greater impact on the environment. The United States, for instance, houses about 5% of the world's population but consumes 25% of the world's energy. All told, the more developed nations, which represent 30% of the world's people, consume 75% of the world's resources! Our numbers are few, but our impact is enormous.

In general, people living in the industrial nations of the world consume 20 to 40 times more resources per capita and produce 20 to 40 times as much pollution per capita as their counterparts in the less developed nations. Consequently, the 8.4 million new residents of the industrial world added to the world population have the environmental impact of 170 to 340 million people in the less developed nations. Because of this, many experts argue that measures to slow growth in the industrial nations are just as important, if not more important, than efforts to slow growth in the less developed nations. Thoughtful analysts realize the importance of growth stabilization everywhere—in the rich as well as the poor nations of the world.

One of the critical thinking rules you learned in Chapter 2 is to question conclusions. In this case, and in others, it is necessary to scrutinize the conclusions of the speaker, which sound valid at first, but fall apart when analyzed more carefully. They may not be based on adequate information.

CRITICAL THINKING AND CONCEPTS REVIEW

1. What is the population of the world? What is the population of the United States?
2. Using your critical thinking skills, analyze the following statement: "Population growth is not the cause of environmental problems; technology is."
3. Analyze the following statement using your critical thinking skills and knowledge you have gained in your readings and in lecture presentations: "Population growth in the United States is of little concern to the future of the planet. It is population growth in the developing world that is out of control and needs to be addressed."
4. You are asked to give a lecture on population growth in the United States and why it should be of concern. Write an outline with sufficient detail to make your case. Be sure to include projections of future growth and the impact they could have on wildlife, open space, air pollution, and resource demand.
5. How many years did it take the world to reach a population of one billion people? How quickly did we reach the second, third, and fourth billion?
6. What factors kept world population in check for so many years? Discuss the advances that have unleashed population growth in the last 200 years.
7. Some individuals think that technology, especially energy-efficient machines and appliances, will be sufficient to solve many of the growing environmental problems. Substantial reductions in energy consumption, for instance, could dramatically cut air pollution. Analyze this point of view. Hint: You may want to do some research on the gains that can be made through energy conservation and compare them to rising demand.
8. Define the term *exponential growth rate*.
9. Using the knowledge you have gained in this course, discuss this statement: "The world population is growing at a relatively slow rate—only about 1.3% per year. Growth at this rate is of little concern. If I had a bank account growing at this rate, I'd be frightened at how slowly it was growing."
10. What is a population histogram? Describe the three general profiles. Why are histograms useful?
11. How is the world population growth rate calculated?
12. How is doubling time calculated? What is the difference in doubling times of populations growing at 0.4% per year and 0.2% per year?
13. Define *replacement-level fertility* and *zero population growth*.
14. Discuss the pros and cons of a lenient policy toward legal and illegal immigration in the United States. Do you favor strong immigration quotas?

KEY TERMS

carrying capacity
doubling time
emigration
exponential growth
histogram
immigration
inner city syndrome

internal migration
lag effect
less developed countries
migration
more developed countries
net migration

overpopulation
overshoot
postwar baby boom
replacement-level fertility
total fertility rate
zero population growth

REFERENCES AND FURTHER READING

The References and Further Reading section at the end of this book contains a list of sources for the information discussed in this chapter and recommendations for further reading.

Connect to this book's website:
http://environment.jbpub.com/
The site features eLearning, an online review area that provides quizzes, chapter outlines, and other tools to help you study for your class. You can also follow useful links for in-depth information, research the differing views in the Point/Counterpoints, or keep up on the latest environmental news.

Human population growth touches all aspects of our lives, even a day at the beach.

CHAPTER 9

Stabilizing the Human Population: Strategies for Sustainability

To rebuild our civilization we must first rebuild ourselves according to the pattern laid down by life.

—*Alex Carrel*

Throughout the world, many countries are taking steps to slow their population growth. For example, in Kenya, which has one of the fastest growing populations in the world, large families have been the norm for many years. Because economic conditions are worsening, however, many parents are worried about supporting large families and are taking action. Between 1984 and 2005, the average number of children a Kenyan woman bore dropped from 7.7 to 4.9. Even though only 39% of women currently practice family planning, 75% of those polled would like to.

Even more encouraging news comes from the northern African nation of Morocco. In 1980, women in this country had an average of 7 children. In 1995, that number dropped to 4. By 2005, it had dropped

Exercise

Many advocates of sustainable development put their faith in technological improvements: new energy-efficient homes, new high-mileage vehicles, pollution-free factories, water-efficient fixtures, energy-efficient appliances, and others. They contend that such innovations can allow both our economy and our population to grow and still permit us to reach a sustainable state. They often use the term *sustainable growth*, especially in an economic arena, to convey their belief in this idea. Their argument boils down to this: By conserving and managing resources properly, we can provide access to energy and food indefinitely. In short, all consumption patterns can be made sustainable simply through conservation.

Using your critical thinking skills and the knowledge you have gained so far, analyze this viewpoint. Is it valid? If not, what principles of sustainability does it violate? Provide some hard evidence to back up your arguments.

Hint: To explore this issue, you will very likely want to gather some additional information on the subject. Two sources that may be of assistance are "Using Less and Still Running Out" by Mark W. Nowak, published in the *National Association of Environmental Professionals News* (November/December 1995) and "Can Technology Spare the Earth?" by Jesse Ausubel, published in *American Scientist* (March/April 1996).

to 2.5. In Morocco the use of **contraceptive measures**, devices or techniques that reduce the chance of fertilization, has increased to 55% of all married couples. This progress is even more encouraging when one considers Morocco's social system, in which women are relegated to an inferior role and gain status primarily by bearing children.

9.1 Achieving a Sustainable Human Population: The Challenges

Achieving a sustainable human population involves two basic challenges. The most immediate is finding acceptable means of slowing the growth of the human population, in order to create a stable population size. Many proponents of sustainability, however, argue that slowing growth and reaching a stable number may be insufficient in the long run. These individuals maintain that the human population already exceeds the Earth's carrying capacity. Further additions would result in even more severe environmental deterioration, especially if people sought higher standards of living. The second, long-term challenge, then, may be to find ways for the human population to decrease in size naturally.

> Nearly half of the world's population lives in extreme poverty or on the edge of poverty with inadequate food and shelter and few amenities. About 940 million people lack access to safe water for drinking, and 1.7 billion live in areas without adequate sanitation.

Is the human population already unsustainable? Chapters 1 and 8 outlined many symptoms of the crisis of unsustainability—facts demonstrating that humans are living beyond the carrying capacity of the Earth. Consider a few statistics regarding the state of the world's people to add to the list. Today, about one sixth of the world's people—approximately one billion people—live in extreme poverty. They are inadequately fed and sheltered. Many of them live in cities, where they wander the streets begging for food or stealing what they can. At night, they sleep in alleyways in makeshift cardboard shelters. Another two billion of the world's people live on the edge, with inadequate food and shelter and few amenities. In many cities, four families live in a two-room apartment and share a water tap with 25 other families. Many have no sewage systems and defecate in the street.

All told, nearly half of the world's population is in bad shape. Strenuous efforts to improve the economic condition of the world's poor, in hopes of increasing personal wealth, have failed to keep up with population growth. More and more people fall into poverty each year. Because of these facts and the deteriorating condition of the environment, many observers believe that stopping human population growth now is essential to reduce further suffering, environmental pollution, and resource depletion.

Stopping population growth, reducing population size, and pursuing sustainable economic plans, all discussed in this chapter, can help break the vicious cycle of poverty and environmental destruction. However, the road ahead will be long and difficult. Even if we could miraculously reach replacement-level fertility today, the world population would continue to swell, adding two to four billion people before it stopped growing, in large part because of the lag effect described in the last chapter. Human conditions would very likely deteriorate, as would the condition of the environment.

KEY CONCEPTS

Most people consider the main challenge of achieving a sustainable human population to be finding acceptable means of reducing the population growth rate, to allow the world population to stabilize. Many others, however, argue that in order to live sustainably on the Earth, we must eventually reduce human numbers through humane, socially acceptable means.

9.2 Stabilizing the Human Population: Some Strategies

Over the last few years, experts on population have come to realize that slowing growth and achieving a stable population will require more than education on birth control and easy access to contraceptives and family planning. Most experts agree that we must apply a number of additional remedies; and the closer they are to the roots of the problem, such as poverty, lack of education, the inequality of women, and poor health care, the better.

KEY CONCEPTS

Stabilizing the human population will require a number of measures besides access to contraception and family planning that attack the root causes of rampant population growth.

Economic Development and the Demographic Transition

For many years, economic growth was viewed as one of the most powerful forces for slowing or even stopping population growth. How does economic development slow growth? Economic growth creates jobs and increases personal wealth. People can then afford decent housing, food, and education. Poverty and disease are often reduced or nearly eliminated. As incomes rise, fertility rates decline. This scenario has been repeated dozens of times over the past couple hundred years in what are now the more developed countries of the world. In all of these countries, population growth was brought under control as economic conditions improved. This phenomenon is called the **demographic transition.**

The demographic transition takes place in four stages (**FIGURE 9-1**). The present-day industrial nations have progressed through all four. In Stage 1, birth rates and death rates are high. In this phase, the population is stable because high birth rates and death rates cancel each other out. In Stage 2, improvements in health care and sanitation (brought about by technological development, including advances in medicine and improving economic conditions) cause death rates to begin to

fall. However, birth rates in Stage 2 tend to remain high. The discrepancy between the birth rates and death rates results in a period of rapid population growth. In Stage 3, as the country continues to develop economically, birth rates begin to decrease, and population growth slows. Finally, over time, birth rates and death rates come into balance. Population growth is stopped in Stage 4.

The decrease in birth rates in Stage 3 can be attributed to several factors. Perhaps the most important is the shift in people's attitudes toward children. Preindustrial farmers view children as an asset because they help on the farm and often support their parents in old age. With industrialization and the inevitable migration of families to the city, however, children become an economic liability.

Because of competition for living space, each child means that more money has to be devoted to necessities. If the children do not work, they create an additional financial drain on the family. As a result, smaller families generally prevail.

If economic development has brought about a demographic transition in so many countries, helping them to stabilize or reach a state of extremely slow growth, it should also work for less developed nations in Asia, Africa, and Latin America. Right?

Unfortunately, the evidence suggests that it will not. At least four reasons for the failure of economic development to effect a timely demographic transition become apparent.

First, the economic resources of many of the less developed countries are too limited to build the type of industry needed for a demographic transition. In some instances, resources were heavily exploited during the era of

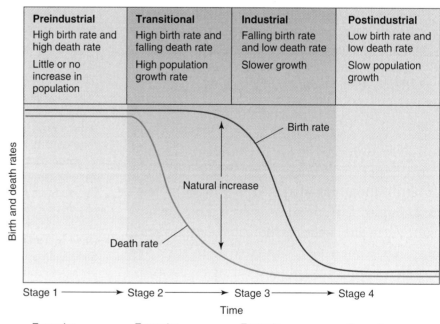

FIGURE 9-1 Demographic transition. This graph of birth rates and death rates in Finland illustrates the demographic transition. As you can see, the transition results in a shift from high birth and death rates to low birth and death rates, a change that is brought about by industrialization and an overall increase in wealth of people in a given population. Industrial nations have all experienced the demographic transition.

Preindustrial	Transitional	Industrial	Postindustrial
High birth rate and high death rate	High birth rate and falling death rate	Falling birth rate and low death rate	Low birth rate and low death rate
Little or no increase in population	High population growth rate	Slower growth	Slow population growth

Stage 1 → Stage 2 → Stage 3 → Stage 4
Time

Example:	Example:	Example:	Example:
Finland 1785–1790	Finland 1825–1830	Finland 1910–1915	Finland 1970–1976
Crude birth rate = 38/1000	Crude birth rate = 38/1000	Crude birth rate = 29/1000	Crude birth rate = 13/1000
Crude death rate = 32/1000	Crude death rate = 24/1000	Crude death rate = 17/1000	Crude death rate = 10/1000
Natural increase = 0.6%	Natural increase = 1.4%	Natural increase = 1.2%	Natural increase = 0.3%

colonialism, which contributed to the wealth and demographic transition of the more developed nations.

Second, the demographic transition in the more developed nations of the world did not take place overnight. For example, it took Finland over 200 years to approach a balance between birth rate and death rate (Figure 9-1). The less developed countries, with rapid doubling times, do not have the luxury of this kind of time.

Third, population growth in many LDCs outstrips economic growth. Recent studies show that a 1% growth in the labor force requires a 3% economic growth rate. Think of the economic growth needed to sustain populations growing at 2.5 to 3% per year. For many LDCs it has been hard enough to keep up with population growth; getting ahead has proven to be impossible.

The fourth reason is that the fossil fuel energy sources that were essential to the demographic transition in the more developed countries are diminishing and becoming ever more costly. Without the rich mines of England or the great oil deposits of Arabia and North America, the less developed nations will probably never experience similar economic growth.

Many of the world's less developed nations are in Stage 2 or early Stage 3 of the demographic transition, with high birth rates and low death rates. Because a demographic transition brought on by industrialization is unlikely, the transition to stable populations must come about in other ways, most likely through family planning and small-scale sustainable economic development (Chapter 26).

> Most of the less developed nations have entered Stage 2 of the demographic transition and are experiencing high birth rates and low death rates—conditions that result in rapid growth of their populations.

KEY CONCEPTS

Economic development can be a powerful force for reducing population growth. Although economic development caused a shift in population growth in the more developed nations, this change took many decades and substantial resources, which the less developed nations of the world do not have.

Family Planning and Population Stabilization

To help usher a transition to Stage 4, most population experts believe that less developed nations must institute aggressive family planning programs. **Family planning** allows couples to determine the number and spacing of offspring. For these countries, family planning programs can accelerate the

> Researchers believe that family planning is responsible for 40 to 50% of the decline in fertility in the less developed nations since the 1960s.

decline in the birth rate that occurs slowly through economic growth and the demographic transition. Some researchers believe that family planning programs are responsible for a 40 to 50% decline in fertility rates in the less developed nations since the 1960s.

Family planning programs offer various means of birth control. **Birth control** includes any device, method, or chemical designed to reduce births in a population. Generally, birth control measures fall into three broad categories: (1) **abstinence**, refraining from intercourse; (2) **contraceptives**, chemicals, devices, or methods that prevent sperm and egg from uniting; and (3) the most controversial, **induced abortion**, the intentional interruption of pregnancy through surgical means or drug treatments.

Family planning programs vary considerably from one nation to another. Most of them offer education on birth control along with clinics that dispense contraceptives or perform abortions (**FIGURE 9-2**). Family planning programs may be either privately run or state sponsored. Planned Parenthood in the United States, for example, is a private, nonprofit organization with clinics in many cities. It offers low-cost medical care, contraceptives, and abortions.

FIGURE 9-2 **Family planning in India.** A doctor lectures to a group of Indian women on various methods of birth control at a family planning clinic in New Delhi. These and other clinics throughout the world are vital to efforts to educate women on family planning.

State-sponsored family planning programs lie on a continuum, from the purely voluntary to the compulsory. **Voluntary programs** as a rule make birth control information and methods available to the public at low cost. There is no promotion on the part of the government. People choose the type of birth control and family size they want.

Family planning programs promoted by governments are called **extended voluntary programs.** In these cases, governmental agencies may hand out information on birth control and sterilization or sponsor posters, newspaper ads, television and radio announcements, and billboards (**FIGURE 9-3**). In Egypt, for instance, a song promoting birth control was played on a government-sponsored commercial. The song was so popular that it became a national hit.

In extended voluntary programs, payments or other incentives are often offered by governments to couples practicing birth control or undergoing sterilization. Transistor radios, for example, were once handed out in rural Egypt to couples who adopted some form of birth control. Special loan programs were used in Thailand (Spotlight on Sustainable Development 8-1). Some government programs offer reminders. In Jakarta, Indonesia, for example, music extolling the virtues of small families plays at intersections. In Bali, church bells ring at 5 P.M., reminding women to take their birth control pills.

Governmental programs often attempt to change people's thinking about family size. Posters in Vietnam, for instance, extol the virtues of a one-child family. Murals and posters in India feature a smiling family of four with the slogan "Two or three children is enough." These efforts strike at one of the roots of the population problem: people's mindsets. Clearly, as Varinda Vittachi, a writer from Sri Lanka, notes, "The world's population problem will be solved in the mind and not in the uterus."

Informational campaigns may also be designed to change stereotypical sex roles. They may try to persuade men that masculinity and self-importance are not related to the number of children their partners have, or they may try to convince men and women alike that childbearing is not the only way for a woman to be a valuable community member.

Forced family planning programs involve strict governmental limitations on family size and punishment for those who exceed quotas. They are extremely rare. They may impose sterilization after a family reaches the allotted size, limit food rations for "excess" children, or tax couples who exceed the allowed number of children. China, which has a forced family planning program, has a one-child policy that is designed to reduce China's population of 1.3 billion to 700 or 800 million. Female workers meet regularly in small groups at work to discuss birth control. Government workers reportedly urge men and women with one child to undergo sterilization. According to unsubstantiated reports, pregnant women who already had one child have been forced to have abortions. The Chinese government also supplements the monthly income of couples who pledge to have only one child. If a couple that has pledged to have one child has a second one, however, the government often requires repayment of the monthly bonus and cuts the monthly salary by

FIGURE 9-3 Family planning poster in China. This poster, extolling the virtues of the one-child family, is a familiar sight in China, a nation of 1.3 billion people. China's government has a strong interest in reducing the nation's birth rate and eventually reducing the overall population size in an effort to temper resource demand and reduce environmental destruction. Special health care workers travel to villages offering advice on birth control. Similar posters are found in India, Thailand, and other Asian countries.

15%. For a discussion of another aspect of population growth control, see Spotlight on Sustainable Development 9-1.

> **KEY CONCEPTS**
>
> Family planning measures permit couples to determine the number and spacing of children to determine family size and are vital to global efforts to reach a sustainable human population.

Small-Scale, Sustainable Economic Development, Jobs for Women, Better Health Care, and Improvements in the Status of Women

Although less developed nations may lack time, funds, and resources needed to promote economic development to reduce fertility, many now understand that small-scale, community-level economic development can raise individual wealth and can be nondetrimental to the environment if fashioned in a sustainable manner. This subject is discussed at length in Chapter 26.

Recognizing that family planning works only if people want fewer children, the **United Nations Fund for Population Activities (UNFPA)**, a leader in promoting family planning throughout the world, invests money in economic programs, especially those that provide employment for women. In Egypt, for instance, the UNFPA recently made substantial investments in clothing factories that employ Egyptian women. The logic behind these and other activities is that working women often delay marriage and childbearing—and thus have fewer children.

In 1994, the United Nations sponsored the International Conference on Population and Development (ICPD),

9-1   Controlling Urban Sprawl: The Other Kind of Population Control

Controlling population means more than slowing the rate of growth. It also entails efforts to halt the spread of human settlements onto farmland, forest, fields, and wildlife habitat, a subject we explore in more detail in Chapter 17. This goal is as vital in the more developed nations of the world as it is in the less developed nations.

A world leader in growth management is the state of Oregon. In the early 1970s, Oregon's legislature passed a law that required all cities and towns to establish urban growth boundaries designed to protect open space, forests, wildlife habitat, and farmland (FIGURE 1). Within the growth zone, city officials are permitted to plan for an annual population growth of 2% over a period of 20 years. At the end of the 20-year period, they can revisit the boundaries and set new ones, if necessary.

Urban boundaries have had many positive effects in Oregon. They have helped reduce the leapfrogging of subdivisions, a leading cause of the loss of open space and agricultural land. Furthermore, the concentrating effect of urban growth boundaries has helped to reduce taxes by reducing the cost of providing basic services such as roads, sewer lines, and schools to new housing developments. Urban sprawl, on the other hand, makes services much more costly to provide.

By concentrating growth in a more confined region, urban growth boundaries can enhance the success of mass transit such as light rail systems. In Portland, the number of passengers who rode the light rail system in its first year was twice what planners had projected. In 1995, the light rail system logged 8 million trips—about 24,000 per day. Because of the popularity, light rail has been greatly expanded. Today, Portland's light rail system provides about 97,000 trips each weekday.

Concentrating growth has been good for business and the environment. Downtown Portland, for example, has seen its share of the regional retail market increase from 7% to nearly 30%, and it has added 30,000 jobs without any increase in the number of cars. The number of days that air quality doesn't meet health standards has dropped from about 100 per year to zero.

Urban growth boundaries are possible in virtually all cities of the world because considerable amounts of land are usually available within cities for growth. In 1989, in fact,

the amount of vacant and underused land in Portland was estimated to be nine times the space needed to accommodate Portland's projected growth for the next 20 years. Portland has contained 95% of growth within urban growth boundaries since the 1970s.

Growth management has been successfully implemented in several other states, including Maine, Florida, Vermont, New Jersey, Rhode Island, Washington, Hawaii, and Georgia. Worldwide, many cities and nations have launched programs to control urban growth, among them Belgium, France, the Netherlands, and Germany. Several Canadian cities, including Toronto and Vancouver, have taken the lead in providing compact urban centers with outlying, high-density suburban areas. Because of efforts to concentrate housing and business, public transportation became a feasible way of linking the areas, decreasing automobile usage, energy consumption, and pollution. Canadians walk, cycle, or use local public transportation within a given area and rely on rapid light rail or express buses to reach other areas. One of the world's best success stories is that of Curitiba, Brazil, featured in Spotlight on Sustainable Development 17-1.

FIGURE 1 View of Portland, Oregon's suburb, showing the developed and protected lands side by side.

which explored many ways to reduce population growth, including steps to improve the status of women and health care. Although controversial, many efforts are underway to raise the social status of women, give them more say in family decisions, and decouple self-worth and child production. Improvements in health care for women are also being widely promoted. Part of the reason behind this strategy is that improvements in health care may help

lower infant death rates and thus will reduce the desire to have more children.

KEY CONCEPTS

Small-scale sustainable economic development, jobs for women, efforts to promote equality, and improvements in health care for women are all vital components of a global strategy to reduce fertility and population growth.

Sustainable Populations in the More Developed Countries

Many people think that curbing population growth applies only to the less developed countries, where 90% of the present growth occurs. Observers point out, however, that the high per capita consumption of the more developed countries puts enormous strains on the Earth's environment. The environmental impact of a nation is often approximated by this equation:

$$\text{Environmental Impact} = \text{Population Size} \times$$
$$\text{Per Capita Consumption} \times \text{Pollution and}$$
$$\text{Resource Use per Unit of Consumption}$$

Population size is a major determinant of impact. If lifestyles are similar in two populations, 1 of 10 million and 1 of 1 million, the larger population will have 10 times the environmental impact of the smaller one. The larger population uses more resources and creates more pollution.

Per capita consumption is also a major determinant in environmental impact. In the more developed countries, especially the United States and Canada, the per capita consumption (the amount of resources each of us uses) far exceeds consumption in the less developed world. As you learned in Chapter 8, a single American or Canadian uses 20 to 40 times more resources than a citizen of the less developed world and has 20 to 40 times the environmental impact.

> Citizens of the more developed countries consume 20 to 40 times as many resources as people of the less developed nations.

To understand the implications of this statistic, let's look at it in another way. Each year, approximately 84 million people join the world population, 75.6 million in the LDCs and 8.4 million in the MDCs. The paltry 8.4 million new residents of the more developed nations, however, will consume as many resources and produce as much waste as 168 to 336 million residents of the less developed nations. Widespread pollution, species extinction, and global resource depletion are three results of the more developed nations' higher standards of living. Because of this, population stabilization is as important (perhaps even more important) in the MDCs as it is in the less developed nations. As the Population, Resources, and Pollution model (Chapter 7) tells us, population stabilization strikes at the root of environmental destruction.

KEY CONCEPTS

Although most of the attention on curbing population growth is focused on the less developed nations—the largest sector of the global population—the more developed nations have an important role because of their high level of per capita consumption and environmental impact.

The third element of the equation, pollution and resource use per unit of consumption, also affects the impact of a given population. The amount of energy required to meet human needs varies. Providing electrical energy from solar panels uses far fewer resources and produces much less pollution than burning coal. To understand this component of the equation, imagine two populations, each with one million people. If population A relies on strategies outlined in this book such as recycling, renewable energy use, mass transit, energy efficiency, and so on, its use of resources and production of waste could be 50 to 90% lower than population B, which is on the traditional, unsustainable path.

KEY CONCEPTS

The impact of a population depends on many factors, most importantly, the size of the population, per capita consumption (how much citizens consume on average), and the resources used and pollution produced to meet needs.

Sharing Knowledge: Another Role to Play By using fewer resources (being frugal), using resources more efficiently, developing technologies and methods that reduce or eliminate pollution, and reducing population growth, the more developed countries can also make enormous progress toward a sustainable future. Frugality and efficiency are often viewed as ways of freeing up resources for the less fortunate people of the world. The logic behind this belief is that if the wealthy nations use less, there would be more for people of the LDCs. This, in turn, would greatly reduce current disparities and growing tensions between the haves and have-nots.

Garrett Hardin, author of numerous books and noted environmental thinker, however, contends that global sharing of resources is not the answer to global sustainability. First, the residents of the less developed nations generally lack the financial resources to purchase surpluses. Hardin argues that the rate of growth and needs of the 5.3 billion residents of the less developed nations far exceed the capacity of the more developed nations to help—food donations could not feed the world's hungry. Even a 10% reduction in the demand for food in the United States, if it could be effected, would not come close to meeting the needs of the world's hungry and poor.

According to Hardin, the more developed countries have an important role to play in bringing about population stabilization. They can assist the LDCs by sharing their knowledge of such things as birth control, sustainable agriculture, health care, and sustainable development. Financial assistance to help achieve sustainable development might also go a long way. Table 9-1 lists some additional suggestions for the more developed countries.

KEY CONCEPTS

The industrial nations can do many things to help build a sustainable future—for example, reducing their own growth, use of resources, and pollution. They can also assist less developed nations through financial aid, especially for family planning and sustainable development, and by sharing information and technical expertise.

Table 9-1

Population Control Strategies for More Developed Countries

Strategy	Rationale	Benefits
Stabilize population growth by restricting immigration and by spending more money and time on sex education and population awareness in public schools.	High use of resources taxes the environment. Immigrants create a serious strain on the economy and create social tension in conditions of high unemployment. Education helps citizens realize the importance of population control.	Limiting resource use leaves more for future generations and less developed countries.
Provide financial assistance to less developed countries for agriculture and appropriate industry. Aid should come from government and private sources.	Economic growth in less developed countries will raise the standard of living and aid in population control.	The rich–poor gap would narrow. A decrease in sociopolitical tension and resource shortages would result.
Provide assistance to population control programs.	Better funded population programs can afford the increased technical assistance and community outreach programs necessary to provide information to the public.	This could result in faster decrease in population growth.
Make trade with less developed countries equitable and freer.	Freer trade will increase per capita income and raise standards of living with little effect on home economy.	A higher standard of living and increased job opportunities could result.
Concentrate research on social, cultural, and psychological aspects of reproduction.	Techniques available today are effective and reliable. What is needed is more motivation for population control, especially among poor countries.	Money will be better spent; research of this nature may help facilitate family planning in less developed countries.

Creating Sustainable Populations in the Less Developed Countries

Most of the LDCs recognize the need for population control. Today, in fact, at least 93% of the world's population lives in countries with some form of population control policy. Support for these programs comes from a variety of sources. One of the major players in the international effort to promote family planning in the less developed nations is the International Planned Parenthood Federation (IPPF), a nonprofit organization established in the 1950s. The IPPF disseminates information on family planning and provides various kinds of assistance to countries.

Another major source of financial support for family planning is the UNFPA. The UNFPA doles out about $140 million a year to less developed nations for family planning programs and is largely responsible for much of the progress in global population control in recent years. While the U.S. helps support this program in theory, the U.S. government withheld its $34 million contribution in 2002–2005 because the program operates in China where forced abortions and sterilizations may occur even though no money from UNFPA supports such activities. Foreign assistance programs sponsored by more developed countries also play a key role in meeting population goals. Germany, Japan, the United States, and other countries all donate huge sums of money for various programs, including sustainable development and family planning. The United States donates money through the U.S. **Agency for International Development (US AID)**. USAID currently spends $441 million on family and reproductive health programs, which includes family planning.

Even international lending agencies help out. For example, since the 1960s, the World Bank, which is largely financed by the United States, has offered financial aid to less developed countries with population control policies; it has recently stepped up its support.

All told, approximately $600 million is donated to the less developed nations from outside sources each year. Although that may sound like a lot of money, it is far from sufficient. In addition, spending on family planning has actually fallen in the past decade or so because of inflation. Today, many countries with active family planning programs provide most of their own financing. In fact, for every $1 of foreign aid received to promote family planning, less developed nations now spend about $4 of their own money. Few of these governments, though, spend more than 1% of their national budget on family planning services—far below what is needed, according to some proponents. Moreover, the director of the U.S. AID's family planning program estimates that the cost of providing family planning services in the less developed countries in the year 2010 will reach $9 to $10 billion. Many believe that the major international donors such as the United States, Japan, and Germany will not be able to meet the projected rise in costs. As a result, many LDCs will have to shoulder

> In the less developed nations, each birth averted by family planning saves between $15 and $200 per year in social services. In the United States, each dollar invested in family planning saves an estimated $3 per year in health and welfare costs.

Table 9-2

Population Control Strategies for Less Developed Countries

Strategy	Rationale	Benefits
Develop effective national plan to ensure better dissemination of information and availability of contraception and other methods of population control. Do not rely on one type of control.	Each country better understands its people and thus can design better programs to spread population control information and devices.	More effective dissemination of information and, probably, a higher rate of success.
Finance education in rural regions, emphasizing population control and benefits of reduced population growth.	Education can help make population control a reality.	Slower population growth, more effective use of contraceptives, and more incentive.
Seek to change cultural taboos against birth control and cultural incentives for large families.	Changes in culture and psychology may be needed to make population control programs effective.	Such changes will help programs succeed.
Develop appropriate industry and agriculture, especially in rural areas to reduce or eliminate the movement of people from the country to the city.	Appropriate agriculture and industry will create jobs and better economic conditions for families. A higher standard of living could translate into better health care and greater survival of young, thus destroying the need for large families.	This will result in a higher standard of living, better health care, and impetus for control of family size.
Seek programs of development that attain a maximum spread of wealth among the people.	Development must not just help a select few because benefits may not trickle down to the needy.	Plans of this nature yield good distribution of income and help the needy rather than a select few.
Integrate population policy with economic, resource, food, and land-use policy to achieve a stable state.	Finite resources require wise allocation and use; success in the long run depends on attempts to achieve a sustainable future.	Longevity and permanence are attainable if policies are integrated and take into account the requirements of a sustainable society.
Seek funding from the United Nations and more developed countries.	More developed countries have a stake in stabilizing world population growth.	More developed countries could provide significant financial support.

an even larger share of the cost—or let their family planning programs languish.

Despite the obstacles facing the less developed countries, family planning makes good sense from environmental and economic standpoints. Depending on the country and the program, each birth averted by family planning yields a savings of between $15 and $200 per year in social services. Estimates in the United States suggest that every dollar invested in family planning saves the LDCs at least $3 in health and welfare costs in the first year. Table 9-2 lists population control strategies for less developed countries.

KEY CONCEPTS

Numerous private and governmental organizations such as the International Planned Parenthood Federation and the United Nations Fund for Population Activities spend millions of dollars a year to support programs to slow the growth of the population in LDCs through family planning and sustainable economic development.

9.3 Overcoming Barriers

Achieving a sustainable human population will not be easy. Many barriers lie in the way. This section describes three major obstacles: (1) psychological and cultural, (2) educational, and (3) religious barriers.

KEY CONCEPTS

Three primary barriers lie in the way of achieving a sustainable human population: psychological and cultural, educational, and religious.

Psychological Barriers

In the less developed countries, children are generally valued by their parents, especially in rural areas, in part because they assist with chores and provide a degree of social security for parents in their old age. Because mortality rates are high in most LDCs, having many children also increases the likelihood that some will survive, but with decreases in death rate as a result of improvements in health care such as vaccination, nutrition, and sanitation, the reason for childbearing is becoming less significant.

Traditional views of family size often change slowly, even in light of changing conditions. In India, for example, larger families are still the norm even in economically developed regions. Although India's government officially promotes families of two or three children, many parents still see the ideal family as two sons and a daughter. The problem is that in trying to reach this goal, couples have, on average, 3.4 children. "One son is no sons," one Indian argues. "To be sterilized is to tempt fate," another asserts. Until people begin to realize that one son is enough and that the son will probably survive, India's population will continue to grow, swelling the second largest population in the world, which reached 1.1 billion in 2005.

Having children is a psychologically enriching activity for many people the world over. Sociologists report that in many LDCs men and women are admired for the number of children they have. As a result, social acceptance and other psychological factors result in the birth of many children that may never have adequate food, clothing, shelter, and education.

Children are psychologically enriching in the MDCs too. However, many citizens of MDCs, even those who love children and put stock in family life, realize that children are an economic drain. In 2004, the U.S. Department of Agriculture estimated that it will cost a family in the United States earning $42,000–70,000 about $184,000 to raise a child, including education through a publicly subsidized college. Families making $70,000 a year or more will spend about $270,000 per child.

> **KEY CONCEPTS**
>
> In the less developed countries of the world, children are often seen as an asset to their parents, and childbearing enhances a woman's social status. Having many children seems desirable because mortality tends to be high, too. These psychological factors contribute to larger family size and rapid growth.
>
> In more developed nations, children are valued but are viewed as a bit of an economic drain, and a woman's status is not so heavily dependent on childbearing.

Educational Barriers

Lack of education is also a barrier to lowering fertility and reducing population growth. As a rule, the lower the level of education, the higher the fertility (**FIGURE 9-4**). Education influences fertility in all nations, rich and poor, in one of several ways. First of all, because education takes considerable time and energy, women often wait till they are finished to get married. The longer they postpone marriage, the fewer children they will have. Second, educated men and women often pursue careers. This also causes couples to postpone marriage and childbearing. In most women, the childbearing years occur between ages 15 and 44; thus, a woman who graduates from college at age 21, marries, but delays having children until she is 30, has decreased her childbearing years by half. She will have half the number of years to produce offspring. Men and women who lack educational opportunities or who choose not to go to school generally marry younger and pursue careers that do not interfere with childbearing. Thus, the period of childbearing is much longer than for couples that pursue higher levels of education. Third, educational programs can also help raise the literacy rate, making it easier for people to understand the benefits of family planning as well as the instructions that accompany contraceptives. Fourth, education improves economic opportunities and, as noted previously in this chapter, shifts attitudes toward larger families. Because of these and other

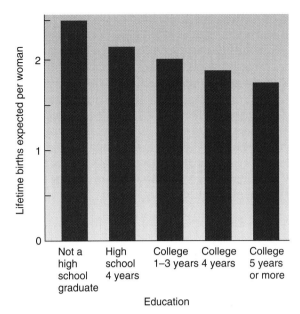

FIGURE 9-4 Total fertility rates of U.S. women, by level of education. In the United States and abroad, education opens up a world of employment opportunities for men and women and lowers fertility rates. As such, education is vital to all efforts to reduce fertility rates, needed to build a sustainable world community.

factors, improvements in public education may be an essential component of a successful population stabilization program.

> **KEY CONCEPTS**
>
> As a general rule, the higher the level of education in a population, the lower its fertility rate. Education and careers that follow decrease the number of childbearing years and open up many options besides childbearing.

Religious Barriers

Religion may also be a powerful force in reproduction. Buddhism promotes family planning in Thailand. The Catholic religion, on the other hand, forbids all "unnatural" methods of birth control, such as the pill, the condom, the diaphragm, and abortion. Theoretically, the Catholic Church guides the sexual practices of approximately 1 billion people. Recent surveys show, however, that the use of contraceptives by Catholic women, especially in Western nations, is nearly as high as that among non-Catholics. In Latin America, many priests speak out against the Vatican's official policy.

Birth control is a generally undiscussed subject among other religions. However, in many of the Eastern and Middle Eastern religions, which compete with one another for followers, birth control is not advocated. The total fertility rate in Iraq is 5.1.

If family planning programs are going to work, they must address these three obstacles. Consider some examples of ways to surmount the psychological and cultural obstacles. In many Middle Eastern countries where family planning is culturally unacceptable, birth control can be promoted as a means of spacing births to improve maternal and child health. This new direction appeals to more people and broadens the base of support for family planning programs. Thus, a couple who may not relate to lofty goals of slowing the growth rate of their country can surely understand the benefits family planning brings to them in improved economic conditions and maternal and infant health.

Attitudes toward family size and childbearing can be changed, especially from within the culture by women themselves. Especially important are ways to improve the status of women and to give women a say in the economic decisions of a village, described earlier.

9.4 Ethics and Population Stabilization

To many people, reproduction is a basic human right. To deny the right to reproduce is to deny one of the most fundamental and important of all human activities. To other people, population control measures violate deep religious beliefs or cultural norms. For various ethnic groups, family planning may even have overtones of genocide. To help you sort out the debate and make up your own mind about the ethics of population stabilization, this section addresses an important ethical question: Is reproduction a personal right? Arguments from both sides of the issue are given to help you formulate your own view. We begin with advocates of family planning.

Some proponents argue that the right to reproduce at will should be curtailed when the rights of the individual interfere with the welfare of society—that is, the **collective rights**. More people means more suffering for everyone and more environmental destruction. Others who argue in favor of population control focus on the welfare of the unborn. Ecologist Paul Ehrlich, for example, notes that we "must take the side of the hungry billions of living human beings today and tomorrow, not the side of potential human beings. . . . If those potential human beings are born, they will at best lead miserable lives and die young." He argues that we cannot let humanity be destroyed by a doctrine of individual freedoms conceived in isolation from the biological facts of life.

Along these lines, ecologist Garrett Hardin argues that the integrity of the biosphere and the Earth's carrying capacity should be the guiding principle in the debate. We must recognize that the condition of the planet determines the well-being of all living things, including humans. Hardin suggests that we have a moral obligation to future generations to protect the biosphere, the life support system of the planet. Human population control is viewed as a way of honoring this obligation.

Proponents of this viewpoint do not ask whether it is ethical to control population growth, but rather whether it is ethical to let the human population continue to grow unabated.

They note that if we permit population to grow, it will very likely cause a massive deterioration of the environment and rob future generations of the opportunities many of us now enjoy. It will also cause the extinction of many species.

Opponents say that individual rights are paramount. To limit family size takes away personal freedom. No one has a right to tinker with such a basic freedom. Moreover, to counter arguments about suffering and deterioration of conditions, they say that we need to find ways to accommodate new people—improve economies, health care, food supplies, and other areas to make conditions better. We can, they say, have more people and not ruin the planet. We need to grow smarter. For a debate on the subject see the Point/Counterpoint in Chapter 8.

Although debate over population stabilization continues today, most countries favor the sustainable viewpoint—that population stabilization is necessary and that the collective rights of present and future generations take precedence over individual rights. While few countries put strict limitations on family size, most encourage smaller families, leaving the decision to the parents. Acting out of this belief, many nations have mounted programs to promote family planning and make contraceptives available to those who want them—even those who cannot afford them. Many of the countries that promote family planning are striving to slow the rate of growth, reach a stable state, or, in some instances, decrease their population size.

> The world population growth rate has fallen from 2% per year in the early 1960s to 1.3% in 2004.

9.5 Status Report: Progress and Setbacks

One of the most encouraging trends in environmental science these days is the decline in population growth rates. As shown in **FIGURE 9-5**, world population growth rates have declined impressively since the early 1960s, falling from 2% (doubling time 35 years) to 1.2% per year in 2005. The decline in growth rates has occurred in many countries, both rich and poor. Many less developed countries such as China, Korea, Taiwan, and Costa Rica, for example, have contributed significantly to this decrease. China's growth rate fell from 2.5% in the 1960s to 1% in 1999. By 2005, it had fallen to 0.6%. Growth rates in the more developed countries have also fallen in the past two decades (Figure 9-5). Today, nearly 80 MDCs have nearly stationary or shrinking populations.

Two of the most encouraging steps were the International Conference on Population and Development held in

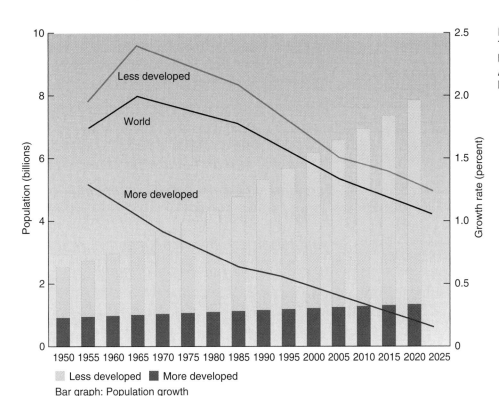

FIGURE 9-5 **Encouraging trends.** The growth rate of the human population has been declining steadily. Although it has declined, the population continues to increase.

Less developed

World

More developed

1950 1955 1960 1965 1970 1975 1980 1985 1990 1995 2000 2005 2010 2015 2020 2025

▢ Less developed ■ More developed
Bar graph: Population growth
Line graph: Growth rate

Cairo, Egypt, in 1994, and the Fourth World Conference on Women held in Beijing, China, in 1995. The Cairo Conference set forth a global strategy to improve the status of women and reduce fertility. The World Conference on Women addressed many of the cultural and psychological factors that stand in the way of progress. Its plan of action calls for measures to enhance women's rights and responsibilities in all levels of society and to improve their social standing. It also calls for measures to increase educational opportunities for women.

Despite this good news, there are some areas of grave concern. One of them is that population growth has resulted in a massive buildup of people in urban areas, who become susceptible to the spread of infectious disease. This is especially troublesome in less developed nations. Poor air quality in LDCs may lower immune system function and unsanitary conditions increase the risk of exposure to new and deadly disease-causing organisms. Increased mobility, largely via jet airliners, may also serve as a means of spreading disease internationally.

Another discouraging trend is the increase in death rates. Researchers estimate that approximately one third of the decline in the annual global population growth rate is the result of rising death rates, largely from AIDS, acquired immune deficiency syndrome. AIDS is a disease caused by the sexual transmission of a virus (HIV) that attacks the immune system. Although HIV infection rates are under 1% in the more developed nations, the incidence in populations in sub-Saharan Africa are astounding, ranging from 22% in South Africa, a country of 47 million people, to 39% of the adult population in Swaziland. Further declines in the growth rate are projected by the United Nations because of AIDS, but

also because of declining groundwater and food supplies, which are setting the stage for a potentially massive increase in death rates.

Another discouraging bit of news is that today, nearly one third of the world's population is under the age of 15 and is soon to enter its reproductive years. Worldwide, contraceptive use is substantial, but demand still greatly exceeds availability (**FIGURE 9-6**). In addition, population growth in Africa, Latin America, and Asia is still fairly rapid. Even

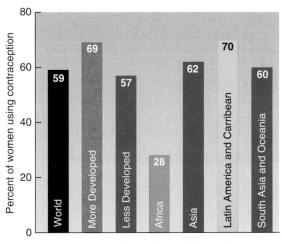

FIGURE 9-6 **Contraceptive use.** This graph shows the percentage of women using contraception in the world, in developed versus developing countries, and on various continents.

though population growth has declined, a 1.2% growth rate still results in the addition of 83 million people every year. Most of the new growth in the coming decades will occur in the less developed nations, which are ill equipped to feed and house the new residents. Widespread hunger, poverty, and environmental decay are likely to occur.

Although the growth of the human population has slowed, about a third of the decline is the result of an increase in the death rate in large part from AIDS.

Human civilization need not end in an explosion of people, but many experts agree that we can improve only if we hold our numbers within the carrying capacity of the Earth. This means stopping the growth of the human population and very likely achieving a smaller population size. To do so will require a multitude of approaches from family planning to sustainable economic development to improvements in education, the status of women, and health care. By attacking the problem at all fronts, we have a better chance of achieving success and creating a sustainable future.

I cannot believe that the principal objective of humanity is to establish experimentally how many human beings the planet can just barely sustain. But I can imagine a remarkable world in which a limited population can live in abundance, free to explore the full extent of man's imagination and spirit.

—*Philip Handler*

CRITICAL THINKING

Exercise Analysis

As noted in previous chapters, population growth is one of the driving forces of environmental deterioration. In other words, it is a root cause. Even with new ways of doing things, more people generally means more pollution, more resource use, and more environmental disturbance. No matter how carefully we conduct our affairs or how efficiently we use resources, human society will always have an impact on the environment. The more of us there are, the greater the impact. However, the impact depends on per capita consumption, resource demand per unit of production, and pollution output per unit of production.

Here's where the issue gets confusing to many people. Being more efficient is absolutely essential to creating a sustainable society. So are efforts to reduce our output of pollution. Technological innovation can have a major beneficial impact on our future. These savings, however, can be easily offset by population growth. Energy efficiency measures can substantially cut energy demand, but those gains can be offset by growth. Recall from the last chapter that the U.S. population may increase by nearly 50% by the year 2050. So, if we cut energy demand and the output of pollution by 50%, population growth would result in no net gain.

This kind of reasoning helps us understand the need for sustainable technologies coupled with population stabilization—and perhaps, as some advocates suggest, efforts to eventually reduce the size of the human population by means that are acceptable and, of course, humane.

Is the notion that all consumption patterns can be made sustainable through conservation and other means valid? Probably not. What principles of sustainability does it ignore? It ignores the fact that the Earth is a finite source of resources. Limits are very real and must be reckoned with, if we are to create a sustainable existence.

CRITICAL THINKING AND CONCEPTS REVIEW

1. Critically evaluate this statement: "The world cannot support the people it currently has at a decent standard of living. Thus, efforts should focus on helping the less developed nations industrialize. Population growth will decline as a result, so family planning programs are not necessary."

2. Define *family planning*. Make a list of the three major types of family planning programs. Give examples of each. What are the weaknesses and strengths of each?

3. The United Nations appoints you as head of family planning programs. Your first assignment is to devise a family planning program for a less developed country with rapid population growth, high illiteracy, widespread poverty, and a predominantly rural population. Outline your program in detail, justifying each major feature. What problems might you expect to encounter?

4. Describe ways in which more developed countries might aid less developed countries in stabilizing population growth.

5. Discuss attitudes about the value of children in less developed countries. How do these views differ from those in more developed countries?

6. If you are considering having children, what factors influence the desirable size of your future family?

7. Discuss reasons why the total fertility rate tends to be lower among more educated women.

8. Discuss general ways to ensure a high rate of success in family planning programs.

9. Do we have the right to have as many children as we want? Should that right be curtailed? Explain your answer.

10. Make a list of the pros and cons of population growth. In other words, what are the benefits, and what are the adverse impacts? You may want to look at U.S. population growth and global population growth separately.

KEY TERMS

abstinence
birth control
collective rights
contraceptive measures
contraceptives
demographic transition

extended voluntary programs
family planning
forced family planning programs
United Nations Fund for Population
  Activities (UNFPA)

U.S. Agency for International
  Development (U.S. AID)
voluntary programs

REFERENCES AND FURTHER READING

The References and Further Reading section at the end of this book contains a list of sources for the information discussed in this chapter and recommendations for further reading.

Connect to this book's website:
http://environment.jbpub.com/
The site features eLearning, an online review area that provides quizzes, chapter outlines, and other tools to help you study for your class. You can also follow useful links for in-depth information, research the differing views in the Point/Counterpoints, or keep up on the latest environmental news.

Strip cropping in Wisconsin improves yield, reduces pest damage, and creates a more sustainable system of agriculture.

CHAPTER 10 | Creating a Sustainable System of Agriculture to Feed the World's People

Civilization itself rests upon the soil.

—*Thomas Jefferson*

Modern farmers typically plant huge expanses of corn and soybeans to produce food in quantities sufficient to meet human needs. Unfortunately, large fields containing a single crop tend to be highly vulnerable to insects, plant diseases, and hail. An outbreak of insects can cause enormous damage. In response to this problem, some progressive Nebraskan farmers are breaking with the traditions of modern agriculture. Instead of planting their fields in one crop, they are planting two crops (such as corn and soybeans) on the same land but in alternating strips. This simple technique called *intercropping* increases the productivity of soybeans and corn dramatically. Why? In one study, researchers found that in this planting arrangement the corn protects the soybeans from the drying effects of the wind. This, in turn, increased soybean output by 11%. They also

Exercise

Some agricultural interest groups object to the phrase *sustainable agriculture*. They argue that it implies that current agricultural practices are not sustainable. In support of their case, they point out that agricultural productivity (the output of food per hectare or acre) has been rising steadily in many countries—such as Canada, Australia, and the United States—thanks to widespread use of insecticides, herbicides, fertilizer, irrigation, and other modern practices. They also point out that farmers in many modern agricultural countries are feeding more and more people every year. Is there anything wrong with this line of reasoning? What critical thinking rules are essential to analyze this issue?

found that the stands of corn are less dense than in a field planted from fencerow to fencerow. This results in better sunlight penetration, which increases corn production by a remarkable 150%. Intercropping not only increases production, it reduces the need for chemical pesticides, for reasons explained later. Reductions in pesticide use save farmers considerable amounts of money, reduce pesticide exposure to farmers and their families, and reduce environmental problems.

This is one example of many efforts aimed at building a more sustainable system of agriculture. **Sustainable agriculture** is a system that produces high-quality foods while maintaining or improving the soil and protecting the environment—the air, water, soil, and wealth of wild species. It is a system that can endure, providing benefits for centuries. This chapter tackles the subject of food and agriculture, beginning with a look at hunger and malnutrition.

10.1 Hunger, Malnutrition, Food Supplies, and the Environment

In Chapter 8, you learned that hunger and starvation are two consequences of overpopulation. As you may already know, hunger and starvation are huge problems. According to estimates from the United Nations, 800 million people living in less developed countries are chronically undernourished—that is, they fail to get enough protein, calories, or both. Thus, about 12% of the world's people are chronically malnourished. Hunger is greatest in sub-Saharan Africa and southern Asia, notably India and Bangladesh. In sub-Saharan Africa, for instance, more than 40% of the population is chronically undernourished, according to a 2005 report by the UN Task Force on Hunger. Many of those affected by this problem are children.

Scientists recognize two types of undernutrition. The first, **kwashiorkor** (KWASH-ee-OAR-core), results from a lack of protein. The second, **marasmus** (meh-RAZ-mess), results from an insufficient intake of protein and calories (food that provides energy). In reality, kwashiorkor and marasmus are two extremes of protein–calorie deficiency and most individuals who are malnourished exhibit symptoms of both. People may also suffer from specific deficiencies such as a lack of vitamin A, which may cause serious damage to the eyes, leading to blindness.

> An estimated 12 million people die each year, mostly in the less developed nations, as a result of malnutrition and diseases worsened by it.

FIGURE 10-1 shows children suffering from kwashiorkor, extreme protein deficiency. As illustrated, the legs and arms of victims of this disease are thin, and their abdomens are swollen with fluids. Victims are weak and passive. Kwashiorkor is most common in children 1 to 3 years of age and generally begins after they are weaned (thus losing the protein-rich milk of their mothers) and have been switched to a low-protein, starchy adult diet.

FIGURE 10-1 Kwashiorkor. This protein deficiency leads to swelling of the abdomen. The loss of muscle protein results in thin arms and legs. Children are physically and mentally stunted, apathetic, and anemic.

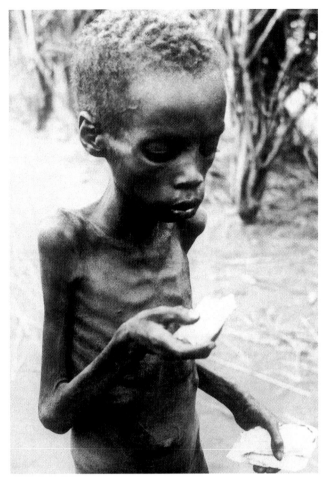

FIGURE 10-2 Marasmus. Victims of marasmus (protein and calorie deficiency) are thin but alert and active. Survivors of malnutrition, however, may be left with stunted bodies and minds.

Victims of marasmus are thin and wasted (**FIGURE 10-2**). Their ribs stick out through wrinkled skin. They often suck on their hands and clothes to appease a gnawing hunger. Unlike victims of kwashiorkor, however, children suffering from marasmus are alert and active. Marasmus often occurs in infants separated from their mothers during breast-feeding as a result of maternal death, a failure of milk production (lactation), or the improper use of milk substitutes. Marasmus often occurs in infants as a result of maternal death or a failure of milk production (lactation) if the infant were breast-feeding, or the improper use of milk substitutes. For example, the distribution of baby formula by food manufacturers to LDCs to introduce mothers to this product can have disastrous results in areas where safe drinking water is unavailable or when poverty-stricken mothers dilute the formula to make it go farther, thus, reducing their infants' intake of protein and calories.

Public outcry led to rational measures to curb such activities. However, according to the International Baby Food Action Network, "Where rational measures are inadequate or have expired . . . Nestle and other companies are quick to return to aggressive and competitive marketing tactics, including free formula supplies to hospitals, samples to mothers and media advertising."

Sadly, for every clinically diagnosed case of marasmus and kwashiorkor in LDCs, there are hundreds of children with mild to moderate forms of undernutrition, a condition much more difficult to detect. Undernourishment is not restricted to the LDCs of the world, however. In the United States, for example, an estimated 10 to 15% of the population is undernourished. Hunger is most prevalent in Mississippi, Arkansas, Alabama, New Mexico, and the District of Columbia.

Although many people suffer from undernourishment—insufficient protein and caloric intake—experts note that 2 billion of the world's people receive inadequate vitamins and minerals. That is, although they may consume a sufficient number of calories and enough protein, their diets may lack essential vitamins. Vitamin A deficiency, for instance, is quite prevalent in LDCs where it causes blindness.

Undernutrition causes considerable human suffering. Imagine the pain of going to bed hungry night after night. Malnutrition can lead to death. For example, mild cases make people more susceptible to **infectious diseases**, ailments caused by bacteria and viruses that can be spread from one person to another. A person weakened by a lack of food is more likely to die from an ordinarily nonfatal disease than a person who is well nourished. In fact, undernutrition so weakens the immune system that normally nonthreatening diseases such as measles and diarrhea can be fatal among children. Making matters worse, crowding in urban centers facilitates the spread of disease. According to the United Nations, an estimated 12 million people die prematurely each year from undernutrition, malnutrition, and nonfatal diseases worsened by poor nutrition. Many of the victims are children—about 19,000 per day!

KEY CONCEPTS

A large segment of the world's people (most of whom live in Asia, Africa, and Latin America) either do not get enough to eat or fail to get all of the nutrients and vitamins they need—or both. Nutritional deficiencies make people more susceptible to infectious disease and, if they are severe enough, can cause death.

Hunger, Poverty, and Environmental Decay

Many students ask why they should study malnutrition in a course on environmental issues. One reason is that food problems are a direct result of population growth, climate change, and loss of farmland caused by growing human population. But there's another reason. Nutritional deficiencies early in life often lead to mental and physical retardation. The more severe the deficiency, the more severe the impairment. Mental retardation occurs because 80% of the brain's growth occurs before the age of two. Malnourished children who survive to adulthood remain mentally impaired. Typically plagued by malnutrition their whole lives, they are often prone to infectious diseases and provide little hope for improving their own or their nation's prospects. Hunger and malnutrition, therefore, may contribute to growing poverty, rising population, and worsening environmental conditions.

Declining Food Supplies

Malnutrition abounds. But what are the prospects for the future? Will we be able to feed the world's people and accommodate the growing population, which expands by about 230,000 people per day?

For many years, hope has been pinned on our ability to grow more food. The use of modern agricultural methods and increases in farmland caused grain production per capita to rise approximately 30% from 1950 to 1970. This resulted in a substantial improvement in the diet of many of the world's people. From 1971 to 1984, however, world grain production barely kept pace with population growth. Between 1984 and 2003, grain production per capita fell nearly 13%.

The decline in food production per capita over the past decade results from numerous factors, among them the warming global climate, population growth, soil erosion, and soil deterioration (from causes described in the next section). Many experts think that food production per capita will continue to decline throughout the next decade as a result of these problems. If global warming, population growth, soil erosion, and other problems worsen, then hunger, poverty, and environmental destruction will become even more widespread.

Because of declining per capita food production, many countries have lost the ability to feed their people and have become dependent on imports from the more developed countries such as Canada, Australia, and the United States. In 1950, grain imports by the LDCs amounted to only a few million tons a year; in the 1980s, they rose to 500 million tons. In 1995, they were over 1.2 billion metric tons per year. By 2004, they had climbed to nearly 1.6 billion tons per year.

Some experts believe that the safety net provided by major food-producing nations is in danger. One of the most serious threats is global warming (Chapter 20). In 1988, after record high temperatures throughout the Midwest, U.S. grain production fell by 35%, plummeting to 190 million metric tons—barely enough to satisfy American needs, let alone foreign demand. Thanks to previous surpluses, domestic and foreign demands were satisfied. In the future, droughts and other factors described in this chapter could slow exports to a trickle, threatening the food supplies of the LDCs.

Many experts believe that unless decisive steps are taken—and soon—millions of people in the less developed nations could perish from hunger and diseases. The famines in Africa and Southeast Asia, in which hundreds of thousands of people died in recent years, may be a portent of what is to come.

The Challenge Facing World Agriculture: Feeding People/Protecting the Planet

Three interrelated challenges face the world today: First, we must find ways to feed the malnourished people alive today (an immediate challenge). Second, we must find ways to meet future needs for food (a long-term challenge). Third, we have to find ways to produce food to meet present and future needs while protecting the soil and water upon which agriculture depends (both an immediate and a long-term challenge). In other words, we have to ensure that current agricultural practices are sustainable.

10.2 Understanding Soils

Malnutrition is only one of a handful of serious food and agricultural problems facing the world today. The next section outlines many problems, such as soil erosion, that are responsible for the deteriorating condition of the world's cropland and the decline in food production. Before we examine this set of challenges and discuss ways to build a sustainable agriculture, however, a brief study of soils is in order. This information will provide you with some of the scientific knowledge you need to assess various solutions and their sustainability.

What Is Soil?

High-quality soils promote plant growth, both in natural ecosystems and on rangeland and farmland. Because of this, soils are vital to our long-term health and our economic well-being.

Soil is a complex mixture of inorganic and organic materials with variable amounts of air and water. The inorganic material includes clay, silt, sand, gravel, and rocks. The organic component consists of living and nonliving plant and animal materials. Living organisms include insects, earthworms, and microorganisms. The nonliving matter includes plant and animal waste and residues (remains of dead bodies) in various stages of decomposition.

Soils are described according to six general features: texture, structure, acidity, gas content, water content, and biotic composition. These components and characteristics combine to form many different soil types throughout the world. A detailed discussion of each soil type is well beyond the scope of this book. As you shall soon see, some soils are better suited for agriculture than others.

How Is Soil Formed?

Soil formation is a complex and slow process, even under the best of conditions. It results from an interaction between the **parent material**, the underlying substrate from which soil is formed, and the organisms. The time it takes soil to develop depends partly on the type of parent material. To form 2.5 centimeters (1 inch) of topsoil from hard rock may take 200 to 1200 years, depending on the climate. Softer parent materials such as shale, volcanic ash, sandstone, sand dunes, and gravel beds are converted to soil at a faster rate—in 20 years or so under very favorable conditions. Because of the time it takes to replenish soil and because soils are so important to society, they should be protected and carefully managed.

As discussed later in this chapter, the cornerstone of sustainable agriculture is prevention—measures that preserve and protect topsoil so that it can remain productive indefinitely. The food we grow on soils is much like interest from a bank account. We can draw on it forever, as long as we do not deplete the soil itself. Soil is, therefore, one form of **natural capital** on which society depends.

Four factors are responsible for the type of soil that forms, the thickness of the soil layer, and the rate of soil formation. They are climate, parent material, biological organisms, and topography.

The **climate** is the average weather conditions, notably temperature and precipitation. The parent material, the underlying rock or sand or gravel from which the mineral matter is derived, is acted on by climate and biological organisms and converted into soil. For example, heating and cooling (both elements of climate) cause barren rock to split and fragment, especially in the desert biome, where daily temperatures vary widely. Water entering cracks in rocks expands when it freezes, causing the rock to fragment further. The roots of trees and large plants reach into small cracks and fracture the rock. Over time, rock fragments produced by these processes are slowly pulverized by streams or landslides, by hooves of animals, or by wind and rain.

Soil formation is facilitated by numerous organisms. Chapter 6 described how lichens erode the rock surface by secreting carbonic acid. Lichens also capture dust, seeds, excrement, and dead plant matter, which help form soil. The roots of plants also help build soil by serving as nutrient pumps, drawing up inorganic nutrients from deeper soil layers. These chemicals are first used to make leaves and branches, which can fall and decay; thus, they become part of the uppermost layer of soil, the **topsoil.**

Grazing animals drop excrement on the ground, adding to the soil's organic matter. The white rhinoceros, for example, produces about 27 metric tons of manure each year, which is deposited in its habitat. A variety of insects and other creatures, such as earthworms, also participate in soil formation.

Topography, the final soil-forming force, is the shape or contour of the land surface. It determines how water moves and how quickly soil erodes. Steeply sloping land surfaces, for instance, are more susceptible to soil erosion. As a result, thin soils tend to form on them. Relatively flat terrain, on the other hand, suffers little from erosion. Soils tend to be thicker in such regions. Valley floors benefit from their flatness and their proximity to steeper terrain whose soil washes away and is deposited on the floor. Many a valley, even in the arid West, has been converted to valuable cropland because of the rich soil that has built up in them over many years. Today, however, valley floors are also desirable land for development, and farmers and ranchers are squeezed out as the population expands and new subdivisions pop up.

KEY CONCEPTS

Soil formation is a complex process involving an interaction among climate; the parent material, which contributes the mineral components of soil; biological organisms; and topography. Because soil is so valuable and because it takes so long to form, we should take care to protect and manage soils carefully.

The Soil Profile

Most of us have seen a road cut or excavation for a building and noticed that soil is composed of layers. The layers, called *horizons,* differ in color and composition. Not all horizons are present in all soils; in some, the layering may be missing altogether.

Soil scientists recognize five major horizons (**FIGURE 10-3**). The uppermost region of the soil is the **O horizon**, or **litter layer.** This relatively thin layer of organic waste from animals and detritus is the zone of decomposition and is characterized by a dark, rich color. Plowing mixes it in with the next layer.

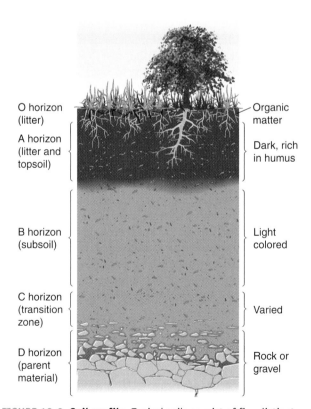

FIGURE 10-3 Soil profile. Typical soils consist of five distinct layers. The topsoil is the most important to agriculture because it contains organic matter and nutrients essential for plant growth.

The **A horizon**, or **topsoil**, is the next layer. It varies in thickness from 2.5 centimeters (1 inch) in some regions to 60 centimeters (2 feet) in the rich farmland of Iowa. Topsoil is generally rich in inorganic and organic materials and is important because it supports crops. It is darker and looser than the deeper layers. The organic matter of topsoil, called **humus**, acts like a sponge, holding moisture. Grasslands have the deepest A horizons. The thinnest A horizons are found in deserts, coniferous forests, tundra, and tropical rain forests. As you shall soon see, most of the land that is viewed as potentially arable lies in the tropics, where the A horizon is practically nonexistent.

The **B horizon**, or **subsoil**, is also known as the *zone of accumulation* because it collects minerals and nutrients leached from above. This layer is lightly colored and much denser than the topsoil because it lacks organic matter. The next layer, the **C horizon**, is a transition zone between the parent material below and the soil layers above. The **D horizon** is the parent material from which soils are derived.

The soil profile is determined by the climate (especially rainfall and temperature), parent material, biology, and topography—that is, the same soil-building forces discussed in the previous section. Soil profiles are histories of the interaction among these factors.

Soil scientists have identified 11 major soil types or orders. These are defined primarily by their *diagnostic horizons*—that is, as having a specific **soil profile**. Soil profiles tell soil scientists whether land is best for agriculture, wildlife habitat, forestry, pasture, rangeland, or recreation. They also tell us how suitable soil might be for various other uses such as home building, landfills, and highway construction.

> **KEY CONCEPTS**
>
> Soils are typically arranged in layers. For agriculture, the most important are the upper two layers: the 0 horizon, which accumulates organic waste from plants and animals, and the A horizon, the topsoil.

10.3 Barriers to a Sustainable Agricultural System

With this brief introduction to soils, we now turn our attention to problems facing world agriculture. We examine the challenges that lie ahead as we attempt to forge a sustainable system of agriculture, beginning with the decline in food supplies.

As already mentioned, as the world population continues to expand, per capita food supplies—the amount of food available per person—are on the decline. This problem is especially acute in the less developed nations. In the more developed nations, in contrast, food surpluses are common and obesity is on the rise. Part of the reason for the unsustainable decline in food production is that agricultural soils are eroding or deteriorating in quality. In addition, many farms are being lost to development.

Soil Erosion

Thomas Jefferson wrote that "civilization itself rests upon the soil." The first towns, early empires, and powerful nations can all trace their origins to the deliberate use of the soil for agriculture (Chapter 7). Agricultural expert R. Neil Sampson wrote that in most places on Earth, "We stand only six inches from desolation, for that is the thickness of the topsoil layer upon which the entire life of the planet depends."

In many places, soil is being washed or blown away—that is, eroded by wind and water. **Erosion** occurs when rock and soil particles are detached by wind or water, transported away, and deposited in another location, often in lakes and streams. **Soil erosion**, the loss of soil from land, is one of the most critical problems facing agriculture today. It is a problem in the MDCs as well as the LDCs.

Soil erosion is classified as either natural or accelerated. As the name implies, **natural erosion** occurs in areas in the absence of human intervention. It generally occurs at such a slow rate that new soil is generated fast enough to replace what is lost. In other words, natural erosion generally occurs at a sustainable rate. In contrast, **accelerated erosion** largely results from human activities such as overgrazing, and it occurs at a rate that outstrips the formation of new soil (**FIGURE 10-4**). Accelerated erosion is dangerous not just because it removes productive topsoil, but because it decreases soil fertility that may cause declines in productivity. Studies on corn and wheat indicate that each inch of topsoil lost to erosion results in a 6% decline in productivity. Severe soil erosion may also result in the formation of deep gullies that make farmland unworkable.

Soil erosion is such a pressing problem because soil washes away quickly but forms very slowly. Soil erosion also has a number of serious environmental impacts. For in-

FIGURE 10-4 Soil erosion on rangeland. All soil erosion above replacement level bodes poorly for farmers and the world's people. This rangeland is suffering from extreme erosion, which not only robs the land of topsoil but also greatly reduces its productive capacity.

stance, pesticides attach to soil particles. Transported to nearby waterways, pesticide-laden particles may be ingested by fish and other aquatic organisms. They may then be passed to birds and human consumers in the food chain. Sediment deposited in waterways fills streams and rivers and reduces their capacity to hold water. This, in turn, increases flooding. Furthermore, sediment destroys breeding grounds of fish and other wildlife and increases the need for dredging harbors and rivers. The World Resources Institute estimates that the off-site damage from soil erosion in the United States is over $10 billion a year.

Since agriculture began in the United States, one third of the nation's topsoil has been lost to erosion, according to the Soil Conservation Service. Unfortunately, soil erosion continues today. According to U.S. Department of Agriculture estimates, about 1.6 billion metric tons (1.8 billion tons) of topsoil were lost annually from U.S. farmland from 1997 to 2001—800 million tons from wind erosion and 1.06 billion tons from water erosion. Although soil erosion is down from earlier years (FIGURE 10-5), the average rate of erosion on U.S. farmland is approximately seven times greater than soil formation, a situation that is clearly unsustainable. Should erosion continue, the U.S. agricultural system could experience substantial declines in productivity.

Unfortunately, little information is available on soil erosion rates throughout the world. Scientists currently estimate that approximately one third of the world's cropland topsoil is being eroded faster than it is regenerated. Soil erosion is especially rapid in many of the less developed nations. In China, for example, the Yellow River annually transports 1.6 billion tons of soil from badly eroded farmland to the sea. The Ganges in India carries twice that amount. The Worldwatch Institute (a nonprofit organization based in Washington, D.C.) estimates that 22.5 billion metric tons (25 billion tons) of topsoil is eroded from the world's croplands each year! At this rate, the world loses about 7% of its cropland topsoil every 10 years! To put this into perspective: If erosion continues at this rate, 225 billion metric tons will be lost in the next decade; this is equivalent to more than half of the topsoil on U.S. farms.

Soil erosion above the natural rate of soil formation is unsustainable. Year after year, it depletes a valuable resource needed to feed the world's people. It also will make it more difficult to feed the growing population. Without efforts to halt soil erosion, malnutrition and starvation could increase in the coming century.

KEY CONCEPTS

Soil is vital to the success of a nation, indeed the world, but agricultural soils are being lost at record rates in many countries—a trend that is clearly unsustainable.

Desertification: Turning Cropland into Desert

Each year millions of acres of cropland, pasture, and rangeland are becoming too arid to be farmed. This phenomenon, called **desertification**, occurs most often in highly vulnerable semiarid lands—that is, lands that are already fairly dry. In such areas, even small changes in rainfall or in agricultural practices can have a profound effect on the ability of land to support livestock and crops.

Numerous factors can be blamed for this problem. One of the leading causes is drought: long, dry periods. Drought may result from both natural climatic changes and changes brought about by humans. Global warming, overgrazing, and deforestation are three human causes. Overcropping semiarid lands—that is, planting them too frequently—also contributes to desertification.

How do global warming, overgrazing, and deforestation cause drought? Global warming is discussed in detail in Chapter 20. Global warming is caused by certain pollutants such as carbon dioxide, which trap heat in the atmosphere. Rising temperature increases evaporation rates and thus tends to dry the soil in various regions. Global warming also appears to be changing climate, especially rainfall patterns. Some areas, scientists predict, will become hotter and drier. Many atmospheric scientists believe that if global warming is not stopped, changes in rainfall and average daily temperature could be severe enough to render entire agricultural regions such as the midwestern United States unsuitable for farming.

Removing vegetation from the land also tends to alter local climate. Deforestation, for example, may decrease rainfall downwind from a site. Why? A well-vegetated surface acts like a sponge, absorbing moisture that supports plants and replenishes groundwater. Some water evaporates, only to fall on downwind sites, creating a cycle of precipitation and evaporation that continues on down the line. In denuded areas, however, water tends to run off the surface of the land; so less is available for replenishing groundwater and for nourishing plants. The drier the landscape, the less water there is to evaporate. This, in turn, tends to decrease rainfall downwind.

The loss of farmland to desertification is a serious problem, especially when combined with other factors, including soil erosion and climate change.

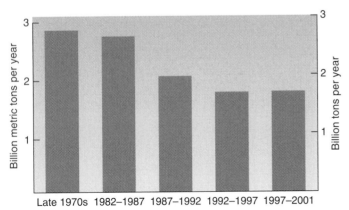

FIGURE 10-5 **Soil erosion on the decline.** Over the past 3 decades soil erosion from wind and water has declined dramatically. Although it still exceeds replacement level in many areas, progress has been substantial.

Throughout the world, cropland, rangeland, and pastures are becoming too dry to use because of climate change (natural and human-induced) and poor land management practices such as overgrazing. This phenomenon is called *desertification*. Desertification destroys millions of hectares of farmland and rangeland each year, further decreasing our ability to produce food.

How Serious Is the Problem? Desertification is a problem rooted in overpopulation and unsustainable land-use patterns. It afflicts numerous countries and regions, including the United States, Africa, Australia, Brazil, Iran, Afghanistan, China, and India. To begin with, a very large portion of the world's productive agricultural land is already experiencing desertification. According to the United Nations Environment Programme (UNEP), desertification is occurring on 70% of all drylands, or one fourth of the world's total land area; much of it used for farming or raising livestock. Desertification is worst in Africa, Asia, and Latin America. Making matters worse, according to UNEP, approximately 9 to 11 million hectares (22 to 27 million acres) of cropland and rangeland become desertified each year.

Desertification is not just an environmental problem, it causes extreme hardship currently. Over 250 million people are directly affected by it. Many of them may soon need to leave their homes because of the loss of cropland and rangeland they depend on. According to the UNEP, desertification costs the world about $40 billion a year in lost productivity.

Desertification is not new to humankind. In the ancient Middle East, for instance, the destruction of forests, overgrazing, and poor agricultural practices caused a deterioration of the water-absorbing capacity of the land and reduced the amount of rainfall. Coupled with a long-term regional warming trend, the decline in rainfall turned once-productive pastureland and farmland in much of the Fertile Crescent (where agriculture had its roots) into desert.

A more recent example occurred in the United States in the infamous dust bowl era of the 1930s. This disaster resulted from prolonged drought combined with fencepost-to-fencepost cultivation of fields in part to supply Europe with food in the early years of World War II. During the prolonged drought, crops withered and died. Field after field turned into an arid tract of dry dirt. Winds swept the parched topsoil into huge dust storms and carried the topsoil away. Only through extensive conservation measures in the postwar years were farmers able to slowly rebuild their soils. Today, however, some of these gains have been lost as farmers attempt to raise food production to increase their earnings. Small dust bowls are occurring in southern California and Texas. Colorado loses about 90 million tons of topsoil a year to wind erosion alone.

Desertification is especially severe in parts of Africa, especially the sub-Saharan region known as the Sahel. Beginning in 1968, a long-term drought in the Sahel (coupled with overpopulation, overgrazing, and poor land management) began the rapid southward expansion of the desert in Ethiopia, Mauritania, Mali, Niger, Chad, and Sudan. The Sahara is also spreading northward. An estimated 100,000 hectares (250,000 acres) of rangeland and cropland are lost in Africa each year.

Desertification and erosion are taking a huge toll on world food production. In Africa, a continent straining under the pressures of 906 million people in 2005, well over 200 million do not have enough food to eat. In Ethiopia, nearly one of every three people is malnourished. In Chad, Mozambique, Somalia, and Uganda, 4 of every 10 people are malnourished. Food supplies are declining in Latin America as well. The number of malnourished preschool children in Peru now stands at nearly 70%. Infant mortality in Brazil continues to rise.

Desertification and soil erosion are destroying agricultural land worldwide, contributing to present-day food shortages and reducing our ability to meet future demands caused by expanding human population.

Farmland Conversion

Besides soil erosion and desertification, valuable farmland is being lost by the conversion to other nonagricultural uses, a phenomenon called **farmland conversion**. Expanding cities, new highways, shopping malls, and other nonfarm uses require millions of acres of farmland each year in the United States and abroad (**FIGURE 10-6**). In Canada, farmland near urban centers has sometimes been paved over and built on. Farmland lost to human development has often been replaced by lower quality farmland. According to the Natural

FIGURE 10-6 Farmland conversion. Urban sprawl, as shown here in Des Moines, Iowa, swallows up farmland at an alarming rate throughout the world. Although home construction and sales contributes to the economy, once houses and other structures are built, the land is lost from agricultural production, a trend with serious consequences.

Resources Conservation Service, between 1997 and 2001, the United States lost an average of 0.9 million hectares or 2.2 million acres of farmland, rangeland, pastureland, and forest each year. This is equivalent to 6,000 acres (2,500 hectares) per year. About 46% of the losses were forest; most of the remainder were cropland and pasture. If this continues, one third of the United States' rural farmland will be gone within the next 100 years.

Farmland conversion is a worldwide phenomenon. The former West Germany, for example, loses about 1% of its agricultural land by conversion every 4 years, and France and the United Kingdom lose about 1% every 5 years. Little is known about the rate of agricultural land conversion in the LDCs of the world, but it is believed to be substantial. The loss of productive farmland is clearly an unsustainable trend that is made all the more troubling by the continual expansion of the world population and losses from soil erosion and desertification.

KEY CONCEPTS

Each year, millions of hectares of productive farmland are lost to human development—roads, airports, shopping centers, subdivisions, and so on—a phenomenon called *farmland conversion*.

Declines in Irrigated Cropland Per Capita

Water is as essential to agriculture as soil and sunlight. Plants need water to grow. Water plays an important role in photosynthesis—the sunlight-driven conversion of carbon dioxide into organic food molecules.

Most of the world's cropland is nourished by rainfall. However, a growing percentage of the world's cropland is irrigated—supplied with water from streams and lakes (surface waters) or from wells that draw water from the ground (groundwater). In the United States, 13% of all cropland is now irrigated. This land produces approximately one third of the nation's food. Globally, about 18% of the world's cropland is irrigated, and that farmland produces about 33% of the world's food.

According to the Worldwatch Institute, further increases in irrigation are likely to be modest. Groundwater depletion and intense competition for surface water supplies between farms and cities, for instance, are the main reasons for such a prediction. Groundwater is already being pumped faster than it can be replaced in several major agricultural areas, including regions of China, India, the Middle East, Northern Africa, and the United States. Water shortages are also evident in many other areas. Although such trends bode poorly for the long-term prospects of world agriculture, there is hope. Numerous sustainable solutions exist, the most important being water conservation (a topic discussed in the next section).

KEY CONCEPTS

Irrigated cropland supplies enormous amounts of food to the world's people, but the amount of irrigated cropland per capita is on the decline—a trend that bodes poorly for world food production. Measures that increase the efficiency of water use may prove helpful in providing an adequate supply of irrigation water.

Waterlogging and Salinization

Although irrigation greatly increases food production, in many semiarid regions, it has created some serious problems that could also significantly reduce food production in the coming decades, further adding to declines caused by soil erosion, desertification, and farmland conversion. Two such problems are waterlogging and salinization.

Waterlogging occurs when too much water is applied to farmland. Irrigating poorly drained fields, for example, often raises the **water table**, the upper level of the groundwater (**FIGURE 10-7**). If the water table rises too near the surface, it can fill the air spaces in the soil and suffocate the roots of plants. It also makes soil difficult to cultivate. Worldwide, about one tenth of the irrigated cropland (an area slightly smaller than Idaho) suffers from waterlogging. Productivity on this land has fallen approximately 20%.

Salinization occurs when irrigation water that has accumulated in the soil evaporates, leaving behind salts and minerals once dissolved in it (Figure 10-7). If not flushed from the soil, enormous quantities of salts and minerals can accumulate in the soil, greatly reducing crop production and making some soils impenetrable.

Worldwide, about one tenth of the irrigated farmland is seriously affected by salinization. Another 33% is affected to some degree. In the United States, the productivity of cropland suffering from salinization is believed to be reduced up to 25%. Moreover, when saline buildup reaches a critical level, soil becomes unproductive and must be abandoned. One expert estimates that 1.5 million hectares (2.5 to 3.75 million acres) of land are abandoned worldwide each year because of salinization and waterlogging. Salinization, for example, continues to be a problem in prairie soils. Although

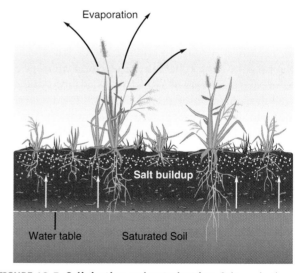

FIGURE 10-7 Salinization and waterlogging. Salts and other minerals accumulate in the upper layers of poorly drained soil (salinization) when irrigation waters raise the water table and water begins to evaporate through the surface. The rising water table also saturates the soil and kills plant roots (waterlogging). Lower arrows indicate the movement of water from groundwater into the topsoil. Upper arrows indicate evaporation.

it affects only about 2% of these farmlands, losses are estimated to be as high as $260 million a year.

Declining Genetic Diversity in Crops and Livestock

Before the advent of modern agriculture, grains and vegetables existed in thousands of varieties. Now, only a few of these varieties are commonly used. In Sri Lanka, for example, farmers once planted 2000 varieties of rice; today, however, only 5 varieties are in use. In India, 30,000 strains of rice were once grown; today, 10 varieties are responsible for about 75% of the nation's rice production.

In most cases, new varieties are chosen because they are more suitable for machine harvesting and because they respond favorably to fertilizer and irrigation water. They also produce higher yields. A similar trend is occurring on ranches throughout the world as ranchers adopt livestock breeds developed for maximum yield. The problem, as you shall soon see, is that huge expanses planted in one species are extremely vulnerable to pests, disease, and adverse weather. To combat disease and pests, farmers often turn to potentially harmful chemical pesticides.

The Green Revolution   The trend toward reduced genetic diversity began with development of high-yield varieties in the 1960s as part of a worldwide agricultural movement called the **Green Revolution.** Research began in 1944, when the Rockefeller Foundation and the Mexican government established a plant-breeding station in northwestern Mexico. The program was headed by Norman Borlaug, a University of Minnesota plant geneticist who developed a high-yield wheat plant for which he was later awarded a Nobel prize. Before the program began, Mexico imported half of the wheat it consumed each year. By 1956, it had become self-sufficient in wheat production, and by 1964 it was exporting half a million tons per year (**FIGURE 10-8**).

The success in Mexico led to the establishment of a second plant-breeding center in the Philippines. High-yielding rice strains were developed at this center and introduced into India in the mid-1960s. Again, the results were spectacular. India more than doubled its rice production in less than a decade.

Important as it was, the Green Revolution contributed greatly to the decrease in species diversity in cultivated crops.

FIGURE 10-8 The Green Revolution continues. Genetic research produced high-yield varieties like the plant on the left, which replaced lower-yield varieties (not shown). The wheat plant on the right is an even higher yield variety now under development. Although it produces a lot of grain, most seeds are shriveled, and plants tend to topple over under their own weight. The new variety is also vulnerable to two serious wheat diseases (rusts). Research is underway to correct these problems.

One of the most important concerns was that the new crops were not as resistant to diseases and insects. Local varieties of plants are acclimated to their environment. New varieties, on the other hand, often have little such resistance.

Moreover, as diversity dwindles, huge monocultures become more and more common. Expansive fields of one genetic strain facilitate the spread of disease and insects. The potato famine in Ireland in the 1840s is one example of the effects of reducing crop diversity. At that time, only a few varieties of potatoes were planted in Ireland. When a fungus (*Phytophthora infestans*) began to spread among the plants, there was little to stop it. Within a few years, two million Irish perished from hunger and disease, and another two million left the country.

The decline in genetic diversity resulting from the Green Revolution and other developments adds to the unsustainability of modern agriculture. But it also presents a focal point for change, a way we can reverse the trend and ensure a more environmentally sustainable system of food production.

The Green Revolution was a worldwide effort to improve the productivity of important food crops: wheat and rice. It succeeded in its primary objectives but created a steady decline in genetic diversity, which makes world food production more vulnerable to insects, plant pathogens, and other factors.

Habitat Destruction: Contributing to the Loss of Genetic Diversity The loss of genetic diversity among crop species is paralleled by an equally troublesome extinction of wild species throughout the world, especially in the tropics. The loss of species in the tropics could adversely affect modern agriculture. Why? Many modern crops came from tropical regions. Many of their relatives remain there today, growing as they have for centuries. These wild relatives (as well as other species) serve as a source of new genetic information for domestic crops to combat drought, disease, and insects. Keeping modern crops vital and successful, therefore, depends on keeping their relatives alive and well.

The loss of wild plant species that gave rise to modern crop species throughout the world, especially in the tropics, is eroding our capacity to improve crops and make them more resistant to pests, disease, and drought.

Politics, Agriculture, and Sustainability

Governments also add to agriculture's growing list of problems, sometimes fostering unsustainable practices. Consider some examples. **Subsidies,** payments made to farmers from their governments, can contribute to an unsustainable system of farming. In the United States, for example, the federal government subsidizes farmers through price supports. In this program, farmers are given a guaranteed price for certain crops such as wheat and soybeans. This helps keep them in business during bad years—that is, in years when prices fall. However, this program has many unforeseen consequences. First, it encourages farmers to plant crops that are insured by the federal government through price support. As a result, farmers tend to plant one or a few crops. It also encourages farmers to plant all of the land they can, even marginal land that may be easily eroded by wind and water. Why not plant every acre if you know the federal government will pay for your product? This practice encourages huge monocultures that are susceptible to insects and other pests. To combat them, farmers rely on an arsenal of toxic insecticides and other chemicals. Many of these chemicals end up in groundwater and in lakes and rivers, where they poison many species.

Government lending policies can also encourage unsustainable practices. In Mexico, for instance, most credit for irrigation systems and roads is given to farmers who produce cash crops such as tomatoes and cattle, both for export to the United States. Cash crop farms and cattle pastures usurp farmland once used to produce crops for domestic consumption.

Governments may also dictate policy. In Ethiopia, farmers have traditionally left semiarid lands fallow for 7-year periods so that nutrients from the highly weathered, poor soil could be replenished by natural processes. This practice, however, has been condemned by the Ethiopian government, which is interested in increasing farm production. If land is not cultivated within 3 years, it is confiscated. Unfortunately, bypassing the fallow period results in rapid soil deterioration.

Numerous other examples could be cited here. The important point in all of this is that to create a sustainable system of agriculture, laws and policies must be systematically re-examined and carefully revised with global sustainability in mind.

The problems facing world agriculture are not all technical. Some result from inadequate or self-defeating policies and governmental intervention. Lawmakers throughout the world have unwittingly facilitated the creation of an unsustainable system of agriculture.

10.4 Solutions: Building a Sustainable Agricultural System

Providing food for the growing human population will require a variety of policies and actions—both private and governmental. As is the case with other social and environmental issues, one of the most important solutions to famine is family planning to slow the growth and perhaps reduce the size of the human population. Efforts must also be made to increase food supplies, and such measures must be sustainable. That is, they must not undermine the long-term health of the global food production system. In other words, they must protect and improve the soil, water, and other resources upon which farming is dependent. Most analysts recommend a multifaceted approach that includes efforts to (1) protect existing soil and water resources; (2) increase the amount of land in production; (3) raise output per hectare of farmland—that is, increase productivity; (4) develop alternative foods; (5) eat lower on the food chain, (6) reduce food losses to pests; (7) increase the agricultural self-sufficiency of less developed nations; (8) enact legislation and policies that ensure a better distribution of food and more sustainable production methods; and (9) end wars. This section describes each part of this multifaceted strategy.

A sustainable system of agriculture consists of practices that produce high-quality food in ways that protect the long-term health and productivity of soils. Creating such a system will require a multifaceted approach, including measures to slow and perhaps stop the growth of the human population.

Protecting Existing Soil and Water Resources

The old adage that "an ounce of prevention is worth a pound of cure" applies to many aspects of our lives. It also applies to the task of building a sustainable system of agriculture. In fact, few measures are as important to creating a sustainable agricultural system as preventive ones: preventing soil erosion, desertification, salinization, waterlogging, and farmland conversion.

> ### KEY CONCEPTS
>
> Protecting soil and water resources is the first line of defense in meeting present and future needs for food.

Soil Conservation: Reducing Soil Erosion Protecting soil from erosion by water and wind is one of the most important steps we can take to ensure adequate food supplies both now and in the future—and protect the environment, too. Fortunately, soil erosion can be minimized and even halted by a variety of simple, often cost-effective techniques. This section describes six strategies: minimum tillage, contour farming, strip cropping, terracing, gully reclamation, and shelterbelts.

> ### KEY CONCEPTS
>
> One of the highest priorities in making the transition to a sustainable system of agriculture is putting an end to excessive soil erosion. Fortunately, there are many simple yet effective measures that can ensure a sustainable erosion rate.

Minimum Tillage Farmers typically plow their fields before planting new crops. They then break up the clumps of soil with a device called a *disc,* making the soil suitable for sowing seeds. In areas where soil is too wet in the spring, farmers plow and disc in the fall, leaving their land barren and subject to wind erosion during the winter.

With special implements, however, farmers can forgo these costly, time-consuming, and energy-intensive steps and plant right over the previous year's crop residue (**FIGURE 10-9**). This technique is one form of **minimum tillage**, or **conservation tillage**, a strategy that reduces the physical disruption of the soil. According to the U.S. Department of Agriculture in 2002, minimum tillage was practiced on about 42 million hectares (103 million acres), 36% of all U.S. farmland (**FIGURE 10-10**). Unfortunately, this practice is not widely used in other countries.

Because fields are protected much of the year by crops or crop residues, soil erosion can be decreased substantially—in some cases by as much as 90%. Minimum tillage also reduces energy consumption by as much as 80%, saving farmers money, and conserves soil moisture by reducing evaporation. Crop residues can increase habitat for predatory insects that prey on pests, reducing the need for pesticides and subsequent contamination of the environment (Chapter 22). Curtailing the use of heavy machinery on farm fields also has the benefit of reducing soil compaction, which makes soils harder, increases surface runoff, and impairs

FIGURE 10-9 Minimum tillage planter. This device is designed to dig furrows in the presence of crop residue, avoiding plowing and discing. Leaving the previous crop residue on the land over the fallow period greatly reduces soil erosion.

root growth. Thus, by cutting back on machinery use, farmers may find that crops actually grow more rapidly and produce more food.

Despite its benefits, minimum tillage has several drawbacks. For example, **herbicides**, chemicals that kill weeds, are often used in place of mechanical cultivation to control weeds. In addition, crop residues may harbor insects that damage crops. Minimum tillage also requires new and costly farm equipment. Already strapped for cash, many farmers can't afford the new equipment. In Canada, this problem has been partially solved by the Provincial government of Manitoba, which purchased or leased

> Terracing and contour farming each reduce soil erosion by about 50%. When combined, they reduce erosion by 75%. Minimum tillage reduces soil erosion by 90%. When combined with terracing and contour farming, soil loss can be slashed by 98%.

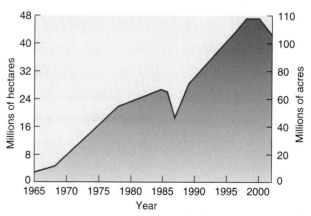

FIGURE 10-10 Growth in minimum tillage in the United States.

conservation seeding equipment, then made it available to farmers for a nominal fee. As a result, hundreds of thousands of acres have been brought into minimum tillage.

KEY CONCEPTS

Reducing the amount of land disturbance by minimizing tillage protects the soil from the erosive forces of wind and rain. This technique, although effective in reducing energy demand and erosion, often requires additional chemical herbicides to control weeds.

Contour Farming   On hilly terrain, crops can be planted along lines that follow the contour of the land, a technique called **contour farming** (**FIGURE 10-11**). Planting crops across the direction of water flow on hilly terrain reduces the rate at which water flows across the land, resulting in a 50 to 80% reduction in soil erosion and a marked increase in water retention. This technique therefore also reduces demand for irrigation water.

KEY CONCEPTS

Planting crops perpendicular to the slope—that is, along the land contour lines—reduces soil erosion and increases water retention.

Strip Cropping   As **FIGURE 10-12** shows, **strip cropping** is a measure in which farmers alternate strips of two or more crops in single fields on flat or hilly terrain. Strip cropping reduces wind and water erosion and increases productivity. For example, farmers may alternate row crops such as corn with cover crops such as alfalfa on hilly terrain. Water flows more easily through row crops and begins to gain momentum, but when it reaches the cover crop, its flow is nearly stopped. Strip cropping can be combined with contour farming to further reduce erosion.

KEY CONCEPTS

Crops can be planted in alternating strips, a practice called *strip cropping*. When combined with contour farming, this technique greatly reduces soil erosion.

Terracing   For thousands of years many peoples have grown crops in mountainous regions using **terraces**, small earthen embankments placed across the slope to check water flow and minimize erosion. Terraces have also been used in the United States for many years on land with less pronounced slope (**FIGURE 10-13**).

A University of Nebraska study showed that terracing and contour farming each reduce erosion by 50%. Together, they can reduce soil loss by 75%. Minimum tillage is even more effective. By itself it can reduce erosion rates by 90%. When it is combined with terracing and contour farming, soil erosion rates can be reduced by 98%. Unfortunately, terraces are expensive to construct and may interfere with the operation of large farm equipment.

KEY CONCEPTS

Terraces, small earthen embankments that run across the slope of the land, greatly reduce soil erosion.

Gully Reclamation   Gullies are a danger sign indicating rapid soil erosion. In fact, some gullies can work their way up hills at a rate of 4.5 meters (15 feet) a year. Gullies also make it difficult to farm land by impairing the use of machinery.

To prevent gullies from forming, farmers must reduce water flow over their land. Contour farming, strip crops, and terraces all help. Already formed gullies can be stopped by seeding them with rapidly growing plants. Small earthen dams can be built across gullies to reduce water flow, retain

FIGURE 10-11 Contour farming. This land is farmed along the contour lines to reduce soil erosion and surface runoff, thus saving soil and moisture.

FIGURE 10-12 Strip cropping. Alternating strips of alfalfa with corn on the contour protects this crop field in northeast Iowa from soil erosion.

FIGURE 10-13 **Terracing. (a)** This Iowa corn is grown on sloping land with the aid of terraces, small earthen embankments (not visible) that reduce water flow across the surface. The corn is planted in the stubble of last year's crop. **(b)** In Asia, terracing has been used in mountainous regions to grow rice.

(a)

(b)

moisture for plant growth, and capture sediment, which will eventually support vegetation as erosion is reduced. Too often, land with severe gullies is abandoned or haphazardly reclaimed, only to suffer worse erosion.

KEY CONCEPTS

Gullies form quickly on hilly terrain and grow rapidly. Preventing them is absolutely essential to protect farmland. If they form, they can be regraded and replanted with fast-growing species to prevent their expansion.

Shelterbelts Wind erosion is a major problem, accounting for considerable amounts of soil erosion in many parts of the world. Wind erosion can be controlled by minimum tillage. Shelterbelts, rows of trees planted along the perimeter of crops, are another effective measure aimed at reducing wind erosion. In 1935, the U.S. government mounted a campaign to prevent the recurrence of the disastrous dust bowl days. This program involved planting long rows of trees as windbreaks, or **shelterbelts**, on farms in the Great Plains to slow the winds that carry soil away (**FIGURE 10-14**). Thousands of kilometers of shelterbelts were planted from Texas to North Dakota.

Besides decreasing soil erosion from wind, shelterbelts prevent snow in fields from being blown away, thus increasing soil moisture. This in turn reduces the demand for irrigation water during the growing season and helps replenish groundwater supplies. Shelterbelts can also improve irrigation efficiency by reducing the amount of water carried away from sprinklers by wind. In addition, shelterbelts provide habitat for animals, pest-eating insects, and pollinators. They also protect citrus groves from wind that blows fruit

from trees. Planted around farmhouses, shelterbelts also reduce heat loss on windy winter days, increasing comfort and saving farmers money.

KEY CONCEPTS

Shelterbelts are rows of trees planted along the perimeter of fields to block wind and reduce soil erosion. Shelterbelts have the added benefit of preventing snow from blowing away from fields, thus increasing soil moisture content. In addition, shelterbelts provide habitat for useful species, such as insect-eating birds that help control crop pests.

Overcoming the Economic Obstacles to Soil Erosion Controls Soil erosion control is one of the most important measures needed to build a sustainable system of agriculture.

FIGURE 10-14 **Shelterbelts.** Rows of trees planted along the margin of farm fields protect farmland in Michigan from the erosive effects of wind.

In many less developed nations, however, farmers struggle to meet their basic needs and claim that they have neither the time nor the means to care properly for the land. Few can see the benefits of soil conservation because the gains tend to materialize slowly.

Economics and short-term thinking impair soil-erosion control efforts in the more developed nations as well. Caught between high production costs and low prices for grains, farmers often ignore soil erosion and offset any losses in production by applying synthetic fertilizers. These additives artificially help farmers maintain yield, despite the loss in soil and valuable soil nutrients. Despite the importance of soil erosion controls to sustainable agriculture, many farmers are reluctant to invest in these practices. Such a view, while understandable, ignores the long-term cost of permitting unsustainable loss of soil—an erosion of the productive capacity of farmland.

Governments can promote soil conservation in a variety of ways. One landmark attempt to end the devastating loss of U.S. topsoil was the **1985 Farm Bill.** This law created a land conservation program (the **Conservation Reserve Program**) that directs the federal government to pay farmers to retire their most highly erodible cropland from production for 10 years and plant trees, grasses, or cover crops to stabilize and rebuild the soil. By 1993, farmers had retired an estimated 14.7 million hectares (36.4 million acres), cutting erosion by an estimated 900 million tons per year. In 1996, the U.S. Congress passed a farm bill that extended the Conservation Reserve Program to 2002, maintaining the protected acreage at slightly less than 15 million hectares (36.4 million acres). In 2004, the USDA announced that it would implement a directive from President Bush that would offer farmers early re-enrollment and contract extensions for acres that begin expiring in 2007. Their goal is to expand enrollment in the program to 39.2 million acres.

The 1985 Farm Bill also established a federal program that called on farmers to develop soil erosion plans in exchange for eligibility for federal crop insurance, subsidies, and other benefits. So far, 1.5 million farmers have signed up for the program and have worked with the U.S. Soil Conservation Service to develop plans for 134 million acres, about 25% of U.S. farmland. These efforts could cut soil erosion on some of the most productive soils in the country. The 1996 Farm Bill provided additional support for these efforts by providing $200 million annually to assist farmers technically and financially with conservation measures. Half of the money is earmarked for livestock operations. This program is known as the **Environmental Quality Incentives Program.** Cost–benefit calculations showed that for every dollar invested in the program, the nation saved two dollars in reduced erosion and pollution.

KEY CONCEPTS

Farmers are sometimes reluctant to take measures to control erosion because of their costs. Carefully crafted government policies can provide economic incentives to protect soils from the erosive forces of wind and water.

Preventing Desertification   Protecting soil from desertification is also needed to preserve cropland and rangeland. Several of the measures designed to prevent soil erosion will be helpful in this vital goal. For example, shelterbelts protect soil from the drying effects of wind. More direct actions are also possible. In China, for instance, agricultural officials have embarked on an ambitious program to plant a 6900-kilometer (4300-mile) "green wall" of vegetation to stop the spread of desert in the northern region (see Spotlight on Sustainable Development 10-1). Contour farming and terracing both increase soil moisture content and combat desertification. In Australia, huge semicircular banks of soil are created in the windswept plains to catch seeds and encourage regrowth in areas denuded by livestock. Better land management—for example, controls on the number of livestock on rangeland—is also necessary (Chapter 12). Pollution controls to slow down global warming are also vital to this effort (Chapter 20).

KEY CONCEPTS

Many measures that protect soil from erosion also make it less susceptible to desertification. When combined with measures to reduce global warming, these steps could help to slow desertification.

Reducing Farmland Conversion   Efforts to prevent the spread of cities, the proliferation of highways, and other nonfarm uses of arable land are also vital to protecting farmland and creating a sustainable future. Slowing the growth of the population is essential. Careful city planning and new zoning laws could help reduce farmland conversion by ensuring that homes, roads, airports, and businesses are not built on agricultural land. (For more on this topic, see the Spotlight on Sustainable Development and the discussion on urban growth controls in Chapter 17.)

KEY CONCEPTS

Numerous techniques are available to prevent farmland conversion, the loss of arable land to highways, airports, subdivisions, and other nonfarm uses.

Saving Irrigated Cropland/Using Water More Efficiently
As you learned earlier in this chapter, irrigated cropland produces a large portion of the world's food supplies, but huge expanses of irrigated land are being salinized and waterlogged. Groundwater supplies are also declining very seriously in key agricultural areas such as the midwestern United States. You also learned that the growth in irrigated cropland is not keeping pace with population growth. One of the reasons for this is a lack of available surface water and groundwater.

One way to solve all of these problems is to use existing water more efficiently. The efficient use of resources, of course, is a key to building a sustainable society. Farmers, for instance, can improve irrigation efficiency through many simple and cost-effective measures. Lining irrigation ditches with cement or plastic can cut water losses by 30 to 50%, thus freeing up tremendous amounts of water to irrigate other land. Transporting water in pipes can result in even greater

10-1   The Green Wall of China: Stopping the Spread of Desert

China is a nation in trouble. With a growing population of over 1.3 billion people, China's land is falling into ruin. Centuries of overgrazing, poor agricultural practices, and deforestation have resulted in severe erosion and rapidly spreading deserts that gobble up the once-productive countryside.

The spread of deserts affects the lives of millions of peasants in China. In the highlands of northern China, for example, the land is cut with gullies, some hundreds of meters deep. Erosion from the raw, parched Earth is an astounding 30–40 metric tons per hectare (12–16 tons of topsoil per acre) per year—far above replacement level. In all, some 1.6 billion metric tons are carried into the Yellow River annually, making it one of the muddiest rivers in the world.

According to one estimate, an area larger than Italy has turned into desert or semidesert in China in the last 30 years. Although most of the desertification is occurring in northern China, few areas are immune to this problem.

In 1978, the Chinese government launched a 73-year reforestation project to stem the tide. By planting trees, shrubs, and grasses, they hope to form a giant green "wall" across the northern reaches of the nation. When completed, the wall will extend more than 4480 kilometers (2600 miles) and will be 400 to 1700 kilometers (250 to 1000 miles) wide. This ambitious project is designed to return much of the land now falling into ruin to productive use.

Between 1979 and 2005, 24 million hectares (59 million acres) were replanted, and 6 million hectares (15 million acres) were naturally regenerated in mountainous and sandy areas. More than 11 million hectares (27 million acres) of farmland have been protected from desertification. In 2005, international experts began an evaluation of the program. Their recommendations will help China draft a plan for phase 2, which will extend to 2050.

The Yulin district is one of China's success stories. Before 1949, more than 400 villages and 6 towns had been invaded or completely covered by sand. Today, four major tree belts have been planted in the area, decreasing the southward push of the desert by 80% (FIGURE 1). Towering sand dunes now peep through poplar trees, and rice paddies sparkle in the sunshine. Grain production has been replaced by a diversified agricultural system, including animal husbandry, forestry, and crop production. The trees provide shade and help reduce wind erosion that causes the sand dunes to shift. Shrubs and grasses now thrive on land once stripped of its rich vegetation. Trees and shrubs grow in gullies, and grasses carpet slopes. Revegetation helps to hold the soil in place and reverse the local climate change.

Local residents have built a diversified economy in what was once an unproductive desert. Juice from the cherry trees, which thrive in the desert climate, is rich in vitamin C and amino acids; it is used to produce soft drinks, preserves, and beer. Twigs of the desert willow are used to make wicker baskets and trunks, which earn local residents $2 million in U.S. currency every year.

Despite the encouraging signs in China, a report by the Shanghai-based World Economic Tribune says that while nearly 10 million hectares (4000 square miles) were planted each year, twice that amount is still being lost. In the northern province of Heilongjiang, home of China's largest concentration of virgin forest, loggers have reduced the tree cover from 50 to 35% in just 30 years. Government pricing policies promote overcutting.

The reforestation project has been plagued by a shortage of money and technical expertise. So far, China has invested over $3 billion in U.S. currency. Because of short-sighted land-use practices, China's Yangtze River, the nation's longest watercourse, could become a second Yellow River. Each year, its tributaries turn muddier from erosion.

Reforestation is an attempt to restore the Earth's vegetative surface, which is vital to building a sustainable future. Efforts such as China's are needed on most continents to help reverse centuries of land abuse that have spawned the spread of deserts. Replanting forests not only slows the spread of deserts; it also helps restore wildlife populations and can help reduce global warming, which now threatens the world climate.

Adapted from: L. Ming (1988). Fighting China's Sea of Sand. *International Wildlife* 18(6):38–45, with permission.

FIGURE 1 **Greenbelt.** These trees planted in China are part of the effort to stop the spread of desert.

FIGURE 10-15 Increasing the efficiency of irrigation. Trickle systems deliver water to roots, cutting evaporation losses substantially. Trickle systems can be used only for certain crops, among them strawberries (shown here) and fruit trees.

savings. Subsidies that help support these costly efforts can assist farmers all over the world. It is taxpayer money well spent, say supporters. Farmers can also use drip irrigation systems to deliver water directly to the roots of fruit trees and a few other crops (**FIGURE 10-15**). Conventional center pivot irrigation systems that once sprayed water into the air (with tremendous losses to evaporation) are modified to spray water downward, creating considerable savings (**FIGURE 10-16**). Computer systems can help farmers monitor soil moisture so

they apply water only when it is needed and in the amount required by crops. Applying irrigation water at night or early and late in the day when evaporation is lowest can reduce demand by 50% or more. Improvements in soil organic matter through application of manure and compost can also help because organic matter acts as a sponge, holding water in the soil. Efforts to protect aquifer recharge zones, areas in which groundwater supplies are recharged, can also ensure farmers a reliable supply of water (Chapter 13).

KEY CONCEPTS

Water efficiency measures help free up water to expand irrigated cropland.

Preventing Salinization and Waterlogging   All water efficiency measures help increase irrigation water and expand cropland under irrigation, but certain measures can reduce salinization and waterlogging. This, in turn, could reduce the annual loss of farmland. By using computerized sensors that measure soil moisture, for example, farmers can apply only the amount of water needed by crops. This not only frees up water for other crops, it reduces salinization and waterlogging. Special drainage systems can also be installed to draw off excess water and prevent the buildup of salt and the potential for waterlogging, although such measures are quite costly. As you will see in Chapter 11, water drained from farm fields may carry high levels of potentially toxic substances. Special care must be taken to avoid solving one problem (salinization) while creating another (surface water pollution). Government programs could also discourage irrigation in soils susceptible to these problems.

KEY CONCEPTS

More frugal application of irrigation water to crops and special drainage systems can reduce salinization and waterlogging in soils susceptible to the problem.

(a)

(b)

FIGURE 10-16 Center pivot irrigation. (a) The standard device sprays water into the air, but much of the water evaporates before it hits the ground on hot days. **(b)** By turning the spray nozzles downward, much more water actually makes it to the plant.

Soil Enrichment Programs

Soil erosion controls help preserve farmland from being washed or blown away. Because soil contains nutrients that are essential to plant growth, soil-erosion control methods also protect soil fertility. As any good farmer will tell you, protecting or even enhancing the fertility of the soil is vital to maintaining or increasing its productivity.

Soil fertility can be enhanced by the use of fertilizers and crop rotation. In many locations throughout the world, agricultural soil fertility is maintained by applying human wastes from homes or sewage treatment plants—an activity that returns the nutrients of foods we eat to their origin.

KEY CONCEPTS
Farming mines the soil, robbing it of valuable nutrients, but numerous methods such as applying organic fertilizer and rotating crops can replenish nutrients and maintain the health of the soil over the long term.

Organic Fertilizers   One of the most effective means of replenishing soil fertility is to apply organic fertilizers to cropland and pasture. **Organic fertilizers** can be waste materials such as cow, chicken, and hog manure and human sewage. All of these replenish the soil's organic matter and add important soil nutrients such as nitrogen and phosphorus.

There are many sources of organic fertilizer. Currently, there are hundreds of millions of sheep, cattle, hogs, and other animals that produce billions of tons of waste. Putting it to good use—making cropland more productive—makes great sense. It helps recycle nutrients and prevents pollution of water supplies.

Organic fertilizers also include **green manure**—plants grown in a field that, rather than being harvested, are plowed under. Especially valuable are the leguminous plants such as alfalfa and clover, which are often grown during the off-season and plowed under before food crops are planted.

Soil enrichment with organic fertilizers of all sorts provides many benefits vital to building a sustainable agricultural system. First and foremost, it increases or helps to maintain soil fertility and crop yield. Because organic matter acts like a sponge in the soil, organic fertilizers increase the water-retention capabilities of the soil. Organic matter also provides an environment conducive to the growth of bacteria necessary for nitrogen fixation. Organic fertilizers help prevent shifts in soil acidity, and they tend to prevent the leaching of minerals from the soil by rain and snowmelt. In addition, careful application of human wastes on farmland helps to reduce water pollution by municipal sewage treatment plants (Chapter 21).

Organic wastes have been successfully applied in some countries, such as China and India, for many years, but this practice is not without its problems; one of these is the cost of transporting waste to farms by pipelines or trucks, for many cities are situated many miles from arable land. Another problem is that organic waste from municipal sewage treatment plants may be contaminated with pathogenic (disease-causing) organisms such as bacteria, viruses, and parasites. Theoretically, some of these organisms could be taken up by crops and therefore enter the human food chain. In industrialized nations, municipal waste may be contaminated with toxic heavy metals such as mercury and lead coming from factories connected to the plant. Better controls at sewage treatment plants or at the factories that produce these materials could alleviate the problem.

You can help promote the use of organic fertilizer by buying organically produced food from your grocery store or by purchasing from local growers either through farmer's markets or through a community-supported agricultural program that is described in Spotlight on Sustainability 10-2.

KEY CONCEPTS
Use of organic fertilizers helps farmers maintain or even improve soil conditions and boost crop production. This strategy also returns nutrients to the soil, thus helping to close nutrient cycles and prevent pollution of waterways.

Synthetic Fertilizers   In the more developed countries such as Canada, Australia, and the United States, farmers apply millions of tons of **synthetic fertilizer** to their land each year to boost crop production. Without these fertilizers, world food production would fall 40% or more, according to the Worldwatch Institute.

Synthetic fertilizers contain three nutrients: nitrogen, phosphorous, and potassium. Because of this, synthetic fertilizers only partially restore soil fertility. They do nothing to replenish organic matter or micronutrients (nutrients required by plants in very small quantity) necessary for proper plant growth and human nutrition. On land that is fertilized solely with synthetic fertilizers, soil fertility slowly declines over time. Moreover, excess fertilizer may be washed from the land by rains and end up in streams, causing a number of problems (Chapter 21). To prevent the gradual depletion of nutrients and to help develop a more sustainable society, many agricultural experts call for much wider use of organic fertilizers.

KEY CONCEPTS
Synthetic fertilizers help boost soil fertility, but they only partially replenish agricultural soils because they contain just three of many nutrients needed for healthy soil: nitrogen, phosphorus, and potassium.

Crop Rotation   In modern agriculture, synthetic fertilizers and pesticides have allowed farmers to grow the same crops year after year on the same plots. This way, farmers can concentrate their efforts on crops that they know well. However, this process is generally viewed as unsustainable because it often depletes soil nutrients, increases soil erosion, and creates serious problems with pests and crop pathogens.

Crop rotation is a practice in which farmers alternate the crops they plant in their fields from one season to the next. For example, corn may be planted for 1 or 2 years, followed by alfalfa, a cover crop. The cover crop reduces soil erosion and replenishes soil nitrogen. If the cover crop is plowed under, it helps to replenish organic matter and return a variety of valuable nutrients to the soil. Therefore, this simple practice helps

10-2   Community-Supported Agriculture

Community-supported agriculture (CSA) "is an innovative and resourceful strategy to connect local farmers with local consumers to develop a regional food supply and strong local economy," according to the University of Massachusetts Extension service, whose website contains a wealth of information on this idea.

The idea began in Japan in the very early 1970s and has since spread to other parts of the globe, including North America, with well over 600 groups and 100,000 members in the United States alone. Although each group is unique, community-supported agriculture generally involves two parties: a local farmer—typically a small farmer who produces food organically—and a group of residents in a nearby city or town who constitute the members. The members purchase produce directly from the farmer. In most groups, says author Sarah Milstein in an article in *Mother Earth News,* "Members pay ahead of time for a full season with the understanding that they will accept some of the risks of production." If the cucumbers produce poorly, so be it. You don't get cukes from the CSA. If zucchini grows particularly well, which is almost a given, you'd better learn how to make zucchini bread or dry the things and use them as firewood in the winter.

"In other groups, members subscribe on a monthly basis and receive a predetermined amount of produce each week," she adds.

Produce is either picked up at the farm by members of the group or delivered to a central location by the farmer. Members of each group generally receive eight or more different types of vegetables each week—starting in the spring and continuing well into the fall. However, "Some groups

offer fruit, herbs, flowers, bread, cheese, eggs, yogurt, beef, honey, maple syrup, and most anything else you can produce on a farm," says Milstein.

CSA programs vary in size. Slack Hollow Farm in Argyle, New York, has a dozen members. They cover only a small percentage of the farm's operating budget. The rest of the produce from their 7-acre farm is sold through a local food co-op. Pachamamma farm in Colorado just north of Boulder grows organic produce on 11 acres of farmland and has 115 members. Across the Hudson River from Slack Hollow Farm in New York is Roxbury Farm, where growers cultivate 25 acres and sell to 700 members whose purchases cover about 90% of the farm's annual operating expenses (FIGURE 2).

FIGURE 2 A community-supported agriculture farm in operation.

build the soil. Crop rotation has two additional benefits: It reduces pest damage and reduces the need for costly and potentially harmful chemical **pesticides** (substances that kill damaging insects, weeds, and other species). The reasons for this benefit are explained in Chapter 22. Properly planned and executed, crop rotation can boost yields by 10 to 15%.

Crop rotation, alternating crops planted in a field one season after another, offers many benefits. Planting the proper crops can help replenish soil nutrients. It also helps reduce erosion, pest damage, and the need for costly and potentially harmful pesticides.

Increasing the Amount of Land in Production

There are several ways of meeting increasing demand for food. The previous section described a preventive approach—strategies such as erosion control and organic fertilizers. The other strategy is to develop new supplies. In the past, virtu-

ally all nations solved the problem of rising food demand by opening up new lands to the plow. Today, however, that option is quickly vanishing. In most parts of the world, potentially arable land is in short supply. In most of the major industrial nations, for example, farmland reserves are relatively small. In Canada, less than 5% of the land is capable of supporting crops, and virtually all of that land is in production. In Southeast Asia, 92% of the potential agricultural land is being farmed. In southwestern Asia, more land is currently being used than is considered suitable for rain-fed agriculture. Consequently, per capita food production in Asia has begun to fall as the population continues to grow.

For those countries that have little farmland reserve, efforts to protect soils, manage urban growth, and reduce population growth offer the greatest hope for meeting future food demand.

Africa and South America have large surpluses of land that could be farmed. Some experts believe that the nations of these continents should develop this land. However, much of this land is currently covered by tropical rain forests. As noted in Chapter 5, although the tropical rain forests are rich in

Community-supported agriculture is a win-win situation. Farmers acquire upfront funding, so vitally needed before the growing season starts, and they cultivate relatively secure markets. By growing a diverse array of crops, farmers reduce pest problems, increase soil fertility, and ensure a decent harvest. You, the consumer, receive an abundance of healthy food at a good price—provided the weather cooperates.

The environment benefits, too. Food produced without pesticides ensures healthy soil and healthy neighboring ecosystems. The birds that frequent the hedge rows surrounding farm fields live healthy lives, gobbling down tons of insects they harvest from the field and feeding the rest to their offspring.

In addition, because food is grown only a few miles from members' homes, rather than on distant farms hundreds or thousands of miles from the dinner table, very little energy is required to transport food to market. The less energy, the less pollution. Production at an oil well in the Mid-East may decline slightly, but the billionaire sheik who owns it or the millionaire executives who run the big oil companies won't notice the difference. They are too busy trying to decide which model Lear jet they want.

Most, if not all, community-supported agriculture operations are initiated by farmers looking for a secure local market for their products. They recruit members through word-of-mouth, brochures, flyers, media coverage, and other methods. If you're interested in such a program, ask around, and do some research on the web. Attend local farmer's markets and ask whether any of the participating farmers also engage in CSA or know farmers in the area who do.

Americans interested in finding a local CSA program can also check with the local U.S. Department of Agriculture Extension Offices in their states or can contact some of the nonprofit organizations, such as the CSA Farm Network (which lists farms in the northeastern United States), for listings of local CSA farms. They are included in the Resource Guide.

If you can't find a local program, you may be able to start one yourself by contacting local farmers. Ads in rural newspapers might help you make contact. Visits to local farm supply stores could prove helpful in identifying potential farmers. You can also contact local and statewide organic farm organizations, too.

Be sure that you don't enter into this venture expecting grocerystore–perfect vegetables. Because organic farmers don't use pesticides, some organic produce may be slightly blemished. And local farmers battle weather, too. In bad years, production can plummet. Some crops may fail entirely. It's not easy producing food; so, help the farms you support, and understand that they're struggling to do the best they can.

Adapted with permission from Dan Chiras and Dave Wann. *Superbia! 31 Ways to Create Sustainable Neighborhoods.* Gabriola Island, B.C.: New Society Publishers, 2003.

plant and animal species, the soils they grow on are poor in nutrients. When stripped of vegetation, these soils also are prone to erosion in the intense tropical rains and may become hardened when exposed to sun (Chapter 5). The potential for expansion on these continents is therefore not as great as some would lead you to believe.

Tapping unfarmed grasslands may be an option in some areas. These soils are rich and productive, but this strategy could severely deplete wildlife populations and disrupt many of the free services provided by nature—for example, flood control and local climate control.

KEY CONCEPTS

Grasslands and forests can be converted to farmland to meet the rising demand for food. In many parts of the world, though, and especially in the more developed nations, farmland reserves are small. Even in countries where there is an abundance of reserve land, much of this land is covered with poor soils. Furthermore, the ecological cost of converting wild land to farmland would be enormous.

Increasing the Productivity of Existing Land: Developing Higher Yield Plants and Animals

Another way to increase food supplies is to develop new, higher yield plants and animals. New, high-yield varieties of rice and wheat developed during the Green Revolution, for example, produce three to five times as much grain as their predecessors. New varieties of plants are created by breeding closely related plants to combine the best features of the parents; these are called **hybrids.**

As the new hybrids were introduced into many poor nations, the hopes of the Green Revolution dimmed, however. Farmers soon found that the hybrids required large amounts of water and fertilizer, which were unavailable in many areas. Without these inputs, farm yields were not much higher than those of local varieties. In some cases, they were actually lower. Another problem was the high cost of the new varieties, which prevented many small farmers from buying them. Moreover, new plants were often more susceptible to insects and disease.

These setbacks stimulated new research to produce varieties that would increase productivity without huge inputs such as fertilizer. Today, plant breeders throughout the world are attempting to develop crops with a high nutritional value and greater resistance to drought, insects, disease, and wind. Plants with a higher photosynthetic efficiency are also in the offing. Efforts are even under way to incorporate the nitrogen-fixing capability of legumes (Chapter 4) into cereal plants such as wheat—a change that would decrease the need for fertilizers and reduce nitrogen depletion.

One exciting improvement is a new variety of corn, a staple for 200 million people worldwide. Because corn is such an important source of calories and protein, researchers spent nearly 2 decades developing a new variety, quality-protein maize (QPM). Studies show that only about 40% of the protein in common corn is digested and used by humans. In contrast, roughly 90% of QPM's protein can be digested and used. In addition, QPM also produces 10% more grain. In areas where corn is a staple, such as Africa and Mexico, QPM could help curb hunger. QPM could help the residents of these areas become more self-sufficient in food production. Efforts are already under way to increase dramatically the use of this type of corn.

Some researchers are also exploring the use of perennial crops for agriculture. A **perennial** is a plant that grows from the same root system year after year—such as grasses. Today, most agricultural crops are **annuals,** plants that only last one season. Annuals are grown from seeds each year. Preliminary research suggests that productivity from perennials may be equal to or slightly lower than conventional annuals such as wheat, but the benefits from soil conservation, soil-nutrient retention, and energy savings may overwhelmingly favor them. Just as new varieties of plants help increase yield, so do fast-growing varieties of fowl and livestock.

KEY CONCEPTS

Numerous efforts are under way to increase the yield of plants and the growth rate of animals to increase food production.

Selective Breeding and Genetic Engineering   Efforts are being made to improve plants and livestock by **selective breeding.** In selective breeding, organisms containing valuable characteristics are bred in hopes of acquiring offspring with these characteristics. This technique, used for hundreds of years, is effective but rather slow. Another more recent development being used to improve livestock and crop species is genetic engineering.

Genetic engineering is a complex process designed to mechanically transfer desirable genes into the genetic material of an organism. Genes that make cattle grow faster, for example, can be transferred into the ova of cattle. If the genes are incorporated into the DNA of the ova, the offspring would then carry this gene and pass it on in turn to their offspring.

Genetic engineering may be used in other ways that improve agriculture. A new strain of bacteria developed by scientists at the University of California, for example, inhibits

the formation of frost on plants, which may provide farmers with a way of reducing crop damage from early frost. Other scientists are working on genes that give plants resistance to herbicides used to control weeds. A group of scientists at the Monsanto Company has developed a strain of bacteria that grows on the roots of corn and other plants. When eaten by insects, the bacteria release a toxic protein that kills the pest. Geneticists have also introduced certain genes that allow oats to thrive in salty soils, and researchers are experimenting with genetically engineered bacteria that could help plants absorb nutrients more efficiently, thus increasing crop yields. In Canada, genetic engineers have developed a new strain of potatoes that poison Colorado potato beetles. Animal geneticists are now working on ways to improve livestock, combining genes from one species with those of another to improve efficiency of digestion, weight gain, and resistance to disease.

Genetic engineering was once touted as a savior for world agriculture. However, researchers are finding that it is more difficult to apply to agriculture and much slower than proponents once thought. Nevertheless, the successes of genetic engineering have fostered extraordinary enthusiasm in the business community. Dozens of new companies have formed in recent years, and billions of dollars have been invested in the fledgling industry. Still, safety questions remain. Will the genetically engineered bacteria escape into the environment, upsetting the ecological balance? Experts agree that, once unleashed, a new form of bacterium or virus would be impossible to retrieve. Controlling it could prove costly and damaging.

Some individuals have criticized genetic engineering as a means of tinkering with the evolutionary process. Deliberate genetic manipulations, such as the transfer of chromosomes from one species to another, are different from anything that ordinarily occurs during evolution. Is it right, critics ask, to interfere with the genetic makeup of living organisms? Will these intrusions alter the evolution of life on Earth?

Recent research suggests that the dangers of genetic engineering have been blown out of proportion and that genetically engineered bacteria are not generally a threat to ecosystem stability. Most scientists agree.

At least two studies now indicate that genetically engineered bacteria that are applied to seeds and take up residence on the roots migrate very little from the site of application in the short term. Critics, however, are concerned with long-term consequences.

Many ecologists want to see the industry properly monitored and advocate careful testing before release. Genetically altered species, they say, are analogous to alien species introduced into new environments. The history of such introductions has been fraught with difficulties (see Chapter 6). In Canada, the **Food Inspection Agency** has assumed responsibility for screening genetically modified foods. They review applications and determine whether new products can be used or subjected to testing.

Genetic engineering stands at a threshold. Still in its infancy, it offers an unparalleled opportunity for improving agriculture and animal husbandry. At the same time, this revolutionary science carries with it an unknown potential for environmental harm that concerns many of its critics.

Protecting Wild Plant Species: Habitat Protection and Germ Plasm Repositories Tropical rain forests and other valuable habitats are being destroyed in record numbers to accommodate growing human populations. The loss of wildlands wipes out wild species, including the ancient relatives of modern crop species such as corn. The loss is not only aesthetic but also economic, for wild species from which modern crop species were derived possess genes that confer resistance to drought, insects, and disease. These genes can be transferred from ancient plants to modern crop species with relative ease. Such genetic boosts are extremely important to the success of modern agriculture. In fact, in a period of 60 years corn harvests climbed from 20 bushels per hectare to 100 to 250 bushels—because of genetic improvements. Consequently, plant breeders are always looking for species that could help produce hardier, more resistant plants.

In 1979, a team of researchers from the United States came across a weedy grass species in Mexico that gave rise to modern corn. Only a few patches were left. This species contains genes that could provide modern corn plants with resistance to several costly diseases. Some researchers are also using the plant to develop a strain of corn that grows in fields much like grass—year after year from the same root structure. As noted earlier, plants such as this are called *perennials*. If successful, this perennial would save farmers the cost of plowing, sowing, and cultivating—and would reduce soil erosion. The potential economic and environmental benefits are enormous. This example points out the importance of protecting rain forests and other valuable habitats—and their immense genetic reservoir.

Because much of the habitat of many organisms is bound to be destroyed by bulldozers and chain saws, biologists have been searching through forests and fields to gather seeds for cold storage in genetic repositories (**FIGURE 10-17**). There they can be held for future study and possible use. The U.S. Department of Agriculture currently runs the National Germ Plasm System, which has (as of June 2005) over 10,000 species in stock with 450,000 genetically different samples. They're adding approximately 10,000 to 20,000 new samples per year.

In 1985, the less developed nations belonging to the UN Food and Agriculture Organization (FAO) voted to establish a worldwide system of storing seeds, cuttings, and roots from native plants, to be available to all countries. This

FIGURE 10-17 Genetic repository. This room in a government facility is home to cuttings and seeds from thousands of plant species that are being preserved for future use.

system was created in large part to thwart genetic imperialism by the more developed nations, whose companies collect plants and seeds from LDCs and extract genes and important medicinal drugs, sometimes reaping huge profits while the less developed nations receive little or no benefit.

However important genetic repositories are to the immediate goal of protecting the world's species, this strategy has some major drawbacks. First, despite storage at low temperature and humidity, many seeds decompose and must be replaced. Others undergo mutations when stored for long periods and are no longer useful. Finally, storage systems will not work for potatoes, fruit trees, and a variety of other plants.

Developing Alternative Foods

Another strategy that may help expand food supplies is the development of alternative food sources—that is, new plants and animals taken from the wild, cultivated on farmlands, or grown in captivity.

Native Species as Sustainable Food Sources Many native plants and animals not currently eaten by humans could help meet the rising demand for food. The winged bean of the tropics, for example, could become a valuable source of food because the entire plant is edible: Its pods are similar to green beans. Its leaves taste like spinach. Its roots are much like potatoes, and its flowers taste like mushrooms. Food scientists are looking for other plants with similar potential.

Native animals may also provide an important, sustainable food source in years to come. In Africa, for instance, native grazers are already becoming a major source of protein. Native grazers are far superior to cattle introduced from Europe and America because they carry genetic resistance to local diseases and rarely overgraze grasslands, unlike cattle. Native grazers also generally convert a higher percentage of the plant biomass into meat and may be cheaper to raise.

KEY CONCEPTS

Many native plant and animal species could be used to provide food. Native animals offer many benefits over domestic livestock, including their resistance to disease-causing organisms.

Fish from the Sea and Aquaculture   Fish provide about 5% of the total animal protein consumed by the world's population. Although three quarters of the fish catch is consumed in the more developed nations, fish protein is important to the people of many poorer countries, in many cases supplying 40% of the total animal protein consumed.

During the 1970s, the world catch stabilized between 66 and 74 million metric tons a year. In the 1980s, it began to climb again, peaking at 96 million metric tons in 2000. In 2002, it was at 93 million tons. Because the total fish catch has remained fairly constant since the 1980s, the amount of fish available per capita has begun to decline. Further declines are very likely. Why?

Of 15 major oceanic fisheries, 11 are in decline. Continued fishing will inevitably result in dramatic population decreases in many commercially important fish species. Depletion of stocks, called **overfishing**, results when commercial fishing interests deplete the breeding stock so that a natural fishery cannot be maintained. Protecting current stocks—so vital to maintaining a sustainable harvest—will require global cooperation.

Another strategy that the world's fishing industry has engaged in is switching to alternative fish species. Table 10-1 shows the shift in major commercial fish species. These changes don't represent changes in preference, but changes in availability as a result of overfishing.

Each year, millions of tons of fish that are netted by commercial fishing interests are thrown overboard because they are the wrong species or sex or are undersize. International fishing regulations require this to protect various species and populations. Undersize fish represent the future reproductive stock. Nontarget species are called *by-catch* and can amount to 15% of the total catch.

Unfortunately, most of the fish are dead before they hit the water. Because of this, efforts are under way to allow commercial fishing interests to keep by-catch and put it to good use. Fish discarded by Alaskan fishing companies would provide 50 million meals a year.

Although this may sound like a reasonable policy, consider the other side of the issue. The regulations were drawn up in an attempt to prevent trawlers from circumventing fishery management regulations designed to help sustain and rebuild fisheries. If by-catch is permitted to be kept, unethical companies could fish with abandon, pulling in whatever they wanted, potentially wiping out important fisheries. The answer may be in the development of better fishing gear that eliminates undersize fish.

Barring any substantial increases in fish from the sea, many observers believe that one of the world's greatest hopes for increasing fish and shellfish production is **fish farms**, commercial endeavors where fish are raised in ponds and natural bodies of water. Fish farms are already common in many parts of the world, and new ones might help increase protein supplies. Fish farms are forms of aquatic agriculture, called **aquaculture** in freshwater and **mariculture** in salt or brackish water.

Table 10-1

Top 10 Species by Weight, 1970, 1980, 1992, and 2000 (catch in million tons)

1970		1980		1992		2000	
1. Peruvian anchovy	13.1	**1.** Alaska pollock	4.0	**1.** Peruvian anchovy	5.5	**1.** Peruvian anchovy	9.7
2. Atlantic cod	3.1	**2.** South American pilchard	3.3	**2.** Alaskan pollock	5.0	**2.** Alaskan pollock	2.7
3. Alaska pollock	3.1	**3.** Chub mackerel	2.7	**3.** Chilean jack mackerel	3.4	**3.** Skipjack tuna	2.0
4. Atlantic herring	2.3	**4.** Japanese pilchard	2.6	**4.** South American pilchard	3.1	**4.** Capelin	1.96
5. Chub mackerel	2.0	**5.** Capelin	2.6	**5.** Japanese pilchard	2.5	**5.** Atlantic herring	1.87
6. Capelin	1.5	**6.** Atlantic cod	2.2	**6.** Capelin	2.1	**6.** Japanese anchovy	1.85
7. Haddock	0.9	**7.** Chilean jack mackerel	1.3	**7.** Silver carp[1]	1.6	**7.** Chilean jack mackerel	1.75
8. Cape hake	0.8	**8.** Blue whiting	1.1	**8.** Atlantic herring	1.5	**8.** Blue whiting	1.6
9. Atlantic mackerel	0.7	**9.** European pilchard	0.9	**9.** Skipjack tuna	1.4	**9.** Chub mackerel	1.47
10. Saithe	0.6	**10.** Atlantic herring	0.9	**10.** Grass carp[2]	1.3	**10.** Largehead hairtail	1.45

Source: U.N. Food and Agriculture Organization.
[1]Raised on freshwater fish farms; all others are wild marine species.
[2]Raised on freshwater fish farms; all others are wild marine species.

Two basic strategies are employed in fish farming. In the first, fish are grown in ponds; population density is high and is maintained by intensive feeding, which is costly and therefore not always suited to the less developed countries. Fish (and shellfish) can also be maintained in enclosures or ponds where they feed on algae, zooplankton, and other fish that occur naturally in the aquatic ecosystem. This system requires little food and energy and is quite suitable for less developed countries.

Worldwide, fish farms produce about 22 million metric tons of food a year. Intensified efforts could double or triple this amount, providing additional food for the growing population.

KEY CONCEPTS
Most of the world's commercially important saltwater fish populations are in decline and in danger of being seriously depleted. The decline in wild fish populations has forced many countries to grow fish commercially in ponds, lagoons, and other water bodies.

Eating Lower on the Food Chain

According to one estimate, current food supplies would feed only approximately 2.5 billion people—less than one half of the world's population—if everyone ate like Americans. If, however, everyone ate a subsistence diet, getting only as many calories as needed, the annual world food production could supply an estimated 6 billion people.

Armed with statistics such as these, some people propose that wealthier citizens of the world can contribute to solving world hunger by eating less and eating lower on the food chain. The logic behind this idea is that by eating less and consuming more grains, vegetables, and fruits—and less meat—citizens of the more developed nations would free grain for the less developed nations. A 10% decrease in beef consumption in the United States, for instance, would release enough grain to feed 60 million people in the less developed countries. (Chapter 4 explained the biological reason why many more people could be sustained on a vegetarian diet.)

Although such a diet would be healthier and would help reduce obesity, which now afflicts a large portion of American adults, the problem with this answer to world hunger, say critics, is that sacrifices on the part of the wealthy would most likely not translate into gains abroad. Another problem is that land suitable for grazing is often not arable.

This is not to say that a vegetarian or a meat-conservative diet isn't desirable. It is. It is healthier and consumes far fewer resources. If there is a lesson to be learned from this issue, it is that to feed their people, the LDCs should concentrate on grain production rather than meat.

KEY CONCEPTS
Efforts to feed the world's people should focus on food sources low on the food chain—plants and plant products. Far more people can be fed on a vegetarian diet than on a meat-based one.

Reducing Pest Damage and Spoilage

Rats, insects, and birds attack crops in the field, in transit, and in storage. Conservatively, about 30% of all agricultural output is destroyed by pests, spoilage, and diseases. In the less developed nations, this figure may be much higher, especially in warm, humid climates where crops are grown year-round in conditions that are conducive to crop-damaging insects and disease.

Reducing the heavy toll of pests in the field could help increase the global food supply (Chapter 22). To create a sustainable system of agriculture, however, pest control measures must be safe and sustainable. Numerous strategies for pest management that fit these requirements are detailed in Chapter 22.

Another measure of great importance is the improvement of storage and transportation. Inefficient transportation can delay food shipments while rats, insects, birds, or spoilage take their share. Spoilage can be prevented by improving transportation networks, refrigeration, and other steps. In the less developed nations, the supply of fish could be increased by 40% by improving refrigeration on ships, in transit, and in stores. Grain supplies in the less developed nations could be stored in dry silos or sheds to prevent the growth of mold and mildew and reduce rodent problems. Technical and financial assistance from the more developed countries could go a long way toward improving food storage and transportation, potentially increasing world food supply by 10% or more.

KEY CONCEPTS
Much of the world's food production is consumed by pests or rots in storage or in transit. Improvements in transit and storage, such as refrigeration, can greatly boost food supplies.

Creating Agricultural Self-Sufficiency in Less Developed Nations

Over the years, many less developed nations have gone from being self-sufficient in food production to becoming major importers of food. The reasons for this are many. Population growth, poor land management, and economics top the list. Continued growth of population raises the demand for food while poor land management decreases a nation's ability to produce food. As for economics, the main problem here seems to be that many people simply can't afford food. Poverty in the less developed nations is therefore one of the key causes of hunger and malnutrition. Trade policies also affect food supplies. Cheap grain from more developed nations, for example, makes it difficult for indigenous farmers to compete. Many have gone out of business as a result of cheap grain from countries such as Canada, Australia, and the United States, grain that was either donated as food aid or sold cheaply because of subsidies provided by the exporting nation's government. Staple crops in less developed nations have also been replaced by export crops, for example, tea, coffee, and bananas, reducing food supplies for local citizens. The force behind this shift has been an economic one. In an effort to pay back debts to more developed nations such as the United States, many governments of the LDCs have encouraged the production of export crops. This generates tax income for the LDCs so they can pay back their debts. Unfortunately, it decreases food supplies for their people.

Improvements in all three arenas could go a long way toward increasing food supplies and helping countries become more self-reliant. Many of the provisions discussed earlier in this section can help improve land management. Population control strategies discussed in Chapter 9 can help reduce pressure on limited food supplies. Changes in economic policies in both LDCs and MDCs and trade policies may be required to promote self-sufficiency as well.

KEY CONCEPTS

Many LDCs have lost their ability to produce food as a result of overpopulation, farmland deterioration, and economic and trade policies. Reversing these trends could help nations become more self-reliant, which is vital for building a more sustainable future.

Legislation and New Policies: Political and Economic Solutions

Laws and policies that promote unsustainable farming and harvesting practices must be changed or abandoned—in all nations, rich and poor. The world's commercial fishing industry spends over $125 billion to catch $70 billion worth of fish. The remainder, $55 billion, came from governments (and hence, taxpayers) in the form of subsidies and low-interest loans. Such policies encourage overfishing and a depletion of this vital food source. Although in the short term, efforts to boost food production may be alluring, in the long run this artificial economic support may lead to ecological impoverishment and a sharp decline in food availability. This phenomenon of short-term production and profit at the expense of long-term production and profit crops up time and time again in environmental issues, and it must be carefully analyzed. Farmers, for instance, may find it prohibitively costly to invest in soil conservation measures. By not investing in them, though, they systematically destroy the productive capacity of their land. Sacrificing long-term sustainable production for short-term profit is a trade-off with serious social, economic, and environmental ramifications. Some advocates of sustainability who are opposed to subsidies argue nonetheless that there are times when they can be used to our benefit to prevent such trade-offs. The economics of the short-term/long-term puzzle are examined in Chapter 25.

To help ensure water supplies, laws that encourage waste should also be modified or replaced. In Colorado, for example, farmers or ranchers who conserve water they are allocated stand to lose the rights to that water, thus eliminating the incentive to be frugal. Ranchers and farmers must use it or lose it. Simple changes in water laws could free up enormous amounts of water to meet future demand as the world population grows and food requirements increase and for wildlife and recreational use.

Voluntary efforts are also needed. For example, some U.S. farmers may find it advantageous to abandon crops that use large amounts of irrigation water in water-short regions. Shifting water-thirsty crops such as cotton and rice from the desert of California's Central Valley to more suitable climates could free up enormous amounts of water for crops requiring less water. Government policies that promote the cultivation of environmentally incongruous crops such as these should be examined and modified.

As noted earlier, changes in public policy in the less developed nations are also vital to creating a sustainable system of agriculture. For example, programs and policies that encourage the cultivation of cash crops such as coffee and bananas for export in place of staples needed to feed the people of a country deserve special attention.

Efforts are needed to encourage small-scale farming and better soil management in the less developed nations. Because many LDCs are poor, international aid can help promote sustainable agricultural practices including the many soil conservation measures described in this chapter.

Ending War

Throughout the world, dozens of civil conflicts are now raging, many of them in Africa. Besides the heavy toll war takes on people, these often bitter conflicts can seriously disrupt the production and distribution of food. Food supplies can be cut off when bridges are bombed or railroad yards are destroyed. Soldiers also damage crops or farmland in an effort to weaken an enemy's resolve. What is more, damage to farm fields and the distribution system can last for many years after a conflict has ended. Mines in farm fields, for instance, may prevent farmers from going back into them to begin planting. It goes without saying that putting an end to war is vital to helping provide not just a peaceful life, but adequate food required for health.

KEY CONCEPTS

Violent conflicts among peoples can greatly disrupt the production and distribution of food, often long after war has ended.

This chapter began by listing three challenges in agriculture: the near-term need to feed malnourished people, the long-term need to provide food for future generations, and the continuing need to protect and enhance the soil and the environment. It should be clear that there are many ways to meet these challenges. No one idea will work. Rather, these ideas must be integrated into a comprehensive policy to build a sustainable agricultural system. (Related personal actions are listed in the Individual Actions Count! table.)

Individual Actions Count!
- Buy organically grown produce. (Try to purchase at least one organically produced vegetable every time you shop.)
- Learn to cook more meals with vegetables. Substitute tofu for meat.
- Support local farmers by buying at farmer's markets and roadside stands. (Farmer's markets can even be found in cities.)
- Plant a garden if you have the space. If you don't, a window planter or a large flowerpot with lettuce, spinach, or tomatoes will work in a sunny location.
- If you don't have room to plant a garden, start a community garden or join a community garden in your neighborhood.
- When you buy a home, make sure there's a good garden space.
- Use natural pesticides and organic fertilizer in your garden.
- Compost leaves and kitchen and yard wastes to build the soil in your garden.
- Contribute to municipal local composting efforts if you don't want to have your own compost pile

As we strive to create a system of food production that provides high-quality food at a reasonable price while protecting the environment, we must remember that even if we take care of the land and manage it sustainably, human population could very well outstrip our maximum sustainable food production. Population stabilization and long-term efforts to reduce the size of the human population, discussed in Chapter 9, are vital components of the long-term strategy to create a sustainable human presence.

To a man with an empty stomach, food is God.

—*Gandhi*

CRITICAL THINKING

Exercise Analysis

There is a problem with the line of reasoning that says that because food production per hectare—or productivity—is on the rise, we shouldn't be worried about agriculture. As you may recall from Chapter 2, just because a system appears to be functioning well does not mean that it can be sustained over the long haul. World fish catch is at an all-time high, but the majority of the world's fisheries are in trouble.

To assess the sustainability of agriculture or any other system that relies on renewable resources, we must look beyond measures of current performance such as productivity to fundamentally more important indicators such as the amount of topsoil on farms, the nutrient levels in the topsoil, the amount of land taken out of production because of salinization and waterlogging, and the supply of water available for irrigation. These are better indicators of agriculture's long-term sustainability. As this chapter points out, when you examine the cold, hard statistics of agriculture, it becomes clear that current agricultural practices are unsustainable.

Why then do we see gains in agricultural productivity if the system is unsustainable? The answer is that gains in productivity largely result from offsetting measures such as irrigation and fertilizer use. Although the soil may be eroding away, farmers are able to maintain or increase productivity by adding more fertilizer. Sooner or later, this approach will fail.

This exercise requires four critical thinking rules. First, it requires us to question the conclusions—notably, that rising productivity means that agriculture is sustainable. It also requires us to define terms such as sustainability and productivity very carefully so we understand what we're talking about and think critically about it. Third, it calls on us to look for hidden assumptions. In this case, it is assumed that rising productivity signifies a healthy system, but what it really reflects is the use of offsetting measures. Finally, we must examine the big picture—that is, look at all of the factors that contribute to the sustainability of farming.

CRITICAL THINKING AND CONCEPTS REVIEW

1. What percentage of the world's population is malnourished? What are the short- and long-range effects of malnutrition?
2. Using your knowledge of environmental issues and agriculture and your critical thinking skills, analyze the following statement: "Hunger is not an environmental issue."
3. What is desertification, what factors create it, and how can it be prevented?
4. Critically analyze this statement: "Soil erosion control is too expensive. We can't afford to pay for it because our crops don't bring in enough money."
5. What justifications can be devised for the expense of soil erosion control and other sustainable farming practices?
6. Critically analyze this statement: "All subsidies to the agriculture and fishing industries should be banned."
7. Describe waterlogging and salinization of soils. How can they be prevented? Which of your proposed solutions might result in further problems?
8. Describe the decline in agricultural species diversity. How could this trend affect world agriculture? Give some examples.
9. Critically analyze this statement: "Technology can solve all of our food problems, so there is no need to slow population growth."
10. List and discuss the major strategies for solving world food shortages. Which are the most important? How would you implement them in this and foreign countries?
11. Critically analyze the statement, "Simply by practicing better soil conservation and replenishing soil nutrients, we can reduce the need for new farmland."
12. Describe the successes and failures of the Green Revolution. What improvements might be made?
13. Critically analyze the following statement: "The world fish catch climbed nicely during the 1980s, and there is no reason to believe that the oceans won't be a major source of new food."
14. Describe the pros and cons of policies designed to eliminate the discard of by-catch (fish thrown back into the sea because they are the wrong sex or species).
15. You have been appointed head of a UN task force. Your project is to develop an agricultural system in a poor African nation that imports more than 50% of its grain but still suffers from widespread hunger. Outline your plan, giving general principles you would follow and specific recommendations for achieving self-sufficiency.
16. Make a list of things you can do to help solve the world hunger problem and create a sustainable system of agriculture.
17. In Chapter 2, Figure 2-3 presented a model that can be used to apply the principles of sustainability to human systems. Look up that model and fill out the chart using the agricultural system as an example.
18. Can you determine ways that sustainable development principles applied to other human systems might influence agriculture and improve our prospects for feeding the world's people?

KEY TERMS

A horizon
accelerated erosion
annuals
aquaculture
B horizon
C horizon
climate
Conservation Reserve Program
conservation tillage
contour farming
crop rotation
D horizon
desertification
Environmental Quality Incentives
    Program
erosion
farmland conversion
fish farms
Food Inspection Agency

genetic engineering
green manure
Green Revolution
herbicides
humus
hybrids
infectious diseases
kwashiorkor
litter layer
marasmus
mariculture
minimum tillage
natural capital
natural erosion
1985 Farm Bill
O horizon
organic fertilizers
overfishing
parent material

perennial
pesticides
salinization
selective breeding
shelterbelts
soil
soil erosion
soil profile
strip cropping
subsidies
subsoil
sustainable agriculture
synthetic fertilizer
terraces
topsoil
water table
waterlogging

REFERENCES AND FURTHER READING

The References and Further Reading section at the end of this book contains a list of sources for the information discussed in this chapter and recommendations for further reading.

Human activities can threaten other species that share the planet with us.

CHAPTER 11

Preserving Biological Diversity

The worst sin toward our fellow creatures is not to hate them, but to be indifferent to them.
—George Bernard Shaw

In 1996, a consortium of research institutes, including the Smithsonian Institution, received funding for a research project in the tropical rain forest of Nigeria and Cameroon. The researchers studied the local abundance, distribution, and life cycles of trees and shrubs with medicinal properties in a large segment of forest in Cameroon and in several smaller plots in Nigeria. Researchers searched for beneficial drugs by examining traditional medicines used by local villagers and by a mass screening of trees and shrubs. This project, funded by the U.S. Agency for International Development and several other U.S. government agencies, was an attempt to help develop alternatives to deforestation. It is believed that cultivation of medicinal trees could provide an important source of income to

You own a guest ranch in western Colorado. Your ranch attracts people from all over the world who come to ride horses and view the abundant wildlife such as deer, elk, eagles, hawks, and an occasional black bear. You also raise sheep to supplement your family's income. One day, a neighboring rancher calls on you to ask for your help in killing coyotes, which he says are killing off his sheep and costing him lots of money. What he wants is permission to spread chunks of meat, containing a lethal poison called Compound 1080 on your property. What he is proposing is not legal, he freely admits, but it is done from time to time in remote parts of the West. He and other ranchers who want to control coyotes extract the poison from sheep collars—devices placed around the necks of sheep to selectively kill coyotes that attack them. Sheep collars are legal. Your neighbor will get the poison and even spread the scraps of meat on your property if you give him permission. He says it will also help protect your sheep herd.

What concerns would you have about this proposal? Make a list of questions you will need to answer before deciding whether to join your neighbor. What critical thinking rules will you use?

local communities. The field work, which ended in 2004, yielded numerous natural chemical compounds that could some day be used to treat tropical diseases, mental disorders, and chronic diseases.

This project is one of many efforts designed to save forests and the species that live within them. It is part of an effort to protect the natural environment and provide for the economic needs of people. As such, it is an example of the new philosophy of sustainable development.

This chapter examines biodiversity—the rich biological world—and new ways of protecting the planet's wealth of wild species. In the following pages, you will examine the causes of extinction and the countless benefits of plants and wildlife to human societies to help you fully understand the importance of saving species.

11.1 Biodiversity: Signs of Decline

By some estimates, as many as 500 million kinds of plants, animals, and microorganisms have made this planet home since the beginning of time. Today, scientists estimate that the world contains between 10 and 80 million species, although only about 1.5 million have been identified and named. Thus, 420 to 490 million species have become extinct over a period of 3.5 billion years.

This tells us that extinction is an evolutionary fact of life. Scientists refer to this as **natural extinction**. The natural occurrence of extinction is viewed by many as a justification for plant and animal extinctions occurring today as a result of human activities, known as **accelerated extinction**. Why not?

Natural extinction differs markedly from accelerated extinction for at least two important reasons. First, natural extinction represents a kind of evolutionary passing. That is to say, many millions of species have become extinct, but they did not disappear. They evolved into new species (**FIGURE 11-1**). Their kind may have vanished, but they gave rise to new species. Today, they are represented by their descendants. Modern extinctions, on the other hand, eliminate species entirely. They represent a dead end in evolution.

Second, the rate of extinction varies considerably. Even though some species did vanish in mass extinctions because of severe climatic changes or for other reasons, the rate of natural extinction—about one species every 1000 years—is slow compared with today's accelerated extinction. Although no one can tell exactly how many species become extinct today, estimates based on loss of habitat and species diversity within these ecosystems suggest that it may be as many as 40 to 100 species every day. Harvard biologist E. O. Wilson thinks the number may be higher. Some biologists put the number at about 140 per day! As a result, many believe that we have entered into an era of extinction unparalleled in the history of the Earth. (For more on the acceleration of extinction and the importance of species, see Point/Counterpoint 4-1.)

Today, thousands of species are endangered or threatened. An **endangered species** is one that is in imminent danger of becoming extinct. A **threatened species** is one that is still abundant in its natural range but, because its numbers are declining, is likely to become extinct in time. Without concerted efforts to protect them, many threatened species will become extinct. How serious is the problem? Today, three fourths of the world's bird species are declining in number or threatened with extinction. Moreover, more than two thirds of the world's 150 species of primates are threatened. Today, approximately 25,000 species of plants are threatened with extinction—1 of every 10 plant species on Earth. In the United States, approximately 3000 species of plants are in danger of extinction.

Scientists around the world are disturbed by an alarming disappearance of amphibians. Many species of frogs, toads, and salamanders are either experiencing steep declines in population size or have vanished. Amphibians are disappearing from a wide variety of habitats—from the jun-

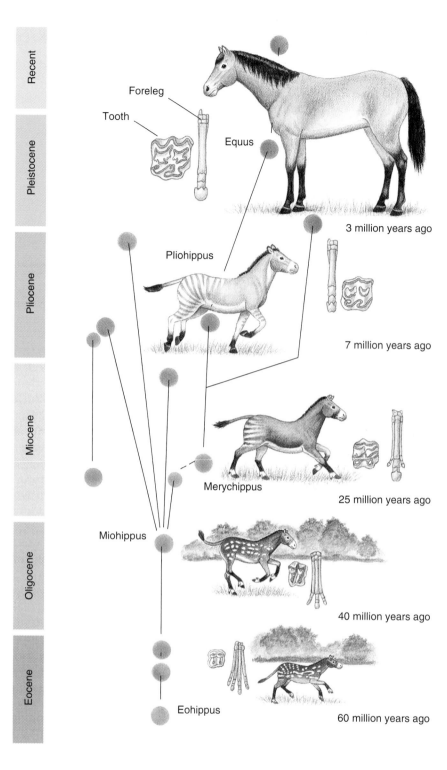

FIGURE 11-1 Gone but not forgotten. This drawing illustrates the stages in the evolutionary history of the horse. All distinct species are extinct except for the modern horse. Like other species, horses evolved through intermediate stages that no longer exist.

Recent

Pleistocene

Pliocene

Miocene

Oligocene

Eocene

Tooth

Foreleg

Equus

3 million years ago

Pliohippus

7 million years ago

Merychippus

25 million years ago

Miohippus

40 million years ago

Eohippus

60 million years ago

gles of Brazil to the suburbs of New York City. The cause is still unknown, but most scientists believe that the decline is the result of a wide range of factors including habitat loss; environmental pollution, such as pesticides; exposure to ultraviolet radiation due to ozone depletion; disease; parasites; global warming; and others.

Unless we curb population growth, reduce pollution and habitat destruction, and manage our resources better, vast expanses of forests, wetlands, and grasslands will vanish

over your lifetime. Millions of species may vanish as a result. The impact of habitat destruction and mass extinction will be felt worldwide.

KEY CONCEPTS

Many species of plants and animals face extinction today as a result of human activities. Although extinction has occurred since the dawn of time, modern extinctions are occurring at a rate much faster than is biologically sustainable.

11.2 Causes of Extinction and the Decline in Biodiversity

Plant and animal extinction, like many other environmental problems, results from many factors. This section examines the two most important factors today: habitat alteration and commercial hunting/harvesting. It also examines other important factors, including (1) the introduction of alien and domesticated species, (2) pest and predator control, (3) the collection of animals for pets and research, (4) pollution, (5) ecological factors, and (6) the loss of keystone species.

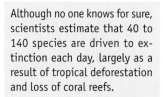

Although no one knows for sure, scientists estimate that 40 to 140 species are driven to extinction each day, largely as a result of tropical deforestation and loss of coral reefs.

KEY CONCEPTS

Many factors contribute to the loss of species, but the two most important are the destruction and alteration of habitat and commercial harvesting.

Physical Alteration of Habitat

We humans have always altered the environment to meet our needs. Virtually every activity we undertake changes the biotic and abiotic conditions of the environment and, hence, the biological communities of aquatic and terrestrial ecosystems (Chapter 6). Every product we buy comes with an environmental price tag. Extraction of the raw materials and production of finished products, even the transportation to stores or our homes, for example, causes damage. Some of the most dramatic environmental changes come from activities such as food production, timber harvesting, mining, road building, the manufacture and operation of automobiles, and the construction and maintenance of homes.

Human activities tend to fragment the habitat of plants and animals. Human settlements, for example, become islands of human activity in the natural environment. As the human population grows, however, and as roads, highways, farms, and buildings increase in number, the pattern changes: All that is left are a few islands of natural habitat in the human-dominated landscape (**FIGURE 11-2**). As the natural habitat fragments, species diversity declines.

Habitat alteration ranges from moderate to extreme and ranks as *the* most significant factor in the extinction of plants and animals the world over. It is no surprise to find that those areas of the world that are at risk are regions of the most intense human activity.

Approximately three fourths of the world's bird species are declining in number. More than two thirds of the world's 150 species of primates (apes and monkeys) and 1 of every 10 plant species is threatened.

FIGURE 11-2 Habitat fragmentation. Humans carve up the natural landscape to make room for cities, towns, farms, and other uses.

Nowhere is the loss of habitat more pronounced than in the tropical rain forests. Tropical forests house at least half of the Earth's species, perhaps more. Once covering an area the size of the United States, the tropical rain forests have been cut by about one-half. Countless species have perished as a result. (Chapter 12 discusses tropical rain forests in more detail.)

Coral reefs, wetlands, and estuaries (the mouths of rivers) are other critical habitats already greatly reduced and rapidly declining because of human development. Wetlands and estuaries, for example, are the home of many species but are also highly prized by humans for development. Damage to wetlands has been particularly severe in the industrial nations. New Zealand and Australia, for example, have lost more than 90% of their wetlands and countless species that lived in them. India, Pakistan, and Thailand have lost at least 75% of their mangrove swamps, a type of coastal wetland. Their ongoing destruction threatens the future of fish and waterfowl throughout the world. (For more on wetland destruction, see Chapter 13.)

Another ecosystem that has experienced serious losses is the tallgrass prairie of North America, which today is virtually nonexistent. Temperate rain forests, such as those in the Pacific Northwest, have also been subjected to intense harvesting. Worldwide, 56% of the temperate rain forests have been logged or cleared. No more than 10% of the original old-growth temperate rain forests remain in the United States. Canadian temperate rain forests have also been heavily cut. One of the many victims of this loss is a bird known as the *marbled murrelet*. The murrelet feeds on schooling fish that congregate near the shores of Canada, but nests on moss-covered branches of the temperate rain forest within 18 miles (30 kilometers) of the shore. Logging of old growth forests in British Columbia has dramatically reduced the habitat of this bird, causing equally dramatic declines in its numbers.

Virtually all human activities alter the environment, changing the biotic and abiotic conditions and fragmenting habitat. Habitat alteration is the number one cause of species extinction. The most dramatic changes occur in biologically rich areas: tropical rain forests, wetlands, estuaries, and coral reefs.

Commercial Hunting and Harvesting

Humans have hunted and killed animals throughout history to provide food, fur, and other products. Several different types of hunting occur today. **Sport hunting** occurs for enjoyment and is generally well regulated. In fact, in many cases sport hunting benefits wild populations by helping to control numbers—so that they remain within the carrying capacity of the environment.

Another form of hunting is **subsistence hunting**, killing animals to provide food for indigenous people such as those who live in the tropical rain forests. Generally, such activities are carried out on a sustainable level and do not pose a threat to animals, although there are notable exceptions.

The third form of hunting is **commercial hunting**, or **harvesting**. This consists of large-scale efforts, such as the whale hunting of years past, and smaller operations, such as hunting African rhinos for their horns. Smaller operations may be legal or, in some cases, illegal.

Whale hunting is one of the most familiar examples of commercial hunting. In the 1700s and 1800s, commercial whalers hunted one species after another to the brink of extinction to provide oil, meat, and other products. The result has been a severe reduction in whale populations (Table 11-1). Thanks to efforts by the International Whaling Commission, commercial whaling has been greatly reduced. In its place is a new industry: whale watching, with annual revenues that exceed those of the commercial whaling industry itself.

Large-scale commercial hunting also doomed the passenger pigeon and greatly reduced the size of the bison herds of North America. The world's fisheries (fishing grounds) have also been heavily harvested for commercial gain, and many important fisheries have been eliminated. Overharvesting continues today in many of the world's remaining fisheries.

Commercial hunting may also occur illegally, on a much smaller scale. Illegal hunting is called **poaching**. Today, poachers continue to slaughter elephants, rhinos, tigers, and a variety of other endangered species because the economic benefits outweigh the risks of small fines or light jail sentences. A Bengal tiger coat, for example, sells for $100,000 in Japan. Poaching of ivory from tusks caused the near demise of elephants, from an estimated 2.5 million in 1970 to an estimated 440,000 to 600,000 today according to the World Wildlife Foundation.

Commercial hunting and harvesting of wild species have occurred for centuries and represent the second largest threat to the world's animal species. This includes past activities, such as whale hunting, and present activities, such as commercial fish harvesting and poaching of endangered species.

The Introduction of Foreign Species

Foreign, or alien, species introduced accidentally or intentionally into new territories often do well in their new homes because conditions may be highly favorable to their growth and reproduction. One reason for this is that these species may face little environmental resistance—for example, there may be no natural predators to hold their populations in check—in their new home. As their populations expand, alien species often outcompete and eliminate native species. The English sparrow is an example. Deliberately introduced into this country in the 1850s, the sparrow quickly spread throughout the continent. It now competes for nesting sites and food once used by bluebirds, wrens, and swallows. The starling is another intentional import from England that has had a similar effect on native North American bird species. The barred owl of North America is yet another example. This native species has spread across the continent, some scientists think, hopscotching from one patch of trees to another on the Great Plains, which was once a vast sea of grass that blocked the westward movement of the owl. It is now a resident of the forests of the Pacific Northwest. The barred owl harasses and preys on spotted owls and takes over their nesting sites. It may even interbreed and could further threaten this species, which is suffering because of the loss of vital old growth habitat.

Alien plants and animals have taken their toll on native species of Canada, too. European starlings have displaced many species. Raccoons that were released on Queen Charlotte Islands to provide additional income for local trappers have caused significant decreases in a type of sea bird known as an *alcid*. These burrow-nesting birds form huge vulnerable colonies. Raccoons raid the nests for food.

Table 11-1		
Whale Populations—Then and Now		
Species	**Number Before Commercial Whaling**	**Current Estimate**
Blue	166,000	400–1400
Bowhead	54,680	8000
Fin	450,000	47,300
Gray	15,000–20,000	26,400
Humpback	119,000	21,570
Minke	250,000	761,000
Right	50,000	3000*
Sei (includes Bryde's)	108,000	36,800–54,700*
Sperm	1,377,000	982,300*
*Population estimates to be updated soon		
Source: International Whaling Commission		

Nonnative plants also outcompete native species. Scotch broom, purple loosestrife, and crested wheatgrass have all been introduced into Canada and now compete aggressively with native species, sometimes wiping them out altogether. Today, 23% of Canada's wild plants are nonnative species. Islands are especially vulnerable to new species. In Hawaii, for example, 90% of all bird species have been wiped out by human inhabitants and by organisms (such as rats) introduced by humans.

Alien species often do well in warm climates such as Florida. In fact, Florida is a showcase of alien species gone wild. Australian pines, for example, were introduced as ornamentals and have spread rapidly along coastal beaches and canals. Their shallow roots are so dense that they destroy sandy beaches, where many sea turtles lay their eggs. Worst of all is a species called the *punk tree*, which grows in swamps, creating a dense tangle of vegetation impassable to many animal species.

The Great Lakes of North America have also been subject to numerous alien invasions, often with devastating economic and ecological consequences. The latest are the zebra mussel and fishook water fleas, both accidentally introduced by ships carrying cargo from Europe. The zebra mussel was described in Chapter 6. The fishook water flea feeds on phytoplankton, the main food source of small fish. Besides posing a threat to fisheries, the flea also has become a nuisance to commercial and sports fishers, clogging nets and lines. It is spreading rapidly throughout the Great Lakes and into surrounding waterways, just like the zebra mussel.

The alien invasion is nothing to take lightly. In the United States, a recent study that took 200 scientists 4 years to prepare estimates that 6500 alien species of plants, animals, insects and arachnids (spiders), and disease organisms have gained a foothold in the United States. They have caused 315 native species to become endangered or threatened. Although no one knows for sure how much damage alien species cause, one study by researchers at Cornell University suggests that the damage in the United States caused by alien plants and animals may be over $120 billion per year! The main problems are predation on native species and habitat destruction.

> By one estimate, there are over 6500 alien species in the United States, which cause an estimated $123 billion in damage each year.

KEY CONCEPTS
Plant and animal species introduced into new regions may thrive because of the favorable conditions and low environmental resistance. Therefore, they often outcompete and eliminate native species. Islands are especially vulnerable to foreign species.

Pest and Predator Control

Well-intentioned efforts to control species considered to be pests also affect wild populations of plants and animals, and they contribute to the steady decline in biodiversity on Earth.

DDT and other pesticides, for example, have taken a huge toll on American wildlife (Chapter 22). The peregrine falcon had disappeared in the eastern United States by the 1960s as a result of DDT. This pesticide caused eggshell thinning, which made eggs fragile and susceptible to breakage. The entire population of falcons east of the Mississippi River was wiped out and dramatic declines occurred in the West, too. Eagles and brown pelicans met a similar fate. Even the California condor, a massive vulture, suffered from eggshell thinning. Especially harmful to migratory birds that spend their summer in North America are those pesticides (such as DDT and related compounds) that have been banned in the United States but are still used in Latin America.

Studies suggest that nonylphenols, a nonactive ingredient of some pesticides, may have a deleterious effect on salmon. This near-ubiquitous water pollutant, also found in detergents, cosmetics, and plastics, enters waterways. In the Pacific Northwest, salmon migrating to the sea that are exposed to minute quantities of this chemical perish when exposed to salt water. Researchers believe that nonylphenols affect the endocrine system, making salmon unable to produce hormones needed to get rid of excess salt, an adaptation vital to the transition to oceanic life.

Predator control, once the cornerstone of wildlife management, has had a profound impact on native species such as wolves, bears, and mountain lions in North America. Killing off predators can also create severe habitat destruction, as the prey populations once controlled by these predators grow beyond the carrying capacity of their environment.

KEY CONCEPTS
Chemical pesticides, sprayed on farms and other areas to control insect pests, and predator control programs have had a profound impact on native species.

The Collection of Animals and Plants for Human Enjoyment, Research, and Other Purposes

Each year, millions of animals and plants are gathered from their native habitat throughout the world and exported to zoos, private collectors, pet shops, research institutes, and other places. The numbers are staggering. In 2003, for instance, more than 85 million fish, 721,000 live reptiles, and 257,000 reptile skins were legally imported into the United States. That same year, nearly 6200 live birds were imported into the United States. Tens of thousands of birds are also imported by Canada and Great Britain. For each bird that makes it into someone's home, though, 10 to 50 may die along the way (**FIGURE 11-3**) Many perish after they are taken home, too.

Scientists throughout the world use a variety of wild caught primates such as monkeys for research on pressing medical problems such as AIDS. Taken from their homeland in Africa, as many as five chimpanzees die for every one that enters a laboratory. In 1975, the United States banned the importation of all primates for pets but allowed contin-

FIGURE 11-3 **Nature for sale.** In the developing countries, local residents sell birds and animals they catch in nearby forests to dealers who export them, sometimes illegally, to the industrial nations. Many animals die along the way.

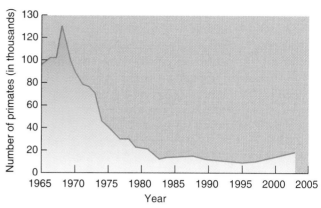

FIGURE 11-4 Decline in the number of primates imported into the United States.

KEY CONCEPTS

Millions of plants and animals are taken from the wild and imported into developed countries for zoos, private collections, pet shops, and research, contributing to the worldwide loss of species.

ued importation for zoos and research. Partly as a result, primate imports have dropped rapidly in the United States and elsewhere (**FIGURE 11-4**). Nonetheless, in 2003, the latest year for which data were available, the United States imported over 18,613 primates.

Some rare and endangered primates are still captured and sent to the United States. Because research animals often do not breed in captivity and because they have a high mortality rate, continual replenishment from wild populations is likely to continue. As noted earlier in the chapter, two thirds of the world's 150 species of primates are threatened with extinction.

Plants are also popular imports. In 2003, over 3.8 million live cacti (including some artificially propagated) were imported into the United States from more than 22 countries. International trade in exotic flowers, especially those originating in the Mediterranean, is also a booming business and currently threatens many native plant species.

At home, collectors pillage the deserts in Texas and Arizona in search of salable cacti to adorn the lawns of customers. To reduce the ravaging impact of commercial cactus rustling, Arizona has made it illegal to remove 222 different plant species. With penalties up to $1000 and jail sentences up to 1 year, Arizona has taken a small step to protect its native plants. With only two "cactus cops" to patrol the state, though, little progress can be made.

Pollution

Pollution plays an important role in the decrease in the planet's biodiversity and is bound to play a larger role as the human population and economy expand. Today, pollution-caused problems such as global warming, acid deposition, and ozone depletion are creating major ecological changes. Pollution could combine with other extinction forces described in this chapter to greatly increase the threat to biodiversity. For example, global warming caused by carbon dioxide pollution from the burning of fossil fuels combined with several other pollutants may be responsible for a massive die-off of the world's coral reefs, already besieged by sediment from onshore development, damage from ships, and chemical pollutants. Coral reefs contain an estimated one million species. By one estimate, 60% of the world's 230,000 square miles of coral reef could be lost by around 2040 if threats continue.

Pollution is also of concern on a local level. In the semiarid farmland of California's San Joaquin Valley, the heavy metal selenium—contained in waters draining from irrigated farm fields into specially built evaporation ponds (**FIGURE 11-5**) in the Kesterson National Wildlife Refuge—has caused massive biological effects. In 1985, for example, biologists found physical abnormalities in 42% of the chicks of waterfowl and wading birds at Kesterson—compared with an expected deformity rate of 1%. Problems included chicks without eyes, beaks, wings, and legs. Many embryos died. Adult birds and many other animals were also affected—as were crayfish, snakes, raccoons, and muskrats, which once flourished in the rich biological community but now have vanished. As one journalist put it, "The Kesterson Refuge had become a place that killed the animals it was

FIGURE 11-5 **Polluted evaporation ponds.** Selenium and other toxic metals, leached from irrigated farmland and carried to nearby evaporation ponds (now closed), have resulted in a number of embryonic defects.

FIGURE 11-6 **Museum specimens.** The passenger pigeon, a once abundant species whose flocks darkened the skies, is extinct in large part because of commercial harvesting.

supposed to protect." (In 1986, the evaporation ponds at Kesterson were filled with dirt.) Federal officials note similar problems in refuges in Utah, Wyoming, Texas, Nevada, Arizona, and other parts of California. In nonrefuge sites, they're proposing costly water treatment facilities to reduce selenium levels, followed by periodic removal of contaminated sediments from the bottom of the ponds in some locations.

Pollution affects many species the world over. One of them is the already-threatened beluga whales of Canada's St. Lawrence estuary. Scientists have found that older whales are dying of cancer and that young calves are ingesting poisons from their mothers' milk. Autopsies of dead whales show they are heavily contaminated with 25 different toxic chemicals. Although this is not the only threat to the beluga whale, whose population numbers slightly more than 1000, it could be a major one.

KEY CONCEPTS

Pollution alters the physical and chemical nature of the environment in ways that may impair the survival of many species. Pollution and climate change (caused largely by pollution) may be altering the health of the world's coral reefs and may cause widespread losses if something is not done to reverse the trend.

Biological Factors That Contribute to Extinction

Not all species are created equal when it comes to extinction. Some species, for instance, are quite adaptable and seem to do well in the human-dominated world. Unfortunately, they're not always the most desirable species. Pigeons and cockroaches are two examples. Others do not fare so well. Some of these species have peculiarities in behavior or lifestyle that make them more vulnerable to extinction. Others produce so few young that they are not very resilient.

Consider the passenger pigeon (**FIGURE 11-6**). At one time, the passenger pigeon inhabited the eastern half of the United States in flocks so large they darkened the sky. Probably the most abundant bird species to ever live, the passenger pigeon is now extinct, in large part because of widespread commercial hunting and habitat destruction. Between 1860 and 1890, countless pigeons were killed and shipped to the cities for food. In 1878, the last colonial nesting site in Michigan was invaded by hunters. When the guns finally fell silent, over one billion birds had been killed. By this time, only about 2000 remained. Broken into flocks too small to hunt economically, the birds were finally left alone. Did they recover? No. The number of birds dwindled year after year, until in 1914, the last bird held captive in the Cincinnati Zoo died.

This story illustrates an important point: Some species have a critical population size below which survival may be impossible. The passenger pigeon needed large colonies for successful social interaction and propagation of the species. Two thousand birds were simply not enough.

Scientists know very little about the critical population size for many species. Unfortunately, overhunting, habitat destruction, and other activities discussed in this chapter can do irreparable damage before we realize it.

Another trait that influences an organism's survivability is its degree of specialization. Animals may be categorized as specialists or generalists. Specialists tend to be more vulnerable than generalists. Why? Generalists can exploit more food sources and can live in more habitats than specialists. They are more adaptable and, consequently, less vulnerable to habitat destruction and other forces.

Animal size also contributes to extinction. Larger animals such as the rhinoceros are easier (and often more desirable) prey for human hunters. Moreover, they are more likely to compete with humans for desirable resources such as grazing land. Larger animals also generally produce fewer offspring, making it more difficult for reduced populations to recover. The California condor, for example, lays a single egg every other year. Young condors remain dependent on their parents for about 1 year but are not sexually mature until age 6 or 7. Combined, these factors give the condor little resiliency to withstand pressures from human populations or natural disasters.

Another factor in extinction is the size of an organism's range. The smaller the range, the more prone the organism is to extinction.

KEY CONCEPTS

Many biological characteristics of organisms determine how vulnerable they are to human impacts on the environment, such as the number of offspring they produce, the size of their range, their tolerance for people, and their degree of specialization.

The Loss of Keystone Species

According to ecological theory, in some ecosystems the extinction of one critical species, known as a **keystone species**, may lead to the loss of many other species. For example, the gopher tortoise in the southeastern United States is a keystone species (**FIGURE 11-7**). Gopher tortoises are important be-

cause they dig long burrows in the sand, up to 400 feet long and 30 feet deep, that many other species share. The Florida mouse becomes a permanent resident, as does the gopher frog. Opossums, gray foxes, and indigo snakes frequent the burrows as well. The gopher tortoise is so important, in fact, that in areas where it has been eliminated, 37 species of invertebrates have disappeared.

Fig trees are a keystone in the tropical rain forests of Peru. Studies of plant communities in these forests show that three fourths of the birds and mammals of the Amazonian rain forest rely on fruits. Most fruits, however, are available only 9 months of the year. During the remaining period, monkeys, peccaries, parrots, and toucans live on figs. Their loss would be devastating.

In the marine ecosystem in the U.S. Pacific, the sea otter is a keystone species. Sea otters are voracious eaters, feeding on sea urchins, abalone, crabs, and molluscs that inhabit kelp beds. The otter helps control sea urchin populations, which feed on kelp and seaweed. In locations where sea otters have been eliminated, kelp beds have disappeared because sea urchin populations increase in the otter's absence. When sea otters repopulated these regions, urchin populations fell and the kelp and seaweed recovered. Moreover, the seaweed provides a habitat for fish and other fish-eating creatures such as harbor seals. Thus, when the sea otter returns to an area, so do other species.

Scientists believe that many little-known animals may be keystone species and that their loss could have profound effects on natural systems and crops. Bats, for example, are a keystone species in the tropical rain forest because they pollinate flowers and disperse seeds of many plants. In fact, many tropical plant species are entirely dependent on bats. So are many important food crops such as avocados, bananas, cloves, cashews, dates, figs, and mangos.

Ironically, the lion's share of conservation funds is dedicated to nonkeystone species. The loss of these species, although tragic, would not be as devastating as the loss of unobtrusive, rare, or little-known plants and animals whose survival is vital to many other species.

KEY CONCEPTS

Keystone species are organisms on which many other species in an ecosystem depend. The loss of a single keystone species may have a devastating effect on other organisms.

A Multiplicity of Factors

The foregoing discussion shows that there are many factors that contribute to the loss of plants, animals, and other organisms. Several factors may operate simultaneously on a single species. California condor populations, for instance, were greatly affected by DDT, which caused eggshell thinning and a devastating decline in offspring. Habitat destruction caused by road and housing construction also played a major role in the decline of this species—taking away the condors' habitat. Even well-intentioned fire control efforts affected this giant bird, which has a wingspan up to 3 meters (nearly 10 feet). Fire suppression in the hills of southern

FIGURE 11-7 Gopher tortoises. This slow-moving tortoise is a keystone species in the southeastern United States. Its burrow is home to many other species.

California permitted brush to grow in the condors' habitat. This, in turn, eliminated takeoff and landing areas needed by these giant scavengers.

The loss of salmon populations in the Pacific Northwest is another example of the way in which many factors combine to eliminate a species. Most readers have heard about the many dams along the rivers that halted the upstream migration of adult salmon and the downstream migration of young salmon. Even with special salmon ladders, migrations are less than optimum. Young salmon, for instance, who migrate to the sea to mature, often get lost in reservoirs and never locate the specially constructed ladders. Salmon have also been the victim of years of overfishing. Logging and agriculture have increased sediment deposition in salmon spawning grounds. Agriculture has also taken its toll on salmon by reducing stream flows resulting from irrigation withdrawals. Agricultural chemicals and sediment eroded from farm fields have both added to the salmon's decline from historical levels of 10 to 16 million to 1.5 million today—three quarters of which are raised in and released from hatcheries.

Not only do many factors contribute to the loss of species, but some observers are concerned that these factors may combine to produce an effect much greater than currently anticipated.

KEY CONCEPTS

Many factors acting together contribute to the loss of biodiversity. These factors may synergize to produce a level of devastation far greater than anticipated.

11.3 Why Protect Biodiversity?

Why should disappearing beetles, plants, or birds concern us? The answer to this question varies, depending on one's ethics and one's grasp of ecology and ecological economics—more specifically, on one's understanding of the importance of other living things to the functioning of the planet and the human economy. In this section, we examine four reasons: (1) aesthetics and economics; (2) food, pharmaceuticals, scientific information, and products; (3) protecting free services and saving money; and (4) ethics. The first three categories can be categorized as *instrumental or utilitarian values*—that is, they are based on humans' obtaining some form of benefit. The fourth, based on ethical reasons to save endangered species and biodiversity, could be categorized as a nonutilitarian or an *inherent value* argument, for humans obtain no direct benefit.

KEY CONCEPTS

Arguments for protecting endangered species and preserving biodiversity can be made on both utilitarian and nonutilitarian grounds.

Aesthetics and Economics

Wildlife expert Norman Myers once wrote, "Every time a species goes extinct, we are irreversibly impoverished." Wildlife and their habitats are a rich aesthetic resource. The sight of a trumpeter swan on a pond in Montana, the eerie cry of the loon at night, and the graceful dive of the humpback whale enrich our lives. Wild places provide solace to battle the stress of the modern world.

Saving wild places and other species protects the beauty around us. It can also be an intelligent economic choice. Tourist towns and resorts the world over depend on natural beauty to attract visitors. Natural beauty also attracts residents to an area and can be a major benefit to those who want to relocate.

Another by-product of the preservation of biodiversity is a relatively new and rapidly expanding industry called **ecotourism**, which has cropped up in recent years to cater to the demands of bird-watchers, wildlife lovers, kayakers, and others who want to travel the globe discovering the beauties that abound in distant countries. Ecotourists spend thousands of dollars each to glimpse a lion or cheetah in Africa or a toucan in Central America or to kayak a remote river in Costa Rica (**FIGURE 11-8**). They sometimes stay in lodges that were built with little impact on the environment. The importance of ecotourism is that it provides a relatively sustainable alternative for nations and local villagers—people who once viewed wild species as competition for food and space or who regarded undeveloped land solely as a resource to provide wood products and other materials for human benefit. The problem with ecotourism is that to travel to the far reaches of the globe requires substantial amounts of energy. And people often crowd ecologically sensitive areas. Trucks carrying sight-seers in Africa are cutting up the savannah to offer a glimpse of wildlife.

KEY CONCEPTS

Some people believe that we should save other species because they are a source of beauty and pleasure. In addition, this can provide an economic benefit through activities such as ecotourism and bird watching.

FIGURE 11-8 Ecotourism. Tourists spend billions of dollars each year to visit faraway places, creating an economic boon to many developing nations and considerable incentive to protect native species and undeveloped land. Ecotourism has a downside, though: Hordes of tourists in tour vehicles can also cause severe environmental damage.

Food, Pharmaceuticals, Scientific Information, and Products

"From morning coffee to evening nightcap," writes Norman Myers, "we benefit in our daily life-styles from the fellow species that share our One Earth home. Without knowing it, we utilize hundreds of products each day that owe their origin to wild animals and plants." Virtually everything in our lives comes from the Earth's crust or the plants and animals that inhabit grasslands, oceans, lakes, rivers, and forests. From the oceans, we harvest millions of tons of fish worth more than $70 billion a year. From rain forests, we harvest a wide assortment of edible fruits and nuts and numerous plant products, such as rattan, which is used to make wicker furniture and baskets. Tropical rain forest plants also yield medicines to fight disease. As noted in Chapter 10, important plant and animal genes needed to improve domestic crops and livestock come from nature. Wildlife provides a wealth of enjoyment for hunters, anglers, and nature lovers, and wild species are a huge source of scientific information that provides valuable insights into our world.

The economic benefits of wild species are huge. By some estimates, 40% of all prescription and nonprescription drugs are made with chemicals derived from or originally extracted from wild plants. The commercial value of the 7000 or so drugs prescribed by Western doctors was estimated to be about $43 billion in 1985, according to the Rain Forest Action Network. The USDA estimates that each year genes introduced into commercial crop species yield over $1 billion worth of food. Similar gains can be documented for other major agricultural nations. Internationally, the most widely traded nontimber product is rattan, which is estimated to be worth $3 billion a year.

Wild species provide a wealth of scientific information that could be of great practical value. Norman Myers, in fact, asserts that "wild species rank among the most valuable raw materials with which society can meet the unknown challenges of the future." Many products from wild species now loom on the horizon and may offer us considerable financial gains and healthier lives. For example, the adhesive that barnacles use to adhere to rocks and ships may provide humankind with a new glue to cement fillings into teeth. A chemical derived from the skeletons of shrimps, crabs, and lobsters may help prevent fungal infections in human beings.

Scientists are also studying a wide range of species to learn new, economical, and environmentally friendly ways of producing a wide range of useful products. The abalone, for instance, produces a ceramic-like material much harder than those humans produce—and they do it in sea water using naturally occurring chemicals. Human ceramic production requires toxic chemicals and high-temperature furnaces. This effort is part of a movement called biomimicry.

KEY CONCEPTS

Wild plants and animals are a valuable economic resource. They could provide new food sources to feed the growing human population, genes that could improve crop species, new medicines to combat disease, scientific knowledge, and an assortment of products useful to us.

Protecting Free Services and Saving Money

Ecosystems provide us with many invaluable services free of charge. For example, birds control insect pests. Plants produce oxygen and absorb carbon dioxide. Forests help maintain local climate. Vegetated land helps to replenish groundwater supplies, reduce flooding, and control erosion. Wetlands help purify water.

Although these services are free to us, they are extremely costly to replace with engineered designs. Consider an example. In 1989, a study of the economic impact of clear-cutting in the watershed of the Cedar River, from which the city of Seattle, Washington, acquires its drinking water, showed that continued harvesting would increase sediment in the river. To purify the water, the city would have to build a $120 million water treatment plant that would cost millions of dollars a year to operate. It would replace a service that is provided free by the relatively intact forest ecosystem.

Collectively, wetlands, forests, grasslands, and other natural systems provide billions of dollars worth of services that most of us take for granted. But this does not mean that humans must create a hands-off policy. Many wildlands can sustain some level of harvest of natural resources without threatening biodiversity and environmental services. Mangrove swamps, for instance, are a nursery ground for commercially important fish and shrimp, but they also reduce coastal flood and storm damage and help filter sediment from waterways (**FIGURE 11-9**). If properly managed, mangroves can provide timber for construction, pulpwood for paper, and charcoal for energy. They can also provide food for livestock, shellfish for human consumption, and a number of other products. In Matang, Malaysia, for example, well-managed mangroves produce fish and wood products worth more than $1000 per hectare ($400 per acre) per year. One job is provided for every

FIGURE 11-9 **Mangrove swamp.** This wonderfully dense thicket of trees along many coastlines provides many ecological services (including protection from storm surges) and can be sustainably harvested without loss of these services.

3 hectares (7.5 acres). If all of Southeast Asia's mangroves were managed sustainably, they would provide about $25 billion annually to the economy and create about 8 million new jobs. If exploited carelessly, this valuable economic and ecological asset, part of our natural capital, could be lost forever.

Ethics—Doing the Right Thing

To some people, protecting plants and animals is simply the right thing to do. What right, some individuals ask, do humans have to drive another species to extinction? Other organisms have a right to live, too. To these people, preserving other forms of life is part of our responsibility. In fact, most major religions instruct their followers to protect other species. Of course, not all people agree with this view. To them, human beings are the most important life-form, and thus, the needs and rights of other species take a backseat to ours.

11.4 How to Save Endangered Species and Protect Biodiversity—A Sustainable Approach

Protecting endangered species and biodiversity, like all environmental challenges, requires a number of strategies. Some will be stopgap in nature. These measures help us protect species in immediate danger of extinction. Others are more long-term and preventive in nature. Both types of effort are essential to sustainability—as vital as both emergency rooms and preventive medicine are to protecting human health. Before we examine these measures, let us consider the question of priorities.

A Question of Priorities: Which Species Should We Protect?

Endangered pandas, blue whales, rhinos, and chimpanzees generally make the headlines because they are the most appealing or visible victims. Most preservation money is spent on these species. Interest in less appealing species is often dif-ficult to stir, but many less conspicuous species are important components of natural systems, even keystone species. Many inconspicuous species are vital to human welfare. An adult frog, for example, can eat its weight in insects every day. In India, sharp declines in the frog populations may be partly responsible for higher rates of insect damage on crops and for an increase in malaria, a debilitating disease transmitted by mosquitos, a main component of the frog's diet. Losing species, therefore, is not just an aesthetic tragedy. It can have profound environmental, economic, and health consequences. Protecting species, regardless of how appealing they are, is vital to sustainability.

Stopgap Measures: First Aid for an Ailing Planet

This section describes two stopgap measures designed to protect endangered species and preserve biodiversity: the Endangered Species Act and captive breeding programs.

The Endangered Species Act   In 1973, the United States Congress passed the **Endangered Species Act**. This act, which has become a model for many other countries, requires the U.S. Fish and Wildlife Service to designate and list endangered and threatened species in the United States. Endangered and threatened species are classified on the basis of biological criteria, primarily their population size and rate of population decrease. A species that is officially listed by the agency is then protected by law. It cannot be legally hunted or trapped or harassed. Violators face prosecution, fines, and jail sentences. The presence of an endangered species on federally owned land or private property slated for development with federal monies is enough to stop a project—unless a plan can be worked out to relocate the species or develop the land without affecting the species.

The Endangered Species Act also requires the U.S. Fish and Wildlife Service to develop recovery plans for endangered species, including measures to protect their habitat. It also provides money to purchase this habitat. Finally, the law enables the United States to help other nations protect their endangered and threatened species by banning the importation of these species and by providing technical assistance.

Protection formally begins with the listing of an endangered or threatened species. In 2005, 389 animals and 599 plants were designated as endangered in the United States. An additional 129 animals and 147 plants were listed as threatened. However, only 21 species are currently awaiting listing, down from 314 only a few years earlier, which some see as a sign of the U.S. Government's waning interest in protecting endangered species. In the time it will take wildlife officials to consider their cases, several of these species will inevitably become extinct.

Since the Endangered Species Act went into effect, thousands of projects on federally owned land or privately owned land with federal funding have been reviewed for their impact on endangered species. In most cases, differences have been worked out amicably. Thus, the Endangered Species Act is not an impediment to development as some critics assert.

The most renowned exception was the case of the snail darter and the Tennessee Valley Authority's (TVA) Tellico Dam on the Little Tennessee River, but it ended with a successful resolution that benefited both people and this endangered fish (FIGURE 11-10). Problems began in 1975, when a federal court ordered a halt to the construction of a multimillion-dollar dam (already 90% completed) that would have flooded what was thought to be the fish's only breeding habitat. The order was upheld in the U.S. Supreme Court, and Congress established a committee to review a request for an exemption to the law. In 1979, the committee refused to grant an exemption, saying that the project was of questionable merit. The TVA applied more pressure on Congress,

however, and later that year Congress authorized the completion of the dam. The story did not end in tragedy for the snail darter, though, for it was transplanted to several neighboring streams where it is doing well. Additional populations were discovered in several nearby streams.

A more recent example has been the battle in Oregon and Washington over protecting old-growth forests and the spotted owl, which was listed as threatened in 1989 (see Point/Counterpoint 12-1).

The Endangered Species Act is one of the toughest and most successful environmental laws in the United States. The success of the act, says Bob Davison, a National Wildlife Federation biologist, is best demonstrated by the fact "that there are species around today that would not have survived if the law had not forced agencies to consider the impacts of what they're doing while allowing development to proceed. . . . To a large extent, the law has succeeded in continually juggling those two competing interests."

As noted earlier, the U.S. Fish and Wildlife Service lists endangered species in other countries as well and forbids their importation. Governments throughout the world have enacted similar measures. An international ban on ivory trade, for instance, has nearly halted the illegal poaching of African elephants. In many countries, however, inadequate funding makes enforcement a joke. Inspectors can be paid off by illegal traffickers in endangered species. Governmental agents patrol only a small fraction of the poachers' range, and the courts have routinely been lenient toward poachers.

At this writing the Endangered Species Act is under severe scrutiny. Critics want economic factors taken into account. They say that the economic impacts of listing a species as endangered should be considered during the process. It is their contention that if it is too costly to protect a species or if it would eliminate jobs, a species should not be listed. Critics also point out that efforts to save species often intrude on private property rights. They ask for compensation when, in order to protect an endangered species, a person's land can no longer be used. Restriction of land use is often called a "taking." For a debate on the takings issue, see Point/Counterpoint 11-1.

A report by the National Research Council, which provides scientific advice to Congress, outlined numerous changes in the act to make it more scientifically sound and economically responsible. For example, it called for faster development of recovery plans and guidelines to avoid provisions that are scientifically and economically unjustified. Such plans, they said, should spell out which human activities are likely to harm recovery and which ones are not—a step that would allow for better economic planning in and around protected areas.

The committee that prepared the report also recommended that a core of "survival habitat" be established as an emergency, stopgap measure when a species is first listed as endangered. This habitat would be able to support the population for 25 to 50 years. After more careful study, scientists could determine the exact dimensions of the critical habitat needed for the species to recover. This might result in either a downsizing or an increase in the protected habitat.

At this writing, (June 2005), the Endangered Species Act is expected to undergo major changes that, conserva-

(a)

(b)

FIGURE 11-10 **The snail darter. (a)** Measuring only 8 centimeters (3 inches), the snail darter created a great controversy between environmentalists and the TVA. **(b)** The impending destruction of the snail darter by the Tellico Dam brought the multimillion-dollar project to a standstill. After years of debate, Congress ordered the dam to be completed.

tionists fear, could dramatically reduce the effectiveness of this important legislation. As Congress prepares to change the law, the Bush Administration has been weakening the Act through regulatory decisions. For example, the administration approved a regulatory change that allows the EPA to approve new pesticides without considering their impact on wildlife. The administration also relaxed restrictions on the trade of endangered species to allow U.S. trophy hunters and wildlife traders to import more endangered species and body parts such as horns, antlers, and skins.

KEY CONCEPTS

The U.S. Endangered Species Act is a model of species protection legislation, but it is essentially an emergency measure aimed at saving species already endangered or threatened with extinction.

The Convention on International Trade in Endangered Species International wildlife trade is big business made possible in part by improvements in transportation that make it feasible to ship live plants and animals from their homeland to almost anywhere in the world to satisfy the pet trade market. Today, the sale of animals, skins, and live plants is worth billions of dollars to those who engage in the legal and illegal trade of these items.

Unfortunately, legal and illegal wildlife trade has also caused massive declines in the numbers of many species of plants and animals. Recognizing the threat trade posed to world's flora and fauna, the United Nations Environment Programme, The World Conservation Union, and the World Wide Fund for Nature launched an effort to ban commercial international trade of endangered or threatened species. The result was an agreement known as the **Convention on International Trade in Endangered Species of Wild Fauna and Flora**, also known as **CITES.** It went into effect on July 1, 1975, and currently has a membership of 150 nations.

CITES, the organization that administers the agreement, keeps a list of species (known as Appendix 1) threatened with extinction by international trade. CITES bans hunting, capturing, and selling of endangered and threatened species. In addition, the signatories meet every 2 to 3 years to assess successes and failures of the convention. They also address new concerns and pass resolutions to protect listed species of wildlife and plants. Meetings are attended by both member nations (signatories) and nongovernmental organizations and nonmember states.

Important as it is, CITES has its shortcomings. Endangered and threatened species continue to be smuggled out of countries, especially in the less developed nations where wildlife populations are disappearing rapidly. Documents are falsified by traders so that it appears as if species were captured or collected in areas where species are still relatively abundant. Species may be mislabeled. Enforcement and investigational staff responsible for monitoring international wildlife trade are small in number and sometimes corrupt. Enforcement is left to each country and is often lax. Where enforcement exists, it is often weak. Penalties are mild. The result: Illegal trade in plants and wildlife whose future is in question continues.

FIGURE 11-11 Gray wolf. U.S. Fish and Wildlife Service Biologist Alice Whitelaw carries a sedated gray wolf, soon to be released in Yellowstone National Park, to an awaiting veterinarian for a checkup.

Captive Breeding and Release Programs Zoos are another frontline component of the global effort to save species from extinction. Today, many of the 1200 zoos and aquariums throughout the world are now involved in breeding endangered species in captivity to prevent them from disappearing from the face of the Earth. Canada's zoos are home to a dozen endangered or threatened animal species such as the whooping crane and ferruginous hawk. Botanical gardens are also lending a hand. In fact, today, about 1000 of the world's 2400 botanical gardens have developed programs to conserve rare plant species. Canada's 60 botanical gardens house 11 endangered, rare, or threatened plant species. More recently, zoos have been trying to establish captive-bred populations to provide animals for release into suitable habitat once the captive population was well established. One recent project was launched by zoos in the mainland United States to rescue three rare species of birds threatened by tree snakes on the island of Guam. The survival of the California condors lies in the hands of biologists and personnel of the San Diego Wild Animal Park and the Los Angeles Zoo.

Many primates used for research are also being raised in captivity by commercial animal supply houses in an effort to reduce the drain on wild populations. Yet another way of reducing demand on wild populations is the use of laboratory tests—for example, cell cultures—rather than tests on live animals. Toxicity tests, for example, could be performed on cells grown in tissue culture. A surprising number of substitutes are already available. Actions such as these could stop the flow of animals from their natural habitat and help prevent the extinction of many primates.

The release of animals into the wild to re-establish populations that have been driven to extinction is an essential component of captive breeding programs. After all, what good is species preservation if the species must remain in captivity for eternity? Gray wolf populations, for instance, are being re-established in and around Yellowstone National Park in Wyoming and in the Northern Rockies (**FIGURE 11-11**). The Mexican wolf is being reintroduced into Arizona and New Mexico. Both programs have been sparked by major

Compensate Landowners for Restrictions on Property Rights

Richard L. Stroup

Richard L. Stroup is a Professor of Economics at Montana State University and a Senior Associate of the Political Economy Research Center (PERC) in Bozeman, Montana.

To understand the takings issue, a good starting point is the Fifth Amendment to the United States Constitution, which includes the clause: ". . . nor shall private property be taken for public use, without just compensation." For most of the nation's history, this "takings" clause referred to eminent domain, the right of a government agency to take property to construct a road or a public building or for some other public purpose. The clause confirms the government's right to take property but requires the government to compensate the former owner.

Compensation is important for two reasons. First, when resources are taken to provide a public good or service, the cost should be shared by the public. Taking a tract of land, a crop, or a tree from an owner without compensation would put an unjust burden on that owner. Second, compensation forces politicians and government officials to recognize the cost of what they are doing. The cost of a school, for example, includes resources such as the bricks used and the land on which it stands. To ignore the cost of these resources is to invite their overuse and waste. Compensation puts the resources into the budget process where, quite properly, it competes for access to tax dollars against other uses.

What has changed is that today's takings may occur not just through eminent domain but through regulations. Of course, regulation has long been used to protect the property and the person of individuals from harmful behavior. Preventing a polluter from wrongfully harming a downwind neighbor, for example, is an exercise of the government's police power. Courts agree that this type of regulation is not a taking, and compensation is not appropriate.

However, the expanding scope of environmental regulations has led the courts, including the Supreme Court, to consider the purpose of a regulation more carefully. If the purpose is to produce a public good, such as public access to a beach or habitat for wildlife, rather than to protect someone downwind from pollution, a court may rule that forcing a landowner to provide the service is a taking, and when a taking reduces the value of property, the government must compensate the property owner.

A 1992 decision by the Supreme Court made clear that when a public good is the purpose of the regulation, rather than preventing the violation of the rights of others, courts may find that a taking has occurred, and compensation is due. David Lucas, a Charleston, South Carolina developer, had purchased two oceanside lots in 1986 with the intention of building homes on them. In 1988, before Lucas began construction, the government of South Carolina enacted the Beachfront Management Act, which prohibited building on beachfront lots. Although his neighbors had built homes on their beachfront property, Lucas was not allowed to build on his. His property, for which he had paid $975,000, lost virtually all its value. Lucas was awarded compensation for the near-total taking.

Unfortunately, some activists seeking preservation regardless of cost tend to confuse the two kinds of regulation—stopping a harm and providing a public service. Another term that they have championed, "givings," also causes confusion. A regulatory taking may actually give value to the owner. For example, when a portion of a field is taken for a highway exit, the value of the remaining property may end up higher than the value of the entire tract prior to the taking. In this case, a portion of the property is taken by the state, but no compensation is required. The owner gains, and the government gains not only from the land use, but also by receiving more revenue from property taxes, higher taxes paid on income from the land, and, if the property is later sold, higher capital gains taxes.

Appropriate compensation encourages good management and cooperation. Today, the Fish and Wildlife Service simply dictates the use of land to carry out its Endangered Species Act mandates. If the agency paid the cost of those restrictions, its staff would seek cheaper land and use economical habitat enrichment techniques. Responding to these restrictions, landowners tend to be uncooperative. Some pre-emptively modify their land to keep endangered animals away. Case studies and statistical evidence showing the importance of this problem are accumulating.

Both logic and evidence suggest that treating the rights of landowners with greater respect, while continuing to hold them responsible for any wrongful harms they cause, would lead to more effective environmental protection. Until these rights are respected, the "takings" issue will remain.

Critical Thinking Questions

1. After you have read both essays, summarize the main points of each author in this debate; then list supporting information.
2. Identify any arguments you would have liked expanded and not why.
3. Which viewpoint best corresponds to yours? Why?

 You can link to websites that represent both sides through Point/Counterpoint: Furthering the Debate at this book's internet site, http://environment.jbpub.com/. Evaluate each side's argument more fully and clarify your own opinion.

In the early 19th century, British textile workers wrecked newly introduced machines out of fear that technological advances threatened their employment. Ever since, these so-called "luddites" have been a symbol of misguided opposition to human progress.

Modern "takings"—or "property rights"—advocates are best understood as latter day luddites. In the face of overwhelming scientific proof of growing environmental stresses, and of the necessity of a new system of property rights and responsibilities in order to protect our environment, takings advocates tenaciously cling to an outmoded, environmentally destructive conception of property rights.

The central contribution of 20th century environmental science has been a new recognition that, in a literal sense, everything *is* connected to everything else. We now know that burning of fossil fuels, once viewed as relatively benign, contributes to higher levels of atmospheric carbon and threatens to produce world-wide warming. We now know that wetlands destruction, the promotion of which was once government policy, increases flooding and pollution levels scores if not hundreds of miles downstream. And we know now that traditional logging practices, unless modified, threaten to wipe certain species out of existence.

Until only a few decades ago, it was hardly farfetched to think that a property owner's rights—as well as an owner's responsibilities—ended at the boundary of his or her property. Today we know better, that the security of the environment, and the long-term value of property itself, depends on careful attention to the effects of an owner's activities on the surrounding neighborhood and society as a whole.

Takings advocates dispute the idea that society can and should evolve to reflect this new scientific understanding. They reject the idea of new responsibilities to protect the environment as both unwanted and unfair. They assert that if society as a whole insists on these changes, the public should pay property owners for shouldering the new responsibilities. There is simply no legitimate support for this position.

Takings advocates seek to ground their argument in the Constitution of the United States, but it is a myth that anyone has a constitutional right not to follow public regulations designed to protect the environment. The Takings Clause of the Fifth Amendment provides that the public must pay "just compensation" when it seizes private property for use in building a road or a school, for example. But, as the U.S. Supreme Court has affirmed, this provision of the Constitution was never intended to prevent regulation of uses of private property to protect society as a whole.

There is nothing surprising nor unprecedented about the idea that society's thinking about private property changes over time. In colonial Massachusetts, a law prohibited the building of a home more than easy walking distance from the meeting house; while this restriction probably strikes most modern American citizens as onerous, the rule apparently suited the society of that time and place. Likewise, in the early 20th century, the idea that government can set the minimum wage or fix maximum hours seemed an affront to private property rights, but the practice is well accepted now.

Rethinking Property Rights and Wildlife Protection

John Echeverria
John Echeverria is director of the Environmental Policy Project, Georgetown University Law Center.

Also, in demanding public payment as a condition of refraining from harming our environment, takings advocates seek unfair windfalls at public expense. After all, reasonable environmental standards benefit everyone, including private property owners. In addition, much of the value of private property is created by nearby public investments, and it is hardly fair for owners to demand payments from the public based on speculative value created in part by the public in the first place.

This argument for the legitimacy of public regulation of the uses of private property suggests no disrespect for private property as an institution or its importance to the proper functioning of our economy. One of the key purposes of government is to safeguard property owners against theft and fraud. But the public's role in safeguarding private property rights goes hand in hand with the public's authority to ensure that those rights are exercised responsibly.

Nor does this viewpoint disparage the value of our market economy in producing an efficient mix of products and services. Quite the opposite. The familiar principle of "polluter pays" reflects a recognition that, in order for the market economy to function efficiently, firms whose activities produce external costs must be required to internalize these costs. In a competitive market, competing timber firms, for example, are driven to seek to maximize the return from the trees on their land, without regard to the negative costs of logging in terms of degraded fisheries, wildlife, or public drinking supplies. In the absence of regulation, these external costs would go unaddressed, and the companies would operate at less than an optimal level of efficiency from the standpoint of society as a whole.

Takings advocates would, in effect, exacerbate the problem of "market failure" by encouraging firms and individuals to ignore the external costs of their actions or, what amounts to the same thing, by forcing the public to pay them not to produce these external costs. In either case, the ultimate result is inefficient from an economic standpoint.

Fortunately, the passage of time solves many problems, and this will likely be the case with the takings issue. At bottom, the controversy over property rights reflects the painful process of change as we adapt our property concepts to reflect our new scientific knowledge. There is ground for hope that, as environmental education advances, firms and individuals will make investment decisions in harmony with the new environmental realities rather than in opposition to them. If so, today's takings luddites can be expected to pass into history along with their 19th century counterparts.

disagreements between environmentalists who seek to re-establish populations and ranchers who are concerned about predation by wolves on sheep and cattle. To address this problem, the Defenders of Wildlife established a Wolf Compensation Trust donated by private sources to pay for losses. Between 1987 and February 2005, they have compensated 408 ranchers for loss of cattle (mostly calves), horses (colts), sheep, and dogs. Total compensation has amounted to $518,000. (For a discussion of boths sides of the wolf introduction, see Point/Counterpoint 11-2.)

KEY CONCEPTS

Zoos are an important player in a global effort to protect endangered species. They not only house many endangered species, protecting them from extinction, they are breeding many species for eventual release into protected habitat.

Long-Term Preventive Measures

Captive breeding programs and laws such as the Endangered Species Act are a kind of first aid measure in global efforts to protect species. They can be thought of as emergency measures needed to preserve the planet's biodiversity. They're similar to the kinds of treatment a heart attack patient receives in an emergency room of a hospital. An emergency room physician may save a patient suffering from a heart attack, but unless the patient diets, exercises, quits smoking, and learns to handle stress, he or she is not likely to survive very long. By the same token, emergency measures are not enough to ensure a diverse, biologically rich world. To achieve this goal, preventive measures are needed. Like other environmental problems, this one can be addressed by slowing, even halting, the

>> SPOTLIGHT ON SUSTAINABLE DEVELOPMENT

11-1   Predator Friendly Wool and Wolf Country Beef: You Must Have Read the Label Wrong

Imagine going to the store to the mall to buy a new sweater. One particular brand catches your eye. It's made from Predator Friendly Wool. After you pay for your purchase, you stop in at the restaurant next door to buy a salad made from Salmon Safe vegetables, while the gentleman at the next table enjoys a hamburger made from Wolf Country Beef.

Thanks to the efforts of far-seeing environmentalists, farmers, and ranchers, this scenario is becoming a reality. In Oregon, the Pacific Rivers Council is spearheading an innovative program to encourage farmers to grow crops in ways that help protect salmon fisheries. Farmers who join the program are seeking ways to reduce runoff from farm fields into streams, which brings with it sediment and pollutants that destroy salmon habitat. Runoff can be reduced by planting buffer zones along streams to trap sediment running off from nearby land or by contour farming and other techniques discussed in Chapter 10. Cover crops, such as clover or rye, may be planted on cropland between seasons or in fallow fields, to protect them from the pernicious effects of rain.

Although the Salmon Safe program doesn't make pesticide prohibitions part of their certification, it does recognize the importance of reducing pesticide use in protecting fish and wildlife. The program's goal is to complement organic certification programs that focus on chemical use. Combined, the two programs provide optimal levels of protection for the nation's fish and wildlife.

In Montana and Idaho, sheep ranchers are trying their own brand of animal protection thanks to the efforts of the Growers' Wool Cooperative. Its Predator Friendly Wool Certification program seeks the coexistence of sheep and native predators, rather than eradication of predators (FIGURE 1). Ranchers use natural methods to keep coyotes away from their sheep, including llamas, which are quite protective, and guard dogs. These nonlethal measures reportedly make Predator Friendly Wool a popular item among some buyers.

In Arizona and New Mexico, the Defenders of Wildlife, a national organization, is working with cattle ranchers to help promote reintroduction of the Mexican wolf. The Mexican wolf or "El Lobo" is a subspecies of the gray wolf. It once roamed freely through much of Arizona and New Mexico. As with other predators, ranchers and government agents took a dim view of this predator, shooting them, poisoning them, and catching them in traps. By 1950, the Mexican wolf population was nearly gone. The last remaining wild wolf was shot in 1970.

Defenders of Wildlife, which has taken an active role in the reintroduction of the gray wolf in Montana, Idaho, and Wyoming, has crusaded hard to reintroduce the Mexican wolf to its native habitat using captive-bred animals. In 1998, the first wolves were introduced into the wild in Arizona. Released by the U.S. Fish and Wildlife Service in the Apache National Forest bordering New Mexico and Arizona, the wolves had a rough go of it. Although breeding pairs stayed together and animals were feeding on natural prey species, a number of wolves were shot, and two disappeared and were presumed to be dead, killed by people unsympathetic to the program. Wolves have left the National Forest and preyed on some livestock as well. Additional releases have brought the wild population to about 50 in 2004.

Because wolves cannot be expected to remain in the National Forest, Defenders initiated a fund, known as the Wolf Compensation Trust, to reimburse ranchers in nearby areas for any losses incurred by wolves. The Fund compensates ranchers in Montana, Wyoming, and Idaho, too, and has so far paid over 400 ranchers for loss of sheep, cattle (mostly calves), colts, and guard dogs killed or presumed killed by gray wolves released in Yellowstone and the northern Rockies. It also covers losses from wolves that have naturally migrated from Canada into Northwestern Montana. It will be used to reimburse livestock losses in Maine, New York, and

growth of the human population and a fundamental redesign of human systems such as agriculture, energy, industry, and waste management.

KEY CONCEPTS

Many stopgap measures are required to save species from immediate extinction, but in the long run, preserving biodiversity requires preventive actions, including steps to help restructure human systems for sustainability.

Restructuring Human Systems for Sustainability To ensure a biologically rich world, we must address the root causes of species extinction and the loss of biodiversity. We need to tackle such issues as habitat destruction, overharvesting, and pollution—and the fundamental driving forces behind these problems, notably unsustainable economic and population growth. To create sustainable solutions that confront these causes will require changes in human systems, from economics to transportation to waste management. This book outlines the need to restructure such systems by applying the operating principles of sustainability—especially conservation, recycling, renewable resource use, and restoration—to human systems (Chapter 2). Such efforts go a long way toward eliminating many of the pressures that endanger our planet and the many species that live on it with us. Spotlight on Sustainable Development 11-1 gives an example of what some ranchers and environmentalists are doing to change practices to help protect wildlife.

More specifically, the sustainable strategy calls on people the world over to use only those resources they need and to use them with efficiency. All systems should be re-

the Olympic Peninsula of Washington, should the proposed wolf reintroductions occur in them. The goal of Defenders of Wildlife is "to shift economic responsibility for wolf recovery away from the individual rancher and toward the millions of people who want to see wolf populations restored." Money for the fund comes from private donations. "When ranchers alone are forced to bear the cost of wolf recovery, it creates animosity and ill will toward the wolf," say defenders.

But Defenders has taken the idea one step further, thanks to a brilliant idea from their Southwest Representative, Craig Miller. Miller proposed that Defenders launch a Wolf Country Beef certification program to provide further incentives to ranchers. For ranchers neighboring the National Forest who agree to allow wolves to roam freely on their land, Defenders provides special stickers for meat products

sold in stores. The Wolf Country Beef stickers "brand" their beef in the market as one that comes from ranches that support wolf recovery. To date, the Wolf Country Beef certification program includes 28,300 hectares (70,000 acres) of private ranch land in Arizona and New Mexico. Wolf Country Beef is now available in grocery stores in Arizona.

Ranchers in the program agree to use nonlethal means to keep wolves away from livestock, such as guard dogs.

FIGURE 1 Label from one of several animal-friendly products on the market in the United States.

They also agree not to set out traps, poisons, or explosive devices that not only kill wolves, but lots of other predators as well. This program, therefore, provides umbrella protection for a great many species including foxes, coyotes, mountain lions, raccoons, and eagles that might be tempted by a baited trap or baited poison. It also protects scavengers who might feed on carcasses of animals poisoned by ranchers.

Wolf Country Beef is a program that demonstrates that endangered species can be an economic asset to communities. It fosters a sense of ownership and pride in wolf recovery. Ranchers become an active part in the process and become allies and economic benefactors. Given the choice of a steak from a Wolf Country Beef producer or a rancher less sympathetic to these wonderful animals, with both products priced the same, which would you choose?

The program, in conjunction with reimbursement for lost livestock, takes the economic sting out of wolf reintroduction. It demonstrates that conservation can be achieved through cooperation and that people need not be hurt economically by efforts to create a sustainable future. Defenders of Wildlife helps ranchers find markets for their beef, by contacting grocery stores and other outlets. It also assists in marketing and educating the public through brochures and articles, further sweetening the deal for ranchers. Retailers win, too, by being able to capitalize on public sentiment for wildlife protection. And, lest we forget, the wolves will benefit, expanding their range into lands adjacent to designated recovery areas.

Exciting as these ideas are, they are not new. The idea of certifying food sources started in 1990 with the Earth Island Institute's Dolphin Safe Tuna program. They later launched a Turtle Safe Shrimp program. Participants got bragging rights and marketing advantages for acting responsibly. It's an arrangement that works for people, the environment, and the economy.

structured to be maximally efficient. Such efforts would greatly reduce both our demand for resources and the habitat destruction caused by resource extraction (such as timber harvesting and mining). Recycling similarly reduces our need for both resource extraction and landfills, which destroy habitat. Recycling, as you will learn in Chapter 16, also uses less energy than making products from raw materials, thus reducing the amount of pollution we produce. All systems should maximize recycling by ensuring the return of usable waste to the production–consumption cycle and by incorporating recycled materials in new products and system components. Turning to re-newable energy resources such as sunlight and wind would have an equally beneficial impact on wild species and natural habitats. Transportation, housing, industry, and other sectors can increase their dependence on renewable energy, which would help reduce some of our most serious pollution problems such as global warming, acid deposition, and oil spills. The restoration of natural habitats is also needed both to save wild species and to protect the free services nature provides. In the Pacific Northwest, for instance, five dams on the Columbia/Snake River watershed, which collectively produce only 5% of the region's electricity, are responsible for wiping out 90% of the wild

POINT 11-2 Controversy Over Wolf Reintroduction

Mexican Wolf Reintroduction Debacle

Laura Schneberger
Laura and her husband Matt own a working cattle ranch that depends on a federal land grazing permit in the Gila National Forest in Southwest New Mexico. She has three children and writes editorials and feature articles that have been published in many New Mexico newspapers and a few national publications.

The Mexican Wolf reintroduction into Southwest Mexico and Southeast Arizona has been in progress for 6 years. In those years, the project has had more than it's share of failures. Those failures never seem to affect the project or the forces that drive it.

Agencies in charge of the Mexican Wolf restoration project and environmental organizations that support it will undoubtedly claim that restoring lobos into the southwest will renew biodiversity in a region that supposedly has none.

What is biodiversity? The word is not in the dictionary. Is it necessary to have a predator, as lethal to other species as wolves have historically been proven to be, in order to have biodiversity? Or is biodiversity just another buzzword brought into the midst of land management policies to further a political agenda?

Environmental organizations across the U.S. have given their support to wolf reintroduction, but only in agricultural communities. Wolves are never restored where people do not inhabit the area for the purpose of raising market livestock.

Supporters of wolf restoration efforts back up the U.S. Fish and Wildlife agency by saying the wolves were here first, and if necessary, those dependent on the land for livestock production can move aside for the animals. Naturally, ranchers see that as fighting words. When agencies and organizations cannot show scientific data that substantiates the need for another predator, supporters explain their compulsion to accomplish wolf reintroduction as the "it's the right thing to do." Who is it right for, certainly not the communities surrounding the wolf release sites, certainly not the game species, certainly not the ranchers and area hunters. How many times have we as individuals made an absolutely immoral decision and altered our lives or hurt someone else because, at the time, the bad decision seemed like the right thing to do?

In reality, the U.S. Fish and Wildlife agency is mandated by law to restore species to their former habitats. The Endangered Species Act (ESA), does not place upon the agency the requirement of good science to list a species. Science seldom has anything to do with it at all. A species can be listed as endangered by licking a stamp, putting it on an envelope, and mailing out a request. The Endangered Species Act has been up for reauthorization for 6 years but has never been reauthorized due to political pressure from environmental extremists.

The ESA has become inefficient and burdensome. The agencies that carry out the act seldom comply with the intent of the act to recover species. Very few species are ever removed from the list. The ESA supports enormous bureaucratic agencies, has given these agencies the power to manipulate people, whole communities and states, without having to show that any species they choose to recover is ever recovered sufficiently. Some grassroots organizations even argue that the ESA is so inefficient and out of date, it is no longer legal as it is now written.

salmon in 25 years. Accordingly, some environmentalists are calling for the destruction of these dams to ensure the salmon's future. In instances where dams have become obsolete and were torn down, salmon populations have rebounded nicely.

As noted earlier, essential to these efforts are measures to stabilize human population growth and, perhaps, reduce human numbers over the long haul. Finally, growth management strategies are needed to stop the spread of human populations. These efforts, when combined, could make tremendous inroads into the problem of species extinction, and they could offer many other benefits to society as well.

KEY CONCEPTS

Protecting biodiversity will be best achieved by efforts that address the root causes of the crisis of unsustainability—our inefficient use of resources, continued population growth, our reliance on fossil fuels, our failure to recycle extensively, and our lack of attention to restoration. Addressing these issues will protect plants and animals and bring many other benefits to society.

Setting Aside Biologically Rich Regions   Buying habitats of endangered species and instituting growth management strategies, described in Chapter 17, to help reduce urban

The areas in the Southwest where government agencies and environmentalists have demanded the lobo be returned have diverse game populations that have not only survived intact, but have grown enough to provide tremendous hunting and tourism income for the states that they occupy. If lobos are successfully introduced into these regions, it is expected that prey species numbers will plummet to pre-1920s levels. The excess game that are now harvested and provide jobs and income in these states will no longer exist. The deer and elk herds in the state belong to the state to manage, not to the federal government.

Since the removal of the Mexican wolf from the Southwest, herds of elk, antelope, and deer have improved, increased, and been managed for the benefit of the herds and the communities. When lobos roamed the Southwest, very few deer could be found for game to support growing communities, livestock were brought in to supply meat, and wolf packs turned to livestock for their prey base.

Cattle still inhabit the areas that are being put to use as wolf habitat. Cattle and other prey species for wolves exist together in a complementary manner and cannot be separated as distinct prey species though wolf proponents try to do so. All ungulate herds in the Southwest are prey for wolves though none are owned by those hoping to introduce wolves. Either individuals or the state own the prey being set aside for reintroduced lobos. The ESA has no provisions for reimbursing individuals or communities for losses to endangered animals. Wolves will migrate into areas occupied by cattle when they follow elk and deer to the fresh grasses brought up by grazing cattle. The wolves will stay once they learn cattle are more easily preyed upon than deer or elk who have the ability to strike back at attackers with their forefeet.

Historically, Mexican lobos were responsible for the demise of hundreds of thousands of dollars in livestock in Arizona, New Mexico, and Texas. One study conducted from 1920 to 1922 showed that the stomach contents of 200 lobos were either empty, or contained beef, burro, horse, pig, or sheep. Only 17% contained antelope, deer, or rabbit.

In another historical account, a 3-year-old girl disappeared from a mining camp in Hillsboro, New Mexico for 4 days. She had been playing with a pack of semiwild burros living in the area of the camp. Trackers followed the burro herd for days trying to find the child. Her tracks were interspersed with those of the burros and those of a wolf pack. When found, she was alive, hungry and exhausted, the burros had surrounded her and kept her moving while defending her from the attack of the wolves. The searchers found areas where tracks showed the burros had surrounded the girl and fought off the lobos. They even left one wolf carcass in the trail. Ironically, the wolves and the burros saved the child's life. The burros kept her moving and kept wolves off her, and the wolves kept the burros moving; if they hadn't, she would have suffered from hypothermia and died.

Just the fact that this story is still told sheds some light on how the people of the Southwest felt about wolves and why they were removed from the area. It isn't the only story left over from a bygone era that shows the Mexican lobo reintroduction to be a senseless endeavor.

Wolf reintroduction is seen by inhabitants of the rural Southwest as an attempt to control the use of land in poor states. It is being orchestrated by organizations that subsist on government and foundation money and bloated government bureaucracies. The welfare of wolves and their species is subordinate to the goal of land and people control.

Controversy Over Wolf Reintroduction

Perhaps no animal has been shrouded in more controversy, misinformation, and emotion than the wolf. While just fifty years ago a federal campaign sought to eliminate wolves, a popular federal program now seeks to reestablish them. This shift reflects society's emerging understanding of the diverse benefits of wolves and our growing sense of responsibility to right the wrongs of the past when possible. Ensuring the survival of wolves is not only possible, but desirable and necessary. Wolf recovery is important to conserve the wolf itself, to maintain important ecological processes and to provide spiritual, recreational, economic, and esthetic benefits to people. While wolf eradication was encouraged by myth, misinformation, and misunderstanding, wolf conservation is supported by ecology, evolution, economics, education, and ethics.

The influence of wolves runs deep. For tens of thousands of years they have been an important evolutionary force shaping and maintaining the landscape of North America and its inhabitants. The alertness of deer, the herding of ungulates and the lodge-building of beaver are all, in part, attributed to the wolf. Wolves have similarly influenced humankind. Evolutionary biologist John Allman asserts that homo-sapiens, through their close alliance with wolves, were able to out-compete neanderthals. It is also believed that humans adopted aspects of wolf family structure and behavior to improve our own efficiency and chances for survival. It is ironic that man is responsible for the recent campaign to exterminate the animal which through centuries of co-existence became "man's best friend."

The systematic destruction of wolves began in the early 17th century with the arrival of British Colonists and their livestock to North America. As settlements were established on top of territories already occupied by native people and native wildlife, policies of assimilation and extermination quickly became the law of the land. Domestic livestock rapidly displaced native populations of deer and other preferred prey of wolves. Over-hunting by humans also severely reduced and in some cases entirely eliminated prey populations. Wolves turned to the larger, slower moving and more abundant livestock. In the West Stockmen faced many new challenges. Not knowing how to deal with drought, overgrazing, overhunting or the general diminished carrying capacity of the land, ranchers targeted wolves with a vengeance. Wolf killing soon became a federal campaign carried out by professional trappers encouraged with bounties, and wolves were slaughtered by the millions. They were shot, trapped, poisoned, burned, and demonized to near extinction. By the mid-1950s wolf populations had been completely extirpated from 90% of their historic range in the lower 48 states.

The removal of wolves brought on abrupt and drastic changes in ecosystems. Animals such as deer and elk now have few natural checks and balances and die by the thousands during unnatural population explosions. As their numbers grow out of control they overgraze the landscape, destroy natural vegetation and contribute to erosion, watershed degradation, and other problems. It is ironic that wolves were extirpated because of a fear of economic loss, because the ecological instability caused by their removal is now severely undermining the profitability of ranching and other resource-dependent industries. The landscape without wolves is biologically impoverished. Learning from history, a few western ranchers are now finding it possible to reinvigorate land productivity and profitability by mimicking the herding and regulatory role of wolves. These ranchers, however, agree that such management is labor intensive and that wolves would be less expensive and more effective (Jim Winder, personal communication).

Wolves also check other predators. Since the reintroduction of wolves in Yellowstone's Lamar Valley, biologist Robert Crabtree has found that wolves have reduced the coyote population by 50%. Conversely, in areas where wolves have been eliminated coyote populations have exploded. The greater density of coyotes has resulted in increased predator problems for livestock producers, as well as for other small to mid-size animals.

By looking for weaknesses in their prey, wolves usually kill vulnerable, sick, or starving animals. This process bolsters the genetic vigor and overall health of prey populations. Additionally, wolves create ecological hotspots when they leave behind carcasses. These carcasses attract a variety of

sprawl are vital to the protection of endangered species and the maintenance of biodiversity—but they're not enough. Another preventive effort of extreme importance is the establishment of biodiversity reserves—protected areas characterized by a high biodiversity. The world map in **FIGURE 11-12** shows numerous high-priority areas in need of such protection. Advocates of this approach argue that protecting these regions is the best economic investment in plant and wildlife diversity we can make, for it safeguards areas with extremely rich plant and animal life.

As you can see from the map, most of these regions are located in LDCs that lack the financial resources needed to protect them. Ironically, though, the wealthy, biologically poor MDCs will probably benefit the most from preserving these genetic resources. Therefore, many people argue that the more developed nations of the world should share the cost of preserving these areas. A 0.1% tax on internationally traded oil, say proponent, would net $1 billion a year and would go a long way toward establishing and maintaining large reserves in high-priority areas. An innovative way more

species in high numbers, including eagles, wolverines, fishers, ravens and more. Beetles and other invertebrates also prosper and attract other birds often carrying fleas, ticks, lice, etc. Carcass decomposition then benefits plant life with renewed fertilization, yielding thicker greener grasses and forbs to support the next generation of ungulates. Simply put, wolves are engineers of biodiversity. Their ability to maintain productive landscapes is the result of over 12,000 years of experience. The wise will embrace this understanding; others will reject it.

Critics of wolf restoration claim that the goal of wolf recovery is to control land use and to force ranchers out of business, but such claims ignore important facts. First and foremost, wolves kill livestock infrequently, especially when native prey is available. In wolf-occupied territories across the U.S. and Canada, less than one half of one percent of livestock present in wolf-occupied regions has been lost to wolves. To share the economic responsibility of wolves, Defenders of Wildlife reimburses ranchers at fair-market value for all verified losses. Since 1987, $115,000 has been paid to 112 ranchers. Defenders has also helped to minimize wolf–livestock conflicts by purchasing portable electric fencing, hiring herdsmen, purchasing hay, providing guard dogs and disposing of carcasses. Defenders also developed a "wolf country" labeling program that creates a market premium for beef products produced in a manner compatible with wolf conservation. It is no coincidence that Arizona "wolf country" ranchers Will and Jan Holder have turned their first profit in 25 years in the presence of wolves. Despite rhetoric from traditional antagonists, opportunities exist to heal the wounds between conservation and agricultural communities as well as between humans and wildlife.

The crusade to exterminate wolves reflected the values of a culture then obsessed with taming wilderness. Today that mindset seems primitive considering that scientists are now predicting that within the next 50 years the only large mammals likely to remain on earth will be those which humans have decided to keep around. Wolves and other wildlife were placed here by a power higher than the livestock industry or the federal government. Right or wrong, humans are now

Why Wolves?

Craig Miller
Southwest Director of Defenders of Wildlife, lives in Tucson, Arizona, with his wife and two sons. All are eager to share their world with the big, good wolf.

determining the existence or non-existence of these and other life forms. This serious responsibility warrants the thoughtful consideration of facts and the generous use of caution. Moreover, the overwhelming majority of the public recognizes the importance of maintaining a broad diversity of life and believes that species like the wolf are worth keeping. This sentiment is reflected in our national policy, the Endangered Species Act. It is our legal, biological, and moral obligation to prevent the senseless loss of wolves and other imperiled wildlife. Failing this would be a reckless abandonment of our responsibility to society, to our children, and to ourselves.

Critical Thinking Questions

1. List the main arguments of each author.
2. Does each author do a good job of supporting his or her arguments?
3. Which viewpoint corresponds to yours?

You can link to websites that represent both sides through Point/Counterpoint: Furthering the Debate at this book's internet site, http://environment.jbpub.com/. Evaluate each side's argument more fully and clarify your own opinion.

developed countries can help in this effort is presented in Spotlight on Sustainable Development 11-2.

Many less developed countries have made important efforts to protect their lands. China, which houses approximately 10% of the world's species, recently set aside 44 million hectares (110 million acres) of forest and tundra—or about 4.5% of its land area—as nature reserves to protect biodiversity. The government of Colombia has turned over 18 million hectares (45 million acres) of rain forest land, an area approximately three fourths the size of Great Britain, to

tribal peoples who have lived on the land sustainably for hundreds of years. The lands will be held in common by the people and cannot be sold unless they have the agreement of three fourths of the adults in the affected tribes.

The Colombian government is also purchasing mountain land, where settlers of European descent have lived for almost 200 years, and is turning it back to its indigenous people. Combined with rain forests, some 26 million hectares (65 million acres) are now in the hands of local tribes. The success of this program has prompted other developing

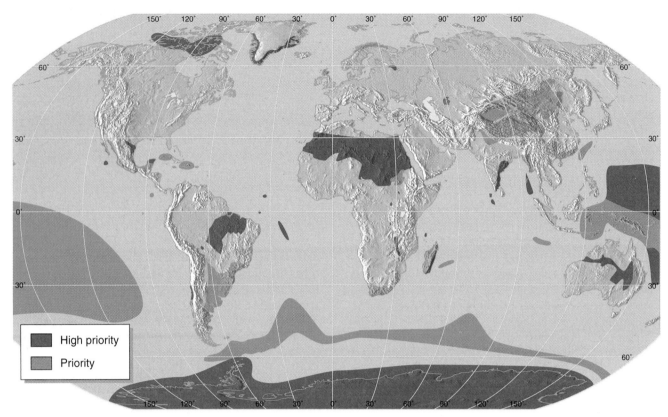

FIGURE 11-12 **Biological hot spots.** High-priority (dark red) and priority (light red) regions in need of protection to preserve important plants, animals, and other species.

countries to follow suit. In 1991, Venezuela gave permanent title to a region of forest about the size of Austria to a native tribe.

Many areas in the more developed nations also need permanent protection. Although efforts are underway by private, nonprofit organizations such as the Nature Conservancy and by various governmental agencies, much more can be done. A study of protected and biologically rich areas in Idaho showed that most of the protected areas—for example, areas set aside as federally designated wilderness—were rather low in biodiversity, whereas the high-biodiversity regions were in private hands and were not in any kind of program to ensure their protection. This type of analysis is called *gap analysis*, because it shows a gap between protection efforts and areas needing protection. Many other states are finding the same result.

KEY CONCEPTS

Setting aside high-biodiversity areas for permanent protection will help to protect species from extinction and will help preserve biodiversity. Unfortunately, the majority of the most biologically diverse areas are located in the less developed nations, which lack the financial resources to protect them.

Buffer Zones and Wildlife Corridors: Protecting and Connecting Vital Areas Existing amid crops, cities, pastures, towns, and mines, protected areas (set aside for wildlife)

have been only marginally successful in safeguarding biodiversity, much to the surprise of many. American ecologist William Newmark studied the loss of mammal species in national parks and reported an alarming decline in the number of species in all but the very largest parks. Utah's Bryce Canyon National Park, which is one of the nation's smallest parks, lost 36% of its species. California's Yosemite, which is nearly 20 times larger than Bryce, lost 23% of its species. Ecologists refer to such isolated patches of protected habitat as **ecological islands**.

The decline in species in ecological islands occurs because they often do not contain enough habitat for all of the species that live within them. Wide-ranging animals are especially affected. Small islands of habitat also reduce populations, sometimes below levels needed for successful reproduction. Tiny fragments of habitat often lead to interbreeding among members of a small population. This may result in inferior offspring, which are less fit. Small, isolated patches of habitat may also have different abiotic conditions from larger patches, making them less hospitable to species. Finally, human activities on the margins may affect the biotic and abiotic conditions.

Nonetheless, isolated patches can offer many benefits. Central Park in New York City, for example, is listed as one of the nation's 14 best bird watching sites. Why? Because it contains a variety of habitats and is strategically located along East Coast migration routes. It is one of the few places

11-2   Debt-for-Nature Swaps

In 1984, Dr. Thomas E. Lovejoy, then with the World Wildlife Federation, proposed a new idea to help less developed nations set aside biologically rich areas. It is called the **debt-for-nature swap.** What is this?

As you will learn in Chapter 26, many LDCs have enormous debt—money they owe to the wealthier nations of the world who helped them finance oil development, agricultural equipment, dams, and other projects. Many of these nations are having trouble paying off their debt. What Lovejoy proposed was a trade of sorts. He suggested that the nations who owed money might agree to set aside land in exchange for some debt relief (**FIGURE 1**). By reducing debt, countries could free up money badly needed for other goals—among them literacy, health care, and sustainable development.

In a debt-for-nature swap, some of a less developed country's foreign debt can be exchanged for a commitment to invest in local conservation programs. A debt-for-nature swap typically involves three parties: a country in debt, a nation or bank to which the debt is owed, and an entity (an international nongovernmental organization, or NGO) that purchases the country's debt, usually at a significant discount. In a typical debt-for-nature swap, Country A, which owes a bank or a government a large sum of money (say, $10 million), sells its debt to an NGO at a considerable discount (for example, 10 to 20 cents on the dollar). In this scenario, the NGO would pay the creditor $1 to $2 million, and the entire debt would be forgiven. In exchange for the NGO's purchasing the debt, the debtor

nation would enter into an agreement with the NGO to invest a certain amount of money in wildlife parks or other conservation measures.

Three international environmental NGOs based in Washington, D.C., have played a major role in promoting these swaps: Conservation International, the Nature Conservancy, and the World Wildlife Fund. The first ever debt-for-nature swap initiated by Conservation International occurred in 1987 in Bolivia. The agreement between Bolivia and Conservation International resulted in a commitment on the part of the Bolivian government to protect an area of biological importance in return for a small debt reduction. Conservation International purchased $650,000 of debt at an 85% discount. In exchange for the debt, the Bolivian government agreed to provide $250,000 in local currency to help manage the newly established Beni Biosphere Reserve, to create several new protected areas nearby, and to ensure that the forest in a vast tract of adjacent land would be managed sustainably.

Since the first debt-for-nature swaps in Bolivia, similar deals have been struck in Ecuador, Costa Rica, Mexico, Madagascar, and the Philippines. Over $1.1 billion has been spent so far by various governments and nonprofit organizations to protect land. In 1998, the U.S. Congress allocated $325 million for debt-for-nature swaps, giving the president authority to enter into such agreements.

Although debt-for-nature swaps help debtor nations reduce debt payments and allow them to convert foreign debt into local currency obligations, such deals are not without problems. First, there is concern for the sovereignty of debtor nations. Understandably, many countries do not want outsiders dictating their environmental and monetary policy priorities any more than they already are. Second, some governments and peoples are particularly sensitive to foreign investment in their countries.

Creditors also experience benefits and costs. Commercial banks, for example, receive immediate cash payment for selling a debt. This is money that they may have never seen had the debtor nations defaulted on their obligations. Some governments give tax credits to the banks for such swaps, helping reduce their losses. NGOs and environmentalists benefit because they help to both reduce debt and poverty and protect the environment.

Debt-for-nature swaps are not a panacea. They represent only one way of protecting biodiversity, but they're a step in the right direction. Existing programs have sparked a great deal of enthusiasm and hope, and they have infused conservation and environmental efforts with badly needed money.

FIGURE 1 Debt-for-nature swaps are another means of setting aside land to protect biodiversity. Photo taken along the Manú River in Peru.

Do We Have Any Other Choice?

Norman Myers
The author, a fellow at Oxford University and wildlife expert, has spent 25 years advising governments on park management. He has written *The Sinking Ark* and *A Wealth of Wild Species*.

I still hurt to recall the first time I went out on an elephant-culling foray. [A culling is a selective harvesting of elephants to reduce herd size.] It's hard to forget the screams of terror, the fountains of blood, and the sudden silence, broken only by the clinical talk of the technicians and scientists. In South Africa's Kruger Park, where elephant culling is a fact of life, officials work under the seemingly arrogant notion that only humans can keep a park wild in this human-dominated world. I spent months agonizing that there must be a better way to deal with nature.

Eventually, I came to the conclusion that Kruger's approach is the right one. It is our only choice, but we must do it with great caution. Despite my realization that such management is necessary, it sticks in my throat.

When I first went to Africa almost 30 years ago, human communities tended to be islands of settlement among a sea of wildlands. Today, it is the wildlands that are islands. Africa is bursting at the seams with people; population pressures threaten parks from Ethiopia to Zimbabwe, from Kenya to Senegal, making intensive management a bitter but necessary reality. The huge Kruger Park, two and a half times the size of Yellowstone National Park, is no exception. A 950-kilometer (600-mile) fence surrounds the park, turning it into an island in an otherwise crowded land. That's where the trouble begins. Biologists theorize that ecological islands tend to have fewer species than a similar-sized portion of contiguous habitat. They are probably more susceptible to environmental change.

In fenced-off islands such as Kruger Park, wardens and scientists have concluded that they cannot allow the rich diversity of species to dwindle. To ensure diversity they have chosen to manage the park to the hilt, which is where this Viewpoint began.

The thought of the flesh of wild elephants ending up in cans in a supermarket may be repellent. But that's what is happening, and for good reason—to prevent overpopulation and habitat destruction within the park. The alternative to control can be dangerous to parks, as I saw in Kenya. During the early 1970s, the country was hit by drought. The elephants in Tsavo Park, already suffering from overcrowding, started to die in the thousands as food supplies shrank. People were starving, too. Yet park officials refused to allow anyone to touch the meat; they were aghast at the idea that the park's wildlife might be used to meet human needs.

Several years ago I returned to Tsavo and found that the local people were trying to acquire sections of the park for cultivation and grazing. One group of elders told me that their overall aim was to have Tsavo abolished altogether. "That park is an insult to us," one of them said. "We have a score to settle."

For their own survival, Africa's parks must take a lesson from Tsavo. Desperate, hungry people do not take pleasure in the pristine wilderness; they need food in their stomachs. Park managers must realize this. In protecting wildlife from the encroachment of humanity they can serve dual needs. Culling herds to eliminate overpopulation preserves habitat and ensures a sustainable ecosystem. In the process they can provide meat, mountains of it for neighboring peoples—reducing the animosity and making them more aware of the benefits of wildlife.

But let us tread the path toward a human-dominated future very carefully. Once we have accepted such total management of a vast area, how long will it be before we go to extremes? Will we choose to eliminate species that aren't of any direct value to us? Will we begin to seed the savanna with "improved" grasses to increase yield, destroying native species uniquely adapted to the environment?

We established our parks as arks against a rising tide of humanity. Now we discover that it isn't enough to play Noah; willy-nilly, we are playing God. Let's do it carefully.

a migrating bird can stop and rest along the heavily populated Northeast coast.

To date, more than 10,000 sites similar to this in 167 nations have been classified as Important Bird Areas in countries such as Russia, Paraguay, South Africa, and the United States. Because of this designation, many have received increased protection.

The success of protected areas can be enhanced by establishing buffer zones around them. A **buffer zone** is a region in which limited human activity is allowed—for example, timber cutting or cattle grazing.

Another potentially important new idea in wildlife protection is the **wildlife corridor**, a strip of land that connects habitats set aside to protect species. Wildlife corridors allow species to migrate from one habitat to another, breeding with members of other subpopulations. This, in turn, increases genetic diversity within a species, ensuring a better chance of survival. Connections between islands also open up new habitat. Food shortages encountered in one region, for example, can be offset by migrating to another area.

The state of Florida and the Nature Conservancy are currently developing a series of wildlife corridors to protect the endangered Everglades panther, which has been relegated to a few patches of land widely dispersed throughout the state. If efforts to connect them are successful, the panther may someday roam much more freely in search of food and mates, hopefully living in relative harmony with the human population.

KEY CONCEPTS

Islands of habitat are vital to protect species, but they may not be enough to prevent species loss. Buffer zones between human activities and protected areas may provide an additional measure of protection. Wildlife corridors, areas that permit wildlife to move from one protected area to another, are also proving vital in the efforts to protect species diversity.

Extractive Reserves   Another way of protecting vital habitat in less developed nations is the **extractive reserve**, land set aside for native people to use on a sustainable basis (**FIGURE 11-13**). Huge tracts of rain forest in tropical countries, for example, are being preserved for sustainable harvesting of rubber, nuts, fruits, and other products. While providing a sustainable income, the reserves also help protect native species (see Spotlight on Sustainable Development 12-1).

Interestingly, several studies show that income from extractive reserves actually exceeds revenues generated from agriculture, grazing, and lumber on the same land. Officials halted a project to create a plantation in the rain forest of the African nation of Niger. They decided to let the rain forest

FIGURE 11-13 Harvesting the rain forest sustainably. Many products like fruit, rubber, and nuts can be harvested from the world's remaining rain forests without creating any damage to the plants, wildlife, and its long-term productivity. This helps to protect natural resources and preserve indigenous populations.

regrow after economic analysis showed that the economic and social benefits of a sustainable harvest from the intact forest outstripped those expected from the proposed project.

The world's very first extractive reserve was established in the Brazilian Amazon, and since that time, 45 additional reserves have been established, protecting 5.8 million hectares (14.3 million acres) of rain forest. Proponents eventually hope to establish reserves on 25% of the Amazon's rain forest, about 100 million hectares (250 million acres). Encouraged by Brazil's success, several other less developed countries have established forest and lake reserves.

Although extractive reserves are an important component of the global plan to protect native species, it should be underscored that sustainable harvesting is vital to their success. Interestingly, scientists are finding huge tracts of jungle throughout the world that are inhabited by hunters and gatherers that are depleted or nearly depleted of animal species by native peoples in search for food, giving rise to a new term, **defaunation** (loss of fauna in an otherwise healthy ecosystem).

KEY CONCEPTS

Protected lands can be sustainably harvested by indigenous peoples to protect biodiversity. These extractive reserves create a long-term source of food and income for native peoples that often exceed the economic benefits of timber harvesting and other short-term, environmentally destructive measures. The key to the success of this approach, however, is sustainable harvesting.

Improving Wildlife Management   Besides establishing protected areas, buffer zones, and connecting corridors, efforts are needed to improve the way we manage fish and wildlife and the ecosystems they depend on. Most important are efforts to regulate harvests of fish and other commercially important species to avoid depletion. Steps are also needed to restore damaged wildlife habitat. Ideas on proper management of forests and rangelands are presented in the next chapter.

One of the most important ideas to come along in this field in recent years is the notion of **ecosystem management**. This calls for management of entire ecosystems—not just single species. It also instructs us in the practice of protecting all of the vital habitat a species requires—not just isolated sections of it. (See also the Viewpoint "Playing God with Nature," in this chapter.)

KEY CONCEPTS

Saving species and protecting biodiversity will require many improvements in wildlife management—especially the adoption of ecosystem management, which takes a broader view of species protection.

The Biodiversity Treaty   In 1992, the nations of the world met in Rio de Janeiro to sign a biological diversity treaty. Unlike other international treaties, this one focuses on biodiversity, not just threatened or endangered species. Currently ratified by more than 175 nations, the treaty calls on each member nation to develop a **national conservation strategy**. This is a detailed plan to manage and protect biological diversity.

Personal Solutions

As in all environmental issues, individuals can make important contributions. By practicing the principles of sustainability, you can become an important part of the solution. Here are some guidelines:

1. Use only what you need, and use all resources efficiently.

2. Recycle and buy recycled products.

3. Increase your reliance on renewable resources like wind energy and economical forms of solar energy and support government programs aimed at increasing their use.

4. Help restore ecosystems; support groups that take an active role in these efforts.

5. Limit your family size; support private and government efforts throughout the world to provide family planning services and other means to help slow the growth of the human population.

You can also help educate others. Join groups and spread the word through educational campaigns, lobbying, television ads, posters, books, and pamphlets. Support organizations and politicians that address population growth, habitat destruction, overharvesting of fish and other resources, and other environmental problems.

Progress in protecting biodiversity has been considerable. However, for most species, the situation is growing worse. The expanding human population and our growing demand for resources threaten to destroy hundreds of thousands of species in short order. The time for action is now. The next decade is extremely crucial—not just to the species that share this planet with us, but for ourselves.

KEY CONCEPTS

Saving species and protecting biodiversity require personal actions. We cannot wait for government or business to solve the problems for us.

To the conservationist, what is civilized in us is not music, literature, cinema, or architecture, although they represent tremendous accomplishment, but a compassion for all living things and a willingness to do more than simply care.
—Daniel D. Chiras

CRITICAL THINKING

Exercise Analysis

Before joining your neighbor, you would probably want to examine his basic premise—that coyotes are killing a large number of his sheep. Can you accept his assertion? Is he exaggerating? Is he biased by a hatred for coyotes? What is the true extent of the damage? Is he a reliable source? Do his losses merit the economic and environmental costs—or the legal risk—of the control measures he is proposing?

Next, you might want to know if poisoning coyotes would decrease wildlife populations on your ranch—and hence, your future income. Poisons that kill a coyote may be transferred to other animals that feed on the carcass. Furthermore, you would want to consider whether the killing of coyotes would increase the number of rabbits (which coyotes prey on) on your property—and thus decrease the amount of grass your sheep can eat. Moreover, do you really need to kill all of the coyotes to make the sheep safe? Are there other ways to control coyote predation that are less damaging to the environment and more economically sound?

CRITICAL THINKING AND CONCEPTS REVIEW

1. Using your critical thinking skills, analyze the following statement: "Extinction is a natural process. Animals and plants become extinct whether or not humans are present. Therefore, we have little to be concerned about."

2. List and describe the factors that contribute directly to animal and plant extinction. Which ones are the most important?

3. Trophy hunters generally try to shoot the dominant males in a population. Natural predators, on the other hand, remove the sick, weak, and aged members of the population. Using your knowledge of ecology and evolution, in what ways are trophy hunting and natural predation different in their effects on the prey population?

4. Why are islands particularly susceptible to introduced species?

5. Discuss the ecological factors that contribute to species extinction.

6. Describe the concept of *keystone species*. What are its implications for the modern conservation movement?

7. You are placed in a high government position and must convince your fellow bureaucrats of the importance of preserving other species. How would you do this?

8. Outline a general plan for preserving species diversity. What measures are short-term in nature, and what measures are long-term? What measures are stopgap, and what measures are preventive?

9. Using your critical thinking skills and your understanding of ecology and human systems, analyze the following statement: "Setting aside land is all well and good, but aside from leaving pristine lands for the elite few who will be allowed to enjoy them, the benefits of this strategy are minimal, even in the long term. The loss of taxable land in this country impacts many, whereas the increase in biodiversity is barely measurable."

10. In Chapter 2, Figure 2-3 presented a model to show how principles of sustainable development can be applied to human systems. How will the application of that model to systems such as agriculture, energy, waste management, and housing affect the planet's wealth of wild species?

KEY TERMS

accelerated extinction
buffer zone
commercial hunting
Convention on International Trade in Endangered Species of Wild Fauna and Flora (CITES)
defaunation
ecological islands

ecosystem management
ecotourism
endangered species
Endangered Species Act
extractive reserve
harvesting
keystone species
national conservation strategy

natural extinction
poaching
sport hunting
subsistence hunting
threatened species
wildlife corridor

REFERENCES AND FURTHER READING

The References and Further Reading section at the end of this book contains a list of sources for the information discussed in this chapter and recommendations for further reading.

Connect to this book's website:
http://environment.jbpub.com/
The site features eLearning, an online review area that provides quizzes, chapter outlines, and other tools to help you study for your class. You can also follow useful links for in-depth information, research the differing views in the Point/Counterpoints, or keep up on the latest environmental news.

Natural systems such as this birch forest provide numerous benefits to human beings.

CHAPTER 12

Grasslands, Forests, and Wilderness: Sustainable Management Strategies

Our duty to the whole, including the unborn generations, bids us restrain an unprincipled present-day minority from wasting the heritage of these unborn generations.

—*Theodore Roosevelt*

In the early 1970s, scientists pored over satellite photographs of the drought-stricken African Sahel, a band of semiarid land that borders the southern Sahara. One of them noticed an unusually green patch of land amid the desert. Curious to find out the reason, Norman MacLeod, an American agronomist, flew to the site. There, surrounded by newly formed desert, was a privately owned ranch of 100,000 hectares (250,000 acres). Its grasses grew rich and thick even though vegetation in the surrounding fields had long since died, leaving the sandy soil unprotected. Why?

Exercise

One of the most dramatic changes on the planet in the last 2 decades has been the steady march of the world's largest desert, Africa's Sahara. In the 1970s and 1980s, researchers estimated that the desert spread southward at a rate of 5 kilometers (3 miles) per year. They attributed this expansion to drought, overgrazing, and agricultural land abuse in the semiarid grasslands bordering the desert. The 5 km/yr projection, however, was based primarily on measurements in a few locations, which researchers assumed were representative of the entire continent.

Using satellite observations of vegetation, however, scientists have found that the Sahara has advanced *and* retreated—largely in response to rainfall. From 1980 to 1984, for example, the desert's southern boundary moved 240 km (150 mi) south. Between 1984 and 1985, it moved north by 110 km (69 mi). In 1987 and 1988, it shifted northward again by 155 km (97 mi). In 1989 and 1990, however, the desert boundary shifted southward 77 km (48 mi).

Although the southern border of the desert in 1990 was 130 km (81 mi) farther south than in 1980, the researchers believe that the shift does not reflect a long-term trend, but rather differences in year-to-year rainfall.

Some critics of global warming use this data to argue that desertification caused by climatic shift is not occurring. They say that the shift of the desert is a natural phenomenon. How would you answer this claim? What critical thinking rules will you use?

Stretching around the perimeter of the ranch was a fence that held out the cattle of the nomadic tribes that had overgrazed the surrounding communal property for decades. The ranch was divided into five sections. The ranch's cattle grazed each section on a 5-year rotation—a sustainable strategy in the semiarid land. This example shows us the devastation wrought by mismanagement but also the possibilities created by sustainable management.

This chapter examines the state of the world's grasslands and forests—sources of food, fiber, and numerous ecological services—and offers suggestions on sustainable management. It also discusses wilderness areas—vital sources of recreation that, if managed sustainably, could also become important means of preserving biodiversity. Before we begin, we must examine one of the reasons why lands are deteriorating.

12.1 The Tragedy of the Commons

The desert surrounding the ranch in the chapter opening was at one time lush communal grazing land. Tribespeople of the Sahel grazed their livestock on the land with abandon, eventually causing its destruction. As far back as ancient Greece, Aristotle recognized that communal property, shared by many people, often deteriorates severely. Early civilizations, for example, clear-cut communal forests and overgrazed their cattle on grasslands. History, however, shows that these civilizations paid dearly for their disregard. The skeletons of buildings from ancient cities stand out in deserts that were once the rich forest and grassland ecosystems of the Fertile Crescent. Much of Iran and Iraq, which is now barren desert, once supported cattle, farms, and rich forests (FIGURE 12-1). Historians believe that the decline of Greece and Rome may be partly the result of the rampant misuse of their lands.

Economists have debated the fate of other common resources such as air, water, and land for decades. It was not until 1968, however, that professor Garrett Hardin exposed the cycle of destruction in a paper entitled "The Tragedy of the Commons."

FIGURE 12-1 **Legacy of past abuse.** View of barren hillsides of Iran. Once covered by rich forests, these lands have been ruined by centuries of abuse, starting with deforestation.

In England, Hardin noted, cattle growers grazed their livestock freely on fields called **commons**, which fell into ruin as the users became caught in a blind cycle of self-fulfillment. Families found that they could enhance their personal wealth by increasing their herd size. Each additional cow meant more income at little if any additional cost because the farmer did not have to buy new land or feed. The commons provided them.

Hardin argues that the cattle growers were rewarded for doing wrong. They realized that increasing their herd would lead to overgrazing and deterioration of the pasture, but they also recognized that the negative effects of overgrazing would be shared by all of the community. Thus, each herdsman arrived at the same conclusion: He had more to gain than to lose by expanding his herd. This shortsighted thinking resulted in a spiraling decay of the commons. As each pursued what was best for himself, the whole was pushed toward disaster. "Freedom in a commons brings ruin to all," wrote Hardin.

The logic that compels people to abuse communal holdings has been with humankind as long as common property has. Today, however, the process has reached epic proportions. The overgrazed communal property of the Sahel is one of the most recent examples. This tragedy has been worsened by unsustainable government policies. In the 1960s, for example, loans were given to nomadic people to drill for water. With a steady supply of water, previously nomadic tribes no longer migrated south during the dry season. The land around the wells and human settlements deteriorated as herders exceeded the carrying capacity.

The tragedy of the commons provides partial explanation for many of the world's problems, from air pollution to the rapid increase in human population. In the rich, industrial nations as well as in the poor, developing countries, for example, fertility is driven largely by personal desires. In India, children provide security in old age or improve the social status of women. In Mexico, childbearing may be motivated by male pride; in the United States, by personal satisfaction. Individual self-serving acts of procreation create a tragedy that befalls all of humanity.

Hardin's analysis of the tragedy of the commons, although important, gives the impression that common resources are unique in their mismanagement. In many places, privately held resources are also abused (**FIGURE 12-2**). Erroneous frontier notions of the Earth as an unlimited supply of resources, ignorance of the impacts of human actions, and a short-term view of economics dictate unsustainable management strategies on a wide variety of privately held resources (Chapter 20).

Today, many lands are gripped by the tragedy of overexploitation. Short-term exploitation may have been permissible at one time, when the human population was small in relation to the Earth's resources. Today, too many people depend on the biosphere for food, water, and other resources, and the cumulative effect of many small insults has become staggering. Local problems have spawned regional calamities. Regional problems are now spreading to create global concerns.

FIGURE 12-2 Private abuse. This privately owned farmland is deteriorating because of poor land management, which has led to severe soil erosion.

KEY CONCEPTS

Communal resources—that is, resources held in common by people, such as land, air, and oceans—often deteriorate as individuals become caught up in a cycle of self-gratification. Personal gain dictates actions that have negative effects shared by all who use communal property. Privately owned lands also deteriorate as a result of ignorance, greed, and other factors.

12.2 Rangelands and Range Management: Protecting the World's Grasslands

Cattle and other livestock range over half the Earth's surface, feeding on grassland vegetation. So important are they to human nutrition that in many parts of the world domesticated animals outnumber humans three to one.

The grasslands on which many of the world's livestock depend are known as **rangelands** and are a vital component of global food, leather, and wool production. When properly managed, grasslands and the livestock they support can provide useful products *ad infinitum*. Regrettably, the vast majority of the world's rangeland is unsustainably utilized. This section examines the problems facing rangelands and shows how they can be managed sustainably to ensure a steady supply of food and other resources. Before we begin, a few words on rangelands are in order to broaden your understanding.

KEY CONCEPTS

Rangelands, grasslands on which livestock graze, are an important element of the global agricultural system. When properly managed, they can be a sustainable food source.

An Introduction to Rangeland Ecology

Dozens of plants, mostly grasses, inhabit the grassland biomes of the world. Grasses, like all other plant species, capture sunlight energy and use it to produce organic matter. Livestock and native grazers feed on grasses, acquiring the nutrients, both organic and inorganic, they need to survive. People, in turn, consume the meat of these animals or drink their milk. Thus, organic plant matter and its energy are passed up the food chain.

Grasses are unique among plants in that they can withstand a considerable amount of grazing without suffering. This is possible because they grow from the **basal zone**, the lowermost portion of the leaf. (It's for this reason that our lawns withstand periodic cutting.) Most other plants grow from the top. Periodic grazing would kill them.

Even though grasses are adapted to grazing pressure, there are limits to how much grazing a grass species can withstand. As a general rule, it is safe to remove the top half of the exposed part of a grass plant. The rest constitutes a **metabolic reserve**—so named because it provides the photosynthetic capability needed to provide food for the roots of the plant, which may extend 2 meters (6 feet) or more into the ground (**FIGURE 12-3**). Without this reserve, the plant is likely to die or be severely weakened.

Rangelands start to deteriorate when there is little or no recovery time between grazings. As a result, plants are so heavily grazed that their metabolic reserve is removed. Otherwise, grasses are rather hardy plants that are adapted to the sometimes harsh conditions of the world's grassland biomes—for example, periodic drought and even occasional fire. Grasses can withstand drought because their roots extend deep into underground water supplies. They also store enormous amounts of nutrients that permit the plant to regrow after drought or periodic fires.

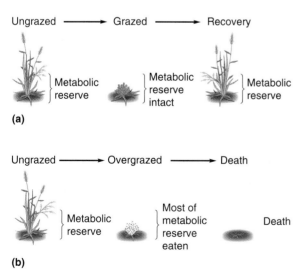

(a)

(b)

FIGURE 12-3 The anatomy of grass. (a) Grasses are generally hardy species adapted to periodic grazing. **(b)** The lowermost portion, the metabolic reserve, however, cannot be removed without weakening or even killing the plant.

KEY CONCEPTS

Grasses form the base of the food chain on rangelands. These hardy species are adapted to periodic drought, fire, and grazing as long as care is taken to protect the metabolic reserve of the plant.

The Condition of the World's Rangeland

In the United States, 85% or over 110 million hectares (276 million acres) of grassland, much of it public land, have undergone slight to moderate desertification in the last 200 years—equivalent to an area nearly twice the size of Texas! The Navajo Indians, for example, live on a 6 million-hectare (15 million-acre) reservation in Arizona, New Mexico, and Utah. Theirs is a sun-parched land, dusty and dry. To feed and clothe their people, the Indians have gradually increased the size of their sheep herds. Today, the herd size exceeds the carrying capacity by at least four times. Baked by the hot summer sun and swept by fierce winter winds, the overgrazed reservation is quickly becoming an arid dust bowl.

Northwest of Albuquerque, New Mexico, is another tract of parched desert land, the Rio Puerco Basin. In the late 1800s, its lush grasslands supported huge herds of cattle. A century later, however, the land has deteriorated under the strain of overgrazing. Erosion has formed deep gullies. Wind and rain erode the soil 5 to 10 times faster than it can be replenished. According to the **Bureau of Land Management (BLM)**, which manages most of the federally owned rangeland, 6.3 million of 12 million acres or 53% of the total rangeland acreage assessed in 2003 meet all standards or are making significant progress toward meeting standards. Some conservationists argue that current standards for good rangeland may not represent truly sustainable conditions. The remaining 47% do not meet standards of good rangeland. On much of this land, conditions appear to be deteriorating.

Worldwide, the condition of rangeland also appears to be less than optimum. The United Nations Environment Programme estimates that nearly 75% of the world's rangeland is at least moderately desertified, with a reduction in carrying capacity of at least 25%. The major culprit in this deterioration is overgrazing. In addition to desertification, this all-too-common practice results in severe erosion, declines in groundwater, loss of wildlife, and invasions of weeds.

Forests are also damaged by livestock. In some regions, branches are cut from trees for animal food. Entire forests are removed to make way for cattle ranches. In India, for instance, grasslands do not provide enough food for the nation's 196 million cattle, which forces people to turn to state forests and other lands.

KEY CONCEPTS

A large percentage of publicly and privately owned rangeland in the United States and other countries has been degraded because of unsustainable land management practices such as overgrazing.

Rangeland Management: A Sustainable Approach

Grasslands have been grazed by native species for thousands of years with no apparent ill effect. In fact, grazing may benefit grasslands, for grazers such as deer and cattle help to spread seeds and fertilize the soil. The hooves of these animals pulverize rock and make soil. They also drive seeds into the ground.

Sustainable grazing management can, if properly carried out, protect or even improve the quality of rangeland. The island within the Sahel, discussed in the introduction to this chapter, provides living proof that humans can manage rangeland sustainably. What are the components of sustainable rangeland management?

Range management involves a number of basic techniques that are often employed simultaneously: (1) controlling the number of livestock on land, (2) deferred grazing, (3) controlling the distribution of livestock, (4) restoration, and (5) rangeland improvements.

KEY CONCEPTS

Grasses are well adapted to grazing pressure, and grasslands and herbivores can coexist in a sustainable relationship that is beneficial to both.

Controlling Livestock Numbers The first and most important step in managing rangeland sustainably is to control the number of animals and the duration of grazing on a piece of land. This measure honors one of the major principles of sustainability: living within the limits (Chapter 2).

The key to proper range management is maintaining livestock populations within the carrying capacity of the ecosystem. The most important variable in determining the carrying capacity is the weather—especially the rainfall and temperature. In dry periods, the carrying capacity may be half that of a normal year. Accordingly, the number of livestock allowed to graze on a piece of land should be cut in half. In wet years, it may be double the normal year's carrying capacity.

Truly effective range management requires a willingness to cooperate with nature, one of the key principles of sustainable ethics. Many ranchers, however, are unwilling or unable to adjust their herd size with the weather. The cost they pay may be a gradual deterioration of their land. The irony of this is that in the long term they may be lowering the carrying capacity and ultimately destroying the land's productive capacity.

KEY CONCEPTS

Rangeland and pasture use must be adjusted according to the carrying capacity of the land, which varies from one year to the next with the weather. Those who cannot adjust their herd size run the risk of lowering the carrying capacity of their land and even destroying grazing opportunities.

Deferred Grazing The ranch mentioned in the opening paragraph of this chapter was successful because its owners shifted cattle from one field to the next each year. This process, called **rotational grazing**, is beneficial because it reduces the grazing pressure on each section and ensures a healthy crop of grass. This technique is growing in popularity and need not just be used in semiarid or arid grasslands. A large percentage of the dairy farmers in Wisconsin practice rotational grazing. This technique reduces costs for feed, equipment, and veterinary care and is especially helpful to smaller operations that would be unable to compete with larger dairy herds. And, in many instances, rotating cattle from one field to the next appears to be promoting a healthier pasture with a diverse array of species. This, in turn, seems to benefit native bird species.

Rotational grazing patterns must be tailored to the local conditions and ecosystem. Some farmers practice short rotational schemes; others may require more complicated schemes as in the deferred grazing scheme showed in **FIGURE 12-4**. As you can see, in this instance the rancher divides his land into three sections: pastures A, B, and C. During year 1, pasture B is grazed first. Pasture C is grazed second. Pasture A is ungrazed. In year 2, pasture C is grazed first; pasture B is grazed second. Pasture A is grazed last. In this rotational scheme, pasture A is kept out of use for nearly 2 years while Pastures B and C are grazed. This gives grasses in pasture A a chance to grow and produce seeds. During the last part of year 2, pasture A is grazed, but only after the plants have reached maturity and dropped their seeds. The cattle spread the seeds in their feces and help drive them into the soil as they move about on the land. This helps to maintain a healthy plant community. During the next 2-year period, Pasture B is withheld from grazing, giving it time to recover.

Rotational grazing in its various forms is a simple yet highly effective and sustainable strategy that benefits ranchers, the economy, and the environment. It increases the vigor of the plants and enhances their nutritional value. It can even increase the carrying capacity of the land over the long run, permitting even greater profit. Furthermore, it helps to ensure a more diverse ecosystem beneficial to native species. Rotational grazing is a system of food production that helps human society meet its current needs in ways that ensure future generations a plentiful supply of food.

KEY CONCEPTS

Cattle can be shifted from one pasture to another to permit grasses to mature and produce seeds. This method enhances the condition of rangeland and may increase the carrying capacity in the long run.

Controlling the Distribution of Livestock Livestock often aggregate around water sources—stock ponds and natural water sources. Because of this, parts of the range may be severely overgrazed while other parts are ignored. In heavily grazed areas, more palatable and nutritious plants may perish, leaving behind less palatable and less nutritious species. In some cases, grasses may be destroyed entirely. Denuded areas, especially stream banks, often suffer serious erosion.

To promote more uniform grazing and protect the land from overgrazing, ranchers can construct fences around ar-

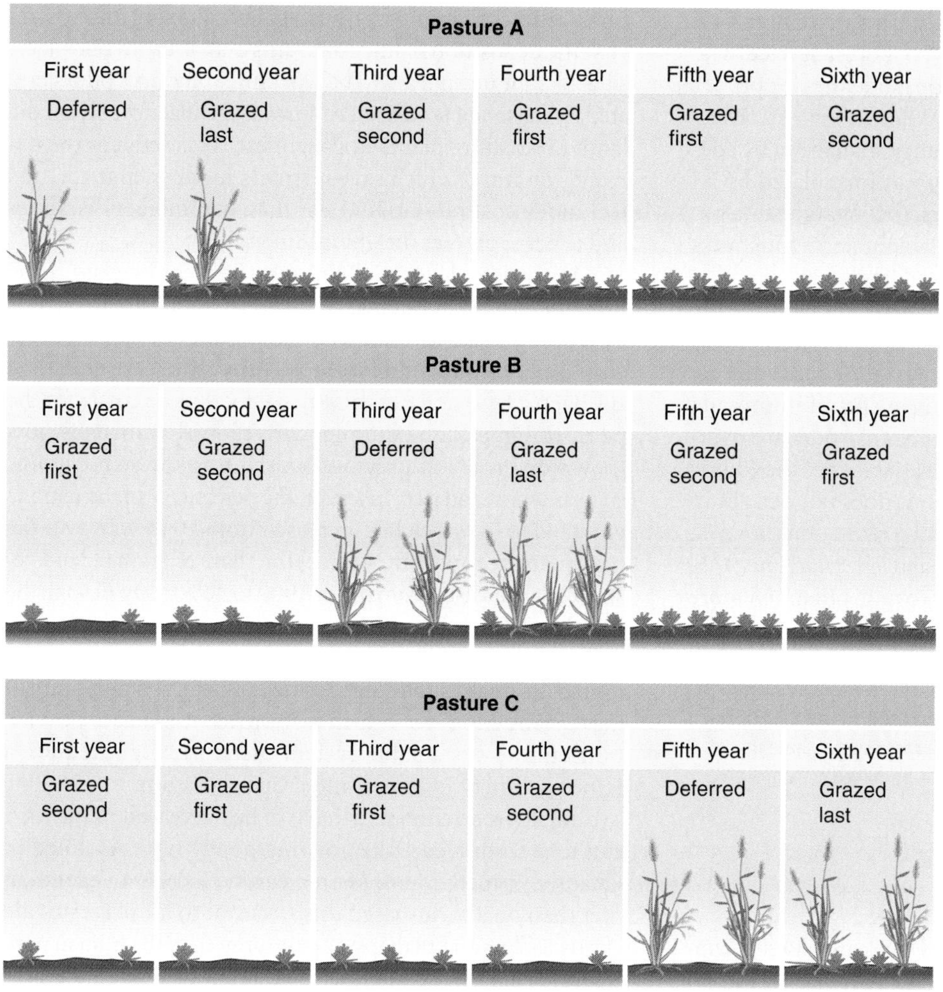

Pasture A

First year	Second year	Third year	Fourth year	Fifth year	Sixth year
Deferred	Grazed last	Grazed second	Grazed first	Grazed first	Grazed second

Pasture B

First year	Second year	Third year	Fourth year	Fifth year	Sixth year
Grazed first	Grazed second	Deferred	Grazed last	Grazed second	Grazed first

Pasture C

First year	Second year	Third year	Fourth year	Fifth year	Sixth year
Grazed second	Grazed first	Grazed first	Grazed second	Deferred	Grazed last

FIGURE 12-4 Deferred grazing scheme. Each field gets nearly a 2-year rest during a 6-year rotation cycle.

eas they want to protect. A more economical way of ensuring a more uniform grazing pattern is to locate water sources and salt licks strategically in areas that are underused.

KEY CONCEPTS

Fencing and careful distribution of water sources and salt licks can help promote a more uniform use of rangeland and protect some areas from serious degradation.

Restoring and Improving the Quality of Rangeland Sustainable rangeland management also requires steps to restore deteriorated land. Restoration includes such measures as fertilizing land and replanting barren areas with native grass seeds. To improve the success of this measure, ranchers often run cattle over the land to force the seeds into the soil. Reseeded areas reduce soil erosion and often support a greater number of cattle than unseeded areas. They also help restore groundwater and protect streams from sediment, by reducing erosion.

The productivity of the land can be improved by periodic controlled burns. This returns nutrients to the soil and thus promotes the growth of grasses. The healthier the grass crop, the faster cattle grow to marketable size. In addition, controlled burns help kill undesirable plant species—for example, woody shrubs and prickly pear cacti.

Restoration and general improvement measures that increase productivity help to create a healthier ecosystem. Besides providing food, such lands are more stable and less prone to erosion and pests. Interestingly, numerous studies have shown that grasshoppers, rabbits, and even weedy plant species such as sagebrush are much more numerous in overgrazed areas than in healthy rangeland.

KEY CONCEPTS

Restoration of degraded grasslands—for example, by applying fertilizer or reseeding the land—is an essential element of building a sustainable system of livestock production. Efforts to boost the productivity of land, including periodic burns, are also helpful in this regard.

Revamping Government Policies

As in other sectors of the agricultural community, government policies that are not based on sound scientific understanding of rangeland carrying capacity can facilitate the rapid deterioration of this valuable resource. In Australia's Northern Territory, for example, ranchers who fail to keep a minimum number of cattle on public land are fined by the government. No maximum limits are set. Such policies promote the deterioration of the land and need to be replaced with sustainable grazing limits based on carrying capacity.

In 1978, the **Public Rangelands Improvement Act** was passed by the U.S. Congress as a means of promoting better range management. This law calls for improvements on lands managed by the BLM and the U.S. Forest Service. The law required these agencies to develop guidelines for proper range management on public lands. Whereas the Forest Service guidelines were designed by range management specialists and others, the BLM's policy was formulated by an advisory board composed of ranchers. Critics argue that because of this, the BLM policy is too lenient and contributes to the deterioration of federally owned lands.

The Public Rangelands Improvement Act requires the BLM and Forest Service to reduce grazing on public land where damage is evident. This strategy is unpopular among ranchers, who either cannot see the benefits of improving range conditions or dispute the claims that they are overgrazing the land. One of the chief weaknesses of the Public Rangelands Improvement Act is that it does not pertain to Indian lands in the West, where grazing reductions are desperately needed. As a result of private and federal actions, U.S. rangelands seem to be gradually improving, but much more work is needed.

> **KEY CONCEPTS**
>
> Many ill-conceived government policies result in the deterioration of publicly owned rangeland. To promote sustainable use of grasslands, government policies should be based on objective scientific criteria.

Sustainable Livestock Production

In many countries, cattle and other livestock are raised in confined quarters called **feedlots** for at least part of their life cycle. Although this practice does not affect grasslands directly, it does have a tremendous environmental impact. For example, pen-raised cattle produce huge amounts of manure in limited spaces. Much of it is dumped in landfills. At one time, farmers used the manure their animals produced to fertilize their cropland. Today, however, many farmers produce only livestock and no longer grow crops—so they have less access to farmland on which to apply manure.

Another problem with livestock raised in enclosures is that they require enormous quantities of grain, mostly corn and sorghum. Ironically, most farmers who supply grain to feed cattle use artificial fertilizer on their land. If applied in excess, the nutrients may wash into nearby waterways, causing pollution. Artificial nutrients also fail to replace all of the nutrients removed by plants.

Hogs also tend to be raised in confined quarters where they are sustained on a steady supply of grain. Such operations produce huge amounts of waste, which is often stored in waste lagoons, where the liquid evaporates or percolates into the ground, polluting groundwater. When waste lagoons break or flood, they can cause severe water pollution problems. Realizing the enormous economic and environmental costs of this technique, many hog farmers in the midwestern United States have found a way around this problem. They're allowing their hogs to range freely in pastures, feeding on natural vegetation. The farmers provide Quanset hut-like shelters to protect the hogs, but move the houses from

time to time, to more evenly distribute grazing and to prevent the deposition of huge amounts of manure in one place. This technique appears to be working. It reduces odors created by most hog farms, and it spreads fertilizer evenly on the land, nourishing plants hogs will eat. This reduces the demand for grain, which saves farmers money. All in all, this technique costs about 70% less than confinement facilities and it helps protect the environment.

In the less developed nations, the wealthy class is the primary consumer of meat. Because livestock are fed grains or are sometimes produced on land that could grow food crops, meat production for the wealthy reduces overall food supplies. These practices also make food more costly for the poor. In Egypt, for example, corn to feed animals is now grown on cropland previously used to grow staple grains such as wheat and rice. In Egypt, the percentage of the nation's grain fed to livestock has increased from 10 to 36% over the past quarter century. In Mexico, the share of grain fed to livestock has increased from 5% in 1960 to 30% today, despite the fact that 22% of the nation's people are malnourished.

Livestock production in confined spaces can be made more sustainable. One step to improve the sustainability of meat production is to return manure to cropland to provide nutrients. A sustainable system therefore might require a reintegration of livestock and crop production.

Meat production may have to be decreased in the long run to accommodate the growing population. As noted in Chapter 4, many more people can be fed on a vegetarian diet than on a meat-based diet. Central to all strategies are efforts to slow and perhaps stop the growth of the human population. Personal actions are listed in the Individual Actions Count! table.

> **KEY CONCEPTS**
>
> In many countries, livestock are raised in pens and fed grains that could be used to feed large numbers of people.

12.3 Forests and Forest Management

Covering slightly more than one quarter of the Earth's land surface, forests are a valuable asset to humankind. The most notable direct benefits are an estimated 5000 commercial products—such as lumber, paper, turpentine, and others—worth tens of billions of dollars a year. Each year, primary and secondary products from U.S. forests sell for more than $378 billion, according to the USFS Forest Products Laboratory. Primary products are roundwood (logs), and secondary products are lumber and paper. Tertiary products such as furniture and turpentine add billions more to the value of the nation's forests. The U.S. forestry and timber products industry employs an estimated 2.7 million people. Forests are important to Canada's economy as well. In Canada, more than 330 communities and nearly 900,000 people are economically dependent on forests. One of every 15 jobs in Canada is derived from timber, wood, paper products, and related industries. Moreover, forest products are the single

individual Actions Count

Protecting Rangelands
- Learn more about rangelands.
- Contact your state's agricultural college and talk with range experts.
- Reduce your meat consumption.
- Write your congressional representative in support of stronger laws to improve rangeland conditions.

Protecting Forests
- Buy recycled paper products—notebooks, paper, toilet paper, paper towels.
- Recycle cardboard and paper.
- Get off junk mail lists by writing to manufacturers.
- Help nonprofit organizations plant trees in national forests or other sites.
- When the time comes to build or buy a house, select a smaller one or use alternative materials such as straw bales and tires.

Protecting Wilderness
- Join nonprofit organizations such as the Nature Conservancy, Sierra Club, and the Wilderness Society.
- If you are a camper, learn how to camp without an impact.
- Write your congressional representatives in support of wilderness designation.

most important export product the nation has to offer. In many poorer nations, forests are a source of wood and charcoal for cooking and heating.

Forests also provide refuge from hectic urban life and opportunities for many forms of recreation. They are home to many of the world's species. Forests protect watersheds from soil erosion, thus keeping rivers and reservoirs free of silt.

They also reduce the severity of floods and facilitate aquifer recharge. In addition, forested lands assist in the cycling of water, oxygen, carbon, and other nutrients.

The United States has about 300 million hectares (740 million acres) of forestland (FIGURE 12-5). Canada has 1 billion acres (418 million hectares) of forest land, about 10% of the world's total. According to estimates of the Worldwatch In-

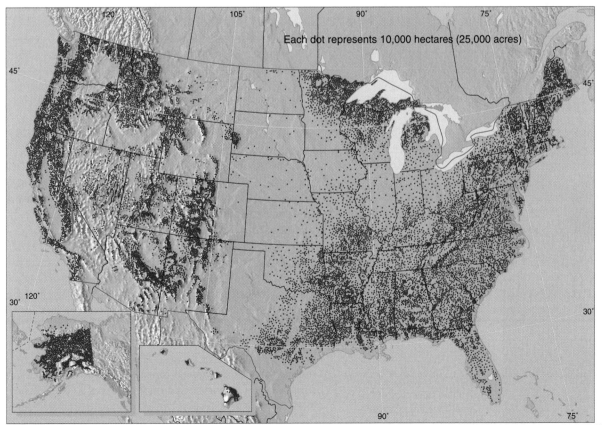

Each dot represents 10,000 hectares (25,000 acres)

FIGURE 12-5 **U.S. forestland.** This map shows the distribution of forestland in the United States.

stitute, world forests cover an area 15 times greater than that of the United States—approximately 4.2 billion hectares (10.6 billion acres). About two thirds of these forests have already been logged or disturbed in some way by human activities. In the United States, 95 to 98% of all forests have been cut at least once since the settlement by Europeans.

Despite the great benefits of forests, only about 13% of the world's forestland is under any kind of management. In addition, only about 2% of the world's forests are protected in forest reserves.

KEY CONCEPTS

The world's forests provide many social, economic, and environmental benefits. A large portion of the world's forests have been logged or disturbed. Very little forested land is under permanent protection.

Status Report on the World's Forests

Since the advent of agriculture, nearly one half of the world's forests or about 3 billion hectares (7.5 billion acres) have been cleared and converted to other uses, mostly farms and human settlements. In East Africa, over 90% of the original moist forest is gone. In Brazil, a country with one of the world's largest forests, 38% of its trees have vanished. In the Philippines, 97% of its forest are gone. To date, the United States has lost about one third of its forests. Europe has lost 70%. Much of this land has been converted to other uses such as farming, or has deteriorated severely after timber harvesting.

Deforestation continues today. According to the Rainforest Action Network, an environmental organization, we deforest and alter 78 million acres, or 31 million hectares of tropical rain forest each year, an area larger than Poland. The World Resources Institute claims that the rate of tropi-

---

>> SPOTLIGHT ON SUSTAINABLE DEVELOPMENT

12-1   Saving the World's Tropical Rain Forests

Rain forests are found in the tropics. Forming a thick, lush carpet of vegetation with a stunningly diverse array of species, the tropical rain forests of the world are being logged at a feverish pace. In 2004, the annual rate of deforestation was estimated to be about 16 million hectares (40 million acres) per year with disturbance to forests caused by road building and other activities bringing the figure to around 32 million hectares. The rates of destruction in specific regions were discussed in Chapter 11. Some experts think that tropical rain forests in all but a few places could be virtually obliterated early in this century if we do not enact strict measures to protect them (FIGURE 1).

FIGURE 1 This boy in Nepal is holding trees that will be planted to restore lost rain forest.

The loss of tropical rain forests is one of the most serious problems facing the world today. It is a major cause of extinction because the tropical forests contain about half to two thirds of the world's species.

The late Al Gentry, a researcher who studied tropical forests before his untimely death in an airplane crash, said that the "loss of so many species is not only a tragic squandering of the Earth's evolutionary heritage but also represents depletion of a significant part of the planet's genetic reservoir, a resource of immense economic potential." Genes from the tropical rain forest help boost agricultural production, as discussed in Chapter 10. Wild species are also a source of new medicines and a host of other new products.

Tropical forests play an important role in global cycling of oxygen and carbon dioxide. By various estimates, global deforestation accounts for about 13% of the world's annual increase in carbon dioxide. It may, therefore, be a major factor in global warming (Chapter 20).

Tropical rain forests are home to indigenous tribes that have lived sustainably for thousands of years. They are vast reservoirs of plants that may have medicinal value.

Clear-cutting tropical rain forests exposes the soil to intense rains that wash away the soil and fill nearby streams and rivers with sediment. These problems are commonplace in Pakistan, India, Thailand, and the Amazon Basin of South America. In Thailand, a devastating flood in 1988 was caused by massive deforestation by commercial timber companies, which ultimately led to a nationwide ban on timber cutting.

Much of the rain falling on intact tropical rain forests evaporates to become clouds again, only to rain down on the neighboring downwind forest. When forests are cut, surface runoff increases, and rainfall decreases by 50% or more downwind from clear-cut regions. As the area of deforestation spreads, rainfall declines over large regions,

cal deforestation is about 16 million hectares per year. For reference, 16 million hectares (40 million acres) of forests covers an area about the size of the state of Washington.

Although most of the news regarding deforestation describes the rapid deforestation occurring in the rain forests in South America, Central America, Asia, and Africa, deforestation is also occurring in temperate deciduous forests and northern coniferous forests. In fact, Canada, Russia, and China are centers of timber cutting that many believe is occurring at unsustainable rates.

The heavy use of forests might not be so bad if efforts were made to replant trees at a rate commensurate with cutting. In less developed countries, for every 10 trees cut down, only 1 tree is replanted. In Africa, the ratio is 29 to 1.

Besides destroying habitat for many species, deforestation decreases sustainable fuel supplies needed for cooking and home heating. According to estimates of the United Nations Food and Agriculture Organization (UN FAO), 100 million people in 26 countries now face acute firewood shortages. In rural Kenya, shortages mean that some women must spend up to 24 hours a week in search of wood. (Spotlight on Sustainable Development 12-1 discusses deforestation in tropical rain forests.)

KEY CONCEPTS

Worldwide, about one half of the world's forests have been cut. The land they once occupied has already been converted to other uses, mostly farming, or undergone severe deterioration. Deforestation continues today in tropical rain forests, northern coniferous forests, and temperate deciduous forests at a rapid pace and threatens the long-term sustainability of human civilization.

and once lush areas turn into barren tropical deserts. Forest fires, almost unheard of in tropical forests, become a problem due to the drying effect. Some tropical soils bake to a hard, brick-like consistency when exposed to sunlight.

Ironically, tropical soils are generally nutrient poor. In fact, approximately 95% of the nutrients are in the biomass (live plants) and only 5% are in the soil, just the opposite of a temperate forest. Cutting and burning the forests to make room for ranches and farms releases enough plant nutrients to support crops for a few years, but because the soil has little reserve and nutrients are quickly taken up by crops or washed away, the land quickly deteriorates, becoming barren and unproductive.

Although clear-cutting of large tracts is widespread in the tropics, efforts to harvest trees by selective cutting often alter the forest so much that they destroy as many species as clear-cutting. Twenty years ago, the International Tropical Timber Organization estimated that less than 0.1% of tropical logging operations were carried out on a sustainable basis, although experts expect that this number has increased in recent years. (A study is currently underway to make a more accurate assessment.)

Because of these problems, many nations have begun to enact laws to protect their forests. In 1988, Brazil established its first extractive reserve in the Amazon Basin, in the state of Acre. Set aside for harvesting rubber, nuts, fruit, oils, and other products, the 40,000-hectare (100,000-acre) reserve allows people to reap sustainable economic benefits while protecting the biodiversity of the forest. All told, 46 extractive reserves have been established in the Amazon, by Brazil and other countries, with hopes of more soon. (Spotlight on Sustainable Development 27-1 discusses efforts to help indigenous people market their products.)

As noted in Chapter 11, extractive reserves can be managed sustainably and represent a long-term solution to economic and environmental problems facing less developed countries. To understand why, contrast cattle ranching with the sustainable harvest of rain forest products. In the state of Rondonia, Brazil, ranchers and farmers have cut down or disturbed at least 35% of the rain forest, creating millions of hectares of wasteland. On most soils, ranchers can graze one cow per hectare for 5 years. In the next 2 years, because of declining soil fertility, they can graze only one cow per four hectares. After that, the land is often destroyed.

In contrast, large tracts of rain forest set aside for sustainable harvest can produce a variety of products (among them latex, nuts, and fruits) indefinitely. Studies show that the income potential of sustainable harvest far exceeds cattle ranching and plantations.

One of the most important strategies for protecting existing tropical rain forests and providing wood for present and future generations is replanting the land already harvested and abandoned by timber companies or settlers. Small teak plantations are cropping up in Costa Rica in an effort to reestablish forests and produce valuable wood sustainably.

Although replanting does not duplicate the incredible diversity of forests, commercial plantations with a modest degree of diversity could be developed to provide wood and other products. This could dramatically reduce pressure on primary forests.

Although positive steps are being taken, rain forests are being destroyed much faster than they are being protected. Individuals can help by joining groups that are working to protect the forests; by avoiding wood products derived from rain forests (teak, disposable chop-sticks, parquet floors, and many hardwoods); and by writing governmental representatives and urging action.

Root Causes of Global Deforestation

Deforestation throughout the world stems from numerous factors. As noted in Chapter 2, a frontier mentality has pervaded many cultures. In the United States, colonists cleared eastern deciduous forests to make room for farms and settlements. Often the land eroded under poor management, so settlers picked up and moved into new territory. Seemingly unlimited forests merited little protection in the frontier days. Forests seemed an impediment to human progress, too. There was, of course, no understanding of the ecological importance of forests and the free services they provide to humans.

Unfortunately, many nations still view vast stands of timber as valuable only if it is harvested or cut down and converted to some other use. And, as nations run out of lumber, corporations expand into new territories, cutting trees in remote locations. Huge tracts are laid barren as a result of this activity.

In less developed countries, deforestation occurs for several main reasons. Forests are leveled to provide fuel wood and charcoal for cooking and heating homes. According to the UN FAO, nearly half of all the wood (tropical and nontropical) cut each year is harvested to supply these needs.

Deforestation also results from poverty, unsustainable population growth, and landlessness. Countries often open up forests to resettlement to ease pressure in cities. The poor people who move into the forest cut trees and try to eke out a living. The faster the population grows, the faster the forest falls. Today, approximately 25% of the annual tropical deforestation is attributed to humans in search of sustenance.

Land ownership patterns in less developed nations also contribute to the loss of forests. In El Salvador, 2% of the landowners have title to 60% of the land. With the good land in the hands of a few, poor rural peasants often enter forests and cut down trees to carve out plots to farm. Many of these people exploit fragile and hilly terrain, semiarid grasslands, or rain forests. Because they are relegated to marginal lands, their farms often fail, and they must move on—often going deeper into the forest to repeat the cycle.

Industrial wood production, wood cut for paper making, lumber, and wood products, accounts for much of the remaining deforestation. This wood is destined for markets in MDCs and LDCs. About half of all industrial timber harvesting is carried out to supply the needs of Europe, North America, and Japan.

> ### KEY CONCEPTS
> Deforestation results from many factors, including frontierism, a lack of knowledge of the importance of forests, population growth, poverty, and inequitable land ownership.

Unsustainable Government Policies Robert Repetto, an economist at the World Resources Institute, puts much of the blame for global deforestation on ill-advised government policies that influence how a nation's forests will be used. This is especially noticeable in nations where governments control the majority of the nation's forest reserves. The government of Canada, for instance, owns 94% of its forest land. The Indonesian government controls 74% of its forest land. Even governments that are committed to conservation often have policies that promote unsustainable forestry practices. Consider some examples.

Governments the world over have typically sold timber below market value to logging companies. Studies have shown that the U.S. Forest Service has lost money on most, if not all, of its national forests for years because it routinely auctions off timber rights below the cost of building roads (which the Forest Service—the taxpayer—pays for), surveying, paperwork, and conducting auctions. Below-cost timber sales are a form of public subsidy to the lumber industry; they amount to about $100 million a year and discourage conservation by companies, builders, and individuals.

The United States is not alone in selling off timber below cost. Many other countries, among them Canada, Ghana, Indonesia, and the Philippines, let their timber go cheaply. Canada, for instance, routinely sells timber at rates 50% lower than those of the United States. Much to the chagrin of environmentalists, and often without public consultation, provincial governments have leased huge tracts of land to private corporations from the United States, Korea, Hong Kong, Japan, and other countries. According to one source, nearly 100% of Canada's boreal (deciduous) forest has been leased over the next 20 years for logging. Some of this land is in provincial parks and wildlife reserves.

Many economic policies encourage an unsustainable exploitation of forest resources. Numerous less developed countries, for example, restrict the export of raw wood by international companies to create jobs at home and encourage economic development of domestic wood-processing industries. Bans or taxes that limit raw wood exports, they think, will increase the export of finished wood products (for example, furniture), netting higher revenues than the sale of raw wood. Unfortunately, says Repetto, many of the small mills are highly inefficient and use 50% more logs than the industry standard to achieve a given output of milled products. Such policies consequently result in a higher rate of deforestation.

Another problem is short-term contracts. In many less developed nations, 35 years or more are required for a stand of trees to recover from logging, but contracts are written for 20 years. Because companies have no long-term interest in their concession, such contracts discourage them from protecting forests. Long-term contracts might encourage them to harvest forests sustainably.

As noted in Chapter 26, heavy borrowing from international banks and industrialized nations has created enormous debt in the developing world. To pay back the loans, countries often lease their land at bargain rates to timber companies from other countries, which come in, clear the forests, then move on. Some countries promote timber harvesting and conversion of forest land to grow export crops. Export crops are also grown on prime farmland, forcing peasants to turn to forests and other fragile ecosystems to make a living.

Government tax policies also encourage deforestation. For instance, in Brazil, the government once offered huge income tax credits to investors in cattle ranches, once a leading cause of deforestation in the Amazon. Although tax credits have ended, Brazil is still aggressively promoting the conversion of its rain forests to agriculture.

In the developing world, governments often promote settlement of forests by giving away land or by building roads into forests. The urban dispossessed then move in, clear the land, and attempt to eke out a living there.

The Canadian government has long encouraged deforestation with little regard for the environment. In British Columbia, where forests are falling at a rate far greater than the estimated annual sustainable yield, there are practically no institutional channels by which citizens can influence forest management on public lands. Citizen participation in decisions to log is extremely limited. As a result of recent deregulation, logging decisions are left largely to logging companies. In addition, forest laws that do exist expressly prioritize logging above all other considerations, making it difficult if not impossible to challenge environmentally harmful logging. Interestingly, although the United States has had policies that allow citizen input on forest management, especially proposed timber sales on Federal land, the Bush administration recently gave Forest Service officials the right to approve logging in federal forests without the usual environmental reviews, a change reportedly made at the request of lumber and paper companies.

Canada's logging industry is a powerful force in its economy and political scene. Here, and elsewhere, logging companies reportedly violate rules and regulations regarding logging, cutting trees right to the banks of rivers; cutting on steep, erosion-prone slopes; making clear-cuts much larger than allowed by law; and failing to protect areas set aside for wildlife. Two thirds of the coastal temperate rainforest has been degraded because of logging and development. In British Columbia, 140 stocks of salmon have been driven to extinction and 624 are at high risk, largely as a result of poor logging practices. Because of this, Canada is often considered the "Brazil of the North." What is more, the government seems to do little to stop lumber companies from such practices.

When conservationists argue for controls on deforestation, the timber industry responds with the threat of lost jobs. In British Columbia, however, as in the American Pacific Northwest, wood products jobs have been steadily declining because of automation for years, while the annual cut has risen sharply. The timber companies wield an enormous amount of power and use the job issue as a smokescreen, say critics. Their power and influence throughout the world are additional factors responsible for widespread deforestation.

In less developed nations, violations and illegal cutting are also major problems. The Brazilian government estimates that 80% of the timber harvesting in the tropical rain forests is illegal. The nations have few resources to monitor logging practices and enforce laws. Logging companies harvest beyond legal boundaries of logging concessions, cut in sensitive areas, and falsify records, taking more timber than allowed.

Especially troublesome is the invasion of reserves set aside to protect native cultures and biodiversity, known as **extractive reserves**. Miners and loggers have moved in, despite official designation. In many less developed nations, governments have stripped rights of native peoples to the land, opening it up to exploitation.

KEY CONCEPTS

Ill-advised government policies including below-cost timber sales contribute to widespread deforestation and unsustainable forest management. These policies are often promoted by powerful economic interests that stand to gain from lenient timber-harvesting practices.

An Introduction to Forest Harvesting and Management

To understand the kinds of reform necessary, we must first look at the way trees are currently harvested. Trees are commercially harvested by four basic methods: clear-cutting, strip-cutting, selective cutting, and shelter-wood cutting.

KEY CONCEPTS

Trees are harvested primarily in four ways: clear-cutting, strip-cutting, selective cutting, and shelter-wood cutting.

Clear-Cutting **Clear-cutting** is a standard practice that is used primarily for softwoods (conifers) that grow in large stands containing relatively few tree species. Those species best suited to clear-cutting have seedlings that grow in open, sunny plots. Clear-cutting is also the method of choice on hardwoods such as aspen. It is the preferred method of harvest in tropical rain forests, which typically have many different tree species.

In clear-cutting operations, loggers remove all the trees in 16- to 80-hectare (40- to 200-acre) plots. They then burn the remaining material. This returns nutrients to the soil, facilitates regrowth, and reduces the threat of fires that could damage the regenerating forest. As the new stand grows, trees are thinned to prevent overcrowding.

Clear-cutting is one of the fastest and cheapest methods of harvesting trees. Clear-cutting may also increase **surface runoff**, the flow of water over the ground's surface, which enhances stream flow. This may increase the supply of water to cities, farms, and industry. Clear-cutting increases suitable habitat for some species, such as deer and elk, which benefits hunters. However, small clearings are better for elk than larger ones because they generally avoid open spaces larger than 8 hectares (20 acres). Elk prefer to remain at the edge of meadows, so they can escape into nearby forests should a predator arrive. Thus, large square or rectangular blocks are less advantageous for elk than smaller, irregular cuts. Another factor that determines whether a clear-cut increases or decreases elk habitat is the location of the cut. Winter range is a limiting factor in elk populations. Thus, clear-cuts in winter range, which make more food available, are more beneficial than cuts in the more abundant summer range. In the

FIGURE 12-6 **Clear-cutting.** This clear-cut, on steep slopes in the South Tongass National Forest of Alaska, is not only an eyesore but also increases soil erosion, impairing forest regrowth and polluting nearby lakes and streams.

Rocky Mountain states, however, clear-cuts are generally made in elk summer range, high in the mountains.

For a long time, researchers have thought that clear-cuts were beneficial to deer and elk because they permit herbs, grasses, and bushes to grow, thus providing additional food. However, a Washington State University wildlife biologist, Charles Robins, recently discovered that although plants grow faster in clear-cuts, they contain lower levels of important nutrients. Robins found that the available protein content of huckleberry growing in clear-cuts was less than half that of huckleberry found in old-growth forests. Plants in clear-cuts also produce more defensive compounds such as tannins, which lower the plants' nutritional value. Robins thinks that these changes in food value could hinder reproduction in wildlife.

Clear-cuts create unsightly scars that may take years to heal (FIGURE 12-6). If not replanted or reseeded naturally, soil erosion may become severe, especially on steep terrain. Eroded sediment fills streams and lakes, destroying fish habitat and increasing the cost of water treatment. Sediment also reduces the water-holding capacity of lakes and streams, which increases flooding, already more likely because of the elevated surface runoff. Erosion in clear-cut areas may deplete the soil of nutrients, thus impairing or even preventing revegetation.

Large open patches in mountainous terrain may accelerate a process called **sublimation,** the conversion of snow to water vapor. When this occurs, snowmelts actually decrease and stream flow declines.

Routine burning in clear-cuts can damage soils by destroying nutrient-cycling bacteria. Burning also volatilizes soil nitrogen, robbing nutrients from the soil itself. Burning may also destroy mychorrizal fungi (MIE-koe-RYE-zl FUN-gee), which grow in soils and attach to plant roots, greatly increasing their uptake of water and nutrients.

Clear-cutting fragments wildlife habitat, creating ecological islands. These islands are exposed to wind and more pronounced changes in temperature, humidity, and light that greatly affect indigenous species. In the Pacific Northwest, studies suggest that for each 10-hectare (25-acre) clear-cut, an additional 14 hectares (35 acres) will be degraded.

Clear-cutting destroys habitat and can contribute to the decline of many species. In the Pacific Northwest, heavy cutting of old-growth forests—ancient forests more than 250 years old, with many sections from 500 to 800 years old—threatens the spotted owl and dozens of other species dependent on this habitat (FIGURE 12-7). Excessive cutting of old-growth forests in the past century has devastated valuable salmon runs in Washington, Oregon, and California, as explained in Chapter 11.

Canadian forests are also threatened. Old-growth forests are especially threatened by timber cutting. One of the hard-

(a)

(b)

FIGURE 12-7 **A bird and its forest. (a)** The spotted owl is just one of many species that are adapted to **(b)** old-growth forests. When its habitat is destroyed, the owl disappears.

est hit areas is Vancouver Island in British Columbia. According to one analysis, Canada's coastal old-growth forests are being clear cut at a rate of approximately 500,000 acres (200,000 hectares) per year. Besides decimating native plants and animals, clear cutting in Canada impinges on indigenous peoples who have hunted and harvested the forest sustainably for many years. (For a debate on old-growth forests, see Point/Counterpoint 12-1.)

In tropical forests, clear-cutting often has devastating results. Soils become baked in the sun and too hard to support growth; others wash away in torrential rains. (For more on this subject, see Spotlight on Sustainable Development 12-1.)

New regulations by the U.S. Forest Service are helping to reduce the impact of clear-cutting in national forests, but on private lands it is largely unregulated. There are 34,000 privately owned tree farms in the United States, covering approximately 30 million hectares (75 million acres). Large commercial tree farms operate much like agribusinesses. Seedlings are planted, fertilized from airplanes, doused with herbicides to control less desirable species, and sprayed with pesticides to reduce losses. When the trees reach the desirable size, they are cut down, and the cycle begins again.

KEY CONCEPTS

Clear-cutting removes entire forests quickly and efficiently. Some tree species such as pines, which grow in open sunny fields, are best harvested in clear-cuts. Clear-cuts benefit certain wildlife but tend to destroy and fragment the habitat of others. Clear-cutting creates ugly scars and can cause considerable environmental damage such as increased soil erosion.

Strip-Cutting   Arguing for smaller clear-cuts on private and federal land, E. M. Sterling, an expert on forest management, notes that Austria harvests as much wood from its forests as does the Pacific Northwest. Yet Austrian forests show little evidence of clear-cutting because of strict forestry laws that apply to public as well as private lands. Austrian law, for instance, forbids clear-cutting on all steep, erodible land. It also limits the size of clear-cuts. A private landowner may cut 0.6 hectare (1.5 acres) without permission but must obtain a permit for larger clear-cuts. Seldom do clear-cuts exceed 2 hectares (5 acres). Most clear-cuts are narrow strips that blend in with the terrain and reseed on their own. This technique is called **strip-cutting**, and it leaves intact forests that will naturally reseed the small clear-cuts. It also reduces erosion and other environmental impacts of clear-cutting. U.S. clear-cutting can be improved to reduce erosion and the visual impact—for instance, by making cuts smaller and by blending them with the terrain.

KEY CONCEPTS

Clear-cutting can be carried out on a smaller scale to minimize visual and environmental impacts. One technique is known as strip-cutting.

Selective Timber Cutting   **Selective cutting**, as its name implies, is the removal of a limited number of trees from a forest. It takes place in multispecies (diverse) forests such as those of the northeastern United States. In selective cutting, foresters remove desirable tree species such as maple or beech trees. They generally remove the mature trees but will cut down deformed trees to get rid of them.

The object of selective cutting is to reduce visual scarring and to preserve species diversity in forests, which helps to protect forests from disease and insects. This method also reduces fire hazards and minimizes soil erosion and the destruction of wildlife habitat.

Selective cutting may sound like the answer to clear-cutting. Unfortunately, it has several major disadvantages, cost and time among them. This practice also removes the genetically superior trees, whose seed is needed to keep the forest healthy. As a result, a forest may slowly degenerate, producing lower quality wood.

Selective cutting in tropical forests can also be quite damaging if done with heavy equipment. A study in the eastern Amazon showed that when only 3% of the trees were removed, 54% had been uprooted, crushed, or damaged during the construction of roads and logging operations. Road building can also accelerate soil erosion, and—because larger tracts of forest must be harvested to achieve the same output as a clear-cut—more roads may need to be built. Selective cutting is also not suited to trees whose seedlings grow in sunny locations. If properly carried out, however, this technique produces little scarring, causes little or no erosion, and does little damage to wildlife habitat.

KEY CONCEPTS

Selective cutting takes place in multispecies forests with species whose seedlings grow best in shade. It reduces visual scarring but is expensive and time-consuming and can cause considerable damage to unharvested trees.

Shelter-Wood Cutting   **Shelter-wood cutting** is a kind of selective harvesting that addresses some of the concerns of critics. In this technique, poor quality trees are first removed from mixed timber stands, leaving the healthiest trees intact. These trees reseed the forest and provide shade for their seedlings. Once seedlings become established, loggers remove a portion of the commercially valuable mature trees, leaving enough in place to provide shade for the seedlings. Finally, when the seedlings become saplings, the remaining mature trees are harvested.

Shelter-wood cutting has many of the advantages of selective cutting. It leaves no unvegetated land except for roads, minimizes erosion, and increases the likelihood that the forest will regenerate. It also reduces habitat destruction and ensures a healthy seed source. However, it is more costly than either clear-cutting or selective cutting.

Shelter-wood and selective cutting can be economically competitive with clear-cutting in **second-growth forests**, ones that have been cut previously, if logging roads are already present. Even in low-diversity forests containing only one or two tree species, these techniques can be economically competitive. Because shelter-wood and selective cutting prevent the scarring of the land, they provide additional economic and aesthetic advantages to regions that rely heavily on tourism.

Old Growth, Spotted Owls, and the Economy of the Pacific Northwest

Ralph Saperstein
Ralph Saperstein is the Public Policy Manager for Boise Cascade Corporation. He has spent the past 18 years tracking forest issues for the forest products industry.

The public debate over national forests in the Pacific Northwest focuses on the volatile passions generated by the images of ancient forests and endangered wildlife. Unfortunately, these images are far from reality. The forests of the Pacific Northwest have always been a mosaic of ages and ecosystems. Natural fires, volcanoes, floods, windstorms, insects, and disease have all played a major role in the evolution of our present forests. These forests were never all old growth, and the wildlife that they support have evolved in a wide variety of age classes and habitat types.

Congress designated national forests to provide the American people a stable source of timber, water, grazing, minerals, recreation, fish, and wildlife. The principle of multiple use has guided the management of national forests for nearly 100 years. Comprehensive land-use management plans developed under congressional directive and presidential administrations have assigned millions of acres of forestland to nontimber uses. In fact, before the listing of the northern spotted owl as a threatened species, only 30% of national forests in the Pacific Northwest were available to grow trees to produce lumber and paper products for millions of people. Protection plans for the spotted owl and other species set aside nearly 90% of our federal forest lands for uses that preclude timber harvesting.

Restrictions on forest management to protect old growth and other forests have drastically reduced the supply of timber from our national forests. Literally dozens of companies, and the communities they support, were established on the promise of a sustained yield of timber from our national forests. Close to 100 towns in the Pacific Northwest are dependent on a lumber mill for their economic lifeblood. When the forest products industry gets a cold, the local communities catch pneumonia.

Trees are a valuable resource that can be planted and grown to produce lumber and paper products for future generations. Over 40% of the lumber used in the United States to build homes is manufactured in the Pacific Northwest. Stopping the scientific management of our national forest will lead consumers to turn to countries with little or no environmental restrictions for their forest products. All forests harvested in the Pacific Northwest are promptly replanted to grow trees for the future, as they have been for over 50 years. Federal and state laws require that streams are protected, wildlife habitat is created, and productive soils are cared for.

The forest products industry supports protecting old growth and wildlife habitat. The issue that must be addressed is one of balance. Over 1.6 million hectares (4 million acres) of old-growth forest are already preserved and will never be harvested. One must recognize that preserving forestland does not assure that it will never be susceptible to fire, wind, insects, and disease. These natural forces and the normal progression of forest growth will eventually drastically alter today's old-growth forests.

The spotted owl, followed by several fish species, has grabbed the attention of the entire nation. Special interest groups use the spotted owl and other fish and wildlife species as surrogates to eliminate needed management on millions of acres of forestland. Despite sympathetic court rulings and administrative actions, special interest groups have not demonstrated that scientifically based forest management cannot be compatible with fish and wildlife habitat enhancement or restoration. In fact, neither the Government nor special interest groups have proven in court that harvesting trees harms a species that enjoys Endangered Species Act protection.

The spotted owl and fish issues have raised important philosophical issues about humans' role in the natural environment. The Endangered Species Act requires protection of all threatened or endangered species without regard to the social and economic costs of such protection. What once was a universally acceptable premise is now being scrutinized at all levels of society. There are literally thousands of species that threaten to stop development, agriculture, fishing, transportation, and power generation in all 50 states. The elevation of wildlife above people's needs is a philosophical change that will shake the foundations of modern society. The answer cannot be nature versus people. The solution to these conflicts must come from the wise management of our natural resources in balance with social and economic values.

You can link to websites that represent both sides through Point/Counterpoint: Furthering the Debate at this book's internet site, http://environment.jbpub.com/. Evaluate each side's argument more fully and clarify your own opinion.

For years, the tobacco companies would trot out their research scientists who announced the results of their latest studies "conclusively proving" that cigarette smoking was not linked to lung cancer. Today, against all evidence that logging the public's national forests is an economic and an ecological disaster, the timber industry continues to indulge in its own version of chronic denial. To listen to industry spokespeople, there are no problems in our national forests that more cutting will not solve.

Let's be clear: 95% of the original native forests that once covered most of our nation are gone. The 5% that remains is primarily in the Northwest, much of it badly fragmented by clear-cuts—the practice of cutting, then burning every living thing in 40- to 80-acre increments. What is left resides almost exclusively on public lands as part of the system of national forests. These forests belong to present and future generations of Americans, not to the timber industry.

The mere fact that in 150 years—a short time in the life of a forest—we have managed to dispatch all but 5% of this once dominant ecosystem attests to unsustainable forestry practices. We have, in fact, been hacking down our native forests at twice Brazil's rate. Ironically, everyone seems to be in agreement that Brazil, which still has 80% of its original forests intact, should stop cutting.

Typically, the industry argument is framed in terms of jobs versus owls: a narrow and inaccurate characterization. The two primary reasons for decreased timber employment are automation and exports. Tellingly, in Oregon, during the decade *before* the emergence of the spotted owl, while the total cut from national forests *increased* 15%, employment in the timber industry *decreased* 16%. The industry exports *more timber annually than the entire cut from all national forests*. According to Oregon Congressman Peter DeFazio's office, one fourth of all the trees cut in the Northwest are exported. If you add minimally processed timber, exports account for up to 60% of all timber cut. It is clear we do not need federal timber for domestic consumption. Like a third-world colony, we export our raw materials—and our jobs— to foreign nations who then sell us back finished goods. If we simply stopped exports, we could cease logging national forests and experience no timber shortage.

Standing national forests offer much wider and more essential values than timber alone. Forests provide us with clean air and pure water. They abate flooding, moderate the climate, and deter desertification. Forests provide wildlife habitat and abundant fisheries. They are a source of medicines. The bark of the Pacific yew tree contains a chemical called taxol, a potent anticancer substance remarkably effective against ovarian cancer. Tragically, for decades, the yew has been cut and burned as a weed species.

Standing forests also act as a vast carbon storehouse. Once cut, they release enormous amounts of carbon into the atmosphere, hastening global warming. And, of course, forests offer us recreation and inspiration. Yet, incredibly, when preparing timber sales, the government attributes no

Owls, Lies, and Taxpayer Waste

Victor Rozek
Victor Rozek is the general manager of the Native Forest Council.

value to standing trees. In Alaska, the U.S. Forest Service sold 400-year-old trees for $1.48 each!

A congressional study showed that the U.S. Forest Service lost $5.6 billion in direct taxpayer subsidies over the past decade on timber sales. We are asked to pay for the destruction of our own heritage for the sake of temporary employment and short-term profits that will disappear with the last native forests.

As for the spotted owl, the disregard for existing federal laws has been so blatant that in May of 1991, U.S. District Judge William Dwyer issued a temporary injunction against timber sales in the Northwest. Judge Dwyer observed: "More is involved here than a simple failure by an agency to comply with its governing statute. The most recent violation of the National Forest Management Act exemplifies a deliberate and systematic refusal by the Forest Service and the Fish and Wildlife Service to comply with the laws protecting wildlife."

It is absurd to suggest that endangered species are "threatening" to stop development or that the Endangered Species Act (ESA) represents an "elevation of wildlife against people's needs." To the contrary, it is an act of last resort. It is precisely a century-long *imbalance* of placing human needs and economics ahead of everything else that has brought the forests to the point of near total ruin. The owl is an indicator species by which scientists judge the health of the entire forest ecosystem. And science tells us that the ecosystem is in deep trouble. As to balance, there are some 1500 pairs of spotted owls and 5.5 billion people. Clearly, humans are not endangered. They can even be restrained, while owls cannot. Are we so impoverished that we need to kill the last handful?

The time has long passed for "compromise" or "balance." It's time to stop managing our national forests like a private social welfare program and return them to present and future generations of Americans. We would not think of hiring displaced quarry workers to fill in the Grand Canyon. We would be far more foolish to sacrifice the enormous value of standing national forests for the benefit of wasteful employment that consumes much more than it produces.

Critical Thinking Questions

1. In your view, which author makes the most compelling case? Why?
2. What factors influence your view on this matter? Do you have certain biases or philosophies that affect your viewpoint?

Selective harvesting can be modified to correct its problems. This method, while more expensive, helps preserve multispecies forests.

Four measures are required to create a sustainable system of wood production: (1) reductions in demand for wood and wood products, (2) sustainable management, (3) establishment of forest preserves, and (4) restoration of forest land.

Creating a Sustainable System of Forestry

With the world population growing by about 83 million people per year and the demand for wood for fuel and wood products rising with it, steps are needed to create a sustainable wood and wood products production system. At least four major strategies are needed. First, efforts are needed to reduce demand for wood and wood products. Second, measures must be implemented to ensure a more sustainable harvest of timber. **Sustainable forestry** optimizes yield on forest land while protecting the long-term health and diversity of forest ecosystems. Third, efforts are needed to protect and perhaps even expand untapped forests throughout the world—for example, by establishing forest preserves that protect native species and provide ecological services such as flood control. Fourth, efforts can be made to replant forests.

Reducing Demand   Reducing the growth of the human population could help to decrease pressure on the world's forests and protect them from further destruction. Demand can also be reduced through efficiency measures. In Brazil, researchers found that about two thirds of each harvested log turned into sawn wood (wood used to make finished products) is wasted. In Southeast Asia, about 60% of the raw wood entering a plywood mill is wasted. In Japan, the amount wasted is 30%. Reducing waste in saw mills and other wood processing facilities could significantly reduce deforestation. How?

Wood can be saved by using thinner saw blades, which reduce the kerf (wood removed by the blade), and improved machines that do a better job of processing logs for plywood along with a host of other technologies. The U.S. Forest Service estimates that the amount of wood required to make a sheet of plywood or plank of lumber could be reduced by 33% through such measures. Special training for workers could also help reduce wood waste. According to the Worldwatch Institute, by training workers, improving equipment maintenance, and managing forests better, companies could produce the same amount of wood from one third less land.

According to the U.S. Forest Service, new homes could be built with about 10% less lumber by spacing studs (vertical framing members) a little wider and by other efficiency measures that would not compromise the structural integrity of the house. Building smaller homes that use 20 to 30% less wood could also reduce pressure on the world's forests.

Individuals can also reduce consumption by using fewer paper products, avoiding overpackaged items, and using the backs of scrap paper rather than new paper for homework and notes. The constant bombardment of printed advertising material can be stopped by writing companies and asking them to take you off their lists.

Another way of reducing the demand for trees is by recycling paper. Paper production is growing dramatically, much faster than other wood products. Increased recycling of newsprint and other paper products could reduce the amount of forest cut down each year. Individuals, companies, colleges, and governments can help by purchasing paper products made from recycled paper (Chapter 23).

The use of alternative building products can also ease pressure on the world's forests. One interesting innovation is the wooden I-beam, which is used to build ceilings and floors (**FIGURE 12-8**). These are made of plywood and oriented strand board, both of which are made from small-diameter, fast-growing trees. All in all, 40 to 60% less wood is needed to make a wooden I-beam to provide the same service as a 2″ × 10″ or 2″ × 12″ board cut from an old-growth tree. Alternative building materials such as straw bales can be used to build homes and can reduce our demand for wood. Many other options are available to us at work and at home.

FIGURE 12-8 **The wooden I-beam.** This product, made from wood scraps and small trees, uses 40% to 60% less wood than a solid beam and thus greatly reduces our impact on old-growth forests.

Demand for wood and wood products can be greatly reduced by controlling growth of the human population, using wood and wood products more efficiently, finding alternatives, and recycling paper and wood materials.

Managing Forests and Tree Farms Sustainably   Protecting forests and the soils that support them is essential to creating a sustainable society. In the United States, forest management began in the late 1800s. In 1905, Republican President Theodore Roosevelt established the U.S. Forest Service. Its first head, Gifford Pinchot, promoted careful use of forests over strict preservation. He promoted the multiple-use concept and sustained yield—using the forest for many purposes, from timber harvesting to recreation, but managing it scientifically to sustain its valuable services.

Pinchot's notion of multiple and sustained use has persisted. Scientifically based management has long been viewed as a key to managing forests to ensure a sustained yield. However, studies show that past knowledge of forest ecology was limited and that some practices were not sustainable.

Sustainable forest management (SFM) is an evolving science. In a sustainable forest management program, forests are managed in ways to protect and preserve a diverse set of goods and services, including recreation and ecological services (flood control). In addition, in sustainable forest management, forest managers seek to support the forest ecosystem in its entirety, paying attention to the health of the soil and the condition of streams and lakes. In more traditional forest management, foresters tend to focus solely on yield. Timber production becomes paramount and forests are treated more like farms but not all forest farming practices are sustainable. In SFM, forests are managed as complex ecosystems, not as simplified farm production designed to produce one species maximally at the expense of the health of the soil and water. To achieve this goal, sustainable forest management often seeks to mimic natural cycles. Clear-cuts, for instance, are patterned after natural fires, which hopscotch across the landscape leaving patches of green between areas of devastation. They provide trees to reseed the area and furnish habitat for species dependent on the forest. Managers may protect dead trees, too, which are home to many bird species. Sustainable forest managers may also attempt to promote forests containing a mix of species, ages, and/or sizes of trees. Tree cuttings are done on longer cycles, rather than shorter cycles dictated by financial considerations. SFM can be applied to all types of forests, although it must be adjusted to climate, terrain, and type of tree.

Many other steps can be taken to ensure the health of forest ecosystems. One of the most important is to eliminate clear-cuts on steep terrain. This decreases erosion and protects streams and other surface waters. In the tropics, smaller clear-cuts help protect biodiversity and ensure reseeding. In the United States and other countries, reseeding by logging companies should be monitored more carefully, especially on public lands.

Protecting forests from natural hazards—including diseases, insects, fires, droughts, storms, and floods—is also important to creating a sustainable system of forestry. Diseases, insects, and fires account for most of the damage. Sound management that seeks to maintain trees and forests in a healthy state can minimize these problems. Just as in rangeland, the healthier the forest ecosystem, the fewer the pest problems. Maintaining a healthy forest may require periodic thinning of trees to reduce competition for water, which weakens trees and makes them more susceptible to pest damage. Maintaining genetic diversity in forests also reduces damage from insects and disease organisms. Insect- and disease-resistant trees could be developed for reseeding on public and private land. Finally, all imported trees and lumber should be carefully inspected to avoid accidentally introducing pests.

Fire accounts for 17% of U.S. forest destruction, leveling 0.8 to 2.8 million hectares (2 to 7 million acres) of forest each year. According to the Forest Service, 85% of all forest fires are caused by human beings—either accidentally or deliberately. The remaining 15% are ignited by lightning, but these fires account for about half of the annual forest damage.

To protect watersheds, timber, and recreational opportunities, the Forest Service and state governments attempt to reduce forest fires by posting fire danger warnings and sponsoring television and radio announcements. In addition, the Forest Service has an active fire surveillance program that seeks to pinpoint fires and, if necessary, put them out as quickly as possible.

In the United States, efforts to protect forests from fire began in the early 1900s. Although strict fire control has saved billions of dollars worth of timber, ecologists and foresters now realize that it can be detrimental to forests. As our understanding of forest ecology has improved, it has become clear that some fires benefit forests. Small fires, for instance, burn dead branches that have accumulated on the ground and return nutrients to the soil. Most animals can escape minor ground fires, and larger living trees are generally unharmed by them. Periodic ground fires also prevent intense, destructive fires that occur if ground litter accumulates.

In forests protected from fires for long periods, the ample fuel supply caused by the buildup of dead branches may cause a fire to burn uncontrollably once it starts, spreading from treetop to treetop as a crown fire. Huge areas are destroyed in firestorms so hot that the soil is charred and wildlife perishes. Trees may be so severely burned that they die.

Periodic fires not only protect forests from devastating fires, they also foster their renewal. Why? Some forest species require occasional fires for optimal growth. The cones of the jack pine, for instance, open up and release their seeds during fires, as do those of the Douglas fir, sequoia, and lodgepole pine. Fires remove brush that shades seedlings and help replenish soils. Fires also help reduce disease and control potentially harmful insects.

Recognizing the benefits of periodic ground fires, forest managers now let many naturally occurring forest fires burn, provided they are not a threat to human settlements.

(a)

(b)

(c)

FIGURE 12-9 **Benefits of forest fires. (a)** Dense undergrowth in an Oregon pine stand results from the control of forest fires. **(b)** Controlled burning removes the undergrowth. **(c)** Periodic burning prevents disastrous fires, returns nutrients to the soil, and increases forage and wildlife.

The Forest Service also deliberately sets hundreds of fires each year (**prescribed fires**) to remove underbrush and litter (**FIGURE 12-9**). This practice reduces the chances of potentially harmful crown fires. It also improves wildlife habitat, soil fertility, timber production, and livestock forage.

Another way to ensure sustainable forestry is through certification. Timber companies practicing sustainable forest management can now be certified as sustainable by independent organizations that inspect their operations. Certification ensures retailers and customers that the wood was produced and processed in an environmentally acceptable manner. To ensure uniform standards, a consortium of environmental groups, foresters, timber producers, indigenous groups, and various independent certifiers formed the Forest Stewardship Council, which drew up guidelines for forest management and wood production. They now accredit independent certifiers, creating a uniform globally consistent system to ensure sustainable forest management and wood production. The Forest Stewardship Council has certified timber operations on 50 million hectares (123 million acres) in more than 60 countries. Today, several thousand wood products are sold each year using Forest Stewardship Council-certified wood. Expect that to change, however, in the near future. In 1999, Home Depot, a retailer that is responsible for selling a huge amount of wood in the United States, announced that it would eliminate all noncertified wood by the end of 2002. Although they were unable to meet this ambitious goal, they sell more Forest Stewardship Council certified wood than any other major supplier in the United States.

One promising development in the management of forests is the **Model Forest Program** launched by Canada in 1991. It seeks to involve a large number of stakeholders (organizations and individuals) in the management of Canada's forests, 94% of which are on public land—that is, land owned by provincial, territorial, or the federal governments. Management plans for 10 large forest areas, ranging from 100,000 to 2.5 million hectares, are based on consensus among loggers, conservationists, scientists, and government officials. They seek to protect valuable ecological resources while setting forth sound harvesting strategies. Other nations, including Mexico and Russia, have adopted similar programs.

Private efforts are also needed. In Alberta, conservation groups started a program to protect the grizzly bear, whose population has declined from 6000 to 800. This species, now in danger of extinction, may be saved by voluntary efforts of these organizations and timber companies. Spray Lakes Sawmills in Alberta, for example, has become a financial sponsor of the project and is voluntarily developing ways to harvest trees that have less impact on grizzly habitat.

Black bears in British Columbia could benefit from wide-ranging efforts to protect and better manage old-growth forests. Studies show that most black bears den in hollows in trees that are at least 500 years old. But much of British Columbia's old-growth forests are being cut and will be harvested every 80 to 100 years, seriously jeopardizing the bear's habitat.

Better management of existing forests, based on sound scientific principles, including tree thinning, prescribed burns, and replanting, helps to create a more diverse and healthier forest that is less susceptible to disease and insects. Certification programs can help promote sustainable forest management.

Saving Primary Forests/Creating Forest Preserves Some forestry experts believe that we need to protect many of the remaining uncut forests, **primary forests**. To do so will require states and nations to set aside large tracts of primary forest throughout the world. Such measures, say proponents, will ensure continued delivery of the many free services forests provide, such as flood control. They will also help protect biodiversity and, in the tropics, will safeguard the homes of many indigenous peoples who still depend on forests.

In 1986, the U.S. Congress passed a law that prevents the U.S. Agency for International Development (AID) from funding projects such as dams and roadways that will destroy tropical forests in less developed nations. The law also directs the agency to help countries find alternatives to forest colonization and requires it to support preserves and other measures to save forests and promote biological diversity.

Primary forest preserves house insect predators (birds and other insects) and thus act as physical barriers to the spread of pests from harvested forests. Old-growth forests in the Pacific Northwest, for instance, house 100 times more insect predators than neighboring tree plantations—and thus tend to limit pest outbreaks.

KEY CONCEPTS

Saving uncut or primary forests helps preserve biodiversity but also protects nearby harvested forests from outbreaks of pests.

Restoring Forestland Restoration is a key principle of sustainability and is vital to efforts to create a sustainable system of timber production. In the tropics, for instance, an estimated 1.2 million hectares (3 million square miles) of tropical rain forest have been cleared—an area approximately half the size of the United States. Although some of the land has been converted to human use, at least two thirds of it is unoccupied and available for replanting. These lands could be replanted with a variety of trees and managed sustainably. This would greatly reduce pressure on the world's remaining tropical rain forests. Planting only 5% of the rain forest land already cleared could nearly double the supply of commercially harvested wood.

Tree farms that support a mixture of species and different-age trees can supply wood and nonwood products while providing wildlife habitat. The Individual Actions Count! table lists some things you can do to help protect the world's forests.

> Planting only 5% of the rain forest land already cleared could nearly double the supply of commercially harvested wood.

KEY CONCEPTS

Building a sustainable system of forestry will require efforts to replant millions of acres of forestland that has been cut and never replanted.

12.4 Wilderness and Wilderness Management

Wilderness, as defined by U.S. law, is "an area where the Earth and its community of life are untrammeled by man, where man is himself a visitor who does not remain." Why is it important? Why should we set aside large tracts of wilderness?

Why Save Wilderness?

Wilderness provides an escape from modern society (**FIGURE 12-10**). Joseph Sax, author of *Mountains without Handrails*, writes that nature "seems to have a peculiar power to stimulate us to reflectiveness by its awesomeness and grandeur." It helps us understand ourselves and the world we live in, awakening us to the forgotten interdependence of living things. "Our initial response to nature," Sax writes, "is often awe and wonderment: trees that have survived for millennia; a profusion of flowers in the seeming sterility of the desert. . . . [It] is also a successful model of many things that human communities seek: continuity, stability and sustenance, adaptation, sustained productivity, diversity, and evolutionary change."

Wilderness, especially forested regions, is vital to preserving biodiversity and nature's free services. Many of the reasons for protecting wildlife, described in Chapter 11, also pertain to wilderness protection.

Historically, however, wilderness has been viewed by many people as something to exploit for short-term gain. In early colonial and postcolonial times, American lands represented untapped wealth—an unequaled opportunity to sustain a young, growing nation. To some frontierspeople and particularly early farmers, many natural resources were perceived more as obstacles than as assets. Forests needed to

FIGURE 12-10 **More than just a pretty place.** Wilderness restores us. It also offers numerous free ecological services, from climate control to watershed protection.

be cleared to permit farming. Marshes needed to be drained. Today many people view wilderness as simply a playground for an upper-middle-class elite—people who fight to protect these lands to the detriment of others who could reap economic benefits from mining and timber harvesting.

Differing views create enormous controversy. Today, many environmentalists lobby for more wilderness to be set aside. The mining and timber industries, both powerful lobbying forces, generally oppose wilderness designation, fearing that it will lock up valuable resources and hinder the economic development of the nation. Because only slightly over 16% of all government land holdings (excluding military bases and land on which public buildings are built), or about 4% of all land in the United States, has been afforded wilderness protection, environmentalists argue that the mining and timber industries' claims that wilderness is locking up the Earth's riches are unfounded. Locally, however, wilderness tracts can tie up huge parcels of land, threatening the economic well-being of communities that have long made a living by timber harvesting and mining. Wilderness advocates, though, note that wilderness designation can create a more sustainable economy by promoting businesses in rural areas that service hunters, anglers, hikers, backpackers, cross-country skiers, and others.

Should the United States set aside wilderness if it contains oil, natural gas, or minerals that could be used today? Many environmentalists believe that wildlands are more valuable than these finite resources because there are no substitutes for wilderness once it has been destroyed (Figure 12-10). They also note that there are many more sustainable ways to meet our demands for resources such as water, energy, minerals, and wood.

KEY CONCEPTS

Wilderness offers many benefits to humans. It provides refuge from urban life, offers valuable ecological services, and is home to many species of plants and animals. Historically, however, wilderness has largely been viewed as either a source of resources or an impediment to human progress. These opposing views are at the root of the controversy over wilderness protection.

Preservation: The Wilderness Act

The earliest efforts at wilderness preservation in the United States began in the 1860s. John Muir, founder of the Sierra Club and a longtime wilderness advocate, is credited with much of the early interest in saving wilderness for future generations. Further advances came in the 1930s, when the U.S. Forest Service began to set aside large tracts of forestland, called **primitive areas**, for protection. Between 1930 and 1964, the Forest Service established over 3.7 million hectares (9.1 million acres) of primitive areas in the national forests.

In 1964, Congress passed the **Wilderness Act**, establishing the **National Wilderness Preservation System**. Under this law, the Forest Service's primitive areas were renamed **wilderness areas**. The Wilderness Act directed the Forest Service, the Fish and Wildlife Service, and the National Park Service to recommend additional land within their jurisdictions for wilderness designation. In 2005, 43.1 million hectares (106.5 million acres) of land were protected as wilderness. Fifty-four percent of the wilderness is in Alaska.

In addition to setting aside land for wilderness designation, the Wilderness Act established rules and regulations for human activities in wilderness. For example, it forbids timber cutting, motorized vehicles, motorboats, aircraft landings, and other motorized equipment—except to control fire, insects, and diseases or where their use was already established.

Although the Wilderness Act sought to create an "enduring wilderness," many unwildernesslike activities were allowed to continue—notably livestock grazing and mining for metals and energy fuels—if claims were filed before the end of 1983. Wilderness areas throughout the United States are also riddled with private inholdings, property owned by individuals and companies who control the mineral and water rights on the property.

Although it has largely been a success, the Wilderness Act has some weaknesses. For example, it did not provide a means of designating wilderness on the 180 million hectares (450 million acres) of land held by the Bureau of Land Management, mostly in the western states and Alaska. To correct this, the U.S. Congress passed the **Federal Land Policy and Management Act** (1976), which calls on the BLM to submit recommendations on the wilderness suitability of its land. In 2005, BLM Wilderness Areas contained 2.9 million hectares (7.3 million acres), 6.8% of the nation's total wilderness.

Since President Bush took office in 2000, however, wilderness designation has virtually stopped. In fact, the Secretary of the Interior Gail Norton has proclaimed a "no new wilderness policy." What is more, in 2003, the Bush administration denied the Tongas National Forest in Alaska protection under the roadless rule. This will open it—and very likely nearly 60 million acres of national forest throughout the United States—to timber and mineral development. Roadless areas are potential wilderness, but once roads are built in them they are no longer able to be designated as wilderness. Timber companies have been lobbying hard and pursuing legal action to kill the roadless rule that would effectively open up all of America's untapped wild areas putting an end to further wilderness designation.

KEY CONCEPTS

The United States has a long history of wilderness preservation that continues today through the Wilderness Act. This law directs federal agencies to establish wilderness areas and stipulates the type of human activities that are permitted on these lands.

Sustainable Wilderness Management

Lured by the thought of quiet and solitude, backpackers pour into some U.S. wilderness areas only to be dismayed by crowds and special camping restrictions aimed at protecting lakes and streams from pollution. Wilderness crowding can result in severe environmental degradation. Grasses near favorite camping spots get trampled and eventually die, leaving behind only topsoil. Human waste deposited around campsites washes into streams. Streams are polluted by soap from dishwashing. Garbage often litters favorite sites and trails. Fragile areas get trampled. Wildlife are displaced. Hiking trails become deeply rutted. Horses soil trails and overgraze popular areas.

Overcrowding and the environmental decay that accompanies it often occur near large metropolitan areas—in Colorado and California, for example. To wilderness proponents, these are signs of the need for more wilderness—especially if the U.S. population is to increase by more than 125 million between now and 2050.

Crowding and the environmental degradation from overuse can be reduced or eliminated by better management: (1) educating campers on ways to lessen their impact, (2) restricting access to overused areas, (3) issuing permits to control the number of users, (4) designating campsites, (5) increasing the number of wilderness rangers to monitor use, (6) disseminating information about infrequently used areas to divert campers from overused areas, and (7) improving trails to promote use of underutilized areas. Personal actions are listed in the Individual Actions Count! table.

Globally, interest in wilderness protection is growing. Chapter 11 cited examples of extractive reserves, which if properly managed could help protect biodiversity. Although some countries—such as Costa Rica, Brazil, and Colombia—have set aside large parcels of land for protection, many others have little or no protection whatsoever. Where they do exist, protected areas are often understaffed and overused.

Countries struggling to feed their people and meet the demands of rapidly growing populations often have little concern for protecting wilderness, even though it may be in their best long-term interest. Ecotourism, briefly described in Chapter 11, provides an economic incentive to protect wilderness in less developed nations. Interest in potential pharmaceuticals from plants and other sources, both discussed in the previous chapter, also provides an economic rationale for protecting lands. Finally, the growing realization that extractive reserves can be much more lucrative and environmentally sound than traditional land-use patterns (such as plantations and ranches) has provided additional impetus for saving land from development. Although wilderness management is generally based on the preservation ideology, these compatible uses bode well for the fate of the wilderness in less developed nations.

Wilderness, grasslands, and forests are vital to our future. They cater to many different needs: eating, shelter, relaxation, and escape. A world without them is almost unimaginable to those of us who love nature. A group of scientists peering through the glass of their space shuttle 100 years from now will see the evidence of our actions today. Whether they see patches of ancient desert within rich, productive land or just the opposite depends on actions and decisions we make today.

We're not so poor that we have to spend our wilderness or so rich that we can afford to.
—*Newton Drury*

CRITICAL THINKING

Exercise Analysis

Critical thinking rules instruct us to question the methods by which information is obtained. In this example, early estimates of desertification were based on studies that looked only at isolated regions; that is, their sample size was quite small. This is a common error in scientific research. Conclusions based on small sample size—a few isolated observations or small numbers of test animals—should always be regarded with caution.

This is not to say that the Sahara may not be expanding. It may be; it may not be. Looking at the new evidence suggests that although the desert may ebb and flow, overall it is marching southward for reasons noted in Chapters 10 and 20. The study cited in this exercise suggests that the desert marches southward in low-precipitation years, then recedes in years when precipitation is heavier. However, the trend has been an expansion of the desert. Clearly, further research is needed.

At this point, it is necessary to exercise one of the most paradoxical of all critical thinking rules—to tolerate uncertainty. Only time will tell whether the desert is indeed marching southward as a result of global warming and other problems mentioned in the introduction. Many global issues such as ozone depletion and global warming take place amid natural cycles. To discern real trends requires a decade or more of data.

CRITICAL THINKING AND CONCEPTS REVIEW

1. What is the tragedy of the commons? Have you seen any examples of it? Is reproductive freedom—the right to have as many children as one wants—contributing to the tragedy of the commons?

2. Critically analyze the concept of the tragedy of the commons. Is it a valid phenomenon? Why or why not? Does the inadequacy of resource management on privately owned land negate the validity of this concept?

3. How can the operating principles of sustainability discussed earlier in the book help reshape the livestock and forestry industries? Give specific examples of ways these principles can be applied.

4. What are the major problems facing the world's rangelands? How can we create a sustainable system of providing meat for human society?

5. Define the following terms as they relate to forest management: *sustained yield*, *multiple use*, and *clear-cutting*.

6. What is meant by *sustainable forestry*? How is it different from the present method of cutting trees, which is based principally on sustained yield?

7. List and discuss ways to satisfy the growing need for wood and wood products in the coming years. Which of your ideas are the most ecologically sound?

8. In what ways can you reduce paper and wood waste and increase recycling?

9. Critically analyze this statement: "Wilderness is not essential to humanity. We should not be concerned with preserving wild areas. For the vast majority of the world's people, they are of no value."

10. Do you agree or disagree with the following statement? "The expanding U.S. population suggests the need for more wilderness designation." State reasons for your position.

KEY TERMS

basal zone
Bureau of Land Management (BLM)
clear-cutting
commons
extractive reserves
Federal Land Policy and Management Act
feedlots
metabolic reserve
Model Forest Program
National Wilderness Preservation System
prescribed fires
primary forests
primitive areas
Public Rangelands Improvement Act
range management
rangelands
rotational grazing
second-growth forests
selective cutting
shelter-wood cutting
strip-cutting
sublimation
surface runoff
sustainable forestry
wilderness
Wilderness Act
wilderness areas

REFERENCES AND FURTHER READING

The References and Further Reading section at the end of this book contains a list of sources for the information discussed in this chapter and recommendations for further reading.

Connect to this book's website:
http://environment.jbpub.com/
The site features eLearning, an online review area that provides quizzes, chapter outlines, and other tools to help you study for your class. You can also follow useful links for in-depth information, research the differing views in the Point/Counterpoints, or keep up on the latest environmental news.

A river in Nature's Valley, South Africa.

CHAPTER 13

Water Resources: Preserving Our Liquid Assets and Protecting Aquatic Ecosystems

A river is more than an amenity–it is a treasure.

—*Oliver Wendell Holmes*

Water. We drink it. We wash with it. We cook with it. We play in it. We irrigate our crops with it, and we use it in our factories. The abundance or lack of water often determines where we live and how well off we are.

Despite its importance to humans and other organisms, water is squandered and polluted by industry, agriculture, and many other systems. To live sustainably on the Earth, we must do a much better job of protecting and managing water resources.

This chapter discusses three components of sustainable water resource management: It examines water supply problems and ways to solve them. It explores flooding and ways to reduce it. Finally, it examines aquatic ecosystems such as wetlands, rivers, and oceans and ways to protect and restore them. Chapter 21 addresses water pollution. The goal in all of this is to find new ways to create a more effective and economically viable ways to meet our needs for water now and in the long haul and to ensure a better future. Before looking at the problems of water supply, however, let's examine the water cycle.

13.1 The Hydrological Cycle

Water is part of a global recycling network known as the **hydrological cycle**, or **water cycle**. The hydrological cycle runs day and night, free of charge, collecting, purifying, and distributing water. Along its way, it serves humans and other living creatures in a multitude of ways.

The water cycle is driven by evaporation and precipitation. **Evaporation** occurs when water molecules escape from surface waters, soil, and plants, becoming suspended in air (**FIGURE 13-1**). When water molecules depart, they leave behind impurities. Thus, evaporated water is free of contamination until it mixes with atmospheric pollutants from human and natural sources.

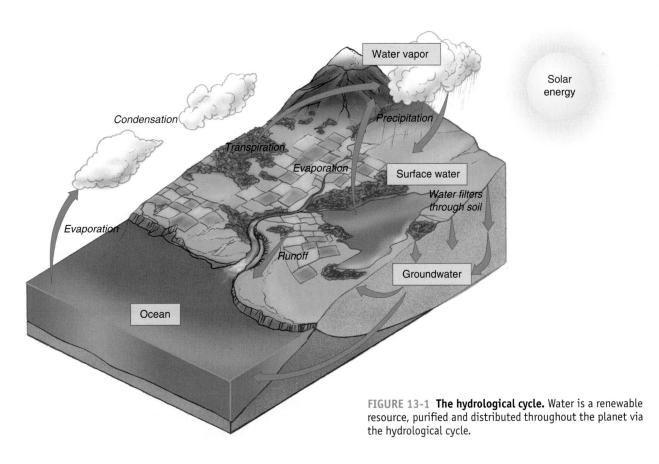

FIGURE 13-1 The hydrological cycle. Water is a renewable resource, purified and distributed throughout the planet via the hydrological cycle.

Evaporation not only ensures the continuation of the water cycle, it also serves important biological functions. In plants, for example, the evaporation of water from leaves draws water and dissolved nutrients up from the roots through stems, helping to nourish plant cells. Water evaporating at the surface of leaves draws water up through tiny vessels in plants, much like sucking on a straw draws water from a glass. The evaporative loss of water from leaves is called **transpiration**. The loss of water from the soil and leaves is called **evapotranspiration**. The evaporation of water is also helpful for animals. In humans, for instance, water evaporating from the skin helps keep our bodies from overheating on hot days.

In the atmosphere, water is suspended as fine droplets known as *water vapor*. Although we're not usually aware of it, you can detect its presence on a cold day when moisture beads up on cold windows or on a warm day when moisture collects on the outside of a glass of ice-cold lemonade.

The amount of moisture (water vapor) air can hold depends on the temperature of the air. The warmer the air, the more moisture it can hold. Atmospheric moisture content can be expressed in one of two ways. The first is the **absolute humidity**, which is the number of grams of water in a kilogram of air. If you extracted a sample of air and removed its water, this would be its absolute humidity. Another measure is the relative humidity. The **relative humidity** is the amount of moisture present in air compared with the amount it could hold if fully saturated, at any given temperature. If the local TV station's meteorologist says that the relative humidity is 50%, she means that the air contains 50% of the water vapor it can hold at the present temperature. If the relative humidity is 90%, the air has 90% of the moisture it can hold at that particular temperature. If the relative humidity is 100%, the air is said to be **saturated.** When the moisture content of a mass of air exceeds the saturation point, clouds, mist, and fog form—that is, water vapor becomes visible to us!

Water that has evaporated eventually returns to Earth through precipitation, thus continuing the hydrological cycle. Here's how it happens. On a warm summer day, sunlight warms the Earth. This causes water to evaporate from water bodies, the land, and plants. Warm, moisture-laden air then rises. As it rises, the air expands because atmospheric pressure decreases. As it expands, it cools. As it cools, its relative humidity of the air mass increases. When the air mass exceeds its saturation point, clouds begin to form. This same phenomenon occurs when moisture-laden air is pushed upward by mountain ranges (see Figure 5-3).

Clouds form when the relative humidity becomes slightly greater than 100%. When this occurs, water molecules begin to attach onto various small suspended particles, called **condensation nuclei**, in the air. Condensation nuclei are always present in the atmo-

> Although the Earth has an enormous quantity of water, only a small fraction (3%) of it is freshwater, and only a small fraction of this water is accessible to humans.

sphere; they may be salts from the sea, dusts, or particulates from factories, power plants, and vehicles.

Clouds and fog contain several hundred water-laden condensation nuclei per cubic centimeter of air. In tropical and semitropical regions, fine cloud droplets collide and coalesce, forming raindrops. Over a million fine water droplets must come together or **condense** to make a single drop of rain. Some evidence suggests that electrical activity in clouds may increase the rate of collision and thus enhance the formation of raindrops.

In temperate climates and at the poles, the temperature of the air in clouds is often well below the freezing point, even in the warmer months of the year. As a result, minute ice crystals tend to form in the clouds. When the crystals reach a certain critical mass, they fall from the sky as snowflakes. In the spring, summer, and fall, the snowflakes generally melt as they fall, producing rain.

Clouds move about on the winds, generated by solar energy (Chapter 5), and deposit their moisture throughout the globe as rain, drizzle, snow, hail, or sleet. This process, called **precipitation**, returns water to lakes, rivers, oceans, and land.

Water that falls on the land may evaporate again, or it may flow into lakes, rivers, streams, or groundwater. Eventually the water returns to the ocean, from which it may once again evaporate. The vast majority of the water in clouds and rain comes from surface waters, primarily the oceans.

On average, over 15 trillion liters (4 trillion gallons) of precipitation fall on the United States every day. Two thirds of all this precipitation evaporates. Thirty-one percent finds its way into streams, lakes, and rivers, and 3% recharges groundwater (**FIGURE 13-2a**).

At any single moment 97% of the Earth's water is in the oceans. The remaining 3% is freshwater. Of this, most (99%) is locked up in polar ice, in glaciers, and in deep, inaccessible aquifers, underground reservoirs usually in some permeable material such as sandstone. This leaves a paltry 0.003% available for use by humans, but most of this water is hard to reach and much too costly to be of any practical value. At present, human civilization and all terrestrial forms of life are maintained by only a tiny fraction of the Earth's water supply, a fact that underscores the importance of treating our freshwater supplies with care. (For a discussion of groundwater, see Chapter 10.)

KEY CONCEPTS

Water is a renewable resource, purified and distributed in the hydrological cycle, which is driven by solar energy. Although the Earth is endowed with an enormous quantity of water, only a small fraction of the planet's water is available for human use, a fact that underscores the importance of managing it wisely.

13.2 Water Issues Related to Supply and Demand

Water issues are of great importance in the study of our environment. In this section, we examine issues related to supply and demand. We look at water shortages and impacts of supplying water to the world's people. Before we begin, however, let's take a look at where our water comes from and who uses it.

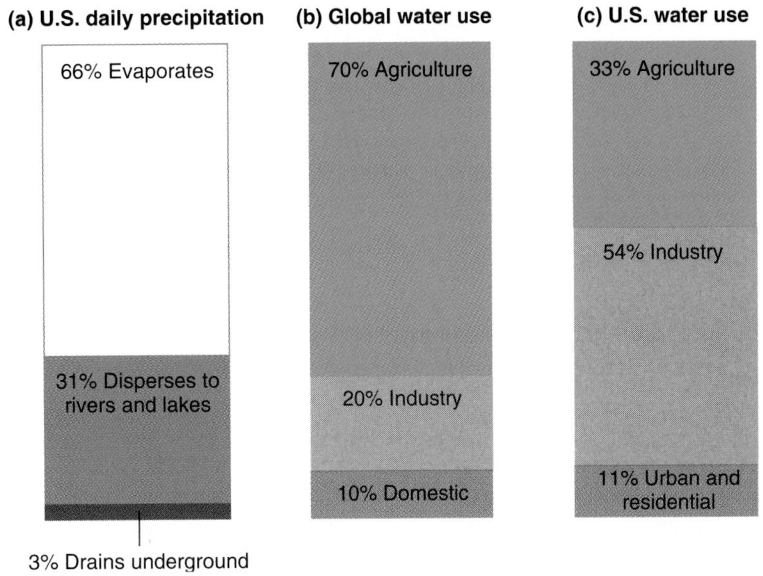

Where Does Water Come from and Who Uses It?

The world population withdraws enormous quantities of freshwater from lakes, streams, and rivers each year. Approximately 70% of this water is used for crop and livestock production. Industrial use accounts for about 20%, and domestic use makes up the remaining 10%, according to the Worldwatch Institute (FIGURE 13-2b). Of course, these averages vary from country to country. In the United States, for example, 33% of the freshwater is withdrawn for agriculture and 54% is for industry. Eleven percent is withdrawn for residential use (FIGURE 13-2c). In Asia, 85% is used for irrigation. Where does this water come from?

In the United States, surface waters supply the bulk of the freshwater we need each day. In fact, about three fourths of our daily water supply comes from lakes, rivers, and streams, known as **surface waters**; the remainder comes from **groundwater**.

> Agriculture is a major consumer of water, using about 70% globally. Industry comes in second, consuming about 20%, and residential use accounts for 10% of all water use.

KEY CONCEPTS

Globally, agriculture and industry are the major users of water. Most water comes from surface water supplies—rivers, streams, and lakes.

Water Shortages

Although the Earth is endowed with a generous supply of freshwater, water shortages are becoming more and more common. Virtually every nation in the world experiences them. And the situation is predicted to get worse. By 2025, the number of people experiencing water shortages year-round or for part of the year is predicted to increase to 2.6 to 3.1 billion, up from about 434 million today. The consequences can be devastating. In many less developed countries, women spend a large part of their waking hours fetching water, often walking 15 to 25 kilometers (10 to 15 miles) a day to get it—frequently from polluted streams and rivers. According to the World Health Organization, three out of every five people in the less developed nations do not have access to clean, disease-free drinking water.

To many, it may seem as if the world is running out of water. Far from it. Today's freshwater supply is the same as it was when civilization began. As you just learned, water is a renewable resource, constantly recycled by the hydrological cycle. Interestingly, enough drinkable water falls on the land each year to flood it to a depth of 86 centimeters (33 inches). On first glance, one would think that this amount should provide sufficient quantities to meet our needs several times over. Unfortunately, precipitation is not evenly distributed across the face of the Earth. Tropical rain forests are drenched with rain, receiving 250 centimeters or more (100 inches). Deserts receive under 25 centimeters (10 inches) a year. People living in areas that receive small amounts of precipitation, such as the desert, often suffer severe water shortages.

Water shortages result from lack of rain and are therefore most common in dry, arid climates, less so in semiarid regions, and uncommon in rainy areas. But rainfall patterns are just part of the equation. Demand also plays a big role. Any time demand exceeds supply, water shortages can occur. A stream and groundwater may be sufficient for a small town, but when that town burgeons to become a city, shortages may result. Water shortages occur even in areas with seemingly abundant supplies when demand exceeds supplies. Canada, for example, has 9% of the world's freshwater and ranks third among the world's nations in freshwater resources. Yet water shortages are common in many parts of Canada, especially the southern parts of the prairie provinces. Part of the reason for this is that Canada's indoor water use is second highest in the world, about 88 gallons (340 liters) per person per day. The other reason is that 90% of Canada's population is concentrated in the lower portion of this vast country and much of the nation's water resources are located in the North.

One thing is certain. As population increases, water shortages will become even more prevalent. Demand for agriculture and industry and personal use will rise significantly, causing even greater shortages than now exist. According to the Worldwatch Institute, between 1950 and 2050, the amount of water available per person will fall by 74%. They and others believe that water scarcity may be the most underrated resource issue in the world today.

(a) U.S. daily precipitation

66% Evaporates

31% Disperses to rivers and lakes

3% Drains underground

(b) Global water use

70% Agriculture

20% Industry

10% Domestic

(c) U.S. water use

33% Agriculture

54% Industry

11% Urban and residential

FIGURE 13-2 **Where does the water go? (a)** Fate of precipitation in the United States, **(b)** global water use, and **(c)** water use in the United States.

KEY CONCEPTS

Water shortages occur virtually everywhere. They are most prevalent in areas that receive small amounts of precipitation but can occur in any region in which demand exceeds existing water supplies. Population growth, agricultural expansion, and demand for water by the industrial sector are likely to make water shortages even more prevalent in coming years.

Drought and Water Shortages

Drought also contributes mightily to water shortages. A **drought** exists when rainfall is 70% below average for a period of 21 days or longer. A severe drought results in a decrease in stream flow, sometimes a complete cessation of flow, and a drop in the level of lakes, streams, and reservoirs. This, in turn, reduces water available for irrigation, domestic use, and industrial production. Drought also results in a drop in the **water table**, the upper level of the groundwater. As anyone who has viewed the evening news can attest, droughts cause devastating losses to agricultural crops. They may cause a loss of wildlife, especially aquatic organisms, and reduction in range production and increased stress on livestock. Drought years are also characterized by an increasing number of forest and grass fires.

Whether you live in New York, Toronto, Tulsa, or Bird City, Kansas, you've probably felt the impacts of water shortage at least once in your lifetime. Most people think of droughts as natural events. Although this is true, droughts can also be caused by human activities. Overgrazing and deforestation, for instance, decrease rainfall in downwind areas, as described in the previous chapter. The devastating fires in Indonesia and neighboring countries in 1999 were caused in large part by deforestation that led to a drying of neighboring intact forests. As you shall see in Chapter 20, global climate change caused by certain forms of pollution also appears to affect rainfall patterns, creating excessive flooding in some areas and drought in others.

KEY CONCEPTS

Droughts occur naturally, but may also result from human actions such as deforestation, overgrazing, and pollution. Droughts reduce water supplies and create significant social, economic, and environmental problems.

Impacts of the Water Supply System

There is no doubt that water shortages cause severe problems, but before we rush headlong into a plan to develop more water supplies to meet our needs, we must carefully study the water supply system and its impacts. Such an understanding could help us find better—more sustainable—ways to meet our needs.

Water is withdrawn from lakes, rivers, and streams to serve human needs. In the case of lakes, water is generally removed by offshore pipes and pumped to water treatment plants. There it is filtered and chlorinated and then distributed through underground pipes to users (FIGURE 13-3). Rivers and streams, however, are generally dammed to produce reservoirs that store water for year-round use. The water from the reservoir is drawn off as needed, filtered, and distributed to users. In some areas, groundwater provides part or all of the water needed by humans. It is pumped from wells.

Although most of us take for granted the water that comes out of the tap when we turn the faucet, this water comes to us through an elaborate and costly system. The system has an enormous impact on our environment. The following section will examine some of these impacts.

FIGURE 13-3 **Water systems.** This diagram shows the components of the water supply system of a city.

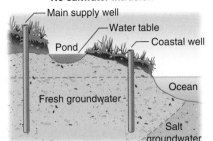

No saltwater intrusion

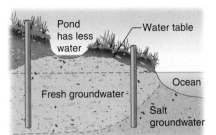

Saltwater intrusion

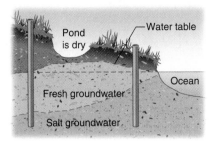

Further saltwater intrusion

FIGURE 13-4 **Saltwater intrusion into groundwater.** Removal of excess groundwater in coastal regions causes freshwater aquifers to shrink, allowing saltwater to penetrate deeper inland underground. Groundwater overdraft not only causes saltwater to enter the freshwater aquifer but also causes surface water in the pond to decline.

KEY CONCEPTS

Water comes from an elaborate and costly system that has a tremendous impact on the environment. Understanding the system and its impacts may help us design more sustainable systems of water supply.

Impacts of Excessive Water Withdrawals   As a rule, small amounts of water can be removed from water supplies without any noticeable impact. When demand becomes excessive, however, impacts start to become evident. One of the reasons for this is that much of the water removed from streams and aquifers to serve humans never returns to its source. For example, over 80% of the agricultural water applied to crops evaporates. Consequently, many rivers flow at a fraction of their natural rates during high-use seasons. Some are completely dried up. The Colorado River in the United States, for example, is siphoned off by farms and cities so that by the time it reaches the sea, it is reduced to a mere trickle. So much for the mighty Colorado! The Yellow River in northern China has been running dry since 1985, and each year the dry period becomes longer. In 1997, it failed to deliver any water to the sea for 226 days. Even the mighty Nile, which flows into the Mediterranean, rarely has any water left in it by the time it reaches the sea.

Reductions in water flow and depletion reduce populations of aquatic species and recreation activities. It can cause neighboring habitat, especially wetlands, to disappear. The vast wetlands at the mouth of the Colorado River, for example, have been reduced to a tiny remnant of its former self. Water that is returned to surface water supplies is often polluted with agricultural pesticides or wastes from homes and factories.

Overuse also causes aquifers to dry up completely, driving farmers out of business (Chapter 10). Groundwater is a vast resource. In Canada, which is home to many large rivers and seven of the world's largest lakes, groundwater resources are approximately 37 times greater than all the water contained in rivers and lakes. Despite the enormity of groundwater, supplies are on the decline worldwide. Groundwater overdraft—the removal of groundwater at a rate faster than it is replenished—has severe long-term implications. In coastal regions, for example, groundwater overdraft can lead to **saltwater intrusion,** the movement of saltwater from ma-

rine aquifers into freshwater aquifers (**FIGURE 13-4**). Saltwater then contaminates previously freshwater wells, destroying this resource.

Ponds, bogs, and streams mark the intersection of aquifers and the land surface. Many think of a pond as "exposed groundwater." Because of this link, groundwater overdraft also drains swamps and ponds, at times drying them up completely. Fish, wildlife, and recreation are often adversely affected.

Chapter 10 noted that excessive withdrawal of groundwater threatens the long-term prospects for irrigated agriculture in the United States (**FIGURE 13-5**). At the very least,

FIGURE 13-5 **Declining groundwater.** This teacher and his student are measuring the water levels in a well sunk into the Ogallala aquifer and are finding that water levels are dropping rapidly due to groundwater overdraft.

FIGURE 13-6 One dramatic effect of groundwater overdraft. This large sinkhole in Florida developed quickly, swallowing a street. It was caused by depletion of groundwater, as explained in the text.

it causes farmers, cities, towns, and homeowners to drill deeper wells, a costly process. It also results in the retirement of irrigated cropland. Excessive groundwater withdrawals are causing the water table to drop in many less developed nations, including parts of China and India, the two most populous countries in the world.

Groundwater fills pores in the soil and thus supports the ground above the aquifer; therefore, when the water is withdrawn, the soil compacts and sinks, a process called **subsidence.** The most dramatic examples of subsidence have occurred in Florida and other southern states, where groundwater depletion has created huge **sinkholes** that may measure 100 meters (330 feet) across and 50 meters (165 feet) deep (**FIGURE 13-6**). Subsidence has occurred over large areas in the San Joaquin Valley of California, damaging pipelines, railroads, highways, homes, factories, and canals. Southeast of Phoenix, over 300 square kilometers (120 square miles) of land has subsided more than 2 meters (6 feet) because of groundwater overdraft. Huge cracks have formed, some 3 meters wide, 3 meters deep, and 300 meters long.

> **KEY CONCEPTS**
>
> Excessive withdrawal of surface water can cause streams to dry up, with profound effects on fish and wildlife. Groundwater overdraft results in many problems, too, among them depletion of water levels in connected surface waters and subsidence.

Impacts of Dams and Reservoirs   Because water flow in streams and rivers varies with the seasons, dams are often constructed on streams and rivers to retain water needed by cities, towns, farms, and industry. This ensures a steady supply year-round. In some cases, dams are built to generate electricity (Chapter 15). Reservoirs behind dams provide recreational opportunities such as fishing, waterskiing, and boating. Dams help to control floods. Dam construction and

accompanying development by the Tennessee Valley Authority (a federal agency created by the U.S. Congress in 1933 to develop the Tennessee River and its tributaries) raised the economic standing of an area of abject poverty. This project, covering parts of seven states, provided hydroelectric power, flood control, and a 650-mile inland waterway. No major floods have occurred in the region since the completion of the major dams, thus averting many millions of dollars in damage.

Despite their social and economic benefits, dams can have a profound impact on fish and wildlife. They flood streams and eliminate fish species that live and reproduce there. They interrupt migration. The once-plentiful salmon of the Pacific Northwest have been greatly reduced in number in large part because of dams that impair migration to the sea. Young salmon (called *smolts*) migrating to the ocean often lose their way in reservoirs. Why? Reservoirs eliminate current, which helps them orient. Thus, many young salmon get lost in reservoirs and never reach the sea, where they would normally mature.

Dams also interfere with the migration of salmon returning from the sea to spawning grounds. Salmon runs on many streams in the Pacific Northwest and on the East Coast have been decimated by dams built without the "ladders" that allow fish to bypass them. Even when ladders are built, there's no guarantee that fish populations will thrive. Some salmon are killed in turbines that produce electricity. By one estimate, 8 to 15 percent of the young deep swimming smolts are killed at each dam.

Dams also flood the habitat of other species, such as bear and deer. Construction roads and noise also disturb wildlife.

Dams impact human activities and economies as well. New dams, for example, threaten recreational rivers, and they place towns that depend on rafting and kayaking in direct conflict with sprawling cities and suburbs that want the water.

In less developed countries, dams built for irrigation, hydropower, and drinking water often flood valuable farmland on which people have lived sustainably for hundreds of years. Pakistan's recently completed Tarbela Dam displaced 85,000 people. China's mammoth Three Gorges Dam and reservoir on the Chang River (formerly the Yangtze River), when completed in 2009, will displace 1.4 million people in 20 towns. It will destroy nearly 100 square kilometers (40 square miles) and inundate about 41,000 hectares (100,000 acres) of farmland in the world's most heavily populated and agriculturally productive river valley. People displaced by dams and reservoirs must often move onto marginal land that deteriorates rapidly under heavy use.

Water withdrawals from reservoirs have many downstream impacts. For example, reduced water flows may allow saltwater to intrude into estuaries during high tide, upsetting the ecological balance in these important life zones. This is a serious problem in Everglades National Park in Florida. (See Spotlight on Sustainable Development 13-1.)

Reduced water flow also decreases the flow of nutrient-rich sediments to estuaries. This interrupts the natural flow of nutrients to floodplains, river deltas, and coastal waters. Rich farmland along riverbanks once replenished by periodic flooding must be abandoned or fertilized, often at a high cost.

Sediment that collects in reservoirs slowly fill them, making them useless. In the United States, well over 2000 small reservoirs are totally clogged with sediment. Many larger reservoirs throughout the world will meet a similar fate. The $1.3 billion Tarbela Dam on the Indus River in Pakistan took 9 years to build, but because of upstream soil erosion, the reservoir could fill with sediment in 20 years. Lake Powell on the Colorado River will be filled in 100 to 300 years.

> In the United States, more than 2000 reservoirs have been rendered useless by sediment.

Dams generally release the coldest water from the bottom of the reservoir. This makes stream water exceedingly cold year-round and decreases the spawning of native species. The Glen Canyon Dam on the Colorado River, for instance, releases chilly water from the reservoir's bottom and has converted what was once a warm-water fishery to a rainbow trout fishery. The cold water now threatens several native species, such as the humpback chub (FIGURE 13-7). To avoid changing stream temperature, some existing dams have been retrofitted with devices that combine warm water from the reservoir's surface with cooler deep waters. Newer dams are often fitted with multiple gates to ensure proper downstream water temperature.

Because up to 80% of the annual runoff of some streams may be withdrawn, wildlife habitat and good recreational sites may be dewatered by dams. In addition, these projects usually transfer water from rural areas of low population density (often agricultural regions) to cities. This practice creates bitter conflicts between urban and rural residents.

Water withdrawals in reservoirs can also affect water quality. Diversion of clean, high mountain water in tributaries of the Colorado River, for instance, diminishes the flow and quality of the river, a source of water for well over 25 million

people. The lower sections of the river now carry a heavy burden of sediment and dissolved salts because of the loss of clean upstream water. By the time the river reaches Mexico, its salt concentration is over 800 parts per million (violating a U.S.–Mexico treaty), compared with 40 ppm at its headwaters. Why is this of concern?

Water with a salt concentration over 700 parts per million (ppm) is toxic to many plants.[1] It cannot be used for agriculture unless it is diluted or desalinated (has the salt removed). Drinking water must have a salt concentration of 500 ppm or less. A bill passed by the U.S. Congress in 1973 authorized the construction of three desalination plants along the Colorado River to remove over 400,000 tons of salt from the river each year. Only one plant was built. This facility in Yuma, Arizona, cost $250 million to build and costs about $15 to $20 million per year to operate.

Sediment and dissolved solids are a problem in agricultural runoff in Canada, too. In the South Saskatchewan River basin, irrigation water returning to streams contains levels of these pollutants that are sometimes twice what they were in the original source. They're also polluted by a host of other contaminants such as pesticides and nitrates.

The diversion of water can also have impacts on people and local economies. For example, excessive water withdrawal from two rivers that once fed the Aral Sea in the former Soviet Union has completely dried up one river and reduced the other to a mere trickle. The sea has dropped 12 meters (40 feet) in 30 years. Two thirds of its water has evaporated now that it is no longer replenished. Towns once situated on the shore are now 30 miles inland. Because freshwater inflow has all but stopped, the concentration of salts in the sea has risen, killing off 20 of the 24 commercial fish species in the sea. Dozens of fishing towns have been abandoned (FIGURE 13-8).

[1]For comparison, saltwater in the ocean has a concentration of 35 parts per *thousand*.

FIGURE 13-7 **The humpback chub.** This is one of several fish species of the Colorado River threatened by dams that release cold water into what was once a fairly warm-water fishery.

FIGURE 13-8 **The Aral Sea.** This body of salt water in the former Soviet Union has been reduced by massive withdrawal of water to irrigate farmland. Most species of commercial fish have disappeared, and fishing villages have been abandoned. These fishing vessels and the fishing village have been left high and dry.

13-1 Undoing the Damage to Florida's Kissimmee River

Between 1964 and 1970, the Army Corps of Engineers hacked away at the winding Kissimmee River in Florida, hauling up mud and dumping it along the banks. When they had finished this huge flood control project, a river that had once lazily meandered nearly 150 kilometers (100 miles) through the Florida marshes had been reduced to a canal 64 kilometers (40 miles) long, 60 meters (200 feet) wide, and 10 meters (30 feet) deep. The canal was designed to drain water quickly from the northern reaches of the watershed. On the heels of the huge dredgers that had converted the river into a canal came contractors who threw up concrete and earthen dams and locks every 10 miles, creating huge reservoirs along the river's previous course. These were designed to control flooding downstream.

Once a rich habitat for bald eagles, deer, fish, waterfowl, and alligators, the Kissimmee River became a sterile tribute to our tireless efforts to control flooding. Most scientists condemned the channelization as a major environmental catastrophe, which destroyed three fourths of the original 16,000 hectares (40,000 acres) of marsh, once a major habitat for dozens of species of waterbirds. Secondary canals built by landowners along the main canal drained another 80,000 hectares (200,000 acres). Soon after the canal's completion, the vast flocks of ducks that once rained down from the skies were gone. Gone, too, were the wading birds. By Florida Game and Fresh Water Fish Commission estimates, 90% of the waterfowl and 75% of the bald eagles vanished from the region, as did the largemouth bass that once attracted anglers from all over the nation.

Two years after this enormous project had been completed, Florida biologists began noticing changes in Lake Okeechobee (pronounced OAK-eh-CHO-bee), into which the

Kissimmee's clean waters once flowed. Dead fish and dying vegetation were the most blatant signs that something was awry in the lake, which provides drinking water for Miami and coastal cities. It didn't take biologists long to determine that the loss of marshlands, which purify waters and hold back sediment, was the reason for Lake Okeechobee's sudden deterioration. The loss of the natural cleansing provided by wetlands—plus a heavy load of pesticides, fertilizer, animal wastes, and sediment from cattle ranches and farms that sprang up along the river's banks—created a monumental water quality problem for the lake.

The waters of the Kissimmee River (which flow south) once fanned out across southern Florida to nourish the huge, multimillion-hectare wetlands known as the Everglades. On the southernmost tip of Florida is Everglades National Park. To make room for farms, much of the Everglades has been drained, and the water from Lake Okeechobee and the Kissimmee River basin has been shunted via canals to the coast. Reduced flows have disrupted the ecology of the Everglades, seriously threatening many species, some already endangered. Reduced water flows also produced an ironic backlash. Because of a lowered water table, farmland once prized because of its rich soil began to sink at a rate that could hinder farming in the region. Lower water flows have also resulted in saltwater intrusion into surface water and groundwater.

Ironically, studies made after the canal was completed indicate that it provides little or none of the expected flood control. Making matters worse, the canal is now seen as a major threat to downstream areas. After heavy rains in central Florida, for instance, a slug of water travels rapidly southward along the canal, wiping out nesting waterfowl and drowning unsuspecting wildlife.

KEY CONCEPTS

Dams and reservoirs have many benefits, such as flood control and water storage, but they also have profound impacts on people, wildlife, and the economy. They can eliminate native fishes, block fish migration routes, flood wildlife habitat and farmland, displace people, destroy recreational opportunities, reduce nutrient flow to estuaries, and alter water quality in streams. Ultimately, all dams have a finite life span because their reservoirs eventually fill with sediment.

Creating a Sustainable Water Supply System

Fortunately, as in so many areas, there are numerous ways to make the present system of water supply and use sustainable. They require changes in both habits and technology. Many water-saving technologies are inexpensive and easy to install, and they provide enormous social, economic,

and environmental benefits. The operating principles of sustainability are a useful guideline for revamping and rebuilding the water supply system.

KEY CONCEPTS

To avert future water shortages, we need a sustainable water policy and management strategy based on four of the operating principles of sustainability: conservation, recycling, restoration, and population control.

Population Stabilization   Given the limited amount of freshwater available to humanity and the importance of free-flowing streams to maintaining life, creating a sustainable system of water supply will require measures to stabilize and perhaps stop the growth of the human population. Especially important are measures to control growth in water-short areas, such as the desert Southwest of the United States, northern Africa, the Middle East, and parts of the Indian

Less than 2 years after the Army Corps of Engineers had trucked in the last load of cement, a special governor's conference committee released a report calling on the state to reflood the marshes that it had just drained. The report concluded that channelizing the river had been a big mistake. With a price tag of $30 million, it had also been a costly one. Even the Army Corps of Engineers commissioned a study to re-evaluate the project and prepare recommendations for returning the river to its original state.

In 1983, Governor Robert Graham and supporters took steps to reverse the damage. In 1984, the Kissimmee River restoration began. However, because no federal funds were available to reclaim the river, funding had to come from another source, property taxes on residents in southern Florida. Financed by property taxes, the South Florida Water Management District (which was put in charge of the project) built three small dams to divert water from the canal back into the old river channel (FIGURE 1). This flooded the marshes along 20 kilometers (12 miles) of the river and cost $1.5 million.

These steps helped wetland vegetation, waterfowl, and fish populations recover, but full recovery will require other measures—especially steps to restore more normal water flows. In 1992, the U.S. Congress approved legislation that began a project that will restore over 40 square miles of river and associated wetlands. This will be accomplished by filling in 22 miles of the 56-mile-long canal and removing two of the five water control structures. The water district will continue to buy up land along the river and will take an ecological approach to restoration. Their goal is not to optimize one or a few valuable species, such as bass, but to restore the ecosystems badly damaged by channelization. They are also work-

ing upstream to help prevent runoff from farms and dairies. Restoring the Kissimmee River is part of a major ecological experiment aimed at saving Florida's fast-vanishing wetlands. The restoration project is expected to take 20 years.

Someday, the complex wetlands of Florida will function more like they did a hundred years ago. But at this very moment, engineers and construction companies are hard at work draining wetlands the world over. One of the largest lies along the Nile River. In the home of countless birds and wildlife, huge dredgers are now busily sucking up the mud and straightening the channel.

FIGURE 1 Small dam on the channelized portion of the Kissimmee River in Florida helps to restore wetlands.

subcontinent. Special attention must be paid to the placement of new housing and other development so that it is not built over aquifer recharge zones. See Spotlight on Sustainable Development 13-2 for a stellar example of ecologically sound development.

Efforts to slow and perhaps stop the growth of the human population are vital to living within the planet's capacity to supply freshwater.

Water Conservation: Using What We Need and Using It Efficiently For years, water shortages were addressed by expanding supply: damming rivers and streams and pumping water out of lakes. As the cost of these options rose and concern over the environment increased, however, more and more attention has been focused on ways to use water more efficiently. One of the main benefits of water conservation is

that it involves measures that can be brought online fairly quickly and at costs well below those incurred by building new dams and reservoirs. Conservation measures also offer numerous environmental benefits.

Because agriculture and industry are the biggest water users in all countries, efforts to conserve water in these two systems should be given the highest priority. Substantial gains in agricultural water efficiency can be achieved by lining the open, dirt-lined ditches that deliver water to crops (and that lose up to 50% of their water) with concrete or plastic. Lined ditches are 80 to 90% efficient. Pipes are even more efficient. Sprinklers, which waste up to half of their water, can be replaced with drip irrigation systems, which lose only 5%. However, drip irrigation can be used for only a limited number of crops, such as fruit trees, grapevines, and some vegetables. For the vast fields of wheat and corn, more conventional methods must be used. Modifications in sprinkler systems for these crops, discussed in Chapter 10, can cut

13-2   Ecological Solutions to Flooding and Water Supply Problems

What do The Woodlands, Texas, and Boston, Massachusetts, have in common, besides the distinguishing accents of their people? The answer is that both places have made important decisions in land-use planning that will save millions of dollars by reducing flood damage.

Boston boasts a fine city park system that stretches from the center of the city into the outlying suburbs. Few realize that the park system, with its meandering streams, was built in large part to help control flooding (FIGURE 1). After heavy rains, excess water flows into this basin, where it is slowly released to the sea, reducing flooding and property damage.

The city also purchased large wetlands in an outlying area, rather than letting developers drain and build on them. The wetlands were set aside to reduce flooding and also to provide valuable habitat for fish and wildlife. The project costs one tenth as much as a dam to control flooding.

In 1971, landscape architect Ian McHarg wrote a landmark book entitled *Design with Nature*. In it, he suggests that builders consider wildlife needs, natural flooding, soil stability, and a half dozen other factors when constructing homes. By careful site analysis and design with nature, builders lessen their impact on the land, air, and water.

Texas developer George Mitchell decided to build a new town called The Woodlands using McHarg's ideas. On his forested tract north of Houston, Mitchell envisioned a city in harmony with the forces of nature. He and his staff of planners first analyzed the region and found that they could leave the natural drainage system as open space, which would carry water away more effectively and more cheaply than a storm sewer system. That step alone saved $14 million in construction costs. Roads were built on high ground, and buildings were restricted from aquifer recharge zones, thus protecting the groundwater that supplies Houston. In 1979, rainwaters drenched the site. The streams swelled by 55%. In neighboring towns built with little regard for nature, water flows increased 180%. Those towns suffered considerable flood damage, whereas The Woodlands managed well.

The Woodlands is an attractive community (FIGURE 2). Most of its trees still stand. The floodplains that were set aside for natural drainage and aquifer recharge harbor numerous birds and mammals, including bobcats and white-tailed deer. This community stands as a testament to the benefits of designing with nature. This approach to development permits nature to direct the design of human settlement and helps people to live in harmony with it—so important in building a sustainable future.

FIGURE 1 Aerial view of the park system of Boston, showing flood control aspects.

FIGURE 2 **The Woodlands, Texas.** Note the large amount of wooded area surrounding the houses.

water losses by 30 to 50%. Farmers can irrigate at night, early in the morning, or late in the day, when evaporation is lowest. In warmer climates, however, nighttime irrigation promotes fungal growth that reduces productivity, so this strategy won't work everywhere.

For many years, there has been little incentive to conserve water on farms in many countries, including the United States, because water projects were subsidized by taxpayers. In the words of one observer, the public often subsidizes the private use of water to grow taxpayer-subsidized crops. In the San Joaquin Valley of California, for example, irrigation water costs farmers about $5 per acre-foot (an acre of water one foot deep) thanks to publicly funded projects. Water is also cheap because the economic costs of the damage created by water projects (for example, the loss of fish) are not included in the cost. California's once-lucrative salmon fishing industry has collapsed, in large part because of dams. Subsidized water projects are not unique to more developed nations. In less developed countries 80 to 90% of the cost of irrigation projects is subsidized by either the government or MDCs.

> Loss of water in irrigation ditches can be cut in half by lining ditches with cement. Pipes that deliver water to fields can cut losses by 80 to 90%.

Water conservation today makes great sense. According to the nonprofit group Environmental Defense, recycling water and scheduling water application according to crop and soil needs costs about $10 per acre-foot. Switching from irrigated crops, such as cotton, to dryland crops costs farmers about $40 per acre-foot in lost crop yield. Water conservation by more efficient drip and sprinkler systems costs about $175 per acre-foot. Water from new dams, however, costs up to $500 per acre-foot.

> Water from new dams costs up to $500 per acre-foot. Conservation measures that free up water cost as little as $10 per acre-foot of water saved.

By investing in water conservation, farmers can cut water demand and save money. They can also sell excess water to willing buyers. California cities and industries, for instance, currently purchase water for about $200 per acre-foot per year. If a farmer could invest $50 per acre-foot to save water and sell the excess for $200, he would reap a nice profit. Because farmers would be making a profit on publicly subsidized water, some critics say that it seems only fair that the public be reimbursed as well; farmers should not be allowed to reap the full profit of a publicly subsidized resource. No matter what the outcome of the debate, water conservation makes good sense in the long run as well as the short run. Worldwide, a 10% decrease in agricultural water use could double the amount of water available for cities and industries.

In Southern California, the Metropolitan Water District pays for water conservation measures in the nearby Imperial Irrigation District. By lining irrigation canals and other measures, farmers free up substantial amounts of water for use in Los Angeles, San Diego, and neighboring cities—at a cost far below that of new dams and diversion projects.

Many sustainable farming practices reduce the demand for water (Chapter 10). Increasing the organic content of the soil, for example, reduces water requirements because organic matter retains soil water. In addition, measures that reduce surface runoff (contour farming, terracing, and strip cropping) or that reduce wind blowing across farm fields (shelterbelts) result in a substantial reduction in irrigation water demands. In India, farmers plant rows of grass along the contours of hilly terrain to retard surface runoff, reduce soil erosion, and increase soil moisture content. Yields can increase by 50%. Such simple measures can be extremely cost-effective in large regions, making costly irrigation dams unnecessary. Shifting water-intensive crops such as cotton from regions of low moisture (California) to areas of higher precipitation (Alabama) could also result in substantial water savings.

Water conservation by industry can be achieved by redesigning existing processes and facilities to make them more efficient. The use of water by water-cooled electric power plants, one of the most water-consumptive industries in the world, could be cut one fourth by using dry cooling towers (FIGURE 13-9), although these require more energy and are more expensive to operate than wet cooling towers.

Improvements can be made in cities and towns as well. Municipal water systems throughout the United States lose on average 12% of

> Municipal water systems lose an average of about 12% of their water each year through leaky pipes.

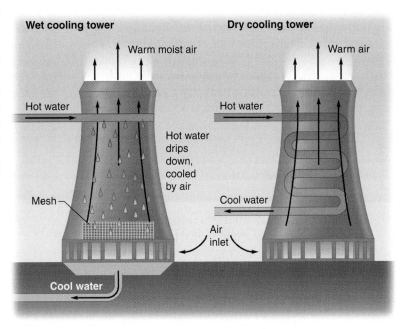

FIGURE 13-9 **Wet and dry cooling towers.** Water from electric power plants is cooled before being reused. Dry towers cost more to operate but conserve water.

their water. New pipes can reduce water loss enormously, saving streams and the wildlife that depend on them. Such measures also provide important employment opportunities.

You and I can do many things to cut down on water use. One very simple and cost-effective solution is the high-efficiency showerhead, which uses about 2 to 2.5 gallons per minute. Using 70% less water than a traditional showerhead, the new models also cut down on energy demand and can save a family of four $200 to $250 a year in energy bills—all for a $12 investment in a new showerhead and five minutes of installation time.

Individual savings can add up. The installation of water-efficient showerheads in 1300 homes would save 22,000 liters (85,000 gallons) of water a day, or about 8 million liters (31 million gallons) a year! Imagine the savings if everyone in the United States used one. The Individual Actions Count! table summarizes some things you can do to conserve water.

KEY CONCEPTS

Water conservation is a relatively fast and highly cost-effective means of meeting demands for water. Some of the largest gains can be made in agriculture and industry, although domestic water conservation measures can also save enormous amounts of water.

Restoring Watersheds and Wetlands to Protect and Enhance Our Water Supplies Slowing, even stopping population growth, and water conservation are the cornerstones of a sustainable water system. But there are other methods that can help us maintain, even expand, our water supply while protecting the environment. One of those is the restoration of vegetation in watersheds—replanting forests and grasslands, for instance. Revegetation reduces sedimentation in streams and reservoirs, extending their lifetime. Reductions in sediment load not only protect reservoirs, they reduce costs for water purification—providing an economic benefit—and protect fish and aquatic organisms—an ecological benefit. Revegetation also helps reduce surface runoff and thus enhances the infiltration of water into the soil, increasing groundwater supplies. Restoring wetlands increases natural filtering.

Another benefit of revegetation of denuded landscapes is that it may help restore normal rainfall patterns in areas where they've been disrupted by deforestation. In such regions, streams that have dried up may once again flow with clean, potable water, benefitting people and wildlife alike.

Preventive measures are also needed. Much greater care should be taken by lumber companies, farmers, and ranchers. Measures to protect streams during logging operations, for instance, help to maintain water quality and quantity (Chapter 12). Erosion controls and sustainable rangeland practices reduce sedimentation and protect vital water supplies. These measures also encourage groundwater recharge.

KEY CONCEPTS

Restoring and protecting watersheds help to prevent siltation in reservoirs and enhance groundwater recharge.

Water Recycling In nature, water is recycled over and over in large part because of evaporation and precipitation. In the process, it is also purified and readied for reuse. Humans can also recycle water. In fact, in the coming decades, water recycling may become an integral part of a sustainable system of water supply. Wastewater from factories, for example, can be purified and returned to productive use instead of being released into surface waters.

Elaborate biological, chemical, and mechanical filtration systems can remove a wide assortment of wastes, from toxic metals such as lead to toxic organic compounds to acids. In Tokyo, for example, Mitsubishi's 60-story office building has a fully automated recycling system that purifies all of the building's wastewater to drinking-water purity. Cities could also convert their existing sewage treatment plants to water recycling facilities. As detailed in Chapter 21, sewage treatment plants remove many impurities from wastewater from homes, offices, and factories, but the final product in most plants, which is released in surface waters, is hardly potable (drinkable). However, sewage treatment plants can be designed to purify water so that it is safe to drink.

Another form of recycling, which lets nature do the work, is the release of wastewater onto **aquifer recharge zones**, regions where rain and snowmelt replenish aquifers. Nutrients in the wastewater are taken up by plants and by bacteria in the soil so that as wastewater percolates through the soil it is cleansed. It then enters the aquifer in a clean state and can be reused. Effluents from sewage treatment plants, irrigation runoff, storm water runoff, and cooling water from industry could all be used to replenish water supplies in existing aquifers. Although aquifer recharge can help increase groundwater supplies, it is costly because many cities are far from aquifer recharge zones.

individual Actions Count

- Take shorter showers.
- Turn the water off while you soap up.
- Install a low-flow showerhead if possible.
- Don't brush your teeth or shave with the water running.
- Keep cold water in the refrigerator so you don't have to run the faucet when you want a drink of cold water.
- Water lawns and gardens conservatively. Watering early in the day or late in the afternoon can save enormous amounts of water.

Water recycling not only helps reduce our demand on freshwater supplies, it also eliminates discharge into streams and thus helps cut water pollution—also vital to a sustainable future. Although building and maintaining large-scale recycling plants can be expensive, they are often environmentally and economically cheaper than new water diversion projects and dams. Many relatively inexpensive, low-tech, biological solutions are also available. In rural settings, for example, sewage from individual homes can be piped into artificial wetlands. The wetlands clean the water and make it usable for flushing toilets or watering gardens or fruit trees. Several artificial wetlands used to purify water from ski resorts and other large sources are highlighted in Spotlight on Sustainable Development 4-1.

In some cases, it may be advantageous not to fully purify water. Nutrient-rich urban wastewater, for example, can be applied to pastures and cropland (Chapter 10). This practice returns valuable nutrients to the soil and reduces pollution of surface waters. In Israel, where freshwater is limited, 35% of the municipal wastewater is used to irrigate cotton. Israel hopes eventually to use 80% of its wastewater for irrigation.

KEY CONCEPTS

Water can be purified and reused in industry, on farms, and even in our own homes. Water recycling reduces pressure on surface water and groundwater supplies and reduces water pollution. Wastewater can even be used to recharge groundwater supplies or can be used to irrigate fields.

Changing Government Policies to Create a Sustainable Water Supply System Revamping government policy can have a substantial impact on efforts to create a sustainable water supply system. One of the most important policy changes is economic: eliminating subsidized water projects (described earlier). Many observers argue that federal and state governments should require those who benefit from water projects to pay their full costs. Making users pay the full cost might encourage them to focus on least cost options, particularly conservation. The cost of a water project should also include the costs of correcting environmental problems, a step that would make conservation and recycling even more attractive.

Governments also establish building codes that include requirements for the type of plumbing fixtures in commercial and residential settings. Revised building codes that favor conservation by requiring water-conserving toilets and fixtures can promote substantial reductions in domestic and commercial water consumption in new and remodeled buildings. Mandating water-efficient toilets and showerheads as well as water-efficient lawns could cut water use in new homes by half. Today, as a result of federal regulation that went into effect in 1996,

> Toilets account for 30 to 40% of a family's indoor water use. Modern low-flush toilets use half as much water as standard models—about 1.6 gallons per flush.

all states require that toilets installed in new and remodeled homes use no more than 5.8 liters (1.6 gallons) of water per flush—about one half as much as current models and one third as much water as toilets in older homes. Toilets use 30 to 40% of the indoor water consumed by a family, so water-conserving models can generate significant savings.

Considerable efforts are needed in arid regions to replace water-thirsty grasses in lawns with water-thrifty species of grass, trees, and shrubs. In Marin County in northern California, for example, one third of the county's water supply is used to irrigate lawns. One town's government offers sizable subsidies to homeowners who **xeriscape** (ZER-eh-scape) their lawns—that is, remove water-intensive vegetation and replace it with water-conserving species. Each converted lawn saves about 460 liters (120 gallons) of water per day during peak months.

Smaller lawns planted with low-water grasses and water-conserving drip and root-zone irrigation systems can be encouraged. Government programs can also encourage the installation of water-conserving measures in existing homes. Some water departments, for example, now offer free water audits, free water-efficient showerheads, and sizable rebates to homeowners who replace existing toilets with low-water models.

Another highly overlooked but potentially important way to help expand water supplies is the **rain catchment system**, also known as a **catchwater system**. Water and snowmelt collected from a roof can be stored in large tanks and used to irrigate gardens and lawns. Passed through filters, this water is even clean enough to drink.

Gray water systems could also play a major role in water-short areas. Gray water is water from showers, sinks, and washing machines. It constitutes about 80% of the wastewater from a home. Gray water is suitable for lawn and garden irrigation. It can also be purified and reused to flush toilets (**FIGURE 13-10**). However, gray water requires extra plumbing because it must be kept separate from **black water** (from toilets). California recently adopted the United States' first gray water codes. Other states will soon follow.

Governments can control water prices, too. In most locations, the more water one uses, the cheaper it is. This, of course, does not encourage water conservation. If users are charged more, not less, for water over a certain sustainable level, water users often respond favorably, cutting back on demand.

Finally, municipal water agencies could adjust water rates by time of day, charging users more for water used when evaporation is greatest and less when evaporation is low. This incentive could help cut down on water evaporation from lawns and save huge amounts of water. (In warmer, moist climates, however, nighttime lawn irrigation may enhance fungal growth that damages lawns.)

KEY CONCEPTS

Many government policies, especially subsidies, contribute to the unsustainable use of water. Changes in these policies can greatly increase the efficiency of water use and promote alternative water supply strategies, such as conservation, rain catchment, recycling, and gray water systems.

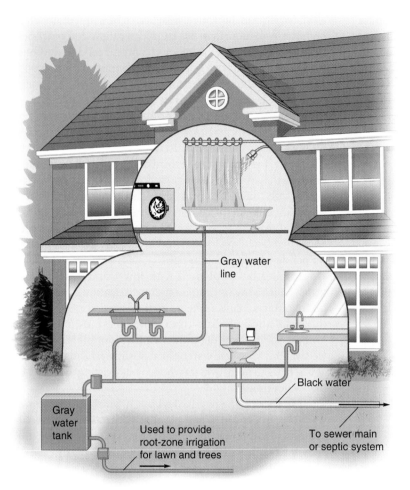

FIGURE 13-10 **Gray water recovery.** Gray water from sinks, showers, and washing machines can be collected by a separate set of pipes and reused for a variety of purposes. Gray water systems such as this one help to reduce household water demand.

Gray water line

Black water

Gray water tank

Used to provide root-zone irrigation for lawn and trees

To sewer main or septic system

Education: Learning to Use Water Wisely   The battle to ensure an adequate water supply can be aided by massive public education. Teaching adults and children throughout the world ways to use this precious resource efficiently may be as important as the math or history they learn. A study of water conservation programs in the western United States, performed by the Sustainable Futures Society in Evergreen, Colorado, found that water conservation programs in cities are most successful if changes in pricing policies (which increase the cost of water the more someone uses it) are complemented by educational programs.

> KEY CONCEPTS
>
> Education is a key to creating individual responsibility and sustainable water-use patterns among citizens.

Tapping New Sources of Water: Dams and Desalination Plants   The measures outlined previously represent the first line of attack. They are generally economical and easy to effect, and many are good for the environment, too. Conservation, for example, reduces environmental impact while supplying us with water at a lower cost than most other options. But there may be times when these measures won't provide enough water. Further supplies may be required. This means more dams and diversion projects or drilling more wells into groundwater and building water purification plants.

Because the economic and environmental costs of dams and reservoirs are substantial, discretion is advised when undertaking such projects. In less developed nations, small-scale irrigation projects combined with efforts to use water more efficiently can eliminate the need for large and costly dams. Smaller projects help build local self-reliance so vital to sustainability.

Another way to increase water supply in communities near the ocean is by **desalination**, removing salts from seawater. Through various methods, seawater can be purified for drinking and irrigation.

Because 97% of the water on Earth is in the oceans, desalination might seem like the best answer to water shortages. Unfortunately, water produced by desalination is 4 to 10 times more expensive than water from conventional sources. Although the world's desalination capacity has increased dramatically, desalination produces only a tiny proportion of the freshwater consumed by humans. In the United States, 3900 desalination plants now produce in excess of 6.1 billion liters (1.6 billion gallons) per day. Although that sounds like a lot, it is only about 0.001% of the daily freshwater requirement. The majority of the plants are located in California, Texas, Florida, and the Northeast. Desalination plants are also in operation in Saudi Arabia, Israel, Malta, and a few other countries.

Desalination plants may not be advisable from an environmental standpoint. For example, they expand the Earth's carrying capacity, with potentially serious ecological impacts. Population growth in the Florida Keys, permitted in part by a new desalination plant, threatens coral reefs and wildlife such as the crocodile. Construction of houses and condominiums causes erosion that pollutes coastal waters. New residents produce an increasing amount of sewage and other pollutants that decrease water quality.

> KEY CONCEPTS
>
> As a last resort, rising demand can be met by developing new water supplies. New dams and diversion are one option, though their costs are enormous. Desalination plants are a likely candidate for coastal communities but they are costly to operate.

13.3 Flooding: Problems and Solutions

Ironically, after shortages, the next major water problem encountered in many nations is flooding. Despite years of flood control work, floods cause damage valued between

$1 to $2 billion a year in the United States. The 1993 flood along the Mississippi River caused $16.4 billion in damage. Floods in 1999 along the southeast coastline also caused billions of dollars of damage. Record floods in England in 2000 also caused enormous damage. The summer of 2002 will long be remembered as the summer of the hundred-year flood in central Europe. In fact, several floods during this period are considered the worst flood-catastrophes since the Middle Ages. Flooding appears to be on the rise in many places because of shifting rainfall patterns and more violent and more frequent storms such as hurricanes Katrina and Rita, which struck the southern United States in September 2005, causing massive loss of life, economic damage, and property loss. These storms result in part from global climate change and alteration of natural systems by human activities that make the land more prone to flooding.

KEY CONCEPTS

Flooding is a major problem in many areas of the world and appears to be on the rise, in part as a result of human activities.

Causes of Flooding

A simple correlation can be drawn between floods and their apparent causes: heavy rainfall and snowfall. The Red River in Canada, for example, has a long history of natural flooding. The river, which flows from south to north, often floods when winter snows in the southern reaches of the watershed melt before ice in the northern portion of the basin melts. Ice forms a natural dam. Closer examination, however, reveals many causes that go unnoticed. As shown in FIGURE 13-11, precipitation that does not evaporate must either run off or percolate into the soil. Whether it flows across the land and empties into rivers or sinks quietly into

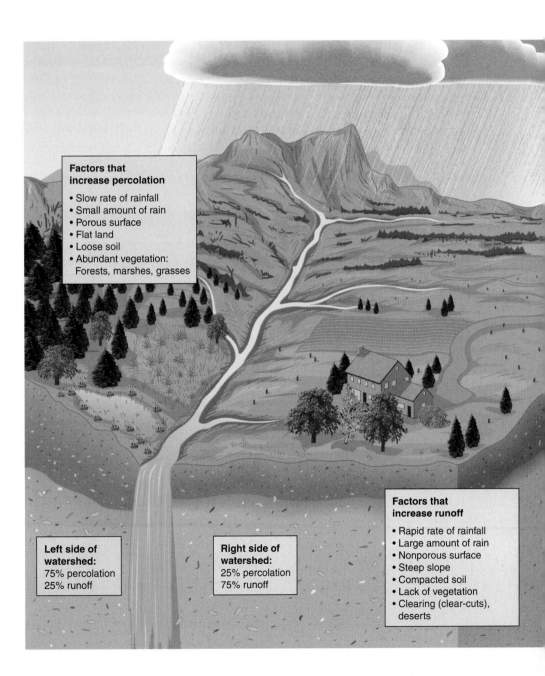

Factors that increase percolation

- Slow rate of rainfall
- Small amount of rain
- Porous surface
- Flat land
- Loose soil
- Abundant vegetation: Forests, marshes, grasses

Left side of watershed:
75% percolation
25% runoff

Right side of watershed:
25% percolation
75% runoff

Factors that increase runoff

- Rapid rate of rainfall
- Large amount of rain
- Nonporous surface
- Steep slope
- Compacted soil
- Lack of vegetation
- Clearing (clear-cuts), deserts

FIGURE 13-11 Percolation–runoff ratio. Numerous factors increase and decrease surface runoff and absorption (percolation).

the soil to become groundwater is largely a matter of the surface features, especially the vegetative cover. For example, a well-vegetated landscape covered by forests and grasses reduces surface runoff and promotes percolation into the soil. Put another way, heavily vegetated watersheds act as sponges. On the other hand, sparsely vegetated areas (for instance, deserts, overgrazed pastures, or clear-cuts) experience increased surface runoff and, hence, flooding.

The fate of rainfall in many cases rests not so much in the hands of nature as in the hands of farmers, urban planners, developers, ranchers, and logging companies, who often strip vegetation from the land. This practice increases surface runoff. Water flowing rapidly over the surface of denuded land flows into streams and often spills over their banks, flooding nearby areas. Such water also often transports a substantial amount of soil from the land. Eroded sediment fills rivers and lakes and reduces their holding capacity. This makes flooding more likely, even after moderate rainfall.

Flooding in urban areas often results from highways, airports, shopping centers, office buildings, and homes, which greatly increase the amount of impermeable surface. Instead of soaking into the ground, rainwater washes off the surface in torrents.

Destruction of wetlands, swamps along rivers, is also responsible for increased flooding. Wetlands act as sponges, too, holding water and releasing it slowly into nearby waterways. Their loss eliminates this benefit and is thus partly responsible for the growing severity of floods in many countries.

FIGURE 13-12 shows the regions of the United States that are susceptible to flooding. Many of these areas are located along the Mississippi River and its tributaries. Several large rivers in Canada (not shown on the map) also experience periodic flooding, among them the Red River (in Manitoba), the Oldman and South Saskatchewan Rivers (in Southern Alberta), and the Saguenay River (in Quebec). The **floodplains** of these rivers, regions along the flanks of rivers naturally subject to flooding, are popular sites for cities, towns, and farms. Thus, living along floodplains, combined with activities that increase surface runoff, ensures humankind a future of flooded basements, costly damage, and lost lives. Sometimes the results can be devastating. A flood in Quebec in 1996, for instance, caused the evacuation of 12,000 people in more than 50 towns. Seven people died in the flood, 100 homes were swept off their foundations and destroyed, and another 1000 homes were badly damaged.

KEY CONCEPTS

Floods result from natural events—for example, too much rain—but are also the result of changes in the environment caused by humans, notably changes to rainfall patterns and changes in the land surface that increase runoff.

Controlling Flooding

Flood prevention measures in most countries have typically taken the form of dams and levees. **Levees** are embankments constructed along the banks of rivers to hold floodwaters

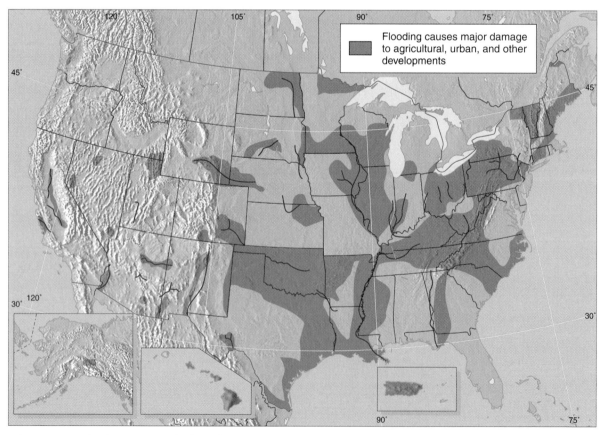

FIGURE 13-12 Flood-prone areas in the United States. Boundaries of the states are shown in black.

back. As is so often the case, dams and levees are stopgap measures: They often treat the symptoms of a serious and costly problem, not the underlying causes.

Another technique for reducing flooding that is falling out of favor is **streambed channelization**, which is a kind of streamlining of rivers and streams, carried out by cutting or bulldozing vegetation along a stream's banks and deepening and straightening river channels. This creates a glorified ditch that may be lined with concrete or rock.

Streambed channelization in the United States was begun seriously in 1954, when the **Watershed Protection and Flood Prevention Act** authorized the Soil Conservation Service to drain wetlands along rivers to make more farmland and to reduce flooding. Between 1940 and 1971, 35,000 kilometers (22,000 miles) of U.S. streams were channelized by the Army Corps of Engineers and the Soil Conservation Service (according to American Rivers, a national river protection organization). More recent data are not available from any of the major players in this arena, although channelization is still occurring. Critics argue that many projects are of dubious merit and should not be undertaken. Why?

Streambed channelization generally eases flooding in the immediate vicinity, but this minor benefit is frequently outweighed by habitat destruction (loss of streambank vegetation), increased erosion of stream banks, loss of river recreation, and, most importantly, increased flooding downstream. (For more on this, see Spotlight on Sustainable Development 13-2.) Like dams and levees, streambed channelization is an unsustainable approach to a serious problem. What can be done to reduce flooding?

One of the most promising and sustainable approaches used to reduce flooding is **watershed management**. Watershed management includes steps to reduce deforestation and overgrazing. It also includes measures to protect and restore wetlands in watersheds. Wetlands act as giant sponges, reducing surface flows and flooding. In fact, the devastating floods along the Mississippi River in 1993, which caused billions of dollars worth of damage to crops and homes, could have been averted had a relatively small percentage of the natural wetlands in the river's massive watershed been left intact. (Wetlands are discussed in more detail later in the chapter.) Wetlands along rivers absorb water and help reduce flooding, but as you learned earlier in this chapter, vegetation within a watershed increases the percolation of water into the ground, thus greatly reducing surface runoff and helping to control floods. This strategy also ensures more groundwater. A revegetated landscape also has streams that flow year-round, rather than drying up in dry months, as water is released from soils in a more controlled fashion.

Two general approaches can be taken to tap into this ecological service. First, efforts can be made to protect existing watersheds by better regulating development. Mining, farming, and housing, for example, can be regulated to ensure minimal runoff. Land can be developed in ways that protect forests and meadows needed to absorb storm water. Second, in areas where the landscape has been altered by human activities, efforts can be made to restore vegeta-

tion. Replanting trees, shrubs, and grasses on denuded hillsides can greatly reduce flooding, while also restoring wildlife habitat and making a more aesthetically appealing environment.

Watershed management also includes measures to redesign urban environments to slow down the rush of water during rainstorms. One effective means of achieving this important goal is the construction of holding ponds that absorb storm water runoff. Lakes in city parks can be situated in such a way as to trap much of the surface flow in a rainstorm. This water can then be released slowly to rivers to prevent flooding, or it could be used to water lawns or recharge aquifers. Holding ponds can also be developed near large parking lots or shopping malls. Some developers are even developing holding ponds under parking lots to absorb runoff and slowly release it into nearby streams. They use a special high-strength porous plastic call Rainstone to create a chamber to temporarily store water after rain storms (**FIGURE 13-13a**).

Storm sewers can also be designed so as to divert water to special holding ponds or water tanks that serve factories. Individual homeowners can divert gutter water to underground holding tanks to be used later for watering lawns and gardens and for washing cars. (For examples of ways to avoid flooding, see Spotlight on Sustainable Development 13-2.) Finally, floodplain zoning laws can help restrict building and other human activities in flood-prone areas. Driveways and walkways can be constructed of materials called porous pavers that permit water to soak into the ground, thus reducing surface runoff (**FIGURE 13-13b**).

As in other areas, efforts to slow or even stop the growth of the human population are vital to solving this growing problem. More people mean more habitat destruction and more flooding. Zoning regulations can be helpful as well, restricting development in floodplains and sensitive areas of watersheds where home construction and roadways are ill-advised.

After years of devastating floods, some cities and towns are taking a new approach: They're "depopulating" river banks. After repeated devastating floods, many cities and towns are relocating to higher ground, out of harm's way. The federal government, which has spent billions of dollars rebuilding flooded towns through flood insurance programs and emergency aid, is helping in these new ventures, secure in the notion that aggressive buyouts of flood-prone land and relocation of homes saves money in the long term. The Federal Emergency Management Agency, which assists after floods, and the U.S. Army Corps of Engineers, a long-time flood fighter and dam builder, have both engaged in an aggressive program to buy property from owners who have filed multiple disaster claims and return the lands to natural uses. The depopulated areas are cleared of human structures and returned to nature, creating ecological and recreation benefits.

As noted in Chapter 20, flood control can also be achieved by reducing our dependence on fossil fuels and curtailing other activities that produce greenhouse gases, which are believed to be changing rainfall and snowfall patterns that may be leading to more frequent and severe flooding in Texas and other southern states.

(a)

(b)

FIGURE 13-13 **(a)** A worker unrolls sod over Rainstore, a porous plastic product that holds runoff from a nearby parking lot, slowly releasing it into the groundwater or into nearby streams to prevent flooding. **(b)** Workers install a plastic product called Grasspave, manufactured by a Denver-based company, Invisible Structures, Inc., in a new parking lot. Grass can be grown over the plastic substrate, thus reducing surface runoff, pollution of nearby waterways, and reducing heat accumulation.

KEY CONCEPTS

Many solutions to flooding—including dams, levees, and streambed channelization—are unsustainable; they treat the symptoms, not the root causes of flooding. Preventive measures—controlling population growth, protecting watersheds, and reducing global climate change—combined with restorative measures such as replanting trees and restoring wetlands, are far more effective in the long run.

13.4 Wetlands, Estuaries, Coastlines, and Rivers

From certain vantage points, Chesapeake Bay on the eastern seaboard of the United States resembles a vast ocean. It is not. It is an estuary, a region where freshwater from inland streams meets the ocean FIGURE 13-14. For years Americans have

FIGURE 13-14 The Chesapeake Bay receives pollutants from the neighboring watershed, which have damaged ecologically and economically important life forms including fish and shellfish.

treated this estuary and its vast wetlands with considerable disrespect. As a result, the rich abundance of organisms is diminishing, threatened by pollution, overfishing, and other activities.

The bay and its surrounding wetlands are home to a variety of economically important fish and shellfish, including blue crabs, oysters, and striped bass. Properly managed, the bay could provide enough food for nearly half the U.S. population. Chesapeake Bay is much more than a food source for the more than 16 million people who live near it, though. It is a popular habitat for all sorts of water fowl, too. It is also a source of recreation for hunters, anglers, boaters, and nature enthusiasts, all of which benefit the local economy.

Although it is only 310 kilometers (195 miles) long, the bay's shoreline measures nearly 13,000 kilometers (8100 miles)—twice the entire U.S. Pacific coastline! The bay contains an abundance of species—nearly as many as the entire Atlantic Ocean. The abundance of fish and wildlife is a result of many factors. The bay's shallowness, for instance, ensures adequate sunlight penetration, which enhances biological productivity. The presence of enormous expanses of wetlands also helps to explain its diversity and productivity. The nutrients it receives from rivers that empty into it and the presence of numerous microhabitats (very distinct habitats within the bay) also contribute to its biological richness and diversity.

Old-timers claim that the bay's waters once contained wall-to-wall oysters. Oyster populations have declined markedly, though, falling by 99% since 1870, with most of the decline occurring in more recent times. Striped bass populations have also fallen off alarmingly. Various strategies have been tried to reverse the downward trend in the bass population, among them a moratorium on striped bass fishing in January 1985. In recent years, striped bass were produced by two federal fish hatcheries to supplement the declining populations. These efforts have resulted in an increase in the striped bass population. Because bass populations have recovered somewhat, limited commercial fishing has been permitted since 1990. In the late 1960s, the total commercial harvest was about 2.7 to 3.2 million kilograms (6 to 7 million pounds). It is currently about 1.4 million kilograms (3 million pounds).

EPA studies show that the bay's submerged vegetation, so vital to fish, has dropped by 76% in the late 1900s, partly as a result of large algal blooms (bursts of algal growth resulting from inorganic pollutants) that block sunlight and impair the growth of submerged aquatic vegetation. Further damage to the bay is caused by aerobic (oxygen-requiring) bacteria, which deplete the oxygen supply when they decompose dead algae and organic pollutants in the water.

Population growth and commercial development—and the pollution they produce—are the biggest threats to the bay. The population is expected to reach 19 million by 2030. Most of the pollution comes from oil spills, sewage, toxic chemicals, heavy metals, and runoff from the bay's extensive drainage system, an area slightly smaller than Missouri. One of the worst pollutants is nitrogen from commercial duck farms, municipal sewage, and farms.

The story of Chesapeake Bay is an example of what is happening to many important surface waters in the United States and elsewhere. It shows us how we are destroying a vital resource and ruining the habitat of other species that are often a valuable source of food for humankind. It reminds us that living sustainably on the planet requires much more than protecting human civilization from floods and providing adequate amounts of water. It requires efforts to protect wetlands, estuaries, coastlines, and rivers. Such actions protect biodiversity, but they also protect valuable food, income, and water sources for human populations.

> **KEY CONCEPTS**
> Surface waters—lakes, rivers, and bays—are under assault. Their destruction affects available water supplies, but also destroys habitat for fish and other organisms that are important food sources for humans and other species.

Wetlands

One of the most endangered ecosystems on the planet is its wetlands. **Wetlands** come in many forms. Swamps are a good example. But so are wet meadows and certain woodlands along streams. What do these have in common? One thing is that the soils are typically saturated with water, either part of the year—as in the case of flooded woodlands (called *bottomlands*)—or year round—as in the case of swamps.

Wetlands are supplied by surface and groundwater and support plants that are adapted for life in saturated soils. Biologists recognized two types of wetlands: inland and coastal.

Inland wetlands are found along streams, lakes, rivers, and ponds; they include bogs, marshes, swamps, and river overflow lands (like bottomlands) that are wet at least part of the year.

Coastal wetlands are wet or flooded regions along coastlines, typically associated with estuaries, the mouths of rivers where freshwater mixes with saltwater. Coastal wetlands include mangrove swamps, salt marshes, bays, and lagoons. Mangrove swamps in the United States are restricted to southernmost biomes of Florida, Louisiana, and Puerto Rico. Hawaii also has mangrove swamps. Wetlands along the rim of the Great Lakes are also considered coastal wetlands. They're typically associated with the mouths of rivers. **FIGURE 13-15** shows where you'll find many of the nation's wetlands.

Wetlands are an extremely valuable and productive habitat for many animal and plant species, including many rare and endangered species. In fact, 45% of the species listed as threatened or endangered by the U.S. Fish and Wildlife Service rely directly or indirectly on wetlands for their long-term survival. In addition to mink, beaver, muskrats, and otters, wetlands are home to many shellfish, amphibians, reptiles, birds, and fish. While they serve as **primary habitat** (year-round) for many species, wetlands also provide important **seasonal habitat** (part-time) for migrating species such as ducks and geese. These species obtain food, shelter, and water during their perilous migrations.

The loss of wetlands can have a devastating effect on species. The Dusky Seaside Sparrow became extinct in 1987. The main reason for its disappearance was the destruction of wetlands around Merritt Island and St. John's Island in Florida. The Ivory Billed Woodpecker was thought to have met a similar fate, largely because of clearing bottomland hardwood forests. In 2005, however, more than 60 years after the bird was believed to be extinct, field biologists caught the bird on video in eastern Arkansas. The Louisiana Black bear, also known as the swamp bear, is now listed as threatened, primarily because of the loss of its wetland habitat. The list goes on.

Aside from their importance to wildlife, wetlands also play an important role in regulating stream flow, as noted earlier. A study in Wisconsin showed that wetlands act like sponges, holding back rainwater and reducing flooding. Wetlands within and upstream from urban areas are especially important in flood control. Loss of wetlands decreases their ability to absorb peak flows and reduce flooding. The sponge effect has the added benefit of recharging groundwater supplies. Because streams are fed by groundwater, wetlands help to maintain year-round base flows. Wetlands also remove sediment in surface runoff and thus reduce sedimentation in streams. This helps keep streams and rivers clean. In addition, wetland plants absorb nitrogen and phosphorus, two common pollutants derived from heavily fertilized land, thus reducing water pollution. According to one study, an acre of coastal wetland is the equivalent of an $85,000 sewage treatment plant. Coastal wetlands also help protect human settlements by absorbing **storm surges**, high waves accompanying high winds during hurricanes and other violent storms. Wetlands are used to grow certain cash crops,

> Wetlands help purify water. Each acre of coastal wetland does the equivalent job of an $85,000 investment in a sewage treatment plant.

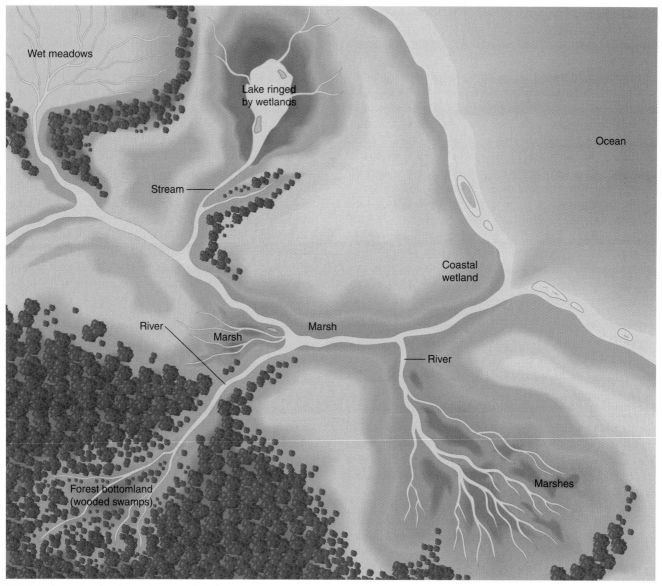

FIGURE 13-15 Wetlands. Wetlands are found in many locations, along rivers or at the mouths of rivers. They are also found in low depressions where groundwater intersects with the land surface or around lakes and ponds.

including rice, cranberries, and blueberries. Wetlands also have other economic values. Canada's remaining wetlands, for instance, annually generate between $5 and $10 billion to the nation's economy from such activities and natural services as commercial fishing, sport fishing, hunting, trapping, recreation, groundwater recharge, and flood control, according to Environment Canada. Because their usefulness is not always apparent, they are often filled in or dredged to make way for farms, housing, recreation, and industry.

Wetlands are highly endangered ecosystems. Wetlands are habitat for aquatic organisms, birds, and mammals. They help control flooding, purify water, and buffer humans from storm surges. Unfortunately, many people fail to understand their economic and ecological importance.

Declining Wetlands In the lower 48 states, wetlands once covered an area twice the size of California (90 million hectares, or 220 million acres). Today, half the wetlands are gone (FIGURE 13-16a). The greatest losses have occurred in California (91%), Ohio (90%), and Iowa (89%). FIGURE 13-16b shows the distribution of existing wetlands and losses by state.

> The United States has lost more than half of its wetlands.

Wetland destruction has slowed in recent times, but the rates are still high (FIGURE 13-17). Between 1954 and 1974, wetlands fell at a rate of 185,000 hectares (458,000 acres) per year. Agricultural conversion accounted for 87% of the loss. In the next ten-year period, the rate of decline dropped to

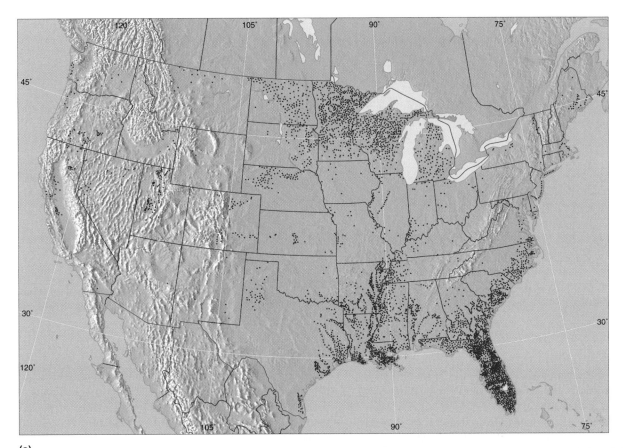

(a)

FIGURE 13-16 Map of wetlands in the United States. These maps show **(a)** where remaining wetlands are found in the lower 48 states and **(b)** the percentage loss in each state.

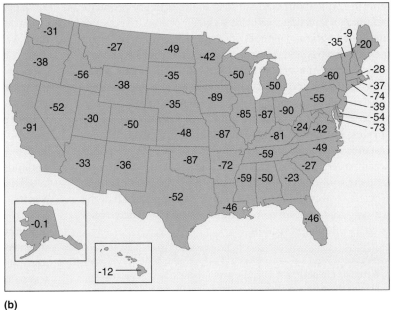

(b)

about 117,000 hectares (290,000 acres) per year. Agriculture still figured prominently in the losses, accounting for 54% of them. Between 1986 and 1997, the average rate of loss fell again to 28,000 to 23,780 hectares (58,500 acres) per year. In more recent times, residential development and urban growth have become the major contributors to wetland losses (**FIGURE 13-18**).

The remarkable decrease in the loss of wetlands is due to many factors. For one, many prime wetlands in agricultural states have been filled in and are now being farmed. Public pressure and new state and federal laws (discussed shortly) resulting from it have also played a key role. Although gains have been significant, the nation's goal of no net loss of wetlands still remains to be met.

Wetland loss is a global phenomenon. Some areas have suffered greater losses than others, however. In California, New Zealand, and Australia, development has claimed more than 90% of the wetlands. Canada has lost huge amounts of wetlands largely to agriculture. In southern Ontario, in the prairie provinces, along the Atlantic seaboard, and along the Frazier River, 65 to 80% of the wetlands have been destroyed.

Coastal wetlands are victims of dredgers that scoop up muck from streams and bays to increase navigability. Dredging drains adjoining swamps. Cities often fill in swamps to make room for homes, recreational facilities, roadways, and factories. Farmers fill in swamps on their land to expand their arable land. About half of the world's coastal wetlands have been destroyed.

Mangrove swamps (located along the coastlines in semitropical and tropical regions) are one of the most valuable forms of wetland and the most threatened, suffering heavy losses in Asia, Latin America, and West Africa. In Ecuador,

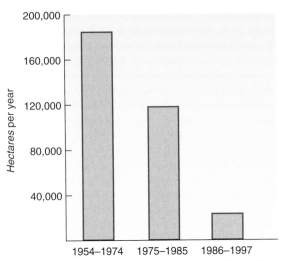

FIGURE 13-17 **Declining wetland losses.** Wetland losses have declined dramatically over the past 50 years.

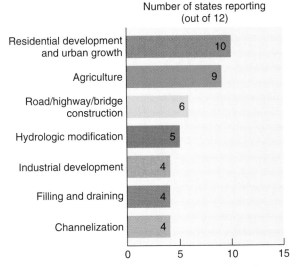

FIGURE 13-18 **Source of wetland losses.** These data show states reporting major causes of wetland losses. The numbers are the number of states reporting the problem.

nearly half of the mangrove swamps have been cleared to build shrimp ponds. India, Pakistan, and Thailand have lost three fourths of their mangroves.

Wetlands are also suffering from declines in environmental quality, primarily caused by pollution. Although we tend to think about wetland loss as the primary problem, pollution of remaining wetlands is equally serious. Excess sediment deposition, chemical pollutants that stimulate plant growth, and pesticide contamination are key problems, as is the reduction of water flow into wetlands. Today, many states have begun to survey and study their wetlands and are finding that a large percentage of them are impaired because of these or other problems including the invasion of exotic species. The Hoopa Valley Tribe in northern California, for instance, found that all of its 3200 acres of surveyed wetlands are impaired. The state of Kansas found that 91% of the wetlands it studied are impaired. In other areas, for example Nevada, however, the picture is just the opposite. All of the wetlands the state surveyed fully supported designated uses such as wildlife protection.

KEY CONCEPTS

Wetlands are declining virtually everywhere, and massive numbers of them have already been destroyed in the United States and abroad. Fortunately, the rate of destruction in the United States has slowed considerably. Nonetheless, many remaining wetlands are impaired because of pollution, invasion of exotic species, and lack of adequate water flow.

Protecting Wetlands   Concern for the loss of wetlands has stirred many governments into action. For example, Florida passed legislation in 1972 to regulate all wetland development. In 1988, New Jersey passed a law requiring buffer zones of 30 meters (100 feet) around important wetlands. Although a plan does not ensure protection, it is a step in the right direction.

The U.S. government has also assumed an increasing role in wetland protection through executive orders that prohibit federal agencies from supporting construction in wetlands when a practical alternative is available. The federal **Coastal Zone Management Act** (1972) calls on states to develop plans to protect coastal wetlands. Virtually all states have either federally approved plans or state plans to regulate their wetlands. However, only 16 states have measures to protect inland, freshwater wetlands.

The federal government can purchase wetlands and set them aside as part of the **National Wildlife Refuge System.** The system today contains more than 4 million hectares (10 million acres) of wetlands. The U.S. Fish and Wildlife Service also buys wetlands to set aside. States own additional wetland acreage. All told, a little over one fourth of our existing wetlands are protected.

Another important action aimed at protecting wetlands was the 1985 Farm Bill (Chapter 10). It included important *swampbuster* provisions—rules that deny federal benefits (low-interest loans and crop insurance) to farmers who drain and farm their wetlands. In 1996, the swampbuster provisions were tightened. The law also established a **Wetlands Reserve Program**, a voluntary program to restore and protect wetlands on private farmland. Farmers are financially reimbursed for the land they enroll in the program.

Unfortunately, many wetland protection laws and regulations are like toothless watchdogs. In many instances, little is done to enforce them. Individuals can write their state representatives to find out what laws exist and how they are being enforced. Local action groups can help stimulate stronger enforcement when necessary. Personal action is needed more than ever in the face of deep cuts in the federal budget.

On an international scale, nations can cooperate to preserve wetlands. In 1971, representatives of many nations met in Iran to discuss the plight of wetlands and agreed to protect those lands within their jurisdiction. In 1986, the United States ratified the agreement. Four U.S. wildlife refuges were added to a list of wetlands of international importance.

Unfortunately, U.S. wetlands are facing a new round of threats—too numerous to cover in detail here. For example, the Bush administration has withdrawn support from a major Everglades restoration project. A 2001 ruling by the U.S. Supreme Court now allows certain types of wetlands (isolated wetlands) to be filled in—without any requirement to enhance wetlands elsewhere or create new wetlands. This ruling has already allowed companies to destroy valuable wetlands. In and around Chicago, for instance, wetlands are being drained and filled as fast as they (builders) can.

> ### KEY CONCEPTS
> Numerous efforts are under way at local, state, national, and even international levels to preserve and protect existing wetlands. Although these measures have not ended the destruction, they have greatly slowed it down.

Estuaries

Estuaries are the mouths of rivers, where saltwater and freshwater mix. Chesapeake Bay, described earlier in the chapter, is a prime example of an estuary. Like wetlands, estuaries are critical habitat

> " Two thirds of all fish or shellfish depend on the estuarine zone at some time of their life cycle. "

for fish and shellfish. Together, coastal wetlands and estuaries make up the **estuarine zone**. Two thirds of all fish and shellfish depend on this zone during some part of their life cycle. Because fish are an important source of food, protection of the estuarine zone is vital to feeding the world's people. It has an economic benefit as well.

The estuarine zone gets its richness from the land—eroded land. Eroded sediment, rich in nutrients, is carried in streams to the ocean, where it supports an abundance of aquatic organisms—especially algae, the base of the aquatic food web.

> ### KEY CONCEPTS
> Estuaries are the mouths of rivers and are biologically important life zones of great economic importance to humans.

Damaging This Important Zone The estuarine zone is vulnerable to a variety of assaults: pollutants from sewage treatment plants or industries; sediment from erosion that buries rooted estuarine vegetation; oil spills; and dams that cut off the life-giving flow of nutrients from the land. Cities may withdraw so much freshwater upstream that rivers and estuaries run dry. In Texas, for example, drought and heavy water demands in past years have critically reduced water flow into estuaries. Further west, the Mexican delta of the Colorado River is a remnant of its former self because of excessive water withdrawal along its course (**FIGURE 13-19**). The Nile

(a)

(b)

FIGURE 13-19 A delta dies. (a) Not more than 80 years ago the Colorado River flowed unhindered from northern Colorado through Utah, the Grand Canyon, Arizona, and Mexico before pouring out into the Gulf of California. But as one can see in this satellite image of the Colorado River Delta taken on September 8, 2000, all of its water is removed by farmers and cites before the river reaches its former destination. The Colorado River can be seen in dark blue at the topmost central part of this image. **(b)** A nine-year-old member of the Cocopah Indians of Northern Mexico plays in dried up mud in an area his people fished and farmed for about 2000 years.

River estuary has been similarly affected. Freshwater inflows are critical to maintaining the proper salt concentration in coastal wetlands, where molluscs and other organisms dwell. Salinity may be one important factor determining shellfish productivity.

New research suggests that pollutants from the ocean can also concentrate in estuaries. Particles can carry heavy metals, organic pollutants such as PCB, and insecticides such as DDT.

Many organisms inhabit the estuarine zone, but the most sensitive to pollution are clams, oysters, and mussels. Pollution is generally thought to be one of the major factors responsible for the decline in mollusc harvests in the last three decades of the 1990s. Molluscs concentrate toxic heavy metals, chlorinated hydrocarbons, and many pathogenic organisms, including those that cause typhus and hepatitis. These pathogens may not affect the molluscs' survival, but they make them unsafe for human consumption.

The estuarine zone is subject to pollution, water loss, sedimentation, dredging, and filling, among other things. Compounding the damage is the widespread problem of overharvesting of fish and shellfish. It is generally agreed that the decline in U.S. oyster production after 1950 was largely the result of overharvesting. Clams in the Northeast have likewise been severely overharvested.

Over 40% of the U.S. estuarine zone has been destroyed. The most severe damage has occurred in California, along the Atlantic coast from North Carolina to Florida, and along the entire Gulf of Mexico. Despite state and federal laws to protect this zone, destruction continues.

KEY CONCEPTS

Because of its location downstream from many human activities, the estuarine zone is seriously endangered. Overharvesting of commercially important species, especially shellfish, in this zone carries enormous ecological and economic costs.

Protecting the Estuarine Zone Protecting estuaries and coastal wetlands is part of an overall sustainable management strategy whose beneficial effects ripple through the biosphere just as the adverse effects do now. Many efforts to solve other problems will also help protect the estuarine zone. Population stabilization, for example, which eases pressure on natural systems and reduces pollution, is vital to the long-term future of estuaries. Improved water pollution control is also essential (Chapter 21). Pollution prevention efforts by businesses can help enormously. Watershed management, including revegetation and other erosion control measures, is also important. Cities and towns can establish water conservation programs to reduce water withdrawals from streams and enhance freshwater flow into estuaries. Restrictions on dredging and filling and ending the overharvesting of fish and shellfish will also help.

Protecting the estuarine zone presents a unique challenge in the United States because 90% of the coastal land in the 48 coterminous states is privately owned. In 1972, Congress responded to the plight of the estuaries and coastal wetlands by passing the Coastal Zone Management Act (described earlier). This law set up a fund to provide the 35 coastal and Great Lakes states with assistance in developing their own laws and programs. It also provides them with money to purchase estuarine zones and estuarylike areas in the Great Lakes states. These regions are to be set aside for scientific study; as of December 2005, 26 national estuarine research reserves had been established covering 0.4 million hectares (1 million acres) of estuary and coastal land.

The Coastal Zone Management Act also allows for the establishment of national marine sanctuaries. Thirteen have been set aside off the coasts of Washington, California, Hawaii, Maine, Massachusetts, North Carolina, Georgia, Florida, Texas, and American Samoa. Two are currently pending approval. These sanctuaries are established to protect vital habitat including coral reefs, open ocean, mangrove swamps, and islands to protect a variety of marine mammals and fish. Several are also home to historic ship wrecks and are thus popular diving spots. Some also support whale watching, sports fishing, and commercial fishing.

Half of the U.S. population lives in counties that are at least partly within an hour's drive of a coast, and most of the major cities are coastal. Discoveries of offshore oil, gas, and minerals pose new problems that require immediate answers. An abundant supply of cooling water makes the coastal zones prime candidates for new power plants and oil refineries. Because of current and potential problems, the Coastal Zone Management Act is an important step in preserving U.S. coasts.

Unfortunately, some states, either have not adopted programs or enforce their programs poorly, leaving their coastal waters vulnerable to further damage.

Fifteen additional federal laws have been passed to promote coastal zone management, but they are only as good as their enforcement. Lackadaisical enforcement is almost as bad as no enforcement at all.

The estuarine zone is being destroyed and severely damaged throughout the world, too. Today, over half the world's people (3.3 billion people) live and work within 120 miles of a coastline, occupying 10% of the Earth's land mass. Two thirds, about 4 billion, live within 250 miles of the coastline. Without further, more serious efforts to improve coastal zone management in the United States and abroad, we will almost certainly lose more of this ecologically and economically important habitat.

> More than half of the world's population lives within 120 miles of the coast.

KEY CONCEPTS

Protecting the estuarine zone requires many efforts; these include establishing protected areas and carrying out numerous preventive measures—such as pollution control and soil erosion controls on watersheds of rivers and their tributaries that empty into estuaries.

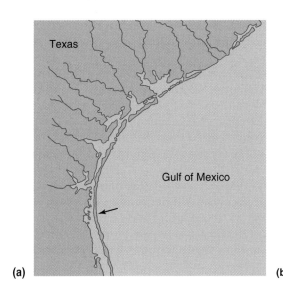

FIGURE 13-20 **Barrier islands. (a)** Map showing location of barrier islands off the coast of Texas. **(b)** Barrier island of Westerly, Rhode Island.

(a)

Texas

Gulf of Mexico

(b)

Barrier Islands

The Atlantic and Gulf coast states are skirted by a chain of **barrier islands**, narrow sand islands separated from the mainland by lagoons and bays (**FIGURE 13-20**). An estimated 250 barrier islands lie along the Atlantic and Gulf coasts. Many of them are popular sites for recreation. Some have been designated as national seashores by the U.S. government and are used for recreation and wildlife habitat; however, most barrier islands are under private ownership.

In the last 50 to 60 years, many of these islands have been developed for vacationers. Summer homes, roads, stores, and other structures have replaced the grass-covered dunes. According to estimates of the National Park Service, in 1950,

only 36,000 hectares (90,000 acres) of barrier islands had been developed, but by 1980, 113,000 hectares (280,000 acres) had been developed. In 1992, the U.S. Fish and Wildlife Service estimates put the number at about 170,000 hectares (420,000 acres).

Barrier islands and their beaches are part of a river of sand that migrates down these coastlines. The islands grow and shrink from season to season and year to year in response to two main forces: First, waves, which tend to arrive at an angle to the beach, erode the beaches (**FIGURE 13-21a**). Waves create **beach drift**, a gradual movement of sand along the beach. In addition, the wind creates **longshore currents** parallel to the beach, which also move sand along it (**FIGURE 13-21b**). Combined, beach drift and **longshore drift** (movement of sand by longshore currents) cause the barrier islands to move parallel to the main shoreline, shortening on one end and elongating on the other. Homes built on the upcurrent side of the island may collapse into the sea (**FIGURE 13-22**). Second, winter storms tend to wash over the barrier islands, robbing them of sand and destroying houses, roads, and other structures.

KEY CONCEPTS

Barrier islands are offshore islands made of sand that are popular sites for homes and other structures, although they are very unstable and subject to violent storms that destroy homes and other buildings.

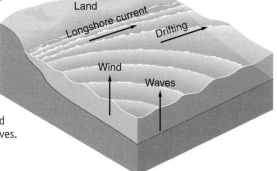

(a)

Beach drift   Backwash

Swash

FIGURE 13-21
Forces that shape a beach. (a) Beach drift is caused by waves approaching obliquely. **(b)** Longshore currents are formed by offshore oblique winds and waves.

(b)

Land

Longshore current   Drifting

Wind   Waves

FIGURE 13-22 **Losing ground.** This home built on a shifting barrier island will soon be engulfed by the sea as the island "migrates" south.

Protecting Barrier Islands—Policy Options Federal actions and various relief programs have, in the past, encouraged development on barrier islands. When erosion and storm damage occurred, for example, the government provided disaster relief funds to rebuild. Federal flood insurance paid for the damage, and federally subsidized construction projects often helped rebuild roads and stabilize the islands. A few years later, storms would devastate the islands again, starting the rebuilding cycle all over at considerable expense. Recognizing the economic unsustainability of this act, Congress passed the **Coastal Barrier Resources Act** in 1982 to prohibit the expenditure of federal money for highway construction and other development on barrier islands. Despite denial of federal subsidies, development continues on many islands.

KEY CONCEPTS
U.S. government policy has promoted the development of barrier islands until quite recently.

Coastal Beaches

Coastal beaches, like barrier islands, are eroded by longshore currents. In fact, U.S. beaches are like great rivers of sand kept in constant motion by the major coastal currents. Sand lost in one area is replaced by sediment carried to the sea by rivers. Consequently, dams that trap sediment diminish the natural replacement of sand on coastal beaches. According to one estimate, dams hold back nearly 40% of the sediment that once reached the mouth of the Santa Clara River north of Los Angeles. This robs California beaches of 15 million metric tons of sand each year. From New Jersey

to Texas, the story is the same. According to one source, 70% of the world's beaches are eroding at a rapid rate.

Some communities erect barriers called **jetties** to prevent erosion by longshore currents. These structures only slow down the process. Jetties are sometimes built to maintain navigable passageways in coastal harbors. In 1911, for instance, two 300-meter (1000-foot) jetties were built on the New Jersey coast north of Cape May to prevent sand from filling in the harbor. Although they performed admirably, the jetties have had a disastrous effect on downcurrent beaches. The beaches at Cape May grew thinner and thinner. By the 1920s, the town was fighting back by building small jetties to keep the remaining sand from being washed away and to trap the sand flowing in the longshore currents. To the townspeople's dismay, their efforts were fruitless. Beaches retreated by 6 meters (20 feet) a year. Lighthouses fell into the sea. The ocean threatened to swallow the airport. After years of anguish the town has turned to an expensive pumping system that draws sand from above the two large jetties and moves it down to the beaches, a costly replacement for a service nature used to provide for free.

KEY CONCEPTS
Beaches are ever changing and rely on sediment eroded from the land to replenish sand lost by coastal currents. Dams often trap the sediment and rob beaches of their natural replenishing sand.

Protecting Coastlines To protect coastlines, the delicate balance between sediment flow and erosion must be maintained. Dams should be built only as a last resort. Other sustainable options to acquire water or control flooding (discussed earlier in this chapter) can eliminate the need for new dams.

Some scientists argue against the construction of jetties so that beaches can be left to grow and shrink with natural cycles. Realizing that it makes more sense to cooperate with nature, in fact, some government officials on the Atlantic coast argue against any repair work on jetties and buildings destroyed after hurricanes, thus returning shorelines to their natural state. This step would begin a retreat from the shorelines, where nature probably intended humankind to be only a visitor in any case.

KEY CONCEPTS
Protecting coastlines may require measures to ensure a steady stream of sediment from rivers and a hands-off policy toward building levees in vain attempts to prevent beach erosion.

Wild and Scenic Rivers

More and more people are flocking to rivers to embark on a variety of sports: fly fishing, kayaking, rafting, inner tubing, and canoeing. Undammed rivers offer much more than excitement for the river runner. Like wilderness, they provide opportunities for relaxation and reflection. They are home to fish and other forms of wildlife. Their canyons offer opportunity for geological study. "Rivers are more than an amenity," said former Supreme Court Justice Oliver

Wendell Holmes, "they are a treasure." Unfortunately, for much of modern history rivers have been viewed in a utilitarian light. They've been perceived as sources of water and repositories for waste. They've been viewed and used as a means of transporting goods. Accordingly, many rivers have been dammed and diverted. Many others are slated for development.

KEY CONCEPTS

Rivers are a great source of recreation, but they provide many other benefits. Unfortunately, for many years people have viewed them solely as a valuable source of water that should be dammed.

Protecting Rivers   Foreseeing the need to protect rivers for recreation and habitat for fish and other species, the U.S. Congress passed the **Wild and Scenic Rivers Act** in 1968. This law made it possible to protect segments of rivers from dams, water diversion projects, and other forms of undesirable development.

At present, 165 river segments totaling 17,490 kilometers (10,931 miles) of rivers have been protected by the Wild and Scenic Rivers Act.

Congress established a three-tiered classification scheme: **Wild rivers** are rivers or sections of rivers that are relatively inaccessible and "untamed." They are free of dams and relatively unpolluted. **Scenic rivers** are free of dams, largely undeveloped, and of great scenic value. They are accessible in places by roads. **Recreational rivers** are rivers or sections of rivers that are readily accessible by roads and that may have some dams along their course or development along their shores. They offer important recreational opportunities. As of September, 1999, 165 river segments had been included in the system, totaling 17,490 kilometers (10,931 miles). No matter what their classification, rivers protected by the "wild and scenic" designation are administered with the goals of ensuring nondegradation and of enhancing the values that caused them to be designated in the first place.

Wild or scenic river designation is often marked by controversy because so many interests vie for a river's benefits: municipal water consumers, paper manufacturers, farmers, anglers, and white-water boaters. Competing interests make compromise difficult or impossible. A dammed river provides water for a new paper mill, waterskiing, and boating, but it also irretrievably floods the kayakers' rapids and the fly fishers' favorite pools.

KEY CONCEPTS

Many rivers or segments of rivers in the United States have been protected from development by the Wild and Scenic Rivers Act. Designating a river for protection is often fraught with controversy.

Sustainable Approaches to River Protection   The fight to keep free-flowing U.S. rivers safe from development continues today. Much of the pressure to dam American rivers, however, has been reduced because of economic and legal forces. After years of paying for questionable dams, Congress has found that many projects return only a few cents for every dollar invested. As a result, federally subsidized water projects, often handed out as political favors, have fallen into disfavor.

Further conflicts can be avoided by finding alternative means of supplying water, as discussed earlier in this chapter. Conservation is essential to this effort.

Many conservationists argue that valuable recreational rivers must be preserved just as endangered species are. When a scenic river gorge is dammed, it is gone forever.

A river is a vital resource to people and to a host of other species. Dammed and diverted to water-hungry, often wasteful consumers, a river becomes a symbol of unsustainable systems design. Our goal should be to manage our rivers and all other water resources wisely and efficiently, minimizing waste and damage, ensuring future generations the use of treasures we now enjoy and too often take for granted.

KEY CONCEPTS

Fortunately, in the United States the demand for new water projects has declined, in large part because of a cutback in federal subsidies for such projects and the realization that alternative sources of water, such as water conservation, are far cheaper and much better from an environmental perspective.

Let him who would enjoy a good future waste none of his present.

—Roger Babson

REFERENCES AND FURTHER READING

The References and Further Reading section at the end of this book contains a list of sources for the information discussed in this chapter and recommendations for further reading.

Connect to this book's website:
http://environment.jbpub.com/
The site features eLearning, an online review area that provides quizzes, chapter outlines, and other tools to help you study for your class. You can also follow useful links for in-depth information, research the differing views in the Point/Counterpoints, or keep up on the latest environmental news.

CRITICAL THINKING

Exercise Analysis

My list of costs for this project includes the following: (1) construction costs for the pipeline, canal system, and pumping stations; (2) pumping costs; (3) canal and pipeline maintenance; (4) interest on loans; and (5) maintenance and repair. Environmental costs would include (1) reduced water flow in the Missouri River, (2) possible effects on fish and wildlife, and (3) increasing pollution levels in the river as water is withdrawn.

The benefits of this project are many. The project would create jobs and boost the agricultural economy. It would help promote a continuation of farming in affected areas, which provides jobs and supports local economies. As farms prosper, so do their communities.

Although we can't examine all of the costs and benefits, let's take a look at the cost of pumping water and see how that compares to the value of the crops grown on the irrigated land. To begin with, it costs about $25 to pump one acre-foot of water 57 meters (188 feet) in elevation from Missouri to western Kansas. If 14 million acre-feet of water were pumped each year, how much would it cost?

To make this calculation, note that you'll be pumping water from 260 meters above sea level to about 900 meters above sea level, or about 640 meters.

Thus, a single acre-foot of water will cost about $280 to pump. Now multiply that by 14 million acre-feet. By my calculation, the annual price tag is nearly $4 billion. That's just electrical cost, though. In reality, repayment of loans for construction, interest, maintenance, and personnel will cost another $4 billion a year, according to my sources.

Now compare this to one economic benefit. Fourteen million acre-feet of water will irrigate about 3.4 million hectares (8.4 million acres) of corn, yielding about 754 million bushels of corn that would sell at $3 per bushel. That's about $2.3 billion worth of corn.

Thus, it would cost about $8 billion to transport water to produce $2.3 billion worth of corn. Remember that the $2.3 billion price tag is gross sales—not profit. It does not take into account costs for labor, machinery, fuel, fertilizer, pesticides, and so on. The profit would be considerably lower than $2.3 billion if these factors were taken into account.

This exercise shows one way of determining the feasibility of a project. We didn't have to work very hard to see that this project, although appealing in many respects, would be a very costly venture indeed.

CRITICAL THINKING AND CONCEPTS REVIEW

1. What is the hydrological cycle? Draw a diagram showing how the water moves through the cycle. Why is this cycle important to you? How does human society alter the cycle?

2. *Define transpiration, evaporation, relative humidity, absolute humidity, saturation,* and *condensation nuclei.*

3. Which sector is the largest water user in the world? In the United States?

4. Define the following terms: *groundwater, water table, saltwater intrusion,* and *subsidence.*

5. Discuss the problems caused by overexploitation of ground and surface water.

6. List the pros and cons of dams.

7. A dam is being proposed for a popular fishing and canoeing river near your home. The dam is designed to provide flood control and water for your community. You are a member of a local organization that opposes the dam. Prepare a list of options—ways to control floods and to supply drinking water to meet additional demand—that will not require a costly dam. What criteria would you use to evaluate each option?

8. Using your critical thinking skills and the knowledge you have gained, analyze the following statement: "Desalination of seawater is an important way of helping to meet future water demands."

9. Describe ways in which you and your family can help conserve water. Calculate how much water your efforts will save each day. How much will they save in a year?

10. Using your critical thinking skills, analyze the following statement: "Water is a renewable resource. Therefore, we can never run out."

11. Chapter 2 (Figure 2-3) described a method of applying the operating principles of sustainability to various human systems. Using that approach, draw up a plan to revamp the water supply system in your area. (You may need to do some research to find out where water comes from in the first place.)

12. What are the connections between flooding and water shortages? How can people solve both problems simultaneously?

KEY TERMS

absolute humidity
aquifer recharge zones
barrier islands
beach drift
black water
catchwater systems
Coastal Barrier Resources Act
coastal beaches
coastal wetlands
Coastal Zone Management Act
condensation nuclei
condense
desalination
drought
estuaries
estuarine zone
evaporation

evapotranspiration
floodplains
groundwater
hydrological cycle
inland wetlands
jetties
levees
longshore currents
longshore drift
National Wildlife Refuge System
precipitation
primary habitat
rain catchment system
recreational rivers
relative humidity
saltwater intrusion
saturated

scenic rivers
seasonal habitat
sinkholes
storm surges
streambed channelization
subsidence
surface waters
transpiration
water cycle
water table
Watershed Protection and Flood
    Prevention Act
wetlands
Wetlands Reserve Program
Wild and Scenic Rivers Act
wild rivers
xeriscape

Finite supplies of coal, oil, and natural gas fuel our economies and our lives.

CHAPTER 14

Nonrenewable Energy Sources

That human beings are fallible has been known since the beginning of time, but modern technology adds new urgency to the recognition.

—*Garrett Hardin*

Energy comes in many forms. In most industrial nations, coal, oil, and natural gas are the predominant fuels. These nonrenewable forms of energy are the lifeblood of modern industrial societies, but they are also a potential Achilles' heel. If you cut the supply off for even a brief moment, industry would come to a standstill. Agriculture and mining would halt. Millions would be out of work. Automobiles would vanish from city streets. Almost everything in our homes would cease to operate.

Many of the less developed nations have pinned their hopes for economic progress on their ability to tap into oil, coal, natural gas, and to a lesser extent nuclear power, which have fueled the industrial transformation of the wealthy nations. China, for example, is hoping to get much of its future energy from its abundant supplies of coal and from oil it imports from other countries.

Is the industrial world's dependence on coal, oil, natural gas, and nuclear energy sustainable? Can less developed nations achieve success by following in our footsteps?

This chapter examines the sustainability of the predominant nonrenewable energy fuels—coal, oil, natural gas, and nuclear energy. It looks at their impacts and their abundance. It ends with some guidelines on creating a sustainable energy future.

14.1 Energy Use: Our Growing Dependence on Nonrenewable Fuels

U.S. and Canadian Energy Consumption

One hundred years ago, Americans had few choices for energy (FIGURE 14-1). Wood, a renewable resource, was the main form of energy. Today, the nation's options are many: coal, oil, natural gas, hydropower, geothermal energy, solar power, nuclear power, and wind.

American energy options began to expand in the late 1800s as wood, which once fueled the nation's factories, became depleted. Coal began to be used in factories, but coal was a dirty, bulky fuel that was expensive to mine and transport. When oil and natural gas were made available in the early 1900s, coal use began to fall. The new, cleaner-burning fuels were easier and cheaper to transport.

Today, despite numerous energy options, the United States depends primarily on three fossil fuels: oil, natural gas, and coal. In 2004, oil accounted for nearly 40% of total energy consumption (FIGURE 14-2a). Natural gas provided about 23%, and coal provided a little over 23% of the energy. All told, fossil fuels account for about 86% of our energy use. Nuclear power, another nonrenewable fuel, provided just under 8%. Renewable sources—solar, geothermal, and hydropower—supplied about 6%.

Canada is also heavily dependent on fossil fuels. Altogether, coal, oil, natural gas, and other fossil fuels account for 72% of the nation's total energy consumption. Canada relies heavily on nuclear energy, which meets 10% of its total demand. Wood makes up the remaining supplies along with renewable energy, primarily hydroelectricity, which supplies 12% of Canada's annual energy demand. Renewable energy from solar and wind sources provide negligible amounts of power.

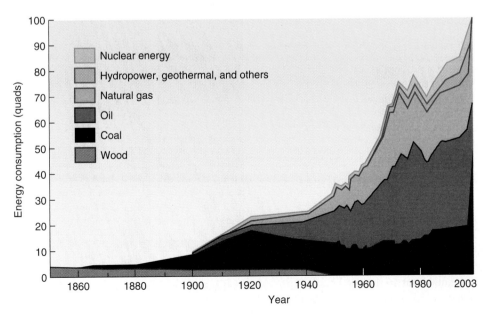

FIGURE 14-1 **Changing options.** Energy consumption in the United States by fuel type from 1850 to the present. As this graph shows, U.S. energy dependence has shifted over the years from wood to oil, coal, and natural gas. (Quad = quadrillion BTUs.)

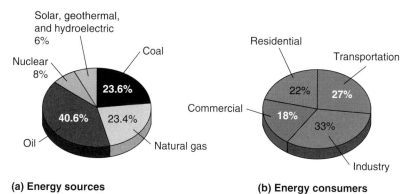

FIGURE 14-2 **The U.S. energy profile. (a)** This pie chart shows the major energy sources in the United States. Oil, natural gas, and coal are the three most commonly used fuels. **(b)** Major energy consumers in the United States. Industry and transportation lead the pack.

(a) Energy sources

(b) Energy consumers

FIGURE 14-2b breaks down energy consumption by user in the United States. As it shows, industry and business consume over 60% of the nation's energy. Transportation consumes about 27%, and residential use accounts for about 11%.

KEY CONCEPTS

Energy use in the United States has shifted considerably over the years. Today, the United States depends on a variety of fuel sources. Fossil fuels provide the bulk of the energy. Industry and business consume the majority of the fuel. Transportation is another major energy consumer.

Global Energy Consumption

Virtually all industrial nations get the energy they need from nonrenewable energy sources. On average, they receive about 85% of their energy from fossil fuels, 5% from nuclear power, a type of nonrenewable energy, and 10% from renewables such as solar and wind energy (FIGURE 14-3a). In the less developed countries, renewable energy sources such as biomass (wood and cow dung, for example) play a much more important role in supplying demand, satisfying about 40% of their energy requirements (FIGURE 14-3b). Nonrenewable fossil fuels supply about 60% of the total energy. Of nonrenewable energy fuels used in these countries, oil supplies the largest share. Coal and natural gas supply the rest. Nuclear power contributes only a tiny fraction of their energy demand, in large part, because of the high cost of this option.

> In the United States, fossil fuels provide 85% of our energy. Nuclear power and renewable energy sources supply 7.5% each.

> Renewable energy supplies a large portion of the energy demands of people in less developed nations, about 40%. In more developed nations, it supplies only about 10%.

Worldwide, the biggest users of energy are Americans, who make up about 4.6% of the world's population but consume about 25% of its primary energy. On a per capita basis, Americans consume more than twice as much energy as the people of Japan and Western Europe and about 16 times more per capita than the people of less developed nations.

Canada is also a major consumer of energy, using more per capita than any other nation except for Luxembourg. With only slightly less than 0.6% of the world's people, it uses 2.5% of the world's energy. The reasons for this are many. It is a large country, situated in a cold climate. Canada has an energy-intensive industrial economy with logging, mining, agriculture, and energy production as the chief sources of income. The extraction and processing of energy resources alone contributes 7% to the nation's Gross Domestic Product. Prices are low, too, and Canadians tend to use energy inefficiently.

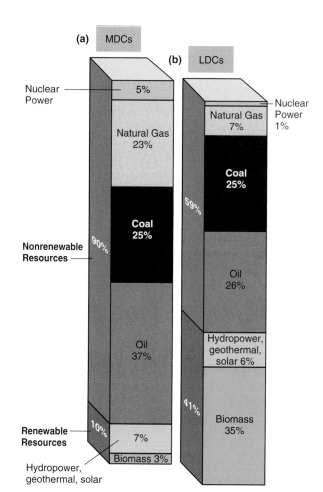

FIGURE 14-3 **Global energy use. (a)** More developed countries. **(b)** Less developed countries.

14.2 What Is Energy?

Energy is all around us, but it is sometimes difficult to define. But what exactly is it?

Energy Comes in Many Forms

Let's begin by making a simple observation as a way to help define this term: energy comes in many forms. For example, humans in many countries rely today on fossil fuels such as coal, oil, and natural gas. Some use lots of nuclear energy to generate electricity. In other countries, wood and other forms of biomass are primary forms of energy. (Biomass includes a wide assortment of solid fuels, such as wood, and liquid fuels, such as ethanol derived from corn, and biodiesel, a diesel fuel made from vegetable oils.) And don't forget sunlight, wind, hydropower, and the geothermal energy—energy produced in the Earth's interior. Even a cube of sugar contains energy! Touch a match to it, and it will burn, giving off heat and light energy, two additional forms of energy.

Energy Can Be Renewable or Nonrenewable Energy in its various forms can be broadly classified as either renewable or nonrenewable. Renewable energy, as noted earlier, is any form of energy that's regenerated by natural forces. Wind, for instance, is a renewable form of energy. It is available to us year after year, thanks in large part to the unequal heating of the Earth's surface. When one area is warmed by the sun, for instance, hot air is produced. Hot air rises, and as it does, cooler air moves in from neighboring areas. As the cool air moves in, it creates winds of varying intensity. Renewable energy is everywhere and is replenished year after year, providing humankind with a potentially enormous supply . . . if only we're smart enough to tap into it!

Nonrenewable energy, on the other hand, is finite. It cannot be regenerated in a timely fashion by natural processes. Coal, oil, natural gas, tar sands, oil shale, and nuclear energy are all nonrenewable forms of energy. Ironically, although most of these sources of energy were produced by natural biological and geological process early in the Earth's history, and, although these processes continue today in some parts of the world, these fuels are not being produced at a rate even remotely close to our consumption. Coal, for instance, may be forming in some swamplands around the world. But its regeneration is taking place at such a painfully slow rate that it is meaningless. Put another way, contemporary production can never replenish the massive supplies that were produced over long periods of time many millions of years ago. Because of this, coal, oil, natural gas, and others are essentially finite.

When they're gone, they're gone.

So, now you know two basic facts about energy: energy comes in many forms, and all forms of energy broadly fit into two general categories: renewable and nonrenewable.

Energy Can Be Converted from One Form to Another

Yet there's more to energy than this. For example, even the casual observer can tell you that energy can be converted from one form to another. Natural gas, for example, when burned is converted to heat and light. Coal, oil, wood, biodiesel, and other fuels are also converted to other forms of energy during combustion. Heat, light, and electricity are the most common by-products of these conversions, but the possibilities don't end here. Visible light contained in the sun's energy can be converted to heat. It can also be converted to electrical energy. Even wind can be converted to electricity or to mechanical energy to drive a pump to draw water from the ground.

Energy Conversions Allow Us to Put Energy to Good Use

Not only can energy be converted to other forms, it has to derive benefit for us. Coal, by itself, is of little value to us. It's a sedimentary rock and fun to behold, but it is the heat and electricity produced when coal is burned in power plants that are of value to us. Sunlight is pretty, too, and it feeds the plants we eat; but in our homes and factories, however, it is the heat that the sun produces and the electricity we can generate from it that is of primary value to us.

In summary, then, it is not raw forms of energy that we need. Not at all. It is the by-products of energy that are unleashed when we "process" them in various energy-liberating technologies that meet the complex needs of society.

Energy Can Neither Be Created nor Destroyed

Another thing you need to know to deepen your understanding of energy is that energy cannot be created nor can it be destroyed. Physicists call this the First Law of Thermodynamics or, simply, the First Law.

The First Law says that all energy comes from pre-existing forms. Even though you may think you are "creating" energy when you burn a piece of firewood in a woodstove, all you are doing is unleashing energy contained in the wood—specifically, the energy locked in the chemical bonds in the molecules that make up wood. It, in turn, came from sunlight. The sun's energy came from the fusion of hydrogen atoms in the sun's interior.

Energy Is Degraded When It Is Converted from One Form to Another

More important to us, however, is the Second Law of Thermodynamics. The Second Law, says, quite simply, that anytime one converts a form of energy to another form—for example, when you convert natural gas to heat—it is degraded. Translated, that means energy conversions transform high-quality energy resource to low-quality energy. Natural gas, for instance, contains a huge amount of energy in a small volume; it's locked up in the simple chemical bonds that attach the carbon atom to the four hydrogen atoms of the methane molecules. When these bonds are broken, the stored chemical energy is released. Light and heat are the products. Both light and heat are less concentrated forms of energy, or lower quality forms of energy. Hence, we say that natural gas, a concentrated form of energy, is "degraded." In electric power plants, only about 50% of the energy contained in natural gas is converted to electrical energy. The rest is "lost" as heat and is dissipated into the environment.

No Energy Conversion Is 100% Efficient, Not Even Close to It!

This leads us to another important fact about energy: no energy conversion is 100% efficient. When coal is burned in an electrical power plant, only about one third of the energy contained in the coal is converted to useful energy, in this case, electricity. The rest is lost as heat and light. The same goes for renewable energy technologies. One hundred units of solar energy beaming down on a solar electric module won't produce the equivalent of 100 units of electricity. You'll only get around 12% to 15% conversion on the most popular modules on the market today.

Energy is lost in all conversions. One hundred units of electrical energy won't produce 100 units of light energy in a standard incandescent lightbulb. In fact, most conventional lightbulbs in our homes (incandescent lights) convert only about 5% of the electrical energy that runs through them into light. The rest comes out as heat!

Each conversion in a chain of energy conversions loses useful energy, as shown in FIGURE 14-4. Don't forget that. To get the most out of our primary energy sources, we must reduce the number of conversions along the path.

But let's get something straight. Some of you may be wondering whether all of this discussion of "energy losses"

is violation of the First Law, which states that energy cannot be created nor destroyed.

The truth be known, the "energy losses" I've been talking about during energy conversion are not really losses in the true sense of the word. Energy is not really destroyed; it is released in various forms, some useful and others, such as heat, not so useful. Chemical energy in gasoline that runs a car, for instance, is converted to mechanical energy of moving parts that propel us forward along the highways. Some is also lost as heat that radiates off the engine. This waste heat is of little value except to use on cold winter days when captured, at least in part, to warm the car's interior.

Eventually, however, all heat produced by a motorized vehicle escapes into outer space. It is not destroyed, per se; it escapes into space and is no longer available to us. Hence, the conversion results in a net loss of useful energy.

By now you know that there are many forms of energy. You know that energy can be renewable or nonrenewable. You understand that raw energy is not as important to us in our homes as is the useful by-products such as electricity, light, or heat. You now also know that energy can neither be created, nor destroyed. It can only be converted from one form to another, and you're privy to the fact that no energy conversion is 100% efficient, not even close.

You also understand that during conversions useful energy decreases. Put another way, all conversions lose energy as heat that is dissipated into outer space. That fact, in turn, is important for nonrenewable fuels; once they've burned or reacted in the case of nuclear fuels their energy is gone forever. The heat radiates endlessly into outer space, heating the universe as it were.

Renewable energy resources, on the other hand, can be regenerated year after year after year. If we're going to persist as a society in the long term, it is renewable energy resources we'll need to rely on. Unlike fossil fuel energy and nuclear energy, renewable resources can return again and again, making our lives bright and cheery and comfortable so long as the sun continues to illuminate the daytime sky.

With these important points in mind, let's define this thing we call energy.

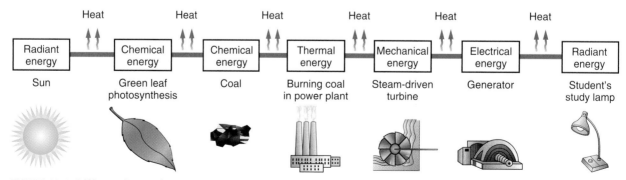

FIGURE 14-4 **Different forms of energy.** Energy can be changed from one form to another; however, with each change a certain amount of energy is lost as heat. Source: Chiras, Daniel D., Reganold, John P., Owen, Oliver S., *Natural Resource Conservation: Management for a Sustainable Future,* 9th Edition, © 2005, p. 50. Reprinted by permission of Pearson Education, Inc., Upper Saddle River, NJ.

Energy Is the Ability to Do Work

To a physicist, energy is defined as "the ability to do work." More accurately, says engineer John Howe, "Energy is that elusive something that allows us to do work." We and our machines, that is.

Any time you lift an object, for example, or slide an object across the floor, you are performing work. The same holds for our machines. Anytime a machine lifts something or moves it from one place to another, it performs work.

Energy, quite simply, is valuable because it allows us to perform work. It powers our bodies. It powers our homes. It powers our society. We cannot exist without energy.

According to physicists, work is also performed when the temperature of a substance, for example, water, is raised. Therefore, your stove or microwave is working when it boils water for hot tea or soup.

14.3 Fossil Fuels: Analyzing Our Options

Energy is the lifeblood of modern society, but it does not come cheaply. In addition to its economic cost, huge environmental costs are posed by many forms of energy. As you shall soon see, these impacts lead many to conclude that the current energy system is unsustainable. When analyzing the sustainability of the world's energy system, one must take into account available supplies as well as the impacts to the environment, climate, and human health. This section examines those impacts. Before we can understand them, though, we must first understand the many steps required to deliver energy to our homes, factories, and gas stations.

FIGURE 14-5 presents a diagram of some of the major steps involved in energy production and consumption. This chain

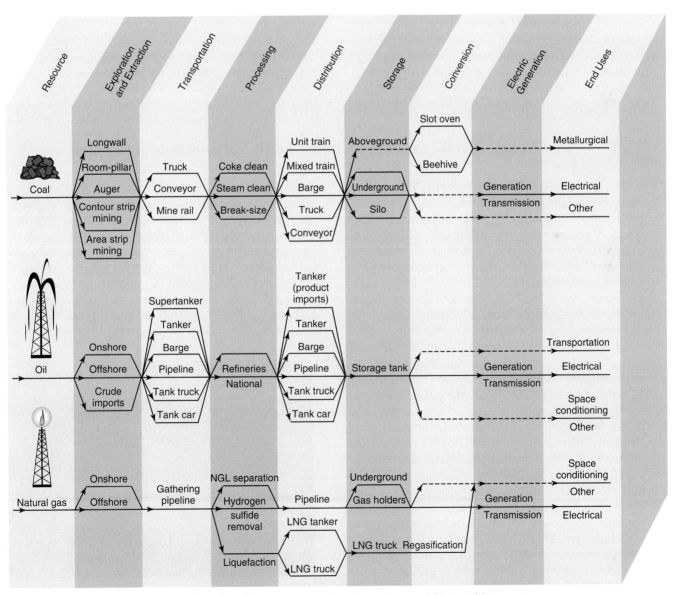

FIGURE 14-5 Energy systems. Most of us are unaware of the many steps involved in providing us with energy.

of events constitutes an **energy system** and is composed of six major phases: exploration, extraction, processing, distribution, storage, and end use. Take a moment to familiarize yourself with these steps. As a rule, the most notable environmental impacts occur at the extraction and end-use phases.

Understanding energy systems and the impacts along the way makes it clear that a simple flick of a light switch or a press on the gas pedal of an automobile creates a trail of environmental damage. It shows our personal connection to the long list of environmental problems facing the world.

KEY CONCEPTS

Energy does not come cheaply. In addition to the economic costs, society pays a huge environmental price for its use of nonrenewable energy in damage to the health of its people and to the environment. These impacts arise at every phase of energy production. The most significant impacts arise from extraction and end use.

Crude Oil

Crude oil or **petroleum** is a thick liquid containing many combustible hydrocarbons (organic compounds made of hydrogen and carbon). Found in deep deposits in the seafloor and on land, crude oil often occurs with natural gas. Geologists locate oil deposits through a careful study of geology—

that is, the type of rocks found in various regions—and by other fairly low-impact means. However, to explore for oil in remote areas, roads must often be built so that exploration crews can enter them. In Canada, plans are in the offing to move into pristine northern areas as far north as the Arctic where oil companies, supported by government subsidies, hope to explore and eventually drill for oil. Even with newer techniques of oil extraction, the impacts could be devastating to the area.

Once located, crude oil is extracted via wells drilled into deposits. Oil flows into the well and is pumped to the surface. Oil companies once thought that the yield from oil wells would be greatly increased by injecting water, steam, and carbon dioxide into the ground, which forces additional crude oil to the surface. Today, geologists realize that these techniques do not increase yield.

Once the crude oil is extracted, it is transported by pipeline, truck, or ship to refineries for processing. Transportation of oil by ships and pipelines is a source of considerable environmental damage. Ships sometimes run aground and spill their contents into lakes and oceans, with devastating environmental consequences (Chapter 21). Pipelines may leak or break, causing additional problems. For a discussion of the potential impacts of oil development in the Arctic, see Spotlight on Sustainable Development 14-1.

>> SPOTLIGHT ON SUSTAINABLE DEVELOPMENT

14-1   Controversy Over Oil Exploration in the Arctic National Wildlife Refuge

In the far northeastern corner of Alaska lies the Arctic National Wildlife Refuge, or ANWR. Set aside to protect wildlife, this refuge encompasses 7.7 million hectares (19 million acres). It has been described as the last great American wilderness. With strong support of the Bush administration, a portion of ANWR has been opened to development. If oil is found, many suspect that widespread drilling will occur in the coastal plain, a 0.6-million-hectare (1.5-million-acre) region—an area almost as large as Yellowstone Park.

The delicate Arctic tundra of the coastal plain, a region 210 kilometers (115 miles) long and 50 kilometers (30 miles) wide, is the only section of land along the entire 1800-kilometer (1100-mile) north coast of Alaska that is closed to oil exploration and drilling. The coastal plain is the annual calving ground of several large caribou herds. It is also home to polar bears, grizzlies, wolves, moose, wolverines, and numerous small mammals. Many thousands of birds spend the summer there, raising their young and feeding on insects.

In 1991, after public protest over the tragic 1989 oil spill in Prince William Sound subsided, President George Bush and the oil companies introduced legislation that would allow drilling in the ANWR. Because of a massive outpouring of public protest, the bill was defeated, but the oil companies continued to push legislation that would permit them to explore for oil within this magnificent wildlife refuge.

In 2005, a couple of major oil companies reportedly withdrew support, after having conducted secret exploration in the region. Despite their having concluded that oil de-

velopment in the area would be uneconomic, in 2005, the Bush Administration and Republican leaders pushed through legislation opening up a small but significant portion of ANWR to oil development. Although they hope to minimize damage by special drilling techniques, many conservationists fear that the pristine region will eventually be turned into a maze of roads, oil platforms, waste ponds, buildings, and gravel pits (**FIGURE 1**). Full-scale development would lead to construction of four airfields and 50 to 60 oil platforms. A power plant and several oil-processing plants would be built on the

FIGURE 1 An oil drilling pad in Prudhoe Bay, Alaska. The delicate Arctic tundra will be damaged to supply more oil to satisfy America's needs.

In a refinery, the oil is heated and distilled. This separates out many of the different hydrocarbon molecules in crude oil—including those that make up gasoline, diesel fuel, heating oil, and asphalt. Many small organic compounds are also extracted during this process and are used to make medicines, plastics, paints, and pesticides (FIGURE 14–6). Refineries use considerable amounts of energy and are major sources of air and water pollution.

The combustion of crude oil derivatives such as gasoline and diesel fuel produces enormous quantities of carbon dioxide (a greenhouse gas) and other pollutants such as sulfur dioxide and nitrogen dioxide, both of which are converted into acids in the atmosphere (Chapter 20). In fact, many of the most significant local and global environmental problems stem from the combustion of oil and its by-products.

KEY CONCEPTS

Oil is extracted from deep wells on the seafloor and on land; it is often found in association with natural gas. After it is extracted, crude oil is heated and distilled, a process that separates the components of oil, which produces useful fuel and nonfuel by-products. The major impacts of the oil energy system come from oil spills and from combustion of oil and its by-products.

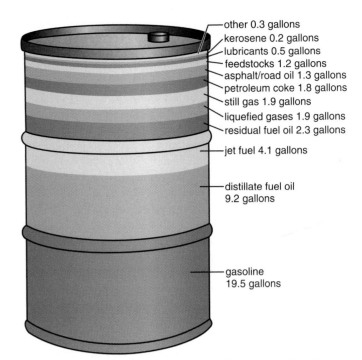

other 0.3 gallons
kerosene 0.2 gallons
lubricants 0.5 gallons
feedstocks 1.2 gallons
asphalt/road oil 1.3 gallons
petroleum coke 1.8 gallons
still gas 1.9 gallons
liquefied gases 1.9 gallons
residual fuel oil 2.3 gallons

jet fuel 4.1 gallons

distillate fuel oil 9.2 gallons

gasoline 19.5 gallons

FIGURE 14-6 **Products from a 44-gallon barrel of oil.** Total is more than 44 because of "processing gain." Source: Data from the American Petroleum Institute.

tundra, and hundreds of kilometers of roads and pipelines would crisscross this delicate land. The rising temperatures discussed in Chapter 10 are threatening the Arctic tundra, which could dramatically alter Bush administration plans to drill for oil in the Arctic. Currently, oil operations are permitted only when the Arctic tundra is frozen at least 12 inches and covered with at least 6 inches of snow. Because of global warming, the "window of opportunity" has shrunk from 200 days per year to only 100. The Bush administration hopes to change more regulations to allow oil companies to operate even when the tundra is thawed.

While proponents think that oil development and wildlife can coexist in harmony, critics predict a 20 to 40% decline in one of the major caribou herds. They expect over half of the musk oxen to perish. Populations of grizzlies, polar bears, wolverines, and other animals are also likely to decrease substantially. Air pollution, water pollution, and hazardous wastes are expected to have major impacts on wildlife and the long-term ecological health of the region. One exploratory well alone requires 35,000 cubic meters of gravel, which would be excavated from streambeds to make pads and roads. Oil spills on the tundra and careless waste disposal, common in nearby Prudhoe Bay, would change this delicate landscape, whose short growing seasons and harsh winters greatly impair the natural healing that normally takes place in the wake of human interference.

Proponents argue that we need the oil to cut reliance on foreign sources and that we need to drill for oil in ANWR now

because of the 10- to 15-year lead time required to fully develop the area. Geologic evidence, some say, indicates a high potential for oil discovery, and oil is needed to help bolster Alaska's economy. They say that environmental regulations will ensure minimal impacts on wildlife and the environment.

Those in favor of preserving the refuge intact argue that the environmental costs are too high. Experience in nearby Prudhoe Bay shows that oil companies often ignore environmental regulations. Frequent violations have been cited. The Department of Interior, which generally favors oil development, has recorded over 17,000 oil spills since 1973 in the Arctic. Where the oil saturates the soil, vegetation fails to recover. Opponents of exploration also argue that because the entire north coast of Alaska is already open to oil exploration, the ANWR should not be—now or ever. If there's oil there, let it lie. Let us find alternatives and let the wildlife live in peace. Furthermore, they point out that slight improvements in automobile gas mileage could easily "provide" as much oil as the ANWR could generate over its lifetime—and at a much lower cost. Moreover, critics say that the proponents have exaggerated the potential for finding oil that is economically feasible to recover. Based on the oil industry's own reports, they say, there's only a one in five chance of finding economically recoverable oil.

Few of us will ever visit the ANWR, but many of us take comfort in knowing there are places where wildlife are free of the intrusion of modern industrial society. Those places, say opponents, may not last.

Natural Gas

Natural gas is a mixture of low molecular weight hydrocarbons, mostly methane. Burned in homes, factories, and electric utilities, natural gas is often described as an ideal fuel because it contains few contaminants and burns cleanly. Substituting natural gas for coal in electric generating power plants, for instance, reduces carbon dioxide emissions by 50 to 70%. But precautions must be taken so methane in natural gas doesn't leak. Methane is 30 times more powerful as a greenhouse gas, so losses of only 3 to 4% of the methane could offset any benefit. Natural gas plants also frequently burn off excess gas, a process called **flaring**. This process releases hydrogen sulfide gas and a variety of pollutants, including carbon dioxide, carbon monoxide, and volatile organic compounds that are harmful to people and the environment, as you will see in Chapters 18 and 19. The Alberta Research Council identified 200 chemical compounds in flare gas. In Alberta, 1.4% of the natural gas summoned from wells is actually flared.

Like oil, it is easy to transport within countries via pipelines. It is much more difficult to transport overseas, however. To do so, it must be compressed and liquified. It is then transported by ship. Natural gas is also fairly economical and requires relatively little energy to extract.

Natural gas comes from wells as deep as 10 kilometers (6 miles) and, as noted earlier, is often found in association with oil. On land, drilling rigs generally have minimal impact unless they are located in wilderness areas, where operations destroy valuable land and roads and noise from heavy machinery and construction camps can disturb wildlife. However, natural gas extraction can cause subsidence in the vicinity of the well. One notable example is in the Los Angeles-Long Beach harbor area, where extensive oil and gas extraction began in 1928 and has caused the ground to drop 9 meters (30 feet) in some areas.

Natural gas is generally safe to transport in the gaseous form in pipelines, but as noted earlier, to transport it across oceans, it must be liquefied. In the liquid form, natural gas is unstable and highly flammable. A ship containing liquefied natural gas could burn intensely.

Coal

Coal is a solid organic fuel extracted from mines. It is the most abundant fossil fuel on Earth and is therefore relatively inexpensive. However, the production and consumption of coal are more environmentally disruptive than for any other major fossil fuel. Significant impacts occur at virtually every step in the process. (Table 14-1 summarizes many of these problems.)

Impacts of Coal Production: Surface Mining   Coal is extracted by surface and underground mines, depending on the terrain and the depth of the seam. In hilly terrain where coal seams lie near the surface, as in the eastern United States, bulldozers strip the rock and dirt lying over them (the **overburden**). The overburden is hauled away and eventually replaced after the coal supply has been depleted.

This type of mining is called **surface mining**, a term that refers to any mine that accesses its mineral from the surface. Several types of surface mines exist. Surface mines that follow the contour of the land are appropriately called **contour strip mines** (**FIGURE 14-7**). Contour strip mining defaces the landscape. Exposed hillsides can increase soil erosion during rainy periods if proper precautions are not taken. Contour strip mining in the hilly terrain of Kentucky, for instance, has been shown to increase erosion from 0.4 in undisturbed watersheds to nearly 150 metric tons per hectare (67 tons/acre). Access roads can also erode. During heavy rains, sediment from the roadways washes into nearby streams. Sediment also fills streams, reducing their

Table 14-1

Major Environmental Impacts of Fossil Fuels

Fuel	Extraction	Transportation	End Uses
Coal	Destruction of wildlife habitat, soil erosion from mine sites, sedimentation, aquifer depletion and pollution, acid mine drainage, subsidence, black lung disease, accidental death	Air pollution and noise from diesel trains	Air pollution from power plants and factories, especially acid pollutants and carbon dioxide; thermal pollution of waterways
Oil	Offshore leaks and blowouts causing water pollution and damage to fish, shellfish, birds, and beaches; subsidence near wells	Oil spills from ships or pipelines	Air pollution similar to that from coal
Natural gas	Subsidence and explosions	Explosions, land disturbance from pipelines	Fewer air pollutants than coal and oil

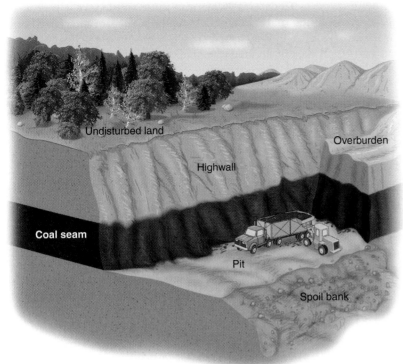

<figure>Undisturbed land
Highwall
Overburden
Coal seam
Pit
Spoil bank</figure>

FIGURE 14-7 **Contour strip mining.** Contour strip mining is common in hilly terrain—for example, in the eastern U.S. coal fields. For many years, overburden was dumped on the downslope, destroying vegetation and creating serious erosion problems. Today, overburden is hauled away and stored. When the coal seam is exhausted, the overburden is replaced, graded, and planted.

water-carrying capacity. When rain falls, water spills over the streams' banks, flooding farms and communities.

Another mining technique in hilly or mountainous terrain is mountaintop removal. In this technique, entire mountain tops are removed by dynamite and bulldozers to expose coal seams. The overlying material is bulldozed into neighboring valleys, causing considerable environmental damage. Although it is prohibited by law, this technique was given a boost by the Bush administration in 2002, which rewrote Clean Water Act regulations. The changes make it legal to fill in valleys and streams by such operations. Further changes in law are expected to make this form of mining fully legal.

In the relatively flat terrain of the West and Midwest, coal is usually mined in **area strip mines** (**FIGURE 14-8a**). In this type of surface mine, the topsoil is first removed by scrapers and is set aside for reapplication (**FIGURE 14-8b**). Next, the overburden is dynamited and removed by huge shovels called **draglines** to expose the coal seam. The coal is removed and hauled away, and another parallel strip is cut. The overburden from the new cut is then placed in the previous one (Figure 14-8b). As required by the **Surface Mining Control and Reclamation Act** (1977), the overburden must be regraded to the approximate original contour and the topsoil replaced. Seeds are sown, increasing the likelihood that the area will revegetate.

Area strip mines create eyesores, destroy wildlife habitat and grazing land, and may increase erosion. Proper reclamation can restore wildlife habitat and grazing land and eliminate the visual impact. Surface mines can also disrupt and pollute groundwater supplies in the West because many aquifers are located in or near coal seams. In Decker, Montana, extraction of coal from an aquifer seam resulted in a drop in the water table of 3 meters (10 feet) or more within a 3-kilometer (2-mile) radius of the mine. Some of the residents who depended on the aquifer were forced to drill deeper wells or find new water supplies.

(a) FIGURE 14-8 **Area strip mine. (a)** Aerial view of a strip mine. **(b)** In flat terrain, coal is extracted by strip mining. A dragline removes the overburden to expose the coal seam. Overburden is placed on the previously excavated strip and eventually regraded and replanted. If land is not carefully recontoured and replanted, it could be permanently ruined.

<figure>Undisturbed land
Reclaimed area
Original ground surface
Highwall
Overburden
Spoil bank
Coal seam
Strip bench</figure>

(b)

Coal is removed by surface and underground mines, both of which create many environmental impacts. Surface mining is especially damaging to the environment; proper controls can greatly reduce impacts, though, and mine reclamation can help companies to return lands to their original use.

Impacts of Coal Production: Underground Mines In mountainous terrain where coal seams lie deep below the surface, coal is extracted via underground mines. Underground coal mines are notorious for explosions and cave-ins, and underground coal mining is ranked as the most hazardous of the major occupations in the United States. Between 1900 and 2003, 104,524 Americans were killed in underground coal mines, and approximately 1.7 million were permanently disabled. Thanks to stricter safety regulations, deaths from coal mining have dropped substantially in the last 70 years.

> In the past 100 years, underground coal mining has killed over 100,000 Americans and permanently disabled nearly two million others. The U.S. government spends about $1.6 billion a year in benefits to coal miners with black lung disease.

Underground coal mines also cause **black lung disease**, or **pneumoconiosis** (NEW-moe-cone-ee-OH-sis), a progressive, debilitating disease caused by breathing coal dust and dirt particles (**FIGURE 14-9**). Victims have difficulty getting enough oxygen because the tiny air sacs (alveoli) in the lungs break down. Exercise becomes difficult, and death is slow and painful. Despite safety improvements, one third of all U.S. underground coal mines still have conditions conducive to black lung, a problem that costs U.S. taxpayers $740 million in 2004 in federal worker disability benefits.

Collapsing mines also cause subsidence, a sinking of the surface. Cracks form on the surface, ruining good farmland. In some cases, streams vanish into the fissures. Over 800,000 hectares (2 million acres) of land has subsided in the United States from underground coal mining. For every hectare of coal mined in central Appalachia, over 5 hectares (12.5 acres) of surface becomes vulnerable to subsidence.

Another problem with underground mines is fires. In Pennsylvania, for example, some fires have burned for years, producing considerable amounts of pollution.

Many abandoned coal mines in the East leak sulfuric acid into streams. **Acid mine drainage**, as it is called, consists of sulfuric acid formed from water, air, and sulfides (iron pyrite) in the mine. A bacterium (*Thiobacillus thioxidans*) facilitates the conversion, making the waters highly acidic. Acid mine drainage kills plants and animals and inhibits bacterial decay of organic matter in water, allowing large quantities of organic matter to build up in streams. Sulfuric acid also leaches toxic elements such as aluminum, copper, zinc, and magnesium from the soil and carries them to streams.

Acid can render water unfit for drinking and swimming. Municipal and industrial water must be chemically neutralized before use. Acid also corrodes iron and steel pumps, bridges, locks, barges, and ships, causing millions of dollars of damage each year.

Active and abandoned U.S. mines produce several million metric tons of acid a year. Acid mine drainage pollutes over 15,530 kilometers (9,709 miles) of U.S. streams, most of which are in Appalachia (**FIGURE 14-10**). Cleaning up the abandoned mines could take decades

> Acid mine drainage currently contaminates about 15,530 kilometers (9709 miles) of streams in the United States.

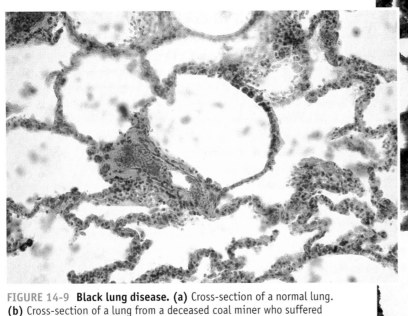

(a)

(b)

FIGURE 14-9 **Black lung disease. (a)** Cross-section of a normal lung. **(b)** Cross-section of a lung from a deceased coal miner who suffered from black lung disease. The black material is carbon from coal dust.

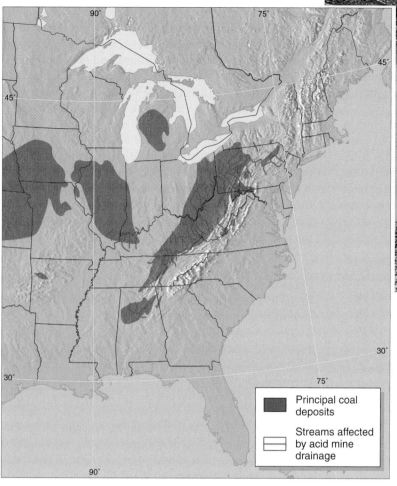

(b)

FIGURE 14-10 **Acid mine drainage. (a)** Map showing streams affected by acid mine drainage in the eastern United States. Acid mine drainage is prevalent in regions where rock contains the sulfur-containing compound known as iron pyrite. **(b)** Stream damaged by acid mine drainage.

(a)

and billions of dollars, and further increases in coal production could increase acid mine drainage although awareness of the problem has resulted in efforts to reduce it.

Underground mines produce enormous quantities of wastes, which are transported to the surface and dumped around the mouth of the mine. These wastes, called **mine tailings**, contain heavy metals, acids, and other pollutants that wash into streams during rains. Coal-cleaning plants, which are designed to crush the coal and wash away impurities, also produce enormous quantities of waste. These wastes are often stored in holding ponds that sometimes leak and pollute nearby streams.

After the coal is extracted, it is loaded onto trucks or rail cars and shipped to power plants. Diesel trains, which move most of the nation's coal, pollute the air with particulates and other potentially harmful pollutants such as sulfur dioxide.

KEY CONCEPTS

Underground coal mines present health and safety hazards for miners. Many miners have been killed or injured by accidents. Many others suffer from black lung disease, a crippling degenerative lung disease, as a result of exposure to coal dust. One of the biggest environmental problems from coal mining results from the release of sulfuric acid from abandoned underground mines, which poisons thousands of miles of streams in the eastern United States.

Impacts of Coal Combustion About 90% of U.S. coal is burned to generate electricity; most of the rest is burned by industry to generate heat or to produce steel and other metals. Power plants require enormous amounts of coal. For example, a 1000-megawatt coal-fired power plant, which produces electricity for about one million people, burns about 2.7 million metric tons (3 million tons) of coal each year. Tens of thousands of tons of pollutants are released into the atmosphere during coal combustion, even with pollution control devices in place. These include particulates, sulfur oxides, nitrogen oxides, carbon monoxide, and carbon dioxide. These pollutants are largely responsible for two of the world's most significant environmental problems, acid deposition and global climate change. They cause billions of dollars in damage to fish, lakes, buildings, and human health, as discussed in Chapters 19 and 20.

Coal combustion also produces enormous quantities of solid waste. During combustion, a fine dust known as fly ash is produced. **Fly ash** is mineral matter that makes up 10 to 30% of the weight of uncleaned coal. It is carried up the smokestack with the escaping gases and, if captured by pollution control devices, becomes a hazardous solid waste. Sulfur dioxide gas is also emitted from the stack but can be removed by a pollution control device known as a **smokestack scrubber** (Chapter 19). Scrubbers produce a toxic sludge, containing fly ash and sulfur compounds. Some mineral matter that is too heavy to form fly ash remains at the bottom of the coal-burning furnace as **bottom ash**. It, too, is a hazardous waste that must be disposed of. Wastes from coal-burning furnaces and their pollution control devices are generally buried in landfills, but toxic substances from these sites may leak into groundwater supplies, polluting them.

A 1000-megawatt power plant may produce 180,000 to 680,000 metric tons (200,000 to 750,000 tons) of solid waste each year, including fly ash, bottom ash, and sludge from scrubbers. Coal-fired power plants in the United States annually produce about 109 million metric tons (121 million tons) of solid waste. Approximately 40% of the waste was used to make road surfaces and other products; the rest ended up in landfills.

KEY CONCEPTS

Electric power plants are the major consumer of coal in the United States. Although pollution controls reduce the amount entering the atmosphere, the large number of power plants and the enormous quantities of coal burned in them result in the production of massive amounts of air pollution. Several air pollutants contribute to two of the world's most pressing environmental problems—acid deposition and global climate change (global warming). Moreover, pollutants captured by pollution control devices end up as solid waste, much of which is buried in landfills.

Oil Shale

Oil shale is a sedimentary rock containing a combustible organic material known as **kerogen**. Oil shale is found in large deposits underlying much of the continental United States, with the richest ones in Colorado, Utah, and Wyoming. Similar deposits are also found in Canada, Russia, and China.

Kerogen can be extracted from oil shale by heat. This process produces a thick, oily substance known as **shale oil**. Like crude oil, it can be refined to produce a variety of fuels and nonfuel materials.

Oil shale is rather abundant, but its extraction and processing are energy intensive and therefore rather costly. **Net energy analysis**, which determines how much energy it takes to extract and process an energy resource in relation to the amount produced, shows that shale oil production creates only about one eighth as much energy as conventional crude oil extraction for the same energy investment. Its **net energy yield** (energy produced − energy invested = net energy yield) is therefore rather low.

Oil shale development could result in serious environmental impacts as well. Strip mining to remove oil shale would disturb large tracts, causing erosion and reducing wildlife habitat. Mined shale is then crushed and heated in a large vessel, called a **surface retort**. Surface retorting produces enormous amounts of solid waste, spent shale. A small operation producing 50,000 barrels per day would generate about 19 million metric tons (21 million tons) of waste per year. Since the shale expands by about 12% on heating, not

all of it could be disposed of in the mines from which the raw ore came. Dumped elsewhere, the spent shale may be leached by water, producing an assortment of toxic organic pollutants that could contaminate underground and surface waters.

Oil shale retorts require large quantities of water as well, but oil shale deposits are typically found in arid country. Retorts also produce significant amounts of air pollutants unless carefully controlled.

To bypass the solid waste problem, oil shale companies once experimented with *in situ* (IN SEE-two) **retorting**. In this process, shale deposits are first fractured by explosives. A fire is then started underground in the oily rock and is forced through the deposit. The heat from the fire forces some of the oil out of the rock. The oil is then collected and pumped to the surface.

In situ retorts eliminate many impacts caused by surface mining, but they have not worked well because it is difficult to fracture shale evenly (a step needed to promote uniform combustion) and because groundwater often seeps into them and extinguishes the fires.

In the early 1990s, the only oil shale plant operating in the United States, which cost $1 billion to build, closed its doors because it was barely covering operating costs and was producing oil at twice the cost of conventional crude oil.

Today, Shell oil company is experimenting with a new process that they believe could make oil shale development cost-effective. They drill holes in the oil shale and insert heaters into them. Heat liberates a gasoline-like substance that rises to the surface. To contain the oil, they also freeze a large (20 to 30 foot) region around the shale. Both processes require lots of energy and time. Although Shell predicts that the final cost will be low, only time will tell.

KEY CONCEPTS

Oil shale is a sedimentary rock containing an organic material (kerogen) that can be extracted from the rock by heating. In a liquid state, this thick, oily substance, called *shale oil*, can be refined to make gasoline and a host of other chemical by-products. Although it is abundant, oil shale is economically and environmentally costly to produce and is currently an insignificant source of fuel.

Tar Sands

Tar sands or oil sands, are sand deposits containing a combustible organic material called **bitumen**. They are another fossil fuel. Although they are found throughout the world, the largest deposits are in Alberta, Canada; Venezuela; and Russia. In the United States, six states have large deposits.

Tar sands are strip-mined and then treated with hot water to extract the bitumen. *In situ* methods similar to those used in oil shale extraction are also being tested.

Tar sands are plagued with many of the problems that face oil shale. They are expensive to mine. The sand sticks to machinery, gums up moving parts, and eats away at tires and conveyor belts. Tar sand expands by 30% after processing. As with oil shale, its production requires large amounts of water, which becomes badly polluted in the process.

The most significant barrier to tar sand development is a low net energy yield, and thus high cost. At least 0.6 of a barrel of energy is consumed per barrel produced. Worse, world reserves of tar sand oil are insignificant compared with world oil demands. Locally, however, tar sands can make a huge difference. Canada's oil sands, which is the largest deposit in the world, contain an estimated 1.7 trillion barrels of oil, of which some experts say 300 billion could be recovered. That's enough to supply Canada at its current oil demand for 200 years. But as noted earlier, oil sands are costly and environmentally damaging. The Canadian government has poured billions of dollars into research and development. By 2020, total investment is expected to be about $30 billion. Environmentalists argue that the federal and provincial governments have paid little attention to the potential impacts of tar sand development.

KEY CONCEPTS
Tar sands are sand deposits impregnated with a petroleumlike substance known as *bitumen*. Although there are large deposits, they are economically and environmentally costly to extract.

Coal Gasification and Liquefaction

The abundance of coal in the United States and abroad has stirred interest in coal gasification and liquefication. In both technologies, coal reacts with hydrogen to form **synfuels**, combustible organic fuels. **Coal gasification** produces combustible gases that can supplement natural gas. **Coal liquefaction** produces an oily substance that can be refined to make gasoline.

Coal gasification produces numerous air pollutants and requires large quantities of water, making it a dirty alternative to natural gas. The cost of synthetic natural gas is high. An additional problem is the low net energy production. Synthetic gas produced from surface mines is about 1.5 times more expensive than natural gas, and synthetic gas from coal taken from underground mines is 3.5 times more expensive.

Coal liquefaction, the production of synthetic oil from coal, is similar to coal gasification. Although there are at least four ways of making a synthetic oil from coal, each involves the same general process: adding hydrogen to the coal. The oil produced by liquefaction must be purified to remove ash and coal particles.

Coal liquefaction could provide liquid fuels, but it would be costly. It produces air and water pollutants and requires large amounts of energy. Like coal gasification, it might be preempted by cheaper, cleaner, and renewable energy sources.

KEY CONCEPTS
Coal can be converted to gaseous and liquid fuels to replace oil and natural gas, but the processes are energy-intensive and produce much pollution.

14.4 Fossil Fuels: Meeting Future Demand

The sustainability of a fuel source must be judged on the basis of its social and environmental costs as well as its economic impact. An analysis of supplies is equally important. However, determining how much fossil fuel is available and how long it will last is fraught with difficulties. First, no one knows how much economically recoverable fossil fuel lies within the Earth's crust. Moreover, some countries have been known to exaggerate supplies, especially oil reserves. Second, future demand is not entirely predictable. Will the former Soviet Union's transformation to a capitalistic economic system dramatically increase its fossil fuel consumption? Will efforts in less developed nations like China and India, which together house over one third of the world's people, have a similar effect, greatly increasing the depletion of oil and other fossil fuels? Or will the nations of the world adopt energy-efficient, renewable technologies on a large scale, cutting energy demand?

This section examines the projected lifetime of three fossil fuels: oil, natural gas, and coal. When examining the prospects for energy supplies, keep in mind the uncertainties outlined previously.

Oil: The End Is Near

Some experts predict that when the last drop of crude oil has been burned, the world's oil fields will have yielded between 1800 and 2200 billion barrels of oil. This figure is called the **ultimate production**. About half of this has already been used. So what's the problem? At the current rate of consumption, all remaining oil would last about 50 years.

At the current rate of consumption, current known oil reserves will last about 50 years. At an accelerated rate of consumption, global supplies could be gone within two decades.

The problem is that oil consumption is increasing approximately 2.5% per year. If oil consumption continues to grow at this rate—and we have every reason to believe that it will—the remaining oil will be gone much sooner.

Long before the last drop is burned, though, signs of shortage will be evident. You can understand this by studying **FIGURE 14-11**, which shows graphs of petroleum production in the United States and the world. As illustrated, world oil production is expected to rise to a peak and then begin to fall as reserves are depleted. When? Most experts predict a peak in oil production between

Global oil production is expected to peak between 2004 and 2010.

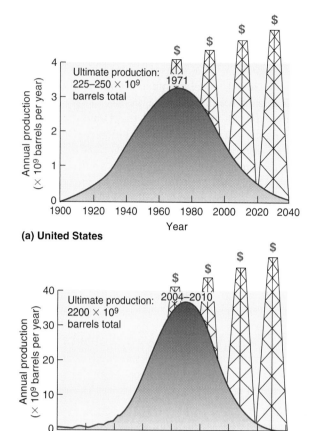

(a) United States

(b) World

FIGURE 14-11 **The fate of U.S. and global oil supplies.** Petroleum production curves for **(a)** the United States and **(b)** the world. U.S. oil production has already peaked. The oil derricks represent the increasing cost of drilling to extract declining supplies.

industrial world to its knees. The 1991 Persian Gulf War was another unfortunate consequence. This war cost several more developed nations $61 billion, a price paid primarily by the United States, Japan, and Saudi Arabia. In 1999, oil-producing nations conspired once again to cut production, sending the price of crude oil skyrocketing again, after years of relatively low prices.

> About two thirds of the world's proven oil reserves are located in the Middle East.

In 2004 and 2005, oil and gasoline prices shot upward again as a result of many factors, including rising demand in the United States and China, a lack of oil refinery capacity, and insufficient supply. Some oil analysts pinned much of the blame on insufficient supply—a peak in global oil production.

If demand for oil remains high and continues to outstrip supplies worldwide, crippling shortages and exorbitant oil prices and potentially inflation could all occur. If alternative energy sources are not available, many countries could experience severe economic turmoil: inflation, economic stagnation, and widespread unemployment. By 2050, oil production will be a small fraction of current levels.

Can't we simply find more oil? Although there's still more oil to find, it's unlikely that oil companies will make any major new finds to relieve the situation. In fact, oil companies haven't found a really large oil deposit since 1962. Of the 40,000 oil fields, only about 40 have been huge. Today, the world consumes about 22 billion barrels of oil per year, yet discovers about 4 billion new barrels of oil. Our prospects are not good. Many of the world's large oil fields are on the decline. Even the mammoth oil fields of Saudi Arabia are thought to have peaked.

2004 and 2010. As oil reserves decline, oil companies will have to work much harder to maintain production, which will raise the price of oil considerably. In addition, as domestic oil reserves fall, many nations may turn to the politically unstable Middle East, which currently houses about two thirds of the world's proven oil reserves (**FIGURE 14-12**). Christopher Flavin and Nicholas Lenssen of the Worldwatch Institute liken the developed world's dependency on oil to an addiction, note, "Not only is the world addicted to cheap oil, but the largest liquor store is in a very dangerous neighborhood."

Increasing our dependence on a fuel from a politically volatile region of the world can have devastating consequences. For example, the 1973 oil embargo imposed by the **Organization of Petroleum Exporting Countries or OPEC** and a second embargo by Iran in 1979 drove the price of oil from $3 a barrel to $34 in a short period, stimulating rapid inflation that brought the

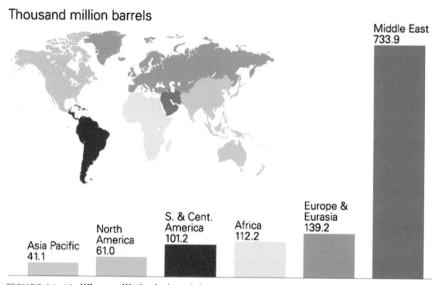

Thousand million barrels

FIGURE 14-12 **Where will the industrial countries get their oil?** This map shows where the world's oil lies. Source: Figure 3 in BP Statistical Review of World Energy 2005; courtesy of British Petroleum, www.bp.com.

FIGURE 14-13 **Toyota Prius, Gen II.** This five-passenger car owned by the author is the second generation of the world's first mass-produced gasoline-electric hybrid vehicle. This car is powered by a gasoline-fueled internal combustion engine and a clean, quiet electric motor. Its low emissions earn its title as a super ultra-low-emission vehicle.

Rising fuel prices could stimulate a massive shift toward energy efficiency—for example, new more efficient vehicles such as the Toyota Prius, which gets 60 miles per gallon in the city and 51 miles per gallon on the highway (FIGURE 14-13). To prevent widespread economic turmoil, however, nations must act soon to use oil more efficiently and to find replacements for oil—which powers our transportation system, heats some of our homes, and provides the raw materials for plastics, medicines, cosmetics, and much more. Some energy options are discussed in Chapter 15.

KEY CONCEPTS

Global oil supplies appear to be quite large; however, when one calculates how long they will last (based on the historical increases in energy consumption), it appears that oil supplies may last only 20 to 40 years. Long before the last drop of oil is removed, shortages will appear, with serious social and economic repercussions.

Natural Gas: A Better Outlook

The outlook for natural gas is brighter than the outlook for oil. For the time being, it appears that global supplies are adequate. In the United States, however, natural gas supplies are running low. (FIGURE 14-14). As illustrated in Figure 14-14, natural gas production peaked in 1973, declined, then plateaued for many years. Some experts believe that natural gas production will decline precipitously around 2008.

> Proven global natural gas reserves will last about 55 years at the current rate of consumption. Undiscovered reserves would last another 40 years. Although an increase in consumption will reduce the life span of natural gas, supplies are still relatively plentiful.

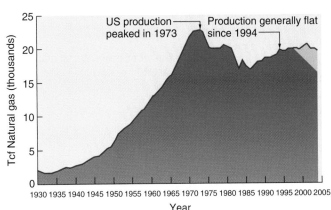

FIGURE 14-14 **U.S. marketed natural gas production, 1930–2004.** Data from U.S. Department of Energy, Energy Information Administration.

Because global natural gas supplies are relatively large and because it is a clean-burning fuel that can be used in cars, power plants, industry, and homes, many believed that natural gas could replace the more environmentally damaging fossil fuels, oil and coal. In fact, sustainable energy proponents believe that natural gas is a great transition fuel. It will help wean us from coal and oil, the most environmentally unsustainable fossil fuels, and provide us time to develop a sustainable energy system based on conservation and renewable energy sources (Chapter 15).

There are problems however, with this strategy, as much of the world's remaining natural gas supplies are in the Middle East. Domestic supplies are on the decline, causing prices to skyrocket. (They tripled from 2000 to 2004.) Can't we merely import more?

Natural gas must be compressed and shipped by tanker. Today, the United States imports only 2% of its natural gas this way and is not equipped to handle much more. Increasing imports would require numerous ports to develop facilities to handle ships and highly explosive gas, and many cities have rejected plans to do so. Even if we could, some experts predict that global natural gas production will peak around 2015.

KEY CONCEPTS

Natural gas is a clean and rather abundant fossil fuel that could provide a source of energy as the world makes the transition to a more sustainable energy supply system.

Coal: The Brightest Outlook, the Dirtiest Fuel

Coal is the world's most abundant fossil fuel. Total resources (all the coal in the ground) are estimated at 12,600 billion metric tons; half are believed to be recoverable.[1] The recoverable reserves would last 1700 years at the current rate of consumption. World proven reserves—the amount of coal that has already been located—are estimated to be 786 billion metric tons and will last 200 years at the current rate of consumption.

[1]Some energy analysts think that this estimate exaggerates the reserves by five or six times.

The United States has about 30% of the world's proven coal reserves, or about 225 billion metric tons (250 billion tons) that can be recovered. This could last the country 100 to 200 years, even at increased rates of consumption. It is believed that an additional 360 billion metric tons (400 billion tons) of coal can be recovered in the United States, giving Americans a supply of coal that could last several hundred years.

Global reserves of coal could last 200 years at the current rate of consumption.

Although growth rates in use would greatly reduce the life span of coal, the large supply suggests that coal could be with us for many years to come. Unfortunately, evidence strongly suggests that coal is an environmentally unsustainable fuel. In fact, with the exception of oil shale, coal is the dirtiest of all fossil fuels. Its combustion produces enormous amounts of ash, toxic sludge, particulates, and gaseous air pollutants. Although cleaner and more efficient ways to burn coal are being developed, progress has been slow. As you will learn in Chapter 19, air pollution control technologies and new clean-coal technologies do little to prevent one of the world's most potentially troubling environmental problems: global climate change caused in large part by carbon dioxide emissions from fossil fuel burning facilities, especially coal-fired power plants.

KEY CONCEPTS

Coal supplies in the United States and many other countries are abundant, but evidence suggests that coal combustion is an environmentally unsustainable fuel.

14.5 Nuclear Energy

Nuclear energy provides a growing percentage of the world's energy demands. In Canada, for instance, nuclear power provides 11% of the total energy. In the United States, it satisfies just under 8% of our total energy demand. The nuclear reactors in use today in most nuclear power plants are fueled by naturally occurring uranium-235 (U-235). To understand how nuclear fuel works and some of its dangers, you must understand the atom and radioactivity.

Atoms: The Building Blocks of Matter

All matter is composed of tiny particles called *atoms*. Atoms, in turn, join to form molecules. The **atom** is composed of two parts: a centrally placed nucleus and an electron cloud (**FIGURE 14-15**). The **nucleus** contains two tiny subatomic particles, protons and neutrons, and constitutes 99.9% of the mass of an atom. Protons are positively charged particles. Neutrons have no charge.

Encircling the nucleus is a comparatively large area, the **electron cloud**. It contains much lighter particles of an atom, the **electrons**. These negatively charged particles orbit around

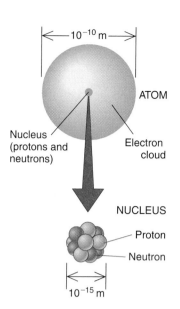

FIGURE 14-15 **Structure of the atom.** Atoms consist of a dense central region, the nucleus, containing protons and neutrons. Electrons are found in the much larger electron cloud, which surrounds the nucleus.

the positively charged nucleus. To give you an idea of the relative size of the electron cloud and nucleus, a simple analogy will suffice. If an atom were the size of Mt. Everest (about 8800 meters or 29,000 feet), the nucleus would be about the size of a football. The rest of the atom would be made of electron cloud.

KEY CONCEPTS

Atoms are the building blocks of all elements. All atoms consist of positively charged nuclei, containing protons and neutrons, surrounded by large electron clouds in which the negatively charged electrons are found.

Radioactive Atoms

If you have ever studied chemistry, even a brief course in high school, you learned that atoms form elements, such as carbon, nitrogen, gold, and uranium. An **element** is the purest form of matter. Each element contains one type of atom. Pure carbon, for instance, contains only carbon atoms. Pure gold contains only gold atoms. Even so, not all atoms in an element are identical. Although carbon contains only carbon atoms, some carbon atoms are slightly different from others. These differences are small. They don't affect the chemical properties of the atoms, that is, how they react, but they can have profound consequences that cause some atoms to emit radioactivity.

How do atoms differ? Although all atoms of a given element contain the same number of protons in their nuclei, some contain slightly different numbers of neutrons. For example, all uranium atoms contain 92 protons, but some have 146 neutrons and others 143. These alternate forms are called **isotopes**. To distinguish one isotope from another, scientists add up the protons and neutrons and tack the sum of these two onto the name of the element. For example, the isotope of uranium that contains 146 neutrons is called uranium-238 because it contains 92 protons and 146 neutrons.

Excess neutrons in some isotopes make them unstable. To reach a more stable state, they emit **radiation**, high-energy particles or bursts of energy. Unstable, radioactive nuclei are called **radionuclides** (ray-de-oh-NEW-klides). They occur naturally or can be produced by various physical means. For the most part, the naturally occurring radionuclides are isotopes of heavy elements, from lead (82 protons in the nucleus) to uranium (92 protons in the nucleus).

Radiation

Radionuclides emit three types of radiation: alpha particles, beta particles, and gamma rays. **Alpha particles** are the heaviest of the three forms of radiation. Each alpha particle consists of two protons and two neutrons, the same as a helium nucleus. Alpha particles are positively charged because they contain protons.

Because alpha particles have the largest mass of all forms of radiation, they travel only a few centimeters in air. They can easily be stopped by a thick sheet of paper, so it is easy to shield people from this form of radiation. In the body, alpha particles can travel only about 30 micrometers (30 millionths of a meter, or about the width of three cells) in tissues. They cannot penetrate skin and are therefore often erroneously assumed to pose little harm to humans. But if alpha emitters enter body tissues—say, through inhalation—they can do serious, irreparable damage to nearby cells and their chromosomes.

Beta particles are negatively charged particles emitted from nuclei. They are equivalent to electrons, except they are more energetic. Beta particles are ejected from nuclei when neutrons in the nucleus are converted into protons, a process that helps stabilize radionuclides.

The beta particle is much lighter than the alpha particle and can travel much farther. It can penetrate a 1-millimeter lead plate and can travel up to 8 meters (27 feet) in air, but only 1 centimeter (0.4 inch) in tissue. Beta particles from some radionuclides have enough energy to penetrate clothing and skin but generally do not reach underlying tissues. They can, however, damage the skin and eyes (causing skin cancer and cataracts).

Gamma rays are a high-energy form of radiation with no mass and no charge, much like visible light but with much more energy. Gamma rays are emitted by nuclei to achieve a lower-energy, more stable state. They are often emitted after a nucleus has ejected an alpha or beta particle because even the loss of these particles does not always allow the nucleus to reach its most stable state. Some gamma rays can travel hundreds of meters in the air and can easily penetrate the body. Some can penetrate walls of cement and plaster or a few centimeters of lead.

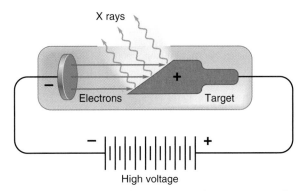

FIGURE 14-16 Anatomy of an X-ray machine. Electrons strike a tungsten filament (red), causing it to release high-energy X rays.

Another common form of radiation is the X-ray. Unlike the three previously discussed forms, X-rays do not originate from naturally occurring unstable nuclei. Rather, X-rays are produced in X-ray machines. An X-ray machine contains a source of electrons, which are emitted when the machine is turned on, and a tungsten collecting terminal in a vacuum tube (**FIGURE 14-16**). Electrons strike the collecting terminal or target, collide with tungsten atoms, and are then rapidly brought to rest. The energy they carried in is released in the form of X-rays, which behave like gamma rays but have considerably less energy. Because of this they cannot penetrate lead.

All the forms of radiation described above are called **ionizing radiation** because they possess enough energy to rip electrons away from atoms, leaving charged atoms called **ions**. Ions are the primary cause of damage in tissues. Ionizing radiation differs from electromagnetic radiation such as radiowaves and magnetic fields, which are weak and unable to ionize materials.

Understanding Nuclear Fission

With this background material, we continue our exploration of nuclear energy. As noted earlier, nuclear fuel contains U-235. Nuclear fuel does not burn like fossil fuels, but rather undergoes fission. **Fission** is the splitting of atoms. Atoms don't split spontaneously, though. Rather, inside a nuclear reactor U-235 atoms undergo fission when they are struck by neutrons. When they split, uranium atoms give off enormous amounts of energy (**FIGURE 14-17**). In fact, the complete fission of 1 kilogram (2.2 pounds) of U-235 could yield as much energy as 2000 metric tons (2200 tons) of coal! Atomic fission takes place inside special devices, **fission reactors**, designed to contain the energy and radioactivity

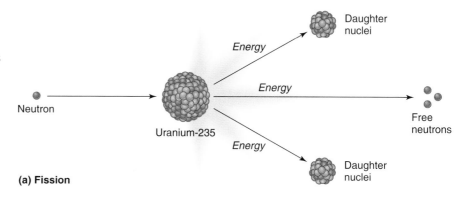

FIGURE 14-17 Nuclear fission.
(a) In a fission reaction a uranium-235 nucleus struck by a neutron splits into two smaller nuclei. Neutrons and enormous amounts of energy are also released.
(b) A chain reaction is brought on by placing fissile uranium-235 into a nuclear reactor. Neutrons liberated during the fission of one nucleus stimulate fission in neighboring nuclei, which in turn release more neutrons. Thus, the chain reaction can be sustained.

(a) Fission

released during fission. As illustrated in **FIGURE 14-18**, U-235 is housed in long **fuel rods** that are located in the **reactor core**. U-235 atoms naturally emit neutrons that bombard other uranium atoms, causing them to split. The heat released during fission is then transferred to water that bathes the fuel rods in the reactor core. As Figure 14-18 shows, the heated water around the reactor core heats water in another closed system. In the latter, hot water is converted to steam, which drives a turbine that generates electricity. The steam is then cooled and the water is used again. Most nuclear plants are cooled by water and are called **light water reactors** (**LWRs**). Other reactors use coolants such as liquid sodium but operate on the same principle.

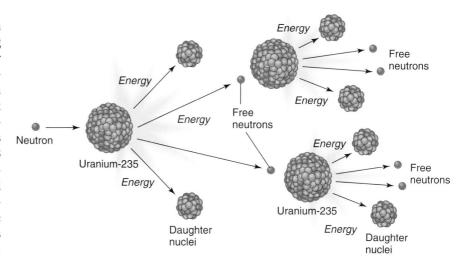

(b) Chain reaction

Fission By-Products When a U-235 nucleus splits, it produces two smaller nuclei, called **daughter nuclei** or **fission fragments** (Figure 14-17). Over 400 different fragments can form during uranium fission, many of them radioactive. When U-235 nuclei split, they also release neutrons, which strike other nuclei in the fuel rods, creating a chain reaction. However, the chain reaction is controlled to keep it from producing so much heat that the reactor core melts. (An atomic explosion in such a case would be unlikely because the fuel is not sufficiently concentrated.)

Fission reactions are held in check by water that bathes the fuel rods. Water absorbs some of the neutrons, thus reducing the rate of fission. Fission reactions are also held in check by the **control rods**, slender rods of neutron-absorbing materials (boron) that lie between the fuel rods. Raising or lowering the control rods regulates the rate of fission. When the control rods are completely lowered, the reactor is shut off.

The entire assemblage of fuel and control rods and their liquid coolant are housed in a 20-centimeter- (8-inch-) thick steel **reactor vessel**. It is surrounded by a huge shield, and the entire unit is contained in a 1.2-meter (4-foot) thick cement shell, the **reactor containment building**.

KEY CONCEPTS

Uranium atoms that undergo fission release additional neutrons, causing additional fission and heat. The chain reaction in a nuclear reactor is kept from running rampant by bathing the reactor core with water, by using control rods, and by maintaining the proper concentration of the fuel in the fuel rods.

Nuclear Power: The Benefits

When nuclear power was first proposed, proponents claimed that its energy would be so cheap that it wouldn't be cost-efficient to meter houses. That dream has failed to materialize. In the United States, for instance, nuclear power costs about 10 to 15 cents per kilowatt-hour, two to three times the cost of electricity from coal and wind farms (Chapter 15). Building a nuclear power plant is two to six times more expensive than building an equivalent coal-fired power plant. Nevertheless, in 2004,

> In the United States, nuclear power supplies about 20% of the total electrical demand.

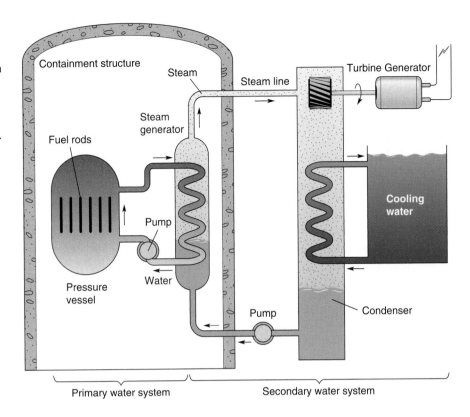

FIGURE 14-18 Anatomy of a nuclear power plant. As shown here, nuclear fission reactions in the reactor core heat up water to generate steam. The steam runs a turbine that generates electricity, as in conventional power plants. The water surrounding the reactor core circulates in a closed system. Heat is transferred to another closed system that generates steam. This double water-heating system helps prevent the escape of radioactivity from the system.

Labels in figure: Containment structure; Steam; Steam line; Turbine Generator; Steam generator; Fuel rods; Cooling water; Pump; Water; Pressure vessel; Pump; Condenser; Primary water system; Secondary water system

the United States' 104 nuclear power plants provided approximately 764 billion kilowatt-hours of electricity—about 20% of the nation's total. Nuclear power also fits well into an electrical grid system that provides electricity to large numbers of people.

Perhaps the most convincing argument for using nuclear power, as opposed to coal and oil, is that it produces very little air pollution. Studies show that the release of radioactive materials into the atmosphere from nuclear power plants is insignificant under normal operating conditions. In fact, a coal-fired power plant may release more radioactivity than a normally operating nuclear power plant. Moreover, nuclear plants do not produce sulfur dioxide and nitrogen dioxide, which are released from fossil fuel combustion sources and are converted into strong acids in the air. Carried to the Earth in rain and snow, and on particulates, these acids can cause considerable damage to the environment (Chapter 20). Nuclear power plants also produce no carbon dioxide. Another advantage of nuclear power is that it requires less strip mining than coal because the fuel is a much more concentrated form of energy. This results in less land disturbance and fewer impacts on groundwater, wildlife habitat, and so on to produce a given amount of electricity. Also, the cost of transporting nuclear fuel is lower than that for an equivalent amount of coal.

> **KEY CONCEPTS**
> Nuclear power has many redeeming qualities. Although it is the most expensive of the major sources of electricity, it fits well into the established electrical grid and produces very little air pollution.

Nuclear Power: The Drawbacks

Despite its many advantages, nuclear power has some substantial drawbacks, including (1) waste disposal problems; (2) contamination of the environment with long-lasting radioactive materials from accidents at plants and during transportation; (3) thermal pollution from power plants; (4) health impacts; (5) limited supplies of uranium ore; (6) low social acceptability; (7) high construction costs; (8) a lack of support from insurance companies and the financial community; (9) vulnerability to sabotage; (10) proliferation of nuclear weaponry from high-level reactor wastes; and (11) questions about what to do with nuclear plants after their useful life of 20 to 25 years is over. This section examines four major areas of concern: reactor safety, waste disposal, social acceptability and cost, and the proliferation of nuclear weapons. Before we examine these, however, a few words on the effects of radiation are in order.

> **KEY CONCEPTS**
> Interest in nuclear power has declined substantially because of major problems, among them questions over reactor safety, unresolved waste disposal issues, low social acceptability, and high costs.

Health Effects of Radiation   All forms of radiation strip electrons from atoms in biologically important molecules in tissues, creating ions. It is these charged molecules (ions) that damage cells. Radiation is like many other harmful substances. Its effect is determined by the dose—how much one receives—and the length of exposure. The higher the dose, the greater the effect. The longer the exposure, the greater the potential for damage.

Numerous studies have revealed some interesting generalizations about radiation: First, fetuses are more sensitive to radiation than children, who are in turn more sensitive than adults. Second, cells undergoing rapid division appear to be more sensitive to radiation than those that divide infrequently or not at all. Some of the most active—and thus most susceptible—tissues are those involved in the formation of blood cells and in immune protection. Epithelial (ep-eh-THEEL-ee-el) cells—those that line body organs such as the intestines—also undergo frequent division and are highly sensitive to radiation. In sharp contrast, nerve and muscle cells, which do not divide, have a very low sensitivity and rarely become cancerous. Third, most if not all forms of cancer can be increased by ionizing radiation.

Health experts divide radiation exposure into two categories, high and low.

Health Impacts of High-Level Radiation   The highest exposures to radiation cause death within a few hours or days. High-level doses that do not result in immediate death cause **radiation sickness**. Nausea and vomiting are the very first symptoms and develop soon after exposure. Several days later, diarrhea, hair loss, sore throat, reduction in blood platelets (needed for clotting), hemorrhaging, and bone marrow damage occur as a result of damage to fast-dividing cells. Although people survive sublethal exposures, many serious delayed effects result, including cancer, leukemia (a form of blood cancer), cataracts, sterility, and decreased life span. Pregnant women who are exposed to sublethal doses suffer from increased miscarriages, stillbirths, and early infant deaths.

Fortunately, high-level radiation exposures are rare. Such exposures would occur only in individuals working at or living very near nuclear power plants or munitions factories where a major accident has occurred. Nuclear war, even on a limited scale, however, would also expose large numbers of people to dangerously high levels of radiation.

> Estimates suggest that a major accident will occur at a nuclear power plant every 20 years or so.

Health Impacts of Low-Level Radiation   The effects of low-level radiation are often delayed and are therefore harder to determine with certainty. A growing body of evidence shows that low-level radiation increases the likelihood of developing several types of cancer, such as leukemia (FIGURE 14-19). Studies of uranium miners, workers who painted watch dials with radium early in this century, employees at nuclear weapons factories, radiologists, and X-ray technicians show that low-level exposure increases the incidence of many forms of cancer.

Studies suggest that low-level radiation is much more harmful than many scientists thought a decade and a half ago.

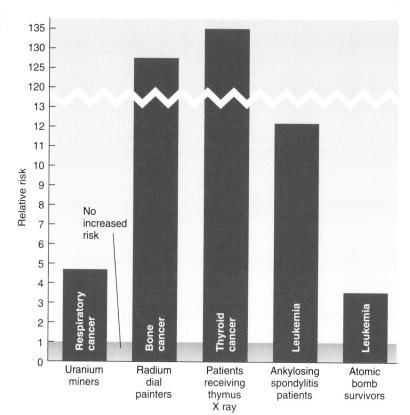

FIGURE 14-19 **Cancer risk.** This graph shows the relative risk of people exposed to radiation. Relative risk is a measure of the probability of developing cancer. As shown here, a uranium miner is four times more likely to develop respiratory tract cancer than someone who is not a miner. Atomic bomb survivors, in general, are 3.5 times more likely to develop leukemia than nonirradiated people.

Noted radiation biologist Dr. Irwin Boss, for example, estimates that low-level exposure is about 10 times more harmful than previously calculated. He and others have called for major revisions of the maximum allowable doses for workers.

Bioaccumulation   One problem with radionuclides is that they are often concentrated in certain tissues, where they can cause considerable damage. For example, iodine-131, which is released from nuclear power plants both during normal operations and in accidents, is deposited on the ground around a power plant. Fallout may be incorporated into grass eaten by dairy cows and passed to humans via milk. It is then selectively taken up by the human thyroid gland, where it irradiates cells and may produce tumors. Milk contaminated with I-131 is especially harmful to children.

Strontium-90 is released during atomic bomb blasts. It may also be released from reactors—in small amounts under normal operating conditions but in large quantities in accidents. Strontium-90 is readily absorbed by plants and may also be passed to humans through cow's milk. It seeks out bone, where it is deposited like calcium. With a half-life of 28 years, it irradiates the bone and can cause leukemia and bone cancer.[2]

Some radionuclides may be biologically magnified in the higher trophic levels of food chains, a phenomenon discussed in Chapter 18.

Reactor Safety   With this background information in mind, we turn to the issue of reactor safety. An accident at a nuclear power plant at Three Mile Island, Pennsylvania, in 1979 heightened awareness of the dangers inherent in this form of energy. This incident was caused by a malfunctioning valve in the cooling system, which triggered the worst commercial reactor accident in U.S. history. Radioactive steam poured into the containment building. Pipes in the system burst, releasing radioactive water that spilled onto the floors of two buildings. Some radiation escaped into the atmosphere, and some was dumped into the nearby Susquehanna River. The accident then took a turn for the worse. Hydrogen gas began to build up inside the reactor vessel, threatening to expose the reactor core, which would cause a meltdown. Although the gas buildup was slowly eliminated, thousands of area residents had to be evacuated. Photographs of the core showed that a partial meltdown had occurred.

[2]*Half-life* is the time it takes for a radioactive material to decrease to half its original mass, a consequence of radioactive decay.

The accident at Three Mile Island had many long-term effects. It cost the utility (and its customers) many millions of dollars to replace the electricity the plant would have generated. Even more money (over $1 billion) was needed for the cleanup.

Although utility officials claimed that the accident would cause few cancers, John Gofman and Arthur Tamplin, radiation health experts, estimated that the exposure to low-level radiation that residents received for 100 hours or longer would cause at least 300, and possibly as many as 900, fatal cases of cancer or leukemia. So far, evidence of elevated cancer rates has not been seen.

A 1975 study on reactor safety called the Rasmussen report (after its principal author, Norman Rasmussen of MIT) showed that the probability of a major accident at a nuclear power plant was about 1 in 10,000 reactor-years. (A reactor-year is a nuclear reactor operating for 1 year; for example, 10 reactors operating for 20 years are 200 reactor-years.) If 5000 reactors were operating worldwide, as advocates once proposed for the year 2050, Rasmussen's prediction means that we could expect a major accident every other year! Each accident might cause between 825 and 13,000 immediate deaths, depending on the location of the plant. In addition, 7500 to 180,000 cancer deaths would follow in the years after the accident. Radiation sickness would afflict 12,000 to 198,000 people, and 5000 to 170,000 genetic defects would occur in infants. Property damage could range from $2.8 billion to $28 billion.

In 2002, 442 nuclear reactors were operating worldwide, and with dozens more under construction or soon to be built. If the Rasmussen report estimate is correct, we can expect a major meltdown every 20 years or so.

Because the Rasmussen study failed to include such possibilities as sabotage and human error, it has been discredited by the Nuclear Regulatory Commission, which argues that it underestimates the real risk. Human error could result in an even higher incidence of accidents. As a case in point, many experts believe that the accident at Three Mile Island was worsened by operator confusion, which turned a small problem into a disaster. Many people argue that misjudgment and performance errors among personnel could negate technological improvements designed to make plants safer. Human errors and oversight can also occur during construction.

The accident at Chernobyl in the former Soviet Union in 1986 reinforced concerns about the possibility of human error (**FIGURE 14-20**). Releasing large amounts of radiation onto farms and cities throughout Europe, the accident was caused by plant operators who were running some tests on the reactor. The amount of cooling water flowing through the reactor core fell rapidly, and—because the operators had deactivated several backup systems—the temperature of the 200 tons of uranium in the reactor's fuel rods soared as high as 2800°C (5000°F), twice the temperature required to melt steel. An enormous steam explosion blew the roof off the building. Flames from 1700 tons of burning graphite (a neutron-absorbing agent in the core) shot 30 meters (100 feet) into the air. While firefighters risked their lives to contain the

FIGURE 14-20 The remnants of human error. This photograph shows the Chernobyl nuclear reactor soon after the 1986 explosion that blew off the top of the reactor building. This reactor and many others in the former Soviet Union were not built with suitable containment.

disaster by spraying water down on the reactor core from the roofs of nearby buildings, the uranium fuel melted, spewing radioactive isotopes into the atmosphere.

Soon after the accident, 135,000 people living within a 30-kilometer (18-mile) radius of the plant were relocated. An additional 65,000 residents were evacuated in the months following the accident from areas as far as 330 kilometers (200 miles) from Chernobyl. Many of those living closest to the nuclear power plant will never be able to return to their homes.

Besides losing their homes and their possessions, tens of thousands of Soviets may have been exposed to high levels of radiation. Estimates of human exposure near Chernobyl indicate that whole-body radiation for persons in the immediate area of the plant was 4 to 20 times higher than the allowable exposure for U.S. nuclear workers. Although no one knows for certain, it is predicted that about 1 of every 10 people, or about 15,000 people exposed to radiation from Chernobyl in the 30-kilometer radius, will die from cancer. The government of Ukraine asserts that Chernobyl has already caused 6000 to 8000 deaths, suggesting that the esti-

mate of 15,000 deaths may be too low. The accident at Chernobyl also caused a number of immediate deaths from radiation poisoning—31 within the first 4 months. Workers who were saved by heroic procedures are likely candidates for cancer. New studies show that the incidence of thyroid cancer has increased dramatically in the Ukraine as a result of exposure to radioactive iodine.

Radiation also spread throughout Europe. One study shows that at least 13,000 people outside of the former USSR will die from cancer in the 50 years following the accident. A study by Dr. John Gofman of the Committee for Nuclear Responsibility suggests that the number may be much higher—about 500,000 to one million. The radiation was greatly diluted by the time it hit the United States, and estimates suggest that only 10 to 20 Americans will die from cancer from the Chernobyl accident. That's not a lot—unless, of course, you're one of the victims.

Besides exposing large numbers of people to potentially harmful radiation, the Chernobyl accident threatened crops, farmland, and livestock in Ukraine, the breadbasket of the former USSR. By some estimates, up to 150 square kilometers (60 square miles) of land is so contaminated that it cannot be farmed for decades. Agriculture outside the former USSR also suffered from the ill effects of radiation. Soon after the accident, for example, Italian officials turned back 32 freight cars loaded with cattle, sheep, and horses from neighboring Austria and Poland because of abnormally high levels of radiation. Radioactive fallout from the accident also settled on Lapland, an expanse of land encompassing northern Sweden, Norway, Finland, and the northwestern part of the former USSR. Lapland is occupied by a seminomadic people who raise reindeer for food. The reindeer feed on lichen, which was contaminated by radiation after the accident. So heavily contaminated was the meat that Laplanders were forced to round up and slaughter their herds.

Critics are also concerned with unforeseen technical difficulties. A hydrogen bubble at the Three Mile Island plant, for example, took the experts by surprise. Numerous backup systems in nuclear power plants are designed to prevent a meltdown and the release of radiation, but as the accident at Three Mile Island showed, plants are not invulnerable.

Clouding the issue of reactor safety is the possibility of terrorism. In 1975, two French reactors were bombed. Nuclear power plants could become targets of similar attacks. Damage to the cooling system could result in a meltdown with radiation leakage. Most plants are vulnerable to attack. Protection from ground and air assaults may be impossible. Even though security has been improved at many plants, the threat of well-planned terrorist actions cannot be ignored.

To address the question of reactor safety, the nuclear industry is promoting new, smaller designs that they claim are much safer than present models. Skeptics question whether this is true and point out that even if it is, making plants safer addresses only one of a dozen problems—among them, what to do with the enormous amounts of radioactive waste produced by a nuclear power plant.

Accidents at nuclear power plants are a major concern to those involved in the nuclear debate, but so are accidents at fuel processing plants—facilities that make nuclear fuel. In September, 1999, Japanese workers set off an uncontrolled chain reaction at a processing plant in Tokaimura when they accidentally poured seven times more uranium into a purification tank than they were supposed to. The reaction lasted about 20 hours and released large quantities of radioactivity into the plant and the neighboring environment. Sixty-three individuals were exposed to high levels of radiation and two are expected to die because of high-level exposure.

Accidents involving trucks and trains that transport uranium ore, yellow cake (processed ore), and uranium wastes also occur. Surprisingly common, accidents in the transportation sector generally result in the release of contaminants over a relatively small area. Special transport containers and other precautions are taken to minimize the potential for the release of nuclear wastes. As an added precaution, some cities actually ban trucks containing such wastes.

KEY CONCEPTS

Several major accidents at nuclear power plants have raised awareness of the potential damage a small mechanical or human error might cause. Estimates suggest that many additional accidents are bound to occur in the future, with costly social, economic, and environmental impacts.

Waste Disposal   Nuclear power plants, nuclear weapons facilities, mines that extract uranium ore, and processing facilities that purify it produce enormous amounts of radioactive waste (**FIGURE 14-21**). For years, much of this waste has been improperly disposed of. In the early days of nuclear power, for example, millions of tons of radioactive waste were dumped on or near uranium mills, sometimes on the banks of rivers. Low-level wastes from power plants and other facilities (hospitals, laboratories) were buried in shallow landfills. In some instances, they were buried in boxes or steel drums that leaked radioactive materials into groundwater and surface waters. Until the 1960s, low- and medium-level radioactive wastes were often mixed with concrete, poured into barrels, and dumped at sea—a practice that continues today off the coast of England. In at least one location— Richland, Washington, home of Hanford nuclear weapons facility—wastes were stored in underground tanks that have begun to leak high-level radioactive materials into the ground. Since it began operating, the plant has produced several hundred million liters of highly radioactive liquid waste. Since 1958, at least 2 million liters (500,000 gallons) of waste have leaked out of tanks into the soil and, possibly, the groundwater. Approximately 150 million liters (38 million gallons) of waste is still in storage tanks that could leak.

Low-level wastes are hazardous for about 300 years, and high-level wastes can be dangerous for tens of thousands of years. The half-life of one highly radioactive waste product of nuclear reactors, plutonium-239 (Pu-239), is 24,000 years. As a rule, it generally takes about eight half-lives for a material to be reduced to 0.1% of its original mass,

FIGURE 14-21 Inactive uranium mine. To acquire large amounts of uranium for nuclear power plants and weapons production, companies extract uranium ore from huge mines like this one. Enormous amounts of material must be removed. Radioactive waste can be washed from the site by rainwater or blown away by wind.

at which point it is considered safe. For Pu-239, this is about 200,000 years. By comparison, the Egyptian pyramids are only about 4500 years old!

High-level nuclear wastes are increasing rapidly. In 1995, there were 28,000 metric tons (31,000 tons) of high-level radioactive waste being held at nuclear power plants, awaiting a permanent disposal site. By 2005, the amount had increased to approximately 52,000 metric tons (55,000 tons). In 1982, the U.S. Congress passed a law requiring the Department of Energy (DOE) to create a suitable site to build a disposal facility for such waste. After eliminating a number of candidates, DOE selected a remote site in Nevada (Yucca Mountain). To date, DOE has spent over $2 billion studying the area. The agency may ultimately spend another $5 to $6 billion to finish the project, which it hopes will open around 2010—well past the original opening date.

Meanwhile, high-level wastes are building up at nuclear power plants and weapons facilities throughout the United States. By 2015, this high-level nuclear waste is expected to total nearly 75,000 metric tons (83,000 tons). Clearly, something needs to be done about these wastes and the radioactive waste currently held at nuclear power plants. (For a discussion of radioactive waste, see Chapter 23.) The Point/Counterpoint in this chapter presents both sides of the current debate over the Yucca Mountain site.

Yucca Mountain Repository: No Place in Nevada

Dina Titus

Dina Titus is the senate minority leader in Nevada and is also a professor of political science at the University of Nevada, Las Vegas. She is author of *Bombs in the Backyard: Atomic Testing and American Politics.*

After World War II, the U.S. government needed a continental site for testing atomic weapons which was remote, sparsely populated, meteorologically predictable, and preferably under federal jurisdiction. They soon found such a setting in the southern Nevada desert, some 65 miles northwest of Las Vegas. Now almost 5 decades later, the federal government needs a similar home for another, even more controversial nuclear project—storing high-level radioactive waste—and once again Washington is looking toward Nevada.

Unlike the selection of Nevada for atomic weapons testing, the decision to store nuclear waste at Yucca Mountain has been a protracted battle in which politics tends to triumph over science. Through a series of legislative actions, including the 1989 "Screw Nevada" bill, which singled Nevada out as the only site for further study, and S.104, originally introduced by Senator Murkowski in 1997 to create a temporary storage site in Nevada pending final approval of the Yucca Mountain facility, Congress has attempted to force high-level nuclear waste down Nevada's throat. In response, a great majority of the citizens and elected officials have voiced vehement opposition to the Yucca Mountain repository, which they feel poses great public health and safety risk.

The arguments are straightforward. First, having dealt with the AEC/DOE for almost 50 years, Nevada is skeptical of the agency's priorities. Good scientific knowledge consistently takes a backseat to expedience, expense, and political extortion. No geologic fault, water table, earthquake, or wind pattern seems problematic enough to disqualify the site; instead, accommodations are simply made to cope with each new "challenge" as it is discovered.

Furthermore, with only four representatives in Congress, Nevada lacks the political clout to fight the NIMBY syndrome exhibited by the other states that either want to get rid of their own waste or avoid being chosen as the site for a future repository. For this reason, and because the DOE sees the people of Nevada as being malleable because they are "accustomed to living with radiation," the state makes an easy political target.

Nevadans may indeed be accustomed to "bombs in their backyards," but we have not been won over by the DOE's extensive PR campaign. On the contrary, we feel that Nevada

has done its share for the nation in the area of nuclear policy. What's more, Nevada has no nuclear reactors and does not benefit from nuclear power. Therefore, we feel it is unfair to force the state to accept the nation's nuclear waste.

Resistance to the repository also stems from a need to protect the state's economy, which is based primarily on tourism and gaming. Las Vegas is the playground of the world and is fast becoming a destination resort for American families. The negative impact of a "dump site," with highly hazardous materials being shipped through the city and stored nearby, could be devastating to this vital source of revenue, which contributes over 60% of the state's general fund.

All of Nevada's opposition to the repository is not so egocentric, however. We believe that a central storage facility is not necessary at this time. Nuclear waste is currently stored on site at 109 reactors, and dry cask technology is available for additional storage at these locales. The Nuclear Waste Technical Review Board, created to oversee DOE, agrees that there is no need to rush forward with a central repository that is lacking in solid scientific underpinnings.

Second, other states besides Nevada will be adversely affected by the establishment of a central repository because of the need to transport the dangerous waste, primarily from the East Coast, to Nevada. Nearly 18,000 shipments of thousands of tons of waste will have to travel through 43 states to reach Nevada. The probability of an accident is astronomical, and there is no way to prepare for the resulting disaster. Such a potentially devastating public health hazard prompted Senator Richard Bryan to dub the plan a "mobile Chernobyl."

Finally, Yucca Mountain is being studied on the basis of a 10,000-year scenario even though the half-lives of the substances involved are much longer. Clearly, there is no scientific basis for such a limitation; and the proposed provision which requires meeting safety standards for only 1000 years is even more outrageous. We cannot afford to be so callous. We have a moral obligation to protect the health and safety of present and future populations on this planet, and we must resist efforts by the nuclear power industry and its sympathizers to focus only on today's bottom line.

You can link to websites that represent both sides through Point/Counterpoint: Furthering the Debate at this book's internet site, http://environment.jbpub.com/. Evaluate each side's argument more fully and clarify your own opinion.

What is the better policy—allowing high-level radioactive spent fuel and waste from electricity generation and defense to pile up near homes and schools at 129 sites in 39 states, or containing materials in a more remote, less populated location?

On April 15, 1997, senators from both political parties chose the latter option. They voted 65–34 to approve bipartisan legislation calling for safe interim storage at a secured federal facility in the Nevada desert.

This decision came only after years of professional scrutiny. Following regional studies in 36 states, scientists focused on three sites most likely to be suitable for an underground repository. Each was located in relatively arid, remote locations—but one, the Nevada Test Site's Yucca Mountain, was deemed the best choice for further geologic characterization and intensive scientific study.

At the western edge of a site used for nearly 1000 nuclear bomb tests, Yucca Mountain has been subjected to billions of dollars worth of tunneling and testing. With much of this testing complete, it is likely scientists will soon judge Yucca Mountain to be suitable as a permanent repository. If so, an interim waste storage site will be needed at the Nevada Test Site as a staging area for eventual waste placement.

Even if Yucca is judged to be unsuitable, something the Clinton Administration admitted is highly unlikely, we will *still* need a central interim storage facility. By 2010, 65 reactors in 29 states will run out of space for used fuel. Since Americans have already paid $13 billion through their electricity bills into a fund in exchange for the government's guarantee it would take the used fuel by 1998, it's unfair to ask them to pay more for onsite interim storage. And Nevadans, like most other Americans, rely to some extent on nuclear power.

We need interim storage, and we need it soon—regardless of what happens at Yucca Mountain. It could take decades to locate and test a new site, and those materials must be safely contained in the meantime.

Understandably, nobody wants to host an interim facility or geologic repository, regardless of how well conceived or designed it may be. The Nevada Congressional Delegation, in a hearing before my committee, admitted it was politically necessary for them to fight storage and disposal in Nevada *even if Yucca Mountain was determined by scientists to be fully suitable for such a facility*. Other politicians in their shoes would probably do the same.

Putting politics aside, however, the U.S. Nuclear Regulatory Commission contends it is far safer to manage and store spent nuclear fuel and other high-level radioactive wastes—currently located at nearly 129 sites in 39 states—at a single, centralized, monitored location. Whether we like it or not, the waste must go somewhere. Even if the portion of this nation's electricity generated by nuclear plants—almost one quarter—vanished tomorrow, we would still be left with

Yucca Mountain: The Scientists' Choice

Frank H. Murkowski
Senator Frank H. Murkowski, a Republican from Alaska, was Chairman of the Senate Committee on Energy and Natural Resources. Elected in 1980, Murkowski is now the governor of Alaska.

spent fuel and defense waste. And the waste in shut-down reactors is more expensive and problematic to maintain.

Critics of interim storage and the geologic repository have worked to incite fear over transportation of used fuel by raising the specter of a "mobile Chernobyl," as if transportation of radioactive wastes and other hazardous materials along our highways and railroads was something new and dangerous.

In truth, the nuclear industry has shipped more than 2400 shipments of used nuclear fuel over the past 30 years *without a single accident resulting in the release of radioactive materials*. Moreover, tests conducted at Sandia National Laboratory have confirmed that transportation casks certified to carry used fuel will safely contain its radioactive cargo in *any conceivable* accident scenario. By comparison, the chemical industry responds to 20 toxic chemical spills, explosions, and releases *each day* according to the U.S. Public Interest Research Group.

The Department of Energy is importing radioactive spent fuel from foreign countries to be transported and stored in the United States. At least one major environmental lobby supports that move. The *Contra Costa Times* in California quoted a physicist with the Natural Resources Defense Council as saying those shipments would not pose a significant safety risk and could not explode like Chernobyl. What is the difference between foreign and domestic waste?

In confronting the need to safely dispose of nuclear waste, there are no insurmountable scientific, technical, or financial issues—the roadblocks are purely political. Legislation currently being considered by Congress would break this stalemate, provide for the continued safe transportation of nuclear materials, enact new safeguards, and allow the Environmental Protection Agency and the Nuclear Regulatory Commission to protect public health, safety, and the environment of present and future generations.

Critical Thinking Questions

1. List the main arguments for and against the use of Yucca Mountain as a nuclear waste repository.
2. Do you sympathize with one viewpoint more than the other? Why?

Social Acceptability and Cost   Two of the most important factors controlling the future of nuclear energy are its social acceptability and its cost. In 1989, for example, voters in California passed a public referendum calling for the closure of the Rancho Seco nuclear power plant. Internationally, several countries, including Sweden and Italy, have voted to phase out nuclear power. Why is nuclear power so unpopular these days?

Besides safety concerns and concern over waste disposal, nuclear power plants are expensive to build and operate. A new nuclear power plant today costs between $2 and $8 billion, compared with $1 billion for an equivalent coal-fired power plant. Costs are high because of strict building standards, expensive labor, construction delays, and special materials needed to ensure plant safety. Because of the high cost and risk of damage, U.S. banks refuse to finance nuclear power plants, and utilities must turn to foreign investors.

The cost of operating and maintaining a nuclear power plant is also high. Repair costs, for example, are many times higher than those for conventional coal-fired plants. For example, saltwater corrosion of the cooling system in a reactor owned by the Florida Power and Light Company cost over $100 million to repair and $800,000 a day to make up for the lost electricity. A similar problem in a coal-fired power plant would cost a fraction of this amount.

Another factor in the cost equation is retirement costs. According to the Worldwatch Institute, one of six reactors already built has been retired. Their average life span was 17 years, compared with an industry estimate of 20 to 30 years. Dismantling these plants at the end of their life span may cost utilities about $0.5 to $1 million per megawatt, bringing the cost of decommissioning a 1000-megawatt plant to $500 million to $1 billion. This cost will inevitably add to electric bills.

A rash of cancellations has also driven costs up. Even though utilities halt construction of nuclear plants, they must pay back the billions of dollars they borrowed to start construction. Customers inevitably foot most of the bill. Bondholders, who often finance such projects, can also be left in financial ruin, as was the case when the Washington Public Power Supply System defaulted on $2.25 billion in bonds.

High construction and maintenance costs, combined with other factors, make nuclear energy the most costly form of electricity currently available on a large scale. In fact, France, which gets 60 to 70% of its electricity from nuclear plants, is thought to be paying 35 to 60% more for that electricity than it would have for electricity produced from coal.

A lack of public support, exorbitant costs, and a rash of canceled plants in recent years, among other problems, have slowed the growth of the nuclear industry. In fact, the World-watch Institute argues that unless there is an immediate turnaround in orders, nuclear power may be declining soon. Although some countries like Canada still rely heavily on nuclear power and export technology and expertise through a government-owned business, the problems outlined in this section may be so insurmountable that even these nations will eventually phase out of nuclear energy.

The Proliferation of Nuclear Weapons   At least 11 countries have or are suspected of having nuclear weapons. Some of these countries like North Korea and Israel are politically unstable or are in volatile regions where war could easily erupt. Seven countries—the United States, China, the Soviet Union, Britain, France, India, and Pakistan—have already tested nuclear weapons and several dozen more countries have the technical capability to produce nuclear weapons. The plutonium used in atomic bombs comes from special nuclear reactors as well as from conventional reactors. Critics of nuclear power argue that the spread of nuclear power throughout the world will make bomb-grade plutonium more widely available.

Nuclear fuels could be stolen by terrorist groups and easily fashioned into a crude but effective atomic bomb that would fit comfortably into the trunk of an automobile. Strategically planted, a terrorist bomb could prove disastrous. Concern for this danger further dampens public and political support for nuclear power.

Breeder Reactors

The world's supply of the uranium-235 used in light water reactors will last about a hundred years at the current rate of use. Increases in the production of electricity by nuclear power plants, however, could greatly reduce the life span of uranium reserves. Because uranium supplies are limited, the nuclear industry has developed an alternative called the **breeder reactor.** Breeder reactors are fission reactors that are similar in most respects to the light water fission reactor described earlier. However, besides producing electricity, breeder reactors convert uranium-238 (an isotope of uranium not used as a fuel in light water reactors) into reactor fuel, plutonium-239.

In the breeder reactor, fast-moving neutrons from small amounts of plutonium-239 located in the fuel rods strike nonfissionable U-238 placed around the reactor core. This converts U-238 into fissile (fissionable) Pu-239 (**FIGURE 14-22**). Theoretically, for every 100 atoms of Pu-239 consumed in fission reactions, 130 atoms of Pu-239 are produced—hence, the name *breeder*.

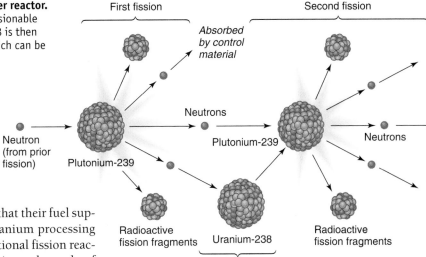

FIGURE 14-22 Nuclear reactions in a breeder reactor. Neutrons produced during fission strike nonfissionable "fertile" materials such as uranium-238. U-238 is then converted into fissionable plutonium-239, which can be used as the reactor in the fuel.

The attraction of breeder reactors is that their fuel supply, U-238, is found in the wastes of uranium processing plants and in the spent fuel from conventional fission reactors. In the United States alone, the estimated supply of U-238 would last 1000 years or more. In addition, breeder reactors could reduce the need for mining, processing, and milling of uranium ore. Moreover, fuel prices might remain stable because of the abundance of uranium-238. The breeder technology does not create chemical air pollution (if we ignore problems from mining and milling).

Breeder reactors have been under intensive development in the United States for well over 30 years. The most popular design is the **liquid metal fast breeder reactor**. It uses liquid sodium as a coolant rather than water. Heat produced by nuclear fission in the reactor core is transferred to the liquid sodium coolant, which transfers this heat to water. The water is then converted to steam that is used to generate electricity.

On the surface, the breeder reactor sounds like the answer to the world's electrical energy needs. Unfortunately, it has numerous problems, some so great that they may make the technology impractical. The most significant problem is that it takes about 30 years for the reactor to break even—that is, to produce as much Pu-239 as it consumes. The fate of breeder reactors hinges on drastically shortening the payback period. The second major problem is the cost: $4 to $8 billion. In addition, breeder reactors have many of the same problems as light water reactors.

Another problem would be the large quantities of plutonium located at breeder reactors. Plutonium is long lived and extremely toxic if inhaled. It burns when exposed to air. Liquid sodium coolant is also dangerous. It reacts violently with water and burns spontaneously when exposed to air. A leak in the coolant system could trigger a catastrophic accident. Should the core melt down, a small nuclear explosion equivalent to several hundred tons of TNT might occur. Rupturing the containment building, the explosion would send a cloud of radioactive gas into the surrounding area.

In 1983, the U.S. Congress canceled funding for an experimental breeder reactor, the Clinch River project in Tennessee. Having spent nearly $2 billion for planning, Congress decided that further investments were unmerited. The prospects of the breeder reactor are currently being assessed in France, which has one fast breeder reactor, and Japan, which has two in operation and two in design or construction phases.

KEY CONCEPTS

Breeder reactors are designed to produce energy and radioactive fuel from abundant nonfissile (nonfissionable) uranium-238 and could greatly increase the supply of nuclear fuels. Unfortunately, breeder reactors are costly to build and take 30 years to produce as much radioactive fuel as they consume.

Nuclear Fusion

The sun is the source of virtually all forms of energy, even the fossil fuels. Although they consist of organic matter, they contain energy that was once sunlight. This energy is captured by plants that are later converted to coal. Oil comes from microorganisms as well. The energy of oil is ancient solar energy captured by plants and passed to animals that fed on them. The sun generates its energy by **nuclear fusion**, which occurs when two hydrogen nuclei join—or fuse—to form a slightly larger helium nucleus. Fusion requires extremely high temperatures to overcome the electrostatic repulsion of the positively charged nuclei. When the hydrogen nuclei do fuse, though, they emit large quantities of energy. Scientists hope to harness fusion reactions and capture this energy to generate electricity (**FIGURE 14-23**).

If scientists can create fusion reactors, nuclear fusion could supply enormous amounts of energy because its fuel supply is also enormous: It's seawater. Energy analyst John Holdren estimates that at current rates of energy consumption in the United States, fusion would meet energy needs for up to 10 million years!

Unfortunately for proponents of this futuristic energy source, fusion has significant drawbacks that may make it commercially unviable. The first of these is that fusion reactions take place at temperatures measured in the hundreds of millions of degrees Celsius. The main obstacle, then, is finding a way to contain such an extremely hot reaction. No known alloy can withstand these temperatures; in fact, metals would vaporize.

To contain fusion reactions, scientists have devised reactors that suspend tiny amounts of fuel in air within a metal

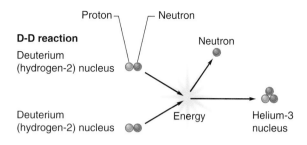

D-D reaction

Deuterium
(hydrogen-2) nucleus

Deuterium
(hydrogen-2) nucleus

Proton — Neutron

Neutron

Energy

Helium-3
nucleus

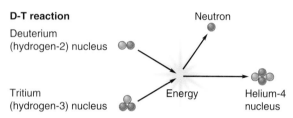

D-T reaction

Deuterium
(hydrogen-2) nucleus

Tritium
(hydrogen-3) nucleus

Neutron

Energy

Helium-4
nucleus

FIGURE 14-23 Two potentially useful reactions. Deuterium and tritium are isotopes of hydrogen atoms.

reactor vessel. The most popular technique in experimental fusion reactors is **magnetic confinement** (FIGURE 14-24). In these reactors, a magnetic field suspends the superheated fuel. Heat given off by the fusion reaction is captured and used to boil water to make steam. Small-scale experimental reactors of this type have been developed and operated in the United States, Europe, the former Soviet Union, and Japan. Despite more than 40 years of research, researchers have failed to reach the break-even point—that is, the point at which they get as much energy out of the system as they put into it. Achieving this goal would be just the first step in a long, costly climb to commercialization.

In the proposed designs, heat released from the fusion reaction would be drawn off by a liquid lithium blanket, which transfers the heat to water. The lithium blanket would also capture neutrons, creating tritium (a hydrogen nucleus containing a proton and two neutrons), which could be extracted and used as additional fuel.

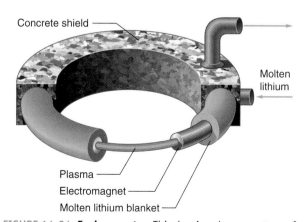

Concrete shield

Molten
lithium

Plasma

Electromagnet

Molten lithium blanket

FIGURE 14-24 Fusion reactor. This drawing shows one type of fusion reactor, in which the fusion reactions occur suspended in an electromagnetic field.

Scientists estimate that a demonstration fusion reactor could be running by the year 2025. Because the U.S. Congress has cut funding, however, this goal is no longer attainable. The cost of a commercial fusion reactor cannot be accurately assessed at this time, but it could cost $12 to $20 billion, making its electricity unaffordable.

Another problem with this technology is the release of radioactivity. The deuterium–tritium fusion reactor is the most feasible type, but tritium is radioactive and difficult to contain. Because of the high temperatures in fusion reactors, tritium can penetrate metals and escape into the environment.

Fusion reactors would produce enormous amounts of waste heat, the ramifications of which are discussed in Chapter 21. Fusion reactions would also emit highly energetic neutrons, which would strike the vessel walls and weaken the metal, necessitating replacement every 2 to 10 years. Metal fatigue could lead to the rupture of the vessel and the release of tritium and molten lithium, which burns spontaneously in air. A leak might destroy the reaction vessel and the containment facilities. Neutrons emitted from the fusion reaction would also convert metals in the reactor into radioactive materials. Periodic maintenance and repair of reactor vessels would be a health hazard to workers, and radioactive components removed from the reactor would have to be disposed of properly.

KEY CONCEPTS

Enormous amounts of energy are produced when small atoms fuse. Fusion reactors could be powered by plentiful fuel sources and could supply energy for many millions of years. However, technical problems, costs, and environmental hazards present major obstacles to the commercial development of this form of energy.

14.6 General Guidelines for Creating a Sustainable Energy System

The previous sections show the wide range of nonrenewable options that exist for forging an energy future. But in choosing options we must select those that are socially, economically, and environmentally sustainable. Creating a sustainable energy future requires us to consider a number of factors—many of which were introduced in the material presented in this chapter. The following rules will help you judge various energy options that you have already learned about and also ones you will encounter in the next chapter.

First, energy resources should have a positive net energy yield. That is, energy we get out of an energy resource must exceed the energy we invest in its exploration, extraction, transportation, and so on. The higher the yield, the better.

Second, energy supplies should be matched with energy demand. Energy comes in a variety of forms. Each of these forms has a different **energy quality**, a measure of the

Table 14-2

Energy Quality of Different Forms of Energy

Quality of Energy	Form of Energy
Very high	Electricity, nuclear fission, nuclear fusion[1]
High	Natural gas, synthetic natural gas (from coal gasification), gasoline, petroleum, liquefied natural gas, coal, synthetic oil (from coal liquefaction)
Moderate	Geothermal, hydropower, biomass (wood, crop residues, manure, burnable municipal refuse), oil shale, tar sands

[1]Workable nuclear fusion reactors do not yet exist. Even if they were technically operable by the end of the first quarter of the 21st century, they would probably remain economically unfeasible.

amount of available work one can get out of it (Table 14-2). Oil and natural gas, for instance, are highly concentrated energy resources. When burned, they produce large amounts of heat useful for certain tasks, such as melting steel. Sunlight is a lower quality form. Streaming through the south-facing windows of a house, it produces heat appropriate for warming homes and offices. Using energy wisely means employing low-quality energy sources (sunlight) for tasks that call for it (home heating) and high-quality energy (oil) for appropriate tasks (running cars).

Third, options should be chosen that minimize pollution and maximize public and environmental safety.

Fourth, options should be selected that are abundant and renewable. Abundant, renewable resources maximize returns on money invested in research, development, and commercialization.

Fifth, options should be pursued that are the most affordable both now and in the future.

KEY CONCEPTS

Although there are many energy options at our disposal, not all are sustainable. Creating a sustainable energy future will require a careful analysis of options for such factors as net energy yield, specific needs, efficiency, environmental impacts, abundance and renewability, and affordability.

REFERENCES AND FURTHER READING

The References and Further Reading section at the end of this book contains a list of sources for the information discussed in this chapter and recommendations for further reading.

14.7 Establishing Priorities

With the energy trends described in this chapter and the guidelines for a sustainable energy strategy just listed, we can establish a list of goals and actions needed to meet our future energy demands.

- **Action 1:** Greatly improve the energy efficiency of all machines, homes, appliances, buildings, factories, motor vehicles, airplanes, and so on. This action stretches fossil fuel supplies and gives us more time to make a transition to pursue other options.
- **Action 2:** Find clean, renewable replacements for oil (which provides transportation fuel and raw materials to make plastics) because its supplies are limited. Also find replacements for coal (primarily used to generate electricity) because it is such an environmentally costly fuel.
- **Action 3:** Find a replacement for natural gas (which is primarily used for heating and industrial processes). Its supplies will begin to decline.

To create an energy system that is economically and environmentally sustainable is an enormously complicated task, made more difficult by current controversies and vested interest in maintaining the status quo, and the enormous political power of the 15 or so companies that control global energy production. To make wise choices, we must base our decisions on what is socially, economically, and environmentally sustainable. Our goal is to design a sustainable energy system that makes sense given the world's limited resources, one that fits into the economy of nature. Wise choices could help create broader prosperity, a better future, a stable economy, and a stronger, less vulnerable nation.

KEY CONCEPTS

Many experts on sustainable energy systems believe that, in the near term, efforts are needed to improve the efficiency of all energy-consuming technologies and to find sustainable alternatives to coal, crude oil, and their derivatives. In the long term, we must find sustainable replacements for natural gas.

Most of our so-called reasoning consists in finding arguments for going on believing as we already do.

—James Harvey Robinson

Connect to this book's website:
http://environment.jbpub.com/
The site features eLearning, an online review area that provides quizzes, chapter outlines, and other tools to help you study for your class. You can also follow useful links for in-depth information, research the differing views in the Point/Counterpoints, or keep up on the latest environmental news.

CRITICAL THINKING

Exercise Analysis

To begin, you should first look at what nuclear power and oil are used for. In other words, be sure to dig deeply and define and understand all terms. As you learned in the chapter, nuclear power is a source of electricity. In the United States and most other nations, however, oil is used primarily to produce gasoline, diesel fuel, home heating oil, and jet fuel. It is also used to produce chemicals needed to make plastics, medicines, and other products. Very rarely is oil burned to generate electricity.

Increasing our reliance on nuclear power will do little if anything to reduce our dependence on oil. Some see this effort, which continues today in an expensive media campaign sponsored by the nuclear industry, as a ploy to dupe the public. Proponents of nuclear power are promoting a desirable goal (energy independence) but are offering a false solution (nuclear power). By defining terms more carefully, you can see that they're consciously or unconsciously hoodwinking the public.

On the issue of nuclear power and global warming, it is quite clear that what proponents say is true. A nuclear power plant produces no carbon dioxide and does not contribute to global warming. What about nuclear power's impacts, though? Would we simply be trading one problem for another?

CRITICAL THINKING AND CONCEPTS REVIEW

1. Critically analyze the statement, "There is enough oil to go around for at least 65 more years."
2. Describe the term *energy system*. List all of the steps you can think of that are needed to produce gasoline from oil, starting with exploration; discuss some of the impacts on the environment associated with each step.
3. Critically analyze this statement: "Coal is our most abundant fossil fuel and, therefore, should be a major source of energy in the next hundred years."
4. What is oil shale? Discuss the benefits and risks of oil shale development.
5. Define *coal gasification* and *coal liquefaction*.
6. Describe how a light water fission reactor works. What is the fuel, and how is the chain reaction controlled?
7. What are the advantages and disadvantages of nuclear power?
8. Critically analyze this statement: "Nuclear energy is a clean source of energy and should be actively promoted."
9. What is a breeder reactor? How is it similar to a conventional fission reactor? How is it different? Discuss the advantages and disadvantages of the breeder reactor.
10. What is nuclear fusion? Discuss the advantages and disadvantages of fusion energy.
11. You are studying future energy demands for your state. List and discuss the factors that affect how much energy it will be using in the year 2025.
12. What principles of sustainability pertain to designing a sustainable energy system?

KEY TERMS

acid mine drainage
alpha particles
area strip mines
atom
beta particles
bitumen
black lung disease
bottom ash
breeder reactor
coal gasification
coal liquefaction
contour strip mines
control rods
crude oil
daughter nuclei
drag lines
electron cloud
electrons
element
energy quality
energy system
fission

fission fragments
fission reactors
flaring
fly ash
fuel rods
gamma rays
in situ retorting
ionizing radiation
ions
isotopes
kerogen
light water reactors (LWRs)
liquid metal fast breeder reactor
magnetic confinement
mine tailings
natural gas
net energy analysis
net energy yield
nuclear energy
nuclear fusion
nucleus
oil sands

oil shale
Organization of Petroleum Exporting Countries (OPEC)
overburden
petroleum
pneumoconiosis
radiation
radiation sickness
radionuclides
reactor containment building
reactor core
reactor vessel
shale oil
smokestack scrubber
surface mining
surface retort
Surface Mining Control and Reclamation Act
synfuels
tar sands
ultimate production

Windy coastlines, with wind machines like these, could provide a significant amount of the world's electricity.

CHAPTER 15 Foundations of a Sustainable Energy System: Conservation and Renewable Energy

I'd put my money on the sun and solar energy. What a source of power! I hope we don't have to wait 'til oil and coal run out before we tackle that.

—*Thomas Edison*

Michael Reynolds is a maverick in the construction industry. Based in Taos, New Mexico, Reynolds builds homes that embody many of the principles of sustainability. The walls, for example, are constructed of used automobile tires that otherwise would have ended up in landfills. The recycled tires are packed with dirt from the construction site, using a local resource. They're laid on top of one another like bricks to build thick walls. Cement stucco or earthen plaster is then applied to the tire walls, creating an appealing

An internationally known expert from the oil industry argues to a group of congressional representatives that solar energy is not an economically sound option. It's a great idea, he says, but the economics of solar energy aren't good enough to merit widespread use in the United States. He argues that the conventional fuels such as oil, coal, and nuclear energy are more economical and, therefore, more desirable options. If you were one of the senators on the panel listening to this testimony, what questions would you ask? How would you evaluate this advice? What critical thinking rules does this examination require?

design (FIGURE 15-1a). Reynolds's houses, called *Earthships*, are generally built into the sides of hills, taking advantage of the Earth to shelter the house from heat and cold. With their thick walls and well-insulated ceilings, Earthships are extremely energy efficient. They stay cool in the summer and warm in the winter.

Reynolds's homes are heated by the sun in the winter and are generally designed with interior planters that line the south wall, permitting resi-

dents to grow a variety of vegetables year round (FIGURE 15-1b). The plants are watered with wastewater from sinks and showers—commonly called *gray water*. In his most recent designs, Reynolds has devised a system to filter toilet water so that it is not released into the environment. The waste is fed into specially lined outdoor planters, where it is broken down by bacteria and other microorganisms. The nutrients are used by plants.

Besides being heated by the sun, Reynolds's homes generate their own electricity from sunlight and are equipped with efficient lighting systems and appliances. His homes even capture and purify rainwater and snowmelt off the roof for cooking, drinking, bathing, washing dishes, and other uses.

Earthships are designed for self-sufficiency and environmental responsibility. They are unlike conventional homes, which Reynolds likens to patients in intensive care units that depend on outside support in the form of food, water, and energy.

FIGURE 15-1 **A house for all seasons.** Radial, that is. **(a)** Architect and builder Michael Reynolds builds homes out of used tires. Tire walls are covered by stucco and appear quite attractive. The interior walls absorb sunlight that streams in through south-facing glass in the winter months and keep the home very warm. **(b)** Planters can be used to grow food and flowers year-round.

Without these inputs, a patient could not survive. Homes are equally as vulnerable. In many ways, the typical modern home is a microcosm of cities and towns. Cut off from oil, food, and water, a city or town would face serious difficulties.

Reynolds and a handful of other builders are part of a growing legion of citizens working to help make the transition to a sustainable society, one that supplies human needs while protecting and even enhancing the Earth's life-support systems. A key element of his design is the use of sunlight, a form of renewable energy. Conservation is also vital to the success of his design. This chapter examines two major "sources" of energy needed to create a sustainable global energy system: conservation and renewable energy. The final section of the chapter shows how these alternative and environmentally sustainable forms of energy could eventually replace oil, coal, natural gas, and nuclear energy.

15.1 Energy Conservation: Foundation of a Sustainable Energy System

We humans need energy, and we need it badly. Yet at least half—perhaps as much as three quarters—of the energy consumed in the world is wasted, largely because of inefficient technologies and wasteful practices. The **second law of thermodynamics** states that when energy is converted from one form to another, some energy is lost as heat. In other words, no energy conversion is 100% efficient; some waste is inevitable. The amount of energy wasted in the United States, however, far exceeds the inevitable loss.

One half to three quarters of the energy used in the United States is wasted.

Given our heavy dependence on nonrenewable energy supplies—and given their economic and environmental costs and eventual depletion—waste is economically, environmentally, and socially unacceptable. It is also ultimately unsustainable. Writer Bruce Hannon put it best when he said, "A country that runs on energy cannot afford to waste it." Lest we forget, waste is lost opportunity. Reducing waste opens up new opportunities.

KEY CONCEPTS

Energy is used wastefully in virtually all nations. Excessive waste is a sign of an unsustainable technology.

Economic and Environmental Benefits of Energy Conservation

Conservation is one of the biological principles of sustainability discussed in Chapter 2 and is essential to the sustainable transition now under way. As noted in that chapter, *conservation* is used here to include frugality (using what you need) and efficiency (using it efficiently). Energy conservation offers numerous social, economic, and environmental benefits. In industry, energy conservation can reduce the cost of producing goods, giving companies a decided economic advantage in the marketplace. An Alcoa plant in Iowa, for example, found that with a minimum investment in energy-efficiency measures it could cut its $6 million annual energy bill by 25%, thus reducing energy costs by $1.5 million a year—a tidy sum that makes them more competitive and more profitable. Energy conservation also results in substantial reductions in pollution. Many factories, for example, produce their own electricity by burning coal. To reduce pollution, they've installed pollution control devices such as smokestack scrubbers on their power plants. These devices capture pollutants but produce a considerable amount of hazardous solid waste. By using energy more efficiently, companies can reduce energy they need. Energy savings therefore not only save fuel costs, they result in economic savings by reducing the amount of hazardous waste that must be disposed of.

The economic savings from energy efficiency can be illustrated by comparing the cost of this strategy with the cost of supplying energy from conventional sources. For example, measures that improve the efficiency of machines and appliances cost about 1 to 2 cents per kilowatt-hour of energy saved. In contrast, electricity from coal-fired power plants costs 5 to 7 cents per kilowatt-hour. Electricity from nuclear power plants, on average, costs 8 to 12 cents per kilowatt-hour. What this means is that for every penny a company invests in energy efficiency it will save 5 to 12 cents in electrical bills, depending on the cost of energy in the area. Similar calculations can be made for other fuels. (For more on this subject, see the Viewpoint in this chapter.)

Every dollar invested in reducing oil consumption by a factory currently saves about $25.

Not surprisingly, the most energy-efficient companies and nations are also the most successful economically (Table 15-1). In the United States, for example, 10 cents of every dollar of the **gross national product** (the total output of goods and services) is spent on energy. Japan, on the other hand, spends about 4 cents per dollar of GNP. Because of this inefficiency, U.S. industries spend $220 billion a year more than is necessary on energy to produce the same goods and services as the Japanese. The cost differential gives Japan a slight economic

Table 15-1	
Oil Consumed per Unit of National Output, Selected Countries, 2000	
Country	**Tons of Oil Equivalent per $1000 (US)**
Canada	0.40
United States	0.34
Germany	0.26
France	0.25
United Kingdom	0.24
Japan	0.19
Switzerland	0.17
Italy	0.16

edge on everything it sells in the United States and in foreign markets.

Besides saving money, energy efficiency addresses a number of environmental problems simultaneously. For example, reducing the combustion of fossil fuels reduces acid deposition, global warming, and urban air pollution. Cutting our demand for oil reduces the number of oil spills, thus saving aquatic species as well as the birds and mammals that depend on clean water. In addition, using less fossil fuel reduces the destruction of wildlife habitat by decreasing the extraction of fossil fuels. Besides saving money, efficiency measures can be rapidly deployed, so they begin saving individuals and businesses money very quickly. Finally, energy efficiency can help us stretch limited supplies of fossil fuel, giving us more time to make a transition to a more sustainable system of energy.

KEY CONCEPTS

Energy-efficiency measures can reduce the world's dependence on fossil fuels. Energy-efficiency measures are easy to implement and cost a fraction of what new energy supplies do, while providing many environmental benefits.

Energy-Efficiency Options

The range of energy-efficient options available to businesses and homeowners is enormous. In some cases, efficiency can be achieved through minor changes in our actions. The Denver Marriott hotel, for instance, made substantial cuts in energy demand by asking its cleaning staff to raise the temperature setting in unoccupied rooms during the summer (so the air conditioning system didn't have to work so hard) and to lower it during the winter (to reduce heating costs). Around the home, shutting off lights, turning off televisions when they're not in use, and lowering the thermostat a little in the winter and raising it in the summer can result in significant reductions in energy demand. For commuters, driving within the speed limit, keeping cars tuned up, and being sure that tires are properly inflated could translate into millions of gallons of gasoline saved each year. Such measures require little or no initial in-

vestment and can save individuals a lot of money. When millions of people contribute to the cause, savings can be enormous.

Many devices that save energy are readily available for homes, businesses, and factories. Water-saving devices—especially shower-heads, dishwashers, and washing machines—can save substantial amounts of energy in homes by reducing demand for hot water. High-efficiency lighting systems can cut electrical demand for lighting in homes and businesses by three fourths. The compact fluorescent lightbulb, available in many stores today, can be substituted for standard incandescent bulbs and uses only one fourth as much energy to produce the same amount of light (**FIGURE 15-2**). Although high-quality compact fluorescents cost more initially ($10 to $12), they outlast 9 or 10 ordinary bulbs and save $30 to $50 over their lifetimes in reduced electrical demand. They also reduce carbon dioxide pollution by about one ton and reduce summer cooling costs because they produce little waste heat, unlike an incandescent bulb. Homeowners are recommended to install them in high-use areas.

Substantial energy savings can also be achieved by installing better insulation in homes, businesses, and factories. Doubling the insulation of a new home, for example, adds about 5% to the cost but pays for itself in 5 years in reduced energy costs.

Planting shade trees around homes can reduce summer cooling costs. One study showed that three trees planted near light-colored houses in a residential neighborhood in Phoenix could cut cooling demand by 18%. In Sacramento and Los Angeles, the savings are even greater—34% and 44%, respectively. Planting conifers on the north and west sides of homes can

> Compact fluorescent light bulbs use one fourth as much electricity as a standard incandescent light bulb to produce the same amount of light. They also outlast 9 to 10 ordinary bulbs. Although they cost more initially, they can save $30 to $50 over their lifetime, more than paying their initial cost.

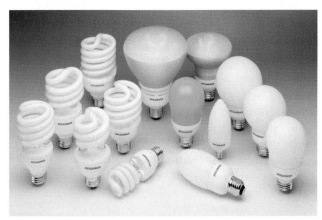

FIGURE 15-2 The compact fluorescent lightbulb is one of many cost-effective means of saving energy. Over its lifetime, a $12 bulb can save $30 to $50 in electricity.

VIEWPOINT The Best Energy Buys

Amory B. Lovins and L. Hunter Lovins
Amory Lovins is a consulting physicist; his former wife and colleague, Hunter, is a lawyer, sociologist, and political scientist. They have worked as a team on energy policy in over 20 countries and were principals in the nonprofit Rocky Mountain Institute (RMI) in Old Snowmass, Colorado, which explores the links among energy, water, agriculture, security, and economic development. Most of RMI's income comes from advising utilities on how to produce electricity more efficiently and more economically. Hunter currently runs a nonprofit organization of her own called Natural Capitalism.

Raw kilowatt-hours, lumps of coal, and barrels of sticky black goo are more messy than useful. Energy is only a means of providing services: comfort, light, mobility, the ability to make steel or bake bread. We don't necessarily want or need more energy of any kind at any price; we just want the amount, type, and source of energy that will do each desired task at least cost.

In the United States about 58% of all delivered energy is needed as heat; 34% as liquid fuels for vehicles; and 8% as electricity (for motors, lights, electronics, smelters). Electricity is a premium form of energy. It is able to do difficult kinds of work but is extremely expensive—far too costly to provide economical heat or mobility. Yet over the next 20 years utilities want to build another trillion dollars' worth of plants—causing spiraling rates, bankrupt utilities, and unaffordable energy.

What's the alternative? First, we ought to use energy just as efficiently as is worthwhile. We're starting to do this. Making a dollar of gross national product (GNP) takes a quarter less energy now than it did 10 years ago. Since 1979, the United States has gotten seven times as much new energy from savings (conservation and improving efficiency) as from all the new oil and gas wells, power plants, and coal mines opened in the same period. Yet it's worth saving even more. With our current technology we could double the energy efficiency of industrial motors or jet aircraft, triple that of steel mills, quadruple that of household appliances, quin-

tuple that of cars, and improve that of buildings by tenfold to a hundredfold. Such increased "energy productivity" gives us the same services as now, just as conveniently and reliably, but at less cost to ourselves and the Earth.

Electricity, being expensive, is especially worth saving. We are writing this under a new kind of lightbulb that gives better light than the old kind, lasts thirteen times as long, uses a quarter as much electricity, and repays its $15 price in a year or two. Better motors and drivetrains in factories could save more electricity than all nuclear plants produce. The best refrigerator we can make uses one-twenty-seventh as much electricity as the one in use today. If Americans used all the best technologies now on the market, they could be using less than a quarter as much electricity as they use now. And they could be paying, for each kilowatt-hour saved, less than it would cost just to run a coal or nuclear plant to generate that kilowatt-hour. Thus, even if a new nuclear plant cost nothing to build, were perfectly safe, and didn't produce radioactive wastes or bomb materials, it would still save the country money to write it off, never run it, and buy efficiency instead.

Indeed, neither new fossil-fueled nor nuclear power plants would be necessary or economically feasible if we used electricity efficiently. Existing and small hydroelectric plants plus a bit of wind power (or, optionally, solar cells or industrial cogeneration—the simultaneous production and use of electricity and heat) would be enough. A government study showed that even if by the year 2000 the United States' GNP had increased by 66%, by buying the cheapest energy options Americans would use a quarter less energy and electricity than now and nearly 50% less nonrenewable fuel. The net savings: several trillion dollars and about a million jobs.

After efficiency, the next best energy buys are, as a Harvard Business School study found, the appropriate renewable sources. Such "soft technologies" include passive and active solar heating, passive cooling, high-temperature solar heat for industry, converting sustainably grown farm and forestry wastes to liquid fuel for efficient vehicles, present and small hydroelectric power, and wind power. Solar cells, too, will soon

cut wind and reduce winter heating costs. Tree planting near homes and businesses in the United States could cut energy demand by nearly 1 quad, or about 1/100th of our total energy demand.[1] Although it is not a lot, individual savings can be substantial.

Shade trees planted around a house can cut summer cooling costs by 18 to 44%.

[1]One quad is a quadrillion BTUs—about 170 million barrels of oil.

Conservation does not mean freezing in the dark, as some would have us believe. For homeowners, adding attic insulation or installing storm windows reduces heat loss in the winter and retains cool air in the summer, resulting in a significant savings on utility bills and a more comfortable life. Both insulation and storm windows require small investments that are paid back in short periods by reduced energy bills. (For additional suggestions, see the Individual Actions Count! table.)

Energy efficiency can be designed into new products—for example, cars, appliances, buildings, and factories. Conservation measures in new office buildings, for example, can be generally cost-effective and join the list. Soft technologies tend to be smaller than huge power plants, so they can provide the cheapest energy where it's needed. Small isn't necessarily beautiful, but it usually saves money by matching the relatively small scale of most energy uses.

Careful studies in 15 countries show that the best soft technologies now available are cheaper than new power plants and could meet essentially all our long-term global energy needs.

Already, efficiency and renewables are sweeping the market, not because we say they should but because millions of people are choosing them as the best buys. The United States since 1979

- has gotten more new energy from renewable sources than from any or all of the nonrenewables.
- has ordered more new electric generating capacity from small hydroelectric plants and wind power than from coal or nuclear plants or both.
- is now getting nearly twice as much delivered energy from wood as from nuclear power, which had a 30-year head start and direct subsidies of well over $100 billion.

Wood burning, solar heat, and the like aren't always done well. People need much better information and quality control to choose and use the cheapest, most effective opportunities. But it is faster to build many small, simple technologies that anybody can use than a few huge, complex projects that take 10 years and cost billions of dollars each. And that's what Americans are doing, to the tune of $15 billion in 1980 alone.

Every time you buy weatherstripping instead of electricity—because you can get comfort cheaper that way—you're part of the transition. And your part matters. The United States could eliminate oil imports in less than a decade, just by making buildings and cars more efficient. Conversely, each dollar spent on reactors can't be spent on faster, cheaper ways to save oil, and hence delays energy independence. Power plants also provide fewer jobs per dollar than any other investment. Thus, every big plant built loses the economy, directly and indirectly, about 4000 net jobs, by starving all other sectors for capital.

In the United States, 64% of the capital charge for new reactors is subsidized via taxes. The taxpayer also picks up much of the tab for nuclear fuel, decommissioning the worn-out plants, developing and regulating them, exporting them, coping with their hazards, and trying to fend off the nuclear bombs they spread. Despite the enormous government intervention, nuclear power is dying of an incurable attack of market forces throughout the world's market economies (and is in deep trouble even in the centrally planned economies, notably France and Russia). Wall Street won't pay for more reactors; over a hundred have been canceled, and most of the industry's best people have already left.

Fortunately, the same best energy buys that are vital to a healthy economy are also keys to national security. Centralized, complex, computer-controlled nuclear plants are sitting ducks for terrorists, accidents, or natural disasters. In contrast, a more efficient, diverse, dispersed, renewable energy system could be resilient. Major failures of energy supply simply couldn't happen. Hooking together decentralized electrical sources via the existing power grid so that they could back up one another just as giant power stations do now, would actually make electrical supplies more reliable, move supplies nearer users, and reduce dependence on fragile transmission lines.

People and communities are starting to solve their own energy problems. They've discovered that the problem isn't where to get 80 quadrillion BTUs a year, but how to seal the cracks around their windows. People are finding more to trust in local weatherization programs, community greenhouses, and municipal solar utilities. The energy transition is happening from the bottom up, not from the top down: Washington will be the last to know.

It's not for "the experts" to choose whether you need caulk or electricity. Pick your own best buys. The energy future is your choice.

Critical Thinking Question

1. Make a list of the major points made by the authors. List the supporting evidence under each one. Using your critical thinking skills, critically analyze each point.

cut energy for lighting, heating, and cooling by 80% or more! Energy-efficient homes can be built with heating bills of $100 a year or lower, even in cold climates!

One extraordinary means of reducing energy waste is **cogeneration**. In industrial cogeneration, waste heat from one process is captured and used in another, thus reducing waste and improving the energy efficiency of a plant. For example, for years, many American industries produced their own steam on site for use in various processes. Steam generation requires natural gas or oil, which companies purchased along with electricity from local utilities. Steam generation in factories was 50 to 70% efficient. To-

day, many companies use waste heat from steam generation to produce electricity, thus boosting the efficiency of the system to 80 or 90%—and driving down the cost of electricity.

Because of its favorable economics, cogeneration is an emerging source of energy. Growth in generating capacity has increased dramatically in the past 25 years—and continues to rise. In Germany, the United States, and other developed countries, small cogeneration technologies are being installed in restaurants, hotels, and apartment buildings. They supply space heat, hot water, and electricity at a lower cost than conventional systems. In Chula Vista, California, a Mc-

1. Water Heating
- Turn down thermostat on water heater.
- Use less hot water (dishwashing, laundry, showers).
- Install flow reducers on faucets.
- Coordinate and concentrate time hot water is used.
- Install an insulated water heater blanket.
- Do full loads of laundry and use cooler water.
- Hang clothes outside to dry.
- Periodically drain 3 to 4 gallons from water heater.
- Repair leaky faucets.

2. Space Heating
- Lower thermostat setting.
- Insulate ceilings and walls.
- Install storm windows, curtains, or window quilts.
- Caulk cracks and use weatherstripping.
- Use fans to distribute heat.
- Dress more warmly.
- Heat only used areas.
- Humidify the air.
- Install an electronic ignition system in furnace.
- Replace or clean air filters in furnace.
- Have furnace adjusted periodically.

3. Cooling and Air Conditioning
- Increase thermostat setting.
- Use fans.

- Cook at night or outside.
- Dehumidify air.
- Close drapes during the day.
- Open windows at night.

4. Cooking
- Cover pots and cook one-pot meals.
- Turn off the pilot lights on stove.
- Don't overcook and don't open oven unnecessarily.
- Double-up pots (use one as a lid for the other).
- Boil less water (only the amount you need).
- Use energy-efficient appliances (Crock-Pots).

5. Lighting
- Cut the wattage of bulbs.
- Turn off lights when not in use.
- Use fluorescent bulbs whenever possible.
- Use natural lighting whenever possible.

6. Transportation
- Carpool, walk, ride a bike, or take the bus to work.
- Use your car only when necessary.
- Group your trips with the car.
- Keep your car tuned and tire pressure at the recommended level.
- Buy energy-efficient cars.
- Recycle gas guzzlers.
- On long trips, take the train or bus (not a jet).

Donald's restaurant produces its own electricity and hot water using a small cogeneration system.

> **KEY CONCEPTS**
>
> There's no shortage of easy, economical ways to save energy. These include simple behavioral changes, such as turning off lights when one leaves a room, as well as many energy-efficient technologies, such as compact fluorescent lightbulbs and cogeneration facilities. Energy savings result in substantial social, economic, and environmental benefits.

The Potential of Energy Efficiency

The United States and other industrial nations have made some modest improvements in energy conservation, but the conservation potential has hardly begun to be tapped. The World Resources Institute, in fact, estimates that the world could meet 90% of its new energy needs between 1987 and 2020 simply by making more efficient use of the energy we now generate. Even though the world population is expected to double between 1980 and 2020, the Institute says, only a 10% increase in energy production would be needed if existing energy-efficiency technologies were put in place. Most experts agree: Fossil fuels will eventually be phased out as

supplies run out and renewable energy supplies will take up the slack. Energy efficiency will also help us get the most from renewable supplies.

> **KEY CONCEPTS**
>
> Because energy is used so inefficiently, huge cuts in energy demand can be made by applying efficiency measures. This should not affect the level of services we receive. Much of our future energy demand can be met by freeing up energy currently wasted in three areas: transportation, buildings, and industry/business.

Saving Energy in the Transportation Sector Over one fourth (27%) of the energy consumed in the United States is used in transportation. Energy consumption in the U.S. transportation sector has grown steadily since the 1960s and shows no sign of declining, a trend that many experts believe bodes poorly for our nation's prospects for creating a sustainable future (FIGURE 15-3).

Fortunately, dramatic improvements in the efficiency of vehicles are possible. Automobiles, the single largest source of fuel consumption in this sector, represent the greatest potential for reductions. The Rocky Mountain Institute's Amory Lovins has completed work on the design of an automobile,

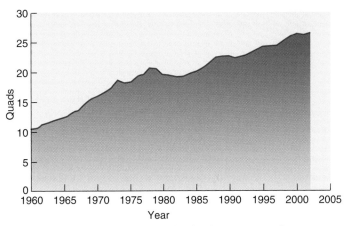

FIGURE 15-3 Energy consumption in the transportation sector. U.S. energy consumption in the transportation sector has increased consistently over the years, an unsustainable trend.

called a **hybrid car**, that he predicts will get 150 to 200 miles per gallon. Honda (Insight) and Toyota (Prius) are now producing offshoots of Lovins's design that get around 70 and 60 miles per gallon, respectively (Figure 14-11). Both went on sale in the United States in the year 2000. Fiat, BMW, Audi, Ford, VW, and virtually every other auto manufacturer, are also marketing hybrid cars. For a discussion of this new breed of energy-efficient car, see Spotlight on Sustainable Development 15-1.

> Hybrid cars could theoretically get 150 to 200 miles per gallon. Earliest versions are getting around 70 miles per gallon.

In contrast, the average gas mileage of cars on the road in the United States (both domestic and imported) in 2004 was 20.4 mpg, one of the lowest averages in the more developed nations. By increasing the average mileage to 60 miles per gallon, Americans could reduce automobile emissions by nearly half. In September 1993, President Clinton challenged the Big Three U.S. auto manufacturers to develop a clean car with triple the efficiency of current vehicles within a decade. They pledged their best efforts to meet this goal and formed a Partnership for a New Generation of Vehicles. Ford, General Motors, and Daimler-Chrysler all have plans to produce hybrid cars. Efforts are also under way to produce cars powered by hydrogen derived from fossil fuels, such as gasoline or natural gas.

> The average gas mileage of cars on the road in the United States is 20.4 miles per gallon.

Energy-efficient vehicles are vital to building a sustainable future. Declining oil supplies, rising population, and increasing congestion on highways in cities beg for additional measures, though—ones that deal with a range of issues. One of these is mass transit—energy-efficient buses and trains that move large numbers of people with a fraction

of the resources. Washington, D.C.'s Metro is a clean, efficient, and convenient means of moving large numbers of people. (This topic is discussed in Chapter 17.)

KEY CONCEPTS

Extremely energy-efficient vehicles are currently available. Many improvements in vehicles could increase efficiency even more, greatly cutting transportation energy consumption.

Saving Energy in Buildings About 40% of the energy consumed in the United States is used in buildings—office buildings, factories, and homes. The modern home and office building make many demands on the energy system. They require lighting, heating, and cooling. Many also contain various machines, appliances, and electronic equipment. Fortunately, very impressive savings can be made in all areas.

Unfortunately, efficiency improvements in buildings have lagged behind other areas. There are many reasons for this lack of progress. For example, new houses and commercial buildings are built by contractors who often seek to minimize initial expenses. To do this, they often select less costly and less energy-efficient appliances, heating, and lighting systems. They inadvertently strap the building owner/operator with a lifetime of high energy bills—all to save a few dollars on the front end. Today, more than 30% of the nation's housing is occupied by renters, who are often responsible for paying utility bills. The owner has very little incentive to invest in energy efficiency.

One encouraging development is the emergence of energy conservation companies, which provide innovative financing schemes for building owners who want to invest in energy efficiency. These companies have been successful in Europe for years and are becoming more common in the United States. They advise clients on ways to cut energy consumption and install various devices to achieve their goals. Some companies work out plans to be paid by their clients through energy savings accrued by the latter.

Some experts believe that cost-effective energy-efficiency measures could cut total energy consumption in U.S. buildings by as much as 30% over the next two decades, despite a 15 to 20% increase in the number of buildings.

KEY CONCEPTS

Energy-efficiency measures in buildings can result in substantial energy savings in heating, cooling, lighting, and appliances and electronic equipment. Unfortunately, short-term thinking and economics often prevent investments in measures to cut energy demand in these areas.

Saving Energy in Industry In the United States, industries consume one third (33%) of the nation's energy. Over the past 30 years, industrial energy savings have been impressive. Energy required to generate one dollar of economic output has

> In the United States, the energy required to produce one dollar of economic output has fallen by 38%.

15-1   Reinventing the Automobile

Imagine a car that gets 150, maybe 200 miles per gallon, and can travel from the West Coast to the East Coast on one tank of gas. Imagine a car that comfortably seats five passengers but weighs only about 400 kilograms (900 pounds)—one half to one fourth the weight of a typical vehicle. Imagine also that this superefficient, lightweight car is also just as safe as a conventional vehicle and consists of a body and chassis that contain only 6 to 20 parts, compared with 250 or so in a conventional automobile.

Is this some creation of a science fiction writer? No. Far from it. This car is the brainchild of Amory Lovins of the Rocky Mountain Institute. Renowned for his work on energy efficiency, Lovins received funds from several foundations to develop designs for a superefficient car of the future to help drastically reduce the world's demand for fossil fuels and cut pollution in a major way.

Shown in FIGURE 1, this vehicle, called a *hypercar*, has a smooth underbody, and an incredibly aerodynamic form. These and other features could dramatically increase the car's efficiency without sacrificing either interior space, safety, or performance.

Lightweight composites (mixtures of materials) and other space-age materials would be used to construct the car. These materials are stronger and much lighter than those currently used by car manufacturers. Because composites absorb a lot more energy in crashes than the steel used in the bodies of a typical car, the lightweight hypercar would actually be safer. The vehicle's small engine, discussed shortly, provides for an efficient performance and also leaves more "crumple zone"—area for the car to cave in during an accident. Force is absorbed by the materials, not the occupants.

FIGURE 1 **The hypercar.**

The hypercar is a hybrid of sorts. Its propulsive unit, its engine, might consist of a small, fossil fuel-powered engine—only 20 horsepower or so—that is used to generate electricity. Other electricity-producing engines are also possible, for example, fuel cells (described in this chapter). Whatever the source, the electricity is sent to four small electric motors in the wheels. During braking, electric motors that drive the wheels actually function as small generators themselves. They could theoretically capture up to 70% of the braking energy—energy that is lost in a typical automobile. This energy is stored in the battery for later use.

Further adding to the car's efficiency are special tires made of materials that reduce the friction on the road surface. This could, according to Lovins and his team at RMI, reduce the rolling drag by 65 to 80%. Together, the ultra-lightweight design, superb aerodynamics, and efficient hybrid propulsion system could translate into a 150 to 200 miles per gallon fuel rating—considerably higher than the 20.4 miles per gallon average achieved by cars on the road in the United States. Air emissions from the vehicle would be 100 times lower than today's cars. In other words, 100 hypercars would produce as much pollution as one conventional vehicle.

Is this a wild dream? Apparently one student at Western Washington University didn't think so. He built a Corvette-sized two-seater light hybrid. In April 1994, the vehicle was tested by the Department of Energy and found to get 202 miles per gallon. In October 1994, a Swiss-built car that seats four achieved 150 miles per gallon.

As noted in the chapter, virtually all automobile manufacturers are now producing cars that begin to resemble the hypercar with gas mileage between 40 and 70 miles per gallon. Toyota was the first to sell its car (Prius) in Japan, starting in 1999, and then in the United States in 2000. Honda came next with a model known as the Insight. Like the Toyota, it was first sold in the United States in 2000. Numerous American manufacturers are now offering hybrid cars, trucks, SUVs, and vans. And all major U.S. auto manufacturers also have diesel hybrid cars that get around 80 miles per gallon (at this writing, there are no plans to manufacture and sell these models).

How does a hybrid work? Hybrids have an electric motor and a small gas engine like the hypercar. In some models, like Toyota's Prius, the car initially operates on electric power. That is, when the car is turned on and begins to move, its electric motor provides the main propulsion. The gas engine kicks in at around 7 or 10 miles per hour. On the highway, however, the gas engine provides much of the thrust. The electric motor provides additional boost to pass or climb hills. When slowing down or going down a hill, however, the vehicle uses very little power at all. In fact, the gas engine may shut off entirely. The instantaneous gas mileage readout in the Prius, located on a computer screen in the center of the dashboard, records gas mileages of 100 miles per gallon, indicating that the car is coasting.

When the vehicle comes to a stop at a stop light, the engine shuts off. When the operator presses on the gas pedal, however, the electric motor kicks in, followed by the gas motor. The car is off and running without hesitation.

Honda's three hybrids (Insight, Civic, and Accord) are similar to the Toyota Prius in many respects. That is, they contain a gas engine and an electric motor, but the electric motor is much smaller than that of the Prius. As a result, it can't be used to start the car from a standstill.

When a Honda Civic Hybrid is turned on, the gas engine starts up and provides the power needed to move the car and most of the propulsion from that point on. The electric motor kicks in when additional power is needed, for example, when climbing a hill or passing another vehicle.

What about the batteries of a hybrid car? Do they need to be charged each night? No, they're charged continuously by the car during normal operation through an alternator, a device that provides electricity for lights and radios and the battery bank. As in the hypercar, batteries are also charged during braking. Electricity flows smoothly in and out of the lightweight battery bank (usually located behind the back seat) during various modes of operation, giving one a clean, energy-efficient ride! Like other functions in the car, it is all computer controlled. You drive; the car figures all of this out automatically. Are you going to have to pay an arm and a leg to purchase a hybrid?

Hybrids cost two to three thousand dollars more than their less frugal cousins. However, Federal and state incentives currently help reduce the costs dramatically, and savings on fuel will help offset what's left very quickly.

Expect to see a lot of hybrids on the road in coming years. Sales are going extremely well. The cars are pretty attractive. They're fast and relatively inexpensive, too. Moreover, California law required 10% of the cars on the road by 2003 be the zero emission variety. Massachusetts, New York, and Arizona have followed suit. Although not reaching the full potential of the hypercar, they represent a dramatic shift in automobile technology with enormous benefit to the environment.

Another vehicle that holds promise is the plug-in hybrid electric vehicle, now under development. What's a plug-in hybrid?

A plug-in hybrid electric vehicle are hybrids, much like those just discussed, with a couple differences. First, they contain extra batteries. Unlike the hybrids just discussed, these cars are plugged in at night into a 120 or 220-volt outlet (like a dryer outlet). This charges the extra battery. The electrical charge, in turn, powers the vehicle in most situations. According to the Institute for Analysis of Global Security, which researches transportation options that could help us free our dependence on foreign oil, "Plug-ins run on the stored (electrical) energy for much of a typical day's driving—depending on the size of the battery up to 60 miles per charge, far beyond the commute of an average American." They go on to say that when the electrical charge is used up, the car "automatically keeps running on the fuel in the fuel tank. A person who drives every day a distance shorter than the car's electric range would never have to dip into the fuel tank.

That of course leads to the second difference. Such cars need larger electric engines than hybrids currently on the market.

Because most of the energy used by plug-ins comes from electricity and not from gasoline and because electricity can be generated efficiently and cleanly from America's abundant renewable energy resources, especially solar and wind power, hybrid electrics may help us combat the coming shortages of oil without increasing global warming and a host of other environmental problems associated with fossil fuel use.

According to the Institute for the Analysis of Global Security, "The plug-in hybrid drive system is compatible with all vehicle models and does not entail any sacrifice of vehicle performance or driver amenities. A mid size plug-in can accelerate from 0 to 60 miles per hour at less than 9 seconds, sustain a top speed of 97 mph and maintain 120 mph for about two minutes even with a low battery."

The Rocky Mountain Institute (RMI) thinks you may see a true hypercar on the road very soon. They've even spun off a for-profit business to help develop the hypercar. RMI predicts that the hypercar will dominate the market by around 2020. The technology to produce them is currently available. Creating a prototype and gearing up for production would, RMI says, take less time than a conventional automobile because of the simpler design and fewer parts.

RMI admits that the hypercar won't solve all of the world's transportation problems. If it became a popular alternative to conventional automobiles, it could drastically reduce fossil fuel consumption and air pollution, both impediments to the goal of creating a sustainable future. The hypercar will not, however, solve such problems as urban sprawl and highway congestion during all hours of the day—and especially that time we mistakenly call "rush hour." Chapter 17 describes options to get people out of their cars to help ease these other transportation problems.

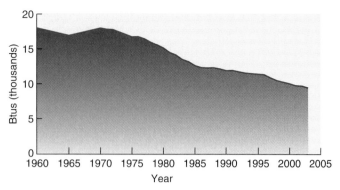

FIGURE 15-4 **Energy vs. GNP.** This graph shows that the energy required to generate a dollar of GNP (in year-2000 dollars) by the U.S. economy has declined over the years, largely as a result of an increase in efficiency. Source: Data from *Annual Energy Review 2003*, Table 1.5. Washington, DC: Energy Information Administration, 2005.

fallen by about 50% (**FIGURE 15-4**). Nevertheless, industries have many additional opportunities to cut energy consumption while increasing their economic strength and profitability. Many corporate executives now realize that an investment in energy efficiency is one of the most cost-effective ways of reducing expenses and boosting profits. Energy efficiency is not an impediment but a keystone to economic success.

Energy can be saved by installing more efficient lighting systems and insulating buildings to reduce heat loss and cooling costs. Energy use can also be cut dramatically by installing more efficient boilers, pumps, and motors. Many pumps, for example, operate at one speed. To control the flow of liquids, valves are used. Installation of variable speed motors has the same effect while dramatically cutting power demand.

Recycling is also an important strategy for reducing industrial energy consumption (Chapter 16). Roughly three quarters of the energy consumed by industry is used to extract and process raw materials. Making metals from recycled scrap uses a fraction of the energy needed to make them from virgin ore.

> **KEY CONCEPTS**
> Industrial energy consumption accounts for a major portion of the world's energy demand but could be cut substantially, making companies more profitable and competitive.

Promoting Energy Efficiency

Energy efficiency is needed in all sectors of society. How can individuals, businesses, and nations be encouraged or compelled to use energy more efficiently?

Once again, many options are available. This section briefly discusses seven options: (1) education, (2) taxes, (3) feebate systems, (4) government-mandated efficiency standards, (5) voluntary programs, (6) changes in pricing, and (7) least cost planning.

One of the most important steps is education. Educating people about the economic and environmental benefits

of energy conservation is vital to the task at hand. Education can occur at many levels, for example, in schools, colleges, and on television. Government agencies such as the U.S. Environmental Protection Agency (EPA) can help educate businesses on the advantages of energy efficiency—and currently take an active role in this arena.

National taxes on fossil fuels—especially oil and coal—could also help promote efficiency. By raising the taxes on fossil fuels, making them more costly to consumers, governments could play an important role in stimulating energy conservation in all sectors of society.

In the United States, the average consumer pays about 38 cents per gallon in state and federal taxes for gasoline. In Denmark, the tax is nearly $3 per gallon, which encourages the efficient use of gasoline. Increased taxes on gasoline and other fuels in the United States and Canada could help cut energy consumption and generate revenues to build mass transit systems, promote conservation, and fund sustainable alternative transportation technologies. Because government currently subsidizes many fossil fuels through tax dollars from general revenues, this step would help it recover the money it already invests in providing us with artificially inexpensive fuel.

Another innovative measure is the *feebate* system. It consists of a fee (tax) paid by those who buy gas-guzzling cars (which would create a disincentive to consumers), and a rebate given to those who purchase energy-efficient autos. The rebate would be paid by the fees and would be an additional incentive to individuals. The United States had a gas guzzler tax for many years, but President Clinton abolished it in 1999.

A fourth means of stimulating conservation is through government-mandated efficiency standards. The **National Appliance Conservation Act** passed by the U.S. Congress in 1987, for instance, established energy-efficiency standards for all new appliances. It doesn't spell out what changes need to be made, only the amount of energy that appliances should be consuming. For example, the act required manufacturers of all major household appliances, such as refrigerators, to produce models that consume 20% less electricity than 1987 models. California passed a similar law, calling for a 50% cut in electrical usage by new appliances. The results were impressive. Efficiency standards could be applied to all new homes and factories as well. Automobile mileage standards, already in place, could call for much greater savings. In the United States, however, improvements in gas mileage standards for new cars have been vigorously opposed and have been frozen for over a decade and continue to face opposition from the Bush administration and Congress, although modest increases in gas mileage were proposed in 2005 for SUVs, vans, and light trucks after months of high gasoline prices.

Voluntary efforts are also vital. Many governments are starting to take a more active role in promoting energy efficiency in buildings through voluntary programs. In Canada, **Natural Resources Canada**, the federal agency once called the Department of Energy, Mines, and Resources, launched its **Super Energy-Efficient Home program** with the Canadian Home Builder's Association. Their goal is to encourage builders to build homes that use half as much energy as conventional ones. Although slightly higher construction costs and red tape have hindered the program, many builders have adopted the ideas and are now building much more energy-efficient homes. More recently, the Canadian federal government also entered into a program with Home Hardware stores to offer videos and publications on ways to improve the energy efficiency of Canadian homes.

In the United States, the U.S. EPA and many states are promoting home energy efficiency through a variety of voluntary programs. EPA has its **Energy Star Homes program.** This program relies on builders to produce homes that use 30% less energy for heating, cooling, and so on than stipulated in the **National Model Energy Code.** When this goal has been verified by an independent certifier, the home receives an **Energy Star label.** Funding for such programs is being dramatically reduced by Congress at this time.

Many cities and states have **green building programs** that promote energy efficiency. By meeting certain criteria, home builders can apply for a green building certificate. This, in turn, helps builders market a home. A leader in commercial green building is the nonprofit group known as the U.S. Green Building Council. This organization created a commercial building rating system known as LEED—Leadership in Energy and Environmental Design—for new and existing buildings. Architects and builders must meet certain criteria to qualify for LEED certification, which dramatically reduces the energy and evironmental impacts of new and existing buildings.

Another encouraging development is the National Association of Home Builders' active role in researching and promoting energy-efficient homes. Some lending institutions are also providing loans for energy-efficient homes, although they tend to simply allow the lender to borrow more money, based on the belief that they will have more money in their pockets. Far better, say some critics, would be a program that gives lower interest loans to buyers of green built homes.

Conservation can also be stimulated by changes in pricing. For example, some utilities now charge customers more for electricity used during peak hours—times when demand is highest. Why? Meeting peak demand is very costly for utilities. It often requires construction of additional power plants that are needed for only a few hours a day. If utility companies can reduce peak demand through pricing, they won't have to build expensive new facilities.

Furthermore, efficiency can also be stimulated by a process called **least cost planning.** In the United States, utilities are regulated by the states. Thus, when a power company decides that it needs to build a new power plant to meet rising demand, it must first receive state approval. Currently, most states have requirements for least cost planning or integrated resource planning (choosing a variety of technologies, including renewables, and measures such as efficiency to meet energy needs). Least cost planning requires power companies to select the least costly way of providing electricity, which almost always entails energy conservation programs. Some cost-saving options include improvements in generating efficiency, peak-pricing schemes, purchasing electricity from other companies (rather than building new power plants), cogeneration, and promoting energy con-

servation in homes, factories, and businesses. These steps are not only cheaper for utilities, they save customers money. In general, efficiency measures are two to three times cheaper than building new power plants. Moreover, the payback period of conservation is only 2 or 3 years compared to 20 for a power plant. Unfortunately, efforts are currently under way in many states to eliminate requirements for least cost planning, and many states are not enforcing their requirements, rendering regulations meaningless.

Finally, energy conservation can be stimulated by a combination of the measures listed previously here. The Canadian government, for instance, is trying to promote conservation through **The Efficiency and Alternative Energy program,** which entails 33 different initiatives to improve efficiency in all end-use sectors, including buildings, industry, transportation, and equipment. One of the keys of this program was the 1993 **Energy Efficiency Act,** which sets minimum efficiency standards for equipment. The **Motor Vehicle Safety Act** places stringent emissions standards on cars. All of these efforts are part of a strategy aimed at becoming more efficient and reducing greenhouse gas emissions to avert global warming.

KEY CONCEPTS

Not only are there many ways to use energy much more efficiently, there are also many ways to promote this strategy, including (1) education, (2) taxes on fossil fuels, (3) feebate systems (which levy a tax on those who choose energy-inefficient options and give rebates to those who opt for energy-efficient technologies), (4) government-mandated efficiency programs, (5) voluntary programs, (6) changes in pricing, and (7) least cost planning.

Roadblocks to Energy Conservation

Despite the economic and environmental advantages of conservation, it still is not as widely practiced as many would like. Why not? Part of the reason is that Americans, Canadians, and others have been blessed for many years with abundant energy and have had little incentive to use it efficiently.

Another reason is that fossil fuels and nuclear energy have been subsidized by federal programs, which reduces the direct cost to the consumer. When federal subsidies paid by taxpayers are included, the cost of U.S. oil is estimated to be about $100 to $200 per barrel compared to the current rate of about $50 per barrel. In other words, if oil companies paid the real cost of producing oil, a barrel of oil would cost 4 to 5 times its current market value. At this rate, gasoline would cost about $5.00 to $10.00 a gallon. Lovins also claims that Middle Eastern oil costs about $500 per barrel when one takes into account the cost of the military escort of tankers moving in and out of the Persian Gulf. This doesn't include the fi-

> Oil is currently selling for about $50 per barrel. When government subsidies are included, the true cost is estimated to be at least $100 to $200 per barrel.

nancial or human costs of the Persian Gulf War in 1991 or the war in Iraq.

Still another reason is that high-efficiency products such as compact fluorescent lightbulbs sometimes cost more than less energy-efficient ones, and many people neither calculate the long-term savings nor think about the environmental consequences of waste.

In addition, many governments have not been committed to energy conservation programs. The U.S. government currently (2005) spends about $868 million a year on energy conservation, down from $950 million per year in 1999. Although this is a lot of money, critics point out that the current outlay for conservation is equal to only about 20 hours of military spending. This minimal investment is hard to reconcile, say some critics, when the government investment-to-savings ratio for conservation can be as high as 1:1000—a thousand dollars saved for every dollar invested.

The lack of interest in energy conservation in the United States and other countries can also be attributed to the influence and power of energy companies, which are a dominant force on the political scene. Nonetheless, as Christopher Flavin and Nicholas Lenssen noted in a recent article, "Powerful economic, environmental, and social forces are pushing the world toward a very different energy system." One of the cornerstones of that system will be energy efficiency. Its potential is great, and many people, businesses, and nations are beginning to recognize that one of the greatest future sources of energy is our waste.

KEY CONCEPTS
Many roadblocks stand in the way of energy efficiency, including the illusion of abundance, federal subsidies that underwrite fossil fuels' true costs, higher initial costs for some energy-efficient products, and powerful political forces. Despite this, energy efficiency is becoming a popular strategy.

15.2 Renewable Energy Sources

Imagine a world powered by the sun, the winds, and other clean, renewable forms of energy. Although this may sound like a dream, 30 years from now you may well live in a world powered by a diverse mixture of renewable energy resources.

The shift to a renewable energy future has already begun. Brazil, for example, is turning to ethanol produced from sugarcane to power trucks and cars. California is turning to wind and geothermal energy. Many other states are following suit. Several European nations, including Germany and Spain, are currently generating a substantial portion of their electricity from wind. Great Britain is following suit and Germany is gearing up for a 100% renewable energy future. Israel and other Middle Eastern countries are increasing their dependence on solar energy. Many Pacific Rim nations now get substantial amounts of energy from geothermal sources (the Earth's heat) and plan to obtain substantially more in the future.

As world oil and natural gas reserves decline, as environmental problems caused by fossil fuels intensify, and as

population spirals upward, more and more countries will shift to renewable energy resources. Each one will find many different clean and abundant renewable fuels that are locally available. By increasing the use of these resources, nations can protect the environment and create more regionally self-reliant economies. By increasing their dependence on clean, affordable, and reliable renewable energy, they create a better future, greater economic and political security, and broader prosperity. This section looks at many of the renewable energy options, discussing the pros and cons of each one.

KEY CONCEPTS
Renewable energy will very likely become a major source of energy in the future; the transition to a renewable energy future has already begun in some nations.

Solar Energy Options

Oil, natural gas, oil shale, and coal all have limits. So does the sun, the origin of solar energy. In the sun's case, though, the supply of energy is expected to last at least 4 to 5 billion years. Although it is finite, solar energy is typically called a renewable resource.

Each day, about two billionths of the sun's energy strikes the Earth. Although this is a small fraction, it adds up to an impressive total. In fact, if all of the sunlight striking an area the size of Connecticut each year were captured and converted into useful energy, it could power the entire United States, including all homes, factories, and vehicles. Despite its enormous potential, solar energy provides only a fraction of U.S. energy needs. Contrary to popular misconception, this poor showing is not because solar energy is limited to a few areas. In fact, significant sources of solar radiation are available throughout the world, although some are better than others (FIGURE 15-5).

> Sunlight striking an area the size of Connecticut each year, if captured and converted to useful energy, would supply all of the power demands of the United States.

Four major solar technologies are in use worldwide: passive, active, photovoltaics, and solar thermal electric. Understanding each one can help us assess the potential of this largely untapped energy source.

KEY CONCEPTS
Solar energy is considered a renewable energy source, but it is really finite. Nonetheless, because it is so abundant and clean, it will very likely be a major contributor to future world energy supplies.

Passive Solar Heating   FIGURE 15-6 shows a house designed to capture sunlight energy. In this structure, sunlight streams through south facing windows and is absorbed by interior walls and floors of brick, tiles, or concrete. The heat stored in these structures radiates into the rooms, heating the air day and night. On cloudy days, solar homes are kept warm by residual

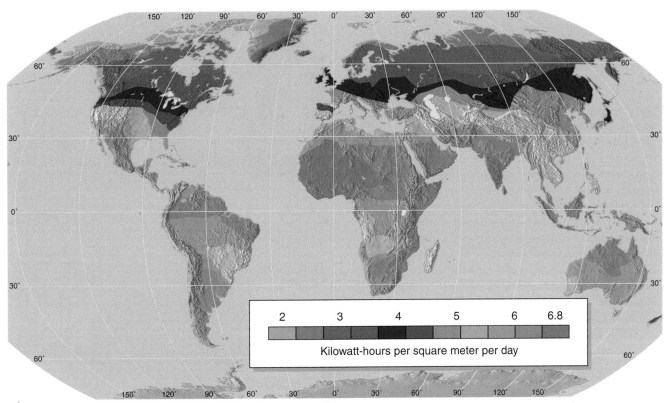

FIGURE 15-5 **Map of global solar energy availability.** The planet is bathed with sunlight energy, although some areas have considerably more than others.

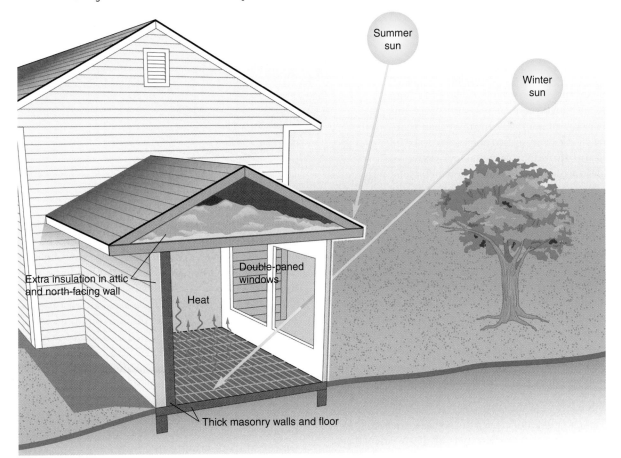

FIGURE 15-6 **A passive solar house.** Sunlight streams into the house during the winter months, when the sun is low in the sky, and is kept out by the overhang during the summer months, when the sun is higher. Interior walls absorb sunlight energy and emit heat, keeping the house warm and cozy at a fraction of the cost of conventional heating systems.

FIGURE 15-7 **Solar retrofit. (a)** Homeowners can add solar green-houses to existing houses. Greenhouses can be used to grow vegetables much of the year. **(b)** Sunlight penetrates the greenhouse glass and is stored in the floor or in water-filled drums. Warm air passes into the house. Fans often help facilitate air movement.

(a)

heat that continues to radiate from heat-absorbent materials (thermal mass) and by backup systems. This is an example of a **passive solar heating system**. Properly designed passive solar homes and buildings also require good insulation and insulated curtains or shades to block the outflow of heat at night. Overhangs block out the summer sun.

Passive solar energy is sometimes described as a system with only one moving part, the sun.[2] It can be added to an existing home by building a greenhouse. In addition to supplying winter heat, a greenhouse can provide a source of food most of the year (**FIGURE 15-7**).

Well-designed passive systems can provide 80 to 90% of a home's space heating. One passive solar home in Canada, built by the mechanical engineering department of the University of Saskatchewan, for instance, had an annual fuel bill of $40, compared with $1400 for an average American home. The house was so airtight and well insulated that heat from sunlight, room lights, appliances, and occupants provided sufficient energy to maintain a comfortable interior temperature. The cost of this house was only slightly more than that of a tract home.

As a rule of thumb, solar houses cost up to 3% more to build than conventional houses of similar size, although there are ways to reduce this cost so that a solar home costs no more than a conventional home. Because used home sales are based primarily on square footage, preowned passive solar homes are often a good buy. My previous solar house cost about the same as conventional used homes, with a fraction of the utility bills! My new solar home was about 5 to 10% cheaper than a conventional home.

Thousands of American homeowners have selected another solar option, the **earth-sheltered house**. Built partly or entirely underground to take advantage of the insulative properties of soil, properly designed earth-sheltered homes are well lighted, dry, and comfortable. They also require less external maintenance. Because they are sheltered by the earth, they stay warm in the winter and cool in the summer. They can easily be designed for passive solar heating. Combined, passive solar heating and earth sheltering can virtually eliminate the need for conventional heat.

KEY CONCEPTS

Buildings can be designed to capture solar energy to provide space heat. Properly designed structures can derive 80 to 90% of their heat from the sun.

[2]Actually, the sun is stationary within our system. The Earth rotates around it.

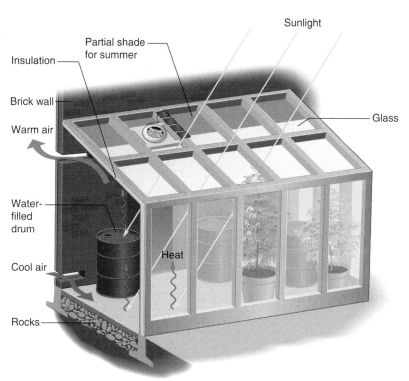

(b)

Active Solar Systems Active heating systems employ **solar collectors**, generally mounted on rooftops. Most collectors are insulated boxes with a double layer of glass on the sunny side (**FIGURE 15-8**). These are called **flat plate collectors**. The inside of the box is painted black to absorb sunlight and convert it into heat. The heat is carried away, by water or some other fluid flowing through pipes in the collector. The heated water is then carried to a storage medium, usually water, in a superinsulated water storage tank. After transferring its heat to the storage medium, the water is returned to the collectors, thus completing the cycle.

The hot water in the storage tank can be used for showers, baths, and washing dishes and can also be used to heat homes. In some parts of the country, active solar water and space heating are competitive with electric heating. In Cyprus, Israel, and Jordan, solar panels supply 25% to 65% of the domestic hot water. Israel eventually hopes to heat 90% of its domestic hot water via rooftop collec-

(a)

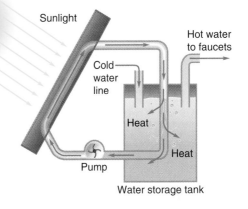

(b)

tors. Specially designed active solar systems can be used in industry to provide hot water and steam, a major consumer of energy.

Solar collectors can be expensive to purchase, mount, and maintain. Leaks in systems can be costly to repair. In fact, a single repair can negate all energy savings for a 6- to 12-month period. Solar collectors are costly to add to a house after it is completed, too. And the extreme temperatures to which they are subjected can be hard on systems. However, some new and inexpensive models are available with few moving parts and requiring little maintenance (FIGURE 15-9).

KEY CONCEPTS

Active solar systems generally employ rooftop panels that collect heat from sunlight and store it in water or some other medium. This solar energy can then be used to heat domestic hot water or to heat the interior of the building.

FIGURE 15-9 **Thermomax Solar Water Heater.** Photo of one of the most ingenious and most efficient solar hot water heaters on the market.

Photovoltaics Photovoltaics provide a way of generating electricity from sunlight. A photovoltaic (PV) cell consists of a thin wafer of silicon or some other material that emits electrons when struck by sunlight. Electrons liberated from the material then flow out of the wafer, forming an electrical current (FIGURE 15-10).

PV cells are assembled onto modules, which are mounted on roofs or poles. Some systems are designed to track the sun across the sky, optimizing electrical production. Solar manufacturers are also making roofing materials, shingles and a laminated solar material that is applied to metal roofing. Some manufacturers even apply PV material to glass for windows and skylights—so they perform double duty: they let light in and also produce electricity (FIGURE 15-11).

Electricity from photovoltaics currently costs about 24–27 cents per kilowatt-hour, or about three to six times more than electricity from conventional sources. Fortunately, costs have fallen rapidly in the last two decades (about 95% since the 1970s) and is expected to decline substantially (around 75%) in the next decade. Experts predict that improvements in production could soon make photovoltaics competitive with electricity from other sources.

> Solar electricity from PVs currently costs about three times more than conventional electrical power.

Photovoltaics are already in wide use in remote villages in the less developed nations. In such locations, photo-

FIGURE 15-10 **Photovoltaic panels.** **(a)** Photovoltaic cells are made of silicon and other materials. When sunlight strikes the silicon atoms, it causes electrons to be ejected. Electrons can flow out of the photovoltaic cell through electrical wires, where they can do useful work. Electron vacancies in the silicon atoms are filled as electrons complete the circuit. **(b)** Array of photovoltaic cells.

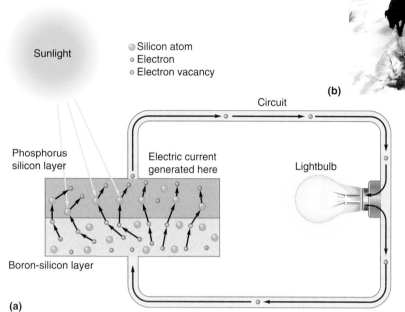

(a)

(b)

voltaics are far cheaper than transmission lines that carry electricity from distant power plants: Today, more than 6000 rural villages in India rely on them. The governments of Sri Lanka and Indonesia have launched ambitious programs to install photovoltaics in remote areas. In the United States, it is often cheaper for a homeowner to install photovoltaics than to pay to string electric lines if the home is more than a few tenths of a mile from a power line. Moreover, several states, including California, New York, New Jersey, Arizona, and Colorado, offer financial incentives that lower the initial cost of PV systems dramatically.

Today, photovoltaics provide only a tiny fraction of the world's electrical power, but sales are growing steadily, averaging a 16% per year increase in the 1990s. Growth has increased even more in the 2000s. Japan, the United States, and the European Union have all adopted million roof policies, which seek to promote installation of solar electric and solar thermal systems on homes and businesses. For a case study on the use of solar electricity in business, see Spotlight on Sustainable Development 15-2.

Another exciting development is the increase in the efficiency of solar cells. Although they are not yet commercially available, solar cells with efficiencies over 30% have been produced by researchers at the National Renewable Energy Labs in Golden, Colorado. Models currently on the market are about half as efficient.

KEY CONCEPTS

Photovoltaics are thin wafers of material such as silicon that emits electrons when struck by sunlight, creating electricity. Although photovoltaics are costly, prices are falling.

FIGURE 15-11 **Solar Shingles.** When reroofing your house, why not replace your ordinary shingles with solar shingles or metal roofing. They protect the house and provide electricity.

15-2   A Solar Giant Grows Taller

In 1998, the Japanese Corporation Kyocera became the world's leading manufacturer of solar electric panels or photovoltaics. That year, the solar giant, long known for its commitment to environmental protection, became even taller with the completion of its new world headquarters, a state of the art office building (FIGURE 1). This 20-story building in Kyoto, Japan was designed to be the world's most environmentally friendly headquarters. What makes it so special?

As you can see from Figure 1, the building is equipped with nearly 2000 PVs, on the roof and on the south side of the structure. They produce about 12.5% of the company's electricity. This, in turn, reduces annual carbon dioxide emissions by 97 metric tons (107 tons), reducing global greenhouse gas emissions. The use of solar electricity also cuts sulfur and nitrogen oxide emissions, two pollutants that are responsible for acid deposition.

The Kyocera headquarters building also acquires much of its electricity from a natural gas cogeneration unit. This system, which burns natural gas to make electricity, operates at about twice the efficiency of a centralized power plant.

Kyocera is air conditioned by ice—that is, by passing cool air over ice. The company makes ice at night using utility power, when demand for electricity is low. Spreading the load out over a 24-hour period makes optimal use of utility power and avoids costly expansion of power production to meet peak demands. It also helps to reduce brownouts during the day, when system-wide demand is high.

Special sensors in the building monitor outdoor and indoor temperature. When appropriate, outdoor air assists in cooling hot spots that develop around windows when the sun shines on them. In addition, the air flow is carefully monitored and controlled in each room to minimize cooling demand. Even the escalator is controlled by a sensor that monitors demand. When a person approaches, it turns on. When there's no activity, the escalator shuts down, saving energy.

During weekends and holidays, excess power from the Kyocera's office building is pumped onto the local electrical grid and is distributed to other customers. Furthermore, Kyocera uses groundwater and rainwater captured on site to irrigate the landscape. High-efficiency lighting and heat-reflecting glass round out the energy features of this building.

Kyocera's headquarters is an environmentally friendly structure, yet aesthetically appealing. Designed to place the highest priority on environmental protection, it is a model for other companies, as is Kyocera itself, most of whose facilities worldwide are environmentally responsible operations. Their whole line of products—from ceramic engines that burn hotter and thus more efficiently to cartridge-free laser printers—shows that companies can prosper while helping build a sustainable future.

FIGURE 1 **Kyocera's headquarters.** This company is the world's leading manufacturer of PVs. This office complex is fitted with thousands of PV panels, which help provide energy.

Solar Thermal Electric   Scientists and engineers have experimented for many years with ways to heat water with sunlight to generate steam that turns a turbine (a special device that drives an electrical generator). Most of their early schemes were large, costly structures. In southern California, one company has pioneered a less expensive alternative. It consists of a series of aluminum troughs that reflect sunlight onto small oil-filled tubes (FIGURE 15-12). The hot oil heats water, which is turned into steam that drives a turbine to generate electric-

ity. This system produces enough electricity to supply 170,000 homes at a cost of 8 cents per kilowatt-hour—cheaper than nuclear power but more costly than coal (if you ignore the costly environmental impacts of electricity generated from coal).

KEY CONCEPTS

Solar thermal electric facilities heat water using sunlight. Steam from this fairly cost-competitive process is used to generate electricity.

FIGURE 15-12 **Solar thermal electric.** These parabolic aluminum reflectors direct sunlight onto an oil-filled tube. The heat is then transferred to water, which boils and produces steam that drives small turbines to generate electricity.

(a)

(b)

FIGURE 15-13 **Wind farm. (a)** These windmill generators in Wyoming produce electricity costing about 5 cents per kilowatt-hour, about the cost of coal-generated electricity but without the adverse environmental impacts. The land can also be used for grazing. **(b)** Wind farm near Canestota, New York.

The Pros and Cons of Solar Energy One of the most notable advantages of solar energy is that the fuel is free. All we pay for are devices to capture and store it. Solar energy is a huge energy resource available as long as the sun continues to shine. While the construction of solar technologies (flat plate collectors and photovoltaics) creates pollution and solid wastes, as does any manufacturing process, once a solar system is operating it is a very clean form of energy. It does not add to global warming, urban air pollution, and other environmental problems. Over their lifetimes, solar systems produce much more energy than is needed to make them. Years of pollution-free operation offset the pollution created by production. Also, most solar systems can be integrated with building designs and therefore do not take up valuable land.

Solar energy has many applications. It can, for example, be used to meet the low-temperature heat demands of homes or the intermediate- or high-temperature demands of factories. Solar energy can be used to cool buildings and generate electricity to power radios, lights, watches, road signs, stream flow monitors, space satellites, automobiles, and industrial motors.

No major technical breakthroughs are required either. Active solar water heating and passive solar space heating are well developed, although some improvements in design and lowered costs could enhance the appeal of others, such as photovoltaics and active solar space heating.

The major limitation of solar energy is that the source, the sun, is intermittent: It goes away at night and is blocked on cloudy days. Consequently, solar energy must be collected and stored. Photovoltaic systems, for example, require storage batteries, although some home owners simply send excess electricity onto the grid during the day. They then draw electricity off at night. (The electrical grid becomes their storage medium.) Passive solar energy stores heat in thermal mass, but most solar users must install a backup system to provide heat during long cloudy periods.

Another disadvantage is cost. Some forms of solar energy do not compete well economically with conventional sources. However, this comparison ignores the massive environmental and economic damage caused by conventional fuels and the huge subsidies that prop them up. When these two factors are taken into account, the economics of solar energy are quite good.

KEY CONCEPTS

Solar energy technologies are well developed. Their advantage over other forms of energy production (such as nuclear or coal) is that they rely on a free, abundant fuel and are relatively clean systems to operate. Although some systems are economically competitive, others are still fairly costly. Storing energy from intermittent sunlight remains one of their major drawbacks.

Wind Energy

About 2% of the sun's energy striking the Earth is converted into wind. Wind can be tapped to generate electricity, pump water, and perform mechanical work (grinding grain, for example). Electricity is produced by generators driven by propeller blades (**FIGURE 15-13**). Large turbines or wind machines are used for commercial energy production. Smaller units can be used to power farms and households. Home owners who rely on wind often install PVs to supplement the wind-generated power. My home, for instance, is powered by PVs and a small wind generator.

Winds are produced by solar energy and can be used to generate electricity or to perform work directly, such as pumping water.

The Pros and Cons of Wind Energy Wind energy offers many of the advantages of solar energy: First, wind energy is an enormous resource. Wind resources are abundant in certain parts of every continent. Tapping the globe's windiest spots could provide 13 times the electricity now produced worldwide. The Worldwatch Institute estimates that wind energy could easily provide 20 to 30% of the electricity needed by many countries. In the United States, North Dakota, South Dakota, and Texas have enough wind to supply all the electrical demands of the entire country.

Wind energy is clean and renewable, uses only a small amount of land (Table 15-2), and is safe to operate. On a typical wind farm, only 10% of the land is used for roads and windmills; the remaining land can be grazed and even planted (Figure 15-13a). Today, many farmers are leasing small portions of their land to companies to install wind generators. Large, commercial generators net them $3000 to $5000 *each* in rent and royalties, greatly boosting farm income.

Windmill technology is well developed, reliable, and efficient, and the fuel is free. Wind-generated electricity (from large windmills) is now cost-competitive with other sources of electricity (Table 15-3). Today, large wind farms are being built all over the world to provide electricity. Spain, Germany, Denmark, India, and the United States are leaders in wind power. Despite good wind resources, Canada has lagged behind other nations in developing wind energy. In the United States, new wind energy now costs on average about 5 cents per kilowatt-hour, clearly cost-competitive

Table 15-3

Present and Estimated Electrical Generation Costs[1]

Source	Cents per Kilowatt-Hour
Nuclear	8–12
Coal	5–7
Gas and oil	6–9
Hydroelectric	3–6
Wind	5–8
Geothermal	4.5–5.5
Photovoltaic	22
Solar thermal	8–12
Biomass	5

Sources: Public Citizen Critical Mass Energy Product and Worldwatch Institute.
[1]Generation costs are costs to companies.

with coal. When one takes into account the environmental benefits, the economics of wind energy (like solar energy) become even more favorable. These factors, combined with the environmental benefit, make wind the fastest growing source of energy in the world with approximately 7000 megawatts of wind electricity being added each year (**FIGURE 15-14**). That's as much electrical energy as two huge coal-fired power plants produce.

> Wind energy is currently cost-competitive with electricity from coal-fired power plants and about half the cost of electricity from nuclear plants.

There are, of course, disadvantages to wind. The wind does not blow all of the time, so backup systems and storage are needed. Second is the visual impact. Individual windmills and wind farms can be eyesores. Third, large wind generators may impair television reception (although fiberglass blades reduce interference by half). Some generators may also impair the microwave communications used by telephone companies.

Although winds do not blow all of the time in all locations, wind still could become a major source of electricity. It could supplement solar and hydroelectricity and conventional power sources as well. Surpluses produced in one area could be transferred elsewhere. Consistently windy locations like coastlines could also provide a steady stream of electricity to the electrical grid.

Wind energy is clean, abundant, and fairly inexpensive, especially when one includes its low environmental costs. Wind energy could provide a significant percentage of our future energy demand. However, because winds are often intermittent, backup systems and storage are necessary.

Table 15-2

Land Use of Selected Electricity-Generating Technologies, United States

Technology	Land Occupied (square meters per gigawatt-hour, for 30 years)
Coal[1]	3642
Solar thermal	3561
Photovoltaics	3237
Wind[2]	1335
Geothermal	404

Source: From L.R. Brown, C. Flavin, and S. Postel (1991). *Saving the Planet: How to Shape an Environmentally Sustainable Global Economy.* New York: Norton.
[1]Includes coal mining.
[2]Land actually occupied by turbines and service roads.

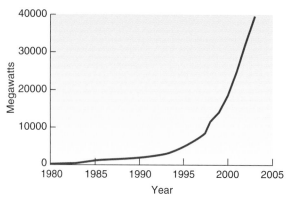

(a) World wind energy generating capacity

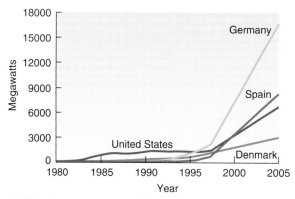

(b) Wind generating capacity of various countries

FIGURE 15-14 **Wind energy. (a)** Wind has become the fastest growing source of energy in the world. **(b)** Some countries are leading the world in the production of electricity from wind.

Biomass

The organic matter contained in plants is called **biomass.** The energy found in this organic plant matter comes from sunlight. Useful biomass includes wood, wood residues left over from the timber industry, crop residues, charcoal, manure, urban waste, industrial wastes, and municipal sewage. Some of these fuels can be burned directly. Others are converted to methane (a gas) and ethanol (a liquid). The simplest way to get energy from biomass is to burn it. In some countries, though, it may make more sense to convert it to gaseous and liquid fuels such as ethanol. These can be burned in motor vehicles or used to produce raw materials for the chemical industry, to manufacture drugs and plastics as oil and natural gas supplies decline.

Biomass (wood, for instance) supplies about 20% of the world's energy. In fact, it is the primary source of energy for about half of the world's population, primarily those in the less developed countries. In sub-Saharan Africa, three fourths of the energy comes from burning wood. In contrast, in the United States and other developed countries, biomass supplies only about 3% of the energy needs. Wood-rich Canada currently supplies about 6% of its total energy demand from biomass, mostly wood.

Certain nonfood crops could also be grown to produce liquid fuel. For example, a desert shrub (*Euphorbia lathyris*) found in Mexico and the southwestern United States produces an oily substance that could be refined to make liquid fuel. In arid climates, the shrub could yield 16 barrels of oil per hectare (6 barrels per acre) on a sustainable basis. The co-paiba tree of the Amazon yields a substance that can be substituted for diesel fuel without processing. Sunflower oil can also be used in place of diesel. Farmers could convert 10% of their cropland to sunflowers to produce all the diesel fuel needed to run their machinery. Eventually, the entire transportation system could be powered by renewable fuels.

Another potential source of energy is manure from livestock operations. Manure is fed into vats where it decomposes, giving off methane, a combustible gas. All over the world, plans are underway to use manure to produce a combustible fuel. This not only helps put a waste to good use, it helps reduce pollution.

KEY CONCEPTS

Biomass is organic matter such as wood or crop wastes that can be burned or converted into gaseous or liquid fuels. It is a common fuel source in most developing nations but supplies only a fraction of the needs of people in the developed nations.

The Pros and Cons of Biomass   Biomass offers many benefits. It can help less developed nations reduce their dependence on nonrenewable energy resources. It also has a high net energy efficiency when it is collected and burned close to the source of production, and it has a wide range of applications. Unlike fossil fuels, biomass does not pollute the atmosphere with carbon dioxide, long implicated in the greenhouse effect—as long as the amount of plant matter burned equals the amount of plant matter produced each year. Burning some forms of biomass, such as urban refuse, reduces the need for land disposal.

Several recent technological and scientific developments could improve the prospects for biomass. First is the use of high-efficiency gas turbines, similar to those in jet aircraft, to generate electricity from hot gases produced by the combustion of various forms of biomass. Second is the development of an enzymatic process that improves the efficiency of ethanol production from wood wastes—a procedure that has decreased the cost of ethanol from wood from $4 a gallon to $1.35 in the last decade. Further improvements could decrease costs to 60 cents a gallon. Wood wastes now dumped in landfills, urban tree trimmings, and sustainably managed tree farms could eventually form the base of a sustainable ethanol production for cars and buses.

Biomass is not a panacea, however. Although the world's reliance on this form of renewable energy will probably increase in coming years, it is doubtful that its use will increase as much as some proponents suggest except perhaps for biodiesel (discussed shortly). Why not? Global warming could reduce water supplies and agricultural output in many areas, thus constraining biomass production. Further, some people believe that rising food demand may limit the amount of farmland available to produce biomass fuels. Furthermore, removing crop and forestry residues may reduce soil nutrient replenishment. Biomass can create large amounts of air pollution—for example, smoke from woodstoves. This smoke contains several toxic air pollutants, such as dioxins, and creosote. Studies show that children living in homes with wood-burning stoves have more respiratory problems than children in homes heated by other means. Finally, transportation costs for biomass are higher than traditional fossil fuels because biomass has a lower energy content per unit of weight.

Biodiesel

Another biomass solution that holds great promise is biodiesel. **Biodiesel** is a renewable fuel made from an assortment of vegetable oils—for example, corn oil, canola oil, and soy oil. The liquid is a combustible, clean-burning fuel that could eventually power many of the nation's trucks, cars, buses, vans, ships, and trains, perhaps even jets. (There are already over 100,000 gas stations in the United States that sell biodiesel.)

Biodiesel is made by mixing vegetable oil with a solution of methanol containing sodium hydroxide (lye). The oil is usually heated. The methanol-lye mixture is then added to the vegetable oil. This solution is heated some more and then stirred for a period of time, usually about an hour. When the chemical reaction is complete, out comes biodiesel—long chain fatty acids that burn very nicely in diesel engines. The only waste product is glycerol, a dark thin oily substance that can be purified and turned into to soap.

The Pros and Cons of Biodiesel Biodiesel is a renewable fuel that could help us meet our needs for transportation fuels as the age of oil winds down. Biodiesel can help stimulate our nation's economy, in part, by reducing its reliance on foreign oil. It could also help stimulate rural economies, too. In the coming years, many farmers will be enlisted to produce "fuel grains" for the biodiesel market. Local biodiesel manufacturers could convert the grains to vegetable oil. Crop production and local manufacturing not only create stronger local and regional economies, they help forge the path to a more decentralized and sustainable system of fuel production.

According to Marc Franke, an Iowa-based proponent of biodiesel, the net energy efficiency of biodiesel production from soy oil is 3.2. The net energy production of biodiesel from canola or rapeseed oil is 4.3. What these figures mean is that for every unit of energy invested in biodiesel production, you get 3.2 units of energy output for soy oil and 4.3 units of energy from canola oil. According to Franke, it would take 7.0 acres of soybeans to supply soybeans to extract the oil needed to make biodiesel for a diesel car that travels 15,000 miles per year and gets 44 miles per gallon. It would take 2.7 acres of canola oil to do the same thing.

Seed crops are not the only potential source of biodiesel, however. Biodiesel can also be produced from vegetable oils discarded by area restaurants—reducing disposal costs for

them while providing a valuable renewable liquid transportation fuel. U.S. restaurants, including all the fast-food chains, produce an estimated 3 billion gallons of waste vegetable oil per year, according to Franke!

There are a lot of other sources, too. For example, biodiesel can be manufactured from algae grown in ponds associated with sewage treatment plants, helping reduce pollution while generating liquid fuel for North America's transportation system. Diesel fuel can also be produced from the organic waste such as that generated at turkey farms using a process known as Thermal De-Polymerization (TDP). Agricultural, organic wastes, says Franke, produce enough material to make 4 billion barrels of biodiesel each year!

According to Franke, biodiesel offers the same performance as regular diesel, but dramatically reduces tailpipe emissions. For example, biodiesel contains no sulfur; therefore, combustion of this fuel source eliminates sulfur oxide emissions that contribute to acid rain and snow—a big problem with conventional diesel vehicles.

There's no black smoke spewing from tailpipes of cars or trucks powered by biodiesel as they pass or power up a hill. Complete cycle carbon dioxide production from the manufacture of the fuel and use of biodiesel cars and trucks is 78% lower than vehicles powered by standard diesel fuel (it would be even lower if biodiesel is made with methanol, most of which comes from natural gas). According to the U.S. Department of Energy, biodiesel is the fastest growing alternative fuel on the market today.

Yet another advantage of biodiesel is that the transition to this fuel won't require major changes in distribution systems. Biodiesel could be produced locally and sold at area filling stations using the same facilities that are used to dispense regular diesel fuel.

Biodiesel can also be run in space heaters and oil furnaces, even forklifts, tractors, and electric generators that currently use diesel. Your home someday could be heated with biodiesel.

Despite these benefits, biodiesel does have a few disadvantages. For one, it is not yet widely available. Second, in the United States, commercially manufactured biodiesel costs a bit more than standard petroleum-derived diesel. This, of course, is largely the result of the fact that biodiesel is only produced by small-scale production facilities. Increasing the scale of production could bring the cost down substantially. As oil prices rise and biodiesel production ramps up, costs could come down even more, making biodiesel the fuel of the future. Improving matters, starting in January 2004, producers will receive a federal tax incentive of up to $1 per gallon of biodiesel. With this incentive, biodiesel may now cost the same or even less than petroleum diesel.

You can also make your own biodiesel at bargain basement prices in your basement. Biodiesel Solutions, a company in Fremont, California sells equipment you can set up in your basement or garage to manufacture biodiesel. They claim that it currently costs only about 70 cents per gallon to make, far lower than standard diesel, which is currently running about $2.50 per gallon.

Another problem with biodiesel is that soy biodiesel gels around 32° F, and, therefore can't be used at full strength when the weather gets cold (rape seed biodiesel gels at a lower temperature). To get around this, biodiesel is ordinarily mixed with conventional diesel to which manufacturers add an antigelling agent to winter petroleum diesel.

Finally, large-scale biodiesel production will put increasing demand on farmland. Production could, if large enough, even reduce the exports of food crops. Although farmers won't suffer, those who rely on food from North America could find themselves in a bind. One way to lessen the impact of widespread biodiesel production would be to make it from algae and waste from turkey and hog farms.

Vegetable Oil as a Fuel

Some people are burning vegetable oil directly in diesel cars, trucks, and buses. Some individuals acquire their fuel from unused vegetable oil. Others salvage vegetable oil from fast-food chains and other restaurants. How can vegetable oil be burned in a diesel car?

It may surprise readers to learn that the German inventor of the diesel engine, Rudolph Diesel, designed his long-lasting, energy-efficient engine to first run on peanut oil. Modern diesel engines, however, can't run directly on vegetable oils. To run a vehicle on straight vegetable oil, which is thicker than biodiesel or conventional diesel fuel, car and truck owners must make some modifications to their vehicles. Several companies now manufacture conversion kits costing $300 to $800. These kits include a separate tank for vegetable oil and heating elements for the tank, the fuel line, and the fuel filter. They heat the vegetable oil, which makes if flow more readily.

Vegetable oil is an ideal fuel, but one needs to understand the pros and cons of this approach to compare it with other options, especially biodiesel.

Let's start on the up side.

One of the biggest advantages of vegetable oil is that the fuel is widely available. You can stop and fill up anywhere in North America as long as there's a restaurant that deep fat fries its food in vegetable oil. (You'll have to filter the gunk out of the deep fryer oil first, however.)

Not only is the fuel abundant and widely available, so are diesel engine conversion kits. You can purchase them on the Internet.

Like other options discussed in this chapter, the fuel is renewable and can be picked up for free—for example, from fast-food restaurants whose owners are typically pleased to give it away, rather than pay for disposal!

Vegetable oil burns cleanly and thus helps solve another thorny issue—air pollution from conventional fossil fuels. Like biodiesel, vegetable oil use reduces carbon dioxide emissions. The fuel itself is often called carbon neutral but that's not entirely true. Although the amount produced during combustion equals the amount the plants take up during photosynthesis, remember that it takes energy to make this fuel (gasoline to power a tractor, for instance). The consumption of energy, in turn, produces carbon dioxide. Even

so, the fuel is light years ahead of conventional fossil fuels in the greenhouse gas production department.

Vegetable oil fuels have some disadvantages. The conversion kits and installation of the kits costs money. In addition, the additional fuel tank takes up a considerable amount of trunk space.

Hydroelectric Power

Humankind has tapped the power of flowing rivers and streams for thousands of years to run flour mills and, more recently, to produce electricity. River flow is made possible by two factors: sunlight energy, which drives the hydrological cycle (Chapter 13), and gravity, which is responsible for the movement of water in streambeds.

Hydroelectric power is a renewable resource usually tapped by damming streams and rivers. Water in the reservoirs behind dams is released through special pipes. As it flows out, it turns the vanes of electric generators, producing electricity.

Hydropower supplies about one fifth of the world's electrical demand. In Canada, it supplies about 12% of the nation's total annual energy demand. Hydropower creates no air pollution and is relatively inexpensive. Furthermore, the technology is well developed. Although dams and hydroelectric power plants are expensive to build, they often have lifespans 2 to 10 times greater than coal or nuclear powered plants.

There are some problems associated with hydropower. As noted in Chapter 13, reservoirs behind dams often fill with sediment, shortening the life span of a hydropower facility to 50 to 100 years. However, large projects with enormous reservoirs may last 200 to 300 years. Once a good site is destroyed by sediment, it is gone forever. Dams and reservoirs flood productive land, displace people, destroy stream fisheries, eliminate certain forms of recreation, upset nutrient replenishment in estuaries, and create many additional problems discussed in Chapter 13.

Many countries have a large untapped hydroelectric potential. In South America, for instance, hydroelectric generating potential is estimated at 600,000 megawatts. By comparison, the United States, the world's leader in hydroelectric production, has a present capacity of about 77,000 megawatts and an additional capacity of about 150,000 to 160,000 megawatts.

As pointed out in Chapter 13, estimates of hydroelectric potential can be deceiving because they include all possible sites, regardless of their economic or environmental costs. For example, half of the U.S. hydroelectric potential is in Alaska, far from places that need power. The potential for additional large projects in the continental United States is small because the most favorable sites have already been developed. Canada faces a similar shortage. Moreover, those developing sites that are available would result in massive environmental impacts and might be strongly opposed by the citizenry. In addition, the high cost of constructing large dams and reservoirs has increased the cost of hydroelectric energy by from 3 to 20 times since the early 1970s. Because of these factors, hydropower is projected to increase very slowly in the coming decades.

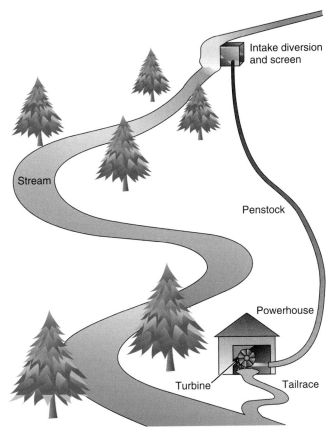

FIGURE 15-15 Small-scale hydropower. Flowing water from a nearby stream turns a turbine to generate electricity.

Two of the most sensible strategies for increasing hydropower may be to (1) increase the capacity of existing hydroelectric facilities—that is, add more turbines—and (2) install turbines on the many dams already built for flood control, recreation, and water supply. In appropriate locations, small dams could provide energy needed by farms, small businesses, and small communities. All projects, however, must be weighed against impacts on wildlife habitat, stream quality, estuarine destruction, and recreational uses.

In LDCs, small-scale hydroelectric generation may fit in well with the demand. In China tens of thousands of medium and small hydroelectric generators account for about one fourth of the country's electrical output. Small instream generators that do not impede the flow of water can also provide power to individual homes located near waterways. Water can also be removed, transported downstream in small pipes (to pick up speed) and then run through small

generators (FIGURE 15-15). These can provide a remarkable amount of electrical energy 24 hours a day, 365 days a year! However, there aren't many suitable sites in most parts of the country for small-scale hydros, called microhydro.

KEY CONCEPTS

Hydroelectric power is renewable and operates relatively cleanly, but dams and reservoirs have an enormous impact on the environment. Although the potential hydropower sources are enormous, they are often far from settlements, and developing them would cause serious environmental damage.

Geothermal Energy

The Earth harbors an enormous amount of heat, or **geothermal energy**, which comes primarily from magma, molten rock beneath the Earth's surface. Geothermal energy is constantly regenerated, but because the rate of renewal is slow, overexploitation could deplete this resource. Geothermal resources fall into three major categories.

Hydrothermal convection zones are places where magma intrudes into the Earth's crust and heats rock that contains large amounts of groundwater. The heat drives the groundwater to the Earth's surface through fissures, where it may emerge as steam (geysers) or as a liquid (hot springs).

Geopressurized zones are aquifers that are trapped by impermeable rock strata and heated by underlying magma. This superheated, pressurized water can be tapped by deep wells. Some geopressurized zones also contain methane gas.

Hot-rock zones, the most widespread but also the most expensive geothermal resource, are regions where bedrock is heated by magma. To reap the vast amounts of heat, wells are drilled, and the bedrock is fractured with explosives. Water is pumped into the fractured bedrock, heated, and then pumped out.

Geothermal energy is heavily concentrated in a so-called *ring of fire* encircling the Pacific Ocean and in the great mountain belts stretching from the Alps to China (FIGURE 15-16). It is also prevalent around the Mediterranean Sea and in East Africa's Great Rift Valley, which extends along the eastern part of the African continent. Within these areas, hydrothermal convection zones are the easiest and least expensive to tap.

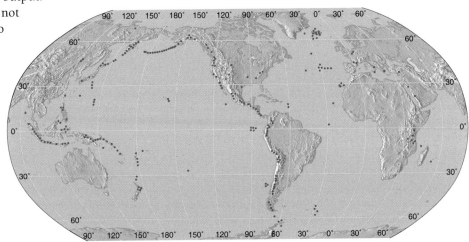

FIGURE 15-16 Global geothermal resources. Note the concentration of geothermal resources on the Pacific coastline—the so-called Ring of Fire.

Hot water or steam from them can heat homes, factories, and greenhouses. In Iceland, for example, 65% of the homes are heated by the Earth's heat. Iceland's geothermally heated greenhouses produce nearly all of its vegetables; Russia and Hungary also heat many of their greenhouses in this way.

Steam from geothermal sources can be used to run turbines that produce electricity. Geothermal plants can produce electricity day and night and can provide electricity when wind or solar systems are not operating. They can also provide electricity in areas without sizable wind or solar resources.

Although still in the early stages of development in most countries, geothermal electric production is growing quickly in the United States, Italy, New Zealand, and Japan. El Salvador in Central America, however, currently generates 40% of its electricity from geothermal sources, and Kenya and Nicaragua acquire 11% and 10%, respectively.

Hydrothermal convection systems have several drawbacks. The steam and hot water they produce are often laden with minerals, salts, toxic metals, and hydrogen sulfide gas. Many of these chemicals corrode pipes and metal. Steam systems may emit an ear-shattering hiss and release large amounts of heat into the air. Pollution control devices are necessary to cut down on air and water pollutants. Engineers have also proposed building closed systems that pump the steam or hot water out and then inject it back into the ground to be reheated. Finally, because heat cannot be transported long distances, industries might have to be built at the source of energy.

> **KEY CONCEPTS**
>
> Geothermal energy is a renewable resource created primarily from magma, molten rock beneath the crust. Geothermal energy is used to generate electricity and to heat structures; it is a major source of energy in some countries.

Hydrogen Fuel

Hydrogen is another renewable fuel that could be used to help replace oil, gasoline, and natural gas. It could someday be burned in our stoves, water heaters, furnaces, factories, and cars.

Hydrogen gas is produced by heating water or passing electricity through it in the presence of a catalyst (a chemical that facilitates the breakdown of water into oxygen and hydrogen without being changed itself). The device used to generate hydrogen is known as an **electrolyzer.**

Hydrogen is a relatively clean-burning fuel. In fact, when hydrogen burns it produces only water and small amounts of nitrogen oxide (formed by the combination of atmospheric oxygen and nitrogen). Nitrogen oxides, which contribute to acid deposition, can be minimized by controlling combustion temperature and by installing special pollution control devices. Unlike fossil fuels, hydrogen produces no carbon dioxide when burned.

Because hydrogen could be generated from seawater, it is an essentially limitless and renewable energy resource. Hydrogen is also easy to transport and has a wide range of uses. Unfortunately, at the present time it takes consider-

able energy to produce hydrogen. This low net energy yield makes it an expensive form of energy.

Hydrogen may also serve as a way of storing energy from hydroelectric, wind, solar, and other renewable energy sources. For example, when demand for these sources is low, the electricity they generate could be used to produce hydrogen from water. It would be stored for later use. When renewable sources are not available (say, on a calm day or during the evening), the hydrogen could be burned to produce electricity.

Although the immediate prospects for hydrogen are not spectacular, efforts to produce hydrogen more efficiently could go a long way to make this energy source more cost effective. Some energy analysts predict that one day areas rich in sunlight, wind, and water could become major sources of hydrogen fuel, which could be piped around the world in existing natural gas pipelines.

In 1996, the U.S. Congress passed the **Hydrogen Future Act.** This law supports research and development of hydrogen fuels and provides approximately $165 million to support these activities.

Fuel Cells Hydrogen can also be used to produce electricity in a **fuel cell.** Many different types of fuel cells have been developed. FIGURE 15-17 shows a simplified view of how a fuel cell works. As illustrated, this particular type of fuel cell looks a lot like a battery. It consists of an anode and a cathode separated by a membrane. Hydrogen is fed into the cathode where it reacts to form hydrogen ions and electrons. The electrons are drawn off by a wire, forming electricity, which powers lights, electronic equipment, and so on. Oxygen is introduced into the cathode, where it reacts with hydrogen ions that pass through the membrane from the anode and electrons returning from the circuit. The product is water.

Electricity produced by a fuel cell can be used to power electric motors in cars. In fact, in 1999 Daimler-Chrysler unveiled a hydrogen-powered car that can travel as fast as 90 miles per hour. It uses hydrogen and oxygen to run a fuel cell, which runs an electric motor to power the car. One company in Canada, Ballard Power Systems, has built hydrogen fuel cell powered buses, which are now being used experimentally in Chicago and British Columbia. During the public ceremony in Chicago, then Mayor Richard Daley, celebrated by drinking a glass of exhaust water collected from the tail pipe of an idling bus to underscore how safe the emissions are. If tests prove successful, Ballard hopes to enter full-scale production of fuel cell bus engines.

Iceland has committed itself to a hydrogen economy and will soon receive all of its energy from hydrogen and geothermal energy. Although hydrogen seems like a promising fuel, remember that it takes energy (electricity) to make hydrogen. Studies show that it is far more efficient to use electricity directly to power an electric car than it is to use that electricity to generate hydrogen to run a fuel cell to produce electricity to run a car. Fuel cells use hydrogen derived from electrolyzers and oxygen, but both hydrogen and oxygen burn. Furthermore, hydrogen is difficult to store. It must be stored in an explosion-proof tank. Today, most fuel cells'

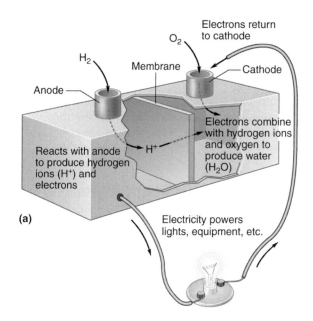

(b)

FIGURE 15-17 Fuel cell. Many people believe that fuel cells are the wave of the future. **(a)** They resemble batteries and produce electricity from hydrogen and oxygen. **(b)** Photo of a real fuel cell.

work is focused on using hydrogen-containing fuels such as methanol, natural gas, ethanol, or gasoline. The fuels will be fed into an onboard fuel processor that strips the hydrogen from the molecules. The hydrogen is then fed into the fuel cell. Although most fuel cells under development are powered by fossil fuels, the process is reportedly very clean and represents a major improvement over standard gasoline- and diesel-powered motor vehicles. They still produce significant amounts of carbon dioxide. At this writing, Daimler-Chrysler, Toyota, Honda, Ford, and General Motors have fuel cell vehicles that could be brought to market soon.

> Renewable energy in the United States could provide the equivalent of 70 to 80 billion barrels of oil per year. There are currently about 9 billion barrels worth of fossil fuel reserves in the United States.

KEY CONCEPTS

Hydrogen may become an important fuel in the future. Hydrogen can be produced by passing electricity through water, a renewable resource. When hydrogen burns, it produces water vapor. Fuel cells use hydrogen, either from water or organic fuels, to produce electricity. The electricity can be used to power cars, and several manufacturers are actively pursuing this option.

15.3 Is a Renewable Energy Supply System Possible?

Enormous amounts of renewable energy are available and accessible with current technologies—far more than are available from fossil fuel reserves. This has led some scientists and environmentalists to propose a sustainable energy supply system far different from the one in existence today. They see energy efficiency and renewable energy resources as the cornerstone of this new, environmentally sustainable system. **FIGURE 15-18** shows one scenario for the projected shift in the world's energy supply system. As illustrated, renewables could take over for all of the nonrenewables. This transition will take many years, and thus, fossil fuels will be around for a long time.

But how will renewable energy supplant the nonrenewables that currently power much of the world? Where will elec-

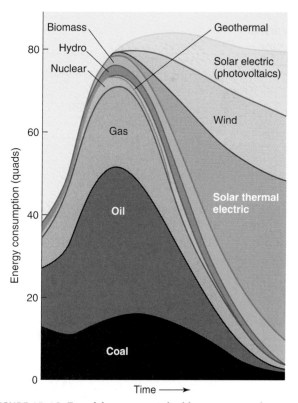

FIGURE 15-18 Transition to a sustainable energy supply system. One projection of possible energy resources in the United States. These figures are based on full commitment to renewable resources.

tricity come from? Where will the fuel come from to power the transportation system? What fuels will power industry?

Let's take these questions one at a time, remembering that energy efficiency will be a key to the success of any future energy system.

Electricity could come from conservation efforts, wind farms, photovoltaics, solar thermal electric systems, fuel cells powered by hydrogen, small-scale hydropower, and geothermal facilities. Biomass facilities and cogeneration could provide additional electricity. One form of biomass, animal manure, could become a major source of energy in the future, helping power farms and rural homes. Eventually, hydrogen may replace natural gas in our kitchens and at power plants. Thus, electricity could come from dispersed as well as centralized sources.

Low- and intermediate-temperature thermal energy (heat) is currently used to heat buildings and power industrial processes. It could be generated from solar energy—or from the combustion of various renewable fuels such as biomass or hydrogen. Solar sources could be used for home space heating, water heating, and many industrial processes.

Table 15-4 gives a further breakdown of energy demand by sector, showing what kind of energy is needed and how it would be used. The point of this table is that there are options—many of them. They may seem farfetched right now, but as the world turns away from fossil fuels, many alternative energy advocates believe that the renewable alternatives will become increasingly popular.

Economic and Employment Potential of the Sustainable Energy Strategy

Numerous studies show that conservation and certain renewable energy technologies actually produce energy more cheaply than conventional sources and eliminate many of the environmental impacts as well. They do this while employing more people than conventional energy supply strategies. The chief reason for this is that conservation and alternative technologies tend to be more labor-intensive (requiring more people) than oil, coal, and nuclear energy, which are more capital-intensive (requiring huge investments in machines and fuels). Consider some examples.

Table 15-4

Meeting Energy Needs of a Solar-Powered Society

Demand Sector	Sources	Application	Percentage of Total Energy Use
Residential and commercial	Passive and active solar systems, district heating systems	Space heating, water heating, air conditioning	20–25
	Active solar heating with concentrating solar collectors		
	Solar thermal, thermochemical, or electrolytic generation	Cooking and drying	~5
	Photovoltaic, wind, solar thermal, total energy systems	Lighting, appliances, refrigeration	~10
			Subtotal ~35
Industrial	Active solar heating with flat plate collectors, and tracking solar concentrators	Industrial and agricultural process heat and steam	~7.5
	Tracking, concentrating solar collector systems	Industrial process heat and steam	~17.5
	Solar thermal, thermochemical, or electrolytic generation		
	Solar thermal, photovoltaic, cogeneration, wind systems	Cogeneration, electric, drive, electrolytic, and electrochemical processes	~10
	Biomass residues and wastes	Supply carbon sources to chemical industries	~5
			Subtotal ~40
Transportation	Photovoltaic, wind, solar thermal	Electric vehicles, electric rail	10–20
	Solar thermal, thermochemical, or electrolytic generation	Aircraft fuel, land and water vehicles	
	Biomass residues and wastes	Long-distance land and water vehicles	5–15
			Subtotal ~25
			Total 100

Source: H.W. Kendall and S.J. Nadis (1980). *Energy Strategies: Toward a Solar Future.* Cambridge, MA: Ballinger, p. 262.

A California-style wind farm generates electricity for about 5 cents per kilowatt-hour, one half the cost of electricity from a nuclear power plant and without the serious environmental impacts. In addition, a wind farm that produces the same amount of energy as a nuclear power plant will employ over 540 workers, compared to 100 workers in the nuclear facility (FIGURE 15-19).

Similar gains can be made through energy conservation. A study in Alaska, for instance, found that state expenditures on weatherization—home energy conservation—would create more jobs per dollar of investment than the construction of hospitals, highways, or new power plants. Conservation spending, for instance, would create three times as many jobs as highway construction.

Weatherization of all homes in the United States would create 6 to 7 million job-years, according to Worldwatch Institute projections. (A job-year is one job for 1 year.) Six million job-years is the equivalent of 300,000 jobs over a period of 20 years! Not all jobs would be low-wage, either. Energy-efficiency companies would be run by well-paid personnel and would employ salespeople, managers, accountants, engineers, and installers.

A wise energy future is economically, environmentally, and socially sustainable. The current system fails on all of these criteria. Nonetheless, great barriers lie on the path to a sustainable system. One of the most significant is that industrial nations have been built on fossil fuels. Billions upon billions of dollars have been spent on the present fossil fuel-based system. Huge investments have been made in mining and drilling equipment, transportation networks, processing facilities, and power plants. Although the system is unsustainable, enormous resistance to change will come from many powerful political and economic interests: legislators from energy-producing states, coal and oil companies, and the power companies, which expect to profit from their investments.

Creating a sustainable system of energy will take many years. As fossil fuels and nuclear plants become obsolete, they can be replaced by renewable energy sources such as wind or solar energy. Some parts of the current energy system, such as electric lines, can be used to transport solar- and wind-generated electricity. Natural gas pipelines could be used to transport hydrogen gas.

Shifting to renewable energy resources and conservation will also require massive amounts of investment capital to finance wind farms, geothermal plants, and so on. Although the shift to a sustainable energy strategy could create many new economic opportunities, jobs will be lost in some sectors. If proponents are right, many more jobs will open up than will be lost. Not only will there be more jobs, but they will undoubtedly be safer. Some workers will be able to turn their old skills to similar activities. Petroleum geologists and oil well crews, for example, might use their expertise to drill for geothermal resources.

The transition to a renewable energy system is occurring, slowly but surely in many parts of the world. Individuals

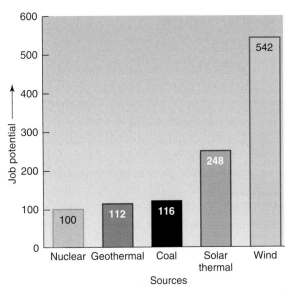

FIGURE 15-19 **Comparison of employment opportunities of various energy sources.** This graph shows the difference in employment opportunities of different energy strategies, all producing the same amount of energy per year. Renewable resources turn out to be stellar job creators.

and businesses can help by buying green power, when it is available. Some utilities, for instance, offer electricity from wind sources to customers at a slightly higher cost. Governments can help, too. In 1996, the Canadian government introduced a program to promote renewable energy that includes a variety of measures, including tax breaks, grants for private research, and steps to increase demand for green power in the marketplace. The government is even using more renewable energy in its own facilities. The U.S. government has taken similar steps. The National Renewable Energy Lab, for instance, and many other government labs and facilities are now getting power from solar electricity and wind. Even the White House has been retrofitted with solar electric panels. Many states now have passed laws calling for an increase in renewable energy (mostly wind) in the next decade.

The length of the transition to renewable energy resources will depend on our political will and our willingness to change for the sake of the planet's future.

KEY CONCEPTS

Shifting to a sustainable system of energy will take many years. Several renewable energy technologies provide competitively priced electricity while creating more jobs than fossil fuels and nuclear energy.

The great end of living is to harmonize man with the order of things.

—Oliver Wendell Holmes

CRITICAL THINKING

Critical thinking warns us that broad statements—generalizations—such as those made by the supposed energy expert often fall apart on closer examination. A careful analysis of the issue reveals a somewhat different picture of reality—one much more favorable for solar energy.

Solar energy, as noted in this chapter, consists of at least four different technologies. One of them, solar thermal electric, is currently cost competitive with nuclear power. Solar thermal electric plants use sunlight to heat water or some other fluid, which is then converted into steam to run an electric turbine.

Another form of solar energy is solar electricity generated from photovoltaics (panels on rooftops that convert sunlight energy into electricity). Photovoltaics are not currently cost competitive with conventional fossil fuel resources in most applications in developed countries. It would cost a fortune for most homeowners, for instance, to add panels to their roofs to supply their homes with electricity.

In rural villages in the developing world, however, photovoltaics are quite cost-competitive because the expense incurred by running power lines to remote sites is astronomical. Even in developed countries such as the United States, though, photovoltaics can make sense. If a house is only half a mile from a power line, it is often more economical to install photovoltaics than to run a power line to the house.

Another solar technology is passive solar heating—that is, heating buildings with sunlight energy that streams in through south-facing windows. Passive solar often costs more, but it can be built economically. The author's house cost no more than a conventional home and is heated almost entirely by the sun, saving several thousand dollars a year.

This exercise also shows that it is important to define what is meant by various terms, such as *uneconomic*. Does the term refer only to conventional economics, which ignores costs such as damage to the environment? In this case, the oil industry spokesperson is talking only about the cost of producing conventional fuels. He's ignoring the enormous external costs from planetary warming, acid deposition, and urban air pollution.

Along that same line, does the cost of production include subsidies? As this chapter points out, oil in the United States is very heavily subsidized by the federal government. The nuclear and coal industries are also heavily subsidized. If subsidies were removed, they would be far less affordable, and many solar technologies would appear quite competitive. Thus, this exercise also shows the importance of considering the source of information that may be biased by economic self-interest.

CRITICAL THINKING AND CONCEPTS REVIEW

1. In your view, is it imperative that we change to a sustainable energy system? Why or why not?
2. Describe the types of solar energy technologies available. How does each one operate? What is it used for? Which ones are already cost competitive with current energy technologies?
3. What are the advantages and disadvantages of solar energy? How can the problems with the technologies be solved?
4. A person living in the Pacific Northwest argues that renewable energy is useless. The sun rarely shines in his neck of the woods. How would you answer this?
5. Describe the difference between passive and active solar systems. What features are needed in a home to make passive solar energy work?
6. What are photovoltaic cells? Why are they economical to use in rural villages in the developing nations and in semiremote sites in developed countries?
7. Wind energy is cost competitive with conventional electricity from coal and cheaper than nuclear energy. Should we develop this energy resource in preference to nuclear power, coal, or shale oil? Why or why not?
8. What is biomass? How can useful energy be acquired from biomass?
9. How is geothermal energy formed? How can it be tapped? Describe the benefits and risks of geothermal energy.
10. Using your critical thinking skills, debate the following statement: "Hydroelectric power is an immensely untapped resource in the United States and could provide an enormous amount of energy."
11. What are the major problems facing hydrogen power? How could these be solved?
12. Using your critical thinking skills, discuss the following statement: "Conservation is our best and cheapest energy resource."
13. Discuss ways in which you could conserve more energy at home, at work, and in transit. Draw up a reasonable energy conservation plan for yourself and your family.
14. Using your critical thinking skills, debate this statement: "A sustainable energy strategy won't work. It will cost money and lose jobs."
15. List and describe the barriers to creating a sustainable energy supply system. How could they be overcome?

KEY TERMS

biodiesel
biomass
cogeneration
conservation
earth-sheltered house
The Efficiency and Alternative Energy program
electrolyzer
Energy Efficiency Act
Energy Star Homes program
Energy Star label

flat plate collectors
fuel cell
geopressurized zones
geothermal energy
green building programs
gross national product
hot-rock zones
hybrid car
hydroelectric power
hydrogen
Hydrogen Future Act

hydrothermal convection zones
least cost planning
Motor Vehicle Safety Act
National Appliance Conservation Act
National Model Energy Code
Natural Resources Canada
passive solar heating system
second law of thermodynamics
solar collectors
Super Energy-Efficient Home program

REFERENCES AND FURTHER READING

The References and Further Reading section at the end of this book contains a list of sources for the information discussed in this chapter and recommendations for further reading.

Connect to this book's website:
http://environment.jbpub.com/
The site features eLearning, an online review area that provides quizzes, chapter outlines, and other tools to help you study for your class. You can also follow useful links for in-depth information, research the differing views in the Point/Counterpoints, or keep up on the latest environmental news.

The Blacklake open pit asbestos mine in Wyoming.

CHAPTER 16 The Earth and Its Mineral Resources

Conservation is humanity caring for the future.

—*Nancy Newhall*

When the Earth first formed, scientists believe it was a mass of rock and ice. But over time, intense solar radiation, radioactive decay within rocks, and other sources of heat caused the Earth to melt, forming a huge molten mass. Gradually, the surface of the Earth cooled and formed a thick, rocky crust rich in minerals (FIGURE 16-1). A molten core remains. As this cooling took place, water vapor in the atmosphere condensed and fell as rain, forming oceans, lakes, and rivers. Today 29% of the Earth's surface is land, and 71% is water, prompting some observers to wonder about the appropriateness of the title *Earth* for what is really a watery planet.

This chapter examines minerals extracted from the Earth's crust, looking at the impacts of this activity and ways to meet our demand for minerals and metals more sustainably.

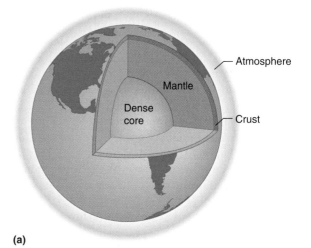

(a)

(b)

FIGURE 16-1 **The Earth in cross section. (a)** The Earth's crust floats on the mantle. Beneath that is the fiery core. **(b)** Lava pouring out of the Earth is evidence of the planet's molten interior.

16.1 The Earth's Mineral Riches

The Earth's surface waters and crust supports life. The crust also contains many important minerals. **Minerals** are largely inorganic substances such as sand, gravel, aluminum ore, and iron ore. The term *minerals* also refers to a few organic substances such as coal and oil; these are technically referred to as *fuel minerals*. The rest are known as **nonfuel minerals**. Fuel minerals were described in Chapter 14.

Nonfuel minerals are grouped into three basic categories: (1) **metal-yielding minerals**, such as aluminum and copper ore; (2) **industrial minerals**, such as lime; and (3) **construction minerals**, such as gravel and sand. Metal-yielding minerals must be processed, usually crushed and smelted (heated), to produce pure metals such as aluminum and copper. Industrial minerals and construction materials are used directly. They typically involve little processing.

Nonfuel minerals are made up of elements such as silicon and oxygen. The most abundant elements in the Earth's crust are oxygen, silicon, aluminum, iron, and magnesium. Other elements, such as gold and platinum, are also present, but are relatively rare.

Minerals typically occur in **rocks**, solid aggregates that usually contain two or more different types of mineral. Geologists divide rocks into three major classes: (1) **Igneous rocks**, such as basalt and granite, are those formed when molten minerals cool. (2) **Sedimentary rocks**, such as shale and sandstone, are formed from particles eroded from other types of rock. These particles are deposited elsewhere (for example, on the floor of the ocean) in huge quantities, where they build up in thick layers. (3) **Metamorphic rocks**, such as schist, are formed when igneous or sedimentary rocks are transformed by heat and pressure during mountain building.

Most metal-yielding minerals come from igneous rocks. These minerals are often concentrated in igneous rocks by geological processes. A concentrated deposit of minerals that can be mined and refined economically is called an **ore**. Most ores are mined and then crushed and processed (for example, heated) to yield their final product—for example, metals such as aluminum and zinc.

KEY CONCEPTS

The Earth contains many valuable nonfuel minerals. Those metal-yielding minerals found in concentrated deposits called *ores* are mined and processed to produce metals.

Mineral Resources and Society

Minerals are extremely important to our lives. Metals derived from some ores, for example, are in many products—among them buildings, computers, bicycles, glasses, and automobiles. Construction minerals are used to make roadbeds, schools, office buildings, and homes. Industrial minerals are used in fertilizers and concrete.

Minerals are so important to our welfare and our cultural evolution that scholars delineate the ages of human history by the chief minerals in use at the time: Stone, Bronze, and Iron. Although the Industrial Revolution is described in reference to the growing use of fossil fuels, minerals also played a key role in this transformation. In fact, fossil fuels were often used to process minerals into useful products.

Minerals are also vital to national economies. Canadian companies, for example, export more than $40 billion (Canadian dollars) worth of minerals each year. Nationally, the mining industry's share of the gross domestic product from exports and domestic sales was about 4.5%, an amount over $225 billion. In the United States, mineral mining is an even bigger business. Nonfuel minerals produced in 2004 were valued at approximately $44 billion. Processing and refining those ores boosted their value nearly 10 times—to $418 billion. Even recycling boosted the economy. In 2004, for instance it netted the economy $13 billion. Products made from minerals and recycled metals were worth nearly $2,000 billion—about 17% of the nation's GDP.

Today, more than a hundred nonfuel minerals are traded in the world market. These materials, worth billions of dollars to the world economy, are vital to industry, agriculture, and our own lives. The major minerals used primarily in the more developed countries are shown in Table 16-1. As shown, global production of construction materials far exceeds that of metals.

Among metals, the most important are iron, aluminum, and copper. Iron is mainly used to make steel for automobiles, bridges, buildings, and a host of other products. Aluminum is used to build jet aircraft and to make beverage cans. Copper is mostly used to make electrical wire and water pipes. So important are the metal-yielding minerals that if any one of several dozen key minerals were suddenly no longer available at a reasonable price, industry and agriculture would suffer.

KEY CONCEPTS

More than a hundred nonfuel minerals are traded in the world market. These materials, worth billions of dollars to the world economy, are vital to industry, agriculture, and our own lives.

Who Consumes the World's Minerals?

The more developed countries are the major consumers of minerals. With approximately 20% of the world's population, they consume about 75% of its mineral resources. The United States, with about 4.5% of the world's population, consumes about 20% of its minerals. Japan, Canada, Europe,

Table 16-1

Estimated World Production of Selected Minerals, 2003

Mineral	Production* (thousand tons)
Metals	
Aluminum	27,700
Copper	13,600
Manganese	8,200
Zinc	9,010
Chromium	15,500
Lead	2,950
Nickel	1,400,000
Tin	207,000
Molybdenum	125,000
Titanium	5,300
Silver	18,800
Mercury	1,530
Platinum-group metals	205,000
Gold	2,590
Nonmetals	
Sand and gravel	110,000,000
Clays	36,000
Salt	210,000
Phosphate rock	137,000
Lime	120,000
Gypsum	104,000
Soda ash	38,000
Potash	28,000

*All data exclude recycling.

and Russia also consume large quantities of minerals.

In the past few years, mineral consumption by the industrial nations has leveled off and in some cases declined. One reason for this is that industrial nations have largely completed their infrastructure—bridges, highways, and buildings—

> The more developed nations constitute one fifth of the world's population but use 75% of the world's mineral resources.

and no longer need massive inputs of minerals. Another is that heavy manufacturing is on the decline in many countries, whereas non–mineral-intensive industries (among them high technology and services) are on the rise. Because the less developed nations are becoming industrialized, their share of the world's mineral consumption is on the rise.

KEY CONCEPTS

The more developed nations consume the bulk of the world's minerals, but as the less developed nations' economies expand, they will use a larger share of the world's mineral supplies.

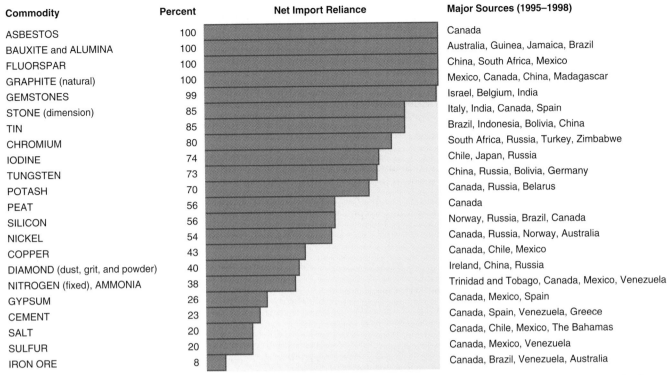

Commodity	Percent	Net Import Reliance	Major Sources (1995–1998)
ASBESTOS	100		Canada
BAUXITE and ALUMINA	100		Australia, Guinea, Jamaica, Brazil
FLUORSPAR	100		China, South Africa, Mexico
GRAPHITE (natural)	100		Mexico, Canada, China, Madagascar
GEMSTONES	99		Israel, Belgium, India
STONE (dimension)	85		Italy, India, Canada, Spain
TIN	85		Brazil, Indonesia, Bolivia, China
CHROMIUM	80		South Africa, Russia, Turkey, Zimbabwe
IODINE	74		Chile, Japan, Russia
TUNGSTEN	73		China, Russia, Bolivia, Germany
POTASH	70		Canada, Russia, Belarus
PEAT	56		Canada
SILICON	56		Norway, Russia, Brazil, Canada
NICKEL	54		Canada, Russia, Norway, Australia
COPPER	43		Canada, Chile, Mexico
DIAMOND (dust, grit, and powder)	40		Ireland, China, Russia
NITROGEN (fixed), AMMONIA	38		Trinidad and Tobago, Canada, Mexico, Venezuela
GYPSUM	26		Canada, Mexico, Spain
CEMENT	23		Canada, Spain, Venezuela, Greece
SALT	20		Canada, Chile, Mexico, The Bahamas
SULFUR	20		Canada, Mexico, Venezuela
IRON ORE	8		Canada, Brazil, Venezuela, Australia

FIGURE 16-2 **U.S. net import reliance for selected nonfuel mineral materials (1999).** Many developed countries such as the United States are heavily dependent on other nations to supply minerals.

Import Reliance

World mineral production is widely dispersed. However, some minerals are found only in commercially valuable quantities only in specific countries. As a result, most nations are highly dependent on the supplies of others (**FIGURE 16-2**). The vast majority of the United States's mineral resources come from politically stable countries. However, some vital minerals such as chromium and platinum come from politically volatile regions (**FIGURE 16-3**). To protect against embargoes or sudden declines in export resulting from political upheaval in such nations, the United States stockpiles a 3-year supply of **strategic minerals** such as bauxite (aluminum ore), cadmium, and graphite, hoping that if political troubles arise in nations that supply minerals they will be resolved well before stockpiles are exhausted.

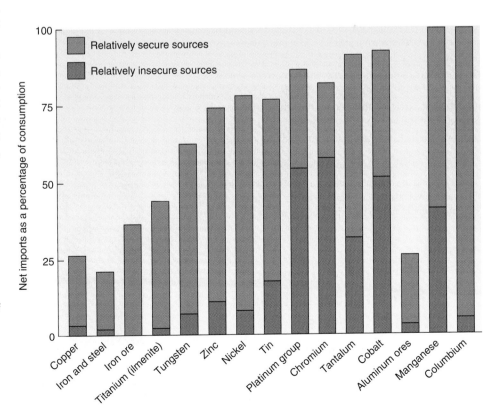

FIGURE 16-3 **Risky business.** The United States imports some minerals from politically unstable countries. Each bar in this graph shows the total imports as a percentage of U.S. consumption. The colored part (maroon) of the bar indicates the percentage of the minerals that comes from insecure sources. The difference between the total and the colored bar (gray) represents imports from secure sources.

Industrial nations have for many years exploited deposits in the less developed countries, taking huge quantities of minerals at extraordinarily low prices. Less developed countries understandably feel cheated by the Western world, which converts the raw materials it acquires at a bargain into products that reap them trillions of dollars a year. To offset this problem, many governments of the less developed nations have taken steps to increase their share of profits—for example, by charging royalties, taxing raw ore exports, or encouraging the production of finished metals. The success of these efforts is reflected by statistics from the U.S. Bureau of Mines, showing that in 1982, the United States imported $4 billion worth of raw minerals and $25 billion worth of finished metals from foreign sources, mostly developing countries. In 1999, the imported material was still $4 billion, but the finished materials had risen to $62 billion.

Will There Be Enough?

The long-term outlook for minerals is mixed. Current estimates of world mineral reserves and projections of consumption suggest that about 75% of the economically vital minerals are abundant enough to meet our needs for many years—or, if they are not, they have adequate substitutes. However, approximately 18 economically essential minerals will fall into short supply—some within a decade or two. Gold, silver, mercury, lead, sulfur, tin, tungsten, and zinc are among them.

Some experts warn us not to count on new discoveries and improved extraction technologies to save the day. Even if we could extract five times the currently known reserves of these materials, the 18 "endangered" minerals will be 80% depleted by or before 2040. Combined with declining oil supplies, the depletion of these important minerals could bring the world economy to its knees unless we make changes soon.

16.2 Environmental Impacts of Mineral Exploitation: A Brief Overview

Like fossil fuels, minerals are part of a production–consumption system that involves exploration, mining, processing, transportation, and end use. Many of these activities can cause significant environmental damage. In fact, mining

and **smelting** (melting ore to extract metal) have created enormous amounts of environmental damage the world over. Copper production is a case in point. Copper is first extracted from the Earth in **open pit mines**, one type of surface mine (FIGURE 16-4). The rock (overburden) above the deposit is first excavated, and then the ore is removed. Next, the ore is crushed and run through a concentrator that removes impurities. The concentrated ore is then heated at high temperatures in a smelter to extract a crude metal, which is later resmelted to produce pure copper.

Each stage in this system produces major environmental impacts, even in the best regulated countries. For example, mining disturbs large areas. In the tropics, forests are leveled to make room for mines. Rock wastes from open pit mines are typically deposited on the land near the mine—burying vegetated areas (Figure 16-4). Mine wastes, called *tailings*, may

> Three fourths of the world's economically important minerals are abundant enough to last many years or have substitutes. Eighteen minerals could fall into short supply within a decade or two.

FIGURE 16-4 **Open pit mine.** This aerial view of the Bingham copper mine in Utah shows the enormous impact of this activity. Waste from the mine is stockpiled around the site, creating further land disturbance.

also erode into nearby leakes and streams, clogging channels and causing a number of problems outlined in Chapter 14, such as flooding. Toxic metals in the mine wastes can contaminate nearby streams and reservoirs, killing aquatic life. Sulfur present in tailings may combine with water to form sulfuric acid, creating acid mine drainage, which is lethal to fish and numerous aquatic organisms (Chapter 14).

Smelting produces enormous quantities of air pollution. Globally, copper and other nonferrous (noniron) smelters produce about 8% of the world's sulfur dioxide emissions. Sulfur dioxide combines with water in the atmosphere to produce sulfuric acid, which rains down on the land, creating severe ecological and economic problems (Chapter 20). Smelters also release enormous amounts of toxic metals in the stack gases, among them arsenic, lead, and cadmium. Toxic metals and acids from smelters are responsible for huge *dead zones*—places where all vegetation has perished. Around the Sudbury smelter in Canada, 10,400 hectares (26,000 acres) have been turned into a barren moonscape because of air pollution from the smelter (FIGURE 16-5).

Other mineral extraction efforts rely on equally devastating techniques. In the Amazon, for example, gold is extracted by **hydraulic mining**. In this process, hillsides are blasted with high-pressure streams of water that wash away the soil and rock containing gold (FIGURE 16-6). The runoff is directed into sluices that separate out the gold. The sediment runs off into streams, killing fish and other aquatic life.

In North America, gold is sometimes extracted from piles of crushed ore or old tailings by a process called **heap leaching**. In this technique, miners spray a cyanide solution on piles, letting it percolate down through the material. During its sojourn, the cyanide solution dissolves gold. The liquid is then collected, and gold is extracted. Cyanide collection reservoirs and contaminated piles endanger wildlife and groundwater. In 1990, for instance, 39 million liters (10 million gallons) of cyanide solution broke through a containment dam in South Carolina, flushing into a stream and killing at least 10,000 fish. Ducks and other waterfowl die by the thousands each year when they land in cyanide ponds near mines.

In summary, mineral exploitation creates environmental damage on a scale matched by few other human activities. It is responsible for deforestation, soil erosion, water pollution, and air pollution. The environmental impacts are particularly severe in less developed countries, which produce a large portion of the world's minerals.

FIGURE 16-5 **Destruction from the Sudbury smelter.** This aerial photograph shows the enormous environmental damage caused by emissions from the copper smelter in Sudbury, Ontario, in Canada.

FIGURE 16-6 **Hydraulic mining.** Water sprayed on this hillside removes the soil and exposes subsoil containing gold, which is also removed by high-pressure water.

KEY CONCEPTS

The mineral production–consumption cycle produces extraordinary environmental impacts. The most noticeable occur in the mining and smelting phases.

16.3 Supplying Mineral Needs Sustainably

Humans need minerals, but they also need clean air, water, and healthy ecosystems. Can these two needs be met simultaneously?

Creating a Sustainable System of Mineral Production

Future demands for minerals and metals can be met in a more sustainable fashion in part by putting into practice the operating principles of sustainability, especially conservation, recycling, and restoration. Let's consider each one.

KEY CONCEPTS

Implementing the operating principles of sustainability—conservation, recycling, and restoration—can help human society create a more sustainable system of mineral production.

Recycling—Closing the Loop Although recycling is discussed in Chapter 23 in detail, it is important to examine a few key concepts that explain why this strategy is so important in the effort to create a more sustainable system of mineral production. **Recycling** is a process in which valuable products such as metals are collected and returned to factories, where they are melted down and used to manufacture new products.

Recycling helps increase the time a mineral or metal remains in use—that is, it increases its **residence time**. Recycling of materials helps to stretch limited mineral supplies, but it has many other environmental benefits. For example, it greatly reduces energy demand, slashes pollution, and reduces water use. Manufacturing an aluminum can from recycled aluminum, for example, uses only 5% of the energy required to make it from aluminum ore (bauxite). Because less energy is involved, less air pollution is produced. Fewer tons of coal need to be mined, and that reduces the impact on the land.

In the United States and many other countries, recycling efforts have fallen short of their full potential, except in a few instances such as the automobile. Approximately 90% of all American cars are recycled. Nearly 100% of all lead acid batteries are also recycled. The recycling of aluminum, steel, and many other metals, however, could easily be doubled.

> In the United States, approximately 90% of all cars and nearly 100% of all lead acid car batteries are recycled.

Meeting future demands sustainably will require a massive shift to recycling. Important as it is, recycling will not permit global exponential growth in mineral use to continue indefinitely, for several reasons. First, during the production–consumption cycle, some minerals and metals enter into long-term uses—for example, aluminum used for wiring. Second, some materials are lost through processing inefficiencies. Others are lost accidentally or are thrown away on purpose. Thus, it is impossible to recycle 100% of any given material. A practical goal would be 60 to 80%. Therefore, recycling can slow down the depletion of mineral supplies but cannot halt it. In theory, recycling can double our mineral resource base, but continually rising demand and inevitable losses will eventually deplete reserves.

KEY CONCEPTS

Recycling metals provides materials needed to manufacture goods, but at a fraction of the environmental cost of producing them from raw materials—in large part because of energy savings created by using recycled (already processed and refined) materials. Recycling can help to extend mineral supplies.

Conservation—Decreasing Product Size, Increasing Product Durability

Conservation (using only what we need and using it efficiently) is another key element of a sustainable system of mineral supply. As in energy supply, conservation is often the cheapest, easiest, and quickest means of stretching mineral resources.

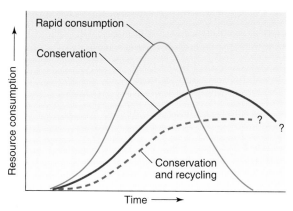

FIGURE 16-7 The future of the world's minerals. This graph shows three possible scenarios on a hypothetical time scale. The current path leads to the depletion of the world's mineral reserves. Conservation and recycling can greatly lengthen the life span of valuable minerals, helping us create a more sustainable economy. Which path will we take?

Conserving minerals may mean making smaller automobiles and finding ways to design many other products using less material. It also means making products more durable so that they last longer. Combined with recycling, conservation measures can greatly extend the lifetime of many valuable mineral resources. As **FIGURE 16-7** shows, continued exponential growth in mineral demand—the track we are now on—is the fastest route to depletion. Recycling will slow down the depletion. Recycling and conservation measures combined will slow it down even more, giving us more time to develop new mining technologies and find substitutes. Conservation and recycling also help us to minimize our impact on the environment.

Individuals can also play a role by purchasing more durable products and by consuming less, meeting needs in other ways—for example, by borrowing or renting a tool rather than buying one. (More on this is in Chapter 23.)

KEY CONCEPTS

Using less material by downsizing products or by making them more durable will help stretch limited supplies and minimize human impacts on the environment. Conservation and recycling combine to produce enormous savings.

Promoting Conservation and Recycling Changes in existing laws are needed to stimulate recycling and conservation. For example, the **General Mining Act** (1872) permits miners to purchase federally owned lands for $5 per acre or less. All that is necessary is to file a claim and perform $100 worth of work on that claim each year. Because of this law, which was designed to promote mineral development, millions of acres of land have been transferred into private ownership for a fraction of their real value. Even foreign companies can purchase U.S. land. A Danish mining company, for instance, paid $275 for 108 acres of federally owned land that is estimated to contain $1 billion in minerals. The U.S. government does not collect royalties on minerals extracted from these lands—as it does for coal, oil, and natural gas. Because of the inexpensive price of this land and the lack of royalties, mineral prices are artificially low. The law of supply and demand correctly predicts that artificially low prices encourage demand. Put another way, low prices discourage conservation and recycling efforts. (For a debate on this law, see the Point/Counterpoint in this chapter.)

U.S. mining companies also receive a sizable tax exemption of about $500 million a year. Known as a **depletion allowance**, this exemption allows companies to deduct 5 to 22% of their gross earnings (depending on the mineral) from their taxes each year, even if the ore deposit is still producing. This law was intended to compensate mining companies for declining ore reserves. While it may do that, it also makes the price of minerals artificially low, again discouraging conservation and recycling. Many countries subsidize mineral development through tax laws or even through direct investment of government funds (tax dollars) in mineral development projects. Eliminating financial incentives for producing raw materials could go a long way toward promoting recycling and efficiency.

Governments could also help promote recycling and conservation by offering tax breaks or other incentives to companies that use recycled materials or that offer longer warranties. Deposit and refund systems for products could also stimulate recycling. Changes in these policies and others are badly needed to promote a more sustainable use of the world's mineral resources.

KEY CONCEPTS

Reforming unsustainable laws, removing subsidies on raw materials, and giving financial incentives to companies that use resources efficiently and incorporate recycled materials in their products—all of these will help promote efficiency and conservation.

Restoration and Environmental Protection

New laws and tighter enforcement of existing laws could improve mining practices and reduce pollution from smelters in both industrial and nonindustrial nations. In the United States, mining and smelting operations are often exempt from existing laws. For example, mine wastes, which contain toxic metals that leach into nearby waterways, are currently not regulated by the **Resource Conservation and Recovery Act** (1976), a law that among other things places tight controls on hazardous wastes. This exemption, insisted on by the mining companies at the time the legislation was being considered, exists despite the fact that mines are the single largest producer of wastes in the United States. Moreover, smelters are not required to report their emissions to the EPA, as other industries are. Efforts are currently under way to close these loopholes, but many observers are skeptical about their outcome.

After years of neglect, abandoned mines, excessive pollution, and environmental destruction from mines, the Canadian government has begun to take steps to ensure the restoration and environmental protection of mined areas.

Canadian mining companies are required to prepare an environmental impact statement for projects, which, among other things, outlines how federal and provincial regulations on mining will be honored. Although previous problems are enormous—for example, cleaning up abandoned mines in Canada is estimated to cost $9 billion—many companies have shown a willingness to cooperate to protect air, water, and wildlife. Dramatic improvements have been especially evident at Canada's large smelters.

KEY CONCEPTS

Meeting mineral demands sustainably will require efforts to minimize environmental damage from mining and other operations. Tougher laws may be needed. Laws that exempt the mining industry from environmental protection may need to be changed, too. Better efforts to restore the damage to natural systems caused by mineral production are also needed in many countries.

16.4 Expanding Reserves

Given the continued growth of the world population, rising expectations of the residents of the less developed nations, economic expansion, and the inevitable loss of metals, recycling and conservation will very likely not satisfy 100% of our future demands. New deposits need to be discovered and mined.

Geologists use the term *reserves* to indicate deposits that they are fairly certain exist and that are feasible to mine at current prices. Several factors determine the size of available reserves and are worth studying as we consider ways of expanding reserves to meet future demand.

KEY CONCEPTS

Future demand cannot all be satisfied by recycling and conservation efforts. Some new minerals must be mined.

Rising Prices, Rising Supplies

Many factors determine the size of mineral reserves. Economics is one of the most important. To understand this, we must look first at the law of supply and demand (discussed in detail in Chapter 25).

The law of supply and demand explains how both supply and demand affect the price we pay for raw materials, goods, and services. Basically, it says that when the demand for a raw material, good, or service exceeds its supply, prices tend to rise. A popular new car, for example, can be sold for more because demand is so high. In contrast, when demand is low in relation to supply, prices tend to be lower. However, that's not the end of the story. Prices also affect demand. As a rule, as prices rise, the demand for a product declines. As prices fall, the demand tends to rise. Supply and demand are elastic. That is, they change over time with changes in the price. Price is elastic as well, for it changes with supply and demand.

The law of supply and demand helps explain how businesses and our economy operate. It also helps to explain mineral reserves. Rising prices, for example, tend to make it economically feasible for companies to search for and produce more minerals. As a result, mineral reserves tend to expand as prices get higher, because it is now more economical to tap into lower grade ore deposits.

Over a decade ago, two rivals, Julian Simon (an economist who supports population growth) and Paul Ehrlich (an ecologist who thinks growth is a prescription for disaster), made a bet. Ehrlich bet $1000 that the price of several key minerals such as copper and nickel would be higher at the end of the decade as a result of scarcity. Simon said they

would not. Who won? Simon, the economist. Reserves of key minerals had expanded because of the many forces described in the previous and succeeding paragraphs.

Many people expect this cycle of rising price and expanding reserve to repeat itself *ad infinitum*, thus creating an endless supply of materials to meet human needs. While this line of reasoning may work in the short term, in the long term resource scientists say that it is bound to fail. Why? Mineral resources are finite. At some point, resources will simply run out or become so costly to mine that they will be unaffordable.

POINT Reforming the 1872 Mining Act

The General Mining Law of 1872: An Idea Whose Time Has Passed

Darrell Knuffke
Darrell Knuffke is the western outreach coordinator for the Wilderness Society. Knuffke joined the Wilderness Society in 1985 and for 11 years directed its Central Rockies regional office in Denver. He lives in International Falls, MN.

Alan Septoff
Alan Septoff is Mineral Policy Center's Reform Campaign Director. Septoff joined MPC in 1997. He spends most of his time fighting anti-environmental initiatives supported by mining industry advocates in the U.S. Senate.

As latte shops thrive in the shadows of feed stores, no one disputes the notion that the American West is settled, the frontier closed.

As frontier disappeared, so did most of the 19th-century land laws meant to spur settlement. The lone statutory artifact from our expansionist youth is the General Mining Law of 1872. Congress enacted the law to promote mineral exploitation and development and to provide incentives to settlers by giving away mineral rights. Today, well over 125 years later, the law remains in full force and effect, unchanged in a new millennium. And there's nothing quaint about it.

Just by staking a claim, anyone can obtain the right to mine hard rock minerals—gold, silver, copper, and the like—on 590 million acres of the national land that all Americans hold in common. Once a claim is posted and registered (no limit to the numbers), the claimant can hold it indefinitely by paying $100 per claim per year to the federal government.

A claimant who discovers a "commercially viable" deposit can buy the land from the government through a process

called patenting for $2.50 to $5.00 per acre. (There is currently a moratorium on patenting, which must be renewed by Congress each year.) More land, contiguous or not, is available for such things as milling and processing at $5 an acre, about which more will be said later.

If the claimant meets the doddering old law's terms, the government can't refuse the sale. After the sale, ownership of the land and its minerals is absolute. Owners may sell the land for any price or purpose—whether related to mining or not. Outside Phoenix stands a golf resort worth an estimated $60 million. The owner paid American taxpayers $170 for the mining claims it stands on.

Whether the land covered by a mining claim is patented or not, anyone who stakes a claim has a right to mine that claim. And that becomes a right protected by the takings clause of the Constitution. Because of this, the mining law is also a zoning law—all public lands not reserved for another purpose, like a National Park, are essentially zoned for mining. It doesn't matter if mining a claim would destroy a nearby town's water supply or if the land were better used for grazing livestock. . . . Once a valid claim is staked, the only foolproof way for the federal government to prevent mining is to buy back the claim.

The 1872 Mining Law offers the ultimate subsidy: We *give away* the minerals and the land with them. The Mineral Policy Center, a leading advocate for U.S. mining law reform, estimates that if the hard rock industry paid U.S. taxpayers an 8% royalty for minerals—the same rate that some coal and oil and gas industries pay for public resources—the Treasury would gain nearly half a billion dollars over 5 years. The industry claims it can't pay the royalty and still mine. But it pays that much and more on state and private land, where much of the nation's mining actually occurs. Of course, the industry claims it is *not* subsidized because it pays taxes on profits. By that lame reckoning, the public should provide at no cost the

Many factors determine the size of mineral reserves. One of the most important is the price. Reserves tend to expand as prices rise because companies are willing to spend more to develop lower grade ores, but ultimately, all mineral resources are finite.

Technological Advances Expand Reserves

Technological improvements in the mining and processing of ore—especially those that make it more economical to mine reserves or tap into lower quality ores—have increased global reserves of several key minerals since the 1970s. Scientists and engineers are currently developing techniques to improve mining efficiency more, thus allowing companies to exploit even lower grade ores. One group of researchers, for example, discovered that certain algae bind gold ions. These algae could permit companies to extract gold from mine wastes or perhaps even from natural waters.

Although technological advances can increase mining efficiency, remember that technology is cost driven. When mineral supplies fall below a certain point, the costs of technologies required to access and mine the lower grade ores

merchandise that every taxpaying business in the country sells.

One of the world's most valuable mines is the Stillwater, located on national forest land in Montana. Its platinum and palladium ore body is an estimated 26 miles long and contains over 225 million troy ounces of minerals worth $32 billion. What will taxpayers get for this monumental treasure? $10,180 in payment for the mining claims.

There are no environmental standards in the 1872 law. According to the U.S. Bureau of Mines, hard rock mining has contaminated more than 12,000 miles of our rivers and streams. The Environmental Protection Agency (EPA) lists over 60 mining related sites among its Superfund cleanup priorities. The industry argues that these are unfortunate relics of another time, that modern mining is different, responsible. Modern mining—especially gold mining—is, if anything, increasingly toxic and increasingly dangerous. Cyanide heap leach mining is responsible for the new gold rush in the American West—and for much else. The U.S. Fish and Wildlife Service says that in Nevada alone, the mining industry reported 6700 wildlife deaths between 1986 and 1990, and the actual number could be 5 to 10 times as high. In Colorado, taxpayers will have spent more than $100 million to clean up the poisonous stew a Canadian company left at the Summitville Mine when, after depleting a gold mine, it declared bankruptcy and walked away from its operation.

Although the 1872 Mining Law doesn't contain any explicit environmental protection provisions, it does effectively limit the amount of space that can be used to dump mining waste. For every 20 acres of mining claims, the text of the law allows one 5-acre claim (called a millsite claim) for ore processing and mine waste dumping. This wasn't an issue in 1872, when gold ore was rich—it didn't take much ore to get an ounce of gold. Nowadays, however, all of the rich ores have been mined out, and it takes many tons of ore to produce an ounce of gold. All of the ore that isn't gold has to be dumped as waste—on the millsite claims. In many cases, the law doesn't provide enough millsite acreage to dump all the waste produced by a modern mine.

Although clearly written into the law, it wasn't until March 1999 that the federal government began enforcing the millsite provision. The mining industry, which has for decades fought reasonable reform and defended as sacrosanct every provision of the old law, now finds it can't possibly live with the plain terms of *this* provision, and it is seeking relief where it has historically found it: in the U.S. Congress. Conservationists are girding to block a legislative rider that is sure to come, probably in the form of an appendage to an appropriations bill, that would deny the Interior Department any money to enforce the millsite limit that the law clearly mandates.

Reform advocates don't seek to bar mining on public lands. We do demand

- That federal land managers be able to prevent mining if it threatens resources of higher public value. The industry's mantra is "gold is where you find it." So are clean drinking water, biological richness, and healthy landscapes.
- Strong environmental safeguards where mining does occur and tough, enforceable standards for reclamation.
- Payment of a reasonable royalty to the public whose mineral resources these are.
- An end to patenting.

The 1872 Mining Law is a fiscal and environmental atrocity, yet it has survived every effort at reform. Its apologists are a few western members of Congress who think the region still lives on livestock, logging, and the mother lode. It doesn't and it hasn't for a very long time. The frontier is closed. So should be the most destructive provisions of this 128-year-old law.

Reforming the 1872 Mining Act

Those who espouse repeal of the Mining Law contend that it amounts to a giveaway of valuable mineral resources. If that were so and if gold, silver, and other valuable mineral commodities were available in big, ready to mine "nuggets" at bargain basement prices, then why isn't the federal government completely "sold out?" Why don't you and I take advantage of the "free gold and other minerals" available? The facts belie such a conclusion. Of the more than 2.2 billion acres of public lands in the United States, only 3 million acres (less than one quarter of one percent) have been transferred to private ownership through mineral patent during the entire 128 year history of the Mining Law. By comparison, more than 94 million acres have been transferred to railroads and 288 million acres conveyed to private ownership under agricultural homesteads. In addition, the mining industry has disturbed an amount of land that's less than the size of a county in Nevada.

"Minerals are where you find them." And minerals can be difficult to locate in commercial quantities; once they are found, it is very costly to extract and process them into a final product. That is why the Mining Law, a law that has been amended more than 20 times in its history, still provides appropriate incentives to reward those individuals who invest the time, capital, and resources to locate and produce minerals on public lands, minerals that are worth nothing in the ground unless someone assumes the risk of exploring for and developing them.

Mining is a capital-intensive industry, and the costs of mine development often run into hundreds of millions of dollars before the company even begins to recoup its investment. True, once the mining company has perfected a claim after location of a valuable mineral deposit, the company may apply for a mineral patent. But, as the Bureau of Land Management (BLM) points out, "the patenting of land under the [Mining Law] is not as easy as simply writing a check . . . In Nevada, mineral patent applications have contained from 1 to 500 claims per application. At a minimum, *it will cost the applicant $37,900 in direct costs to process a single claim or millsite."* And these costs are based on the assumption that the claim will be uncontested. Companies are further required to pay annual maintenance fees in the amount of $100 per claim.

Contrary to what the opponents of mining have been saying, patent fees are not the only costs that the mining industry will incur in connection with the development of a claim and extraction of the mineral resource. In addition to the costs of mine development, mining companies add millions of dollars each year to local economies in the form of sales, property, and other taxes. More than $27 billion of the revenues received by state and local governments throughout the country in 1995 (the most recent year for which such information is available) were generated by mining. The federal government also receives a large amount of revenue from mining, nearly $57 billion in taxes and other revenues. The Mining Law is no giveaway; it actually has generated billions of dollars in revenues to support local communities, schools, and economies. The mining industry shares its wealth for the benefit of its most valuable resource, the mine workers, who earn the highest average annual wages of all industrial workers, about $60,000 annually in Colorado. More than 320,000 people owe their livelihoods to the mining industry.

Opponents of mining falsely state that the Mining Law allows claimants to obtain patents on the cheap and convert such lands to residential, resort or other valuable uses, reaping a windfall in the process. Federal regulations preclude such a practice and already require the forfeiture of a mining claim in the event of an unauthorized use. One often cited example of alleged abuse of the Mining Law involves land near Phoenix, Arizona, on which the claimant had obtained claims for sand and gravel. In the mid-1950s, the Congress excluded such common variety minerals from the operation of the Mining Law. Meanwhile, the city itself had grown and reached the proposed boundaries of the mining operation and the city government precluded the mine operator from developing the claim. Only then was the land later sold for commercial hotel and golf course development. Now you know the rest of the story. . . .

These opponents also criticize the Stillwater Mine in Montana, which produces platinum and palladium. Antimining lobbyists argue that the operator's sole financial contribution is the amount paid for mineral patents. They conveniently ignore the massive costs the company has incurred in mine development, as well as the taxes and other economic contributions to the Montana's economy. Stillwater is the largest private employer in Montana. According to the company's annual report, it paid its employees (95% of whom reside in the state) $57 million in salary and benefits in 1999. The company has spent millions developing the claim and pays on an annual basis more than $24 million payroll, property, and other taxes.

Does the mining industry support reform of the Mining Law? The answer is an emphatic "yes!" The industry has supported legislation in Congress to require the payment of a royalty. Industry is also willing to pay fair market value for the surface of the land in addition to any royalty for the minerals in the subsurface. Any unauthorized use would result in forfeiture of the mining claim. We have also supported legislation that would impose a permanent claim maintenance fee and would further earmark revenues for the cleanup of abandoned mines.

Any royalty imposed, however, must be reasonable and not discourage mineral production. Yet antimining special interests urge the imposition of a 12-$\frac{1}{2}$ percent royalty on all minerals, the same royalty paid by the coal industry. Such an argument, however, fails to account for the fact that the Mining Law applies to many different varieties of minerals, i.e., gold, silver, copper, palladium, and the like. Unlike coal, a one-size-fits-all approach would impose an inequitable royalty burden on these minerals, despite the fact that gold, copper, molybdenum, and other locatable minerals are priced differently and compete in different markets. Moreover, such a royalty would have a devastating impact on mineral production and would actually reduce the amount of revenues the federal and state governments receive in taxes and other fees. The accounting firm Coopers & Lybrand confirmed in a 1993 study that the imposition of even an 8 percent gross royalty would result in a minimum $422 million annual net loss to the Federal treasury. By contrast, the royalty proposed by the mining industry that would only tax net proceeds would actually increase revenues by $53 million.

Opponents of the Mining Law allege that additional one-size-fits-all federal reclamation standards are needed because the Mining Law contains no environmental standards. What they fail to point out is that mining activity is regulated by literally dozens of federal environmental laws, including the National Environmental Policy Act, the Clean Water Act, the Safe Drinking Water Act, the Endangered Species Act, and numerous others, not to mention hundreds of regulations. Mining operations on federal lands are required to meet the reclamation standards of the Federal Land Policy and Management Act, which direct mining operators to prevent degradation of such lands. Moreover, the states in which mining operations occur contain very comprehensive regulatory programs that govern the permitting of mining operations, and impose strict reclamation standards that require mine operators to return lands to a productive, post-mining land use. Colorado has one of the toughest such laws in the nation.

The idea that we need yet another set of federal reclamation standards is contrary to sound science and to the recommendations contained within the most recent study by the National Academy of Sciences. In a 1999 report entitled "Hardrock Mining on Public Lands," the Academy reviewed all existing federal and state regulations and concluded they were adequate, comprehensive, and well coordinated. The study further concluded that the federal government would be better served by improving the manner in which it implements and enforces existing laws, given the array of federal laws and remedies already on the books. We believe that the states, because of their greater familiarity with

The Case for Reasonable Reform

Stuart A. Sanderson
Stuart A. Sanderson is president of the Colorado Mining Association and is an attorney with nearly 25 years experience in mining and related issues, in both government and private practice.

the terrain and other unique conditions within their borders, are best equipped to manage the environmental and reclamation plans for mineral activities. A new set of federal regulations would cost the taxpayer more and is no substitute for adequate budget, support, and enforcement of existing laws.

Mining is important to all Americans, adding nearly $523 billion annually to our nation's economy. The average American consumes more than 40,000 pounds of minerals each year, according to the National Mining Association. Your telephone contains more than 40 different minerals. Minerals are the salt in our food, the limestone in the sidewalk under our feet, the gypsum used in fire resistant wallboard, the clay for the brick in our homes, the gold used in medical equipment and computers (not to mention jewelry and currency), and the molybdenum used in alloy steels. Modern life would not be possible without minerals. It is important, therefore, to fashion laws and regulatory programs in a manner that protect the environment without discouraging legitimate and responsible mineral resource development.

Critical Thinking Questions

1. Summarize each author's viewpoint, noting supporting data.
2. How do their positions overlap?
3. What are the points of difference?

 You can link to websites that represent both sides through Point/Counterpoint: Furthering the Debate at this book's internet site, http://environment.jbpub.com/. Evaluate each side's argument more fully and clarify your own opinion.

could conceivably become prohibitive—too great for consumers to bear. Mining would be curtailed. Thus, although technology can help, it is only part of the solution.

Factors That Reduce Supplies

Higher prices and new technologies can expand reserves (economically recoverable minerals) within limits. However, many factors have an opposite effect. Rising labor costs, for example, tend to reduce reserves. Higher interest rates on money that companies borrow to operate also increase costs and make exploration, mining, and production more costly. Some economists predict that competition for investment capital (money provided by investors to start new companies or expand operations) will escalate in coming years, driving up interest rates and slowing the expansion of global mineral reserves.

Energy prices also play an important role in determining when mining a particular mineral becomes uneconomical. As the concentration of an ore decreases, the amount of energy required for mining and refining is fairly constant up to a certain point, beyond which it increases dramatically (FIGURE 16-8). At this point, reserves become too expensive to mine; their costs exceed what the market will bear.

Rising energy prices combined with declining ore deposits could very well bring an end to some of our important mineral resources within 30 to 40 years, say some analysts. Runaway inflation could grip the world economy unless something is done.

A final factor is environmental costs. Mining lower grade ores produces greater environmental damage than mining higher grade ores. Why? Larger surface mines are needed to produce the same amount of ore. More material will be transported to smelters for processing. More waste will be produced at the mines and smelters, and more air and water pollution will result. This could cause environmental protection costs to escalate, adding to the cost of production and accelerating the economic depletion of some minerals.

The prospects for expanding mineral reserves are mixed. As noted earlier, reserves for some minerals appear to be quite limited. Although supplies of others may be huge, environmental constraints and energy costs could put a damper on expansion, making conservation and recycling even more attractive.

Minerals from the Sea

The mineral deposits on land are finite, and on many continents, they have been heavily exploited. As a result, some companies are looking toward two untapped frontiers—Antarctica and the floor of the world's oceans—as potential sources for new minerals. Superficially promising, these options face serious economic, environmental, and social barriers.

Spotlight on Sustainable Development 16-1 examines the mineral potential of Antarctica. This section describes the seafloor deposits in the ocean.

The ocean is a vast resource of minerals, many of which are dissolved in the water itself. However, the concentrations of most dissolved minerals are generally too low to be of economic importance. More important are mineral deposits on the seafloor. The most important of these minerals are small lumps called manganese nodules (FIGURE 16-9). Abundant in the Pacific Ocean, manganese nodules contain several vital minerals, notably manganese (24%) and iron (14%), with smaller amounts of copper (1%) and cobalt (0.25%).

Several mining companies have explored the possibility of dredging the seabed to mine nodules. Preliminary studies show that seafloor mining may be technically feasible as well as profitable, but little is known about its environmental impact. At this point, the biggest impediment appears to

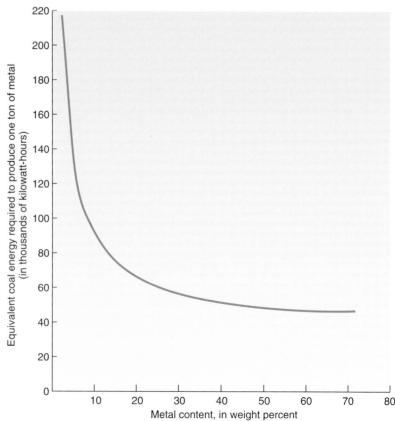

FIGURE 16-8 **Energy and mineral concentration in ores.** This graph shows that as the concentration in an ore body declines, the energy required to recover the mineral increases dramatically.

16-1   Antarctica: Protecting the Last Frontier

Antarctica makes up one tenth of the Earth's surface and is home to many animals, among them penguins, seals, and whales. Relatively untouched by humans, Antarctica has been the center of a huge controversy over minerals (FIGURE 1).

Between 1908 and 1943, Great Britain, France, Norway, Australia, New Zealand, Chile, and Argentina claimed 85% of Antarctica. The United States and the former Soviet Union, which had carried out their share of South Pole exploration and research, did not establish claims and refused to acknowledge the territorial claims of other nations. It was the position of these nations that it was preferable to retain an interest in all of the continent rather than to stake out a claim on one part.

In 1959, 16 nations signed a cooperative agreement, the **Antarctic Treaty System,** to maintain the continent for peaceful uses. The Treaty prohibited all military activities and weapons testing. The treaty also set aside arguments over land claims and permitted the nations that had signed it to carry out research unencumbered.

Since then, hints of Antarctic minerals and oil have threatened to destabilize the peace, pitting one nation against another. Part of the conflict arose between the haves—the seven claimant nations—and the have-nots—those who signed the treaty and have participated in Antarctic research and management but have no legal claim to the land or resources. The have-nots also include a growing number of less developed nations, who would like a share in the potential wealth of Antarctica. Added to the growing controversy was the involvement of several international environmental organizations, such as Greenpeace International and the International Union for the Conservation of Nature and Natural Resources. They are seeking to prevent development so that this vast, pristine wilderness will remain so.

The first real push for Antarctic mineral exploration came after the perilous climb in oil prices in the 1970s. Reports from the U.S. Geological Survey indicated that huge

FIGURE 1 **Antarctica.** This vast land may house coal, oil, and minerals. But is it worth risking this pristine area to boost natural resource supplies?

oil reserves might lie beneath Antarctica's continental shelf, but the evidence is sketchy. Sizable nonfuel mineral deposits may also lie beneath the ice and snow of Antarctica.

In 1988, the signatories of the treaty proposed regulations that would permit oil and mineral exploration. If approved, oil and mineral prospecting by seismic testing and other techniques with relatively minor environmental impacts could have begun. Full-scale development, such as mining and drilling, however, would have been barred until a new agreement was reached.

Many critics were concerned that development would threaten Antarctica's environment. Offshore oil rigs, for instance, would operate in some of the roughest seas in the world. Huge icebergs could rip apart drilling rigs. Ice floes could trap and crush ships. Mining would require huge amounts of energy to drill through the mile-thick ice cap.

Environmentalists were also concerned about the potential impacts of resource development on the rich marine ecosystem. Oil spills could be carried ashore, wiping out huge breeding colonies of penguins, birds, and seals. Oil slicks might also kill algae, the base of the aquatic food chain. Caught under the ice and in the chilly waters, oil could persist a hundred times longer than in warmer waters. Finally, cleaning up an oil spill would be impaired by the short "warm" season. If a well blew out, winter might set in before workers could plug it up.

Nonfuel mineral extraction would require enormous amounts of energy, create pollution, damage the ice pack, and severely disrupt the land surface. Most damaging to wildlife would be the use of ice-free coastal areas for processing and shipping ores.

Experience had shown that research, generally a relatively benign activity, had already created significant air and water pollution. Further human activity on the scale required to extract minerals and oil, critics argued, would create an environmental disaster.

In October 1991, though, after considerable political pressure from environmentalists, 24 industrial nations signed an agreement that banned environmentally damaging oil and mineral exploration in Antarctica for at least 50 years. Japan signed the agreement in 1991 but has not ratified it. South Korea signed in 1992.

Besides prohibiting oil, gas, and mineral development, this treaty includes provisions that will protect the environment from the impacts of other human activities. Although the agreement still leaves the possibility of oil and mineral development open many years down the road, for now, Antarctica is saved from development. Protecting areas of the world from human development—especially environmentally fragile areas with slow recovery rates—is an essential element of global efforts to create a more sustainable human presence.

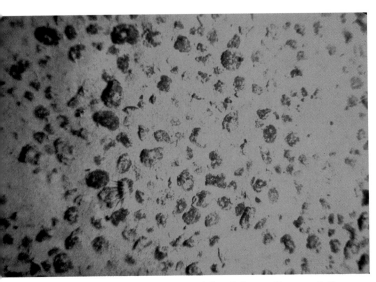

FIGURE 16-9 **Manganese nodules.** Lying on the ocean's floor, these nodules range from the size of peas to that of oranges.

For example, fiber-optic cables are made of thin glass fibers made from silica (sand). They are currently replacing many uses of copper, a mineral whose extraction and processing are environmentally harmful, even under the best of circumstances. Digital cameras are replacing film cameras that rely on silver, a product used to make photosensitive emulsion on film. But is substitution a crutch we can lean on forever?

Critics argue that substitutions have created unreasonable faith among the public in the ability of scientists to come up with new resources to replace those that are being depleted. Some resources may not have substitutes. For instance, it may be impossible to find substitutes for the manganese used in desulfurizing steel, the nickel and chromium used in stainless steel, the tin in solder, the helium used in low-temperature refrigeration, or the tungsten used in high-speed tools. Scientists also note that many substitutes have limits themselves. Plastics have replaced many metals, for example, but the oil from which plastics are made is a limited resource.

Finding substitutes is a race against time. The time of economic depletion of some minerals is fast approaching, perhaps more quickly than substitutes can be found. A wise strategy would be to identify those resources that are nearest to economic depletion and then promote widespread conservation and recycling. Research to find substitutes should begin immediately, or if research is already under way, it should be greatly accelerated.

KEY CONCEPTS

Substitution of one resource for another that has become economically depleted has been a useful strategy in the past, but it may not work in all cases. Substitutes have limits, and some materials have no suitable alternatives.

be a political one. To date, the world is in a quandary over the issue of ownership. Mining companies in the West have the wealth to exploit the seafloor and argue that they should be allowed to do so. However, the less developed nations contend that the seabed belongs to all nations and that they should receive a portion of the proceeds.

At a United Nations Conference on the Law of the Sea, the less developed countries proposed an international tax on seabed minerals that could provide them with millions of dollars a year for agricultural and economic development. A comprehensive Law of the Sea treaty, worked out with U.S. negotiators during the Carter administration, would have included this plan, but the United States under Presidents Reagan, Bush, and Clinton refused to ratify the treaty—as did Great Britain and West Germany. Today, many people believe that a seabed taxation scheme is fair and necessary. Others see it as an unjust way of cutting into corporate profits. Consequently, progress toward mining the seabed has come to a halt.

KEY CONCEPTS

Mineral-rich nodules are found on the ocean's floor. Although they appear to be economically feasible to mine, little is known about the environmental impact. Questions of ownership also plague their exploitation.

16.6 Personal Actions

Tolstoy wrote that "everyone thinks of changing the world, but no one thinks of changing himself." We're quick to point out all of the things government or business or other people can do but we are often remiss when it comes to taking personal action. As noted in previous chapters, personal actions are essential to building a sustainable future (see the Individual Actions Count! table). They go a long way toward reducing demand and the impacts caused by meeting those demands, and they have the added advantage of saving money.

KEY CONCEPTS

Individual action is vital to building a sustainable future. Buying durable products, recycling, and choosing recycled materials are three steps people can take.

16.5 Finding Substitutes

Historically, the substitution of one resource for another that has been depleted has been a useful strategy for industrialized nations. Shortages of cotton, wool, and natural rubber, for example, have been eased by synthetic materials made from oil.

Substitution will unquestionably play an important role in the future, too. It could help us find alternatives to some minerals that fall into short supply. It could help us replace materials whose production is environmentally damaging.

Tomorrow's growth depends on the use we make of today's materials and experiences.
—Elmer Wheeler

Conservation
- Buy durable products.
- If you're going to buy a car, purchase a compact or subcompact.

Recycle
- Recycle steel, aluminum, and copper.
- Purchase soft drinks in recyclable containers.
- Buy recycled products.
- Sell or donate old bicycles and other items made of metal—so that they can be reused.

Support
- Support nonprofit organizations working on recycling and reform of mining laws.

Write
- Write to state and federal representatives supporting recycling and mining reform.

CRITICAL THINKING

Exercise Analysis

In the late 1980s and the early 1990s, a dramatic increase in the number of U.S. recycling programs glutted the market with recycled newspapers and other products such as steel cans and plastic pop bottles. Consider paper recycling as an example: Recycling mills in the United States were flooded with newspaper, but their capacity to recycle this material increased only slightly during that time. This created a huge bottleneck in the system. Because of the glut of recyclables, prices for newsprint dropped considerably, sometimes making this endeavor rather uneconomical. Similar problems arose in plastic bottle and steel can recycling.

During this dark era for recycling, analysts predicted that these gluts would be transient. New plants would open up to take advantage of these materials. Were they right?

Absolutely. By 1993, paper recycling was back on its feet. New mill capacity had eased the paper glut. Plastic pop bottle recycling followed a similar pattern, as did steel can recycling. The problem with the critics was that they assumed a static condition—that the market would remain the same. Gluts were viewed as permanent. The critics failed to recognize that with millions of tons of relatively cheap material (recycled newspaper, plastic, and steel) on hand, businesses were bound to notice and avail themselves of the opportunities.

Assumptions can be dangerous, especially in dynamic systems that respond to abundance and price, like the economy. Critical thinking reminds us to examine assumptions carefully—our own and those of others.

Critics also point out that recycling is a dirty process. Indeed, recycling produces pollution. But digging a little deeper, you find that as a rule recycling produces much less pollution than manufacturing goods from raw materials. The energy savings alone can be staggering, and as you learned in Chapter 14, fossil fuel energy is a leading producer of air pollution. Cut back on energy consumption, and you automatically reduce air pollution.

CRITICAL THINKING AND CONCEPTS REVIEW

1. Critically analyze this statement: "Economic forces will ensure a continual supply of mineral resources. As prices rise, we'll find new resources, develop new technologies, and find substitutes for minerals currently used. So there is nothing to worry about."

2. What are reserves? Describe the factors that cause mineral reserves to expand and contract.

3. Outline a plan to meet the future mineral needs of our society sustainably. Describe each element of your plan. What are the most important steps? What barriers would you expect to the implementation of each idea?

4. What legislative changes would help promote a sustainable system of mineral production and consumption?

5. Make a list of ways in which you can cut down your resource consumption.

6. Describe ways in which recycling and resource conservation can be promoted in your community. Be sure to list a variety of options, ranging from compulsory to voluntary. Which ideas do you like best? Why?

7. Using your critical thinking skills, analyze the following statement: "Outer space offers a vast supply of mineral resources. We will never run out of resources as long as we have access to these minerals." Do you agree with it or not? Why?

KEY TERMS

Antarctic Treaty System
construction minerals
depletion allowance
General Mining Act
heap leaching
hydraulic mining
igneous rocks
industrial minerals

infrastructure
metal-yielding minerals
metamorphic rocks
minerals
nonfuel minerals
open pit mines
recycling

residence time
Resource Conservation and Recovery Act
rocks
sedimentary rocks
smelting
strategic minerals

REFERENCES AND FURTHER READING

The References and Further Reading section at the end of this book contains a list of sources for the information discussed in this chapter and recommendations for further reading.

Connect to this book's website:
http://environment.jbpub.com/
The site features eLearning, an online review area that provides quizzes, chapter outlines, and other tools to help you study for your class. You can also follow useful links for in-depth information, research the differing views in the Point/Counterpoints, or keep up on the latest environmental news.

Wellington, New Zealand, a clean livable city.

CHAPTER 17

Creating Sustainable Cities, Suburbs, and Towns: Principles and Practices of Sustainable Community Development

The optimist proclaims that we live in the best of all possible worlds, and the pessimist fears this is true.

—*J.B. Cabell*

This book presents a systems approach to sustainability. One of its goals is to show how principles of sustainability derived from the study of ecology and other areas can be used to restructure human systems such as energy, transportation, waste management, and housing to make them compatible with the natural systems that support our lives and to ensure a long, prosperous human presence. Chapter 2 made the case for this new approach to environmental protection.

(If you haven't read them, you may want to do so now.)

This book also presents many ideas on ways to create nature-compatible designs to permit humans to thrive within the limits of the natural world. These new designs could allow human civilization to prosper without disrupting nutrient cycles, climate, wildlife, and natural environments.

Table 17-1 lists the human systems that were discussed in Chapter 2. In Part IV, the chapters have reviewed natural resource issues and solutions, highlighting ways to restructure systems such as energy, water supply, and agriculture to resolve resource issues using the operating principles of sustainability. Although restructuring human systems and the massive global economy are essential to protect the environment and build a sustainable future, the task will not be easy, inexpensive, or quickly executed. The chapter you are about to read presents some additional ideas on sustainable systems. It addresses land use and transportation.

Table 17-1
Human Systems
Energy
Transportation
Waste management
Water
Industry
Agriculture

17.1 Cities and Towns as Networks of Systems

Think of the city or town you live in. It consists of people and the built environment—streets, shopping centers, office buildings, and schools. Look a little harder, though, and you will see the city or town as a network of interdependent human systems. There's a transportation system, consisting of roads and highways, gas stations, automobiles, buses, trucks, and airports. There's an energy system, consisting of gas stations, power plants, underground pipes that carry natural gas, and power lines that transmit electricity. There's most likely an elaborate system of water supply, consisting of deep wells or dams and reservoirs, water treatment plants, and an extensive set of pipes under the streets to transport water to homes, factories, businesses, schools, government offices, car washes, and parks. There's a waste management system as well. It consists of thousands of generators of waste—homes, factories, office buildings, and copy shops that produce millions of tons of waste each year. It also consists of ways to deal with waste, such as recycling centers, waste-to-energy plants that burn garbage, landfills in which waste is buried, sewage treatment plants, and hazardous waste facilities.

These networks of systems are designed to meet our needs for raw materials, finished goods, and services. They also get rid of the mountains of waste produced by human society. Some futurists believe that making cities sustainable will require a restructuring of these systems. Virtually every aspect of city or town life will need to be rethought and redesigned, although not everyone agrees that this is necessary or, if they do, about how this should be accomplished.

KEY CONCEPTS

Cities and towns consist of numerous systems, such as energy, housing, and transportation, that many experts think are largely unsustainable. Making our living environment sustainable will require us to redesign human systems to better fit within the natural systems that support us.

The Invisibility of Human Systems

Although there has been a lot of progress in the last three or four decades in environmental protection, many problems are worsening. As pointed out earlier, the vast majority of the trends are leading us away from a sustainable existence. It may be hard for people to accept the assertion that human society and the systems it depends on are unsustainable. Throughout this book you have seen many statistics that support this conclusion. As has been pointed out in previous chapters, pollution, species extinction, global warming, and a host of other environmental problems aren't mere surface wounds that can be fixed with Band-Aids; they're symptoms of deeper problems, most notably overpopulation and unsustainable human systems.

Why is it hard to grasp the problem with human systems? For much of the past 50 years, environmentalists, policymakers, teachers, and researchers have focused most of their attention on solutions that treat the symptoms of the envi-

FIGURE 17-1 **The kitchen is more than a place to cook.** The kitchen is an integral part of several key systems. Can you name them? How do natural systems support the human systems?

ronmental crisis. Very little attention has been focused on human systems and their fundamental unsustainability. The basic assumption has been that if we solve the immediate problems, we'll be all right. A little change here and there will cut pollution and reduce the negative effects on people and other living creatures.

Another important factor is that for most people systems are invisible. That is, they were designed to operate with minimal bother to us (**FIGURE 17-1**). The fact that we don't think about systems very much is a tribute to the engineers who designed them. In fact, it's generally only when a system breaks down that we notice it exists.

We turn on a light switch, and electricity surges through the wires. We turn on the faucet, and out comes drinkable water. We go to the grocery store, and the shelves are packed with food. We pull up to the gas station, and there's plenty of fuel. We're aware of light switches, faucets, and gas pumps, but we're fairly ignorant about the rest of the systems. If we barely recognize their existence, how can we be concerned about them?

KEY CONCEPTS

Most efforts to solve environmental problems have focused on treating symptoms rather than on rethinking and revamping the systems that are at the root of the problems. Most people are unaware of the systems that support our lives until they break down.

Performance versus Sustainability: Understanding a Crucial Difference

Another obstacle in the way of understanding the premise that human systems are unsustainable has to do with the distinction between performance and endurance. For most of us, the systems work well. The important distinction here is that just because a system is supplying us with the services we need doesn't mean that it is sustainable. For example, just because the world's fishing fleet is producing about 93 million tons of fish each year does not mean that this amount is a sustainable harvest level. In fact, this level of fish catch is severely depleting key fisheries, and we are headed for major shortages in the near future. The same is true with virtually all human systems.

KEY CONCEPTS

Just because a system such as energy or manufacturing appears to be functioning well does not mean it is sustainable in the long run.

Why Are Human Systems Unsustainable?

Although it is difficult to determine the carrying capacity of the planet—that is, how many people it can support—it is easy to find evidence that we are exceeding it. Global warming, species extinction, land devastation, soil erosion, desertification, food shortages, and other problems are signs that we are exceeding the capacity of the Earth to support human life—and other life-forms as well.

These problems are partly a result of the massive size of the world population, but also a result of the design of the systems that support our lives. A system of energy based on efficiency and clean, renewable sources, for example, would have a fraction of the environmental impact of the present system (Chapters 14 and 15). A system of agriculture based on minimum tillage, crop rotation, soil conservation measures, and natural pesticides would be able to feed the world's people with much less impact on water, wildlife, and soils. Both could endure for many centuries, too.

In short, the present systems are unsustainable because they produce waste and pollution in excess of the Earth's capacity to absorb them. They end up poisoning us and other species. They change the climate. They change the chemistry of rainfall. They've contributed to the present depletion of the ozone layer. They use up the resources they need to function. They deplete both renewable and nonrenewable resources upon which their and our future depend.

Why do they do all of these things? As noted in Chapter 3, most systems are inefficient. They do not recycle or use recycled materials. They depend primarily on nonrenewable energy. They destroy but do not restore.

Human systems are unsustainable because they exceed the carrying capacity of the Earth. They produce pollution in excess of the planet's ability to absorb it, use renewable resources faster than they can be replenished, and deplete nonrenewable resources.

The Challenge of Creating Sustainable Cities and Towns

Cities and towns are a lot like the bird colonies that dot the rocky coastlines of many continents (FIGURE 17-2). Bird colonies are nesting and resting sites where large numbers of birds of the same species aggregate. They are also sites of enormous waste production and enormous food consumption—just like cities. But food for the colony and the city does not come from the immediate vicinity. It comes from neighboring ecosystems.

To be sustainable, cities and towns must have a lasting supply of resources. Proper resource management is therefore essential to the survival and sustainability of this pattern of habitation. Many changes can also be made in cities to ensure their sustainability. These changes are all designed to lessen resource demand and waste production while ensuring we meet our many needs.

The challenge facing humankind is twofold. First, we must revamp existing infrastructure. Second, we must build new infrastructure in a sustainable fashion. Consider the task of revamping existing infrastructure first—the so-called *redevelopment strategy*.

One of the first steps in reducing resource demand and pollution is to make all buildings much more efficient in their use of energy—electricity, fuel oil, and natural gas (Chapter 15). We can make heating systems much more efficient by installing better insulation and by replacing worn-out systems with newer models. We can also find ways to increase their dependence on renewable energy. For example, we can add photovoltaic panels to generate electricity or add hot water panels for domestic hot water and space heating (FIGURE 17-3). We can add low-flush toilets and water-efficient showerheads, faucets, and appliances. We can use recycled paint or low-toxicity paints. The list goes on.

Similar actions must be taken in transportation, housing, agriculture, waste management, and other systems—and are detailed in this book. Especially important are efforts

(a)

(b)

FIGURE 17-2 **Colonies and cities. (a)** Bird colony. **(b)** City. Cities resemble bird colonies in many ways. They are areas of intense activity where resources from outlying areas are consumed. They are also areas of concentrated waste production.

FIGURE 17-3 **Challenges of sustainable development.** This solar home utilizes three solar technologies: solar electric panels on the roofs of the garage and house to generate electricity; solar hot water panels on the roof of the garage (left) to provide domestic hot water; and passive solar, south-facing windows that emit the low-angled winter sun for interior space heat.

to reduce **urban sprawl**, the continual expansion of human communities into farmlands and wildlands that provide vital resources required for our long-term economic health and survival. This topic is discussed in Spotlight on Sustainable Development 9-1 and in Chapter 10 on agriculture and Chapter 11 on preserving biological diversity.

Redevelopment is a vital but often overlooked strategy for sustainable development, in large part because so much infrastructure is already in place and because billions and billions of dollars have been invested in it. It is imperative that we make this infrastructure as sustainable as possible. Tens of millions of single-family homes and apartment buildings and millions of businesses are currently in use and are prime targets for sustainable refurbishment.

The second challenge—designing and building new infrastructure to be as sustainable as possible—takes place when systems need replacement or require expansion. To expand a transportation system in an urban environment, for example, cities can install light rail systems rather than adding new lanes to existing highways. New housing would be constructed from recycled products and would be equipped with a host of resource-saving devices so that new houses use only a quarter of the raw materials of existing houses. Rather than adding a landfill, a city could expand its recycling facilities to handle waste.

Specifics of these two prescribed actions are outlined in chapters on energy, waste, water, forestry, mining, and air pollution. The reader should refer to those chapters for specifics. In this chapter, we look primarily at land-use planning and transportation, which are not covered elsewhere.

KEY CONCEPTS

Two challenges face existing communities: revamping existing infrastructure and building new infrastructure in as sustainable a manner as possible.

17.2 Land-Use Planning and Sustainability

A city or town is home to many different activities, some that conflict with one another: for example, housing and factories. For many years, cities and towns the world over have engaged in **land-use planning** to ensure that incompatible activities are kept apart. Land-use plans, for example, set up industrial zones and residential zones. Besides determining where people can live and do business, land-use plans establish sites for water pipes, electrical lines, roads, and shopping malls. In some countries, land-use planning has been fashioned in a way to prevent or reduce sprawl.

KEY CONCEPTS

Land-use planning helps cities establish the locations of various structures and activities and keep incompatible uses apart. As conceived and practiced in most places, it doesn't do much for sustainability.

Sustainable Land-Use Planning: Ending Sprawl

Sustainable land-use planning seeks to accomplish the same goals as land-use planning, notably separating humans from unsightly, noisy, and potentially dangerous activities. But it also attempts to achieve more efficient use of the land—that is, to create patterns of land use that minimize the conversion of farmland and wildlands to asphalt, concrete, and lawn. That is, it strives to prevent or control sprawl. This helps to preserve farmland, recreational areas, wetlands, scenic views, watersheds, aquifer recharge zones, and wildlife habitat—retaining aesthetic values and ecological services. The rationale for saving these lands has been discussed in Chapters 9 through 11.

Sustainable land-use planning also entails efforts to coordinate key uses such as housing, business, and transportation development. Ultimately, land-use planning seeks the best for people, the economy, and the environment. It can in fact improve economies by reducing the loss of productive farmlands, by reducing air pollution and costly health problems, and by making the provision of government services such as police and fire protection more efficient. It can, therefore, create more efficient government.

FIGURE 17-4 shows four major types of development: dispersed, compact, satellite, and corridor. Let's examine each one very briefly and assess its potential for promoting sustainable development.

KEY CONCEPTS

Sustainable land-use planning and development seek to optimize land use and minimize the loss of economically and ecologically important lands. They offer other benefits as well, including more efficient mass transit, reduced air pollution, and reductions in the cost of providing water, sewage, and other services.

Dispersed Development **Dispersed development** occurs in many cities and is commonly referred to as *urban sprawl*. Urban sprawl is the steady outward expansion of urban/suburban areas that occurs as new housing subdivisions, highways, shopping malls, and other forms of development spring up on the perimeter of existing development, taking over farmland, forest, and grassland (Figure 17-4a).

In many cities, land is already zoned for this type of development. Dispersed development, however, ranks low on the sustainability scale. In fact, it is the least desirable of all of the alternatives because it consumes lots of land. Farmland, forests, wetlands, and grasslands are all victims of sprawl. Because cities are often located near prime farmland, sprawl decreases a nation's long-term ability to produce food (Chapter 10) and displaces wildlife (Chapter 11). Loss of natural habitat and replacement with paved surfaces often increases flooding (Chapter 12). Furthermore, sprawl results in a rather haphazard pattern of settlement, with poor aesthetic appeal.

Because housing spreads out inefficiently on the landscape, dispersed development increases vehicle travel, adding

FIGURE 17-4 **Four development patterns.** These maps of the Denver metropolitan area show four different development paths. (a) Dispersed development, or urban sprawl; (b) compact development, the most sustainable alternative; (c) satellite development; and (d) corridor development. Each one has its benefits and its costs.

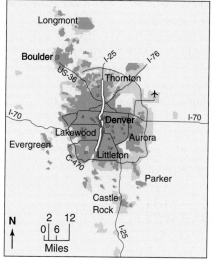

(a) Dispersed development

(b) Compact development

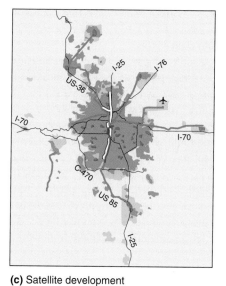

(c) Satellite development

(d) Corridor development

	Commuter rail transit		Light rail transit		HOV or bus lane		Major highways and freeways

Urban center	1990 Urban area	2020 Urban area

to commuting time and increasing energy consumption and air pollution. Each of these has a cost. Increased commuting time means more time away from families and friends and less free time. Increased energy consumption, caused by the need to travel farther to work each day, costs more in gas and wear and tear on the automobile. Air pollution takes its toll on people and the environment.

Sprawl has other costs. Providing highways, mass transit, sewers, water, and other forms of infrastructure generally costs more than the alternative forms of development because more miles of water and sewer lines, telephone lines, and power lines must be run to supply homes and businesses strewn out inefficiently across the landscape. Police and fire protection in a sprawling urban environment is often more costly.

KEY CONCEPTS

Dispersed development or urban sprawl is the most environmentally and economically unsustainable form of urban/suburban development.

Compact Development Figure 17-4b shows an alternative development pattern, compact development. Ranking highest on the sustainability scale, **compact development** is a denser or more compact form of development. It requires the least amount of land to accommodate people and the services they require. Because of this, it helps to preserve open space, farmland, forests, and other ecologically and economically important lands surrounding a city.

Many measures can be implemented to ensure compact development: smaller lots, more efficient placement of houses in subdivisions, an increase in the number of multifamily dwellings such as apartments and condominiums, and more compact placement of services (shops and the like). Smaller lots mean that more houses can be located on a given piece of property.

Some people argue that compact development means crowding and unpleasant living conditions, but others note that this doesn't have to be so. For example, in a suburb, placing houses closer to the street so that there is less front lawn and more back lawn gives people more usable outdoor space. Backyards can open into a common area that serves many homes: Community swing sets and community gardens could be added to create a sense of community and save resources. Commons could make suburbs more human. Narrowing the streets and using the saved space to create larger parks can provide playgrounds and ball diamonds for neighborhood children. Existing zoning laws could be changed to allow homeowners to add small apartments in the tradition of the carriage house to existing homes or to convert unused space into apartments for renters. Condominiums and townhouses can accommodate far more people per hectare of land than single-dwelling units. These are ideal for singles and married couples without children or with small families. Paying attention to soundproofing, views, privacy, access to lawns/playspace to make such dwellings more livable can help make them a desirable option for people.

One promising development, which began in Denmark in the 1970s, is **cohousing**, a residential living arrangement that strives to create a community. Cohousing developments are small neighborhoods that offer many social and economic benefits, described in Spotlight on Sustainable Development 17-1.

In many cities and towns, businesses are strewn along highways in highly dispersed fashion. Aggregating services close to residences can cut down on driving time and make them accessible by foot or bicycle.

>> SPOTLIGHT ON SUSTAINABLE DEVELOPMENT

17-1   Cohousing: Building a Community

In the past, many people lived in villages or at least in tightly knit urban neighborhoods. Families were familiar with people's past histories, their talents, and weaknesses. There was a sense of community that is often lacking in modern neighborhoods. Community provided a sense of belonging, a sense of security. It was a more practical way to live, too. People helped each other, sharing the burden of everyday life. For example, when a barn needed to be built, neighbors chipped in to assist. They built one-room school houses together, too. They also shared in planting and harvesting.

In modern society, community has withered and died. Designers of modern suburbs where many of us live take special care to ensure privacy, rather than promote closeness. In many neighborhoods, for example, fences line our back yards, walling us off from contact with neighbors. In fact, few of us see our neighbors any more. If we do, it is usually just to glimpse them coming and going—driving in and out of the garage. At night, many people returning from work simply slip into the garage without a trace. They hit the garage door opener, then disappear. Although people do socialize, it is often with friends who live far from their homes. Given the distance that separates us, many people complain that they have to make appointments to see their friends. No one just drops in to say hello. Because there isn't much of a community in the neighborhoods we live in, many people begin to feel isolated. Life becomes a chore.

Disenfranchised with the lack of community in modern society, people throughout the world are banding together to recreate a sense of belonging through cohousing. What exactly is cohousing?

Cohousing usually consists of houses or apartments clustered around a common area and a common house. Usually consisting of around 30 homes or apartment dwellings, cohousing is occupied by people who planned their "development" and now manage the community they've developed. Individuals own or rent their homes, but share in the ownership of the commons. They also share in some of the work. One person, for example, might lead a committee that plans social events. Another might organize weekend work projects. Still another might coordinate day care, while someone else manages the laundry facility.

In cohousing, many resources are shared, too. For example, in Golden, Colorado, Harmony Village's 27 families share a single lawn mower. The common house also offers a tremendous opportunity to share. In many common houses, there's a shared laundry facility. Two washers often serve an entire community. A television room and places for kids to hang out, to play without disturbing adults, are features of many common houses. The common house also typically has a large kitchen where community meals are prepared, sometimes every night of the week. The preparation of community meals is a shared responsibility. Imagine coming home from work to have a meal on the table and only being responsible for cooking once or twice a month.

Because of the sharing of resources, individual units are typically downsized. Kitchens, for example, tend to be much smaller, more like those found in apartments. If you are planning a big meal for friends, you simply book the common house and use the larger kitchen facility.

The common house may also have a room or two for guests, so you don't need a spare bedroom for that once-a-year visit by mom and dad. Mechanic shops, greenhouses, gardens, and playgrounds are also shared by residents, greatly cutting down on expenses and reducing personal living space.

Common meals and activities help knit a community together. But the initial weaving of the social fabric comes during the development of the community. In most instances, a core group of people does the planning from day 1 and see the project through. This involves endless meetings and discussions to reach consensus. Although this is difficult at first, with practice, residents say it becomes much easier. In fact, many people claim that the skills they learned in cohousing benefit them in both their home and work lives.

FIGURE 1 shows the layout of cohousing. As you can see, houses open onto a central area, a common area, where kids can play freely. Adults socialize in the commons, too.

Compact development not only minimizes land use, it also reduces vehicle miles traveled and can substantially reduce energy consumption and air pollution. Perhaps more important, dense settlement patterns are much more amenable to mass transit systems, which, as discussed in the next section, are far easier and more efficient to implement if people live closer together. Mass transit systems use much less energy and require much less space to move people than automobiles do.

Compact development is also more efficient from an economic perspective. First and foremost, it is much cheaper to provide people living in a more compactly designed city with services such as fire protection, water, sewers, and other amenities than it is to provide them if homes and businesses are dispersed over a wider area.

Compact development, very common in Europe, creates a clear line between city and suburb and the outlying lands. In many U.S. cities, making compact development occur will require changes in existing zoning and other tools such as urban growth boundaries, described shortly.

Cars are typically relegated to a peripheral position, usually garages, parking lots, or carports along the outside of the community. Although the car is never more than a block or so away from your home, you generally can't park next to your home, except perhaps to deliver groceries. The walkways are made for people, not cars. As a result, cohousing is rather kid and pedestrian friendly.

The common house is usually in view of all the units in cohousing. It is the heart of the community. Some cohousers have organized day care in the common house so they can leave for work and not have to transport their children to another location, saving time and money.

Privacy is designed into all cohousing so that people can satisfy this need too. In many instances, private patios or gardens are located outside the back door. You're not forced to socialize if you need alone time.

Cohousing is found in rural and urban settings. It is a great place for kids to grow up, and adults, too. It provides a rich social atmosphere.

Cohousing is not a new idea by any stretch of the imagination. It began in Denmark in the 1970s. The pioneers in cohousing sought to create a child-friendly atmosphere and opportunity to share some daily functions. Since that time, hundreds of new communities have sprung up in Europe, Canada, and the United States. Most are organized around the idea of community. In other words, most were developed for social reasons, but a growing number of communities have been formed to create a more environmentally responsible way of living. In Nyland Cohousing in Lafayette, Colorado, the homes were designed and built to conserve energy and use solar energy for heating. Special care was taken to use some environmentally friendly building materials, too. In Denmark, several cohousing developments have incorporated passive solar design and wind energy to generate electricity.

Cohousing is growing rapidly and is bound to increase in years to come. Being a more compact form of development, in which some space is often devoted to habitat preservation, cohousing is not only good for people, it is good for the planet.

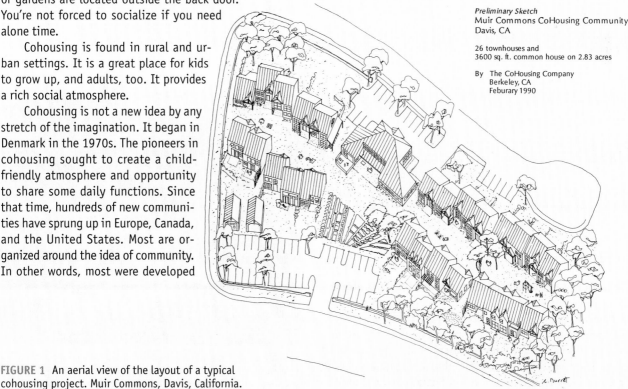

Preliminary Sketch
Muir Commons CoHousing Community
Davis, CA

26 townhouses and
3600 sq. ft. common house on 2.83 acres

By   The CoHousing Company
     Berkeley, CA
     Feburary 1990

FIGURE 1 An aerial view of the layout of a typical cohousing project. Muir Commons, Davis, California.

Corridor and Satellite Development The last two options are satellite development and corridor development. **Satellite development,** shown in Figure 17-4c, involves the development of outlying communities connected to the metropolitan area (the city and the surrounding suburbs) by highways and rail lines. Although this permits people to live outside the city and suburbs in small towns—and can create a very desirable quality of life—it is not as advantageous from many perspectives as compact development. It results in the conversion of a fair amount of open space and farmland. Although satellite development does not require as much land as dispersed development, it increases commuting time, energy consumption, and air pollution. Of the four options, it ranks number two in desirability.

Corridor development, concentrating housing and business growth along major transportation corridors, is even less desirable (Figure 17-4d). It requires much more vehicle use to get to work and to services. It uses more energy and produces more pollution than compact development. It is, however, amenable to mass transit.

Clearly, of all of the options described in this section, compact development offers the best benefits and is the most sustainable form. When properly executed, land-use planning can help us achieve a more sustainable relationship with the Earth. Land-use plans can protect the foundation of tomorrow's civilization: renewable resources such as farmland, pastures, forests, fisheries, and wild species.

Land-Use Planning and Building

Proper land-use planning can also be carried out on a smaller scale—for example, the development of a subdivision. When applied to specific sites, land-use plans take into account the slope of the land, soil quality, water drainage, location of wildlife habitat, and many other features. This permits planners and developers to design with nature, rather than redesigning nature. In The Woodlands, Texas, for example, a developer placed the homes outside the natural drainage areas, saving millions of dollars that would otherwise have been required to construct storm sewers. This also left an attractive open space rich in wildlife for local residents to enjoy. Protecting natural drainage, wildlife areas, aquifer recharge zones, and other important land features can reduce development costs, increase the quality of life, and help protect valuable ecological services. For more on this, see Spotlight on Sustainable Development 13-1.

Statewide and Nationwide Sustainable Land-Use Planning

The Japanese provide a model that many countries could adopt to protect their land. In 1968, the entire country was placed under a nationwide land-use planning program. Lands were divided into urban, agricultural, and "other" classes. Several years later, the zoning classifications were expanded to include forests, natural parks, and nature reserves. The success of the Japanese plan lies in protecting land from the market system, which, left on its own, appropriates land irrespective of its long-term ecological value.

Many European nations have adopted similar programs. In Belgium, France, the Netherlands, and the former West Germany, national guidelines for land-use planning were established in the 1960s. Administered by local governments, they protect farmland, prevent urban sprawl, and help to establish **greenbelts,** undeveloped areas in or around cities and towns (**FIGURE 17-5**). The Netherlands has one of the best programs of all. Its national planning program also governs water and energy use.

Land-use planning at the federal level in the United States is rudimentary at best. Except for establishing national parks, wilderness areas, national forests, and wildlife preserves, the federal government has done little to systematically protect its land. Most zoning occurs on the community level. On the local level, planners concern themselves primarily with restrictions on land use for commercial purposes—housing developments and industrial development. Because of the United States' reliance on community-level planning, states are

FIGURE 17-5 The greenbelt. This aerial photograph of Boulder, Colorado shows the greenbelt that surrounds the town, preserving valuable farms and open space and creating a more aesthetic living environment.

a patchwork quilt of conflicting rules and regulations. Some people believe that statewide land-use planning is needed.

Statewide land-use planning is an idea that is slow in coming. Oregon passed such a program in the early 1970s, discussed in Spotlight on Sustainable Development 9-1. Noteworthy programs now exist in Maine, New Jersey, Florida, Vermont, Washington, and Hawaii.

KEY CONCEPTS

Many states and nations have land-use planning that minimizes urban sprawl.

Beyond Zoning

The main tool of land-use planning for years has been **zoning regulations**, which classify land according to use. Zoning can protect farmland and other lands from urban development. In rural Black Hawk County in Iowa, for instance, zoning laws prohibit housing developments on prime farmland, but permit them on lands with lower productivity.

New approaches are also being adopted to protect valuable land, especially farmland. One public policy tool is the **differential tax rate.** This allows city officials to tax different lands at different rates. Farmland is taxed at a lower rate than housing developments, making it more economical to farm—and encouraging farmers to hold onto their lands rather than sell them to developers.

Another technique that helps farmers keep their land instead of selling it to developers is the purchase of development rights. A **development right** is a fee paid to a farmer to prevent the land from being "developed," that is, being bulldozed, paved, and built on. To determine the cost of a development right, two assessments of the land are made, one of its value as farmland and one of its value for development. The difference between the two is the development right. State or local governments may buy the development rights from the farmer and hold them in perpetuity. From then on, the land must be used for farming, no matter how many times it changes hands.

Still another way to reduce the spread of human populations onto valuable land is by **making growth pay its own way.** This idea, partially practiced in Boulder, Colorado, calls on developers and new businesses to pay the cost of new schools, highways, water lines, sewer lines, police protection, and other forms of infrastructure needed as communities expand. The rationale is that the cost of new development should be passed on to those who profit from it, not to existing residents. A new home thus comes with a development fee attached, which the new homeowner pays.

This not only keeps local taxes from rising to subsidize new development, it encourages builders to locate new housing projects closer to existing schools, water lines, and highways. This, in turn, reduces sprawl and reduces the destruction of farmland and other ecologically important sites. Development fees might also encourage developers to install water-efficient fixtures and pay for water efficiency measures in existing homes and businesses, thus preventing an increase in demand.

Open space—fields, forests, and other valuable lands—can also be purchased and set aside for wildlife or designed for mixed use—that is, as parks in which people and wildlife can coexist. Open space acquisition is practiced in many areas. In some counties, a small sales tax raises millions of dollars to purchase land. In others, real estate transfer fees (a tax paid when houses and land are bought and sold) are used to buy open space. Although these methods raise much money, open-space lands often command a premium price.

KEY CONCEPTS

Many methods can help promote sustainable land-use patterns, including zoning, differential tax rates, purchase of development rights, making growth pay its own way, and open space acquisition.

Land-Use Planning in the Less Developed Nations

Urban sprawl is a major problem in the less developed nations. In fact, millions of hectares of farmland are destroyed each year by expanding urban centers. Land-use planning is therefore as essential in the LDCs as in the industrial nations.

In some areas, land reform is badly needed. Wealthy landowners in many Latin American countries, for example, graze their cattle in rich valleys while peasants scratch out a living on the erodible hillsides. Hilly terrain that should be protected from erosion is being torn up by plows and washed away by rainfall. Some argue that sensible land use hinges on reform of these outdated landholding systems.

These are just a handful of ideas that could help reshape government policy to foster sustainability. Combined with many other ideas given in previous chapters—to promote sustainable ethics, revamp economics, and reshape unsustainable systems—they could form a national framework for dramatically realigning human systems to steer us onto a sustainable course.

KEY CONCEPTS

Land-use planning and land reform are also essential to creating sustainable land-use patterns in the developing nations.

17.3 Shifting to a Sustainable Transportation System

In 1950, the global automobile fleet numbered only 50 million; in 1998, the number was 508 million. In 2005, it had climbed to around 550 million. Americans are among the most avid automobile users on the planet. Each year, Americans travel nearly 2.6 trillion miles in their automobiles—the equiva-

In 2005, the global automobile fleet reached 550 million vehicles, approximately 200 million of which are in the United States. Automobiles account for 90% of the motorized passenger transport in the United States.

lent of more than 13 round trips to the sun, 93 million miles away.

In the United States, an estimated 200 million automobiles are currently on the road. This impressive total results from a number of factors, among them inefficient land-use planning (that results in urban sprawl) and inadequate mass transit systems. Affluence and the expansive nature of the country also contribute to America's having the highest per capita auto ownership on the planet. In the United States, automobile travel accounts for 90% of the motorized passenger transport. In Europe, where mass transit is much better developed and where cities are more compact, the auto still accounts for 78% of the passenger transport.

Air travel is another major component of modern transportation. World air travel has grown rapidly in the past five decades from 28 billion passenger kilometers per year to nearly 3000 billion in 2003, the latest year for which data are available (FIGURE 17-6). Besides carrying people, jets also carry huge amounts of freight. Both trends are expected to continue. The Worldwatch Institute, however, points out that the rising demand for air travel comes at an increasingly high environmental cost. Aviation consumes at least 5% of the world's oil each year. It is also the most polluting form of transport per kilometer traveled. According to the Worldwatch Institute, a single DC 10 flight from Los Angeles to Tokyo emits 266 tons of carbon dioxide, which appear to be altering global climate. The Worldwatch Institute points out that aviation currently accounts for 3% of global human carbon dioxide emissions and 2% of global nitrogen dioxide emissions, but these figures could rise to 11% and 6%, respectively, by the year 2050 if global air travel continues to increase as projected.

Today, nearly 30% of the energy Americans consume is used by the transportation sector, and much of that powers our automobiles and jet aircraft. But are automobiles and jets sustainable forms of transportation?

Many observers believe that the answer to this question is no. The planet cannot absorb the pollutants produced by them, especially cars. As the number of cars and jets expands and as the number of miles traveled each year increases, air pollution is bound to worsen. Oil spills could increase as the amount of oil being shipped increases. Atmospheric carbon dioxide levels are bound to rise as the combustion of fossil fuels increases. Acid deposition and urban pollution, both caused in large part by pollutants in automobile exhaust, could expand and threaten the health of people and ecosystems. Urban congestion, already at headache levels in many cities, is bound to worsen. Declining oil supplies also make the automobile and jet aircraft unsustainable. Road construction and maintenance are also quite costly. And tens of thousands of people are killed and injured each year in automobile accidents.

Clearly, even though they provide us with great joy and mobility and are an important aspect of our economy, automobiles come with a huge price tag. So do jets. What can be done to shift to a more sustainable transportation system? Will the automobile continue to play a predominant role? Will jet aircraft travel continue to rise? The shift to a sustainable system of transportation will very likely occur in phases.

KEY CONCEPTS

Automobiles are a major component of the global transportation system. Declining oil supplies, congestion in urban areas, regional air pollution problems, and global climate change are problems associated with their use that are likely to help stimulate a shift to a more sustainable transportation system.

Phase 1: The Move toward Efficient Vehicles and Alternative Fuels

The first step in the transition to a sustainable transportation system will be an improvement in efficiency, a change that has already begun. Improving fuel efficiency has several benefits essential to sustainability. First, it helps stretch oil supplies, providing time for the development of alternative fuels and alternative modes of transportation. Second, it reduces air pollution. The more efficient a vehicle is, the less pollution it produces. Third, it reduces the need to drill for and transport oil. Important as they are, efficiency gains could easily be offset by both the rising number of vehicles on the road and the increasing number of miles people are driving each year.

What is happening to fuel efficiency in the United States? In 1982, the average new American automobile got about 9 kilometers per liter of gasoline (22 miles per gallon). By 1992, the average fleet mileage had edged up to 11.9 kilometers per liter (about 28 mpg). But progress stagnated in the 1990s and 2000s because of heavy pressure on Congress from auto manufacturers interested in selling minivans, trucks, and sport-utility vehicles. Sales of these relatively inefficient, but highly popular vehicles, are more profitable than energy-efficient vehicles. Today, minivans, trucks, and sport-utility vehicles account for about two thirds of all new car sales in the country. Because of this trend, in 2001, the average gas mileage of new vehicles in the United States fell to 20.4 mpg.

The best gas mileage achieved by a standard gasoline engine comes from a Japanese-made vehicle, the Chevy

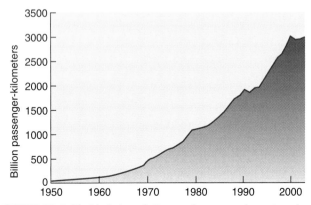

FIGURE 17-6 **World air travel.** More and more people are traveling by air these days, one of the most energy-inefficient modes of transportation.

Metro, which averages 24 kilometers per liter (58 miles per gallon) on the highway. It's production was recently halted. The British Leyland, a four-passenger prototype vehicle, leaves the Chevy Metro in the dust, getting 34 kilometers per liter (83 miles per gallon). A Japanese vehicle not imported to this country currently gets the same mileage.

Creating More Fuel-Efficient Cars and Better Highway Systems Improvements in gasoline mileage can be achieved by many different techniques that can be combined to produce safe, lightweight, and extremely efficient vehicles. Engine redesign, for instance, can improve fuel efficiency. Improvements in aerodynamics—how the air flows over a moving vehicle—can also improve efficiency, as can lightweight materials such as the new foams and plastics. Space-age materials and air bags can increase the safety of the smaller, more energy-efficient vehicles. The alleged dangers of smaller cars could also be mitigated by tougher drunk-driving laws, enforcement of speed limits, and better driver education.

California is a catalyst for many of the changes occurring in automobile design. Los Angeles, which is plagued by traffic congestion and air pollution, has mandated *ultraclean* cars. Ultraclean cars include automobiles that run on a mixture of methanol and gasoline. They also include cars with advanced catalytic converters, cars that burn natural gas or propane, hybrid cars, and electric cars. By 2003, state law requires that 10% of all cars on the road be ultraclean. In 2002, California enacted legislation to reduce emissions even more but is being sued by a consortium of auto makers who oppose even more efficient vehicles designed to low CO_2 emissions to combat global warming. The law requires that 10% sold in the state by 2009 would be electric zero emission.

Auto manufacturers have responded admirably to the challenge. In 1998, for example, Honda released a natural gas version of the Civic whose emissions are so clean they don't register on emission test equipment. As noted in Chapter 15, hybrid cars are now being sold by many auto manufacturers, domestic and foreign. These vehicles have small gasoline-powered piston engines combined with electric motors, which are controlled by an on-board computer. In the Toyota Prius, the electric motors are used to power the vehicle at slow speeds and when idling (**FIGURE 17-7**). The gas engine is used for higher speeds. These new vehicles are getting around 40 to 70 miles per gallon, but further improvements could greatly increase the gasoline mileage.

Another exciting development is the plug-in hybrid, discussed in Chapter 15. Besides increasing efficiency of vehicles by improving engines and automobile aerodynamics, engineers are studying ways to change the way cars are driven. Imagine climbing into your car, punching your destination into the computer, and then sitting back to read the newspaper. This is the dream of some technologists who are working on cars that drive themselves. The experimental models are either equipped with on-board computers that sense the sides of the road and traffic and keep the car in its lane or sensors that pick up a signal from a wire embedded

FIGURE 17-7 This five-passenger hybrid vehicle boasts outstanding gas mileage and low emissions. It runs on gasoline and electricity but never needs to be plugged in. Hybrid trucks, sedans, vans, and SUVs are gaining in popularity and could help us bide time to develop alternative transportation fuels and modes of transportation.

in the highway. Early studies have shown some promise. One of the chief benefits of this strategy would be that traffic would move more smoothly and more efficiently, reducing commuting time and pollution.

Another innovation is the use of computers to unjam congestion. Currently, engineers are using computers to understand and eliminate some of the factors that cause traffic congestion. These analyses are also helping cities to design highways to reduce congestion. Better controlled traffic lights and electronic signs that provide information to drivers to guide them toward better choices could reduce commuting time and prevent congestion.

> **KEY CONCEPTS**
> More efficient cars are part of the first phase of the transition to a sustainable energy system. Improvements in engines and automobile aerodynamics are key elements of this effort. Computer systems that operate cars automatically, monitor traffic and signal congestion, and permit designers to create more efficient highways are also key to the success of these efforts.

Creating More Efficient Aircraft Aircraft manufacturers have made significant strides in improving fuel efficiency. Concerns over the impact of aviation and desires to reduce costs have led manufacturers such as Boeing/McDonnell Douglas and General Electric to take measures to improve efficiency in jet engines. In large part because of their efforts, today's new jets use about half as much fuel as those manufactured in the mid-1970s. Even with these advances, air travel is growing so quickly that other means may be needed to reduce environmental impacts.

> **KEY CONCEPTS**
> Aircraft manufacturers have made much more impressive strides in improving fuel efficiency.

Alternative, Clean-burning Fuels Alternative fuels are also part of the immediate transition to a more sustainable system of transportation. Cars, trucks, and buses can be powered by a variety of fuels, including hydrogen and ethanol, an alcohol produced from renewable sources such as corn and wood. Besides being renewable, these fuels burn very cleanly. Interestingly, many new cars are now designed to operate on gasoline and a 85% ethanol–gasoline blend (called E-85). Known as flex fuel cars, they're growing in popularity.

Another technology that holds some promise is the electric vehicle. Although, as you may have learned by doing the critical thinking exercise, electric vehicles are not much cleaner than standard gasoline-powered engines if the electricity comes from coal-fired power plants, solar and wind-derived electricity make the electric vehicle a clean one.

Another technology for electric cars that holds tremendous promise is the fuel cell, described in Chapter 15. This device produces electricity used to run electric motors in cars, home generators, and other applications. Fuel cells are powered by hydrogen that can be derived from renewable resources (such as water) and nonrenewable fuels (gasoline, natural gas, and methanol).

Yet another fuel that holds great promise is biodiesel. Derived from vegetable oil, biodiesel burns in conventional diesel engines and produces a fraction of the air pollution of a conventional oil-based diesel. Biodiesel was discussed in Chapter 15. Also covered in Chapter 15 is vegetable oil, a fuel that can be burned in diesel engines with only minor modifications.

KEY CONCEPTS

Alternative fuels that burn cleanly and are renewable could also help reduce many problems created by the gasoline-powered automobile.

Phase 2: From Road and Airports to Rails, Buses, and Bicycles

The shift to clean-burning cars is essential, but like improvements in efficiency, it may be only a stopgap measure. By the year 2010, over 500 cities in the world will contain over a million people, and more than 26 of these will house over 10 million people. As cities grow, large numbers of commuters will very likely have to shift to **mass transit**—buses, commuter trains, and light rail (single-car trains). Recognizing this, Los Angeles's program calls on employers with 100 or more workers to develop programs to reduce vehicle miles traveled by workers. Some companies will offer free bus passes, for example, to encourage workers to get out of their cars. Companies can also offer economic incentives (small bonuses) to employees who commute via mass transit or with fellow workers in car or van pools. Employers may also meet their requirement by per-

> By 2010, 500 cities will house more than one million people and more than 26 cities will have populations over 10 million.

Table 17-2

Relative Efficiencies of Various Modes of Transportation

Mode of Transportation	Kilojoules per Passenger Kilometer (passenger mile)[1]
Van pool	400 (640)
Rail	400 (640)
Bus	450 (720)
Car pool	650 (1040)
Automobile	1800 (2800)
Airline	3800 (6080)

Source: Worldwatch Institute.
[1]A kilojoule is 1000 joules, a unit of energy.

mitting workers to **telecommute**—to work at home, linked to the office by phone lines. Others may shift to 4-day work weeks.

For a case study on one city's successful use of mass transit, see Spotlight on Sustainable Development 17-2 .

No matter how much automobile fuel economy improves, the car does not compare favorably with bus and train transportation (Table 17-2). In urban centers, buses and trains achieve a fuel efficiency of about 62 passenger kilometers per liter (150 passenger miles per gallon) of fuel. Intercity train and bus transport increases efficiency to 82 passenger kilometers per liter of fuel (200 passenger miles per gallon)—seven times better than the average new car today.

> In cities, buses and trains achieve a fuel efficiency of about 150 passenger miles per gallon. In travel between cities, it climbs to about 200 passenger miles per gallon.

Given the relative efficiency of mass transit—and given declining fuel supplies, congestion, and pollution—many urban residents may within the next few decades give up their second and third automobiles. They will turn to more efficient and less polluting forms of transportation, among them commuter trains, light rail, and buses. Some mass transit users may even join car-share programs. Participants in commercial car-share programs lease cars by the hour for special trips but rely on mass transit for most of their daily trips. To learn more about car-sharing programs, see Spotlight on Sustainable Development 17-3.

In Stockholm, the Office of Future Studies has proposed that the city work to phase out the private automobile. It recommends considerable expansion of the existing mass transit system and expansion of the fleet of rental vehicles for vacations and other special occasions.

The shift to mass transit is inevitable over the coming decades, but cities will have to improve their present systems, making them much more rapid and convenient. With declining automobile traffic, cities may be able to convert high-

17-2 Curitiba, Brazil—A City with a Sustainable Vision

Most modern cities have grown up around the automobile. City planners have literally shaped their cities around major transit corridors. Thus, the location of subdivisions, industrial facilities, and services has been largely determined by roads and highways, access routes primarily traveled by people in automobiles.

Proving that there is an alternative path is a city that's gaining wide recognition: Curitiba, Brazil. Lying near the east coast in the southernmost part of Brazil about 500 miles south of Rio de Janeiro, Curitiba is a showcase of wise planning and sustainable design principles that have served the city and the planet well.

Since 1950, Curitiba's population has grown from 300,000 to over 2.1 million. For most cities, this rate of growth, combined with poor planning, would have been a prescription for disaster. Poverty, pollution, crime, and highway congestion would have been the inevitable results—not so in Curitiba.

The city was blessed with a visionary mayor in the 1970s, Jaime Lerner, who adopted proposals first made in the 1960s to plot a future for the city based on mass transit, ecological design, appropriate technology, and public participation—all essential elements of sustainability. Lerner was an architect and planner of extraordinary vision.

As witness to his foresight, consider this: Today, 1.3 million commuters travel into Curitiba each day to go to work. Three quarters of these people travel by bus. This remarkable feat is made possible by an extensive, privately operated bus system that transports people in and out of the city with remarkable speed.

In most cities, bus systems are notoriously slow. Although they move large numbers of people, they bog down in heavy traffic on city streets. In Curitiba, however, buses move passengers into and out of the city at such rapid speeds for several reasons. First, the city has constructed five major roadways that penetrate into the heart of the city. Each of these roadways has two lanes designated for buses. In addition, bus stops are equipped with special devices called raised tubes, which allow passengers to pay before they get on the bus (FIGURE 1). This greatly speeds up the boarding process that slows down many a bus in the United States and other countries. Extra-wide doors also contribute to the speed of boarding. Double- and triple-length buses increase the system's capacity. Together, these innovations reduce the transit time by one third.

Curitiba has made it possible to move in other directions as well. Many smaller bus routes connect residential areas with the main transit corridors so that one can travel about freely. Commuters can take a bus to the main route, hop on an express bus, and be downtown in record speed.

This system of transit not only makes good sense from an environmental standpoint—because buses transport people with fewer resources and much less pollution than the automobile—it also makes sense from an economic standpoint. It's much cheaper than a subway system.

So that the less fortunate, Curitiba's poor, can gain access to the system, the city purchased land along major corridors, which was developed for low-income families.

Curitiba has more to boast about than its mass transit system. It has established an extensive network of parks along natural drainages. This not only provides residents with someplace to escape from the buildings and roadways, it reduces damage from flooding. Prior to the establishment of this system, developers often built homes and other structures in drainage areas. When floods came, many a home was damaged. Today, this system of parks with specially constructed ponds has nearly eliminated flooding and saved the city millions of dollars in engineering and construction costs. Low-tech solutions work and save money.

Curitiba promotes participation and cooperation, too. The city recognizes that solutions require the participation of many sectors, including business, government, community groups, and others. Fortunately, other cities are beginning to learn from Curitiba.

FIGURE 1 Raised tube bus stop in Curitiba.

17-3 Car Sharing: On the Road—Cheaper and Greener

Do you own your car—or does it own you? If you're like most people, you'll answer "yes" to both questions. You own a car that gets you where you want to go, but you're a slave to it, spending considerable amounts of money (and time) supporting your driving habit.

According to the American Automobile Association, each month Americans pay an average of US$700 to own and operate a car. Payments for a new car can easily run US$300 to $500 per month. Insurance adds from US$75 to $150 per month to the cost of car ownership, and then there's gasoline, costing another US$100 to $200, and maintenance expenses. . .

Fortunately, if you want the convenience of a car without the expense of ownership, there's an option for you—it's called car sharing.

Popular in Europe for almost 3 decades, car sharing has finally begun to gain momentum in the United States. Today, two commercial ventures—Zipcar and Flexcar—offer car share programs in cities across the country, including Boston, New York, Chicago, Denver, Los Angeles, Seattle, Washington, D.C., and Portland, Oregon. Even universities, including the University of North Carolina–Chapel Hill, UCLA, and the University of Washington, are jumping on the wagon, saving students money, and helping curb campus traffic congestion and parking problems.

Most car share programs require an application and membership fee, which average about US$75. Once you join, vehicles are available for one hour to several days. (Special arrangements can be made to rent cars for long trips.) Businesses, families, and individuals can all participate. Car share programs screen applicants using age and traffic violation criteria to eliminate risky clientele. Qualified members are covered by comprehensive and liability insurance when behind the wheel.

Urban car share programs place their cars conveniently throughout the city, usually in reserved parking lots or spaces. Members pay a small hourly fee to use the car, typically under US$10, and a per-mile charge. San Francisco's nonprofit City Car Share program charges users US$4 per hour and 44 cents per mile. Some programs give members a certain number of free miles before charging for mileage.

Both Zipcar and Flexcar use online and phone reservation systems, which allow you to reserve a car in a few quick keystrokes or with a phone call. A computer tells you where the car is and its license plate number. You show up at the site, hold your membership card next to the windshield, where it is read by a scanner, and if you've reserved the car and the time is correct, the doors unlock. The car's onboard computer sends an electric signal to company headquarters, indicating your rental period has begun, and activates a billing record.

When you're done, you return the car to its parking space, lock it, and leave. Your credit card is billed monthly for any usage you incur. The program pays for gas, although members are responsible for filling the tank when the gauge drops below the one-quarter mark—using a company credit card. And what if you have a fender-bender? Neither Zipcar nor Flexcar charge a damage deposit. According to Flexcar spokesperson John Williams, members pay half of the deductible ($500) if they caused the accident; otherwise, there's no penalty. (Policies vary among different organizations; be sure to inquire first.)

Car sharing programs make it easy to choose a vehicle to meet your needs by offering a wide range of vehicles, from small, efficient commuter cars such as the hybrid-electric Honda Civic or Toyota Prius to larger vehicles for special uses, such as Ford pickups and SUVs. According to Zipcar, more

way lanes to light rail lines. Median strips could be converted to light rail systems serving surrounding suburbs. Fast, efficient buses could carry commuters from their homes to outlying rail stations, where people board high-speed trains that transport them rapidly to urban centers.

To be profitable, high-speed rail requires high-participation, high-density population in outlying areas, and a large central business district. In order to achieve this, it may be necessary to *densify* new suburbs—that is, to foster a more compact form of development, as described in the previous section.

Another idea gaining interest is in making new subdivisions more like small towns—that is, making them more self-contained. Known as *neotraditional towns*, these new communities are being built so that residents are within

walking or biking distance of shops, stores, and small office buildings—which are often part of the new development itself. At this writing, nearly 100 are in various stages of development in the United States. They are also called new urbanist communities.

Urban centers can also densify by converting empty parking lots (produced by the decline in automobile use in cities) into office buildings. This makes mass transit more efficient and affordable. Unfortunately, the economics of mass transit is currently skewed by massive subsidies to the automobile and gasoline industry. According to national statistics, the automobile is subsidized to the tune of about $300 billion a year—or about $1500 per car. This subsidy includes expenses for police protection, traffic control, city-paid parking, and other things—expenses that are paid out of general tax rev-

than half of their members say that they tried the service for the opportunity to get behind the wheel of many different makes and models of cars, including hybrid-electric vehicles.

How much you save depends on your driving habits and needs. According to a recent customer survey, Zipcar members state they save an average of US$435 per month when compared with car ownership. Williams says that most of Flexcar's customers report savings of at least a couple hundred dollars each month.

To determine whether a car sharing program makes economic sense for you, Zipcar offers an online savings calculator. Just click on "Run the Numbers" on their website and enter the current or projected costs of your vehicle, including monthly payments, insurance premiums, fuel costs, parking, and maintenance costs. The program calculates the monthly and annual costs of car ownership and computes car sharing savings.

Beyond your pocketbook, car sharing also offers some environmental savings. Personal vehicles produce a large portion of the nation's annual emissions of carbon dioxide and other pollutants, contributing to global warming and localized smog. Accommodating the growing fleet of vehicles as we pave the planet also results in the loss of huge amounts of open space, farmland, and wildlife habitat.

According to Zipcar research, the company alone has been responsible for removing more than 10,000 cars from U.S. city streets and highways. Their survey also showed that car share members drive an average of 4000 fewer miles (6437 km) each year, compared with their habits before joining the program. Most people drive less when they have to pay a per-hour or per-mile fee, and studies have shown that car sharing folks are more likely to combine trips, take mass transit, bike, or walk when it's convenient. Combined with cities that have good mass transit systems, car share programs are highly effective at reducing congestion and pollution in urban settings.

If car sharing hasn't come to your town, consider setting up a program yourself. Across the country, groups of friends, neighbors, and colleagues have established their own car share programs by using cars already owned by individual members of the group or by purchasing cars together.

To make a community car sharing program run smoothly, members should consider providing convenient locations to park the vehicles, and draft agreements on buying fuel, accessing keys, and servicing and insuring the vehicles. A booking system should also be established. The Eugene, Oregon, BioCarShare program uses Online Resource Scheduler free software as their scheduling tool. Members log on at the website with their user name and password to reserve a car. Set up as a cooperative, BioCarShare requires members pay a joiner's fee, which is refundable, and a small monthly membership fee. Like most programs, they also require that drivers pay in proportion to their use, per hour or per mile. Car clubs such as these usually operate with standard insurance coverage—as long as the group or any of its members makes no profit. In most states, car clubs apply for insurance in the club's name and can list four or five people on a single policy.

Car sharing is an idea that offers the best of both worlds. It ensures people access to transportation while saving them considerable sums of money, and, it is good for the environment.

Adapted with permission from Dan Chiras, *EcoKids: Raising Children Who Care for the Earth*, Gabriola Island, British Columbia: New Society, 2004

enues. As you learned in Chapter 14, oil is also heavily subsidized. Because of this, sustainable strategies such as mass transit generally have a hard time competing with the automobile. Some critics say that the playing field is tilted heavily in favor of the automobile. Removing the hidden subsidies from oil and automobiles would clearly make mass transit compete more favorably with the automobile.

Jet travel could shift in the coming decades as well. The Swiss have, for example, levied an extra tax on aircraft that do not meet air pollution standards. In 1997, Denmark introduced a $15 fee on all domestic flights that has encouraged people to travel by train instead. High-speed rail, which is growing in popularity in Europe and Japan, could help reduce air travel and shift passengers to a much more efficient form of transit.

KEY CONCEPTS

Mass transit is much more efficient than automobiles and produces much less pollution per passenger mile traveled. Congestion, fuel concerns, and interest in cutting pollution will all stimulate the shift to efficient mass transit in urban areas. More compact development patterns will help complement the move to mass transit.

The Bicycle For many urban dwellers, the bicycle is not a viable commuting option. Streets are too hilly, or workers live too far from work for practical commuting. In other places, cold winter climates prevent commuting except during the spring, summer, and early fall.

In some cities, however, the bicycle could play a significant role in transporting people. Where climates are mild, streets are not too crowded, and people live relatively close to work, the bicycle can transport surprisingly large numbers of people. Investments that promote bicycle commuting represent one of the cheapest options available to cities and towns.

For decades, the bicycle has been a major means of transportation in many European and Asian countries. In Beijing and other major Chinese cities, bicycles outnumber cars many times over, although bikes are being squeezed out by automobiles as China's economy improves. In some Asian cities, half of all trips are made by bike. Europe is also relatively bicycle friendly. In some cities, bicycles account for 20 to 30% of all trips. In the United States, in contrast, bicycles account for less than 1% of all trips. In Portland, Oregon, Austin, Texas, and other cities, however, bikes are becoming more popular thanks to a "fleet" of city-owned bikes that are left on the sidewalk for anyone who wants them (**FIGURE 17-8**). If you need to travel five blocks and don't want to walk, you hop on a bike and pedal to your next meeting or to lunch. Anyone who wants to use the bike you leave by the side of the road can have it.

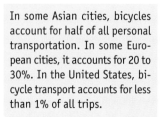

In some Asian cities, bicycles account for half of all personal transportation. In some European cities, it accounts for 20 to 30%. In the United States, bicycle transport accounts for less than 1% of all trips.

Following the example set by Europe and Asia, some cities in the United States have laid out extensive bike paths for commuters. Davis, California, is a leader in promoting bicycle transportation. Today, 30% of all commuter transport within the city is by bicycle. Some streets are closed entirely to automobile transport, and 65 kilometers (40 miles) of bike lanes and paths have been established.

Bicycles won't replace cars, buses, and trains, but they can augment them—in some places more than others. Unfortunately, bicycle sales are beginning to decline in Asia in part because of an increased interest in motorized transport, mostly motorbikes and cars. In some countries, official policies are discouraging bicycle transport in an effort to reduce congestion to make travel by car more feasible. It may be only a matter of time before congestion from auto traffic becomes a problem as it is in many more developed countries.

KEY CONCEPTS

In many cities, bicycles already carry a significant number of commuters. The bicycle could help supplement the mass transit systems of cities in the future.

Economic Changes Accompanying a Shift to Mass Transit

The automobile industry is the world's largest manufacturing endeavor and supports a number of other economically important industries. Manufacturers of rubber, glass, steel, radios, and numerous automobile parts will also feel the impacts of the shrinking automobile market. So will the service

FIGURE 17-8 The yellow bikes of Austin, Texas. These bicycles are left on the street for anyone who wants to use them. Riders take them to their destination and leave them for someone else to use.

sector: gas stations, automobile dealerships, and repair services. The shift to more efficient forms of transportation is likely to lead to significant shifts in the world's economy.

Today, 20 cents of every dollar spent in the United States is directly or indirectly connected to the automobile industry and its suppliers. Eighteen cents of every tax dollar the federal government collects comes from automobile manufacturers and their suppliers.

Although shifting toward a sustainable transportation system could result in a dramatic shift in our economy, experts point out that some of the steel and glass now destined for autos will be used for buses and trains. Many automobile workers will very likely find jobs in plants that produce buses and commuter trains. Many mechanics could shift as well to service the new fleet of more efficient vehicles. Some workers, however, may be forced to find employment in new areas. Helping them adjust to the changes is an important task.

Studies suggest that the employment potential of mass transit, like other sustainable strategies, exceeds that of the current automobile-based economy. A study in Germany showed that spending $1 billion on highways yields 24,000 to 33,000 (direct and indirect) jobs. The same amount spent on mass transit produces 38,000 to 40,000 jobs.

A sustainable transportation system is possible, but it will require a massive restructuring of the current, unsustainable system. Making that transition will require foresight and considerable political will.

Creating sustainable cities and towns is part of the challenge outlined in this text. It will take many years. The technology and knowledge required to make the changes are not barriers, but the political will and the costs of such endeavors surely stand in the way of such a massive shift. Incrementally, however, cities, businesses, and citizens are already beginning to make the changes required to create a more sustainable human presence. Automobile manufacturers are making and selling tens of thousands of more efficient hybrid cars and trucks. The hypercar featured in Chapter 15 is currently under development. Biodiesel is currently sold at 100,000 gas stations in the United States, and numerous cities such as Denver are dramatically expanding their light-rail systems. Growth management strategies are popping up in the most unlikely places. The Individual Actions Count! table lists personal actions that you can take to do your part in creating a more sustainable system of transportation and a better future.

KEY CONCEPTS

A shift away from the automobile will have serious repercussions on the global economy, but much of the slack could be taken up by a shift to the manufacture and maintenance of alternative transportation modes such as buses.

The world of politics is always twenty years behind the world of thought.

—John Jay Chapman

REFERENCES AND FURTHER READING
The References and Further Reading section at the end of this book contains a list of sources for the information discussed in this chapter and recommendations for further reading.

Connect to this book's website:
http://environment.jbpub.com/
The site features eLearning, an online review area that provides quizzes, chapter outlines, and other tools to help you study for your class. You can also follow useful links for in-depth information, research the differing views in the Point/Counterpoints, or keep up on the latest environmental news.

CRITICAL THINKING

Exercise Analysis

Two of the very first things you must do in analyzing the assertion that electric cars help to clean up urban air are to understand what an electric car is and where its energy comes from. Most electric cars are specially built subcompact vehicles. They're powered by an electric motor supplied with electricity from a large and heavy battery bank. The batteries of typical electric cars must be recharged every 60 to 100 miles the vehicle travels. The electricity to power them is typically generated by burning coal. Regular automobiles are powered by gasoline derived from oil.

Side by side, the comparison is pretty startling. Electric cars are quiet and clean. They have no tailpipes; thus, unlike the standard gasoline-powered automobile, they're fairly clean. To make a fair comparison of the two technologies, though, one must assess the total amount of energy consumed and pollution produced—not just by the cars, but by the entire systems that they depend on. Thus, we must analyze the energy demands and pollution output of the electric car *and* the power plants and mines and all of the transportation and processing facilities associated with producing electricity from coal. We must then compare it with the energy and pollution output of the standard automobile and the oil-gasoline automobile energy system.

When you do so, you find that electric vehicles use slightly less energy overall. When you compare the two with regard to the production of the greenhouse gas carbon dioxide, you find that the electric vehicle supplied by electricity from coal-fired power plants actually produces a little more than the gas-powered vehicle. Electric vehicles also produce more particulates, nitrogen oxides, and sulfur oxides, the last two of which contribute to acid deposition (Chapter 20). Electric vehicles do better than gas-powered vehicles with respect to carbon monoxide and hydrocarbons.[1]

The difference between electric- and gasoline-powered vehicles isn't as huge as proponents would lead you to believe. That doesn't mean, though, that electric vehicles don't offer advantages over gasoline-powered cars. The major difference between the two is that the pollution from gasoline-powered vehicles is dispersed throughout an entire urban area so that many people are exposed to potentially harmful pollutants. Pollution from power plants that supply electricity for electric vehicles is produced at concentrated sources that are easier to control. If these facilities are sited properly, for example, downwind of cities, they present less harm than gas-powered autos. We mustn't forget, however, that power plants contribute significantly to global pollution levels, and pollution will be carried downwind to neighboring cities and towns and could have effects there.

Electric vehicles could provide an enormous advantage over gas-powered cars if the electricity were generated by sunlight or wind energy (Chapter 15). Another consideration to bear in mind is that electric cars may be quieter than their fossil fuel cousins. Finally, electric cars may alleviate urban air pollution, but they won't solve another perplexing problem facing cities: traffic congestion.

This exercise required you to dig deeper and to define terms—to learn a little more about electric cars and their sources of fuel. It also required you to look at the big picture—that is, to step back and consider the entire cost and impact of running different types of vehicles.

---

[1]These data come from a study by the U.S. EPA entitled *Compilation of Air Pollutant Emissions Factors,* which was published in 1991.

CRITICAL THINKING AND CONCEPTS REVIEW

1. List all the human systems you have relied on since you woke up this morning.

2. Where does the water you drink come from? Where does your garbage go? Is it recycled? Is it landfilled? Is it incinerated? Where does the wastewater from your domicile go?

3. Using your critical thinking skills and the knowledge you have gained in your coursework and reading, critically analyze the following statement: "We must restructure human systems to make them compatible with natural systems."

4. What is meant by the statement "Just because a system is functioning well doesn't mean that it is sustainable"? Give some examples.

5. List and describe several reasons why most if not all human systems are currently unsustainable.

6. When most people think about creating a sustainable future, they think about designing anew—that is, creating new superefficient homes and autos. Is this sufficient?

7. Critically analyze the following comment: "Our transportation system is just fine. My commute hasn't changed very much. Air pollution levels are down because of greater efficiency in automobiles and pollution control devices. What's everyone so concerned about?"

8. What are the traditional functions of land-use planning? How is sustainable land-use planning different?

9. Compare and contrast compact development and dispersed development patterns according to the following criteria: use of land, cost of infrastructure (roads, bridges, and so on), feasibility of mass transit, and air pollution.

10. Describe corridor and satellite development. What are the advantages and disadvantages of each?

11. Imagine that you are a developer. You are about to draw plans to develop a 200-acre piece of property. Make a list of ways to make the development as environmentally sustainable as possible. Try doing this exercise by addressing one system at a time—for example, energy, waste, water, and transportation.

12. How can differential tax rates, the purchase of development rights, and making growth pay its own way (ending the subsidy for growth) be used to promote more sustainable land use?

13. Using your critical thinking skills, analyze the following assertion: "Land reform in the developing nations will help create a sustainable future."

14. The text outlined a simplified version of a plan to shift the developed nations such as the United States to a more sustainable system of transportation. Describe this plan and point out its strengths and weaknesses. How could the weaknesses be eliminated?

15. Debate the following statement: "A shift to a transportation system based in large part on mass transit will devastate the global economy and put thousands of people out of work."

KEY TERMS

cohousing
compact development
corridor development
development right
differential tax rate

dispersed development
greenbelts
land-use planning
making growth pay its own way
mass transit

satellite development
telecommute
urban sprawl
zoning regulations

Illegally disposed hazardous waste.

CHAPTER 18 Principles of Toxicology and Risk Assessment

CHAPTER OUTLINE

Life is a perpetual instruction in cause and effect.
—*Ralph Waldo Emerson*

All across the globe, companies are taking a new approach to the problem of hazardous chemical wastes. For example, numerous progressive companies are finding ways to reduce the amount of hazardous chemical wastes they produce. Others are finding ways to recycle hazardous wastes, generating profit at the same time. This approach, called **pollution prevention**, is like energy efficiency. It's a preventive action that is not only good for the environment, but also great for a company's bottom line. This encouraging shift has been stimulated in part by costly regulations, but also by business leaders who realize that industry is partly responsible for the deteriorating condition of the planet and that they can play a significant role in solving the problems we face. The fruits of their labor can be seen in FIGURE 18-1, which shows how dramatically toxic chemical release has declined in the United States.

CRITICAL THINKING

Exercise

You are an environmental manager for a major corporation. Your boss tells you that the company needs to cut back on toxic pollutants emitted into the air and water. If you don't, the company could face heavy fines. You turn to your staff and ask them to come up with a plan to reduce pollution. In a week, they come back with a proposal that calls for the addition of a $20 million pollution control device. You present the proposal to your boss, who says that the company can't afford it. He argues that it would be cheaper to continue releasing toxic pollutants and be fined than to comply with the law. What would you do? Would critical thinking help you out of this mess?

This chapter deals with toxic substances, laying the groundwork for the rest of Part V, on pollution. It examines principles of **toxicology**, the study of toxic chemicals or **toxicants**, chemicals that adversely affect living organisms. It discusses the effects of toxic chemicals and several means of reducing exposure of people and the environment to these substances. It concludes with an introduction to a growing field, risk assessment, which helps us analyze the risks that various activities, technologies, and toxic substances have on us and the environment.

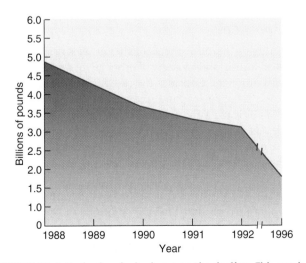

FIGURE 18-1 Toxic chemical release on the decline. This graph shows that the release of toxic chemicals from industry is on the decline, thanks in part to pollution prevention programs that seek ways to reduce the production of toxic chemicals.

18.1 Principles of Toxicology

Nearly 85,000 chemical substances are sold commercially in the United States. These chemicals are used to produce a long list of products, including cosmetics, food additives, and pesticides to which humans are routinely exposed. In the United States, chemical production and use skyrocketed after World War II (**FIGURE 18-2**). Of the commercially important chemicals, however, only a small number—perhaps 2%—are known to be harmful. Nonetheless, this small percentage amounts to hundreds of potentially dangerous chemicals that pose a threat to workers and to the general public.

Perhaps the biggest problem with toxic substances is our lack of knowledge about their effects. The National Academy of Sciences notes that fewer than 10% of U.S. agricultural chemicals (mostly pesticides) and 5% of food additives have been fully tested to assess long-term health effects. Testing potentially harmful substances is a costly and time-consuming task, made more difficult by the 700 to 1000 new chemicals entering the marketplace each year.

> **KEY CONCEPTS**
>
> Many thousands of chemicals are produced each year in industrialized nations, but only a small percentage pose a risk to humans. Vast gaps exist in our knowledge of the effects of toxic substances, in large part because of the sheer number of chemicals that need to be tested and the expense of thorough testing.

The Biological Effects of Toxicants

People are exposed to toxic substances at home (Table 18-1), at work, or in the out-of-doors. Toxicants are present in the food we eat, the water we drink, the air we breathe, the furniture we buy, and even in the clothes we wear. These substances come from natural and human sources. Toxic substances also affect a wide range of plants and animals. (Many effects are discussed in subsequent chapters.) This chapter deals primarily with effects on humans, although the principles you will learn apply to all living things.

In many cases, people have little control over exposure. Polluted air from nearby power plants or highways, for in-

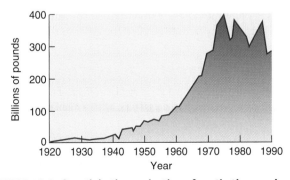

FIGURE 18-2 Growth in the production of synthetic organic chemicals in the United States. Production has more or less leveled off since 1970.

Table 18-1

Common Toxicants Around the Home

Ibuprofen and Tylenol
Gasoline
Rubbing alcohol
Bleach
Mineral spirits
Many cleaning agents

stance, exposes them to dozens of potentially harmful substances. In other cases, however, we intentionally expose ourselves to harmful substances, such as toxic fumes from cleaning agents.

Bear in mind, though, that exposure to a toxic substance does not necessarily mean that one will be adversely affected. The effect a toxic substance has on us, if any, depends on many factors. Before we examine those factors, let's study some key characteristics of toxic effects.

KEY CONCEPTS

Humans are exposed to potentially toxic substances in virtually every aspect of our lives, but exposure does not necessarily mean that we will be adversely affected.

Immediate and Delayed Toxicity   Toxicants produce a wide variety of effects. Some stimulate immediate effects. Some immediate effects are subtle, such as a slight cough or headache from urban air pollution. Others can be pronounced, such as the violent convulsions induced by exposure to certain insecticides. As a rule, immediate effects disappear shortly after the exposure ends; they are generally caused by fairly high concentrations of chemicals that result from short-term exposures.

Other toxic substances produce delayed effects such as cancer or birth defects. They may occur months to years after exposure and usually persist for years, as in the case of emphysema caused by cigarette smoke and air pollution. Delayed effects often result from low-level exposure over long periods (chronic exposures). It is important to note, though, that short-term exposures may also have delayed effects. A one-time exposure to certain cancer-causing agents, for example, may actually cause the disease that appears many years later.

KEY CONCEPTS

Toxic substances may produce immediate effects, ranging from slight to severe, or delayed effects, with a similar range of severity.

Local and Systemic Toxic Effects   Still another important distinction regarding toxic chemicals is their site of action. Some toxicants exert local effects, most often at the site of contact. Certain industrial chemicals, for instance, cause skin rashes. Other toxic substances get into the body and cir-

culate to many different sites, where they exert their effects, often on entire organs or organ systems, such as the brain and spinal cord (central nervous system). These effects are described as **systemic** and are the most common type. Most systemic toxic substances act on the central nervous system. But how do toxicants cause their effects?

KEY CONCEPTS

Toxic substances can exert their effects locally, often at the site of exposure, or systemically—that is, in an organ system or even throughout much of the body.

How Toxicants Work

Toxic substances exert their effects at the cellular level in three major ways: First, some toxicants bind to enzymes, the cellular proteins that regulate many biochemical reactions. A disturbance of enzymatic activity can seriously alter the functioning of an organ or tissue. As examples, mercury and arsenic both bind to certain enzymes and block their activity. Second, some toxic chemical substances bind directly to nonenzyme molecules in the body, upsetting the chemical balance. Carbon monoxide, a pollutant from automobiles and other sources, for example, binds to hemoglobin in the red blood cells, blocking their ability to bind to oxygen. This decreases the transport of oxygen throughout the body and can lead to death if levels are high enough (Chapter 19). Third, some toxicants interact with the genetic material of cells, causing mutations, potentially harmful changes in the structure of the DNA (described shortly). As you will learn in a later section, mutations can lead to cell death, cancer, and birth defects.

KEY CONCEPTS

Toxic substances may bind to enzymes and other important molecules, including the genetic material, DNA. Binding to these chemicals alters cellular function, sometimes in profound ways.

Factors That Affect the Toxicity of Chemicals

Predicting the harmful effects of chemicals is no easy task. Age, sex, health, and a variety of other factors contribute to the final outcome. Consider the case of a family of six living near a Canadian lead and zinc smelter that released large amounts of lead. Although each member was exposed to high levels of lead, the symptoms varied. For example, the father and a 4-year-old boy suffered from acute abdominal pain and pancreatitis. The mother developed a neural disorder. Two other children experienced convulsions, and the last developed diabetes. Differences in age, sex, and genetic composition lead to such variations in toxic effect. They also make studies of the health effects of toxic agents very difficult to perform and can make the results difficult to interpret.

Complicating matters further, individuals can develop tolerance to certain toxicants. In other cases, they can become **hypersensitized** to them—that is, extremely sensitive to

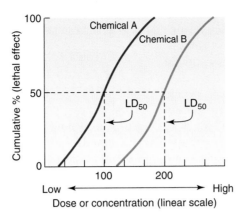

FIGURE 18-3 Dose-response graph for two chemicals with differing toxicities. The LD_{50} is the amount of chemical that kills one half of the experimental animals within a given time. The higher the LD_{50} value, the less toxic the chemical is. See the chart at right.

Chemical A
Chemical B
LD_{50}
LD_{50}

Cumulative % (lethal effect)

100

50

0

100   200

Low ←——————→ High
Dose or concentration (linear scale)

Approximate Acute LD_{50}s of a Variety of Chemical Agents	
Agent	LD_{50} (mg/kg)
Ferrous sulfate	1500
Morphine sulfate	900
Phenobarbital sodium	150
DDT	100
Picrotoxin	5
Strychnine sulfate	2
Nicotine	1
d-Tubocurarine	0.5
Hemicholinium-3	0.2
Tetrodotoxin	0.10
Botulinus toxin	0.00001

tiny doses. People exposed to formaldehyde, a chemical released from furniture, plywood, and other products, are often tolerant of the chemical at high levels at first, but over time, they become sensitive to extremely low levels.

Numerous factors influence the effects of a given chemical, the four most important being (1) dose, (2) duration of exposure, (3) biological reactivity of the chemical in question, and (4) route of exposure.

Chemical Reactivity, Dose, and Duration of Exposure

Chemicals vary considerably in their reactivity—how readily they react with biological molecules. Highly reactive chemicals generally have a greater effect on the body than those that are less reactive. In addition, the effect is also determined by the **dose**, the amount to which an individual or animal or plant is exposed. In general, the greater the dose, the greater the effect.

To study the effect of dosage, toxicologists often expose laboratory animals to varying amounts of a substance, determining the response at each level. The resulting graph is called a **dose–response curve** (**FIGURE 18-3**). To compare one chemical with another, toxicologists often determine the dose that kills half of the test animals—that is, the lethal dose for 50% of the test animals or LD_{50}. By comparing LD_{50} values, scientists can judge the relative toxicity of two chemicals. For example, a chemical with an LD_{50} of 200 milligrams per kilogram of body weight is half as toxic as one with an LD_{50} of 100 milligrams. Thus, the lower the LD_{50}, the more toxic a chemical.

The **duration of exposure** is the amount of time an individual or laboratory animal is exposed to a toxic substance. Exposures generally fall into two categories: **acute**, or short-term exposures, generally last less than 24 hours; **chronic**, or long-term exposures, last more than 3 months. Obviously, these are very broad categories, and many intermediate exposure possibilities exist. Studies show that for many toxicants, acute exposures result in very different effects than those resulting from repeated (chronic) exposure. An acute

exposure to benzene, for instance, may result in a transient bout of depression. Repeated, chronic exposure may result in leukemia, a cancer of the white blood cells.

Routes of Exposure The toxicity of a chemical is also dependent on the route of exposure—that is, how it gets into a body. Three routes of exposure are most common: inhalation, ingestion, and dermal exposure (**FIGURE 18-4**).

Inhalation exposure, as the name implies, results from breathing a chemical. Smokers inhale many potentially toxic

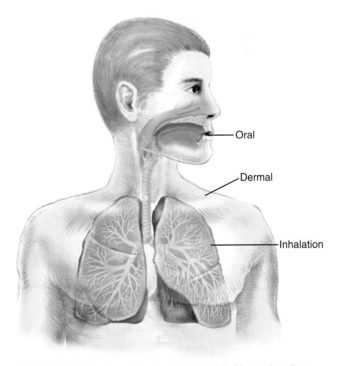

Oral
Dermal
Inhalation

FIGURE 18-4 Three routes of exposure. This illustration shows the three major ways that toxic chemicals can enter the human body.

substances, as do workers in chemical factories or commuters in heavy traffic. Toxic substances that are inhaled enter the bloodstream via the lungs and quickly become dispersed throughout the body. Inhalation is the most rapid route for chemicals to enter the body.

Ingestion, the next major route of exposure, occurs when toxic substances enter our bodies or the bodies of other animals in the food we eat and the water or liquids we drink. Ingestion is the next most rapid form of exposure. Toxicants, however, can be neutralized by acids in the stomach and enzymes in the small intestine. Those that survive this harsh environment can be absorbed into the blood vessels in the lining of the intestinal tract and may be distributed throughout the body.

The slowest and least effective route of exposure is the skin. Toxic substances spilled on the skin may penetrate this rather thick layer and enter the blood vessels. As a general rule, the more readily a toxic substance is absorbed, the more effect it has.

Age and Health   Although we think of youngsters of all species as resilient, young, growing organisms are generally more susceptible to toxic chemicals than are adults. For example, two common air pollutants, ozone and sulfur dioxide, affect young laboratory animals two to three times more severely than they affect adults. Among humans, infants and children are more susceptible to lead and mercury poisoning than adults because their nervous systems are still developing (**FIGURE 18-5**).

Health is determined by many factors, among them one's nutrition, level of stress, and personal habits such as smoking. As a rule, the poorer a person's health is, the more susceptible he or she is to a toxicant.

Chemical Interactions   Modern society is dependent on thousands of chemicals, and most of us are exposed to many different substances in many different ways. Predicting the effect of these chemicals is therefore very difficult. Further adding to the difficulty of this task is the fact that chemical substances interact in a variety of ways. Some chemical substances, for example, team up to produce an **additive response**—an effect that is the sum of the individual responses (for example, $2 + 2 = 4$). Others may produce a superadditive effect, also known as a **synergistic response**—that is, a response that is greater than the sum of the individual ones ($2 + 2 = 6$). One of the most familiar examples of synergism is the combination of barbiturate tranquilizers and alcohol; although neither taken alone in small amounts is dangerous, the combination can be deadly. Pollutants can also synergize. For instance, sulfur dioxide gas and particulates (minute airborne particles) inhaled together can reduce airflow through the lung's tiny passages. The combined response is much greater than the sum of the individual responses. The synergistic effect of smoking and asbestos is discussed later.

Another fascinating interaction is **potentiation**, which occurs when a chemical with no toxic effect combines with a toxic chemical and makes the toxicant even more harmful. This response can be represented by the equation $0 + 2 = 6$. Isopropyl alcohol (rubbing alcohol), for instance, has no effect on the liver, but when combined with carbon tetrachloride, it greatly boosts the toxicity of the latter.

Certain chemicals can also negate each other's effects, a phenomenon called **antagonism**. In these cases, a harmful effect is reduced by certain combinations of potentially toxic chemicals. This can be represented by the equation $2 + 4 = 3$. In mice exposed to nitrous oxide gas, mortality is greatly reduced when particulates are present. Scientists are uncertain of the reasons for this phenomenon.

The many variables mentioned in the previous sections—such as age, nutrition, personal habits, and chemical interactions—often make it difficult to study the effects of pollutants on people. **Epidemiologists**, scientists who study the effects of chemicals and other disease-causing agents on people, must be careful to eliminate as many of these variables as possible so that they don't cloud the interpretation of the results (see Chapter 1 for a review of epidemiology).

Bioaccumulation and Biological Magnification   Two additional factors that profoundly influence toxicity are bioaccumulation and biological magnification. **Bioaccumulation** is the buildup of chemicals within the tissues and organs of

FIGURE 18-5 **Children are more susceptible to poisoning.**

FIGURE 18-6 **Bioaccumulation.** Radioactive iodine released from accidents at nuclear power plants is taken up by cows and passed to humans in cow's milk. In humans, it accumulates in the thyroid gland in the neck.

the body. It results from at least three factors: selective uptake, a resistance to chemical breakdown, and long-term storage. Consider each one, beginning with selective update.

Iodine is actively absorbed from the bloodstream by the thyroid gland of humans. This is called *selective uptake*. What's the problem? Radioactive iodine, present in the milk of cows who feed on grass contaminated by an accident at a nuclear power plant, is selectively concentrated in the thyroids of people who consume the milk (FIGURE 18-6). The radioactive iodine remains in the gland and subsequently irradiates the cells, sometimes producing tumors.

Heavy metals bioaccumulate because they resist chemical breakdown in the body. Further aiding the bioaccumulation of heavy metals, some of them bind strongly to proteins. Chlorinated hydrocarbons such as DDT are also resistant to breakdown and, because they are fat soluble, tend to concentrate in body fat. Here, they remain for many years.

The accumulation of toxic substances within an organism can have serious effects on the organism itself, as in the case of radioactive iodine. They may also have adverse effects on the organisms that eat them. Consider an example. Scallops and other molluscs feed on material suspended in water. They also selectively take up certain toxic elements (heavy metals) from seawater, such as zinc, copper, cadmium, and chromium. The level of cadmium in scallops in polluted waters may be 2.3 million times that of seawater. While such concentrations may not always cause problems to the organism, they can be toxic to organisms that eat them.

Biological magnification refers to the progressive increase in the concentration of a chemical substance in the organisms of a food chain. As shown in FIGURE 18-7, the pesticide DDT is present in water and is taken up by **zooplankton**, single-celled organisms that live in water. Small fish ingest the fat-soluble DDT when they feed on zoo-

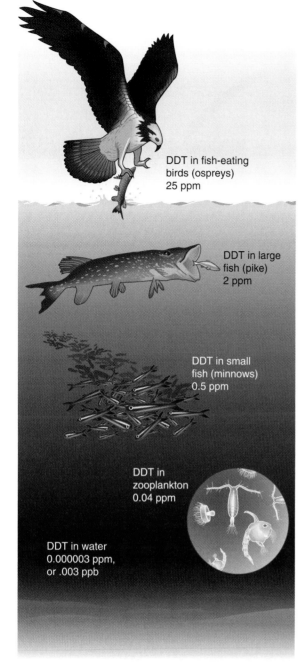

FIGURE 18-7 Biological magnification. Concentrations of some toxic metals and fat-soluble organic compounds such as the pesticide DDT increase at higher levels in a food chain. This phenomenon is known as *biological magnification* and results from the fact that fat-soluble molecules tend to be stored in the fat of organisms, rather than being broken down. As a result, they are passed with nearly 100% efficiency from one level of the food chain to the next.

[Labels within figure:]
DDT in fish-eating birds (ospreys) 25 ppm

DDT in large fish (pike) 2 ppm

DDT in small fish (minnows) 0.5 ppm

DDT in zooplankton 0.04 ppm

DDT in water 0.000003 ppm, or .003 ppb

plankton, and the persistent insecticide is then concentrated in their body fat. DDT also accumulates in the fat of fish-eating birds such as the osprey. As illustrated, tissue concentrations increase substantially at each level of the food chain.

Biological magnification occurs because DDT is a fat-soluble chemical that is stored in body fat; it is not readily broken down and is not excreted. Because of this, the more fish an osprey eats, the higher its DDT levels become. The concentration of DDT, in fact, may be several million times greater in fish-eating birds than it is in water (Figure 18-7). (Chapter 22, on pesticides, describes the effects DDT has had on several bird species.) For humans, magnification that occurs in our food chain may be as much as 75,000 to 150,000 times. Synthetic chemicals such as DDT, some lead and mercury compounds, and even some radioactive substances are all biomagnified.

Biological magnification exposes organisms high on the food chain to potentially dangerous levels of persistent toxicants. Therefore, the presence of this phenomenon is important to keep in mind when judging the risk that a chemical poses to people and to the many species that live among us. Levels that may appear safe can actually be quite dangerous.

KEY CONCEPTS

Toxic substances may accumulate in certain tissues and organs, causing local effects. Some may also increase in concentration in the food chain, being the highest in top-level consumers. Because of these phenomena, ambient concentrations of a toxicant may be an insufficient means of predicting its toxic effects.

18.2 Mutations, Cancer, and Birth Defects

Much of the concern about toxic chemicals today stems from their effects on the genetic material and from two likely offshoots of these effects: cancer and birth defects.

Mutations

The hereditary material of the cell is contained in a molecule called **DNA**, short for **deoxyribonucleic acid**. This material is housed in **chromosomes** found in each cell in the body. The DNA not only passes on traits from parents to offspring, it also controls how cells grow and develop and how they function. DNA can be a target for various chemical and physical agents. Those agents that cause changes, or **mutations**, in the hereditary material are known as **mutagens**. The term *mutation* actually refers to three possible alterations in the hereditary material: (1) changes in the DNA molecule itself, (2) alterations of chromosomes that are visible by microscope (for example, deletion or rearrangement of parts of the chromosome), and (3) missing or extra chromosomes.

Mutations can be caused by chemical substances, such as benzene, or by physical agents, such as ultraviolet radiation and other high-energy radiation. In humans, mutations can occur in normal body cells, or **somatic cells**, such as skin and bone. Such mutations occur quite frequently, but they are usually repaired by cellular enzymes. If a mutation is not repaired, and if it affects important genes, the cell may die or become cancerous (described shortly).

The reproductive cells, or **germ cells**, in the male and female gonads are also susceptible to mutagens. Unrepaired germ cell mutations may be passed to offspring. If an ovum with a genetic mutation, for example, is fertilized by a normal sperm, the mutation may be passed on to the embryo.

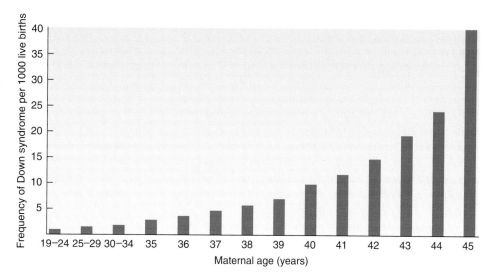

FIGURE 18-8 **Down syndrome.** The incidence of Down syndrome and several other abnormalities caused by abnormal chromosome numbers is related to the mother's age. The incidence of Down syndrome is about 5 in 1000 at age 35, but 40 in 1000 by age 45—about the time of menopause, when women stop ovulating (releasing ova).

The defective gene may prove lethal, killing the embryo, or it may manifest itself as a birth defect (described shortly), a metabolic disease (a biochemical disorder), or childhood cancer. Some germ cell mutations may not be expressed in the first generation but may appear in the second and third generations. This delayed effect makes it hard for scientists to pinpoint the causes of some diseases.

Genetic mutations are present in about 2 of every 100 newborns. The causes of mutations in humans are not well understood. Abnormal chromosome numbers, responsible for some diseases such as Down syndrome, are related to maternal age (**FIGURE 18-8**). Broken and rearranged chromosomes are also related to maternal age. As women enter their 30s, their chances of having a baby with an abnormal number of chromosomes increase; after age 40, the chances skyrocket. Geneticists once hypothesized that the older a woman is, the greater the chance that she has been exposed to mutagens; hence, the greater the chance that her child will have a mutation. New research suggests that the increased incidence of birth defects in babies of older women may result from the fact that older mothers are, for unknown reasons, more likely to carry a defective fetus to term than younger women.

Other diseases, associated with structural defects in the DNA molecule itself, seem to increase in incidence as the *father* gets older and are not related to the mother's age (**FIGURE 18-9**). These defects may be caused by mutagens.

KEY CONCEPTS

Chemical and physical agents can alter the hereditary material in a number of ways. Such changes, called *mutations,* can occur in body cells or in reproductive cells. Those occurring in body cells may kill the cells or lead to uncontrolled growth, a cancer. Nonlethal changes in reproductive cells can be passed on to one's offspring.

Cancer

One of the most serious consequences of mutations in cells is cancer (defined shortly). In the United States, recent statistics indicate that one of every two men and one of every three women will develop some form of cancer. In the United States, cancer kills about 560,000 people a year. Worldwide, the number of deaths is estimated to be about six million per year, according to the World Health Organization.

According to two Oxford University scientists, 8000 Americans die each year of cancer caused by environmental factors such as air pollution. Another 8000 cancer deaths are attributed to food additives and industrial products, such as pesticides used around the house; 16,000 deaths result each year from occupational exposure to harmful substances. The researchers note, by comparison, that tobacco causes more than 90% of the lung cancer deaths in the United States each year, or at least 144,000 deaths. The Point/Counterpoint in Chapter 23 looks at the presumed epidemic of cancer in the United States.

What is cancer, and how does it form? **Cancer** is a disease produced when a single cell or a group of cells somehow "escape" from mechanisms that control their growth. The unchecked division of these cells results in the formation of

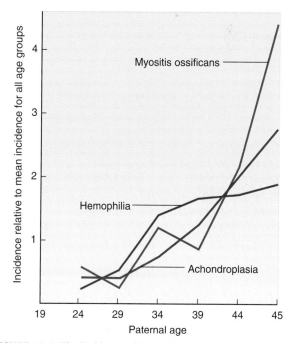

FIGURE 18-9 The incidence of three diseases in newborns: myositis ossificans, affecting bone; hemophilia, affecting blood; achondroplasia, affecting cartilage. All are caused by DNA damage related to the father's age.

a **tumor** or **neoplasm**. In some instances, the tumor is limited to a single expanding mass of cells. This is referred to as a **benign tumor**. Benign tumors can create problems, for example, by compressing nerves or other vital tissues. If a tumor continues to grow, spreading into tissues and organs surrounding it, it is called a **malignant tumor**. Cells may break off from the site of formation, the **primary tumor**, and travel in the blood and other body fluids to other sites. The spread of cancerous cells is called **metastasis** (meh-TASS-tah-siss). In distant sites the cancerous cells may form **secondary tumors**. Certain types of cancers spread (metastasize) in characteristic ways. Breast cancer, for example, tends to spread to the bones. Lung cancer typically spreads to the brain.

Malignant tumors are dangerous because they continue to enlarge, demanding huge amounts of nutrients. They also often invade neighboring areas and destroy vital tissues and organs.

Cancers occur most commonly in the rapidly dividing cells of the body, such as those of the skin, bone marrow, lungs, and lining of the intestines. As a rule, nondividing cells such as nerve cells and muscle cells rarely become cancerous. Nonetheless, many different types of cells can form tumors; there are hundreds of different types of cancer. Not only are there different types, but many types behave differently, a fact that has complicated efforts to find a cure for this disease. The most likely scenario is that we will discover numerous cures, many of them specific to a particular type of cancer.

KEY CONCEPTS

Cancer is the uncontrolled proliferation of cells of the body. Tumor cells typically develop in rapidly dividing cells such as those of the skin and often spread to other parts of the body. Hundreds of different types of cancer exist, and many behave differently, a fact that has complicated efforts to find a cure for the disease.

What Causes Cancer?   Cancer can be caused by a variety of factors—chemicals in the food we eat, physical agents such as X-rays, and biological agents such as viruses. Of these carcinogenic (cancer-causing) agents, the ones of greatest concern to most people are toxic chemicals. Chemicals that increase the chances of getting cancer are called **carcinogens** (car-SIN-oh-gins).

Carcinogens usually require repeated exposures for many years to stimulate tumor production. Thus, smoking a single cigarette (or even a whole pack) will not increase your chances of contracting lung cancer very much. However, smoking a pack a day for several years will greatly increase the likelihood that you will develop lung cancer. Moreover, the longer and more you smoke, the greater your chances are of getting cancer.

KEY CONCEPTS

Cancers can be caused by chemical, physical, and biological agents. Chemicals that cause cancer, called *carcinogens*, generally require repeated exposures over many years.

DNA-Reactive and Non–DNA-Reactive Carcinogens Chemicals that induce cancer by altering the DNA of cells are called **DNA-reactive**, or **genotoxic**, (jean-oh-TOX-ick)

carcinogens. A growing body of evidence, however, indicates that some chemical carcinogens may cause cancer without altering DNA directly. Such substances are referred to as **epigenetic** (ep-eh-gin-ET-ick) **carcinogens.**

Some of the epigenetic carcinogens, for instance, are thought to cause hormonal imbalances that lead to rapid cell proliferation. For example, toxicologists have discovered natural chemicals (certain plant-derived substances) and synthetic chemicals (for example, the pesticide DDT) that mimic the actions of the sex hormone estrogen. Estrogen stimulates cell division in the breasts and wombs (uteri) of women. It is widely believed to play an important role in the development of some types of cancer in women, particularly cancer of the breast and uterus. Toxicologists speculate that exposure to estrogen-like chemicals in our diet and the environment may also increase cancer risk.

Epigenetic carcinogens may act in other ways, too. For example, some chemicals may alter the function of the immune system, severely impairing its ability to recognize and destroy precancerous cells. Other epigenetic carcinogens result in persistent tissue injury. Asbestos fibers, for instance, may end up in cells, where they slice dividing chromosomes.

KEY CONCEPTS

Most carcinogens react with the growth control genes of cells, causing mutations that lead to cancer. Evidence suggests that some other chemical carcinogens cause cancer through mechanisms that don't directly involve the DNA.

Biotransformation, Emotions, and Intrinsic Factors Interestingly, many cancer-causing chemicals must be chemically modified by enzymes in the body in order to be able to react with the DNA. Nitrites, for example, are found in processed meats such as bologna (as a preservative) and are fairly benign chemicals. However, the liver converts nitrites into carcinogenic nitrosamines. Although the liver usually detoxifies chemical substances and protects us from harm, in this instance it converts a relatively harmless substance into a carcinogen with potentially lethal consequences. This process is called **biotransformation**.

New studies indicate that emotions may also play an important role in the development of cancer (and other diseases), possibly by acting through the immune system. Researchers at the Johns Hopkins University, for example, studied the incidence of cancer in medical students who took a personality test between 1948 and 1964. The research showed that students who suppressed emotions were 16 times more likely to develop cancer later in life than students who vented their emotions. More research is needed to determine if the cause-and-effect relationship between mental health and cancer is real.

Other factors may also influence one's susceptibility to cancer. For example, nutritional deficiencies may promote cancer of the liver and esophagus. Excessive alcohol intake may promote cancer of the larynx, especially in heavy smokers.

Not every individual who is exposed to a potentially carcinogenic agent develops cancer. Scientists think that the reason for this is that certain intrinsic factors—for example, one's genetic makeup, immune system, and sex—may influence one's susceptibility to carcinogens.

Birth Defects

A **birth defect** is a physical (structural), biochemical, or functional abnormality. The most obvious defects are the physical abnormalities such as cleft palate, lack of limbs, or spina bifida (incomplete development of the spinal cord, often resulting in paralysis). According to some estimates, about 7% of all U.S. newborns have a birth defect. Others believe that the incidence of birth defects may actually be higher, about 10 to 12%. The reason for this discrepancy is that many minor defects escape detection at birth, among them mental retardation and certain enzyme deficiencies.

Agents that cause birth defects are called **teratogens**; the study of birth defects is **teratology** (from *teratos*, Greek for "monster"). Teratogenic agents include drugs, physical agents such as radiation, or biological agents such as the rubella virus, which causes German measles (Table 18-2). No one knows for sure what percentage of birth defects are caused by chemicals in the environment.

Timing of Exposure Embryonic development can be divided into three parts: (1) a period of early development immediately after fertilization; (2) a period when the organs are developing (**organogenesis**); and (3) a growth phase, a period during which the organs have formed and the fetus primarily increases in size (**FIGURE 18-10**). Teratogens exert their effects during organogenesis. As Figure 18-10 shows, organs are most sensitive early in organogenesis, and it is during this period that the most noticeable effects occur. Thus, the effects of a teratogen are related to the time of exposure.

FIGURE 18-10 Embryonic development and teratogenesis. Schematic representation of human development, showing when some organ systems develop. Sensitive periods are early in development, indicated by the blue part of the bar. Exposure to teratogens during these times will almost certainly cause birth defects. Exposure to potentially harmful chemicals after organogenesis may result in physiological defects, minor anatomical changes, or death, if levels are sufficiently high.

Table 18-2

Some Known and Suspected Teratogens in Humans

Known Agents	Possible or Suspected Agents
Progesterone	Aspirin
Thalidomide	Certain antibiotics
Rubella (German measles)	Insulin
	Antitubercular drugs
Alcohol	Antihistamines
Radiation	Barbiturates
	Iron
	Tobacco
	Antacids
	Excess vitamins A and D
	Certain antitumor drugs
	Certain insecticides
	Certain fungicides
	Certain herbicides
	Dioxin
	Cortisone
	Lead

Teratogenic effects are also determined by the type of chemical involved. Certain chemicals affect certain organs; for example, methyl mercury damages the developing brains of embryos. Other chemicals such as ethyl alcohol can affect several systems; children born to alcoholic mothers exhibit numerous defects, including growth failure, facial disfigurement, heart defects, and skeletal defects. Like most tox-

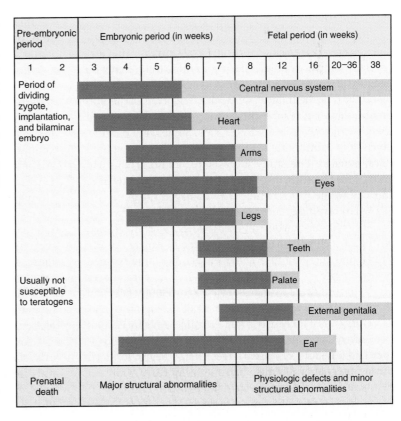

icants, teratogens usually exhibit a dose–response relationship: the greater the dose, the greater the effect.

Fetal development, occurring after organogenesis, may also be affected by physical and chemical agents in our homes and places of work. Some toxic chemicals, for instance, stunt fetal growth and in higher quantities may even kill fetuses, resulting in a **stillbirth** or **spontaneous abortion.**

Humans are not the only species whose offspring are altered by environmental chemicals. Numerous studies have shown that environmental pollutants such as selenium and PCBs alter development in bird embryos (Chapter 11). Chapters 19 through 23 describe many examples of the effects of chemical pollutants on plants and wildlife.

> KEY CONCEPTS
> Birth defects may occur when an embryo is exposed to a teratogen during the critical period of organ development, organogenesis. Chemicals vary in their effects. Some affect specific organ systems. Others affect a number of systems.

18.3 Reproductive Toxicity

The effects of environmental chemicals on reproduction in humans and other species have become a major concern in recent years. Studies have shown that male factory workers become sterile when exposed on the job to DBCP (1,2-dibromo-3-chloropropane), some permanently. Men who routinely handle various organic solvents often have abnormal sperm, unusually low sperm counts, and varying levels of infertility. A wide number of chemicals such as borax, cadmium, diethylstilbestrol (DES), methyl mercury, and many cancer drugs are toxic to the reproductive systems of males and females.

Reproductive toxicants may exert their effects long after exposure. Two researchers from Laval University in Quebec, for example, examined the records of 386 children who had died of cancer before the age of five. Their study showed that at the time the children were conceived, many of the children's fathers had been working in occupations that exposed them to high levels of hydrocarbons. Some were painters exposed to paint thinners; others were mechanics exposed to car exhaust. This study suggests that hydrocarbons had entered the bloodstream of the fathers at work and may have traveled to the testes, where they damaged the germ cell DNA. The resulting mutations were passed to the offspring.

In 1970, seven women (ages 15 to 22) were diagnosed with clear-cell adenocarcinoma (a type of cancer) of the vagina and cervix, a disease never seen before in women below the age of 30 and usually only seen in women over 50. A study of these women showed a correlation between their cancer and maternal ingestion of a drug known as DES. DES was widely used from the mid-1940s to 1970 in the United States and was administered to women who either had a history of miscarriages or had begun to bleed during pregnancy. Because bleeding is an early symptom of miscarriage, DES was given in hopes of preventing it. (It is now known that DES cannot prevent miscarriage.) Years later, vaginal cancers and other abnormalities began to appear in the daughters of DES-treated women. Additional research has uncovered reproductive abnormalities in approximately one fourth of male offspring exposed to DES through their mothers. Problems include small testes, cysts, and low ejaculate volume. No evidence of testicular tumors has been uncovered.

> KEY CONCEPTS
> Reproduction is a complex process, involving many steps. Chemical and physical agents may interrupt any of these complex processes, interfering with reproduction.

18.4 Environmental Hormones

Hormones are chemicals produced by certain cells in the body. They travel in the bloodstream to other parts and regulate numerous body functions, from reproduction to body temperature to growth. Hormones help maintain internal constancy, or homeostasis.

The secretion of various hormones can be influenced by certain pollutants such as dioxins. Dioxins are a group of structurally similar chemicals found in some herbicides (chemicals used to control weeds) and in paper products. At high levels, dioxins are thought to be carcinogenic. At low levels, however, dioxins may suppress the immune system by binding to hormone receptors, proteins in the plasma membranes of body cells. As a result, some researchers are calling dioxins **environmental hormones.** Environmental hormones include a growing list of chemical toxicants. They can alter a variety of hormonal systems, creating a wide range of effects. For example, studies suggest that a number of common herbicides (the thiocarbamates) may produce thyroid tumors by altering normal hormonal balance. The result is a goiter-like condition and even cancer.

Environmental hormones alter reproduction and other functions in nonhuman animals, too. Nonylphenols or chemicals that break down to form them, for example, alter the endocrine (hormonal) systems of salmon. Nonylphenols are present in dishwashing detergent, pesticides, and even some contraceptives (spermicides) and end up in waterways. New studies suggest that they can evaporate and travel long distances in the atmosphere, depositing via rain in other areas. Nonylphenols alter fish reproduction and cause other abnormalities. One major problem is that they appear to impair internal hormonal changes that are required by salmon as they migrate from freshwater to saltwater. Unable to adapt to the high concentrations of salt, many of these fish weaken and die. Some ecologists suspect that hundreds of environmental hormones may be responsible, in part, for the decline in numerous species of wildlife—fish, birds, and mammals. Effects may be subtle and difficult to demonstrate.

Another environmental hormone is a group of chemicals called phthalates (pronounced thall-lates). These chemicals are softening agents added to plastics—IV bags, beach balls, plastic milk jugs, and pop bottles, among others. Phthalates act as estrogens and may be partly responsible for the earlier onset of sexual maturity, including early breast development in girls. Because of problems such as this, the EPA now requires companies to screen all new chemicals for potential hormonal effects.

18.5 Case Studies: A Closer Look

A look at two case studies illustrates how complex and controversial some problems are and how supposed experts can examine the same data and sometimes come up with different opinions on the risk posed by a chemical substance.

Asbestos: How Great a Danger?

Asbestos is the generic name for several naturally occurring silicate mineral fibers. Asbestos is useful because of its flexibility, its great tensile strength, and its resistance to heat, friction, and acid. In the United States and other industrial countries, asbestos was once added to cement to make it more resistant to weather. Asbestos was also applied as a heat insulator on ceilings and pipes in factories, schools, and other buildings. In addition, asbestos was sprayed on walls and ceilings to make them more sound proof and fire proof. It was also once used in the manufacture of brake pads, brake linings, hair driers, patching plaster, and a multitude of other products.

Asbestos fibers can be easily dislodged. Floating in the air, these fine particles may be inhaled into the lungs, where they are neither broken down nor expelled. Remaining in the lungs for life, they produce three disorders: pulmonary fibrosis, lung cancer, and mesothelioma.

Pulmonary fibrosis, or **asbestosis**, is a buildup of scar tissue in the lungs that occurs in people who inhale asbestos on the job or in buildings with exposed asbestos insulation. This disease makes breathing nearly impossible and takes 10 to 20 years to develop.

Exposures to asbestos at low levels, even for short periods, can cause lung cancer. The death rate from lung cancer among asbestos insulation workers is four times the expected rate (FIGURE 18-11). Interestingly, the incidence of lung cancer in asbestos workers who smoke is 92 times greater than in asbestos workers who don't, providing a striking example of synergism.

Asbestos is the only known cause of **mesothelioma** (mez-oh-theel-ee-OH-ma), a cancer that develops in the lining of the lungs. This highly malignant cancer spreads rapidly and kills victims within a year from the time of diagnosis. Although the incidence of other asbestos-related diseases is clearly related to dose, mesotheliomas have been observed in individuals with only brief exposure.

Scientists have long wondered how asbestos causes cancer. Studies have shown measurable amounts of DNA in tissue fluids surrounding cells of the body. Research suggests that asbestos fibers may attach to this DNA and then pierce the cell membrane, carrying the DNA inside. Inside the cell this DNA may disrupt or turn off the genes that control a cell's growth. With the control mechanism altered or completely shut down, the cell begins to duplicate wildly. Another possibility is that the DNA may carry cancer-causing genes into the cell. Once inside, the genes become activated, triggering the cell to divide. Still other research suggests that asbestos fibers inside cells may damage chromosomes during cell division. This, in turn, may cause harmful mutations.

An estimated 8 to 11 million American workers have been exposed to asbestos since World War II. Studies of these workers show that over a third had lung cancer, mesothelioma, or gastrointestinal cancer. The expected death rate in the population for these diseases is about 8%. Between 1990 and 2020, some health experts expect about 2 million people, mostly workers, to die from exposure to asbestos.

Based on these and other statistics, the use of asbestos in the United States for insulation, fireproofing, and decorative purposes was banned in 1974. In 1979, the EPA began to help states and local school districts identify asbestos crumbling from pipes and ceilings. In 1989, the EPA ordered bans on almost all other applications, which went into effect in 1997 and will, with the previous ban, eliminate almost 95% of all asbestos use in the United States.

Companies that produce asbestos argue that removing asbestos from school buildings will not protect public health. Workers, they point out, have been the main victims of asbestos-related diseases. Furthermore, workers who smoke are at a much higher risk. Because young children do not smoke, their risk is very small. A far more economical way

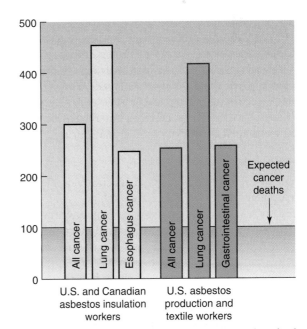

FIGURE 18-11 **Incidence of cancer in asbestos workers in the United States and Canada.** (Ratio of the number of observed to the number of expected deaths times 100.) Asbestos workers are much more likely to develop cancer than the general public.

of dealing with the problem would be to stabilize the asbestos layer around pipes and in ceilings—say, by painting it.

Asbestos is a naturally occurring silicate mineral fiber with many practical uses. Unfortunately, it produces three disorders: pulmonary fibrosis (buildup of scar tissue), lung cancer, and mesothelioma. It is especially dangerous to asbestos workers and to individuals who also smoke.

Electromagnetic Radiation: A Hazard to Our Health?

In 1979, two researchers from the University of Colorado reported a link between high-current electric power lines and the incidence of childhood leukemia. Their study suggested that extremely low-frequency (ELF) magnetic fields produced when electricity flows through wires may be the cause of the increased incidence of cancer in children living nearby. The researchers found that the death rate from cancer in these children was twice what was expected in the general public.

ELF magnetic fields are virtually everywhere. Moreover, they easily penetrate building walls and the human body. Magnetic fields are also found around power stations, welding equipment, subways, and movie projectors.

In 1986, researchers from the University of North Carolina announced the results of a study that supported the Colorado research. The study showed a fivefold increase in childhood cancer (particularly leukemia) in residents living near the highest magnetic fields, 7 to 15 meters (25 to 50 feet) from wires carrying electricity from power substations to neighborhood transformers. Adding to the concern, a researcher from Texas recently found that magnetic fields increase the growth rate of cancer cells. In addition, cancer cells exposed to magnetic fields are 60 to 70% more resistant to the body's naturally occurring killer cells.

In 1986, another group of scientists reported that magnetic fields from electric waterbed heaters and electric blankets increased the likelihood of miscarriage. In the group that used electric heaters in their waterbeds, 61% of the miscarriages occurred from September to the end of January. By comparison, women using neither a waterbed heater nor an electric blanket had a 44% miscarriage rate during this same period.

ELF magnetic fields may also be a cause of birth defects in humans. Scientists know that magnetic fields affect fetal development in pigs, chickens, and rabbits.

Many scientists and industry representatives are skeptical about the link between cancer and magnetic fields. Eleanor Adair, environmental physiologist and senior research scientist at Yale University, notes that of the more than 30 epidemiological studies, about half show a weak correlation between magnetic fields—and the other half do not. She finds many serious problems with the design of virtually all epidemiological studies, further shedding doubt on their validity. In most studies, she notes, scientists did not measure the strength of the magnetic fields. Instead, they relied on measures of the proximity of homes to power lines

or other sources and measured the size of the wires carrying electric current to homes. She also notes the failure of such studies to take into account other variables, such as smoking and exposure to pollution, that might be responsible for reported cancers. Other scientists, including David Carpenter from the University of Albany School of Public Health, disagree and feel that the studies probably underestimate the risk of cancer. A recent analysis of over 500 studies on the subject led the prestigious National Research Council to conclude that the bulk of the evidence suggests that exposure to ELF magnetic fields does not cause cancer, neurological or behavioral problems, or damage to reproductive cells or the developing fetus. They and others point out that electromagnetic radiation is not an ionizing radiation, like that emitted from radioactive materials, and thus probably can't cause cancer. Still others say that although ELF magnetic fields may not cause cancer, they may accelerate the growth of cancer cells.

Some studies suggest that ELF magnetic fields produced when electricity flows through wires may increase the incidence of cancer in children living nearby. To date, however, results of studies on the effects of ELF magnetic fields have yielded inconclusive results.

18.6 Controlling Toxic Substances: Toward a Sustainable Solution

The United States produces about 125 million metric tons (280 billion pounds) of synthetic chemicals and creates 200 million metric tons of hazardous wastes each year. Many laws have been passed to regulate them but most have been end-of-pipe controls (Table 18-3). Although important, many experts realize that we also need long-term approaches, for example, efforts to prevent hazardous waste production in the first place. This section will look at both stop-gap and long-term, sustainable solutions. See Spotlight on Sustainable Development 18-1 for a discussion of efforts to reduce exposure to lead.

Many laws have been passed in the United States and other countries to reduce the release of toxic chemicals into the environment, but most of these relied on end-of-pipe controls. More recent efforts are aimed at preventing pollution—that is, eliminating hazardous waste production in the first place.

Toxic Substances Control Act

In 1976, the U.S. Congress passed the **Toxic Substances Control Act (TSCA)**. The TSCA consists of three parts. The first is **premanufacture notification**, which requires all companies to inform the EPA 90 days before they import or manufacture a chemical substance not currently in commercial use. The agency then has 90 days to approve the chemical or, if necessary, place restrictions on its use to reduce exposure to peo-

Table 18-3

Federal Laws and Agencies Regulating Toxic Chemicals

Statute	Year Enacted	Responsible Agency	Sources Covered
Toxic Substances Control Act	1976	EPA	All new chemicals (other than food additives, drugs, pesticides, alcohol, tobacco); existing chemical hazards not covered by other laws
Clean Air Act	1963, 1967, 1970, amended 1977, 1990, 1997	EPA	Hazardous air pollutants
Federal Water Pollution Control Act	1972, amended 1977, 1978, 1987	EPA	Toxic water pollutants
Safe Drinking Water Act	1974, amended 1977, 1996	EPA	Drinking water contaminants
Federal Insecticide, Fungicide, and Rodenticide Act	1948, amended 1972, 1973, 1988	EPA	Pesticides
Act of July 22, 1954 (codified as § 346(a) of the Food, Drug and Cosmetic Act)	1954, amended 1972	EPA	Tolerances for pesticide residues in food
Resource Conservation and Recovery Act	1976	EPA	Hazardous wastes
Marine Protection, Research and Sanctuaries Act	1972	EPA	Ocean dumping
Food, Drug and Cosmetic Act	1938	FDA	Basic coverage of food, drugs, and cosmetics
Food additives amendment	1958	FDA	Food additives
Color additive amendments	1960	FDA	Color additives
New drug amendments	1962	FDA	Drugs
New animal drug amendments	1968	FDA	Animal drugs and feed additives
Medical device amendments	1976	FDA	Medical devices
Wholesome Meat Act	1967	USDA	Food, feed, and color additives; pesticide residues in meat, poultry
Occupational Safety and Health Act	1970	OSHA	Workplace toxic chemicals
Federal Hazardous Substances Act	1966	CPSC	Household products
Consumer Product Safety Act	1972	CPSC	Dangerous consumer products
Poison Prevention Packaging Act	1970	CPSC	Packaging of dangerous children's products
Lead Based Paint Poison Prevention Act	1973, amended 1976	CPSC	Use of lead paint in federally assisted housing
Hazardous Materials Transportation Act	1970	DOT (Materials Transportation Bureau)	Transportation of toxic substances generally
Federal Railroad Safety Act	1970	DOT (Federal Railroad Administration)	Railroad safety
Ports and Waterways Safety Act	1972	DOT (Coast Guard)	Shipment of toxic materials by water
Dangerous Cargo Act	1952	DOT (Coast Guard)	Shipment of toxic materials by water

NOTE: CPSC = Consumer Product Safety Commission
DOT = U.S. Department of Transportation
EPA = U.S. Environmental Protection Agency
FDA = Food and Drug Administration
OSHA = Occupational Safety and Health Administration
USDA = U.S. Department of Agriculture
Source: Council on Environmental Quality.

ple and the environment. To do this, EPA scientists review existing toxicity data on the chemical or data on chemicals with a similar structure. If the new chemical is believed to pose insignificant risk, it is approved. If it could be hazardous, the agency usually asks the manufacturer to test its toxicity and

report back. At this point, many manufacturers drop the candidate because of the cost involved in toxicity testing.

The second part of the TSCA requires the EPA to examine chemicals that were in commercial use before the law passed. Those the agency deems potentially hazardous are

18-1 Getting the Lead Out: Steps to Reduce Lead Exposure in the United States

Lead is one of the most useful metals in modern industrial societies. It is found in ceramic glazes, batteries, fishing sinkers, solder, and old pipe. Lead was once added to gasoline to help reduce engine knocking.

Unfortunately, lead is also a highly toxic poison, entering the body primarily through inhalation and ingestion. Although it affects many organs, lead has a special affinity for bone, brain, and kidneys.

High-level exposure in certain factory workers has been linked to neurological symptoms: fatigue, headache, muscular tremor, clumsiness, and loss of memory. If exposure is discontinued, patients may slowly recover, but residual damage such as epilepsy, idiocy, and hydrocephalus (fluid accumulation in the brain) often lingers. Continued high exposure may lead to convulsions, coma, and death. Some scientists believe that lead drinking vessels and lead pipes in water systems may have caused a decline in birth rates and increased psychosis in ancient Rome's ruling class, contributing to the fall of the Roman Empire.

Organic lead (alkyl lead in gasoline, for example) has been linked to a number of psychological disorders—including hallucinations, delusions, and excitement—and may lead to death.

Lead also damages the kidneys, causing a disturbance in the mechanisms that help us conserve valuable nutrients (such as glucose and amino acids) that might otherwise be lost in the urine. Prolonged, high-level exposure causes a progressive degeneration of the filtering mechanism that removes wastes from the bloodstream.

Lead has a profound effect on reproduction. Numerous reports show that the rate of spontaneous abortion is much higher in couples when one or both has been exposed to high levels of lead in the workplace. Recent studies show decreased fertility and damaged sperm in male workers with high-to-medium levels of lead in their blood. According to one study, exposure of a pregnant woman to high levels of lead in household drinking water nearly doubles her risk of having a retarded child.

Today, because of better controls on lead in the workplace and in commercial products, acute lead poisoning is rare in most countries. Nonetheless, many people throughout the world are regularly exposed to low levels of lead, with potentially serious consequences. Blood levels in humans in industrial nations today are about a hundred times greater than before the advent of lead pollution. However, even residents of Nepal, a nonindustrialized country, have levels 10 times higher than those estimated to be present before the widespread use of lead—attesting to the global distribution of lead in the atmosphere.

Recently, scientists have begun to learn about the effects of low-level exposure, especially on the central nervous system. Herbert Needleman and his colleagues studied over 3000 first-and second-grade children with different levels of lead in their bodies. Children with high lead levels (caused by exposure to low levels of lead over time) had significantly lower IQ scores than those with low levels. Attention span and classroom behavior were also significantly impaired. Several other studies have shown similar results. A recent study in England showed that at an early age even marginally elevated levels of lead may have lasting adverse effects on intelligence and behavior.

In 1990, Needleman reported that childhood exposure to lead levels once considered moderate or low can seriously and permanently alter the intelligence of adults. Studies show that elevated lead levels in the blood of men cause hypertension and increase the risk of heart disease.

Lead is found in our food, water, air, and soils. For most Americans, food is the number one source of lead exposure. Lead emitted by power plants, smelters, and boilers that burn used motor oil is frequently deposited in the soil, where it is taken up by crops.

Although food is the major source of lead, most of the concern for lead exposure in humans has focused on automobile exhausts. Regulations to eliminate lead from gasoline have markedly decreased concentrations in and around our cities.

In 1986, a major study by the EPA revealed that lead levels in drinking water in many cities exceeded federal

required to undergo toxicity testing. If they prove too risky, they can be banned from use or restricted.

Finally, the act calls for controls on chemicals believed to be hazardous to humans and the environment. The most radical controls were placed on **polychlorinated biphenyls (PCBs)**, an insulating fluid used in electrical transformers. PCBs are stable in the environment because they resist biodegradation. They are also widely dispersed in the environment, can biomagnify, and are fairly toxic to laboratory

animals. Because of these factors, in May 1979, Congress banned their manufacture and distribution except in a few limited cases.

KEY CONCEPTS

The Toxic Substances Control Act seeks both to prevent the introduction of chemicals that will be harmful to people and the environment and to eliminate those already in use that pose an unacceptable risk.

standards, potentially threatening the health of millions of Americans. Lead in drinking water comes from lead-based solder and from lead pipes often found in older homes.

Lead victims are primarily children and, among them, mostly children of poor, black families who tend to live in regions where pollution from automobile exhaust is prevalent (FIGURE 1). Children may also eat dirt contaminated with lead from vehicles in previous years (before lead was removed from gasoline). Children living in old, neglected buildings often ingest flakes of lead-based paint, which was applied before a ban was enacted in the 1940s.

Children are more susceptible to lead in part because they absorb more in the intestines. Adults absorb about 8 to 10% of the lead they ingest, whereas the absorption rate in children may be as high as 40%. In addition, children are more sensitive to lead than adults. The developing brain seems to be the most sensitive organ. Furthermore, the toxicity of lead is increased in malnourished and iron-deficient children, who often come from poor, urban families.

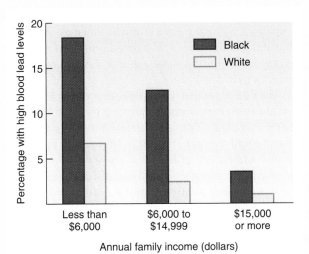

FIGURE 1 High lead levels in U.S. children from 6 months to 5 years old, according to parents' income and race (1976–1980).

A new study suggests that lead poisoning may be responsible for the higher incidence of cavities in children in economically depressed areas. A slight increase in lead concentrations in the blood (of just 5 micrograms per deciliter of blood) was attributed to an 80% increase in cavities.

Alarmed by the mounting evidence regarding the effects of lead on children, the U.S. EPA in 1973 began a progressive restriction of the lead content of gasoline. Today, it has declined to nearly zero (there's a nearly undetectable amount in unleaded gas). Blood lead levels have fallen as well. Other countries have also taken an active role in reducing lead emissions.

Cities in less developed nations are farther behind. In these countries, the lead content of the gasoline is, on the average, twice that of the more developed countries. Malnutrition and high lead levels in the air will almost certainly have serious effects on the children.

As noted previously, food is the major source of lead in the United States. In 1979, the Food and Drug Administration took steps to reduce the intake of lead from lead-soldered cans. Lead has been virtually eliminated in this application.

In 1986, Congress also banned the use of lead solder in pipes. Because drinking water accounts for about 20% of the lead exposure to Americans, the EPA adopted regulations that will reduce lead in public drinking water supplies, which could lower lead exposure in drinking water for about 138 million people. One regulation requires public water suppliers to treat their water with alkaline additives if lead is a problem or if the water is slightly acidic. This measure is expected to reduce lead leaching from pipes. The new EPA regulations also lower allowable levels of lead in the drinking water to about 5 parts per billion (ppb), down from 50 ppb. This measure is in all likelihood not going to result in a 10-fold decrease because the current limit of 50 ppb is for water measured at the tap, whereas the new standards measure lead levels in water leaving treatment plants. Because most lead enters water *after* it leaves the treatment plant, the new ruling may not lower public exposure very much. Only time will tell.

Market Incentives to Control Toxic Chemicals

In 1988, one of the most controversial environmental laws in the United States went into effect. The state of California's **Safe Drinking Water and Toxics Enforcement Act** set in motion a market strategy to reduce dramatically the exposure to toxic chemicals in foods and consumer products. Created as a result of a citizen-sponsored initiative, Proposition 65, which was passed overwhelmingly by voters in 1986, this law directs the state to set standards (acceptable levels) for potential toxicants in a variety of consumer products and foods. The law requires that manufacturers who violate these standards print warnings on their products, noting that the amount of the chemical in the product exceeds the state's safe level. Manufacturers are not required to meet standards, only to print the fact on the packages. The proponents of this law believe that consumers will shun products such as these, which will encourage manufacturers to reduce levels of toxicants in their products.

- Select nontoxic alternatives to traditional cleaning products and chemical pesticides.
- Use nontoxic paints and finishes.
- Reduce your consumption because production of virtually all products results in the production of toxic substances.
- Support nonprofit organizations involved in promoting pollution prevention and in improving regulations to reduce or eliminate hazardous wastes.
- Properly dispose of all potentially toxic chemicals—cleaning agents, paints, finishes, and paint thinners. Many counties have some form of household hazardous waste pickup services.

Numerous other market-based incentives for controlling toxic pollution are also available. Two of the more promising are pollution taxes and tradable permits. Basically, **pollution taxes**, or **green taxes**, are fees levied on environmentally unsound activities—for example, the combustion of fossil fuels or the use of toxic chemicals or the production of hazardous wastes. Taxes make it more costly to do things wrong and thus provide an incentive to switch to more environmentally sustainable practices.

Tradable permits are discussed in Chapter 19. For now, suffice it to say that each company is given permits to release certain amounts of pollution. If a company can find a way to reduce emissions economically, it can sell its permit to pollute to others, making a profit on the transaction. This system encourages innovation and cost-effective means of reducing or even eliminating the production and release of toxic substances.

Other market-based measures will be discussed in Chapters 19 through 23 and Chapter 25. In summary, market-based policies are designed to produce monetary incentives for companies to reduce pollution and regulatory burden. They also provide more freedom of choice for businesses to select methods that suit their unique situations. Accordingly, market-based actions are gaining popularity among businesses, government regulators, and environmentalists, but they are still a long way from reaching their full potential.

KEY CONCEPTS

Many efforts are under way to harness market forces, rather than impose regulations on companies, to encourage the use of nontoxic or less toxic products. These mechanisms promote innovation, freedom of choice, and cost-saving solutions that many businesses view as an acceptable way to reduce pollution.

The Multimedia Approach to Pollution Control: An Integrated Approach

The U.S. EPA has historically regulated toxic chemicals by controlling the emissions into various media (air, water, and soil). In addition, each medium has been monitored and regulated by a separate branch of the agency. Wastewater inspectors looked only at the pollutants in the wastewater. The air pollution inspectors examined what goes out of the smokestacks. So long as a company was obeying the necessary regulations, the inspectors were happy.

The shortsightedness of this method became evident in the early 1990s when EPA officials found that companies often disposed of wastes in the medium with the weakest regulations. Refinery wastes that would normally be shipped out in barrels and thus were subject to hazardous waste rules, for instance, might end up in the wastewater stream, which is less strictly regulated.

In 1991, William Reilly, then administrator of the EPA, called for a **multimedia approach** to prevent this environmentally costly toxic shell game. In the regional offices of the EPA, Reilly set up multimedia branches that trained inspectors in several media—for example, air and water releases. Their job is to determine if companies are in compliance with all pertinent environmental regulations, not just regulations in a specific medium—hence, the term *multimedia* (not to be confused with modern audiovisual productions).

The EPA hopes that multimedia inspections will help the agency identify companies that are illegally or improperly disposing of wastes. Such a tightening in the regulatory approach could promote pollution prevention (preventive actions) and reduce emissions to the environment. Why? Increased agency awareness of the activities in a facility decreases a company's ability to hide pollution and avoid regulation by transferring pollution from one medium to another.

Controlling toxic substances also requires individual actions. For a list of ways you can help, see the Individual Actions Count! table.

KEY CONCEPTS

Pollutants from factories exit via one of several avenues, such as wastewater or air pollution. For years, each medium has been regulated separately. Efforts are now under way to regulate and monitor several media simultaneously to keep companies from dumping potentially toxic substances in the least regulated medium.

18.7 Risk and Risk Assessment

Ralph Waldo Emerson once wrote, "As soon as there is life there is danger." Every day of our lives we face many dangers—some obvious, some hidden. The assessment of the risks (potential harm) of modern technological societies has become important in policymaking and is clearly vital to efforts to build a sustainable society. This section looks at risk and how it is assessed.

Risks and Hazards: Overlapping Boundaries

Two types of hazard are broadly defined by risk assessors: anthropogenic and natural. **Anthropogenic** (an-throw-poe-GEN-ick) **hazards** are those created by human beings. **Natural hazards** include events such as tornadoes, hurricanes, floods, droughts, volcanoes, and landslides. Interestingly, events that are typically classified as natural hazards may actually be caused or worsened by human actions. Take flooding as an example. Although most of us think of floods as natural phenomena, the severity and frequency of flooding is often the result of such actions as channelizing streambeds or deforestation, as described in Chapter 13.

Three Steps in Risk Assessment

Hazards befalling human society exact an enormous price. They may kill people or destroy property. They may cause a deterioration in our health or lead to the extinction of plants and animals. How do we go about assessing the hazards and creating policies that minimize or eliminate them in socially and economically acceptable ways?

In the mid-1970s, a new science called **risk assessment** began to take root. Its goal was to help people understand and quantify risks posed by technology, our lifestyles, and our personal habits (smoking, drinking, and diet).

Risk assessment involves two interlocked steps: hazard identification and estimation of risk. **Hazard identification** involves steps to identify potential and real dangers. **Estimation of risk** generally involves two processes (**FIGURE 18-12**). The first determines the probability of an event or occurrence. This process answers the question, "How likely is the event?" The second stage determines the severity of an event, answering the question, "How much damage could be caused?" To assess the risk of a toxic chemical, we must estimate the number of people (or other organisms) exposed, the levels and duration of exposure, and other complicating factors such as age, sex, health status, personal habits, and chemical interactions (described earlier). This information helps us determine what the impact might be—that is, how severe the effects might be.

Why Assessing the Risk of Toxic Substances Is So Difficult
Estimating risk is fraught with uncertainty, especially in the area of toxic substances. Several factors contribute to the difficulties. First, our knowledge of the effects of toxic sub-

FOUNDATION TOOL

FIGURE 18-12 Determining the acceptability of a risk. After the probability and severity of a risk are determined, the next step is to determine its acceptability. Cost-benefit analysis is the most common method of determining risk acceptability, but three other methods can also be used, as shown here.

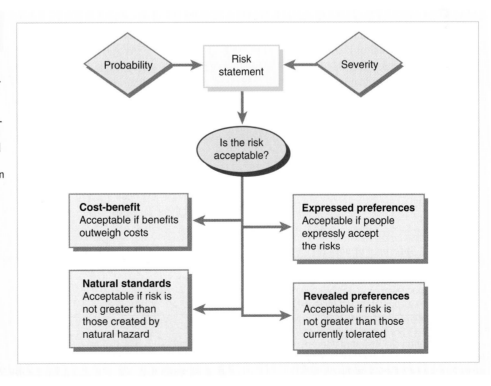

stances on humans is incomplete. As noted earlier in the chapter, modern societies use thousands of potentially toxic chemicals. Testing each one is expensive and time consuming. In addition, our knowledge of the toxic effects of pollution on humans is limited because it is neither practical nor ethical to test toxic chemicals on human beings. As a result, toxicologists must often rely on tests on laboratory animals (rats, mice, and rabbits) to estimate human toxicity. The results of experiments on laboratory animals cannot always be extrapolated to humans, though. As my graduate advisor once noted, "Contrary to popular belief, the human is not a large rat." Lab animals often react differently to chemicals than humans do; they may be able to break them down better, or they may not be able to break them down as well. Physiological differences between humans and lab animals therefore make it difficult to predict if a chemical harmful to an animal will be injurious to us.

Second, our lack of knowledge about toxic effects also stems from the fact that we humans are often exposed to many potentially harmful chemicals. For practical reasons, most toxicity tests are performed on one substance at a time. Because of chemical interactions described earlier, extrapolating the results from these tests to the real world can be misleading.

A third problem is that most animal tests of toxicity, especially those for cancer, are performed at high exposure levels. It is assumed that if a chemical is harmful at high levels, it will also be harmful at the lower levels. However, the fact that a large dose of a chemical induces cancer in a lab animal does not necessarily mean that the chemical will cause cancer at the low doses to which humans are typically exposed. Most toxicants have a **threshold level**—that is, a level below which no effects occur (**FIGURE 18-13**). Thus, extremely low levels of certain chemicals may be completely safe. The reason for this may be that at low levels, protective mechanisms in the body can cope with the chemical—either inactivating it or excreting it fast enough to prevent damage. Tissues may also be able to repair damage at low levels. At levels above the threshold, these mechanisms may not be able to keep up, and noticeable effects may then appear.

Although thresholds may exist for most toxicants, there are some exceptions, among them asbestos and radiation, discussed in Chapter 14. Many scientists believe that even the smallest exposure has an effect and that damage caused by repeated low-level exposures accumulates over time.

If scientists can't make accurate extrapolations from high doses to low doses, why do they perform their experiments that way? Researchers use high doses to speed up their experiments. As a general rule, the entire process from exposure to manifestation takes about one eighth of the life span of an animal. In humans, the time required to develop a noticeable cancer is 5 to 30 years after exposure. Any measure that speeds up the process, such as a high dose, helps cut costs. To test for low-level effects, scientists would need very large numbers of experimental animals to generate statistically valid results. High-dose studies, therefore, reduce the number of lab animals needed and can cut time and costs—a significant factor because cancer studies can cost $500,000 to $1 million per chemical. (For a debate on the validity of testing animals for cancer, see the Point/Counterpoint essays in this chapter.)

Assessing the risk of a toxic substance is difficult because so little is known about the thousands of chemicals and because much of the work is done on laboratory animals, the results of which may not be applicable to humans.

Risk Acceptability

No matter what we do—whether it is screwing in a light-bulb or flying cross-country in a jet—we put ourselves (and possibly our environment) at risk. Nothing is safe—that is, entirely free from harm. The science of risk assessment recognizes that human life is haunted by hazards. Therefore, rather than speaking in terms of safety, which is absolute, the risk assessor speaks in terms of risk, which is relative. Activities that we commonly consider safe are referred to as *low-risk functions*. "Unsafe" activities are better labeled *high-risk functions*.

Knowing the relative risk of a technology is one thing. Knowing whether the risk is acceptable to the general public is another story. **Risk acceptability**, in fact, is one of the trickiest issues facing modern society. Why? One reason is that our awareness and perception change from time to time. Thus, what appears safe one day becomes suspect the next, after a widely publicized accident. Irrational fears crop up and frighten us away from relatively low-risk activities. Jet crashes, for example, are widely publicized and may give the im-

FIGURE 18-13 **Threshold or not? (a)** Hypothetical dose-response curves indicating the absence of a threshold level, a level below which no effect occurs. Some responses are linear: For every increase in dose, there is a corresponding increase in the response. In others, the response is nonlinear; that is, there is not a 1:1 correlation between dose and response, although the response does increase with increasing dose. **(b)** A hypothetical dose-response curve showing a threshold level.

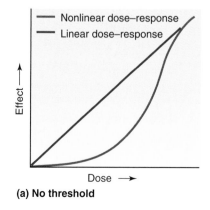

(a) No threshold

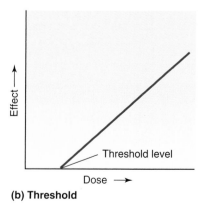

(b) Threshold

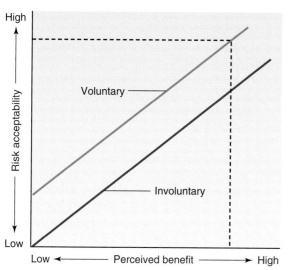

FIGURE 18-14 **The acceptability of a risk increases as the perceived benefit rises.** Voluntary risks—that is, ones people agree to—are generally more acceptable than risk imposed upon us without consent.

pression that this form of travel is quite risky. The fact of the matter is, you have a one in 5000 chance of being killed in an automobile accident in a given year and only a one in 7 million chance of being killed in a jet crash.

The **perceived harm**, the damage people think will occur, heavily influences our assessment of risk acceptability. In general, the more harmful a technology and its by-product are *perceived* to be, the less acceptable they are to society.

The acceptability of risks is also influenced by **perceived benefit**—how much benefit people think they will get from something. In general, the higher the perceived benefit, the greater the risk acceptability (**FIGURE 18-14**). As an example, the risks of a new steel mill might be acceptable to a community with high unemployment. Automobile travel provides the most telling example of the way in which perceived benefits affect our decisions. The risk of dying in an automobile accident in the United States is 1 in 5000 in any given year. Over your lifetime, the risk of dying in a car accident is higher if you don't wear seat belts, about 2 in 100, than if you do, 1 in 100. Meanwhile, substances believed to be far less hazardous than driving are banned from public use because their benefit is not so highly valued. Chemical substances that pose a risk of one cancer death in a million people, for example, are banned.

The acceptability of different forms of risk is determined by many factors, especially perceptions.

How Do We Decide If a Risk Is Acceptable?

Decisions on modern sources of risk—technologies, personal habits, and pollution—are becoming more and more commonplace. One of the most commonly used techniques is **cost–benefit analysis**, a process in which one analyzes

the many costs and benefits and weighs one against the other (Figure 18-12).

One of the most important benefits of a sustainable strategy—including efforts such as energy efficiency, pollution prevention, recycling, and use of renewable resources—is that it offers many ways of having both a clean, healthy environment and economic vitality. It makes the environment and economy complementary, not antagonistic, forces. For now, however, many decisions must be made about risk acceptability. In this technique, one enumerates costs and benefits and then tries to determine if the benefits outweigh the costs. Although that may sound simple, it isn't. As a rule, benefits are generally easily measured: financial gain, business opportunities, jobs, and other tangible items. Many of the costs, however, are less tangible. External costs, costs incurred to society by pollution and other forms of environmental damage, are among the most difficult to quantify. In addition, researchers find it difficult to assign a dollar value to human health effects, environmental damage, and lost species. Moreover, the costs may be long term and borne by future generations.

Another problem with cost–benefit analysis is that the benefits, most importantly economic gain, often accrue to a select few who have inordinate power to influence the political system and sway people's opinions. American companies that design and build hazardous-waste incinerators, for instance, spend extraordinary sums of money trying to convince people in rural areas of the benefits of siting an incinerator near them.

In sum, cost–benefit analysis suffers because many costs are poorly documented, spread out, and unquantifiable, whereas the benefits are often obvious and readily quantifiable. Recent efforts by economists to assign a dollar value to environmental and health costs may help improve the process. In addition, the efforts of scientists to measure the impacts of technologies and their by-products on wildlife, the environment, recreation, health, and society in general may also help.

Several techniques are used to determine if the risk(s) posed by a technology or activity are acceptable. The most common is cost–benefit analysis, which weighs the costs against the benefits. Sustainable development strategies minimize social, economic, and environmental costs and maximize the benefits.

Actual versus Perceived Risk The main purpose of risk assessment is to help policymakers formulate laws and regulations that protect human health, the environment, and other organisms. Ideally, good lawmaking requires that the **actual risk**, or the amount of risk a hazard really poses, be equal to the **perceived risk**, the risk perceived by the public. When actual and perceived risk are equal, public policy yields cost-effective protection.

When the perceived risk is much larger than the actual risk, costly overprotection occurs. In contrast, when the perceived risk is much smaller than the actual risk, under-

Animal Testing for Cancer Is Flawed

Philip H. Abelson
Philip H. Abelson is the Deputy Editor of *Science*. This article is copyright (© 1990) by the American Association for the Advancement of Science.

The principal method of determining potential carcinogenicity of substances is based on studies of daily administration of huge doses of chemicals to inbred rodents for a lifetime. Then by questionable models, which include large safety factors, the results are extrapolated to effects of minuscule doses in humans. Resultant stringent regulations and attendant frightening publicity have led to public anxiety and chemophobia. If current ill-based regulatory levels continue to be imposed, the cost of cleaning up phantom hazards will be in the hundreds of billions of dollars with minimal benefit to human health. In the meantime, real hazards are not receiving adequate attention.

The current procedures for gauging carcinogenicity are coming under increasing scrutiny and criticism. A leader in the examination is Bruce Ames, who with others has amassed an impressive body of evidence and arguments. Ames and Gold summarized some of their recent data and conclusions in *Science* (31 August 1990, p. 970). Three articles in the *Proceedings of the National Academy of Sciences* provide an elaboration of the information with extensive bibliographies. The articles also provide data about other pathologic effects of natural chemicals.

A limited number of chemicals tested, both natural and synthetic, react with DNA to cause mutations. Most chemicals are not mutagens, but when the maximum tolerated dose (MTD) is administered daily to rodents over a lifetime, about half of the chemicals give rise to excess cancer, usually late in the normal life span of the animals. Experiments in which synthetic industrial chemicals were administered in the MTD to both rats and mice resulted in 212 of 350 chemicals being labeled as carcinogens. Similar experiments with chemicals naturally present in food resulted in 27 of 52

tested being designated as carcinogens. These 27 rodent carcinogens have been found in 57 different foods, including apples, bananas, carrots, celery, coffee, lettuce, orange juice, peas, potatoes, and tomatoes. They are commonly present in quantities thousands of times as great as are the synthetic pesticides.

The plant chemicals that have been tested represent only a tiny fraction of the natural pesticides. As a defense against predators and parasites, plants have evolved a large number of chemicals that have pathologic effects on their attackers and consumers. Ames and Gold estimate that plant foods contain 5,000 to 10,000 natural pesticides and breakdown products. In cabbage alone some 49 natural pesticides have been found. The typical plant contains a total of 1% or more of such substances. Compared with the amount of synthetic pesticides we consume, we eat about 10,000 times more of the plant pesticides.

It has long been known that virtually all chemicals are toxic if ingested in sufficiently high doses. Common table salt can cause stomach cancer. Ames and others have pointed out that high levels of chemicals cause large-scale cell death and replacement by division. Dividing cells are much more subject to mutations than quiescent cells. Much of the activity of cells involves oxidation, including formation of highly reactive free radicals that can react with and damage DNA. Repair mechanisms exist, but they are not perfect. Ames has stated that oxidative DNA damage is a major contributor to aging and to cancer. He points out that any agent causing chronic cell division can be indirectly mutagenic because it increases the probability of DNA damage being converted to mutations. If chemicals are administered at doses substantially lower than MTD, they are not likely to cause elevated rates of cell death and cell division and hence would not increase mutations. Thus, a chemical that produces cell death and cancer at the MTD could be harmless at lower dose levels.

Diets rich in fruits and vegetables tend to reduce human cancer. The rodent MTD test that labels plant chemicals as cancer causing in humans is misleading. The test is likewise of limited value for synthetic chemicals. The standard carcinogen tests that use rodents are an obsolescent relic of the ignorance of past decades. At that time, extreme caution made sense, but now tremendous improvements of analytical and other procedures make possible a new toxicology and far more realistic evaluation of the dose levels at which pathological effects occur.

The vast majority of the scientific community endorses the conduct of experimental studies in order to try to identify those materials that could cause disease and prevent harmful exposures. In the typical toxicologic study of 50 rodents, each animal is a stand-in for 50,000 people. Because rodents live about 2 years on average, they are usually exposed to amounts of the suspect agent that approximate what a human would encounter in an average lifetime of 70 years.

Some have argued that the use of the maximum tolerated dose (MTD) in these studies produces tissue damage and cell proliferation, which in turn lead to cancer. This is biological nonsense that ignores a fundamental characteristic of cancer biology that has been known for more than 50 years: Cancer is a multistage disease, with multiple causes, which arises by a stepwise evolution that involves progressive genetic changes, cell proliferation, and clonal expansion. Thus, swamping of tissues with high doses alone may well kill an animal or damage its tissues, but high doses alone are not sufficient to cause cancer.

A 2-year study at the National Institute of Environmental Health Sciences (NIEHS) by David Hoel and colleagues provides good evidence on this point. They looked for signs of damage in tissues taken from rats and mice used in cancer studies. Cancers occurred in organs that did not show apparent damage, and some damaged organs were completely free of tumors.

In addition, most compounds tested do not cause cancer only in the highest dose group tested but typically produce a dose–response relationship, where the amount of cancer developed is proportional to the dose administered. Some chemicals cause toxicity only; others cause only cancer. Not all those that cause cancer do so through organ toxicity. In fact, almost 90% of the substances shown to cause cancer in the National Toxicology Program do so without producing any increased cellular toxicity, and they also cause cancer at both lower and higher doses.

Animal studies are evolving and being further refined, as is our understanding of differences between species that need to be taken into account in conducting these studies. Every compound known to cause cancer in humans also produces cancer in animals, when adequately tested. For 8 of the 54 known human cancer-causing agents, evidence of carcinogenicity was first obtained in laboratory animals; in many cases, the same target organs and doses have been involved in producing cancer in both animals and humans.

The NIEHS has carried out nearly 400 long-term rodent studies, which have been published following public peer review by specialists in the field. Chemicals nominated for testing usually represent a sample of potentially "problematic" materials. About 40% of the "suspect" pesticides eval-

Current Methods of Testing Cancer Are Valid

Devra Davis
Devra Davis is the author of National Book Award Finalist, *When Smoke Ran Like Water* (www.whensmokeranlikewater.com), and Director, Center for Environmental Oncology, University of Pittsburgh Cancer Institute, and Professor of Epidemiology, Graduate School of Public Health.

uated to date have been found to cause cancer. As to the role of so-called natural pesticides, rodent diets are also loaded with many of these materials. Nevertheless, a number of test compounds added to these diets markedly increase the amount of tumors produced. Thus, animals are more sensitive to certain synthetic compounds than to the background level of natural materials. Humans are also likely to have acquired some resistance to these persisting, natural materials throughout our evolution.

To date, only about 20% of all synthetic organic chemicals have been adequately tested for their potential human toxicity. Those who must set public policy on the use of chemicals need a rational basis on which to stake their actions. Continued advances in animal testing provide an important contribution to environmental health sciences and to public health efforts to predict, and then prevent, the development of environmentally caused disease.

In summary, the current system should be used until it can be replaced by a demonstrably better one. There is no scientific basis for rejecting the MTD as capable of inducing cancer.

Critical Thinking Questions

1. Summarize each author's main points and supporting data. Do you see any inconsistencies or examples of faulty logic in either essay? Where?
2. Why is it that two scientists can disagree on an issue such as this?
3. Given the disagreement, what course of action would you recommend?

 You can link to websites that represent both sides through Point/Counterpoint: Furthering the Debate at this book's internet site, http://environment.jbpub.com/. Evaluate each side's argument more fully and clarify your own opinion.

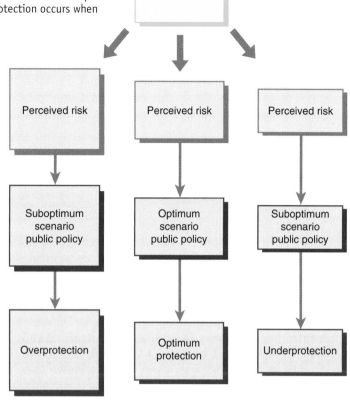

FIGURE 18-15 **Matching risk.** Matching the actual risk and the risk that a society perceives is essential to the formulation of good public policy. Perceived risk and actual risk do not always match, however, as shown. Policy decisions may result in overprotection or underprotection when the actual and perceived risk do not match. Optimal protection occurs when actual and perceived risk are equal.

protection occurs (FIGURE 18-15). It, too, can be costly to present and future generations.

KEY CONCEPTS

Risk assessment is designed to facilitate decision making by ensuring that the risk we perceive to be posed by any factor is equal to the actual risk.

The Final Filter: Ethics and Sustainability   We hear over and over again that if it doesn't make economic sense, people won't do it. This is not always the case. Many decisions we make are based on our ethics, the values we hold—or, more simply, what we view as right and wrong. Values that affect our decisions come from our parents, relatives, friends, enemies, teachers, religious leaders, and politicians. These values shift over time, sometimes subtly, sometimes dramatically. They often change as we become older and as our priorities shift. Although our ethics are often never explicitly stated, they play an important role in our lives. They determine how we think, how we vote, how we treat one another, how we carry on business affairs, how we treat the Earth, and how we treat future generations.

Because values influence the way we think and the decisions we make, they often play an important role in decisions regarding risk acceptability. Most of us, however, assign different ethical priorities to different things. Some people care more about the environment than about the economic health of their community. Others care more about economics than the environment. In sustainability, the goal is to place economics, the environment, and society on an equal level and to make decisions that serve all three. This requires creative thinking, but the options are many.

KEY CONCEPTS

Our values, what we perceive as right or wrong, come from many sources and often affect our decisions.

Space-Time Values and Sustainability   Sustainable development requires a long-term perspective. Unfortunately, most of us are rather narrow in our outlook. FIGURE 18-16 is a hypothetical graph of where our values lie. One axis plots our concern in space, the other in time. Space refers to self, family, friends, community, state, and so on, expanding outward. Time, obviously, ranges from the present to the long-term future.

FIGURE 18-16 **Space-time values.** Graph of people's hypothetical spatial and temporal interests, indicated by the points. Most individuals tend toward the lower end of the scales, being concerned primarily with self and the present. This is a natural tendency, but the rapid pace of global environmental change makes it necessary to shift space-time values to reflect long-term needs and to stimulate actions needed to build a sustainable society. Unfortunately, most political leaders are mired in the immediate—given their human nature, the short terms of office, and reelection concerns.

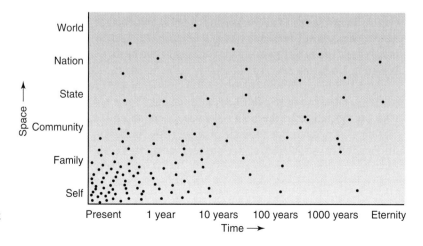

As the scatter diagram shows, individual interest can be identified by a single point that denotes one's space and time concerns. These are called **space-time values**. Most people's interests lie toward the lower end of the scales, tending toward self-interest and immediate concerns. Some people call this selfishness, but it can also be considered a natural biological tendency to be concerned with the self. Among animals, awareness of the needs of others is a feature only of social creatures such as monkeys and lions; however, concern for the uppermost end of the space-time graph is a distinguishing feature of the human animal. The unique human ability to ponder the consequences of actions is fortunate because humans have reached a position of unprecedented power as molders of the biosphere. Our power to change the world to our liking has never been greater, nor has our power to destroy ever attained such heights.

One of the things you have learned about sustainable development is that it takes a long-range view. It requires us to push the limits in space and time.

KEY CONCEPTS

People's values tend to be somewhat narrow with respect to time and space. A long-range view that encompasses the entire planet is essential for sustainability.

Chance fights ever on the side of the prudent.

—*Euripides*

CRITICAL THINKING

Exercise Analysis

Pollution control devices cost money, and they usually remove pollutants from a waste stream and create a concentrated waste that must be disposed of elsewhere. Pollution prevention, on the other hand, involves adjustments to the production system that reduce or even eliminate hazardous emissions. Several options are available under the pollution prevention strategy: modifying production processes, finding substitutes for toxic chemicals, recycling wastes, and a number of other cost-effective measures. Not only will these help you comply with the emissions standards and save your company huge sums of money, but you may actually exceed the standards—that is, reduce pollution emissions below those stipulated by state and federal laws and regulations.

Pollution has been dealt with after the fact for many years. In this exercise, all it took was a look at the bigger picture (alternatives to pollution control) and a little digging into the various options to find some viable solutions. This exercise is a good reminder to you to beware of conventional strategies for solving problems—not just in the environmental arena, but in all areas. Often, solutions that people propose reflect old ways of thinking—and therefore overlook creative and cost-effective measures that could help save money and build a sustainable society.

REFERENCES AND FURTHER READING

The References and Further Reading section at the end of this book contains a list of sources for the information discussed in this chapter and recommendations for further reading.

Connect to this book's website:
http://environment.jbpub.com/
The site features eLearning, an online review area that provides quizzes, chapter outlines, and other tools to help you study for your class. You can also follow useful links for in-depth information, research the differing views in the Point/Counterpoints, or keep up on the latest environmental news.

CRITICAL THINKING AND CONCEPTS REVIEW

1. Define the terms *toxicant, carcinogen, teratogen*, and *mutagen*.
2. Using your critical thinking skills and your knowledge of toxicology, analyze the following statement: "All substances are toxic. It's the dose that makes the poison."
3. Describe how toxicants affect the cells of our bodies.
4. What is cancer? Discuss the steps in its formation.
5. Using your critical thinking skills, analyze the following statement: "If a chemical is found to be carcinogenic in a laboratory animal, it is likely to be carcinogenic in humans."
6. List some of the possible consequences of somatic and germ cell mutations in humans.
7. What is teratology? Do teratogenic chemicals always create birth defects when given during pregnancy? Why or why not?
8. Make a list of factors that influence the toxicity of a chemical in a given individual.
9. Define the terms *synergism, antagonism*, and *potentiation*. What is the difference between synergism and an additive effect?
10. A research study shows that a certain chemical is toxic to laboratory mice. The researchers believe that it may also be harmful to humans. What information would you need to support controls on human exposure?
11. Define the terms *bioaccumulation* and *biological magnification*. Based on your knowledge gained in this chapter, what factor(s) can be used to predict whether a chemical will be biologically magnified?
12. What is meant by a threshold level? Explain the reasons why threshold levels exist for many toxicants.
13. Describe the major provisions of the Toxic Substances Control Act.
14. What are the two major types of risk? Give examples.
15. Describe the steps involved in determining the level of risk posed by technology.
16. What factors determine whether a risk is acceptable to a population? What is the difference between voluntary and involuntary risks? What is the difference between actual and perceived risks?
17. Many more people die in Montana and Wyoming from falls while hiking than are killed by grizzly bears. Why, then, are people so concerned about being killed by a bear when their chances of being killed in a fall are much greater?
18. What are space-time values? In general, where do your concerns lie in space and time?
19. Critically analyze the market-based incentives for controlling toxic substances that were discussed in this chapter.
20. The following statement was made at a public hearing regarding the contamination of a lake by effluents containing hazardous wastes generated by a chemical manufacturing plant. Using your critical thinking skills and your knowledge of toxicology, analyze the statement: "Our facility will produce hazardous substances, but because we are using state-of-the-art pollution control devices, releases of mercury and other toxicants will be minimal. Levels in the lake receiving our treated wastewater will be rather small and should pose no threat to people drinking the water."

KEY TERMS

actual risk
acute
additive response
antagonism
anthropogenic hazards
asbestosis
benign tumor
bioaccumulation
biological magnification
biotransformation
birth defect
cancer
carcinogens
chromosomes
chronic
cost-benefit analysis
deoxyribonucleic acid (DNA)
DNA-reactive carcinogens
dose
dose-reponse curve
duration of exposure
environmental hormones
epidemiologists

epigenetic carcinogens
estimation of risk
genotoxic carcinogens
germ cells
hazard identification
hypersensitized
LD$_{50}$
malignant tumor
mesothelioma
metastasis
multimedia approach
mutagens
mutations
natural hazards
neoplasm
organogenesis
perceived benefit
perceived harm
perceived risk
polychlorinated biphenyls (PCBs)
pollution prevention
pollution (green) taxes
potentiation

premanufacture notification
primary tumor
pulmonary fibrosis
risk acceptability
risk assessment
Safe Drinking Water and Toxics
    Enforcement Act
secondary tumors
somatic cells
space-time values
spontaneous abortion
stillbirth
synergistic response
systemic
teratogens
teratology
threshold level
Toxic Substances Control Act (TSCA)
toxicants
toxicology
tradable permits
tumor
zooplankton

Photochemical smog in Vancouver damages the area's human health, environment, and economy.

<table>
</table>

CHAPTER 19

Air Pollution and Noise: Living and Working in a Healthy Environment

Not life, but a good life, is to be chiefly valued.

—Socrates

At the Ford automobile manufacturing plant in Cologne, Germany, plant managers modernized the paint-spray line, cutting pollution generated by their operation by 70%. This change reduced the cost of painting a car by about $60 and will net the company hundreds of thousands of dollars in the next decade. Another German company that makes plastic films cut emissions by 70% by instituting controls that capture 90% of solvents previously lost to the atmosphere. The savings will pay for the technological improvements and are expected to save the company considerable sums of money in years to come. These two examples are part of a trend among industries designed to reduce operating costs while protecting the air we breathe.

This chapter discusses ambient air pollution—that is, pollution in the outside environment—and indoor air pollution. It looks at traditional strategies and sustainable approaches that make sense economically and environmentally. It also examines noise pollution.

19.1 Air: The Endangered Global Commons

Air is a mixture of gases, primarily nitrogen (78%) and oxygen (21%). It also contains small amounts of carbon dioxide (0.04%) and even smaller amounts of several inert substances, among them argon (almost 1%), helium, xenon, neon, and krypton. Suspended in this gaseous mixture is water vapor, which exists in varying amounts (depending on the atmospheric humidity). Air also contains various amounts of pollutants, chemicals that can have adverse effects on animals, plants, and materials.

The air we breathe is a renewable resource that is replenished by natural processes. Oxygen, for example, is replenished by plants. Air is also cleansed by other natural processes—for example, rain. Rain scours many pollutants from the sky, washing them into lakes and into our soils, sometimes with adverse effects.

Transparent and nurturing, air is a global resource, akin to the commons described in Chapter 12. Owned by no one and used by all of us, it is a source of oxygen vital to animals and carbon dioxide needed by plants. Industrial societies, however, have traditionally used the air as a waste dump for hundreds of pollutants. Like other commons, no one has sole responsibility for protecting our air, and even those who do not pollute it suffer from the disregard of others.

KEY CONCEPTS

Air is a renewable resource cleansed by natural processes and regenerated by living things. This global resource is used by many and protected by few. It suffers from the tragedy that befalls many commons.

Table 19-1

Natural Air Pollutants

Source	Pollutants
Volcanoes	Sulfur oxides, particulates
Forest fires	Carbon monoxide, carbon dioxide, nitrogen oxides, particulates
Wind storms	Dust
Plants (live)	Hydrocarbons, pollen
Plants (decaying)	Methane, hydrogen sulfide
Soil	Viruses, dust
Sea	Salt particulates

Sources of Air Pollution

Pollutants in the atmosphere arise from two major sources: natural and anthropogenic. It may be surprising to learn that globally, the largest sources of many air pollutants are natural. Natural events such as volcanoes, dust storms, and forest fires produce huge quantities of air pollution each day (Table 19-1). Thus, in sheer quantity, natural pollutants often outweigh the products of human activities, the **anthropogenic pollutants**, pollutants from human sources. Nevertheless, anthropogenic pollutants generally create the most significant long-term threat to the biosphere. Why? Natural pollutants, except those from volcanoes, come from widely dispersed sources or infrequent events and therefore generally do not raise the ambient pollutant concentration very much. As a result, they have little effect on biological systems. In contrast, power plants, automobiles, factories, and other human sources emit large quantities in restricted areas, making a significant contribution to local pollution levels. The higher the concentration, the greater the effect.

KEY CONCEPTS

Pollutants arise from natural and human or anthropogenic sources. Human-generated pollution is generally of greatest concern because it is produced in localized regions so that concentrations reach potentially dangerous levels.

Anthropogenic Air Pollutants and Their Sources

Take a deep breath. If you live in a city or even a large town, chances are you have just inhaled tiny amounts of dozens of different air pollutants, most in concentrations too small (we think) to be harmful. If you are indoors, you may have inhaled even greater amounts because indoor air is often more contaminated than outdoor air, a subject we discuss shortly.

This chapter concerns itself primarily with six major pollutants: carbon monoxide, sulfur oxides, nitrogen oxides, particulates, hydrocarbons, and ozone, called **criteria air pollutants** by the U.S. EPA. Lead, another important air

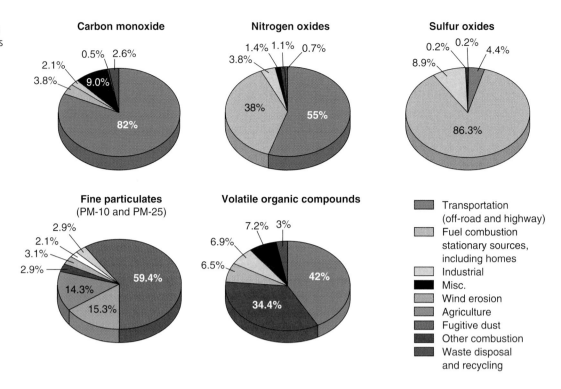

FIGURE 19-1 **Criteria air pollutants.** Sources of the five regulated air pollutants in the United States.

Carbon monoxide

2.6%
0.5%
2.1%
3.8%
9.0%
82%

Nitrogen oxides

1.4% 1.1% 0.7%
3.8%
38%
55%

Sulfur oxides

0.2% 0.2% 4.4%
8.9%
86.3%

Fine particulates
(PM-10 and PM-25)

2.9%
2.1%
3.1%
2.9%
14.3%
15.3%
59.4%

Volatile organic compounds

7.2% 3%
6.9%
6.5%
42%
34.4%

■ Transportation
(off-road and highway)
■ Fuel combustion
stationary sources,
including homes
■ Industrial
■ Misc.
■ Wind erosion
■ Agriculture
■ Fugitive dust
■ Other combustion
■ Waste disposal
and recycling

pollutant, was discussed in Chapter 18. Carbon dioxide and chlorofluorocarbons are discussed in Chapter 20.

In 2004, the U.S. produced 140 million tons of criteria air pollutants, down from 188 million tons in 1995 (**FIGURE 19-1**). The criteria air pollutants are generated by three principal sources: transportation (autos, trucks, jets, and trains), fuel combustion at stationary sources (mostly power plants and factories), and various industrial processes.

Air pollutants arise from vaporization (or evaporation), attrition (or friction), and combustion. Combustion is by far the major producer. Among combustion sources, the burning of fossil fuels (coal, oil, and natural gas) and products refined from them (gasoline, diesel, jet fuel, and home heating oil) pose the most significant threat to the environment. In fact, the more developed nations' heavy reliance on fossil fuels, which have fueled most progress, is considered by many scientists to be one of the root causes of the crisis of unsustainability.

Fossil fuels consist primarily of carbon and hydrogen atoms linked by chemical bonds. When they are ignited, an interesting thing happens. The initial source of heat—say, a match—breaks some of the chemical bonds holding the atoms together. This releases energy in two forms: light and heat. Heat released in the process

breaks other bonds, permitting combustion to occur until the fuel runs out. During combustion, oxygen in the air reacts with the carbon and hydrogen atoms of the fossil fuel to form carbon dioxide (CO_2) and water (H_2O). More commonly, combustion is not 100% complete. Incomplete combustion produces carbon dioxide, carbon monoxide (CO) gas, and unburned hydrocarbons (**FIGURE 19-2**).

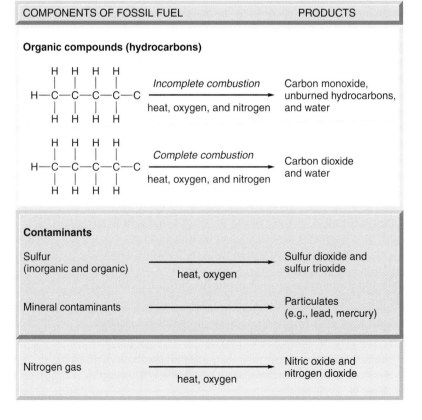

FIGURE 19-2 **The chemistry of pollution.** Products of fossil fuel combustion.

Many fuels are contaminated by heavy metals, such as mercury or lead. These unburnable contaminants may be carried off by hot combustion gases, escaping into the air as fine particulates. Other contaminants are also present. One of them is sulfur. It is especially prevalent in coal. Sulfur reacts with oxygen at high combustion temperatures, forming sulfur oxide gases, notably sulfur dioxide (SO_2) and sulfur trioxide (SO_3) (Figure 19-2). In the absence of pollution control devices, these gases escape with the other smokestack gases.

Combustion must take place in the presence of air, which provides oxygen. But air also contains nitrogen. During high-temperature combustion, nitrogen (N_2) reacts with oxygen to form nitric oxide (NO). NO is quickly converted to nitrogen dioxide (NO_2), a brownish orange gas seen in the air of many cities. Nitrogen dioxide is a key reactant in the formation of photochemical smog, described shortly. Nitrogen dioxide can also react in the atmosphere to produce nitric acid. For a review of the sources and effects of the criteria air pollutants, see Table 19-2.

Primary and Secondary Pollutants

The atmosphere contains hundreds of air pollutants from natural and anthropogenic sources. These pollutants are called **primary pollutants** because they are produced directly by various sources. Primary pollutants often react with one another or with water vapor. These reactions, which are often powered by energy from the sun, produce a whole new set of pollutants called **secondary pollutants**. For example, sulfur dioxide gas is a primary pollutant released from a variety of sources, such as coal-fired power plants, diesel trucks, and automobiles. In the atmosphere, SO_2 reacts with oxygen and water to produce sulfuric acid (H_2SO_4), a toxic and corrosive secondary pollutant with far-reaching effects (Chapter 20).

Toxic Air Pollutants

Health officials and environmental activists have long been concerned about the effects of criteria air pollutants. Many are also concerned about hundreds of other potentially toxic pollutants released into the atmosphere each year from factories and other sources. By some estimates, approximately 400 toxic air pollutants are released into the air in the United States. Toxic air pollutants can cause mutations, cancer, and possibly birth defects. Although generally emitted in much smaller quantities than the criteria pollutants, toxic air pollutants may be responsible for numerous cancer deaths each year. The U.S. EPA estimates that 45 of the toxic air pollutants may cause as many as 1700 cases of cancer each year.

Table 19-2

Major Air Pollutants: Their Sources and Health Effects

Pollutant	Major Anthropogenic Sources	Health Effects
Carbon monoxide	Transportation (cars, trucks, vans, SUVs, buses, and planes)	Acute exposure: headache, dizziness, decreased physical performance, death Chronic exposure: stress on cardiovascular system, deceased tolerance to exercise, heart attack
Sulfur oxides	Stationary combustion sources such as coal-fired power plants, industry	Acute exposure: inflammation of respiratory tract, aggravation of asthma Chronic exposure: emphysema, bronchitis
Nitrogen oxides	Transportation, stationary combustion sources such as coal-fired power plants	Acute exposure: lung irritation Chronic exposure: bronchitis
Particulates	Stationary combustion sources such as coal-fired power plants, industry	Irritation of respiratory system, cancer
Hydrocarbons	Transportation (cars, trucks, vans, SUVs, buses, and planes)	Unknown
Photochemical oxidants	Transportation, stationary combustion sources (indirectly through hydrocarbons and nitrogen oxides)	Acute exposure: respiratory irritation, eye irritation Chronic exposure: emphysema

Industrial and Photochemical Smog

Air pollution in cities generally falls into one of two categories, based on the climate and the type of air pollution. For years, older industrial cities such as New York, Philadelphia, St. Louis, Pittsburgh, and London belonged to a group of **gray-air cities** (FIGURE 19-3a). Newer, relatively nonindustrialized cities such as Denver, Los Angeles, Salt Lake City, Albuquerque, and Vancouver, Canada, belong to the group of **brown-air cities** (FIGURE 19-3b).

Gray-air cities were generally located in cold, moist climates. The major pollutants are sulfur oxides and particulates from factories. These pollutants combine with atmospheric moisture to form the grayish haze called **smog**, a term coined in 1905 to describe the mixture of smoke and fog that plagued industrial England. The gray-air cities depend greatly on coal and oil and are usually heavily industrialized. The air in these cities is especially bad during cold, wet winters, when both the demand for home heating oil and electricity and the atmospheric moisture content are high.

Brown-air cities are typically located in warm, dry, and sunny climates and are generally newer cities with few polluting industries. The major sources of pollution in them are automobiles and electric power plants; the major primary pollutants are carbon monoxide, hydrocarbons, and nitrogen oxides.

In brown-air cities, unburned hydrocarbons and nitrogen oxides from automobiles and power plants react in the presence of sunlight to form a witch's brew of harmful secondary pollutants. Some of the worst are ozone, formaldehyde (the same chemical used in embalming fluid), and peroxyacylnitrate (PAN). The reactions that form these and other potentially toxic chemicals are called *photochemical reactions* because they involve both sunlight and chemical pollutants. The resulting brownish orange shroud of air pollution is known as **photochemical smog**. Ozone (O_3) is one of the most prevalent chemicals in photochemical smog. This molecule is highly reactive. It erodes rubber, irritates the respiratory system, and damages crops, trees, and other plants.

In brown-air cities, early morning traffic provides the ingredients for photochemical smog, which usually reaches the highest levels in the early afternoon (FIGURE 19-4). Because the air laden with photochemical smog often drifts out of the city and because the reactions require sunlight and take time to occur, the suburbs and surrounding rural areas usually have higher levels of photochemical smog than the cities themselves. Major pollution episodes in brown-air cities usually occur during the summer months, when the sun is most intense. Pollution may extend for many miles. In Vancouver, British Columbia, photochemical smog may cover an area 30 miles (50 kilometers) downwind.

Today, the distinction between gray- and brown-air cities is disappearing. Most cities have brown air in the summer when sunlight and automobile pollutants are prevalent and gray air in the winter when pollution from wood stoves and oil burners conspires with the moist, wet air to darken the skies.

Although anthropogenic pollutants are the most significant sources of air pollution, researchers have found some instances in which naturally occurring pollutants noticeably affect air quality. In Atlanta, for example, the trees emit a number of highly reactive hydrocarbons. These chem-

FIGURE 19-3 **A tale of two cities. (a)** Gray-air smog in New York City. **(b)** Brown-air smog in Los Angeles.

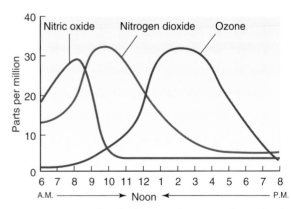

FIGURE 19-4 **Photochemical smog.** Nitrogen oxides and hydrocarbons (not shown here) react to form ozone and other photochemical oxidants. Because sunlight and time are required for the reactions to occur, maximum ozone concentration occurs in the early afternoon. Hydrocarbon levels would follow the same pattern as nitric oxide levels.

icals react with nitrogen dioxide from automobiles and other combustion sources to produce ozone. New research shows that hydrocarbons from trees are 50 to 100 times more reactive than hydrocarbons from human sources.

Pollution control often requires an analysis of both natural and anthropogenic sources. Such studies can keep us from making costly mistakes. For instance, the EPA once thought that the city of Atlanta could meet federal air quality standards for ozone by reducing the level of hydrocarbons from human sources, mostly automobiles, by 30%. New data, however, suggest that when the contribution from trees is added, the human-related sources would have to be cut by 70 to 100%. Thus efforts to cut nitrogen oxide would probably be more effective and more feasible.

> **KEY CONCEPTS**
>
> *Smog* is a term applied to two different types of air pollution. In cities with drier, sunnier climates and minimal industrial activity, hydrocarbons and nitrogen oxides from motor vehicles react to form a brownish haze called *photochemical smog*. In older industrial cities with moister climates, particulates and sulfur oxides form *industrial smog*.

Air Pollution—A Symptom of Unsustainable Systems?

Criteria air pollutants come from a variety of sources—the main ones being motorized vehicles, factories, power plants, and homes. These, in turn, are components of basic human systems of transportation, industry, energy, and housing. To some observers, air pollution is viewed as an indication that human systems are unsustainable. This book focuses on ways to make these systems sustainable through ecological design, energy conservation, alternative fuels, and a host of other measures. Such efforts are an attempt to create human systems that operate in ways that do not threaten the climate and other natural systems upon which we depend. They are, in short, a way to improve our prospects, create a more prosperous society, a healthier population, and a better future.

> **KEY CONCEPTS**
>
> Air pollution, like other forms of environmental deterioration, is a symptom of unsustainable systems of transportation, industry, housing, and energy production.

19.2 The Effects of Climate and Topography on Air Pollution

Many factors affect ambient air pollution levels. The following section examines some of the most important ones.

> **KEY CONCEPTS**
>
> Air pollution levels in a region are affected by a number of factors, among them temperature, sunlight, wind, and other climate factors. They are also affected by the topography.

The Cleansing Effects of Wind and Rain: Don't Be Fooled

Pollution is something of an enigma to many people. One day is clear. The next, the skies fill with brown haze. Why?

Numerous factors contribute to this puzzle. Wind, for example, sweeps dirty air out of cities so that on a windy day the air above a city appears fairly clean. Rain also cleanses the air in and around our cities. The folks in Seattle, for example, may brag about their clean air. The fact is, they're producing as much pollution as any other city of similar size; it just tends to be blown away or washed from the sky.

In contrast to what many people think, the pollutants do not disappear. They are merely shifted elsewhere or deposited onto another medium—for example, surface waters or the soil or onto buildings. This transfer is known as **cross-media contamination**. Airborne pollutants can travel hundreds, sometimes thousands, of kilometers to other cities or to unpolluted wilderness, where they are deposited. In Canada, for example, ground-level ozone pollution in the Windsor-Quebec corridor and southern New Brunswick and western Nova Scotia comes not from Canadian sources, but largely from the United States.

> **KEY CONCEPTS**
>
> Wind and rain both tend to cleanse the air above our cities, but pollutants do not disappear. They're either blown elsewhere or fall from the sky, ending up in waterways or in our soil or on buildings and other structures.

Mountains and Hills

Salt Lake City stretches out below a giant mountain range. Granite peaks and jagged cliffs make this one of the most scenic American cities—that is, when the mountains are visible through the air pollution. Residents of Salt Lake City would tell you that mountain ranges can be an asset, but they can also be a curse for they often block the flow of winds and trap pollutants for days on end. Mountains also block the sun, which helps disperse pollutants, as explained shortly. In Colorado and other mountainous states, ski resort towns nestled in the valleys among towering peaks often experience elevated levels of pollution as a result of their protected locations.

> **KEY CONCEPTS**
>
> Mountains and hilly terrain can impede the flow of air, resulting in the buildup of pollutants in cities and industrialized areas.

Temperature Inversions

If you rode a hot-air balloon into the sky on a normal summer day, you would find that warm ground air rises. As it rises, it expands and cools. Thus, as you lifted into the sky, you'd find that the temperature decreases. If you had proper equipment to measure pollution, you would also find that pollution rises with the warm ground air. Because of this, the atmosphere is gently stirred, and ground-level pollution is reduced.

Atmospheric mixing is caused in part by sunlight. As sunlight strikes the Earth, it heats the rock and soil. This heat is transferred to the air immediately above the ground. The warm air then rises, mixing with cooler air above it. The temperature profile is shown in FIGURE 19-5a.

Under certain atmospheric conditions, a slightly different temperature profile is encountered. As FIGURE 19-5b shows, on some days air temperature drops as one gains altitude, but only to a certain point. After that point, the temperature begins to increase. This inverted temperature profile is called a **temperature inversion**. This bizarre temperature profile results in the creation of a warm-air lid over cooler air (Figure 19-5b). Because the cool, dense ground air cannot mix vertically, pollutants become trapped in the lower layers, sometimes reaching dangerous levels. How do temperature inversions form?

Temperature inversions occur in two ways. A **subsidence inversion** occurs when a high-pressure air mass (warm front) slides over a colder air mass that failed to move out when the warm-air front arrived. Subsidence inversions may extend over many thousands of square kilometers. A **radiation inversion**, the second type, is typically local and usually short lived. It is a phenomenon many of us witness on cold winter days. Radiation inversions begin to form a few hours before the sun sets. As the day ends, the air near the ground cools faster than the air above it. Thus, warm air lies over the cooler ground air, preventing it from rising and causing pollutants to accumulate. Making matters worse, radiation inversions also correspond with the evening commute, when traffic and pollution emissions peak. Fortunately, radiation inversions usually break up in the morning when the sun strikes the Earth and vertical mixing begins. Radiation inversions are common in mountainous regions—especially in the winter, when the sun is obstructed by the mountains and is therefore unable to warm the ground enough to stimulate vertical mixing.

KEY CONCEPTS

Temperature inversions result in the formation of warm-air lids that form over cities and even large regions, impeding the vertical mixing of air. This, in turn, traps cooler pollutant-laden air below.

19.3 **The Effects of Air Pollution**

According to EPA estimates, although air pollution is decreasing, over 150 million Americans—about one in two people—are breathing unhealthy air (air that violates federal air pollution standards for ozone) at some point during the year. According to the U.S. EPA, 474 counties in 31 states failed to meet federal air quality standards. Most counties are in the eastern third of the United States, although heavily populated California tops the list. The Los Angeles basin had the worst photochemical smog problem. Ozone poses the biggest problem. Particulates

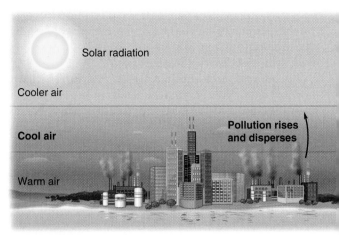

(a) Normal pattern

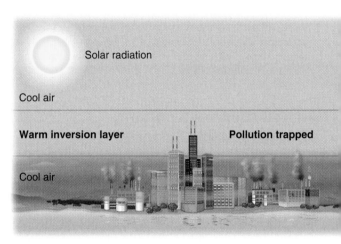

(b) Thermal inversion

FIGURE 19-5 **Temperature profiles. (a)** During normal conditions, air temperature decreases with altitude; thus, pollutants ascend and mix with atmospheric gases. **(b)** In a temperature inversion, however, warm air forms a lid over cooler air, thus trapping air pollution.

rank second, despite dramatic declines since 1970. All told, about 50 million Americans breathe air containing potentially harmful levels of particulates. According to a recent government report, approximately 60,000 Americans a year will die prematurely because of exposure to air pollution.

KEY CONCEPTS

Although 474 counties failed to meet federal air pollution standards, 2668 counties met all standards. Nineteen states had all counties in compliance: Alaska, Florida, Hawaii, Idaho, Iowa, Kansas, Minnesota, Mississippi, Montana, Nebraska, New Mexico, North Dakota, Oklahoma, Oregon, South Dakota, Utah, Vermont, Washington, and Wyoming.

The American Lung Association (conservatively) estimates that air pollution costs Americans $50 billion a year in health costs, or about $200 per year for every man, woman, and child. Air pollution damages crops and buildings, costing us billions each year. Estimates put ozone crop damage at $5 to $10 billion a year. Many other costs cannot be calculated: the loss of scenic views, the destruction of a favorite fishing spot, the erosion of a valuable and important statue.

Air pollution causes extraordinary environmental and economic damage worldwide. In Europe, three fourths of the forests are affected by air pollution, and the cost from damage is estimated to be about $30 billion a year, according to the International Institute of Applied Systems Analysis in Austria.

Air pollution causes considerable damage in the less developed countries, which (with the exception of China) produce only a fraction of the world's air pollutants. Although no statistics are available, damage to health, forests, buildings, and statuary surely carries with it an enormous price tag.

> ### KEY CONCEPTS
> Air pollution adversely affects human health, damages the environment and the organisms that live in it, and damages buildings and a wide assortment of materials, costing billions of dollars a year.

The Health Effects of Air Pollution

This section examines the impacts of pollution on human health, other organisms, and materials. These costs are typically referred to as *external costs*, as noted in previous chapters, because they are almost universally not factored into the costs of the activities (such as driving an automobile) that produce them.

> ### KEY CONCEPTS
> Air pollution has a variety of health effects, ranging from immediate to delayed and from slight irritation to potentially life-threatening conditions.

Immediate Health Effects   Air pollutants may cause many immediate effects on human health, from shortness of breath to eye irritation to death (Table 19-2). Visitors to Los Angeles, for example, often complain of burning or itching eyes and irritated throats caused by photochemical smog. Residents in other cities report similar problems, especially in less developed nations where pollution laws are lax or nonexistent. In Mexico City, for example, the air is badly polluted with exhausts from vehicles, factories, and homes.

Some problems are not very obvious. Commuters in heavy traffic, for example, sometimes complain of headaches. Although they are often attributed to stresses and strains of work or highway congestion, they may also be due to carbon monoxide from automobile fumes.

In 1966, researchers found that a sudden increase in levels of sulfur dioxide in New York City caused by adverse weather conditions resulted in a dramatic increase in the incidence of colds, coughs, rhinitis (nose irritation), and other symptoms. When the air cleared, the symptoms disappeared. More recent studies have shown that pollution episodes occurring in the summer months in the northeastern United States and Canada, which are characterized by an increase in sulfate and ozone, increase the incidence of hospital admissions for respiratory disease. These effects, however, are generally ignored or simply seen as part of the price we pay for city living.

Far more serious health problems have been observed in industrial nations during pollution episodes—periods when atmospheric concentrations reach dangerous levels for short periods because of changes in the weather that impede the cleansing of the air. Several notorious episodes have been documented. The very first occurred in the Meuse Valley in Belgium in 1930; the second in Donora, Pennsylvania, in 1948; and the third in London in 1952.

In each incident, pollution rose to dangerous levels as a result of temperature inversions that lasted 3 to 4 days. In addition, in each episode, the pollution came primarily from the combustion of coal in homes and factories. In Belgium, 65 people died. In Pennsylvania, 20 succumbed, and in London, 4000 people may have died as a result of toxic levels of particulates and sulfur dioxide. In all locations, many other people became ill. Those who died or became ill were usually old or suffering from cardiovascular disease—and were therefore unable to cope with the added stress caused by the heavily polluted air.

Another serious episode occurred in Bhopal, India, in December 1984, when approximately 40 tons of methyl isocyanate were accidentally released into the atmosphere from a pesticide facility. At least 2000 people died in the incident.

> ### KEY CONCEPTS
> Air pollution causes many immediate effects, such as shortness of breath, eye irritation, or upper respiratory tract irritation. Few people are aware of the source of these problems. In extreme cases, pollutants can become lethal.

Chronic Health Effects   Chapter 18 noted that exposure to some toxic substances may take place over many years and at low levels. This exposure regime may result in a variety of health effects.

In 2002, 9.1 million Americans suffered from **chronic bronchitis**, a persistent inflammation of the bronchial tubes, which carry air into the lungs. Symptoms include a persistent cough, mucus buildup, and difficulty breathing. Cigarette smoking is a major cause of this disease, but urban air pollution is also a major contributing factor. Studies of children ages 10 to 12 in four U.S. cities show a noticeable increase in the prevalence of chronic bronchitis as pollution levels rise. Sulfur dioxide, nitrogen dioxide, and ozone are believed to be the major causative agents.

Emphysema is another chronic effect. Currently, nearly three million Americans suffer from this incurable disease, according to the American Lung Association. As these people become older, the small air sacs, or **alveoli**, in their lungs break down. This reduces the surface area for the exchange of oxygen with the blood. Breathing becomes more and more labored. When lung surface area is reduced by about 40%, victims suffer shortness of breath even when exercising lightly.

Emphysema is caused primarily by cigarette smoking (about 80% of all cases) but may be caused by urban air pollution as well. One study, for instance, showed that the incidence of emphysema was higher in relatively polluted St. Louis than in relatively unpolluted Winnipeg (FIGURE 19-6). Studies in Great Britain showed that mail carriers who worked in polluted urban areas had a substantially higher death rate from emphysema than those who worked in unpolluted rural areas. Ozone, nitrogen dioxide, and sulfur oxides are the chemical agents believed responsible for this disease.

More recently, several studies have shown that **bronchial asthma** attacks may be associated with elevated levels of pollution. Asthma is a chronic disorder, marked by periodic episodes of wheezing and difficult breathing. Most cases of asthma are caused by allergic reactions to common stimulants such as dust, pollen, and skin cells (dander) from pets. In some individuals, pollution may trigger asthma attacks. During such an attack, the passageways that carry air to the lungs (bronchi and bronchioles) fill with mucus, making breathing difficult. Irritants also stimulate the contraction of the smooth muscle cells in the walls of the smallest air-carrying ducts, the bronchioles, making it even more difficult to breathe. Periodic attacks can be quite disabling and may even lead to death. By one estimate, several thousand Americans die each year from severe asthma attacks. Victims are generally older individuals who are suffering from other diseases. Recent studies suggest that fine particulates are to blame for asthmatic attacks caused by air pollution.

A number of studies have shown that lung cancer rates are slightly higher among urban residents than among rural residents (even after the influence of cigarette smoking has been ruled out). Critical thinking recommends caution because health studies of human populations, or epidemiological studies, rely on statistical methods with some unavoidable shortcomings (described in Chapter 1). In lung cancer studies, for example, researchers usually compare death certificates of urbanites with death certificates from rural residents. If the urban population shows a higher incidence of lung cancer, it is tempting to conclude that the cause was urban air pollution. But researchers must be careful to eliminate other causative agents, such as smoking and occupation. Most researchers eliminate smokers from the study, but not all take into account urban workplaces (factories), where men and women may have been exposed to high levels of cancer-causing pollutants. Thus, the slightly higher incidence of lung cancer in some urban settings may result from occupational exposure or some other factor. It could also result from pollution. Only time and further research will tell.

KEY CONCEPTS

Long-term exposure to air pollution may result in a number of diseases, including bronchitis, emphysema, asthma, and lung cancer.

High-Risk Populations Not all individuals are affected equally by air pollution. Particularly susceptible are the old and infirm, especially people with preexisting lung and heart

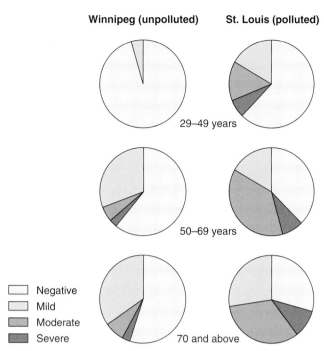

FIGURE 19-6 **Urban air pollution and emphysema.** Incidence of emphysema in Winnipeg and St. Louis. Note the increased incidence of emphysema in all three age groups in the more polluted urban environment of St. Louis.

disorders. Carbon monoxide is especially dangerous to people with heart disease because CO binds strongly to hemoglobin, a protein in the red blood cells (RBCs). Hemoglobin normally binds to oxygen, which is transported through the bloodstream in the RBCs. The binding of CO to hemoglobin reduces the oxygen-carrying capacity of the blood. For sufficient oxygen to be delivered to the body's cells, the heart must pump more blood during a given period. This puts a strain on the heart and may trigger heart attacks in those whose hearts are already weakened.

As you may recall from Chapter 18, toxic substances often affect the young more severely than adults. Air pollution is no exception. In fact, some researchers estimate that the health risk from air pollution is six times greater for children than for adults. Several reasons may account for this difference. First, children may be more susceptible than healthy adults because they are more active and therefore breathe more. As a result, they may be exposed to more pollution. In addition, children typically suffer from more frequent colds and nasal congestion and thus tend to breathe more through their mouths. Air bypasses the normal filtering mechanism of the nose; thus, more pollutants enter the lungs.

KEY CONCEPTS

Three groups are generally the most susceptible to air pollution: the young, the old, and the infirm (sick).

Effects on Plants and Nonhuman Animals

Air pollutants at levels people are exposed to have a wide range of effects on experimental animals. Unfortunately, very little research has been performed to study the effects

FIGURE 19-7 **Dying forest.** These trees in the Appalachian Mountains have been killed by acids in rain and snow, which are produced by pollution from automobiles, factories, and coal-fired power plants.

FIGURE 19-8 **Lichens.** This organism lives on rock and erodes away its surface. It is extremely sensitive to air pollution.

of air pollution on wild species. We do know, however, that fluoride and arsenic released from smelters can seriously poison cattle grazing downwind. Acids produced from pollutants released by power plants, smelters, industrial boilers, and motor vehicles have been shown to be extremely harmful to wildlife, especially fish (Chapter 20). In southern California, millions of ponderosa pines have been damaged by air pollution (mostly ozone) from Los Angeles.

Ozone, sulfur dioxide, and sulfuric acid are the pollutants most hazardous to plants (**FIGURE 19-7**). Ozone, for instance, makes plants more brittle. Farms in southern California and on the East Coast report significant damage to important vegetable crops. City gardeners also report damage to flowers and ornamental plants. Urban landscapers must plant pollution-resistant trees, notably locusts, along heavily traveled roads, but even these live only a fraction of their normal life span.

Sulfur dioxide damages plants directly, causing spotting of leaves. Recent studies show that air pollutants may also make some plant species more desirable to leaf-eating insects. According to botanists from Cornell University, air pollution and other stresses cause plants to produce a chemical called *glutathione*, which protects leaves from pollution but also attracts insects that normally have no interest in these plant species.

One group of organisms that is extremely sensitive to air pollution is the lichens, which grow on rock, wood, or soil (**FIGURE 19-8**). Lichens are composite organisms, consisting of a fungus and an internal alga. Because they are amazingly hardy and resilient, lichens can endure a wide variety of climates, often living where few other life-forms exist. However, lichens obtain nourishment from the air, rain, and snow, which makes them extremely vulnerable to air pollution.

In the 1800s, scientists began to notice that lichens were being killed by pollution in many European cities. In the 1860s,

32 species of lichens were collected in the Jardin du Luxembourg in Paris. By 1896, not a single species survived. In the town of Mendlesham, England, 129 species were identified between 1912 and 1921. In 1973, only 67 remained. In England, researchers found that lichens were absorbing heavy metals and pollution; wherever pollution increased, the lichens died.

So sensitive are lichens to air pollution that air quality and the spread of pollution from an industrial source can be determined by mapping the presence or absence of lichens or by chemically analyzing the lichens in the area.

KEY CONCEPTS

Air pollution adversely affects plants, animals, and their habitat.

Effects on Materials

Air pollutants severely damage metals, building materials (stone and concrete), paint, textiles, plastics, rubber, leather, paper, clothing, and ceramics (Table 19-3). The four most corrosive and harmful pollutants are sulfur dioxide, sulfuric acid, ozone, and nitric acid.

The Statue of Liberty, for example, was recently given a costly ($35 million) face-lift after years of exposure to two secondary pollutants, sulfuric and nitric acids (which are produced from sulfur and nitrogen oxides). The Taj Mahal in India, like many other buildings in the world, is being defaced by sulfur oxide air pollution from local power plants. Sulfuric and nitric acids not only cause cosmetic damage to metals, they also reduce their strength. In the Netherlands, bells that had been ringing true for 3 or 4 centuries have recently gone out of tune because of acid pollutants that have eaten away at them, lowering their pitch and rendering once familiar tunes indecipherable. Ozone cracks rubber windshield wipers, tires, and other rubber products, necessitating costly antioxidant additives. Particulates also cause damage. For example, particulates blown in the wind erode the surfaces of stone, doing significant damage.

Table 19-3

Damage to Materials from Air Pollution

Material	Damage	Principal Pollutants
Metals	Corrosion or tarnishing of surfaces, loss of strength	Sulfur dioxide, hydrogen sulfide, particulates
Stone and concrete	Discoloration, erosion of surfaces, leaching	Sulfur dioxide, hydrogen sulfide, particulates
Paint	Discoloration, reduced gloss, pitting	Sulfur dioxide, hydrogen sulfide, particulates, ozone
Rubber	Weakening, cracking	Ozone, other photochemical oxidants
Leather	Weakening, deterioration of surface	Sulfur dioxide
Paper	Embrittlement	Sulfur dioxide
Textiles	Soiling, fading, deterioration of fabric	Sulfur dioxide, ozone, particulates, nitrogen dioxide
Ceramics	Altered surface appearance	Hydrogen fluoride, particulates

Society pays a huge price tag for cleaning sooty buildings, repainting pitted houses and automobiles, and replacing damaged rubber products and clothing. The damage can also be tragic, for air pollution attacks irreplaceable works of art such as the marble of the Parthenon in Athens, which has deteriorated more in the last 60 years than in the previous 2000 because of air pollution.

KEY CONCEPTS

Air pollution damages many human-made materials, from metal to concrete and stone.

19.4 Air Pollution Control: Toward a Sustainable Strategy

Air pollution exacts an enormous cost and air pollution control can result in many benefits. The economic benefits are surprisingly large. For example, in a report by the U.S. Environmental Protection Agency entitled *Human Health Benefits from Sulfate Reductions*, EPA scientists estimated that reducing sulfate pollution alone would result in an annual health benefit of $12 to $78 billion. Another EPA report, *The Benefits and Costs of the Clean Air Act 1970 to 1990*, says that in 1990 Americans received about $20 benefit for every dollar spent to control air pollution—and this included only health benefits.

> The U.S. EPA estimates that every dollar spent on air pollution control saves $20 in health care costs.

How can air pollution be reduced to sustainable levels, though? We begin our study of pollution control with a look at some of the more traditional approaches included in the end-of-pipe strategy, which typically captures pollutants from smokestacks or converts them into other less harmful substances. We examine the pros and cons of this strategy. Then we turn to the alternative—and potentially more sustainable—strategies.

KEY CONCEPTS

Air pollution control is costly but appears to reap huge economic benefits—far in excess of the cost of controls.

Cleaner Air through Better Laws

Society's laws constantly evolve to fit the conditions of the present. This is true with U.S. clean-air legislation—now in its fifth decade of existence. Early clean-air laws were fairly weak and ineffective, but they laid the foundation for some of the most progressive environmental legislation in the world.

The federal **Clean Air Act** (1963) was our nation's first real attempt to reduce air pollution. But the law was weak and ineffectual. The first significant advances in protection came in 1970 when the Congress amended the law, adding (1) emissions standards for automobiles, (2) emissions standards for new industries, and (3) ambient air quality standards for urban areas. The ambient air quality standards established by the EPA covered six criteria air pollutants. These standards were designed to protect human health and the environment.

The 1970 amendments have reduced air pollution from automobiles, factories, and power plants. In addition, they stimulated many states to pass their own air pollution laws, some with regulations even tighter than federal ones. Despite these gains, the amendments created some problems. For instance, in regions that exceeded ambient air quality standards, the law prohibited the construction of new factories or the expansion of existing ones. Predictably, the business community objected. In addition, some of the wording of the 1970 amendments was vague and required clarification. Of special interest were provisions dealing with the deterioration of air quality in areas that had already met federal standards.

Because of these and other problems, the Clean Air Act was amended again in 1977. To address the limits on industrial growth in areas that were violating the air quality standards, called **nonattainment areas**, lawmakers devised a strategy that allowed factories to expand and new ones to be built, but *only* if they met three provisions: (1) The new sources achieved the lowest possible emission rates.

(2) Other sources of pollution under the same ownership in that state complied with emissions-control provisions, and (3) unavoidable emissions were offset by pollution reductions by the same or other companies in the region. In 2003, the Bush Administration changed regulations, allowing thousands of older power plants, oil refineries, and factories to expand operations without having to install pollution-control devices on existing (heavy polluting) facilities.

The last provision, known as the **emissions offset policy**, forces newcomers to cut emissions in existing facilities or to request other companies in noncompliance areas to reduce their emissions. In most cases, the newcomers pay the cost of air-pollution control devices. The emissions offset policy has been used to reduce regional air pollution—because the air pollution emissions permitted from both the new and existing facilities are set below preconstruction levels.

The issue of how to protect air quality in areas that were already meeting federal air quality standards sparked considerable debate in Congress. Environmentalists felt that the ambient air quality standards, in effect, gave industries a license to pollute up to permitted levels. That is to say, standards for the quality of air meant that clean-air regions could be polluted up to the standard. The 1977 amendments set forth rules for the **prevention of significant deterioration (PSD)** of air quality in **attainment regions**, regions where air quality meets federal standards. However, PSD requirements apply only to sulfur oxides and particulates. Many air pollution experts think that the PSD requirements should be expanded to include other pollutants, such as ozone.

The 1977 amendments also strengthened the enforcement power of the EPA. In previous years, when the EPA wanted to stop a polluter, it had to initiate a criminal lawsuit. Violators would often engage in protracted legal battles because legal costs were often lower than the cost of installing pollution control devices. Thanks to the 1977 amendments, the EPA is allowed to levy **noncompliance penalties** without going to court. These penalties are assessed on the ground that violators have an unfair business advantage over competitors that comply with the law. Penalties equal to the estimated cost of pollution control devices eliminate the cost incentive to pollute.

The Clean Air Act and additional regulations have helped reduce the release of six major pollutants, excluding carbon dioxide, despite a significant increase in economic output, vehicle miles traveled, energy consumption, and population (FIGURE 19-9). Additional gains in urban air quality may result from new regulations on wood-burning stoves. In 1990 and 1992, for instance, the EPA put into effect regulations requiring all new fireplace inserts and freestanding wood stoves to produce far less pollution. In fact, EPA-certified wood stoves emit 70 to 90% fewer particulates than older models. Many cities and towns have adopted regulations that prohibit wood burning during high-pollution days. Clean air legislation in Canada and other actions have resulted in significant declines in urban pollution as well. Canada's National Air Pollution Surveillance network showed marked declines in all pollutants except ground-level ozone since 1979.

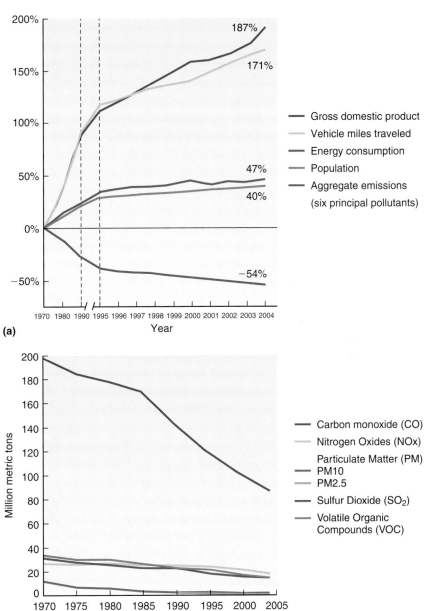

FIGURE 19-9 **Signs of success. (a)** Despite dramatic increases in the U.S. gross domestic product, vehicle miles traveled, energy consumption, and population, emissions of six major pollutants (not carbon dioxide, however) fell thanks to more fuel-efficient vehicles and pollution control devices on factories, cars, and power plants. **(b)** Total pollution emissions of six criteria pollutants in the United States since 1970. Source: www.epa.gov/airtrends/2005/econ-emissions.html.

In 1990, the Clean Air Act was amended once again to address inadequacies in the previous amendments. The new law, for example, set deadlines for establishing emission standards for 190 toxic chemicals, which had not been previously addressed. More important, it established a system of **pollution taxes** (market-based incentives) on toxic chemical emissions. These provide a powerful incentive for companies to reduce their releases.

The 1990 Clean Air Act amendments tightened emission standards for automobiles and raised the average mileage standards for new cars, a step that will improve automobile efficiency and help attack the pollution problem at its roots.

The 1990 amendments also established a market-based incentive program to reduce nitrogen and sulfur oxides, primary contributors to **acid deposition** (Chapter 20). According to the law, the EPA will issue **tradable permits** to companies throughout the United States. These permits stipulate allowable emissions for nitrogen and sulfur oxides— lower than present emission rates—to improve air quality. Companies that find innovative and cost-effective ways to reduce pollutants below their permitted allowance can sell their unused credits at a profit. (These were bought and sold on the commodities market starting in 1995.) This system encourages companies to develop cost-effective ways to prevent pollution. Thus, if a company can find inexpensive means of reducing its output of pollutants, it could benefit economically from the sale of its pollution credits. Finally, the 1990 amendments called for a phaseout of ozone-depleting chemicals (more on this in Chapter 20).

In 2003, the Bush Administration introduced the Clear Skies Act. According to the administration, this bill, if passed, would create a mandatory program that would dramatically reduce power plant emissions of sulfur dioxide (SO_2), nitrogen oxides (NO_x), and mercury by setting a national cap on each pollutant. The law would primarily affect power plants. If enacted and if successful, the act would cut sulfur dioxide emissions by 73%, from year 2000 emissions of 11 million tons to 4.5 million tons in 2010 and to 3 million tons in 2018. It would also cut emissions of nitrogen oxides by 67% from year 2000 emissions of 5 million tons to 2.1 million tons in 2008 and to 1.7 million tons in 2018. In addition, it would also cut mercury emissions by 69%—the first-ever national cap on mercury emissions. Emissions would be cut from 1999 emissions of 48 tons to 26 tons in 2010 and to 15 tons in 2018.

This law, if passed, would set national caps and would allow power companies to find the most economical solutions and would increase emissions trading, discussed in this chapter. If successful, it could cut pollution and result in substantial reductions in death, lung disease, and costs to companies according to the EPA.

Unfortunately, critics say this law is fundamentally flawed. For one, although it would, if successful, reduce pollution, its goals are less ambitious than the existing Clean Air Act. That is, it actually reduces pollution much less than the existing Clean Air Act. "Americans don't have to settle for only a 70 percent cut in air pollution when existing laws and existing technology mean that we can do better," argues the

Sierra Club, a nonprofit environmental group. "By the 15th year of the Bush plan," they argue that the Clear Skies Act would actually result in annual emissions of "450,000 more tons of NO_x, one million more tons of SO_2, and 9.5 more tons of mercury" than "would be allowed than under strong enforcement of existing Clean Air Act programs."

Eliminating mercury from coal-fired industrial emissions may be one of the greatest challenges in the next 25 years. If new technologies that are currently being tested at coal-fired power plants are successful, electric utilities could possibly achieve upward of 90% reductions in this toxic heavy metal. These technologies being developed by the Department of Energy in conjunction with the nation's largest private utilities are designed to remove mercury from smokestack gasses by injecting fine carbon particles into them. Mercury binds to the carbon particles.

Although this technology is relatively new and has not (at this writing) been tested at full-scale power plants, experts say carbon-based sorbents promise to help the industry meet new mercury regulations imposed by the U.S. EPA and a growing number of state governments. Although such technologies may help, critics point out that mercury would become a solid waste that would need to be disposed of in a safe fashion.

Another potential problem is that the Bush administration's energy proposal calls for the addition of 1400 coal-fired power plants over the next decade, a measure that would dramatically increase coal combustion, making it difficult to meet the goals of the new, apparently weaker legislation. In addition, continued expansion of the coal-fired power plant will only result in dramatic increases in carbon dioxide emissions, a potent greenhouse gas discussed in Chapter 20. For a truly effective reduction in pollution, they argue that the United States and other countries should dramatically increase their reliance on clean, reliable, and affordable renewable energy technologies such as wind.

KEY CONCEPTS

The Clean Air Act has been strengthened over the years to provide a comprehensive means of reducing air pollution. It consists of many parts, including standards for emissions from various sources, air quality standards, and several market-based incentives designed to reduce the emissions of pollutants.

Cleaner Air through Technology: End-of-Pipe Solutions

Until recently, clean air laws and regulations have typically prescribed many technological controls to cut down on air pollution. These devices either (1) remove harmful substances from emissions gases or (2) convert them into "harmless" substances. Such controls are called *end-of-pipe strategies*, for they address a pollutant after it has been produced.

End-of-pipe controls are much like first aid. They're essential in the short term and are thus a critical part of a sustainable strategy. Ultimately, however, many advocates promote preventive strategies, hoping that we can restructure our industrial and energy systems so that they produce very little air pollution in the first place. Before we examine such ideas, let's look at some of the end-of-pipe technologies.

KEY CONCEPTS

To date, many efforts to control pollution have relied on end-of-pipe solutions, mostly pollution control devices that capture pollutants or convert them into (supposedly) less harmful substances.

Pollution Control at Stationary Sources In some electric power plants, **bag filters** separate particulate matter from the stack gases (FIGURE 19-10a). In these devices, smoke passes through a series of cloth bags that filter out particulates. Filters often remove over 99% of the particulates, but they do not remove gaseous pollutants such as sulfur dioxide.

Cyclones are also used to remove particulates, generally in smaller operations (FIGURE 19-10b). In the cyclone,

particulate-laden air is passed through a metal cylinder. Centrifugal force causes the particulates to strike the walls and fall to the bottom of the cyclone, where they are removed. Cyclones remove 50 to 90% of the large particulates but few of the small- and medium-sized ones. The small particulates are the ones that cause the most harm because they can enter the lungs. Like bag filters, cyclones have no effect on gaseous pollutants.

Electrostatic precipitators, also used to remove particulates, are about 99% efficient and are installed on many U.S. coal-fired power plants (FIGURE 19-10c). In electrostatic precipitators, particulates first pass through an electric field, which charges the particles. The charged particles then attach to the wall of the device, which is oppositely charged. The current is periodically turned off, allowing the particulates to fall to the bottom.

The **scrubber,** unlike the other methods, removes particulates *and* gases such as sulfur dioxide (FIGURE 19-10d). In

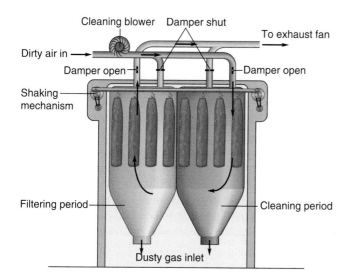

(a) Typical bag filter

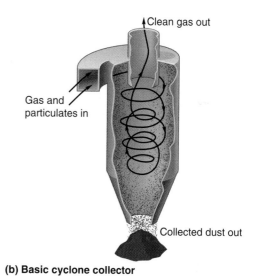

(b) Basic cyclone collector

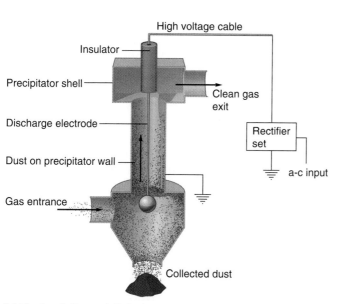

(c) Electrostatic precipitator

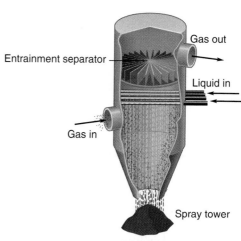

(d) Spray collector (scrubber)

FIGURE 19-10 Four pollution control devices used for stationary combustion sources.

scrubbers, pollutant-laden air is passed through a fine mist of water and lime, which traps over 99% of the particulates and 80 to 95% of the sulfur oxide gases. Nitrogen oxides and carbon dioxide are not removed by these devices.

> Smokestack scrubbers remove 99% of the particulates and 80 to 95% of the sulfur dioxide from smokestack gases.

Removing air pollution from stack gases helps clean up the air, but it creates a problem: hazardous wastes (Chapter 23). Particulates from power plants trapped by the various pollution control devices, for instance, contain harmful trace elements such as mercury. Scrubbers produce a toxic sludge rich in sulfur compounds. Improper disposal can create serious pollution problems elsewhere.

Mobile Sources Reducing pollution from mobile sources (cars, trucks, and planes) can be achieved by changes in engine design, especially those that improve efficiency. Such preventive measures are essential to sustainability. However, most new engine designs do not reduce emissions of carbon monoxide, nitrogen oxides, and hydrocarbons to acceptable levels. This makes it necessary to pass the exhaust gases through **catalytic converters**, special devices that are attached to the exhaust system (**FIGURE 19-11**). Catalytic converters transform carbon monoxide and hydrocarbons into water and carbon dioxide.

Catalytic converters once seemed like a logical strategy to reduce automobile emissions. Today, however, many people realize that they merely reduce one pollution problem (urban air pollution caused by carbon monoxide and hydrocarbons) while contributing to another (global warming caused by carbon dioxide). In addition, in many cities improvements in air quality brought about by catalytic converters are being negated by increasing numbers of motor vehicles on the road and increases in vehicle miles traveled. Catalytic converters in widespread use today also do nothing to reduce nitrogen oxides, a major contributor to acid deposition and photochemical smog. American auto manufacturers once argued that an affordable converter to do this job could not be developed. However, Volvo, the Swedish automobile manufacturer, introduced a catalytic converter in 1977 that lowered nitrogen oxide emissions to well below current U.S. automobile standards.

New Combustion Technologies

Another approach to pollution control is to develop new technologies that help us burn fossil fuels more efficiently. The more efficient the combustion chamber, the more energy one acquires from a given fuel source—and the less pollution is produced in meeting our energy demands.

One technology for burning coal more efficiently is called **magnetohydrodynamics (MHD)**. As shown in **FIGURE 19-12**, coal is first crushed and mixed with potassium carbonate or cesium, substances that are easily ionized (stripped of electrons). The mixture is burned at extremely high temperatures and produces a hot ionized gas—a *plasma*—containing electrons. The plasma is passed through a nozzle into a magnetic field, generating an electrical current. The heat of the gas creates steam, which powers a turbine.

Magnetohydrodynamics is about 60% efficient, compared with the 30 to 40% efficiency of a conventional coal-burning power plant. MHD systems remove 95% of the sulfur contaminants in coal, have lower nitrogen oxide emissions, and produce fewer particulates than conventional coal plants; however, they release more fine particulates.

Coal may also be burned in **fluidized bed combustion (FBC)**, a technology that is also more efficient and cleaner than conventional coal-fired burners. In FBC, finely powdered coal is mixed with sand and limestone and then fed into the boiler. Hot air, fed from underneath, suspends the mixture while it burns, thus increasing the efficiency of combustion. The limestone reacts with sulfur, forming calcium sulfate and reducing sulfur oxide emissions. Lower combustion temperatures in FBC reduce nitrogen oxide formation.

The coal industry and Bush Administration are actively seeking ways to make coal combustion more environmentally acceptable. These techniques—referred to as **clean coal technologies**—are designed to enhance the efficiency of coal

FIGURE 19-11 The catalytic converter. In this device, exhaust gases pass over a hot catalytic surface that converts several pollutants into less harmful substances.

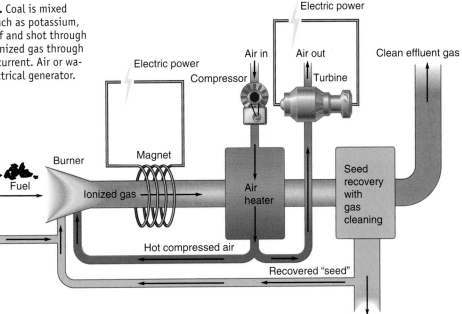

FIGURE 19-12 Magnetohydrodynamics. Coal is mixed with an ion-producing seed substance, such as potassium, and burned. A hot ionized gas is given off and shot through a magnetic field. The movement of the ionized gas through the magnetic field creates the electrical current. Air or water is also heated and used to run an electrical generator.

combustion, a desirable goal that results in fewer pollutants per unit of energy extracted from coal.

Clean coal technologies also include pollution control devices like the mercury-capturing technique described earlier in the chapter. In addition, they include ideas on ways to capture carbon dioxide emissions from coal plants and sequester them, that is, remove them from the carbon cycle. For example, some proponents hope to capture carbon dioxide gas released from power plants and inject it into the deep ocean waters, a costly and difficult process, that may not effectively remove them from the carbon cycle in the long run because deep waters often emerge at the surface in upwellings where they could release carbon dioxide into the atmosphere.

Clean coal technologies could play a key role in improving coal's environmental acceptability and performance, say proponents. However, critics still point out that this approach is a stopgap measure that could easily be offset by increasing dependence on coal.

Chapters 14 and 15 described the importance of shifting to natural gas and using more efficient combustion technologies—in particular, natural gas turbines similar to jet engines. These efforts could dramatically reduce the emission of sulfur dioxide, particulates, and carbon dioxide from power plants and other facilities.

Economics and Air Pollution Control

The previous section describes some new technologies to burn fossil fuels more efficiently and a number of air-pollution control devices that are end-of-pipe solutions.

The end-of-pipe strategies can be very costly. Smokestack scrubbers and electrostatic precipitators, for instance, cost millions of dollars to install and operate. Catalytic converters on automobiles add $300 to $500 to the sticker price. Although air pollution controls generally save money, some controls may reap benefits far lower than their actual costs. One study, for instance, suggests that pollution controls on automobiles and other mobile sources cost about $15 billion a year but create benefits of under $1 billion a year. Although such analyses often grossly underestimate the economic savings of benefits (such as reducing the cost of global warming), they do suggest the need for more cost-effective alternatives. More important, what if one could reduce pollutants at little or no cost—or even better, while enhancing one's profits?

Another problem with end-of-pipe solutions is that they often result in the production of hazardous wastes that must be disposed of in another medium—the ground, for instance. When pollution control devices are used, creative thinking and ingenuity can lead to strategies to capitalize on the waste. The Long Island Lighting Company, for example, began recovering vanadium from particulates collected at its power plants. In 1976, when the company began its program, it sold 362 tons of vanadium a year—about 9% of the total U.S. vanadium production—for $1.2 million. The company continues to extract vanadium today but makes only $10,000 to $20,000 a year because it is using a cleaner fuel. The chemical division of the Sherwin-Williams Company installed pollution control systems that capture solvents at a Chicago plant, saving that company $60,000 a year. In Germany and England, scrubber waste is used to make drywall (see Spotlight on Sustainable Development 19-1).

Pollution control strategies are often overwhelmed by increasing activity. For example, catalytic converters cut pollution, but as the number of cars increases and as people

19-1   Germany's Sustainable Approach Pays Huge Dividends

In Iphofen, Germany, the Knauf gypsum manufacturing plant produces drywall from scrubber sludge from a nearby coal-fired power plant (FIGURE 1). This plant has been so successful that Knauf recently opened a second facility in England, which generates $48 million per year in revenue.

This example is just one of many environmentally sustainable economic projects in Germany, indisputably the world's leader in environmental protection.

In an article in *International Wildlife*, U.S. Senate Committee on Environment and Public Works counsel Curtis Moore writes that "the world's future begins in Germany." He goes on to say, "More than anywhere else on Earth Germany is demonstrating that the greening of industry, far from being an impediment to commerce, is . . . a stimulus." Germany is finding that efforts to use energy more efficiently (conserve), to recycle, and to tap into the Earth's generous supply of renewable resources—all vital components of the sustainable strategy—are creating enormous economic opportunities. Consider a few examples.

FIGURE 1   Dried sludge ready to be converted to drywall.

Germany passed legislation that requires all car manufacturers to take back and recycle old cars. To facilitate this process, new cars rolling off the assembly line contain barcoded parts that identify the material used in their manufacture. German auto manufacturers have also made cars so that they can be more easily disassembled. Disassembly plants will be able to take a car completely apart in 20 minutes.

By 1994, take-back requirements went into effect for virtually all products, helping close the loop and reduce waste. By 1995, 72% of all glass and metals and 64% of all paperboard and plastic had to be recycled. Even old refrigerators are hauled free of charge to recycling centers, where ozone-depleting CFCs are extracted.

In the village of Neunburg stands a $38 million plant owned by a consortium of governments and industries. This facility, which produces hydrogen from solar-generated electricity, is really an experimental testing ground for new technologies. The Germans hope to be in the forefront when fossil fuel supplies run out or as nations look for alternatives to reduce global warming and other air pollution problems (Chapter 15). If successful, Germany could power its entire economy by solar electricity and hydrogen.

Knowing that the future depends on clean, renewable energy, the German government initiated the Thousand Roofs program, which offers a 75% tax credit to homeowners who install photovoltaics on their roofs. The hope is that government funding will drive the costs downward and also provide valuable hands-on experience in manufacturing—experience that could position Germany well in ensuing years as photovoltaics come into wider use.

The Germans have found that even conventional pollution control strategies can pay off. For example, Germany adopted rules that required all power plants within its borders to cut sulfur oxide emissions by 90%, a goal achieved in 6 years. These rules stimulated considerable innovation on the part of Germany's industrial technologists. Today, technological innovators stand to make huge sums of money selling pollution control technology and know-how to Americans and others.

Adapted from: Curtis A. Moore (1992). "Down Germany's Road to a Clean Tomorrow." *International Wildlife* 22 (5): 24–28.

drive more miles, the savings from the catalytic converter are often negated. From an economic standpoint, pollution control strategies cost business far more than is necessary. Pollution prevention strategies often achieve better reductions at a fraction of the cost.

KEY CONCEPTS

Air pollution control via end-of-pipe methods often adds substantial economic costs to various processes. It also produces waste products that, if not used for other purposes, end up in landfills.

To Reduce Air Pollution
- Walk or ride a bike.
- Use mass transit whenever you can.
- Drive an energy-efficient vehicle.
- Buy only what you need.
- Use energy efficiently at home.
- Recycle and buy recycled products.
- Support nonprofit organizations that promote energy efficiency and renewable energy.

Toward a Sustainable Strategy

Air-pollution control strategies have clearly improved the quality of the environment. Although important in addressing immediate pollution problems, end-of-pipe solutions are falling into disfavor for reasons cited earlier. These and other problems with the conventional air-pollution control strategy have led many people the world over to seek alternatives—a sustainable approach that is based on several of the key principles of sustainability: conservation, recycling, renewable resources, restoration, and population control.

As you learned in Chapter 15, conservation, recycling, and the use of renewable energy sources can produce substantial reductions in air pollution. Cutting energy demand by half through conservation measures, for instance, usually reduces air pollution output by half. Recycling aluminum rather than making it out of raw ore, however, cuts air pollution by 95%. Paper recycling reduces air pollution by 75%. Making steel from recycled scrap cuts air pollution by 30%.

Through individual, corporate, and government efforts in these areas, global demand for fossil fuel consumption can be substantially lowered—and along with it global pollution. Individuals can also help. Each unnecessary product we buy, each bottle or can we toss out, and each gallon of gas we waste contribute to air pollution. Individual responsibility is an effective means of cutting down on resource demand and pollution. When added together, individual contributions can make a big difference. For a list of ideas on ways you can help to reduce air pollution, see the Individual Actions Count! table.

> **KEY CONCEPTS**
>
> Pollution prevention is a key element of a sustainable society. Pollution prevention results from enacting measures that promote conservation (frugality and efficiency), recycling, renewable resource use, and restoration of habitats.

19.5 Noise: The Forgotten Pollutant

Noise is one of the most widespread environmental pollutants. Few of us can escape it: Even when we are enjoying a backcountry camping trip, the silence is often broken by the roar of jets, chain saws, and off-road vehicles. Especially noisy are the cities and factories where many of us live and work. To understand noise pollution and its effects, let us first understand sound.

What Is Sound?

Sound is transmitted through the air as a series of waves. For a simple example, take the bass speaker element of a loudspeaker. When the music is loud, the speaker cone can actually be seen to vibrate, and a hand placed in front of it can feel the air move. If we could slow down the speaker cone to observe what was happening, we'd see that as it moves outward, it compresses the air molecules in front of it (FIGURE 19-13). As the speaker cone returns inward, the air it just compressed expands in both directions, much as a coiled spring would expand when released. This expansion compresses neighboring air molecules slightly farther away. These have the same effect that the outward-pushing speaker cone had (Figure 19-13). Thus, waves of expansion and compression are set up, transmitting the sound. Air molecules do not travel with the sound but only oscillate back and forth in the direction the sound is traveling. Sound is a train of high-pressure regions following one another through the air at about 340 meters per second (760 miles per hour).

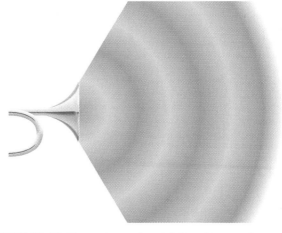

FIGURE 19-13 **The anatomy of sound.** Sound waves, shown here, are compression waves. Sound vibrates molecules in the air, which in turn vibrate neighboring molecules, transmitting the sound from its source.

Sound can be described in terms of its loudness and its pitch, or frequency. Loudness is measured in **decibels (dB)**. The decibel scale encompasses a wide range of volume (Table 19-4). The lowest sound the human ear can detect is set at 0 dB. This is the threshold of hearing. In the decibel scale, each 10 dB increase results in a 10-fold increase in sound intensity. Thus, a 10 dB sound is 10 times louder than 0 dB sound. A 20 dB sound is 10 times louder than a 10 dB sound and 100 times louder than a 0 dB sound.

Pitch, or frequency, is a measure of how high or low a sound is. Bass notes played by a tuba have a low pitch, and a violin's treble notes have a high pitch. Pitch is measured in cycles (that is, waves) per second. This is the number of compression waves passing a given point each second. The higher the pitch, the larger the number of cycles per second. Cycles per second are commonly called **hertz (Hz)**, after the German physicist Heinrich Rudolf Hertz.

The human ear is sensitive to sounds in the range of 20 to 20,000 hertz. Sounds below 20 hertz are not detected by the human ear and are described as *infrasonic*. Sounds above the audible range are called *ultrasonic*.

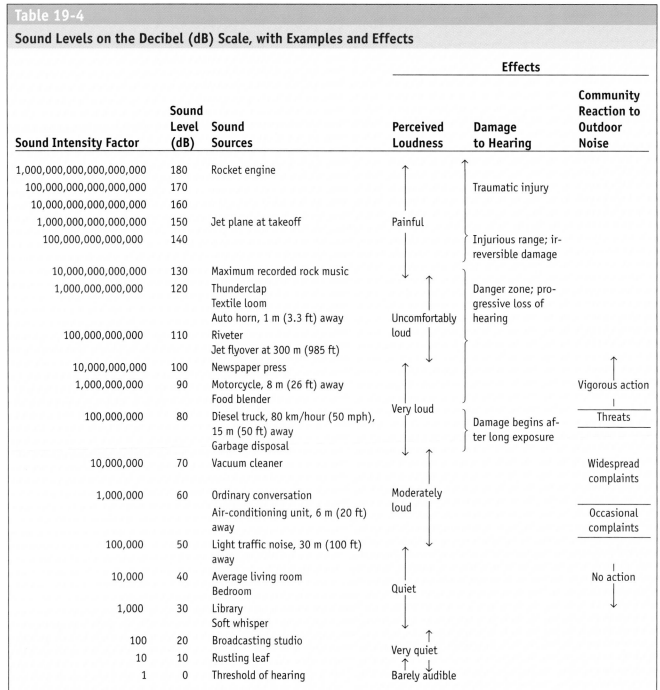

Table 19-4

Sound Levels on the Decibel (dB) Scale, with Examples and Effects

Sound Intensity Factor	Sound Level (dB)	Sound Sources	Effects		
			Perceived Loudness	Damage to Hearing	Community Reaction to Outdoor Noise
1,000,000,000,000,000,000	180	Rocket engine	Painful	Traumatic injury	
100,000,000,000,000,000	170				
10,000,000,000,000,000	160				
1,000,000,000,000,000	150	Jet plane at takeoff			
100,000,000,000,000	140			Injurious range; irreversible damage	
10,000,000,000,000	130	Maximum recorded rock music	Uncomfortably loud	Danger zone; progressive loss of hearing	
1,000,000,000,000	120	Thunderclap / Textile loom / Auto horn, 1 m (3.3 ft) away			
100,000,000,000	110	Riveter / Jet flyover at 300 m (985 ft)			
10,000,000,000	100	Newspaper press	Very loud		Vigorous action
1,000,000,000	90	Motorcycle, 8 m (26 ft) away / Food blender			Threats
100,000,000	80	Diesel truck, 80 km/hour (50 mph), 15 m (50 ft) away / Garbage disposal		Damage begins after long exposure	
10,000,000	70	Vacuum cleaner	Moderately loud		Widespread complaints
1,000,000	60	Ordinary conversation / Air-conditioning unit, 6 m (20 ft) away			Occasional complaints
100,000	50	Light traffic noise, 30 m (100 ft) away	Quiet		No action
10,000	40	Average living room / Bedroom			
1,000	30	Library / Soft whisper			
100	20	Broadcasting studio	Very quiet		
10	10	Rustling leaf			
1	0	Threshold of hearing	Barely audible		

Source: Table from Jonathan Turk and Amos Turk (1987). *Physical Science with Environmental and Other Practical Applications*, 3rd ed. Copyright © 1987 by Saunders College Publishing. Reproduced by permission of the publisher.

KEY CONCEPTS

Sound waves are compression waves that travel through the air. Sound is characterized by loudness (measured in decibels) and pitch (how high or low it is).

What Is Noise?

Noise is any unwanted, unpleasant sound. What any individual considers noise depends on many variables, among them one's background, mood, and occupation. Location and hearing ability are also important determinants, as are the time of day, the duration and volume of a sound, and other factors. A sound may be pleasant when it is soft but noisy when it is loud, or it may be acceptable when you generate it but obnoxious when someone else does. On this people generally agree: The louder a sound is, the more annoying it becomes and the more likely people are to describe it as noise.

Sources of noise fall into four categories: transportation, industrial, household, and military. As the world becomes more dependent on technology, noise pollution will grow worse. Of greatest concern are noises from off-road vehicles, construction, air traffic, home appliances, and surface transportation.

KEY CONCEPTS

Noise is an unwanted, unpleasant sound. What individuals consider to be a noise depends on many variables, such as the time of day or loudness of the sound.

Impacts of Noise

As we grow older, our hearing often declines. Several factors seem to be at work here. The loss of hearing results from occasional infections in the inner ear and from the natural aging of the sound receptors. Exposure to noise also causes a deterioration of hearing. Men, for instance, generally suffer a greater loss of hearing with age than women do. This difference probably results from exposure to noise at work. If loud and persistent enough, noise can cause premature degeneration of the sensory cells inside the ear. According to one estimate, 1 in every 10 to 20 people suffers some hearing loss from anthropogenic noise.

Exposure to extremely loud noises, such as rock music, gunfire, and noisy machinery, can cause a momentary decrease in our ability to hear, which is called a **temporary threshold shift**. A permanent loss of hearing, or a **permanent threshold shift**, occurs after continued exposure to loud noise. Studies suggest that continuous, long-term exposure to noise levels as low as 55 dB can permanently damage hearing. Noise of this level is common in many factories and jobs, especially in construction and mining. Even city traffic noise can damage the hearing of people exposed on a regular basis. This explains why the average 20-year-old resident of New York City hears as well as a 70-year-old tribal resident of Africa's grasslands.

As a general rule, the higher the sound level, the less time it takes to induce a loss in hearing. Different frequencies

produce differing amounts of damage. The lower frequencies, for example, do less damage than the higher-pitched sounds at the same loudness. Explosions (130 dB) and other extremely loud noises can cause instantaneous damage to the hair cells, resulting in deafness. Noise levels over 150 dB can severely damage the sensory cells, rupture the eardrum, and displace the tiny bones in the ear (the ossicles), which convey sound waves from the eardrum to the hair cells.

Workplace exposure to noise is slowly deafening millions of Americans. Large numbers of Americans, in fact, work at jobs where the noise levels are over 80 dB. Military personnel fall victim to noise from tanks, jets, helicopters, artillery, and rifles. Studies show that about half of the soldiers who complete combat training suffer so much hearing loss that they no longer meet the enlistment requirements for combat units. Bars, nightclubs, and traffic noise (especially diesel trucks and buses) are other important contributors to the deafening of America.

Noise interferes with family, friends, and coworkers, causing increased tension—not only because the sufferer can't understand what's being said, but also because deafened individuals may talk annoyingly loudly.

Noise also has profound effects on sleep. It may prevent us from falling asleep as soon as desired, or it may keep us from sleeping at all. Noise may wake us during the night or may alter the quality of sleep, leaving us irritable.

KEY CONCEPTS

Noise affects us in many ways. It damages hearing, disrupts our sleep, and annoys us in our everyday lives. It interferes with conversation, concentration, relaxation, and leisure.

Controlling Noise

Noise pollution receives little attention worldwide compared with other forms of air pollution—in part because hearing loss is generally progressive. Because of this, victims are generally unaware of the gradual loss in auditory acuity. In fact, workers who are exposed to loud noises eventually become accustomed to them, partly because their hearing declines.

Noise control can be carried out on many different levels (Table 19-5). Individuals can take actions, even simple ones such as wearing ear plugs when operating chain saws or circular saws or other machinery. A nonprofit organization called Hearing Education Awareness for Rockers (H.E.A.R.), works with radio stations in the United States and Canada to protect audiences from loud rock music. In 1996, they gave out over 60,000 earplugs. Even some rock bands, among them Motley Crue, now sell earprotectors at concerts. Changing the design of machinery and other products can cut turbulence and vibration. Better urban planning can isolate people from noisy railroads, highways, factories, and airports. Legal controls on noise emissions can also be helpful.

Many countries have regulations to control noise from motorized vehicles, including Japan, Norway, the Netherlands, Sweden, Switzerland, the United Kingdom, Denmark,

Table 19-5

Controlling Noise

Design and Planning

Minimize air turbulence created by vehicles.

Minimize vibration from vehicles, machines, and home appliances.

Build railroads and highways away from densely populated areas.

Build better insulated subways, railroad cars, trucks, and buses.

Eliminate noisy two-stroke motorcycle engines.

Install better mufflers on motorized vehicles and equipment.

Build airports away from high-density population centers.

Restrict growth in the vicinity of existing airports.

Require low-noise home appliances, tools, and office equipment.

Pass laws to eliminate or block noise.

Control traffic flow to eliminate stop-and-start driving.

Barriers and Sound Absorbers

Build embankments along noisy streets.

Use smooth road surfaces.

Build level streets to cut down on engine noise generated on steep inclines.

Plant dense rows of trees along highways and around sources of noise.

Construct artificial noise barriers along streets and around factories.

Install better insulation in houses.

Use double- and triple-paned windows.

Add sound-absorbent materials in factories, offices, and homes.

Provide earplugs for ground personnel at airports and workers in factories and construction.

Personal Solutions

Use fewer power tools and appliances.

Time activities to minimize disturbance of others.

Use mass transit.

Refuse to buy noisy vehicles, tools, office equipment, and appliances.

Maintain vehicles to eliminate noise.

Wear ear guards when engaged in noisy activities.

Work with employer to reduce noise.

---

must be moved regularly from noisy jobs to quieter jobs to reduce their overall exposure. Personal protective equipment is a less desirable solution because it is uncomfortable to wear and also blocks out important sounds or signals necessary for worker safety. Many workers find it uncomfortable and refuse to wear it.

The U.S. Environmental Protection Agency also once played a role in noise control and abatement. Unfortunately, the EPA's noise control office, with a staff of about 800 people, closed its doors in 1982 when Congress refused to appropriate funds to support their work any longer. Before the EPA's noise office was eliminated, however, regulations on noise controls on interstate motor carriers, new trucks, motorcycles, and railroads had been created. These regulations are still on the books, as stipulated by the Noise Control Act, but nothing new has been added since then. New trucks generally comply with these regulations, as do cars, but motorcycles often fail to meet standards. Without a watchdog agency, there is little that can be done about this.

KEY CONCEPTS

Noise levels can be controlled by redesigning machinery and other sources, by sound-insulating buildings, by separating noise generators from people, and by other measures.

19.6 Indoor Air Pollution

Americans and other residents of industrial nations spend, on average, 50 to 90% of their time indoors, away from the harmful pollutants in the outside air. Well, not exactly. Studies show that the air in our homes and offices contains high levels of potentially harmful substances known as **indoor air pollutants**. The U.S. EPA identifies four indoor air pollutants as the most dangerous: cigarette smoke, radon (a radioactive substance), formaldehyde, and asbestos.

> Americans spend, on average, about 90% of their lives indoors. If you have an outdoor job, you still spend about 50% of your life inside.

Indoor air pollutants come from a surprising number of sources, among them stoves, tobacco smokers, and heating systems. Furniture, wood paneling, plywood, and carpet also contribute to indoor pollution. Naturally occurring radioactive materials in the ground beneath a building may also release a potentially dangerous gas known as radon, which seeps into buildings. Outdoor air pollutants also enter buildings. Making buildings more airtight to reduce energy loss keeps these pollutants out, but it can also trap pollutants generated internally, resulting in higher levels. Special precautions must be taken when retrofitting a building for energy efficiency or constructing a new energy-efficient one to ensure proper air exchange (more on this later). Table 19-6 lists the four major indoor air pollutants, their sources, and their major effects.

France, Italy, and Canada. In the United States, the **Noise Control Act** (1972) authorized the EPA to establish maximum permissible noise levels for motor vehicles and other sources. Individual cities and towns have also passed noise ordinances.

Noise in U.S. workplaces is controlled by the **Occupational Safety and Health Administration**. OSHA's workplace standard is 90 dB for an 8-hour exposure, a level many authorities believe is too high. The agency can issue abatement orders for noise violations. These orders require the employer to first tackle the problem through engineering and design changes in the equipment. Should these prove infeasible, ear guards must be worn by workers, or workers

Table 19-6

Major Indoor Air Pollutants

Pollutant	Source(s)	Effects on Human Health
Tobacco smoke—carbon monoxide, particulates, sulfur dioxide, nitrogen dioxide	Cigarettes, pipes, and cigars	Carbon monoxide—shortness of breath and heart attacks; particulates—lung cancer; sulfur and nitrogen dioxide—emphysema and chronic bronchitis. Effects seen in smokers and nonsmokers.
Radon	Naturally occurring radioactive material (radium) in the soil. Radon leaks through cracks in the foundation	Lung cancer
Formaldehyde	New furniture, particleboard, plywood, chipboard, and wood paneling	Irritation of the eyes, nose, and throat; nasal cancer and lung cancer
Asbestos	Insulation around pipes in old homes, school buildings, and factories	Lung cancer and buildup of fibrous tissue in the lungs, which makes breathing difficult

KEY CONCEPTS

Indoor air pollutants come from combustion sources, from a variety of products that release potentially harmful substances, and even from naturally occurring radioactive materials in the ground beneath a building.

How Serious Is Indoor Air Pollution?

The EPA estimates that indoor air pollutants may cause as many as 2000 to 20,000 cases of lung cancer each year. Indoor air pollution, in fact, may cause more lung cancer than ambient air pollutants. EPA officials believe that toxic substances in our homes and offices are much more likely to cause cancer than outdoor air pollutants, for two main reasons. First, indoor levels are often much higher than outdoor levels, in some cases up to 100 times higher. Even in pristine rural areas, indoor air can be more polluted than outside air next to a chemical plant. Second, as noted earlier, people spend an enormous amount of time indoors.

> The U.S. EPA estimates that indoor pollutants cause somewhere between 2000 to 20,000 cases of lung cancer per year. The agency also estimates that one fifth to one third of all office buildings in the United States contain unhealthy levels of indoor air pollution and that 10 to 20 million Americans are adversely affected by exposure to these pollutants.

As you might expect, the people with the highest risk are the young, elderly, and infirm—that is, individuals who are sick or who have heart and lung disease. Pregnant women are also at greater risk, as are office or factory workers who spend inordinate amounts of time indoors in polluted buildings.

The EPA estimates that one fifth to one third of all U.S. office buildings, including its own headquarters in Washington, D.C., contain unhealthy levels of indoor air pollutants. These are often classified as **sick buildings**. The agency also

estimates that 100,000 to 200,000 workers die prematurely each year as a result of exposure to indoor air pollution at work—either in offices or, more likely, in factories.

The EPA also estimates that 10 to 20 million Americans suffer from **sick building syndrome**, an assortment of debilitating health effects caused by polluted indoor air. These individuals display a variety of health problems, among them chronic respiratory problems, sinus infections, sore throats, and headaches. People may also complain of dizziness, rashes, eye irritation, and nausea. These symptoms are largely caused by exposure to formaldehyde from furniture, plywood, and other sources in buildings.

Although estimates of the effects of indoor air pollution are crude, they do suggest the need to take action. In order to understand our options, we examine three of the major indoor air pollutants: products of combustion, radon, and formaldehyde.

KEY CONCEPTS

Indoor air pollution is present in millions of homes and offices and adversely affects millions of Americans.

Why Is Indoor Air in Buildings So Polluted?

Indoor air pollution is not a recent phenomenon. It has been around since houses were built. Early homes, heated by fires, for example, could become choked with smoke. Indeed, this is the case in many less developed nations where people cook on open fires inside their homes.

Today, indoor air pollution exists because of efforts to tighten houses to reduce air infiltration and save energy. Cracks are sealed and many buildings are wrapped with a wind-proof material to reduce leakage. But the problem is worsened by the introduction of many new building materials containing pollutants. Carpets, furniture, plywood, oriented strand board, and a host of other products are made with glues that contain formaldehyde or other potentially toxic substances. These are trapped inside homes by weather-proofing measures.

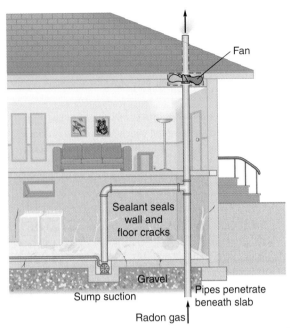

FIGURE 19-14 **Radon protection.** This house is protected from radon by pipes that draw radon from underneath the slab.

FIGURE 19-15 **Milk paint?** Yes, you read correctly. Paints were made out of milk long before pigments started being used. These paints, while expensive, are environmentally safe.

KEY CONCEPTS

Indoor air pollution has been around for centuries. Today, problems are created by efforts to reduce air infiltration which traps pollutants and the use of many new building materials and products that contain glues and other chemicals that are harmful to human health.

Controlling Indoor Air Pollutants

Indoor air pollutants can be reduced or eliminated by a variety of techniques. Let's take each of the major indoor air pollutants and consider some strategies, starting with tobacco smoke.

Tobacco smoke is difficult to remove from a room, so one of the easiest solutions is to ask smokers to carry on outside. Better ventilation systems can remove the smoke-laden air and replace it with cleaner outside air. Bans on smoking in offices or special zones set up for smokers can also help reduce overall exposure.

Radon can be removed by special devices called *air-to-air heat exchangers,* which pump air out of a building and draw new air in. During the process, much of the heat from the outgoing air is transferred to the incoming air to reduce heat loss. Radon can also be controlled by placing plastic on the ground under the cement slab a house is built on and by installing pipes under the slab to draw radon out of the earth immediately below it—before it can seep into a building (FIGURE 19-14). Pipes can be installed in new construction and can be retrofitted in existing structures.

Formaldehyde is much more difficult to control because it is found in so many products. Fortunately, new building products such as formaldehyde-free oriented strand board are being made available for builders. One manufacturer is producing a wood sheathing for making floors and exterior walls out of straw. It performs well and is glued with a nontoxic material. Some builders are using natural materials such as adobe or straw bales to build walls, avoiding the use of glue-laden wood products altogether. Formaldehyde is also present in furniture, especially inexpensive products, which are often made with particleboard or other formaldehyde-containing wood products. These products should be avoided. Natural carpeting can help, too.

Asbestos can be removed from buildings, although it is costly. Alternatively, asbestos in pipes and ceilings can be stabilized with special sprays that coat the product and prevent fine fibers from flaking off.

Paints, stains, and finishes can also add lots of indoor air pollution. Latex paints contain volatile organic compounds (VOCs), substances such as benzene, toluene, ethanol, and other compounds, that evaporate as the paint dries. These can cause a variety of health problems. Stains and finishes often have even higher concentrations of VOCs and outgass (give off) these materials for months after application.

To avoid these problems, some builders are using low toxic or nontoxic paints, stains, and finishes, containing reduced amounts of VOCs. Some manufacturers have even produced no-VOC paints. One company, for example, makes a water-soluble paint from milk protein, which is popular in hospitals and schools (FIGURE 19-15).

Solving indoor air pollution is not easy, but it can be done. Ways to build more healthful homes are becoming more widely known and many builders are adopting these practices.

The power of man has grown in every sphere, except over himself.

—Winston Churchill

CRITICAL THINKING

Exercise Analysis

The first statement made by the person cited in this exercise is true. That is, natural sources of pollution such as volcanoes often do exceed human sources of pollution. There is an important distinction, though. Natural sources are often dispersed. Although natural sources produce more pollution than human sources overall, the fact that the pollution from natural sources is spread out over a large area greatly dilutes its effect on ambient levels. As a rule, most natural sources do little to increase concentrations, and they rarely, except in isolated events, produce levels of pollution that are harmful to humans and other species. Oil, for instance, seeps from the ocean floor in rather large quantity overall. However, because the sources are many and widely dispersed, large oil slicks such as those produced when an oil tanker runs aground are not produced by natural seepage.

In contrast, human sources tend to be fairly concentrated. The Ohio River valley, for instance, is home to numerous power plants that burn coal. The sulfur dioxide they produce results in a substantial increase in atmospheric levels, with effects on downwind sites. Because of the concentrated nature of human sources, controlling air pollution from them is vital to protecting human health and the quality of the environment.

The second part of the speaker's argument, that environmental protection reduces productivity, is true. The extent to which it does so is rather small, though. Moreover, environmental protection efforts have long-term economic benefits. They lower economic externalities—social and health costs passed on to you and me. An accurate financial reckoning of the costs of pollution control should, many people argue, take into account economic externalities (Chapter 25). As noted in this chapter, air pollution reductions can have enormous economic benefits in reduced health care costs alone.

On this topic, it is important to note that many laws and regulations could be revamped to encourage more creative and cost-effective solutions. Rather than weakening environmental legislation and regulation, we might do well to strengthen them through provisions that encourage pollution prevention and other measures that enable us to both cut pollution and boost productivity.

CRITICAL THINKING AND CONCEPTS REVIEW

1. Describe the pollutants produced by burning fossil fuels. How is each product formed?
2. Define the terms *primary* and *secondary pollutant*.
3. In what ways are brown-air and gray-air cities different? What are the major air pollutants in each city? Why does the distinction break down?
4. What is photochemical smog? How is it formed?
5. Describe some factors that affect air pollution levels in your community.
6. What is a temperature inversion?
7. What are the health effects of (1) carbon monoxide, (2) sulfur oxides and particulates, and (3) photochemical oxidants?
8. List the major chronic health effects of air pollution. Which pollutants are thought to be the cause of these effects?
9. Of all the impacts of air pollution, which are of the most concern to you? How have your education, home life, and religious beliefs affected your answer to this question?
10. Discuss the air-pollution control legislation enacted by the U.S. Congress, highlighting important features of the acts and amendments. How would you characterize the conventional approach to air pollution? How has this approach contributed to the prevalent notion that environmental regulation is bad for the economy?
11. Make a list of ways you can help reduce air pollution.
12. Critically analyze this statement: "Coal is an abundant fuel with a potentially long life span. Efficient technologies can be used to burn coal more efficiently and represent a sustainable solution to air pollution."
13. Why do you think noise pollution is often overlooked in our everyday concerns for our health and well-being?
14. Describe a sound wave and define these terms: *decibel*, *pitch*, and *frequency*.
15. List and describe reasons why what we consider to be noise is so subjective.
16. Critically analyze this statement: "Noise is a subjective thing. No one agrees on it, so why bother regulating it? Besides, no one's dying of noise pollution."
17. Define the terms *temporary threshold shift* and *permanent threshold shift* and the reasons each one occurs.
18. List the major indoor air pollutants and describe the effects of each one.
19. How would you go about estimating the number of people in the United States who are exposed to unhealthy air in their office or workplace? Describe an experimental design to derive fairly accurate numbers.

KEY TERMS

acid deposition
alveoli
anthropogenic pollutants
attainment regions
bag filters
bronchial asthma
brown-air cities
catalytic converters
chronic bronchitis
Clean Air Act
clean coal technologies
criteria air pollutants
cross-media contamination
cyclones
decibel (dB)

electrostatic precipitators
emissions offset policy
emphysema
fluidized bed combustion (FBC)
gray-air cities
hertz (Hz)
indoor air pollutants
magnetohydrodynamics (MHD)
noise
Noise Control Act
nonattainment areas
noncompliance penalties
Occupational Safety and Health
      Administration (OSHA)
permanent threshold shift

photochemical smog
pollution taxes
prevention of significant
      deterioration (PSD)
primary pollutants
radiation inversion
scrubber
secondary pollutants
sick building syndrome
sick buildings
smog
subsidence inversion
temperature inversion
temporary threshold shift
tradable permits

REFERENCES AND FURTHER READING

The References and Further Reading section at the end of this book contains a list of sources for the information discussed in this chapter and recommendations for further reading.

Connect to this book's website:
http://environment.jbpub.com/
The site features eLearning, an online review area that provides quizzes, chapter outlines, and other tools to help you study for your class. You can also follow useful links for in-depth information, research the differing views in the Point/Counterpoints, or keep up on the latest environmental news.

Acid rain from pollutants produced by a Manitoba smelter destroyed the vegetation around the facility.

CHAPTER 20

Global Air Pollution: Ozone Depletion, Acid Deposition, and Global Warming

For 200 years we've been conquering nature. Now we're beating it to death.

—Tim McMillan

In June 1992, representatives from nearly 180 nations met in Rio de Janeiro to hammer out the final language of Agenda 21, a massive global blueprint for sustainable development. They also negotiated final language for various agreements on global climate change, biodiversity, forest protection, and other subjects. (The outcome of the Earth Summit is described in Chapter 27.) This meeting was held in large part because of widespread recognition that human society is destroying key life-support systems for the entire planet and that without global coop-

Exercise

A business magazine article notes that "on the issue of global warming the scientific community is divided." In support of this assertion, it quotes two prominent scientists. One says that he's "convinced the world is in a human-induced warming phase," and another claims that "there's simply not enough evidence to support such a conclusion." The author of the article goes on to say that because of the uncertainty among the scientific community, it makes no sense to launch a global effort to reduce carbon dioxide emissions. This position is supported by some segments of the business community (especially the oil and coal companies) but not by others (such as the insurance and natural gas industries). Can you detect any problem in this reportage? After you have finished, list the critical thinking rules that were helpful to you in this exercise.

intense influx of ultraviolet radiation, many plants would perish—and with them, millions of species that depend on them for nutrition. How does the ozone layer work?

The ozone layer blocks out 99% of the sun's ultraviolet radiation.

When ultraviolet radiation strikes ozone molecules, it causes them to split:

$$UV + O_3 \rightarrow O + O_2$$

The products, however, quickly reunite, reforming ozone and giving off heat:

$$O + O_2 \rightarrow O_3 + heat$$

Thus, the ozone layer is a renewable form of protection that converts harmful ultraviolet radiation into heat.

> **KEY CONCEPTS**
> The ozone layer is a portion of the stratosphere with a slightly higher concentration of ozone molecules. It forms a protective shield that filters out harmful ultraviolet radiation.

Activities That Deplete the Ozone Layer

This section examines two primary threats to the ozone layer: the release of chlorofluorocarbons and jet travel in the stratosphere.

> **KEY CONCEPTS**
> Human civilization threatens the ozone layer through two principal activities: (1) the use of a class of chemical compounds called *chlorofluorocarbons* and (2) jet travel through the stratosphere.

The Use of Chlorofluorocarbons In 1951, manufacturers introduced a promising new product on the market in the United States, spray cans that contained a chemical substance known as *freon-11*. It served as a propellant. Freon-11 and a similar compound, freon-12, belong to a group of chemicals commonly referred to as **chlorofluorocarbons (CFCs)**. Table 20-1 provides a few details on the two major

eration we could greatly decrease the habitability of the planet for humans and all other life-forms.

Twenty-five years ago, most environmental issues were local or regional in scale. Today, however, many problems have reached global proportions. Solving them will require global cooperation such as that witnessed during the Earth Summit. This chapter looks at three global issues—ozone depletion, acid deposition, and global climate change or global warming—and sustainable solutions.

20.1 Stratospheric Ozone Depletion

Encircling the Earth is a layer of ozone gas (O_3), which screens out 99% of the sun's harmful ultraviolet radiation (**FIGURE 20-1**). Called the ozone layer, this protective zone occupies the inner two thirds of the stratosphere, which extends 15 to 45 kilometers (10 to 30 miles) above the Earth's surface. The screening effect of the ozone layer protects all organisms from ultraviolet radiation. Without the ozone layer, terrestrial life would all but vanish. Some animals would suffer serious burns and would develop cancer and lethal mutations. Humans would be especially vulnerable to these changes. Plants would suffer as well. Unable to cope with the

Table 20-1

Commonly Used Freons

Generic Name	Use	Chemical Name	Chemical Formula
Freon-11	Spray can propellant	Trichloromonofluoromethane	Cl | Cl–C–Cl | F
Freon-12	Coolant in refrigerators, freezers, and air conditioners	Dichlorodifluoromethane	Cl | F–C–Cl | F

FIGURE 20-1 The ozone layer.
(a) Ozone molecules absorb ultraviolet radiation striking the Earth's atmosphere and convert it into heat. **(b)** Ozone concentrations in the atmosphere and stratosphere.

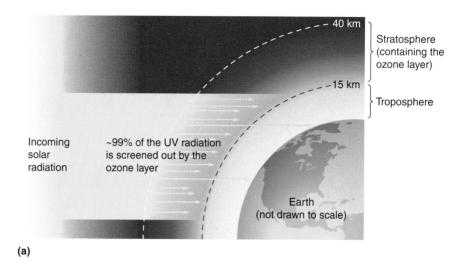

(a)

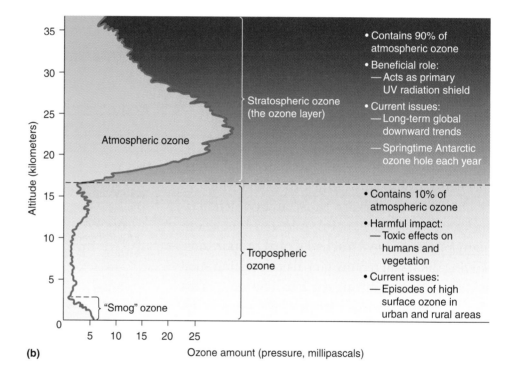

(b)

CFCs: Freon-11 and Freon-12. As shown, **Freon-11** is a propellant used in spray cans; **Freon-12** is used in refrigerators, air conditioners, and freezers. As you can see from the chemical formulas, CFCs contain two to three chlorine atoms (indicated by the symbol Cl).

Chlorofluorocarbons have been used in other ways as well. They were once widely used as foam *blowing agents*. A blowing agent is a gas mixed with a liquid polymer such as polystyrene and blown or extruded into molds, to produce styrofoam cups or foam insulation boards. CFCs form bubbles in the foam, making it a good insulator. One class of CFCs was also used to clean electronic equipment, precision ball bearings, and medical equipment.

CFCs were attractive substances because they were chemically unreactive—or inert. Until the early 1970s, chemists believed that CFCs released from spray cans or es-

caping from refrigerators simply diffused into the upper layers of the atmosphere. There, they were presumed to be partially broken down by sunlight, a process that liberated some of the chlorine atoms. It was thought that this process, called *photodissociation* (breakdown in sunlight), would have little if any effect on the upper atmosphere.

In the early 1970s, two U.S. scientists, Mario J. Molina and F. Sherwood Rowland, began to question this assumption. They pointed out that **chlorine free radicals** (highly reactive chlorine atoms produced by the breakdown of CFCs) might react with stratospheric ozone.

> A single chlorine free radical will destroy up to 100,000 molecules of ozone before it is removed from the stratosphere.

FIGURE 20-2 The chemistry of CFCs and ozone depletion. (a) Chlorofluorocarbons are dissociated by ultraviolet radiation in the stratosphere. This produces a highly reactive chlorine free radical. **(b)** The free radical can react with ozone in the ozone layer, forming chlorine oxide. This reaction reduces the ozone concentration. **(c)** A single molecule of Freon gas can eliminate many thousands of molecules of ozone because chlorine oxide breaks down, re-forming the chlorine free radical. Note too that chlorine oxide also reacts with ozone molecules, destroying them.

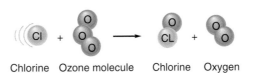

Ultraviolet radiation from the sun strikes the CFC molecule and causes a chlorine atom to break away.

(a)

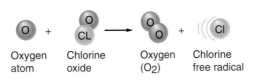

The chlorine atom reacts with an ozone molecule to form chlorine oxide and diatomic oxygen.

(b)

Shortly after their announcement, three research teams reported that a single chlorine free radical could react with and destroy many molecules of ozone. The reactions are shown in **FIGURE 20-2**. Because of the nature of these reactions, a single chlorine free radical from a CFC molecule can destroy 100,000 molecules of ozone.

In addition to CFCs, there are several chlorine and bromine compounds that deplete the ozone layer. Carbon tetrachloride, for example, is an ozone-depleting solvent that was once widely used. Methyl chloroform is a cleaning agent for clothes and metals. It is also used as a solvent in such common products as typewriter correction fluid and was used as a propellant in many sprays such as adhesives and cleaning agents. Methyl bromide is a soil fumigant, designed to kill pests in soil.

Like chlorine-containing compounds, bromine substances also diffuse into the stratosphere, where they photodissociate. Bromine then reacts with ozone, causing its concentration to fall. However, bromine free radicals are far more destructive of ozone than are chlorine free radicals and destroy hundreds of times more ozone molecules than chlorine atoms do.

When a free atom of oxygen reacts with a chlorine oxide molecule, diatomic oxygen is formed, and the chlorine atom is released to destroy more ozone.

(c)

High-Altitude Jets High-flying aircraft, such as military jets that routinely travel through the stratosphere, also destroy ozone. How? Jet engines release a pollutant called nitric oxide (NO). Nitric oxide gas, in turn, reacts with ozone to form nitrogen dioxide and oxygen:

$$NO \text{ (nitric oxide)} + O_3 \text{ (ozone)} \rightarrow NO_2$$
$$\text{(nitrogen dioxide)} + O_2 \text{ (oxygen)}$$

The threat to the ozone layer from high-flying jets is much lower than that from CFCs. The Concorde, a supersonic jet that once traveled between the United States and Europe in the stratosphere, was taken out of service in 2004 **(FIGURE 20-3)**. In 1971, the U.S. Congress killed a proposal to subsidize the construction of 300 to 400 SSTs. Had they

KEY CONCEPTS

Chlorofluorocarbon molecules were once used as propellants, refrigerants, blowing agents, and cleaning agents. CFCs are stable molecules that diffuse into the stratosphere, where they break down, releasing chlorine atoms that react with ozone molecules, destroying them. Other chlorine and bromine-containing compounds have also been used widely and are known as *ozone depleters*.

FIGURE 20-3 Supersonic speed: at what price? This supersonic transport once flew through the stratosphere, releasing nitric oxide, destroying ozone.

approved this proposal, jets would have become a major depleter of stratospheric ozone. Ordinary commercial jets also produce nitric oxide, but their effect on the ozone layer is still in question.

KEY CONCEPTS

All jets release nitric oxide gas, a pollutant that reacts with ozone. Jets, such as SST, that travel through the stratosphere have the greatest impact. Because few high-flying jets are in use today other than in the military, jet travel poses a lower risk than the use of CFCs.

Ozone Depletion: The History of a Scientific Discovery

Concern for the ozone layer originally began with projections of the effects of high-altitude jets. It reached a fever pitch, though, when in 1974 the chemists Rowland and Molina announced that CFCs, previously considered safe, could react with and remove ozone molecules from the stratosphere. Their initial projections indicated that CFCs could eventually destroy 20 to 30% of the ozone layer. Their work on CFCs and ozone depletion eventually won them a Nobel prize.

Evidence in support of the claim that CFCs were depleting the ozone layer began to be published in scientific journals in the 1980s. In May 1985, for example, British scientists reported a large decrease in the ozone layer above Antarctica, which has since been dubbed an *ozone hole* because of the severity of the decrease. In 1986 and 1987, intense study of the ozone hole strongly suggested that it was caused in large part by CFCs. Studies showed that the Antarctic ozone hole was also caused by several natural climatic conditions. One such weather phenomenon was a vortex of wind (a whirlpool in the atmosphere) that circles the south pole during the winter months, blocking warmer air from penetrating the region. Studies showed that the vortex contributes to the formation of polar stratospheric clouds. CFCs and other ozone-depleting chemicals adhere to the surface of the ice crystals. After winter, when sunlight returns to this frigid region, CFCs are released and ozone levels plummet. As the vortex breaks down, huge masses of ozone-depleted air then move northward, where they hover over nearby land masses such as Australia, New Zealand, and the southern tips of South America and Africa.

In 1988, researchers discovered that a similar hole was forming over the Arctic. Ozone levels there, however, were not as severely depleted because the vortex in the Arctic is not as strong as it is in the south. Invading winds can break through, keeping the Arctic air much warmer and reducing ozone destruction. Although the wintertime ozone depletion is only 10 to 25%, less than in Antarctica, the northern hemisphere is more heavily populated. When the Arctic ozone hole breaks up, it also releases huge masses of ozone-depleted air, which may linger over highly populated areas of North America and Europe.

In December 1990 (Antarctica's summer), studies of ultraviolet radiation reaching the ground at Palmer Station, a U.S. base on Antarctica, showed that levels had reached a record high—twice their normal value. The researchers suggested that the high levels of ultraviolet radiation resulted from the longer-than-normal persistence of the springtime ozone hole. In October 1991, studies showed that ozone depletion over Antarctica was even greater, marking the third year in a row that a severe ozone hole had formed over Antarctica.

Concern over ozone depletion was further galvanized by the report of an international panel of more than a hundred atmospheric scientists convened by several U.S. government agencies, including NASA. This group studied satellite data that had been gathered since 1962; in March of 1988, they published a report claiming that ozone concentrations had declined 1.7 to 3% over the northern hemisphere since 1969 (FIGURE 20-4). Over the heavily populated areas of North America and Europe, ozone levels had fallen 3%. The greatest declines, ranging from 5 to 10%, were recorded over Antarctica and the southern tip of Argentina. More recent studies have uncovered an even greater decrease. Data showed that ozone levels over the northern hemisphere were 13 to 14% lower in the late 1990s than in the early 1970s. Globally, the average decline was 3%. Since that time, ozone depletion has continued at a steady pace, falling at a rate of about 5% per decade at midlatitudes.

Ozone-depleting chemicals released on the Earth's surface take many years to rise into the stratosphere; therefore, scientists predict that ozone depletion will grow worse—eventually reaching 10 to 30% over North America.

KEY CONCEPTS

Studies of the ozone layer show substantial declines over the globe, with the highest level of depletion in the southern hemisphere and Antarctica.

The Many Effects of Ozone Depletion

Do the declines in the concentration of ozone gas in the ozone layer result in an increase in ultraviolet radiation striking the Earth?

Although ozone depletion has not resulted in the predicted increase in ultraviolet radiation striking the Earth's surface in all areas, there are some places where declines have been noticeable. As noted earlier, in Antarctica, ozone depletion has resulted in a substantial increase in the amount of ultraviolet radiation striking the ground. In mid- and high-latitude regions, away from the polluted air of cities (which block ultraviolet radiation), decreases in ozone continue to occur in the spring.

What impacts could ozone depletion have? Like many natural components of our environment, ultraviolet light is beneficial. In small amounts, ultraviolet radiation tans light skin and stimulates vitamin D production in the skin. However, excess ultraviolet exposure can cause problems. In humans, it can cause serious skin burns, cataracts (clouding of the eye's lens), skin cancer, and premature aging. Increased exposure to ultraviolet light may also suppress the human immune system, making us more susceptible to infectious diseases.

EPA researchers estimate that a 1% depletion of the ozone layer would lead to a 0.7 to 2% increase in ultraviolet light striking the Earth. This, in turn, would lead to an increase in skin cancer of about 4%. Over the next 5 decades,

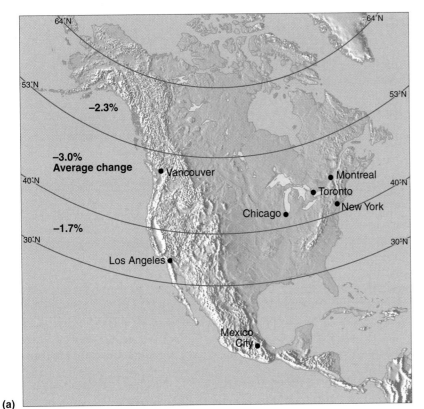

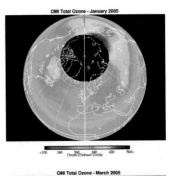

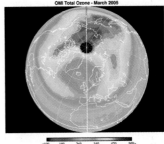

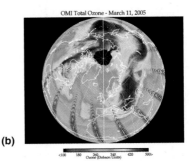

(a)

(b)

FIGURE 20-4 **Mapping ozone depletion. (a)** This map of North America shows the ozone depletion at different latitudes between 1969 and 1988. **(b)** Computer-generated images of changing size of ozone hole over the Arctic.

the EPA estimates that ozone depletion will result in approximately 200,000 cases of skin cancer in the United States. Worldwide, the number will be much higher. Especially hard hit will be countries such as New Zealand and Australia. In November 1991, a panel of scientists convened by the United Nations released a report on the predicted effects of global ozone depletion. They estimated that a 10% decrease in stratospheric ozone concentrations will cause 300,000 additional cases of skin cancer and an additional 1.6 million cases of cataracts (loss of opacity) per year worldwide.

Studies of skin cancer show that light-skinned people are much more sensitive to ultraviolet radiation than more heavily pigmented individuals. In addition, studies show that some chemicals that are commonly found in drugs, soaps, cosmetics, and detergents may sensitize the skin to ultraviolet radiation. Thus, exposure to sunlight may increase the incidence of skin cancers among light-skinned people and users of many commercial products.

Land- and water-dwelling plants could also suffer from increasing ultraviolet radiation. Intense ultraviolet radiation is usually lethal to plants; smaller, nonfatal doses damage leaves, inhibit photosynthesis, cause mutations, or stunt growth. Declining ozone and increasing ultraviolet radia-

> EPA scientists estimate that each 1% decrease in ozone layer could lead to a 4% increase in skin cancer.

tion could also cause dramatic declines in commercial crops such as corn, rice, and wheat, costing billions of dollars a year. It may also affect certain commercially valuable tree species.

In a hearing before the U.S. Senate in November 1991, Susan Weiler, head of the American Society of Limnology and Oceanography, testified that studies in Antarctica by other scientists have shown that phytoplankton (algae and other free-floating photosynthetic organisms) populations decrease about 6 to 12% when stratospheric ozone concentrations over the region drop by 40%. Because phytoplankton form the base of the aquatic food chain, damage to them could cause widespread ecological problems. One scientist thinks that ozone depletion and subsequent effects on the food chain may be the reason why two species of penguin are declining in Antarctica.

Finally, ultraviolet light is harmful to many products. Paints, plastics, and other materials deteriorate when exposed to ultraviolet light. Losses from further decreases in the ozone layer could cost society enormous amounts of money.

KEY CONCEPTS

Ozone depletion is resulting in an increase in ultraviolet radiation striking the Earth, especially in unpolluted areas. Ultraviolet radiation causes skin cancer, cataracts, and premature aging. It could also seriously damage ecosystems, crops, materials, and finishes.

Banning CFCs and Other Ozone-Depleting Chemicals: A Global Success Story

In the 1970s, fears caused by early projections of ozone depletion moved several nations, including the United States, Sweden, Finland, Norway, and Canada, to cut back on CFC emissions. In 1978, for example, the United States banned CFC use in spray cans. Freon-12—the refrigerant, coolant, and blowing agent—was not affected by the ban.

Additional scientific evidence on the decline of the ozone layer made it evident a decade later that worldwide cooperation was needed. As a result, in 1987, the United Nations sponsored negotiations aimed at reducing global CFC production. In September of that year, 24 nations signed a treaty called the **Montreal Protocol**, which would cut production of five CFCs in half by 1999 and freeze production of halons (used in fire prevention systems) at 1986 levels. (In halons, bromine atoms replace some or all of the chlorine atoms.) Although halons are used in much smaller quantities, as noted earlier, bromine is far more effective in destroying ozone than chlorine from CFCs and other sources.

This agreement paved the way for a gradual decline in CFC production in the industrial nations, but critics argued that it had too many loopholes. Like so many other pollution control strategies, it would only slow the rate of destruction, not stop it. EPA computer projections showed that an 85% reduction in CFC emissions was needed to stabilize CFC levels in the atmosphere.

Before the Montreal Protocol went into effect, something unusual happened. In March of 1988, an international panel (described earlier) announced that ozone levels had fallen throughout the world. Two weeks later, DuPont, a major producer of CFCs, called for a total worldwide ban on CFC production—when only 2 weeks earlier it had said that it would not support a ban.

Continuing bad news about ozone depletion brought negotiators to the table once again, this time in London, where in June 1990 they reached a new agreement. This treaty was signed by 93 nations and called for the complete elimination of CFCs and halons by the year 2000, if substitutes were available by then. The signatories also agreed to phase out other ozone-depleting chemicals. Among these were carbon tetrachloride, methyl chloride, and even the hydrochlorofluorocarbons (HCFCs), a class of chemicals (described shortly) once thought to be an excellent substitute for CFCs.

The news about the ozone layer continued to worsen. In 1992, a team of 40 scientists announced record-high concentrations of chlorine monoxide (ClO is produced when chlorine free radicals from the breakdown of CFCs react with ozone molecules) in the air above New England and Canada. Concentrations such as these had never been seen before, even in the Antarctic ozone hole. If chlorine levels continued to climb, chances were good that the Arctic ozone hole would begin to appear with great regularity, exposing Canada and parts of the United States, Europe, and Asia to dangerous levels of UV radiation.

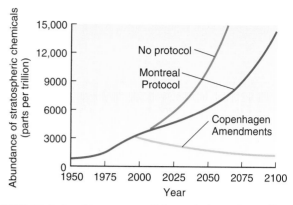

FIGURE 20-5 Graphing success. This graph shows the predicted concentrations of ozone-depleting chemicals in the atmosphere under various scenarios—no action, the Montreal Protocol, and the Copenhagen Amendments.

Aircraft measurements in 1992 also produced rather disturbing findings about global ozone outside the Arctic. In flights as far south as the Caribbean, scientists detected ClO concentrations of up to five times the amount they had anticipated.

In 1992, the nations of the world met in Copenhagen to sign another agreement calling for an acceleration of the phaseout of CFCs, carbon tetrachloride, and other ozone-depleting chemicals within 4 to 9 years. The success of the global efforts to phase out CFCs is shown in **FIGURE 20-5**, a graph of projected concentrations of ozone-depleting compounds. Although progress has been impressive, police are finding that CFCs are being illegally imported into the United States (and presumably other countries as well) in massive quantities. Stopping this flow will be required to reach the goals of the various international treaties.

KEY CONCEPTS

As scientific evidence on ozone depletion accumulated, the nations of the world tightened restrictions on the production of ozone-depleting chemicals. Three international treaties have already been signed to eliminate the production of ozone-depleting compounds, and progress toward meeting these goals has been very impressive.

Substitutes for Ozone-Destroying CFCs

As the previous section showed, discouraging scientific reports led to the signing of the three treaties to ban CFCs and other ozone-depleting chemicals. The development of replacement chemicals has also played a big role in the dramatic change. It gave industry options and, in some cases, opportunities to profit from the shift to less harmful means of refrigerating and of propelling liquids from spray cans. Many companies also led the way. AT&T, for example, was the first U.S. company to set a goal of phasing out the use of CFCs, which they succeeded in achieving in 1993.

CFCs are extremely important to modern society. Every refrigerator, freezer, and air conditioner uses them. To continue to provide these desirable services *and* protect the ozone layer, manufacturers have pursued two basic options: the use of less stable CFC compounds, which break down before they reach the stratosphere, and the production and use of non-CFC chemicals as substitutes.

Consider the first option and some of the roadblocks to bringing it about: By adding a hydrogen atom to the stable CFC molecule, which results in the formation of a substance known as HCFC, researchers can make CFCs that break up in the lower atmosphere. In theory, the chlorine atoms released during this process are less likely to reach the stratosphere. In practice, some do, but they cause less damage. Because some of the chlorine from HCFCs reacts with and destroys ozone molecules, HCFCs are viewed as interim solutions.

Several less stable CFCs are already on the market. One of these, HCFC-22, is now used as a coolant in new home air conditioners. By 1998, over 40% of the existing cooling systems in American businesses had been converted or replaced with equipment that uses non-CFC refrigerants such as HCFC-22. Using the opportunity to save energy, too, many companies installed more efficient systems. According to the Air Conditioning and Refrigeration Institute, these companies collectively save about $480 million a year with their new energy-efficient equipment. Incidentally, this also reduces carbon dioxide emissions by 4 million tons per year and cuts sulfur dioxide emissions by about 34,000 tons per year.

HCFC-22 is 20 times less destructive than the CFC-12. However, because a single HCFC molecule will destroy 5000 ozone molecules, the HCFCs are slated to be phased out by 2030.

The second most widely used CFC is CFC-11. Until recently, it was primarily used as a blowing agent for foam and, outside of the United States and a few other countries, as a spray can propellant. HCFC-123 has been touted as a possible replacement. Some manufacturers are using steam and carbon dioxide as blowing agents, eliminating the use of CFC and HCFC, for example, in rigid foam insulation used to insulate foundations of new homes.

Perhaps one of the most difficult challenges is finding a replacement for CFC-113. This compound is an all-purpose cleaner for circuit boards produced for the computer industry. Because CFC-113 was not being considered for banning initially, industry had not actively pursued replacements. At the signing of the Montreal Protocol, in fact, work to find a substitute had not even begun.

In January 1988, researchers announced the development of a compound called BIOCAT EC-7 that may be a partial replacement for CFC-113. This substance, isolated from orange peels, is very similar to kerosene and turpentine. EC-7 has replaced a sizable share of the CFC-113 market, but it has its limitations. It is not versatile and is flammable. Industry representatives believe that no single compound will replace CFC-113 completely. Many companies making sophisticated electronic equipment such as computers and medical equipment that was once cleaned with CFC-113 have switched to water-based cleaning agents—while maintaining product quality and lowering cost. At this writing, however, printed circuit board cleaning still requires some use of CFC, although companies typically collect vapors and recycle the cleaning agent. Although options are available, intense research is under way to find ozone-friendly substitutes.

KEY CONCEPTS

Most CFCs have been being replaced by a class of compounds called HCFCs. These less stable compounds deplete the ozone layer, but not as rapidly as their predecessors. Because of this, they are viewed mainly as an interim solution. Efforts are under way to replace CFCs used for cleaning agents with environmentally friendly water-soluble agents.

The Good News and Bad News about Ozone

The ozone story is one of humankind's greatest success stories. It not only illustrates how scientific knowledge can be used by society for the common good of humankind and all other species; it also shows how quickly changes can be made to bring about an end to the destruction of the global environment. It further illustrates the power of corporate responsibility. Many of the changes were brought about voluntarily without the need for governmental actions.

Already, studies are showing that the rate of increase in the concentration of ozone-depleting chemicals is slowing. A word of caution, however. Many millions of tons of CFCs have already been released into the atmosphere. CFCs take 15 years or so to migrate into the stratosphere, and some CFCs last 100 years in the atmosphere. Because of this, scientists predict that the ozone layer will continue to thin for many years to come. It will then slowly start to recover. Estimates on full recovery vary. National Oceanic and Atmospheric Administration (NOAA) scientists predict that if international agreements are adhered to, recovery to pre-1980 levels will be achieved by 2050. Others' predictions are less optimistic. They believe that at least 100 years will be required to return the ozone layer to 1985 levels. Another 100 to 200 years may be needed for full recovery.

Many people will contract skin cancer in the interim. The EPA notes that skin cancer in the United States has already reached epidemic proportions. One in five Americans will develop skin cancer in their lifetime and one American currently dies from skin cancer every hour. Melanoma, the most serious form of skin cancer, which has been linked to early exposure to excess ultraviolet light, is one of the fastest growing forms of cancer in the United States.

Nonetheless, many will be spared from skin cancer by the phaseout. The EPA estimates that the successful phaseout of CFCs will result in 295 million fewer cases of skin cancer over the next 100 years than would have occurred without these important changes.

Finally, many ozone-depleting chemicals (and their replacements) are greenhouse gases and thus contribute to global warming and climate change. They will accelerate the warming of the Earth's atmosphere, but their eventual decline may, when combined with other efforts, help us lower the Earth's surface temperature.

KEY CONCEPTS

The accumulation of CFCs and other ozone-depleting compounds in the atmosphere has begun to slow. Despite this progress, the ozone layer will take many years to recover. Many people will contract and die from skin cancer, but the phaseout will also spare many people as well.

20.2 Acid Deposition

In the 1960s, forest ranger Bill Marleau built the cabin of his boyhood dreams on Woods Lake in the western part of New York's Adirondack Mountains. Isolated in a dense forest of birch, hemlock, and maple, the lake offered Marleau excellent fishing. Ten years after Marleau finished his cabin, however, something bizarre happened: Woods Lake, once a murky green suspension of microscopic algae and zooplankton, teeming with trout, began to turn clear. As the lake went through this mysterious transformation, the trout stopped biting and soon disappeared altogether. Then the lily pads began to turn brown and die; soon afterward, the bullfrogs, otters, and loons disappeared.

What had happened to Woods Lake? What destroyed the web of life at this small, isolated lake, far from any sources of pollution? Scientists from the New York Department of Environmental Conservation say that Woods Lake is "critically acidified." As a result, virtually all forms of life in and around it have perished or moved elsewhere. The lake became acidified from acids and acid precursors deposited from the skies in several forms.

Acid deposition—the deposition of acidic substances generated from pollution largely human in origin—is commonplace today, as are lakes like Marleau's. In fact, Woods Lake is only one of about 375 lakes and ponds in the western Adirondacks turned acidic and hazardous to virtually all forms of life by acid deposition (FIGURE 20-6). In eastern

Canada, 100 lakes have met a similar fate. In Scandinavia, the death count is 10,000. Across the globe thousands of lakes will die unless something is done, and quickly.

Acid deposition is a phenomenon not just of environmental interest, but of grave social and economic importance. Studies show that acid deposition turns lakes acidic, kills fish and other aquatic organisms, damages crops, destroys forests, alters soil fertility, and destroys statues and buildings. Moreover, scientists are finding that acid precipitation is more widespread than once thought and is taking a large toll on our environment and pocketbooks. Recent reports indicate that it poses a universal threat, affecting the more developed countries as well as many less developed nations. Earthscan, an international environmental group, reports that acid deposition is already damaging soils, crops, and buildings in much of the developing world. Rapidly growing urban centers, with their poorly regulated industry and traffic congestion, are largely the culprits. Ironically, tough pollution laws in MDCs have given multinational corporations incentives to set up operations in LDCs, whose pollution laws (if they have any) are certainly much weaker.

KEY CONCEPTS

Acid deposition from pollutants is a global problem with serious social, economic, and environmental impacts.

What Is an Acid?

Understanding acid deposition requires a brief explanation of acids. An acid is a chemical substance that adds hydrogen ions to a solution. Hydrochloric acid (HCl), for example, dissociates into hydrogen and chloride ions and is therefore an acid.

Acids come in varying strengths. Strong acids add many hydrogen ions; weak acids add fewer. The degree of acidity of a solution, then, is related to the concentration of hydrogen ions in it. Chemists and others measure acidity on the **pH (potential hydrogen)** scale, which ranges from 0 to 14 (FIGURE 20-7). Substances that are acidic, such as vinegar and lemon juice, have low pH values—that is, less than 7. Basic (alkaline) substances, such as baking soda and lime, have high pH values—greater than 7. Basic substances have very

FIGURE 20-6 This crystal clear lake in the Adirondacks is too acidified to support aquatic life.

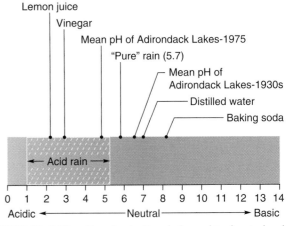

FIGURE 20-7 **The pH scale.** A pH scale is used to denote levels of acidity.

low hydrogen ion concentrations. Neutral substances, such as pure water, have a pH of 7.

The pH scale is logarithmic, like the decibel scale discussed in Chapter 19 (which measures loudness). This means that a change of one unit on the scale—in this instance, 1 pH unit—represents a 10-fold change. Therefore, rain with a pH of 4 is 10 times more acidic than rain with a pH of 5, 100 times more acidic than rain with a pH of 6, and 1000 times more acidic than rain with a pH of 7.

> **KEY CONCEPTS**
>
> Acids are chemical substances that add hydrogen ions to a solution. Acids are measured on the pH scale, which ranges from 0 to 14, with 7 being neutral—neither acidic nor basic.

What Is Acid Deposition?

In an unpolluted environment, rainwater is slightly acidic, having a pH of about 5.7. The normal acidity of rainwater is created when atmospheric CO_2 is dissolved in water in clouds, mist, or fog and is converted into a mild acid (carbonic acid). Acid precipitation is rain and snow with a pH below 5.7.

Acid deposition includes two broad categories: wet deposition and dry deposition. **Wet deposition** refers to acids deposited in rain and snow. These acids are formed when two air pollutants, the sulfur and nitrogen oxides (which are both gases), combine with water in the atmosphere. Sulfur oxides form sulfuric acid; nitrogen oxide gases react with water to form nitric acid. Sulfuric and nitric acid are two of the three strongest acids known to science.

Produced in the atmosphere, these acids may accumulate in clouds and fall from the sky in rain and snow. Even coastal fogs may contain droplets of acid that, when deposited on buildings or plants, can cause noticeable damage. One study shows that moisture droplets in low-lying clouds (fog) tend to contain higher concentrations of acid than rain or snow that falls from them. Therefore, fog and clouds may bathe trees in highly acidic water. Making matters worse, recent studies suggest that the evaporation of recently deposited cloud water from forest canopies may result in acid concentrations on leaf surfaces much higher than those found in the cloud droplets themselves.

Sulfate and nitrate particulates are also present in the atmosphere. These pollutants may settle out of the atmosphere much like fine dust particles. This process is one form of **dry deposition**. Settling onto surfaces, these particulates can combine with water to form acids. Sulfur and nitrogen oxide gases may also be adsorbed onto the surfaces of plants or solid surfaces, where they, too, combine with water to form acids. This is another type of dry deposition.

> **KEY CONCEPTS**
>
> Rainfall in unpolluted areas has a pH of about 5.7 and is just slightly acidic. Acid deposition refers to rain and snow with a pH of less than 5.7 and the deposition of acid particles and gases. Acids reach the surface of the Earth either as wet deposition (rain or snow) or dry deposition (particulates and gases).

Where Do Acids Come From?

Acid precursors, the primary pollutants that give rise to acids, come from natural and anthropogenic sources. Before we look at them, let us review how these pollutants are generated. As noted in Chapter 19, fossil fuels such as coal and oil contain sulfur impurities. When these fuels are burned, sulfur reacts with oxygen in the intense heat and forms sulfur oxide gases. These gases are released into the atmosphere where they may be converted into sulfuric acid. Chapter 19 also noted that the combustion of any organic matter in air, with its high concentration of nitrogen gas, produces nitrogen oxide gases. These gases are converted into nitric acid in the atmosphere after reacting with water.

Several natural events can produce significant amounts of sulfur dioxide. The natural sources of sulfur oxides include volcanoes and forest fires (**FIGURE 20-8a**). Bacterial decay of organic material such as plants also produces hydrogen sulfide gas, which can be converted into sulfuric acid in the atmosphere. Anthropogenic sources of this pollutant are of major concern, however, because they are often concentrated in urban and industrialized regions, causing local levels to be quite high. About 70% of all anthropogenic sulfur dioxide comes from electric power plants, most of which burn coal. Smelters, devices that melt mineral ores to extract pure minerals, also release huge quantities of sulfur dioxide (**FIGURE 20-8b**).

Like the sulfur oxides, the nitrogen oxides arise from a wide variety of sources. Forest fires, for example, can release millions of tons. The two most important anthropogenic sources are electric power plants and motor vehicles. American factories, cars, and power plants currently produce approximately 16.4 million metric tons (18.2 million tons) of sulfur dioxide and about 22.4 million metric tons (24.9 million tons) of nitrogen oxides a year.

> **KEY CONCEPTS**
>
> Acid precursors come from natural and anthropogenic sources, the latter being the most important. Of the anthropogenic sources, the combustion of fossil fuels is the most significant.

The Transport of Acid Precursors

Acid precursors and acids formed in the atmosphere can remain airborne for 2 to 5 days and may travel hundreds, perhaps even thousands, of kilometers before being deposited. Studies have shown that acids falling from the sky in southern Norway and Sweden largely come from England and industrialized Europe. In the United States, acids falling in the Northeast come from the industrialized Midwest, primarily the Upper Mississippi and Ohio River valleys. Indiana and Ohio are the two major producers. Moving eastward, the mass of pollutants tends to converge on New York State

> Acid precursors and acids can travel 2 to 5 days in the atmosphere, often being deposited hundreds, even thousands, of miles downwind from their site of production.

(a)

(b)

FIGURE 20-8 **Sources of sulfur dioxide. (a)** Volcanic eruptions, such as the Mount St. Helens blast shown here, is one natural source of sulfur dioxide. **(b)** Coal-fired power plants like this one produce most of the anthropogenic sulfur dioxide.

and New England, where 50% of the lakes are endangered because of low acid-neutralizing capacities.

As noted earlier, acid deposition is a global phenomenon. It is found in the United States, Canada, the Amazon Basin, Europe, and the Netherlands, among other places. All of these places are downwind of heavily polluted areas. In Europe and Scandinavia, rain and snow samples frequently have pH values between 3 and 5. In the White Mountains of New Hampshire, for example, the average annual pH of rainfall is about 4 to 4.21, nearly a hundred times more acidic than normal precipitation. Rain samples collected in Pasadena, California, in the 1970s had an average pH of 3.9. One of the low-

est pH measurements was made in Kane, Pennsylvania, where a rainfall sample with a pH of 2.7 was recorded—rain as acidic as vinegar. The grand prize for acidic rainfall, however, goes to Wheeling, West Virginia, where a rainfall sample had a pH of 2—stronger than lemon juice. More recent studies of the pH of fog downwind from the Los Angeles basin, however, revealed levels as low as 1.7.

In southern Norway and Sweden and in the northeastern United States, two ominous trends have been observed. First, acid precipitation is falling over a wider area than it was 60 years ago; second, the areas over which the strongest acids are falling are expanding (FIGURE 20-9).

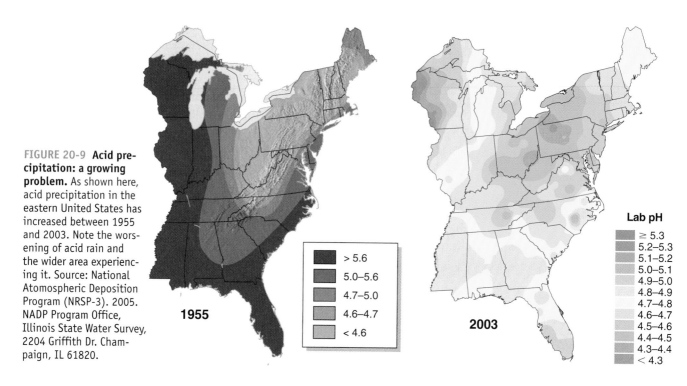

FIGURE 20-9 **Acid precipitation: a growing problem.** As shown here, acid precipitation in the eastern United States has increased between 1955 and 2003. Note the worsening of acid rain and the wider area experiencing it. Source: National Atomospheric Deposition Program (NRSP-3). 2005. NADP Program Office, Illinois State Water Survey, 2204 Griffith Dr. Champaign, IL 61820.

1955

2003

> 5.6
5.0–5.6
4.7–5.0
4.6–4.7
< 4.6

Lab pH

≥ 5.3
5.2–5.3
5.1–5.2
5.0–5.1
4.9–5.0
4.8–4.9
4.7–4.8
4.6–4.7
4.5–4.6
4.4–4.5
4.3–4.4
< 4.3

Acid precursors can be transported hundreds of kilometers from their site of production to their site of deposition. Acid deposition occurs downwind from virtually all major industrial and urban centers. Acid deposition is increasing in strength (acidity) and expanding in geographic range.

The Social, Economic, and Environmental Impacts of Acid Deposition

Acid deposition has many impacts on the environment, our economy, and society. This section describes some very real impacts and some possible ones.

Acidification of Lakes   Throughout the world, thousands of lakes and rivers and their fish have fallen victim to acid deposition. In the 1930s, for example, scientists surveyed lakes in the western part of the Adirondacks. Sampling the pH of 320 lakes, they found that most had pHs ranging from 6 to 7.5. In a 1975 survey of 216 lakes in the same area, a large number of them had pH values below 5, a level at which most aquatic life perishes (**FIGURE 20-10**). Of the acidified lakes, 82% were devoid of fish life. A more recent study of 1500 lakes in New York's Adirondack Park found that 25% of the lakes are so acidic that fish no longer live in them. Another 20% of lakes are acidic enough to be endangered.

In 1988, the National Wildlife Federation (NWF) published a list of U.S. lakes that had become acidified. The study showed that eastern lakes had been particularly hard hit. One of every five lakes in Massachusetts, New Hampshire, New York, and Rhode Island was acidic enough to be harmful to aquatic life. As acid deposition continues to fall, these lakes could become lethal to virtually all forms of life.

More recent data show that 4% of the lakes and 8% of the streams in acid-sensitive areas in the United States are chronically acidic. This includes lakes in the mid-Appalachians, the Adirondack Mountains, New England, the Atlantic coast, Florida, and the upper Midwest. Although the percentages may seem small, they translate into thousands of lakes and streams.

Recent studies have shown that acid deposition also occurs widely over the northern and central portions of Florida, with precipitation 10 times more acidic than normal. One third of the lakes in Florida are now acidic enough to be harmful to aquatic life.

In the mountains of southern Scandinavia, acidification of surface waters has occurred at a rapid rate for more than 50 years. In Sweden, approximately 20,000 lakes are without or soon to be without fish. In a recent study, Swedish authorities estimated that about 17% of the nation's lakes are acidified from acid deposition. Salmon runs in Norway have been eliminated because of the impact of acid precipitation on egg development, putting an end to inland commercial fishing in some areas.

In Canada, nine of Nova Scotia's famous salmon-fishing rivers have already lost their fish populations because of acidity. Eleven more are teetering on the brink of destruction. In a survey of 8500 lakes in southeastern Canada, 31% of them (2635 lakes) had pHs lower than 6. Thirteen percent (1100 lakes) had pH values below 5.5.

Acid deposition has acidified lakes throughout the world. Hundreds of lakes no longer support aquatic life, and thousands are on the verge of ecological collapse.

Why Are Some Lakes More Susceptible Than Others?   All lakes are not created equal. Those that lie downwind from major industrial centers are most vulnerable, but location is not a sufficient predictor of a lake's future. Those found in watersheds with thin soils with little **buffering capacity** fall victim to this insidious form of pollution. Some lakes also contain **buffers**, chemical substances that allow aquatic systems to resist changes in pH. When hydrogen ion (H^+) levels increase, buffers combine with the free ions and eliminate them. When levels fall, they release them, thus maintaining a constant pH. In general, the higher the buffering capacity of a lake and its surrounding watershed, the less vulnerable it is. **FIGURE 20-11** shows acid-sensitive areas in the United States and Canada.

The buffering capacity of the soil and surface waters—their ability to resist changes in pH—plays an important role in determining if a lake will be damaged by acid deposition.

How Does Acid Deposition Affect Aquatic Ecosystems? Many species of fish die when the pH drops below 4.5 to 5.0 (**FIGURE 20-12**). Falling pH is only part of the reason fish die. Scientists have found that acidic rainwater or snowmelt dissolves toxic elements such as aluminum from soil and rocks. The acidic waters carry the metals to streams and lakes. Aluminum irritates the gills of brook trout, causing a buildup of mucus and, ultimately, death by asphyxiation (**FIGURE 20-13**).

Spring poses a special threat to fish and other aquatic organisms. When the snow begins to melt, the surface melts first. This water drains through the unmelted snowpack and

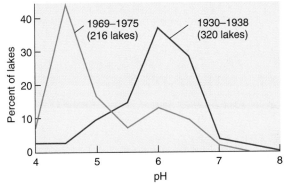

FIGURE 20-10 Changing pH. These graphs show the pH level in Adirondack lakes. In the 1930s most lakes had a pH of 6.0 or higher, but in the 1970s a large percentage had a pH of less than 5.

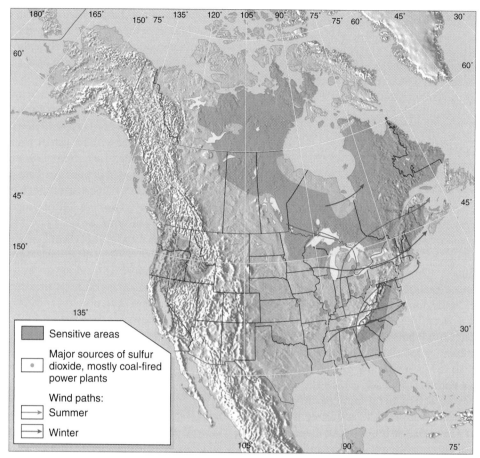

FIGURE 20-11 **Sensitive areas.** This map of North America shows areas experiencing acid deposition and the most geologically vulnerable regions.

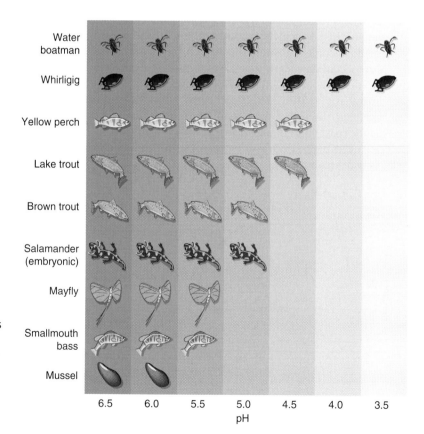

FIGURE 20-12 **Differing sensitivity.** The sensitivity of fish and other aquatic organisms to acid levels varies. The figure indicates the lowest pH (highest acidity) at which the various organisms can survive. The yellow perch, for example, can withstand a pH of 4.5, but populations plummet if the pH falls any further. Smallmouth bass and mussels are more acid sensitive, perishing at levels below 5.0 and 5.5, respectively.

FIGURE 20-13 Death by asphyxiation. These fish were confined to a cage in a stream affected by acid rain. They died of asphyxiation caused by acid leaching aluminum from the soil. Aluminum irritates the gills and causes mucus buildup, which blocks oxygen influx and kills the fish.

leaches out most of the acids. In fact, the first 30% of the meltwater contains virtually all of the acid and typically has a pH of 3 to 3.5, which is toxic to eggs, fry, and adult fish as well. Thus, when snow begins to melt, the concentration of acid in nearby lakes and streams rises rapidly. This surge of acids coincides with the sensitive reproductive period for many species of fish.

Widening the Circle of Destruction Many other species are affected by acid deposition. Songbirds living near acid-contaminated lakes in Scandinavia, for instance, lay eggs with softer shells than birds feeding near unaffected lakes. Scientists have found elevated levels of aluminum in the bones of affected birds and hypothesize that this comes from eating aquatic insects living in acidified waters. The aluminum interferes with normal calcium deposition, resulting in defective (soft) eggshells and fewer offspring.

Acidification of surface waters may also be partly responsible for a nearly 50% decline in the population of black ducks on the East Coast from the 1950s to the early 1980s when populations reached an all-time low. Today, populations remain steady. Although other factors have accounted for part of this decline, it appears that acids are killing the aquatic insects needed by female ducks and their offspring.

Spotted salamanders are also adversely affected by acidity. In the laboratory, exposure to water with pH of 5 prevents normal embryonic development and results in gross deformities that are usually fatal. In one study, it was found that the mortality of fertilized eggs was 60% at pH 6 but only 1% at pH 7.

In the wild, spotted salamanders breed in temporary ponds created by melted snow. These ponds are likely to be highly acidic in regions where acid precipitation is prevalent; as a result, the fate of the spotted salamander is bleak.

The spotted salamander is as important as birds and small mammals in the food chain. A drastic change in its population would very likely have serious repercussions in the entire ecosystem. Many other amphibian species have disappeared or severely declined in their natural habitat, largely as a result of acid deposition, as pointed out in Chapter 7.

Forest and Crop Damage Researchers estimate that 120,000 hectares (300,000 acres) of forest has been destroyed in the former Czechoslovakia by pollution, mostly acid precipitation. In Germany, 500,000 hectares (1.25 million acres) of forest is dead or dying (**FIGURE 20-14**). Even the famous Black Forest is now severely damaged by acidic pollutants from industry and automobiles. In Vermont's Green Mountains, half the red spruce, a high-elevation tree, have died from acid precipitation and acid fog. Lower-elevation sugar maples are also on the decline. The death of trees exacts huge ecological and economic costs. Forest-dwelling species perish as their forests die. The loss of timber reduces revenues. Swiss scientists believe that damage to trees on mountain slopes may increase the likelihood of avalanches because trees help retain snow on steep mountainsides. Further loss of the "barrier forests" endangers the safety of mountain residents, skiers, and highway travelers.

FIGURE 20-14 Forest die-off. This forest in Germany is now largely dead because of many years of acid deposition.

Numerous studies show that acid deposition damages forests both directly and indirectly. Consider direct damage first. Acid deposition damages leaves of birch and needles of pine trees. It also impairs germination of spruce seeds. It erodes protective waxes from oak leaves and leaches nutrients from plant leaves. In 1988, Robert Brock, a forest epidemiologist, reported findings from studies on Mt. Mitchell in North Carolina. In his study, he found that low-lying clouds containing acids that often bathe spruce and fir trees on the mountain are considerably more acidic than vinegar. Two days after a 2-day cloudy period, Brock found that needle tips looked singed. The needles contained 7 to 11 times more sulfate than healthy ones. Needles and leaves damaged by acid develop brown spots. Photosynthesis, the process by which needles and leaves make foodstuffs, declines—causing the tree to suffer nutritionally. Badly damaged leaves and needles actually fall off.

Indirect effects can also be important. Acids, for example, dissolve nutrients and valuable minerals from the soil trees grow in. These important substances are flushed from the soil, thus depriving plants of vital components essential for growth and reproduction. Acids also cause the release of toxic substances in soil, such as naturally occurring aluminum, which is normally chemically bound within the soil and thus not a problem. Recent evidence shows that aluminum damages cells in the water-transporting tubules of trees, closing off water transport. Trees die from thirst. The release of these substances can damage trees and other plants.

In 1988, Professor Lee Klinger from the University of Colorado proposed another hypothesis to help explain why the world's forests are dying. One of the chief culprits, he says, may be an acid-loving moss that grows on the forest floor. Klinger has studied 100 regions in 30 states where forests are dying. In each one he has found a thick layer of moss carpeting the forest floor.

These mosses act as sponges, holding so much water that the surface soils become saturated. In affected areas, the feeder roots of the trees and the trees themselves die, for the same reason that a houseplant dies when it is overwatered: Water eliminates air from the soil. Plants literally suffocate. Mosses may also kill mycorrhizal fungi that help trees absorb nutrients. Mosses also acidify the water passing through them. Acidic water dissolves toxic trace metals such as aluminum found in the soil, which can also kill the root system. It is likely that direct foliar damage and root damage may combine forces.

Acid deposition does not kill trees directly, according to the U.S. EPA. It is more likely to weaken them, by damaging their leaves, limiting the nutrients available to them, or poisoning them with toxic substances slowly released into the soil. Weakened tress are more susceptible to diseases, insects, drought, and other pollutants such as ozone that ultimately kill them. Combined, these forces may be responsible for the massive reduction in forest growth noticed by foresters in the eastern United States. It reminds us how important it is to consider the interaction of many changes we have wrought.

Just like trees, crops are affected directly and indirectly by acids. The direct effects include damage to leaves and buds. Acids falling on crop plants in the spring may impair growth at a very important time of year. In addition, acids may inhibit photosynthesis, the process by which plants produce carbohydrates and other important chemicals.

Acids damage plants indirectly—by altering the soil. For example, they may leach important elements from the soil, resulting in reduced growth. Acids impair soil bacteria and fungi that play an important role in nutrient cycling and nitrogen fixation, both essential to normal plant growth.

Concern for agriculture has also been raised by numerous researchers, but the results of many studies are inconclusive. Some researchers have reported that simulated acid precipitation decreases crop productivity, but others have found increases. Still others have found no effect. On balance, the EPA concludes that food crops are not seriously affected by acid deposition.

KEY CONCEPTS

Acid deposition damages forests in many parts of the world and may affect crops as well. Trees and other plants are damaged directly by acids but also indirectly, through changes in the soil chemistry and soil-dwelling organisms.

Acids: Fertilizing Effect   Aquatic and terrestrial plants require sulfur and nitrogen to grow. In some instances, acid rain may enhance soil fertility and improve crop growth.

On balance, acids probably do more damage than good. Direct damage to growing plants and damage to the soil offset the fertilizing effect. Studies have also shown that nitrogen in the form of nitrates or nitric acid deposited in Chesapeake Bay actually stimulate the growth of algae and aquatic plants.

Plant overgrowth in Chesapeake Bay caused by nitric acid impairs navigation. In addition, plants growing on the surface block sunlight needed for photosynthesis in plants and algae in deeper layers. These deeper plants help maintain oxygen levels; without them, other forms of aquatic life may perish. In addition, when aquatic vegetation dies in the autumn, it decays. The bacteria that break down this organic matter rob the water of oxygen, killing many aquatic organisms.

In a recent study of the effects of nitrogen fertilization on the prairies of Minnesota (published in *Science*), researchers found that additional nitrogen benefited weedy species that overtook native prairie grasses. Native species require less nitrogen and were choked out by weedy species. This not only altered the species composition, it resulted in a greater emission of carbon dioxide, with implications for global warming. The excess carbon dioxide emissions result from the fact that weedy species decompose much more rapidly than native prairie grass species.

KEY CONCEPTS

The sulfur and nitrogen in sulfuric and nitric acid promote plant growth, but their negative effects (such as direct damage and changes in the soil chemistry) typically outweigh any benefits resulting from their fertilizing effect.

FIGURE 20-15 **Lady Liberty gets a face-lift.** The Statue of Liberty was recently given a multimillion-dollar face-lift because of damage from air pollution, including acid deposition. The repair work cost well over $35 million.

FIGURE 20-16 **Liming lakes.** Millions of dollars are spent by the Swedish government to lime lakes each year to offset the influx of acids from the sky.

Damage to Materials

Acid precipitation also corrodes humanmade structures. It has taken its toll on some of special importance, such as the Statue of Liberty, the Canadian Parliament Building in Ottawa, Egypt's temples of Karnak, and the caryatids of the Acropolis—not just architectural works but works of art, priceless treasures (**FIGURE 20-15**). Acid rain may also damage house paint and etch the surfaces of automobiles and trucks. A U.S. report claims that acid precipitation causes an estimated $5 billion a year in damage to buildings in 17 northeastern and midwestern states. The price tag includes the cost of repairing mortar, galvanized steel, and stone structures as well as the cost of repainting. It does not include damage to automobile paint, roofing materials, and concrete, potentially adding billions of dollars to the cost.

KEY CONCEPTS
Acids cause billions of dollars of damage to priceless statues, buildings, and materials.

Solving a Growing Problem— Short-Term Solutions

In 1984, the New York State legislature passed a bill that required utilities to reduce sulfur emissions by 30% by 1991. Minnesota also passed legislation to curb the growing problem. In 1984, nine European nations and Canada signed an agreement to make similar reductions, but over a 10-year period. Canada's program was finalized in 1985 with seven provinces agreeing to reduce their combined sulfur dioxide emissions by 2.3 million metric tons per year by 1994, a goal they met in 1993 primarily by switching to low-sulfur coal.

The first significant U.S. governmental action came with the passage of 1990 amendments to the Clean Air Act (Chapter 19), which contained provisions for substantial reductions in sulfur oxide emissions but more modest cuts in nitrogen oxides by the year 2010. According to the law, by the year 2010, sulfur dioxide emissions are to be 40%—or 9 million metric tons (10 million tons)—below 1980 releases. Nitrogen oxides, which are much more difficult to control, are supposed to be only 1.8 million metric tons (2 million tons) below 1980 levels. Most of the sulfur dioxide reductions will be achieved by burning low-sulfur coal or by installing pollution control devices on 110 power plants, mostly coal-burning facilities, in eastern and midwestern states.

In 1991, the United States and Canada signed an agreement to reduce emissions of sulfur dioxide and nitrogen dioxide from power plants, smelters, and other stationary sources. Canada, highly motivated to protect its lakes from further deterioration, met its goal by 1993. Huge reductions occurred at some facilities such as the world famous Inco nickel smelter in Sudbury, Ontario, where 90% reductions were achieved. The United States has not yet achieved either of its goals. Although sulfur dioxide emissions have been reduced, they're still over 2 million metric tons higher than the goal of the agreement. Nitrogen dioxide emissions have been even harder to bring under control and have increased by over a million tons per year in the past 5 years.

International efforts to control sulfur oxide emissions generally rely on three strategies: (1) the installation of scrubbers on new and existing coal-fired power plants, (2) the combustion of low-sulfur coal or natural gas in utilities, and (3) the combustion of desulfurized coal—that is, coal that has had most of its sulfur removed. Nitrogen dioxide levels are more difficult to control because nitrogen oxide gases come from the nitrogen in air that reacts with oxygen in high-temperature furnaces. Changes in the temperature of combustion, however, can help reduce nitrogen oxide emissions.

Another approach to the problem involves treating the lakes themselves. In 1977, for example, the Swedish government embarked on an expensive program to neutralize acidic lakes by applying lime to thousands of lakes and rivers (**FIGURE 20-16**). These actions improved the water quality in many lakes, saving fish populations, but cost tens of millions of dollars per year.

Critics argue that liming is a short-term, stopgap solution, a little like CPR administered to a heart attack victim. In Canada liming costs $120 per hectare ($50 per acre). Treating a single lake can cost between $4000 and $40,000. In 5 years, however, treated lakes turn acidic again.

Stopgap measures are essential, as noted earlier in the book. They are the first line of defense against acid deposition, but many much less costly—indeed, profit-generating—options are available to us.

KEY CONCEPTS

Many stopgap measures have been initiated to help reduce the threat of acid deposition, including the installation of smokestack scrubbers, combustion of low-sulfur or desulfurized coal, and liming lakes to neutralize acidity. Such measures are necessary in the short term but must eventually be replaced by long-term, preventive actions.

Long-Term Sustainable Strategies

Liming lakes and other approaches outlined in the previous section generally treat the symptoms of the problem while ignoring the root causes—in this case, our heavy dependence on and inefficient use of fossil fuels. To solve this problem sustainably, we must find ways to address the root causes. Increasing energy efficiency and reducing our dependence on fossil fuels by using solar and wind energy are key elements of a sustainable strategy, which are discussed in Chapter 15.

Recognizing the importance of these alternative measures, several countries—including Germany, Great Britain, Japan, Norway, and Sweden—are actively developing alternative fuel sources of energy and increasing the efficiency with which they use fossil fuels. In Mexico, utilities, government, and industry have teamed up in hopes of making energy efficiency a cornerstone of the country's development. In India, private businesses have launched a program to install thousands of energy-efficient compact fluorescent lightbulbs in homes and businesses.

Population stabilization, growth management, and recycling are also essential elements of our prevention strategy. All of these reduce our demand for energy and, therefore, our production of pollutants. These strategies help solve more than acid deposition—they'll help reduce habitat destruction, the loss of species, urban air pollution, water pollution, and other environmental ills.

KEY CONCEPTS

Fuel efficiency, renewable fuels, recycling, population stabilization, and growth management are key elements of a sustainable design strategy to help prevent the production of acid precursors—and hence reduce acid deposition.

Are Controls on Acid Deposition Working?

In 1980, the U.S. Congress passed the **National Acid Precipitation Act**. Among other things, this law authorized the formation of the **National Acid Precipitation Assessment Program** (NAPAP)—a program staffed by people from a variety of federal agencies including the EPA, Department of Energy, Department of Agriculture, and National Oceanic and Atmospheric Administration. Their goal was to first assess the state of the environment—how much damage acid deposition has caused to determine how severe the problem is. Their findings showed that although it is a problem in some areas, it has not reached crisis stages in the United States. Some critics take issue with this statement, noting that huge numbers of lakes have been destroyed in the northeastern United States with a huge impact on the tourism economy.

The Acid Deposition provisions of the 1990 Clean Air Act (discussed in Chapter 19) called for a dramatic reduction of sulfur dioxide emissions to 9 million metric tons (10 million tons) per year by 2010—over a 50% reduction. It also required reduction, albeit a modest one (about 10%), of 1.8 million metric tons (2 million tons) per year in nitrogen oxide emissions.

The 1990 Clean Air Act also directed the NAPAP to develop market-based controls on acid deposition—that is, strategies that encouraged electric utilities to cut emissions through innovation and other methods that made more sense economically than old command and control regulations. One idea was the tradable permit, discussed in Chapter 19. Furthermore, the NAPAP was directed to monitor progress and economic costs and benefits of controls.

Their early findings were that market-based controls were very effective in cutting sulfur emissions. In the first year alone, emissions of sulfur dioxide from the worst polluters fell by nearly 40%. Even more remarkably, costs of compliance have been far less than originally projected—only about 15% of the estimates. Have sulfur dioxide and nitrogen oxide emissions fallen?

From 1990 to 2003, the most recent year for which data are available, sulfur dioxide emissions declined by 32%. Nitrogen oxide emissions decreased very slightly but appear to be on the rise over the past 5 years. Has the decline in acid precursors affected acid deposition and other problems associated with it?

In response to the decreases in sulfur dioxide emissions, concentrations of sulfate and sulfuric acid in precipitation have shown a significant decline in many parts of the United States. Nitrate concentrations in rain, as expected, have shown no change. Many lakes and streams also reflect changes in the chemical makeup of the rain. There is now evidence of recovery of acidified lakes in New England, but the most badly damaged lakes of the Adirondacks have shown no change. The NAPAP concluded that most forests in the nation are not adversely affected by acid deposition, but "if deposition levels are not reduced in areas where they are presently high, adverse effects may develop due to chronic multiple decade exposure."

KEY CONCEPTS

Market-based strategies, in particular tradable permits, have proven successful in reducing sulfur dioxide emissions in the United States, with corresponding changes being seen in the acidity of rainfall as well as many lakes and streams. Still, Americans are a long way from the goals of reducing sulfur and nitrogen dioxide set out in the 1990 Clean Air Act.

20.3 Global Warming/Global Climate Change

Scientists have long known that air pollution can affect local weather. For example, smoke from factories can substantially increase rainfall in areas downwind. In 1995, a group of 2500 atmospheric scientists concluded that the bulk of the evidence showed that human activities (pollution and deforestation) were having a discernible effect on global climate that could have profound effects on people, the economy, and the environment. In December 1999, the head of the U.S. National Oceanic and Atmospheric Administration and the head of the British Meteorological office, agencies that normally do not venture into the political arena, said in letters to London's *Independent* newspaper, "Our climate is changing rapidly and it's important that we take action now. We're now coming clean and saying we believe the evidence is almost incontrovertible, that man has an effect (on climate) and therefore we need to act accordingly." Some oil companies, such as BP PLC (formerly British Petroleum) have accepted the scientific consensus that fossil fuels are a main contributor to the problem and are investing in alternatives to fossil fuels and decreasing their output of CO_2. Exxon Mobil, however, does not agree and is not rushing to spend money on alternatives. Quite the opposite is true. The company is spending over a million dollars to convince the American public that global warming is not a threat. To understand this issue, and why so many scientists are concerned about it, we must first look at the global energy balance, the basis for the planet's climate.

Global Energy Balance and the Greenhouse Effect

Each day the Earth is bathed in sunlight. Approximately one third of the sunlight striking the Earth and its atmosphere is reflected back into space. The rest is absorbed by the air, water, land, and plants. As anyone who has sat in the sun knows, sunlight is absorbed by surfaces and converted into heat or, more technically, infrared radiation. As shown in **FIGURE 20-17**, this heat is slowly radiated back into the atmosphere and eventually escapes into outer space. As a result, energy input is balanced by energy output.

Scientists have long known that certain chemical substances in the atmosphere alter this balance. These include water vapor, carbon dioxide (CO_2), nitrous oxide (N_2O), methane (CH_4), and chlorofluorocarbons (Table 20-2). How do these gases affect the Earth's energy balance?

Consider carbon dioxide. Carbon dioxide molecules in the air absorb infrared radiation escaping from the Earth's surface and radiate it back to Earth, acting much like the glass in a greenhouse (Figure 20-17). Other molecules act similarly.

This phenomenon, known as the **greenhouse effect**, is essential to life on planet Earth, for it helps maintain the Earth's surface temperature. Without it, the Earth would be at least 30°C (55°F) cooler than it is today—and inhospitable to more life-forms.

The Earth's surface temperature can be altered by changes in the concentrations of CO_2 and other chemical substances noted above. An increase in carbon dioxide levels, for instance, causes more heat to be radiated back to the Earth, which warms the planet. Decreases in carbon dioxide or other similar compounds have the opposite effect.

The chemical substances that increase the Earth's surface temperature are called **greenhouse gases** because they act somewhat like the glass in a greenhouse. You've observed the greenhouse effect in a car or house heated by the sun. The glass permits sunlight to enter the car's interior, where it is absorbed by interior surfaces and converted to heat. The glass also prevents heat from escaping. On Earth, sunlight is absorbed by plants, dirt, buildings, parking lots, and roadways and is converted into heat. The heat or infrared radiation then rises, but some of it is trapped in the atmosphere by greenhouse gases.

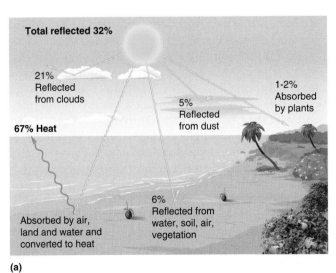

(a)

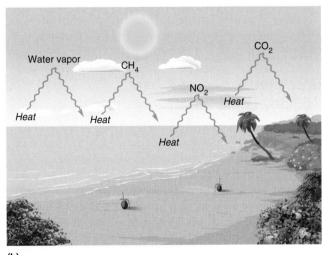

(b)

FIGURE 20-17 Global energy balance. (a) This drawing shows the influx of solar radiation and its fate. **(b)** Carbon dioxide and other greenhouse gases radiate heat back to Earth, causing the Earth's surface temperature to increase.

Table 20-2

Major Greenhouse Gases and Their Characteristics

Gas	Atmospheric Concentration (ppm)	Annual Increase (%)	Life Span (Years)	Relative Greenhouse Efficiency ($CO_2 = 1$)	Current Greenhouse Contribution (%)	Principal Sources of Gas
Carbon dioxide (CO_2) (from fossil fuels)	351.3	0.4	x^1	1	57 (44)	Coal, oil, natural gas, deforestation
Carbon dioxide (from biological sources)					(13)	
Chlorofluorocarbons (CFCs)	0.000225	5	75–111	15,000	25	Foams, aerosols, refrigerants, solvents
Methane (CH_4)	1.675	1	11	25	12	Wetlands, rice, fossil fuels, livestock
Nitrous oxide (N_2O)	0.31	0.2	150	230	6	Fossil fuels, fertilizers, deforestation

Sources: Data from World Watch Institute, U.S. EPA, and *Journal of Geophysical Research*.
[1]Carbon dioxide is a stable molecule with a 2- to 4-year average residence time in the atmosphere.

Studies of the global energy balance show that the Earth's surface temperature is influenced by two key factors: those that affect the amount of sunlight striking the Earth and those that alter the amount of heat lost or retained. In other words, the Earth's temperature is affected by the influx of solar energy and loss of heat. Both natural and anthropogenic factors influence these processes. For example, the sun is not a constant source of energy. It undergoes periods in which it produces more energy. During such periods the Earth's temperature tends to increase. Pollution from factories increases the amount of heat radiated back to Earth and warms the planet.

KEY CONCEPTS

Much of the sunlight striking the Earth and its atmosphere is converted into heat and is eventually radiated back into space. Natural and anthropogenic factors affect the amount of solar radiation striking the Earth and the rate at which heat escapes— and thus influence the temperature of the Earth's atmosphere.

Greenhouse Gases: Where Do They Come From?

Greenhouse gases typically come from natural and anthropogenic sources. Carbon dioxide, nitrous oxide, and methane, for example, are emitted by natural as well as human sources. Although the release of these gases from natural sources has remained fairly constant over the past 100 years, emissions from human sources have increased dramatically. Annual carbon dioxide emissions, for instance, have climbed from 534 million tons (of carbon) to about 7000 million tons per year in 2004 (FIGURE 20-18a) in the past 100 years. Not all greenhouse gases have a natural source. Chlorofluorocarbons and their replacements, the hydrochlorofluorocarbons, for example, are greenhouse gases but they have no natural source. Their production and release have risen rapidly between 1950 and the mid-1990s.

The rise in carbon dioxide emissions and the release of other greenhouse gases from anthropogenic sources is believed to be the primary cause of the increase in average global temperature in the past 30 years (FIGURE 20-18b and c).

KEY CONCEPTS

Greenhouse gases come from natural and anthropogenic sources, the latter of which have been increasing dramatically in the past 50 years.

Upsetting the Balance: Global Warming and Global Climate Change

Many scientists believe that the increase in greenhouse gases is responsible for rising global temperatures, known as **global warming.** (See the Point/Counterpoint for a discussion of the issue of global warming.) Numerous studies suggest that the increase in the Earth's surface temperature could have a profound effect on the planet. It could, for instance, cause changes in rainfall patterns. It may be responsible for an increase in the frequency and severity of storms. In other words, it could affect global climate. Consequently, most scientists prefer the term **global climate change** over *global warming.* Global climate change may affect ecosystems, agriculture, sea level, insurance rates, and our economy.

Table 20-2 lists the four greenhouse gases and several important facts about each one, including their sources, the rates at which they are increasing in the atmosphere, and their life spans. It also lists their relative greenhouse efficiency—that is, how they compare as a greenhouse gas to carbon dioxide. As shown, one molecule of CFC is equivalent to 15,000 molecules of carbon dioxide, partly because CFCs last so long. Pay special attention to the contribution

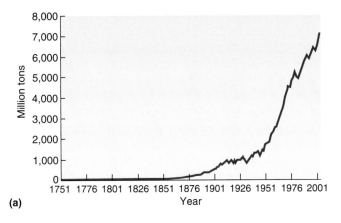

(a)

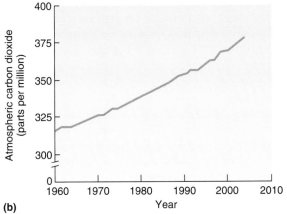

(b)

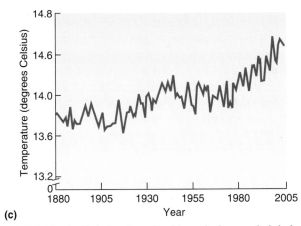

(c)

FIGURE 20-18 **Global carbon dioxide emissions and global temperature.** (a) Annual carbon emissions from fossil fuel burning, 1751–2004. Data source: UN, BP, DOE, IEA. (b) Carbon dioxide levels in the atmosphere have risen dramatically since 1960. Data source: GISS, BP, IEA, CDIAC, DOE, and Scripps Institute of Oceanography. (c) Graph of average global temperature since 1880. Data source: GISS, BP, IEA, CDIAC, DOE, and Scripps Institute of Oceanography.

of each to the greenhouse effect in column 6. This takes into account the relative effect of each one and the amount in the atmosphere. As you can see, carbon dioxide is the major greenhouse gas. Not all carbon dioxide comes from the combustion of fossil fuels. Deforestation and torching of cleared forests, for example, also produce a significant amount of carbon dioxide. How does deforestation result in an increase in atmospheric carbon dioxide? As you may recall

from studying the ecology chapters, all plants, including trees, take up carbon dioxide from the atmosphere, which they convert into plant matter. As trees are stripped from the land, the Earth's capacity to absorb carbon dioxide decreases. Atmospheric levels increase.

KEY CONCEPTS

The accumulation of greenhouse gases is very likely responsible for the increase in average daily temperatures, or global warming. This, in turn, is linked to changes in other aspects of climate, such as rainfall patterns and the frequency and severity of storms. Many of these changes affect ecosystems and could have profound effects on human society.

Predicting the Effects of Greenhouse Gases

Many factors will determine the level of greenhouse gases in the future. Population growth and the rate of industrialization will be primary determinants. Industrialization alone is not a sufficient predictor, though. What is more important is the amount of fossil fuel burned to spur economic growth and industrial development. Public policy will also have a profound effect on future concentrations of greenhouse gases. Measures to promote renewable energy and efficient energy use will both reduce the growth in the emissions of carbon dioxide. International treaties are also important. The treaties discussed earlier in this chapter that call for the phaseout of CFCs, for example, will reduce at least one type of greenhouse gas. Corporate actions and individual actions will also influence how much carbon dioxide is released into the atmosphere each year.

To predict the effects of different levels of atmospheric greenhouse gases, scientists use **global climate models**, computer programs that contain mathematical equations that attempt to simulate the various aspects of our climate and how they change. Scientists feed data into the computer regarding projected greenhouse gas levels, and the computers use this information to predict possible climatic effects.

KEY CONCEPTS

Scientists predict the effect of greenhouse gases by using computer programs that simulate global climate. They use information on projected levels of greenhouse gases to determine future temperature and other climatic effects.

Rising Temperatures, Rising Sea Levels   In 1995, scientists from the UN-sponsored Intergovernmental Panel on Climate Change (IPCC) predicted a 1.8°F to 6.3°F increase in average global temperature from 1990 to the year 2100. In 2000, the same group issued a report saying that the Earth is likely to get a lot hotter than previously predicted, rising 2.7°F to 11°F over the next 100 years.

Although a 2.7°F increase may seem insignificant, it could drastically alter global climate and sea level. IPCC scientists predict that sea level will increase 50 centimeters (20 inches) from 1990 to 2100. Sea level has already increased 10 to 12.5 centimeters (4 to 6 inches) in the past

Global Warming Is Real

Stephen H. Schneider
Stephen H. Schneider is a senior fellow at the Center for Environmental Science and Policy, Stanford University, and editor of the journal *Climate Change*. He has done pioneering modeling work in the fields of atmospheric science and global climatology. His research interests include climate change, global warming, and modeling of human impacts on climate.

Observations have already established beyond doubt that atmospheric constituents, such as water vapor, clouds, carbon dioxide (CO_2), methane, (CH_4), nitrous oxide (N_2), and chlorofluorocarbons (CFCs), trap heat escaping from the Earth's surface, causing the greenhouse effect. Likewise, it is virtually certain that an unprecedented 25% increase in CO_2 and 100% increase in CH_4 over the past 150 years have resulted from increased use of fossil fuels and expanded deforestation. It is also well accepted that the buildup of these gases has trapped 2 extra watts of radiative energy averaged over every square meter of Earth. What then is the basis of the debate over global warming?

First of all, translating 2 watts per square meter of heating into X degrees of temperature rises requires calculations that are based on not yet verified assumptions about how clouds, soils, forests, ice, and oceans will respond to this heating. These factors could change in ways that feed back on the heating, either reducing it, as global warming critics like to point out, or enhancing it, as most present climate models project. Such models predict that the past hundred years should have experienced some 1°C of warming from the extra 2 watts per square meter of heating, provided all other factors were constant, a dubious prospect. Already, half a dozen national and international assessment bodies have suggested that if the CO_2, CH_4, and CFC trends continue, global warming of some 1.5° to 4.5°C can be expected in the next century. Already, 0.5° $\pm$ 0.2°C of global warming has been observed from 1890 to 1990.

Some critics note that there is not a perfect match between decade-to-decade fluctuations in climate and the buildup of greenhouse gases. Unfortunately, such critics often fail to mention that no knowledgeable scientist would ever expect such agreement, because fluctuations of several 10ths of a degree Celsius per decade occur naturally. Thus, it is illogical to look for a decade-by-decade match. Only long-term, global trends can verify that a warming trend due to the buildup of greenhouse gases has been detected. A high level of certainty will take 10 to 20 years more to establish, not a few years as some critics of immediate danger argue. Moreover, waiting for such scientific certainty is not cost free, because the Earth could then be forced to adapt to a greater amount and rate of climate change than if we act now.

Over the past 10 years, half a dozen government-sponsored assessments have agreed that there is a better than 50% chance that current trends in population growth, fossil-fuel use, and land utilization practices will cause climatic changes of 2°C or more in the next century. Moreover, the rate of projected human-induced changes is some 10 times greater than the long-term rate of natural, global climate changes.

Critics suggest that 3 to 5 more years of testing is needed before taking action. Although this sounds prudent, such testing will not provide definitive answers on climatic change or its implications for ecosystems, forestry, agriculture, water supplies, human health, sea level, and severe storms.

I'm not a planetary gambler. I'd prefer to slow down the rate of buildup of greenhouse gases rather than gamble that things may work out all right in the end. Study after study has shown that the best way to start to reduce the buildup rate of greenhouse gases is to make cost-effective improvements, such as controlling population growth, increasing the efficiency of energy use, and virtually eliminating the production of CFCs. The Yale economist William Nordhaus, a critic of severe cuts in CO_2 emissions, has argued nevertheless that modest cuts in CO_2 emissions would, at present, yield economic benefits in excess of the costs, and his calculations did not even include the free extras: less acid rain, less air pollution, a lower balance-payments deficit from importing foreign oil, and lower long-term energy costs of manufactured goods.

To me, reducing CO_2 and other measures is a kind of climate change "insurance" that pays other dividends. We need to eliminate the billions of dollars of government subsidies to inefficient current fossil-fuel uses and deforestation practices and move toward lower real costs and an environmentally more stable society. Political rhetoric about uncertainty only commits the future to greater risks.

Greenhouse warming (GW) has emerged as the issue of the Millennium. Wide acceptance of the Montreal Protocol, which reduces the manufacture of chlorofluorocarbons, considered a threat to the stratospheric ozone layer, has encouraged environmental activists to call for similar controls on carbon dioxide. They have expressed disappointment with the White House for not supporting immediate action on CO_2. Should the United States assume "leadership" in this campaign, or would it be more prudent to first assure through scientific research that the problem is both real and urgent?

The scientific base for GW includes some facts, lots of uncertainty, and just plain ignorance. What is needed are more observations, better theories, and more extensive calculations. There is consensus about an increase in greenhouse gases (CO_2, CFCs, methane, nitrous oxide, ozone) in the Earth's atmosphere. There is some uncertainty about their rate of generation and their rate of removal. There is major uncertainty and disagreement about whether this increase has caused a change in the climate during the last hundred years; many observations do not fit the theory. There is also major disagreement in the scientific community about predicted changes from further increases in greenhouse gases; the models used to calculate future climate are not yet refined enough to simulate nature. As a consequence, we cannot be sure whether the next century will bring a warming that is negligible or significant. Finally, even if there is a warming and associated climate changes, it is debatable whether the consequences will be good or bad; likely, we will get some of each.

Has the observed increase of greenhouse gases in the last decades had an effect on climate? The data are ambiguous to say the least. Advocates of immediate action profess to see a global warming of about 0.5°C since 1880 and point to record temperatures experienced in the 1980s. Others tend to be more cautious; they call attention to the fact that the strongest increase occurred *before* the major rise in greenhouse gas concentration; it was followed by a 35-year decrease, between 1940 and 1975. Most researchers consider the warming observed before 1940 to be a recovery from the "Little Ice Age" that prevailed from 1600 to about 1850.

Weather satellites, in continuous operation since 1979, have shown negligible warming in recent decades. These results are independently verified by radiosondes carried on weather balloons. Human effects on climate seem to be minor.

We can sum up our conclusions in a simple message: *the scientific base for greenhouse warming is too uncertain to justify drastic action at this time*. There is little risk in delaying policy responses to this century-old problem, because there

Too Early to Tell

S. Fred Singer
S. Fred Singer, professor of environmental sciences at the University of Virginia, has served as deputy assistant administrator of the Environmental Protection Agency and as the first director of the U.S. weather satellite program in the Department of Commerce. An atmospheric and space physicist, he predicted the increase of atmospheric methane due to human activities.

is every expectation that scientific understanding will be substantially improved within a few years. Instead of panicky and premature actions, we will then be able to apply specific remedies as necessary. That is not to say that steps cannot be taken now; indeed, many kinds of energy conservation and efficiency increases make economic sense even *without* the threat of greenhouse warming.

Drastic, precipitous, and especially unilateral steps to delay the putative greenhouse impacts can cost jobs and prosperity without being effective. The Yale economist William Nordhaus, one of the few who has been trying to deal quantitatively with the economics of the greenhouse effect, has pointed out that "those who argue for strong measures to slow greenhouse warmings have reached their conclusion without any discernible analysis of the costs and benefits." Economists now tend to agree that warmer climates are beneficial for agriculture and for humanity. It would be prudent to complete the ongoing and recently expanded research so that we will know what we are doing before we act. "Look before you leap" may still be good advice.

Critical Thinking Questions

1. Summarize the major points made by each author. Are there areas where they are looking at nearly the same data but reaching different conclusions? If so, how is this possible?
2. Using your critical thinking skills, analyze each essay. What flaws do you see in the logic, if any?

 You can link to websites that represent both sides through Point/Counterpoint: Furthering the Debate at this book's internet site, http://environment.jbpub.com/. Evaluate each side's argument more fully and clarify your own opinion.

FIGURE 20-19 **Paradise lost.** Small islands like these (foreground) in the blue-green waters off Bora-Bora in the Pacific Ocean could vanish as sea levels rise as a result of global warming.

coast of Bangladesh, killing an estimated 60,000 to 140,000 people and flooding many rice fields, destroying crops. By some estimates, 17% of the land area of Bangladesh could be under water by 2030, worsening crowding inland.

Low-lying island nations, many of them tropical resort places, would also be dramatically affected by rising sea level. Many island nations in the South Pacific are already beginning to feel the effect as ocean waters rise and rob them of valuable beaches and shoreline (FIGURE 20-19). In fact, a coalition of island nations is one of the leading proponents in global climate change negotiations.

KEY CONCEPTS

A small but climatically significant increase in global temperature is expected by the end of the next century as a result of increasing emission of greenhouse gases. Scientists predict that this increase will very likely result in a rise in sea level with potentially devastating effects on coastal human population centers.

50 years. The rise in sea level would result from melting of glaciers and the land-based Antarctic ice pack and an expansion of the seas resulting from warmer temperatures.

Rising sea levels would threaten coastal cities throughout the world. Today, over half the world's population lives in coastal cities and towns. In fact, over 40 of the world's largest cities are in coastal regions. Among cities most at risk are Miami, New Orleans, Bangkok, Hamburg, London, St. Petersburg, Shanghai, Sydney, Alexandria, and Dhaka. In the United States, more than half of the population lives within 83 kilometers (50 miles) of the ocean. Even a modest increase in sea level would flood coastal wetlands, low-lying fields, and cities. The rise in sea level would also worsen the damage from storms. Waves produced during hurricanes and other storms would sweep further inland, damaging more homes and cities than they do today. Many people would have to relocate. Cities may be forced to build levees to hold back the seas or gradually move to higher ground as buildings are retired.

The inland creep of the ocean would usurp farmland and wildlife habitat and create more crowding as people compete for a limited land base. The changes would not come cheaply, and they would not be easy.

Less developed nations would also suffer enormously if the oceans rise. In Asia, for instance, rice is produced in many low-lying regions that might be reclaimed by the sea. Storm surges could carry saltwater onto some of the remaining fields, killing crops and poisoning the soil. In 1991, a devastating storm ravaged the

Rising Temperatures, Changing Rainfall Patterns Global temperature is expected to continue to increase after 2100 even if the emission of greenhouse gases ceases. According to some computer models, not only will temperature and sea level increase, rainfall patterns will change (FIGURE 20-20). Much of Europe, Africa, and South America, for example, are already experiencing decreased rainfall. Such decreases reduce agricultural production, causing food shortages. Many areas, however, are experiencing increased rainfall. Although that may seem beneficial, it too can disrupt farming and lead to decreased food production. In 2000, heavy spring rains in western New York, for example, flooded fields and prevented farmers from planting their crops on time. By the Fourth of July, corn was only ankle high.

KEY CONCEPTS

Global temperature increases could shift rainfall patterns, increasing precipitation in some areas and decreasing it in others. Too little rain in some areas and too much in others could have a profound effect on food production and economic output.

FIGURE 20-20 **Changing rainfall patterns.** This map shows measured changes in global rainfall over the past 100 years. Source: Intergovernmental Panel on Climate Change; http://www.ipcc.ch/present/graphics/2001syr/small/07.17.jpg.

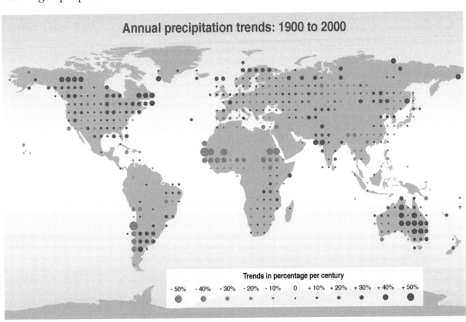

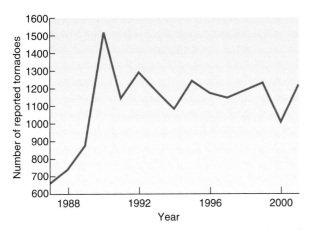

FIGURE 20-21 Tornadoes. The number of tornadoes each year is on the rise.

Rising Temperatures, Violent Storms Computer models predict that global warming could spawn bizarre and violent weather, notably hurricanes, such as those witnessed in growing numbers in recent years. Warming seas impart more energy to the atmosphere, which can greatly intensify a hurricane as it did in the case of hurricane Katrina, which struck Louisiana and Mississippi in 2005. An increase in devastating storms could have potentially serious economic impacts as well, by destroying crops and making food more expensive. Property insurance would rise.

A recent study of storm severity in the United States showed that violent downpours are indeed on the rise. Studies also show that the occurrence of tornadoes is on the rise in the United States. In fact, 1999 was the fourth "busiest" year for tornadoes in U.S. history with 1225 reported tornadoes, approximately twice as many as in the 1980s (**FIGURE 20-21**). In 2004, the United States experienced over 1400 tornadoes. Worldwide, the incidence of major weather events has also increased. In 1999, there were four times as many major storms and floods as there were in the 1960s, and damage was seven times higher. Plagued by devastating floods in Europe in recent years that many experts think were caused by global climate change, the European insurance industry has become a major supporter of international efforts to curb the emission of greenhouse gases.

KEY CONCEPTS

Preliminary studies suggest that global warming may be responsible for an increase in the number and severity of storms in the United States and other countries.

The Ecological and Health Impacts of Global Climate Change

In 2000, a report by the National Research Council predicted an increase in North America of 3° to 6°C (5° to 10°F) over the next 100 years. In this report, scientists predicted a major shift in climate zones in North America. The authors suggest that the tropical climate of equatorial regions will shift northward into the lower tier states. Their climate, in turn, will shift to the mid-tier states and the climate of the top-tier states would move into Canada. The effect of such a rapid shift on vegetation and wildlife could be devastating, with many trees simply dying as temperatures climb outside of their range of tolerance.

A great many plants and animals could face difficult times as the planet warms and its climate shifts. In fact, if the change in temperature continues at the predicted rate, many species could become extinct. Others could suffer enormous declines in their populations. A limited number will be able to adapt or migrate to suitable habitat.

Interestingly, most species and habitats have dealt with changing conditions for eons, so warming itself is less a concern than the rate at which it is likely to occur. Professor Margaret Davis of the University of Minnesota shows why, in a computer simulation she performed to predict the effects of a global temperature increase on several eastern tree species. She found that if global atmospheric carbon dioxide concentrations double by 2050 and temperatures rise as predicted, hardwood trees east of the Mississippi would face tremendous losses. Her studies predict that beech trees would probably disappear from the southeastern United States, except in a few mountainous regions. Suitable beech tree habitat would shift north to New England and southeastern Canada, which is now the extreme northern limit of the tree's range.

Although species can shift their range, most would not be able to move anywhere as fast as required. At the end of the last ice age, for instance, beech trees "migrated" northward by dispersing seeds at a rate of about 10 kilometers per 50 years—far slower than the 500-kilometer migration needed to avoid destruction from global warming taking place within the next 50 years. Many other trees in the United States would face a similar fate. Over time, these trees will die out, possibly being replaced by less desirable heat-resistant species.

Practically every ecosystem on Earth would be affected by global warming. Some of the most important and most threatened are coastal ecosystems, notably mangrove swamps and coastal marshes. The future of these areas and the services they provide—among them protecting coastal regions from erosion and providing habitat for commercially important food species—would not be bright. As sea level rises, most mangrove swamps would be lost.

Some studies suggest that plants might thrive in a CO_2-rich world. After all, carbon dioxide is essential to plant growth. A closer examination of the issue, however, suggests that the benefits of increasing levels of CO_2 are overstated. In fact, studies have shown that elevated CO_2 levels benefit certain plants but harm others—among the latter corn, sugar cane, and many grasses. In addition, studies show trees in tropical regions grow more slowly when exposed to higher temperatures. Moreover, photosynthesis grinds to a halt when temperatures exceed 100° F. Making matters worse, elevated levels of CO_2 result in a reduction in nitrogen levels in virtually all plants, for reasons not understood. Changes in the nutritional quality of plants could affect the entire food web. Studies show that to compensate for the lower nutritional value of plants grown in elevated carbon dioxide, insects eat more. If insects cannot get enough to eat, declines in insect populations might occur. This, of course, would have devastating effects on birds and other insect-eating species.

Animals, like plants, respond to warming trends by shifting to new habitats, but not all animals are capable of moving as far as necessary. Stanford University researcher Dennis Murphy studied the Great Basin Mountains, lying between the Cascades and Sierra Nevadas on the west and the Rockies on the east. His studies indicated that 44% of the mammals, 23% of the butterflies, and a smaller percentage of birds would be destroyed by a 3°C (5.4°F) increase in global temperature. A study published in the prestigious journal *Nature* says that global climate change could drive as many as one million species to extinction by 2050.

Some species may actually thrive amid global warming—for example, organisms that are responsible for infectious diseases. Robert Shope of the Yale University School of Medicine predicts that some diseases now restricted to tropical areas could invade new territory as the planet warms. For example, one form of rabies, now transmitted by vampire bats, could spread northward from Mexico and could result in damage of about $1 billion a year to the Texas cattle industry. Some scientists believe that the recent outbreaks of the deadly Hanta virus, which is transmitted by rodents in the desert Southwest, may be a result of rising global temperature.

Warmer climates are already causing the spread of insects that carry malaria and dengue fever to higher altitudes and higher latitudes. In Africa, studies show that malaria is spreading to higher altitude regions in five countries, which scientists attribute to the warming climate. In Mexico and Costa Rica, dengue fever is spreading into the highlands as well. "The control issue looms largest in the developing world, where resources for prevention and treatment can be scarce," writes Dr. Paul Epstein, Associate director of the Center for Health and the Global Environment at Harvard Medical School in an article in *Scientific American.* "But the technologically advanced nations, too, can fall victim to surprise attacks—as happened last year (1999) when the West Nile virus broke out for the first time in North America, killing seven New Yorkers. In these days of international commerce and travel, an infectious disorder that appears in one part of the world can quickly become a problem continents away if the disease-causing agent, or pathogen, finds itself in a hospitable environment."

KEY CONCEPTS

Organisms and ecosystems could be profoundly influenced by global climate change, especially if the rate of change occurs faster than their ability to adapt (which seems inevitable).

Changing Ocean Currents

Ironically, global warming may plunge some areas into a period of intense and catastrophic cooling? In particular, global warming could bring on ice age conditions in Great Britain, Europe, and Russia. How?

As noted in Chapter 5, seawater circulates throughout the oceans. For example, warm water from tropical regions circulates northward in the Atlantic Ocean via the Gulf Stream (Figure 5-4). This makes Great Britain and Europe (located on the same latitude as northern Canada) much warmer than it would otherwise be. Warm saltwater circulating northward cools as it moves toward the North Pole. The cooler, denser water then sinks and flows back toward the equator, only to rise again. This gigantic current forms a huge conveyor shown in **FIGURE 20-22** that distributes heat from the equator to the poles. It makes the poles warmer and the equator cooler than they would be otherwise, but global climate change could alter all of this, and abruptly. How?

Scientists speculate that melting glaciers and increased rainfall in the northern latitudes could infuse the ocean with less dense, cool fresh water that could shut down the conveyer belt. The cool less-dense fresh water would essentially put a halt to an essential driving force of the global conveyor belt—the sinking of dense, cool, salt water. Studies of ocean currents and climate suggest that such changes could occur abruptly, and soon, and may take decades, even centuries, to reverse. The result? Areas like Europe could become much colder, while the equator becomes hotter. As you shall soon see, scientists have already detected significant changes in this system, which has caused considerable concern in Europe and stimulated major efforts to curb greenhouse gases.

National and International Security

As the previous material illustrates, climate change could have extremely serious economic and environmental consequences throughout the world. Over the next 20 years, the U.S. Department of Defense noted in a classified report, it could result in a global catastrophe that would cost millions of lives through warfare and recurrent natural disasters. The report, for example, warns that major European cities could sink beneath rising seas as Great Britain is plunged into a Siberian climate as early as 2020. It also warns of nuclear conflict, serious droughts, famine, and widespread civil unrest that could erupt throughout the world.

The document, suppressed by the Bush Administration, predicts that abrupt climate change could cause many nations to develop nuclear weapons to defend and secure dwindling food, water, and energy supplies. Some experts who have viewed the classified document argue that the threat of terrorism is currently eclipsed by the potential global instability that could result from global climate change. "Disruption and conflict will be endemic features of life," concludes the Pentagon. "Once again, warfare would define human life." Unfortunately, the Bush Administration has repeatedly denied that climate change even exists. Climate change "should be elevated beyond a scientific debate to a US national security concern," say the authors, Peter Schwartz, CIA consultant and former head of planning at Royal Dutch/Shell Group, and Doug Randall of the California-based Global Business Network.

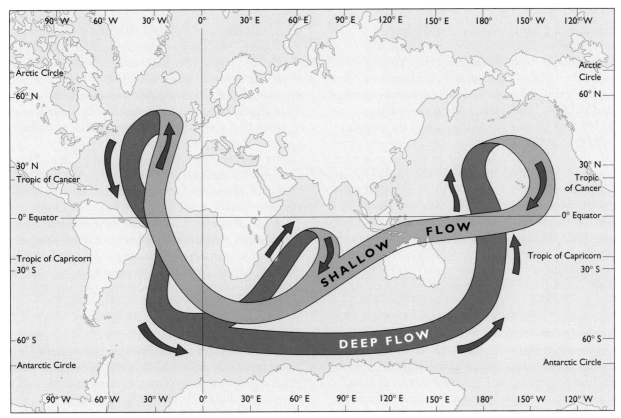

FIGURE 20-22 Surface and deep water exchange.

Is Global Climate Change Occurring?

Many of the predictions of the global climate models and scientists are coming true. As noted earlier, sea level is on the rise, in part due to melting of glaciers in mountainous areas and in Antarctica. Huge blocks of ice, often the size of small New England states, are breaking off the Antarctic ice sheet, signaling a melting and break-up of this massive expanse of ice. Recent studies of Arctic polar ice show a dramatic decrease as well. A research ship once took 2 years to traverse the North Pole; in 2000, the ship made the trip in a few months because of the lack of ice. Over a period of 2 years, a huge ice shelf that has extended into the Arctic ocean for the past 3000 years has broken up, threatening polar bear habitat. Sea temperatures are rising, too. And violent storms are on the rise. Studies of global carbon dioxide levels show a definite increase, along with it an increase in global temperature. Record temperatures are now becoming very common. Seventeen of the hottest years in the past 100 years have occurred since 1980. Although atmospheric temperatures vary from year to year and heat waves are not uncommon, scientists point out that the probability of this many record-setting years in such a short period is very small.

Heat and drought are taking their toll on agriculture worldwide, causing severe economic problems in some areas. In 1999, severe drought plagued many parts of the United States. Heat waves are also responsible for numerous deaths.

In 1998, 455 Americans died as a result of sweltering summer heat waves. Droughts and record-breaking heat waves continued into the first half of the 2000s throughout the world. In 2003, 20,000 people died in France alone as a result of a record-breaking heat wave. Scientists who have studied ocean currents for decades are reporting major changes that indicate that the global conveyor belt in the Atlantic is slowing down, reducing the distribution of heat from warm equatorial regions to northern climates. This could plunge Great Britian into an ice age. Studies show that this slowdown is caused by a dramatic increase in fresh water flowing into the North Atlantic as a result of melting glaciers and increased rainfall in northern Russia.

These and other problems have prompted many people to take serious action to prevent, even reverse, the build-up of greenhouse gases. In June 1999, for example, President Clinton signed an executive order committing the federal government to cut greenhouse gas emissions by 30% through energy-efficiency and other conservation measures. Because the U.S. government is the single largest energy consumer in the United States, such actions could have a profound effect on efforts to slash greenhouse gases. Concern for global climate change has also spurred two major oil companies to voluntarily pledge to cut their emissions of carbon dioxide. The Bush administration, however, and many influential corporations like Exxon Mobil steadfastly oppose any actions to halt global warming.

Uncertainties: What We Don't Know

Although there is growing evidence that the Earth is warming up and that it is affecting us in many ways, there are some uncertainties worth considering. For example, studies of global climate models show that the improved computer programs do a very good job of predicting current temperature and weather. That is, when scientists feed in data on current levels of various greenhouse gases, the computer programs predict the global temperature and weather patterns fairly accurately. The models lose their accuracy on a regional scale, though. In regions smaller than 2400 kilometers (1500 miles), they are not able to predict with great accuracy because of local factors that influence climate.

Some scientists believe that the projections of global warming by the IPCC are somewhat conservative. They do not take into account some factors that could result in a rapid deterioration of climatic conditions. They believe that conditions could deteriorate more quickly than anticipated because changes now apparently in motion may stimulate dangerous positive feedbacks (FIGURE 20-23). A **positive feedback** occurs when one factor leads to an increase in another. In nature, positive feedback mechanisms are rare. Homeostatic mechanisms, discussed in Chapter 6, tend to operate through negative feedback mechanisms that maintain more or less constant conditions. In natural systems altered by humans, positive feedback mechanisms are dangerously common. To understand this phenomenon, consider some examples.

The oceans currently serve as a major reservoir for carbon dioxide. In fact, they store 60% more CO_2 than the atmosphere does. Without the oceans, carbon dioxide levels in the air would be considerably higher than they are today. As the Earth's temperature rises, however, the ocean's ability to dissolve and hold carbon dioxide may decline (Figure 20-23). If this happened, the oceans would release some of the CO_2 they have already absorbed. This would cause atmospheric carbon dioxide levels to increase more rapidly, possibly accelerating the warming trend. This, in turn, would further reduce the amount of carbon dioxide held in ocean waters, creating a dangerous positive feedback cycle. New research suggests that rising temperatures may be reducing the ability of oceans to hold carbon dioxide.

Several additional positive feedback mechanisms may also be initiated. Melting glaciers, for instance, decrease the reflective surface of the Earth, resulting in an increase in the absorption of sunlight. This would result in a further warming of the atmosphere, for reasons explained at the beginning of this section. Rising temperatures would accelerate the melting of glaciers and would also increase the amount of energy people use to cool their houses—which adds more CO_2 to the atmosphere, once again accelerating the rise in temperature.

As noted earlier, the rise in atmospheric carbon dioxide concentrations is partly the result of deforestation—the rapid loss of trees, especially in the tropics, but also in parts of the United States, such as the Pacific Northwest. Trees absorb carbon dioxide, which they use to produce food molecules and structural tissues. Worldwide, forests are being cut much faster than they regrow. As a result, deforestation "contributes" approximately one fourth of the annual global increase in carbon dioxide and, as noted in Table 20-2, about 13% of the annual rise in global temperature. Trees may also be destroyed by climatic change (rising temperature and drought) and by fires. This loss could also become part of a wildly accelerating positive feedback mechanism that causes a rapid increase in CO_2 in the atmosphere and a more rapid shift in global temperature.

Several factors may counteract these forces, however, either negating the climb in temperature or lessening it. The oceans, for instance, tend to act as a temperature buffer, keeping temperatures from rising as rapidly. In addition, rising temperatures may result in an increase in cloudiness. This could reduce sunlight penetration and hence reduce the climb in global temperature. Increased particulates from denuded, parched lands could also reflect incident solar energy.

Current evidence suggests that the depletion of the ozone layer will have a cooling effect on global climate. Other information suggests an adjustment in the other direction. A study of 17 computer models published in *Science* in 1991 also suggests that global warming may affect snowfall and average temperature in unanticipated ways. As a rule, climate modelers think that global warming will result in a net decrease in planetary snow cover. This, they predict, would increase absorption and increase global warming. However, five of the world's best climate models show that a reduction in snow cover would cool the atmosphere and counteract a small part of the greenhouse warming. The cooling effect may result from an increase in cloud cover or

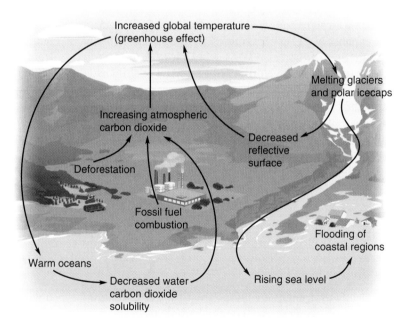

FIGURE 20-23 **Dangerous feedback.** If the planet continues to warm, many changes could occur. Some could accelerate the increase in atmospheric carbon dioxide levels and result in an acceleration in the rate of global warming.

an increase in heat radiation into space. In sum, many climate modelers acknowledge that current models don't do a very good job of simulating cloud behavior.

KEY CONCEPTS

Warming of the world's oceans, melting of land-based ice, and loss of forests may result in a rapid increase in global carbon dioxide levels, causing an accelerated planetary warming. Other factors, however, may offset these changes.

Solving a Problem in a Climate of Uncertainty: Weighing Risks and Benefits

Global climate is a very complex system, and our understanding, while growing, is still less than complete. Many uncertainties still exist with regard to the behavior of the CO_2 sinks such as the oceans and forests. Future levels of greenhouse gases and global warming activities such as deforestation are unknown.

With this much uncertainty, some people prefer to take a wait-and-see approach. Many oil companies, coal companies, and the Bush Administration oppose measures to cut greenhouse gas emissions, arguing that the threat is not real and that such changes will hurt the nation's economy. Critics of this viewpoint argue that the stakes are too high and that the economic and environmental benefits of cutting fossil fuel combustion are so significant that we should make the conversion despite the uncertainty. In short, the cost of accepting the greenhouse hypothesis and acting accordingly is far lower than the potential cost of rejecting it. Today, many nations and progressive companies like British Petroleum are taking steps to reduce CO_2 emissions. Germany recently announced plans to develop a 100% renewable energy economy. Great Britain has pledged to reduce its CO_2 emissions by 60%. Meanwhile, the U.S. government remains steadfastly opposed to such actions. In fact, the Bush Administration ruled that CO_2 cannot be regulated, which will make any measures to reduce this greenhouse gas difficult at best. President Bush established a voluntary CO_2 emissions program, but as of 2004, only a handful of companies (54) have joined.

KEY CONCEPTS

Considerable uncertainty exists on the global climate change issue, which has tended to slow progress toward solutions. Even though there are many unanswered questions, many scientists believe that the costs of reducing or even eliminating greenhouse gases are outweighed by the potential social, economic, and environmental costs of global climate change.

Solving the Problem Sustainably

Humans produce more than 7.0 billion tons of carbon (in the form of carbon dioxide) a year—40% from the less developed nations and 60% from the more developed industrial nations. To create a sustainable society, free from the impacts of global warming, one prestigious group of scientists known as the Intergovernmental Panel on Climate Change, believes that global carbon dioxide emissions will need to be cut by about two thirds, down to about 2 billion tons per year. Unfortunately, the world is poised to increase emissions. Emissions from the less developed nations are projected to quadruple in the next half century. In fact, by 2020, the less developed nations could be producing as much CO_2 as the industrial nations. Emissions from the more developed nations are projected to increase by 30%.

> Globally, humans produce more than 7.0 billion tons of carbon (in the form of carbon dioxide) a year. To create a sustainable society, the Intergovernmental Panel on Climate Change believes that global carbon dioxide emissions will need to be cut by two thirds, down to about 2 billion tons per year.

Further adding to the dilemma, tropical deforestation and burning of forests could greatly accelerate carbon dioxide emissions and global climate change. The fires occurring in Asia in 1997 and 1998 released more carbon dioxide than western Europe produces in a year. These fires were intentionally ignited in many cases. In others, however, extraordinarily dry rain forests caught fire in lightning storms. The dry conditions were caused by deforestation in upwind areas, which tends to reduce rainfall, a phenomenon discussed in Chapter 12.

Global warming is ultimately a symptom of overpopulation and unsustainable human systems and technologies that are designed to meet the needs of people. Especially noteworthy in the group of human systems are transportation, housing, industry, and timber production. The energy system, for example, relies heavily on fossil fuel, the combustion of which produces carbon dioxide in excess of the planet's ability to absorb it.

The good news is that numerous solutions are available to reduce or even eliminate greenhouse gases. Bans on ozone-depleting and greenhouse-enhancing CFCs, discussed earlier in the chapter, are one example. Other sustainable solutions, those that strike at the roots of the problems, include measures to reduce population growth and eliminate the harmful emissions of carbon dioxide, methane, CFCs, and nitrous oxide. This section discusses some of the most effective measures. Not surprisingly, the most effective approaches turn out to be applications of the operating principles of sustainability discussed in earlier chapters: conservation, recycling, renewable resource use, restoration, and population control.

KEY CONCEPTS

Redesigning human systems according to sound principles of sustainability could help alleviate the problem of global warming.

Population Stabilization and Restoration To reduce atmospheric carbon dioxide levels and stave off or stop the possible increase in global temperature will require a massive reforestation of the Earth. Australia recently announced that it was embarking on an ambitious program to plant one billion trees, partly to offset global warming. A few other countries are following suit, among them China, although efforts are well below that required to have any appreciable effect.

Norman Myers, an international expert on tropical forests, argues that replanting 2.6 million square kilometers (1 million square miles) of tropical rain forest would reduce annual emissions of carbon dioxide by 2.25 billion metric tons (2.5 billion tons), or about 41%. Although this ambitious project would cost approximately $100 billion, Myers argues that it is a small price to pay, especially when one takes into account the potential economic damage caused by global climate change. A bad hurricane can easily cause several billion dollars worth of damage if it hits a metropolitan area. According to the EPA, protecting the U.S. East Coast from rising sea waters could cost the nation $75 to $110 billion. Global warming would necessitate costly modifications of irrigation systems and hydroelectric dams estimated at $100 bil-

> A family of four would need to plant 2.5 hectares (6 acres) of fast-growing trees to offset its lifetime carbon dioxide production

lion. Replanting could also bring direct economic benefits to less developed nations through reduced soil erosion and sustainable harvest of forests. (See Spotlight on Sustainable Development 20-1 for a discussion of a company's efforts to replant trees.)

Individuals can help plant trees in clear-cuts, roadsides, abandoned fields, and backyards. If you really want to offset the carbon dioxide produced by your lifestyle, you would have to plant 400 trees. A family of four would need to plant 2.5 hectares (6 acres) of fast-growing trees to offset its lifetime carbon dioxide production. Obviously, not everyone can replant trees, but you can help by supporting public and private reforestation projects.

KEY CONCEPTS

Population stabilization and attrition can go a long way in reducing our demand for fossil fuels and other greenhouse-enhancing activities such as deforestation. Restoring forests, especially in the tropics, and other carbon sinks could have a profound effect on global carbon dioxide levels and global climate.

Recycling, Energy Efficiency, and Renewable Energy   Individuals and businesses throughout the world can also help reduce global carbon dioxide levels by recycling and reusing

---

>> SPOTLIGHT ON SUSTAINABLE DEVELOPMENT

20-1   Offsetting Global Warming: Planting a Seed

Progressive companies throughout the world are taking steps to reduce the rise in greenhouse gases through energy efficiency and other measures. One leader in this effort is Applied Energy Services (AES) of Arlington, Virginia, which in 1988 announced plans to help finance the planting of 50 million trees in Guatemala (FIGURE 1). The company is all

FIGURE 1 Trees being planted in Guatemala.

the more remarkable because it is the first utility company ever known to take direct action to offset carbon dioxide emissions of fossil fuel power plants it planned to build.

AES currently owns or has part ownership in six power plants in the United States and Great Britain. When it decided to build two new ones, its chief executive officer, Roger Sant, decided that they ought to do something about emissions. "Given the scientific consensus on the seriousness of the greenhouse problem, we decided it was time to stop talking and act," Sant remarked. AES also wanted the project to confer additional benefits—for example, improving the economic conditions of communities currently affected by deforestation and preserving endangered plant and animal species.

AES contacted the World Resources Institute (WRI) for suggestions. WRI found an ongoing project in Guatemala sponsored by CARE, Inc., an international relief and development agency. The CARE Guatemala Agroforestry Project was created to replant rain forests to reduce soil erosion and flooding. Over a 10-year period, the agency (now with financial support from AES) hopes to plant enough trees to remove 15 million metric tons of carbon dioxide (16.5 million tons) from the atmosphere, approximately the same amount that AES's planned 180-megawatt coal-fired plant in Connecticut will emit over its 40-year life. The trees will also provide a sus-

all materials to the maximum extent possible. Table 20-3 shows a comparison of energy use in manufacturing cans and bottles. It shows that recycled glass and aluminum require far less energy to produce—and thus produce less carbon dioxide air pollution—than if they were made from raw ore and used only once. Refillable bottles use much less energy even if they are only used 10 times. (For more on recycling and reuse, see Chapter 23.)

Energy efficiency is another means of reducing carbon dioxide and methane emissions. According to energy experts Amory Lovins of the Rocky Mountain Institute and Christopher Flavin of the Worldwatch Institute, global emissions could be cut by 2.7 billion metric tons (3 billion tons) per year within two decades by cost-effective and profitable technologies. No technical breakthroughs are needed, either. Combined with reforestation efforts outlined above, these efforts could nearly eliminate anthropogenic global carbon dioxide emissions.

Individuals can help cut energy demand by walking, bicycling, or riding a bus to school or work; by building smaller, more energy-efficient homes; by insulating existing homes; by recycling; and by using efficient appliances or doing some things by hand (for example, mixing by hand rather than using an electric mixer, or drying clothes on a line rather than

Table 20-3	
Energy Consumption per Use for 12-Ounce Beverage Containers	
Container	**Energy Use (BTUs)**
Aluminum can, used once	7050
Steel can, used once	5950
Recycled steel can	3880
Glass beer bottle, used once	3730
Recycled aluminum can	2550
Recycled glass beer bottle	2530
Refillable glass bottle, used 10 times	610

Source: From L.R. Brown, C. Flavin, and S. Postel (1991). *Saving the Planet: How to Shape an Environmentally Sustainable Economy.* New York: Norton.

using a dryer). When the time comes to buy appliances, choose the most energy-efficient ones available. New energy-efficient refrigerators, the leading user of electricity in our homes, can cut electrical demand from 1200 kilowatt-hours per year to 240. (See Chapter 15 for more ideas on energy efficiency.)

tainable harvest of fruit, lumber, and fuelwood, and will help reduce soil erosion and flooding.

The price tag for this project was substantial. AES provided $2 million. CARE, the U.S. Agency for International Development, and the government of Guatemala each donated $2 million. Peace Corps volunteers planted the trees. Their service and training was estimated to be valued at $7.5 million.

Some critics think that American companies would be more likely to invest in domestic tree planting. One approach might be to plant trees on marginally productive, highly erodible farmland taken out of use in the United States to reduce soil erosion. Instead of the government's subsidizing the planting of trees, power companies could do it, says Daniel Dudek of the Environmental Defense Fund. His proposal calls for utilities to lease the land currently being set aside by farmers (with federal assistance) and plant fast-growing trees on it. The cost would be about 70 cents per ton of carbon dioxide emissions removed. A utility wishing to offset emissions from a 1000-megawatt coal-fired power plant would need to plant a forest with a 48-kilometer (30-mile) diameter.

AES is a model of sustainable business. This innovative power company is values-driven rather than being competition-driven solely for profit. Carrying its mission further, AES has also taken steps to offset pollution from a coal-fired power plant on Oahu, Hawaii, that came on line

in 1992. Instead of planting trees, though, AES has taken actions to protect a forest in Paraguay from destruction, halting the release of millions of tons of carbon sequestered in the trees. AES agreed to donate up to $2 million for the purchase and preservation of the 58,000-hectare (143,000-acre) Mbaracayu forest. Located in eastern Paraguay, the Mbaracayu is one of South America's few remaining tracts of dense, humid sub-tropical forest. The area is vital to the survival of many species. It includes 19 distinct plant communities, at least 300 bird species, and threatened and endangered animals including tapirs, jaguars, giant armadillos, peccaries, the rare bush dog, king vultures, and macaws. The forest is the traditional hunting and gathering area for the Ache tribe. AES is partner in this project with the Nature Conservancy and a Paraguayan group that promotes sustainable development.

AES plans to continue its carbon-offset programs with power plants in Oklahoma and Florida. Tree planting alone will not save us from global climate change. Current emissions from fossil fuel combustion are far too great for this strategy to work by itself. Combined with measures to reduce deforestation and slow the growth of the human population—plus energy efficiency, recycling, and renewable energy technologies—replanting can play an important part in reversing the dangerous global climate change now under way.

Renewable energy technologies can also dramatically reduce carbon dioxide and methane emissions, as pointed out in Chapter 15. Many options are available, and several are quite cost competitive.

Recycling and energy efficiency greatly reduce energy demand and cut greenhouse gas emissions. Renewable energy technologies can provide us with much-needed power, with little or no impact on global climate.

International Cooperation to Halt Global Warming   This chapter opened with a discussion of the Earth Summit, the highly acclaimed global conference that addressed such pressing problems as forest protection and global warming, among others. One outcome of the meeting was an agreement calling on nations to reduce carbon dioxide emissions. The nations also signed an agreement to protect forests. Details of these agreements are outlined in Chapter 27.

Although both agreements are important, they fall seriously short of the job at hand. Further, the global climate agreement is largely voluntary. Making matters worse, many U.S. industries that initially supported the global climate treaty reneged on their support and stonewalled efforts to curb greenhouse gas emissions. By 1997, little progress had been made in reducing CO_2 levels. During that year, the United Nations sponsored a meeting in Kyoto, Japan, to negotiate an international agreement on global climate change. Known as the 1997 **Kyoto Protocol** and signed by 84 nations, this agreement calls for a slow, steady decline in greenhouse gas emissions. It went into effect in 2005 when Russia signed on (the United States did not). This agreement commits the industrial nations and the former Eastern bloc nations to cut carbon dioxide and other greenhouse gas emissions 5.2% below 1990 levels between 2008 and 2012. The signatories have met twice since then, but so far success in achieving these goals has been poor. Many countries have experienced increases in greenhouse gas emissions of 9 to 12%, among them Japan, the United States, and Australia. The fastest growth in greenhouse emissions has occurred in the less developed nations, with growth rates of about 40%, in large part due to industrialization. There is some good news, however, notably in the United Kingdom, France, and Germany, where greenhouse gas emissions are falling as a result of energy-efficiency policies. Russia and many members of the former Soviet Union have also recorded outstanding declines in carbon dioxide emission, due to the collapse of their economies following the dissolution of the USSR.

Many observers believe that, as was the case with the first treaty on ozone protection (the Montreal Protocol), nations will soon see the need to take more drastic steps to curb or even stop the rise in greenhouse gas, which the vast majority of the world's atmospheric scientists believe is responsible for record-hot temperatures in the past two decades and a host of other changes, including a rise in sea level. Continuing bad news—hot years, devastating floods, and violent storms—could spur further agreements that eventually lead us to a more sustainable system of energy production and a climate we can all live with.

Solving the problem of global warming will require the efforts of all sectors of society and virtually every country on Earth.

You cannot escape the responsibility of tomorrow by evading it today.

—Abraham Lincoln

CRITICAL THINKING

Exercise Analysis

Critical thinking rules encourage us to question sources of information and their conclusions. In this example, we find that the author of this article is grossly in error when he claims that the scientific community is divided on the issue. The key word is *divided*, which makes it sound as if there's a 50–50 split of opinion.

In truth, about 99% of the United States' 700 atmospheric scientists believe that global warming is a reality; only a handful embrace the opposite view. The evidence the author introduces to support his assertion—a quote from each side of the issue—is terribly misleading.

This type of reporting is quite common in newspapers, television, and magazines. In journalism, it is customary to get both sides of the issue, but one must not mistake the notion of two sides for an equal split of opinion. Statements such as the one we're analyzing here, although intended to give a balanced view of issues, in reality provide an extremely unbalanced view.

What would have been a more accurate statement? The author might have noted opposing views but also noted that, although not all scientists agree that global warming is occurring, the vast majority do.

That said, it is still important to note that even though the majority of the world's atmospheric scientists agree that global warming is happening, it doesn't mean they're right. For many centuries, scientists held that the Earth was at the center of the solar system.

This exercise shows that digging a little deeper and finding out more often throws conclusions into question.

CRITICAL THINKING AND CONCEPTS REVIEW

1. What is the ozone layer? Why is it important to life on Earth?

2. Describe the major activities and pollutants that deplete the ozone layer.

3. What measures have been taken to reduce the destruction of the ozone layer? Will they reverse the decline quickly? Why or why not?

4. HCFCs, replacements for CFCs, also deplete the ozone layer, but to a lesser degree. Consequently, they are considered interim solutions. What other reason accounts for their interim-solution status?

5. Some critics of the environmental movement suggest that stratospheric ozone depletion is a hoax. Look up some of their writings and, if possible, the work on which they base their claims. What do you find? Are the conclusions based on good scientific evidence? With your knowledge of the issue, debate the main points they make.

6. Critically analyze the following statement: "We must balance the harm done by ozone depletion and maintaining CFCs for refrigeration with the potential loss of refrigeration in developing countries, which could result in many lost lives due to foodborne disease."

7. Define the terms *acid deposition, acid precursor, wet deposition*, and *dry deposition*.

8. How are acid precursors formed?

9. Using your critical thinking skills, analyze the following statement: "Natural events such as volcanoes produce far more acid precursors than human activities, so we shouldn't worry about controlling anthropogenic sources. Such controls won't do anything to affect acid levels."

10. Using your critical thinking skills and your knowledge of biology, how would you assess current efforts aimed at reducing acid deposition or counteracting its impact (for example, smokestack scrubbers, liming lakes, and breeding acid-resistant fish)?

11. Outline key elements of a sustainable strategy for reducing acid deposition.

12. What is the greenhouse effect? What gases are responsible for it? Which ones are natural? Which ones are anthropogenic?

13. Describe the potential social, economic, and environmental impacts of the continued rise of greenhouse gases. Describe how environmental impacts could lead to social and economic impacts.

14. Describe the factors that could accelerate greenhouse warming and those that may lessen it.

15. Outline the short-term and long-term solutions that have been proposed for addressing global climate change and critically analyze each one. Which ones would be the most effective? Which ones would be the most politically acceptable?

16. Compare the difference between heeding warnings about global warming and ignoring them. What action do you recommend? Why?

KEY TERMS

acid deposition
acid precursors
buffers
buffering capacity
chlorine free radicals
chlorofluorocarbons
dry deposition
Freon-11

Freon-12
global climate change
global climate models
global warming
greenhouse effect
greenhouse gases
Kyoto Protocol

Montreal Protocol
National Acid Precipitation Act
National Acid Precipitation
    Assessment Program (NAPAP)
pH (potential hydrogen)
positive feedback
wet deposition

REFERENCES AND FURTHER READING

The References and Further Reading section at the end of this book contains a list of sources for the information discussed in this chapter and recommendations for further reading.

Connect to this book's website:
http://environment.jbpub.com/
The site features eLearning, an online review area that provides quizzes, chapter outlines, and other tools to help you study for your class. You can also follow useful links for in-depth information, research the differing views in the Point/Counterpoints, or keep up on the latest environmental news.

Water pollution in rich and poor countries of the world affects our health and economy.

CHAPTER 21

Water Pollution: Sustainably Managing a Renewable Resource

It is a crime to catch a fish in some lakes, and a miracle in others.

—*Evan Esar*

Author Loren Eiseley once wrote, "If there is magic in this planet, it is in water." Water has a practical side, too. Water covers 70% of the Earth's surface and makes up two thirds or more of the weight of most animals and up to 95% of the weight of plants. It plays an important role in metabolism, the chemical reactions in the cells of the bodies of all organisms. It is vital to agriculture and industry, too. Despite its crucial role in our lives, water is one of the most badly abused resources. Chapter 13, for example, described unsustainable withdrawals of groundwater and surface water. Chapter 13 also showed how pollution of estuaries was destroying food sources for humans and other species. This chapter covers water pollutants: where they come from, how they affect living organisms, and finally, ways to reduce them. Like other chapters, this one outlines some components of a sustainable response to this problem.

> Water covers 70% of the Earth's surface and makes up two thirds of the weight of most animals and 95% of the weight of plants.

21.1 Surface Water Pollution

Water exists in many forms. It can be found in the atmosphere as mist or clouds. It can be locked in glaciers as solid ice. It can be found on the Earth's surface in ponds, lakes, streams, and rivers, which are collectively referred to as **surface waters.** Chapter 5 examined some features of surface waters that will help you understand the effects water pollution has on lakes, rivers, and streams. You may want to review them now before you begin your study of water pollution. Water is also stored underground in aquifers. Such deposits are part of the planet's vast **groundwater** holdings discussed in Chapter 13. This chapter deals with water pollution occurring in groundwater and surface waters.

Water pollution is any physical or chemical change in water that adversely affects organisms. Like many other problems, it is global in scope, but the types of pollution vary according to a country's level of development. In poor, nonindustrialized nations, water pollution is predominantly caused by human and animal wastes, pathogenic organisms from this waste, pesticides, and sediment from unsound farming and timber practices. The rich industrial nations also suffer from these problems, but their more opulent lifestyles and numerous industries create an additional assortment of potentially hazardous pollutants: heat, toxic metals, acids, pesticides, organic chemicals, and an assortment of pharmaceuticals (excreted in peoples' urine). In between these two extremes are numerous countries with various levels of industrialization. They often have inadequate laws—or none at all—to combat water pollution, or if they do have good pollution control laws, they frequently lack ad-

equate funding to enforce them. Their waters are therefore often badly polluted with an assortment of industrial and municipal wastes.

> **KEY CONCEPTS**
>
> Nations produce many different types of water pollutants. The water in all countries is plagued with pollutants from human and animal wastes, but in the industrial nations numerous toxic chemicals also contribute to water pollution.

Sources of Water Pollution

Like air pollutants, water pollutants come from numerous sources, both natural and anthropogenic. As a rule, anthropogenic sources are the most important because they tend to be localized and thus contribute significantly to the deterioration of local waterways or groundwater.

Like air pollutants, water pollutants respect no boundaries. Pollutants produced in one city, for example, may affect water quality in another city located downstream. They may even flow from one country to another, ending up in the water supply of downstream neighbors, or they may flow into oceans and seas. For example, for many years the rivers flowing into the Mediterranean Sea and the sea itself were viewed as a convenient dump for domestic and industrial wastes. Along Italy's coast, the waters became a cesspool. Fortunately, internal actions and negotiations among the nations that border the Mediterranean have spawned a region-wide plan that is beginning to reverse the years of decline.

Pollutants also enter surface waters from the air and surrounding land. This problem is called **cross-media contamination** and refers specifically to the movement of a pollutant from one medium (such as air) to another (such as water). Pesticides sprayed on crops, for example, may evaporate, becoming airborne. They may then be transported in the air to nearby lakes. They may then flow to the oceans. Volatile organic chemicals from factories may evaporate and later be washed out of the sky by rains. Hazardous wastes buried in the ground may leak into aquifers whose waters replenish streams.

> **KEY CONCEPTS**
>
> Water pollutants arise from natural and anthropogenic sources. They travel freely from one location to another through rivers, streams, and groundwater. They are also transported from one medium (land or air) to another (water).

Point and Nonpoint Sources

When we ponder the sources of water pollution, we generally think of factories, power plants, mines, oil wells, and sewage treatment plants that release tons of potentially toxic chemicals into sewers and lakes and rivers (**FIGURE 21-1a**). These **point sources,** so named because they are discrete locations, are relatively easy to identify and control. Although they are obvious sources for which controls can be effected, pollution control can be costly. Abandoned metal mines in Canada's Arctic region, for example, leak a variety of pollu-

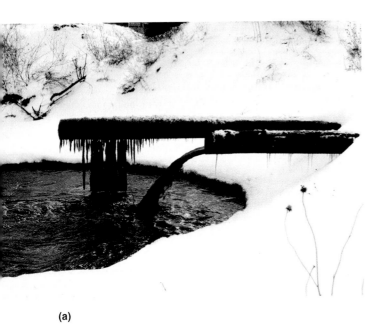

(a)

FIGURE 21-1 **Water pollution sources. (a)** This pipe dumping factory waste into a nearby river is an example of a point source. **(b)** Urban streets, lawns, and gardens are nonpoint sources. Heavy rains often wash pollutants from these sources into nearby water ways.

(b)

tants (mostly heavy metals) into nearby streams and could cost billions of dollars to clean up. Active and inactive oil wells in the Arctic region of the United States and Canada are equally troublesome from an environmental and economic standpoint.

In the past decade, scientists have come to realize that point sources are only half the problem. The other half are the **nonpoint sources**—less discrete sites such as farms, forests, lawns, and urban streets (**FIGURE 21-1b**). Rainwater washes pollutants from these sources into nearby streams, rivers and lakes (Table 21-1).

Together point and nonpoint sources can result in huge problems. Many Canadian waters are still polluted by direct discharge of pollutants from industrial and municipal point sources and by indirect discharge from a long list of nonpoint sources. The Frazer River in Canada, for instance, receives incompletely treated sewage and a variety of toxic chemicals

from wood treatment and wood pulp and paper mills. It is polluted by chemical leaching from nearby landfills. Rivers in the prairie provinces are polluted by agricultural runoff. The Great Lakes are contaminated by industrial and municipal point sources and by a host of nonpoint sources as well.

In the United States, nonpoint sources release nearly two thirds of the pollutants that end up in our waterways (**FIGURE 21-2**). These substances include dust, sediment, pesticides, asbestos, fertilizers, heavy metals, salts, oil, grease, litter, and even air pollutants washed from the sky by rain. Because the sources are many and spread out, control has proven difficult. One reason is that it is far easier for governments to force a company to install a pollution con-

Table 21-1	
Major Nonpoint Pollution Sources in the United States	
Activity	**Explanation**
Silviculture	Growing and harvesting trees for lumber and paper production can produce large quantities of sediment.
Agriculture	Disruption of natural vegetation leads to increased erosion; pesticide and fertilizer use, coupled with poor land management, can pollute neighboring surface water and groundwater.
Mining	Leaching from mine wastes and drainage from mines themselves can pollute surface and groundwater with metals and acids; disruption of natural vegetation accelerates sediment erosion.
Construction	Road and building construction disrupts vegetation and increases sediment erosion.
Salt use and ground-water overuse	Salt from roads and storage piles can pollute groundwater and surface water; saltwater intrusion from groundwater overdraft pollutes ground and surface water.
Drilling and waste disposal	Injection wells for waste disposal, septic tanks, hazardous waste dumps, and landfills for municipal garbage can contaminate groundwater.
Hydrological modification	Dam construction and diversion of water both can pollute surface waters.
Urban runoff	Pesticides, herbicides, and fertilizers applied to lawns, as well as residues from roads, can be washed into surface waters by rain.

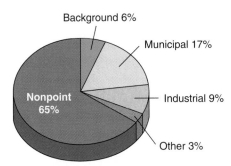

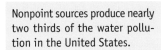

FIGURE 21-2 **Major sources of U.S. stream pollution.** This pie chart shows the contribution of point and nonpoint sources to total water pollution in the United States. Municipal, Industrial, and Other are point sources. Nonpoint sources are by far the greatest polluter of U.S. waters. (Background is natural pollution.)

trol device on a factory than to try to force tens of thousands of city dwellers to get oil leaks in cars fixed or to apply pesticides to lawns more conservatively. Although there are numerous sources of nonpoint water pollution, **FIGURE 21-3** shows that agriculture is the predominant nonpoint source in the United States, affecting nearly 60% of the nation's streams.

> Nonpoint sources produce nearly two thirds of the water pollution in the United States.

The remaining portion of this section covers the major types of point and nonpoint water pollution.

KEY CONCEPTS

Water pollution arises from identifiable point sources, such as factories, and from diffuse nonpoint sources, such as farm fields and streets. Point sources are much easier to control.

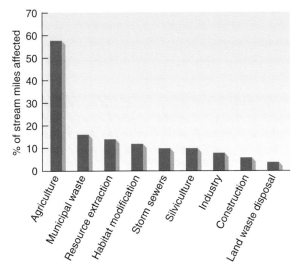

FIGURE 21-3 **Sources of nonpoint water pollution affecting streams.** This bar graph lists nonpoint water pollution sources in the United States by their impact—that is, the percentage of miles of streams they affect in the United States. Agricultural activities and municipal wastes (sewage) turn out to be the number one and two contributors.

Organic Nutrients

Rivers, streams, and lakes contain both organic and inorganic pollutants that are nutrients for plants and other organisms. Let's consider each one.

In more developed countries, large quantities of organic chemicals are released into surface waters from point sources such as feedlots, sewage treatment plants, and various industries such as paper mills and meat-packing plants.[1] In many cases, these wastes are treated prior to release into streams. Even as such, large quantities of organic waste still enter surface waters. Flooding can also cause severe problems. In 1999, for example, a series of hurricanes and tropical storms hit the southeastern United States. In North Carolina, 20 municipal sewage treatment plants were flooded, releasing enormous amounts of untreated human waste into streams and other water bodies. Making matters worse, many of the state's 4000 hog farms were flooded out by heavy persistent rainfall. Waste on the ground and in lagoons, ponds where the waste is held for treatment, flowed into streams (**FIGURE 21-4**). Hundreds of millions of gallons of hog wastes poured into wetlands, creeks, and rivers.

[1]An organic compound consists primarily of carbon and hydrogen atoms.

FIGURE 21-4 Hundreds of hog farms, many built in flood plains like this one were flooded by Hurricane Floyd in 1999, drowning tens of thousands of animals. Sewage lagoons overflowed and waste from the thousands of farms flowed into rivers, lakes, and the ocean.

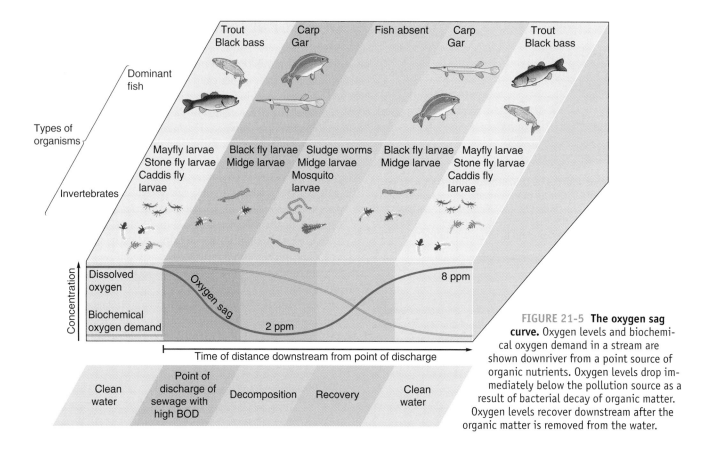

FIGURE 21-5 **The oxygen sag curve.** Oxygen levels and biochemical oxygen demand in a stream are shown downriver from a point source of organic nutrients. Oxygen levels drop immediately below the pollution source as a result of bacterial decay of organic matter. Oxygen levels recover downstream after the organic matter is removed from the water.

In less developed countries, organic pollutants come primarily from human sewage and animal wastes. Raw sewage often enters water without treatment.

In surface waters, organic pollutants are consumed by naturally occurring bacteria. With an abundant food source, the bacterial populations proliferate. As bacteria consume the organic matter, they help to purify the waters and are therefore part of nature's system of restoration. Unfortunately, there's a catch.

The degradation of organic pollutants by bacteria utilizes oxygen, which is dissolved in water (**FIGURE 21-5**). Excess organic matter therefore causes oxygen levels to drop; as a result, fish and other aquatic organisms may perish. When oxygen levels become very low, anaerobic (non–oxygen-requiring) bacteria take over, breaking down what's left. In the process, though, they produce foul-smelling and toxic gases (methane and hydrogen sulfide).

Oxygen depletion in rivers and streams occurs more readily in the hot summer months because stream flow is generally low and organic pollutant concentrations are higher. In addition, increased water temperatures speed up bacterial decay.

Because streams are flowing systems, oxygen levels may return to normal if the pollution is a one-time event. However, when numerous sources of organic pollutants are found along the course of a river—or when organic input exceeds the ability of the aquatic system to restore conditions through bacterial decomposition and water flow—recovery may be impossible. Lakes can also recover from organic pollutants,

but usually much more slowly than rivers, for reasons explained earlier.

The organic nutrient concentration in streams is estimated by a test that determines the rate at which oxygen is depleted from a test sample. In this test, polluted water is saturated with oxygen and held in a closed bottle for 5 days; during this period, bacteria in the water degrade the organic matter and consume the oxygen that was added to it. The amount of oxygen remaining after 5 days gives an indication of the organic matter present; the more polluted the sample, the less oxygen is left. This standard measurement is called the **biochemical oxygen demand**, or **BOD**.

KEY CONCEPTS

Organic nutrients come from a variety of sources, primarily treated and untreated waste (human and other animals) accidentally and intentionally released into waterways. Organic compounds stimulate bacterial growth, which depletes oxygen levels in water bodies, killing off oxygen-dependent species. Oxygen levels can return to normal levels, but only if the influx of organic materials ceases.

Inorganic Nutrients—Nitrates and Phosphates

Inorganic nutrients are chemical compounds produced by a variety of sources, many of which also release organic nutrients. Two of the most common inorganic nutrients are nitrogen and phosphorus. Unlike organic nutrients, which stimulate the proliferation of algae and aquatic plants, inorganic nutrients tend to promote the growth of aquatic plants.

In terrestrial and aquatic ecosystems, nitrogen is contained primarily in the form of ammonia and nitrates. Phosphorus is found in phosphates. In aquatic ecosystems, nitrogen and phosphorus are often limiting factors for populations of algae and other aquatic plants (Chapter 6). Consequently, if levels of these pollutants increase, plant growth may increase dramatically, choking lakes and rivers with thick mats of algae or dense growths of aquatic plants. In freshwater lakes and reservoirs, phosphate is usually the limiting nutrient for plant growth; marine waters are usually nitrate limited.

Excessive plant growth impacts people and the environment. For example, during the summer the proliferation of plants often impairs fishing, swimming, navigation, and recreational boating. In the fall, most of these plants die and are degraded by bacteria. When large amounts of plant matter decay, dissolved oxygen levels fall, killing other aquatic organisms. As oxygen levels drop, anaerobic bacteria resume the breakdown and produce the noxious products noted earlier. Thus, inorganic nutrients create many of the same problems that organic nutrients do. Where do inorganic nutrients come from?

Inorganic fertilizer from croplands is the major anthropogenic source of plant nutrients in fresh waters. When highly soluble fertilizers are used in excess, as much as 25% may be washed into streams and lakes by the rain. Applying fertilizer in smaller amounts and implementing measures to reduce surface runoff of farmland, such as terracing and strip cropping, could greatly reduce the amount of phosphates and nitrates entering surface waters (Chapter 10).

Fertilizers also pollute groundwater with nitrates. In Canada's Lower Frazer River Valley and in southern Ontario, nitrate pollution in groundwater has been traced to the application of manure and artificial fertilizer and from septic tanks.

Laundry detergents are the second most important anthropogenic source of inorganic nutrient pollution in the United States and other developed countries. Many detergents contain synthetic phosphates called tripoly-phosphates (TPPs). These chemicals cling to dirt particles and grease, keeping them in suspension until the wash water is flushed out of the washing machine.

Low and no-phosphate detergents are currently widely available, but many people don't need phosphate-based detergents at all. In fact, nearly 60% of the U.S. population lives in soft-water regions where soap-based cleansing agents work as well as detergents. In the hard-water regions, harmless substitutes for TPPs can be used. For example, lime soap-dispersing agents have been used in bar soaps for years and could be used for laundry detergents.

Nitrates and nitric acid can also enter surface waters from the atmosphere, as described in Chapter 20. According to the Environmental Defense Fund, about 25% of the nitrogen polluting the Chesapeake Bay comes from wet and dry acid deposition. Nitrates and phosphates are also contained in animal waste—humans and livestock being the principal sources.

KEY CONCEPTS

Inorganic nutrients stimulate excess plant growth, which impairs navigation and swimming and disrupts the aquatic environment. When the plants die, they decompose, causing oxygen levels to decline precipitously—an effect that can be harmful to a host of organisms.

Eutrophication and Natural Succession

In undisturbed ecosystems, lakes naturally pick up plant nutrients from surface runoff and rainfall. The natural accumulation of nutrients in lakes is called **natural eutrophication** (u-TROH-feh-KAY-shun). Given sufficient time, natural eutrophication and natural soil erosion can transform shallow lakes into swampland and then into dry land, a process called **natural succession** (discussed in Chapter 6). In this process, inorganic nutrients stimulate plant growth. The plants eventually die and contribute organic sediment to the lake's bottom (**FIGURE 21-6**). This sediment combines with silt from erosion, gradually filling in a lake. Of course, shallow lakes are many times more vulnerable than deeper lakes.

Humans can accelerate the decay of lakes and ponds. Accelerated erosion, caused by human activities, and the accumulation of nutrients from farms, feedlots, and sewage treatment plants—called **cultural eutrophication**—both contribute to the deterioration of lakes. Inorganic nutrients

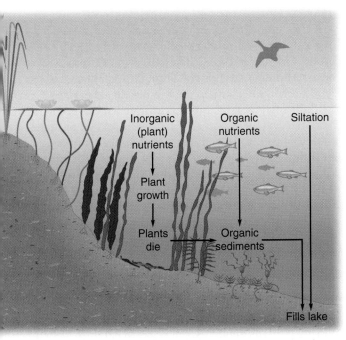

FIGURE 21-6 Eutrophication and succession. This drawing shows the contributions of inorganic and organic nutrients and sediment to the succession of a lake into swampland. Inorganic nutrients contribute to eutrophication. Combined with sediment from natural or anthropogenic sources, they can cause a lake to fill in.

may cause good, productive lakes to become choked with vegetation. In temperate climates, the plants rot in the fall, which depletes dissolved oxygen. Sediment from erosion and organic debris caused by excessive plant growth can fill the lake, destroying it.

The fate of lakes overfed with nutrients from sewage treatment plants and farms, however, is not as dim as scientists once believed; if nutrient inflow is greatly reduced or stopped, a lake may make a comeback. Lake Washington near Seattle, for example, became a foul-smelling, eutrophic eyesore after decades of abuse during which millions of gallons of sewage were dumped into its waters. In 1968, local communities began to divert their wastes to Puget Sound, an arm of the sea with a greater capacity to assimilate the wastes. Lake Washington began a slow recovery.

Of course, the diversion of sewage to Puget Sound has had some negative effects. Although the sound cleanses itself more quickly than the lake, certain toxic substances in the waste and in surface runoff (nonpoint pollution) are having a harmful effect on marine life. The effects are especially noticeable in areas where industries have discharged their wastes for decades. Today, sediments in the sound contain high levels of toxic organic wastes and heavy metals, which may explain the high incidence of tumors in sole, a fish that lives on the ocean's bottom. Shellfish beds have been closed to prevent people from harvesting contaminated shellfish. Efforts are now under way to minimize sewage and industrial waste discharged into the sound.

The story of Puget Sound reminds us that when it comes to pollution there is no "away." The diversion of waste from one location to another is often not a remedy, just a quick fix that transfers a problem from one location to another. It also illustrates the importance of preventing problems in the first place. By minimizing or eliminating the discharge of pollutants into the environment, we stand a better chance of living sustainably on the planet. The Puget Sound case also illustrates the economic impact of human actions. Contamination of shellfish beds by human wastes not only threatens natural systems and organisms, it threatens human health and the economic well-being of those who make a living from the sea.

Eutrophication is the most widespread problem in U.S. lakes. According to a report by the EPA that summarized the results of water-quality assessments of lakes, reservoirs, and ponds by states and Native American tribes, 22% of the water bodies assessed were polluted by nutrients that cause eutrophication. (The states only assessed 43% of their ponds, lakes, and reservoirs acreage.) Nearly all receive wastes from industry and municipalities, but even if these sources were eliminated, many lakes would probably not improve significantly because of continued pollution from nonpoint sources.

KEY CONCEPTS

The accumulation of nutrients in lakes, from both natural and human sources, is called *natural* and *cultural eutrophication*, respectively. Combined with the deposition of sediment from human activities, cultural eutrophication causes lakes to age prematurely.

Infectious Agents

Water may be polluted by pathogenic (disease-causing) organisms, including bacteria, viruses, and protozoans. The major infectious diseases transmitted via water include viral hepatitis, polio (viral), typhoid fever (bacterial), amoebic dysentery (protozoan), cholera (bacterial), schistosomiasis (parasitic worm), and salmonellosis (bacterial). These diseases are especially harmful to the young, old, and infirm. The major sources of infectious agents are (1) untreated or improperly treated sewage, (2) animal wastes in fields and feedlots located near waterways, (3) meat-packing and tanning plants that release untreated animal wastes into water, and (4) some wildlife species that transmit waterborne diseases.

Waterborne infectious diseases are a problem of immense proportions in the less developed nations of Africa, Asia, and Latin America, as witnessed by the outbreak of cholera in the 1990s in many Central and South American nations. Safe drinking water is very hard to come by in many less developed nations.

Infectious agents were once the major water pollutants of now-developed nations, too, before sewage treatment plants and disinfection of drinking water became commonplace. Despite these improvements, problems remain. In many cities, sewage and storm runoff travel through the same pipes. Both end up in the sewage treatment plant. However, when heavy storms occur, the stormwater fills the system, and excess water containing both sewage and stormwater bypasses the sewage treatment plants, ending up in rivers and lakes. Flooding also causes sewage treatment plants located in flood plains (a common practice) to overflow. According to a U.S. EPA report that summarized the findings of states, 13% of the rivers studied had unacceptable levels of bacteria. (This survey only covered less than half of the nation's waterways.) Although this is high, it is nothing compared to the condition of waterways in the less developed nations.

Still, infectious agents can cause enormous problems in more developed countries. In 1993, 312,000 residents of Milwaukee, Wisconsin, developed flulike symptoms after drinking water contaminated by a protozoan parasite known as *Cryptosporidium* (CRYPT-toe-spore-ID-eum) (FIGURE 21-7). Eight people died. This organism entered the drinking water drawn from Lake Michigan. The parasite lives in the intestinal tract of land-dwelling animals such as cows, pigs, and chickens. Scientists believe that its eggs were washed into the Milwaukee River and then into the lake after heavy spring rains. Water is drawn for the city's treatment plant only 3 miles from the river. A few days before the outbreak of the disease, the treatment plant's filtration system failed to operate at peak efficiency, permitting millions of microscopic eggs to enter the system.

Measuring the level of each pathogenic organism would be costly and time consuming. By measuring levels of a naturally occurring intestinal bacterium, the **coliform bacterium**, water quality personnel can determine how much fecal contamination has occurred (FIGURE 21-8).

FIGURE 21-7 **Cryptosporidium.** This microorganism can cause widespread illness if it contaminates public drinking supplies.

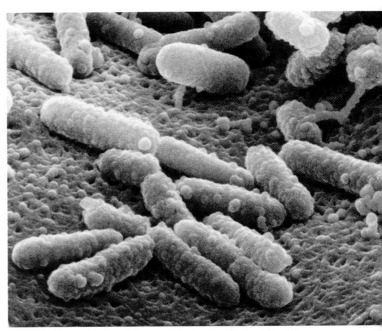

FIGURE 21-8 **Fecal coliform bacteria.** This nontoxic bacterium is an indicator organism. Drinking water quality personnel measure it to determine the level of fecal contamination in water supplies. Its presence suggests the possible presence of pathogenic organisms.

The higher the coliform count, the more likely the water is to contain some pathogenic agent from fecal contamination. As noted previously, more than one third of U.S. rivers now violate standards for coliform bacteria. Rivers in many less developed nations are very heavily polluted with human wastes. A river in New Delhi, India, for example, has a coliform bacteria count of 7500 per 100 milliliters, which is extremely high, before it enters the city. The count jumps to 24,000,000 per 100 milliliters after leaving the city. Disease and mortality rates are understandably very high.

Monitoring coliform bacteria in water is not by itself sufficient to protect human health. Several studies show that swimmers have developed gastrointestinal disorders and flu-like symptoms after swimming in seawater containing levels of coliform bacteria considered safe by government standards. Researchers have found that these symptoms are related to exposure to another noncoliform bacteria, a group known as **fecal streptococcal bacteria**. Some water quality experts believe that they are a more sensitive indicator of human fecal contamination. Others argue that, to protect human health adequately, two or more microorganisms ought to be monitored, including viruses.

KEY CONCEPTS

Numerous infectious agents are found in surface waters, especially in the less developed nations where they cause considerable amounts of suffering and death. Sewage treatment facilities and drinking water purification have greatly reduced the incidence of disease in more developed countries, although outbreaks do occur. To monitor infectious agents, officials generally measure fecal coliform levels, a harmless bacterium itself, but an indicator of the presence of fecal contamination.

Toxic Organic Water Pollutants

About 85,000 synthetic organic compounds are in use today in the industrialized nations. These include solvents, pesticides, and a host of others. Some of these chemicals find their way into surface waters directly from factories.[2] Others, such as pesticides, enter waterways in surface runoff. Still others, such as toxic cleaning agents, enter sewage from homes. Concerns over toxic organic pollutants are many: (1) Many toxic organic compounds are nonbiodegradable or degrade slowly, so they persist in the ecosystem; (2) some are biomagnified in the food web (Chapter 18); (3) some bioaccumulate; (4) some chemicals cause cancer in humans and aquatic organisms, whereas others are converted into carcinogens after reacting with the chlorine used to disinfect water; (5) many are toxic to fish and other aquatic organisms; and (6) some are mere nuisances, giving water and fish an offensive taste or odor.

Unfortunately, our knowledge of the effects of synthetic organics, which are often found in low concentrations, is rudimentary. Reports of diseases traceable to a single chemical are few, but many experts worry that cancer and genetic damage may result from long-term exposure to the wide array of toxic substances found in the waterways of the world.

KEY CONCEPTS

Numerous toxic chemicals enter the waterways from factories, homes, farms, lawns, and gardens. They may have many different effects on people and the environment. As a rule, concentrations in surface waters are low.

[2]Toxic organic pollutants differ from organic nutrients discussed earlier because they do not nourish plants. They are harmful to both plants and animals because they alter metabolism, as discussed in Chapter 18.

Toxic Inorganic Water Pollutants

Inorganic water pollutants encompass a wide range of chemicals, including metals, acids, and salts. Some of them have become major water pollutants. In most states, for example, toxic metals such as mercury and lead are found in levels that are considered unsafe for people and other organisms. Metals come from many sources, among them industrial discharge, urban runoff, sewage effluents, and mining. They are also derived from air pollution fallout. This is another example of cross-media contamination (Chapter 18). Mercury emitted from coal-fired power plants, for instance, is washed from the sky and enters rivers, lakes, and streams. Recent surveys of U.S. drinking water show that some heavy metals such as lead come from pipes; others may come from groundwater supplies.

> **KEY CONCEPTS**
>
> Numerous inorganic pollutants such as acids and heavy metals make their way into the surface and groundwater of industrial nations, usually from industrial sources. Some, such as lead and mercury, are of major concern.

Mercury　One of the more common and potentially most harmful toxic metals is mercury. In the 1950s, mercury was thought to be an innocuous water pollutant, although it was known to have been hazardous to miners and to 19th-century hatmakers exposed to mercury during the manufacture of hat brims. Both miners and hatters frequently developed tremors, or "hatter's shakes," and lost hair and teeth.

In the 1950s, an outbreak of mercury poisonings in Japan raised awareness of the hazard. Residents who ate seafood from Minamata Bay, which was contaminated with methyl mercury, developed numbness of the limbs, lips, and tongue. They lost muscle control and suffered from other neurological defects, among them deafness, blurring of vision, clumsiness, apathy, and mental derangement. Of 52 reported cases, 17 people died and 23 were permanently disabled.

Mercury is a by-product of manufacturing the plastic vinyl chloride, from which beach balls, toys, and other products are made. It is also emitted in aqueous wastes of the chemical industry and from incinerators, coal-fired power plants, research laboratories, and even hospitals. Worldwide, an estimated 5000 to 10,000 metric tons (11,000 tons) of mercury are released into the air and water each year.

> Worldwide, more than 10,000 metric tons of mercury are released into the atmosphere each year.

In streams and lakes, inorganic mercury is converted by bacteria into two organic forms. One of these, dimethyl mercury, evaporates quickly from the water. The other, **methyl mercury**, however, remains in the bottom sediments and is slowly released into the water, where it enters organisms in the food chain and is biologically magnified. The mercury poisoning that affected those eating seafood from Minimata Bay resulted from biological magnification of methyl mercury. In Minnesota and other northern states, mercury levels in fish are so high that wildlife officials publish booklets that warn anglers to limit their intake of fish. Mercury even contaminates salt water fish like swordfish, making some hazardous to eat, especially for young children and pregnant women.

> **KEY CONCEPTS**
>
> Mercury is emitted from many sources, among them vinyl factories and coal-fired power plants, and is one of the most common and most toxic inorganic pollutants, primarily exerting its effect through the nervous system.

Nitrates and Nitrites　Nitrates and nitrites are common inorganic pollutants of water. Nitrates in surface and groundwater come primarily from animal waste and artificial fertilizers. The sources include septic tanks, barnyards, sewage treatment plants, and heavily fertilized crops. Besides stimulating algal growth, described earlier, nitrates are also converted to toxic nitrites in the intestines of humans. Nitrites combine with the hemoglobin in red blood corpuscles and form methemoglobin, which has a reduced oxygen-carrying capacity. Nitrites can be fatal to infants. Although rare today, nitrite poisonings usually occur in rural areas where drinking water is contaminated by septic tanks and farmyards.

> **KEY CONCEPTS**
>
> Nitrates can be converted to nitrites, which bind to hemoglobin and reduce the oxygen-carrying capacity of the blood.

Salts　Sodium chloride and calcium chloride are used on winter roads, driveways, and sidewalks to melt ice and snow. However, meltwater from ice and snow carries salts into streams and groundwater. Salts kill sensitive plants such as the sugar maple. Apple and peach trees are also affected.

In surface waters, salts may kill salt-intolerant organisms, allowing salt-tolerant species to thrive. However, the fluctuations in water flow lead to varying salt concentrations. This is a condition that neither salt-tolerant species (which thrive in high salt concentrations) nor salt-intolerant organisms can survive.

> **KEY CONCEPTS**
>
> Road salts used to remove ice and make driving safe can profoundly affect aquatic ecosystems, forests, and orchards.

Chlorine　Chlorine is a highly reactive inorganic chemical. It is commonly used (1) to kill bacteria in drinking water; (2) to destroy potentially harmful organisms in treated wastewater released from sewage treatment plants into streams; and (3) to kill algae, bacteria, fungi, and other organisms that grow inside and clog the pipes of the cooling systems of power plants. Chlorine and some of the products it forms in water are highly toxic to fish and other organisms.

Chlorine reacts with organic compounds to form **chlorinated organics**. These chemicals may show up in drinking water downstream from sewage treatment plants and other sources. Many of them are known carcinogens and teratogens. However, medical studies indicate that the rates of certain cancers (liver, intestinal tract) are only slightly elevated in populations consuming water contaminated by these compounds.

Sediment

Sediment is the leading water pollutant in the United States in terms of volume. Consisting of dirt particles and sand, sediment is largely a by-product of a number of poorly planned and executed practices such as logging, agriculture, mining, and construction of roads and buildings. Agriculture, for example, increases erosion rates on average four to eight times above normal, sometimes much more. Poor construction and mining may increase the rate of erosion by 10 to 200 times. Sediment is also a major problem in the less developed countries, where poor farming practices, overgrazing, and careless logging practices occur with great regularity.

> Agriculture may increase soil erosion rates four to eight times above normal. Poor construction and mining may increase it 10 to 200 times.

Sedimentation, the deposition of sediment in surface waters, has many social, economic, and environmental impacts. Sediment, for instance, destroys spawning and feeding grounds for fish. It also smothers fish eggs and fry and thus reduces fish and shellfish populations. Suspended sediments decrease light penetration in water bodies, which may impair the growth of aquatic plants, the base of the food chain. Each of these environmental impacts exacts a social and economic cost. As rivers and lakes are destroyed, for instance, they reduce recreational opportunities such as fishing; this not only robs many individuals of satisfying entertainment, it also reduces revenues to local bait shops and other segments of the local economy dependent on fishing.

The deposition of sediment in lakes speeds up natural succession. The filling in of streambeds, or **streambed aggradation** (ah-gra-DAY-shun), results in a gradual widening of the channel, and as streams become wider, they become shallower. Water temperature may rise, lowering the amount of dissolved oxygen and making streams more vulnerable to organic pollutants that deplete oxygen. Lowered dissolved oxygen also wipes out species such as trout that require higher levels. Streambed aggradation also makes streams more susceptible to flooding. In addition, sediment can fill shipping channels, which must then be dredged at considerable expense. As noted in Chapter 13, sediment fills in agricultural and municipal reservoirs throughout the world, reducing their lifespan. Moreover, hydroelectric equipment associated with dams may be worn out by sediments. Finally, some pollutants—such as pesticides, nitrates, phosphates from agricultural fertilizers, and pathogenic organisms—bind to sediment. This extends their lifetime and biological impacts. Sediment pollution can be checked and even eliminated, however, by good land management and restoration (described in Chapters 10 and 12).

Thermal Pollution

Unknown to many people, industry also pollutes water by dumping warm or hot water into cooler surface waters. Rapid or even gradual changes in water temperature can disrupt aquatic ecosystems with profound effects. Where does all this hot water come from? Power plants and factories often must use water to cool liquids or machinery. They obtain this water from nearby lakes, oceans, and rivers. The U.S. electric power industry is a major contributor (**FIGURE 21-9**). It uses about 86% of all cooling water in the United States, or about 730 billion liters (190 billion gallons) per day. Steel mills, oil refineries, and paper mills also use large amounts of water for cooling.

> Electric power plants are a major source of thermal pollution. They're responsible for 86% of all cooling water use in the United States.

Small amounts of heat have no serious effect on the aquatic ecosystem, but large quantities can shift conditions beyond the range of tolerance of aquatic organisms (discussed in Chapter 4). Heat-intolerant organisms perish. Elimination of heat-intolerant species may allow heat-tolerant species to take over. These are usually less desirable species.

Thermal pollution lowers the dissolved oxygen content of water, at the same time increasing the metabolic rate of fish and aquatic organisms. **Metabolism** consists of the chemical reactions occurring in the cells of organisms. Because metabolism requires oxygen, some species may be eliminated entirely if the water temperature rises 10°C (18°F). At the Savannah River nuclear power plant, the number of rooted plant species and turtles was at least 75% lower in ponds receiving hot water than in ponds at normal temperature. The number of fish species was reduced by one third.

> Groundwater supplies about one fifth of the United States's freshwater needs.

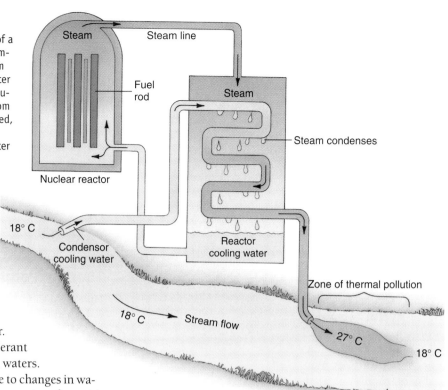

FIGURE 21-9 Thermal pollution. This schematic drawing shows a cooling system of a nuclear power plant and its effect on the temperature of surface waters of a nearby stream into which the heated water is released. Water from the river is used to cool water that circulates through the reactor. Heat picked up from this process by the river water is then dumped, along with the water, back into the stream (or lake), causing a dramatic increase in water temperature that is harmful to aquatic life.

Sudden changes in water temperature can cause **thermal shock,** the sudden death of fish and other organisms that cannot escape into cooler neighboring waters. Thermal shock is frequently experienced when power plants begin operation or when they temporarily shut down for repair. The latter event can devastate heat-tolerant species that inhabit artificially warmed waters.

Fish spawn and migrate in response to changes in water temperature, and thermal pollution may interfere with these processes. Water temperature also influences the survival and early development of aquatic organisms. For instance, trout eggs may not hatch if water is too warm. Thermal pollution can also increase the susceptibility of aquatic organisms to parasites, certain toxic substances, and pathogens.

Thermal pollution can be controlled by constructing ponds for collecting and cooling water before it is released into nearby lakes and streams. Cooling towers are another way to dissipate heat.

KEY CONCEPTS

Water from rivers and lakes is used to cool many industrial processes, electric power production being one of the major ones. Heat generated by these processes is often discharged directly into surface waters, where it kills organisms outright or shifts the composition of the aquatic system.

21.2 Groundwater Pollution

Aquifers supply one fifth of the annual water demand in the United States and drinking water for over half the population (about 150 million people). In rural areas, groundwater supplies 95% of the drinking water. Aquifers are also a major supplier of water in many other countries. In Canada, a country containing seven of the world's largest lakes and with nearly 8% of its land area covered by lakes and rivers, groundwater still contains 37 times more water than surface waters. According to one estimate, approximately one in four Canadians relies on groundwater for domestic uses. In some areas, however, such as Prince Edward Island, groundwater supplies 100% of the domestic demand. Sixty percent of the population of New Brunswick and the Yukon are dependent on groundwater.

Scientists are now reporting that U.S. and Canadian groundwater is increasingly threatened by pollution. In fact, many pollutants are present at much higher concentrations in groundwater than they are in most contaminated surface supplies. Also, many contaminants are tasteless and odorless at concentrations thought to threaten human health.

About 4500 billion liters (1185 billion gallons) of contaminated water seeps into the ground in the United States every day from septic tanks, cesspools, oil wells, landfills, agriculture, and ponds holding hazardous wastes. Unfortunately, very little is known about the extent of groundwater contamination. The Environmental Protection Agency estimates that 1% of the drinking water wells in the United States have contaminants that exceed the standards designed to protect human health. Although that may seem small, 1% of hundreds of thousands of wells is a large number. In fact, one study reported that at least 8000 private, public, and industrial wells in the United States are contaminated.

KEY CONCEPTS

Groundwater is an important source of drinking water in many nations and may be heavily contaminated in numerous industrialized nations by industrial waste pits, septic tanks, oil wells, and landfills. Groundwater in some rural areas may also be contaminated by agricultural chemicals, notably pesticides and fertilizer.

Effects of Groundwater Pollution

Thousands of chemicals, many of them potentially harmful to health, turn up in water samples from polluted wells. The most common chemical pollutants are chlorides, nitrates, heavy metals, and various toxic organics such as pesticides, degreasing agents, and petroleum products. In Canada, many underground storage tanks containing gasoline and other petroleum by-products have begun to leak. The problem is especially severe in the Atlantic provinces where groundwater use is high.

Low-molecular-weight organic compounds are particularly worrisome because many of them are carcinogenic. Concern among medical experts is great because some fear that there is no threshold level for some of these compounds—that is, there is no level free from risk of cancer or other problems. Others fear that many chemicals may act synergistically, making a potentially difficult problem much worse.

With so many different chemicals in groundwater, it is not surprising that the effects vary widely. Nonetheless, some effects are more prevalent, according to Beverly Paigen, an Oakland, California, researcher renowned for her studies at Love Canal in Niagara Falls, New York. In a summary report of health studies of Americans exposed to groundwater pollutants, Paigen noted that the most common problems include miscarriage, low birth weight, birth defects, and premature infant death. Adults and children suffer skin rashes, eye irritation, and a whole host of neurological problems, including dizziness, headaches, seizures, and fainting spells. In a widely publicized case in San Jose, California, pollutants from a leaky underground storage tank owned by the Fairchild Camera and Instrument Company are thought to have doubled the rate of miscarriage in pregnant women and tripled the rate of heart defects in newborns. In Woburn, Massachusetts, contaminated groundwater is blamed for a doubling in the childhood leukemia rate.

KEY CONCEPTS

Thousands of chemicals may be found in a nation's groundwater. Many of them are potentially harmful to human health, causing problems for unborn children (miscarriage, birth defects, and premature infant death) and adults (rashes and neurological problems).

Cleaning Up Groundwater

Many people think of groundwater as fast-flowing underground rivers. Nothing could be farther from the truth. Groundwater typically moves from 5 centimeters (2 inches) to 64 centimeters (2 feet) a day. Because groundwater moves so slowly, it can take years for water polluted in one location to appear in another. Additionally, once an aquifer is contaminated, it can take several hundred years for it to cleanse itself.

Detecting groundwater pollution is expensive and time consuming. Numerous test wells must be drilled to sample water and determine the rate and direction of flow. Despite intensive drilling, health officials can easily miss a tiny stream of pollutants that flows through one portion of a large aquifer. Liquids that do not readily mix with water, for example, may travel along the top or bottom of the aquifer in thin layers and are often difficult to detect.

Preventing groundwater pollution is generally the cheapest way to protect this vital resource and is an essential element in creating a sustainable water supply system. How can we prevent this problem? Cutting down on the use of potentially toxic chemicals and taking measures to reduce or eliminate the production of hazardous wastes are important first steps. Individuals can help by using natural or biodegradable cleaning agents. Hazardous waste recycling and improvements in the ways we dispose of wastes would also help. Ways to achieve these goals are discussed in Chapter 23 on waste management.

Because many aquifers are already contaminated by potentially toxic substances, efforts are needed to reclaim them. It may, for example, be necessary to pump contaminated water from polluted aquifers to the surface and purify it. The water can be pumped back into the ground at that site or used for a variety of purposes. This process, although feasible for some contaminants, is extremely expensive. Unfortunately, it is not feasible for contaminants that are not very water soluble. Many other approaches are being tested. For example, steam, various solvents, and surfactants (substances that reduce surface tension of liquids) can be injected into aquifers to improve recovery rates of chemicals. Laboratory studies of these ideas seem promising.

New techniques are also being developed to use naturally occurring bacteria in soil and groundwater to clean up some contamination. For instance, hydrocarbons (such as crude oil, gasoline, and creosote) that have leaked from storage tanks or are spilled from vehicles have polluted more groundwater used for drinking than any other class of chemicals in the United States. Microbiologists have known since the late 1970s that some bacteria can digest or break down hydrocarbons in the soil and groundwater, converting them into carbon dioxide and methane gases. Bacteria in the soil, however, as a rule can degrade only about 1% of the hydrocarbon pollution flowing past them. Why? Quite simply, they lack key chemical nutrients needed for metabolism. By supplying these nutrients, researchers may be able to accelerate the bacterial decomposition of hydrocarbons.

KEY CONCEPTS

Groundwater moves slowly and takes many years to cleanse itself. Preventing groundwater pollution is essential to creating a sustainable water supply. Equally important are efforts to clean up groundwater supplies already contaminated by potentially toxic chemicals.

21.3 Ocean Pollution

"When we go down to the low-tide line," Rachel Carson wrote, "we enter a world that is as old as the Earth itself—the primeval meeting place of the elements of Earth and water, a place of compromise and conflict and eternal change." Today this compromise, conflict, and eternal change have taken on a new meaning as humankind forges out into the oceans in search of food, fuel, and minerals. In our use of the seas, we spill many toxic substances, such as oil. Ocean pollutants also arise from land-based activities, though. The Kansas farmer and the Minnesota factory, for instance, have an impact on this vast body containing more than 1.3 billion cubic kilometers of water. Many inland pollutants are washed from the land or dumped into rivers, in which they are transported to the world's oceans.

Oceanic life zones and the hazards of pollution in the biologically rich coastal zones were discussed in Chapters 5 and 13. You may want to review this material before reading this section, which deals with five crucial challenges: (1) oil pollution, (2) plastic pollution, (3) medical wastes, (4) red tide (algal blooms), and (5) disposal of sewage in the ocean.

KEY CONCEPTS

The oceans are polluted by chemicals spilled into them directly and by pollutants washed from the lands and transported to them by rivers.

Oil in the Seas

About 3.2 million metric tons of oil enters the world's seas every year. About half of the oil that contaminates the ocean comes from natural seepage from offshore deposits. One fifth comes from well blowouts, breaks in pipelines, and spills. The rest comes from oil disposed of inland and carried to the world's oceans by rivers. The homeowner who dumps oil in the sewer at night, for example, is one of the biggest sources of oceanic oil pollution.

> About 3.2 million metric tons of oil enters the world's seas every year. About half of the oil that contaminates the ocean comes from natural seepage from offshore deposits. One fifth comes from well blowouts, breaks in pipelines, and spills. The rest comes from oil disposed of inland and carried to the world's oceans by rivers.

Some oil entering the ocean comes from the transfer of oil from offshore platforms to on-shore pipelines. Contamination from this source has not captured the public attention, even though its effect on marine life and birds can be quite significant. What generally becomes headline news are the dramatic events such as large spills—well blowouts or wrecked tankers that spill out tons of black, viscous oil. In 2004, a year during which there were relatively few tanker spills accidents accounted for the loss of 15,000 metric tons of oil or 17.8 million liters (4.6 million gallons).

KEY CONCEPTS

Half of the oil polluting the oceans comes from natural seepage; the rest comes from human sources, including tanker accidents and inland disposal. The largest natural source is inland disposal.

Biological Impacts of an Oil Spill   What happens to oil spilled from a tanker or a well? You may be surprised to learn that about 25% of the chemicals making up crude oil are volatile—that is, they evaporate readily. Within 3 months, one fourth of the chemical substances in oil evaporate, becoming air pollutants. Relatively nonvolatile compounds

that are lighter than water float on the surface, where they are broken down by bacteria over the next few months. Nearly 60% of the oil spill is destroyed in this way. The remaining 15% consists of heavier compounds that stick together and sink to the bottom in huge globs. In cold polar waters, oil decomposes very slowly. In some cases, it may become incorporated into sea ice and be released very slowly for years afterward.

Before oil's chemical components can evaporate or be broken down, they can have many harmful effects. Oil kills plants and animals in the estuarine zone. Especially hard hit are the barnacles, mussels, crabs, and rock weed (a type of algae). Their recovery after a major spill may take two to ten years. Oil also settles on beaches and kills organisms that live there. It settles to the ocean floor and kills benthic (bottom dwelling) organisms such as crabs. Those benthic organisms that survive may accumulate oil in their tissue, making them inedible. Oil poisons algae and may disrupt major food chains and decrease the yield of edible fish. It also coats birds, impairing flight or reducing the insulative property of feathers, thus making the birds highly vulnerable to hypothermia (**FIGURE 21-10**). Gallant efforts are often made to rescue birds sullied by oil. After cleaning off the oil, the birds are kept in captivity, nursed back to health, and then released. Two studies published in 1996, however, suggest that such efforts (which are quite costly) are fairly ineffective. One extensive study of more sensitive species of rehabilitated seabirds found that they lasted, on average, only

> About 25% of the chemicals in crude oil are volatile and evaporate within three months of an oil spill. Sixty percent consists of relatively nonvolatile compounds that float on the surface, where they are broken down by bacteria over the next few months. The remaining 15% are heavier compounds that stick together and sink to the bottom in huge globs.

FIGURE 21-10 Oil-covered duck. This bird was beyond the help of volunteers who attempted to save many oil-covered animals after the 1989 oil spill in Prince William Sound (Alaska).

4 days after release. A study of brown pelicans showed that half of the birds released into the wild died within 6 months. Only 12–15% of the rehabilitated birds were still alive after 2 years, far lower than the 80–90% two-year survival rate of birds not exposed to oil. Why?

Birds immersed in oil inhale and ingest (swallow) toxic substances. These birds commonly suffer from anemia, immune system suppression, tissue damage, and hormonal imbalances. Researchers hypothesize that rehabilitated birds, although apparently healthy upon release, fare poorly because their immune systems cannot protect them adequately under the stressful conditions of life in the wild.

Oil also endangers fish hatcheries in coastal waters and can contaminate the flesh of commercially valuable fish, as it did in Prince William Sound in Alaska after the Valdez spill.

The amount of damage caused by oil pollution depends partly on the amount of oil spilled and the direction in which it is carried by the wind and ocean currents. If slicks reach land, they damage beaches and shorelines, recreational areas, and marine organisms. Oil may be driven over portions of the continental shelf, a highly productive marine zone; there it can poison clams, scallops, flounder, haddock, and other important food species. Oil driven out to sea has fewer environmental consequences because there is little life in the open waters of the ocean.

The damage resulting from an oil spill also depends on when it occurs, too. For example, the devastating spill off the coast of Alaska in 1989 occurred only 2 weeks before hundreds of thousands of ducks and other waterfowl migrated through the sound or came to nest. Had the spill occurred after the migration, wildlife losses would have been much lower.

Oil pollution of the oceans poses less of a threat to the overall marine environment than was once feared, a National Research Council committee has concluded. Oil can have serious local effects, the group noted, and these can persist for decades, as in the case of Prince William Sound. Overall, however, the marine environment has not suffered irreversible damage from oil. The committee was quick to point out that scientists have only limited knowledge of the potential damage of oil in tropical and Arctic regions, where much of the current oil development is occurring. Studies of the 1989 oil spill in Alaska, for example, showed that the hydrocarbons in the oil are more stable and more persistent in the cold waters than in warmer seas. This could make their impact even greater.

KEY CONCEPTS

Oil spilled from human sources evaporates or is broken down by naturally occurring bacteria or sinks to the bottom. Before it is eliminated, however, it can cause serious environmental damage. The extent of the damage depends on the amount spilled, the location of the spill, the prevailing weather conditions, and the season.

Reducing the Number of Oil Spills Thanks to public outcry in the United States, the number of oil spills began to fall after 1980 and has remained relatively stable in recent years. Tougher governmental standards for new oil tankers went into effect in 1979. New safety standards for older tankers were phased in between 1981 and 1985. Dual radar systems, backup steering controls, collision avoidance aids, and improved inspection and certification were instrumental in reducing spills. Under these regulations, crude oil must be cleaned before it is pumped on board to eliminate sludge buildup in storage tanks. This sludge was once rinsed out at sea. New regulations also require tankers to have separate ballast tanks. These tanks are filled with salt water to help ships keep their balance when returning after discharging their cargo. In older ships, empty oil tanks were once filled with water for ballast. When the ship arrived at port, the oil-contaminated water was dumped into the sea.

The serious effects of oil washing and combined ballast-oil tanks must not be underestimated. About 1.3 million metric tons (1.4 million tons) of oil were once released each year during tank purging and ballast tank discharge—over six times the amount released by tanker spills.

Although the number of oil spills has decreased worldwide since the 1970s, the quantity of oil released fluctuates wildly. In 2004, for example, 4.6 million gallons were spilled from ships and barges; in 2002, oil spills from these sources was 4.5 greater, about 21 million gallons. Much room for improvement remains.

In 1990 the U.S. Congress passed the **Oil Pollution Act,** which establishes a $1 billion fund to be used to clean up oil spills and pay for damages. The money comes from a 3-cents-per-barrel tax on domestic and imported oil and thus passes the costs of cleanup and damage on to the consumer. Under the bill, oil companies would ultimately be responsible for cleanup costs, but only up to a point because the law sets strict limits on their financial liability. A company responsible for a spill, for instance, would pay only $10 million for cleanup. If the spill occurred near an onshore facility where damage is much greater, the company would pay $350 million for each spill. The fund would be used to pay the additional costs. Many critics are disappointed with the liability limits because the costs of cleaning oil are likely to be far greater. Cleanup in Prince William Sound to date has cost Exxon $2.5 billion and the federal government $154 million. In addition, in 1992, Exxon agreed to pay an additional $1.25 billion in criminal fees, restitution, and civil recovery (to pay damages to salmon fishing companies and other parties).

Although the law requires new tankers to be equipped with double steel hulls or effective double containment features to reduce the likelihood of spills, existing oil tankers are not required to be retrofitted. However, existing tankers must be escorted by two towing vessels in high-risk areas, and single-hull vessels started being phased out in 1995. By 2015 all tankers with single hulls will be banned. As FIGURE 21-11 shows, these and other measures have resulted in a dramatic decline in tanker accidents and spills.

In addition, the law establishes regional oil spill response teams that can be deployed immediately after a spill to coordinate cleanup efforts. It also allows states to set stricter guidelines. A state, for instance, could mandate double hulls for use in its harbors.

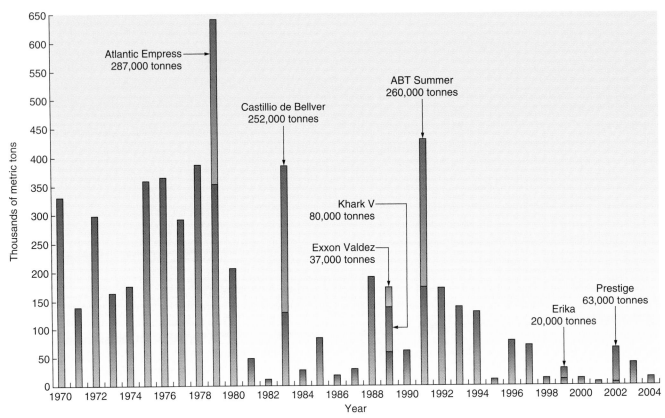

FIGURE 21-11 Oil spills from 1970 to 2004. Note the dramatic decline in recent years thanks to higher safety standards. (Tonnes = metric tons.)

Many efforts are under way in the United States and other countries to reduce oil spills, including new standards for the construction of oil tankers.

Sustainable Solutions to Oil Pollution   Many of the steps outlined in this book to make human systems more sustainable will have a profound impact on water quality. This section describes how key principles of sustainable development—conservation, recycling, renewable energy use, restoration, and population control—can be applied to reduce oil pollution, beginning with conservation.

Conservation is one of the most important principles of sustainability. Energy conservation—using what we need and using it efficiently—is one of the most important areas to practice this principle. For example, energy efficiency in automobiles and increased use of mass transit will decrease the world's dependence on crude oil and will reduce the number of oil wells that leak oil into the ocean. It will also decrease oil spills occurring each year, as noted previously. Because much of the oil entering the oceans comes from disposal of oil on land, recycling of oil, rather than disposing of it in sewers, will also help protect the oceans. Renewable energy resources cut pollution of water and air by reducing our dependence on fossil fuels.

Conservation, recycling, and renewable resource use can all reduce our dependence on oil and reduce oil pollution in the seas.

Plastic Pollution

A young seal swims playfully in the coastal waters of San Diego Bay. Floating in its watery domain is a piece of a plastic fishing net that has drifted with the currents for months. The seal swims around and around curiously and then plunges through an opening in the net, only to be entrapped.

At first, the net is just a mild nuisance, but as the seal grows, the filament begins to tighten around its neck. Eventually, it cuts into the seal's skin, leaving an open ring of raw flesh exposed to bacteria. Unless it is helped, the seal will perish, along with an estimated 400,000 marine mammals each year that die as a result of the estimated 6.4 million metric tons (7 million tons) of plastics (including nets) discarded into the ocean annually by commercial fishers, sailors, and military personnel. Tens of thousands of tons of plastic may also come from private boats, factories, and onshore facilities (**FIGURE 21-12**). No one knows the number of seabirds or fish that die each year from nylon fishing nets, plastic bags, six-pack yokes, plastic straps, and a myriad of other objects made from nonbiodegradable plastic.

> An estimated 400,000 marine mammals each year die as a result of the estimated 6.4 million metric tons (7 million tons) of plastics (including nets) discarded into the ocean annually by commercial fishing interests, sailors, and military personnel.

FIGURE 21-12 **Victim of plastic pollution.** A young hawksbill sea turtle is caught in a plastic fishing net cut loose by commercial fishing interests.

FIGURE 21-13 Trash on a beach in California. National and local initiatives are helping to reduce the amount of plastic and other trash like this on our beaches and in our oceans.

Plastic nets entangle fish, birds, and sea mammals. They may strangle, starve to death, or drown their victims. Plastic bags, looking like jellyfish, are eaten by sea turtles. One scientist pulled enough plastic out of a leatherback turtle's stomach to make a ball several feet in diameter. In such cases, starvation is the usual result because the animal's stomach is packed with indigestible plastic that cannot pass through its digestive tract. Birds and fish gobble plastic beads resembling the tiny crustaceans that are a normal part of their diet. They may become poisoned and die. Discarded plastic eating utensils, when swallowed, may cut into an animal's stomach lining, causing it to bleed to death.

KEY CONCEPTS

Millions of tons of plastic are dumped into the ocean each year, killing hundreds of thousands of marine mammals, fish, and birds.

Controlling Plastic Pollution   As with many other environmental issues, public outcry created a groundswell of activity at many levels. On the local level, concern over plastic spurred the Oregon Fish and Wildlife Department to sponsor annual beach cleanups to remove plastic that might be washed out to sea (**FIGURE 21-13**). Italy placed a ban on all nonbiodegradable plastics, which went into effect in 1991. Following suit, Oregon and Alaska passed laws requiring that all six-pack yokes be biodegradable. On another front, in what may prove to be a precedent-setting move, New Jersey legislators introduced legislation to ban the release of nonbiodegradable balloons. Why? Many balloons released to celebrate national holidays and other occasions eventually end up in the ocean. Here they are often mistaken for food by marine organisms. They accumulate in their digestive tracts and cause starvation.

On a larger scale, the U.S. Congress passed the **Plastic Pollution Control Act** (1988), which makes it unlawful for any U.S. vessel to discard plastic garbage into the ocean. This legislation also requires all manufacturers of six-pack yokes to use degradable plastic if technically feasible. In 1988, the U.S. Senate approved an international treaty banning the disposal of plastics in the ocean. The agreement was signed by 28 nations and went into effect in December 1988.

Despite these steps, millions of fish, birds, and sea mammals will perish in coming years. Without stricter controls, international cooperation, and private efforts, the ever-increasing use of plastic will bring about the unnecessary death of innumerable sea creatures. The use of biodegradable plastics could help, but they are not a panacea. In fact, the term *biodegradable plastic* is a bit of a misnomer. Biodegradable plastics are plastics mixed with starch or some other material that breaks down, dispersing small particles of plastic into the environment.

KEY CONCEPTS

Many steps have been taken to reduce the disposal of plastic into the ocean, but huge amounts are still being disposed of each year, posing a serious threat to sea life.

Medical Wastes and Sewage Sludge

In the summer of 1988, many Americans were shocked to learn that medical wastes were being illegally dumped into the ocean. Bloody bandages, sutures, vials of AIDS-infected blood, and used syringes washed up onto the eastern coast of the United States as well as onto the shores of Lake Erie.

Because there was no way to track the wastes to their source, the U.S. Congress passed the **Medical Waste Tracking Act** (1988). It established a two-year program that covered 10 states and required people generating, storing, treating, and disposing of medical wastes to keep records that were accessible to the public. If wastes showed up on shorelines, they could be traced back to their source so that responsible parties could be brought to justice. Unfortunately, this law expired in 1990 and has not been reauthorized, leaving controls up to the states. All states have some rules regarding medical waste disposal.

A far more important concern has been the dumping of sewage into the ocean from sewage treatment plants. In 1988, Congress passed the **Ocean Dumping Ban Act,** which prohibited the dumping of sewage sludge (organic material from sewage treatment plants, discussed later) into the ocean. This longstanding practice of many coastal cities officially ended on December 31, 1991.

According to the Natural Resources Defense Council, however, 8.9 trillion liters (2.3 trillion gallons) of liquid waste generated from sewage treatment plants are dumped directly into the ocean. Some of the waste receives little or no treatment before it is discharged. Much of it is industrial waste containing toxic organic chemicals and toxic metals.

KEY CONCEPTS

Medical wastes and sewage have long been dumped into the ocean, but steps have been taken to eliminate both practices. The direct dumping of sewage has stopped entirely, but millions of gallons of sewage enter the sea each year from sewage treatment plants located near the coast.

Red Tide

Another growing problem occurring in the oceans of the world is known as *red tide*. **Red tide** is a term that refers to reddish algal blooms in coastal waters. Reports of sometimes harmless, sometimes noxious, and sometimes deadly red tides appear to be on the rise. At one time, they were found mainly in the waters off the coasts of Europe and North America. Today, they are also reported in coastal waters of Asia and South America. Scientists are not certain whether there is an actual increase in the incidence of such outbreaks or simply more awareness of the problem and more frequent reporting.

Red tides are only occasionally red. They may be brown or green or even orange. Some evidence suggests that they may be caused by elevated levels of plant nutrients, such as algal blooms in freshwater. The problem with red tides is that the organisms that discolor the water contain toxic substances that protect them from predators. These organisms may be ingested by shellfish, however, which themselves become toxic to people.

Scientists have also discovered an entirely new family of microorganisms (pfeisteria) that release an extremely toxic substance into the water. The substance is so toxic that scientists must take very special precautions when working with it—similar to those taken when studying the AIDS virus in the laboratory. First identified in coastal waters of North Carolina, this organism causes massive fish kills. In the laboratory, the organism is nourished by levels of phosphate similar to those found in polluted rivers.

Some ecologists believe that toxic red tides represent a serious problem worldwide and are urging states and nations to step up efforts to reduce pollution that may be feeding these organisms.

KEY CONCEPTS

Outbreaks of microscopic and often highly toxic algae appear to be on the rise worldwide and may be caused by an increase in inorganic nutrient pollution from agriculture, industry, and the human population.

The Case of the Dying Seals

In the spring of 1988, harbor seals in the North Sea began to die in record numbers. Adult seals floated aimlessly in the water, too weak to eat. Pregnant females aborted their fetuses. The mysterious epidemic, which began off the coast of Denmark, quickly spread to seal colonies throughout the North and Baltic Seas. By the middle of the summer, seals were dying along hundreds of miles of North Sea coastline. By September, the disease had spread to the Atlantic coast of Ireland.

Some people dubbed this incident the "black death of the sea," for it recalled the epidemics of bubonic plague, or black death, which devastated the human population of Europe in the 1300s. In this tragic turn of events, though, the victims were helpless seals. In a few short months, the population of harbor seals, once numbering 18,000, had fallen to only 6000.

Studies showed that the massive die-off was caused directly by the phocine distemper virus (PDV), a new virus to science. Biologists believe, however, that the virus was only the most immediate cause of death. Pollution in the seas, they say, may have greatly weakened the immune systems of the seals, making them highly susceptible to infection.

Seals living in the waters off the coast of Germany and the Netherlands are heavily contaminated with a toxic chemical called PCB (polychlorinated biphenyl), a substance once used as an insulator in electrical devices. PCBs and other chemical contaminants are believed responsible for the reproductive problems and the suppression of the seals' immune systems. PCBs entered the ocean from either land or air.

In fact, an emergency working group formed in London to deal with the epidemic. Made up of biologists, veterinarians, and toxicologists, the group concluded that "persistent pollutants could not be excluded as an additional factor in the seal deaths." Though there is not a great deal of evidence about what role pollution played, scientists found that the highest seal mortality occurred in waters that were most polluted.

The North and Baltic Seas have been polluted for years. The North Sea alone annually receives 60 billion liters (15 billion gallons) of wastewater from factories and waste treatment facilities in bordering industrial nations. The pollution problem in the North and Baltic Seas, however, is compounded by the fact that both seas are shallow and cleanse themselves very slowly. The North Sea, for instance, purges itself only twice every 10 years. The Baltic Sea is much slower. Because they are cleaned so slowly, pollution levels can reach dangerous levels and have serious impacts on fish, wildlife, and possibly people.

The seal plague may be the latest manifestation of a chronic pollution problem in the North and Baltic Seas. Fortunately, many countries have decided to take action and have entered into cooperative agreements. Most of the countries bordering the North Sea, for instance, agreed to halve the amount of nutrient pollution (nitrates and phosphates) and toxic chemicals flowing into the sea by 1995, a goal not yet met. A similar agreement was reached to cut pollution entering the Baltic Sea.

Seal deaths off the coast of Europe are a symptom of a global problem. Similar tragedies are occurring elsewhere. Since June of 1987, for example, as many as 4 of every 10 dolphins off the Atlantic coast of the United States have perished. Studies of gulls in the Great Lakes have shown an alarming reproductive failure because of PCBs and other organic pollutants.

KEY CONCEPTS

A massive seal die-off in the late 1980s, caused by a virus, may ultimately have resulted from immune system suppression caused by a common pollutant, PCBs.

21.4 Water Pollution Control

Table 21-2 offers a quick summary of the different sources of water pollution and shows the major pollutants from each one. This information is helpful in plotting control strategies. You may want to take a few minutes to go through the table and devise some solutions. The real challenge is to devise solutions that prevent the problem in the first place. Also bear in mind the need to address both point and nonpoint sources.

As with other issues, water pollution control has traditionally focused on end-of-pipe measures: pollution control devices that remove the pollutants from a waste stream. Although these are important, the waste must go somewhere. Ultimately, the most sustainable approaches are those that prevent the pollutant from being generated in the first place.

KEY CONCEPTS

Reducing water pollution requires efforts on two levels—those that capture wastes emitted from various sources (the so-called end-of-pipe solutions) and those that prevent waste production and pollution.

Legislative Controls

The cornerstone of U.S. water pollution policy is the **Clean Water Act**. While important, it has focused primarily on point sources, notably sewage treatment plants and factories. In fact, the Clean Water Act has provided funding to construct and improve thousands of U.S. sewage treatment plants that handle municipal wastes (FIGURE 21-14). It has also required

FIGURE 21-14 **Sewage treatment plant.** This photograph of a sewage treatment plant shows the large, circular trickling filters that are used for secondary treatment.

Table 21-2

Common Water Pollutants

Sources	Bacteria	Nutrients	Ammonia	Total Dissolved Solids	Acids	Toxics
Point sources						
Municipal sewage treatment plants	•	•	•			•
Industrial facilities						•
Combined sewer overflows	•	•	•	•		•
Nonpoint sources						
Agricultural runoff	•	•		•		•
Urban runoff	•	•		•		•
Construction runoff		•				•
Mining runoff				•	•	•
Septic systems	•	•				•
Landfills						•
Forestry runoff		•				•

Source: U.S. Environmental Protection Agency.

Table 21-3

Major Provisions of the U.S. Clean Water Act

Planning

The states receive planning grants from the EPA to review their water pollution problems and to determine ways to solve them by reducing or eliminating point and nonpoint water pollutants.

Standards Development

The states adopt water quality standards for their streams. These standards define a use for each stream and prescribe the water quality needed to achieve that use.

Effluent Standards

The EPA develops limits on how much pollution may be released by industries and municipalities. These limits are developed at the national level based on engineering and economic judgments. The EPA or the states are required to make the discharge limits for individual plants more stringent if necessary to meet the state water quality standards.

Grants and Loans

The EPA provides financial assistance to state water programs for the construction of sewage treatment plants, processing of permit applications, monitoring of water quality, and enforcement.

Dredge and Fill Program

The EPA develops environmental guidelines to protect wetlands from dredge and fill activities. These guidelines are used to assess whether permits should be issued.

Permits and Enforcement

All industries and municipal dischargers receive permits from either the EPA or the states. The EPA and the states regularly inspect these dischargers to determine whether they are in compliance with the permit and take appropriate enforcement actions if necessary.

Source: U.S. Environmental Protection Agency.

factory owners to install water-pollution control equipment to reduce the discharge of industrial waste into surface waters (Table 21-3). For a success story, see Spotlight on Sustainable Development 21-1.

Unfortunately, in many cases nonpoint pollution from expanding cities offsets the gains from sewage treatment plants. Pesticides from lawns; oil on driveways, parking lots, and streets; nitrate fertilizers from farm fields; and dozens of other nonpoint sources all contribute millions of tons of pollution each year to U.S. waterways. Ironically, despite intense efforts to control point sources, in many regions experiencing rapid population growth water quality has not improved or has improved only slightly.

KEY CONCEPTS

Legislation to address water pollution has historically focused on point sources—primarily factories and sewage treatment plants. Gains made in controlling such sources, however, have often been offset by increasing levels of pollution from nonpoint sources such as city streets, lawns, and farm fields.

Controlling Nonpoint Pollution

Recognizing that nonpoint sources are a major source of water pollution, the U.S. Congress amended the Clean Water Act in 1987. Under the amended law, the states were required to identify nonpoint sources and draft plans to control them. The EPA currently provides $100 million per year in grant money to states to address nonpoint sources. States are required to partially match federal funds as well. This program has resulted in numerous successes.

Some state and local governments have taken actions on their own by passing zoning laws aimed at reducing agricultural and urban runoff. Maryland, for instance, limits the amount of pavement accompanying development in watersheds of Chesapeake Bay. In northern California, loggers are prohibited from clogging streams with silt. In addition, local soil conservation districts help to identify trouble spots and work with farmers to decrease erosion and other problems. North Carolina has instituted some innovative approaches to control nonpoint sources. For example, it currently permits industry to work with farmers to implement nonpoint pollution controls—for example, erosion controls. Such measures are often cheaper than building treatment facilities and offer multiple benefits.

 Despite these gains, much work is needed to reduce nonpoint water pollution to sustainable levels. Because agricultural runoff is the largest source of nonpoint water pollution, affecting between 50 and 70% of all surface water and groundwater in the nation, laws that address pollution from farms are important. New regulations that require terracing of steep roadbanks, revegetation of denuded land, and use of mulches to hold soil in place while grasses take hold along newly constructed highways and housing sites, are also essential. Laws that require sedimentation ponds to collect runoff before it can reach streams and that require porous pavements that soak up rainwater could also help reduce erosion and the influx of pollutants into nearby water bodies.

> Agricultural runoff is the largest source of nonpoint water pollution, affecting between 50 and 70% of all surface water and groundwater in the nation.

KEY CONCEPTS

In the United States, efforts to control nonpoint water pollution are still in their infancy but are gaining in popularity because they are often economical solutions that offer other benefits as well. Especially important are efforts to control agricultural runoff.

Preventing Groundwater Pollution Several federal laws give the EPA the authority to prevent and control sources of groundwater pollution and to clean up polluted groundwater. In 1984, the EPA adopted a formal groundwater protection strategy that, among many things, established an Office of Groundwater Protection. This strategy focuses on building state capacity—that is, state regulation and control of groundwater. Under the plan, the EPA provides technical assistance in analyzing problems and gives advice needed by states to establish their own groundwater protection programs. Unfortunately, say critics, states often lack the necessary funds, experience, and expertise. Nonetheless, many states have adopted their own groundwater standards. Some have taken preventive measures by mapping out aquifers and banning industrial development over them. Today, the vast majority of the states have developed programs that at least minimally regulate discharges to groundwater. Clearly, much improvement is needed, especially in groundwater pollution prevention.

In 1996, Congress toughened the U.S. drinking water law by adding regulations that protect groundwater. Congress also recently passed legislation that requires the EPA to set standards for drinking water quality for 85 additional chemical substances, including pesticides and various industrial chemicals.

The new law requires drinking water suppliers to test for these chemicals and to maintain the drinking water standards. It also requires them to monitor drinking water supplies for other substances that might pose a threat to human health. Even though the EPA is now monitoring these chemicals, hundreds more could be monitored.

Water Pollution Control Technologies: End-of-Pipe Approaches

The first sewage treatment plant in the United States was built in Memphis, Tennessee, in 1880; today, there are over 16,255 of them. Sewage entering a treatment plant often contains pollutants from homes, hospitals, schools, and industries. It contains human wastes, paper, soap, detergent, dirt, cloth, food residues, microorganisms, and a variety of household chemicals. In some cases, water from storm drainage systems is mixed with municipal waste drainage systems to save the cost of building separate pipes for each, which can be exorbitant. Combined systems generally work well, but as noted earlier, during storms inflow may exceed plant capacity. Consequently, some untreated storm runoff and sewage pass directly into waterways, raising the coliform count in downstream waters and rendering them unfit for swimming.

Primary Treatment Sewage treatment can take place in three stages: primary, secondary, and tertiary treatment. Primary treatment physically removes large objects by first passing the sewage through a series of grates and screens. Sand, dirt, and other solids settle out in grit chambers (**FIGURE 21-15**). The solid organic matter, or sludge, settles out in a settling tank.

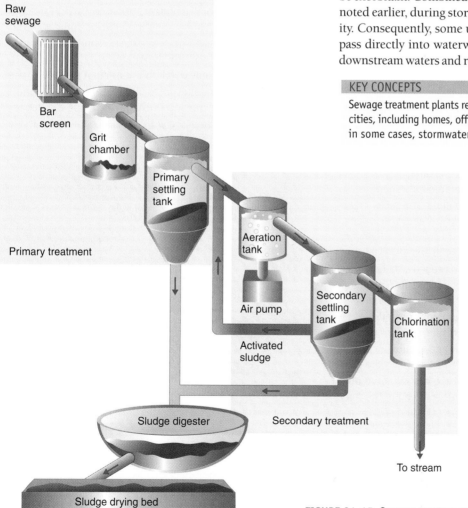

FIGURE 21-15 **Sewage treatment.** Primary and secondary sewage treatment facilities. The secondary treatment facility contains all of the components of the primary system but has an additional aeration tank (or trickling filter) to further decompose organic matter.

21-1   Cleaning Up the Great Lakes

Carved by ancient glaciers, the five mammoth Great Lakes hold one fifth of the world's standing freshwater. Approximately 40 million people live within their drainage basin. Like so many of the continent's waters, they have suffered years of abuse from pollution and poor land management. Especially hard hit were Lakes Ontario and Erie.

Cultural Stress: The Death of a Lake?

The story of Lake Erie serves as a reminder of the immense impact of human civilization and as a lesson on ways to prevent future deterioration of the world's waters. Lake Erie was once surrounded by dense forest. The healthy forests reduced erosion and streams ran clean. Today, more than 13 million people live in the lake's watershed. The dense woodlands that once protected the soil were cut down to make room for farms, homes, industries, and roadways. As a result, large quantities of topsoil washed into the rivers and the lake, clogging navigable channels and destroying spawning areas essential to the lake's once rich fish life. In the early 1900s, many of the wetlands along the lake's shore were drained, and hundreds of small dams were built to provide power to mills, blocking the upstream migration of fish such as the walleye and sturgeon.

By the 1960s, the lake's waters were polluted with a wide assortment of organic and inorganic nutrients. Raw sewage floated on the water's surface, and algal blooms (rapid growth of algae) were commonplace (FIGURE 1). Dissolved oxygen levels frequently dropped to low levels, especially in the profundal zone that occupies the large central basin of the lake. Blue-green algae proliferated in the warm summer months in the shallow western end of the lake, creating a foul-smelling, murky green water. Lead, zinc, nickel, mercury, and other toxins from industry polluted harbors and built up in nearshore sediments. In 1970 and 1971, mercury levels in fish from Lake Erie often exceeded safe levels set by the U.S. Food and Drug Administration. Reductions in mercury discharges in 1975 resulted in a decline in mercury in fish, but violations of health standards continue.

In colonial days, numerous fish species inhabited the lake and its tributaries. Largemouth and smallmouth bass, muskellunge, northern pike, and channel catfish were common in the tributaries of the lake. Lake herring, blue pike, lake whitefish, lake sturgeon, and others lived in the open waters. By the 1940s, however, blue pike and native lake trout had vanished. Sturgeon, lake herring, whitefish, and muskellunge managed to hold on, but in reduced numbers.

Lake Erie suffered from overfishing, introduction of alien species, pollution, and destruction of shorelines and spawning grounds. Algal blooms, beach closings, thick deposits of sludge, oxygen depletion, taste and odor problems, and contaminated fish were the legacy of years of disregard and mismanagement.

By the late 1950s, large areas of Lake Erie's central basin were anoxic (without oxygen) for weeks on end during the summer. Until the late 1970s, anoxia spread cancerously. The lake was pronounced dead; many feared that the other lakes would follow suit.

The Joint Cleanup Program—Not Enough

Alarmed by the condition of Lake Erie and the other lakes, in 1972, the United States and Canada agreed to restrict the discharge of pollution into the Great Lakes. The Great Lakes Water Quality Agreement, updated in 1978, called for controls of point and nonpoint pollution sources. It demanded that releases into the lakes of "any or all persistent toxic substances" be "virtually eliminated." With cooperation from industrial and municipal polluters, the lakes began to show signs of recovery. Lake Erie, the shallowest and fastest flowing of the lakes, made a quick recovery. Today, it is teeming with fish. Gone are the massive

FIGURE 1 Plant nutrients from sewage treatment facilities caused massive algal blooms in the Great Lakes.

algal blooms and the raw sewage discharges that once discolored the waters. The other lakes also show signs of improvement. In a recent report to the U.S. Congress, in fact, Great Lake states reported on a survey that included nearly all of the nearshore waters. This report showed that most of the waters are now safe for swimming and other recreational activities, and much of it is clean enough for drinking water.

Despite these efforts, 43 areas failed to meet the standards established in the U.S.–Canadian accord a decade or so later. In 1987, a Buffalo-based environmental group, Great Lakes United, published a report noting that while highly visible and odoriferous pollutants such as sewage had been significantly reduced, many toxic substances not visible to the naked eye continue to pour into the Lakes. For instance, PCBs and pesticides still persist at unacceptable levels. Officials administering the 1978 agreement conceded that the U.S. and Canadian governments had not fully lived up to its terms.

Toxicants enter the lakes from factories and sewage treatment plants; from nonpoint pollution, including farmland and urban runoff; from toxic fallout—that is, pollutants deposited from the atmosphere; and from resuspension of substances contained in the bottom sediments, caused in part when harbors and river mouths are dredged.

One of the most significant avenues of entry for toxins is atmospheric deposition. Trace metals, pesticides, phosphorus, nitric acid, nitrates, sulfates, sulfuric acid, and organic compounds are all deposited from the air. Approximately 60 to 90% of all PCBs entering Lakes Superior and Michigan come from the atmosphere. An estimated 14,000 metric tons (15,400 tons) of aluminum is deposited in Lake Superior from the atmosphere, and nearly 29,000 metric tons (31,900 tons) annually falls from the skies into Lake Michigan. The atmosphere is also a major source of phosphorus, providing an estimated 59% of the total input to Lake Superior. Without controls on atmospheric deposition, it will be impossible to end the discharge of toxins into the Great Lakes.

What does the continued pollution assault mean? First, commercial fishing, once an economic mainstay in the region, has virtually ceased in all of the lakes except Superior. Even there, the commercial fishers operate under continual uncertainty, never knowing when their catch will exceed safe limits for contaminants and be declared unsafe by the FDA. Second, introduced Pacific salmon and reintroduced lake trout survive in the lakes, but the salmon population must be restocked each year. Lake trout do not reproduce successfully either, except in Superior. Restocking costs millions of dollars a year. Third, continued pollution has forced states to issue warnings advising women who are pregnant, lactating, or of childbearing age not to eat certain fish. Parents are also advised not to feed their children lake-caught fish. This warning came after a scientific study on chronic low-level exposure, which showed that infants of women who had eaten PCB-contaminated fish two to three times a month were smaller, more sluggish, and had slower reflexes than infants of women who had not eaten contaminated fish. Children also accumulate the toxicants, which may impact future fertility. Acceptable levels for most of the 800 pollutants found in the lakes simply have not been established, primarily because of a lack of information on health effects. In its report to Congress, the EPA noted that 97% of the nearshore waters of the Great Lakes is under fish consumption advisory—meaning the fish consumption is inadvisable or should be limited to prevent health problems. PCBs are the main culprit. Aquatic life in these waters is also in jeopardy from both toxic chemicals that accumulate in the food chain and habitat loss or degradation and competition and predation by nonnative species such as the zebra mussel and sea lamprey.

Reducing the problems posed by toxic pollutants is complicated. Eight states, two provinces, numerous tribal councils, and two nations share an interest in managing the waters of the Great Lakes. The result is often conflict that holds up the important steps needed to bring the Great Lakes back to a full, productive, and healthy life. In 2002, 20 years after starting cleanup efforts only 2 of the 43 areas that failed to meet standards have been cleaned up. Two more are said to be in recovery.

In recent talks, Canada and the United States agreed on ways to control airborne pollutants. They also agreed to remove contaminated sediments from the lakes and to better control groundwater pollution that contributes to the degradation of the Great Lakes. At least 15 to 30 years of rehabilitation are needed, but improvements are already beginning to be seen. Clearly, there's a long way to go.

FIGURE 21-16 Trickling filter. A decomposer food chain consisting of bacteria and other microorganisms in the rock or bark bed of this system consumes organic matter, nitrates, and phosphates in the liquid sewage.

Secondary Treatment The secondary stage destroys biodegradable organic matter through biological decay (Figure 21-15). Sludge from primary treatment enters a large tank, where bacteria and other organisms decompose the agitated waste. Another common way is to pass the liquid sludge through a trickling filter (**FIGURE 21-16**). Here, long pipes rotate slowly over a bed of stones (and sometimes bark), dripping wastes on an artificial decomposer food web consisting of bacteria, protozoa, fungi, snails, worms, and insects. The bacteria and fungi consume the organics and, in turn, are consumed by protozoans. Snails and insects feed on the protozoans. Some inorganic nutrients—notably nitrates and phosphorus—are also removed during secondary treatment.

Secondary treatment is completed by a secondary settling basin, or clarifier, which removes residual organic matter. In most municipalities, liquid remaining after secondary treatment is chlorinated to kill potentially pathogenic bacteria and protozoans and is then released into receiving streams, lakes, or bays.

The efficiency of primary and secondary treatment is shown in Table 21-4.

Tertiary Treatment Many methods exist for removing the chemicals that remain after secondary treatment. Most of these tertiary treatments are costly and therefore are rarely used unless water is being released into bodies of water that require a high level of purity—for example, Lake Tahoe in California. Fortunately, some cheaper options are gaining recognition. For example, effluents can be transferred to holding ponds after secondary treatment, where algae and water hyacinths growing in the water consume the remaining nitrates

Table 21-4

Removal of Pollutants by Sewage Treatment Plants

Substance	Percentage Removed by Treatment	
	Primary	Primary and Secondary
Solids	60	90
Organic wastes	30	90
Phosphorus	0	30
Nitrates	0	50
Salts	0	5
Radioisotopes	0	50
Pesticides	0	0

and phosphates. Certain aquatic plants such as duckweed absorb dissolved organic materials directly from the water.

Aquatic plants grown in sewage ponds can be harvested and converted into food for humans or livestock. In Burma, Laos, and Thailand, duckweed has been consumed by farmers for years. The protein yield of a duckweed pond is six times greater than that of an equivalent field of soybeans. One of the problems with this approach, however, is that water hyacinths and duckweed also absorb toxic metals from the water. Therefore, consumption by humans and livestock must be carefully monitored. Another approach is the use of artificial wetlands, either indoors in greenhouses or outdoors. These were discussed in Chapter 13 and Spotlight on Sustainable Development 4-1.

Sustainable Solutions for Water Pollution

Despite an outpouring of laws to control pollution, the United States and other countries have hardly come to grips with the problem. Tens of thousands of hazardous waste sites litter the American landscape. Pollution control laws passed in the 1970s initially decreased water pollution nationwide, but since the early 1980s, pollution levels have remained more or less constant. Making matters worse, regulation and enforcement of hazardous-waste laws have been lax.

Solutions to global problems require new laws and tighter controls. Critical thinking demands a search for additional systemic solutions. New technologies, for example, can help reduce waste.

In making the transition to a sustainable society, considerable effort must be made to find solutions that strike at the root causes. These strategies do not merely capture pollutants bound for waterways and dispose of them in a slightly more acceptable manner, in the process often contaminating another medium; they seek to prevent pollution in the first place.

Table 21-5

Environmental Benefits Derived from Substituting Secondary Materials for Virgin Resources

Environmental Benefit	Reduction of Use (percent)			
	Aluminum	Steel	Paper	Glass
Energy use	90–97	47–74	23–74	4–32
Air pollution	95	85	74	20
Water pollution	97	76	35	—
Mining wastes	—	97	—	80
Water use	—	40	58	50

Source: R. C. Lechter and M. T. Sheil. 1986. "Source Separation and Citizen Recycling." In *The Solid Waste Handbook.* W. D. Robinson, ed. New York: Wiley.

Conservation, recycling, renewable resource use, and population control are four forms of pollution prevention. For example, by using energy more efficiently, individuals and companies can reduce their demand for electricity. This, in turn, decreases the amount of cooling water needed by utilities and cuts back on thermal pollution and pollution of streams and lakes with chlorine. Cooling systems are also treated with chemicals to keep organisms from clogging pipes. The less cooling water that is used, the less chemical pollution of waters will result. Farmers can find many ways to cut back on their use of fertilizers and pesticides (Chapters 10 and 22). Many techniques that are essential to creating a sustainable agriculture help control soil erosion (Chapter 10). These measures therefore reduce sedimentation and the release of pesticides, inorganic plant nutrients, and organic chemical pollutants into surface and groundwaters.

Table 21-5 shows the dramatic reductions in water pollution possible from recycling aluminum, steel, paper, and glass. Because recycling these commodities also reduces air pollution—and because air pollution can become water pollution through cross-media contamination—this strategy pays a double dividend. A third payoff comes from the reduction of mining wastes, which often make their way into streams.

In contrast to fossil fuel power plants, most renewable energy sources such as photovoltaics and wind energy require little if any water. Thus they produce little if any air and water pollution. Population stabilization, of course, is the ultimate pollution prevention strategy. Each avoided birth means that much less waste going into waterways.

The sustainable strategy also seeks to weave human actions into the cycles of nature. That is, it attempts to return nutrients to their site of origin or to utilize them for other purposes—for example, by using municipal sewage to fertilize crops. This approach is known as **land disposal.**

In ancient times, land disposal of human sewage was commonplace; it is still practiced in many LDCs, such as China and India. As the populations of less developed countries grew and became more urbanized, this natural method of recycling wastes was gradually phased out. Land disposal, discussed in Chapter 13 as a means of recharging groundwater, uses the surface vegetation, soil, and soil microorganisms as a natural filter for many potentially harmful chemicals. Sewage can be piped to pastures, fields, and forests (**FIGURE 21-17**). Organic matter in the effluent enriches the soil and improves its organic content and ability to retain water. Nitrates and phosphates serve as fertilizers. The water supports plant growth and helps recharge aquifers. Crops nourished by effluents from treated sewage show a remarkable increase in yield.

Land disposal of sludge from sewage treatment plants is not a perfect solution. First, treated sewage may contain harmful bacteria, protozoans, and viruses that could adhere to plants consumed by humans or livestock or become airborne after the effluent dries. To get around this problem, the Swiss and Germans heat their sludge to destroy such organisms before applying it to pastures and cropland. Alternatively, sewage sludge can be decayed in compost piles

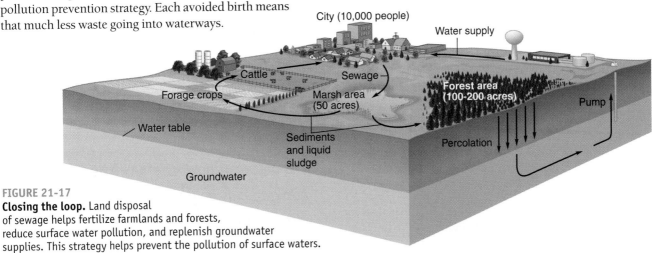

FIGURE 21-17
Closing the loop. Land disposal of sewage helps fertilize farmlands and forests, reduce surface water pollution, and replenish groundwater supplies. This strategy helps prevent the pollution of surface waters.

- Compost organic waste rather than using the garbage disposal.
- Use nontoxic, biodegradable cleaning supplies.
- Don't discard oil or other harmful substances down the drain.
- Use energy efficiently to lower your demand for oil and your personal share of global air pollution.
- Recycle all the waste you can and use recycled materials to reduce water and air pollution.
- Use organic fertilizers and natural pesticides on lawns and gardens.
- Help replant denuded landscapes in your area to prevent soil erosion and nonpoint water pollution.
- Help protect wetlands or support groups that take an active role in this activity.

before application. The heat given off during composting kills virtually all of the viruses, bacteria, and parasite eggs.

The second problem is that toxic metals found in some sewage may accumulate in soils and be taken up by plants and livestock. Metal-contaminated sewage usually comes from industries. By removing metals from their waste stream (an option called **pretreatment**) or by finding ways to prevent them from entering the waste stream entirely (pollution prevention), industries can eliminate the problem.

Third, transporting sludge to fields increases the cost of sewage treatment, limiting some of the incentive to use this method. Experts are quick to point out, however, that land disposal is 10 times cheaper than building and operating a tertiary treatment plant.

Scientists at the University of Maryland recently developed an unusual way to put sludge to good use. By combining it with clay and slate, they formed odorless *biobricks*, which look like ordinary bricks. Washington's Suburban Sanitary Commission recently put the idea to the test by building a 750-square-meter (8300-square-foot) maintenance building in Maryland with 20,000 biobricks. If successful, biobricks could help reduce land disturbance resulting from mining materials used for brick making—and could help cut sewage disposal costs and environmental contamination.

Restoration efforts can also help clean up water supplies. Direct efforts are needed to clean up groundwater, as noted earlier in the chapter. In addition, restoration of wetlands can help reverse the decline of aquatic systems. In Florida, for example, drained swampland that was converted to farms is now pouring vast amounts of plant nutrients into the once massive Everglades. The native sawgrass that covers much of this massive wetland is adapted to a nutrient-poor environment. In the presence of excess nutrients, cattails are taking over, outcompeting the slow-growing sawgrass. To date, about 8100 hectares (20,000 acres) of the Everglades National Park has been destroyed by cattails, which are spreading at a rate of about 1.6 to 2.4 hectares (4 to 6 acres) a day. Debris from the plants rots in the water and creates anaerobic conditions that kill fish and other aquatic species.

To reverse this trend, Florida officials have proposed taking 33,000 acres of drained swampland, now being farmed, out of production. These lands would be flooded and set aside as a marshy area that would filter out nutrients from surrounding farmland, reducing the influx to the Everglades Park by about 85%.

These are just a handful of steps needed to solve the water pollution dilemma sustainably. Of course, personal actions are also needed (see the Individual Actions Count! table for ideas). Limiting family size and reducing the consumption of unnecessary goods are important actions we can each take. By reducing the purchase of goods, we can greatly reduce hazardous waste production at factories. As a general rule, each ton of garbage generated by consumers is preceded by 5 tons of manufacturing waste and 20 tons of waste produced at the source (the mine or forest where the raw materials were acquired). So every savings we make, by buying less or by buying more durable items, can effect large upstream savings in waste and pollutants that might otherwise make their way into our waterways.

> As a general rule, each ton of materials we throw away is preceded by 5 tons of manufacturing waste and 20 tons of waste produced at the source (the mine or forest where the raw materials were acquired).

Another effective way to cut your personal contribution to water pollution is to install a composting toilet. The composted wastes can be added safely to gardens, eliminating the need for synthetic fertilizer. If and when you become a homeowner, restrict your use of synthetic fertilizers, insecticides, herbicides, bleaches, detergents, disinfectants, and other household chemicals—or find safe alternatives. Select low-phosphate or no-phosphate detergents for washing your clothes. You may also contact local and federal officials to support further cleanup efforts.

KEY CONCEPTS

Preventing pollution is a key to creating clean, healthy ground and surface waters, which are a prerequisite to living sustainably on the Earth. Efforts to reduce consumption; recycling materials, industrial waste, and municipal sewage; using renewable resources; and stabilizing population growth—all of these will collectively serve to reduce our production of water pollutants.

It is astonishing with how little wisdom mankind can be governed when that little wisdom is its own.

—*W.R. Inge*

CRITICAL THINKING

Exercise Analysis

Water conservation in the home is designed to save water, but it has other benefits as well. For example, it helps reduce energy demand because it takes energy both to pump water to treatment plants and to heat water for domestic use. The less water you use, the less energy is needed in your home and at the water treatment plant. This in turn results in numerous environmental savings, from reduced habitat destruction from coal mining to less air pollution.

Water conservation in the home also means less water going down the drains to sewage treatment plants. That saves energy at the sewage treatment plant. It also reduces the demand for chlorine, which is used to disinfect the effluent of sewage treatment plants. Can you think of any other benefits?

This exercise clearly shows how important it is to examine the big picture. Creating a sustainable future requires steps with multiple benefits. We need to conserve energy and reduce pollution wherever we can; so look at each solution as an opportunity to solve several problems. Many people think that conservation means sacrifice, when in fact water conservation measures in the home can be quite inexpensive and easy to implement. Simple changes in behavior—such as not flushing the toilet every time or watering your lawn less frequently—actually reduce your work. Conservation measures can save you money as well, which can be used to enhance your life.

CRITICAL THINKING AND CONCEPTS REVIEW

1. List the major types of water pollutants found in developing and developed nations. What are the major differences? Why do these differences exist?
2. Define the terms *point source* and *nonpoint source*. Give some examples of each and explain why nonpoint sources of water pollution are often more difficult to control than point sources.
3. Describe where organic nutrients come from, what effects they have on aquatic ecosystems, and how they can be controlled.
4. What are inorganic plant nutrients, and how do they affect the aquatic environment? List some strategies for eliminating these pollutants.
5. Define the terms *natural eutrophication* and *cultural eutrophication*. Describe how inorganic and organic nutrients accelerate natural succession in a lake.
6. What are the major sources of infectious agents in polluted water? How can they be controlled?
7. What are some of the major inorganic water pollutants? How do they affect the aquatic environment and human populations?
8. Why is chlorine used in the treatment of human sewage and drinking water? What dangers does its use pose? How would you determine if the risks of chlorine use outweigh the benefits?
9. What are the major sources of sediment, and how can they be controlled? What are the costs and benefits of sediment control?
10. A factory that has been polluting a nearby stream with toxic wastes (volatile organic chemicals) has agreed to stop. The owners are proposing the use of evaporation ponds, where the wastes would sit until they evaporated. Using your critical thinking skills, your systems-thinking abilities, and your knowledge of water pollution, critically analyze this proposal.
11. What are the major sources of groundwater pollution? How can groundwater pollution be reduced or eliminated? What can you do?
12. Discuss the sources of oil in the ocean and ways to reduce oil contamination.
13. Critically analyze the following statement: "Oil naturally seeps into the ocean. In fact, about half of the oil in the ocean comes from natural sources. Because of this, it makes little sense to worry about spills from tankers and other sources."
14. Describe primary and secondary wastewater treatment. What happens at each stage, and what pollutants are removed?
15. A housing development is being built in your town. The developers say that this project will not increase pollution in the river that flows into the town because they will pay to expand the town's sewage treatment plant. Do you agree or disagree with their assertion? Why?
16. Pollution control devices are often viewed as stopgap measures. Why are such measures essential for long-term sustainability?
17. List and critically analyze all of the options available to us to prevent pollution, such as population stabilization. Why are these approaches essential to the task of building a sustainable society?

KEY TERMS

biochemical oxygen demand (BOD)
chlorinated organics
Clean Water Act
coliform bacterium
cross-media contamination
cultural eutrophication
fecal streptococcal bacteria
groundwater
inorganic nutrients
land disposal

Medical Waste Tracking Act
metabolism
methyl mercury
natural eutrophication
natural succession
nonpoint sources
Ocean Dumping Ban Act
Oil Pollution Act
Plastic Pollution Control Act

point sources
pretreatment
red tide
sediment
sedimentation
streambed aggradation
surface waters
thermal shock
water pollution

A crop duster leaves a cloud of toxic pesticides that often drifts to neighboring areas.

CHAPTER 22

Pests and Pesticides: Growing Crops Sustainably

What we do for ourselves resides with us. What we do for others and the world remains and is immortal.

—*Albert Pine*

Ron Rosmann is not a revolutionary; he's an Iowa farmer. Like many of his cohorts, though, he is sowing the seeds of a revolution in farming. For years Rosmann, like other corn and soybean farmers, believed that without chemical pesticides on his farm, he'd be out of business. **Pesticides** are chemical substances that kill insects, weeds, and a whole assortment of organisms that reduce crop output.

In the 1980s, Rosmann visited a couple of farmers that grew crops without pesticides. Much to his surprise, their fields looked great. There were few weeds and insects. Encouraged by what seemed an impossibility, Rosmann began to experiment with alternatives to pesticides the next year. On his farm, weeds were the biggest problem, and herbicides (chemicals used to kill them) were a big expense. By using techniques he had seen the year before, Rosmann found that he could virtually eliminate herbicides. In the 9 years that followed, he used herbicides only once, and then in small amounts because he had mistakenly let weeds get out of control. Today, Rosmann saves $4000 to $5000 a year on herbicides. Much to his surprise, his crop yields have gone up. In other words, he's growing more food per acre at a lower cost.

Rosmann's efforts are a prime example of sustainable development. They are good for people, the environment, and the economy. Unlike Rosmann, farmers in the industrial countries collectively spend billions of dollars on chemical pesticides. To them, weeds and other pests are an impediment to efficient farming. They reduce production and profits.

Each year weeds, insects, bacteria, fungi, viruses, birds, rodents, mammals, and other organisms—pest species—consume or destroy an estimated 48% of the world's food production (this includes both preharvest and postharvest losses). The highest rate of destruction occurs in the tropics and subtropics, where because of favorable year-round climate, as many as three crops are grown each year on the same field.

> Each year pest species consume or destroy an estimated 48% of the world's food production.

The warm weather and long growing season create optimal conditions for crop-eating insects.

Crop destruction from pests is high even in the more developed nations, despite elaborate and costly control strategies. In the United States, for instance, preharvest losses are estimated to be about 37%, and postharvest losses amount to 9% of what is left. Together, about 45% of annual U.S. production is lost to various pests. Some of the most common insect pests in the United States and their ranges are shown in FIGURE 22-1.

Similar losses are reported in other more developed countries. In a world where about 20% of the people are malnourished, such losses are tragic. The silver lining behind this ominous cloud, however, is that efforts to reduce losses through pest control could help increase food supplies. As you will see in this chapter, pest control must be embarked on judiciously. It requires a careful analysis of the social, economic, and environmental costs of all strategies—and the use of sustainable strategies.

To help set the stage for the discussion of sustainable pest management, this chapter begins with a discussion of the conventional pest control strategies and the many environmental problems that make them unsustainable. It ends with a discussion of the many options, including those pursued by Ron Rosmann, for controlling pests economically and in a socially and environmentally sustainable manner.

22.1 Chemical Pesticides

In China over 3000 years ago, farmers controlled locusts by burning infested fields. In the ancient Middle East, open ditches were used to trap immature locusts. In 1182, Chinese citizens were required to collect and kill locusts in an effort to control an outbreak. Such measures involving direct human intervention are often called *cultural controls*.

Over the years, many farmers have also tried various toxic chemicals. Early chemical pesticides, known as **first-generation pesticides**, were simple but sometimes highly toxic preparations made of ashes, sulfur, arsenic compounds, ground tobacco, or hydrogen cyanide. Lead, zinc, and mercury compounds were also used. In Greece, for instance, farmers used arsenic, sulfur, caustic soda, and olive oil to control crop pests. Today, few of the first-generation pesticides remain in use. Most proved to be too toxic to people or ineffective. Many were environmentally persistent. Many of the compounds used 50 to 60 years ago still contaminate soils.

Pest control measures have been used for centuries. Earliest techniques included cultural controls, such as burning fields to kill locusts, and chemical controls, the use of toxic chemicals to kill insect pests. Early chemicals have been abandoned because they were toxic, ineffective, or environmentally persistent.

Modern Chemical Pesticides

In 1939, the Swiss chemist Paul Müller discovered the insecticidal properties of a synthetic organic compound called **DDT (dichlorodiphenyltrichloroethane)**. This discovery ushered in a new era of chemical control. DDT became the first in a long line of **second-generation pesticides**, synthetic organic pesticides. For the 25 years following its development, DDT was viewed by many as the savior of humankind, for it was quite lethal to insect pests and thus increased crop yield. Since DDT was relatively inexpensive to produce, its use became widespread. In 1944, in fact, Müller was awarded a Nobel prize.

Over the years, thousands of new chemical pesticides have been synthesized and tested. Today, 1500 different substances are used in over 33,000 commercial formulations of herbicides, insecticides, fungicides, miticides, and rodenticides, with one goal in mind: to reduce pests to tolerable levels. Some pesticides, including DDT, are **broad-spectrum pesticides**, which attack a wide variety of organisms; others are **narrow-spectrum pesticides**, used in controlling a few pests. **FIGURE 22-2** shows the pesticide use in the United States since 1980.

KEY CONCEPTS

The first synthetic organic chemical used as a pesticide was a chemical called DDT, whose discovery ushered in a new era of synthetic chemical pesticides. Chemical pesticides are either broad-spectrum substances that kill a variety of pest species or narrow-spectrum chemicals that are effective against one or a few pests.

Types of Chemical Pesticides Synthetic chemical pesticides fall into three chemical families: (1) chlorinated hydrocarbons (organochlorines), (2) organic phosphates (organophosphates), and (3) carbamates.

The first group to be developed, the **chlorinated hydrocarbons**, is a high-risk group that includes DDT, aldrin, kepone, dieldrin, chlordane, heptachlor, endrin, mirex, toxaphene, and lindane. All of these have been banned or drastically restricted or are being considered for such actions because of their ability to cause cancer, birth defects, neurological disorders, and damage to wildlife and the environment. Chlorinated hydrocarbons are extremely resistant to breakdown, persist in the environment, are passed up the food chain (biomagnified), and remain for long periods in body fat. As you will see later in this chapter, a pesticide banned in the United States may still be found on food imported from countries where no bans are in place.

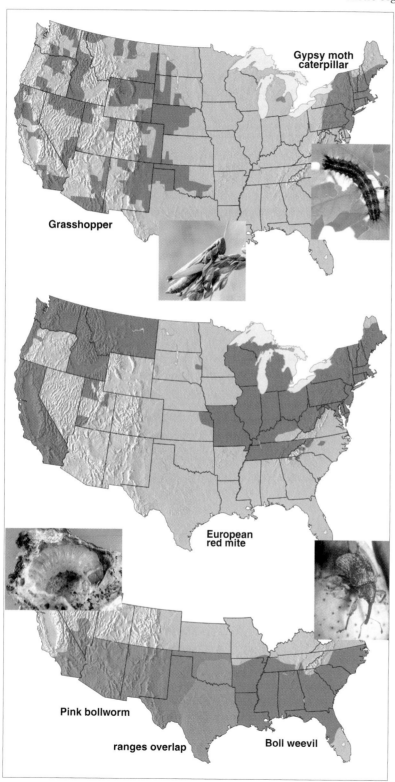

FIGURE 22-1 The geography of major pests. These maps of the United States show the most prominent insect pests and where they are found.

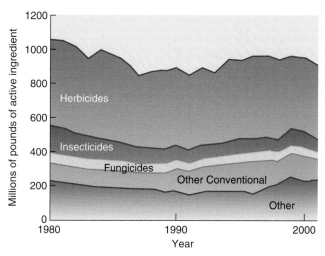

FIGURE 22-2 A profile of pesticide popularity. This graph shows pesticide production in the United States began to decline in the 1980s and early 1990s, in part because of a shift away from environmentally harmful pesticides. (Data from http://epa/gov/oppbead1/pestsales/01pestsales/historical_data2001_3.html.)

The second group, the **organic phosphates**, consists of chemicals such as malathion and parathion. These pesticides, while still toxic, break down much more rapidly than chlorinated hydrocarbons. The chlorinated hydrocarbon DDT, for example, has a half-life in the environment of a couple of years; parathion, an organic phosphate, has a half-life of a couple of days.

The organic phosphates are water soluble and are excreted in the urine. Because they are water soluble, they are less likely to bioaccumulate (Chapter 18). Despite their benefits, the organic phosphates are still of great concern, for humans exposed to even low levels may suffer an assortment of health effects such as drowsiness, confusion, cramps, diarrhea, vomiting, headaches, and difficulty breathing. Higher levels cause severe convulsions, paralysis, tremors, coma, and death.

The third group, the **carbamates**, are widely used today as insecticides, herbicides, and fungicides. One of the most common is carbaryl (commonly known as Sevin). As a group, the carbamates are, like the organic phosphates, less persistent than the chlorinated hydrocarbons, remaining a few days to two weeks after application. They are also water soluble and do not bioaccumulate. Carbamates are nerve poisons, as are the chlorinated hydrocarbons and organic phosphates. Although they are environmentally safer than their predecessors, carbamates have been shown to cause birth defects and genetic damage.

Organophosphates and carbamates have one key advantage over the chlorinated hydrocarbons— their lack of persistence. This means that they kill their pests and then disappear within a few days or weeks. Unfortunately, they are much more toxic to humans than the chlorinated organics. They enter the body rapidly through the skin, digestive system, and lungs and therefore can be harmful to people who apply them to crops and to people who live nearby.

Although chlorinated hydrocarbons, organic phosphates, and carbamates are the three main types of pesticides in use today, they are not the only ones. Two others worth noting are the **triazines** (TRY-ah-zeens), a relatively harmless family of chemical herbicides, and the **pyrethroids** (PIE-rithroids), a group of natural and synthetic insecticides, also relatively nontoxic to humans.

KEY CONCEPTS

Three main types of synthetic organic chemical pesticides have been developed over the years—chlorinated hydrocarbons, organophosphates, and carbamates. These are all neurotoxins. Carbamates are less persistent in the environment or body fat of organisms—and are thus less likely to be biomagnified in the food chain than organochlorines. Unfortunately, they are much more toxic to people.

Growth in the Use of Chemical Pesticides

Approximately 2.5 million metric tons (2.7 million tons) of chemical pesticides are used annually throughout the world, approximately 22% in North America and about 57% in Europe and other more developed nations. The remaining 21% is used primarily in less developed countries. The United States is a leading agricultural nation and a leader in pesticide use. In fact, about one fifth of all pesticides used in the world each year are applied to U.S. crops. Of these pesticides, 10% are insecticides for insect control, 46% are herbicides for weed control, 7% are fungicides, and the remaining 37% include nematicides, fumigants, rodenticides, molluscicides, and fish/bird poisons.

KEY CONCEPTS

The more developed nations, including the United States, are the major consumers of chemical pesticides. In the United States, the bulk of the pesticides in use are herbicides.

Overuse

According to University of Illinois entomologist Robert Metcalf, author of a standard college textbook on pest management, farmers apply more than twice the pesticide they need. Adding to the unnecessary environmental contamination are a cadre of misinformed homeowners who apply about one tenth of all U.S. chemical pesticides on gardens, lawns, and trees in higher amounts per acre than farmers do. Rains wash excess pesticides into sewers, streams, and other water bodies. Soils exposed to heavy applications of pesticide often suffer because beneficial bacteria are destroyed. Without bacteria, grass clippings decay very slowly, creating a layer of undecomposed material

On average, farmers use about two times as much pesticide as is needed. Homeowners tend to use even more and account for one tenth of all pesticides used in the United States.

called *thatch*. Thatch buildup on lawns is believed to be one by-product of excessive use of pesticides.

Individuals not only apply pesticides in excess, they often take poor precautions to protect themselves when applying chemicals to gardens and lawns. That is, they rarely use or wear gloves and face masks. As noted later in the chapter, extremely toxic pesticides are often restricted—that is, not available for homeowners to use.

KEY CONCEPTS

Pesticides are often applied far in excess of what's needed by farmers and especially homeowners. Homeowners also often fail to take precautions to protect themselves from exposure.

Biological Impacts of Pesticides

Pesticides constitute only about 3% of the commercial chemicals commonly used in the United States each year. Nonetheless, because they are released into the environment in large quantities and have the potential to alter ecosystem balance and threaten human health, their use has created widespread and often heated controversy.

In the 1940s, DDT and other new pesticides were applied to a variety of crops, with astonishing results. DDT was even used to delouse soldiers and civilians in World War II. DDT and other chemical pesticides proved to be fast and efficient in controlling insects, weeds, and other pests. In India, for example, before the use of DDT in the 1950s to control malaria-carrying mosquitoes, there were over 100 million cases of malaria a year. By 1961, the annual incidence had been reduced to 50,000. DDT no doubt saved millions of lives.

Pesticides also allowed farmers to respond quickly to pest outbreaks, thus avoiding economic disaster. Pesticides were relatively cheap and easy to apply, and their use resulted in substantial financial gains as yields increased. Some insecticides, such as DDT and dieldrin, persisted long after application, giving extended protection. The persistence of these pesticides, in fact, was thought to be one of their major advantages, for one application could have lasting effects.

The successes of the early phase led to a rapid expansion in pesticide production and use. During this period, researchers developed many new pesticides. In the ensuing years, however, many problems started to emerge.

KEY CONCEPTS

The initial success of DDT led to a rapid increase in pesticide use and in the number of new chemical pesticides. Problems soon began to emerge, though.

Destroying Natural Pest Controls and Beneficial Insects
One of the most troubling problems is that pesticides, especially broad-spectrum chemicals, often destroy natural biological pest controls—that is, insect predators and parasitic insects that naturally help control pests and potential pests. The loss of these beneficial insects (predators, for example) may result in a proliferation of pest species. In fact, some insect species that were previously benign suddenly in-

crease in number and begin creating a need for additional pesticides. Agronomists call this population explosion of new pests an **upset**.

A classic example of such an upset took place in California. Spider mites, once only a minor crop pest, have become a major pest because of the use of pesticides that killed off many of their natural enemies, which were more sensitive to the sprays. Today, mites cause twice as much damage as any other insect pest in California and cost farmers (in damage and control) five times what they cost 35 years ago. Two of modern farming's most costly pests, the cotton bollworm and the corn-root worm, were minor problems 50 to 60 years ago, before widespread pesticide use reduced their natural predators. In the United States, one third of the nation's 300 most destructive insect pests are secondary pests—that is, insects that once caused little or no trouble at all.

> " In the United States, one third of the nation's 300 most destructive insect pests are secondary pests—that is, insects that once caused little or no trouble at all. "

Pesticides also destroy other beneficial insects, such as honeybees, which play an important role in pollination. Honeybees pollinate U.S. crops annually worth an estimated $15 billion a year in 1999, the latest data available. Apple orchards are particularly hard hit. Honey produced by honey bees was valued at $225 million in 2003. Over 400,000 bee colonies are destroyed or severely damaged by pesticides in the United States each year. Honey bee colonies are also being destroyed by parasitic mites that suck the blood from adult bees. Mites are transmitted by Africanized honeybees and other species. Losses are also attributed to habitat loss, and even habitat modifications.

KEY CONCEPTS

One major impact of pesticides is that they often kill natural pest control agents such as predatory insects that are more sensitive to pesticides. This unleashes the growth of pest populations. Pesticides also kill other beneficial insects such as pollinators.

Creating Genetically Resistant Pests   Another unanticipated effect of pesticide use has been the dramatic increase in **genetically resistant insects**. Because of genetic diversity (Chapter 6), a small portion of any insect population (roughly 5%) is genetically resistant to a pesticide. That is, it is not killed by a normal application (**FIGURE 22-3**). Why are they resistant? They may contain enzymes, for example, that destroy or detoxify the poison. The presence of these enzymes is a product of their unique genetic makeup. Therefore, an initial application of pesticides will kill all but the genetically resistant members of a pest population. Although these survivors do little damage at first, over time they reproduce and form a sizable population, which can cause significant crop damage.

As farmers encountered genetic resistance, they began to try new approaches. The first strategy was to increase the amount of pesticide. Because of genetic resistance, this so-

FIGURE 22-3 Pesticide resistance. A tobacco budworm crawls through deadly DDT unaffected. Because of genetic resistance, many other species are unaffected by the toxic chemicals designed to kill them.

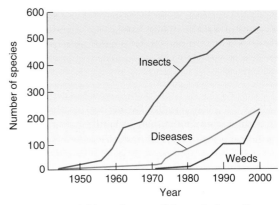

FIGURE 22-4 Pesticide resistance. This graph shows the number of species of insects, plant pathogens (disease organisms), and weeds that are resistant to at least one pesticide.

lution is effective only in the short term. Why? Even though the higher dose wipes out the majority of the insect pests, it invariably leaves behind a small subpopulation that is genetically resistant to the increased dose. They soon proliferate and cause farmers to apply even more, creating a vicious addictive cycle that many observers call the **pesticide treadmill**. When DDT and other insecticides were first introduced into Central America, cotton fields were sprayed eight times each growing season; today, because of genetic resistance, 30 to 40 applications are typical.

The second strategy was the development of new pesticides. However, scientists found that it was expensive to create pesticides and that the insects developed genetic resistance to these new chemicals almost as quickly as they were released.

Genetic resistance to DDT was first reported in 1947 by Italian researchers. Today, approximately 540 insect species are resistant to at least one form of pesticide (**FIGURE 22-4**). Over 20 of the world's worst pests are now resistant to most types of insecticide. Moreover, farmers are finding that crop diseases and weeds are also developing resistance to the very chemicals that are used to control them. Worldwide, 230 plant pathogens (organisms that cause disease) and 220 weeds have developed resistance to at least one pesticide.

Since the introduction of DDT, insects have never met a chemical pesticide they couldn't defeat. What is more, despite the increased application of chemical pesticides, annual losses due to pests have continued to climb. Although insecticide use has increased 10-fold

> Worldwide, over 220 weeds, 230 plant pathogens, and 540 insect species are resistant to at least one form of pesticide. More than 20 of the worst insect pests are now resistant to most types of insecticide.

since World War II, crop damage has doubled. Today, not only do insects take a larger percentage of the U.S. harvest than they did before the introduction of DDT, but damage from fungi and weeds is climbing as well.

Damage to Fish and Wildlife Many chemical pesticides have proved harmful to nontarget species: birds, fish, and other animals. Survival rates in newly hatched trout in upstate New York, for example, were depressed by DDT that had been applied to nearby forests. Insect- and worm-eating birds also perished in areas where aerial spraying of insecticide had occurred. As a result of widespread pesticide use, populations of many birds plummeted.

The manufacturers of pesticides argued that the chemicals were found in only minute concentrations in the environment and could not be the cause of declining populations of fish and wildlife. Numerous experimental studies, however, showed that certain persistent pesticides—even when present in small amounts in the environment—could drastically affect the reproduction and survival rate of birds and other animals through **biomagnification**, described in Chapter 18.

Studies showed that although DDT and DDE (a toxic breakdown product of DDT) levels in aquatic ecosystems were quite low, concentrations were higher in producers and still higher in consumers because of biomagnification. Fish-eating birds, the consumers at the top trophic level, had the highest concentrations of these chemicals. Although these levels were not lethal to adults, they impaired reproduction. DDT and DDE reduced the deposition of calcium in the eggshells of birds that feed on fish and of other birds. This list includes peregrine falcons, brown pelicans, cormorants, bald eagles, gulls, and ospreys. Of the two compounds, biologists discovered that DDE posed the greater physiological threat to birds, in part because it persisted longer. Reduced eggshell calcium levels create a thinner, more fragile shell that cracks easily during incubation. As a result of widespread DDT contamination, many predatory populations were nearly wiped out. In one study of bald eagle reproduction, North Dakota State University zoologist James Grier showed that the number of young per nest in northwestern Ontario declined by about 70% between 1966 and 1974.

The newer pesticides are also harmful to fish and wildlife. In the United States, more than 213 million pounds (97 million kilograms) of chemical pesticides are used on lawns (both private and commercial), gardens, and golf courses. One pesticide, diazinon (die-AS-eh-non), was once commonly used on sod farms and golf courses throughout the United States to control insects. In 1984, three fairways were treated with diazinon in Hempstead, New York. In the 2 days following treatment, 700 Atlantic brant geese, 7% of the state's entire population, died of acute diazinon poisoning.

Diazinon was also used to treat nine fairways at a golf course in Bellingham, Washington. After application, the area was irrigated to decrease pesticide concentrate on the surface. Nevertheless, 85 American wigeon ducks perished after eating grass on one of the treated fairways on the day of application.

In 1988, the EPA banned the use of diazinon on golf courses and sod farms because of similar incidents. It is still available for use in other outdoor areas. Although it continues to kill wildlife, especially birds, mortality rates are down now that it is no longer used on golf courses and sod farms.

Birds have also been dramatically affected by the use of granular carbofuran (car-bow-FUHR-an), which was developed in 1970 by the FMC Corporation of Philadelphia. Farmers throughout the United States use carbofuran to eradicate nematodes and insects from corn, rice, and other crops. Although carbofuran apparently poses no threat to humans when applied to crops, it is lethal to songbirds. In fact, a songbird can die after ingesting a single granule.

In 1989, EPA records showed that about 2 million birds were dying of carbofuran poisoning each year. In 1990, more than 200 songbirds were poisoned in eastern Virginia near the Rappahannock River. FMC Corporation acknowledged a problem but faulted the farmers for misusing carbofuran or for mishandling it by spilling it.

In 1991, FMC issued new instructions for the use of carbofuran. Still suspicious, citizens in Virginia decided to monitor the results of that year's pesticide application on 360 hectares (900 acres) of treated farmland. They found 62 bird carcasses, 10 sick birds, and 47 feather spots (indicating a bird had died but been eaten by a scavenger) on the field. Many more birds could have died outside the study site. Considering that hundreds of thousands of hectares of farmland in the state had been treated with carbofuran that spring, it was estimated that tens of thousands of birds probably died from the pesticide.

In 1991, the state of Virginia banned the use of carbofuran, and since that time, it has been banned nationwide by the EPA. Though the ban will likely have a positive impact on birds, it is not faultless. Export of carbofuran will continue and will probably increase to make up for shrinking domestic sales.

Currently, the U.S. EPA and wildlife agencies are considering banning or restricting use of carbaryl (Sevin™), a likely carcinogen that may also be toxic to certain fish like salmon. To learn more about this and other pesticides, you can log on to Beyond Pesticides' website (www.beyondpesticides.org).

KEY CONCEPTS

Pesticides poison fish, birds, and other species outright and also biomagnify in the food chain. Thus, seemingly low concentrations in the environment can result in very high levels in organisms in the uppermost trophic level, which may impair reproduction.

Human Health Effects   Early studies showed the presence of DDT in fish, beef, and other foods consumed by humans. DDT also appeared in the fatty tissues of seals and Eskimos in the Arctic, far from its point of use, indicating that it was traveling in the atmosphere to remote parts of the globe. DDT in the atmosphere was washed from the sky by rain and passed through the food chain. It was also detected in human breast milk, a discovery that caused considerable alarm although the long-term effects of low levels on humans remain unknown.

Although the general public is exposed to pesticides in a variety of ways, it is farm and chemical workers who are exposed to the highest levels from direct exposure to pesticides on the job as well, indicating widespread misuse. Workers pick up pesticides on their clothing and skin through accidents and negligence or by prematurely entering sprayed fields. In most cases, workers are poorly protected (**FIGURE 22-5**). They are often given few instructions and no protective gear to minimize exposure—especially in the less developed countries, where poor worker safety standards and practices and widespread illiteracy are common. In the less developed countries, workers often complain of being sprayed with insecticides from hel-

FIGURE 22-5 Farm workers at risk. Poorly protected farm workers are exposed to the highest levels of pesticides.

icopters or planes while they are working in fields and on plantations. A study in Palestine uncovered another serious problem: The instructions on pesticide containers were printed in Hebrew, despite the fact that most of the farmers and farm workers read only Arabic.

In less developed countries where farm labor is abundant and worker safety provisions minimal, workers are often viewed as an expendable commodity. Christopher Brady, who is active in development work in Latin America and Africa, writes, "As a result of increasing health risks to the sprayers, Tela (a Honduran subsidiary of Chiquita Brands, Inc., a U.S. company) has changed their hiring policy. Sprayers are now only hired on a six-month contract." They are "let go and new ones hired before any serious health problems may be detected." Workers spraying herbicides on banana plantations are given a thin cotton mask as their only protective gear. Moreover, empty barrels once containing concentrated pesticide are often used by local villagers to hold drinking water.

Symptoms of poisoning are many. They include insomnia, nausea, and loss of sex drive. Some people exposed to pesticides complain of reduced powers of concentration, irritability, and nervous disorders. Spills on one's skin may cause rashes and a burning sensation. In severe cases, death will occur.

In the United States, at least 45,000 workers are seriously poisoned each year, but many experts believe that this figure grossly underestimates the number of serious poisonings. Surveys in California, for instance, show that three fourths of all serious poisonings go unreported. Each year 200 to 1000 people die from pesticide poisoning in the United States. Worldwide, there are at least 500,000 (perhaps as many as two million) pesticide poisonings annually, according to various estimates. These result in somewhere between 4000 and 19,000 deaths each year, according to World Health Organization statistics. Pesticide poisoning also results in numerous chronic and fatal illnesses. What is more, the WHO expects the number of poisonings to increase if pesticide use increases in the less developed countries.

> An estimated 45,000 people, mostly workers, are seriously poisoned by pesticides in the United States each year. Worldwide, the number is at least 500,000. Deaths in the United States resulting from pesticide poisoning are estimated to be 200 to 1000 people each year. Worldwide, the number is estimated to be 4000 to 19,000.

Although farm and chemical workers are the groups most heavily exposed to pesticides, residents of rural and even suburban areas are often exposed to potentially dangerous levels if they live near agricultural lands. Families living near fields sprayed with herbicides and pesticides outside Scottsdale, Arizona, for example, suffered from persistent headaches, cramps, skin rashes, dizziness, high blood pressure, chest pains, persistent coughs, internal bleeding, and leukemia. Health officials are worried the most about possible long-term problems from exposure.

People living near pesticide-treated fields are heavily exposed because one half to three fourths of the sprayed material never reaches the ground but is carried away by light winds (FIGURE 22-6). Recent studies have also shown that pesticide applied to one crop may be able to vaporize under sunlight and drift to neighboring crops. Exposure may also occur through contaminated groundwater.

In an effort to control mosquitoes, many cities in the United States routinely spray insecticides in neighborhoods and nearby breeding areas. In Florida, for instance, various mosquito control agencies argue that controlling the insect is vital to real estate interests and tourism because it minimizes the risk of encephalitis, a potentially deadly brain infection that is caused by an organism carried by mosquitoes. The 1990–1991 Florida encephalitis epidemic resulted in over 130 cases and more than 10 deaths. Encephalitis is still a problem. In 2004, there were numerous reported cases and nine deaths.

To protect residents from West Nile Virus, which is carried by mosquitoes, many cities have started or increased pesticide spraying. To protect against diseases, trucks and aircraft spew out tons of pesticides while residents sleep. What long-term health impact this has, if any, is unknown. Florida wildlife officials, however, think that pesticide use is largely responsible for a 70% decline in the population of snook, a popular sport fish. Adding to the problem, city officials use a variety of pesticides in city parks, and lawn-care

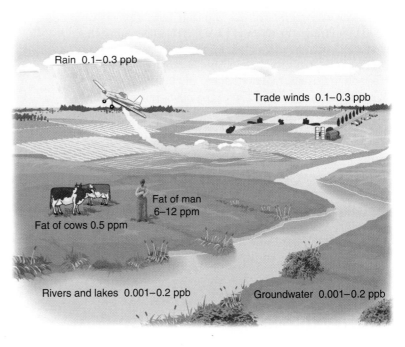

FIGURE 22-6 **Pesticide drift.** Pesticide sprayed from planes contaminates the ecosystem because much of the pesticide drifts away. Various avenues for the dispersal of pesticides are shown. Average values for DDT concentrations are indicated in parts per million and billion. Families living near sprayed fields are exposed to herbicides and pesticides.

companies and individuals douse lawns and trees with a variety of toxic substances, often incorrectly and without warning neighbors.

Consumers may also become the victims of pesticide poisonings. In the summer of 1985, for instance, 1400 people on the West Coast were stricken with nausea, diarrhea, vomiting, and blurred vision after eating watermelons contaminated with the pesticide aldicarb illegally used by farmers. The EPA permits aldicarb for use on cotton and on vegetables that are cooked before consumption (such as beans and potatoes) but not on crops such as watermelon whose produce is eaten without cooking. In 1992, 29 people were poisoned and three hospitalized in Ireland after eating cucumbers sprayed with aldicarb, also not approved for use on this vegetable.

> **KEY CONCEPTS**
> Pesticides contaminate many foods and have been found in human body tissues even in remote areas of the world, indicating that pesticides are globally distributed. Chemical and farm workers are frequently exposed to the highest levels, especially in the less developed nations. The effects of pesticide exposure range from mild neurological problems to death, depending on the exposure level and type of chemical.

The Economic Costs of Pesticide Use

Although pesticide use helps protect crops and saves us billions of dollars, pesticide use has an environmental impact. All environmental impacts have an economic cost. Human pesticide poisonings, described in the previous section, cause illness and death, both of which exact a huge economic cost. Some pesticides have been shown to damage crops. In the 1991 growing season, for instance, a fungicide called *benomyl*, produced by DuPont and sprayed on a variety of crops in the United States and other countries, caused millions of dollars worth of crop damage. Studies showed that in hot, humid climates, benomyl is converted into several phytotoxic (plant toxic) compounds. Florida was one of the worst-hit areas, with damages estimated at $1 billion. To date, more than 1600 U.S. farmers in 40 states have filed claims against the company; three quarters of them have been settled out of court at a cost of over $1 billion. Similar problems have been reported in Costa Rica, Puerto Rico, and Jamaica.

Pesticides may also end up in groundwater in rural communities, causing a potential health risk and necessitating costly cleanup. The Monsanto Company, which produces herbicides, estimated that of 6 million wells in a surveyed area in the United States, 13% (770,000 wells) were contaminated with one or more of five herbicides; 6600 wells contained levels exceeding the EPA's maximum contamination level. Although the company insists that none of the herbicides poses a threat to health, it is offering well owners whose water is contaminated above specified concentrations up to $2000 a well to make corrections.

> **KEY CONCEPTS**
> Pesticide use is not just a threat to human health and the environment, it causes considerable economic damage.

Herbicides in Peace and War

Forty-six percent of all chemical pesticides applied to crops are herbicides. Although there are over 180 types of synthetic herbicides on the market, four products are used in greatest quantity: atrazine, alachlor, butylate, and 2,4-D (2,4-dichlorophenoxyacetic acid). Two herbicides have received most of the attention: 2,4-D and 2,4,5-T. These are nonpersistent synthetic organic compounds similar in function to plant hormones called **auxins**. When sprayed on plants, 2,4-D and 2,4,5-T increase the metabolic rate of cells so much that plants cannot keep up with increased nutrient demands and literally grow to death.

> **KEY CONCEPTS**
> Plant auxins are hormones that stimulate plant growth. Two of the most widely studied herbicides are auxin-like compounds.

Peacetime Uses: Pros and Cons 2,4-D and 2,4,5-T were once widely used to control brush and plants along roadways, power lines, and pipelines. They have also been used to control poison ivy and ragweed, eliminate unwanted trees in commercial tree farms and in National Forests, kill aquatic weeds, and rid rangelands of brush and poisonous plants. Overall, three fourths of these chemicals are used for weed control on farms.

The benefits of these and other herbicides are many:
1. They decrease the amount of mechanical cultivation needed to control weeds—and thus reduce labor costs.
2. They reduce weed damage when soils are too wet to cultivate, because crops can be sprayed by plane.
3. They help farmers reduce water usage because water escapes more rapidly from ground that has been tilled to control weeds.

However beneficial their use may be, herbicides also have many drawbacks:
1. Some weeds have become resistant to herbicides and have become more troublesome.
2. Some nonpest species proliferate after spraying. To get rid of them, additional herbicide or more toxic ones may be needed.
3. When herbicides are used, farmers often reduce tillage; the weeds killed by herbicides remain on the ground and provide food and habitat for insect pests. Herbicide use therefore may actually increase insect pest populations, increasing the need for insecticides. In addition, herbicides may decrease the farmer's incentive to rotate crops, an effective way of reducing insect pests.
4. Herbicides may make some crops more susceptible to insects and disease. For example, some herbicides reduce the waxy coating on plants, change their metabolic rates, and retard or stimulate plant growth; all of these effects may make plants more susceptible to insects and disease.
5. Some herbicides are toxic and may cause birth defects, cancer, and other illnesses in animals.

Critics argue that **integrated weed management** could reduce the use of herbicides. Such management would employ special equipment such as wick applicators that apply herbicide only on weeds between rows. Wick applicators consume much less herbicide than aerial spraying and create less environmental contamination. The reasonable use of herbicides could be complemented by mechanical cultivation, proper spacing of rows for healthy crops, biological weed controls, and crop rotation.

Controversy over Wartime Use of 2,4-D and 2,4,5-T Herbicides were used extensively in the Vietnam War in the 1960s and early 1970s as defoliants: to prevent guerrilla ambushes along roads and waterways; to deter the movement of soldiers through demilitarized zones and across the border of Laos; to destroy crops that might be eaten by the enemy; and to clear areas around camps. Three herbicide preparations were sprayed from planes, helicopters, boats, trucks, and portable units from 1962 to 1970.

The most effective and most controversial of the herbicides was **Agent Orange**, a 50-50 mixture of 2,4-D and 2,4,5-T (**FIGURE 22-7**). During the war, over 42 million kilograms (93 million pounds) of Agent Orange was sprayed on the wetlands and forests of Vietnam, decimating 1.8 million hectares

Agent Orange used in the Vietnam war as a defoliant decimated 1.8 million hectares (4.5 million acres) of countryside, at least 190,000 hectares (470,000 acres) of farmland, and over half of the mangrove vegetation of South Vietnam (1930 square kilometers, or 744 square miles).

(4.5 million acres) of countryside and at least 190,000 hectares (470,000 acres) of farmland. Over half of the mangrove vegetation of South Vietnam (1930 square kilometers, or 744 square miles) and about 5% of the hardwood forests were destroyed. The forests alone represent about $500 million worth of wood, a supply that would last 30 years.

The prospect for these forests seems dim because hardy weed species such as cogon grass and bamboo have invaded the deforested zones. Ecologists fear that these species may greatly delay recovery or prevent it altogether. The National Academy of Sciences estimates that defoliated mangroves may take 100 years to recover because the destruction was so great that few seed sources remain.

Defoliants and numerous insecticides used to control mosquitoes created an ecological disaster in Vietnam, resulting in the death of numerous fish and animals. In one survey of a heavily sprayed forest visited years after the war ended, Harvard University biologist Peter Ashton found 24 species of birds and 5 species of mammals, compared with 145 and 170 bird species and 30 and 55 mammal species in two nearby forests that had not been sprayed. The total impact of such actions will never be known.

Agent Orange may have been the cause of serious medical problems that developed in soldiers and villagers throughout Vietnam. Studies suggest that health effects attributed to Agent Orange probably resulted from contaminants belonging to a group of chemicals known as dioxins. **Dioxins** cause birth defects and cancer in mice and rats. Dioxin, discussed in Chapter 18, is believed to be 100,000 times more potent than the tranquilizer thalidomide, which caused many birth defects in Europe. Soldiers from the United States and Australia who fought in herbicide-defoliated areas and were exposed to substantial amounts of the chemical developed severe headaches, nausea, diarrhea, internal bleeding, chloracne (a severe skin rash similar to acne), and depression.

In 1970, as a result of the public outcry in the United States, the government banned the use of Agent Orange in Vietnam. In 1985, 2,4,5-T was banned altogether. Use of 2,4-D continues today to control hardwood trees that are considered weeds in commercial evergreen forests.

Vietnam veterans returning from the war also began to suffer unusual medical disorders. A high proportion of the men have fathered offspring that were born dead or aborted prior to term, as well as infants with multiple birth defects. Other veterans have developed lymphoma, leukemia, and testicular cancer.

Perhaps the most compelling evidence linking adverse health effects to Agent Orange came from a study carried out by Vietnamese doctors. They examined the rate of birth defects in the offspring of 40,000 Vietnamese couples and found that women whose husbands had fought in South Vietnam, where Agent Orange was sprayed, were 3.5 times more likely to miscarry or give birth to defective children than women whose husbands had remained in the north during the war. American scientists who studied the results, although expressing caution, found the study convincing.

FIGURE 22-7 Spraying Agent Orange. Vietnamese jungles were sprayed with Agent Orange to clear the vegetation.

Data recently released from an independent study of military veterans indicate an increased risk of elevated blood pressure, benign fatty tumors, miscarriage, visual and skin sensitivity to light, and depression. Other evidence has implicated dioxin as the cause of soft-tissue sarcomas (cancers); veterans exposed to dioxin had a rate of soft-tissue sarcomas seven times higher than normal.

In May 1984, manufacturers and Vietnam veterans reached an out-of-court settlement that established a $180-million fund for veterans and their families claiming injury from Agent Orange. In 1989, seven companies that were being sued by veterans for damages caused by Agent Orange agreed to pay them $240 million. Still unconvinced that the data showed a link between Agent Orange and these serious problems, the VA has reluctantly agreed to treat all Vietnam veterans, a change in policy that may be too late for many of the soldiers.

KEY CONCEPTS

Extensive use of chemical defoliants (Agent Orange) during the Vietnam War has resulted in substantial environmental and health impacts in soldiers and their offspring. The health effects are attributed to dioxins that contaminate the herbicide.

The Alar Controversy: Apples, Alar, and Alarmists?

In the spring of 1989, the Natural Resources Defense Council (NRDC), a U.S. environmental group, announced the results of a 2-year health study on children and pesticides. It concluded that U.S. children are exposed to dangerous levels of pesticides in fruits and vegetables and that 5500 to 6200 children alive at the time will develop cancer in their lifetimes from just eight of the many pesticides to which they are exposed in their preschool years.

The NRDC also charged the EPA with routinely neglecting children in setting their standards. Children face a higher risk from pesticides in part because they eat more fruit than adults. Although the EPA has begun to take into account the higher level of fruit consumption in children, the agency still relies on 1977 consumption data that grossly underestimate the amount of fruit children eat. Several researchers who reviewed the NRDC study thought that the group underestimated the cancer risk from pesticide residues because it examined only 27 of the 500 or more pesticides in common use that leave residues on fruit and vegetables. The NRDC calculated the risk for only 8 of the 27 pesticides it examined. The NRDC research team also omitted several foods, such as milk, that are important components of children's diets and a source of further exposure to pesticides. Furthermore, the NRDC did not take into account continued exposure to pesticide residues as children became adults. In addition, some research suggests that children are more sensitive to chemical exposure than adults because rapidly dividing cells are prime targets for carcinogens and the enzymes needed to detoxify chemicals may not have fully developed in children.

The main culprit in pesticide-caused cancer, the NRDC charged, is a chemical called *Alar*, a substance that delays ripening so that apples do not fall off the tree prematurely. Alar also promotes the reddening of apples and delays over-ripening, so that apples stay fresher in storage. Alar penetrates the flesh of apples and cannot be washed off.

Based on what it considered the best available information at the time, the NRDC estimated that 86 to 96% of the total pesticide cancer risk arises from Alar and its breakdown product, UDMH. This conclusion was based on risk assessment derived from toxicology studies performed in the 1970s.

The EPA's safety standards seek to limit cancer risk to one in a million. The estimates used by the NRDC in its Alar campaign suggested that UDMH posed a cancer risk of 1 in 4200—240 times greater than the routinely accepted standard.

Not everyone agreed that Alar presented a health threat. Dr. Bruce Ames of the University of California at Berkeley argued that Alar posed a much lower threat than some naturally occurring chemicals found in some foods. He also argued that the ban on Alar, which followed the NRDC report, would require orchard owners to increase pesticide use. An insect called the *leafminer*, for instance, causes apples to fall prematurely. To control them, apple growers would have to use more pesticides, increasing human exposure to additional carcinogens. Molds that produce their own toxins may also increase in apples on trees and in storage, said Ames, because untreated apples are less firm and more susceptible, exposing people to a greater danger than Alar itself.

The Alar controversy has taken some interesting turns since the chemical was pulled off the market by its manufacturer. Since that time, the EPA has actually twice *lowered* its estimate of cancer risk because of new information. The first reduction resulted in a cancer risk value 10 times lower than the one the NRDC used. Still maintaining that Alar is carcinogenic, the EPA nonetheless halved its estimate again in 1991. Although this won't resurrect Alar, it has invigorated the apple growers of Washington State who were hit hard by the Alar ban.

KEY CONCEPTS

Children may be at higher risk to pesticides than adults, especially pesticides used on fruit.

22.2 Controlling Pesticide Use

Concern over the health and environmental effects of pesticide use has led to a number of actions to minimize impacts, including outright bans of harmful pesticides, registration, and establishing tolerance levels on produce. This section discusses these actions and their strengths and weaknesses.

Bans on Pesticide Production and Use

In the United States, caution regarding pesticide use began in 1962 with the publication of the late Rachel Carson's book *Silent Spring*, which pointed out many of the real and potential impacts of pesticide use (FIGURE 22-8). This widely read book is credited with touching off the controversy that re-

FIGURE 22-8 Antipesticide crusader. Scientist Rachel Carson first drew attention to the dangers of pesticides in her environmental classic *Silent Spring*.

sulted in increased public awareness and research on the effects of the indiscriminate use of pesticides. Concern reached a peak a decade later with the near extinction of peregrine falcons, brown pelicans, cormorants, bald eagles, and other bird species exposed to DDT.

Faced with the growing body of evidence that illustrated the many hazards of using chemical pesticides, environmentalists pushed for bans on certain pesticides. Scientists have found that such bans have greatly benefited U.S. wildlife. The ban on DDT in the United States, for instance, has resulted in increases in the number of endangered bald eagles, brown pelicans, ospreys, and other species. Recent studies show that DDE and DDT levels have dropped in wild populations and that normal reproductive rates have returned. Affected bird populations are recovering nicely and some may be moved off the endangered species list.

Bans on the production and use of pesticides in the United States, however, are only part of the answer. Continued use of pesticides outside the United States, as in the case of carbofuran, poisons migratory species such as songbirds that winter in Central America and Mexico but spend spring and summer in the United States. In addition, much of the produce imported into the United States has recently been found to be contaminated with pesticides, many of which were banned in the United States. Because one fourth of the fruit and vegetables sold in the United States comes

from foreign countries, global bans of harmful pesticides are needed to protect human health.

KEY CONCEPTS

Many environmentally harmful pesticides have been banned in the United States in the past 30 years. Some of these pesticides, however, continue to be used in other countries, where they poison wildlife and migratory birds and contaminate crops destined for local markets as well as markets in the United States and other more developed countries.

Registering Pesticides

Another way of controlling pesticides is the act of registration, described shortly. In the United States, pesticide registration was first required by the **Federal Insecticide, Fungicide, and Rodenticide Act (FIFRA)**, a law currently administered by the Environmental Protection Agency. The FIFRA is designed to protect farmers, farm workers, the general public, and the environment from new chemical pesticides. It does this by requiring manufacturers to test new pesticides for a range of effects before they can be registered for use. Pesticides introduced prior to the passage of the law in 1972, however, were originally not affected.

Registration is a kind of permitting process managed by the EPA. When a company develops a new pesticide that it wants to market, it must perform a number of tests on plants and animals to assess the potential toxicity of the chemical. Test plots are also sprayed with the pesticide to determine residues—that is, how long the chemical remains. These data are then submitted to the EPA scientists for their review. Taking into account the average American diet, residue levels, and toxic effects, the EPA then determines on which crops, if any, the pesticide can be used.

The critical thinking skills you have learned may suggest a problem right away: Is the average American diet accurately determined? If the truth be known, the average American diet includes a great deal of red meat, chicken, and other meat products. That automatically excludes vegetarians, however, who tend to consume large quantities of fruits and vegetables. Ironically, many vegetarians who select a special diet for health reasons are inadvertently exposed to the highest levels of pesticide—unless, of course, they consume organic produce. **Organic produce** includes fruits and vegetables grown without pesticides or synthetic fertilizers. Most grocery stores contain a small selection of organically grown produce. In a number of states large grocery stores such as Wild Oats and Alfalfa's specialize in healthy food, including organically grown fruits and vegetables, as do many smaller health food stores.

Pesticides are registered for general or restricted use. General use means that anyone can purchase and use them. Technically, restricted-use pesticides are to be used by licensed applicators—farmers and lawn-care companies, for instance.

The problem with this system is that pesticide use is largely regulated by an honor system. Labels on restricted-use products describe their legal uses, and only licensed applicators can buy them; other than those limitations, there's very little if anything to stop people from using pesticides any way they please. The aldicarb poisonings cited earlier illustrate one of the common problems.

In 1988, the FIFRA was amended to correct some of its weaknesses. One of the most significant gains was a plan to register many pesticides that had been introduced before the EPA took over the process in 1972. Pesticide registrations in the 1950s and 1960s were made with very little if any sound toxicological data, say critics. Registering the over 300 chemical pesticides will take many years to complete and will be funded by fees paid by the chemical companies. Some experts see this as a weeding-out process. They believe that many pesticides will be canceled as companies that are afraid their product won't be approved for health and environmental reasons won't invest the money needed to register them.

Despite this improvement, critics say that additional reform is necessary. For example, pesticide registration does not currently require manufacturers to test for neurotoxicity—toxic effects on the nervous system, the brain, spinal cord, and nerves. To many critics, this is a glaring omission because about half of the pesticides (especially the organophosphates) are insect neurotoxins that could also affect the human nervous system.

New research also shows that some pesticides damage the immune system, the body's defense against bacteria, viruses, and even cancer. Pesticide registration, however, requires no test of immune system effects—another omission, say critics. In addition, pesticide registration does not take into account possible synergism, the super-additive effect described in Chapter 18.

Perhaps the most glaring problem, though, is that the FIFRA provides virtually no monitoring functions. The end users are free to do as they please. They can apply as much pesticide as they want and can apply it wherever they want. Moreover, there is no one to see that solutions are mixed correctly or that equipment is working properly. As a result, farmers frequently apply much more than is needed.

Licensing and training are the only avenues available to address this problem. To be licensed, most states require farmers to take a test. In Colorado, for example, farmers must study a booklet provided by the EPA and then take an open-book test. If they pass, they're licensed to spray restricted pesticides on their fields. Critics argue that more rigorous education and testing are needed.

Governments, pesticide companies, farmers' groups, and universities could provide additional hands-on training and monitoring to be sure that pesticides are used more safely. As mentioned earlier, programs that teach farmers to become better at identifying pests and monitoring pest populations in the field could decrease insecticide use and costs. Special crop scouts, such as those now in use in Indonesia, could be trained to monitor pest populations and determine when they have reached the threshold level. They could also assist farmers in finding alternative ways of controlling pests. (See Spotlight on Sustainable Development 22-1 for more on Indonesia's program.)

KEY CONCEPTS

The EPA registers newly developed and previously introduced pesticides for general or restricted use and stipulates what crops they can be used on, a process called *registration*. Many improvements are needed in this system to make it more effective, especially more rigorous education and testing of users.

Establishing Tolerance Levels and Monitoring Produce

To help protect public health, the FIFRA also authorizes the EPA to set **tolerance levels** for pesticides on fruits, vegetables, and other foods. Tolerance levels are concentrations in or on foods that are believed to pose an acceptable health risk. For cancer, the EPA sets the concentration at a level it thinks will cause no more than one additional cancer death in one million people. This determination is fraught with difficulty, as explained in Chapter 18.

Although the EPA sets tolerance levels, it is up to the U.S. Food and Drug Administration and state agricultural agencies to monitor the nation's food supply and to enforce tolerance levels. Inspectors examine fruits and vegetables and have the authority to seize and condemn foods containing residues that exceed EPA levels or that contain illegal pesticides—but only when they are shipped from one state to another. Inside the states, the responsibility for monitoring food lies with state agencies.

Both the FDA and state agencies suffer from a common problem: a chronic lack of funds and a shortage of inspectors. Given these problems and the massive amount of food consumed by the American public, it can be no surprise that only a small portion of our food is actually tested. Furthermore, examiners test only for the presence of a handful of the pesticides that could be on our food. Unless you grow your own food, you are probably consuming small amounts of many pesticides. In a recent study, more than 50% of all food in supermarkets was found to contain detectable levels of pesticides.

Similar problems occur in other countries. In Ireland, for instance, the Pesticide Control Service has only 11 scientists to test food for the entire country. Consequently, few farmers are inspected, and only 2000 food samples are tested annually.

In 1987, the EPA reported that at least 55 of the pesticides that leave residues on food are thought to be carcinogenic. In 1987, the National Academy of Sciences issued a report concluding that one million Americans alive today will develop cancer as a result of pesticide contamination of their food—that's one of every 250 Americans. Add to that possible birth defects, miscarriages, mutations, neurological effects, and other milder symptoms, and it is little wonder that the EPA ranks pesticides in food as one of the nation's most serious health concerns.

Despite its recognition of this potential problem, the EPA has a long way to go. Most critics think that the tolerance levels set for pesticide residues are inadequate because they fail to take into account the special diets of vegetarians and children. A law passed by the U.S. Congress in 1996 directed the EPA to consider the increased susceptibility of

22-1 Indonesia Turns to Biological Pest Control

Indonesia is a country of islands—nearly 14,000 of them—in Southeast Asia. In 1983, this rural nation, once the world's leading importer of rice, succeeded in growing enough rice to feed its own people. New strains of rice, fertilizers, and an intricate irrigation system deserved much of the credit for the success. In 1985, however, the notorious brown planthopper threatened the progress of the previous years. This insect causes rice to dry out, rot, and fall in the field. Infestations of the insect can cause enormous damage.

To combat the planthopper, in 1985, the government decided to try integrated pest management. It was advised to do so by an Indonesian entomologist, Dr. Ida Oka, who had received his Ph.D. from Cornell University under David Pimentel, a leading expert on natural insect control. Oka had been in charge of pest management in the late 1970s and early 1980s and had implemented sound IPM for rice and other crops. During that time pesticide use had been greatly reduced, and rice yields had soared. Oka had resigned, however, when a new minister of agriculture was appointed. This official's pro-pesticide policies, in which the government paid about 85% of the cost of pesticides to farmers, resulted in widespread use of chemical pesticides and clearly placed the farmers on the pesticide treadmill. This policy was largely responsible for the outbreak of the brown planthopper in 1985.

Indonesian scientists had found that the use of pesticides to control the planthopper and other insects killed many beneficial insects that preyed on the pest. One of the beneficial insects destroyed in the spraying was the wolf spider, which can devour 5 to 20 brown planthoppers a day. If left alone, the beneficial insects could often control the harmful ones. Researchers also found that farmers sprayed fields regularly, regardless of whether they needed it. The overuse of pesticides actually increased the severity of infestations, so much so that by 1986 the country was in danger of becoming a rice importer once again.

Convinced of the danger of the continuing use of pesticides, the Indonesian government asked the UN Food and Agricultural Organization (FAO) to help it promote an IPM program. In 1986, with the help of the FAO, the government embarked on a program to educate farmers on IPM and the dangers of pesticides. Experts ventured into the rice paddies, where they showed farmers how to diagnose problems, calculate the ratio of good bugs to bad ones, and decide how much damage the crop could stand without a decrease in yield. They then taught farmers ways to reduce spraying and methods to protect beneficial predatory insects.

Early results showed that IPM worked. IPM reduced pesticide use substantially. Trained farmers, for example, apply one ninth as much pesticide as they did before training, with no decrease—sometimes even an increase—in crop yield. Farmers have also learned to discern more carefully insect damage from fungal damage, and this helps to reduce pesticide use.

The average yield on farms using pesticides was 4% lower than on fields controlled by IPM. Despite the high subsidies for insecticides, the farms using IPM proved more profitable than those sprayed more frequently. The government saves an estimated $120 million a year on pesticide subsidies. Indonesia's streams and wildlife also started showing signs of recovery.

The success of the pilot program in Indonesia convinced the government to adopt IPM as a national pest control strategy. The government, in fact, banned 56 of the 57 pesticides previously approved for farming in Indonesia, in order to help protect predatory insects. The government then launched a massive campaign to educate Indonesia's farmers in IPM. In 1992, the country hired 2000 crop scouts to educate farmers on IPM and hoped that all farmers would soon be using this method.

Since then, integrated pest management has spread to Thailand, Bangladesh, Sri Lanka, Malaysia, India, and China. Farmers in Indonesia who were not introduced to the technique raised their voice. They wanted to be included in this effort to help the world move to a more sustainable, environmentally safe form of farming. Agricultural experts believe that IPM could be used on fields that provide 45% of the rice for people living in southern and southeast Asia and could save millions of dollars, preserve wildlife, and protect human health without endangering high crop yields.

---

infants and children when it sets pesticide residue limits. Another problem critics point out is that the EPA rarely revises tolerance levels when new scientific data about risks become available. Also, it rarely bans a pesticide if it is harmful to wildlife but not to people.

Some additional reforms in pesticide controls were also included in the recent legislation. For example, the EPA is directed to consider health risks other than cancer and to consider hormone-mimicking properties of pesticides. As noted in Chapter 18, some pollutants mimic naturally occurring hormones and can dramatically alter reproductive and other functions in fish, wildlife, and perhaps even humans.

Another problem is that the chemicals that pesticides are mixed with or dissolved in may also be toxic. In fact, studies of pesticides used indoors show that some of these substances reach rather high levels—and remain high indoors for a considerable length of time after a house has been sprayed for cockroaches, ants, or other insects.

The next section shows that there is a way to greatly reduce or even eliminate pesticide use. It is called *integrated pest management.* In the interim, individuals can avoid some produce containing pesticides by growing some of their own fruits and vegetables or by purchasing organically grown produce. Washing fruits and vegetables can help but won't eliminate all pesticide residues. Beyond that, individuals can exercise their democratic prerogatives by writing local, state, and federal officials and asking for tougher laws to regulate pesticide registration and use.

Weakening Protection

Federal pesticide regulations have been refined over the years, providing more and more protection to people, wildlife, and the environment. Some states have even increased protection by passing more stringent regulations. Other states, however, are working to weaken pesticide regulations. Congress is currently considering legislation that could result in more pesticide contamination of streams and rivers from commercial timber operations and a host of other actions. The Bush administration is also considering regulations that would allow EPA officials to ignore the toxic effects of pesticides on wildlife when considering approval of a new pesticide.

22.3 Integrated Pest Management: Protecting Crops Sustainably

Integrated pest management (IPM) is a new strategy of pest control gaining popularity throughout the world. It depends on four sustainable means of pest control: environmental, genetic, chemical, and cultural. Two or more of these approaches may be used simultaneously to control pests, with little or no damage to the environment and human health. Studies show that they also offer important economic benefits. Remember, as you read about the various strategies described in the following pages, the central goal of integrated pest control: to reduce pest populations to levels that do not cause economic damage, while protecting human health and the environment. The goal is not to eliminate pest species entirely, which may be impossible anyway.

Environmental Controls

Environmental control is a term used to designate a number of techniques that alter the biotic and abiotic conditions in crops, making them inhospitable to pests. Because they generally rely on knowledge more than on technology, these practices are especially suitable for less developed countries. Still, they can be equally effective if used properly in modern agricultural societies.

Increasing Crop Diversity: Heteroculture and Crop Rotation
In Chapter 10, we saw that monocultures generally promote pest and disease outbreaks. Crop diversity, on the other hand, reduces the amount of food available to any one pest and helps prevent such rapid population growth. Several techniques can increase crop diversity, among them heteroculture and crop rotation.

A farmer who plants several crops side by side in his field, rather than huge expanses of one crop covering, is practicing **heteroculture**. This simple but effective measure works because it provides environmental resistance to pests; that is, pest populations are often much smaller in heterocultures than in monocultures because there is less to eat. In addition, some crops harbor predatory insects that feed on pests in nearby crops. Corn and peanuts grown in adjacent fields, for instance, can reduce corn borers by as much as 80%. Part of the reason for this success may be that predatory insects that feed on the corn borer live in peanut crops.

Chapter 10 discussed one of the most recent and innovative ways of intermixing crops and **strip cropping** (FIGURE 22-9). In this technique, alternative strips of corn and soybeans (or other crops), each with a dozen or more rows, are planted side by side in the same field. This not only decreases insect pests but also increases yield because the corn protects the soybeans from wind while the openness of the field provides more sunlight to the corn.

Heteroculture decreases pesticide use and also provides a means of diversifying farm production from year to year.

FIGURE 22-9 **Strip cropping.** Alternating strips of alfalfa with corn on the contour protect this crop field in northeast Iowa from soil erosion and increase productivity of the crops as explained in text.

In some ways, then, it is a form of insurance. Ron Rosmann, for example, plants half of his 200-hectare (500-acre) farm in corn and soybeans. The rest he devotes to hay, pasture, cattle, hogs, chickens, and a tree nursery that he hopes will help pay for his three sons' college educations. A bad year for corn will not wipe him out.

Heteroculture can be practiced by home gardeners with great success. I have been growing vegetables in a pesticide-free garden for nearly 30 years and have had virtually no trouble with insects, in large part because I intermix species. Small patches of carrots are planted next to peas, which are next to spinach, and so on. I also plant onions, marigolds, and other species that tend to repel pests.

Crop rotation, discussed in Chapter 18 as a means of reducing soil erosion and increasing soil fertility, also helps control pests, for at least two reasons. First, the healthier the soil, the healthier and more resistant the plants are to insects and disease. Second, it also helps hold down pest populations because it reduces food available from one year to the next for specialized insects—those that feed on only one crop. For instance, wireworms feed on potatoes but not alfalfa. Therefore, if potatoes and alfalfa are alternated from year to year in the same field, wireworm offspring that hatch in the alfalfa patch will have little to feed on. Their numbers will decline severely. When potatoes are planted the next year, few wireworms will be around. Although the population may increase during the growing season, it will generally not reach a level that causes harm. The next year alfalfa is planted, and wireworm offspring once again perish. Gardeners can also practice crop rotation on a small scale to hold insects in check.

Altering the Time of Planting   Some plants naturally escape insect pests by sprouting early or late in the growing season. A good example of this adaptation is the wild radish, which sprouts early in the season before the emergence of the troublesome cabbage maggot fly.

Agriculturalists can use their knowledge of an insect's life cycle to their advantage by coordinating plantings with the expected date of hatching. A slight delay in planting of wheat, for example, helps protect it from the destructive Hessian fly. In general, if a pest emerges early in the spring, planting can be delayed to avoid that pest (within the limits of the growing season). Without food, the pest will perish.

Home gardeners can also foil pests by planting seedlings instead of seeds. Certain crops such as lettuce can be mowed down by hungry slugs and other insects when they first emerge from the ground. If larger seedlings are planted, though, the plant has a better chance of surviving.

Altering Plant and Soil Nutrients   The levels of certain nutrients in soil and plants can also affect pest population size. By regulating soil nutrients, then, a farmer may be able to control pests. Nitrogen is one of the important nutrients that insects and parasites derive from plants. Too much or too little nitrogen can alter the population size of various pests. For example, grain aphids reproduce better on grains high in nitrogen. Other insects, such as the greenhouse thrip and mites, do poorly on high-nitrogen spinach and tomatoes, respectively. Once again, knowledge reigns supreme. A little knowledge of pest nutrient requirements, soil nutrient levels, and plant nutrient content can become a useful ally.

Controlling Adjacent Crops and Weeds   All crops are surrounded by forests, meadows, or other crops. Each of these houses harmful as well as beneficial insects. To reduce pest damage, farmers often avoid planting crops that provide food and habitat for harmful pests next to other crops. Those that provide habitat for beneficial insects are often encouraged. Farmers have even used low-value crops to attract pests away from more valuable crops. The former are called **trap crops**. Alfalfa is a good example. When planted adjacent to cotton, it lures the harmful lygus bug away from the cotton plants, which prevents serious damage to the cotton. Some farmers may even spray the alfalfa with pesticide to get rid of the bug, using far less chemical than would be necessary if the entire cotton crop had to be sprayed.

Biological Control: Introducing Predators, Parasites, and Disease Organisms   In nature, thousands of potential insect pests never become real pests because of natural controls exerted by predators, diseases, and parasites. If you recall from your reading of the ecology chapters, these are biotic components of natural environmental resistance. Farmers can capitalize on this knowledge to manage a variety of pests including weeds, insects, and rodents.

To date, scientists have documented well over 300 examples of partial or complete control of crop pests through natural predators and parasites. This technique is generally referred to as **biological control**. One classic example of the effectiveness of biological control is the control of the prickly pear cactus in Australia.

The prickly pear was introduced into Australia from its native Mexico. By 1925, over 24 million hectares (60 million acres) of land had been badly infested. Half of this land was abandoned because of the thick carpet of cactus. Farmers introduced a cactus-eating moth to Australia to eradicate the

FIGURE 22-10
Prickly pear cactus invasion. (a) A prickly pear cactus infestation in Queensland, Australia. **(b)** The same area after the introduction of the cactus-feeding moth.

(a)

(b)

pest, and 7 years later, much of the land had been cleared and could once again be used (**FIGURE 22-10**).

The predatory lady beetle was introduced into California from Australia in the 1880s to control an insect that destroyed citrus trees. Parasitic insects from Iran, Iraq, and Pakistan have been introduced to control the olive scale, an insect that once threatened the state's olive trees. Both lady beetles and the predatory insects now exert complete control on their prey, keeping their populations at manageable levels without the use of pesticides.

Entomologists in the United States are currently experimenting with a new method of controlling mosquitoes using *Toxorhynchites rutilis* (Big Tox, for short), a large, non-biting mosquito whose larvae feed on the larvae of other mosquitoes. Bred in captivity, this predatory mosquito will be released in infested regions to control biting mosquitoes.

A few insect pests can also be controlled by birds, a natural control organism whose potential has been overlooked. Brown thrashers can eat over 6000 insects in one day. A swallow consumes 1000 leafhoppers in 12 hours, and a pair of flickers can snack on 500 ants and go away hungry. In China, thousands of ducklings are driven through rice fields; in some places they reduce the populations of insects by 60 to 75%, allowing farmers to reduce insecticide use considerably. Their droppings also provide fertilizer for the crops.

Bacteria and other microorganisms can be brought to bear on pests. One common example is the bacterium *Bacillus thuringiensis* (bah-SILL-us thur-in-GEEN-siss) or BT, used to control many leaf-eating caterpillars. Cultivated in the lab and sold commercially, it is available as a powder that is either dusted on plants or mixed with water and then sprayed on plants. Caterpillars that eat the bacteria die because BT produces a toxic protein that paralyzes the digestive system of insects. Humans and other organisms are usually unaffected. I use BT routinely in my garden to control cabbage butterflies.

BT is used by organic gardeners with considerable success. It has been sprayed in China to control pine caterpillars and cabbage army worms. In California, it has been used for more than 20 years to control various crop-eating caterpillars, and it is currently applied in the northeastern United States to help control gypsy moths, which devastate forests. Another strain of BT has been employed in the battle against mosquitoes in Colorado and other states. The use of BT and other techniques has resulted in a measurable reduction in insecticide use in certain crops, especially almonds and tomatoes.

Researchers have also successfully inoculated corn plants with genetically altered bacteria containing the BT gene. The bacteria multiply in the corn plant as they grow, and they kill European corn borers that feed on the stalks. Studies show that the bacteria do not migrate into the corn kernels.

Viruses and fungi may be used similarly. In Australia, after years of fruitless efforts to control rabbits, scientists introduced a pathogenic myxoma (mix-OH-ma) virus, which eliminated almost all the rabbits within one year. Unfortunately, the virus has evolved to an nonvirulent form, and the rabbit has evolved resistance. Control is no longer as effective as it once was.

Cabbage loopers can be controlled with 0.5 gram of an experimentally produced virus applied to a hectare of cropland. Other viruses are being used to control pests such as the pink bollworm, which damages cotton, and the gypsy moth.

Biological control agents must be developed with caution to ensure that they do not pose a threat to humans, livestock, and natural ecosystems. One of the major concerns has to do with the introduction of alien species into new environments, a problem discussed in Chapter 11. Careful testing is necessary to be sure that an organism introduced to control a pest does not become a pest itself. In the early 1980s in sub-Saharan Africa, for example, an insect known as the mealybug became a major pest, attacking cassava plants. This plant produces an edible root that is a staple for about 200 million people. To find a control for this troublesome pest, researchers scoured the mealybug's South American homeland for natural enemies. They eventually located a small parasitic wasp that injects its eggs into the larvae of the mealybug. When the eggs hatch, they devour the larvae. Before the researchers could release the wasp, though, they had to perform extensive tests to see if the wasp would survive in its new home and if it would become a pest itself. Convinced that it wouldn't, the bug was first released in 1986 and today is providing protection in 24 African countries. So far, the wasp seems to be working fine.

Another undiscussed problem of biological controls is that target organisms can develop genetic resistance to them, as did the Australian rabbits mentioned earlier. Researchers in Kansas recently found that larvae of the Indian meal moth, which feed on grain stored in sheds and bins, develop genetic resistance to BT. In such cases new controls could be introduced or alternated with BT. In some instances, biological control agents themselves may undergo genetic changes that offset the newly acquired resistance of the pest. This process is called **coevolution**. As yet, there is no record of such changes in biological control agents, but some scientists think that they are inevitable.

Genetic Controls

Integrated pest management includes two major genetic control strategies, the sterile male technique and the breeding of genetically resistant plants and animals. Both are important components and can be used in conjunction with other methods.

Sterile Male Technique   As the previous examples have shown, a little knowledge of biology can go a long way. Another example in which knowledge is brought to bear on pest control is the **sterile male technique**. As the name implies, this technique involves the introduction of sterile males of insect pests into the environment. Males are raised in captivity and sterilized by irradiation or by exposure to certain chemicals. The sterilized males are then released in large numbers in infested areas. The males far outnumber the fertile wild males and thus account for a large percentage of the matings with wild females. Because many insect species mate only once, eggs produced by such a union are infertile. Thus, if the population of sterilized males greatly exceeds that of the wild males, most of the eggs will be infertile. Insect populations can be brought under control swiftly.

The sterile male technique has been used effectively against several species of insect pests, including the screwworm fly in Mexico and the United States, the Mediterranean fruit fly in Capri, the melon fly on the island of Rota (near Guam), and the Oriental fruit fly on Guam.

The sterile male technique has not always succeeded. It has, for instance, proved unsuccessful in mosquito control. Scientists believe that the chief reason for its failure in this instance is the lower sexual activity of sterilized males compared with wild males. Other reasons may include an inadequate number of sterile males, ignorance of the insect pest's breeding cycle, the in-migration of additional pests, and impatience on the part of state agricultural agents. Some researchers also suggest that through natural selection a new strain of insects may evolve that recognizes and avoids sterile males.

Despite these problems, the sterile male technique is an important tool in integrated pest management. It is species specific, can be used with environmental controls, and can be effective in eliminating pests in low-density infestations.

Developing Resistant Crops and Animals   Many species are naturally resistant to pests and disease organisms. However, crop species often have lost their resistance through years of special breeding programs aimed primarily at increasing yield. Many geneticists think that introducing genes that provide genetic resistance to crops can help reduce or eliminate pesticides. For example, researchers have found that certain oils in the skins of oranges, grapefruits, and lemons are highly toxic to the eggs and larvae of the Caribbean fruit fly, which lays its eggs in the skins of these fruits. The flies' larvae destroy the fruit, but scientists may now be able to breed citrus selectively to increase the amount of toxic oils in their peels. By creating natural resistance, this technique eliminates the need for pesticides.

Cornell University scientists are developing a potato plant whose leaves, stems, and sprouts are covered with tiny, sticky hairs that trap insects and immobilize their legs and mouth parts. Field tests show that this plant can reduce by half the infestation of green peach aphids, which (despite their name) also attack potatoes. The new variety was developed by crossing cultivated potatoes with a wild species with sticky hairs, which grows as a weed in Bolivia.

Other genetic research has led to Hessian fly-resistant wheat and leafhopper-resistant soybeans, alfalfa, cotton, and potatoes. Work on chemical factors that attract insects to plants may help scientists selectively remove them to make plants unappealing.

The Monsanto Company is working on another promising weapon in the fight against pests. Robert Kaufman and his colleagues have isolated the gene that gives BT its pesticidal action. The scientists have transferred that gene to another bacterium, *Pseudomonas fluorescens*, which lives on the roots of corn and several other plants. The transplanted gene renders the host bacterium lethal to insects and other organisms, such as the black cutworm, that feed on the roots of commercially important crop species. Simply by planting seeds that have been pretreated with *P. fluorescens* bearing the toxic gene, farmers may be able to provide long-term protection without many of the dangers of pesticides. However, widespread use of BT in corn and other crops may result in the development of resistance and loss of the BT control.

Monsanto hopes that more insecticidal genes can be added to *P. fluorescens* in the years to come, giving corn a wider range of protection, reducing chemical pesticide use and in the process protecting wildlife from the harmful toxic pesticides that have been the mainstay of agriculture for decades.

Root-zone protection is not the only strategy that geneticists are developing. Numerous bacteria colonize aboveground plant parts; fitted with insecticidal genes from BT and other naturally occurring biological agents, these bacteria could create a protective barrier to ward off dozens of insect pests.

Chemical Controls

Chemical controls are also be a part of IPM. These chemicals are conventional pesticides and a new breed of natural (presumably nontoxic) substances. First, though, we look at the conventional pesticides.

Reducing the Use of Second-Generation Pesticides   Even with wider use of biological control agents and other strategies of IPM, second-generation pesticides will likely remain a part of our pest control strategy for many years to come. However, several principles should guide their use: (1) They should be applied sparingly; (2) they should be applied at the most effective time to reduce the number of applications; (3) they should destroy as few natural predators, nonpest species, and biological control agents as possible; (4) they should not be applied near drinking water supplies; (5) they should be carefully tested for toxic effects; (6) they should be avoided if they are persistent and tend to bioaccumulate; (7) they should be used in ways that reduce exposure to workers and nearby families; and (8) they should be used to reduce populations to low levels, with environmental, genetic, and cultural control measures used to maintain populations at low levels.

One way to minimize the use of insecticides is to spray only affected areas. Insects, for example, may infest a small portion of a crop. If possible, only that portion should be sprayed. A technique that is useful for herbicides is the use of special wick applicators, rather than sprayers, which deliver a small dose directly to the target species with minimum environmental contamination (**FIGURE 22-11**).

Timing of application also helps farmers reduce pesticide use. Consider an example. The red spider mite can be kept under control in apple orchards by applying insecticides early in the season, well before the mite's natural predators emerge. Throughout the rest of the season, the natural predators will keep the spider mites under control. No further pesticides are needed. By using a similar approach on

FIGURE 22-11 Wick applicator. Instead of spraying the entire crop with herbicide, farmers can use this device to apply pesticides only to the weeds growing between the rows.

cotton, researchers at Texas A&M have cut pesticide use by 70% while maintaining normal cotton crop production.

Better monitoring of crops also help reduce pesticide use. Some farmers spray their entire crop when they encounter any pests. They don't even bother to see how extensive the infestation is. In many cases, only a tiny portion of the crop needs spraying, a discovery that could result in considerable savings of time and money.

Others spray on a routine schedule, often provided by the pesticide salesperson, without even checking to see if insect pests are about. Schedules may not reflect the needs of farmers' crops. Were farmers to check, they might find that the pests were under control—or weren't even there. A little monitoring might save lots of time, money, and environmental contamination.

Successful IPM requires a better understanding of insect biology, as well as better skills in recognizing and counting insects in a farmer's fields. By knowing what insects are present, and where, farmers can become wiser participants in the ecosystem they manage.

Finally, more developed nations have an important role to play by discouraging companies from exporting banned pesticides to less developed countries, where they often return on the produce we import. The rich can also help the poor develop a sustainable pest management program through technical and financial assistance.

KEY CONCEPTS

Conventional pesticides will remain in use, but they must be screened carefully to avoid harmful ones. Those that are used must be administered with caution—in amounts and at a time when they are most effective, to minimize their environmental impact. This requires better monitoring of crops and infestations.

Third-Generation Pesticides   This section discusses a whole new arsenal of chemicals that are produced in nature and could, with a little ingenuity, be applied to pest control on a large scale. These nontoxic agents, such as natural chemical repellants, could, in conjunction with measures outlined previously, displace potentially harmful second-generation pesticides. This new class of chemical compounds is sometimes referred to as the **third-generation pesticides**.

KEY CONCEPTS

Scientists and farmers are exploring a whole new group of naturally occurring chemicals to control insect pests. These could eventually become the cornerstone of pest management in a sustainable system of agriculture.

Pheromones   Insects and other animals release chemicals called **pheromones** (FAIR-eh-moans). Pheromones provide a chemical means of communication. One well-known group of pheromones is the **sex attractants**, which are emitted by female insects to attract males at the time of breeding. Effective in extraordinarily small concentrations, pheromones draw males to females. This evolutionary adaptation ensures a high rate of reproductive success.

Some sex attractants are now produced commercially and are available for pest control. Some of these substances are

FIGURE 22-12 **Pheromone trap.** This trap contains a sticky substance to immobilize male gypsy moths, which are drawn to it in search of mates by chemical sex attractants known as pheromones.

used in **pheromone traps** that lure males. These traps may contain a pesticide-laden bait or a sticky substance that immobilizes insects (**FIGURE 22-12**). Pheromone traps of various sorts have been used to control at least 25 insect species and can be used with other IPM methods.

Pheromone traps can also be used to pinpoint the time when insect eggs hatch. By knowing precisely when insects emerge, farmers can time their pesticide applications for maximum effectiveness. This technique helps reduce the amounts of pesticide applied.

Another technique for controlling insects with pheromones is known as the **confusion technique**. In this method, pheromones are sprayed on crops at breeding time. Unsuspecting males are drawn by the pheromone in all directions. Fertile females hardly stand a chance and may never find a partner. One modification of this technique involves the release of wood chips treated with sex attractants. Males are attracted to the wood chips and may attempt to breed with them.

Finally, pheromones can be used to lure beneficial insects out of fields. Once they are gone, pesticide sprays can be applied.

The use of pheromones offers several advantages over second-generation pesticides. They are, for example, nontoxic and biodegradable and therefore not expected to have any significant environmental impacts. They can be used at low concentrations. Third, they are highly species-specific. The major disadvantage is the high cost of developing them.

Insect Hormones   The life cycle of many insects is shown in **FIGURE 22-13**. As illustrated, adult insects lay eggs, which develop into larvae (caterpillar stage). The larvae are voracious eaters and are often the most troublesome form of pest species. Eventually, the caterpillar spins a cocoon in which it undergoes an amazing change, transforming into a flying form such as a moth or butterfly—the adult form.

The entire life cycle of insects is regulated by two hormones, **juvenile hormone** and **molting hormone. Hormones** are chemical substances that are produced by specific cells in the body and travel through the bloodstream to distant sites, where they exert some effect. Altering the levels of juvenile and molting hormones disrupts an insect's life cycle, sometimes resulting in death. For example, larvae treated with juvenile hormone are prevented from maturing and eventually die. If given molting hormone, they will enter the pupal stage too early and die. Interestingly, some plants contain chemicals structurally similar to juvenile hormone, an evolutionary adaptation that helps protect them from hungry insects. When ingested by larvae, these chemicals prevent the larvae from pupating. This, in turn, prevents formation of the adult form that produces eggs and additional generations of larvae.

Insect hormones applied to crops offer many of the same advantages that pheromones offer, including biodegradability, lack of toxicity, and low persistence in the environment. Like pheromones, however, they are costly. Insect hormones act rather slowly, sometimes taking a week or two to eliminate a pest, by which time extensive damage may have been done. In addition,

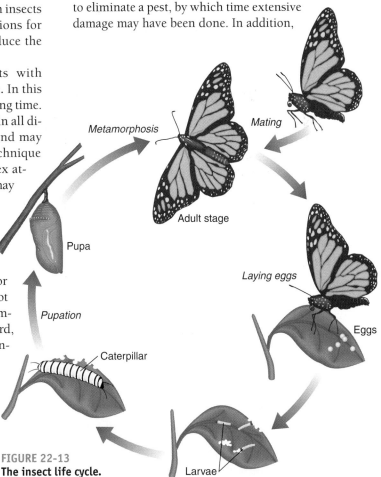

FIGURE 22-13
The insect life cycle.

insect hormones are not as species specific as pheromones and therefore may affect natural predators and other nonpest species. The timing of application is also critical, for hormones are effective only at certain times in an insect's life cycle.

Researchers recently discovered a plant from Malaysia that produces juvenile hormone. They hope that the genes responsible for the production of this hormone can be transferred to commercially important crops, offering another avenue of protection.

> **KEY CONCEPTS**
>
> Two hormones control the life cycles of insects and can be sprayed on crops to alter these cycles and lead to the death of pests. Unfortunately, to be effective, application must be precisely timed and these hormones take a fair amount of time to kill off a pest.

Natural Chemical Pesticides   Natives of the South American tropics have used the seeds and leaves of the neem tree for many years to control pests. Researchers have found that this tree produces chemicals that kill or repel a variety of insects. This extract may become useful in the control of larvae that feed on vegetables and ornamental crops.

Egyptian researchers found that flies ignored a species of brown algae left out on a counter to dry. Curious, researchers extracted a mixture of chemicals from the algae and found that they, too, repel a variety of insects that attack cotton and rice. Natural chemicals such as these may prove useful in years to come. Like other third-generation pesticides, they are biodegradable and nonpersistent.

Researchers have also found that some plants produce chemicals that alter insect metabolism. Petunias, for example, synthesize a chemical that dramatically stunts the growth of corn earworms. Scientists hope that they can transfer the genes to crop plants, through either genetic engineering or more conventional means, to provide a natural protection against pests, thus providing an on-site means of control. Nicotine, caffeine, and citrus oil are all natural insecticides and are under investigation today.

Gardeners often use insecticidal soap and hot pepper to spray their plants to control insects such as pesty white flies and aphids. I've used them for years on vegetables and ornamentals grown in my indoor planters. Gardeners also use pyrethrums, a naturally occurring substance derived from plants. It's an effective pesticide against many insects.

> **KEY CONCEPTS**
>
> Plants have evolved many natural insect repellants that can be commercially produced and sprayed on crops to help control pests.

Cultural Controls

The final components of IPM are the **cultural controls**, any one of a dozen techniques to control pest populations that do not fall under the previous categories—environmental, genetic, and chemical. These methods include cultivation to control weeds, noisemakers to frighten birds, and manual re-

moval of insects from crops (especially suitable for gardeners). Also included in this group are such measures as destroying insect breeding grounds; improved forecasting of insect emergence; quarantines on imported foods, notably fruits and vegetables, to prevent the spread of pests; and water and fertilizer management to ensure optimum crop health and resistance to pests. One of the most badly needed cultural controls is monitoring.

> **KEY CONCEPTS**
>
> Insect pests can be controlled by many techniques that do not require the use of chemicals. These are called *cultural controls* and include measures such as noisemakers to frighten birds from crops, manual removal of insects, and quarantines on imported food.

Educating the World About Alternative Strategies

According to a report by Worldwatch Institute's Peter Weber, "Farmers seem to see in pesticides . . . the illusion of a guarantee, which they can never get from the weather, markets, or politicians." Some banks even require farmers to use pesticides to qualify for crop loans.

With a little imagination and thought, farmers are finding that they can reduce pesticide use by 50% and still not lower harvests. As farmers like Ron Rosmann are finding, they can even find ways to curtail pest damage without chemicals.

Training the world's farmers in integrated pest management is vital to efforts to build a sustainable society. This will require a massive educational effort. Schools and universities, extension services, and even agricultural magazines can help retrain farmers.

For many years, though, in virtually all countries, farmers have received most of their advice on pest control from the sales reps of the chemical manufacturers that produce pesticides—who have an obvious conflict of interest. In some universities, much of the research on pest control is sponsored by pesticide manufacturers, which may bias the system. In the United States, for example, the Department of Agriculture spends under 1% of its $95 billion annual budget on integrated pest management and sustainable agriculture. In contrast, pesticide companies shell out $1.7 billion a year for research and development of chemical controls.

Fortunately, many farmers have become aware of the pitfalls of the indiscriminate use of pesticides. Cost consciousness has led to efforts to seek alternatives on the part of both small farmers and large corporate farms. In addition, several U.S. universities have sustainable agriculture programs that teach students about IPM, among other subjects. Students at Iowa State University and the University of California at Davis, for example, can learn cost-effective ways of controlling pests without chemicals or with minimal use of them. State university extension programs also provide information to farmers and gardeners on alternatives to chemical pesticides.

Universities in the less developed countries are also beginning to train farmers and students. Birzeit University on the West Bank in Palestine, for instance, recently embarked on a program to help reduce the use of pesticides in Palestine and introduce farmers to integrated pest management through an ambitious educational campaign.

In less developed countries, agricultural departments have trained farmers in IPM and encouraged them to educate fellow farmers. In Indonesia, the government trained and hired 2000 crop scouts. These scouts work with farmers to monitor insect populations and also teach the farmers the techniques of IPM. Farmers can also work together without government support. Ron Rosmann, introduced at the beginning of the chapter, is a member of Practical Farmers of Iowa (PFI), a group of more than 500 farmers interested not only in cutting costs but also in preventing soil erosion and using fewer chemicals—ultimately, farming more sustainably. Groups like PFI have been started in several other states as well. Farmers who belong to such groups are pioneering new ways to produce crops that could result in dramatic decreases in the use of pesticides while maintaining or increasing yields. By sharing their ideas with nonmembers, they can spread the word to a broad range of farmers.

Farmer groups are also beginning to form in less developed countries. Bolivian farmers, for instance, recently formed a group called the Association of Ecological Producers. A similar group was formed in Mexico and draws off a long tradition of pesticide-free agriculture. Mexico's 13,000 organic farmers export an estimated $20 million worth of food to the United States and Europe.

Nongovernmental organizations or **NGOs**—among them, consumer and environmental groups—can play a major role in educating farmers and government officials worldwide. A recent ban on several widely used and highly toxic pesticides in the Philippines, for instance, is the result of the efforts of numerous NGOs. Concerted efforts on their part resulted in a dramatic shift in official and public opinion away from using chemical pesticides toward safer alternatives. The success of these groups can be attributed in large part to the efforts of the NGOs to demonstrate alternatives that are available to farmers. NGOs have also publicized research showing that crop yields don't have to fall when pesticides are abandoned. Through their work, they have successfully promoted the idea of IPM, which has gained wider acceptance than the chemical pesticide dogma promoted by chemical companies.

The shift to pesticide-free farming will require attitude changes on the part of consumers, too. To make sustainable agriculture successful will require buyers to choose organic produce. In some cases, organic produce is competitively priced. In most cases, though, it costs more. So, consumers must be willing to pay extra to support this growing industry. Many consumers will have to rethink and change their buying habits. They will have to learn to accept slightly blemished fruits and vegetables rather than demand picture-perfect produce made possible only by the heavy application of chemical pesticides. A few more spots on our oranges could mean many more birds overhead, cleaner waterways, improved health for workers and the general public, and cheaper oranges. They won't change the nutritional value of the produce one bit.

> **KEY CONCEPTS**
>
> Educating farmers and others about alternative strategies is imperative if sustainable pest management is to become widely adopted the world over. Universities, agricultural agencies, farmers' groups, and nonprofit organizations are several of the avenues available for this important task.

Government Actions to Encourage Sustainable Agriculture

Governments can promote sustainable agriculture by providing low-cost crop insurance for farmers who make the transition to IPM. As a farmer shifts from chemical-intensive use to IPM, losses can be severe because the ecology of the farm may be severely out of balance. With insurance, farmers can wean themselves from pesticides and not go bankrupt in the process.

Governments can also help by developing **organic certification programs**. Developed nationally and in several states, such as Colorado and California, these programs set standards farmers must meet to permit them to label their produce "Organically grown." Certification will avoid dishonesty and help consumers determine which produce is truly pesticide-free. It could also stimulate other farmers to think about switching to chemical-free methods of farming. To the delight of many people, numerous U.S. farmers have already made the switch. In 1999, the Organic Farming Research Foundation estimated that there were 6600 certified organic growers in the United States. That's good news to some, but it is only a small fraction of the two million U.S. farmers.

You can help, too by purchasing organic produce and taking other steps listed in the Individual Actions Count! table.

Our doubts are traitors and make us lose the good we oft might win by fearing to attempt.
—*William Shakespeare*

CRITICAL THINKING

As with many other issues, it is important to dig deeper and always consider the big picture. It is crucial to uncover biases that may enter into the arguments of some proponents as well. Scrutinize the experiment reported in the newspaper. Was it performed correctly, and are the results applicable to humans?

Suppose you found that the results were valid. You would then want to look at the next most important question: Should the pesticide be banned? In the public policy arena, this question pivots on another relatively simple question: Are the risks worth the benefits?

In order to answer this, it is necessary to seek more information and viewpoints. If you did, you might find that farmers and pesticide manufacturers would argue that this insecticide helps them prevent crop damage, which saves farmers millions of dollars a year. They also say that pesticides make food cheaper for consumers. The pesticide manufacturing industry also provides thousands of jobs. Farmers provide food for billions of people. In the developing world, this food keeps millions of people alive.

As for the costs, if EPA estimates are correct, 10 to 20 U.S. citizens will contract cancer each year, and many of them will die each year. The pesticide may also contaminate groundwater, kill fish and birds, and have other adverse environmental effects.

When you dig deeper, you would very likely find that there are alternative ways of controlling insects without using potentially harmful pesticides. These methods may have the added benefit of reducing groundwater contamination and mortality in birds, fish, and other wildlife. In some cases, alternative methods lower crop yield, but because of lower input costs, farmers make as much or even more money.

Can you think of any other costs and benefits? After analyzing both sides of the argument, what is your opinion? Should the pesticide be banned? Would your opinion change if you were one of the cancer victims? if you were a farmer?

CRITICAL THINKING AND CONCEPTS REVIEW

1. List and discuss reasons why pest damage is high in developed nations despite the extensive use of chemical pesticides.

2. Critically analyze this statement: "Without pesticides, crop damage in the United States and other major food producers would be much higher."

3. Describe some of the environmental and health problems caused by the use of pesticides. How can they be avoided?

4. How are beneficial species affected by insecticide use? Give some examples.

5. What does the term *pesticide treadmill* refer to?

6. Why do DDT and other chlorinated hydrocarbons persist in the environment? Why do they cause problems even though they are found in low concentration in water?

7. Make a list of the major components of integrated pest management. What role do soil conservation measures discussed in Chapter 10 play in pest control? What advantages does IPM strategy offer over current management techniques?

8. Explain why crop rotation and increasing crop diversity reduce pest populations.

9. Describe some of the biological control methods. Give examples. Using your knowledge of biology and evolution, would you expect pest species to develop resistance to biological controls?

10. Describe why the sterile male technique works.

11. Discuss some ways in which genetic engineering may be used to help cut down on pest damage.

12. You are appointed director of the state agricultural department. Outline a way to encourage farmers to minimize and ultimately eliminate pesticide use.

13. Using critical thinking, analyze this statement: "U.S. agriculture cannot go organic. We cannot eliminate pesticides without greatly reducing productivity and farmers' profits."

KEY TERMS

Agent Orange
auxins
biological control
biomagnification
broad-spectrum pesticide
carbamate
chlorinated hydrocarbon
coevolution
confusion technique
crop rotation
cultural controls
DDT (dichlorodiphenyltrichloroethane)
dioxin
environmental control
Federal Insecticide, Fungicide, and
    Rodenticide Act (FIFRA)

first-generation pesticide
genetically resistant insect
herbicide
heteroculture
hormone
integrated pest management (IPM)
integrated weed management
juvenile hormone
molting hormone
narrow-spectrum pesticide
nongovernmental organization (NGO)
organic certification program
organic phosphate
organic produce
pesticide

pesticide treadmill
pheromone
pheromone trap
pyrethroid
registration
second-generation pesticide
sex attractant
sterile male technique
strip cropping
third-generation pesticide
tolerance level
trap crop
triazine
upset

REFERENCES AND FURTHER READING

The References and Further Reading section at the end of this book contains a list of sources for the information discussed in this chapter and recommendations for further reading.

Connect to this book's website:
http://environment.jbpub.com/
The site features eLearning, an online review area that provides quizzes, chapter outlines, and other tools to help you study for your class. You can also follow useful links for in-depth information, research the differing views in the Point/Counterpoints, or keep up on the latest environmental news.

Developed nations produce mountains of trash, much of which can be recycled.

CHAPTER 23 Hazardous and Solid Wastes: Sustainable Solutions

There is nothing more frightful than ignorance in action.

—*Goethe*

The 3M Corporation, based in Minnesota, is a model of corporate ingenuity and foresight. In 1975, the company started a Pollution Prevention Pays program designed to reduce solid and hazardous waste generated by the company. By substituting less toxic or nontoxic chemicals for solvents, by recycling everything they could, and by modifying manufacturing processes, 3M has made substantial cuts. In the first 15 years, they reduced pollution and waste by 50%. Since 1990, they have made an additional 33% cut.

The company's program *annually* eliminates nearly 90,000 metric tons (100,000 tons) of air pollutants. It reduces their *annual* output of solid waste by 250,000 metric tons (275,000 tons) per year. It also reduces the production of wastewater by 5.8 billion liters (1.5 billion gallons) per year. What is more, these and other actions taken by the

Exercise

As a hazardous-waste manager of a chemical company, you are faced with a dilemma. Your company produces over 400,000 pounds of highly toxic waste annually. The cost of disposal is several million dollars. An official from a new waste disposal firm contacts you and says that he can dispose of the waste at half the cost. When you ask him where it is going, he says that it will be shipped to a developing nation, where it will be incinerated. He tells you that pollution controls on the incinerator where the waste will be burned are not as sophisticated as those on the one owned by the U.S. disposal company you've always used, but it doesn't really matter because there's so little toxic waste burned in the country anyway. What choice do you make? What factors will affect your choice?

FIGURE 23-1 **Toxic hot spot in Stratford, CT.** Inactive industrial sites like this one are often contaminated by a variety of hazardous chemicals. Cleanup of such sites can easily cost several million dollars.

company saved more than $1 billion between 1975 and 2000.

Although interest in reducing waste by recycling and using resources more efficiently is increasing, businesses are a long way from tapping into the full potential of this strategy. Many continue to throw away millions of tons of perfectly usable material, including cardboard, wood, office paper, and metals. Moreover, some companies continue to illegally and irresponsibly dump hazardous waste into the air, water, and soil.

This chapter discusses solid and hazardous wastes. Like other chapters in this book, it shows how individuals, businesses, and governments have traditionally addressed the problem of waste and how limited these strategies are. It will also show more sustainable approaches, measures that make sense from social, economic, and environmental perspectives.

23.1 Hazardous Wastes: Coming to Terms with the Problem

Hazardous wastes are waste products of homes, factories, businesses, military installations, and other facilities that pose a threat to people and the environment. They are toxic,

carcinogenic, or mutagenic. For many years, pollution like hazardous waste was seen as a sign of progress. Today, many individuals view pollution in general and hazardous waste in particular as signs of unsustainable technologies or unsustainable industrial systems (**FIGURE 23-1**).

KEY CONCEPTS

Hazardous wastes come from a variety of sources, among them factories and even our homes. These toxic materials are now viewed as signs of unsustainable practices.

Love Canal: The Awakening

The severity of the U.S. hazardous-waste problem caught the attention of the American public in the 1970s when toxic chemicals began to ooze out of a hazardous-waste dump known as Love Canal in Niagara Falls, New York. The story of Love Canal began in the 1880s, when William T. Love began digging a canal that would run from the Niagara River just above Niagara Falls to a point on the river below the falls. The canal was built to divert water to an electric power plant to supply future industrial facilities that would be built along its banks. Unfortunately for Love, the canal was never completed. Only a small remnant of the canal remained in the early 1900s. In 1942, the Hooker Chemical Company signed an agreement with the canal's owner to dump hazardous wastes into the abandoned canal. In 1946, Hooker bought the site, and from 1947 to 1952 it disposed of over 20,000 metric tons (22,000 tons) of highly toxic and carcinogenic wastes, including dioxin.

In 1952, the story took an ironic twist. In that year the city of Niagara Falls began condemnation proceedings on the property. This legal maneuver would allow the city to acquire the land to build an elementary school and residential community. Hooker sold the land to the city for $1 in exchange for a release from any future liability. Hooker insists that it warned against construction on the dump site itself, but it allegedly never disclosed the real danger of building on it. Before turning the land over to the city, Hooker sealed the pit with a clay cap and topsoil, once thought sufficient to protect hazardous-waste dumps.

Troubles began in January 1954, however, when workers removed the clay cap during the construction of the school. In the late 1950s, rusting and leaking barrels of toxic waste began to surface. Children playing near them suffered chemical burns; some became ill and died. Hooker said that it had warned the school board not to let children play in contaminated areas, but the company apparently made no effort to warn local residents of the potential problems.

The problem continued for years. Chemical fumes took the bark off trees and killed grass and plants in vegetable gardens. Smelly pools of toxins welled up on the surface. In the early 1970s, after a period of heavy rainfall, basements in homes near the dump began to flood with a thick, black sludge of toxic chemicals. The chemical smells in homes around the dump site became intolerable.

Tests in 1978 on water, air, and soil in the area detected 82 different chemical contaminants, a dozen of which were known or suspected carcinogens. In that same year, the state health department found that nearly one of every three pregnant women in the area had miscarried, a rate much higher than expected. Birth defects were observed in 5 of 24 children. Another study, released in 1979 by Dr. Beverly Paigen of the Roswell Cancer Institute, showed that over half of the children born between 1974 and 1978 to families living in areas where groundwater was leaching toxic chemicals from the dump had birth defects. In this study, the overall incidence of birth defects in the Love Canal area was one in five, compared with a normal rate of less than 1 in 10. The miscarriage rate was 25 in 100, compared with 8 in 100 for women moving into the area. Asthma was four times as prevalent in contaminated areas as in uncontaminated areas in the region; the incidence of urinary and convulsive disorders was almost three times higher than expected. The incidence of nasal and sinus infections, respiratory diseases, rashes, and headaches was also elevated.

As a result of public outcry, the school was soon closed. The state fenced off the canal and evacuated several hundred families (**FIGURE 23-2**). President Carter declared the site a disaster area. In May 1980, a new study revealed high levels of genetic damage among residents living near the canal. An additional 780 families were evacuated from outlying areas.

In 1987, the EPA announced plans to clean the sewers and dredge two creeks in the Love Canal area to remove sediments contaminated with toxins. In 1988 and 1989, the creeks were diverted so they would dry. Bulldozers then removed the top 18 inches of mud, which was burned by the company in a special incinerator built especially for this project. All told, about 35,000 cubic meters of sediment will be burned, making this the largest single application of thermal destruction in modern history. The removal of the sediments cost $13 million, and sewer cleanup added another

FIGURE 23-2 **The price of pollution.** Houses awaiting the bulldozer in the Love Canal area of Niagara Falls, New York. This tragedy not only cost the government (and taxpayers) millions of dollars, it also resulted in considerable disruption of people's lives as families were forced from their homes.

$5 million to the price tag. Occidental Petroleum (formerly Hooker) estimates that the incineration will cost an additional $14 million. Love Canal cost the state of New York and the federal government approximately $272 million for cleanup, relocation of residents, and other expenses.

A 1980 study by the EPA showed that chemical contamination was pretty much limited to the canal area (the actual dump), an area immediately south of it, and two rows of houses on either side of the canal (**FIGURE 23-3**). The last group of residents to be evacuated, the report said, were probably moved out unnecessarily. The EPA study also showed that the dump had contaminated shallow groundwater but not deeper aquifers. The EPA concluded that further migration of toxic chemicals was highly unlikely. Based on this study and other work, the EPA and the state of New York declared two thirds of the evacuated Love Canal site "habitable" and proposed to sell the houses. At this writing, nearly all of the homes had been sold. Lois Gibbs, the Love Canal resident largely responsible for drawing public attention to the disaster and getting the state and federal governments to take action, argues that the decision to resettle the area was improperly made. In fact, she claims that in assessing the habitability of the Love Canal site, the New York State Health Department compared it to only two other sites, both badly contaminated by industrial wastes, and deemed it suitable for resettlement. Gibbs warns that resettling Love Canal will put more people at risk. According to Gibbs, 20,000 tons of hazardous waste still remain at the site. It will take 20,000 years for these wastes to decompose. So far, there have been no medical concerns.

KEY CONCEPTS

Love Canal was a hazardous-waste disposal site in Niagara Falls, New York, similar to many others around the world. Leakage from the site caused serious health problems in residents living near it. The incident alerted the public and government officials to the problem of improper hazardous-waste disposal.

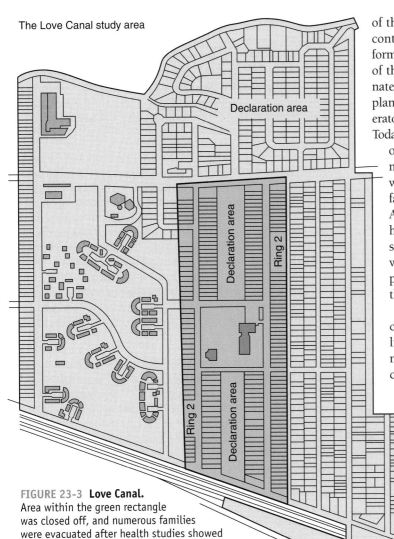

The Love Canal study area

FIGURE 23-3 **Love Canal.**
Area within the green rectangle was closed off, and numerous families were evacuated after health studies showed elevated incidence of birth defects, stillbirths, and a variety of symptoms most likely attributable to toxic wastes from the hazardous-waste dump. Residents were also evacuated from an outer region (the declaration area); but tests have shown that hazardous wastes have not migrated into this area and that these residents may have been unnecessarily evacuated.

The Dimensions of the Problem

In the years following the Love Canal incident, health officials found that Love Canal was not an isolated incident, but just the tip of an enormous hazardous-waste iceberg. In 1989, the EPA announced that the number of hazardous-waste sites was approximately 32,000. The General Accounting Office estimated that the number of hazardous-waste sites could be much higher, perhaps 100,000 to 400,000. These estimates do not include the 17,000 hazardous-waste *hot spots* on U.S. military bases. Hazardous wastes have also been mixed with oils that are sprayed on dirt and gravel roads in rural areas to prevent dust. Tens

> Although no one knows for sure, the number of hazardous waste sites in the United States could be as high as 100,000 to 400,000. Europe may contain a similarly high number.

of thousands, perhaps hundreds of thousands, of badly contaminated sites may exist in Europe, especially in the former east bloc nations and the nations that once were part of the Soviet Union. Canada is home to many contaminated sites, too. One of the worst of all is the Sydney steel plant in Nova Scotia. For nearly 100 years, the plant's operators dumped waste from its operations in a nearby creek. Today, 34 hectares (85 acres) of the tidal flats where the creek opens onto the ocean contain an estimated 500,000 metric tons of toxic coal-tar from the facility. Cleanup was estimated to cost $35 million, but the task proved far more difficult and far more costly than anticipated. At this writing, the federal and provincial governments have spent over $52 million cleaning up the site and are still not finished. Findings such as these have fueled widespread concern about cancer. (For a debate on the presumed epidemic of cancer in the United States, see the Point/Counterpoint in this chapter.)

Making matters worse, each year U.S. factories create an estimated 37 million metric tons (41 million tons) of hazardous waste from large facilities—nearly one metric ton for every man, woman, and child.[1] But the United States is not alone. European countries and many less developed nations also produce tens of millions of tons of hazardous waste each year. In 1985, 90% of the hazardous wastes in the United States were improperly disposed of, according to one estimate. This waste ended up in abandoned warehouses; in rivers, streams, and lakes; in leaky landfills that contaminate groundwater; in fields and forests; and along highways. No current estimates are available, but the percentage is now believed to be much lower.

Nevertheless, improper waste disposal has left behind a long list of costly effects: (1) groundwater contamination, (2) well closures, (3) habitat destruction, (4) human disease, (5) soil contamination, (6) fish kills, (7) livestock disease, (8) sewage treatment plant damage, (9) town closures, and (10) difficult or impossible cleanups. Irresponsible and ill-conceived waste disposal continuing today will create a legacy of polluted groundwater and contaminated land that could persist for decades, perhaps centuries.

Three decades after the United States first awakened to the hazardous-waste issue, most experts believe that the nation is in for a much longer, more difficult battle than once was anticipated. Why? There are many more sites that are far more difficult to deal with than anybody ever anticipated. The price tag could also be much higher than expected. The U.S. Office of Technology Assessment estimates that it will cost $100 billion to clean up the 10,000 sites in the United States that pose a serious threat to health. Researchers at the University of Tennessee estimate the cleanup of all hazardous-

[1] This includes only facilities that produce over 1000 kilograms (450 pounds) of hazardous waste per year. Thousands of smaller facilities also generate hazardous waste but are not required to report it.

America's Epidemic of Chemicals and Cancer

Lewis G. Regenstein
Lewis G. Regenstein, an Atlanta writer, is author of *Cleaning Up America the Poisoned* and is president of the Interfaith Council for the Protection of Animals and Nature.

America is in the throes of an unprecedented cancer epidemic, caused in part by the pervasive presence in our environment and food chain of deadly, cancer-causing pesticides and industrial chemicals.

Today, significant levels of hundreds of toxic chemicals known to cause cancer, miscarriages, birth defects, immune and central nervous system damage, and other health effects are found regularly in our food, our air, our water—and our own bodies. Accompanying this widespread pollution has been a dramatic and alarming rise in the cancer rate in recent decades (which has only now begun to abate for some cancers, with the restricting of some of the most dangerous chemicals).

Over 4.6 billion pounds of active pesticide ingredients are used in the United States each year—mainly on farms—which amounts to some 18 pounds for every man, woman, and child in the country. Some 75 million pounds of pesticides are used on just lawns and turf. Of the 36 most widely used lawn chemicals, 13 can cause cancer; 14 can cause birth defects, 11 can cause reproductive effects, 15 can damage the liver, and 21 can harm the nervous system.

Indeed, pesticides are used almost everywhere in our society: in homes, schools, restaurants, hospitals, parks, offices, hotels—they are virtually inescapable. One EPA study found 23 pesticides in indoor dust and air, many of which had not been used on the premises; another found that most households had at least five pesticides in indoor air, often at levels 10 times higher than outdoors.

So it should not be surprising that by the time restrictions were placed on some of the deadliest compounds, such as DDT, dieldrin, and BHC, these carcinogens were being found in the flesh tissues of literally 99 percent of all Americans tested, including mother's milk.

Numerous studies link exposure to pesticides to incidences of cancer, especially in children. Higher rates of childhood brain cancers, leukemia, and soft tissue sarcomas are found in homes where pesticides are used. Indeed, a child in a household where home and garden pesticides are used has a 6.5-fold increased risk of contracting leukemia.

The U.S. National Research Council and the EPA have estimated that pesticide residues in food may be responsible for 20,000 to 60,000 excess cancer cases in America each year. And the American Academy of Pediatrics has warned that infants and children are at particular risk, since "the government is permitting 100 to 500 times as much chemicals in the food as a health basis number would dictate."

Cigarette smoking is by far the biggest cause of cancer, mainly of the lung, and diet can also be an important risk factor, especially the frequent consumption of meat and other high fat foods. But the constant, unavoidable, lifelong chemical onslaught to which we are subjected, from conception on, also plays a role in this disease, once considered rare, that now strikes 1.2 million Americans a year. (This does not count the rapidly growing number of usually nonfatal skin cancers, now a million cases a year. These are mainly caused by exposure to sunlight and increased ultraviolet radiation resulting from the ongoing depletion of the earth's protective ozone layer, which is also caused by pesticides and other synthetic chemicals.) One American in three can now expect to contract a potentially fatal cancer, which kills some 570,000 of us every year—more Americans than were killed in combat in World War II, Korea, and Vietnam combined.

And cancer has now become a common disease of the young as well as the old, with the actual incidence (and not just detection) of childhood cancers, especially leukemia and brain tumors, mounting sharply in recent years.

The response of the U.S. government, through various administrations, has been largely weak or nonexistent enforcement of the nation's health and environmental protection laws. EPA, under pressure from Congress and chemical and agricultural interests, has, with few exceptions, refused to carry out its legal duty to ban or restrict cancer causing pesticides.

Thus, we are even now acting to ensure that the current cancer epidemic will continue long into the future. Only time will tell what will be the effect on this generation of Americans, and future ones—the chemical industry's ultimate guinea pigs.

You can link to websites that represent both sides through Point/Counterpoint: Furthering the Debate at this book's internet site, http://environment.jbpub.com/. Evaluate each side's argument more fully and clarify your own opinion.

There is no debate that cancer is a devastating and deadly disease. One in three people living in the United States today will contract some form of cancer in his or her lifetime, and one in four will die from it, if current rates continue. Are cancer rates increasing in epidemic proportions? The total number of people and the fraction of all deaths attributable to cancer have increased dramatically in the past 75 years. However, cancer is largely a disease of old age, and thus, it is necessary to adjust such statistics for changes in the age distribution of our population. Annual mortality statistics from both the American Cancer Society and the National Cancer Institute indicate that age-adjusted deaths from all causes of cancer have not increased significantly in the past several decades *if lung cancer mortality is excluded*. There is no question that we have experienced an epidemic of lung cancer in the 20th century. The age-adjusted mortality for lung cancer in both males and females has increased over 50-fold since the early 1900s, peaking in the late 1980s in men and late 1990s in women. About 85–90% of all lung cancer is directly attributable to smoking. Per capita consumption of cigarettes increased 5-fold in men from 1900–1960, and with it concomitant increase in lung cancer. This pattern was repeated, 20 years later, in women. Tobacco products also increase the risk of several other types of common cancers (e.g., cancers of the bladder, kidney, esophagus and mouth). In contrast, there was an equally dramatic decline in the incidence of stomach cancer in the United States, during the first half of the 20th century. This decline was mostly likely due to changes in dietary habits that came with the widespread use of refrigeration and possibly antioxidant food preservatives (versus the older methods of smoking and salting methods used to preserve foods). Although mortality statistics have not risen sharply, the incidence (number of cases per 100,000 population) of certain cancers has been increasing in the past several decades. For example, the incidence of childhood leukemia and brain cancers, and adult cases of non-Hodgkin's lymphoma, have increased significantly. While at least part of these increases are likely to be due to better disease surveillance and reporting, the cause(s) of these increases are not known and may be related to as yet unidentified environmental factors. Infectious agents also cause many cancers. For example, nearly all cases of cervical cancer are thought to result from papilloma virus infection that is spread by sexual activity. Likewise, many cases of liver cancer are from hepatitis virus infections, and stomach cancer risk is increased by certain bacterial infections.

Finally, recent advances in the understanding of the biology of cancer suggest that spontaneous or background alterations in DNA may explain much of the cause of cancer. The use of modern techniques in molecular biology has revealed that DNA is inherently unstable and can be altered by normal errors in DNA replication. Within our life span, our cells undergo about 10 trillion cell divisions. Spontaneous errors in this process, which lead to mutations and cancer, accumulate with age. DNA is also subject to extensive oxidative damage from processes associated with normal cellular metabolism that occurs in each cell every day of our lives. Although most of this damage is repaired in the cells, over many years, small amounts of unrepaired damage to DNA accumulates. It is not surprising then that cancer is a frequent outcome of old age. One reason that diet is an important risk factor for many types of cancer is because of the presence of naturally occurring "antioxidants" found in fruits and vegeta-

"America's Epidemic of Chemicals and Cancer"—Myth or Fact?

David L. Eaton, Ph.D.
David L. Eaton is Professor of environmental health and Associate Dean for Research in the School of Public Health and Community Medicine at the University of Washington. He has an active research program on the molecular mechanisms by which chemicals cause cancer.

bles. These chemicals help protect against DNA damage from normal cellular metabolism and may also offer protection against other DNA damaging chemicals found in cigarette smoke, our diet, and other environmental sources of exposure.

The vast majority of cancer researchers throughout the world do not support the view that we are in an overall cancer epidemic caused largely by industrial chemicals. Unfortunately, it will take some time for the political arena, influenced greatly by public fears, to come to grips with the fact that further reduction in exposure of the general public to trace amounts of synthetic chemicals will not have a measurable impact on overall cancer incidence. Hundreds of billions of dollars are spent each year on pollution reduction in the United States alone, often under the rubric of reducing cancer risk to the public. I believe that much of these expenditures are justified to enhance the quality of our environment, protect valuable natural resources, and ensure the habitability of our planet for our children and our children's children. But these expenditures for pollution prevention are not going to have a significant impact on cancer incidence and mortality in the United States. Certainly we must continue our efforts to identify potentially cancer-causing industrial chemicals and take appropriate actions to minimize occupational and environmental exposures to such chemicals. However, if our society is truly concerned about reducing the human tragedy from cancer, more efforts should be focused on eliminating smoking and alcohol abuse and increased public education about dietary risk factors. More research into the biochemical and molecular events that lead to cancer will ultimately lead to more effective prevention and treatment of cancer.

For up-to-date information on cancer and cancer risks, see the websites for the American Cancer Society (www.cancer.org) and the National Cancer Institute (www. nci. nih.gov).

Critical Thinking Questions

1. Summarize each author's major points and supporting data. In your view, how well has each of the authors supported his contentions?
2. How can the critical thinking skills presented in Chapter 1 help you analyze the arguments presented here?
3. Do you agree with Dr. Eaton that to determine whether we are in a cancer epidemic caused by synthetic chemicals, we should eliminate lung cancer from the calculation? Why or why not?

waste sites in the United States by state and federal programs will cost $750 billion.

LUST—It's Not What You Think

You feel dizzy. Your head spins. Your insides ache. You haven't been yourself for weeks. What may be ailing you is LUST—but not the usual kind. Your symptoms may be caused by the latest in a long list of hazardous chemical problems: groundwater pollution from a leaking underground storage tank, which EPA's top acronymists dubbed LUST. Some time later, no doubt responding to complaints from citizens vigilant for political correctness, they dropped the L, leaving UST: underground storage tanks.

Generally concentrated in heavily developed urban and suburban areas, underground storage tanks are used primarily to store petroleum products such as gasoline, diesel fuel, and fuel oil. They're found at gas stations, factories, and homes. Some are used to store hazardous wastes, too.

The problem with tanks is that moisture and soil acidity gradually corrode the steel they're made of, causing them to deteriorate over time and leak petroleum by-products, toxic chemicals, and hazardous wastes. The main concern is the potential effect on groundwater and human health. Even a small leak can contaminate large quantities of groundwater. For example, 1.5 cups of hazardous liquid leaking out of a tank per hour can contaminate nearly 4 million liters (over 1 million gallons) of groundwater in a day. Contaminated groundwater is very difficult—sometimes impossible—and expensive to clean up.

Contaminated groundwater poses a problem to people and animals (such as livestock) that drink it. Contaminated drinking water used for baths and showers can also be dangerous. Benzene, a component of gasoline that can cause cancer, is absorbed through the skin when bathing. Showering generates dangerous vapors that can cause skin and eye irritation.

Although there are many sources of groundwater contamination, according to the U.S. EPA, leaking underground storage tanks are the number one cause of groundwater pollution, followed by landfills (FIGURE 23-4). A small leak can go undetected for long periods, causing considerable contamination. According to the EPA, the most significant effects are seen in the shallow aquifers, the ones domestic water is typically drawn from. A report by the New York Department of Environmental Conservation suggests that at least half of the state's underground steel tanks containing petroleum products that are over 15 years old may be leaking. Nationwide, 3 to 5 million underground storage tanks containing hazardous materials dot the United States. In 1996, the EPA reported confirmation of leakage in more than 300,000 tanks. They estimated that 60% of these leaks have affected groundwater nationwide, although the percentage may be as high as 90% in certain locations.

Major oil companies have already spent millions

> The EPA confirmed the presence of more than 300,000 leaking underground storage tanks in the United States, 60% of which they estimate are polluting groundwater.

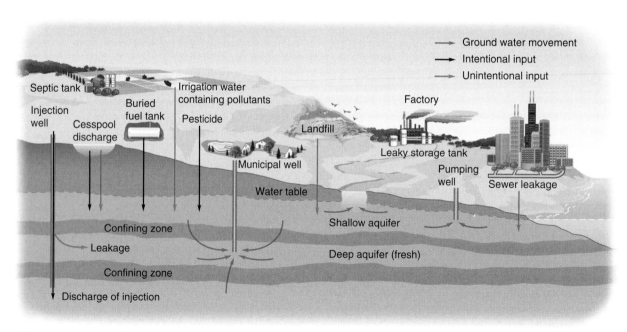

FIGURE 23-4 **Source of groundwater pollution.** Groundwater is contaminated by a variety of sources such as septic tanks, farms, and abandoned industrial sites.

to clean up polluted groundwater and soil and to install new tanks at gas stations. The cost of such actions can be exorbitant. Chevron alone estimates its replacement costs at about $100 million. Unfortunately, half of U.S. service stations are owned by independent dealers who generally are not financially able to replace the leaking tanks.

About 90% of the cleanup and replacement of leaking underground storage tanks is being financed and performed by private industry. The rest falls on the shoulders of the states. The EPA sets guidelines for cleanup and replacement and also provides financial assistance to help states. Today, all 50 states have created funds to pay for their part of the cleanup cost. State and federal funds are derived from taxes on gasoline.

KEY CONCEPTS

Hundreds of thousands of underground storage tanks have been installed in industrial nations and are used to store many potentially toxic substances such as gasoline, diesel fuel, hazardous waste, and heating oil. Over time, steel tanks corrode and begin to leak, contaminating groundwater used for cooking food, drinking, and bathing.

23.2 Managing Hazardous Wastes

Two hazardous-waste problems face virtually all industrial nations and, to a lesser extent, the less developed countries. First, how do they clean up existing hazardous-waste sites, leaking storage tanks, and polluted groundwater? Second, how do they deal with the enormous amounts of hazardous waste produced each year to avoid creating new sites and further contamination of groundwater?

The first problem requires immediate action. It's one place where the Band-Aid approach is appropriate. That said, all solutions that fall into this category must be sustainable; they must not merely shift the problem from one location (a contaminated factory site) to another (a landfill where the hazardous wastes are dumped). The second problem calls for long-term preventive measures that eliminate the production of wastes.

KEY CONCEPTS

Addressing the issue of hazardous waste requires both plans to clean up contaminated sites and preventive actions to greatly reduce or eliminate hazardous-waste production in the first place.

The Superfund Act: Cleaning Up Past Mistakes

In June 1983, the 2400 residents of the town of Times Beach, Missouri, agreed to sell their 800 homes and 30 businesses to the federal government for $35 million. Why? The roads in Times Beach, like those elsewhere, had been sprayed with oil containing hazardous wastes, including dioxin. How did the oil get contaminated?

As noted earlier in the chapter, unscrupulous hazardous-waste disposal companies had mixed toxic chemical wastes with waste oil and then spread it on dirt roads to control dust. In Times Beach, the dioxin levels in the soil were 100 to 300 times higher than levels considered harmful during long-term exposure. The town had to go, and the federal government bought it. Today, Times Beach is a ghost town bordered by a tall chain-link fence. Its only occupants are occasional EPA officials or scientists from companies that are exploring ways to detoxify the soil.

The $35 million purchase price for this contaminated piece of real estate came from a special fund known as the **Superfund**. It was created in 1980 by the **Comprehensive Environmental Response, Compensation and Liability Act, CERCLA** for short. Commonly called the **Superfund Act**, it and its two amendments (1986 and 1990) established a $16.3 billion fund financed by state and federal governments and by taxes on chemical and oil companies. The money is earmarked to clean up both leaking underground storage tanks that are deemed a threat to human health and abandoned and inactive hazardous waste sites, including hazardous waste dumps, landfills, and contaminated factories, mines, and mills.

Under CERCLA, the EPA is authorized to collect fines from parties responsible for the contamination—totaling up to three times the cleanup cost. Thus, this law makes owners and operators of hazardous waste dump sites and contaminated areas, as well as their customers, responsible for cleanup costs and property damage. Under the law, all businesses, hospitals, schools, cities, and other parties that deposited hazardous waste at a site are liable for a portion of the cleanup cost, based on the type and the volume of waste they deposited. CERCLA requires a sharing of costs even at licensed hazardous-waste disposal facilities where hazardous wastes were legally disposed of in previous years. As one industry representative put it, "You are liable for your waste forever." If a company or party is no longer in business, the remaining parties must pay the cost.

Superfund has clearly had an impact. By the end of 1991, for instance, the EPA had surveyed more than 30,000 potential Superfund sites and completed more than 2700 emergency actions—steps to reduce immediate threats, such as removing barrels of waste stored in abandoned warehouses. In 1999, the EPA listed 1529 sites on a National Priority List (NPL) because of their potential health threat (**FIGURE 23-5**). Although some action has been taken on over 90% of all NPL sites, cleanup has proven costly and slow. By June 2004, 263 sites had been cleaned up, and 583 had been largely decontaminated. Under the Bush Administration, the pace of Superfund cleanups has slowed by half that of during the Clinton Administration. Cleanup costs recovered from polluters fell by 13%, and penalties collected from polluters fell 41%. In addition, Congress dropped the Polluter Pays program by letting the Corporate Superfund tax expire. Cleanup funds declined to an all-time low of $159 million, down from $3.8 billion.

Because of public pressure and their own realization that cleanup was going much too slowly, the EPA identified ways to cut the time for a typical cleanup by two years (the current average is seven to ten years).

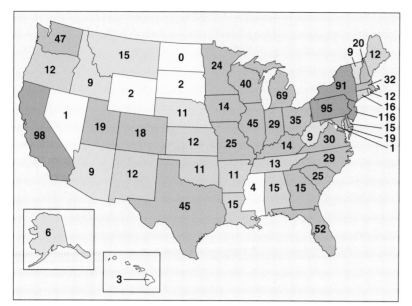

FIGURE 23-5 **U.S. hazardous waste sites.** This map shows the number of hazardous waste sites on the EPA's 2003 Priority List. How does your state rank? (*Source:* U.S. Environmental Protection Agency. Supplementary Materials: National Priorities List, Proposed Rule. December 2003.)

KEY CONCEPTS

The Superfund Act provides money to clean up hazardous waste dumps and other contaminated sites. This money comes from a tax on oil and petrochemicals and is replenished by fees charged to those responsible for the contamination, including the owners and operators of waste sites and the companies that paid to have their waste disposed of in them.

Problems with the Superfund Program Despite these successes, the Superfund program has been riddled with problems. One of the most significant is cost. Stabilizing a leaking pond designed to hold hazardous wastes costs the EPA $500,000. A study to determine what chemicals are leaking from a site can cost $800,000.

A second problem is that it has created a legal nightmare, described as a "massive web of litigation between the EPA, waste depositors, and insurance companies." Of the funds paid in the years the program has been operating, nearly 60% have been for legal fees. This money has largely been spent on identifying liable parties to get them to pay their fair share—not on cleanup.

Initial mismanagement by top EPA officials also delayed serious action by the agency in the early years. Officials negotiated with owners of hazardous-waste dumps to begin private cleanups, but critics say that they let some companies off too easily and required only superficial cleanups. Officials waived future liability in some cases. Thus, if problems develop in the future, companies will bear no responsibility. Investigations conducted in 1983 led the EPA's top leadership to resign or be fired because of the issue. One EPA official went to jail for perjury.

The Superfund Act has also been criticized because, although it provides money for cleanup and financial compensation for property damage, it fails to provide avenues for victims of illegal dumping of hazardous wastes and their families to be compensated for personal injury or death. According to former senator George Mitchell, "Under the legislation, it's all right to hurt people, but not trees." Many people believe that a fund similar to worker's compensation is needed for victims.

According to the Office of Technology Assessment (OTA), the EPA has often opted for quick-fix solutions to clean up areas. Three quarters of the cleanups, they say, are inadequate in the long term. In Love Canal, for instance, the EPA simply put a clay cap over the dump site and dug a ditch around it to collect hazardous wastes that escape. In other sites, contaminated soil was excavated and hauled off to another landfill. Incineration and biological destruction of the wastes might have been more long-lasting solutions.

KEY CONCEPTS

The cleanup of hazardous-waste sites under the Superfund Act has proven extremely slow, costly, and litigious. Much of the money spent has gone to legal fees. The Superfund Act has been criticized because it provides money for cleaning up property but no compensation for health damage. Many of the cleanups are considered inadequate.

Alternative Cleanup Funding Options Some critics of CERCLA argue that the law has created an adversarial relationship among many parties that is counterproductive and inefficient. Too much money is spent on litigation and too little on cleanup. Although critical of CERCLA, these individuals do not suggest we abandon the program, only that we find options that put the money to better use.

One proposal calls for a federal hazardous-waste tax levied on each ton of hazardous waste disposed of, incinerated, or treated. This money would generate revenue for cleanup of the most heavily contaminated sites. There would be no need to try to assign responsibility for contamination and no need for the lengthy litigation that often results.

The American Association of Property and Casualty Insurers has proposed funding cleanup via a small fee on each new commercial insurance policy written in the United States. Their calculations suggest that this would generate more than $4 billion a year that could be used for cleanup. Both this and the previous alternative might allow the EPA to redirect its efforts to cleanup.

KEY CONCEPTS

Rather than spending millions of dollars to identify responsible parties, the Superfund program might work better with a no-fault policy—one that provides funds to clean up sites regardless of who is liable.

What to Do with Today's Waste: Preventing Future Problems

The high cost of cleanup strongly suggests the need for active preventive measures to avoid further contamination. Several possibilities exist.

The Resource Conservation and Recovery Act: Preventing Improper Disposal Cleaning up hazardous-waste sites is an essential first step in protecting health and creating a sustainable future. It should help reduce further contamination of groundwater. Efforts are also needed to prevent illegal and improper waste disposal. In 1976, the U.S. Congress passed the **Resource Conservation and Recovery Act (RCRA)**. This law is designed to monitor hazardous waste to eliminate illegal and improper waste disposal.

Under RCRA, the EPA was designated the nation's hazardous-waste watchdog. The EPA's first role was to determine which wastes were hazardous. RCRA also called on the agency to establish a nationwide reporting system for all companies handling hazardous chemicals. This requirement created a trail of paperwork that follows hazardous wastes from the moment they are generated to the moment they are disposed of—a so-called cradle-to-grave tracking. Congress believed that this requirement would make it difficult for waste generators to dump hazardous wastes improperly. RCRA also directed the EPA to set industry-wide standards for packaging, shipping, and disposing of wastes. Only licensed facilities could receive wastes.

Unfortunately, RCRA's implementation has been slow. It was not until 4 years after Congress adopted the law that the EPA came up with its first hazardous-waste regulations. To the dismay of many, the regulations were full of loopholes, and about 40 million metric tons (44 million tons) of pollutants annually escaped control.

Because of public pressure, in 1984, Congress passed a set of amendments to eliminate RCRA's loopholes and ensure proper waste disposal. For example, under the original law, if a company produced under 1000 kilograms (2200 pounds) of hazardous wastes per month, it could dump them in a local landfill. The amendments changed the rules so that any individual or company that generated more than 100 kilograms (220 pounds) of hazardous waste a year must follow the same guidelines imposed on large waste producers.

The 1984 amendments also declared a national policy to reduce or eliminate land disposal of hazardous waste. Congress made it clear that land disposal technologies must be a last resort. The 1984 amendments gave preference to reuse, recycling, detoxification (such as incineration), and other measures discussed shortly. These approaches are bringing the United States closer to a sustainable waste management system.

The 1984 amendments to RCRA also addressed leaking underground storage tanks. After May 1985, for example, all newly installed underground tanks had to be protected from corrosion for the life of the tank. The lining of the tank must consist of materials compatible with stored substances. Furthermore, owners and operators must have methods for detecting leaks, must take corrective action when leaks occur, and must report all actions.

KEY CONCEPTS

The Resource Conservation and Recovery Act established a nationwide reporting system to monitor hazardous wastes from their production until their disposal. This provision seeks to eliminate illegal and improper hazardous-waste disposal. Recent amendments call for an elimination of land disposal of hazardous wastes and regulations to prevent leaking underground storage tanks.

Weaknesses in RCRA Despite these changes, RCRA still has many loopholes. Some critics believe that its definition of *hazardous wastes* is too narrow. Michael Picker of the National Toxics Campaign, for example, thinks that municipal waste (garbage) should be classified as hazardous waste because it contains toxic chemicals such as pesticides, ore, and heavy metals like lead (from batteries). Leachates (water containing contaminants) seeping from some municipal landfills are as toxic as those coming from regulated hazardous-waste facilities. According to John Young of the Worldwatch Institute, more than one of every five hazardous-waste sites on the U.S. Superfund cleanup list is a municipal landfill.

Picker also thinks that sewage and untreated wastewater handled by publicly owned sewage treatment plants should be considered a hazardous waste. Toxic chemicals in the sewer system, released by factories (legally and illegally) and homeowners, can escape into the air and into waterways. Agricultural wastes, mostly pesticides, are also not regulated. In California, for example, rules *require* that leftover pesticides be diluted and sprayed into the environment. Mill and mine tailings are also excluded from most regulatory control. By expanding the definition of what is toxic and by instituting better controls, the government could greatly cut back on the influx of hazardous materials into the environment.

One lesson environmentalists have learned in the past 4 decades is that passing a law is not a guarantee of protection. Why not? For one thing, agencies responsible for administering and enforcing new laws don't always perform as they are instructed. Some drag their feet because they don't approve of the law. Additionally, agencies may be so underfunded and so overworked that they can't take on new responsibilities or, if they do, they do a shoddy job. The EPA is a case in point. Understaffed and underfunded, the EPA today struggles to implement RCRA and the handful of other laws aimed at protecting public health and the environment. Since the agency was formed, its workload has more than doubled, but funding has until quite recently remained at more or less the same level (adjusted for inflation). EPA's funding was drastically scaled back under the Bush administration.

Additional steps may be needed to further reduce environmental contamination by hazardous wastes, among them broadening the definition of what is hazardous so that municipal waste, sewage, pesticides, and mine waste are included. Passing a law to protect the environment does not always result in immediate or expected gains for a number of reasons, notably a lack of funds or personnel to carry out the work.

Exporting Toxic Troubles   Another lesson we've learned in the last few decades is that tough environmental legislation often has unanticipated effects elsewhere. For example, regulations that add to the cost of disposing of hazardous wastes have caused some companies to find ways to prevent waste production, but the very same regulations have caused unscrupulous companies to illegally dump their wastes or to export toxic wastes (including incinerator ash) abroad. In the 1980s, these wastes often ended up in cash-hungry LDCs or east bloc nations, neither of which had adequate laws requiring proper hazardous-waste disposal. Today, European and U.S. companies are believed to be the most heavily involved in illegally exporting toxic wastes. Because the trade is presumed to take place illicitly, no reliable records exist regarding the quantities of materials being exported. In the United States, the export of hazardous wastes is not only common, it is on the rise, according to Hilary French of the Worldwatch Institute. The problem with exporting waste is that many of the countries that receive waste don't know what is in it, don't know how toxic the materials really are, and don't have facilities to store it or dispose of it properly.

In 1986, Congress amended RCRA by establishing procedures to notify importing countries and obtain prior written consent. However, these regulations may be insufficient and EPA officials think that hundreds of tons of hazardous wastes are still being exported illegally.

Exporting hazardous waste to a nation without its full consent goes against principles of international law. Accordingly, numerous African nations have passed laws banning the import of hazardous wastes. In some countries, importing hazardous wastes is punished by stiff jail terms and multimillion-dollar fines. In Nigeria, an importer can be put to death. The Organization of Eastern Caribbean States and 22 Latin American countries have also joined forces to stop the dumping of hazardous wastes on their soils.

In 1990, the European Community (a coalition of European nations) agreed to ban exports of toxic and radioactive waste to 68 former European colonies. Many of the less developed nations who were part of the accord also agreed not to import hazardous wastes from non-EEC members. Today, 121 nations have signed an agreement (the Basel Convention) that bans the transfer of hazardous wastes to less developed nations, including Canada, Mexico, 13 European nations, but not the United States. Although these are important steps forward, many less developed nations are still open to exports, representing a potentially huge repository for hazardous wastes from the industrial nations. Signing an agreement will also not stop the illegal flow.

Tighter regulations for the disposal of hazardous wastes in the United States and other nations have led to commendable efforts to reduce waste production by some companies, but to illegal dumping by less scrupulous ones. Some companies export hazardous wastes to less developed nations with little or no oversight of such practices. Although many less developed nations ban such activities, many still accept wastes.

Dealing with Today's Wastes: A Variety of Options

In 1983, the National Academy of Sciences, a prestigious body composed of the nation's leading scientists, issued a report outlining options for handling hazardous wastes (FIGURE 23-6). This section discusses the various options and illustrates some of the most effective means of dealing with the problem, notably reducing or eliminating hazardous waste, showing that you don't have waste if you don't make it.

Many options exist for getting rid of waste. The most sustainable approaches involve steps that reduce or eliminate hazardous-waste output. You don't have waste if you don't make it.

Process Manipulation, Recycling, and Reuse   As illustrated in Figure 23-6, at the top of the National Academy of Sciences (NAS) list are in-plant options, generally relatively simple and cost-effective changes that reduce hazardous-waste production. All qualify as highly sustainable practices because they reduce hazardous-waste output—avoiding the need to dispose of waste.

In-plant options include three general actions. The first, **process manipulation**, involves alterations in manufacturing processes to cut waste production. One alteration of manufacturing is known as substitution. **Substitution** involves the use of nontoxic or less toxic substitutes in manufacturing. Cleo Wrap, the world's largest producer of gift wrap, for example, switched to nontoxic inks and cut its annual production of hazardous waste by 140,000 kilograms (300,000 pounds). Industries can also change the chemical composition of their products, eliminating those that are harmful or that might produce harmful by-products during manufacturing. Nontoxic household cleaners are a good example. Another type of process manipulation involves the monitoring of manufacturing processes to locate and fix leaks that are emitting toxic chemicals into the environment. Exxon Chemical, for example, installed floating lids over vats that contained volatile organic compounds, greatly reducing losses from evaporation. According to OTA, U.S. industries could reduce or prevent more than 50% of their hazardous-waste generation through process manipulation.

The second and third in-plant options are the **reuse and recycling strategies.** In some instances, companies can capture toxic wastes and, with little or no purification, reuse them to manufacture other products or sell them to other

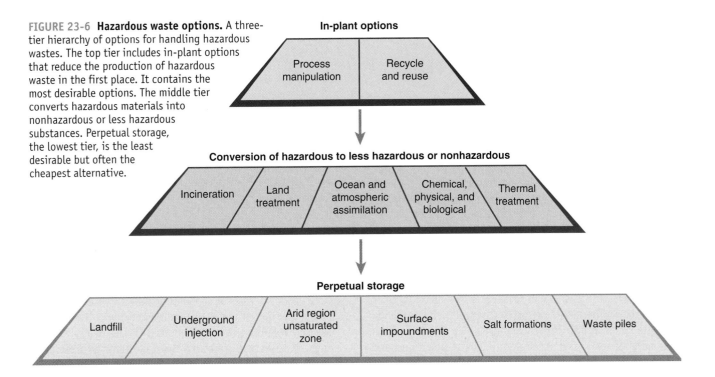

FIGURE 23-6 Hazardous waste options. A three-tier hierarchy of options for handling hazardous wastes. The top tier includes in-plant options that reduce the production of hazardous waste in the first place. It contains the most desirable options. The middle tier converts hazardous materials into nonhazardous or less hazardous substances. Perpetual storage, the lowest tier, is the least desirable but often the cheapest alternative.

companies for reuse. In metal finishing, nickel that's left behind in rinse water can be recovered and reused or sold. For plants that cannot afford onsite technologies, shared facilities or third-party recyclers can provide an economical alternative. Regional waste exchanges—including both private and government-operated facilities that put waste producers in touch with companies that need their waste—can assist in the exchange of potentially hazardous wastes, save companies money, and protect the environment.

The reuse and recycling strategies help cut the output of waste by putting perfectly good materials to use. These strategies may save companies millions of dollars a year in materials costs. In addition, they eliminate the cost of waste disposal, cut down on potential environmental and health damage, and eliminate costly cleanups.

KEY CONCEPTS

Changes in manufacturing processes such as substitution and process manipulation are often the simplest and most cost-effective means of reducing or eliminating hazardous wastes. Waste output can also be dramatically reduced by recycling and reusing wastes.

Conversion to Less Hazardous or Nonhazardous Substances Waste reduction, recycling, and reuse can make a significant dent in waste production. Unfortunately, not all waste can be eliminated, reused, and recycled. Some waste will always be produced. The NAS recommended that, where appropriate, remaining wastes be destroyed or detoxified—that is, converted to less hazardous or nonhazardous materials.

Detoxification can be accomplished for certain types of waste by **land disposal**, applying them to land. When

mixed with the top layer of soil, some waste materials can be broken down by chemical reactions, by oxidation from sunlight, or by bacteria and fungi in the soil. Some nondegradable wastes may be absorbed onto soil particles and held there indefinitely (we think). Others may migrate into deeper layers. A word of caution: Land treatment is an expensive option, requiring care to avoid polluting ecosystems, poisoning cattle and other animals, and contaminating groundwater. Also, studies show that changes in conditions that bind toxins to soil particles may change, causing a sudden and unanticipated release into the environment. Plants can also remove toxic materials from contaminated soil, as explained in Spotlight on Sustainable Development 23-1.

Another option available for organic wastes is **incineration**. High-temperature furnaces at stationary waste disposal sites, on ships that burn wastes at sea, and on mobile incinerators can be used to burn toxic organic wastes such as PCBs, pesticides, and dioxin (**FIGURE 23-7**). In these facilities, oil and natural gas are used as fuels. Hazardous substances are injected into the furnace or mixed with the fuel before combustion.

Mobile hazardous-waste incinerators can be used to clean up contaminated sites or to deal with wastes stored in warehouses. Permanent hazardous-waste incinerators can also be established to deal with wastes from a variety of producers. Such facilities may also be used to provide energy for communities and factories. However, many communities object to hazardous-waste incinerators, fearing the release of toxicants from spills during transport or leaks at the plants. Incinerators may not always perform adequately, and operating personnel may bypass regulations. Low-level releases from smokestacks may also result in a long-term exposure to hazardous chemicals. Recent studies also show that when

23-1   Plants Clean Up Toxic Chemicals

Several small rafts float in a pond near the Chernobyl nuclear power plant, the site of the world's most costly nuclear reactor accident. The raft contains sunflowers, whose roots dangle in the water. These rafts are not intended to make the pond more aesthetically appealing, but to decontaminate the radioactive water.

This is just one of dozens of similar efforts worldwide to use plants to clean up soils and waters contaminated by hazardous materials such as lead, selenium, oil, and radioactive substances.

Cleaning up contaminated sites with plants is technically referred to as *phytoremediation*. It relies on plants' abilities to absorb, store, and in some cases even detoxify toxic substances. The plants are often assisted by bacteria and single-celled fungi. These organisms help to break down some chemicals. The sunflowers in the radioactive pond near Chernobyl act as a kind of biological sponge. The roots absorb the radioactive cesium. Radioactive strontium contained in the water is taken up and is stored in the shoots. After 3 weeks or so, the plants are removed and discarded. This technique costs a fraction of what the more elaborate filtering mechanisms do.

Plants can also be used to remove radioactive cesium and strontium from contaminated soils. One species, known as *Indian mustard*, is particularly effective (FIGURE 1).

Lead-contaminated soil can also be cleansed by plants, but because lead forms strong chemical bonds with soil components, scientists must first apply a special agent to release the lead so that it can be taken up by plant roots. After absorbing all of the lead they can, the plants are removed and properly disposed of. Eventually, scientists would like to see the lead such plants contain recycled. Plants could, for example, be added to lead smelters. The organic matter would burn off, leaving behind the lead for reuse.

Experiments are also under way to use plants to clean up contaminated groundwater. One of the most commonly encountered groundwater pollutants is a chemical known as TCE (trichloroethylene). This dry cleaning agent and degreaser was widely used before being banned.

Scientists have found that TCE can be removed by poplar trees whose roots extend 12 to 15 meters (40 to 50 feet) into the ground. Studies have shown that poplars can remove 95% of the TCE in experimental settings.

Another serious contaminant is selenium, a toxic substance that is leached from irrigated soils in California and other arid regions. Irrigation water draining from farm fields can have extremely high concentrations of this substance, too high to release the water safely into lakes and streams. To handle this waste, special evaporation ponds have been built as a kind of low-tech repository. In these ponds, the selenium-rich water evaporates, leaving behind the toxic metal. Over time, selenium concentrations can become quite high. Wildlife that inhabit the ponds can suffer enormously, as described in Chapter 11.

Scientists are now experimenting with artificial wetlands placed upstream from the evaporation ponds. These wetlands contain plants that remove selenium and could greatly reduce the selenium levels deposited in the evaporation ponds. A 36-hectare (90-acre) wetland built by Chevron Corporation in California currently removes up to 75% of the selenium from the 10 million liters (2.6 million gallons) of wastewater the company pumps through it every day.

Plants aren't the answer to hazardous waste cleanup, but they could be a welcome tool in the gigantic task of cleaning up the hundreds of thousands of contaminated aquifers and sites throughout the world.

FIGURE 1 Indian mustard growing in this field absorbs radioactive contaminants in the soil.

hazardous waste is burned, unburned waste and other chemical compounds in the exhaust may combine to form toxic air pollutants. Incinerators also require fuel and produce carbon dioxide, a greenhouse gas.

Low-temperature decomposition of some wastes, including cyanide and toxic organics such as pesticides, offers some promise. In this technique, wastes are mixed with air and maintained under high pressure while being heated

to 450°C to 600°C (840°F to 1100°F). During this process, organic compounds are broken into smaller, biodegradable molecules. Valuable materials can be extracted and recycled. One advantage of this process is that it uses less energy than incineration.

Chemical, physical, and biological agents can also be used to detoxify or neutralize hazardous wastes. For example, lime can neutralize sulfuric acid. Ozone can be used to

FIGURE 23-7 **Chemical waste incinerator.** This incinerator in the United Kingdom is used to destroy toxic organic wastes from factories. High-temperature combustion and good pollution controls help to reduce emissions from the facility.

break up small organic molecules, nitrogen compounds, and cyanides. Toxic wastes can be encapsulated in waterproof plastic and disposed of in landfills. Many bacteria can degrade or detoxify organic wastes and may prove helpful in the future. New strains capable of destroying a wide variety of organic wastes may be developed through genetic engineering.

KEY CONCEPTS

Hazardous wastes can be converted to nontoxic or less toxic substances by chemical, physical, and biological means, such as neutralization, combustion, low-temperature thermal decomposition, and bacterial decay. Such measures, although effective, present some problems and are clearly not as desirable as preventive measures.

Perpetual Storage   In-plant modifications and conversion techniques that destroy or detoxify wastes cannot rid us of all of our hazardous waste. By various estimates, 25 to 40% of the waste stream will remain even after the best efforts to reduce, reuse, recycle, and destroy it—although some companies have achieved far better reductions.

Residual hazardous waste could be stored by one of a half dozen options (Figure 23-6). For example, residual waste could be dumped in **secured landfills**, excavated pits lined by impermeable synthetic liners and thick, impermeable layers of clay. To lower the risk of leakage, landfills should be placed in arid regions—neither over aquifers nor near major water supplies. Special drains must be installed to catch any liquids that leak out of the site. Groundwater and air should be monitored regularly to detect leaks.

Growing public opposition to this strategy makes it more difficult for companies to find dump sites. Some observers have labeled this the **NIMBY syndrome**: Get rid of the stuff, but Not In My BackYard. Paradoxically, it seems that most of us want the products available in an industrial econ-

omy that inadvertently generates waste, but few of us want the wastes dumped (or even burned) nearby.

Even though the EPA has issued tough regulations for hazardous-waste landfills, critics argue that landfills are only a temporary solution. No matter how well constructed they are, they will eventually leak. In an attempt to reduce problems for future generations, the EPA has drawn up a list of chemicals that cannot be disposed of in landfills.

Landfills are one of the cheapest waste disposal options today and are therefore often favored by industry. The savings they offer today are very likely to be charged to future generations, though.

Other methods of perpetual storage include (1) use of surface impoundments and specially built warehouses that hold wastes in ideal conditions and prevent any material from leaking into the environment; (2) deposition in geologically stable salt formations; and (3) deposition deep in the ground in arid regions where groundwater is absent.

KEY CONCEPTS

Not all waste can be eliminated by prevention, recycling, or detoxification. Perpetual storage remains the final, yet least sustainable, option.

Disposing of Radioactive Wastes

High-level radioactive wastes are some of the most hazardous of all wastes, but they have long been ignored by many countries. High-level radioactive waste is generated by commercial nuclear power plants and weapons production facilities. Lower-level wastes come from research laboratories and hospitals. Many radioactive wastes have a long lifetime. Others can concentrate in animal tissues. Virtually all pose a serious threat to human health, as described in Chapter 14.

High-level radioactive waste is not a problem that will go away, even though the nuclear power industry and nuclear weapons manufacturing are declining (Chapter 14). The waste already produced and now stockpiled at nuclear reactors and the continued operation of the existing power plants, fuel processing facilities, and nuclear weapons facilities necessitates long-term, low-risk storage of nuclear wastes.

Recognizing the need to do something with nuclear wastes rather than continuing to stockpile them at power plants or weapons facilities, Congress passed the **Nuclear Waste Policy Act** in 1982. This law established a timetable for the Department of Energy to select a deep underground disposal site for high-level radioactive wastes. The DOE focused its attention on a site in Nevada called Yucca Mountain. Located on federal land about 160 kilometers (100 miles) northwest of Las Vegas, Yucca Mountain could someday be home to a huge underground storage site excavated in the volcanic rock, costing an estimated $7 to $8 billion (**FIGURE 23-8**).

The DOE had hoped to open the site in 2010. In order for federal decision makers to approve the site, though, they must be relatively sure that earthquakes and volcanic eruptions will not threaten the stability of the repository.

If the Nevada site passes muster, workers will dig a repository 600 meters (2000 feet) below the surface. High-

level nuclear wastes from power plants and defense facilities all over the United States (and possibly from other countries) will be shipped in casks to their (hopefully) permanent home. However, if the Nevada site proves unsatisfactory, the process would have to begin again, delaying construction even more.

Some observers believe that the Nevada site may never be opened because scientists have discovered numerous active earthquake fault lines at the site and a fairly active volcano 11 kilometers (7 miles) away. In 1997, they found that rainwater is leaking through the site faster than predicted and may pose a threat to deep aquifers. The presence of faults and underground fractured rock has some scientists concerned about the intrusion of groundwater into the repository.

Several other options are available. Radioactive waste, for example, can be bombarded with neutrons in special reactors to convert some of it into less harmful substances. However, existing reactors do a poor job of altering cesium-137 and strontium-90, two of the more dangerous by-products of nuclear fission.

Seabed disposal has been used in the past by the United States and European countries, but is now forbidden. Still, some scientists suggest that the seabed may provide a site for radioactive wastes; the effects are difficult to predict.

The problems of disposal suggest to some that nuclear power should be phased out. Cleaner, less costly measures to produce energy should be developed. As noted earlier, though, we need to establish cost-effective and low-risk methods to dispose of (safely store) the vast amounts of waste that have already been generated.

In the United States, low-level radioactive waste from hospitals and research laboratories is packaged and shipped to three disposal sites—in Nevada, Washington, and South Carolina—where it is buried in the ground. Medium-level waste from nuclear power plants and weapons facilities is another matter altogether. In the 1980s, the Department of Energy began construction on a medium-level radioactive waste depository called WIPP (Waste Isolation Pilot Plant). WIPP is the first geological repository for the disposal of radioactive waste produced by the U.S. defense programs—nuclear weapons research and production. It will also house waste from the disassembly of nuclear weapons. The waste slated for disposal in the site consists mostly of protective clothing,

tools, glassware, and equipment contaminated with waste. It will not house high-level radioactive waste or spent nuclear fuels.

WIPP is an experimental project of DOE monitored and regulated by the Environmental Protection Agency. This $700 million facility was carved out of a thick salt deposit 630 meters (2100 feet) below the ground in southeastern New Mexico near Carlsbad. Radioactive waste will be held in steel canisters.

In 1993, the DOE started receiving wastes to test the safety of the site. The facility will eventually hold 6.2 million cubic feet of radioactive waste. In 1998, the EPA gave final approval for the site, saying that it complies with disposal regulations and is safe to contain wastes for 10,000 years.

Although the site has been opened, many citizens are concerned about accidents that could occur as radioactive waste is shipped to WIPP federal weapons facilities in California, Colorado, Idaho, Illinois, Nevada, New Mexico, Ohio, Tennessee, South Carolina, and Washington. Most of the waste to be disposed of at the site will be generated as nuclear weapons are disassembled. To reduce the risk of transportation accidents, DOE requires wastes to be shipped in special casks certified by the Nuclear Regulatory Commission and subjected to a series of tests to demonstrate their ability to survive severe crashes and punctures followed by fires or immersion in water.

Unlike many other hazardous wastes, most nuclear wastes cannot be reused or recycled. Process modification can help reduce waste output, but for the most part the problem has to be addressed at the end of the pipe. The end of the cold war and the decline in the nuclear arsenals of the United States and the former Soviet Union have dramatically reduced production at nuclear weapons facilities. Nuclear waste from power plants could also decline as they are phased out in the United States and elsewhere, as discussed in Chapter 14. Nonetheless, huge volumes of waste already exist and must be dealt with somehow.

Some Obstacles to Sustainable Hazardous-Waste Management

Hazardous-waste production has dropped steadily since 1980 in the United States. This encouraging trend has resulted in large part because of efforts to reduce hazardous-waste production by process modification, recycling, and reuse (**FIGURE 23-9**).

For years, one of the main problems with hazardous waste was that much of it was highly diluted in water released by various industrial processes. This dilute waste stream was typically pumped into deep wells, from which hazardous substances may leak into groundwater. Besides contaminating drinking water, industrial wastes pumped into deep wells can increase the incidence of earthquakes. Two Ohio geologists believe that industrial waste injected into the ground by a chemical company in Ohio may have created underground pressure that triggered the 4.9-Richter-scale earthquake that shook northeastern Ohio and a nearby nuclear power plant in 1986. Their studies suggest that pumping waste into the sandstone increased fluid pressure in the pores of a bed of rock that lies directly above the crystalline bedrock, causing the earthquake along a fault in the bedrock.

The sheer volume of polluted water has made it hard to regulate. Removing the hazardous substances from the water is extremely costly, so most plant owners are unwilling to make the investment in wastewater treatment.

Today, underground injection of hazardous waste continues, but at a greatly reduced level, about 11% of the total release. By far, the most significant release occurs in the air. It accounts for nearly 60% of all toxic chemicals released into the environment.

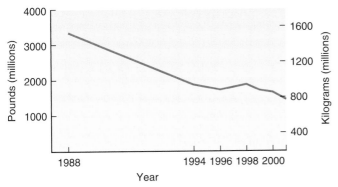

FIGURE 23-9 Reduced but not eliminated. Although hazardous waste released to the environment has declined, substantial amounts are still dumped into our air, water, and land.

The push to reduce hazardous waste release into the environment has stirred considerable interest in incinerators. Incinerators are equipped with pollution control devices and can remove many of the chemical pollutants. They can also be used to generate energy and reduce environmental pollution by hazardous materials. In rural areas, they provide employment. Companies that want to locate them in rural areas often offer help in financing schools, roads, water supply systems, and other needed projects as incentives to local communities.

One person's solution, however, often becomes another person's problem. Incinerators often require landfills to dispose of highly toxic residues. Leaks may pollute the site and seep into groundwater, which is used for irrigation and drinking. Although plants typically have state-of-the-art pollution control equipment, small amounts of toxic substances may be emitted and over time could accumulate in the soil, ecosystems, and crops downwind. An incinerator becomes a hazardous waste magnet, drawing shiploads in by truck and train from neighboring cities and states. Accidents could cause spills, environmental damage, injury, and evacuations.

Although proponents of incineration think that the rural siting strategy may be shrewd business, it ultimately diverts attention from finding permanent solutions, such as process modification, reduction, recycling, and reuse. These are environmentally more acceptable and, in the long run, are more sustainable strategies for dealing with a nation's hazardous wastes.

Individual Actions Count

As with all environmental issues, individuals can participate in many ways. For example, each of us can contribute by recycling oil, not dumping it in sewers or in vacant lots. We can properly dispose of paints, paint thinners, and other potentially toxic wastes. Many cities and counties provide toxic roundups, periodic collection dates during which they will take toxic household products for disposal free of charge. Some of these products, such as paints, can be reclaimed and used again.

One of the best strategies is to avoid toxic chemicals such as pesticides and cleaning agents in the first place. You can cut back on potentially hazardous materials by purchasing environmentally safe cleaning products and insecticides. Reducing your consumption of nonessential goods, whose production invariably creates toxic wastes that poison our land and water, also helps.

You can also learn about waste sites in your area and become active in grassroots organizations working on reduction, recycling, and safer disposal methods. Together, millions of Americans using resources wisely can make significant inroads into the hazardous-waste problem. (See the Individual Actions Count! table for some ideas on ways you can help.)

individual Actions Count

- Recycle oil from your automobile crankcase.
- Take unwanted paint and hazardous materials (pesticides) to county toxic roundups—don't dump either down the drain.
- Buy reclaimed paint.
- Use nontoxic, water-soluble finishes.
- Decrease consumption of all goods, for all production processes result in the release of hazardous wastes.
- Support legislation and nonprofit organizations that promote the reduce, reuse, and recycle strategy.
- Recycle and compost all the waste you can.
- Buy recycled goods.
- Use nontoxic cleaners.

KEY CONCEPTS

Individuals can help reduce the hazardous-waste threat by properly disposing of hazardous materials, avoiding their use whenever possible, using nontoxic alternatives, and cutting down on nonessential consumption—because the production of goods often results in the generation of hazardous waste at factories.

23.3 Solid Wastes: Understanding the Problem

Each year, human society produces mountains of **municipal solid wastes**: discarded paper, metals, leftover food, and other items that come from businesses, hospitals, airports, schools, stores, and homes (**FIGURE 23-10**). The problems are especially acute in the more developed nations.

> Americans produce enough garbage each day to fill the Superdome in New Orleans nearly three times a day.

In 2003, for instance, Americans generated 212 million metric tons (236 million tons) of municipal solid waste—enough garbage to fill the Superdome in New Orleans nearly three times a day. On a more individual level, in 2003 Americans produced on average approximately 727 kilograms (1600 pounds) per person—over 2 kilograms (4.5 pounds) per person per day. A city of 1 million people could fill the Superdome once every year. The composition of the U.S. municipal waste stream is shown in **FIGURE 23-11**.

Municipal solid waste production has increased sharply since 1980, but growth slowed in the 1990s. From 1990 to 2003, for instance, municipal solid waste production only grew 13% while resource recovery (recycling and composting) climbed around 105%. Despite the large quantity of municipal solid waste produced by Americans, municipal solid wastes still make up only about 4 to 5% of the total solid waste discarded in the United States each year. Nonetheless, the continued growth and the sheer volume of waste create major problems in many cities, where land for disposal is growing scarcer and more costly by the day.

FIGURE 23-10 **Mountains of trash.** Environmental activist Lynn Landes standing in front of a landfill in Tullytown, PA. Environmentalists and residents who live near the landfills complain that Pennsylvania's policy of allowing other states easy access to its landfills has made the state a magnet for the trash of others.

Garbage disposal is also of concern to those interested in building a sustainable future because it squanders the Earth's resources. The more that is thrown away, the more minerals that must be mined. The more we throw away, the more trees that must be cut. The more plastic we discard, the more oil wells that must be drilled. Each of these activities produces enormous waste itself and equally impressive amounts of environmental damage.

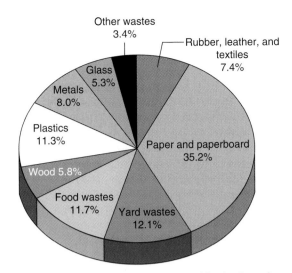

Other wastes
3.4%

Rubber, leather, and textiles
7.4%

Glass
5.3%

Metals
8.0%

Plastics
11.3%

Wood 5.8%

Food wastes
11.7%

Yard wastes
12.1%

Paper and paperboard
35.2%

FIGURE 23-11 **Anatomy of U.S. waste.** This pie chart shows the composition of U.S. municipal solid waste by weight, after recovery of some recyclable and compostable materials. (Data from U.S. Environmental Protection Agency, Municipal Solid Waste Generation, Recycling, and Disposal in the United States: Facts and Figures for 2003. http://www.epa.gov/epaoswer/osw/conserve/priorities/msw.htm. August 9, 2005.)

Science fiction writer Arthur C. Clarke once noted that "solid wastes are the only raw materials we're too stupid to use." In the United States, 55% of the municipal solid waste is buried in landfills. Nearly 30% of America's trash is currently recycled and composted. Another 15% is burned in incinerators. Landfill disposal not only wastes valuable resources, it costs communities millions of dollars each year. Landfilling the disposable diapers used in the United States, for example, costs the nation an estimated $300 million a year. For most local governments, the cost of trash disposal is usually exceeded only by the costs of education and of highway construction and maintenance.

> In the United States, 55% of all municipal solid waste is still buried in landfills and 28% is recycled.

Like so many other problems, municipal solid waste is the product of many interacting factors: (1) large populations; (2) high per capita consumption; (3) low product durability; (4) our heavy dependence on disposable prod-ucts; (5) low reuse and recycling rates; (6) a lack of personal and governmental commitment to reduce waste; (7) widely dispersed populations, in which producers of recyclable and reusable items are separated from those willing to purchase these materials; and (8) relatively cheap energy and abundant land for disposal.

KEY CONCEPTS

More developed nations produce enormous amounts of solid waste each year, much of which is burned or landfilled, squandering precious resources and creating an enormous and costly waste disposal problem in urban areas. Waste production is increasing in many countries such as the United States; recovery rates (recycling and composting) are growing much faster, a trend that bodes well for the future.

23.4 Solving a Growing Problem Sustainably

Actions to reduce our output of solid waste generally fall into three broad categories (**FIGURE 23-12**). The traditional response to solid waste is known as the **output approach**. It consists of ways to deal with trash flowing out of cities and towns. Most often, this means incinerating trash or dumping it in landfills. A more sustainable strategy is known as the **input approach**. This consists of activities that reduce the amount of materials entering the production–consumption cycle—for example, efforts to reduce consumption and waste, say, by increasing product durability. The third approach, also essential to building a sustainable society, is the **throughput approach**. It consists of ways to direct materials back into the production-consumption system, creating a closed-loop (cyclic) system akin to those found in nature. Reuse and recycling fall under this category.

KEY CONCEPTS

Solid waste management strategies fall into one of three categories: those that deal with waste after it has been produced, those that divert waste back into the production–consumption cycle, and those that prevent waste generation in the first place.

FIGURE 23-12 **Strategies for reducing solid waste.** There are many ways of dealing with waste. The strategies fall within one of three groups: output, input, and throughput. In most cases, a combination of all three must be applied to alleviate the solid waste problem. Efforts should concentrate on the input and throughput solutions, which will go a long way in helping to create a sustainable society.

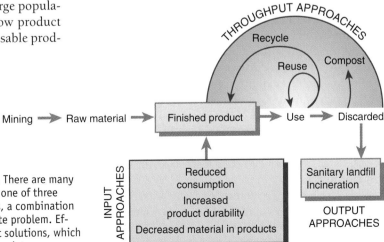

The Traditional Strategy: The Output Approach

The most widely used strategy throughout the world is the output approach, landfilling and incineration. Landfilling and incineration are end-of-pipe controls. Like many others discussed in this book, they are not sustainable in the long run.

KEY CONCEPTS

Worldwide, most trash is still dumped in landfills.

Dumps and Landfills   Until the 1960s, garbage dumps were prevalent features of the American landscape. Public objection, however, to wafting odors, rat- and insect-infested midden heaps, and dark plumes of smoke that billowed out of burning dumps forced cities to look for other ways to deal with their growing trash problem. The federal government contributed to the demise of the dump by passing RCRA, discussed earlier. Besides addressing hazardous waste, RCRA also required all open dumps to be closed or upgraded by 1983.

The open dump has been replaced by the sanitary landfill. A **sanitary landfill** is a natural or humanmade depression into which solid wastes are dumped, compressed, and daily covered with a layer of dirt. Because solid wastes are no longer burned, as they were in many open dumps, air pollution is greatly reduced. Because trash is covered each day with a layer of dirt, odors, flies, insects, rodents, and potential health problems are eliminated or sharply reduced.

Despite their immediate benefits, landfills have some notable problems. First and most important, landfills require land. In the United States, the trash from 10,000 people in a year will cover one hectare (2.47 acres) about 1.2 meters (4 feet) deep. Around many cities, usable land is in short supply or is expensive. Second, landfills, like dumps, require a great deal of energy for excavation, filling, and hauling trash. Third, they can pollute groundwater. Toxic household wastes (paint thinner, pesticides, and other poisons) and feces (from disposable diapers, kitty litter, and backyard cleanup of Rover's messes) are discarded in municipal landfills, where they can leak into groundwater. Landfills are a major source of groundwater contamination and many retired landfills are listed as Superfund sites, as you learned earlier in the chapter. Fourth, they produce methane gas from the decomposition of organic materials. Methane can seep through the ground into buildings built over or near reclaimed landfills. Methane is explosive at relatively low concentrations. Fifth, landfills sink or subside as the organic trash decays, requiring additional regrading and filling. Buildings constructed on top of reclaimed landfills may suffer serious structural damage. Sixth, they have low social acceptability. Quite understandably, most people don't want the noise, traffic, and blowing debris that come with even the best-managed landfills.

Although landfilling is ultimately an unsustainable strategy, there are many ways to make it more environmentally acceptable. Energy requirements, for example, can be cut by new methods of waste collection. Packer trucks now reduce waste volume by 60%, meaning that fewer trucks are needed to haul garbage to landfills. This saves on fuel consumption and pollution.

Vacuum collection systems can also be used to save energy in dense urban settings, especially apartment complexes. Solid waste is dumped into pipes that carry it to a central collection point. One such system is in operation in Sundbyberg, Sweden. Garbage is whisked away from wall chutes to a central collection facility, where the glass and metals are removed by an automated process. The burnables are incinerated, providing heat for the 1100 apartments using the system. A similar system handles 45 metric tons (50 tons) of waste per day at Disney World in Florida. Today, over 400 such systems are in operation in Europe in hospitals, apartment buildings, and housing tracts.

Water pollution problems can be reduced by locating landfills away from streams, lakes, and aquifers. Test wells around the site can be used to monitor the movement of pollutants, if any, away from the site. Special drainage systems and careful landscaping can reduce the flow of water over the surface of a landfill, thus reducing the amount of water penetrating it. Impermeable clay caps and liners can reduce water infiltration and the escape of pollutants. In addition, pollutants leaking from the site can be collected by specially built drainage systems and then detoxified. The toxic seepage is shipped to hazardous-waste facilities.

Methane gas produced in landfills can be drawn off and sold as fuel, supplementing natural gas. Subsidence damage to buildings built on reclaimed sites can be reduced by removing organic wastes before disposal and by allowing organic decay to proceed for a number of years before construction.

KEY CONCEPTS

Open garbage dumps have been replaced by sanitary landfills, pits into which garbage is dumped and then covered daily with a thin layer of dirt to eliminate odors and pests. Although sanitary landfills are unsustainable, they can be made more environmentally acceptable by locating them away from groundwater supplies, collecting and treating toxic leachate, capturing methane gas, and other steps.

Ocean Dumping   For years, many city officials have viewed the world's vast oceans as a huge garbage dump for wastes, including municipal garbage and human sewage. At one time, U.S. municipalities dumped their trash in 126 offshore dump sites. Although most of the solid waste dumped at sea was mud and sediment from dredging of harbors, estuaries, and rivers, considerable amounts of industrial hazardous wastes, radioactive wastes, municipal wastes, and sludge from sewage treatment plants were dumped at sea. Even today, city planners are trying to find ways to justify dumping some of their trash offshore. Some propose building offshore "islands of trash" from discarded automobiles, demolished buildings, and other solid wastes to support hotels and airports. Others talk of (and build) artificial reefs of discarded stone, toilets, cement slabs, and automobiles.

Ocean dumping of many wastes in U.S. waters declined as a result of the **Marine Protection, Research and Sanctu-**

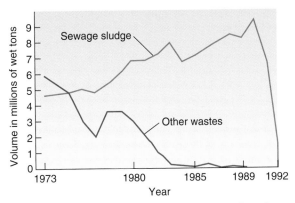

FIGURE 23-13 **Signs of improvement.** Ocean dumping of sewage sludge has declined significantly in recent years. Other waste dumping was largely brought to a halt in the early 1980s.

aries Act passed by Congress in 1972. The long-term goal of this act was to phase out all ocean dumping, especially of sewage and solid wastes. In 1977, the EPA issued regulations calling for an end by 1981 to all dumping of industrial wastes and sludge in the Atlantic Ocean, where 90% of the waste disposal was occurring. Although hazardous-waste and solid waste dumping declined sharply, sewage sludge disposal remained a significant problem because New York and other cities continued to dump sludge in the ocean for many years.

In 1988, the U.S. Congress passed the **Ocean Dumping Ban Act** (1988), which called for an end to sewage sludge disposal at sea by December 21, 1991. The EPA negotiated agreements with local jurisdictions that phased out ocean dumping of sewage sludge by June 1992 (**FIGURE 23-13**). Unfortunately, many other countries continue to view the oceans as a vast repository for their waste.

KEY CONCEPTS

Oceans have long been viewed as a huge waste dump site for a variety of wastes, including radioactive materials. This practice is no longer legal in the United States, but it continues in other countries.

Incineration   Incineration is another output control widely used in many industrial countries. In Denmark, 60% of the municipal solid waste is incinerated. The Netherlands and Sweden both burn about one third of theirs. The United States burns about 15% of its solid waste.

Burning trash can reduce the waste volume by two thirds, cutting requirements for landfill space. Incineration can also be used to produce heat and electricity. Because of this, incinerators are often called **waste-to-energy** (WTE) **plants**. One ton of garbage is equivalent to about one barrel of oil. But even with the energy gain, WTE plants cost more to build and operate than landfills. WTE plants are also much more expensive than recycling programs. Nationwide, recycling projects cost about one third as much as incin-

> Nationwide, recycling projects cost about one third as much as incinerators.

erators. Each incinerator must be individually designed to accommodate the local mixture of burnable (leaves) and nonburnable refuse (steel cans). Operating these incinerators is made more difficult because the mixture varies from season to season. In the spring and fall, for example, yard and garden waste increases dramatically. But all yard waste does not behave the same. Wet leaves, for example, do not burn well. To avoid problems caused by wet leaves and other similar materials, municipalities may require homeowners to separate combustible material (dry branches) from wet organic matter (grass clippings).

Another problem with incinerators is that they may emit toxic pollutants, especially when plastics are burned. Another major problem with garbage incinerators is that the ash they produce may be hazardous to human health. Test data show that the ash from incinerators contains dioxins and toxic metals, such as lead and cadmium, in concentrations that may be harmful to humans. Consequently, some scientists and environmentalists argue that ash from garbage incinerators should be reclassified as a hazardous substance and disposed of in hazardous-waste facilities, which is much more costly than dumping in ordinary landfills.

Ash disposed of in ordinary landfills may, over the years, begin to leak into groundwater, polluting public and private drinking water supplies. Cleaning up these sites could cost many millions of dollars.

Many municipalities are opting for incinerators in which nonburnable materials such as glass and steel cans are removed before combustion, rather than for mass burn facilities, in which the entire waste stream is burned without separation. Separation of noncombustible wastes permits recycling and increases the combustion efficiency of incinerators. It is also much cleaner and is more efficient than mass burn, although it still produces hazardous materials.

WTE plants are a step in the right direction, but they are not as cost-effective and sustainable as waste management programs that rely on waste reduction, composting, and recycling. Incinerators also rank low on the social acceptability scale. Residents of Lowell, Massachusetts, defeated an incinerator that its city council was planning to build because they found, among other things, that Lowell produced only 225 metric tons of waste per day, whereas the plant would require 1350 metric tons to operate. That meant that the city council would have had to enter into agreement with neighboring towns to accept their trash to meet the needs of the plant. Lowell would have become a municipal solid waste magnet. Citizens in Spokane, Washington, were not so lucky. Their city council approved a massive incinerator that requires a considerable amount of outside trash to keep it running.

Incinerators may seem like a good way to solve the growing trash problem, but critical thinking suggests the need for a long-term view of the problem. The most important question today is not just how to reduce landfilling, but how to cut our waste; conserve valuable resources; protect our air, water, and land; and ensure vital wildlife habitat. In short, how do we create a sustainable system of waste management? Clearly, incinerators reduce trash but don't contribute significantly to the goals of sustainability.

Garbage can be burned in incinerators, which greatly reduces trash. This option, however, even when linked to energy production, is viewed by many as an unsustainable way of dealing with municipal solid waste.

Sustainable Options: The Input Approach

In Chapter 3, you learned that sustainable solutions are those that often approach problems by addressing their root causes—that is, by confronting them at their source. In solid waste, as in hazardous waste, the most effective and sustainable strategies are those that seek to reduce the amount of materials entering the production–consumption system in the first place. This strategy is called **source reduction.** The three main source reduction strategies include measures that (1) increase product life span, (2) reduce the amount of materials in goods and their packaging, and (3) reduce consumption (demand for goods).

KEY CONCEPTS

Source reduction techniques reduce the amount of waste entering the waste stream and represent the most sustainable waste management strategy.

Increasing Product Life Span: Making More Durable Products

The trouble with today's products is that, as writer John Ruskin noted, "There is hardly anything in the world that some man cannot make a little worse and sell a little cheaper." In the long run, however, cheaply made goods end up costing consumers more than well-made and slightly more expensive goods. The rapid turnover may be profitable to businesses, but it is unsustainable from an ecological standpoint. Planned obsolescence of products destroys the air, water, and land. More durable toys, garden tools, cars, and clothing require less frequent replacement and thus decrease resource use. It's a great personal strategy and also can be a profitable business strategy because consumers who are fed up with shoddy products will gravitate to well-made ones.

KEY CONCEPTS

Higher quality, more durable goods last longer and consequently reduce the amount of waste. Manufacturers and consumers can play significant roles by making and buying more durable products.

FIGURE 23-14 Aseptic container. Aseptic containers like this one are convenient and energy efficient but are fairly difficult to recycle.

Reducing the Amount of Material in Products and Packaging

In the United States, packaging (including bottles) uses 90% of the glass, 50% of the paper, 11% of the aluminum, and 8% of the steel consumed, according to Concern, Inc., a citizens' group based in Washington, D.C. All told, 32% of the municipal solid waste (by weight) is discarded packaging.

Of course, packaging is necessary, but much of it is superfluous and wasteful. (Consider the package containing the latest software you bought.) The Campbell Soup Company realized this and redesigned its soup cans; today, they use 30% less material than they did in the 1970s. Some beverage companies now package drinks in **aseptic containers,** boxes constructed of several thin layers of polyethylene, foil, and paper (**FIGURE 23-14**). The containers hold milk, juices, and wine and keep them fresh for several months without refrigeration. Aseptic containers require less energy, too. Canned drinks, for instance, must be pasteurized for 45 minutes, whereas the contents of aseptic packages are sterilized out of the package for only one minute. This not only greatly reduces energy demand but preserves flavor. Furthermore, beverages in aseptic containers like soy milk do not require refrigeration during transportation and storage, which also lowers the energy demand. Being lighter than cans also helps cut down on transportation costs. Aseptic containers can be recycled, but access to recycling programs that handle them is fairly limited.

Virtually all products can be redesigned to reduce waste. Many large newspapers, for example, have gone to a more economical design that has cut the use of newsprint by 5%. Smaller cars and trucks have emerged in the 1970s and early 1980s in the United States and are popular in many countries. In the United States, however, the trend in the 1990s and 2000s has been toward larger and larger vehicles such as the Ford Expedition and Chevy Suburban (**FIGURE 23-15**). Not only do they consume more minerals, they are less efficient and consume much more energy to operate. Smaller computers and calculators have also helped save valuable materials. Smaller houses could help as well.

KEY CONCEPTS

Efforts to make products smaller and more compact can significantly reduce resource demand.

Reducing Consumption H. W. Shaw once wrote, "Our necessities are few, but our wants are endless." Critics of modern-day society argue that ceaseless efforts to satisfy our endless wants are a big part of the solid waste dilemma the United States, Canada, and other countries face. By cutting back on consumption, individuals can help reduce solid wastes and many other problems described in this book.

Reductions in consumption will require shifts in our attitudes. The attitude that "new is always better" leads many con-

FIGURE 23-15 **An oversized SUV.** This sport utility vehicle, like many other models on the market today, uses an incredible amount of resources to build and gets about 12 miles per gallon.

The Throughput Approach: Reuse, Recycling, and Composting

In natural systems, all waste is food material for other organisms. In short, there is no waste. Everything is recycled. Creating a sustainable society, say many experts, will require steps to greatly increase recycling efforts to emulate nature.

Recycling is part of the throughput approach. The second part of this approach is reuse. Both strategies remove useful materials from the waste stream and channel them back to manufacturers (in the case of recycling) or to end users (in the case of reuse) (Table 23-1).

KEY CONCEPTS

The throughput approach diverts waste from the waste stream for recycling and reuse.

The Reuse Option Reuse is the return of operable or repairable goods into the market system for someone to use. In most cities, organizations such as Goodwill and the Disabled American Veterans pick up used products, including clothes, shoes, silverware, plates, pans, books, tools, bicycles, and appliances. Many of them provide drop-off stations or pick up goods at your home. These products are cleaned and then resold to the needy and the frugal. Profits go to help the needy. For-profit secondhand stores also provide consumers with options for buying children's clothing and toys, furniture, appliances, plumbing supplies, and a host of other products. Check them out. You can often get great bargains!

Packaging materials—such as cardboard boxes, bottles, and grocery bags—can also be reused, saving both energy and materials. Shopping bags can be reused by individuals. Reusable beverage containers can be sterilized, refilled, and returned to the shelf, sometimes completing the cycle as many as 50 times. Disposable and recyclable bottles and cans have virtually eliminated the reusable container from the market in more developed countries, although this technique is popular in less developed nations.

Table 23-2 compares the energy demand of refillable glass bottles (used 10 times) with various other packaging

sumers to purchase new goods when old ones still work. The fashion industry thrives on its ability to convince the public that new fashions are "in"—and that anyone wearing the old is "out of fashion." Advertisers capitalize on this strategy as well, and many consumers fall into the trap. Chapters 3 and 24 detail other attitudinal factors that contribute to the solid waste dilemma.

Another way to reduce consumption is through outright bans. Some cities, for example, have taken steps to reduce waste by banning objectionable materials—notably plastics. Minneapolis and St. Paul, Minnesota; Berkeley and Palo Alto, California; Newark, New Jersey; and other cities have passed ordinances to ban many plastics, prompting the plastics industry to take measures to develop recycling facilities.

Each of us can make a personal effort to reduce consumption with little noticeable change in lifestyle. There are limitless possibilities; all we have to do is try them.

KEY CONCEPTS

One of the most effective means of reducing solid waste is to reduce consumption—buy what you need.

Table 23-1

Reuse and Recycling of Solid Wastes

Material	Reuse and Recycling
Paper	Repulped and made into cardboard, paper, and a number of paper products. Incinerated to generate heat. Shredded and used as mulch or insulation.
Organic matter	Composted and added to gardens and farms to enrich the soil. Incinerated to generate heat.
Clothing and textiles	Shredded and reused for new fiber products or burned to generate energy. Donated to charities or sold at garage sales.
Glass	Returned and refilled. Crushed and used to make new glass. Crushed and mixed with asphalt. Crushed and added to bricks and cinder blocks.
Metals	Remelted and used to manufacture new metal for containers, buildings, and other uses.

Source: Modified from B. J. Nebel (1981). *Environmental Science.* Englewood Cliffs, NJ: Prentice-Hall, p. 297.

Table 23-2

Energy Consumption per Use for 12-Ounce Beverage Containers

Container	Energy Use (BTUs)
Aluminum can, used once	7050
Steel can, used once	5950
Recycled steel can	3880
Glass beer bottle, used once	3730
Recycled aluminum can	2550
Recycled glass beer bottle	2530
Refillable glass bottle, used 10 times	610

Source: Modified from L. R. Brown, C. Flavin, and S. Postel (1991). *Saving the Planet: How to Shape an Environmentally Sustainable Economy.* New York: Norton, Table 4-1, p. 71.

options. As illustrated, when it comes to energy demand, recycling is generally more efficient than one-time use (compare the energy required to make an aluminum can from raw ore versus recycled scrap). Reusing glass bottles only 10 times, though, is four times more efficient than recycled glass.

Besides saving energy, the reuse option (1) reduces the land area needed for solid waste disposal, (2) provides jobs, (3) provides inexpensive products for the poor and the thrifty, (4) reduces litter, (5) decreases the amount of materials consumed by society, and (6) helps reduce pollution and environmental degradation.

KEY CONCEPTS

Reusing materials and products cuts down on resource demand and offers many other economic and environmental benefits. Citizens can participate in two ways, by donating used products and buying used goods.

The Recycling Option In human societies, **recycling** refers to the return of materials to manufacturers, where they can be melted down or chopped up, refashioned into the original finished material, and then incorporated into products. Paper, for instance, is shredded, immersed in water, de-inked, and then used to make more paper. Recycling conserves resources, alleviates future resource shortages, reduces energy demand, cuts pollution, saves water, and decreases solid waste disposal and incineration.

Consider some examples of the benefits: Each 1.2-meter (4-foot) stack of newspaper you recycle saves a 12-meter (40-foot) Douglas fir tree. Recycling a ton of newspaper saves 17 trees. Paper recycling uses one third to one half as much energy as the conventional process of making paper

> Recycling a 1.2 meter (four-foot) stack of newspapers saves one tree, measuring 12 meters (40 feet).

from wood pulp. Small savings can add up to make huge changes. If the United States, for example, increased paper recycling by 30%, we would save an estimated 350 million trees each year. Paper recycling has an added benefit of reducing the air pollution generated in making paper by 95%.

Aluminum recycling offers great benefits as well. As noted in two previous chapters, aluminum recycling requires 95% less energy than making aluminum from raw ore (bauxite). Thus, a manufacturer can make 20 aluminum cans from recycled metal with the same energy it takes to make one can from bauxite ore. Aluminum recycling produces 95% less air pollution as well. Similar environmental benefits are available from recycling other metals and plastics.

Japan is a world leader in recycling. Currently, about one half of that nation's waste is composted and recycled. In the United States and many other countries, however, recycling efforts are a long way from achieving their full potential, although there are exceptions. One exception is the automobile. In the United States, approximately 90% of all cars are recycled. Lead in car batteries is another exception, with about 95% of all batteries making their way back into the manufacturing process. Despite these impressive statistics, in 1997, nationwide only about 5.2% of the plastic, 24% of the glass, 42% of the paper and cardboard, and 35% of the aluminum was recycled. Recycling rates for many products could easily double or triple. Plastic recycling could increase 10 to 20 times.

> Recycling aluminum is 95% more efficient than making aluminum from raw ore. You can make 20 aluminum cans from recycled scrap with the same amount of energy it takes to make one can out of raw ore.

KEY CONCEPTS

Many products can be returned to recycling facilities, where they are shipped to factories to be used to make new products—a process that offers many social, economic, and environmental benefits. Although recycling is on the rise, most countries have barely tapped the full potential of recycling.

Recycling's Growing Popularity Recycling is widely practiced in Europe and Japan. Many major U.S. cities have developed recycling programs to reduce landfilling. In 2001, there were over 9704 curbside recycling programs, serving more than 139 million people. Seven states—South Carolina, Virginia, New Jersey, New York, Maine, Minnesota, and South Dakota—report total recycling rates of 40% or more. Fifteen other states report recycling rates of 30 to 40% or more (Table 23-3). Impressive as this is, many other countries are doing much better. Japan, for example, currently

> In 2001, there were 9700 cities and towns in the United States with curbside recycling programs, up from 8800 in 1996.

Table 23-3

How's Your State Rate?

State	Recycled (%)	State	Recycled (%)
Minnesota	45	Rhode Island	27
New Jersey	43	New Hampshire	26
New York	43	North Dakota	26
Maine	42	Pennsylvania	26
South Carolina	42	Michigan	25
South Dakota	42	Connecticut	24
Virginia	40	Hawaii	24
Florida	39	Alabama	23
Arkansas	36	Indiana	23
Wisconsin	36	Delaware	22
Tennessee	35	Utah	22
Texas	35	West Virginia	20
Iowa	34	Louisiana	19
Massachusetts	34	Colorado	18
California	33	Arizona	17
Georgia	33	Ohio	17
Washington	33	Mississippi	14
Kentucky	32	Nevada	14
North Carolina	32	Kansas	13
Maryland	30	Oklahoma	12
Missouri	30	District of Columbia	8
Oregon	30	Alaska	7
Vermont	30	Montana	5
Nebraska	29	Wyoming	5
Illinois	28	Idaho	n/a

Source: Environmental Protection Agency.

FIGURE 23-16 Drop-off site. This roadside center is conveniently located so that people can drop off recyclables on the way to nearby shopping centers.

recycles about half of its garbage. One suburb of Tokyo, for instance, currently recycles and composts 90% of its garbage.

KEY CONCEPTS

Many states now recycle 30% or more of their municipal solid waste.

Types of Recycling Programs Cities and towns offer a variety of recycling options. Some use drop-off sites, where residents can deposit their recyclables on the way to work or to the grocery store (**FIGURE 23-16**). In some locations, recycling companies will actually pay for some materials such as aluminum and copper. Drop-off centers can be successful, but generally only if containers are conveniently placed—for example, at train stations, near parks, or near heavy-use intersections. Many colleges and universities also offer recycling programs with recycling containers located in the hallways of classroom buildings.

Curbside recycling, in which recyclables are periodically picked up by trash haulers at the curb, is by far the most successful type of recycling program (**FIGURE 23-17**). Participation rates

FIGURE 23-17 Curbside recycling. Many cities have curbside recycling programs, which make recycling very convenient. Trash and recyclables are often picked up by different trucks.

as high as 60 to 80% can be expected if recyclables and trash are picked up on the same day. One study in Canada showed that curbside recycling requires about 10% less energy than a drop-off program.

Unsorted trash can also be collected and shipped to resource recovery centers, where it is separated by machines or people. This technique, called *end-point separation*, is generally more costly than source separation in which people sort recyclables at home or in the office. Moreover, complete separation may not be possible at these facilities, thus lowering the value of recyclable materials.

A growing number of recycling programs are being run by private haulers. Toronto, for example, is home to a program run by Waste Management, Inc. In my area, all three private trash haulers offer curbside recycling.

KEY CONCEPTS

Recycling programs generally involve drop-off sites or curbside pickup, which may be run by private industry or by city and town governments.

Obstacles to Recycling. From an energy and resource standpoint, recycling is generally not as good an option as reuse, but it is far better than burning materials and infinitely better than throwing them away. If recycling is such a good idea, why don't Americans, Canadians, and other people do more of it? The reasons are complex.

First, some industrial societies grew up with abundant resources. Steeped in the frontier notion that "there's always more," industrialists and political leaders have traditionally seen little need for recycling, except perhaps in times of war when raw materials were in short supply. Given the seemingly inexhaustible supply of materials, factories were primarily set up to handle virgin material. The entire production–consumption system was built without recycling in mind. Corporate empires were built on the profits of extractive industries, the mining companies, and those companies today wield enormous political power. Changing this ingrained and wasteful system will not be easy.

Second, the nation's tax laws today support extraction and discourage recycling. As pointed out in Chapter 16, U.S. laws work against recycling. Even today, for example, mining companies receive generous tax breaks (depletion allowances) that give them an unfair advantage over recyclers. These tax breaks often make virgin materials cheaper than recycled ones. Logging companies that supply wood for paper mills (and other uses) are also heavily subsidized by the federal government—and thus by the taxpayer. Logging roads in national forests, for instance, are built with public money, and timber sales on public land have long been made below their cost (Chapter 12), further benefiting virgin paper industry over recycling. Federal subsidies create unfair economics. Hard-rock mining companies can also purchase federal land at $5.00 an acre, thanks to the Mining Act of 1872. They are also not required to pay any royalty to the government for extracted minerals, which artificially reduces the price of minerals and metals.

The traditional extractive industries receive another hidden subsidy that's far more difficult to quantify. It is called an *economic externality*—a cost that is passed on to the public from pollution and other harmful effects of these activities. Because the traditional ways of making paper, steel cans, and other products produce more pollution, they have a bigger impact on our health and our environment than does recycling, which uses less energy and produces far less pollution. The higher environmental cost of virgin materials, however, is not completely reflected in the price of the product. It is, instead, paid in federal taxes that go to clean up our air and water. It is paid in higher health bills and in dozens of other ways that few of us are aware of.

Another difficulty is the built-in transportation price differential, mandated by law, for scrap metal and ore. For example, ore travels more cheaply than metal bound for a recycling mill.

In some locations, such as the U.S. West, cheap landfill costs were once a barrier to overcome. In Colorado, for example, landfill tipping fees—the cost to dump a ton of trash in a landfill—once ranged from $4 to $7 per ton. Recycling costs $30 to $50 a ton. On the East and West Coasts, landfill tipping fees ranged from $40 to $100, making recycling far more profitable. As landfills were closed and new, more expensive sites were developed, however, tipping fees began to rise. In the northeastern United States, tipping fees now average $71 per ton. In the central portions of the nation, where undeveloped land is more abundant, tipping fees now average $24 to $34 per ton. In addition, many interior states are a long way from the nation's paper recycling mills, making it more expensive and less profitable to recycle paper.

Recycling suffers from an image problem as well. In the 1970s, many recycled paper products were of inferior quality. Many people who used them were dissatisfied and soon returned to virgin materials. Since that time, however, recycled paper products have improved immensely. Recycled office paper, stationery, and computer paper are indistinguishable from virgin stock. Still, the notion that recycled paper is inferior sometimes persists.

Plastics pose a special problem for recycling. Most plastic is perfectly recyclable. The problem is that there are more than 45 different types of plastic commonly used for packaging. Making matters worse, some packages contain two or more types of plastic, making it difficult to recycle. A plastic ketchup bottle, for example, has five layers of plastic, each one different, each one providing a special feature needed to make a perfectly squeezable bottle to deliver our ketchup.

One way plastic manufacturers have helped promote plastic recycling is by placing codes on plastic packaging so that individuals and recyclers can tell exactly what type of plastic it is made of (**FIGURE 23-18**). In most cities, only two or three of the most widely used plastics are recycled.

KEY CONCEPTS

Although recycling is a promising strategy for waste management, many obstacles hinder its full implementation. Some of the most important are federal laws and subsidies that give the raw materials industry an unfair economic advantage over recycling.

FIGURE 23-18 Plastic code. The codes on the bottom of virtually all plastic containers tell the type of plastic the product contains, making it easier for individuals and recyclers to tell one from another.

Overcoming the Obstacles. Despite these barriers, U.S. recycling efforts are on the rise and are bound to increase substantially. In 1988, the U.S. Environmental Protection Agency announced a nationwide goal of reducing the solid waste stream by 25% through waste reduction and recycling by 1994, a goal reached a year early. In 2004, however, the rate had climbed to 29.7%.

Lofty government goals are only part of the equation, though. Rising energy prices, decreasing landfill space, and depletion of high-grade ores will all increase recycling in coming years. Public awareness through education from schools, environmental groups, and the media can also help. Policy changes are also needed to eliminate preferential freight rates and subsidies for timber harvesting and mining.

To be successful, recycling programs must also be convenient and cost-effective. Recycling also requires individual commitment and action. Separating trash before collection requires a little effort, but not much more than the effort we expend to separate our white clothes from our colored clothes when we do our laundry. I like to think of source separation as the small price we have to pay for the riches the Earth gives us. It is a small sacrifice that pays off in a better world and a sustainable future.

To promote recycling among citizens, cities and towns can create several economic incentives. In Seattle, residents who recycle pay $5 per month less if they fill only one trash can each week, and super-recyclers get an even lower rate. The city of High Bridge,

New Jersey, at one time charged residents an annual flat fee of $280 for garbage collection, for which they received virtually unlimited trash pickup. Today, though, residents have to purchase stickers, which they attach to every trash bag left by the curb. Fifty-two stickers cost $140. Additional stickers cost $1.25 each. Since this program began, waste output has dropped by 25%, in large part because of increased efforts on the part of citizens to compost, recycle, and compact their trash.

KEY CONCEPTS

Expanding the amount of garbage that is recycled will require removal of some of the legal and economic barriers and subsidies that give benefits to companies that use virgin materials rather than recycled ones. Commitment on the part of citizens is also vital to making recycling a success.

Procuring Recycled Materials: Closing the Loop. Besides overcoming the barriers just described, to make recycling successful, we must also develop markets for recycled materials. In other words, we must close the loop. What does this mean?

Creating markets for recyclables causes an important shift in the production–consumption system and our impact on the Earth. Most human economies are linear in design, as illustrated in **FIGURE 23-19** (top panel). They are

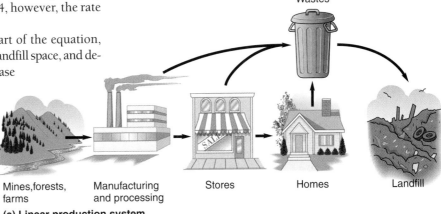

(a) Linear production system

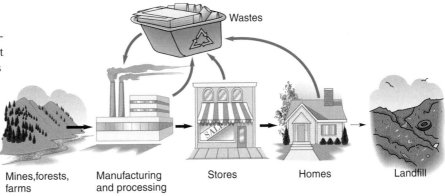

(b) Cyclic production system: The industrial ecosystem

FIGURE 23-19 Systems of production and consumption. By reducing waste, reusing, composting, and recycling, individuals and businesses can convert **(a)** the linear production system into **(b)** a cyclic one. Jobs will shift away from the ends of the linear system to the center, where recycling and composting are done. Jobs at all levels of income will be created in recycling and composting industries.

dependent on resources extracted from the environment, which are then fashioned into products, used, and then disposed of. A sustainable industrial system based on recycling is cyclic. It resembles the nutrient cycles in ecosystems. In such a system, materials flow cyclically through the system—from production facility to consumer and back for reuse a lot like carbon or nitrogen in ecosystems.

Although some raw materials will need to be extracted and some wastes will invariably be produced in a sustainable system of production and consumption, the goal of a sustainable economy is to maximize the amount of material that is recycled within the economy. As you shall soon see, this will mean a shift away from employment at both ends of the system and toward the middle, where the recycling occurs (both pickup and remanufacturing).

Building an industrial ecosystem—that is, a system of commerce that is an analog of the natural ecosystems—means even greater increases in the amount of material that is returned by end users; but this is not enough. To create a truly cyclic economy, companies must be willing to use the materials picked up in curbside programs or drop-off sites, as noted above. Financial incentives and disincentives can be used to promote remanufacturing. For instance, Florida levied a 10% tax on newspaper and book publishers using virgin paper, to discourage its use. Colorado offers a tax credit to companies that purchase equipment that allows them to incorporate recycled materials into the manufacturing process. In addition, people like you and I must be willing to purchase recycled paper and other materials to help support manufacturers. If recyclers cannot find a market for their materials, all the recycling in the world is of no use.

In November 1990, Germany's minister for the environment issued a decree calling for an 80% reduction in the amount of packaging entering the waste stream. To meet these goals, retailers are required to collect packaging materials and recycle them. This puts pressure on manufacturers to reduce excess packaging wherever possible and perhaps to help develop markets to use the recyclables. To promote public participation, the government is also levying a small deposit (equivalent to about 30 cents in U.S. currency) on most packages. Retailers can avoid the deposit and take-back requirements if industry establishes alternative collection and recycling systems that meet the established goals.

Laws requiring deposits on beverage containers, commonly known as *bottle bills*, can also dramatically increase recycling rates. At this writing, nine U.S. states have bottle bills. In such states, over 90% of the glass and aluminum containers are returned for recycling. Container deposit legislation is also on the books in the Netherlands, Scandinavia, Russia, Canada, Japan, parts of Australia, and a handful of less developed nations.

Recycling programs in the United States are currently hindered by small markets for some materials. For example, in the past decade, municipalities starting many new recycling programs have discovered that there are no markets for some materials or that the payments they will receive are

inadequate to support their programs. When New York State began recycling plastic soft drink bottles, it couldn't find a buyer; the recycling program became an expensive trip to the landfill. Over time, though, markets developed, and the soft drink bottles are now ground up and used to make carpeting, filler for pillows, jackets, and other products. In the late 1980s, many cities found that newsprint recycling efforts were crippled by a dramatic decline in newsprint prices. Over the past 2 decades, gluts of recyclable materials have always seemed to end, as markets develop in response to rising secondary material supplies. In the late 1980s and early 1990s, over 70 paper product mills that rely on recycled paper opened in the United States or were modified to handle recycled paper.

Recycled newsprint can be used to make a variety of useful products besides new newsprint. For instance, it can be used to make egg cartons, cereal boxes, map tubes, drywall, ceiling tiles, animal bedding, and insulation. The insulation filling the rafter spaces in my environmentally sustainable home is 100% recycled newsprint and cardboard. Cities and towns can encourage companies to manufacture these products from waste, thus putting locally available wastes to good use and helping build stable, self-reliant regional or state economies.

Individuals can help expand markets by purchasing recycled products. Governments can also help. When added together, local, state, and federal governments account for about 20% of the U.S. **Gross National Product**—that is, the nation's total annual output of goods and services. Government is the single most important purchaser in the U.S. economy. Think of the impact that local, state, and federal governments could have if they all started buying recycled paper. The demand they created would greatly increase production, lower costs, and create supplies for the rest of us.

Today, all 50 states and the District of Columbia have procurement programs, purchasing millions of reams of recycled office paper and paper products. New York, for example, allows state agencies to purchase recycled paper if it comes within 5% of the cost of virgin stock; California allows purchase if it comes within 10%. These are called **price preference policies.** In these two states, about one fourth of all the office paper, tissue, paper towels, and cardboard purchased by state agencies are made from recycled stock. On average, the states pay only about 2% more for office paper. Who knows what they saved in reduced pollution and health bills?

In October 1991, President George Bush signed an executive order that required all federal departments and agencies to purchase products made with recycled materials whenever possible. This order also requires federal agencies to name recycling coordinators to increase recycling of discarded items by the nation's three million federal employees.

Unknown to many, the Resource Conservation and Recovery Act called on federal agencies to purchase recycled materials in 1976, but only if the products were reasonably priced. Unfortunately, two problems thwarted progress. First, the law called on the EPA to publish a list of guidelines for a dozen or so recycled products. However, the EPA did not begin drawing up the guidelines for over a decade—and only then after it was sued by environmentalists. The job was not completed until 1989.

The second problem was interpretation of the "reasonable cost" provisions to mean the lowest cost. Given federal transportation policy, the smaller markets, and the failure of current pricing systems to reflect external costs (among other problems), recycled products are often slightly more expensive. Colorado officials, however, have found that some items are cost-competitive or even cheaper. Furthermore, money saved on one product, can be used to pay the slightly higher price of another, thus keeping state spending constant.

As recycling becomes more common, more and more manufacturers are bound to shift to recycled materials. Recycling facilities are bound to open in many states, creating jobs and economic opportunities. In the not-too-distant future, recycling could well be the chief source of materials to make consumer products.

It may be surprising to many, but recycling is much more prevalent in the less developed countries than in the industrialized nations. The poor raid the dumps for food, clothing, and materials for shelter; they also seek out discarded metals and other goods that they can sell. The more developed nations can help promote recycling as they assist less developed nations in their efforts to increase their economic well-being. Information on the energy and material savings, as well as the technologies for recycling, could help convince these countries and businesses to adopt recycling policies.

Composting.   Another valuable throughput strategy that reduces waste and recycles materials is **composting**, the process in which nutrients from organic wastes such as leaves, grass clippings, cardboard, and paper are returned to the soil. Composting is a form of nutrient recycling.

In most large-scale operations, organic matter is collected from various sources, stockpiled, mixed with some dirt, and then allowed to decompose. The resulting product, **compost**, is a nutrient-rich, organic material that can be used to build soil fertility. Composting may occur in backyards, in neighborhood facilities, or in large municipal operations.

In 1999, the latest year for which data are available, over 3800 municipal composting operations were in business in the United States. In Seattle, for example, zoo officials compost all of the manure produced by the zoo's many animals. When decomposed, the manure forms a rich, organic soil supplement they call *zoodoo*, which is sold to gardeners and homeowners for use around the yard. Yard waste composting programs exist in virtually every state in the United

States. Despite this success, most large-scale composting programs are in Europe—in the Netherlands, Belgium, England, and Italy—and in Israel.

Widespread composting, combined with recycling, can result in substantial reductions in municipal waste. Nationwide, organic wastes such as leaves, grass clippings, and kitchen scraps constitute about 12% of the garbage dumped into landfills or burned in incinerators, During the fall, compostable waste can make up 75% of the waste stream. Composting these wastes would not only reduce landfilling but could save considerable amounts of energy needed to transport wastes to landfills. Composting is often cheaper than landfilling and can help recycle nutrients to farmland, closing the loops of an extremely important nutrient cycle. In New Jersey, composting in Morris County costs municipalities $16 to $32 per ton, much less than landfilling, which costs over $110 per ton. Collecting and composting leaves and yard waste in Minneapolis and St. Paul costs the cities about $65 a ton, compared to landfilling at $90 a ton.

> In the United States, organic waste such as grass clippings, leaves, and kitchen waste constitute about 12% of the garbage dumped in landfills.

Despite their obvious benefits, large-scale composting operations have some drawbacks. First, they require large tracts of land because they can produce odor and create breeding sites for pests. Because of this, composting facilities are usually sited far from homes. This adds significantly to the transportation costs and energy consumption. Large-scale composting facilities are often expensive undertakings, in part because of the need to sort out the noncompostable materials such as plastics, metals, and glass. They also need to invest in machinery to turn the compost regularly to accelerate decay.

To make municipal composting more cost-effective, cities and counties could rely on labor from convicts, welfare recipients, or the unemployed. Citizens could be required to sort out recyclable metals, plastics, and glass, thus eliminating the cost of separating the wastes later.

Composting can be practiced very successfully at home. Gardeners can make their own compost piles of leaves, grass clippings, and vegetable wastes from the kitchen. By mixing these materials with a little soil (which contains the bacteria that do the breakdown) and by watering the pile from time to time to keep it moist, homeowners can produce a nutrient-rich soil supplement for lawns and gardens—eliminating the need for artificial fertilizer. A commercially available container or a simple wooden enclosure helps keep neighbors' dogs or neighborhood raccoons and skunks out of the pile. Home composting also eliminates the need to haul wastes to central facilities.

Another waste product of cities and towns is sewage sludge. As noted in Chapter 21, sludge is typically dumped in landfills, wasting very valuable soil nutrients. Fortunately, sludge can be combined with municipal solid waste (leaves, paper, grass clippings, and so on) and composted. This process, called **co-composting**, destroys viruses and bacteria in the sludge, so the product can be sold for use on farms or on gardens and lawns. Co-composting is expected to become increasingly popular in the United States and other countries because it costs less and is more ecologically sound than landfilling.

The Economic Benefits of Recycling, Reuse, and Composting John Young of the Worldwatch Institute once wrote that "more efficient use of materials could virtually eliminate incineration and dramatically reduce dependence on landfills. It could also substantially lower energy needs. . . . Taken together, source reduction, reuse, and recycling can not only cut waste but also foster more flexible and self-reliant economies. Decentralized collection and processing of secondary materials can create new industries and jobs."

In fact, numerous studies show that recycling creates far more jobs than the traditional strategies of landfilling and incineration. A study in New York showed that recycling 10,000 tons of trash through curbside recycling programs produced, on average, 36 new jobs. Landfilling or incinerating that trash produced only about one job each. While some jobs will be lost in the front end of the production–consumption system—that is, at the mines and in the forests where raw materials are extracted—many more jobs will be opened up in the middle, the recycling portion. As Young points out, while the economic health of nations is often measured by the amount of materials they consume, prosperity doesn't have to be so tightly linked to consumption. We can live well and live sustainably.

We are poisoning ourselves and our posterity.
—*Barry Commoner*

CRITICAL THINKING

Exercise Analysis

This is a clear-cut case of dollars versus environmental protection that hinges on ethics—values. You are fairly sure that the incinerator in the developing country will produce toxic emissions and that spills along the way are also possible. Lax rules may result in considerable environmental contamination. Furthermore, waste from the incinerator will very likely be dumped in an open pit, where it could leak into the groundwater. In essence, your economic savings will be passed on as an economic liability to unsuspecting residents of the receiving nation.

A savings of a million dollars is fairly substantial, however. Your boss will no doubt take notice and may even give you a sizable raise for saving the company so much money. How do you decide? Which is more important: protecting the environment of another country or saving your company money and maybe setting yourself up for a raise? Do you feel a sense of intragenerational equity? What other options are available to you?

CRITICAL THINKING AND CONCEPTS REVIEW

1. Summarize the major events occurring at Love Canal. Who was to blame for this problem? What might have been done to avoid it?

2. You are appointed to head a state agency on hazardous-waste disposal. You and your staff are to make recommendations for a statewide plan to handle hazardous wastes. Draw up a plan for eliminating dumping. Which techniques would have the highest priority? How would you put your plan into effect?

3. Discuss the major provisions of the Resource Conservation and Recovery Act (1976) and the Comprehensive Environmental Response, Compensation, and Liability Act (1980), the so-called Superfund Act. What are the weaknesses of each?

4. Describe the pros and cons of hazardous-waste management strategies, including process modification, recycling and reuse, conversion to nonhazardous or less hazardous materials, and perpetual disposal.

5. Debate the following statement: "All hazardous wastes should be recycled and reused to eliminate disposal."

6. A hazardous-waste site is going to be placed in your community. What information would you want to know about the site? How would you go about getting the information you need? Would you oppose it? Why or why not?

7. List personal ways in which we can each contribute to lessening the hazardous waste problem.

8. Discuss some of the options we have for getting rid of radioactive wastes. Which ones seem the most feasible to you? Why?

9. Debate the following statement: "Victims of improper hazardous-waste disposal practices should be compensated by a victim compensation fund developed by taxing the producers of toxic waste."

10. Describe the three basic approaches to solving the solid waste problem. Give examples of each one. Which is (are) the most sustainable? Why?

11. Describe the pros and cons of landfilling, incineration, source reduction, composting, reuse, and recycling.

12. What shifts in activities would result from changing the linear production–consumption system to a cyclic one? How will this affect jobs? In your opinion, are the proposed shifts necessary? How can we soften the blow of such a change?

13. Critically analyze this statement: "Recycling is not as great as advocates would have you think. Recycling takes energy, for example, to pick up materials and reprocess them. It produces waste and pollution."

KEY TERMS

aseptic container
co-composting
compost
composting
Comprehensive Environmental Response, Compensation and Liability Act (CERCLA)
detoxification
Gross National Product
hazardous waste
input approach
land disposal
low-temperature decomposition

Marine Protection, Research and Sanctuaries Act
municipal solid wastes
NIMBY syndrome
Nuclear Waste Policy Act
Ocean Dumping Ban Act
output approach
price preference policies
process manipulation
recycling
Resource Conservation and Recovery Act (RCRA)

reuse
reuse and recycling strategies
sanitary landfill
secured landfills
source reduction
substitution
Superfund
Superfund Act
throughput approach
waste-to-energy (WTE) plant

REFERENCES AND FURTHER READING

The References and Further Reading section at the end of this book contains a list of sources for the information discussed in this chapter and recommendations for further reading.

Connect to this book's website:
http://environment.jbpub.com/
The site features eLearning, an online review area that provides quizzes, chapter outlines, and other tools to help you study for your class. You can also follow useful links for in-depth information, research the differing views in the Point/Counterpoints, or keep up on the latest environmental news.

Environmental ethics, often passed from parent to child, are key to creating a sustainable world.

CHAPTER 24

Environmental Ethics: The Foundation of a Sustainable Society

Modern man is the victim of the very instruments he value *most. Every gain in power, every mastery of natural forces,* *every scientific addition to knowledge has proved potentia* *dangerous, because it has not been accompanied by equal* *gains in self-understanding and self-discipline.*

—Lewis Mum

This book began with an outline of today's environmental crisi really a crisis of sustainability. Each part has explored a facet and has explained the role of individuals, businesses, and g ernment in creating problems and solving them. Part II, for insta covered the population question. Part III surveyed resource proble and Part IV described the many faces of pollution. Each chapter lined traditional strategies, which address problems but often fai address their root causes. Each chapter also included a set of sustainable remedies—solutions that confront the root causes and thus hold great promise for creating lasting change and long-term prosperity. Included in the sustainable approach are ways to promote conservation, recycling, renewable resource use, restoration of renewable resources, population stabilization, and population growth management.

In order to bring about desired change, we also need a set of values that support a sustainable society. Without them, changes in our practices will be slow in coming. Chapter 3 described such a set of values, called *sustainable ethics*. They are based on the realization that current patterns of existence are not sustainable. This chapter re-examines the subject, highlighting key points already made and expanding the discussion to deepen your appreciation of our ethical options. We begin with a review of frontier ethics. Next, we explore in more depth sustainable ethics and discuss some ways to develop and implement a global sustainable ethic.

24.1 The Frontier Mentality Revisited

To understand sustainable ethics, we first examine the predominant ethic of today's industrial society, the frontier ethic (Table 24-1). As noted in Chapter 3, the **frontier ethic** is characterized by three tenets: First, the Earth has an unlimited supply of resources for exclusive human use. In other words, "There is always more, and it's all for us." Embodied in this notion is the belief that the Earth has an unlimited capacity to assimilate pollution from human activities. Second, humans are apart from nature, rather than a part of it. In other words, we can survive without natural systems and are immune to the natural forces and ecological laws that affect all other organisms. Third, human success is best achieved through the domination and control of nature.

The frontier mentality has been a part of human thinking for tens of thousands of years. It may have begun to emerge in the hunting and gathering societies and was clearly present in agricultural societies. It flourishes today in the industrial world. The damage this outmoded way of thinking creates has reached enormous proportions, in large part

Table 24-1

Frontier and Sustainable Ethics Compared

Frontier Ethics	Sustainable Ethics
* The earth is an unlimited bank of resources for exclusive human use.	* The earth has a limited supply of resources used by all species.
When the supply runs out, move elsewhere.	Recycling and the use of renewable resources will prevent depletion.
Life will be made better if we just continue to add to our material wealth.	Life's value is not simply the sum total of our bank accounts.
The cost of any project is determined by the cost of materials, energy, and labor. Economics is all that matters.	The cost is more than the sum of the energy, labor, and materials. External costs such as damage to health and the environment must be calculated.
* The key to success is through domination and control; nature is to be overcome.	* We must understand and cooperate with nature.
New laws and technologies will solve our environmental problems.	Individual efforts to solve the pressing problems must be combined with tough laws and new technologies.
* Humans are apart from nature; we are above nature, somehow separated from it and superior to it.	* We are a part of nature, ruled by its rules and respectful of its components. We are not superior to nature.
Waste is to be expected in all human endeavors.	Waste is intolerable; every wasted object should have a use.
* Indicates key ethical principles of each set of ethics.	

because human numbers have increased dramatically and many of our technologies create enormous, now life-threatening, changes in the environment.

A prime example of the frontier mentality can be seen in the settling of North America. Early frontierists cut down forests to make room for farms and grew crops on the soil until the nutrients had been depleted or the soil eroded away. They then moved on, forging into new territory to start the cycle over again. Modern society displays the same brand of careless frontierism in its quest for wealth. By cutting down forests without replanting them; by depleting ocean fisheries with little concern for the long-term well-being of fish populations; by digging up minerals as if there will always be more; and by polluting the air, water, and soil, modern society has created the potential for widespread ecological collapse.

The frontier ethic is not unique to the Western world. Similar views are evident in Latin America, Africa, and Asia. In Panama, for example, to promote economic development, past leaders have called for "the conquest of the forest." In 1940, Brazil's president wrote of the need to "conquer the land, tame the waters, and subjugate the jungle." In Asia, Japan has become a world leader in exploiting the global environment. Thailand and Indonesia have suffered mightily under resource development. In the 12th century, the king of Sri Lanka wrote, "Let not a single drop of water that falls on the land go into the sea without serving the people."

Perhaps the most troubling aspect of frontier ethics is the disregard for the Earth it fosters. Why worry about soil erosion or water pollution? Why worry about overfishing? Why worry about tropical forests? There is always more, and it's ours for the taking.

Another troubling aspect of the frontier notion that humans succeed best through the subjugation of nature is that this strategy often backfires on us, creating severe ecological backlashes such as pest resistance, increased flooding, wildlife extinction, global climate change, and severe soil erosion. Ecological backlashes affect our economic well-being and our own survival, underscoring an important but often overlooked fact: The future of people and nature is closely intertwined.

Today, people are learning to cooperate with nature—to weave human activities into the economy of nature. Biological pest control, watershed protection to control flooding, solar energy, growth management to protect valuable farmland, and a host of other measures are good examples of ways we can meet human needs without destroying the biosphere, our life-support system, and foreclosing on future generations. Refraining from building homes on barrier islands or in the floodplains of rivers, grazing cattle at the carrying capacity of grasslands, and letting natural forest fires burn are additional means of living and thriving within natural cycles and the limits inherent in them. They all pay huge dividends in the long term and represent a kind of insurance policy for the future. That is, they will allow us to prosper well into the future. The Spotlights on Sustainable Development included in many chapters in this book illustrate the ways people are learning to fit into nature's grand scheme, abiding by immutable ecological laws.

Despite these gains, the frontier mentality is still the dominant social paradigm of modern society (paradigms are

discussed in Chapter 1). If you listen to politicians or newscasters, you will hear the talk of unlimited possibilities—for continued growth, bold new frontiers, and vast, virtually limitless resources. So deeply imbedded is the frontier mentality that it affects how most of us view our problems and how we go about solving them. What is more, nearly all of the social and political institutions function to maintain the chief goal of frontierism: continued growth. This deep entrenchment makes the frontier mentality difficult to dislodge.

KEY CONCEPTS

The frontier ethic has long historical roots and is deeply embedded in modern civilization. This notion of unlimited resources for exclusive human use and of humans being apart from nature, succeeding by dominating it, profoundly influences the way we meet our needs and how we view and solve environmental problems of great urgency.

24.2 Sustainable Ethics: Making the Transition

As entrenched as our belief system may seem, it can be changed. Colonialism and slavery and the beliefs that supported them, for instance, have fallen by the wayside. Communism fell dramatically and almost without warning, despite (or maybe because of) 60 years of rule. New paradigms have replaced these outmoded systems when it became clear that they did not serve humanity well. Although their demise was not without turmoil, they did fall (FIGURE 24-1). The same may hold true for the frontier ethic, which a growing number of people are recognizing has outlived its welcome. Change may occur as the evidence of global environmental decline and the social and economic repercussions of it become more evident. What will replace frontier ethics?

FIGURE 24-1 Communism fell dramatically despite 60 years of domination.

The frontier ethic may be replaced by a more sustainable view—just as other predominant beliefs have fallen over the years—as evidence of global decline and its wide-ranging effects continues to mount.

Leopold's Land Ethic: Planting the Seed

One of the most influential proponents of sustainable ethics was the late Aldo Leopold, a wildlife ecologist best known for his book *A Sand County Almanac*. Leopold carried on the battle of John Muir, the founder of the Sierra Club and a long-time crusader for wilderness. Leopold described the need to include nature in our ethical concerns—more specifically, to extend our concerns beyond people. He called his ethic a **land ethic**. The land ethic teaches us to respect the land and ecosystems. It instructs us, in Leopold's own words, to enlarge "the boundaries of the community to include soils, water, plants and animals, or collectively: the land." In so doing, Leopold thought that the role of *Homo sapiens* would shift from that of "conqueror . . . to plain member and citizen of it."

Leopold first suggested the need for a land ethic in 1933. His book, which was published in 1949, took the message further. Charles E. Little, author and founder of the American Land Resource Association, calls the land ethic "one of the most important ideas of the century."

Leopold's view was considerably more encompassing than the view of Theodore Roosevelt–style conservationists, who protected resources principally because they were valuable to humans. The land ethic has helped to change the thinking of many people the world over.

Leopold suggested caution and deferred rewards in our use of natural resources. For Leopold, conservation required equal amounts of reflection and action. He called for individual responsibility in maintaining the health of the land, but his guidelines revolved mostly around resource management. The land ethic, while planting the seeds of the ethical revolution needed to build a sustainable society, falls short of the complex challenges facing the world today.

Aldo Leopold described a land ethic that called on people to view themselves as a part of the environment and to discard the notion of humans as conquerors of nature.

A New View to Meet Today's Challenges: Sustainable Ethics

The main tenet of sustainable ethics is that "there is *not* always more." As previous chapters have shown, the Earth has a limited supply of nonrenewable resources such as metals and oil. Even renewable resources can be depleted if harvested at a rate that exceeds the replacement rate. Furthermore, the Earth has a limited capacity to absorb our waste. Truly, there are very real biophysical limits.

To develop a sustainable society, we must recognize that infinite growth of material consumption in a finite world is an impossibility. We must also accept the fact that ever-increasing production and consumption in a world of limits could destroy the life-support systems of the planet, upon which we depend.

Sustainable ethics also holds that humans are not *apart from* but rather *a part of* nature. We are dependent on natural systems for a variety of goods and services. In Chapter 2, this principle was called *dependence*. Sustainable ethics also holds that humans are subject to the laws that govern all life on Earth.

Just as humans depend on natural systems, so too these systems depend on us. Although natural systems would do very well without us, human society will indeed have an enormous impact on the future of the biosphere. As you have seen repeatedly in this book, the damage we create in the biosphere often has profound influences on people. In Chapter 2, this principle was called *interdependence*. The notions of dependence and interdependence are eloquently articulated in a quote once attributed to the Native American Chief Seattle who watched as white people usurped the Indian lands of the Pacific Northwest:

> You must teach your children that the ground beneath their feet is the ashes of our grandfathers. So they will respect the land, tell your children that the Earth is rich with the lives of our kin. Teach your children what we have taught our children—that the Earth is our mother. Whatever befalls the Earth, befalls the sons of the Earth. If men spit upon the ground, they spit upon themselves.
>
> This we know. The Earth does not belong to man; man belongs to the Earth. This we know. All things are connected like the blood which unites one family. All things are connected.
>
> What befalls the Earth befalls the sons of the Earth. Man did not weave the web of life; he is merely a strand in it. Whatever he does to the web, he does to himself.

Sustainable ethics also maintains that the key to success is through cooperation rather than domination and control. In short, we must learn to fit the human economy within the economy of nature. Put another way, we must live within the Earth's carrying capacity. Just as all aircraft must obey certain laws of aerodynamics, human actions must obey the laws of nature, respecting limits and protecting the life-support systems upon which our economy and our lives depend.

The core value of sustainable ethics is respect and care for the community of life. This value reflects a duty to one another and to other life-forms, both now and in the future. The core value implies that human activity should not occur at the expense of other species or other people. Chapter 1 defined these ideas as intergenerational equity (fairness to future generations), intragenerational equity (fairness to others who are alive today), and ecological justice (fairness to other species). As you have seen many times in this book, managing our activities so that they do not threaten the survival of other species or eliminate their habitats is as much a matter of ethics as it is of practicality. Chapter 2 listed other principles of sustainable development. You may want to take a moment to review Table 2-1 on page 24.

One outgrowth of a change to a sustainable ethic is restraint. In regard to technology and development, the ability to say "we can" would not inevitably be followed by "we will." Instead, new questions would be asked: Should we build this dam? Should we introduce this product? Should we build more nuclear weapons? Should we have another child? Thus, "we can" would be followed by two important questions: "What are the environmental consequences?" and "Should we?"

Sustainable ethics turns us away from self-centered thinking and favors good for the whole of society and the Earth. Restraint is exercised because it benefits all people alive today, future generations, and the many species that share the planet with us. Restraint could help create a high-synergy society in which the individual parts function for the good of the whole.

As noted in Chapter 3, sustainable ethics outlines seven principles by which society can operate, thus putting ethical guidelines into action. They are: conservation, recycling, renewable resources, restoration, sustainable management, population control, and adaptability. These operating principles could help us reshape all human activities and systems, from the way we acquire food to the way we transport ourselves to and from work.

Sustainable ethics and practices are growing worldwide, as witnessed in 1992 by the Rio Conference and in 1997 by the follow-up conferences, both of which brought leaders from virtually every nation of the world together for a common goal.[1] Although the success of these efforts is less than expected, they did initiate many important programs.

The sustainable ethical system is a new paradigm that lays the foundation for a sustainable society. However different a sustainable society may be, it does not necessarily renounce all technology, all growth, or all material goods. Instead, it advocates a thoughtful look at the long-term health of the planet and an evaluation of the consequences of technology, population growth, and materialism. All decisions are passed through the sustainability filter to determine if they enhance our present and our future.

Toward a Humane, Sustainable Future

In 1992, Donella Meadows, Dennis Meadows, and Jörgen Randers, three of the four authors of the groundbreaking 1972 book *The Limits to Growth* (which forecast social and environmental disaster as a result of trends in population growth, resource use, and pollution), published a sequel entitled *Beyond the Limits*. In it, the authors noted that many trends have worsened. They wrote, "The human world is beyond its limits. The present way of doing things is unsustainable. The future, to be viable at all, must be one of drawing back, easing down, healing." Sustainable ethics can provide an ethical framework for these changes.

Because ethics are the foundation of human behavior, changes in the way we think could begin to alter the ways we

[1]The 1997 conference was judged by some critics as a limited success.

act. If our actions are based on a set of principles conducive to a humane, sustainable existence, our future could shift similarly. But how does one go about changing the way the world thinks?

24.3 Developing and Implementing Sustainable Ethics

The daunting task of changing the way people think and act worldwide elicits great pessimism. Nonetheless, examples of the shift in ethical values abound, testing the resolve of even the most ardent pessimists. This section describes several ways by which we can change the way the world thinks. The Viewpoint in this chapter discusses the rationale behind a future-centered perspective.

Promoting Models of Sustainability

The many Spotlights on Sustainable Development in this book highlight examples of people, businesses, and governments that have taken significant actions based on their commitment to the Earth and to building a sustainable world. Models such as these are extremely important catalysts in the sustainable transition (FIGURE 24-2). Similar examples

FIGURE 24-2 **Models of sustainability.** What people do has a profound influence, especially if it provides a sound model that others can study and emulate.

"Why should I feel obligated to future generations? We're inevitably separated by time and space. My presence here on Earth now will have no influence on someone living 200 years from now."

You may have heard this opinion expressed by your friends—perhaps you even hold it yourself. But if you have ever explored a wilderness preserve, used a library, or visited a historical monument, you already have some reasons for being responsible. Much of what we value in our family, our society, and our world has been provided by our predecessors, sometimes at considerable cost and effort on their part.

In today's world we face a number of issues that will affect future generations even more profoundly than they affect us now. Exploding world populations, shrinking nonrenewable resources, and plant and animal organisms threatened with extinction all add up to one thing—an ailing environment.

These are not isolated problems. Each stems from a common perception of our relationship to the world and our future. This perception can be characterized by a description of people and things as unique, immediate, individual, and separate from everything else. Any solutions that we might use to resolve our problems would have to start by challenging this perception.

Let me propose four basic considerations that may suggest a new paradigm for understanding our relationship to the world and our future. I believe our moral responsibility to coming generations will follow directly from these.

1. Future generations will be essentially the same as we are. They may have different wants and priorities, but they will manifest the same basic needs for food, water, air, and space. In addition, they will have the same basic physical and mental capacities with which to interact with their environment. Once born, they too will claim a right to life and protection from life-threatening conditions, such as extreme temperatures, toxins, famine, and disease. To give them life without also providing the basic means to sustain and enhance life would be cruel. If we expect the species to continue, we are obliged to leave a hospitable environment for those still to come.

2. One is born into a given generation by historical accident. None of us chose when to be born or to whom. Because we have no special claim to the time and

Why We Should Feel Responsible for Future Generations

Robert Mellert
The author teaches philosophy and future studies at Brookdale Community College in Lincroft, New Jersey. He has published numerous articles on process philosophy, ethics, and future studies.

place of our birth, justice would require that we have no more rights over the world and its resources than anyone else.

3. Our survival as a species is more important than our individual survival. This is confirmed in nature every day; parents, whether they be rabbits, wolves, whales, or humans, spend their energies to reproduce and care for their young before they themselves die. Many will even risk their own lives for their offspring. This is because life is not ours to keep, but to share with others.

4. Even after we die, the effects of our life continue. We will be present in the memories of others and in the habits and traditions we shared with them. Our ideas will continue to enlarge the range of options for others; even when these memories and ideas are no longer consciously a part of the future, they will ripple onward, actively influencing the course of future events and people. What has been can never die. We are what we have been given and what we have chosen to make of these "gifts." In short, we are the product of our ancestors—all that they died for and believed in—and we are the product of our decisions. Future generations will be the result of what we are now and how they use what we leave them.

If we accept these four simple ideas, it is easy to see why we have an obligation to future generations. Our obligation is based on the truth that we are more than unique and separate individuals, living only in the immediacy of the now. We are, rather, parts of a much larger whole, one that transcends space and time. As John Locke, the great English philosopher, once said, we owe the future "enough and as good" as we received from the past.

are often highlighted in books, television news reports, newspaper articles, documentaries, magazine articles, government reports, and other media. They not only inspire other people to take action, they offer practical guidance on the ways to achieve similar goals. In addition to inspiring hope, models illustrate that humans can prosper within limits. The successes of models are also broadcast by word-of-mouth com-

munications and via the Internet. For a look at an inspiring model, see Spotlight on Sustainable Development 24-1.

Publicizing models of sustainable action offers inspiration and practical examples of what individuals, businesses, and governments can do to build a sustainable future.

24-1   First Publisher in North America to Go Carbon Neutral

New Society Publishers is a progressive publisher on Gabriola Island, British Columbia that puts out about 25 books a year. As their name implies, their books are designed to help people understand the many pressing problems of our times and take actions to build a more just and economically and environmentally sustainable future.

In May, 2005, New Society Publishers became the first publisher in North America to become carbon neutral. What does carbon neutral mean?

It means that they have taken steps to ensure that their operations do not add to the carbon dioxide emissions to the atmosphere. New Society Publishers (NSP) achieved this goal, in part, by purchasing 213 metric tons of carbon offsets; these offsets, which I'll explain shortly, have neutralized the effect of the 213 metric tons of emissions that the company released into the atmosphere during 2003 as a result of the publisher's use of paper, fuel, and electricity and their production of garbage.

"This is a great way to play our part, as we try to find ways to conduct our business in a more environmentally friendly manner," said Chris Plant, co-owner of NSP (FIGURE 1). "We see it as a fundamental component of being a socially responsible business and walking our talk."

Becoming "carbon neutral" is a new expression that is gathering momentum around the world. The growing list of companies and organizations that have become carbon neutral includes the HSBC Bank, with 10,000 offices in 76 countries; the City of Newcastle, UK; and Volvo's fleet sales team. If you know where to look, it is possible to buy carbon-neutral flights, cars, holidays, concerts, conferences, and houses.

In 2003, Canada's Premiers' Conference on Prince Edward Island became carbon neutral through the purchase of 850 trees planted by Tree Canada to offset the emissions from the ministers' flights to PEI. The 2006 Commonwealth Games in Victoria, Australia, is planning to be carbon neutral, as well as the National Football League's 2006 Superbowl XL, to be held in Detroit.

FIGURE 1 Chris and Judith Plant, owners of New Society Publishers, were the first North American publishers to go carbon neutral.

Education

Educators—at all levels of the educational system—can play an important role in forging environmentally sustainable attitudes and actions among children, teenagers, and adults the world over. Recognizing the importance of environmental education, some states (such as Wisconsin) have made environmental education mandatory. Some colleges, such as Prescott College in Arizona, include environmental education as part of their mission statement.

Care must be taken in developing an environmental education program. Steve Van Matre, head of the Institute for Earth Education, has proposed an alternative to issues-based environmental education for schoolchildren. He argues that environmental education for school-age children is often disorganized and hit-or-miss. It falls down in three areas: (1) It fails to develop a true understanding of the importance of the environment in supporting our lives. It also tends to overlook (2) how we individually impact natural systems and (3) what we can do personally to change our habits. By focusing on these three areas and developing strong connections with the environment, Van Matre believes, we can make important gains in environmental education. David Orr, director of the environmental studies program at Oberlin College, argues for the need to revamp college curricula and the ways colleges operate, making them models of sustainability. The environmental studies building Orr developed with students and experts—and now teaches in—is a model of sustainable building (FIGURE 24-3). It uses about one fifth as much energy as a conventional teaching facility and is heated by passive solar energy and powered by solar electricity. In fact, the building was designed to produce more electricity than it requires. It can be naturally ventilated and recycles water. It even uses an electrolyzer to produce hydrogen gas for heating and hot water. Numerous other colleges have followed suit, including Colorado College where I teach, building classrooms and offices with many environmentally friendly features (FIGURE 24-4).

KEY CONCEPTS

All levels of education can be enlisted in the effort to foster an understanding of ethics and actions that will contribute to a sustainable future.

New Society's 213 metric tons of carbon dioxide emissions came from its use of paper (140 metric tons), flights and couriers (65 metric tons), trucking and vehicle travel (7 metric tons), fuel for heat and power (0.75 metric tons), and the garbage it produced (0.14 metric tons).

Half of the publisher's emissions were offset by a contribution to Tree Canada, which will plant 182 trees to absorb 106 metric tons of NSP's emissions. The other half (for 107 metric tons) was offset by a contribution to the Solar Electric Light Fund (SELF), based in Washington, D.C. SELF brings solar electric lighting to villagers in countries such as South Africa and Bhutan, replacing the use of kerosene, which is a dangerous and expensive carbon-producing fossil fuel.

NSP's emissions analysis was done by climate specialist Guy Dauncey, author of the award-winning book *Stormy Weather: 101 Solutions to Global Climate Change* (New Society Publishers, 2001). Mr. Dauncey is himself carbon neutral, offsetting his personal emissions each year, a habit he shares with Hollywood stars Leonardo DiCaprio, Brad Pitt, and Jake Gyllenhaal. "It takes a bit of time and effort, and some number crunching," said Mr. Dauncey, "but it's nothing that any other business need be shy of."

Mr. Dauncey has also created a template that NSP can use in future years and a paper called *Going Carbon Neutral— A Guide for Publishers,* which can be used by any publisher in the world to help them do the same. This is available for free downloads on NSP's website at www.newsociety.com.

The 25-year-old Gabriola Island–based publisher has a staff of eight people and, as noted above, publishes over 25 titles a year, including four of my trade books. Its goal as a company, says Plant, is "to contribute in fundamental ways to building a more ecologically sustainable and just society."

NSP has a history of environmental firsts. In 2001, they became the first publisher in North America to print all of their books on ancient forest–free paper (100% postconsumer recycled, nonchorine bleached). Since then, NSP has saved over 6,000 trees with this strategy, receiving awards from both the publishing community (Publisher of the Year) and Canadian Businesses for Social Responsibility (Ethics in Action). Mr. Dauncey's analysis shows that by switching to postconsumer recycled paper NSP prevented 112 metric tons of CO_2 from entering the atmosphere in 2003, because recycled paper has a smaller environmental footprint than virgin paper. Along with other decisions, using forest-free paper has helped NSP's annual emissions to be 119 metric tons fewer than they would otherwise have been.

Judith Plant, NSP's other co-owner, said, "Our staff has been really behind us on this one. As soon as we knew what our emissions were, we sat down and discussed how to offset them. We plan to continue doing this every year."

NSP is the second publisher in the world to become carbon neutral. The first was the small British publisher Snowbooks, which was launched in 2004.

FIGURE 24-3 **Practicing what they preach. (a)** Oberlin College's environmental studies program is taught in this state-of-the art classroom building on campus. **(b)** Workers install photovoltaic panels on the roof of the building during its construction.

(a)

(b)

FIGURE 24-4 The Tutt Science building at Colorado College, where the author is a visiting professor, is LEED-certified with many green features, including energy-efficient design, energy-efficient lighting, water-efficient bathrooms, and a host of environmentally friendly building materials.

Churches

Churches can also become an important force in creating global change needed to foster a sustainable future. Religious leaders who are not already involved, for example, may want to explore a broader sustainable ethic with their parishioners and suggest ways to change. Many churches sponsor adult education classes that deal with social, political, and environmental issues. These also can be an excellent route for raising awareness and promoting a new Earth ethic.

Declarations of Sustainable Ethics and Policy

Nations can help redirect human values and actions on a broad scale by adopting declarations of sustainable development and by ratifying constitutional amendments that make the environment a leading consideration. In the United States, for instance, numerous states have passed laws or adopted official language embracing the notion of sustainable development. Many businesses have done the same. Although talk can be cheap, such acts can have a solidifying effect—galvanizing people around key concepts and goals. In 1992, over 170 nations adopted the Rio Declaration, a kind of international declaration for sustainable development.

A World Organization Dedicated to Sustainable Development

Participants in the 1992 Earth Summit in Rio created an organization known as the UN Commission on Sustainable Development (UNCSD). UNCSD consists of high-level government representatives—heads of governments' environmental protection branches—with the power to make decisions. The commission uses its powers of publicity and peer pressure to encourage the implementation of Agenda 21, a massive blueprint for sustainable action. It also promotes the Rio Declaration, a statement of 27 principles regarding environment and development, and the agreement on forest protection. The commission could become an international watchdog, overseeing promises made at the Rio conference, "shining the spotlight on countries that renege and pushing through stronger commitments as necessary," as Worldwatch Institute's Hilary French says.

Equally important, the commission helps coordinate the UN's many programs and agencies that deal with the environment—among them the Development Program, the World Bank, the Food and Agriculture Organization, and the International Monetary Fund. All UN entities will report to the UNCSD on what they are doing to integrate the sustainable policies adopted at the 1992 Earth Summit.

Although it is too early to judge its success, this effort could go a long way toward instilling sustainable beliefs and practices among important players in the world community. The commission is likely to become the leader in maintaining the momentum of the 1992 Earth Summit. At this writing, however, the commission's success has been rather limited.

A Role for Everyone

One of the recurring themes of this book is that sustainable development will require efforts on the part of all of us. Without widespread participation and cooperation based on a new environmental ethic, sustainable development is unlikely to succeed. Political leaders can play a major role in re-educating people through speeches and television advertisements. Parents can help by teaching their children. Children can help by teaching their parents. Businesses can work with their employees, and employees can help instill a sustainable ethic in businesses they work in. The opportunities are limitless.

Surely, promoting a global sustainable ethic will be slow and uphill work. Many people do not see the need for change, and others will resist change because it threatens personal interests. Clearly, many obstacles exist. Overcoming those obstacles requires an understanding of them.

> **KEY CONCEPTS**
> Participation by all sectors of society is essential to building a sustainable society.

24.4 Overcoming Obstacles to Sustainability

The Industrial Revolution did not occur overnight—nor did the Agricultural Revolution. The Sustainable Revolution will very likely take a century or more. The reason for the slow pace is simple: There are many obstacles to the adoption of sustainable ethics and sustainable practices. This section explores some of these obstacles and ways to overcome them.

> **KEY CONCEPTS**
> Many obstacles lie on the road to sustainability, and the process will likely take a century or more.

Faith in Technological Fixes

Many people express extreme optimism in the power of technology to solve our problems, a phenomenon labeled **technological optimism.** The late economist Julian Simon in the Point/Counterpoint in Chapter 8, for instance, says that there are no limits. As long as we exercise our ingenuity, we can develop new technologies to expand our reserves or find substitutes. So why change our views? The world truly is unlimited!

In their zeal to find high-tech solutions, technological optimists often overlook logical, inexpensive, and Earth-friendly low-tech solutions, such as conservation. As you have seen in previous chapters, conservation is a far more effective way of reducing global carbon dioxide emissions and is far cheaper than nuclear energy or other options. A dollar invested in conservation, in fact, reduces carbon dioxide emissions seven times more than a dollar invested in nuclear power. Recycling, discussed in Chapter 23, is a far cheaper solution than high-tech trash incinerators. Nonetheless, many municipalities are still building incinerators.

Unfortunately, our unqualified optimism in technology blinds us to the obvious solutions at our feet. Spurred by thoughts of limitless resources, many applied scientists, technologists, politicians, and businesspeople promise an unlimited future. Believing that these experts can't be wrong, many people place their faith in their proposed solutions and resist changing their views.

Today, more and more people are beginning to understand that technological fixes won't save the day and are calling for simpler, more effective solutions that strike at the root causes of the crisis of unsustainability. This is not to say that technology has no role in creating a sustainable world. Far from it. Technology will play an important part

in the future. The solar panels that are powering my computer are a good example of the positive benefits of technology. The point is that technology is not enough. A new ethic that fosters personal responsibility and action is a large part of the equation. Education can surely help illustrate the limits of technology and the many positive examples of ways that sustainable ethics and personal actions can help forge an enduring human presence.

> **KEY CONCEPTS**
> The belief that technology can solve all of the world's problems hinders the development of a sustainable ethic and the implementation of simple, cost-effective, sustainable practices.

Apathy, Powerlessness, and Despair

Many people are apathetic about the course of modern society. Involved in their own lives, they see resource limitations and pollution as problems for which someone else must take responsibility. Apathy is effortless, noncontroversial, and cheap.

Where does apathy come from? In part, many of us are taught or conditioned to be apathetic—not to rock the boat or make waves. Someone bulldozes a favorite forest of ours to put up a shopping mall, and someone else calms our indignation by saying, "That's the price of progress."

Apathy may also stem from our early education. Government and history teachers, for example, instruct students about democracy but may fail to drive home the important principle that democracy relies on participation. To many of us, democracy means freedom, but democracy also comes with certain responsibilities. First and foremost, we have a responsibility to participate: to vote, to write congressional representatives, to sit on citizen advisory committees. Many people, however, don't even vote.

Another cause of apathy is powerlessness. In the United States and other countries, people are often paralyzed by a sense of powerlessness—feelings of insignificance. Big corporations influence government decisions, steamrolling citizen interests. If you can't do anything, why bother?

This pervasive feeling has a powerful effect on our own lives and on the environment. Feelings of insignificance create many of the environmental problems facing the world and keep us from solving them. I call this the **Paradox of Inconsequence.** Here's how it goes: Being just one of more than six billion people on Earth is an excuse many of us use for continuing to do what we have always done, living unsustainable lifestyles. What difference, we ask, does it make if you drive 10 miles per hour over the speed limit and waste a little gasoline? You're just 1 of 295 million Americans and more than six billion Earth residents. You're just a small fraction of the problem.

The trouble with this thinking is that millions of people think the same thing and thus go about their affairs as if their roles were of no consequence. Their actions can add up to a great deal of waste and pollution, though. The many problems we have are, in fact, the result of millions of individuals' actions.

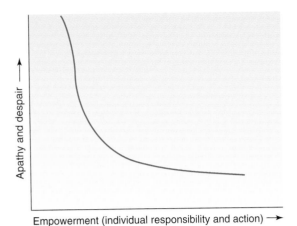

FIGURE 24-5 Apathy and despair. Getting active can help reduce one's feelings of despair. When an individual's actions are combined with those of others, people can make a big difference.

Ironically, feelings of insignificance also keep us from solving many problems. What difference does it make if you drive the speed limit, recycle aluminum and paper, and keep your thermostat at a reasonable temperature? Because you are only 1 of 295 million Americans or six billion world citizens, your contribution to resource depletion and pollution is insignificant. So why bother acting responsibly?

Feelings of insignificance create many of the problems we face and keep us from solving them. But all of this can be turned around. Millions of people acting responsibly can have huge impacts. It is a powerful mathematics. For example: Ascend a marble stairway that tens of thousands have ascended before you. Feel the grooves worn by their feet. Although each one would say he or she had no effect, together all have had an enormous impact. If we can have a negative impact, we can also have a positive impact through our actions. Education can go a long way to convince individuals of the collective benefit of responsible living.

FIGURE 24-5 shows a hypothetical curve that plots apathy and its close cousin despair against empowerment—taking control of your life and taking action. It suggests, and with good reason, that as people take control of their lives, get active, and work toward change, apathy and despair tend to decrease. A good example of what individuals can do is presented in Spotlight on Sustainable Development 24-2.

KEY CONCEPTS
Apathy and feelings of powerlessness and despair hinder progress toward sustainability. Many examples show that individuals can be truly powerful in changing the world.

The Self-Centered View

The average man and woman on the street often take a self-centered view. As a rule, their actions are driven not by concern for sustainability but by their own welfare. Author and social critic Tom Wolfe coined the phrase *the me-generation* to describe the Americans of the 1980s who seemed intent on self-indulgence. The 1990s proved to be much the same.

Many people of the me-generation buy what they can afford, giving little thought to the effects of their consumption on the environment. Replacing the self-centered approach to life with a global environmental perspective is the main thrust of sustainable ethics.

Ironically, for many years, environmental groups failed to promote individual responsibility and action among their members and the general public. Much of their educational outreach was through direct mail. Their appeals generally outlined the problems we faced and presented proposals to solve them through legislative action. Individuals were asked to help, but individual involvement was generally limited to writing a check. Wendell Berry, philosopher and author of *The Unsettling of America*, writes that "the giving of money has . . . become our characteristic virtue. But to give is not to do. The money is given in lieu of action, thought, care, time." I call this phenomenon the *cash conscience*. Fortunately, many groups have realized the failing of this approach and have taken strides to promote more individual action.

Environmental groups have also spent much of their time pointing accusatory fingers at lawmakers, inept government regulators, and business executives. Although these parties are indeed part of the equation, so are individuals. This lopsided approach nurtures the idea that the blame for the environmental crisis lies almost entirely in someone else's hands, a phenomenon I call *blame shifting*. In addition, it fosters the idea that problems can be solved only by regulating someone else's activities or by applying new and more efficient technologies to control someone else's pollution. It nurtures the idea that money alone can solve our problems. The legislative and legal work of environmental organizations such as Environmental Defense, the Natural Resources Defense Council, Greenpeace, the Sierra Club, and others is vital to our efforts to improve the environment, but individual action is equally important.

KEY CONCEPTS
Concern for the self, feelings of insignificance, cash conscience, and blame shifting hinder efforts to foster individual action and responsibility.

Ego Gratification

According to psychologists, as we pass from infancy to adulthood we develop a sense of the self, of our own separate identity. Psychologists call this the *derived self*. Once the derived self becomes established, it also needs reaffirmation. We can get that from our parents, friends, teachers, and ourselves. If not, we may seek it in new clothes, fast cars, and extravagant homes, self-rewarding behavior that builds our egos or reaffirms that we are important (Chapter 3). After all, important people live in big houses and drive expensive cars, don't they? Reaffirmation may be at the roots of materialism and overconsumption. They lead to the accumulation of material possessions, which depletes essential resources, causes pollution, and reduces wildlife habitat. What can a society do to build feelings of self-worth?

24-2   Students Put Rotting Garbage to Use Heating School

Science students in Maryland Heights, Missouri, launched an ambitious program that could start a trend in the United States and abroad. Their project: to capture methane, a combustible gas released from a nearby landfill, to heat their school.

Today, thanks to the efforts of a handful of students and a local businessperson, all 117 classrooms and two gymnasiums at Pattonville High are heated by methane from rotting garbage. Pattonville High is believed to be the first school in the United States that is heated by methane from rotting garbage.

The methane, piped to the school by an 1100-meter (3600-foot) pipeline from the 35-hectare (85-acre) landfill, is burned in the furnace in place of natural gas. What makes this all the more exciting is that the landfill operators were burning off the waste gas. Even more exciting is that this project, which cost $175,000, saves the school $40,000 a year in heating bills. The savings would pay the capital costs in less than 5 years. The principal of the school, Tom Byrnes, said, "The methane gas is there. It's a matter of burning it off or using it productively."

The school is not entirely heated by methane. Some natural gas is needed for the cafeteria and other operations, but the use is minimal.

The U.S. Environmental Protection Agency estimates that 750 landfills in the United States could be tapped to generate gas. If only half of them were tapped, it would cut the emissions of thousands of tons of carbon dioxide each year. According to the EPA, this would be equivalent to removing 12 million cars from the road. Obviously, landfill waste gas is not the answer to creating a sustainable future, but it can help save money and reduce emissions.

New regulations require that landfills with more than 2.5 metric tons (2.75 tons) of trash be tested for gaseous emissions. If emissions exceed a certain level, the new Clean Air Act amendments require the operator to install a system to collect and burn off the gas, to prevent possible explosions caused when the gas reaches certain levels. If methane can be put to good use, say proponents, it should be.

Adult education programs and better training in health classes can help us break patterns of child rearing and create new ones. Religious leaders can help as well.

KEY CONCEPTS
Feelings of inadequacy are often offset by the accumulation of material possessions. Overconsumption is a key factor in the decline of the environment.

Economic Self-Interest and Outmoded Governmental Policies

Most of us live in countries whose socioeconomic systems are out of line with the natural world—that is, whose human systems are operating unsustainably. Greed, ignorance, and a host of other factors drive us further down the road to ecological ruin (Chapter 3). Our governments and the economic systems they often serve are therefore a barrier to sustainability. Although previous chapters have pointed out some of the flaws in both government policy and economics, the next three chapters go into more depth, showing where reform is needed to place these human activities back on a sustainable course.

KEY CONCEPTS
Unsustainable economic systems and governments that support them are major barriers to sustainable development.

24.5 Sustainable Ethics: How Useful Are They?

Each day people are faced with dozens of decisions, many of which influence the environment. For example, should you ride your bicycle or drive your car to school or work? Should you turn up the heat or put on a warm sweater? The sum of the decisions made by many millions of individuals such as yourself adds up. Environmentally sustainable decisions result in positive outcomes. Environmentally unsustainable decisions create negative results.

Corporations and governments face many decisions as well. Single decisions at this level have impacts as profound as many millions of individual decisions. For example, should the government develop nuclear power or push for energy conservation? Should the government use recycled paper or continue buying virgin stock?

Ultimately, most of our decisions are influenced by our values—what we view as right or wrong, desirable or undesirable. Values are learned from our parents, relatives, peers, teachers, religious leaders, politicians, writers, news commentators, reporters, books we read, and articles in magazines and newspapers. Values may shift over the years, sometimes drastically.

Many people today base their judgments on **utilitarianism**, a doctrine by which the worth of things is determined by their usefulness. This view tends to put human needs above most, if not all, others. Economics, discussed in

¥ Read a book on environmental ethics.
¥ Discuss environmental ethics with friends and family.
¥ Write your own code of environmental ethics.
¥ Support political candidates that have similar views.
¥ Make a list of actions you will take to manifest your environmental ethics.

the next chapter, is the yardstick of many utilitarian decisions. Utilitarian resource management, for example, finds the fastest, cheapest ways of acquiring resources. It puts human needs above all else. Thus, forests may be reseeded, but not so much to provide habitat for animals as to provide wood for future generations.

At first glance, frontier ethics appears to be heavily utilitarian. It serves human needs. After closer examination, though, it is clear that it is anything but utilitarian, for it leads to destructive behavior that forecloses on our future. In the long run, frontier ethics is nonutilitarian.

At the opposite end of the spectrum is a new and controversial view of **natural rights**. It says that all living things have rights, irrespective of their value to human society. Wilderness should be set aside, not just so people can use it, but to protect the species that have lived on the land for thousands of years. Animals, plants, and insects should be preserved, not because they are of use to humans, but because they have a right to live.

Another value system is based on **divine law**, the word of God as written, for example, in the Bible. For the most part, divine law dictates personal behavior—interactions between people. The Ten Commandments instruct us on ways to treat one another: Thou shalt not kill. Thou shalt not commit adultery. One religious group has recently added an eleventh commandment: Thou shalt protect the Earth. As you may recall from Chapter 3, divine law does indeed call on us to protect the environment, God's creation. Virtually every religion has some teachings to advance this mandate.

These are some of the principal value systems that affect our decisions. As you think about them, you may realize that no one system determines your actions at all times. In some cases, you may act out of utility. In others, you may make a sacrifice for the good of the whole.

Sustainable ethics is another system of values. It calls on us to consider a new set of parameters in making decisions. It reminds us of our place in the world and implores us to act in cooperation and consideration, rather than in isolation and strict self-interest. At first this may seem anything but util-

itarian. It is utilitarian, however. Sustainable ethics is a means of serving human needs and the needs of nature, which are the same. It serves people in the present and the future. In many ways, sustainable ethics combines aspects of natural rights and utilitarianism. Some religious scholars also contend that it incorporates divine law. Bottom-line, sustainable ethics and practices that manifest these ideas will help all countries ensure brighter futures, stronger economies, healthier populations, and even more efficient government.

KEY CONCEPTS

Values affect the way we act. To some people, the value of things is based on their usefulness. Sustainable ethics may be one of the most useful sets of values because it serves people, the environment, and the economy much better than shortsighted, outmoded systems of belief.

A human being is part of the whole, called by us "Universe." He experiences himself, his thoughts and feelings as something separated from the rest—a kind of optical delusion of his consciousness. This delusion is a kind of prison for us, restricting us to our personal desires and to affection for a few persons nearest to us. Our task must be to free ourselves from this prison by widening our circle of compassion to embrace all living creatures and the whole of nature in its beauty.

—Albert Einstein

CRITICAL THINKING

Exercise Analysis

You, your friends, and your parents probably do many things to help protect the environment that don't benefit you economically. For example, you may recycle even though you probably don't earn enough from the bottles and cans you return to make it worth your while economically. So do millions of other people in the United States and other countries. You and your family are therefore living proof that some people do act for unselfish reasons. (See the Individual Actions Count! table for some ideas on what you can do.)

The cynic who asserts that people respond only when they're financially benefited is making a broad generalization that fails to stand up to scrutiny. A more accurate statement might be that some people will take actions on their own, whereas others seem to respond best when they're financially aided. People who make statements such as the one that launched this exercise may be motivated by money themselves. Consequently, they may think that everyone else is the same. This bias clearly affects one's view of the world. Beware of powerful generalizations; they can become crippling thought stoppers.

CRITICAL THINKING AND CONCEPTS REVIEW

1. Describe the three major tenets of frontier ethics.
2. For one week, when you listen to the news, watch TV, read articles, and listen to people talk, note examples of the frontier mentality. Make a list of them.
3. Is your personal ethic closer to a frontier or a sustainable ethic? Do your actions reflect your personal beliefs? How could you improve?
4. Discuss the tenets of sustainable ethics. Indicate which of these tenets coincide with your personal beliefs. Which ones don't? Why?
5. Critically analyze the tenets of sustainable ethics. Do you see any flaws in them?
6. Describe modes by which sustainable ethics can be promoted worldwide. Would you approve of such an activity?
7. Describe the roles that apathy, self-centeredness, feelings of insignificance, and technological optimism play in creating environmental problems and blocking change. How can each one be changed?
8. "Animals and plants have rights," says a leading philosopher, "irrespective of their value to humans." Do you agree? Why or why not?
9. Critically analyze the following statement: "Sustainable ethics is, in many ways, a utilitarian set of beliefs."
10. Using your critical thinking skills, analyze the following statement: "All of the changes prescribed in this book won't be enough to create a sustainable society. We need to change attitudes and ethics. Without changes at the deep, personal level, we cannot create a sustainable future."

KEY TERMS

Agenda 21
divine law
frontier ethic
land ethic

natural rights
Paradox of Inconsequence
Rio Declaration
technological optimism

UN Commission on Sustainable
   Development (UNCSD)
utilitarianism

REFERENCES AND FURTHER READING

The References and Further Reading section at the end of this book contains a list of sources for the information discussed in this chapter and recommendations for further reading.

Connect to this book's website:
http://environment.jbpub.com/
The site features eLearning, an online review area that provides quizzes, chapter outlines, and other tools to help you study for your class. You can also follow useful links for in-depth information, research the differing views in the Point/Counterpoints, or keep up on the latest environmental news.

Windfarms promise to produce clean, reliable, affordable electricity for many years to come.

CHAPTER 25

Sustainable Economics: Understanding the Economy and Challenges Facing the Industrial Nations

Thirty years of exploration into biology, physics, and human nature have brought me to the realization that humanity itself has, in fact, sharpened the sword that is potentially responsible for piercing it through the heart.

—*Michael Reynolds*

I n southern India, people once made traps for monkeys by drilling small holes in coconuts, filling the shells with rice, and chaining them to trees. The success of this trap was attributed to the fact that the hole was large enough for a monkey to put its empty hand into, but too small for it to pull a handful of rice out. As monkeys clung tenaciously to their rice, villagers threw nets over them. The monkeys were trapped by their own refusal to let go.

Many observers believe that humankind is caught in a similar dilemma. Clinging tenaciously to an unsustainable way of living and doing business, many among us ignore the critical environmental problems such as global warming that undermine our long-term future including the health of our economy. As a result, we often resist changes needed to build a sustainable human economy. This chapter begins with an overview of some key principles of economics, describes the economics of pollution control and resource management, and then discusses major weaknesses in economic systems from the viewpoint of sustainability. It concludes with a discussion of ways to build sustainable economic systems in industrial nations. Chapter 26 tackles the less developed nations.

25.1 Economics, Environment, and Sustainability

As you may recall from your study of the cultural history of humankind (Chapter 7), manufacturing and trade began in earnest with the rise of towns and villages. Towns and villages, in turn, owe their origin to agricultural surpluses made possible by technological advancements. New implements and techniques permitted farmers to produce an abundance of food with less labor. Displaced farm workers and their families moved to cities and towns, where many took up trades to earn a living.

Although the economic system began a long time ago, the science of economics began only 2 centuries ago with the publication of the Scotsman Adam Smith's book *The Wealth of Nations*. But what is economics?

Economics is the study of the production, distribution, and consumption of goods and services. It concerns itself primarily with two factors: inputs and outputs. **Inputs** are the things companies require to manufacture and sell products—primarily labor, land, and commodities, or natural resources such as energy or minerals. The **outputs** are the products of participants in the economy. These include materials, goods, and services that companies produce for use by consumers or other companies. Economics, like ecology and environmental science, is concerned with relationships. It also employs scientific tools to discover the laws that regulate economies.

Economists recognize two broad disciplines within economics. The description of economic facts and relationships falls within the purview of the first discipline, known as **descriptive economics**, so named because it is supposed to be free of judgment. Descriptive economics is a relatively pure science. Its questions can be answered by research and facts.

Economics melds with political science and sociology when it attempts to answer value-laden questions. For example, should companies pay for pollution controls? Should the economy continue to grow? Such questions cannot be answered by empirical facts and figures. There are no right or wrong answers to them, for they are value judgments and are left to the political process and the economic players themselves. This realm of economics is called **normative economics**.

More and more, people are calling for a new kind of economics, an **ecological economics** or **sustainable economics**, one that concerns itself with supplying the needs of people while protecting the environment. In this chapter, we examine **sustainable economics**. As you ponder this new and exciting discipline, remember that it seeks ways to ensure that economic development serves people—and serves them equitably, while protecting, even enhancing the natural world. For example, its proponents attempt to end economic exploitation. Sustainable economics, as envisioned by its many proponents, is a system that would supply goods and services in ways that honor limits and protect the carrying capacity of the planet. It is a system that would enhance human welfare in the present but also in the future, thus supporting the goals of intergenerational and intragenerational equity. Before we examine what a sustainable economic system might look like and study ways to make this vision a reality, we must first examine the current economic system. To do so, we examine some of the basic precepts of economics.

The human economy has been functioning for thousands of years, but only recently have people begun to think seriously about developing a sustainable economy—one that serves people equitably and protects and enhances the environment on which we all depend.

Economic Systems

Economics is a science that helps businesses and societies solve three fundamental problems:

1. *What* commodities should it produce and in what quantity?
2. *How* should it produce its goods?
3. *For whom* should it produce them?

Of course, there are many ways for a society to solve these three basic problems. The type of economy influences the answers. In **command economies**, such as those found in China, Cuba, North Korea and formerly in the Soviet Union and east bloc nations, government officials dictate production and distribution goals. Decisions are often left to bureaucrats. These decisions are supposed to be made to benefit all, not an elite few, but even in these systems based on egalitarian goals, corruption abounds.

In **market economies**, like those of most nations of the world, governments (theoretically) take a back seat to the marketplace. In other words, the market operates on its own without government intervention. Companies produce the goods and services for which there is the highest demand or products that yield the highest profit, thus answering the first question: What goods and services and in what quantity? Profit ultimately dictates how goods are produced. In market economies, the least costly method of production yields the greatest profit.

In a market economy, the question "For whom?" is also determined by money. In general, whoever can afford a good or service can get it.

One of the key principles of economics is the **law of scarcity**, which states that most things that people want are limited (**FIGURE 25-1**). As a result, their sale is rationed. In a market economy, price is the principal rationing mechanism. For instance, few of us drive Porsches, even though many of us might like to. Why? Because the price greatly exceeds the ability of the vast majority of us to pay for this car. In a sense, then, price rations Porsche sales and purchases. In contrast, in command economies, governments ration most of the output, although prices do play a role.

In truth, most nations' economies are mixed—that is, they contain elements of market and command economies. China's economy is becoming more and more a market economy as thousands of government-owned businesses are converted to private ownership. In the market economies of

25-1   A Small Business Park Unpaves the Way to a Sustainable Future

Umeå is a small city in northern Sweden, a country renowned for environmental stewardship. Among its many environmental features, Umeå is home to a remarkable small business park—a model that other cities could emulate. The business park is home to franchises of three multinational corporations: a Ford Motor Company sales and service dealership; a Statoil gas station, car wash, and convenience store; and a McDonald's fast-food restaurant.

Okay, you say, there's no way these three companies possibly contribute to environmental protection and a sustainable future?

Well, let's take a look . . .

To begin, all three businesses have green roofs—not roofs painted green, but rather specially constructed roofs that are planted in grass and other forms of vegetation. These roofs help give back to nature a little more green space, taken up by the development itself. They absorb moisture, which nourishes the plants, rather than letting it run off parking lots, picking up oil and other pollutants that then wash into surface waters. Green roofs also help keep the buildings cooler in the summer.

The parking lot at the Ford dealership is unlike other parking lots you'll find in industrial nations. This one is made from porous pavers, special blocks that allow moisture to seep into the ground where it replenishes groundwater. By letting water seep into the ground, these pavers reduce surface runoff, water pollution, and flooding.

The buildings are made of natural or recycled building materials, too. Moreover, the buildings are built to be efficient—so efficient that they use 60% less energy than similar structures. As an example, motion sensors in the buildings turn off lights when the rooms are unoccupied.

Energy demands are met from renewable energy sources, such as the sun and wind. Electricity comes from a wind generator on the coast about 15 miles away. The Ford dealership also has solar panels on one side that preheat incoming air, reducing energy demand of the heating systems. It also has skylights that reduce daytime electrical consumption by 60 to 70%. At the McDonald's, waste heat from various processes such as cooking grills and deep fryers and from refrigeration systems is captured and piped to parts of the complex that require heat.

The business park also uses water efficiently and captures and reuses all of the storm water onsite. It flows into an onsite water garden. Ninety-nine percent of the water

Great Britain, Canada, and the United States, for example, free enterprise is often touted as the guiding rule. Let the market determine who gets what. Nonetheless, in these and other countries, governments also influence economic behavior. Governments, for example, may subsidize some activities, giving some businesses and some products an economic advantage over others. The subsidies to oil and coal artificially distort their price.

Governments also influence businesses through various laws and regulations—for example, those that require companies to control pollution. Laws that stipulate how much pollution a company can emit affect the price of goods and services. Outright government bans on dangerous products limit product availability and therefore dictate production and consumption, interfering in a free economy to protect health or the environment. (For a look at a creative way to regulate natural resources, see Spotlight on Sustainable Development 25-1.)

Another way of influencing the market is through federally mandated freight rates. As noted in Chapter 23, freight rates are set by the federal government. In the case of raw ore versus recycled materials, raw ore travels more cheaply than scrap for recycling—providing a great benefit to the mining industry and a deterrent to recycling.

As you may recall from Parts III and IV, state and federal governments also subsidize various activities by special tax

FIGURE 25-1 **A treat for the wealthy.** High-end vehicles, such as this Mercedes SLR, are targeted to China's newly rich.

breaks or by sponsoring or funding research. These and other subsidies create an uneven playing field, which benefits some and harms others. Last, but far from least, governments also impose tariffs (taxes on imports) to regulate the flow of goods into a country, thus stifling free international competition.

Regulations, bans, subsidies, and other policy instruments are levers through which governments influence mar-

from the car wash is filtered and reused. Sewage from the facility is used for fertilizer in local farming operations.

In addition, the business park reuses or recycles all of the waste that it produces. Wood used to build the facility comes from local wood lots and was sustainably grown and harvested.

When building the small business park, the developer removed a house located on the site. Rather than demolishing the structure and hauling the waste to the landfill, he lifted it and trucked the building to a new location. Rather than cutting down a large oak tree on the site, it, too, was transplanted elsewhere. The developer left a pine tree that was home to an endangered beetle on the site, too, rather than cutting it down.

The list of environmental attributes of this remarkable facility does not stop here. The gas station, for instance, sells three types of fuels that replace environmentally unfriendly diesel and gasoline. The convenience store sells organic produce and other organic foods. They encourage customers to recycle food containers by including well-labeled receptacles.

The Ford Dealership recycles all waste oils and other waste fluids from cars such as brake fluid and radiator fluid.

They use vegetable oil, rather than a petroleum-based fluid, in their hydraulic lifts. They even work hard to make this the healthiest and safest possible workspace for their employees. All three businesses provide ongoing education for their employees to ensure the continuation of sustainable business practices.

Now, you say, sure this is all well and good, but the businesses must be suffering.

Wrong.

The fact is that during the first year of operation revenue from sales at the car dealership and service shot through the roof. The other businesses are thriving as well, indicating that this is clearly not a case of environmental philanthropy, but a sound, economically profitable business practice that provides a model for the rest of the world. Over a half a million people visit the park every year, and city officials are encouraging other similar development.

Imagine how the world would change if all businesses pursued a similar strategy?

Adapted with permission from Dan Chiras, *EcoKids: Raising Children Who Care for the Earth,* Gabriola Island, British Columbia: New Society Publishers, 2004.

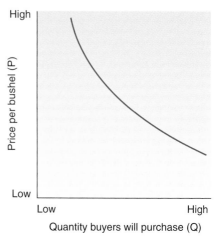

(a) Demand curve

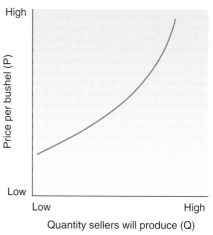

(b) Supply curve

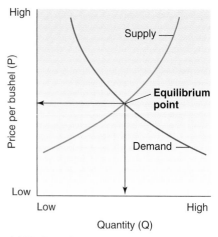

(c) Market price equilibrium point

FIGURE 25-2 **Supply and demand curves for rice. (a)** The demand curve shows the relationship between the price (P) on the vertical axis and demand (Q) on the horizontal axis. This graph shows that a rise in prices reduces the demand. Falling prices increase demand. **(b)** The supply curve shows the relationship between the price and supply, or amount produced. The higher the price, the more farmers will produce. **(c)** The market equilibrium point is the intersection of the supply and demand curves. It's the price people will pay for rice and the amount farmers will produce at that price.

ket economies to protect natural resources, people, and special economic interests, such as the mining industry.

Although they're nearly extinct, command economies also generally consist of a mixture of market and command practices. China, for example, allows some free market enterprise within an economic system tightly controlled by the central government.

Before turning to the principles that govern market economies, it is important to note that both types of economic systems have enormous impact on the environment. The command economies of eastern Europe and the former Soviet Union, for example, produced enormous amounts of air and water pollution and have left behind a trail of toxic hot spots, not unlike those of the Western world, where free enterprise has been the rule.

KEY CONCEPTS

Two types of economic systems exist: command and market economies. Command economies are run in large part by governments. In market economies, decisions about production are largely determined by prices and by people's ability to pay. Nonetheless, governments affect market economies in many ways. Both types of economies have had enormous environmental impact.

The Law of Supply and Demand

In market economies, the three essential questions posed in the previous section are generally solved by price. The price of a good or service is, in turn, determined by two additional factors: supply and demand. **Supply** refers to the amount of a resource, product, or service that's available. **Demand** refers to the amount people want. The predictable interplay of price, supply, and demand constitutes the **law of supply and demand**.

To understand the relationship between these three factors, please take a moment or two to study the graph in **FIGURE 25-2a**, which shows a demand curve for rice. On the vertical axis of the graph is the price (P) of rice per bushel. On the horizontal axis is the quantity (Q) that people will buy at each price. This graph shows that as the price rises, the demand decreases. Conversely, a lowering of price generally increases demand. The relationship between P and Q is inverse—that is, as one increases, the other decreases.

Most of us are familiar with the interplay of price and demand. We all know, for example, that when popular video games first come out, they often command a high price. Popular rock bands sell their tickets for more than struggling bands.

The supply curve is of interest chiefly to producers (**FIGURE 25-2b**). It plots the relationship between price and the quantity that suppliers will produce. The supply curve illustrates an intuitively simple concept: The higher the price, the more producers are generally willing and able to produce. At lower prices, producers often have to reduce production. Numerous examples exist. The fall in oil prices in the early 1980s, for example, put many American oil drillers out of business and caused economic hardship throughout the world. What happened? As oil prices fell because of conservation (which reduced demand) and cheaper foreign oil, many American companies went out of business or shut down wells because they couldn't produce oil profitably at the lower prices.

The economy is a balancing act (offset by some government tampering, as mentioned earlier) with two principal players, supply and demand. Supply and demand interact to determine the price of goods and services. Graphically, this is represented by the intersection of the supply and demand curves and is known as the **market price equilibrium point** (FIGURE 25-2c). The market price equilibrium point represents the price at which both consumers are willing to buy a product and producers can afford to produce it.

Imagine an economy in which literally millions of prices are set by this kind of interaction without interference by governments. That system would be a **free-market system**. However, as noted previously, in most countries, governments do meddle in economic affairs, sometimes benefiting businesses through subsidies and sometimes thwarting free enterprise through regulations.

The free-market system may also be tampered with by business itself. Some businesses, for instance, push for protectionist trade policies that limit or eliminate imports from foreign producers, in an effort to keep markets to themselves. Others buy up competitors, creating monopolies. Monopolies eliminate competition and can result in prices that are set at any level the monopoly wants (provided the public will pay). Antitrust laws in the United States are aimed at protecting individuals from monopolies. Companies may also try to drive others out of business. For example, Browning Ferris Industries (BFI), a major U.S. trash hauler, was found guilty of trying to illegally drive a smaller Vermont trash hauler out of business and was fined over $6 million.

KEY CONCEPTS

The law of supply and demand shows that prices affect both the supply of goods and services and the demand for them. In addition, the prices of goods and services in a market economy are largely determined by the interaction of supply and demand.

Environmental Implications of Supply and Demand

The law of supply and demand has some very real implications for sustainability. First, consider the impact of declining supplies by looking at the ivory trade. For years, elephants were slaughtered by the thousands in Africa to support the profitable ivory trade. When populations began to decline, African countries made it illegal to shoot elephants, and many countries prohibited the importation of ivory. Nonetheless, poaching continued. Why? Bans on ivory export and the dramatic decline in wild elephant populations caused by overhunting increased the price of ivory. That, in turn, gave poachers a considerable economic incentive to continue the illegal slaughter, even at the risk of being killed by game wardens. The supply graph predicts such activity, showing that the higher the price, the more willing someone is to produce a given product. People become rich, and the elephant is pushed toward extinction. It's a simple line of cause and effect, with devastating consequences.

Government regulations that put a stop to the sale of ivory products in many countries have helped to solve the problem. These bans caused the market for raw ivory to evaporate, making poaching unprofitable and brightening the future of the elephant. Now that the elephant populations are recovering, efforts are under way to allow limited ivory production—say, from animals killed to keep populations on game preserves from skyrocketing and ruining the environment.

Supply and demand economics also have considerable impact on conservation efforts. For instance, when the price of oil climbed in the late 1970s because of the artificial shortages created by foreign oil embargoes, many industrial nations found ways to improve the efficiency of factories, automobiles, and homes. Supply and demand can also spawn wasteful behavior, though. The fall in oil prices in the early 1980s, created by conservation and by increases in production by the United Kingdom and other countries, eroded many people's resolve to save energy. The U.S. government, in fact, lost virtually all interest in energy efficiency at that time.

As a general rule, abundant supplies (at least in the near term) lead to low prices, which often foster wasteful practices. "We've got plenty, so why conserve?" seems to be the attitude of many. Section 25.4 discusses some of the weaknesses of supply and demand.

KEY CONCEPTS

Economic considerations have profound impacts on activities that affect the quality of our environment and the sustainability of our society.

Measuring Economic Success: The GNP

Economists need ways to measure economic activity. The most widely used measure of a nation's economy is its gross national product. The **gross national product (GNP)** is the market value of the nation's output—in other words, the value of all goods and services that a nation produces and sells (including government purchases) in a given year. GNP includes domestic and overseas operations of all businesses. Real GNP is the GNP adjusted for inflation. Per capita GNP is defined as follows:

$$\text{Per capita GNP} = \text{Real GNP} \div \text{Total population}$$

Widely used to track economies, the GNP gives a general picture of the wealth of nations and the living standards of their people. As you will soon see, the GNP, like the law of supply and demand, is somewhat flawed and in need of corrections to make it sensitive to environmental concerns and sustainability.

KEY CONCEPTS

The gross national product is a measure of the economic output of a nation, including all goods and services, and is used as a means of tracking the success of economies.

25.2 The Economics of Pollution Control

The free-market economic system in the United States and other countries, as well as the command economies of socialist and communist nations, treated pollution with almost uniform disregard until the late 1960s and early 1970s (FIGURE 25-3). (Some disregard continued even through the 1980s.) Pollution was something that issued from smokestacks and that symbolized progress and prosperity. Where it landed no one seemed to know or care. The costs incurred from pollution are considered an **economic externality**, a cost to society and the environment not paid directly by manufacturers or consumers. Economic externalities include damage to human health, fish and wildlife populations, vegetation, climate, and others. Numerous examples have been pointed out in this text.

In many instances, businesses were simply unaware of the external costs of their activities. Gradually, though, citizens throughout the world began suing polluters (if their system of government allowed it). Governments established pollution standards to protect people and the environment,

and businesses began to curb pollution, usually by installing pollution control devices. Reducing pollution reduces external costs, but pollution controls cost money. Thus, they became part of the cost of doing business—in other words, they became internalized. Such costs are passed on to the consumer.

Chapters 18–23 described U.S. pollution laws, which have forced significant controls on pollution and subsequent cost internalization. Among them are the Clean Air Act, the Clean Water Act, the Resource Conservation and Recovery Act, and the Surface Mine Control and Reclamation Act. Although these laws are important, they have not resulted in a full internalization of costs. In fact, despite billions of dollars spent on controlling pollution, environmental damage (externalities) still occurs at unsustainable levels.

The reason environmental destruction continues is largely one of policy. As explained in Chapter 2, most government policy designed to control pollution has relied on end-of-pipe controls. In most cases, such controls reduce only the output of pollution and lessen problems or slow their rate of development. To a large extent, economics determines the level of control—that is, just how much money we are willing to spend on controlling pollution.

KEY CONCEPTS

Pollution creates outside costs called *externalities*, which are borne by society at large, not by producers. Pollution control technologies reduce the emission of harmful substances and thus reduce external costs, but they add to the cost of producing goods and services.

Cost–Benefit Analysis and Pollution Control

As you may recall from discussions of risk assessment in Chapter 18, the chief goal of pollution control is to reduce pollution in ways that yield the maximum benefit at the lowest cost. This goal is made possible by cost–benefit analysis.

To understand how this process operates, take a look at FIGURE 25-4. FIGURE 25-4a illustrates an intuitive concept: the higher the level of pollution, the higher the cost of damage. Thus, the more acid deposition, the more damage one can expect.

FIGURE 25-4b shows the relationship between control costs and the amount of pollution that can be removed from a smokestack. This graph indicates that a little bit of effort can remove large quantities of pollutants cheaply. As more and more pollutants are removed, however, the cost to remove each additional increment rises. In the initial phase, for example, removing 100 units of pollution may cost only $1— or 1 cent per unit. Later on, the cost of removing pollution increases rapidly—for example, to $1 per unit of pollution. The phenomenon is called the **law of diminishing returns**. It states that for each additional dollar invested, we get a smaller return.

If both graphs are placed together, the point where the two lines intersect is the *break-even point*—theoretically, the level at which costs and benefits are equal (FIGURE 25-4c). If

FIGURE 25-3 Disregard for pollution. In the past, factories such as this one belched out huge amounts of smoke, which was considered a sign of progress. Little thought was given to the damage it created and the subsequent costs.

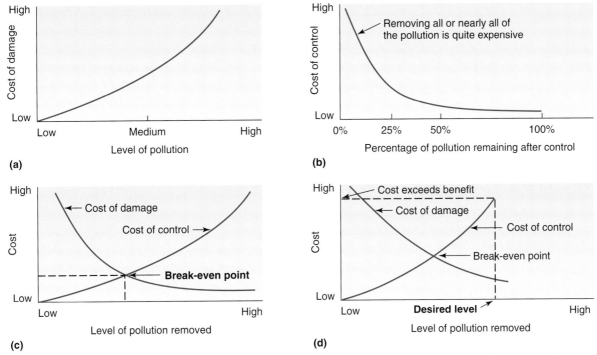

FIGURE 25-4 Cost of damage vs. cost of control. Optimizing pollution control efforts requires balancing the cost of cleanup against the economic benefits of control. **(a)** This graph shows that as pollution levels rise, the amount of damage increases, as reflected in rising costs. **(b)** Removing pollution requires an economic expenditure when pollution control devices are used. Removing 100% of the pollution is quite expensive. Removing 75% costs less per unit removed. This graph shows the relationship between cost of control and percentage of pollution remaining. **(c)** This graph shows the point at which pollution control costs equal economic damage. This is the break-even point. Unfortunately, determining the full costs of pollution is not easy, and estimates may be in serious error. **(d)** Society may wish to lower pollution levels past the break-even point. The desired level of pollution may technically cost society more than it benefits in reducing damage.

a society wants even lower pollution levels, it will have to pay more (**FIGURE 25-4D**).

Although this sounds very easy, it isn't. As Chapter 18 noted, determining the actual cost of pollution is very difficult. One of the first problems in this exercise is determining the amount of damage. For example, how many people will die from air pollution-induced cancer? How many fish will be poisoned? The second problem comes in assigning a value to lost lives, lost wilderness, polluted air, extinct species, or obstructed views. How much is a person's life worth? How much are the many free services (oxygen production and carbon dioxide trapping, for example) provided by nature really worth?

One way around the dilemma of the unpriceable good is to calculate mitigation costs, the cost of offsetting damage. How much would it cost to restore an eroded statue? How much would it cost to move an endangered species to a new habitat? These costs can be surprisingly high. Engineers estimated that the cost of destroying nearly 3400 hectares (8500 acres) of wetlands in the Charles River Basin (near Boston, Massachusetts) was over $17 million a year. The Army Corps of Engineers instead opted to protect the wetlands rather than build expensive flood control facilities along the river. The Minnesota Department of Natural Resources estimates that it costs the public $300 in flood control measures to replace 0.4 hectare (1 acre) of wetland that holds 30 cm (12 inches) of water.

The cost of replacing 2000 hectares (5000 acres) would be $1.5 million, which exceeds the state's annual spending for flood control. Another means is to calculate maintenance costs—that is, how much it costs to maintain and protect natural services.

Reducing all life-forms to dollars and cents bothers many people. It strikes many as morally wrong to consider sacrificing people's lives so that society can have its endless supply of disposable pens, diapers, and razors. Others consider it a cruel form of business logic. How can people or species such as the spotted owl be sacrificed for profit? Clearly, it is not an easy issue.

Because of the difficulties in assessing damage and the elusive nature of determining value, modern society is bound to find itself torn between two factions. On the one side are those who would allow some acceptable damage as a trade-off for the benefits of modern technology. On the other side are those who argue for eliminating harmful pollutants altogether. Torn between polar opposites, we muddle along, using the best scientific information available on environmental and health damage while battling one another in courts and legislatures over the proper levels of control. Interestingly, all this conflict could be greatly reduced by applying sustainable measures: implementing conservation, practicing pollution prevention, recycling, and turning to renewable energy technologies. These alternative solutions eliminate many adverse impacts because they eliminate the source of the problem—and they often do it at a fraction of the cost of the traditional end-of-pipe controls.

KEY CONCEPTS

Cost–benefit analysis is a balancing act that allows one to determine how much should be spent to reduce pollution to a level at which the costs of control equal the benefits (reduced externalities). Sustainable approaches have the advantage over end-of-pipe solutions in that they can reduce pollution emissions at a far lower cost.

Who Should Pay for Pollution Control?

In today's world, pollution control devices remain the dominant means of dealing with pollution. One of the most frequent economic questions is, "Who should pay for pollution control?" Should corporations be required to pay for scrubbers and other devices to clean up the environment? In such cases the costs will ultimately be passed on to the consumers of industry's products. This strategy is known as the **consumer-pays option**. Another option, the **taxpayer-pays option**, places the financial responsibility on taxpayers. Government payments to coal miners who suffer from black lung disease are an example of this option. Government programs to lime lakes to neutralize acids are another. Some government programs, such as those from Superfund, were supported by taxes on business, but those taxes were paid by consumers in the form of higher prices for goods.

Individuals who favor the consumer-pays option argue that the people who use the products that create pollution should bear the cost. The more one buys (and pollutes and depletes), the more one should pay. Frugal individuals should not have to subsidize the cost of cleanup through taxes. Passing the cost directly to the consumer, they add, could create more frugal buying habits—so essential to our long-term sustainability.

Those who support the taxpayer-pays option argue that through their elected officials, voters (and hence taxpayers) have allowed industry to pollute with impunity for years. Today, new standards are imposed on industry that place costly burdens on companies that have been operating lawfully for long periods. A good example of this is the Superfund Act, described in Chapter 23. As pointed out in that chapter, some think that when society changes its rules, society (citizen taxpayers) ought to be required to pay for controls. Advocates of the taxpayer-pays option also argue that society has elected and continues to elect officials who make deals with polluters to entice them and their polluting businesses into the community. In other cases, elected officials have overlooked flagrant violations of environmental laws. If society is responsible for elected officials who permitted pollution and other forms of environmental destruction, then the voting public (taxpayers) must bear at least some of the cost.

As in most controversies, there is validity to both arguments. In the case of old industries suddenly confronted with new laws, taxpayers might bear the economic burden of new controls. However, in new industries, pollution control costs should probably be borne by the corporation and consumers.

KEY CONCEPTS

Pollution controls and other environmental protection strategies are paid for by consumers if the costs are borne by businesses, or by taxpayers if the costs are shouldered by government.

Does Pollution Control Always Cost Money?

The previous discussion may leave the impression that reducing pollution always costs inordinate sums of money. Far from it. In some cases, pollution control devices can be installed to capture useful products that might otherwise be dispersed into the air. These materials can be sold or reused, thus generating considerable profits. Pollution prevention strategies, discussed in Chapter 23, can be even more profitable. Redesigning chemical and industrial processes, for example, can sharply reduce energy and material demand and eliminate hazardous waste disposal. These reductions often save companies considerable sums of money. In other words, companies can profit from such changes.

Nowhere is this case for pollution prevention more obvious than in the 1989 *Exxon Valdez* oil spill in Alaska, where a little prevention could have saved billions of dollars (**FIGURE 25-5**). Accidents such as this have three costs: direct costs, indirect costs, and repercussion costs. **Direct costs** from an oil spill are those that oil companies incur in the weeks and months following an accident. These include the costs of lost oil, cleanup, waste disposal, and ship repairs. The oil lost from the *Valdez* was worth an estimated $4.8 million. Cleanup costs were $2.5 billion the first year. The company also agreed to pay an additional $1.25 billion in fines and penalties.

Indirect costs are those costs that are incurred by state and federal agencies. Oil companies are usually required to

FIGURE 25-5 Alaskan oil spill. This tragedy substantially boosted the GNP of Alaska, but such expenditures are not a true measure of economic welfare.

reimburse the government for some or all of these costs. Another indirect cost results from damage to the local economy: reduction in tourism, fishing revenues, and so on. Yet another indirect cost is damage to wildlife. The Clean Water Act holds companies liable for the cost of damage to natural resources from oil spills. The law permits federal agencies to collect money for lost sea otters, waterfowl, and eagles. All told, indirect costs in the *Exxon Valdez* case could come to several hundred million dollars.

Repercussion costs, the image problems arising from the spill, cause people to boycott the company or reduce their patronage. Adverse publicity may also result in more costly restrictions on oil tankers, and it has certainly damaged the oil companies' plans to explore for oil in the Arctic National Wildlife Refuge (at least for the time being).

Companies invariably balance the costs of control and prevention against the possible costs of an accident. If they don't incorporate the full costs, however, cost–benefit analyses are likely to be flawed and pollution may end up costing them hundreds of millions of dollars.

> **KEY CONCEPTS**
>
> Reducing or eliminating pollution can be a profitable venture that adds to the bottom line of companies. Pollution prevention and other techniques often save companies considerable sums of money, especially if all costs are calculated.

25.3 The Economics of Resource Management

The previous section looked at the economics of pollution control as it is typically practiced. This section examines the economics of resource management—for example, the management of forests and soils—offering additional insights into the problems of modern society as well as some new solutions. Many decisions about natural resources are influenced by basic economic considerations, among them a factor called *time preference*.

> **KEY CONCEPTS**
>
> The management of natural resources is affected by numerous economic factors.

Time Preference

Time preference is a measure of one's willingness to postpone some current income for greater returns in the future. For example, suppose your parents offer you $100 today or promise to invest it so it is worth $200 by the end of the year. If you are short on cash and need to pay for school books, you may take the money now. Your decision to accept the money now is based on your current needs, which in this case outweigh the benefits of waiting the year, even though you would be $100 ahead. Economists would say that your need for current income outweighs greater returns in the future.

Time preference is also influenced by uncertainty. In the previous example, how certain are you that your parents will be able to invest the money successfully so you get the $200? Another factor affecting time preference is the rate of return. The higher the rate of return, the more likely it is that you will wait for the income. For instance, if your parents said they had a sure thing they could invest in and that your money would grow to $1000 a year from now, you'd probably wait. You could borrow $100 from a friend, if you needed the money, at 10% interest and pay her back at the end of the year with the $1000 you earned—and you would have lots of money left over.

Inflation also affects time preference. In times of inflation, people are apt to invest now to avoid higher costs later. But inflation can also drive interest rates up, making savings more appealing.

Time preference applies equally well to the ways in which we manage many of our natural resources such as water, farmland, and forests. Take agriculture as an example. Farmers have two basic choices when it comes to managing their land. They can choose a depletion strategy, which permits them to acquire an immediate high rate of return for a short period. This might involve the use of artificial fertilizers, herbicides, and pesticides to maximize production (Chapter 10). Soil erosion control and other techniques might be ignored. Alternatively, farmers may choose a conservation strategy, which includes techniques to conserve topsoil and maintain soil fertility. These actions require immediate monetary investments. They may cost the farmers a little more in the short run and cut into immediate profits. When one takes into account the possible loss of future income from soil erosion, the conservation strategy may make the most sense.

The choice of strategy in this example depends on the time preference. Will farmers choose the cheapest method of production, which gives the highest profit in the short term, or will they choose a slightly more expensive route, forgoing immediate high profits in favor of sustainable profits in the future? The economic needs of farmers determine, to a large extent, their time preference. For example, a young farmer looking forward to a productive career may opt for the conservation strategy. His immediate needs may be small. He may have no family and few debts. He can sacrifice income now for larger returns in the long run. However, an established farmer may have a family to support and excessive debt. He may, therefore, maximize his short-term profits through a depletion strategy.

Farmers' willingness to give up the potentially higher later income resulting from conservation may also result from uncertainty about future prices, the long-range prospects for farming, and interest rates. If the price of corn is high this year but might drop significantly in coming years, farmers may choose to make their money now. If the bottom falls out of the market in the next few years, they will have made the most of this short-term opportunity. If interest rates are likely to rise, short-term profit making may be the preferable choice. High interest rates on land and machinery that farmers purchase tend to encourage the depletion strategy.

Resource management is influenced by time preference, one's temporal preference for earnings. Time preference is affected by current needs, uncertainty, and inflation.

Opportunity Cost

Another factor that greatly influences economic decisions regarding resource management is the **opportunity cost**—the cost of lost opportunities. For instance, the conservation strategy requires a monetary investment. The money put into conservation could have been invested in the stock market or a new business venture, possibly yielding more profit with less work than the conservation strategy. As a result, when opportunity costs are high, farmers are likely to choose options other than conservation.

Opportunity costs are also incurred when resources are wantonly destroyed. Chapter 11 (on wildlife extinction), for example, describes the economic benefits of medicines derived from plants from the tropical rain forest. Losing them through disregard creates a significant opportunity cost— a loss of profit and potentially life-saving drugs. As another example, many of the world's ocean fisheries have been badly depleted. Salmon runs have also been ruined by dams and water diversion projects, pollution, and outright habitat destruction. The economic loss to commercial fishing interests and the lost recreational opportunities are enormous. These losses suggest that a broader view of opportunities should be considered during resource management decisions and when making new laws and regulations.

Money can be put to many uses. Many people choose options that provide the highest returns. That may not include investment in wise resource management.

Discount Rates

Economists rely on a decision-making tool that reflects one's time preference and opportunity costs. Known as the **discount rate**, it allows investors and economists to determine the present economic value of different profit-making options. Under this complicated practice, immediate profit becomes the main concern. Thus, it is perfectly rational to liquidate a forest or a fishery to achieve the maximum profit. If the interest rate of one's money in a bank is growing faster than a natural resource (which is almost always the case), it makes perfect sense *from an economic standpoint* to liquidate the resource as quickly as possible and put the money in the bank. If the natural resource is a forest, this would lead one to cut down the trees at once and invest the money in ways that yield higher returns.

The World Bank, which lends about $20 billion a year for projects in the less developed countries, uses a discount rate of 10% as its measure of investment wisdom. That is, if a project yields a 10% return on investment, it is deemed suitable. Clearly, a forest growing at a rate of 2 to 3% per

year hasn't much of a chance. Thus, it will be cut, and the money will be put into other money-making ventures.

Discount rates are used to calculate the present value of different options in resource management. Decisions based on the discount rate tend to emphasize immediate returns, which results in the liquidation of natural resources rather than their sustainable harvest.

Ethics

To many people, money is a key driving force in society. You can't convince them that other people will act out of a sense of duty to future generations or other species. Others see this view as hopelessly pessimistic. They point out that many noneconomic factors influence economic decisions. One of the key factors is ethics. In some cases, ethics can be as powerful as—or even more powerful than—profit and other economic factors.

In building a sustainable society, writers, educators, business leaders, and government leaders can play an important role in creating a long-term view that seeks to ensure the survival and well-being of all life. This view, if widely held, could help foster wiser management of the Earth's resources. Most important, it could help shift time preference and encourage us to reconsider opportunity costs and discount rates.

Not all decisions about resource management are based on economics. Ethics can play a big role in determining actions, overriding other immediate concerns such as opportunity costs.

25.4 What's Wrong with Economics: An Ecological Perspective

Herman Daly, who has led the effort to create an Earth-friendly system of economics, studied three leading economics textbooks and found that not one of them mentioned pollution, the environment, or natural resources. Some forward-thinking economists see this almost complete disregard for the environment as a fundamental flaw in their discipline. Worldwatch Institute's Sandra Postel writes, "While the environment and the economy are tightly interwoven in reality, they are almost completely divorced from one another in economic structures and institutions."

In my book *Lessons from Nature*, I outline four major flaws in economic thinking when viewed through the lens of sustainability: It is shortsighted. It is obsessed with growth. It promotes dependency, and it tends to exploit people and the environment. This criticism is not meant to be a denunciation of capitalism or a condemnation of those who are part of the economic system—which is all of us. It is intended to show how we can bring the human economy back into line with the economy of nature.

The economic system has several key flaws when viewed from an ecological perspective. Correcting these flaws can help us create a sustainable human economy.

Economic Shortsightedness

Earlier in this chapter you learned that the law of supply and demand governs modern economic transactions and profoundly influences our thinking. Proponents of sustainable economics, however, point out that this crucial law of economics is shortsighted, for it fails to take into account the finite supplies of many natural resources: oil, natural gas, and minerals. As a rule, supply and demand economics focuses on immediate supplies and is blind to long-term stocks and long-term demand. Current prices reflect short-term supplies. "Why worry about running out?" ask supply and demand proponents. As a resource is depleted, rising prices will stimulate exploration and more discovery, thus opening up new supplies. Falling supplies may also force us to find substitutes, permitting society to continue on the endless treadmill of production and consumption.

At some point, nonrenewable resources become economically depleted—that is, they fall into such short supply that they are no longer affordable. They're too expensive to extract. No one would pay the price. If substitutes are not available—and there appear to be a number of important minerals for which there are no substitutes—hard times are likely. Long before that point, though, the rising prices of declining resources could stimulate crippling global inflation.

Soil, the ozone layer, and the current climate are vital Earth assets. Exploited and abused, these resources have no substitutes. Their exploitation is clearly part of a short-term economic thinking. Only now are we finding that these supplies are limited and that previous indicators of their abundance were misleading.

Supply and demand economics as it is practiced today is a serious impediment to sustainability. To overcome it, supply and demand economics must be adjusted to reflect ecological realities. Key economic players, among them professors of business and economics, can assist by pointing out the limitations of supply and demand theory. Such activities could temper our lust for growth and stimulate efforts to recycle and use resources more efficiently.

Changes are also needed in public policy to help adjust economic activity to honor limits. The basic goal of these policy measures would be to adjust current prices to reflect long-term supplies. Several market tools are available, including a range of incentives and disincentives. One of the most important adjustment tools is **user fees**, or **green taxes**. User fees are taxes on raw materials, paid by producers and ultimately passed on to consumers. One example is the severance tax charged to coal companies on each ton of coal they mine. Chapter 18 described carbon taxes currently used in several European countries; they are another user fee.

User fees artificially increase the cost of raw materials and finished products. This, in turn, helps promote conservation and raises revenues that can be used to develop alternative supplies. For instance, money raised by a tax on coal could be used to promote energy conservation or renewable energy such as solar. Thus, a tax can help ensure future generations that they too will have access to resources needed for a healthy, productive life. User fees help us honor the rights of future generations. Although new taxes are not popular, the impact of user fees or green taxes can be offset by lowering income taxes.

Another means of instilling vision in the economic system is to revamp cost–benefit analysis, described earlier in the chapter. Efforts are needed to identify all environmental, social, and human health impacts—present and future—and to quantify them. As best they can, companies should ensure that all costs are incorporated into the price of their product or service. This is known as **full-cost pricing**. By incorporating the external costs into the cost of a good or service, we can narrow the gap between the market price and the real cost. Spotlight on Sustainable Development 20-1 showed how Applied Energy Services offsets carbon dioxide pollution from their coal-fired power plants and is an example of full-cost pricing in action.

Full-cost pricing is an ideal to work toward. It could stimulate economic change. For instance, it might compel companies to find ways to prevent problems in the first place, through sustainable practices such as pollution prevention, energy efficiency, recycling, and renewable energy use. To make it happen, government agencies such as the EPA and the Department of Energy could widen their research efforts to determine the full costs of various economic activities, thus helping businesses adjust their cost–benefit analyses. Business and economics professors could train future business leaders in the practice of full-cost pricing.

Governments can also play a role by requiring least-cost policies. As Chapter 15 notes, numerous states now require utilities to choose the least costly ways of producing new energy. To reflect the cost of various strategies more accurately, some states require utilities to add 15% to the cost of conventional strategies such as coal, oil, and nuclear power. This helps account for economic externalities and usually makes environmentally sustainable strategies more competitive.

Economic systems and the participants in them often fail to take into account long-term supplies, a dangerous trend that results in an underpricing of many natural resources and that leads to environmentally unsustainable activities. Several mechanisms are available to incorporate such concerns into economic decision making, including user fees.

Economics and Growth

In the 1967 edition of his popular economics textbook, Nobel prize-winning MIT economist Paul Samuelson defined economics as "the science of growth." Subsequent editions have dropped the wording, but the bias remains. In his 1984 edition, for example, Samuelson wrote, "Today, the ultimate measure of economic success is a country's ability to generate a high level of and rapid growth in the output of economic goods and services. Greater output of food and clothing, cars and education, radios and concerts—what else is an economy for if not to produce an appropriate mix of these in high quantity and fidelity?" Although some economists do not subscribe to this view, many do—especially those in business and government.

Our focus on growth is rooted in the frontier notion that "there is always more." For many decades, human civilization has been caught up in it and has been willing to pay almost any price for it.

Economic growth is based on an increasing consumption of goods and services. Such an increase generally arises from (1) an increase in population size and (2) an increase in the amount each of us buys—per capita consumption. Because it means greater production and, presumably, greater economic wealth, population growth has traditionally been viewed as an asset to society, with each new baby viewed as a potential new consumer.

The dangerous aspect of our nearly singular focus on growth is that it tends to equate economic growth with human progress. The faster an economy is growing, many assume, the better off people are. Thus, a rising GNP is taken to mean that a country's health and people's lives are improving. Over the years, so many people have bought into this logic that economic growth has become a way of life. In fact, it is no exaggeration to say that economic growth has become the abiding principle of economics and business and the central focus of political campaigns and government policy. With the growth-is-essential philosophy so deeply imbedded in our society, many people are blind to the outcome of continual growth. Few people today recognize that continuous economic growth is ultimately incompatible with the economy of nature, where very real limits exist.

KEY CONCEPTS

Continual economic growth is the abiding principle of many economies and the measure of success, but it is ultimately incompatible with the finite world in which we live. Economic growth is fueled by population growth and ever-increasing per capita consumption.

Growth and the GNP: Some Fundamental Flaws

Economic growth is measured by tracking the GNP of nations, but the GNP is believed by many observers to be a flawed index of a nation's welfare. Why? The GNP includes many goods and services that make no contribution to the welfare of the people. More to the point, the GNP fails to differentiate between "good" and "bad" expenditures. For example, the GNP includes all expenditures on homes, books, concerts, and food—deemed good because they improve the standard of living. However, it also includes expenditures on oil spill cleanup, cancer treatment, and air pollution cleanup. It includes hospital bills from drive-by shootings and the ever-increasing cost of insurance to protect our homes from theft. Therefore, a country with filthy air and polluted water faced with rising cancer rates and violent crime will register a high GNP.

By carefully defining terms, as is required by critical thinking, we can see the fallacy of overdependence on GNP

as a measure of success. A telling example is the state of Alaska, whose economic output increased by $1 billion in 1989 because of the *Valdez* oil spill.

Another problem with the GNP is that it fails to account for the destruction of natural assets. Suppose that lumber companies in a country cut down the nation's forests, in order to increase exports, and failed to replant them. If one studied the GNP of the country during the cutting phase, the economy would look good. If the destruction of the forests were taken into account, however, the GNP is meaningless. It fails to show the dramatic decline in assets. Unfortunately, many countries the world over are depleting their natural resources. Bolivia, Colombia, Ethiopia, Ghana, Indonesia, and Kenya, for example, depend on primary resources (such as minerals, timber, and crops) for 75% or more of their exports, but they are rapidly depleting them. Their GNPs hide evidence of the decline. Soon there will be no assets and no source of income.

Yet another largely unrecognized problem with the GNP is that it is blind to accumulated wealth. Modern economies today compare one to another on the basis of their annual growth in GNP. This criterion, however, fails to take into account the accumulated wealth of a country—how much people already have. A newly developed nation may have a 5% annual growth in its GNP compared with a 2% growth rate in a developed country such as the United States. Does this mean the newly developed nation is doing better? Not necessarily. It only means that its economy is expanding more rapidly. Most likely, it has more room to expand, but its people are far less wealthy than those of the United States. Keeping in mind their accumulated wealth may help nations temper their seemingly insatiable obsession with growth.

Finally, GNP also pays no attention to the distribution of wealth in a society. Many projects that improve the GNP of a nation fail to improve the lives of the people they are designed to help. A select few benefit, and in some cases, the masses may actually be harmed. (Some people like to call this the *trickle-up theory*.) In developed countries, the GNP hides frightening inequities. In the United States, for example—a country deemed rich by GNP standards—one fifth of all children live in poverty, and nearly 40 million Americans have no health insurance. Looking at the GNP, one would be compelled to think that all was well within U.S. borders.

Author Wendell Berry calls the GNP the "fever chart of our consumption." It is much more than that. It is an indiscriminate measure of our consumption, our waste, and our disregard for environmental conditions. It has become a measure of our unsustainability. Clearly, something else is needed to measure progress.

KEY CONCEPTS

Continual economic growth is the abiding principle of many economies and the GNP is our measure of success. However, the GNP is an inaccurate measure of the welfare of a nation's people because it fails to distinguish economic activity that enhances our welfare from that which results in a decreased quality of life. It also fails to take into account resource depletion, accumulated wealth, and the distribution of wealth.

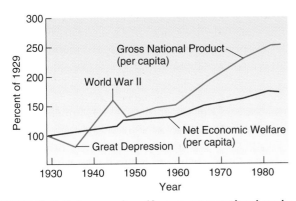

FIGURE 25-6 Net economic welfare vs. gross national product. The GNP is the market value of all goods and services produced by a country. U.S. per capita GNP has risen continuously for many years. The NEW, or net economic welfare, is a measure of beneficial goods and services. It is derived by subtracting the negative aspects of the GNP, which do nothing to improve the quality of life, such as damage from air pollution. As shown here, the per capita NEW is lower than the GNP and rises at a slower rate, suggesting diminishing returns from economic growth.

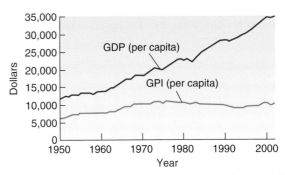

FIGURE 25-7 Alternative indicators of progress. This graph is part of a set of indicators that measure the well-being of the United States. Note that while GNP continues to rise, GPI falls.

Alternative Indicators of Progress One of the most exciting developments of recent years is the push to develop alternative measures of progress. The Yale economists William Nordhaus and James Tobin, for example, devised a measure that adjusts the GNP to make it a more accurate representation of the good that people receive from their nation's economic growth. This measure, called the **net economic welfare (NEW)**, subtracts the *disamenities* of an economy—the cost of pollution, the cost of medical care for victims of lung disease caused by urban air pollution, and other costs made necessary by economic growth—from the GNP. The NEW also *adds* the cost of certain activities, such as household services (cleaning and cooking) provided by men and women, that are not part of the traditional GNP calculations but that improve well-being.

FIGURE 25-6 shows the relationship between NEW and GNP. As you might expect, NEW is lower than the GNP. That is to be expected. Moreover, NEW is growing more slowly than GNP. In other words, as the nation's output grows, the economic benefits of growth fall behind, largely because of rising pollution and environmental destruction. We are spending more money but getting less for it.

Economist Herman Daly proposed an alternative measure, the **Index of Sustainable Economic Welfare (ISEW)**, which is now called the **Genuine Progress Indicator** or **GPI**. It includes factors such as the cost of air and water pollution, cropland and wetland losses, and other forms of environmental decay. It also entails the costs of car accidents and a host of other factors that affect human welfare. As in the case of the GNP and NEW, GPI and GNP do not compare favorably over the past 20 years. In fact, in the 1970s the per capita GNP increased about 2% per year and about 1.8% per year in the 1980s (Figure 25-6). In contrast, the per capita GPI increased only about 0.7% in the 1970s and declined 0.8% per year in the 1980s, a decrease largely attributed to the rapid deterioration of the environment (**FIGURE 25-7**).

New measures of economic success such as the GPI and the NEW are aggregate measures. This means that they are single numbers that indicate the economic output of a nation, taking into account all positive and negative economic factors discussed in the previous section. These measures are needed to help create a more accurate picture of economic performance.

Fortunately, many efforts are under way to recalculate GNPs to take into account the depletion and deterioration of natural resources. At this writing, Australia, Canada, France, the Netherlands, and Norway have initiated programs to inventory their natural resources to determine the loss of natural capital. Germany and the United States have launched programs to calculate alternative GNPs that take into account environmental damage. The United Nations Statistical Commission developed guidelines for nations that want to calculate alternative GNPs. However, it still views the GNP as an acceptable measure. Because this group will not revisit the issue for 20 years, critics think a stronger stand is needed.

Essential as these efforts are, it is important to note that aggregate measures such as the GPI hide important trends needed to assess a nation's progress. To get an even more accurate picture, over 75 U.S. cities, towns, and states are developing or have developed alternative measures of success—often called **benchmarks** or **indicators of sustainability**. These include data on a variety of social, economic, and environmental trends that allow communities to track crime, energy consumption, pollution, and a host of other factors. Indicators of sustainability tell communities where they are and where they're going. They provide valuable information needed to develop policies that address the most critical problems in an era of limited revenues. Trends data also show how effective policies are at solving problems.

In summary, while single measures of the economic, social, and environmental health of a nation may be helpful, they could hide important distinctions within countries. What's needed are new national report cards based on factors such as health, literacy, environmental conditions, resource supplies, and income distribution. Pertinent measures of sustainability could also be included, among them efficiency of resource use, reliance on renewable energy, level of recycling, and expenditures on restoration. Population control measures and public policy supporting sustainable goals could also be included.

Making the Economic System Work Better   Some economists believe that the continued divergence of the GNP and measures of economic welfare and sustainability such as the GPI may be inevitable as the world becomes more congested and more dependent on fossil fuels, large-scale technologies, synthetic chemicals, and disposable goods.

Advocates of a sustainable future argue that nations should strive to reduce the difference between the GNP and GPI, that is, we should make our economic systems work *for* us, not to our detriment. For every gain in economic output, we should receive a similar social benefit, not a shot of pollution that needs to be cleaned up. How can this be done?

One way of maximizing the social benefit of economic activity is to reduce the pollution, waste, and environmental destruction per dollar of GNP. This can be accomplished by applying the operating principles of sustainability, discussed in previous chapters. By becoming more efficient in our use of resources; by recycling and reusing all materials; and by converting to clean, renewable energy supplies, we can reshape our economic system so that economic activity translates into measurable improvements in our lives and the ecosystems upon which we depend.

Corporations, small businesses, and even individuals can promote these activities by adopting new approaches. As many previous examples in this book have shown, such changes can result in substantial economic savings. Governments can also assist by providing a range of incentives and disincentives, many of which have been discussed in previous chapters.

Rethinking Growth: Focusing on Development   Many advocates of sustainable development assert that our economy as presently configured is much like a cancer, eating away at its host, the Earth. They remind us that sooner or later all untreated cancers kill their hosts. Put another way, continual economic growth in a finite world results in environmental degradation and resource depletion. Although some advocates of sustainability think we can grow continually and sustainably, others disagree.

Critics of economic growth argue that the future of industrial nations depends more on development than on continual economic growth. Continued economic growth requires a continual extraction of resources and ever-increasing production and consumption, which many observers believe cannot be sustained.

As Michael Kinsley of the Rocky Mountain Institute notes, "Development is very different from growth. After reaching physical maturity, we humans can continue to develop in many beneficial and interesting new ways—learning new skills, gaining deeper wisdom and much more." Likewise, a nation can develop without increasing (growing) its material consumption. It can still create jobs and expand cultural and educational opportunities without expanding its demand for resources.

In the more developed nations, then, sustainable economic development must strive to create a higher quality of life without depleting resources or causing environmental impacts that undercut future generations and other species. Many of the strategies outlined in this chapter will help us reach this goal. Especially important are efforts to use resources more efficiently and to reduce our demand for unnecessary goods and services. Using renewable resources and recycling also meet our needs, at a fraction of the environmental costs of traditional strategies.

In the less developed nations, development and limited growth within ecological limits are probably required. Why is growth desirable? Limited economic growth is essential to raise the standard of living of the poor. Such efforts are needed to improve the lives of the world's people and will help put an end to the cycle of poverty and environmental decay, say proponents of this viewpoint. Strategies for sustainable development in the LDCs are detailed in Chapter 26.

Fostering Local and Regional Self-Reliance

The economic system of much of the world also tends to foster dependency among individuals, regions, and nations through trade. Economic interdependence, while desirable in many respects, creates very real problems from an environmental perspective. First and foremost, it tends to separate producers of goods and services from consumers. As a result, consumers are often blind to the source of human wealth, the Earth, and oblivious to the environmental costs of satisfying their needs and desires.

Trade and interdependence allow human populations to flourish beyond the local and regional carrying capacity. Water imported from the Colorado River to the cities and towns of southern Arizona, for example, has resulted in explosive population growth. Food imports allow populations in regions with poor agricultural potential to flourish. Without infusions of fuel, food, and water, many regions could not support large human settlements (FIGURE 25-8). If resource supplies dwindle, many of these regions could suffer extreme economic and social hardship. In the U.S. West, for instance, studies of global warming suggest that water flow in the Colorado River, which currently serves an estimated 20 million people, could fall by 30%, with devastating effects on this population. This could cripple agriculture in arid southern California.

Despite these problems, business economists usually refer to the global marketplace (and interdependence in general) as a highly desirable goal. From an ecological viewpoint, however, the global marketplace and the rising interdependence that accompanies its development could be a dangerous bargain.

Some leading thinkers on the subject argue that from a sustainable vantage point, the future of human society lies not in globalization, but rather in local or regional self-reliance—that is, human communities living within the means of a bioregion. Food, water, energy, and other resources would come from local sources, not be imported from sources hundreds, even thousands of miles away. Greater regional and local self-reliance does not eliminate all imports and exports; it seeks *greater* self-sufficiency and reliance on resources that are immediately available—for example, sunlight and wind energy instead of imported oil.

The transition to a sustainable society may ultimately require increasing regional self-reliance. For the United States, regional self-reliance means tapping into local energy supplies, preferably clean and renewable ones. For states, it may mean developing diverse economies that produce many of the goods and services needed by people on a day-to-day basis. In fact, many products could be manufactured locally from recycled materials gleaned from our wastes. The diversification of regional economies could provide an added benefit: It could help make communities and states more recession proof.

Greater local self-reliance, although in opposition to the dominant view of economic success, is beginning to emerge throughout the world. In Brazil, for instance, a program to produce ethanol for cars and trucks was created to reduce the nation's dependence on foreign oil. California's public utilities have developed an energy strategy that relies principally on energy efficiency and local supplies of renewable energy, such as wind and geothermal. Many states have programs to promote the purchase of goods produced within the state. Efforts to promote U.S. or state-produced goods are another sign of emerging self-reliance.

FIGURE 25-8 **The price of interdependence.** The city of Las Vegas, Nevada, is surrounded by desert. If its energy or water were cut off, the city could not exist.

Creating an Ecologically Compatible Society

Many economic practices widen the gap between the rich and the poor, the powerful and the weak. Income statistics support this assertion. In 1980, for instance, the average salaries of the CEOs of the 300 largest companies in the United States were 20 times greater than the average salaries of manufacturing employees; by 1990, the CEOs were earning 93 times more than the average manufacturing employee. As further proof of the widening gap, from 1980 to 1990, the income of the top 1% of the U.S. population (those earning over $250,000 a year) doubled, whereas middle income increased only slightly. These trends continued through the 1990s. Social scientists view this as a socially disruptive phenomenon.

Economic exploitation is part of the reason why less developed nations remain so poor. As in colonial times, many wealthy nations continue to reap the benefits of natural resources in the less developed countries, but they often pay a fraction of their real value. Gold mined in several African countries, for example, provides little economic benefit to miners and their local economies, whereas the middlepersons who purchase and sell it to others reap huge profits. As pointed out in Chapter 16, poverty caused in part by this form of exploitation is a key element in the complex equation of rapid population growth and environmental decay in many less developed countries.

The human economy also exploits the environment. In the 1800s, for example, the U.S. timber industry cut down most of the white pine trees from New England to Minnesota. When the last of the forests were depleted, they moved to the Southeast and the Northwest to continue this unsustainable harvest. Widespread overharvesting of many whale species is another example of economic exploitation. So is commercial fishing, which (as pointed out in Chapter 10) has depleted at least two dozen ocean fisheries in the North Atlantic in the past 40 years and is threatening the majority of the remaining ones.

Making economies less exploitive of nature and of people is an enormous challenge. Virtually all of the practices outlined in this book to promote sustainability and to help correct the flaws in our economy could greatly reduce our exploitation of nature by making society more efficient, less resource intensive, and more ecologically compatible. Some

measures will reduce our exploitation of people as well. For instance, alternative measures of progress that indicate income distribution or education and illness among different socioeconomic strata could help adjust economic activities so that they serve all people, not just a select few. Conservation can also help. Protecting forests, for instance, by reducing demand for materials that are extracted from them (timber) or under them (minerals), protects the large segment of humanity that makes the forest its home. When a rain forest is converted into a coffee plantation, most often it is the owners who reap a sizable profit, while poor rural families and indigenous peoples that depend on the forest for food and fuel suffer. (For a discussion of sustain-

25-2   Cultural Survival: Serving People and Nature Sustainably

Throughout the tropics, efforts are under way to improve the lives of people who live in the rain forests without destroying the rich diversity of their biome. One way to protect a resource is to set it off-limits to people. Such reserves, however, stand little chance of remaining intact if local people are robbed of the resources they need to survive. Recognizing that in some instances land must be used in order to be preserved, some countries have established extractive reserves (Chapter 12).

Extractive reserves are protected regions where local residents harvest the natural products of the forest, among them fruits and nuts, rubber, oils, fibers, and medicines. Because they are harvested in ways that do not harm the for-

est, extractive reserves can be productive sources of food and resources *ad infinitum*. In fact, indigenous peoples of the world have depended on a sustainable harvest of forest resources throughout history (**FIGURE 1**).

Although numerous countries have set aside huge tracts of land, the success of extractive reserves also depends on efforts to market forest products in ways that benefit those who do the work. Jason Clay, an anthropologist with a U.S.-based group known as Cultural Survival, is one of many volunteers helping to make this happen. Clay manages a rain forest marketing project. He helps rain forest inhabitants process and sell the nuts they harvest. Instead of receiving 4 cents per pound, the typical price for unprocessed nuts, processing and marketing raise the price to $1.00 per pound. Cultural Survival charges its commercial clients an additional 5%, which it uses to support the group's activities throughout the world.

One of Cultural Survival's biggest customers is a company called Rainforest Products, which currently markets two cereals in the United States, Rainforest Crisp and Rainforest Granola. Both cereals contain Brazil nuts and cashews harvested by indigenous Amazonians. Ben and Jerry's Ice Cream of Vermont once marketed a popular ice cream flavor called Rainforest Crunch, made with handpicked Brazil nuts and cashews from the Amazon.

"For the forests, it is a question of use it or lose it," says Cultural Survival's Jason Clay. "The value of the rain forest will have to be tested in the marketplace. But the point is to change the market, not the forest."

By that, Clay means that the rain forest needs to be viewed differently—as a source of sustainably harvested fruits, nuts, and other products, rather than as a source of wood and pasture land. Do such nonwood forest products have adequate economic value, though?

The World Bank argues that "the extractive reserves are *the* most promising alternative to land clearing and col-

FIGURE 1 Nonprofit organizations are finding ways to help those closest to the land reap great financial reward from the sustainable harvest of forest products.

able rain forest management, see Spotlight on Sustainable Development 25-2.)

Other changes are also needed to make the economic system fairer. In mineral-exporting countries, for instance, eliminating the middlepersons can bring more income to the people who extract the ore. (For an example see Spotlight on Sustainable Development 20-1.) Furthermore, developing factories in which people can convert raw materials (ore) into intermediate (metal) or finished products (hubcaps) can shift the wealth to those who have the raw materials to begin with. Further ideas on reducing the exploitation of people in less developed countries are presented in the next chapter.

Kenneth Boulding was one of the first to write about changes needed in the U.S. economy. In 1966, he coined the phrase *cowboy economy* to describe the present economic system, characterized by maximum production, consumption, resource use, and profit. Boulding suggested that the cowboy, or frontier, economy be replaced by a *spaceship economy*—an economic system that recognizes that the Earth, much like a spaceship, is a closed system wholly dependent on a fragile life-support system.

onization schemes" (italics added). In 1989, Charles M. Peters of the New York Botanical Gardens' Institute of Economic Botany and his associates proved this to be true. These researchers evaluated a 1-hectare (2.5-acre) plot of Amazonian forest in Peru to determine the economic value of a variety of options. An inventory of plant life turned up 842 trees with diameters greater than or equal to 10 centimeters (4 inches). The trees included specimens from 275 species. Seventy two of the species yield products that are currently marketed by natives. Sixty species produce commercial timber, 1 species produces rubber, and 11 species produce edible fruits. Peters and his associates found that the 1-hectare survey area produced fruit whose annual net worth was about $400 and rubber worth more than $22.

The 1-hectare plot also contained 93.8 cubic meters of merchantable timber, which if clear-cut would be worth about $1000. However, clear-cutting would destroy the fruit and rubber trees and would put an end to future timber production. The $1000 net income would be a one-time profit.

The researchers also examined selective cutting of marketable trees (Chapter 12), which could be compatible with fruit and rubber harvesting. They found that this option by itself yielded about $310 per year.

As pointed out in the chapter, economists often rely on the discount rate to determine the present economic value of different profit-making strategies. Net present value (NPV) is equal to the net revenue produced each year, divided by a discount rate. The equation is $NPV = V/r$, where V is the annual revenue and r is the discount rate. Peters and his colleagues used a 5% inflation-free discount rate to determine the net present value of the fruit and latex resources of all future harvests. They found that the NPV of sustainable fruit and latex harvests was $6330 per hectare, assuming that 25% of the fruit crop was left in the forest for regeneration.

The researchers used another equation to determine the NPV of a perpetual series of sustainable timber harvests. Using that equation, they found that the NPV of sustainable harvest by selective cutting was $490. The fruit, rubber, and sustainable timber harvests of the 1-hectare plot have a combined net worth of $6820. The value would increase even further if the revenues from medicinal plants, small palms, and other plants were included.

Clear-cutting has a much lower net present value because, as noted previously, the practice is not sustainable. Interestingly, even conversion of the forest to tree farms was found to have a lower net present value than sustainable harvest. Timber and pulpwood produced on a 1-hectare plantation of a commercially valuable tree called *Gmelina arborea* in Brazil was estimated at $3184, less than half that of the natural forest. Calculations also showed that a fully stocked cattle ranch in Brazil has a present value of only $2960 per hectare, even if the costs of weeding, fencing, and animal care are excluded.

This economic assessment clearly shows that the value of a standing forest is much higher than the most common alternatives. Except for a few cases, though, decisions about the forests seem to favor economically less productive and environmentally unsustainable options.

The goal of sustainable economics is to serve people and to protect, even enhance the environment, making extractive reserves an essential element of the new economy. Some advocates believe, however, that for such ventures to be truly successful, the people who do the work have to have more control in the market. By eliminating the middle persons, locals can generate more income for local collectors. If extractive reserves are successful, not only will lifestyles improve, but large tracts of rain forest and the species that live in them could be saved.

In this book, an economic system that seeks to meet human needs while protecting or even enhancing the life-support systems of the planet is called a **sustainable economy** (Table 25-1). The notion of a sustainable economy was not widely accepted when Boulding first proposed it. Today, as more and more nations face limits, experience environmental deterioration firsthand, and ponder threats such as global warming, the need to reshape economic systems is becoming clear. The Earth Summit in Rio de Janeiro is a symbol of these global realizations. How do we create a sustainable economy?

The previous sections have outlined numerous ways to make the economy more sustainable—among them user taxes, full-cost pricing, and better cost-benefit analysis. They outlined some efforts to create new measures of progress and described the need for more self-reliance. New laws and private initiatives that promote conservation (efficiency), restoration, and recycling were also discussed, as were efforts that encourage the use of renewable resources and promote population stabilization. These steps could help revamp major sectors of our society, such as agriculture, transportation, housing, and energy. No one of the changes listed in this paragraph or this book will work alone; all are needed, and each one is vital to the effort.

Because many private and government initiatives that promote these activities have been outlined in previous chapters, in this section we examine four major ideas: harnessing market forces, corporate reform, green products and green seals of approval, and appropriate technology. Each of these is an important element in the complex equation of sustainability.

KEY CONCEPTS

Numerous changes can help us forge a sustainable economy, one that meets human needs without foreclosing on future generations by destroying the natural resource base that makes all economic activity possible.

Table 25-1

Characteristics of a Sustainable Economy

Participants take a long-term systems view that

- Recognizes the importance of natural systems to human well-being.
- Ensures that economic activities improve the quality of life of all, not just a select few.
- Seeks ways to ensure that economic activities maintain or improve natural systems.
- Uses all resources extremely efficiently.
- Promotes maximum recycling and reuse.
- Relies heavily on clean, renewable technologies.
- Restores damaged ecosystems.
- Promotes regional self-reliance.
- Relies on appropriate technology.
- Designs and builds using principles of ecological sustainability.

Harnessing Market Forces to Protect the Environment

Companies often argue that, left to their own devices, they could find ways to reduce pollution at a much lower cost than governmentally mandated controls. In response to this complaint and the growing cost of regulation, former Democratic senator Tim Wirth of Colorado and a Republican colleague, the late John Heinz of Pennsylvania, assembled a multidisciplinary, bipartisan team from businesses, colleges and universities, the environmental community, and government. Their goal was to propose ways that government could tap into economic forces to reduce pollution and manage resources sustainably. Such measures could supplement—even supplant—traditional laws aimed at regulating pollution and resource use.

Their report, issued in 1988, outlined a number of "marketplace solutions," including (1) economic disincentives, (2) economic incentives, (3) tradeable and marketable permits, (4) laws that eliminate market barriers to efficient resource use, and (5) laws that remove subsidies for environmentally destructive activities.

KEY CONCEPTS

Many market mechanisms can be brought to bear on environmental problems, allowing businesses to innovate, save money, and reduce the burden of regulations.

Economic Disincentives Governments have many legal means to force companies to control pollution, reclaim damaged lands, or reduce resource consumption. In the United States and many other industrial countries, the mainstay of environmental control is a complex and costly set of rules and regulations. U.S. environmental regulations generally establish standards of conduct for a wide range of activities, from the release of pollutants from smokestacks to the restoration of surface-mined land. Some regulations stipulate how companies will meet standards—that is, what technologies they will use.

Regulations are generally backed by legal recourse, among them fines and prison sentences. Fines are a type of economic disincentive to discourage environmentally destructive activities.

The 1988 report proposed a different set of economic disincentives, among them user fees and pollution taxes, described earlier. **User fees** and **pollution taxes** impose a fee on products that waste resources, pollute the environment, or damage ecosystems. A good example is the **gas guzzler tax**, a tax on cars that get poor gas mileage, which President Clinton rescinded through executive order. Carbon taxes, described in Chapter 20, are a good example of pollution taxes.

Economic disincentives can work. The 1990 Clean Air Act, for instance, imposed a steep tax on the use of ozone-depleting chlorofluorocarbons (CFCs) that provided considerable incentive for companies to find alternatives.

Hazel Henderson, a leader in rethinking economics, calls user fees and pollution charges **green taxes** and argues that they should be applied to a wide variety of products and activities, among them disposable goods, airplane travel, international tourism, and oil consumption. Interestingly, some European nations that embrace the idea that present generations have an obligation to future generations have instituted a number of green taxes. In such cases, green taxes have been found to generate a substantial source of revenue. Henderson notes that many business executives find green taxes more acceptable than bureaucracy and regulation. Surprisingly, most economists approve of the idea. But efforts to institute a carbon tax on coal during the Clinton Administration met with stiff opposition from Western senators and was quickly killed in Congress.

Economic Incentives   In recent years, government officials have also explored a variety of economic incentives that induce companies to comply with pollution laws. Tax credits can also be effective incentives. A tax credit works this way: A government gives a company a tax credit—say 10%—for investing in recycling equipment or for buying recycled material. This lowers the cost of business and encourages responsible practices. The 10% credit for a $100,000 purchase would amount to a $10,000 savings on taxes. Tax credits can be given to individuals who invest in environmentally responsible products, for example, solar energy, wind energy, and conservation. Wisconsin and Colorado offer a 5% tax credit for companies that invest in recycling equipment.

Faced with budget shortfalls, governments are often wary of tax credits because they can lower tax revenues. Therefore, careful analysis is a must when considering offering tax credits—to avoid losing revenues, avoid investing in businesses that could prosper without support, and avoid investing in activities that could not survive even with tax credits. Creative leaders, however, can find ways to help environmentally responsible businesses get started and save the government money. For example, a city that is responsible for its trash pickup might find it advantageous to subsidize private recycling businesses. In the process, the city can save enormous amounts of money by reducing the amount of garbage it needs to landfill. This cuts landfill tipping fees, labor costs, wear and tear on trucks, and fuel consumption in vehicles that transport garbage to landfills.

The Permit System   Another incentive is the **marketable permit.** As Chapter 19 notes, a marketable permit works this way: The EPA grants a permit to Company A, allowing it to emit 10,000 tons of sulfur dioxide each year, but the company finds an inexpensive yet effective way to cut emissions to 5000 tons. It can then sell its permit for the 5000 tons to Company B. For Company B, purchasing the permit may be cheaper than installing pollution control devices. Under this system, the total emission of pollution in the region remains the same. However, permit systems are generally designed to reduce pollution levels. In this example, the EPA would simply lower permitted levels of emissions for all companies in an area. Those that can reduce pollution below their permitted value—say, through pollution prevention—can sell unused permitted emissions.

In Colorado, a tradeable permit system was used to control water pollution entering Dillon Reservoir, located in the Rocky Mountains 70 miles west of Denver. Nitrogen and phosphorus entering the reservoir from farms, sewage treatment plants, and other sources had begun to make the reservoir eutrophic. A study of various options to reduce phosphorus pollution showed that additional controls in treatment plants would cost the towns about $1800 per kilogram ($820/pound) of phosphorus removed. Controlling nonpoint pollution would cost only about $150 per kilogram ($68/pound)—over 10 times less. The legislature and the EPA approved a tradeable permit plan allowing the publicly owned sewage treatment plant to pay for nonpoint pollution controls on farmland, saving an estimated $1 million per year. Several countries now have tradeable permit system pollutants that create acid rain, notably sulfur oxides, and the greenhouse gas carbon dioxide.

Removing Market Barriers   Numerous laws and regulations create market barriers and provide subsidies for environmentally destructive activities, as discussed in chapters on mining, forestry, and solid waste. Depletion allowances, for example, provide tax breaks for fuel and mineral companies (Chapter 16). As these companies deplete their reserves, they are allowed a tax credit. The money is supposed to be used to invest in more exploration, but many companies use it to diversify—that is, to buy other companies unrelated to fuel and mineral production. Chapter 23 also described another barrier, preferential freight rates—mandated by federal regulations—that make it cheaper to haul virgin materials than scrap bound for recycling. Chapter 14 mentioned subsidies to oil, gas, coal, and nuclear industries that give them a considerable advantage over renewable fuels.

By removing the subsidies and shipping regulations, economically inefficient and environmentally unsustainable practices can be eliminated. Removing these barriers can help protect the environment, ensure sustainable practices, and reduce government spending.

Corporate Reform: Greening the Corporation

A sustainable economy depends in large part on the emergence of companies that operate sustainably. Their products will be made from recycled materials and will themselves be recyclable. They will be produced in ways that optimize energy efficiency and minimize or eliminate hazardous waste. They could be powered by renewable energy sources, and so on. Marketplace solutions such as those described earlier and the traditional command-and-control legislation can promote such operations. Corporate change may also come internally, though—from a sense of corporate environmental responsibility.

In 1989, a number of environmental groups and several managers of major pension funds in the United States, including those of California and New York, joined forces to form the Coalition for Environmentally Responsible Economies (CERES). The CERES proposed a set of guidelines for responsible corporate conduct. Called the Valdez Principles, they call on companies to do the following:

1. Protect the biosphere by reducing and eliminating pollution.
2. Promote the sustainable use of natural resources by ensuring sustainable management of land, water, and forests.
3. Reduce the production and disposal of wastes by recycling, waste minimization, and other measures; use safe disposal methods for wastes that cannot be handled otherwise.
4. Employ safe and sustainable energy sources and use energy efficiently.
5. Reduce risk to the environment and to workers.
6. Market safe products and services—those that have minimal environmental impacts—and inform consumers of the impacts of the products and services a company offers.
7. Restore previous environmental damage and provide compensation to persons who have been adversely affected by company actions.
8. Disclose accidents and hazards and protect employees who report them.
9. Employ environmental directors, with at least one member of the board of directors qualified to represent environmental interests; employ environmental managers, with a senior executive responsible for environmental affairs.
10. Assess and audit progress in implementing the Valdez Principles.

Take a moment to look back over them, and you will see that almost all of the principles of sustainability (and many other ideas discussed in this chapter) are embodied in them. Number 9 is particularly important. It instructs companies to appoint high-ranking executives to the task of environmental management. IBM, for example, appointed a vice president of environmental health and safety, with a staff of about 30 people, to ensure that corporate environmental policy is carried out. IBM has also upgraded the status of its environmental staff, creating a more powerful and autonomous group that ensures production goals are met while minimizing pollution and reducing environmental damage. Environmental management staff are responsible for **environmental auditing**, periodic checks to ensure compliance with federal laws and enactment of sustainable practices. Most agree that environmental auditors must have the autonomy of corporate financial auditors. Ciba-Geigy (a manufacturer of drugs, pesticides, and other products) recently placed all environmental auditors into an independent group that reports directly to the chief executive officer (CEO).

The success of corporate programs depends in large part on leadership from CEOs or presidents of corporations who look favorably upon corporate environmental policy. Thus, when DuPont's chairman Edgar S. Woolard, Jr., proclaimed a new policy of "corporate environmentalism," the vice president of safety, health, and environmental affairs (who had faced stiff resistance from plant managers) noticed a sudden change in interest from employees.

Although few businesses and governments adopted the Valdez Principles, many have begun to implement the principles and practices of a **natural capitalism**, described in Spotlight on Sustainable Development 20-3. Natural capitalism incorporates some of the key principles of sustainable economics described in this book and has been popularized by Amory and Hunter Lovins and Paul Hawkens in a book called *Natural Capitalism*. Individuals can help influence the direction of business by investing in socially and environmentally responsible companies and mutual funds. The New Alternatives Fund, for example, is a mutual fund that invests in companies involved in many environmentally sustainable activities, among them renewable energy, cogeneration, insulation, efficient light bulbs, and other forms of energy conservation. Offering a respectable rate of return, it and other environmentally responsible investments illustrate that good business and environmental protection can go hand in hand.

KEY CONCEPTS
Companies can be forced to become sustainable through regulations and market mechanisms, but individual responsibility and action on the part of business owners, members of boards of directors, CEOs, and employees can have a profound effect on the nature of business.

Green Products and Green Seals of Approval

Individuals can also influence corporate policy—and make an enormous impact on the environment—by avoiding environmentally unfriendly products (among them disposable diapers and pens) and purchasing green products, defined as environmentally friendly goods and services. **Green products** may be goods made from recycled materials—for example, books, toilet paper, and paper towels made from recycled paper. They also include nontoxic cleaners and high-efficiency showerheads or light bulbs. They include a wide range of recyclable and reusable products, such as shopping bags. Hun-

25-3   Principles of Natural Capitalism

The principles of natural capitalism provide the basis for a complete rethinking of business. They show how, in contrast to conventional wisdom, far greater profits are achieved through protecting and enhancing nature, culture, and community than by harming them.

Radically increase resource efficiency

Most companies can dramatically increase the productivity of almost any resource they use. Fourfold increases in resource productivity are now the basis of economic development policy for the European Union, and 10-fold increases or greater are possible with existing technologies. Increasing efficiency also encompasses the development of innovative business models that focus on meeting consumer needs in ways that require fewer manufactured products and reward companies for reducing their environmental footprint.

"Biomimicry"

Biomimicry calls for a shift away from conventional "heat, beat, and treat" manufacturing methods, which require enormous energy inputs and usually create toxic byproducts, to production based on models derived from nature. The benign productive processes of living things can guide industrial innovation and teach us how to eliminate waste and toxics. This principle, outlined in the book *Biomimicry*, by Janine Benyus, can be the source of dynamic industrial innovation.

Invest in restorative practices

All good capitalists reinvest in productive capital. Restoring the world's depleted natural and human capital is a critical foundation of sustainable wealth creation. This principle encourages businesses to behave in ways that restore the capacity of both the earth and society to sustain life by investing in human and natural capital.

The goal is

No net loss of natural or social capital. No current balance sheet accurately captures the real economic value of natural or social capital. But no human system that systematically degrades either one can long endure. To achieve genuine prosperity and an economy worth sustaining, it is essential to ensure that neither form of capital is diminished.

You can link to websites that represent both sides through Point/Counterpoint: Furthering the Debate at this book's internet site, http://environment.jbpub.com/. Evaluate each side's argument more fully and clarify your own opinion.

dreds of green building products are now available, and thus, a house can be built from the foundation to the roof using environmentally friendly materials (**FIGURE 25-9**). Many non-toxic, environmentally friendly paints, stains, and finishes are also available. Many of them are cost competitive with conventional products, too. Using the operating principles of sustainability—conservation, recycling, renewable resource use, and restoration—as guidelines for product purchase will help you select the most sustainable products.

Individuals can find out about a company's environmental and social record by consulting the book *Shopping for a Better World*, which is published by the Council on Economic Priorities, a nonprofit consumer/environmental group. It rates hundreds of products that people routinely buy. Companies are scored in 10 categories, including environmental policy, charitable giving, minority advancement, and women's advancement. Green building products have become so popular there are several well-respected books on the subject and at least one website where individuals can learn more about them.

Another means of learning about products is government labeling programs. In the early 1980s, the government of the former West Germany instituted a product-labeling program

FIGURE 25-9 Green building products. Virtually every part of a home can now be made from an environmentally friendly or green building material like bamboo flooring (shown here), a fast-growing species that can be harvested without killing the plant. (Courtesy of Teragren SIGNATURE Bamboo Flooring, Photo by Teragren LLC.)

- Become a more environmentally conscious consumer.
- Reduce consumption.
- Buy durable goods.
- When buying stocks and mutual funds, invest in environmentally responsible companies.
- Support nonprofit organizations that encourage/promote sustainable economic development.

known as the *Blue Angel* program. The Blue Angel label on a product assures the customer that the product is environmentally acceptable and was produced in an environmentally acceptable manner. Since its beginning, the program has scrutinized over 3500 products in 50 different categories.

The Blue Angel program and a German magazine that offers even more complete analysis of the environmental acceptability of products have had a profound influence on German industries. In some cases, companies whose products have been given poor ratings have altered their manufacturing processes or made drastic changes in the products themselves to receive a seal of approval.

Ecolabeling programs have also emerged in Europe, Canada, Japan, and the United States. The Green Seal program in the United States, headed by Earth Day cofounder Denis Hayes, for instance, passes judgment on products using criteria similar to those used in Canada (**FIGURE 25-10**). This program currently certifies about 300 products.

A truly sustainable economy would require that virtually all products be produced in efficient, environmentally sustainable ways. In the meantime, individuals can send a strong signal to the marketplace by purchasing products that are environmentally acceptable.

Although green products and green seals of approval are an important step along the road to sustainability, some critics point out that green consuming is still consuming. To build a sustainable world, we still need to reduce our consumption—that is, learn to do less with less. For more ideas on what you can do, see the Individual Actions Count! table.

FIGURE 25-10 The Green Seal of approval. This seal affixed to a product indicates that it is a green product—one whose purchase causes less harm to the environment than competing products made by other manufacturers.

KEY CONCEPTS

Products vary in the degree to which they contribute to sustainability. Some promote the principles of sustainability, such as recycling, renewable resource use, and restoration. By purchasing such green products, individuals help promote a sustainable economy. Product labeling programs can help individuals select the most environmentally sustainable products.

Appropriate Technology and Sustainable Economic Development

In his classic book *Small Is Beautiful*, the late E. F. Schumacher popularized the term **appropriate technology**. Summarized in Table 25-2, appropriate technology puts people to work in meaningful ways. Compared with many modern forms of technology, it is efficient on a small scale. It uses locally available resources, takes less energy to run, and produces minimal amounts of waste. Some examples of appropriate technologies geared to a less exploitive lifestyle include passive solar heating for homes and businesses, wind generators and photovoltaic panels for making electricity, bicycles for commuting to work when the weather permits, and compost piles or compost worm bins to decompose yard wastes.

The use of appropriate technologies can help the less developed nations reduce their resource demand and impact on

Table 25-2
Characteristics of Appropriate Technology

- Machines are small to medium sized.
- Human labor is favored over automation.
- Machines are easy to understand and repair.
- Production is decentralized.
- Factories use local resources.
- Factories use renewable resources whenever possible.
- Equipment uses energy and materials efficiently.
- Production facilities are relatively free of pollution.
- Production is less capital intensive than conventional technology.
- Management stresses meaningful work, allowing workers to perform a variety of tasks.
- Products are generally for local consumption.
- Products are durable.
- The means of production are compatible with local culture.

the environment. It could also free up resources for other people, making a better life possible for a larger number of people.

In more developed countries, the shift to appropriate technology will eliminate some jobs that have been a part of the economy for decades. Auto and steel workers and miners, for example, may be phased out as we shift to a more efficient way of life. However, as pointed out in the next section, many new jobs will emerge as our priorities shift.

Appropriate technology, like other suggestions in this chapter, is not a panacea. It is one part of the solution. (See Table 25-3 for a summary of the ideas.) The choice for appropriate technologies belongs to us and the business world.

KEY CONCEPTS

Appropriate technologies are an essential part of a sustainable future. They rely on local resources, are efficient, and produce little if any pollution.

A Hopeful Future

"Throughout most of its tenure on Earth," economist Herman Daly points out, "humanity has existed in near steady-state conditions. Only in the past two centuries has growth become the norm." As pointed out earlier, continual economic growth may be the norm today but is very likely not sustainable. If we are to create an enduring human presence, our economic system must conform to the design principles of the ecosystem. The human economy must operate within the limits of the economy of nature. However, that does not mean that the economic future is bleak. The sustainable economy need not be a prescription for dull living and stagnation. In fact, it may entail growth in some sectors, such as solar energy and energy efficiency, while others, such as oil and steel production, are phased out. A sustainable economy based largely on renewable resources does not mean a retreat into the Dark Ages of drafty, cold domiciles. Solar heating and insulation could mean even more comfortable living spaces. Many nonrenewable resources (metals, for instance) would remain in use, recycling through the system many times. Some mining would be necessary to replace what is lost or locked up permanently in structures. Renewable fuels such as ethanol and perhaps even hydrogen could power our transportation system. Photovoltaics (such as the panels on my roof, which supply 100% of my electricity) along with wind energy might provide the electricity we need.

Clearly, today's economy and that proposed by a growing number of people are worlds apart. One of the principal differences is the continued high level of consumption and waste in the former, which is bound to deplete our nonrenewable (oil and aluminum) and renewable (wood from the tropical rain forest) resources and possibly create a population crash. In contrast, a truly sustainable economy would achieve a level of production and consumption that could be sustained forever.

KEY CONCEPTS

A sustainable economy conforms to ecological design principles and is dynamic and full of opportunity.

Table 25-3

Steps Essential to Building a Sustainable Economy

Goals

- Reduce population growth, then gradually reduce population size through attrition.
- Reduce resource consumption and waste by reducing demand, increasing product durability, and increasing efficiency.
- Recycle, reuse, and compost to the maximum extent possible.
- Develop a wide range of renewable energy resources.
- Protect and conserve renewable resources, such as farmland, fisheries, forests, grasslands, air, and water.
- Improve renewable resource management to ensure sustainability.
- Repair past damage to natural resources by replanting forests and grasslands, reseeding roadsides, restoring streams, cleaning groundwater, and reducing overgrazing.
- Increase national and regional self-sufficiency by using renewable resources whenever possible.
- Support sustainable development projects in developing nations.
- Develop sustainable ethics and promote individual and corporate responsibility and action.
- Improve social conditions by promoting democracy, justice, and a more equitable distribution of wealth.
- Work for global peace and cooperation.

Some Policy Tools

- Environmental education programs.
- Green taxes and full-cost pricing requirements for business.
- Alternative measures of progress, such as the NEW and GPI, that look at a wide range of economic, social, and environmental conditions.
- Laws that harness market forces, including (1) economic disincentives, (2) economic incentives, (3) tradeable and marketable permits, (4) laws that eliminate market barriers that promote inefficient resource use, and (5) laws that remove subsidies of environmentally destructive activities.

25.6 Environmental Protection versus Jobs: Problem or Opportunity?

During the 1992 presidential election campaign, then-President George Bush repeated a theme that arose many times during his campaign: We need environmental protection, but not at the cost of jobs. He pointed out that we can't have higher-mileage cars because it would put tens of thousands of American autoworkers out of work. We can't save the spotted owl because it will cost jobs. Ignoring the loss of employment from automation and other factors (which cost the timber industry 12,000 jobs from 1977 to 1987), Bush and others hammered on the point over and over again, while Bill Clinton and Al Gore asserted that the choice between jobs and environmental protection was a false one.[1] Who's right in this debate? Do environmental regulations cost jobs? Do they cripple businesses?

Environmental regulations and permits can delay projects, such as dams and power plants. However, several studies show that delays are often the result of poor planning on the part of the companies and government agencies overseeing the work. Delays could be greatly reduced if corporations and governmental agencies were willing to invite the public to give input early on. As it is, citizens often become involved late in projects, and—because their concerns have not been incorporated—they oppose them.

Business leaders argue that environmental regulations also cut productivity and reduce income, resulting in loss of jobs. **Productivity** is the dollar value of goods per hour of paid employment. This figure is often used to determine how healthy an economy is. Businesspeople argue that environmental regulations divert workers from productive jobs (miner) to nonproductive ones (mine safety inspector), which raises the cost of doing business and lowers productivity. Careful analysis shows that environmental regulations do indeed diminish the output of industry. By various estimates, environmental regulations decrease productivity by an estimated 5 to 15%. However, they decrease output much less than other factors, such as high energy prices and the general shift to a service economy. Efforts to prevent pollution and other problems can actually increase productivity, as pointed out in previous chapters.

In an article in *International Wildlife*, Curtis Moore described the efforts of Germany to protect the environment. He wrote: "More than anywhere else on Earth, Germany is demonstrating that the greening of industry, far from being an impediment to commerce, is in fact a stimulus." He went on to say, "Precisely because it is so environmentally advanced, Germany is lean, competitive, and poised to dominate the global marketplace." Through end-of-pipe controls and numerous sustainable strategies that address the root causes of the problem, Germans have shown that cleaning up industry can stimulate business. How? For one, tough standards stimulate innovation, and nations with the most rigorous requirements for pollution control often become leading exporters of control technologies. In addition, environmentally responsible business can result in better employment opportunities.

Several studies show that far more jobs are created by environmental controls than have been lost. The Clean Air Act and Clean Water Act have created at least 300,000 jobs in pollution control. Studies by the EPA show that very few jobs are lost as a result of environmental regulation. Only companies that are hanging on for dear life are affected. Even the AFL-CIO admits that not one plant shutdown can be attributed to environmental regulations.

This is not to say that tighter controls on pollution, reductions in timber cutting, or other environmental regulations are always benign. They do affect communities, sometimes profoundly. Environmental protection measures, especially the kinds of changes needed to develop a sustainable system, will result in massive shifts in employment. As Chapter 23 notes, employment will very likely shift from the ends of the present linear system of production and consumption (mines, forests, and waste dumps) to the center of an efficient, cyclic system.

Numerous studies suggest that this shift will result in a net increase in the number of jobs. Thus, restructuring the economy and our way of life to rely more on energy efficiency, recycling, mass transit, and alternative fuels will create far more jobs than are lost. As Chapter 15 notes, a large nuclear power plant employs 100 workers. A coal-fired power plant that generates the same amount of electricity employs 116 workers, whereas a solar thermal facility employs more than twice as many workers, about 250. However, a wind farm employs more than twice as many people again, 540 in all. Furthermore, wind and solar thermal energy are less expensive and more environmentally benign than nuclear power.

Studies in the United States show that energy efficiency also creates jobs. Michael Renner of the Worldwatch Institute estimates that weatherizing all U.S. households could create six to seven million job-years—that's 300,000 jobs lasting 20 years each!

A 1985 European study that examined the employment potential of six energy conservation and renewable energy technologies in four countries (Great Britain, Denmark, France, and the former West Germany) found that these technologies could create 142,000 job-years—7000 jobs lasting 20 years. A full-fledged program for all 12 European nations could create 530,000 job-years.

Michael Renner notes that "although a shift from fossil fuels to solar energy entails job losses in the oil, coal, and gas industries, there are overlaps among the kinds of supplies and skills required by the solar industry that will minimize overall job loss." Companies that produce the materials needed for solar panels are already in existence. Many of the skills needed to tap renewable energy supplies are similar to those required for conventional construction and heating system installation. Work opportunities could spring up in a variety of existing occupations, among them carpentry, plumbing, and construction.

Shifting to mass transit may also create more jobs than are created by our current system of automobile transit. A German study showed that spending $1 billion on highway construction yields only 24,000 to 33,000 direct and indirect jobs. Spending the same amount on railway and light-rail construction creates 38,000 to 40,000 jobs. Furthermore, the shift from cars and trucks to railroads, subways, light rail lines, and buses offers alternative job opportunities for much of the workforce.

---

[1] Another 35,000 jobs in the timber industry could be lost in the next 45 years as a result of mechanization, according to the Association of Forest Service Employees for Environmental Ethics.

Recycling offers similar employment opportunities, as pointed out in Chapter 23. Alcoa estimates that at least 30,000 people in the United States are employed in aluminum recycling, nearly twice as many as in aluminum production. Vermont's recycling facilities employ 550 to 3000 people for each one million tons of materials they handle. On the other hand, incinerators equipped to handle the same amount employ only 150 to 1100 people, and landfills employ only 50 to 360.

Environmental protection can help save companies money. Pollution prevention measures, for instance, cut down on the production of hazardous waste and substantially reduce waste disposal costs. They also eliminate future liability for cleanup and may reduce or eliminate employee health effects and future claims for health impacts. Pollution prevention can reduce lawsuits and greatly slash the cost of compliance—installation and operation of pollution control equipment. Many examples cited in this book highlight companies that have saved considerable amounts of money by preventing pollution, using energy more efficiently, and recycling and reusing wastes.

Worldwatch Institute's Cynthia Pollock Shea notes that "businesses that protect the environment can make a healthy profit." As German and Japanese companies are showing us, environmental protection is a precondition for success.

> **KEY CONCEPTS**
>
> Environmental protection, rather than being an impediment to economic progress, may be a stimulant. Many sustainable strategies result in cost-competitive goods and services that create as many or more jobs, with little environmental impact. A sustainable economy, however, will involve a major employment shift.

A penny will hide the brightest star in the Universe if you hold it close enough to your eye.

—*Samuel Grafton*

CRITICAL THINKING

Exercise Analysis

Two essential critical thinking rules are to examine one's biases and to consider the big picture. These rules are important in analyzing the assertions made in this exercise. Consider bias first.

Many people in positions of power are wedded to the notion that all growth is good. As noted in the chapter, however, much of the economic growth occurring in the United States has not improved our well-being. Growth results in extraordinary costs, such as lost wildlife, lost open space, polluted air, crowding, and so on. As pointed out in the chapter, a substantial portion of the growth of the U.S. GNP has resulted from "bad" economics—pollution cleanup, repair of damage from pollution, and unnecessary health care expenses. Statistics also show that economic growth tends to benefit the wealthiest members of societies; that is, economic benefits of the growing economy are not trickling down to the less fortunate, as they're supposed to.

Also noted in this chapter is the assertion that to improve an economy it is not always necessary to manufacture more goods. One of the fastest and most profitable ways of improving an economy is to use resources more efficiently. By using energy, water, and materials more efficiently, for instance, companies can produce goods more cheaply, make a higher profit, and/or compete more effectively. Pollution prevention is another strategy with extraordinary economic benefits. All of these strategies permit companies to produce goods while improving the bottom line and the environment.

CRITICAL THINKING AND CONCEPTS REVIEW

1. Define the following terms: *command economy, market economy,* and *law of scarcity.*
2. Describe the law of supply and demand. What is the market price equilibrium? What are the major weaknesses of the law of supply and demand? How can they be corrected?
3. Define your own economic goals. Would you classify them as consistent with a frontier economy or a sustainable economy?
4. Using your critical thinking skills, analyze the following statement: "Economic growth is essential to a healthy economy."
5. In your view, is continued economic growth in the developed nations possible? If so, why, and for how long? If not, what are the alternatives?
6. Describe the gross national product and its strengths and weaknesses.

7. Define the term *net economic welfare* (NEW). How does it differ from the GNP? Is it greater than the GNP, or less? Does it grow as quickly as or more slowly than the GNP? Is this good or bad?

8. Describe the index of sustainable economic welfare (GPI). How does it differ from the GNP?

9. Using your critical thinking skills, analyze the following statement: "The GNP is a fundamentally sound economic measure and a good indicator of the well-being of a nation's people because economic health means a higher standard of living for all."

10. Discuss how time preference, opportunity costs, and discount rates are related. How do they affect the ways in which people manage natural resources?

11. Define the term *economic externality* and describe how externalities can be internalized. What are the benefits of doing this?

12. How can externalities be avoided?

13. What is the economically optimal level of pollution control? Why is it impractical to consider reducing pollution from factories and other sources to zero via pollution control devices?

14. Describe the law of diminishing returns. How does it apply to pollution control? Can you think of any other examples where the law applies?

15. The analysis of the economic system and of economic thinking suggests a number of weaknesses when viewed through the lens of sustainability. What are they? Describe each one and discuss how each can be corrected.

16. Describe a sustainable economy. What are its main goals? In your view, is it a practical alternative to the current economic system? What are its strengths and weaknesses?

17. Using your critical thinking skills, analyze this statement: "Cleaning up the environment will put thousands of people out of work. We simply can't afford to do it."

KEY TERMS

appropriate technology
benchmarks of sustainability
command economies
consumer-pays option
demand
descriptive economics
direct costs
discount rate
ecological economics
economic externality
economics
environmental auditing
free-market system
full-cost pricing
gas guzzler tax

Genuine Progress Indicator (GPI)
green products
green taxes
gross national product (GNP)
Index of Sustainable Economic Welfare (ISEW)
indicators of sustainability
indirect costs
inputs
law of diminishing returns
law of scarcity
law of supply and demand
market economies
market price equilibrium point
marketable permit

natural capitalism
net economic welfare (NEW)
normative economics
opportunity cost
outputs
pollution taxes
productivity
repercussion costs
supply
sustainable economics
sustainable economy
taxpayer-pays option
time preference
user fees

REFERENCES AND FURTHER READING

The References and Further Reading section at the end of this book contains a list of sources for the information discussed in this chapter and recommendations for further reading.

Connect to this book's website:
http://environment.jbpub.com/
The site features eLearning, an online review area that provides quizzes, chapter outlines, and other tools to help you study for your class. You can also follow useful links for in-depth information, research the differing views in the Point/Counterpoints, or keep up on the latest environmental news.

Sustainable economic development raises the standard of living in less developed countries.

CHAPTER 26 Sustainable Economic Development: Challenges Facing the Developing Nations

He helps others most, who shows them how to help themselves.

—*A.P. Gouthey*

A huge dam constructed with international financial assistance now spans India's Tawa River. The project, like many others in the less developed nations of the world, has proved to be a mixed blessing. Although it did provide electricity as promised, the irrigation canals built to transport water from the reservoir to nearby farms were constructed in porous soils. As water flowed through the canals, much of it drained into nearby farm fields, making it difficult for farmers to work the land. In some instances, water filled the soil pores and suf-

In many less developed nations, people enjoy few amenities. On average, they spend 40% of their income on food and live in marginally adequate dwellings. The less fortunate residents live in substandard conditions with inadequate food and shelter. Hunger and disease run rampant. Many experts believe that sustainable economic development will require massive improvements in the lives of people in the developing countries. They see industrialization (much like that in the West) that leads to much more affluent lifestyles as a key to creating a sustainable society. Critically analyze these ideas. What key questions need to be asked? What critical thinking rules are helpful in this exercise?

focated the plants, killing crops. Tragically, many of the farmers who were supposed to have benefited from the project have experienced a marked decrease in their food production.

Near the village of Sukhomajri, India, livestock overgrazed fields, reducing vegetative cover by 95%. This in turn led to a substantial increase in soil erosion. To combat the problem, local villagers built small earthen dams to capture water from rain, with relatively modest financial assistance. The water was used to irrigate the fields and coax the land back to life. A communitywide agreement to reduce overgrazing is also giving the land a chance to recover.

The first example is typical of many efforts to improve conditions in less developed countries. Unfortunately, many such projects do more harm than good. The second represents **sustainable development**, aimed at meeting the needs of people at an affordable price while protecting the environment on which they depend. Sustainable development is both people centered and conservation based. It is part of a growing list of ventures that could revolutionize the way less developed countries progress. Like other projects, its aim is to enable people to enjoy long, healthy, and fulfilling lives and to promote economic prosperity.

This chapter concerns itself with sustainable economic development in the less developed nations of the world. It first examines traditional strategies, many of which are unsustainable, then offers guidelines on sustainable development strategies, and concludes with a discussion of some of the barriers to sustainable development.

26.1 Conventional Economic Development Strategies and Their Impacts

Many less developed countries are steeped in economic and environmental problems. The sources of their dilemma are many. For example, a large number of countries strain under the burden of rapid population growth. Many suffer from political corruption and inept governance. Others are ruled by oppressive dictators. Still others are enmeshed in civil wars or ongoing military conflicts with neighboring countries, two activities that waste precious resources and destroy valuable land (FIGURE 26-1). Economic and environmental problems also stem from the exploitation of international corporations and foreign governments. As mentioned in the opening paragraph, many well-meaning forms of assistance coming from the more developed countries and international lending agencies add to the environmental and economic troubles.

During the past 50 years, less developed countries have received considerable economic and technical assistance from other countries, especially industrial nations. This aid has been aimed primarily at improving health care, increasing food production, providing energy (particularly electricity), and building an industrial base. Unfortunately, many development

FIGURE 26-1 **Conflicts.** Civil wars and other military conflicts with neighboring countries threaten national security, but so does environmental deterioration. Resolving conflicts peacefully is essential to living sustainably on the Earth.

projects funded by the West have spawned enormous social, economic, and environmental damage in recipient nations. The environmental impacts include massive deforestation, soil erosion, and desertification. Some projects have resulted in widespread poisoning of people and the environment with pesticides. Valuable wilderness and farmland have been flooded by dams or destroyed by mines. Given the importance of the environment to the well-being of people, it is imperative that development projects protect the environment.

KEY CONCEPTS

Economic and environmental problems abound in the less developed countries and are the result of many interacting forces, including overpopulation, political corruption, ineptitude, war, economic exploitation, and misguided development efforts.

What Is Wrong with Western Development Assistance?

In the past, most development strategies have attempted to impose Western ways on less developed nations. Large-scale agricultural projects and dams, such as the one on the Tawa River in India mentioned in the introduction, are examples. These and other Western-style schemes seek to remold nature to benefit humans. As you have seen, however, efforts to reshape nature to our liking often have severe ecological backlashes with tremendous economic repercussions.

As a rule, Western-style development projects tend to centralize wealth in cities and in the hands of a few. In many cases, they benefit a small segment of the population while ignoring or, even worse, harming the people they were meant to serve. Huge dams built for hydropower, for example, often flood productive farmland and displace farmers, forcing them to move to marginal land that cannot withstand heavy use. Dams also inundate forests, destroying the homes and food source of native peoples. In arid regions, irrigation water from reservoirs increases salinization and waterlogging of agricultural soils, both of which decrease the productivity of the land and can render it useless (Chapter 10). Salinization and waterlogging translate into economic disaster for small farmers. Today, over half of the irrigated cropland in less developed countries suffers from salinization.

As pointed out in previous chapters, some development projects negate gains achieved from others. Deforestation projects in Guatemala to increase agricultural output have resulted in massive soil erosion. This tends to increase flooding along streams and increase sedimentation in reservoirs, greatly reducing their useful life span.

Another common form of development involves the introduction of high-yield crops to increase food production. While potentially valuable, these projects tend to create a costly cycle of dependency. In order for many of these crops to survive, they must be doused with pesticides, fertilized, and then irrigated. Pesticides and fertilizers are often imported from the more developed nations. Their costs are often prohibitive, and their impact on the environment and health of the workers is profound. In Central America, chronic and acute pes-

ticide poisoning are among the region's most serious problems (Chapter 22). In order to promote pesticide use, the governments of some less developed nations sometimes subsidize them, again creating a costly economic dependency that drains money needed for education and environmental protection. In the African nation of Nigeria, a factory was built to produce fertilizer for farmers, but developers discovered that the nation's transportation system was inadequate to transport the fertilizer to rural farmers—who didn't have enough money to pay for fertilizer anyway. The factory is now abandoned. Projects such as this may have been the stimulus for one author's comment: "Developing countries are littered with the rusting good intentions of projects that did not achieve social or economic success" (FIGURE 26-2).

Western-style development, with its disregard for limits, can decimate potentially sustainable sources of food, timber, and fiber. For instance, before 1960, most of Thailand's commercial fishing fleet consisted of small-scale family operators. Today, modern fishing vessels dragging trawl nets (a Western technology) cruise the shorelines, catching virtually everything that swims and leaving nothing to sustain fisheries. This development could bring an end to a once lucrative commercial fishing industry.

International development assistance has also helped finance road building to colonize remote regions. In western Brazil, for instance, tropical rain forests were cut to make way for farmers and ranchers, but 80% of the people soon left because the soils quickly washed away or lost their fertility (Chapter 12).

Another problem, some critics say, occurs in rebuilding war-torn nations such as Afghanistan and Iraq using expensive imported materials common in the Western world like concrete and steel. Far better, they argue, would be the use of locally available natural materials that have been the cornerstone of building in these countries for centuries. Such materials can be used to build modern, safe, and durable structures. Finally, Western-style development often un-

FIGURE 26-2 **Inappropriate technology.** This tractor in Niger lies rusting in a field. When it broke down, knowledge, skills, and the parts to repair it were not available.

dermines sustainable practices and sustainable cultures. Some refer to this as *cultural erosion*. In a remote area of northern India, Helena Norberg-Hodge, who has worked in the area for many years, notes that people once lived sustainably and well. Their community was close-knit, its food locally produced, and its clothing made from local materials. The culture was shaped by a close and intimate relationship with the environment.

Because of exposure to Western ideology and Western ways of doing things, the people now often feel inadequate, even inferior. Many have rejected their own culture. Today, they are struggling with the choice of preserving an economy that was once self-reliant and sustainable or turning to the alluring but unsustainable ways of the West.

KEY CONCEPTS

Western-style development efforts often result in a means of production that tends to centralize wealth and destroy sustainable lifestyles. These efforts also radically reshape the environment and, in the process, destroy the soil and other natural resources and promote a costly dependency on Western economies.

Who's Financing International Development?

Financial backing for international development comes from thousands of individual sources, which fall into one of four categories: (1) multilateral development banks (MDBs), (2) private commercial banks, (3) development agencies of the industrial nations, and (4) private foundations. The two key players are the MDBs and development agencies.

Four MDBs provide the bulk of the money: the World Bank, the Inter-American Development Bank, the Asian Development Bank, and the African Development Bank. All receive funding from the more developed nations. The World Bank, for example, is headquartered in New York City and is supported by the United States, France, Germany, the United Kingdom, and others. The World Bank lends about $200 billion a year to support a wide range of projects, from road construction to farming to dam building and family planning.

Additional money comes from private commercial banks that make loans directly to less developed countries. The international development agencies are also key players in development. The U.S. Agency for International Development, for example, provides outright grants for development projects. Although private foundations also help support international development, their contribution is small.

As pointed out in the previous material, many projects sponsored by the various organizations noted previously are unsustainable. Thus, they have not only failed to improve the lives of those they were designed to help, they have worsened them. In fact, volumes of horror stories could be written about well-intended projects gone awry. Fortunately, many funding agencies are beginning to see the need for new practices—for sustainable development. Thankfully, sustainable development does not require a technological revolution or new knowledge. The technology and knowledge already exist. It does, however, require steps to put ideas and technologies into practice.

KEY CONCEPTS

Most financial resources come from international banks and government development agencies. Most of it is uncoordinated and largely unconcerned with achieving a sustainable form of development.

26.2 Sustainable Economic Development Strategies

To create a sustainable future, many experts believe that the less developed nations must first slow the growth of their populations. In many cases, it may be necessary to decrease population size to avoid overshooting the carrying capacity. Dramatic improvements are also needed in education and health care. Freedom from repression and gains in women's rights are also essential. Ultimately, the less developed nations must build sustainable systems of agriculture and commerce. Demand for energy, food, water, waste removal, housing, and transportation must all be met in ways that protect and enhance natural systems.

Some progressive thinkers believe that the less developed nations, like the industrial nations, must become more self-reliant (Chapter 25). In fact, many proponents of sustainable development believe that self-sufficiency is one of the cornerstones of success. As pointed out in Chapter 25, efforts to increase self-reliance oppose the powerful forces shaping an interdependent global economy. Nonetheless, in order to create a sustainable future in which people live within the limits of the environment, it may be necessary to promote ways for less developed nations to become less dependent on others for aid and trade. One way to promote self-reliance is to tap into local resources to meet local needs.

Sustainable economic development also requires appropriate policies—that is, laws, regulations, and subsidies that promote sustainable practices. Inappropriate policies, among them subsidies for pesticide use or tax breaks for companies that destroy natural systems, need to be eliminated.

Sustainable development can also be fostered by routine analyses of social, economic, and environmental impacts of proposed projects. Until very recently, concern over the long-term environmental sustainability of development projects received little if any attention from developers and lending institutions. Surprisingly little thought was given to the social costs and benefits. Impact analyses could help countries avoid projects that are likely to fail or that are destined to create irreparable environmental damage.

KEY CONCEPTS

Creating a sustainable future for the less developed countries will require many steps, including efforts to slow or even halt the growth of the human population, to improve education and health care, and to ensure the protection of basic human rights. Ultimately, the infrastructural systems of a country will need to be rethought and reshaped.

Employing Appropriate Technology

One means of promoting economic development while protecting the environment is appropriate technology. As noted in Chapter 25, appropriate technology refers to an assortment of environmentally compatible technologies, among them hand-operated tools, solar cells (PVs), and methane digesters, devices that produce methane gas from organic waste. Appropriate technologies are generally small-scale solutions to local problems. They are reliant on locally available resources and local knowledge for repair and service (Chapter 25). As a rule, appropriate technologies depend more on people (human labor) to produce goods and services than on machines and fossil fuel energy. In other words, they are labor intensive rather than energy intensive—in sharp contrast to the highly automated factories and machines of the West, which require large amounts of costly fossil fuel and capital and much less labor.

Appropriate technologies are suitable for the less developed nations which typically lack the financial capital and fossil fuel resources required to support modern, Western-style industries equipped with an assortment of labor-saving, energy-intensive devices. Obviously, because appropriate technologies rely more on human labor, they are quite suitable to many heavily populated nations such as India and China.

Appropriate technologies, such as ox-drawn steel plows and solar-powered water pumps, are badly needed on farms throughout the less developed world (FIGURE 26-3). They provide means of improving the lives of people in less developed nations. The Western alternatives, tractors and diesel-powered water pumps, are quite often inappropriate technologies for less developed nations. Although tractors increase food output, they displace farmworkers. In many cases, displaced farmworkers move to already overpopulated cities in search of work, creating serious problems there (Chapter 8). In addition, the costs of tractors and diesel-powered pumps, repair, and energy to run them are prohibitive in poor, rural regions of less developed countries. To learn more about socially and economically sustainable development, readers may want to contact Builders without Borders, Engineers without Borders International, Solar Energy International, or the World

Hands Project. All four groups sponsor projects in LDCs that employ appropriate technology to better peoples' lives.

One inexpensive, appropriate technology is the solar box cooker, which uses sunlight to cook meals (FIGURE 26-4). Constructed from aluminum foil, glass, and cardboard boxes, solar box cookers are suitable in regions where firewood is scarce and sunlight abundant, among them northern Pakistan, Nepal, and parts of Africa. This relatively inexpensive device can reduce deforestation. Because many families cook over wood fires in unventilated or poorly ventilated spaces, solar box cookers can also reduce health problems from smoke inhalation.

Another appropriate technology is the photovoltaic cell (Chapter 15). Photovoltaics are far cheaper than centralized, coal-fired power plants and transmission lines in rural regions of less developed countries. Installed in villages, they can provide electricity for refrigerators and freezers that store medicines. They can also be used to power water pumps.

Mohandas Gandhi summed up the challenge when he said that the poor of the world cannot be helped by mass production, only by production of the masses. Many experts agree that the world cannot support economic development in the less developed nations on a scale similar to that in the more developed nations. By one estimate, if everyone lived like people in the United States, the world would run out of resources in about 3 months. Without some improvement in lifestyles, though, hunger, disease, environmental destruction, and death are bound to continue.

To many people, progress means moving toward labor-saving devices, manufacturing processes, and cutting-edge technology. The shift to appropriate technology, however, will require a change in this mindset.

KEY CONCEPTS

Appropriate technology is environmentally compatible, easily understood and repaired, and relies on local resources, especially people. It is an essential component of sustainable development in the less developed nations. These technologies are as important to food production as they are to the production of goods.

FIGURE 26-3 **Appropriate technology.** Farmers in Asia learn to use and maintain relatively simple, handmade plows, an appropriate technology in many less developed countries.

FIGURE 26-4 Solar box cooker. This solar box cooker is inexpensive and easy to make. It helps reduce the need for firewood in sunny, arid regions and cuts pollution generated by cooking with wood.

Creating Environmentally Compatible Systems of Production

Over the years, Western-style development has been driven by frontier ideology. As you may recall from Chapter 24, the frontier ethic insists that the key to human success lies in the domination and control of nature. Western-style agricultural projects in the rain forest, for instance, first require the removal of trees from huge areas, to create clearings for farm equipment. Such projects introduce equipment and farming practices that may be suitable in Iowa but that often fail miserably in the tropics.

The key to success lies not in trying to modify the environment to fit the production system, but rather in modifying production systems to fit the environment. Developers must design with nature, rather than attempt to redesign nature to fit human needs. Growing crops in small clearings in tropical rain forests or harvesting food and fiber from relatively undisturbed forests, for example, are both sustainable activities that fit within the workings of nature, whereas leveling huge tracts of forest to plant row crops does not.

Sustainable development that seeks to work within the constraints of natural systems may require many regional, small-scale projects that honor limits and avail people of nature's opportunities. Such projects require little if any financial backing and can pay for themselves many times over, in contrast to the large Western-financed projects that often cause tremendous economic and environmental damage and result in a tremendous burden of debt.

> **KEY CONCEPTS**
>
> Production systems custom-designed to fit within the local environment are essential to create a sustainable future. They are also more affordable and tend to benefit participants more than large-scale Western development projects.

Tapping Local Expertise and Encouraging Participation

Convinced of the superiority of Western ways, many international development experts overlook the abundance of knowledge held by indigenous peoples. For instance, the Lancandons of Mexico, believed to have descended from the Mayans, have a profound knowledge of tropical rain forest agriculture. They grew as many as 80 different crops in the rain forest until they were forced off their land by the Mexican government.

Although many native people have abandoned sustainable farming practices for one reason or another, the knowledge still exists. Tapping into this vast reservoir can help countries find sustainable solutions to local problems. A good example of a sustainable agricultural practice is agroforestry, growing crops and trees together, as discussed in Spotlight on Sustainable Development 7-1.

Besides overlooking local knowledge, many development projects have failed because they have addressed issues that are not viewed as problems by local people. By finding out what people want and then involving them in the project design, developers are more likely to succeed. Development is unlikely to be sustained unless the needs of people are identified and local residents support the project.

Finally, development projects must be culturally sensitive. In many countries, rural people value a sense of community and cooperation. Unlike the West, where competition reigns supreme, people in many communities in the less developed world tend to work together for the good of the entire citizenry. Western-style projects that promote individual gain may be viewed as a hindrance to progress. In such instances, developers might be advised to find projects that promote the good of the entire community, rather than enterprises that promote private ownership and competition.

For years, local cultures have been viewed as obstacles to success. Today, more and more people in the development community realize that for their efforts to be successful, they must preserve cultures. Such efforts must be woven into the social fabric of the people they are intended to help, not forced on them in a culturally disruptive manner.

Knowledge of a sustainable means of production often exists within local communities in less developed nations and can provide the basis for much of the new development. Such projects must be deemed important, however, and solutions must enhance the local culture to be successful.

Promoting Flexible Strategies

Another problem with traditional development strategies is that they tend to be organized from the top down. Decisions tend to be made by a select few—those with the training and the money. Relying on the top-down approach, large bureaucracies often attempt to manage projects great distances from their headquarters.

This organizational structure tends to be inflexible. Soon after they begin, many projects face obstacles. Without flexible goals and some flexibility in achieving them, such projects often fail.

To be successful, development programs could be viewed as experiments. Small pilot projects could be run first to determine problems and find solutions to them. At the very least, development projects need the flexibility to change. If troubles arise, managers need to be able to make adjustments to address them. This technique, discussed in Spotlight on Sustainable Development 25-2, is called *adaptive management*.

One way of increasing flexibility is to remove the actual management of projects from the hands of the international development agencies and banks. Who would take over? One candidate is certain **nongovernmental organizations (NGOs)**—private groups involved in sustainable development. With funding from development agencies, NGOs could manage projects to ensure adequate safeguards to protect the environment while improving the lives of the people. Fortunately, more and more MDBs are involving NGOs in project management.

KEY CONCEPTS

Large bureaucracies that attempt to manage development projects lack the flexibility needed to respond to problems and reach a successful conclusion. Creating flexible management structures that are more responsive to emergent needs will help ensure greater success.

Improving the Status and Expanding the Role of Women

In most of the less developed countries, women are important managers of natural resources. They collect firewood and maintain gardens to feed their families. They bear and raise the children. In some communities, women have assumed a major role in restoring the environment and in creating sustainable means of production (FIGURE 26-5). Yet in most countries, women have little access to education, credit, and land. They are poorly trained, and despite the fact that they perform most of the work, their labors account for only a small portion of the total family income.

Sustainable development clearly requires steps to improve the status of women. Many different reforms are needed,

FIGURE 26-5 **The importance of women.** Women play a key role in environmental decision making and restoration. This woman in Africa is part of a movement to replant that continent's forests.

among them measures to eliminate discrimination and cultural traditions that cast women in an inferior role. Proponents of this movement call for legislation to ensure equal pay for equal work, job training, and other benefits. Efforts are also needed to ensure that women have a full voice in decision making. Women also need access to credit at a reasonable rate of interest. In India, for instance, uneducated rural women are often forced to borrow money from loan sharks at 10% per day. Some use the money to buy fish, which they sell. Today, thanks to the Working Women's Forum, an organization that seeks to improve the conditions of women, rural women can receive credit at a reasonable rate to start small businesses to generate additional income for their families. Many of them have proven to be highly successful entrepreneurs. The Working Women's Forum boasts a 95 to 100% payback on loans.

Women also need access to better education, health care, and family planning. Once established in a region, the Working Women's Forum helps women acquire government services for prenatal and child care, immunization, and family planning. As the income and health of the family improve, the number of children born drops drastically.

Groups similar to the Working Women's Forum are forming all over the world. To many observers, this movement represents another ray of hope that population growth and environmental decline may be brought under control.

KEY CONCEPTS

Women play an important role in shaping a country's future, but are often relegated to an inferior position. Increasing the status of women and creating greater opportunities for them in education, employment, credit attainment, and other areas will contribute positively to the creation of a sustainable future.

Preserving Natural Systems and Their Services

Sustainable development also requires efforts to protect natural systems that provide goods and services. Appropriate technology and many efforts to design with nature help protect natural systems. Developers and local people can also protect the environment by establishing guidelines that minimize the destruction of wildlands.

The World Bank adopted a set of guidelines that could serve as a model for all international development activities. The bank's policy strictly prohibits supporting projects that will destroy wildlands of special concern, including wetlands and virgin rain forest. Instead, the bank favors projects on land that has already been disturbed—for example, forestland that has already been cut or grassland that has been farmed. This stipulation minimizes the destruction of undisturbed land. When the Bank feels that undisturbed land is justifiably developed, it favors projects on lands that are less valuable—for instance, forests with lower species diversity or land not essential to protect water supplies. Even in such instances, it calls on developers to offset losses by improving management on similar lands. Finally, the World Bank calls on developers to protect lands surrounding development projects. The Bank might, for instance, provide financial support for a dam project, but only if efforts were made to protect its watershed from deforestation.

This set of guidelines reflects the importance of wildlands and the free services they provide. Replacing free services such as watershed protection can be quite costly. Corrective measures to regain nature's gifts may require massive outlays of money for reforestation, flood control, water purification, and other services.

KEY CONCEPTS

To be sustainable, social and economic development strategies must protect natural systems, which provide many free services such as flood control and water purification.

Improving the Productivity of Existing Lands

The efforts outlined above protect land from unnecessary development. As Chapter 12 pointed out, better management of existing lands can also save them from the ravages of ill-conceived development. Erosion control and grazing management, for instance, maintain and improve the quality of farmland and grassland, thus reducing the need to use undisturbed lands. Efforts to improve the fertility of existing cropland—in order to maintain or even increase crop output—reduce the amount of farmland conversion required to feed burgeoning populations.

International lending agencies and development agencies could support projects to improve the condition of lands already under the plow or lands already being grazed. In addition, they could promote better forest management and efforts to replant tropical rain forests, to reduce the pressure on undisturbed lands (Chapter 12).

KEY CONCEPTS

Measures to restore and sustainably enhance the productivity of previously or currently used land and other resources (such as farmland and forests) are a vital first step in sustainable economic development. They also reduce the pressure on undeveloped land.

26.3 Overcoming Attitudinal and Economic Barriers

The previous section outlined many barriers to sustainable development, among them continuing efforts to impose Western ways on foreign cultures and the failure of developers to involve native people and tap local knowledge. This section discusses some important attitudinal and economic barriers to sustainability.

KEY CONCEPTS

Several substantial attitudinal and economic barriers also lie in the way of sustainable development.

Attitudinal Barriers

The first barrier is the predominant notion that environmental protection is a luxury—that is, a matter governments can address *after* people have achieved prosperity. The prevalent view is that poor people—and their leaders—simply can't be concerned about protecting the environment if people are starving and doing without basic necessities.

Fortunately, many people recognize that environmental protection is essential to the survival and prosperity of all people. Moreover, increasing evidence shows that poverty and environmental deterioration are part of a vicious cycle. Environmental decay drives many people into poverty. As desperate people consume and destroy the resource base on which they depend, poverty worsens and increases environmental destruction, creating a vicious downward spiral that undermines their long-term prospects. Lester Brown, Christopher Flavin, and Sandra Postel note in their book *Saving the Planet*, "Rather than a choice between the alleviation of poverty and the reversal of environmental decline, world leaders now face the reality that neither goal is achievable unless the other is pursued as well."

Environmental protection, therefore, is not an indulgence that can be afforded only by the wealthy. It is a prerequisite for survival and a decent standard of living (FIGURE 26-6). As noted by the World Bank, natural and man-made environmental resources—fresh water, clean air, forests, grasslands, marine resources, and agro-ecosystems—provide sustenance and a foundation for social and economic development. The goal of sustainable development is to meet human needs in ways that protect and enhance natural systems.

> Environmental protection, therefore, is not an indulgence that can be afforded only by the wealthy. It is a prerequisite for survival and a decent standard of living.

FIGURE 26-6 Environmental protection is not a luxury. Protecting the environment is essential to ensuring a sustainable future for all of the world's people because the environment is the source of all our resources and the sink for all our wastes. It is not a luxury that can come later, after people have achieved a comfortable existence.

Another similar belief is that efficiency is a luxury, too. Today, less developed countries spend nearly two thirds of their export earnings on oil imports. Making more efficient use of oil and other fossil fuels is vital to the future of these countries and could reduce the sizable trade deficits that cripple their economies.

Efficiency measures cost far less than new power plants and can improve the lives of many people. For example, if India replaced 20% of its incandescent lightbulbs with high-efficiency compact fluorescent bulbs over the next 20 years, it could avoid building eight power plants at a total cost of $4 to $8 billion. Installing energy-efficient light bulbs would cost less than $0.9 billion and thus would save substantial sums of capital that could be put to other uses.

Considerable economic and environmental gains can be made in the less developed nations by improving energy efficiency. High-efficiency wood cookstoves, for instance, could cut demand for wood and help reduce deforestation. Energy efficiency, combined with projects to replant trees and better manage forests, could go a long way toward helping people meet their needs sustainably. Today, however, energy-efficiency projects constitute less than 1% of all international aid.

KEY CONCEPTS

Environmental protection and efficiency are often seen as luxuries, but they are truly essential to the survival and prosperity of people in less developed nations. Sustainable development strategies must seek ways to meet current human needs while promoting efficiency and protecting the life-support systems on which people rely.

Economic Barriers

Even in the wealthiest nations, shortages of money to carry out the many things we need to do to create a sustainable future seem to be a perennial problem. In less developed nations, money problems are even more intense. Despite the shortages of cash, many less developed nations spend large amounts of their national budgets on military protection. Many observers believe that reduced military expenditures could be a substantial source of money that could be used in such underfunded areas as housing, education, sanitation, clean water, sustainable agriculture, family planning, and sustainable economic development.

Successfully reducing military expenditures depends on convincing governments that their borders are secure from invasion and that internal strife can be managed with less investment. It also rests on efforts to convince countries that their economic and environmental well-being could improve by such steps.

Two international environmental groups and the UN Environment Programme have outlined some measures needed to demilitarize the world, among them an acceleration of agreements to limit weapons production. International agreements to regulate (limit) the arms trade and treaties to reduce financial assistance for military expenditures could also help. International conventions that limit certain types of action, such as the destruction of Kuwait's oil fields in the 1991 Iraqi invasion, are also needed. These groups also call on the world community to work together to redeploy military personnel from wartime activities to peacetime activities, among them disaster relief and conservation.

Wealthy nations can help improve the economic conditions of less developed nations by ending their exploitation, discussed in Chapter 25. As you may recall, exploitive practices tend to rob the less developed nations of their natural resources (For a few ideas, see Spotlight on Sustainable Development 20-1).

Besides reducing economic exploitation, wealthy nations can help by relieving the burden of international debt—money lent to support many unsustainable development projects. Many of the poorest less developed nations are also the most debt-ridden. As described in Chapter 10, servicing the debt has forced many less developed nations to convert food crops for domestic production to cash crops (tea and citrus fruit, for instance) for export. Debt-for-nature swaps, discussed in Chapter 11, are one means of reducing debt.

Many individuals think that a more equitable distribution of world resources could help the poorer nations feed, clothe, and house their people. How that would be done and how successful it would be is debatable. Food handouts, while helping alleviate suffering, don't get at the root causes of the problem: rapid population growth, poverty, ill-conceived agricultural systems, and other unsustainable practices. More effective might be assistance aimed at helping less developed nations become more self-reliant and sustainable.

Although this chapter has emphasized the role of the West in promoting sustainable economic development in the less developed nations, it is important to point out that the less developed nations themselves can play a huge role in this process. Individuals, businesses, and governments can do many things to promote sustainable development. For an example of one nation's efforts, see Spotlight on Sustainable Development 26-1.

The changes and policies outlined in this chapter could help the less developed nations progress along a sustainable path, operating within the limits imposed by the Earth and availing themselves of many opportunities. As astute observers point out, however, they could also serve as a set of guidelines for development in the industrial nations. Widespread use of appropriate technology and ecological design are especially needed in the industrial nations. The World Bank ecosystem policy, which seeks to protect untouched wildlands by diverting development to already disturbed lands, is also essential, as are efforts to sustainably improve the productivity of existing lands.

We in the West can also learn a great deal from the cultures of people in the less developed countries. In fact, many modern writers champion a new worldview based on cooperation and holistic (systems) thinking. The new worldview is *anticipatory* in that it seeks to understand the long-term, systemwide impacts of our actions. It is also *participatory* in that it seeks to involve many people in decision making. These values are common in what people in the West disparagingly call "primitive cultures." On this matter, anthropologists Margaret and William Ellis once wrote, "There are new future concepts already conceived and still practiced by various people around the world that we need to understand, adapt, and adopt." Our future depends in part on learning important life lessons from the people we have inadvertently victimized in the past with our generous development assistance.

KEY CONCEPTS

Less developed nations will require huge sums of money to develop in a sustainable fashion. Several measures could make this possible, including decreases in military expenditures, reduced economic exploitation by the Western world, and debt relief.

To change and change for the better are two different things.

—German proverb

CRITICAL THINKING

Exercise Analysis

Raising the standard of living in the less developed countries is probably essential to creating a sustainable future, for poverty and environmental decay tend to go hand in hand. Numerous examples show that people who are poor and hungry treat the environment poorly. Furthermore, raising the standard of living—and along with it education levels, health care, and job opportunities—will have a positive effect on controlling population growth. However, raising the standard of living also puts additional strain on the environment's source and sink functions. In fact, the developed countries, with their high standard of living, are responsible for enormous environmental damage.

Ultimately, the Earth and global ecological systems cannot support a world population with a standard of living equivalent to that of the United States and Canada. One study, in fact, showed that if the entire world population lived like North Americans, resources would last about three months. Thus, many experts believe that some economic development is necessary, but it would probably have to be limited. Does this seem fair to you? Can you think of ways to allow greater economic development?

One proposal calls on the developed nations to make massive improvements in efficiency, recycling, renewable resource use, restoration, and population control to free up resources for the less developed nations. By greatly reducing our demand for resources and our stress on the environment, we in the developed world could free up resources for the less developed nations and permit them access to global sink functions—pollution assimilation.

The bottom line, though, is that whatever level of development occurs, human society must fit within global ecological constraints. Many experts think we've already pushed beyond the limits. In other words, we're already living unsustainably. This book has presented many statistics that suggest this may indeed be true. If that is the case, there is very little room for improvement of the lives of people in less developed nations without considerable reductions in resource use and environmental pollution by the developed world.

26-1 Bhutan: Pursuing a Sustainable Path

The country of Bhutan, nestled in the majestic Himalayas, is a country in transition. Its air is clean. Its forests are largely intact, and its wildlife are abundant and diverse (**FIGURE 1**). About the size of Switzerland, Bhutan has a population of only 700,000.

In Bhutan, land is distributed relatively equitably so that landlessness and poverty are rare. Today, 90% of the people live in rural areas, mostly on small farms where they grow corn and rice to feed their families, using draft animals to pull their plows. Although most basic needs are met, infant and maternal mortality rates are among the highest in the world, largely because of a lack of medical facilities and clean water. Additionally, 70% of the population is illiterate.

FIGURE 1 Bhutan: A country trying to preserve its natural resources and scenic beauty.

Bhutan's young monarch, King Wanchuk, wants to improve the condition of his people. To do this, he realizes, Bhutan could follow the economic development strategy of nearby countries, which have paid little attention to the environment. In neighboring India and Nepal, the results of this approach are painfully evident: Forests are severely overcut, and much of the soil is eroded. These examples and other factors (discussed shortly) convinced Bhutan's ruler to follow a sustainable path—one that would sustain human progress, not just in a few places for a few years, but for the entire country into the distant future.

In the May 1991 issue of *National Geographic*, the king was quoted as saying, "We would like to develop rapidly, but we would also like to ensure that there is a certain amount of harmony between rapid development and our culture and environment."

Fortunately, culture and environment are closely linked in Bhutan. The religion of Buddhism practiced by many of the nation's people dictates respect for the environment. In fact, some think that the practice of Buddhism is one reason the natural environment is still largely intact in Bhutan.

Today, 64% of Bhutan's land area is covered by forests, largely because of religiously dictated respect for trees. When it came to the government's attention that forest resources were being overcut in some areas, officials responded by nationalizing most logging operations. In 1985, the government enacted a National Forest Policy, which made the conservation of forests a top priority and economic considerations secondary. This policy, designed to ensure that 60% of the country remains forested, requires that the rate of tree harvesting be equal to the rate of replanting. To preserve part of the forests, the government of Bhutan has set aside more than 20% of its land in 10 reserves. The Royal Manas National Park covers 450 square kilometers (165 square miles) and provides sanctuary for many of the endangered species of south Asia. Elephants, golden langur monkeys, tigers, wild buffalo, and more than 500 species of birds can be found in the park.

CRITICAL THINKING AND CONCEPTS REVIEW

1. Describe the goals of sustainable economic development.
2. Traditional development schemes in the poor, nonindustrialized nations often result in increases in the GNPs of the target nations. Why is an increase in GNP not necessarily a sign of progress? Give some examples from this chapter.
3. You are appointed head of development in a poor African nation. Your population is 35 million and is growing at a rate of 3% per year. Many of your rural people are poor. Outline how you would go about improving the lives of the people and creating a sustainable future.
4. List some ways to make current economic development strategies more sustainable. Give examples of each one.
5. List the barriers currently standing in the way of sustainable economic development and ways to address them.
6. Using your critical thinking skills, debate the following statement: "In order to aid the people of the less developed nations, we must help them build factories to make tractors, bulldozers, and other equipment that

In 1990, Bhutan's highest officials passed the Paro Resolution, a call for a national sustainable development strategy. It recognizes that the key to sustainable development is to find a path that will allow the country to meet the needs of the people without undermining the natural resource base of the nation. The agreement states that new industries, new agricultural markets, and new forestry products need to be carefully developed with respect to the environment.

Development in Bhutan is slated to proceed slowly and to be based primarily on renewable resources, among them hydropower and solar energy. Officials see hydropower as the country's largest source of potential earnings. Today, most rivers still run free in Bhutan. The largest dam in the country produces $25 million worth of electricity, which is sold to India. It also supplies power to approximately 23,000 families in and around Bhutan's two largest towns. Bhutan hopes to build more dams in the future with financing from the World Bank. Aimed at increasing export income, these projects will be built only when the King and his advisors are sure that they will cause minimal environmental damage.

If the goal of development is to improve the lives of people, a better alternative might be photovoltaics, which produce electricity from sunlight. Used to power thousands of isolated farms and villages throughout the world, photovoltaics could provide a sustainable supply of energy for local needs without having to dam the nation's rivers. Unfortunately, foreign sources of funding for this approach are scarce, so photovoltaics are not being aggressively pursued by the government at this time.

Bhutan has also begun to develop extractive industries. In the north, the conifer forests are selectively harvested by the government-owned Bhutan Logging Corporation. In the south, a joint government–private company operates a large plywood plant. The company is allowed to clear-cut hardwood forests in the area as long as it replants them.

Bhutan has also opened its borders to tourists, but it limits the number to 1500–2000 a year. Those allowed into the country must follow a strict, government-set itinerary to limit their impact.

To deal with its rapidly growing population and fears that it will have to increase grain imports more than the current 10% to feed its people, the nation hopes to limit the growth rate of its population to 2% a year, in part by limiting families to two children. Contraceptives and family planning advice are available in 70 clinics across the country.

If industrialization is allowed to progress too far or too rapidly, the Bhutanese government knows that it is likely that natural resources will be irrevocably destroyed and the people will be impoverished. Rapid economic growth may be accompanied by growing inequality, which could tear the social fabric of the country. Increasing consumerism may lead to the pursuit of narrow economic goals.

The country of Bhutan, perhaps unique in the world, has the option of creating a prosperous and sustainable society from scratch. A fundamentally sustainable course is already in place. Environmentally damaging infrastructure and polluting industries have not been developed. The country is not hindered by debt, as many other nations are. Its cultural values remain strong and intact. Income is evenly distributed. There are few economic or ethnic tensions. Moreover, the government has strong convictions about sustainability.

Few countries have the option to start from scratch with so many things in their favor. But even with such strong convictions, the road to sustainability will require constant vigilance. Most think that the effort will be well worth it. As Worldwatch Institute's Christopher Flavin notes, "It will involve a process of continual adaptation and change, preserving what is best in Bhutanese society while improving living standards and protecting the natural resource base on which the country depends. Such an endeavor would truly make Bhutan a model for the world."

has been the key to success in the United States and other developed countries."

7. Using your knowledge of environmental issues and solutions and your knowledge of sustainable development, critically analyze the following statement:

"Environmental protection is a luxury. People in the less developed nations are starving and dying of disease. They cannot afford environmental protection efforts. Those can come only when the economic picture in these nations is much brighter."

KEY TERMS
nongovernmental organizations
    (NGOs)
sustainable development

Citizens working with local, state, and national governments are creating a sustainable future.

<table>
</table>

| CHAPTER 27 | # Law, Government, and Society |

Our environmental laws are not ordinary laws, they are laws of survival.

—Edmund Muskie

We need a future we can believe in, one that is neither so optimistic as to be unrealistic nor so grim as to invite apathy or despair. In short, we need a future that is not only hopeful but also attainable. This book has outlined one of many possible futures, a sustainable one in which human society thrives and prospers within the limits of nature. Creating such a system requires an understanding of basic principles of ecology and an appreciation of how vital natural systems are to our lives. It also depends on a decrease in population growth (Chapters 8 and 9) and dramatic changes in the way we use and manage resources (Chapters 10–17). To create a sustainable human presence we must also alter the way we deal with pollution (Chapters 18–23). Building a sustainable future, say

In 1991, many nations began to draft national reports in preparation for the United Nations Conference on Environment and Development in June 1992. The national reports were designed to assess what nations were doing to protect their environments and also to promote sustainable development. In the *U.S. National Report*, the authors frequently pointed to the billions of dollars that the government and the nation's businesses were spending on pollution control as a sign of our commitment to environmental protection and sustainable development. The authors also cited case studies that showed specific actions on the part of businesses and government agencies.

To some critics, the authors of the *National Report* seemed to be saying that because the United States is spending a great deal of money on environmental problems and is putting enormous effort into solving them, the nation is acting responsibly and with sufficient vigor. Is there a fallacy in this kind of reasoning?

many observers, will require fundamental changes in human systems, from agriculture to transportation to industry. The changes needed depend on a shift in people's attitudes (Chapter 24) and dramatic changes in economic systems (Chapters 25 and 26). Such modifications may come about in many ways—for example, through personal change and personal initiative or through market forces (such as the demand for green products). One effective means of social change is through government.

This chapter examines government: the roles of government, the participants in government policy making, and the decision-making process. It then discusses environmental law. Next, it discusses some key obstacles to a sustainable future and ways to overcome them. Finally, it outlines important changes in international governance needed to make a sustainable future a reality.

27.1 The Role of Government in Environmental Protection

Chapter 25 described two basic economic systems: market economies, in which the marketplace determines the availability and price of goods and services, and command economies, in which the government controls these functions. Chapter 25 also noted that most have features of both economies. However, they tend to lean in one direction or the other.

Forms of Government

For the most part, governments and economic systems go hand in hand. Countries with market economies are generally **democratic nations** with representative governments—that is, governments largely staffed by people who are elected by the people. Representative governments operate by rules (laws) that are, technically, agreed on by the majority through their elected officials. In democratic nations, the means of production and distribution of goods and services are, for the most part, privately held and privately managed. Profit is a major driving force of this system, and because of this, many large corporations play a powerful role in influencing policy, often to the detriment of the people. Thus, the economic systems tend to be market driven. However, democratic nations have found that varying degrees of governmental control are needed to reduce an assortment of unlawful activities.

Communist nations are based on the notion that all property is held in common. It is the antithesis of capitalism, with its privately held production capacity. Communist or socialist nations are politically organized around the social ownership of all goods and services. In other words, technically speaking, the people own and operate the production and distribution network. Such nations emphasize the requirements of the whole, rather than individual liberty. Profit is not a motive in such systems. The goal is that goods should be distributed equally, according to needs (although in practice this rarely occurs). In communist nations, the government manages the economy in the name of the betterment of people. History shows, however, that personal oppression is often rampant in such systems. Many thinkers believe that such systems are fundamentally unsustainable, and transformation of them is a key element of creating a sustainable future.

Just as economies are mixed, so are governments. New Zealand, Germany, and Sweden are all democratically governed, but in all three nations, the government provides health care and other services that are paid for through taxes. In China, one of the few remaining communist nations, the private sale of products for profit is currently permissible and growing rapidly. This chapter focuses primarily on democratic nations.

KEY CONCEPTS

Two basic forms of government exist: democratic and communist. Democratic nations are those in which most decisions are made by elected officials. The economies tend to be controlled by market forces. Communist nations are those designed around the notion of public ownership of the means of production and the distribution of goods and services according to need, not ability to pay.

Government Policies and Sustainability

Although democratic nations generally support free-market economies, governments have found it necessary to regulate free enterprise to protect unbridled business interests from exploiting people and the environment. The tools for regulating or influencing development are many, but fall into three general categories: taxes, financial support, and laws and regulations. The President is given a special tool, known as the *executive order*, which we examine shortly.

Taxes are a means of generating funds for public services, among them highways, education, road maintenance, police protection, pollution control, and wildlife programs. Taxes are levied on personal property, commodities, services, and income (both personal and corporate). As Chapter 25 notes, taxes can be used to discourage undesirable activities. For instance, a sizable tax on gasoline in European nations discourages automobile use and promotes the purchase of energy-efficient automobiles. It's not surprising that some of the best mass transit systems are found in Europe. Another example is taxes on ozone-depleting CFCs in the United States, which have discouraged companies from using them.

Two other common tools are tax credits and income tax deductions. A **tax credit** is a dollar amount of the purchase price that a person or company can deduct from income or corporate tax. In the United States in the late 1970s and early 1980s, for example, the federal government offered individuals and businesses sizable tax credits if they invested in energy conservation (insulation, weatherstripping) and renewable energy (solar panels and wind energy). Many states offered similar tax credits. In Colorado, for instance, the combined federal and state tax credit for solar energy came to 70%. Thus, 70% of the cost of a solar system could be deducted from one's tax bills. If you owed $1000 in tax and had a $700 tax credit, your tax bill would be $300. This form of tax policy passes the cost on to taxpayers.

Income tax deductions work similarly. When a government offers a deduction for an activity, the amount of the purchase or some percentage of the purchase price is deducted from one's total income. This lowering of income lowers the amount of tax one pays. If your taxable income was $50,000, but you had a $5000 tax deduction, you would only be taxed on $45,000. The U.S. government currently offers income tax deductions for children. Couples with children receive a small break for every child they have. Interest on loans for second homes was, until recently, also tax deductible. Tax policies that favor undesirable practices can also have significant impacts on our future.

Direct government expenditures can also have a profound effect on the environment and on sustainability. For instance, in the 1980s, government grants to communities for sewage treatment plant construction created thousands of jobs and helped reduce water pollution (Chapter 21). Governments also fund research and development in many activities, such as energy conservation and renewables. Government-funded research helps reduce the cost of commercial development and thus provides an economic advantage to some technologies. Government procurement (purchase) programs for environmentally beneficial technologies and materials (such as photovoltaics and recycled products) help increase production and bring costs down, making new products more attractive to citizens.

Government expenditures on environmentally less desirable technologies (nuclear energy research, for instance) and activities (war and oil development) have the opposite effect. The most obvious reason, say critics, is that such expenditures direct society down a potentially unsustainable path of development. The farther we go along that path, the harder it will be to shift to a sustainable one. Such expenditures also divert money from other important tasks, such as energy efficiency and renewable energy development.

Cutting government expenditures on environmentally destructive activities could free up considerable sums of money, which could be invested in promoting sustainability. Even modest reductions in global military spending could finance a massive campaign to restore the Earth and shift to wind and other renewable energy technologies to build a sustainable world community.

The final governmental tool, and probably the most widely used one, is laws and regulations. In democratic societies, laws emanate from legislative bodies: congresses, parliaments, and town councils. They are generally designed to curb or ban those activities people have deemed undesirable. For example, the **U.S. Corporate Average Fuel Economy Act** is a law passed in 1975. It established a set of goals for gas mileage standards for automobile manufacturers, thus curbing the waste of fuel.

A nation's laws are only as good as their enforcement. Enforcement depends on having adequate interest, personnel, and funds. If an agency lacks funds or human power or even the will to enforce a new law, little will come of it. Chapter 23 noted that the EPA took 12 years to make its first recommendations on the acceptable recycled content of a dozen or so products, an action mandated by the Resource Conservation and Recovery Act (RCRA). It finally complied only because of a lawsuit mounted by a prominent environmental group. It had failed to do so previously because of a shortage of labor and pressing tasks in other areas.

The Corporate Average Fuel Economy Act (just cited) is an example of a law whose utility was greatly reduced by lack of interest on the part of the responsible government agency. The Department of Transportation was given the responsibility for implementing and enforcing mileage standards, but it bowed to heavy pressure from Ford and General Motors, rolling back the mileage standards twice during the Reagan administration. In 1985, the average fleet mileage was far below Congress's goal of 11.9 kilometers/liter (27.5 mpg). When President George H. W. Bush took office, however, he pledged to raise standards, and by 1989, the average new car did achieve this goal. Further improvements called for in the ensuing years, however, have been slow, and efforts to raise the mileage standard in the United States to 17.3 kpl (40 miles per gallon) by 2000 were defeated in 1991. For 5 years in a row in the late 1990s, Congress passed laws forbidding the Department of Transportation from even considering increases in mileage standards! Similar resistance has been met by the Congress, Senate, and present administration.

Like the president or chief operating officer of a corporation, the president of a nation can have a powerful impact on environmental protection. Unsympathetic leaders make it difficult to move forward. The record shows that President Reagan, for instance, who was in office from 1981 to 1989, openly opposed the environmental protection goals of previous administrations. He actively sought to undermine protection by weakening existing laws and agencies, such as the EPA and the Council on Environmental Quality. President George Bush, who laid claim to the title of "Environmental President," improved matters somewhat; behind the scenes, the Council on Competitiveness (chaired by Vice President Dan Quayle) systematically worked to undermine the nation's environmental laws and regulations. The Council proposed, for instance, dramatic changes in wetland designation, which would have eliminated about half of the nation's wetlands from protection. After passage of the 1990 Clean Air Act amendments, the Council unsuccessfully proposed a strategy for companies to avoid compliance. If a company felt it couldn't meet requirements of the law, all it would have had to do was write a letter to the governor of its state. If the governor didn't reply within a week, the company would have been exempt from the law.

President Clinton was much more sympathetic to the environmental issues. He and Vice President Gore took numerous steps to improve environmental protection but were thwarted by a Republican-controlled Congress. (see Spotlight on Sustainable Development 27-1). President George W. Bush has systematically worked to weaken environmental laws and regulations—far more actively and successfully than President Reagan.

The United States is not the only country in which environmental laws fail to accomplish their goals because of lack of funds, apathy, lack of enforcement or political opposition. Before it disintegrated, the Soviet Union had some of the most stringent environmental standards in the world, but enforcement was virtually nonexistent. The country lacked the technical expertise and the financial resources to control pollution.

Environmental lawmakers usually set broad goals in their legislation, but the job of writing specific regulations to put those goals into effect lies in the hands of experts within various government agencies. In the United States, the EPA is the principal environmental rule maker. A host of other agencies also participate in this process when laws pertain to matters under their jurisdiction. The Forest Service, for instance, drafts regulations for management of forests. These agencies propose regulations, based on their interpretation of the law. The regulations are then subject to public comment. (This is where citizens and nonprofit organizations can play a big role in helping to shape final regulations!) After receiving input from various stakeholders, the agency may modify its proposed regulations and put them into effect. The EPA is currently empowered by nine federal laws, among them the Clean Water Act, the Clean Air Act, the Superfund Act, and the Resource Conservation and Recovery Act (RCRA).

The **executive order** is yet another tool for effecting development and protecting the environment, but it is reserved for the President of the United States. An executive order is just that, an order issued by the President of the United States. President Clinton used it effectively and often to reduce the environmental impact of the federal government. Executive Order 12843, for instance, implemented an early phaseout of ozone-depleting chemicals at federal facilities. Executive Order 12844 accelerated the acquisition of vehicles that use alternative (clean-burning) fuels by federal agencies. Executive Order 12845 required the federal agencies to purchase energy-efficient (EPA Energy Star) computers, monitors, and printers. Executive Order 12856 directed federal agencies to cut toxic emissions by 50%. President Clinton also signed an Executive Order 12873 promoting the purchase of green products. Green dot markers in the federal supply catalogues helped officials select environmentally friendly products. The federal government also maintains lists of energy-efficient lighting and energy-efficient computers to make the job easier. President Nixon used an executive order to establish the Environmental Protection Agency. The executive order can also be used to weaken environmental protection, President Bush, for instance, has used it often in this fashion.

Through the mechanisms outlined previously, governments around the world have made tremendous strides in protecting the environment, managing resources, and controlling population. Many chapters in this book have pointed out the most significant advances. Table 18-3 lists nearly two dozen environmental laws and amendments enacted in the United States since 1958. Similar progress has been made in most other industrial nations. Although environmental conditions are better than they would have been in the absence of these measures, most governments have dealt with environmental problems after the fact; that is, they have typically relied on end-of-pipe solutions or regulations that merely slow down the destruction of global ecosystems. This chapter describes policies needed to confront the roots of the problems and create a sustainable way of life.

What about the less developed nations, where over 5 billion people now live? Many less developed nations have adopted policies to control population growth, save wildlife, and reduce pollution. However, progress has been hindered in many cases by a lack of money, rapid growth of population, corruption, civil war, war with neighboring nations, and governmental apathy and ineptitude. Interference by foreign interests and misguided development projects have also worsened environmental quality (Chapter 26). In many cases, hunger and poverty are so severe that environmental matters are relegated to the back seat. Without population stabilization and programs to raise the standard of living and the availability of food, little progress can be expected.

KEY CONCEPTS

Democratic governments regulate activities, such as environmental protection, via three measures: (1) tax policy, (2) direct financial support, and (3) laws and regulations. The President also wields enormous power through the executive order.

27-1   Greening the White House

Shortly after he took office, President Bill Clinton (FIGURE 1) began the task of remodeling the White House. Although Presidents and their wives have long been known to spruce up the Executive mansion to meet their individual tastes, this remodeling project was unlike any other president's in U.S. history. Far from the cosmetic changes of previous administrations, this was to be a "green makeover"—a series of environmental upgrades of the White House complex, consisting of the Executive Residence, where the President and his family live, the Old Executive Office Building (OEOB) where many White House staffers work, and the White House grounds.

FIGURE 1 President Clinton

Spurred on by Vice President Al Gore, long known for his commitment to environmental issues, this project was designed to achieve a number of important goals: to reduce pollution, slash energy demand, protect water quality, conserve water, improve the comfort of the structure, and save the government and taxpayers money. Actions were many and varied from reducing pesticide use to using soaker hoses to water plants to reduce water consumption to installing an energy-efficient refrigerator in the First Family's residence. Here's a quick overview of some of the many measures taken on behalf of the President:

*Installing more efficient cooling units, known as chillers, which provide cool air to rooms.

*Installing chillers that do not use ozone-depleting CFCs in air coolers, using HFC-134a, which contains no chlorine.

*Upgrading heating ventilation and air conditioning systems with energy management control systems that allow more precise control of room temperature to reduce overall heating and cooling demands.

*Capturing waste heat to preheat domestic hot water.

*Replacing window-mounted air conditioners with much more efficient ones—there are 1000 air conditioners and about 10% are replaced every year. New units have timers and occupancy sensors.

*Rezoning steam heating system for better control of heating.

*Cleaning up the White House complex paint shops—they were a source of toxic VOCs (volatile organic solvents). Problems resulted from inadequate ventilation, numerous open cans, pans, and trays of paint, lacquer, thinner, solvents, etc.

*Painting off site at a state of the art painting facility maintained by the National Park Service.

27.2 Political Decision Making: The Players and the Process

Now that we've seen the basic tools by which governments influence the environment and shape the future of a country (indeed, the entire world), let us take a closer look at the players and the process.

Government Officials

In communist and other authoritarian governments, policy is largely determined by a ruling elite. Ironically, as noted earlier in this chapter, in such systems the people whom the government is meant to serve generally have little or no input. In fact, they often live under severe oppression. In contrast, in democratic nations a wide assortment of government of-

ficials make important decisions that affect the environment and the future of the nation. Some government officials (senators, for instance) are elected by the people; others (agency heads) are appointed by the president, governors, or mayors. Their role is to make laws and regulations and enforce them. In other words, they are the makers and guardians of public policy, although in recent years just the opposite is taking place in the United States.

Corruption is known to abound in communist and other authoritarian nations, but representative government, for all its merits, does not always operate as its founding fathers intended. Appointed officials in the EPA and state health departments, for example, have been known to overlook violations of environmental and health laws. Today, many high-ranking employees of federal agencies that once worked for industry are actively weakening environmental

*Implementing a no smoking policy in the Residence, in East and West Wings, and OEOB.
*Reducing pesticide use and training pesticide applicators.
*Reducing fertilizer use.
*Adjusting mowing practices to minimize water demand.
*Using mulching lawn mowers to grind up grass clippings, returning nutrients to soil.
*Irrigating early in day between 6 and 7 am to reduce water use.
*Using soaker hoses whenever possible for watering vegetation.
*Installing moisture sensors for irrigation system.
*Adjusting and replacing faulty sprinkler heads.
*Using integrated pest management to control insect pests.
*Installing compact fluorescent lighting wherever possible—for example in table lamps—to reduce electrical demand and pollution.
*Installing occupancy sensors for lighting.
*Upgrading outdoor lighting, cutting electrical consumption by 51%.
*Rehabilitating existing historical skylights on fifth floor of the OEOB to increase daylight—dirt and paint on glass had reduced light penetration. This will reduce electrical use in these areas by 20 to 75%.
*Purchasing energy-efficient computers and office equipment, cutting energy consumption by 50% or more.
*Installing water conserving devices in restrooms, kitchens, and other areas.
*Recirculating cooling water in chillers, rather than using fresh water to supply them. This eliminates five gallons of water per minute previously discharged into sewage system.
*Instructing staff to use green products as often as possible.

*Establishing a source reduction policy to reduce paper consumption, limit use of disposable products, and conserve office supplies.
*Making special events green by minimizing disposable items, using recycled materials, making bulk purchases, and making recycling containers visible.
*Recycling batteries.
*Increasing composting of yard waste.
*Recycling compact fluorescent light bulbs to reduce flow of heavy metals such as mercury to landfills.
*Recycling polystyrene plastic in the cafeteria.
*Replacing glass in third-floor solarium and greenhouse in Executive Residence with high-performance insulating glass.
*Retrofitting windows of OEOB with double-paned glass.

The greening of the White House is an inspiration to all of us. As President Clinton said at a speech on Earth Day, 1994, "We're going to identify what it takes to make the White House a model for efficiency and waste reduction, and then we're going to get the job done. . . . Before I can ask you to do the best you can in your house, I ought to make sure I'm doing the best I can in my house." That's called leadership by example.

But this project wasn't carried out by the President on weekends. The American Institute of Architects coordinated 100 national experts, who made recommendations. The Environmental Protection Agency and the Department of Energy played a huge role in the success of this project, as well. Tonight when the current President sits down to read in his study, he'll be working under the light of a compact fluorescent light bulb.

protection. Elected officials are often inclined to vote in favor of legislation that is supported by big-money interests that, not coincidentally, donate large sums to their campaigns. This is not to detract from the fact that dedicated, hard-working men and women at all levels of government have also made important contributions to bettering the environment.

KEY CONCEPTS

Government officials are charged with the task of creating laws and regulations and enforcing them for the general good. In democratic nations, government officials are either elected or appointed to office. Theoretically, both types are accountable to the general public. Elected officials are generally more responsive, though, as their reelection depends on satisfying the electorate.

The Public

In a democratic society, voters generally have a say in policy making through their elected representatives (FIGURE 27-1). Letters, phone calls, attendance at town meetings, and responses to surveys are four lines of communication that permit us to let our opinions be known on a variety of issues.

Although citizens are sometimes prone to forget, elected representatives are their employees. Their job is to interpret public preferences and find ways of enacting policies that satisfy the needs of the electorate. In many cases, however, the populace is divided on issues. Elected representatives may even choose to vote against their constituencies' wishes. Public officials are also frequently influenced by special interests—business groups or environmental groups. For this and other reasons, many people have grown cynical about their influence on the government and about their own fu-

FIGURE 27-1 **The many lines of connection among government, individuals, and corporations.** Laws are passed by lawmakers. These, in turn, affect the behavior of individuals and businesses. Individuals, environmental groups, and businesses also influence the laws through lobbyists and the court system.

ture. Many people feel victimized and disempowered by the political process, which leads to apathy and poor voter turnout at elections.

Apathy is counterproductive to the political process. Citizens can make a difference. A letter to a congressional representative may not change a vote, but many letters could. As a rule, for each letter they receive, legislators estimate that there are 5000 to 10,000 people who feel the same way. One thousand letters means there are 5 million people who think similarly, which is very compelling to an elected official.

Letters to state legislators can have special impact because they rarely hear from the public. Twenty or 30 letters are cause for concern. In their minds, 40 or 50 letters indicate a disaster requiring action. So, write and vote. Let your voice be heard.

KEY CONCEPTS

Public policy is forged by elected officials with input from the general public. Although many people are discouraged with the process and are apathetic, letters and other forms of communication to elected officials can help shape public policy.

Special Interest Groups

The **theory of public choice** states that politicians act in ways that maximize their chances of reelection. To do so, they must appease the voters and the special interest groups (Figure 27-1). Special interest groups, defined as organized forces of like-minded individuals with common goals (such as environmental organizations and industry groups), often work through **lobbyists**, individuals who work to educate legislators and their aides and convince them of the merits of their particular views. In some cases, lobbyists even draft legislation. When Harry Truman was president, about 450 lobbyists worked the halls of Congress; today, an estimated 23,000 lobbyists swarm the nation's capital.

Special interests often reward politicians monetarily, providing funds needed to run costly political campaigns. To make an even larger impact, many special interests band together to form **political action committees**, or PACs. A PAC is a consortium of individuals or organizations that pool their financial resources to create a single large donation. In the 1988 elections, PACs donated more than $150 million

to congressional candidates, up from $34 million a decade earlier (**FIGURE 27-2**). In the 1992 election, they contributed an estimated $254 million. In the 2004 election, PAC contributions totaled $916 million, up from $214 million in 1996.

According to Common Cause, a nonprofit citizens' lobbying group, PAC contributions pay off "in billions of dollars worth of government favors for the corporations and other special interest groups that make them." PAC contributions from business, for example, may help to defeat or to water down important environmental legislation. In the

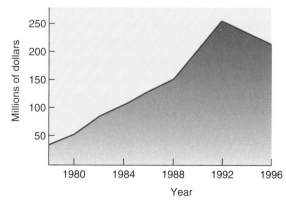

FIGURE 27-2 **The power of PACs.** Political action committee contributions to congressional candidates from 1978 to 1996.

1980s, PAC contributions and coal and utility industry lobbyists allegedly stalled the enactment of legislation to reduce acid deposition, toxic air pollution, and other important problems. In 1984, auto manufacturers successfully lobbied the U.S. Congress to pass import quotas on energy-efficient Japanese automobiles, which yielded the companies $300 million in profit and cost the consumer an estimated $2 billion or more.

Another new political and economic force is the 527 group. These political organizations accept huge donations from wealthy individuals and ordinary citizens. The donations are used to run ads supporting like-minded candidates or political positions. They currently bypass federal election campaign contribution laws.

One of the most insidious political games played by special interests, and especially business interests, is known as the **Double-C/Double-P game**. Double-C/Double-P stands for "Commonize the Costs and Privatize the Profits." Here's how it is played: Business lobbyists support and sometimes even propose legislation that promotes government programs that could generate huge profits for the private sector. Thus, taxpayers end up paying for programs that support business. Publicly supported sports stadiums are a good example. Some people believe the war in Iraq is another, as several large corporations that supported the Bush Administration are now profiting in the rebuilding of the country.

Although democracy can be distorted by special interest groups, especially business PACs, money does not always reign supreme. In the United States, powerful antipollution laws, auto safety standards, hazardous-waste laws, and other important environmental laws have been enacted in the last three decades despite strong opposition from some of the business community. A good measure of this success can be attributed to the efforts of environmental groups.

Environmental Groups

Environmental groups affect public policy in several ways. Some groups, such as Greenpeace, take an active role on the front lines, protesting environmental injustices. Greenpeace members often meet face to face with whalers, seal hunters, and polluters, endangering their own lives to protest actions they oppose (FIGURE 27-3). Such public displays have proved highly successful in raising awareness on issues. In 1980, a radical group of environmentalists came into being, calling themselves **Earth First!** Under the leadership of Dave Foreman, they took more active, frequently illegal steps to protect the environment. They put their bodies on the line to stop bulldozers. They drove spikes into old-growth trees to thwart chain saws. They put dirt in the fuel tanks of heavy equipment. Their tactics were called **ecotage**—sabotage in the name of the environment. (For a debate on the pros and cons of direct action, see the Point/Counterpoint in this chapter.)

Some years later, Paul Watson, one of the original founders of Greenpeace, started the oceanic equivalent of Earth First! Called the Seashepherd Conservation Society, it roams the high seas, ramming ships that are illegally whaling, flying miniature airplanes to disrupt sonar, and generally disrupting illegal activities. Its members have never been arrested or charged with a crime because they stop people engaged in illegal activities, who don't want public attention drawn to them.

Most environmental groups operate within legal means. Some groups, for instance, are involved in education. They may prepare educational materials for schools or deliver talks to public gatherings. Some groups work to preserve land to protect biodiversity. The Nature Conservancy, for instance, purchases land that is then set aside to protect endangered wildlife and plants, to provide future generations with opportunities to enjoy the out-of-doors. Other groups are involved in lobbying efforts, writing new laws and getting them passed, and strengthening existing laws. Still others serve as watchdogs, making sure that polluters obey the rules and that government agencies do their jobs.

Many environmental groups have become involved in proactive planning, devising sustainable alternatives to environmentally harmful projects. Environmental Defense has been a leader in this arena. Its staff of economists, public policy experts, and scientists research issues and offer economically and environmentally sound alternatives. Environmental Defense worked with McDonald's Corporation, for example, to institute the company's recycling and waste reduction program, such as the Trial Lawyers for Public Justice in Washington, D.C.

Some nonprofit groups spend much of their time in court, suing governmental agencies when they are not do-

FIGURE 27-3 Greenpeace activists position themselves between whaling ships and whales in order to thwart the slaughter of whales.

Direct Action Wastes Limited Resources of the Environmental Community

Randal O'Toole
As an economist for Cascade Holistic Economic Consultants, of Portland, Oregon, Randal O'Toole has examined management practices on over 60 national forests. He is the author of *Reforming the Forest Service,* a comprehensive package of reforms based on his research on the national forests.

Civil disobedience, ecotage, and other direct actions aimed at stopping ecological destruction have scored a few spectacular victories. In Australia, civil disobedience successfully halted construction of the economically and environmentally costly Franklin Dam. In the Pacific Northwest, direct action brought national attention to the spotted owl and other old-growth forest-dependent wildlife, which may yet lead to their permanent protection.

For the most part, however, direct action has proven largely ineffective, even counterproductive, at protecting the environment. Moreover, it raises important ethical questions that its advocates often ignore.

Of the two basic types of direct action—civil disobedience and ecotage—ecotage is the least effective because all it does is increase the cost to resource developers. If millions of dollars can be made from cutting old-growth forests or drilling for oil, adding a few thousand dollars to the costs will not stop the cutting or drilling. Ecotage may make the ecosaboteurs feel good, but it does little to protect the environment.

The most successful direct actions have all used civil disobedience—a human blockade in the case of Franklin Dam, tree-sitting in the case of old-growth forests. The goal of such actions is to gain media exposure so that people learn about environmental problems. Yet, civil disobedience suffers from two major weaknesses. First, it only works a few times. The first person who sits in a redwood tree is news. The hundredth person is boring.

More importantly, civil disobedience works because it polarizes people into action—letter writing, testimony, and other lobbying. This seems good at first because it encourages Congress or state legislatures to address environmental problems. In the long run, however, direct actions polarize people on both sides of an issue. They help opponents of environmental protection as much as they help supporters.

This is apparent in the West, which has recently seen the rapid growth of a counterenvironmental crusade known as the Wise Use Movement. Their followers are people upset by environmental tactics, government bureaucracy, taxes, and

other problems. Direct action drives people into this camp as much as it gains environmental supporters.

Beyond the effectiveness of the direct action are questions of ethics. A few environmental problems may be life-and-death issues for the human race. But, to be honest, most are just questions of values: Is a fish worth more than a dam? Is an owl worth more than the houses that can be built from wood?

Environmentalists sometimes turn to direct action out of frustration that the current system does not allow them to express their values. But too often, direct action attempts to forcibly or legislatively impose their values on others, which threatens the stability of our society.

Instead of trying to impose values on others, the environment is better protected by finding ways to express environmental values. Many resources, such as air, wildlife, and outdoor recreation on the public lands, are free or nearly free. As a result, they are overused and abused. When a forest manager compares timber values, which directly add to the manager's income, with recreation values, which may be real but contribute nothing to the manager's income, timber values nearly always win out.

The solution is to find institutions that allow us to express our appreciation for the environment. Some of these institutions already exist. Groups like The Nature Conservancy buy critical wildlife habitat. The Clean Air Act allows polluters to sell their pollution permits, thus giving them an incentive to reduce their pollution. Other institutions must be created. Recreation fees would give forest managers a reason to protect scenic beauty and other resources that recreationists value. Tradeable water rights would allow fishing groups like Trout Unlimited to buy irrigation water and leave it in the stream.

Like anyone, environmentalists have limited resources to spend on actions to protect the environment. Those resources must be divided between lobbying, litigation, research, and other activities. To be as effective—and ethical—as possible, environmentalists should concentrate on promoting institutions that protect our environment as well as the stability of our society.

You can link to websites that represent both sides through Point/Counterpoint: Furthering the Debate at this book's internet site, http://environment.jbpub.com/. Evaluate each side's argument more fully and clarify your own opinion.

Civil disobedience and other direct actions catapulted old-growth forest destruction into the national consciousness. Direct action helped stop the infamous Vegas-Barstow dirt bike race in the fragile California desert.

Recently, the U.S. Forest Service cancelled the remainder of the Cove and Mallard timber sales in central Idaho's Salmon-Selwag Ecosystem. Nearly a decade of direct actions had slowed logging to less than half of the original goal. Now, a new policy to protect roadless areas has given the uncut wildlands a major reprieve. And in Seattle, mass demonstrations and human blockades made the World Trade Organization's ability to veto environmental and labor laws of sovereign nations an international issue. Overnight Ecotage (monkey-wrenching) has also received lots of press, but the subject upsets many, and discussing it tends to obfuscate the biodiversity issue.

Traditional tactics (letters, lobbying, litigation, articles, slide shows, rallies, and boycotts) have also saved many wild places. The Grand Canyon remains undammed and the National Wilderness System is expanding, proving that traditional tactics work. Or do they?

Despite impressive victories, habitat destruction is rampant. Consequently, species, subspecies, and populations become locally, regionally, or entirely extinct as native wildland ecosystems collapse. Dozens of species and subspecies in the United States have become extinct since the Endangered Species Act was enacted in 1973. Since before World War II, the U.S. Forest Service has annually destroyed—mostly via logging and road building—about a million acres of unprotected wilderness, despite the growing National Wilderness System (mostly growing with rocky high-altitude additions).

Wildland ecosystems are under assault everywhere, by loggers, miners, cattlemen, power companies, dam builders, and subdividers. Urban sprawl abuts wilderness. Air pollution kills forests. Fences halt wildlife migrations. Poachers exterminate black bear and black rhino for body parts used for mythical aphrodisiacs. Ignorant all-terrain vehicle enthusiasts crush desert tortoises, bird nests, insects, and small mammals—not to mention the monumental environmental threat of global warming.

Today's global ecological crisis is complex and unprecedented. The late Cretaceous extinction 65 million years ago pales when compared with today's galloping loss of biodiversity. Globally, over one species per hour is wiped out. The obvious concern, of course, is survival. How much biodiversity can we lose before the life-support web crumbles or before humans can no longer live comfortably? Or at all? In truth, we don't know and perhaps can never know in time to avert monumental disaster.

But the question of survival obscures another profound consideration. Life has intrinsic value. Some economists reduce all conflicts to opposing values: costs versus benefits. "Is an owl 'worth' more than the houses derived from wood?" they ask. Such reductionism opposes nearly all enlightened historic human thought. Do we value only human life? Are we appalled only when torture or genocide victims are human? I think not. The value of thriving diverse native life transcends the almighty buck. Those who resort to nonviolent civil disobedience-related direct actions are imposing their values on others no more than corporations that donate money

Direct Action Fills an Empty Niche in Wildland Conservation

Howie Wolke
Howie Wolke is a wilderness backpacking guide, owner of Big Wilderness Adventures, based near Darby, Montana. A longtime wildland activist, he is the author of *Wilderness on the Rocks,* a critique of wildland conservation in America, and coauthor (with Dave Foreman) of *The Big Outside,* a descriptive inventory of the remaining big roadless areas in the United States.

to political campaigns or, for that matter, than those who threw the Boston Tea Party.

Three questions should always be confronted when considering direct action: (1) Is the action nonviolent toward life? (2) Will the action increase public awareness of a problem? (3) Will it slow or delay destruction so that long-term protection remains plausible?

The success of direct action is apparent. Besides some obvious victories, direct action brings new activists into wildland battles. It permits various mainstream environmental groups to take stronger positions without appearing "radical." Despite contrary claims, civil disobedience still draws attention. Yes, the first tree-sitter is news; the hundredth isn't. Unless she stays tree-bound for many months. Civil disobedience is effective when applied strategically: Save it for the worst atrocities. And use your imagination.

The bottom line is this: *Nothing* yet has stopped today's spiraling loss of habitat and biodiversity. We need better laws, better incentives, better news media, better representatives, and better institutions. But, we can't afford to wait. We must seize any opportunity to save habitat. Life is falling through the cracks of the well intentioned. In these complex times, simple solutions won't work. Wildland conservation must be eclectic.

Just as lobbying isn't for everyone, neither is civil disobedience. To expand the realm of effective conservation activism, few avenues of ecological defense should be closed. Specialization works. The Nature Conservancy buys land. The Alliance for the Wild Rockies lobbies and educates. The EarthJustice Legal Defense Fund litigates. Earth First! specializes in direct action. Ultimately, no human or nonhuman society can sustain itself in a wasteland. The challenge is to protect habitat. Sustainability and the joy of unharnessed life will follow.

Critical Thinking Questions

1. There are many forms of direct action to protect the environment. Which ones, if any, do you support? Why?
2. Would you support spiking trees in a forest to prevent it from being cut down? If so, why? If not, why not?
3. What values influence your answer to question 2?

ing their job or are doing it poorly, or suing businesses that violate environmental laws.

Many environmental groups operate on several levels. Environmental Defense (ED), for instance, began as a litigator, suing government agencies and businesses. Today, it works closely with business and government to find alternative practices that serve people and the environment. ED also publishes books and newsletters that influence public opinion and public policy.

Environmental groups can offset the disproportionate influence of business, but they are usually at a disadvantage. Corporations such as Exxon Mobil, which boasts a $25 billion annual income, usually wield more influence in the political process. They are currently funding a massive campaign to convince the American public that global warming does not pose a significant threat.

Unfortunately, a strong antienvironmental movement is cropping up in the United States. Called the **Wise Use Movement,** it is heavily funded by timber companies, mining companies, oil and coal companies, cattle ranchers, and other similar parties. It is spreading half-truths about the environment and is actively trying to dismantle environmental protection.

KEY CONCEPTS

Environmental organizations take many forms and work in many ways to protect the environment.

27.3 Environment and Law: Creating a Sustainable Future

The late Edmund Muskie, a longtime senator from Maine, once noted that "environmental laws are not ordinary laws, they are laws of survival." Today, more and more people are realizing that environmental protection is indeed essential to the long-term future, but they also recognize that environmental protection is not enough to achieve sustainability. The social and economic goals described in this book are also an essential component. They will help us build a better future, more effective and cost-efficient government, broaden prosperity, and a more stable world through mutual responsibility. This section, however, examines U.S. environmental laws.

Evolution of U.S. Environmental Law

How did the U.S. environmental laws and regulations evolve? The progression of laws followed a logical evolutionary route that began with ordinances imposed by local governments. Interested in protecting health and environmental quality, officials of cities and towns passed local laws to limit activities of private citizens for the good of the whole. For example, municipal ordinances regulated trash burning within city limits to reduce air pollution. These efforts began in the 1800s.

In the mid-1900s, however, it became clear that efforts to control problems such as air pollution through local ordinances were often inadequate, especially in densely populated regions. Disputes arose between neighboring municipalities with different laws. Regions with strict laws were hampered in cleaning up their rivers by upstream cities with lax pollution laws. Thus, states began drafting legislation to regulate water pollution.

Soon, state laws proved inadequate, too, because air and rivers flow freely across state borders. Thus, interstate conflicts over pollution replaced the conflicts between neighboring municipalities. In addition, state programs were often inadequately funded and lacked the technical expertise needed to set pollution standards. State agencies also found themselves powerless against large corporate polluters with political influence in the courts and legislatures. Because of these problems and the growing effectiveness of special interest groups, environmental controls shifted to the federal government, gradually in the 1940s and 1950s and then more rapidly.

Initially, the federal government restricted itself to research on health and pollution control technologies. This approach met little opposition from state and local governments. Next, the federal government began offering grants to fund pollution control projects and the formation of state pollution enforcement agencies.

With increasing pressure from environmental groups, citizens, and many businesses, the federal government began to take a larger role in enforcement. Today, much to the dismay of some, the federal government sets ambient pollution standards and standards for emissions from factories, automobiles, and other sources and can take strict enforcement actions if needed. Stiff fines and penalties are frequently used to coerce businesses into cooperating. Corporate executives have even been jailed.

The shift to federal control is based on at least two important principles of American federalism: (1) When it is important to maintain uniform standards, the federal government provides the best means. Uniform policies minimize interstate conflicts and create a fairer economic system for businesses. (2) The power of the federal government to tax is much stronger than a state's. To control pollution effectively requires expensive research, which the federal government can more easily afford. Furthermore, it would be costly, time-consuming, and redundant for each state to carry out this extensive research.

The shift to the federal level has some disadvantages, however. First, the federal government officials may not always understand the problems of the regions it regulates as well as local officials do. Another criticism is that states should have the right to do as they please with their own resources; in other words, federal control diminishes self-determination and self-governance. Without central controls, however, states impinge on each other's quality of life. For example, poor watershed management leading to erosion in the Rocky Mountain states could have long-term adverse impacts on downstream users to the east and west.

One way of addressing these problems is to develop federal standards but to allow the states to manage their own programs. Thus, the Clean Air Act, the Surface Mining Control and Reclamation Act, and the Resource Conservation and Recovery Act all permit the states to run their own programs as long as they are at least as stringent as the one set up by federal law. These acts also provide money to assist the states in setting up their own programs.

KEY CONCEPTS

U.S. environmental law began locally and evolved to higher levels of government as it became apparent that local problems spread to other areas and required a higher political authority to address them.

The National Environmental Policy Act

One of the most significant advances in environmental protection of the past three decades was the **National Environmental Policy Act** (NEPA) (1969). NEPA is a brief, rather general statute with several key goals. First, it declares a national policy to "use all practicable means" to minimize the environmental impact of federal actions. More specifically, NEPA requires that decisions regarding federally controlled or subsidized projects (such as dams, highways, and airports) must be made after considering possible adverse impacts outlined in an **environmental impact statement (EIS)**. Among other things, an EIS must describe (1) what the project is; (2) the need for it; (3) its environmental impact, both in the short term and the long term; and (4) proposals to minimize, or (in the language of bureaucrats) to mitigate, the impact, including alternatives to the project. Drafts of the EIS are available for public comment and review by federal agencies. Written comments from the public must be addressed in the final EIS, which is issued at least 30 days before undertaking the proposed action.

The EIS has been an effective way of getting businesses and governmental agencies to focus on the environmental impacts of their projects. Thus, it has helped promote vision in economic development strategies. The underlying goal of the EIS is that those who become aware of potential impacts of their plans will act responsibly to avert them as much as possible. In this sense, the EIS is a political carrot (an inducement) rather than a political stick (a punishment). Available early in the planning stage, it can help decision makers determine whether they are implementing their policies, programs, and plans in compliance with the national environmental goals expressed in the NEPA and other legislation.

NEPA also established the **Council on Environmental Quality (CEQ)** in the executive branch. The Council once published an annual report (*Environmental Quality*) on the environment and on environmental protection efforts of the federal government. It also developed and recommended new environmental policies to the President. The CEQ was all but abolished by President Clinton, who established in its place the **Office of the Environment** (an environmental advisor to the President) and the **President's Council on Sustainable Development**, a large volunteer organization composed of business representatives, environmentalists, and others who have actively examined various strategies for sustainable development in the United States.

NEPA has significantly affected federal decision making. It has led to hundreds of lawsuits filed by environmental groups against the government, perhaps more than any other environmental statute. In addition, several states and nations throughout the world have passed laws or issued executive or administrative orders patterned after NEPA. France, Canada, Australia, New Zealand, and Sweden all require EISs. California passed an **Environmental Quality Act** in 1970 that requires EISs for all projects—private and public—that will affect the environment in a significant way.

The success of NEPA has been great, but the law is not without flaws. One of the most frequent criticisms is that EISs are too lengthy and deal with too many peripheral subjects. Reports are often ignored; they may show serious adverse impacts, but the project will be approved and carried out anyway, often without ameliorative actions. A common complaint from environmentalists is that reports are often based on inadequate information. Projections of environmental impact are difficult to make and often too subjective. Practically no work has been done to see if the projected impacts actually materialize; thus, Americans continue to be unprepared to make sound projections about impact. EISs may be "doctored" by agencies or private consulting firms that write them for federal agencies, to hide the real impacts. Some agencies can avoid writing EISs by simply stating that there will be no adverse environmental impact; it is then up to others to prove the need in court. Other critics of the EIS contend that it is too costly and often leads to delays in important projects. The paperwork and time involved seem excessive.

To answer some of these complaints, the Council on Environmental Quality issued streamlined procedures for preparing EISs in 1979. They call for (1) a maximum length of 150 pages, except for more involved projects; (2) a summary of no more than 15 pages that describes major findings and conclusions; (3) documentation and referencing of projected impacts; and (4) the use of clear, concise, and understandable language.

EISs are prepared after the fact; that is, they are designed to assess impacts of projects that have already been planned. A more appropriate method would be to require analyses of need and then seek sustainable ways to meet the needs. Some environmentalists believe that NEPA should require agencies to select the most environmentally benign and cost-effective approach, both in the short and the long terms.

Another unanswered problem is that although environmental groups can sue an agency that they believe should have filed an EIS or that has filed an inadequate EIS, they cannot recover attorney's fees from such suits. Recovery of attorney's fees would relieve the costly burden of forcing governmental agencies to heed the law (although critics point out that it might also open the door to numerous costly lawsuits, which would be a burden on the taxpayer).

Although it has its critics and still stands in need of improvement, NEPA is the cornerstone of U.S. environmental policy. Numerous federal agencies have reported important environmental benefits from it as well as economic savings from recently revised rules.

The Environmental Protection Agency

Another major environmental accomplishment in the United States was the establishment of the **Environmental Protection Agency** in 1970. The EPA was founded by an executive order drafted by President Nixon, which called for a major reorganization of 15 existing federal agencies that were already working on important environmental issues.

The EPA was directed to carry out the Federal Water Pollution Control Act and the Clean Air Act. Today, it manages and enforces most of the environmental protection laws that issue from Congress. Current responsibilities of the EPA include research on the health and environmental impacts of a wide range of pollutants, as well as the development and enforcement of health and environmental standards for pollutants outside the workplace. The EPA is concerned with a variety of areas, including pesticides, hazardous wastes, solid wastes, toxic substances, water pollution, air pollution, radiation, and noise pollution.

The EPA can provide incentives to state and local communities through grants that help pay for water pollution control projects. Grants to universities have helped expand the research capability of the agency. The EPA carries on much of its own research at four National Environmental Research Centers, located at Cincinnati, Ohio; Research Triangle Park, North Carolina; Las Vegas, Nevada; and Corvallis, Oregon.

The EPA is often caught in the crossfire between opposing groups—for example, between environmentalists who seek tighter controls and the businesses the EPA regulates, which commonly complain that regulations are too stringent and costly.

Through polls, the American people continue to express a strong concern for a healthy environment; they support maintaining existing laws or even strengthening them. New problems such as acid precipitation and hazardous wastes constantly crop up, demanding the attention of the EPA and other federal bureaucracies. A growing population and an expanding economy create an ever-increasing burden on the environment and on the agencies that regulate environmental issues. Thus, many argue that the EPA and other agencies involved in environmental problems should be expanded, not merely maintained or even reduced. Today, the EPA is facing severe cutbacks in many key programs, including enforcement, pollution prevention, and energy efficiency.

Principles of Environmental Law

In the United States, government's authority to protect the environment is conferred by the U.S. and state constitutions. Environmental protection is conferred by common law, federal and state statutes and local ordinances, and regulations promulgated by state and federal agencies. Statutory law and common law are the mainstays of environmental protection.

Statutory Law   Many examples of state and federal laws for environmental protection and resource management have appeared throughout this book. These are forms of **statutory laws**—that is, laws written by and agreed upon by legislative bodies. Statutory laws generally establish broad goals, among them the protection of health and the environment by reducing air pollution, or the judicious use of natural resources. However, as pointed out in this chapter, Congress and the state legislatures lack the time and expertise to determine specifically how these goals can be met. Thus, Congress assigns the technical details—the setting of standards, pollution control requirements, and resource management programs—to federal agencies such as the EPA.

Besides setting standards of acceptable behavior and calling on agencies to determine regulations, statutory laws also provide authority for various enforcing agencies to take legal action—to fine polluters or to take polluters to court to face criminal charges and possible jail sentences.

Common Law   Many environmental cases are tried on the basis of **common law**, a body of unwritten rules and principles derived from thousands of years of legal decisions. Common law is based on proper or reasonable behavior.

Common law is a rather flexible form of law that attempts to balance competing societal interests. As an example, a company that generates noise may be brought to court by a nearby landowner who argues that the factory is a nuisance. The landowner may sue to have the action stopped through an injunction. In deciding the case, the court relies on common-law principles. It weighs the legitimate interests of the company in doing business (and thus making noise) and the interests of society (which wants its citizens employed and wants to collect taxes from the company) against the interests of the landowner (who is trying to protect the family's rest, health, and enjoyment of property).

The court may favor the **plaintiff** (the one who files the lawsuit) if the damage (loss of sleep, health effects, and inconvenience) is greater than the cost of preventing the risk (costs of noise abatement, loss of jobs, and loss of tax revenues). However, the court might not issue an **injunction**, an order to cause the factory to shut down; instead, it may simply require the **defendant** (the one defending the case or whose actions are being contested) to reduce noise levels within a certain period, striking a balance between competing interests. Cases such as this one illustrate the **balance principle**.

Many cases involving common law are settled on two legal principles: nuisance and negligence.

Nuisance   The most common ground for action in the field of environmental common law is **nuisance**. A nuisance is a class of wrongs that arises from the unreasonable, unwarranted, or unlawful use of a person's own property that obstructs or injures the right of the public or another individual. It may result in annoyance, inconvenience, discomfort, or hurt. This means that one can use one's personal property or land in any way one sees fit, but only in a reasonable manner and as long as that use of the property does not cause material injury or annoyance to the public or another individual.

Generally, two types of remedy are available in a nuisance suit: compensation and injunction. Compensation is a monetary award for damage caused. Injunctions are court orders requiring the nuisance to be stopped.

Nuisances are often characterized as either public or private. Until recently the two were distinctly different concepts. A **public nuisance** is an activity that harms or interferes with the rights of the general public. Typically, public nuisance suits are brought to court by public officials. A **private nuisance** is one that affects only a few people. For example, the pollution of a well affecting only one or two families is considered a private nuisance. A public nuisance would be pollution that affects hundreds, perhaps thousands, of landowners along a river's shores. The most common environmental nuisance is noise (Chapter 19). Water pollutants and air pollutants such as smoke, dust, odors, and other chemicals are other major nuisances.

Historically, the distinction between private and public nuisance has hampered pollution abatement because the courts have traditionally held that an individual could bring a private nuisance suit only if the individual had suffered a unique injury—that is, one different from that suffered by others. For example, an individual would not be permitted to sue a company for polluting a river shared by many others. In this case, the only legal recourse would be a public nuisance suit brought by an official (the local health department, for instance). In such instances, though, public officials are often unwilling to file suit against local businesses that provide important tax dollars for the community (and possibly campaign support as well).

Increasingly, the distinction between private and public nuisance is fading; private persons can bring suit to stop a public nuisance. As a result, the private individual is gradually acquiring more power to stop polluters.

Several common defenses are used to fight nuisance suits. Because most nuisance suits are decided by balancing the rights and interests of the opposing parties, **good-faith efforts** of the polluter may influence the decision. For example, if a small company had installed pollution control devices and had attempted to keep them operating properly but was still creating a nuisance, the court might hold it liable but would probably be more lenient in damages or conditions of abatement. If, on the other hand, the company had made no attempt to eliminate a pollution nuisance, the court would generally be more severe.

The availability of pollution control must also be considered. If a company is using state-of-the-art pollution control and still creates a nuisance, the court may not impose damages or an injunction. In contrast, if the company has failed to keep pace with pollution control equipment, the court may order it to install such controls.

In states that still distinguish between public and private nuisance, a **class action suit** can be filed. Class action suits are brought on behalf of many people and seek remedy for damage caused to the entire group by a nuisance. In order for a federal class action suit for compensation to be permitted, however, each person named in the suit must have suffered at least $10,000 in damage. If not, the suit can be dismissed.

One defense against nuisance suits is that the plaintiff has "come to the nuisance." **Coming to the nuisance** occurs when an individual moves into an area where a nuisance—such as an airport, animal feedlot, or factory—already exists and then begins to complain. An old common-law principle holds that if you voluntarily place yourself in a situation in which you suffer injury, you have no legal right to sue for damages or an injunction. In most courts, however, even though you purchase property and know of the existence of a nuisance, you still have the right to file suit to abate it or recover for damages. This is based on another common-law principle: that clean air and the enjoyment of property are rights that go along with owning the property. Thus, if population expands toward a nuisance, it may be the responsibility of the party creating the nuisance to put an end to it.

According to environmental attorney Thomas Sullivan, "The courts are moving to strict liability for environmental nuisances, so that practically speaking, there are no good defenses. The solution is: do not create nuisances."

Negligence   A second major principle of common law is **negligence**. From a legal viewpoint, a person is negligent if he or she acts in an unreasonable manner and if these actions cause personal or property damage.

Negligence provides a basis for liability, just as nuisance does, but negligence is generally more difficult to prove.

What is reasonable action in one instance may not be reasonable in another. Statutory laws and regulations help the courts determine whether behavior is reasonable. For example, regulations drawn up by the EPA specify how certain hazardous wastes should be treated. Failure to comply with those standards may be evidence of negligence. Negligence may be shown in instances in which a company fails to use common practices in the industry. For example, a company may be found negligent if it fails to transport hazardous wastes in containers like those used by other companies.

In a much broader sense, the courts may decide that a company is negligent simply if it fails to do something that a reasonable person would have done. For example, negligence might be demonstrated if a company failed to test its wastes for the presence of harmful chemicals when a reasonable person would have done so. Likewise, negligence may stem from action that a reasonable person would not have taken. In summary, then, negligence can result from either inaction or action that may be deemed unreasonable considering the circumstances.

The **concept of knowing** is fundamentally important to negligence suits. Briefly, negligence can be determined on the basis of what a defendant knew or should have known about a particular risk. The standard of comparison is what a reasonable person should have known under similar circumstances. For example, a man on trial may argue that he is not negligent because he did not know that a harmful chemical was in the materials that he dumped into a municipal waste dump and that subsequently polluted nearby groundwater. His argument will be valid if a reasonable person in his position could not have known about the wastes.

Interestingly, in some cases, determinations of liability for damage or harm need not be based on proof of negligence. For instance, in cases where the risk is extraordinarily large, one need prove only that injury or damage occurred, not that the operator was negligent or acted unreasonably.

Business interests often try to lessen their liability by getting Congress to pass laws that put ceilings on liability. The Price-Anderson Act, for example, frees utility companies from financial liability incurred by a nuclear power plant accident. The act requires the government to pay for all damages outside the plant, but only up to $560 million. The airlines have a similar law; at this writing, the oil companies have successfully pushed for similar limits on liability resulting from tanker spills.

Problems with Environmental Lawsuits   Legal actions to stop a nuisance or collect damages from a nuisance or act of negligence carry with them a **burden of proof**: In order to win, plaintiffs must prove that they have been harmed in some significant way and that the defendant is responsible for that harm. This is not always easy, for several reasons. First, the cause-and-effect connection between a pollutant and disease may not have been definitely established by medical research. If doubt exists, the case is weakened. Second, diseases such as cancer may occur decades after the exposure, as pointed out in Chapter 18. This makes it extremely difficult to prove causation. As a rule, it is easier to link cause and effect with acute diseases. Third, it is often difficult or impossible to identify the party responsible for damage, especially in areas where there are many industries or where illegal acts such as midnight dumping of hazardous wastes have occurred. Any reasonable doubt about the party responsible for personal or property damage may weaken a lawsuit.

The **statute of limitations** limits the length of time within which a person can sue after a particular event. This provision creates problems in cases of delayed diseases. Statutes of limitations are designed to reduce lawsuits in which evidence is unavailable or memories of potential witnesses may have faded. In latent-disease cases, though, they create a major obstacle to individuals seeking compensation for damage in states that apply the time limitations to the onset of exposure. This essentially makes it impossible for cancer victims to file suit. Other states start the judicial clock from the time the victim learns of the disease. This makes it a little easier to collect compensation for diseases such as cancer, black lung disease, and emphysema.

Resolving Environmental Disputes Out of Court

An increasing number of disputes between environmentalists and businesses are being settled out of court, either privately or by dispute resolution, known as **mediation**. This innovative approach often employs a neutral party who mediates the discussion between opposing parties. The mediator keeps the proceedings on track, encourages rivals to work together, and tries to resolve the dispute in a way that is satisfactory to both groups.

The benefits of mediation over litigation are many. Mediation is much less costly and time consuming. Mediation also tends to create better feelings among disputants, whereas court settlements create winners and losers and often leave bitter feelings. In addition, and perhaps most important of all, mediation may bring about a more satisfactory resolution. For instance, environmental lawsuits often hinge on specific points of law rather than on substantive issues, and thus, they may have little to do with what the plaintiff really wants. An environmental group might bring a suit over the adequacy of an EIS but, in reality, might want the government to ensure protection of a valuable species that would be affected by the project. In mediation, this will be the central issue. Mediation may therefore result in more appropriate solutions.

Mediation also tends to promote a more accurate view of problems. For instance, in lawsuits each party tends to bring up evidence that favors its goal and may ignore or dismiss unfavorable or ambiguous information that might weaken its stand. In mediation, both parties are encouraged to openly discuss the uncertainties of their positions, discovering many points of agreement and building a better understanding of opposing positions.

Mediation has its drawbacks, too. First, funding is inadequate. In the past, mediation has been financed largely by foundations. Federal, state, and local governments need to develop programs to fund mediation. Second, some groups fear they will lose their constituency if they enter into negotiations on certain issues because they will be giving the impression that they are failing at their stated goals and compromising with the "enemy." A third drawback is a lack of faith in the outcome of mediation. Unlike court orders, resolutions drawn up in mediation are not legally enforceable. Thus, months of discussion may produce nothing but a piece of paper that polluters will ignore.

Out-of-court settlements present another legal problem. Such settlements have hindered environmental law. Eager to avoid a costly settlement, a company may pay victims if they agree to dismiss the company from further liability. Out-of-court settlements may also benefit plaintiffs, saving them the time, headaches, and costs of environmental litigation.

Although advantageous to both parties, these settlements provide no precedents for environmental laws. In short, the fewer cases that make it to court, the fewer guidelines courts have to settle future cases. This lack of precedent may discourage attorneys and citizens from filing court cases. Without clear examples from the past, they may simply be unwilling to face costly, time-consuming legal battles.

Effective mediation requires the following: (1) A truly neutral party must serve as mediator, and (2) A formally agreed-on agenda for discussion and a point of focus are also important. Resource and pollution issues should be the focus of discussion; disputes over values should not dominate the proceedings because they cannot be solved by mediation. Although values will surely come out in the debate, they should not be the central point. (3) There must be a willingness to explore new ideas and possibilities on both sides. (4) Disputants must deal honestly with each other. (5) An adequate representation of all interested parties must also be achieved. If someone is not represented, a solution that is unsatisfactory may result. This could lead to a lawsuit. (6) Strict rules should be imposed regarding news releases. The media should not be employed as a lever by either group.

Dispute resolution is growing in the United States, but it will not replace litigation. Still, it can play a valuable role in the future. Eight states have organizations that offer mediation services, and a growing number of private organizations have been formed to provide professional mediators with experience and knowledge in environmental issues. Thus, more and more disputes may be settled by this noncombative approach.

KEY CONCEPTS

Out-of-court settlements through mediation provide a less hostile approach to conflict resolution and can reduce or avoid costs of environmental lawsuits for both defendant and plaintiff.

Creating Governments That Foster Sustainability

From an environmental standpoint, democratic governments display the same weaknesses as the economy: They are generally shortsighted, obsessed with growth, and exploitive of people and nature. They also tend to promote dependency and undermine self-reliance.

Why do governments and economies have the same flaws? In a nutshell, the frontier ethics in many countries is largely responsible for creating mutually supporting systems of government and economics. We create what we believe in.

To build a humane, sustainable society, governments and the economies they serve must be made responsive to the needs of future generations and other species that share this planet with us. First and foremost, governments must adopt the goal of sustainability as a central organizing principle. This section offers additional ideas on ways to correct the flaws in governments and economies. The following section looks at international governance.

KEY CONCEPTS

Most governments promote shortsighted, growth-oriented, exploitive policies and economies. Correcting perceptions and government policies could help us reshape our unsustainable economy.

Creating Government with Vision

Responding to problems requires a consensus for action, which may be translated into appropriate governmental policies. In the United States, one of the key barriers to building a sustainable future is a lack of consensus about the severity of the world's problems and the need for long-term action. As humorist Herbert Prochnow once noted, "How can a government know what the people want when the people don't know? How can we agree on the need to tackle a problem when there is so much conflicting evidence, much of it planted by public relations firms hired by large corporations or conservative think tanks funded by millions of dollars by donations from big business? How can we agree on a course for the future when few agree on what is best?" This section discusses ways of increasing public awareness of problems and the need for long-term sustainable solutions—creating a society and a government with a vision.

KEY CONCEPTS

A long-term perspective is essential to create a sustainable government.

Increasing Public Awareness through Research and Education The lack of consensus on environmental issues results in part from widespread ignorance about these issues among the general public. Often, people who speak out against environmentalist concerns have very little knowledge of the issues. Many of them are compelled by a sense of optimism to dismiss claims of impending environmental disaster.

FIGURE 27-4 Outdoor classroom activity. Environmental education of children of all ages is a key to building a sustainable society.

One solution to the problem of public ignorance is government support for scientific study of environmental problems. Research is also needed to identify the levels of consumption and other human activities that are sustainable. As these factors are being studied, researchers can also explore creative, sustainable solutions to help society reach these goals while achieving economic prosperity. Research is also needed on ways values can be changed to achieve a sustainable society. Support for this work could come from governments and private foundations.

Information is important, but in order to make a difference it must be disseminated to the citizenry. Concerted efforts are needed to inform people the world over of the problems facing the global ecosystem. Well-informed teachers and public officials can help disseminate this knowledge, as can the media (**FIGURE 27-4**). Environmental groups and religious leaders can also play a role. Education can assist in making the search for solutions into immediate concerns of everyday citizens, elected officials, and businesspeople. (For more on education, see Chapter 24.)

KEY CONCEPTS

Research and education are essential to raise the level of awareness of environmental problems and sustainable solutions among the public and elected officials.

Getting Beyond Crisis Management Public ignorance of issues is also a result of a tendency for public policymakers to practice **crisis politics**, which refers to the management of immediate crises at the expense of long-range problems. Henry Kissinger put it best when he said that, in government, the "urgent often displaces the important." In crisis management, immediate problems such as strikes, disaster relief, and pressing economic problems tend to get all of the attention. They reduce the amount of time policymakers spend thinking about and addressing long-range problems, especially environmental ones. When new problems do arise, governments ignore them. Too heavily embroiled in the present crisis, they defer decision making until a small—and probably solvable—matter becomes a monstrous problem that defies solution. In addition, governments are generally unwilling to devote money and resources to problems when they're small because they're already spending so much treating the latest list of crises. In so doing, they end up with a menu of very costly problems that could have been solved at a fraction of the cost. A good example is the savings and loan scandal. Government representatives were warned about the problem in 1984, but ignored it until 6 years later, when a minor problem had grown to a $500 billion headache.

When immediate issues take precedence over important long-term problems, governments lumber from crisis to crisis. It's not an efficient way to run a government or a society. Making matters worse, legislators often apply marginal corrections to the present-day crises. Thus, they often end up treating the symptoms and not the causes. In time, our problems grow worse.

In 1985, then Senator Al Gore introduced a bill to Congress called the **Critical Trends Assessment Act**. Had it passed, it would have required the federal government to establish an **Office of Critical Trends Analysis (OCTA)** to examine long-term trends, especially economic and environmental ones. The OCTA would collect and analyze vast amounts of data from a variety of government agencies. Much of this information currently goes unused. Using knowledge from many sources, the OCTA would then advise the President on policy needs to avert problems and bring government and society in line with economic and ecological realities. State governments could also establish OCTAs, whose central purpose would be to forge a sustainable economic and environmental future.

OCTAs might help convince influential government leaders as well as business interests of the importance of many environmental issues, reducing ignorance and creating a consensus for change that helps improve our prospects socially, economically, and environmentally. Combined with the many other changes described in this book, OCTAs could help shape a sustainable future.

Another approach would be the establishment of special commissions on the future to address critical trends. In 1977, for example, President Carter directed several key agencies in the federal government to study population, resource, and pollution trends and issue a report on their findings, which would serve as a foundation for longer-term planning. The resulting study, *The Global 2000 Report to the Pres-*

ident, was a gold mine of information about the future, on which many decisions could be based. This report could be updated and revised.

Another way of creating vision in government is to establish special study sections in various governmental agencies, such as the Departments of Agriculture, Energy, and the Interior. Reports from these agencies could be distributed to political leaders, teachers, and the public.

Internationally, long-range vision can be facilitated by the United Nations, through the Environmental Programme (UNEP) or specially established commissions. In the mid-1980s, for instance, the UN established the **World Commission on Environment and Development.** Members of the commission were asked to produce a global agenda for sustainable economic development. Their report, *Our Common Future*, called on the General Assembly of the United Nations to find ways to achieve sustainable development; it was the main stimulus for the United Nations Conference on Economic Development in Rio de Janiero in 1992. The UN's role is described more fully later in this chapter.

KEY CONCEPTS

In government, long-range planning and action (which are essential for sustainable development) are often sacrificed in the interest of addressing immediate problems, creating a phenomenon called *crisis politics*. Crisis management tends to apply marginal solutions and allows little problems to become larger and more difficult to solve.

Getting Beyond Limited Planning Horizons Another barrier to sustainability is the limited planning horizon of many elected officials. Limited planning horizons result from several factors. First and foremost, politicians, like most of the rest of us, tend to be shortsighted. Re-election concerns also contribute mightily to this phenomenon. Frequent re-election campaigns tend to force politicians to choose short-term solutions, rather than long-term answers that may be more difficult and painful—and consequently less popular. To get re-elected, politicians need results, even if they treat only the symptoms of a crisis. Members of the U.S. House of Representatives, for instance, serve 2-year terms. Although this may have been appropriate 200 years ago when the nation's problems were few, today some observers think that it is inadequate. Why? During their 2-year terms, some House members spend inordinate amounts of their time collecting money, plotting re-election strategies, and campaigning—time that they could have devoted to their job.

One way to address this problem is to institute longer terms of office for members of the U.S. House of Representatives, and perhaps even for the President. Term limitations could also help. Fourteen states now have some form of term limitation—that is, laws that limit the number of years a state or federal official can serve. Colorado's governor, for instance, can serve no more than three four-year terms. A 6-year nonrenewable term for the President, for example, might reduce the reelection pressures that lead to shortsighted political decisions and take valuable time away from the office. It would also reduce the amount of time spent on the campaign trail while in office and increase the amount of work time. Longer terms of office, of course, could have damaging effects as well.

Shorter campaigns could help make democracies work better, too. In England, campaigning begins 6 weeks before general elections. Thus, incumbents spend less time on the campaign trail than U.S. politicians. In addition, they spend considerably less money. In the United States, campaigns often run a full year and cost tens of millions of dollars. Presidential candidates spend $200 to $300 million in their bids for office. In order to finance such costly campaigns, they frequently take money from PACs or businesses. As was pointed out earlier in the chapter, this distorts the political process, making it bow to the economic interests of business and others, which may be shortsighted.

Overcoming the barrier of limited planning horizons will also require more farsighted leaders, who are willing to implement policies that have implications well past the next election. Such policies may not favor the immediate electorate as much as the future electorate. Citizens must become willing to elect men and women with a vision of the future and with the skills to articulate and popularize their vision. If we expect visionaries, though, we must become more visionary ourselves and make tough choices.

Environmental groups at the national and state levels are helping support responsible leadership. You can help by supporting candidates who promote the efficient use of resources, recycling, renewable resource use, restoration and sustainable management of resources, growth management, and population stabilization. Citizens can write or visit their elected officials. One other effective way for citizens to make their voices heard is to join advocacy groups.

KEY CONCEPTS

Longer terms of office for elected officials could reduce political shortsightedness and encourage solutions that contribute to sustainability, as would term limitations and shorter campaigns.

Becoming Proactive A government that lives and acts for today is by definition a **reactive government.** Its laws and regulations are often ineffective, and in the long run they can complicate matters, making truly effective solutions harder to reach. Many of the laws passed by reactive governments are **retrospective**—that is, they attempt to regulate something that has already gotten out of hand. For example, the Superfund Act provides money to clean toxic hot spots in the United States. Retrospective laws are part of crisis management. They're necessary like emergency rooms in hospitals but less desirable in the long run (as pointed out in many previous chapters) than long-term, sustainable solutions.

Sustainability relies in large part on **proactive laws,** laws that attempt to prevent problems in the first place or to confront them while they're still small and easily solvable. A good example is the Toxic Substances Control Act, with its provisions for screening new chemicals before they are introduced into the marketplace (Chapter 18). Laws that promote solar energy, conservation, recycling, growth management, and population stabilization are additional examples. Ultimately, such laws could help us transition from the wasteful frontier economy to the efficient, sustainable society.

Many examples of future-oriented acts by Congress exist:

1. Establishment in 1946 of the National Science Foundation to promote research
2. Passage in 1969 of the **National Environmental Policy Act**, which requires environmental impact statements for all projects sponsored or supported by the federal government
3. Creation in 1972 of the **Office of Technology Assessment**, to examine the costs and benefits of new technologies
4. Authorization of the Congressional Research Service to create a Futures Research Group
5. Passage of the **National Appliance Energy Conservation Act** (1987), which establishes efficiency standards for appliances

Western democracies today generally have a mix of proactive and retrospective (and reactive) laws, with a strong leaning toward a reactive approach. As we move into the 21st century and as the world population continues to grow exponentially, proactive government becomes more necessary than ever before.

KEY CONCEPTS

Creating a sustainable future requires short-term solutions to address immediate problems, but it ultimately hinges on a long-term, proactive approach that seeks to prevent problems.

Ending Our Obsession with Growth

Some of the steps outlined above could also help nations reduce their obsession with growth. Offices of critical trends analysis, for instance, could help show the potential outcome of current policies that promote continual growth in production and consumption at the expense of the environment. Education and research could have similar effects. As pointed out in Chapter 25, new measures of progress could also temper our obsession with growth. Governments can help by promoting the use of these measures and by publicizing real progress toward improvement—rather than the illusion of progress suggested by the GNP.

In 1949, the U.S. Congress passed the **Full Employment Act**, which calls on the government and its various agencies to promote maximum production and consumption to promote full employment. Although full employment is a desirable goal, the notion that we can achieve it only through maximum production and consumption is erroneous and ecologically ill-advised. Repealing this law and replacing it with a **Sustainable Futures Act** might be a symbolic act, but it could be an important step in building a sustainable future.

The Sustainable Futures Act would call on Congress to actively promote sustainability through conservation (efficiency and pollution prevention), recycling, renewable re-

source use, restoration, growth management, and population measures. Moreover, it would require that all new policies be judged through the lens of sustainability. The act might also set up a special legislative body to review all existing policy and to suggest amendments or new laws that would bring human activities—from international trade to energy to manufacturing—in line with the dictates of sustainability.

KEY CONCEPTS

Many measures could help temper the obsession with growth found in many nations, which is so detrimental to the goal of sustainability.

Reducing Exploitation and Promoting Self-Reliance

Reducing the exploitation of people and of natural resources is another goal of sustainable government. Interestingly enough, many of the changes mentioned above would contribute to this goal. Steps that reduce our obsession with growth and create vision, for instance, tend to make societies less resource-intensive and thus less exploitive. Other steps can also be taken to reduce our exploitive ways, some of which have been discussed in Chapters 25 and 26. Additional ideas are listed later.

KEY CONCEPTS

Countless measures that promote other goals of sustainability could help reduce society's exploitation of natural systems and of other people—and promote greater self-reliance, broader prosperity, and a better future.

Creating Offices of Sustainable Economic Development
One idea that may promote self-reliance and an end to our exploitative economy is the establishment of offices of sustainable economic development (OSED) in state governments or, perhaps, at the national level. All state governments currently have offices of business development or offices of economic development, which promote business by recruiting new business and helping existing business become more profitable. As a rule, little thought is given to matters of sustainability.

OSEDs could serve as a catalyst for change. First, they could help existing businesses become more sustainable—for example, by using energy and other resources more efficiently or by recycling and purchasing recycled goods. Besides promoting the use of locally available waste resources and helping build industries that divert wastes to useful purposes, OSEDs could promote pollution prevention and renewable energy resources. OSEDs could also identify economic opportunities arising from local restoration projects. Second, they could assist in promoting sustainable business opportunities—for example, helping identify locally available resources (wastepaper or aluminum scrap) around which new businesses could be developed. By focusing on sustainable economic progress, OSEDs would help build strong, stable, and regionally self-reliant economies.

FIGURE 27-5 Reinhard Buetikofer, chairman of Germany's Green party (right) and election campaign manager, Fritz Kuhn (left), take a ride in a bike taxi at the opening of the Green election campaign center in Berlin in 2005. The sign on the bike taxi reads "Take Part!"

Working with the state offices of critical trends analysis and elected officials, OSEDs could help create a comprehensive strategy for sustainability. They could also help promote better land-use planning, which is vital to sustainable economic development throughout the world. For a summary of ideas on creating a sustainable government, see Table 27-1.

KEY CONCEPTS

Offices of sustainable economic development in state and national governments could replace existing offices of business development. They could promote sustainable business development and help current businesses become more sustainable, actions that are beneficial to both the Earth and the economy.

Models of Sustainable Development

In any transition, people and nations often become models of desirable action. In the sustainable transition, several European countries have recently emerged as models. One of the best examples is Germany (see the Spotlight on Sustainable Development 19-1).

Germany probably has the highest environmental standards in the world. Part of the reason for this lies in the **Green Party**, a political party that advocates measures to protect the environment. The Greens support reductions in pollution and hazardous wastes, putting an end to nuclear power and nuclear weapons, efficiency measures, and the use of renewable resources. Rallying under the banner, "We are neither right nor left, but in the front," the Greens have been instrumental in transforming German society (FIGURE 27-5). If they get their way, Germany could become a nation of nonpolluting industries that concentrates mainly on socially necessary products.

The Greens represent a major shift in political thinking. They take a long-term view of the future, calling for redirection of policy consistent with sustainable ethics. In western Europe, a dozen green parties now exist, although in recent years the Green Party has lost importance. The Green Party has also spread to the United States. Colorado, Connecticut, Washington, and dozens of other states now have green parties, which promote social justice and a clean and healthy environment. Green Party members write letters and phone political leaders, stage peaceful demonstrations, and work to get their members elected to state legislatures and other offices.

Conventional political wisdom holds that in the United States popular ideas are absorbed by major political parties. They enter the mainstream of political thought and become integrated into public policy. This has happened in Germany. Thus, no matter whether the U.S. Green Party survives or is simply absorbed into the mainstream, it may have begun a movement that could reshape American politics.

Like Germany, Sweden, Norway, and the Netherlands are also models, whose policies and actions could inspire other countries to follow suit—especially when skeptical leaders see the economic benefits of pursuing sustainable options. Besides providing inspiration, model countries also become a source of new technologies and will enjoy a competitive advantage over those countries that are resisting change.

KEY CONCEPTS

The Green Party has been instrumental in encouraging sustainable practices in Germany and other countries, which serve as models for other nations. Taking the lead in sustainable development also positions some nations to become leading exporters of knowledge and technologies.

Global Government: Toward a Sustainable World Community

For many years, environmental problems were largely local, regional, or national issues. Solutions required actions on similar scale. Today, though, many environmental ills affect the entire globe and require international efforts to solve.

This realization and the awareness that ozone depletion, deforestation, global warming, and other problems have serious impacts on the future of a country have led many nations to rethink outdated notions of **national security**. Once equated with the military strength needed to protect a nation from outside aggression, national security today also requires efforts to protect a nation's environment from internal threats (such as deforestation and soil erosion) as well as external threats (such as global warming). Achieving true national security will require many of the changes described in this book, in addition to massive global efforts to address the roots of the current crisis of unsustainability.

KEY CONCEPTS

Many environmental problems threaten the national security of nations and are global in scale. Because of this, they will require international cooperation.

Regional and Global Alliances

Cooperation and participation are key principles of sustainable development. These efforts are often fostered by various alliances. Several environmental alliances have been discussed in this book. Chapter 11 described the International Whaling Commission (IWC). Composed of members from all of the whaling nations, the IWC sets quotas on whale kills and enacts outright bans. The IWC, however, has no enforcement power; it relies principally on cooperation. As noted in that chapter, not all nations comply with the regulations. To help give the IWC a little muscle, the United States and other nations prohibit trade with countries that violate the agreements.

Spotlight on Sustainable Development 21-2 discussed an agreement reached by the European nations that border the Baltic Sea. Chapter 20 described the international treaties aimed at protecting the ozone layer.

Some of the most recent alliances emerged from the Earth Summit in June 1992 and a smaller follow-up conference in 1997. This section examines the various accords and discusses further measures needed to improve them.

KEY CONCEPTS

Because the task of building a sustainable future is global in nature, regional and global alliances are needed to solve the problems that afflict many nations.

The Climate Convention Signed by 154 nations at the 1992 Earth Summit, this agreement calls on the world community to hold greenhouse gas emissions at 1990 levels by the year 2000. Unfortunately, this agreement did not create any stipulations that required actions. In the years that followed the signing of this accord, little was done. So, in 1997, the nations of the world met again in Kyoto, Japan to hammer out a more rigorous agreement. Industrial and former eastern bloc nations agreed to cut carbon dioxide and other greenhouse gas emissions to 5.2% below 1990 levels between 2008 and 2012. In 1997, emission levels from these nations were already 4.7% below 1990 levels, not because of efforts to control pollution, but rather because of the sudden economic downturn in former eastern bloc nations. Decreases of 33% in these nations were offset by an increase of 12% in the United States. During that same period, emissions of carbon dioxide grew by over 30% in the less developed nations.

In 1998, more than 160 countries meeting in Buenos Aires adopted further actions. In a move that stunned the world, British Petroleum, the third largest oil and gas company in the world, announced that it would reduce carbon dioxide emissions at its facilities by 10%. British Petroleum has become a major supplier of photovoltaics, too, after acquiring a major U.S. solar producer.

In November, 2000, the signatories of the Kyoto Accord met in the Netherlands, but were unable to reach an agreement on ways to cut emissions in the United States. In 2004, Russia signed the agreement, thus putting it into effect. The United States still refuses to sign the agreement.

Although the Climate Convention is an important first step forward, it will—if followed—only slow the rate of climate change because greenhouse gas emissions already exceed the Earth's carrying capacity.

KEY CONCEPTS

The Climate Convention is a first step in reducing global greenhouse gases, but success has been limited and much more is needed to solve the problem of global warming.

The Biodiversity Convention This agreement was signed by all participants except the United States, which refused because of language the Bush administration deemed unacceptable. President Clinton signed it in 1993 soon after taking office, however. This agreement asks nations to prepare inventories of species to be preserved and to develop strategies to conserve and use biological resources sustainably. Like the climate treaty, no deadlines were set. In addition, the language of the proposal is quite weak.

KEY CONCEPTS

The Biodiversity Convention calls on nations to take steps to protect species.

Agenda 21 By far the most comprehensive document to emerge from the Earth Summit, **Agenda 21** is a massive work plan for national action and international cooperation aimed at achieving sustainable development. Its 39 chapters of nonbinding recommendations are a blueprint for a sustainable future. For instance, Agenda 21 urges nations to adopt national strategies for sustainable development. It calls for community-based sustainable development. It also urges countries to adopt national strategies that eliminate obstacles to the full participation of women in sustainable development. It recommends that countries eliminate subsidies for environmentally destructive activities and encourages them to ensure public input.

Despite these and other important recommendations, Agenda 21 has some serious weaknesses, according to some critics. Language that would commit more developed nations to donate a certain percentage of their annual GNP to sustainable development overseas was eliminated. In addition, Agenda 21 did not propose new ways of eliminating debt. Nor did it offer new ways to accelerate the transition to sustainable systems of energy (because of pressure from the United States, Saudi Arabia, and Kuwait). All references to full-cost pricing were eliminated. Moreover, the chapter on forestry does not recommend policies of sustainable forest management. The chapter on population fails to underscore the importance of population stabilization to sustainable development and avoids the term *family planning* altogether, reportedly because of pressure from the Vatican.

Forest Principles Another nonbinding agreement adopted by attendees of the Earth Summit was the **Authoritative Statement of Principles on the World's Forests.** Although in this instance the United States government pushed for stronger efforts to preserve native forests, 77 less developed nations, led by India and Malaysia, watered the agreement down. In fact, they eliminated language pointing out the value of forests to the global environment. In addition, Malaysia eliminated wording that encouraged efforts to promote international trade of products from sustainably managed forests. According to Gareth Porter of the Energy and Environment Institute, the result is a set of principles that underscore sovereign rights of nations to exploit their forests, legitimizing existing policies in those countries that are currently endangering the world's forests.

Rio Declaration The final outcome of the Rio Conference was the **Rio Declaration.** It includes recommendations for a number of political principles vital to achieving sustainable development, among them public access to government information on the environment. It also calls on nations to use environmental impact statements and to exercise caution in their development plans. In addition, it encourages timely notification of activities that could exert their effects across a nation's boundaries. The Rio Declaration also recognizes the special responsibility of the more developed countries in achieving global environmental restoration because of their technological and financial capabilities, consumption patterns, and pollution production.

Although the outcome of the 1992 Earth Summit was less than many would have hoped, the progress made there was a good start, especially considering the complexity of negotiations (involving over 150 nations with very different goals and priorities). Because many environmental, economic, and social problems will inevitably worsen in the coming years, progress is bound to occur in areas that were ignored or deleted because of their political sensitivity.

Strengthening International Government

To solve global environmental problems, national sustainable development plans and international agreements are badly needed. Also needed are efforts to strengthen existing international institutions, notably the United Nations (**FIGURE 27-6**). Some observers think that we may need some form of world government to tackle the problems facing humankind.

Strengthening the United Nations' Role in Sustainable Development The United Nations has several key programs, such as the Fund for Population Activities (UNFPA), the Food and Agricultural Organization (FAO), the United Nations Environment Programme (UNEP), and the United

FIGURE 27-6 **The United Nations.** The UN has helped increase global cooperation in pollution control and other important issues. It could provide a forum for more progress on building a sustainable society.

Nations Development Program. Over the years, these programs have helped scores of less developed nations in many ways. Conferences on population and its impact, sponsored by UNEP, have helped convince most less developed nations that something must be done to control rampant population growth. In 1987, UNEP facilitated negotiations that led to the Montreal Accord and two revisions, and in 1990, it sponsored negotiations that produced the groundbreaking London Agreement (Chapter 20). In 1988, the UN Commission for Europe negotiated a treaty that calls for a freeze on nitrogen oxide emissions at their 1987 levels by using the best available control technology on new vehicles and power plants in Europe. The Rio Conference and the follow-up conference in 1997, sponsored by the UNEP, are additional examples. For a discussion of the follow-up conference, see Spotlight on Sustainable Development 27-2.

Despite its successes and its importance to the future of the world, UNEP operates on a meager $60 million annual budget. The 330 professional staff and 350 support staff are, according to most observers, much too small for the task at hand. Increasing the budget and the staff of UNEP could go a long way toward improving its effectiveness. It would allow UNEP to gather more data and disseminate information more widely. In addition, it would allow UNEP to expand efforts to promote alliances on key issues such as population growth, desertification, and deforestation.

KEY CONCEPTS

Several agencies within the United Nations already work diligently toward the goal of environmental protection and sustainable development. Strengthening these organizations could greatly improve these efforts.

A global sustainable society and government may seem like Utopia. As writer Rolf Edberg reminds us, though, "The utopia of one generation may be recognized as a practical necessity by the next." I think that the present generation is beginning to recognize that sustainability is a practical necessity.

Edberg also notes that achieving such goals depends on our ability to free ourselves from "ideas and emotions that

>> SPOTLIGHT ON SUSTAINABLE DEVELOPMENT

27-2   Earth Summit II: The Follow-Up to Rio

Five years after the highly regarded Earth Summit in Rio de Janeiro, the United Nations sponsored a follow-up meeting, commonly called Earth Summit II. This weeklong, follow-up conference held in New York City in June, 1997, was attended by more than a hundred presidents, prime ministers, and other top government officials and was designed to assess progress made since Rio.

What attendees heard was not encouraging. Vice President Al Gore gave the opening speech, noting that deforestation was occurring at rates only slightly less than those of the 1980s. Greenpeace International representative Clif Curtis summed up the grim message of the conference, noting that governments have done far too little to turn around the ongoing environmental decay. He noted that "environmental degradation is worsening at the global level" and pointed out that "we're going backwards on the road to sustainable development." For example, although 150 nations signed the Framework Convention on Climate Change at the 1992 Earth Summit, in which they agreed to reduce or prevent greenhouse gas emissions, in 1996, global carbon emissions reached a record high of 6.25 billion tons. The Framework Convention called on nations to curb emissions of carbon dioxide so that by 2000, emissions of this greenhouse gas would not exceed 1990 levels. However, U.S. carbon dioxide emissions had increased 8% since 1992 and were projected to be 13% above 1990 levels. Emissions in Japan and the nations of the European Union were an estimated 6% higher than 1990 levels. Moreover, many rapidly industrializing nations such as China, Brazil, and Indonesia had experienced 20 to 40% increases in emissions since 1990.

At Rio I, 158 countries signed an agreement to protect plant and animal species by preserving and restoring natural habitats. This agreement, known as the Convention on Biodiversity, went into effect in December 1993. Since that time, an estimated 100,000 species of plants and animals have become extinct, according to estimates of a nonprofit organization, Earth Summit Watch, which monitors progress of the Earth Summit and encourages governments to implement measures they agreed to. The loss of forest habitat, as noted earlier, is a key factor in this decline and is continuing at a rapid pace.

One bright spot is the global phaseout of leaded gasoline. But on balance, Earth Summit Watch notes that "a lot of international institutional development, dialogue, and debate" have taken place, but "not much concrete improvement" in environmental conditions. While many countries are addressing the complex problems, continued growth of the human population and the economy place extraordinary demands on the environment. Progress in environmental protection cannot keep pace with the destruction.

In December 1997, national leaders met to address global climate change, perhaps one of the most pressing issues of our times. Many proposals were offered by the United States, Europe, and small island nations that will suffer the most if the climate continues to warm and oceans rise. Although change will be slow, it is clear that global consensus is emerging on the topic of global climate change: Most nations recognize that something must be done to reduce the emissions of greenhouse gases.

once had a function in our battle for survival but have since become useless." Starving masses and the universal fear of environmental destruction may be the psychological forces that set the stage for a new world society governed by sustainable principles.

Some people think that we need a world government to deal with the many complex global issues confronting human society.

The legitimate object of government is to do for a community of people what ever they need to have done, but cannot do at all, in their separate and individual capacities.

—*Abraham Lincoln*

CRITICAL THINKING

Exercise Analysis

One of the myths we propagate in our society is that results can be measured by the amount of money we spend on a problem. Thus, if we are spending a lot of money, we must be doing a good job. Unfortunately, results don't always parallel expenditures. For example, billions of dollars have been spent on hazardous-waste cleanup in the United States, but much of this money ends up in the pockets of attorneys, not in actual cleanup efforts. Spending on water pollution control at sewage treatment plants keeps sewage out of streams, but sludge ends up in landfills.

The *National Report* also fails to address the most important question of all: Is the money we are spending enough to create a sustainable way of life? In other words, is it spent on stopgap measures, or does it go into systemic solutions that foster a sustainable human presence? Far more important than dollars spent is the outcome—for example, how much pollution is actually removed, and is it enough to ensure sustainability?

The important lesson here is that when you dig a little deeper, you find that simple but appealing logic falls apart. The report fails to answer the most important question of all: Are solutions advancing the transition to a sustainable society or merely making marginal improvements?

CRITICAL THINKING AND CONCEPTS REVIEW

1. What are the major tools by which governments control economic activity and environmental protection? Make a list and describe the pros and cons of each one, as you see it.
2. In what ways are democracies unrepresentative? Can you give some examples?
3. Look around your community. What aspects of it could have been planned better? Why?
4. This chapter lists a number of weaknesses in the current system of government. What are they? What proposals were given to overcome them? Evaluate each of the proposals and make a list of the benefits of each one. Critically analyze them, pointing out problems and ways to solve them.
5. Critically analyze the following statement: "Nations should be left on their own. Others should not meddle in their internal affairs, even in the name of global environmental protection."
6. List the pros and cons of increasing the power of the United Nations in matters of environmental protection.
7. Debate this statement: "We need an international democratic government to address social, economic, and environmental problems."

KEY TERMS

Agenda 21
Authoritative Statement of Principles
    on the World's Forests
balance principle
burden of proof
class action suit
coming to the nuisance
common law
communist nations
concept of knowing
Council on Environmental Quality
    (CEQ)
crisis politics
Critical Trends Assessment Act
defendant
democratic nations
direct government expenditures
Double-C/Double-P game
Earth First!
ecotage
environmental impact statement (EIS)

Environmental Protection Agency
Environmental Quality Act
executive order
Full Employment Act
good-faith efforts
Green Party
income tax deductions
injunction
lobbyists
mediation
National Appliance Energy
    Conservation Act
National Environmental Policy Act
    (NEPA)
national security
negligence
nuisance
Office of Critical Trends Analysis
    (OCTA)
Office of Technology Assessment
Office of the Environment

plaintiff
political action committees
President's Council on Sustainable
    Development
private nuisance
proactive laws
public nuisance
reactive government
retrospective
Rio Declaration
statute of limitations
statutory laws
Sustainable Futures Act
tax credit
theory of public choice
U.S. Corporate Average Fuel Economy
    Act
Wise Use Movement
World Commission on Environment
    and Development

REFERENCES AND FURTHER READING

The References and Further Reading section at the end of this book contains a list of sources for the information discussed in this chapter and recommendations for further reading.

Connect to this book's website:
http://environment.jbpub.com/
The site features eLearning, an online review area that provides quizzes, chapter outlines, and other tools to help you study for your class. You can also follow useful links for in-depth information, research the differing views in the Point/Counterpoints, or keep up on the latest environmental news.

biotic factors Nonliving components of the ecosystem, including chemical and physical factors such as nitrogen, temperature, and rainfall.

accelerated erosion Loss of soil due to wind or water in land disturbed by human activities.

accelerated extinction Elimination of species due to human activities such as habitat destruction, commercial hunting, sport hunting, and pollution.

acid deposition Rain or snow that has a lower pH than precipitation from unpolluted skies; also includes dry forms of deposition such as nitrate and sulfate particles.

acid mine drainage Sulfuric acid that drains from mines, especially abandoned underground coal mines in the East (Appalachia). Created by the chemical reaction among oxygen, water, and iron sulfides found in coal and surrounding rocks.

acid precursor Refers to sulfur oxides and nitrogen oxides, chemical air pollutants that can be converted into sulfuric and nitric acids, respectively, in the atmosphere.

active solar Capture and storage of the sun's energy through special collection devices (solar panels) that absorb heat and transfer it to air, water, or some other medium, which is then pumped to a storage site (usually a water tank) for later use. Contrast with *passive solar*.

actual risk An accurate measure of the hazard posed by a certain technology or action.

acute effects In general, effects that occur shortly after exposure to toxic agents. Contrast with *chronic effects*.

acute exposure Short-term exposure, usually to high levels of one or more agents. Symptoms generally appear soon after exposure.

adaptation A genetically determined structural, functional, or behavioral characteristic of an organism that enhances its chances of reproducing and passing on its genes.

adaptive management Experimental approach to resource management that allows officials to monitor and evaluate practices and change them as needed.

adaptive radiation Evolution of several life-forms from a common ancestor.

advanced industrial society Post-World War II industrial society characterized by a great increase in production and consumption, increased energy demand, and a shift toward synthetics and nonrenewable resources.

age-specific fertility rate Number of live births per 1000 women of a specific age group.

agricultural land conversion Transformation of farmland to other purposes, primarily cities, highways, airports, and the like.

agricultural revolution One of the dominant epochs of human cultural development characterized by the cultivation of plants for food and fiber and the domestication and use of animals for meat and other products.

agricultural society A group of people living in villages or towns and relying on domestic animals and crops grown in nearby fields. Characterized by specialization of work roles.

algal bloom Rapid growth of algae in surface waters due to increase in inorganic nutrients, usually either nitrogen or phosphorus.

alien species (or foreign species) Any species introduced into or living in a new habitat. Also known as an *exotic*.

alpha particles Positively charged particles consisting of two protons and two neutrons, emitted from radioactive nuclei.

alveoli Small sacs in the lungs where exchange of oxygen and carbon dioxide between air and blood occurs.

ambient air quality standard Maximum permissible concentration of a pollutant in the air around us. Contrast with *emissions standard*.

annuals Plants that must be grown from seeds—for example, domestic corn and radishes.

antagonism In toxicology, when two chemical or physical agents (often toxins) counteract each other to produce a lesser response than would be expected if individual effects were added together.

anthropogenic hazard A danger created by humans.

appropriate technology A term coined by the late E. F. Schumacher to refer to technology that is "appropriate" for the economy, resources, and culture of a region. It is characterized by small- to medium-sized machines, maximum human labor, ease of understanding, meaningful employment, use of local resources, decentralized production, production of durable products, emphasis on renewable resources (especially energy), and compatibility with the environment and culture.

Aquaculture Cultivation of fish and other aquatic organisms in freshwater ponds, lakes, irrigation ditches, and other bodies of water.

Aquatic life zone Distinct regions of the ocean akin to terrestrial biomes.

Aquifer recharge zone Region in which water from rain or snow percolates into an aquifer, replenishing the supply of groundwater.

Aquifer Underground stratum of porous material (sandstone) containing water (groundwater) that may be withdrawn from wells for human use.

Artificial selection Selective breeding to create new plant and animal breeds to bring out desirable characteristics.

Asbestos One of several naturally occurring silicate fibers. Useful in society as an insulator but deadly to breathe, even in small amounts. Causes mesothelioma, asbestosis, and lung cancer.

Asbestosis Lung disease characterized by buildup of scar tissue in the lungs. Caused by inhalation of asbestos.

Asthma Lung disorder characterized by constriction and excessive mucus production in the bronchioles, resulting in periodic difficulty in breathing, shortness of breath, coughing. Usually caused by allergy and often aggravated by air pollution.

Atmosphere Layer of air surrounding the Earth.

Atom A basic unit of matter consisting of a nucleus of positively charged protons and uncharged neutrons, and an outer cloud of electrons orbiting the nucleus.

Autotroph Organism, such as a plant, that produces its own food, generally via photosynthesis. See *producer*.

Auxins Plant hormones responsible for stimulating growth.

Bacteria A group of single-celled organisms, each surrounded by a cell wall and containing circular DNA. Responsible for some diseases and many beneficial functions, such as the decay of organic materials and nutrient recycling.

Barrier islands Small, sandy islands off a coast, separated from the mainland by lagoons or bays.

Beach drift Wave-caused movement of sand along a beach.

Beta particles Negatively charged particles emitted from nuclei of radioactive elements when a neutron is converted to a proton.

Big bang Theory of the Universe's formation. States that all matter in the universe was infinitely compressed 15 to 20 billion years ago and then exploded, sending energy and matter out into space. The matter was in the form of subatomic particles, which formed atoms as the Universe cooled over millions of years.

Bioaccumulation The buildup of chemicals within tissues and organs of the body; results in a resistance to breakdown and slow excretion rates.

Biochemical oxygen demand (BOD) Measure of oxygen depletion of water (largely from bacterial decay) due to presence of biodegradable organic pollutants. Gives scientists an indication of how much organic matter is in water.

Biodegradable Material that can be broken down by naturally occurring organisms such as bacteria in air, water, and soil.

Biogas A gas containing methane and carbon dioxide. Produced by anaerobic decay of organic matter, especially manure and crop residues.

Biogeochemical cycle Complex cyclical transfer of nutrients from the environment to organisms and back to the environment. Examples include the carbon, nitrogen, and phosphorus cycles.

Biological community An assemblage of plants, animals, and microorganisms.

Biological control Use of naturally occurring predators, parasites, bacteria, and viruses to control pests.

Biological extinction Disappearance of a species from part or all of its range.

Biological infrastructure Also known as infrainfrastructure. All that supports human life and the human economy: the air, water, soil, and living things.

Biological magnification Buildup of chemical elements or substances in organisms in successively higher trophic levels. Also called *biomagnification*.

Biomass As measured by ecologists, the dried weight of all organic matter in the ecosystem. In the energy field, any form of organic material (from both plants and animals) from which energy can be derived by burning or by bioconversion, such as fermentation. Includes wood, cow dung, agricultural crop residues, forestry residues, scrap paper.

Biomass pyramid See *pyramid of biomass*.

Biome One of several immense terrestrial regions, each characterized throughout its extent by similar plants, animals, climate, and soil type.

Biosphere All the life-supporting regions (ecosystems) of the Earth and all the interactions that occur among organisms and between organisms and the environment.

Biotic (reproductive) potential Maximum reproductive potential of a species.

Biotic factor The biological component of the ecosystem, consisting of populations of plants, animals, and microorganisms in complex communities.

Birth control Any measure designed to reduce births, including contraception and abortion.

Birth defect An anatomical (structural) or physiological (functional) defect in a newborn.

Black water Water from toilets and bidets contaminated with feces. Kitchen sink water containing food scraps is often placed in this category.

Bloom See *algal bloom*.

Breeder reactor Fission reactor that produces electricity and also converts abundant but nonfissile uranium-238 into fissile plutonium-239, which can be used in other fission reactors.

Broad-spectrum pesticide (or biocide) Chemical agent effective in controlling a large number of pests.

Bronchitis Inflammation of the bronchi caused by smoking and air pollutants. Symptoms include mucus buildup, chronic cough, and throat irritation.

Brown-air cities Newer, relatively nonindustrialized cities whose polluted skies contain photochemical oxidants (especially ozone) and nitrogen oxides, largely from automobiles and power plants. Tend to have dry, sunny climates. Contrast with *gray-air cities*.

Buffer zone Region around a protected area in which limited human activity is permitted.

Calorie Amount of heat needed to raise 1 gram of water 1 degree Celsius or 1.8 degrees Fahrenheit. One thousand calories is equal to one kilocalorie, commonly written as *Calorie*.

Cancer Uncontrolled proliferation of cells in humans and other living organisms. In humans, includes more than a hundred different types, afflicting individuals of all races and ages.

Carbon cycle The cycling of carbon between organisms and the environment.

Carcinogen A chemical or physical agent that causes cancer to develop, often decades after the original exposure.

Carrying capacity Maximum population size that a given ecosystem can support for an indefinite period or on a sustainable basis.

Catalyst Substance that accelerates chemical reactions but is not used up in the process. Enzymes are biological catalysts. Also see *catalytic converter*.

Catalytic converter Device attached to the exhaust system of automobiles and trucks to rid the exhaust gases of harmful pollutants.

Cation Any one of many kinds of positively charged ions.

Cellular respiration Process by which a cell breaks down glucose and other organic molecules to acquire energy. Also called *oxidative metabolism*.

Chlorofluorocarbons Organic molecules consisting of chlorine and fluorine covalently bonded to carbon. CCl_3F (Freon-11) and CCl_2F_2 (Freon-12) are common forms. Used as spray can propellants and coolants. Previously thought to be inert, but now known to destroy the stratospheric ozone layer. Also called *chlorofluoromethanes* and *freon gases*.

Chlorophyll Pigment of plant cells that absorbs sunlight, thus allowing plants to capture solar energy.

Chromosomes Genetic material of organisms, containing DNA and protein. Carries the genetic information that controls all cellular activity.

Chronic bronchitis Persistent inflammation of the bronchi due to pollutants in ambient air and tobacco smoke. Characterized by persistent cough.

Chronic effects In general, the delayed health results of toxic agents, for example, emphysema, bronchitis, and cancer. Contrast with *acute effects*.

Chronic exposure Exposure to one or more toxic agents generally over a long period. Symptoms (for example, cancer) usually appear long after exposure.

Chronic obstructive lung disease Any one of several lung diseases characterized by obstruction of breathing. Includes emphysema, bronchitis, and diseases with symptoms of both of these.

Clear-cutting Removal of all trees from a forested area.

Climate Average weather conditions, including temperature, rainfall, and humidity.

Climax community or **ecosystem** See *mature community*.

Closed system A system that can exchange energy, but does not exchange matter, with the surrounding environment. Example: the Earth. Contrast with *open system*.

Coal gasification Production of combustible organic gases (mostly methane) by applying heat and steam to coal in an oxygen-enriched environment. Carried out in surface vessels or *in situ*.

Coal liquefaction Production of synthetic oil from coal.

Coastal wetlands Wet or flooded regions along coastlines, including mangrove swamps and salt marshes. Contrast with *inland wetlands*.

Coevolution Process whereby two species evolve adaptations as a result of extensive interactions with each other.

Cogeneration Production of two or more forms of useful energy from one process. For example, production of electricity and steam heat from combustion of coal. Increases energy efficiency.

Coliform bacterium Common bacterium found in the intestinal tracts of humans and other species. Used in water quality analysis to determine the extent of fecal contamination.

Commensalism Relationship between two organisms that is beneficial to one and neither harmful nor helpful to the other.

Common law Body of rules and principles based on judicial precedent rather than legislative enactments. Founded on an innate sense of justice, good conscience, and reason. Flexible and adaptive. Contrast with *statutory law*.

Commons Any resource used in common by many people, such as air, water, and grazing land.

Community Also called a *biological community*. The populations of plants, animals, and microorganisms living and interacting in a given locality.

Competition Vying for resources among members of the same or different species.

Composting Aerobic decay of organic matter to generate a humus-like substance used to supplement soil.

Confusion technique (of pest control) Release of insect sex-attractant pheromones identical to pheromones released by normal breeding females to attract males for mating. Release in large quantities confuses males as to the location of the females, thus minimizing the chances of males finding females and helping to control pest populations.

Conifers Trees with needles for leaves that remain green (and photosynthetic) year round.

Conservation A strategy to reduce the use of resources, especially through increased efficiency, reuse, recycling, and decreased demand.

Conservation biology See *restoration ecology*.

Consumer (or consumer organism) An organism in the ecosystem that feeds on autotrophs and/or heterotrophs. Synonym: *heterotroph*.

Continental drift Movement of the earth's tectonic plates on a semi-liquid layer of mantle, forcing continents to shift position over hundreds of thousands of years.

Contour farming Soil erosion control technique in which row crops (corn) are planted along the contour lines in sloping or hilly fields rather than up and down the hills.

Contraceptive Any device or chemical substance used to prevent conception.

Control group In scientific experimentation, a group that is untreated and compared with a treated, or experimental, group.

Control rods Special rods containing neutron-absorbing materials. Inserted into a reactor core to control the rate of fission or to shut down fission reactions.

Convergent evolution The independent evolution of similar traits among unrelated organisms resulting from similar selective pressures.

Cosmic radiation High-energy electromagnetic radiation similar to cosmic rays but originating from periodic solar flare-ups. Possesses extraordinary ability to penetrate materials, including cement walls.

Cost-benefit analysis Way of determining the economic, social, and environmental costs and benefits of a proposed action such as construction of a dam or highway. Still a crude analytical tool because of the difficulty of measuring environmental costs.

Crisis of unsustainability Used to define the general state of affairs of our society living, as it were, in an unsustainable fashion.

Critical population size Population level below which a species cannot successfully reproduce.

Critical thinking The most ordered type of reasoning we're capable of. Helps us to discern between beliefs (what people believe to be true) and knowledge based on facts.

Crop rotation Alternating crops in fields to help restore soil fertility and also control pests.

Crossing over Transfer of genetic material from one chromosome to another during the formation of gametes.

Cross-media contamination The movement of pollution from one medium, such as air, to another, such as water.

Crude birth rate Number of births per 1000 people in a population at the midpoint of the year.

Crude death rate Number of deaths per 1000 people in a population at midyear.

Cultural control (of pests) Techniques to control pest populations not involving chemical pesticides, environmental controls, or genetic controls. Examples: cultivation to control weeds and manual removal of insects from crops.

Cultural eutrophication Eutrophication (see definition) due largely to human activities.

Daughter nuclei Atomic nuclei that are produced during fission of uranium.

DDT (dichlorodiphenyltrichloroethane) An organochlorine insecticide used first to control malaria-carrying mosquitoes and lice and later to control a variety of insect pests, but now banned in the United States because of its persistence in the environment and its ability to bioaccumulate.

Debt-for-nature swap Arrangement made between a debtor nation, an intermediary, and a lending institution in which the intermediary buys up a debt of a debtor nation at a discount in exchange for funds or programs to protect vital habitats in the debtor nation.

Decibel (dB) A unit to measure the loudness of sound.

Deciduous trees Trees that lose their leaves each year during the late fall and remain leaveless.

Decomposer An organism that breaks down nonliving organic material. Examples: bacteria and fungi.

Decomposer food chain A specific nutrient and energy pathway in an ecosystem, in which decomposer organisms (bacteria and fungi) consume dead plants and animals as well as animal wastes. Essential for the return of nutrients to soil and carbon dioxide to the atmosphere. Also called *detritus food chain*.

Deforestation Destruction of forests by clear-cutting.

Demographic transition A phenomenon witnessed in populations of industrializing nations. As industrialization proceeds and wealth accumulates, crude birth rate and crude death rate decline, resulting in zero or low population growth. Decline in death rate usually precedes the decline in birth rate, producing a period of rapid growth before stabilization.

Demography The science of population.

Depletion allowance Tax relief given to extractive industries as they deplete reserves. Intended to allow the companies to invest more in exploration. Gives extraction industries unfair advantage over recycling companies.

Desert Biome located throughout the world. Often found on the downwind side of mountain ranges. Characterized by low humidity, high summertime temperatures, and plants and animals especially adapted to lack of water.

Desertification The formation of desert in arid and semi-arid regions from overgrazing, deforestation, poor agricultural practices, and climate change. Found today in Africa, the Middle East, and the southwestern United States.

Detoxification Rendering a substance harmless by causing it to react with another chemical or destroying the molecule through combustion or thermal decomposition.

Detritus Any organic waste from plants and animals.

Detritus feeders Organisms in the decomposer food chain that feed primarily on organic waste (detritus), such as fallen leaves.

Detritus food chain See *decomposer food chain*.

Deuterium An isotope of hydrogen whose nucleus contains one proton and one neutron (a hydrogen atom has only one proton).

Development In this book, development is defined as ways of improving human well being.

Dioxins A large group of highly toxic, carcinogenic compounds containing some herbicides (2,4-D) and 2,4,5-T) and Agent Orange. Once disposed of by mixing with waste crankcase oil that was spread on dirt roads to control dust.

Diversity A measure of the number of different species in an ecosystem.

DNA (deoxyribonucleic acid) A long-chained organic molecule that is found in chromosomes and carries the genetic information that controls cellular function and is the basis of heredity.

Dose-response curve Graphical representation of the effects of varying doses of chemical or physical agents.

Doubling time The length of time it takes some measured entity (population) to double in size at a given growth rate.

Ecological backlashes Ecological effects of seemingly harmless activities, for example, the greenhouse effect.

Ecological equivalents Organisms that occupy similar ecological niches in different regions of the world.

Ecological justice An ethical concept stating that other organisms have a right to live and prosper using the Earth's resources.

Ecological niche See *niche*.

Ecological system See *ecosystem*.

Ecology Study of living organisms and their relationships to one another and the environment.

Economic depletion Reduction in the supply of a resource to the point at which it is no longer economically feasible to continue mining, extracting, or harvesting it.

Economic externality A cost (environmental damage, illness) of manufacturing, road building, or other actions that is not taken into account when determining the total cost of production or construction. A cost generally passed on to the general public and taxpayers; external cost.

Ecosphere See *biosphere*.

Ecosystem management A new approach to managing human and natural landscapes. Attempts to protect entire watersheds and biological communities, rather than small isolated parcels or individual species, and considers a wide range of factors, including abiotic and biotic factors of the environment.

Ecosystem Short for *ecological system*. A community of organisms occupying a given region within a biome. Also, the physical and chemical environment of that community and all the interactions among organisms and between organisms and their environment.

Ecotone Transition zone between adjacent ecosystems.

Element A substance, such as oxygen, gold, or carbon, that is distinguished from all other elements by the number of protons in its atomic nucleus. The atoms of an element cannot be decomposed by chemical means.

Emigration Movement of people out of a country to establish residence elsewhere.

Emissions offset policy Strategy to control air pollution in areas meeting federal ambient air quality standards, whereby new factories must secure emissions reductions from existing factories to begin operation; thus the overall pollution level does not increase.

Emissions standard The maximum amount of a pollutant permitted to be released from a given source.

Emphysema A progressive, debilitating lung disease caused by smoking and pollution at work and in the environment. Characterized by gradual breakdown of the alveoli (see definition) and difficulty in catching one's breath.

Endangered species A plant, animal, or microorganism that is in immediate danger of biological extinction. See *threatened species*.

End-of-pipe controls Refers to strategies and technologies such as sewage treatment plants that attempt to deal with pollution after they are created. Generally, pollutants are removed from smoke stack gases or liquid effluents.

Energy The capacity to do work. Exists in many forms, including heat, light, sound, electricity, coal, oil, and gasoline.

Energy quality The amount of useful work acquired from a given form of energy. High-quality energy forms are concentrated (such as oil and coal); low-quality energy forms are less concentrated (such as solar heat).

Energy system The complete production-consumption process for energy resources, including exploration, mining, refining, transportation, and waste disposal.

Entropy A measure of disorder. The second law of thermodynamics applied to matter says that all systems proceed toward maximum disorder (maximum entropy).

Environment All the biological and nonbiological factors that affect an organism's life.

Environmental control (of pests) Methods designed to alter the abiotic and biotic environments of pests, making them inhospitable or intolerable. Examples include increasing crop diversity, altering time of planting, and altering soil nutrient levels.

Environmental impact statement (EIS or ES) Document prepared primarily to outline potential impacts of projects supported in part or in their entirety by federal funds.

Environmental phase (of the nutrient cycle) Part of the nutrient or biogeochemical cycle in which the nutrient is deposited or cycles through the environment (air, water, and soil).

Environmental resistance Abiotic and biotic factors that can potentially reduce population size.

Environmental science The interdisciplinary study of the complex and interconnected issues of population, resources, and pollution.

EPA Federal agency charged with creating and enforcing environmental regulations.

Epidemiological studies Studies of diseases based on analysis of populations.

Epidemiology Study of disease and death in human populations, which attempts to find links between causes and effects through statistical methods.

Epilimnion Upper, warm waters of a lake. Contrast with *hypolimnion*.

Estuarine zones Coastal wetlands and estuaries.

Estuary Coastal region such as an inlet or mouth of a river where salt and fresh water mix.

Ethanol Grain alcohol, or ethyl alcohol, produced by fermentation of organic matter. Can be used as a fuel for a variety of vehicles and as a chemical feedstock.

Eukaryotes Cells whose genetic material is located in nuclei. Contain distinct organelles that carry out specific functions.

Eutrophication Accumulation of nutrients in a lake or pond due to human intervention (cultural eutrophication) or natural causes (natural eutrophication). Contributes to process of succession (see definition).

Evapotranspiration Evaporation of water from soil and transpiration of water from plants.

Evolution A long-term process of change in organisms caused by random genetic changes that favor the survival and reproduction of the organism possessing the genetic change. Through evolution, organisms become better adapted to their environment.

Exclusion principle Ecological law holding that no two species can occupy the exact same niche.

Experimental group In scientific experimentation, a group that is treated and compared with an untreated, or control, group.

Exponential curve See *J curve*.

Exponential growth Increase in any measurable thing by a fixed percentage. When plotted on graph paper, it forms a J-shaped curve.

Externality A spillover effect that harms others. The source of the effect (say, pollution) does not pay for the effect.

Extinction See *biological extinction*.

Extractive reserve A forest protected for the sustainable harvest of fruit, nuts, and other products, generally by local people.

Fall overturn Annual cycle in deep lakes in temperate climates, in which the warm surface water and cool subsurface water mix.

Fallout Deposition of radioactive materials produced during an atomic detonation.

Family planning Process by which couples determine the number and spacing of children.

Feedlot Fenced area where cattle are raised in close confinement to minimize energy loss and maximize weight gain.

First law of thermodynamics Also called the *law of conservation of energy*. States that energy is neither created nor destroyed; it can only be transformed from one form to another.

First-generation pesticides Earliest known chemical pesticides such as ashes, sulfur, ground tobacco, and hydrogen cyanide. Contrast with *second-* and *third-generation pesticides*.

First-law efficiency A measure of the efficiency of energy use. Total amount of useful work derived from a system divided by the total amount of energy put into a system.

Fission fragments See *daughter nuclei*.

Fission Splitting of atomic nuclei when they are struck by neutron or other subatomic particles.

Floodplain Low-lying region along a river or stream, periodically subject to natural flooding. Common site for human habitation and farming.

Fly ash Mineral matter escaping with smokestack gases from combustion of coal.

Food chain A specific nutrient and energy pathway in ecosystems proceeding from producer to consumer. Part of a bigger network called the *food web*. See *decomposer food chain* and *grazer food chain*.

Food web Complex intermeshing of individual food chains in an ecosystem.

Foreign species See *alien species*.

Fossil fuel Any one of the organic fuels (coal, natural gas, oil, tar sands, and oil shale) derived from once living plants or animals.

Freons See *chlorofluorocarbons*.

Frontier mentality A mind-set that views humans as "above" all other forms of life rather than as an integral part of nature and that sees the world as an unlimited supply of resources for human use regardless of the impacts on other species. Implicit in this view are the notions that bigger is better, continued material wealth will improve life, and nature must be subdued.

Fuel rods Rods packed with small pellets of radioactive fuel (usually a mixture of fissionable uranium-235 and uranium-238) for use in fission reactors.

Full-cost pricing Technique to include all costs, including the cost of ecological damage, into the price of a product or service.

Gaia hypothesis Term coined by James Lovelock to describe the earth's capacity to maintain the physical and chemical conditions necessary for life.

Galaxy Grouping of billions of stars, gas, and dust, such as the Milky Way galaxy.

Gamma rays A high-energy form of radiation given off by certain radionuclides. Can easily penetrate the skin and damage cells.

Gasohol Liquid fuel for vehicles, containing nine parts gasoline and one part ethanol.

Gene pool Sum total of all the genes and their alternate forms in a population.

Gene Segment of the DNA that either codes for proteins produced by the cell (structural gene) or regulates structural genes.

Generalists Organisms that have a broad niche, usually feeding on a variety of food materials and sometimes adapted to a large number of habitats.

Genetic control (of pests) Development of plants and animals genetically resistant to pests, through breeding programs and genetic engineering. Also, introduction of sterilized males of pest species (see *sterile male technique*).

Genetic engineering Isolation and production of genes that are then inserted into bacteria or other organisms. Can be used to produce insulin and other hormones. May someday also be used to treat genetic diseases.

Geopressurized zone Aquifer containing superheated, pressurized water and steam trapped by impermeable rock strata and heated by underlying magma.

Geothermal energy Energy derived from the earth's heat that comes from decay of naturally occurring radioactive materials in the earth's crust, magma, and friction caused by movement of tectonic plates.

Global change General term that refers to all of the changes—climatological, ecological, environmental, perhaps even economic—caused by greenhouse gas accumulation in the atmosphere.

Global climate change Change in the Earth's climate caused by greenhouse gases.

Global warming Warming of the Earth's atmosphere caused by the build up of greenhouse gases.

GNP See *gross national product*.

Gradualism Theory of evolution holding that species evolve over long periods. Contrast with *punctuated equilibrium*.

Grasslands Biome found in both temperate and tropical regions and characterized by periodic drought, flat or slightly rolling terrain, and large grazers that feed off the lush grasses.

Gray-air cities Older industrial cities characterized by predominantly sulfur dioxide and particulate pollution. Contrast with *brown-air cities*.

Grazer food chain A specific nutrient and energy pathway starting with plants that are consumed by grazers (herbivores).

Green product General term referring to environmentally friendly products. They may be made from recycled materials or may be fully recyclable. They may be reusable or nontoxic or may help promote efficient use of resources.

Green Revolution Developments in plant genetics in the late 1950s and early 1960s resulting in high-yield varieties, producing three to five times more grain than previous plants but requiring intensive irrigation and fertilizer use.

Green tax General term applying to user fees, severance taxes, and pollution taxes.

Greenhouse effect Mechanism that explains atmospheric heating caused by increasing carbon dioxide. Carbon dioxide is believed to act like the glass in a greenhouse, permitting visible light to penetrate but impeding the escape of infrared radiation, or heat.

Greenhouse gas Any of several naturally occurring or anthropogenic gases that trap heat escaping from the Earth's surface, causing the atmosphere to heat up.

Gross national product (GNP) Total national output of goods and services valued at market prices, including net exports and private investment.

Gross primary productivity The total amount of sunlight converted into chemical-bond energy by a plant. This measure does not take into account how much energy a plant uses for normal cellular functions. See *net primary productivity*.

Groundwater Water below the earth's surface in the saturated zone.

Growth factor Any one of many biotic and abiotic factors that stimulate growth of populations. Contrast with *reduction factors*.

Habitat The specific region in which an organism lives.

Half-life Time required for one-half of a given amount of radioactive material to decay, producing one-half the original mass. Can also be used to describe the length of residence of chemicals in tissues. *Biological half-life* refers to the time it takes for one-half of a given amount of a substance to be excreted or catabolized.

Hazardous waste Any potentially harmful solid, liquid, or gaseous waste product of manufacturing or other human activities.

Herbicide Chemical agent used to control weeds.

Herbivore Heterotrophic organism that feeds exclusively on plants.

Heteroculture Agriculture in which several plant species are grown simultaneously to reduce insect infestation and disease.

Heterotroph An organism that feeds on other organisms such as plants and animals. It cannot make its own food-stuffs.

Homeostasis State of relative constancy in organisms and ecosystems. A kind of dynamic equilibrium.

Hot-rock zones Most widespread geothermal resource. Regions where bedrock is heated by underlying magma.

Humus Decaying organic matter that increases fertility, aeration, and water retention of soils.

Hunting and gathering society People who lived as nomads or in semipermanent sites from the beginning of human evolution until approximately 5000 B.C. Some remnant populations still survive. They gathered seeds, fruits, roots, and other plant products, ate carrion, and hunted indigenous species for food.

Hybrid car A new type of vehicle powered by gasoline and electricity (which it generates itself).

Hybrid Offspring produced by cross-mating of two different strains or varieties of plants or animals.

Hydrocarbons Organic molecules containing hydrogen and carbon. Released during the incomplete combustion of organic fuels. React with nitrogen oxides and sunlight to form photochemical oxidants in photochemical smog.

Hydroelectric power Electricity produced in turbines powered by running water.

Hydrological cycle The movement of water through the environment, from atmosphere to earth and back again. Major events include evaporation and precipitation. Also called the *water cycle*.

Hydrosphere The watery portion of the planet. Contrast with *atmosphere* and *lithosphere*.

Hydrothermal convection zone Rock strata containing large amounts of water heated by underlying magma and driven to the surface through cracks and fissures in overlying rock layers. Forms hot springs and geysers.

Hypolimnion Deep, cold waters of a lake. Contrast with *epilimnion*.

Hypothesis Tentative explanation for a natural phenomenon.

Immature ecosystem An early successional community characterized by low species diversity and low stability. Contrast with *mature ecosystem*.

Immigration Movement of people into a country to set up residence there.

Income tax deduction An expense by an individual or business that can be subtracted from one's gross earnings, thus lowering one's taxes. Results in a smaller overall reduction in one's taxes than a tax credit.

Indoor air pollution Generally refers to air pollutants in homes from internal sources such as smokers, fireplaces, woodstoves, carpets, paneling, furniture, foam insulation, and cooking stoves.

Induced abortion Surgical procedure to interrupt pregnancy by removing the embryo or fetus from the uterus. In the first trimester, generally carried out by vacuum aspiration. Contrast with *spontaneous abortion*.

Industrial Revolution Period in history marked by a shift in manufacturing from small-scale production by hand to large-scale production by machine. Occurred in England in the 1700s and in the United States in the 1800s.

Industrial smog Air pollution from industrial cities (gray-air cities), consisting mostly of particulates and sulfur oxides. Contrast with *photochemical smog*.

Industrial society Group of people living in urban or rural environments characterized by mechanization of industrial production and agriculture. Widespread machine labor causes high energy demands and pollution. Increasing control over natural processes leads to feelings that humans are apart from nature and superior to it.

Infant mortality rate Number of infants under one year of age dying per 1000 births in any given year.

Infectious disease Generally, a disease caused by a virus, bacterium, or parasite that can be transmitted from one organism to another (example: viral hepatitis).

Infrared radiation Heat, an electromagnetic radiation of wavelength outside the red end of the visible spectrum.

Inland wetlands Wet or flooded regions along inland surface waters. Includes marshes, bogs, and river outflow lands. Contrast with *coastal wetlands*.

In-migration Movement of people into a state or region within a country to set up residence.

Inorganic fertilizer Synthetic plant nutrient added to the soil to replace lost nutrients. Components include nitrogen, phosphorus, and potassium. Also called *artificial fertilizer* or *synthetic fertilizer*.

Input approach A method of solving an environmental problem by reducing the inputs. For example, reducing consumption and increasing product durability can cut production of solid wastes, pollution, or hazardous wastes.

Insecticide One form of pesticide, used specifically to control insect populations.

Integrated pest management Pest control with minimum risk to humans and the environment through use of a variety of control techniques (including pesticides and biological controls).

Integrated wildlife or species management Control of populations through the use of many techniques, including the reintroduction of natural predators, habitat improvement, reduction in habitat destruction, establishment of preserves, reduced pollution, and captive breeding.

Intergenerational equity Key concept of sustainable ethics. A kind of fairness to future generations.

Interspecific competition Competition among members of different species.

Ion A particle formed when an atom loses or gains an electron.

Ionizing radiation Electromagnetic radiation with the capacity to form ions in body tissues and other substances.

Isotopes Atoms of the same element that differ in their atomic weight because of variations in the number of neutrons in their nuclei.

J curve A graphical representation of exponential growth.

Juvenile hormone Chemical substance in insects that stimulates growth through early life stages. Used with some success as an insecticide. When applied to infested fields, JH alters normal growth and development of insect pests, resulting in their death.

Kerogen Solid, insoluble organic material found in oil shale.

Keystone species Critical species in an ecosystem whose loss profoundly affects several or many others.

Kilocalorie One thousand calories. See *Calorie*.

Kilowatt One thousand watts. See *watt*.

Kinetic energy The energy of objects in motion.

Kwashiorkor Dietary deficiency caused by insufficient protein intake; common in children one to three years of age in less developed countries. Characterized by growth retardation, wasting of muscles in limbs, and accumulation of fluids in the body—especially in feet, legs, hands, and face.

Lag effect The tendency for a population to continue growing even after it has reached replacement-level fertility. Caused by an expanding number of women reaching reproductive age.

Land ethic View that extends ethical concerns beyond humans to the ecosystem.

Landfill Depression in the ground or excavated site in which waste is deposited. Sanitary landfills are so named because trash is covered daily with soil to prevent odors and proliferation of rodents and flies. Can be used for hazardous wastes, but preferably as last resort. Must be lined with clay and impermeable liners to accept hazardous materials.

Land-use planning Process whereby land uses are matched with the needs of the community and environmental considerations—for example, need for open space and agricultural land and for control of water and air pollution.

Laterite Soil found in some tropical rain forests. Rich in iron and aluminum but generally of poor fertility. Turns bricklike if exposed to sunlight.

Least-cost planning Process in which demand for a resource (such as electricity) is met in the least costly way. Generally includes only economic costs.

Less developed country Term describing the nonindustrialized nations, generally characterized by low standard of living, high population growth rate, high infant mortality, low material consumption, low per capita energy consumption, low per capita income, rural population, and high illiteracy.

Life expectancy The number of years people in a society live, on average.

Light water reactor Most common fission reactor for generating electricity. Water bathes the core of the reactor and is used to generate steam, which turns the turbines that generate electricity. Contrast with *liquid metal fast breeder*.

Limiting factor A chemical or physical factor that determines whether an organism can survive in a given ecosystem. In most ecosystems, rainfall is the limiting factor.

Limnetic zone Open water zone of lakes through which sunlight penetrates; contains algae and other microscopic organisms that feed on dissolved nutrients.

Linear thinking A mode of thinking that oversimplifies complex issues. Often ignores complex networks of cause and effect and systems impacts.

Linearity Refers to linear thinking and linear systems design.

Liquefaction Production of liquid fuel from coal.

Liquid metal fast breeder Fission reactor that uses liquid metals such as sodium as a coolant.

Lithosphere Crust of the earth. Contrast with *hydrosphere* and *atmosphere*.

Littoral drift Movement of beach sand parallel to the shoreline. Caused by waves and longshore currents parallel to the beach.

Littoral zone Shallow waters along a lakeshore, where rooted vegetation often grows.

Macronutrient A chemical substance needed by living organisms in large quantities (for example, carbon, oxygen, hydrogen, and nitrogen). Contrast with *micronutrient*.

Magma Molten rock beneath the Earth's crust.

Malnourishment A dietary deficiency caused by lack of vital nutrients and vitamins.

Manganese nodules Nodular accumulations of manganese and other minerals such as iron and copper found on the ocean floor at depths of 300 to 6000 meters. Particularly abundant in the Pacific Ocean.

Marasmus A dietary deficiency caused by insufficient intake of protein and calories; occurs primarily in infants under the age of one, usually as the result of discontinuation of breast-feeding.

Mariculture Cultivation of fish and other aquatic organisms in salt water (estuaries and bays).

Mature community A community that remains more or less the same over a long period of time. Climax stage of succession. Also called a *climax community*.

Mature ecosystem An ecosystem in the climax stage of succession, characterized by high species diversity and high stability. Contrast with *immature ecosystem*.

Measure of economic welfare Proposed standard that takes into account the accumulated wealth of a nation.

Megawatt Measure of electrical power equal to a million watts, o 1000 kilowatts. See *watt*.

Mesothelioma A tumor of the lining of the lung (pleura). Cause by asbestos.

Metalimnion See *thermocline*.

Metastasis Movement of cancer cells to another location where nev tumors are formed.

Methyl mercury Water-soluble organic form of mercury, formed b bacteria in aquatic ecosystems from inorganic (insoluble) mer cury pollution. Able to undergo biological magnification.

Micronutrient An element needed by organisms, but only in smal quantities—such as copper, iron, and zinc. Contrast with *macronutrient*.

Migration Movement of people across state and national bound aries to set up new residence. See *immigration, emigration, in migration*, and *out-migration*.

Mill tailings Residue from uranium processing plants. Spent ura nium ore that is contaminated with radioactivity.

Mineral A chemical element (such as gold) or inorganic compoun (such as iron ore) existing naturally.

Minimum tillage Reduced plowing and cultivating of cropland be tween and during growing seasons, to help reduce soil erosio and save energy. Also called *conservation tillage*.

Molecule Particle consisting of two or more atoms bonded together The atoms in a molecule can be of the same element but are usu ally of different elements.

Monoculture Cultivation of one plant species (such as wheat) ove a large area. Highly susceptible to disease and insects.

More developed country A convenient term that describes indus trialized nations, generally characterized by high standard of liv ing, low population growth rate, low infant mortality, excessiv material consumption, high per capita energy consumption, higl per capita income, urban population, and low illiteracy.

Municipal solid waste Refers to garbage from homes and busi nesses in cities and towns. Typically contains high percentage o recyclable and compostable materials.

Mutagen A chemical or physical agent capable of damaging the ge netic material (DNA and chromosomes) of living organisms i both germ cells and somatic cells.

Mutation In general, any damage to the DNA and chromosomes.

Mutualism Relationship between two organisms that is beneficial t both.

Narrow-spectrum pesticide A chemical agent effective in control ling a small number of pests.

Natural erosion Loss of soil occurring at a slow rate, but not cause by human activities. A natural event in all terrestrial ecosystems

Natural eutrophication See *eutrophication*.

Natural gas Gaseous fuel containing 50%–90% methane and lesse amounts of other burnable organic gases such as propane an butane.

Natural hazards Dangers that result from normal meteorologic, at mospheric, oceanic, biological, and geological phenomena.

Natural resource See *resource*.

Natural selection Process in which slight variations in organism (adaptations) are preserved if they are useful and help the organ ism to better respond to its environment.

Negative feedback Control mechanism present in the ecosystem an in all organisms in which signals cause processes to shut down.

Net energy See *net useful energy production*.

Net migration Number of immigrants minus the number of emi grants. Can be expressed as a rate by determining immigratio and emigration rates.

Net primary productivity Gross primary productivity (the tota amount of energy that plants produce) minus the energy plant use during cellular respiration.

Net useful energy production Amount of useful energy extracte from an energy system.

Niche Also called an *ecological niche*. An organism's place in th ecosystem: where it lives, what it consumes, what consumes it and how it interacts with all biotic and abiotic factors.

Nitrate (NO_3) Inorganic anion containing three oxygen atoms an one nitrogen atom linked by covalent bonds.

Nitrite (NO_2) Inorganic anion containing two oxygen atoms an one nitrogen atom. Combines with hemoglobin and may caus serious health impairment and death in children.

Nitrogen cycle The cycling of nitrogen between organisms and the environment.

Nitrogen fixation Conversion of atmospheric nitrogen (a gas) into nitrate and ammonium ions (inorganic form), which can be used by plants.

Nitrogen oxides Nitric oxide (NO) and nitrogen dioxide (NO_2), produced during combustion when atmospheric nitrogen (N_2) combines with oxygen. Can be converted into nitric acid (HNO_3). All are harmful to humans and other organisms.

Noise An unwanted or unpleasant sound.

Nonattainment area Region that violates EPA air pollution standards.

Nonpoint source (of pollution) Diffuse source of pollution such as an eroding farm field, urban and suburban lands, and forests. Contrast with *point source.*

Nonrenewable resource Resource that is not replaced or regenerated naturally within a reasonable period (fossil fuel, mineral).

Northern coniferous, boreal forest Same as taiga.

Nuclear fission Splitting of an atomic nucleus when neutrons strike it. Products are two or more smaller nuclei, neutrons (which can cause further fission reactions), and an enormous amount of heat and radiation energy.

Nuclear fusion Joining of two small atomic nuclei (such as hydrogen and deuterium) to form a new and larger nucleus (such as helium), accompanied by an enormous release of energy. Source of light and heat from the sun.

Nuclear power (or energy) Energy from the fission or fusion of atomic nuclei.

Nutrient Any substance that is required for the survival of an organism.

Nutrient cycle Same as *biogeochemical cycle.*

Off-road vehicle (ORV) Any vehicle used cross country, often in a recreational capacity (four-wheel-drive vehicles, dune buggies, all-terrain vehicles, snowmobiles, and trail bikes).

Oil See *petroleum.*

Oil shale A fine-grained sedimentary rock called *marlstone* containing an organic substance known as kerogen. When heated, it gives off shale oil, which is much like crude oil.

Old-growth forest Ancient forests with trees often 150 to 1000 or more years old. Also called *primary forest.*

Omnivore An organism that eats both plants and animals.

Open system A system that freely exchanges energy and matter with the environment. Example: any living organism. Contrast with *closed system.*

Opportunity costs Costs of lost money-making opportunities (and potentially higher income) incurred when we make a decision to invest our money in a particular way.

Optimum range Biotic and abiotic conditions in which survival and reproduction of a given species is maximized.

Ore deposit A valuable mineral located in high concentration in a given region.

Ore Rock bearing important minerals—for example, uranium ore.

Organic farming Agricultural system in which natural fertilizers (manure and crop residues), crop rotation, contour planting, biological insect control measures, and other techniques are used to ensure soil fertility, erosion control, and pest control.

Organic fertilizer Material such as plant and animal wastes added to cropland and pastures to improve soil. Provides valuable soil nutrients and increases the organic content of soil (thus increasing moisture content).

Organismic phase The part of the nutrient cycle in which nutrients are located in organisms: plants, animals, bacteria, fungi, or others.

Out-migration Movement of people out of a state or region within a country to set up residence elsewhere in that country.

Output approach A method of solving an environmental problem by controlling the outputs. For example, composting or burning trash reduces the land requirements for solid waste disposal. Control devices reduce air and water pollution.

Overgrazing Excessive consumption of producer organisms (plants) by grazers such as deer, rabbits, and domestic livestock. Indication that the ecosystem is out of balance.

Overpopulation A condition resulting when the number of organisms in an ecosystem exceeds its ability to assimilate wastes and provide resources. Creates physical and mental stress on a species as a result of competition for limited resources and deterioration of the environment.

Overshoot The phenomenon occurring when a population of organisms exceeds the carrying capacity of its environment.

Oxidants Oxidizing chemicals (for example, ozone) found in the atmosphere.

Oxygen-demanding wastes Organic wastes that are broken down in water by aerobic bacteria. Aerobic breakdown causes the oxygen levels to drop.

Ozone (O_3) Inorganic molecule found in the atmosphere, where it is a pollutant because of its harmful effects on living tissue and rubber. Also found in the stratosphere, where it helps screen ultraviolet light. Used in some advanced sewage treatment plants.

Ozone layer Thin layer of ozone molecules in the stratosphere. Absorbs ultraviolet light and converts it to infrared radiation. Effectively screens out 99% of the ultraviolet light.

Paradigm A major theoretical construct that is central to a field of study. For example, the theory of evolution and the structure of DNA are two paradigms that are central to biological science.

Parasitism Relationship in which one species lives in or on another, its host.

Particulates Solid particles (dust, pollen, soot) or water droplets in the atmosphere.

Passive solar Capture and retention of the sun's energy within a building through south-facing windows and some form of heat storage in the building (brick or cement floors and walls). Contrast with *active solar.*

PCBs See *polychlorinated biphenyls.*

Perennial A plant that grows from the same root structure year after year (for example, rose bushes).

Permafrost Permanently frozen ground found in the tundra.

Permanent threshold shift Loss of hearing after continued exposure to noise. Contrast with *temporary threshold shift.*

Pesticide A general term referring to a chemical agent that kills organisms we classify as pests, such as insects and rodents. Also called *biocide.*

Petroleum A viscous liquid containing numerous burnable hydrocarbons. Distilled into a variety of useful fuels (fuel oil, gasoline, and diesel) and petrochemicals (chemicals that can be used as a chemical feedstock for the production of drugs, plastics, and other substances).

pH Measure of acidity on a scale from 0 to 14, with pH 7 being neutral, numbers greater than 7 being basic, and numbers less than 7 being acidic.

Pheromone Chemical substance given off by insects and other species. Sex-attractant pheromones released into the atmosphere in small quantity by female insects attract males at breeding time. Can be used in pest control. See *pheromone traps* and *confusion technique.*

Pheromone traps Traps containing pheromones, to attract insect pests, and either pesticide to kill them or a sticky substance to immobilize them. These traps may be used to pinpoint the emergence of insects, allowing conventional pesticides to be used in moderation.

Photochemical oxidants Ozone and a variety of oxygenated organic compounds produced when sunlight, hydrocarbons, and nitrogen oxides react in the atmosphere.

Photochemical reaction A chemical reaction that occurs in the atmosphere involving sunlight or heat, pollutants, and sometimes natural atmospheric chemicals.

Photochemical smog A complex mixture of photochemical oxidants and nitrogen oxides. Usually has a brownish-orange color.

Photosynthesis A two-part process in plants and algae involving (1) the capture of sunlight and its conversion into cellular energy and (2) the production of organic molecules such as glucose and amino acids from carbon dioxide, water, and energy from the sun.

Photovoltaic cell Thin wafer of silicon or other material that emits electrons when struck by sunlight, thus generating an electrical current. Also called *solar cell.*

Physical infrastructure Refers to humanmade environment or humanmade systems: buildings, bridges, factories, and so on.

Phytoplankton Includes single-celled algae and other free-floating photosynthetic organisms.

Pioneer community The first community to become established in a once-lifeless environment during primary succession.

Pitch (or frequency) Measure of the frequency of a sound in cycles per second (cps) hertz (Hz)—compressional sound waves passing a given point per second. The higher the cps, the higher the pitch.

Pneumoconiosis (black lung) A debilitating lung disease caused by prolonged inhalation of coal and other mineral dusts. Results in a decreased elasticity and gradual break-down of alveoli in the lungs. Eventually leads to death.

Point source (of pollution) Easily discernible source of pollution, such as a factory. Contrast with *nonpoint source*.

Pollution Any physical, chemical, or biological alteration of air, water, or land that is harmful to living organisms.

Pollution prevention Any one of several methods to reduce pollution production, such as process modification and substitution.

Polychlorinated biphenyls (PCBs) Group of at least 50 organic compounds, used for many years as insulation in electrical equipment. Capable of biological magnification. Disrupts reproduction in gulls and possibly other organisms high on the food chain.

Population A group of organisms of the same species living within a specified region.

Population control In human populations, all methods of reducing birth rate, primarily through pregnancy prevention and abortion. In an ecological sense, regulation of population size by a myriad of abiotic and biotic factors.

Population crash (dieback) Sudden decrease in population that results when an organism exceeds the carrying capacity of its environment.

Population growth rate Rate at which a population increases on a yearly basis, expressed as a percentage. For world population: GR = (crude birth rate – crude death rate) × 100. For a given country, population growth rate must also take into account the net migration rate.

Population histogram Graphical representation of population by age and sex.

Positive feedback Control mechanism in ecosystems and organisms in which information influences some process, causing it to increase.

Potential energy Stored energy.

Predator An organism that actively hunts its prey.

Prey Organism (such as deer) attacked and killed by predator.

Primary air pollutant A pollutant that has not undergone any chemical transformation; emitted by either a natural or an anthropogenic source.

Primary consumer First consuming organism in a given food chain. A grazer in grazer food chains or a decomposer organism or insect in decomposer food chains. Belongs to the second trophic level.

Primary forest See *old-growth forest*.

Primary pollutant Pollutant produced by combustion or other sources. Can be chemically modified after release, creating a secondary pollutant.

Primary succession The sequential development of biotic communities where none previously existed.

Primary treatment (of sewage) First step in sewage treatment, to remove large solid objects by screens (filters) and sediment and organic matter in settling chambers. See *secondary* and *tertiary treatment*.

Proactive government One that is concerned with long-range problems and lasting solutions. Contrast with *reactive government*.

Producer (autotroph or producer organism) One of the organisms that produces the organic matter cycling through the ecosystem. Producers include plants and photosynthetic algae.

Productivity The rate of conversion of sunlight by plants into chemical-bond energy (covalent bonds in organic molecules). See *gross* and *net primary productivity*.

Profundal zone Deeper lake water, into which sunlight does not penetrate. Below the limnetic zone.

Prospective law Law designed to address future problems and generate long-lasting solutions. Contrast with *retrospective law*.

Punctuated equilibrium A theory of evolution stating that species are fairly stable for long periods and that new species evolve rapidly over short periods of thousands of years that punctuate the equilibrium. Contrast with *gradualism*.

Pyramid of biomass Graphical representation of the amount of biomass (organic matter) at each trophic level in an ecosystem.

Pyramid of numbers Graphical representation of the number of organisms of different species at each trophic level in an ecosystem.

Quad One quadrillion (10^{15}) Btus of heat.

Rad (radiation absorbed dose) Measure of the amount of energy deposited in a tissue or some other medium struck by radiation. One rad = 100 ergs of energy deposited in one gram of tissue.

Radioactive waste Any solid or liquid waste material containing radioactivity. Produced by research labs, hospitals, nuclear weapons factories, and fission reactors.

Radioactivity Radiation released from unstable nuclei. See *alpha* and *beta particles* and *gamma rays*.

Radionuclides Radioactive forms (isotopes) of elements.

Rain shadow Arid downwind (leeward) side of mountain range.

Range of tolerance Range of physical and chemical factors in which an organism can survive. When the upper or lower limits of this range are exceeded, growth, reproduction, and survival are threatened.

Rangeland Grazing land for cattle, sheep, and other domestic livestock.

Reactive government A government that lives and acts for today, addressing present-day problems as they arise. Shows little or no concern for long-term issues and solutions. Contrast with *proactive government*.

Reactor core Assemblage of fuel rods and control rods inside a reactor vessel. Bathed by water to help control the rate of fission and absorb the heat.

Real price (or cost) The price of a commodity or service in fixed dollars—that is, the value of a dollar at an earlier time. Helpful way to determine whether a resource has experienced a real increase in cost or whether higher costs are simply due to inflation.

Reclamation As used here, the process of returning land to its prior use. Common usage: to convert deserts and other areas into habitable, productive land.

Recycling A strategy to reduce resource use by returning used or waste materials from the consumption phase to the production phase of the economy.

Reduction factors Abiotic and biotic factors that tend to decrease population growth and help balance populations and ecosystems, offsetting growth factors.

Relative humidity The amount of moisture in a given quantity of air, divided by the amount the air could hold at that temperature. Expressed as a percentage.

Rem (roentgen equivalent man) Measure that accounts for the damage done by a given type of radiation. One rad = one rem for X rays, gamma rays, and beta particles; but one rad = 10 to 20 rems for alpha particles because they do more damage.

Renewable resource A resource replaced through natural ecological cycles (water, plants, animals) or through natural chemical or physical processes (sunlight, wind).

Replacement-level fertility Number of children a couple must have to replace themselves in the population.

Reproductive age Age during which most women bear their offspring (ages 14–44).

Reproductive isolation Any of many mechanisms that prevent species from interbreeding or producing viable offspring.

Reserve Deposit of energy or minerals that is economically and geologically feasible to remove with current and foreseeable technology.

Residence time Length of time a chemical spends in the environment.

Resilience Ability of an ecosystem to return to normal after a disturbance.

Resource (as a measurement of a mineral or fuel) Total amount of a mineral or fuel on earth. Generally, only a small fraction can be recovered. Compare with *reserve*.

Resource (in general) Anything used by organisms to meet their needs, including air, water, minerals, plants, fuels, and animals.

Restoration ecology Study of restoring ecosystems to their natural state after human interference. Also called *conservation biology*.

Retorting Process of removing kerogen from oil shale, usually by burning or heating the shale. Can be carried out in surface vessels (surface retorting) or underground in fractured shale (*in situ* retorting).

Retrospective law One that attempts to solve a problem without giving much attention to potential future problems. Contrast with *prospective law*.

Reverse osmosis Means of purifying water for pollution control and desalination. Water is forced through porous membranes; pores allow passage of water molecules but not impurities.

Risk acceptability A measure of how acceptable a hazard is to a population.

Risk assessment The science of determining what hazards a society is exposed to from natural and human causes and the probability and severity of those risks.

Risk probability The likelihood that a hazardous event will occur.

Risk severity A measure of the total damage a hazardous event would cause.

Salinization Deposition of salts in irrigated soils, making soil unfit for most crops. Caused by rising water table due to inadequate drainage of irrigated soils.

Saltwater intrusion Movement of salt water from oceans or saltwater aquifers into freshwater aquifers, caused by depletion of the freshwater aquifers or low precipitation or both.

Sanitary landfill Solid waste disposal site where garbage is dumped and covered daily with a layer of dirt to reduce odors, insects, and rats.

Scrubber Pollution control device that removes particulates and sulfur oxides from smokestacks by passing exhaust gases through a fine spray of water containing lime.

Second law of thermodynamics States that when energy is converted from one form to another, it is degraded; that is, it is converted from a concentrated to a less concentrated form. The amount of useful energy decreases during such conversions.

Secondary consumer Second consuming organism in food chain. Belongs to the third trophic level.

Secondary pollutant A chemical pollutant from a natural or anthropogenic source that undergoes chemical change as a result of reacting with another pollutant, sunlight, atmospheric moisture, or some other environmental agent.

Secondary succession The sequential development of biotic communities occurring after the complete or partial destruction of an existing community by natural or anthropogenic forces.

Secondary treatment (of sewage) After primary treatment, removal of biodegradable organic matter from sewage using bacteria and other microconsumers in activated sludge or trickle filters. Also removes some of the phosphorus (30%) and nitrate (50%). See also *tertiary treatment*.

Second-generation pesticides Synthetic organic chemicals such as DDT that replaced older pesticides such as sulfur, ground tobacco, and ashes. Generally resistant to bacterial breakdown.

Second-law efficiency Measure of the efficiency of energy use, taking into account the unavoidable loss (described by the second law of thermodynamics) of energy during energy conversions. Calculated by dividing the minimum amount of energy required to perform a task by the actual amount used.

Secured landfill One lined by clay and synthetic liners in an effort to prevent leakage.

Sediment Soil particles, sand, and other mineral matter eroded from land and carried in surface waters.

Selective advantage An advantage one member of a species has over others by virtue of some adaptation it has acquired.

Selective cutting Selective removal of trees. Especially useful for mixed hardwood stands. Contrast with *clear-cutting* and *shelter-wood cutting*.

Sewage treatment plant Facility where human solid and liquid wastes from homes, hospitals, and industries are treated, primarily to remove organic matter, nitrates, and phosphates.

Shale oil Thick, heavy oil formed when shale is heated (retorted). Can be refined to produce fuel oil, kerosene, diesel fuel, and other petroleum products and petrochemicals.

Shelterbelts Rows of trees and shrubs planted alongside fields to reduce wind erosion and retain snow to increase soil moisture. May also be used to reduce heat loss from wind and thus conserve energy around homes and farms.

Shelter-wood cutting Three-step process spread out over years: (1) removal of poor-quality trees to improve growth of commercially valuable trees and allow new seedlings to become established, (2) removal of commercially valuable trees once seedlings are established, and (3) cutting remaining mature trees grown from seedlings.

Sigmoidal curve An S-shaped curve.

Simplified ecosystem One with lowered species diversity, usually as a result of human intervention.

Sinkhole Hole created by sudden collapse of the Earth's surface due to groundwater overdraft. A form of subsidence.

Slash-and-burn agriculture Farming practice in which small plots are cleared of vegetation by cutting and burning. Crops are grown until the soil is depleted; then the land is abandoned. This allows the natural vegetation and soil to recover. Common practice of early agricultural societies living in the tropics.

Sludge Solid organic material produced during sewage treatment.

Smelter A factory where ores are melted to separate impurities from the valuable minerals.

Smog Originally referred to a grayish haze (combination of smoke and fog) found in industrial cities. Also pertains to pollution called *photochemical smog*, found in newer cities. See *industrial smog*.

Social Darwinism The application (or misapplication) of the theory of evolution to social behavior.

Sociocusis Hearing loss from human activities.

Soil horizons Layers found in most soils.

Solar collector Device to absorb sunlight and convert it into heat.

Solar energy Energy derived from the sun. Used for heating water for domestic use, space heating, and producing electricity.

Solar system Group of planets revolving around a star.

Sonic boom A high-energy wake creating an explosive boom that trails after jets traveling faster than the speed of sound.

Spaceship Earth Metaphor introduced in the 1960s to foster a greater appreciation of the finite nature of Earth's resources and the ecological cycles that replenish oxygen and other important nutrients.

Specialist Organism that has a narrow niche, usually feeding on one or a few food materials and adapted to a particular habitat.

Speciation Formation of new species.

Species A group of plants, animals, or microorganisms that have a high degree of similarity and generally can interbreed only among themselves.

Species diversity Measure of the number of different species in a biological community.

Spontaneous abortion Loss of an embryo or fetus from the uterus, not caused by surgery. Generally the result of chromosomal abnormalities. Contrast with *induced abortion*.

Spring overturn Annual cycle in deep lakes in temperate climates, in which surface and subsurface waters mix.

SST (supersonic transport) Jet that travels faster than the speed of sound.

Stable runoff Amount of surface runoff that can be counted on from year to year.

Star Spherical cloud of hot gas, such as the sun, fueled by nuclear fusion reactions in its core.

Statutory law Law enacted by Congress or a state legislature. Contrast with *common law*.

Sterile male technique Pest control strategy whereby males of the pest species are grown in captivity, sterilized, and then released en masse in infested areas at breeding time. Sterile males far exceed normal wild males and mate with normal females, resulting in infertile matings and control of the pest.

Sterilization A highly successful procedure in males and females to prevent pregnancy. In males the ducts (vas deferens) that carry sperm from the testicles are cut and tied (vasectomy); in females the Fallopian tubes, or oviducts, which transport ova from the ovary to the uterus, are cut and tied (tubal ligation). Sterilization is not to be confused with castration in males (complete removal of the gonads).

Stratosphere Outer region of the earth's atmosphere, found outside the troposphere, extending 11 to 40 kilometers (7 to 25 miles) above the Earth's surface. Innermost layer of the stratosphere contains the ozone layer.

Streambed aggradation Deposition of sediment in streams or rivers, thereby reducing their water-carrying capacity.

Streambed channelization An ecologically unsound way of reducing flooding by deepening and straightening of streams, accompanied by removal of trees and other vegetation along the banks.

Strip cropping Soil conservation technique in which alternating crop varieties are planted in huge strips across fields to reduce wind and water erosion of soil.

Subsidence Sinking of land caused by collapse of underground mines or depletion of groundwater.

Succession The natural replacement of one biotic community by another. See *primary* and *secondary succession*.

Sulfur dioxide (SO_2) Colorless gas produced during combustion of fossil fuels contaminated with organic and inorganic sulfur compounds. Can be converted into sulfuric acid in the atmosphere.

Sulfur oxides (SO_x) Sulfur dioxide and sulfur trioxide, common air pollutants arising from combustion of coal, oil, gasoline, and diesel fuel. Also produced by natural sources such as bacterial decay and hot springs. Sulfur dioxide reacts with oxygen to form sulfur trioxide, which may react with water to form sulfuric acid.

Supply and demand theory Also known as the *law of supply and demand*. Economic theory explaining the price of goods and services. The supply of and demand for goods and services are primary price determinants. High demand diminishes supply, creating competition for existing goods and services, thus driving up prices.

Surface mining Any of several mining techniques in which all the dirt and rock overlying a desirable mineral (coal, for example) are first removed, exposing the mineral.

Surface runoff Water flowing in streams and over the ground during a rainstorm or snowmelt.

Surface water Refers to bodies of water visible from the Earth's surface, such as lakes, rivers, ponds, and marshes.

Sustainable development Economic development that meets current needs without compromising ability of future generations to meet their needs. Relies on appropriate technology, efficient use of resources, recycling, renewable resource use, restoration, growth management, and other measures.

Sustainable economics Economic system that seeks to meet human needs while protecting the life-support systems of the biosphere.

Sustainable ethics (mentality) A mind-set that views humans as a part of nature and Earth as a limited supply of resources, which must be carefully managed to prevent irreparable damage. Obligations to future generations require us to exercise restraint to ensure adequate resources and a clean and healthy environment.

Sustainable society A society based on sustainable ethics. Lives within the limits imposed by nature. Based on maximum use of renewable resources, recycling, conservation, and population control.

Sustained yield concept Use of renewable natural resources, such as forests and grassland, that will not cause their destruction and will ensure continued use.

Sympatric speciation Formation of new species without geographical isolation. Common in plants.

Synergism Phenomenon occurring when two or more agents (often toxicants) together produce an effect larger than expected based on knowledge of the effect of each alone. Sometimes simply refers to two or more components acting together.

Synergistic response Refers to a biological response to two or more chemicals or physical factors such as radiation that is greater than the sum of responses to individual factors.

Synfuel See *synthetic fuel*.

Synthetic fertilizer Same as inorganic fertilizer.

Synthetic fuel Gaseous or liquid organic fuel derived from coal, oil shale, or tar sands.

Systems thinking Mode of thinking that takes into account complex webs of cause and effect. Contrast with *linear thinking*.

Taiga Biome found south of the tundra across North America, Europe, and Asia, characterized by coniferous forests, soil that thaws during the summer months, abundant precipitation, and high species diversity.

Tar sands Also known as *oil sands* or *bituminous sands*. Sand impregnated with a viscous, petroleumlike substance, bitumen, which can be driven off by heat, producing a synthetic oil.

Tax credit An expense by an individual or business that can be subtracted directly from one's annual tax liability.

Technological fix A purely technological answer to a problem. Also called a *technical fix*.

Technological optimism Undying faith in technological fixes.

Tectonic plates Huge segments of the earth's crust that often contain entire continents or parts of them and that float on an underlying semiliquid layer.

Temperate deciduous forest Biome located in the eastern United States, Europe, and northeastern China below the taiga. Characterized by deciduous and nondeciduous trees, warm growing season, abundant rainfall, and a rich species diversity.

Temperature inversion Alteration in the normal atmospheric temperature profile so that air temperature increases rather than decreases with altitude.

Temporary threshold shift Momentary dulling of the sense of hearing after exposure to loud sounds. Can lead to permanent threshold shift.

Teratogen A chemical or physical agent capable of causing birth defects.

Terracing Construction of small earthen embankments on hilly or mountainous terrain to reduce the velocity of water flowing across the soil and thus reduce soil erosion.

Tertiary treatment (of sewage) Removal of nitrates, phosphates, chlorinated compounds, salts, acids, metals, and toxic organics after secondary treatment.

Thermal pollution Heat added to air or water that adversely affects living organisms and may alter climate.

Thermocline Sharp transition between upper, warm waters (epilimnion) and deeper, cold waters (hypolimnion) of a lake. Also called *metalimnion*.

Thermodynamics The study of energy conversions. See the *first and second laws of thermodynamics*.

Third-generation pesticides Newer chemical agents to control pests, such as pheromones and insect hormones.

Threatened species A species whose population is declining and could become extinct.

Threshold level A level of exposure below which no effect is observed or measured.

Throughput approach A method of solving an environmental problem by recycling and reuse. For example, recycling or reusing hazardous wastes reduces their output.

Time preference A measure of the value of an immediate gain in comparison with a long-term gain.

Tolerance level (for pesticides) Level of residue on fruits and vegetables permitted by EPA because it is considered "safe."

Total fertility rate Average number of children that would be born alive to a woman if she were to pass through all her childbearing years conforming to the age-specific fertility rates.

Toxicant A chemical, physical, or biological agent that causes disease or some alteration of the normal structure and function of an organism. Impairments may be slight or severe. Onset of effects may be immediate or delayed.

Tradeable permit Permit issued by government agency that allows companies to release certain amounts of pollution. Companies that find ways to reduce pollution can sell permits to other companies.

Transpiration Escape of water from plants through pores (stomata) in the leaves.

Tree farms Private forests devoted to maximum timber growth and relying heavily on herbicides, insecticides, and fertilizers.

Tritium (hydrogen-3) Radioactive isotope of hydrogen whose nucleus contains two neutrons and one proton. Can be used in fusion reactors.

Trophic level The position of an organism in a food chain.

Tropical rain forest Lush forests near the equator with high annual rainfall, high average temperature, and notoriously nutrient-poor soil. Possibly the richest ecosystem on Earth.

Tumor Any abnormal growth of cells. May or may not spread. Tumors that do not continue to grow or spread to other sites are known as *benign tumors*. Those that continue to grow and spread to other areas are called *malignant tumors*.

Tundra (alpine) Life zone found on mountaintops. Closely resembles the Arctic tundra in terms of precipitation, temperature growing season, plants, and animals. Extraordinarily fragile.

Tundra (Arctic) First major life zone or biome south of the north pole. Vast region on far northern borders of North America, Europe, and Asia. Characterized by lack of trees, low precipitation and low temperatures.

Ultimate production Total amount of a nonrenewable resource that could ultimately be extracted at a reasonable price.

Ultraviolet (UV) light or radiation Electromagnetic radiation from sun and special lamps. Causes sunburn and mutations in bacteria and other living cells.

Undernourishment A lack of calories in the diet. Contrast with *malnourishment*.

Variation Genetically based differences in behavior, structure, or function in a population.

Waste minimization Any one of several strategies to deal with hazardous and solid wastes such as recycling. Designed to minimize or reduce waste output.

Waste-to-energy plant Incinerator for rubbish that produces small amounts of electricity from heat given off by combustion.

Water cycle See *hydrological cycle*.

Water table Top of the zone of saturation.

Waterlogging Saturation of soils with water due to poor soil drainage and irrigation. Decreases soil oxygen and kills plants.

Watershed Land area drained by a given stream or river.

Watt Unit of power indicating rate at which electrical work is being performed.

Wave power Energy derived from sea waves.

Wet cooling tower Device used for cooling water from power plants. Hot water flows through rising air, which draws off heat. Cool water is then returned to the system.

Wetlands Land areas along freshwater (inland wetlands) and salt water (coastal wetlands) that are flooded all or part of the time.

Wilderness An area where the biological community is relatively undisturbed by humans. Seen by developers as an untapped supply of resources such as timber and minerals; seen by environmentalists as a haven from hectic urban life, an area for reflection and solitude.

Wilderness area An area established by the U.S. Congress under the Wilderness Act (1964), where timber cutting and use of motorized vehicles are prohibited. Most are located in national forests.

Wildlife corridor Protected piece of land that connects protected habitat, allowing animals to migrate.

Wind energy Energy captured from the wind to generate electricity or pump water. An indirect form of solar energy.

Wind generators Windmills that produce electrical energy.

Zero population growth A condition in which population is not increasing; the population growth rate is zero.

Zone of intolerance Range of environmental conditions that an organism cannot survive in.

Zone of physiological stress Upper and lower limits of range of tolerance, where organisms have difficulty surviving.

Zooplankton Nonphotosynthetic, single-celled aquatic organisms.

Chapter 1 Critical Thinking, Science, and Scientific Method

Ayres, R. U. (2001). How Economists Have Misjudged Global Warming. *World-Watch* 14 (5): 12–25. Extremely important reading about the myths of economic strength and controls on emissions of carbon dioxide from U.S. industries.

Chiras, D. D. (1992). Teaching Critical Thinking in the Biology and Environmental Science Classrooms. *American Biology Teacher* 54 (8): 464–468. One view of critical thinking that you may find useful.

Eckblad, J. W. (1991). How Many Samples Should Be Taken? *Bioscience* 41 (5): 346-348. Gives examples of how to determine a suitable sample size for biological studies.

Goodstein, E. (1999). *The Trade-Off Myth: Fact and Fiction about Jobs and the Environment.* Washington, DC: Island Press. Examines a deeply held belief, notably that environmental protection threatens jobs.

Hardin, G. (1985). *Filters against Folly: How to Survive Despite Economists, Ecologists, and the Merely Eloquent.* New York: Viking. Eloquent writing on scientific bias. See Chapters 1–7.

Kelley, D. (1988). *The Art of Reasoning.* New York: Norton. Especially good is Part 4, "Inductive Reasoning."

Klemke, E. D., Hollinger, R., Kline, A., eds. (1988). *Introductory Readings in the Philosophy of Science.* New York: Prometheus Books. An excellent introduction to critical thinking and its application in science.

Kuhn, T. (1970). *The Structure of Scientific Revolutions.* Chicago: University of Chicago Press. The original description of paradigms.

Meadows, D. H. (1991). *The Global Citizen.* Washington, DC: Island Press. Excellent discussions of paradigms.

Sarewitz, D., Pielke, Jr., R. A., and Byerly, Jr., R. (2000). *Prediction: Science, Decision Making, and the Future of Nature.* Washington, DC: Island Press. Examines predictive science and how it is used and can be better used in making policy.

Chapter 2 Living Sustainably on the Earth

Bennear, L. S., and Coglianese, C. (2005). Measuring Progress: Program Evaluation of Environmental Policies. *Environment* 46 (2): 22–43. Answers important questions about the effectiveness of different types of environmental policy.

Bright, C. (2000). Anticipating Environmental 'Surprise' in *State of the World 2000,* ed. L. Starke. New York: W.W. Norton. Extremely interesting and thorough discussion of the way environmental disturbances and alterations combine to produce unanticipated or more rapid deterioration than one might expect.

Brown, L. R., Flavin, C., and Postel, S. (1991). *Saving the Planet: How to Shape an Environmentally Sustainable Economy.* New York: Norton. General overview of key areas of sustainable development.

Chiras, D., and Wann, D. (2003). *Superbia! 31 Ways to Create Sustainable Neighborhoods.* Gabriola Island, BC: New Society. One of the author's most recent books, which describes actions we can take to build a more sustainable world right in our own neighborhoods.

Cobb, C., Halstead, T. and Rowe, J. (1995). If the GDP is up, why is America down? *The Atlantic Monthly* 276 (4): 59–78. Takes a good hard look at alternative indicators of success.

Chiras, D. D. (1990). *Beyond the Fray: Reshaping America's Environmental Response.* Boulder, CO: Johnson Books. Describes changes needed to improve our environmental response.

Chiras, D. D. (1992). *Lessons from Nature: Learning to Live Sustainably on the Earth.* Washington, DC: Island Press. Describes the biological principles of sustainability and how they can be applied to systems of ethics, economics, government, agriculture, industry, energy, and so on.

Chiras, D. D. (1992). Eco-Logic: Teaching the Biological Principles of Sustainability. *American Biology Teacher* 55 (2): 71–76. Overview of the biological principles of sustainability.

Chiras, D. D. (1993). Toward a Sustainable Public Policy. *Environmental Carcinogenesis and Ecotoxicology Reviews* C11 (1): 73–114. Good overview of some basic principles of sustainable development.

Chiras, D. D. (1995). Principles of Sustainable Development: A New Paradigm for the Twenty-First Century. *Environmental Carcinogenesis and Ecotoxicology Reviews* C13 (2): 143–178. Much more detailed look at principles of sustainable development.

Chiras, D., and Corson, W. (1997). Indicators of Sustainability and Quality of Life: Translating Vision into Reality. *Environmental Carcinogenesis and Ecotoxicology Reviews* C15 (1): 61–92. Describes the use of alternative measures of progress.

Chiras, D., and Herman, J. (1997). Sustainable Community Development: A Systems Approach. In *Rural Sustainable Development in America,* ed. Ivonne Audirac. New York: Wiley. Presents a more detailed discussion of the systems approach.

Council on Environmental Quality. (1996). *Environmental Quality: 25th Anniversary Report.* Washington, DC: U.S. Government Printing Office. Excellent survey of progress and problems.

Dauncey, G. (1999). *Earthfuture: Stories from a Sustainable World.* Gabriola Island, BC: New Society Publishers. Over 40 thought-provoking and informative case studies showing what people are doing to create a more sustainable world. Uplifting.

Earth Works Group. (1989). *50 Simple Things You Can Do to Save the Earth.* Berkeley, CA: Earthworks Press. Full of important steps you can take to help build a sustainable future that are as important today as they were in the late 1980s.

Elgin, D. (2000). *Promise Ahead: A Vision of Hope and Action for Humanity's Future.* New York: HarperCollins. A powerful book that calls on us to take action to help build a sustainable society.

Ford, A. (1999). *Modeling the Environment: An Introduction to System Dynamics Modeling of Environmental Systems.* Washington, DC: Island Press. Introduction to system dynamics designed to help students learn principles of modeling and develop models themselves.

Gardner, G. (2001). Accelerating the Shift to Sustainability. In *State of the World 2001,* ed. L. Starke. New York: Norton. Looks at the role of business, government and citizens in shaping a sustainable future.

Gore, A. (1992). *Earth in the Balance: Ecology and the Human Spirit.* Boston: Houghton Mifflin. Excellent reading. An in-depth analysis of environmental problems.

Harris, J. M., Wise, T., Gallagher, K., and Goodwin, N. R. (2001). *A Survey of Sustainable Development: Social and Economic Dimensions.* Washington, DC: Island Press. Broad overview of important topics.

Hayes, D. (2000). *The Official Earth Day Guide to Repair the Planet.* Washington, DC: Island Press. Good overview of what individuals can do to help build a sustainable society.

Kates, R., Parris, T. M., and Leiserowitz, A. A. (2005). What is Sustainable Development? *Environment* 46 (3): 8–21. Excellent overview of a complex subject.

Mastiny, L., et al. (2005). *Vital Signs: The Trends That Are Shaping Our Future.* New York: Norton. Extremely valuable resource. Analyzes key trends that affect sustainability.

Meadows, D. (1991). *The Global Citizen.* Washington, DC: Island Press. Collection of short, insightful essays on a variety of environmental topics. Still as timely a read as it was when it first came out.

Meadows, D. H., Meadows, D. L., and Randers, J. (1992). *Beyond the Limits: Confronting Global Collapse, Envisioning a Sustainable Future.* Post Mills, VT: Chelsea Green. Excellent reading.

Orr, D. W. (1992). *Ecological Literacy: Education and the Transition to the Postmodern World.* Albany: SUNY Press. Excellent treatise on sustainability and reform needed in our educational system.

Prugh, T., and Assadourian, E. (2003). What is Sustainability Anyway? *World-Watch* 16 (5): 10–21. Great overview of the challenges that lie ahead.

Speth, G. (2004). *Red Sky at Morning: America and the Crisis of the Global Environment.* New Haven, CT: Yale University Press. A look at global environmental catastrophes that may strike in this century and their effect on the United States.

Starke, L., ed. (2005). *State of the World 2005,* ed. L. Starke. New York: Norton. A great annual publication that covers many important topics, analyzing issues and suggesting viable soltuions.

Starke, L., ed. (2005). *Vital Signs 2005.* New York: Norton. Annual publication that tracks many important social, economic, and environmental trends.

United Nations Environment Programme. (1999). *Global Environment Outlook 2000.* New York: Earthscan. Region-by-region analysis of the state of the world's environment with recommendations for change.

Wann, D. (1990). *Biologic: Environmental Protection by Design*. Boulder, CO: Johnson Books. Calls for a revolution in our way of life to include nature-compatible designs.

Wann, D. (1996). *Deep Design*. Washington, DC: Island Press. Contains many good examples of preventive design for promoting sustainability.

Weisman, A. (1998). *Gaviotas: A Village to Reinvent the World*. White River Junction, VT: Chelsea Green. An uplifting tale about a community in Colombia that is practicing principles of sustainability.

World Commission on Environment and Development. (1987). *Our Common Future*. Oxford: Oxford University Press. Important reading on sustainable development.

Chapter 3 Root Causes and Root-Level Solutions

Chiras, D. D. (1992). An Inquiry into the Root Causes of the Environmental Crisis. *Environmental Carcinogenesis and Ecotoxicology Reviews* C10 (1): 73–119. Detailed look at a network of root causes of the environmental crisis, especially biological and evolutionary roots.

Chiras, D. D. (1992). *Lessons from Nature: Learning to Live Sustainably on the Earth*. Washington, DC: Island Press. Introduces some of the root causes of environmental unsustainability and ways to address them.

DeGraaf, J., Wann, D., and Naylor, T. (2001). *Affluenza: The All-Consuming Epidemic*. San Francisco: Berrett-Koehler. An insightful look into overconsumption—one of the root causes of the environmental crisis—and the effects it has on us and our planet.

Chapters 4–6 Ecology, Biomes, and Aquatic Life Zones

Benyus, J. M. (1997). *Biomicry: Innovation Inspired by Nature*. New York: Perennial. A book that looks at many exciting technological innovations that are inspired by nature's design.

Berger, J. J. (1985). *Restoring the Earth: How Americans Are Working to Renew Our Damaged Environment*. New York: Knopf. A delightful book that's a must.

Berger, J. J., ed. (1989). *Environmental Restoration: Science and Strategies for Restoring the Earth*. Washington, DC: Island Press. Technical discussion of restoration.

Botkin, D. B. (1990). *Discordant Harmonies: A New Ecology for the 21st Century*. New York: Oxford University Press. Dispels common myths about ecology.

Bush, M. B. (2002). *Ecology of a Changing Planet*. 3rd ed. Upper Saddle River, NJ: Prentice-Hall. Good introductory-level coverage of ecology.

Chiras, D. D. (1992). *Lessons from Nature: Learning to Live Sustainably on the Earth*. Washington, DC: Island Press. Shows how we can apply the biological principles of sustainability to modern society.

Dodson, S. I., Allen, T. F. H, Carpenter, S. R., Ives, A. R., Jeanne, R. L., Kitchell, J. F., Langston, N. E., and Turner, M. G. (1998). *Ecology*. New York: Oxford University Press. Great reference for students who are eager to learn more about ecology.

Dodson, S. I, et al., eds. (1999). *Readings in Ecology*. New York: Oxford University Press. Collection of essays including classical studies and new, interesting thinking about ecology.

Ehrlich, P. R. (1996). *The Machinery of Nature*. New York: Simon and Schuster. Great general discussion of ecology. Great reading for those who don't want to tackle a textbook.

Ehrlich, P. R., and Roughgarden, J. (1987). *The Science of Ecology*. New York: Macmillan. Higher-level coverage of ecology.

Hudson, W. E., ed. (1991). *Landscape Linkages and Biodiversity*. Washington, DC: Island Press. Describes the importance of protecting biological diversity.

Morgan, S. (1995). *Ecology and Environment: The Cycles of Life*. New York: Oxford University Press. Very readable and graphic introduction to ecology.

Odum, E. P. (1993). *Ecology and Our Endangered Life-Support Systems*. Sunderland, MA: Sinauer. Excellent introduction to ecology.

Smith, T. M. and Smith, R. L. (2005). *Elements of Ecology*. 6th ed. New York: Harper-Collins. Good overview of ecology and environmental problems.

Southwick, C. H. (1996). *Global Ecology in Human Perspective*. New York: Oxford University Press. Very readable introduction to human ecology.

Stiling, P. D. (2001). *Ecology: Theory and Applications*. 4th ed. Upper Saddle River, NJ: Prentice-Hall. Slightly more advanced coverage of ecology.

Chapter 7 Human Ecology

Benyus, J. M. (1997). *Biomimicry: Innovation Inspired by Nature*. New York: Perennial. A book that looks at many exciting technological innovations that are inspired by nature's design—and could help create the next industrial revolution.

Caufield, C. (1991). *In the Rainforest: Report from a Strange, Beautiful, Imperiled World*. Chicago: University of Chicago Press. Describes efforts to eradicate some indigenous cultures in Latin America.

Commoner, B. (1990). *Making Peace with the Planet*. New York: Pantheon Books. Insightful look into some of the root causes of the environmental crisis.

DeGraaf, J., Wann, D., and Naylor, T. (2001). *Affluenza: The All-Consuming Epidemic*. San Francisco: Berrett-Koehler. An insightful look into overconsumption and the effects it has on us and our planet.

Isaacs, J. (1980). *Australian Dreaming: 40,000 Years of Aboriginal History*. Sydney, Australia: Lansdown Press. A wonderfully illustrated book on the hunters and gatherers of Australia.

McDonough, W., and Braungart, M. (2002). *Cradle to Cradle: Remaking the Way We Make Things*. New York: North Point. An excellent look at a new and much more environmentally sound way of producing goods for modern society.

McPhee, J. (1989). *The Control of Nature*. New York: Farrar, Straus & Giroux. Three case studies showing the extent to which modern society will go to attempt to control natural forces.

Meadows, D. H., Meadows, D. L., and Randers, J. (1992). *Beyond the Limits: Confronting Global Collapse, Envisioning a Sustainable Future*. Post Mills, VT: Chelsea Green. Excellent reading.

Wackernagel, M., and Rees, W. (1995). *Our Ecological Footprint: Reducing Human Impact on Earth*. Gabriola Island, BC: New Society Publishers. One of the most important new books of our times. Explains how much land is dedicated to each of us to meet our needs.

Chapters 8 and 9 Population

Brown, L. R., et al. (1990). *State of the World 1990*, ed. L. Starke. New York: Norton. See Chapter 8 for a discussion of means to end poverty.

Brown, L. R., et al. (1992). *State of the World 1992*, ed. L. Starke. New York: Norton. Excellent information vital to population control.

Brown, L. R., Gardner, G., and Halweil, B. (1998). *Beyond Malthus: Sixteen Dimensions of the Population Problem*. Worldwatch Paper 143. Washington, DC: Worldwatch Institute. A must read for all students. This aptly describes the many impacts of overpopulation.

Brown, L. R., and Halweil, B. (1999). Where Death Rates are Rising, *World-Watch* 12 (5): 20–29. A startling look at rising death rates.

Chiras, D. D. (1992). *Lessons from Nature: Learning to Live Sustainably on the Earth*. Washington, DC: Island Press. Offers insights into development measures useful in curbing population growth in the developing world.

Cincotta, R. P., Engelman, R., and Anastasion, D. (2003). *The Security Demographic: Population and Civil Conflict After the Cold War*. Washington, DC: Population Action International. An insightful look into conflicts that could worsen as the human population grows.

Cole, H. S. D., Freeman, C., Jahoda, M., and Pavitt, K. L. R. (1973). *Models of Doom: A Critique of the Limits to Growth*. New York: Universe Books. A rebuttal of the computer study.

Ehrlich, P. R., and Ehrlich, A. (1990). *The Population Explosion*. New York: Simon & Schuster. Updated version of the 1971 classic *The Population Bomb*.

Frazer, E. (1992). Thailand: A Family Planning Success Story. *In Context* 31: 44–45. Provides additional information on the success of Thailand in promoting family planning.

Gelbard, A., Haub, C., and Kent, M. M. (2000). *World Population Beyond Six Billion*. Washington, DC: Population Reference Bureau. A great survey of major population changes and projected changes to year 2050.

Gupte, P. (1984). *The Crowded Earth: People and the Politics of Population*. New York: Norton. Superb book on family planning successes.

Hardin, G. (1993). *Living Within Limits: Ecology, Economics and Population Taboos*. New York: Oxford University Press. A comprehensive analysis of the current population crisis by a world authority.

Jacobson, J. L. (1988). *Environmental Refuges: A Yardstick of Habitability*. Worldwatch Paper 86. Washington, DC: Worldwatch Institute. Excellent.

Jacobsen, J. L. (1991). *Women's Reproductive Health: The Silent Emergency*. Worldwatch Paper 102. Washington, DC: Worldwatch Institute. Examines the importance of family planning as a means of protecting the health of women.

Kent, M. M., and Mather, M. (2002). What Drives U.S. Population Growth. *Population Bulletin* 57 (4): 1–40. A must-read for anyone interested in understanding U.S. population growth.

Lowe, M. (1992). City Limits. *World-Watch* 5 (1): 18–25. Explains different strategies of controlling urban growth.

MacDonald, M., and Nierenberg, D. (2003). Linking Population, Women and Biodiversity. In *State of the World 2003*, ed. L. Starke. New York: W.W. Norton. Looks at population growth in most ecologically diverse areas and the role women are playing in helping curb growth.

Malthus, R. (1994). *An Essay on the Principle of Population*. New York: Oxford University Press. A classic; must reading for all students.

McFalls, J. A., Jr. (1998). *Population: A Lively Introduction*. Population Bulletin 53 (3). Washington, DC: Population Reference Bureau, 1998. An excellent overview.

McFalls, J. A. (2003). Population: A Lively Introduction. *Population Bulletin* 58 (4): 1–40. Washington, DC: A great introduction to demography.

Meadows, D. H., Meadows, D. L., Randers, J., and Behrens, W. W. (1974). *The Limits to Growth*. 2d ed. New York: Universe Books. Excellent study of population, resources, and pollution.

Meadows, D. H., Meadows, D. L., and Randers, J. (1992). *Beyond the Limits*. Post Mills, VT: Chelsea Green. Sequel to the Limits to Growth study.

Population Reference Bureau. (2005). *World Population Data Sheet*. Washington, DC: Population Reference Bureau. A great source for information on populations of every country in the world. Updated annually.

Tedesko, S. (1992). Family Planning Media: That's Entertainment! *In Context* 31: 42–43. Interesting look at the promotion of family planning through songs, music videos, and television.

United States Bureau of the Census. (2000). *Current Population Reports*. Washington, DC: U.S. Government Printing Office. Published annually. Latest information on births, immigration, and changing age structure of our nation's population.

United States Bureau of the Census. (2000). *Statistical Abstract of the United States*. Washington, DC: U.S. Government Printing Office. Provides latest population data for the United States.

World Commission on Environment and Development. (1987). *Our Common Future*. Oxford: Oxford University Press. Good reference.

Chapter 10 Sustainable Agriculture

Bender, W. H. (1997). How Much Food Will We Need in the 21st Century? *Environment* 39 (2): 6–11, 27–31. Very important reading.

Brown, L. R. (1994). Facing Food Insecurity. In *State of the World 1994*, ed. L. Starke. New York: Norton. An excellent discussion of recent trends in food production.

Brown, L. R. (2001). Eradicating Hunger: A Growing Challenge. In *State of the World 2001*, ed. L. Starke. New York: W. W. Norton. Discusses many strategies covered in this chapter.

Brown, L. R. (2003). *Plan B: Rescuing a Planet Under Stress and a Civilization in Trouble*. New York: Norton. Lots of valuable information on agricultural issues and sustainable solutions.

Brown, L. R., and Young, J. E. (1990). Feeding the World in the Nineties. In *State of the World 1991*, ed. L. Starke. New York: Norton. Excellent study of the prospects for world agriculture.

Chiras, D. D., and Reganold, J. P. (2005). *Natural Resource Conservation: Management for a Sustainable Future*. 7th ed. New York: Prentice-Hall. See sections on agriculture, soils, and land management.

Conway, G. (2000). Food for All in the 21st Century. *Environment* 42 (1): 8–18. Describes the need for a second Green Revolution and ways this can be achieved.

Durning, A. T., and Brough, H. B. (1992). Reforming the Livestock Economy. In *State of the World 1991*. New York: Norton. Important discussion of reforms needed in meat production.

Gardner, G. (1996). *Shrinking Fields: Cropland Loss in a World of Eight Billion*. Worldwatch Paper 131. Washington, DC: Worldwatch Institute. Excellent update on the current loss of cropland and food trends.

Gardner, G., and Halweil, B. (2000). Nourishing the Underfed and Overfed. In *State of the World*, ed. L. Starke. New York: W. W. Norton, pp. 58–78. Describes very aptly the world of haves and have nots.

Gujja, B., and Finger-Stich, A. (1996). What Price Prawn? Shrimp Aquaculture's Impact in Asia. *Environment* 38 (7): 12–15, 33–39. A good overview of the impact of a growing activity on a developing nation.

Gupta, A. (2000). Governing Trade in Genetically Modified Organisms: The Cartagena Protocol on Biosafety. *Environment* 42 (4): 22–33. Describes the provisions of an important international treaty on trade in genetically engineered food.

Halweil, B. (2002). Farming in the Public Interest. In *State of the World 2002*, ed. L. Starke. New York: W. W. Norton, pp 51–74. Critical analysis of farming with sound ideas for creating a sustainable system of agriculture.

Halweil, B. (2002). *Home Grown: The Case for Local Food in a Global Market*. Worldwatch Paper 163. Washington, DC: Worldwatch Institute. Excellent look at local food production—that is, the production of food by local farmers for local markets—and the benefits it can have on the environment and the economy.

Levidow, L. (1999). Regulating BT Maize in the United States and Europe: A Scientific-Cultural Comparison. *Environment* 41 (10): 10–21. Worthwhile reading for those interested in the controversy over genetically engineered foods.

Mortimore, M. (2005). Dryland Development: Success Stories from Africa. *Environment* 47 (1): 8–21. An uplifting piece on successes in Africa to stop the spread of deserts through responsible agricultural practices.

Norse, D. (1992). A New Strategy for Feeding a Crowded Planet. *Environment* 34 (5): 6–11, 32–39. Important reading on sustainable agriculture.

Paarlbert, R. (2000). Genetically Modified Crops in Developing Countries: Promise or Peril? *Environment* 42 (1): 19 –27. Candid look at this important subject.

Postel, S., and Vickers, A. (2004). Boosting Water Productivity. In *State of the World 2004*, ed. L. Starke. New York: Norton. Examines ways we can make better use of existing water supplies for agriculture.

Pretty, J. (2003). Agroecology in Developing Countries. *Environment* 45 (9): 8–20. An interesting look at ways agriculture can be made more sustainable in developing countries.

Reganold, J. P., Papendick, R. I., and Parr, J. F. (1990). Sustainable Agriculture. *Scientific American* (June): 112–120. Excellent overview.

Renner, M. et. al. (2003). *Vital Signs: The Trends that Are Shaping Our Future*. New York: W. W. Norton. Great analysis of trends in agriculture and population.

Rosegrant, M. W., and Livernash, R. (1996). Growing More Food, Doing Less Damage. *Environment* 38 (7): 6–11, 28–32. Good look at policy changes required to promote sustainable agriculture.

Rosson, P. (1997). Will Erosion Threaten Agricultural Productivity? *Environment* 39 (8): 4–9, 29–31. Important reading.

Soule, J. D., and Piper, J. K. (1992). *Farming in Nature's Image: An Ecological Approach to Agriculture*. Washington, DC: Island Press. Important reading.

Taylor, D. A. (1997). Saving the Forest for the Trees: Alternative Products from Woodlands. *Environment* 39 (1): 6–11, 33–36. Important reading.

Timmer, V., and Juma, C. (2005). Taking Root: Biodiversity Conservation and Poverty Reduction Come Together in the Tropics. *Environment* 47 (4): 24–44. Detailed piece on the feasibility of sustainable ways to protect wild species and ecosystems while creating economic prosperity in Third World Nations.

Chapter 11 Preserving Biological Diversity

Achiron, M. (1988). Making Wildlife Pay Its Way. *International Wildlife* 18 (5): 46–51. Elaboration on Myers' Viewpoint essay.

Batisse, M. (1997). Biosphere Reserves: A Challenge for Biodiversity Conservation and Regional Development. *Environment* 39 (5): 6–15, 31–33. Overview of important challenges.

Brandon, K., Redford, K. H., and Sanderson, S. E., eds. (1998). *Parks in Peril: People, Politics, and Protected Areas*. Washington, DC: Island Press. Analyzes trends in park management and implications for protection of biodiversity.

Chester, C. C. (1996). Controversy over Yellowstone's Biological Resources. *Environment* 38 (8): 10–15, 34–36. Interesting insights into the value of biological resources.

Cox, G. W. (1999). *Alien Species in North America and Hawaii*. Washington, DC: Island Press. Valuable resource on invasive species.

DiSilvestro, R. (1996). What's Killing the Swainson's Hawk? *International Wildlife* 26 (3): 38–43. An examination of the effect of pesticide use on migratory birds.

DiSilvestro, R. (2004). Where Would They Be Now? *National Wildlife* 42 (5): 48–57. A look at species that have been saved by the Endangered Species Act.

Domalain, J. (1977). Confessions of an Animal Trafficker. *Natural History* 87 (5): 54–57. Startling account of illegal practices in the animal trade.

Ellis, G. (2005). Making Progress with Pandas. *National Wildlife* 43 (4): 36B–36H. Shows how captive breeding programs can help restore populations of endangered species.

Forsberg, M. (2005). Hovering on the Edge of Existence. *National Wildlife* 43 (6): 22–30. Explains how captive breeding programs may help save sandhill cranes from extinction.

Freese, C. H. (1998). *Wild Species as Commodities: Managing Markets and Ecosystems for Sustainability*. Washington, DC: Island Press. Valuable reading.

Gorke, M. (2003). *The Death of Our Planet's Species: A Challenge to Ecology and Ethics*. Washington, DC: Island Press. In this book, the author makes a case for a holistic environmental ethic to help save imperiled species of the world.

Harris, T. (1991). *Death in the Marsh*. Washington, DC: Island Press. Tells the story of the Kesterson National Wildlife Refuge.

Irwin, P. G. (2000). *Losing Paradise: The Growing Threat to Animals, Our Environment, and Ourselves*. Garden City Park, NY: Square One. Important reading for anyone interested in learning more about the mounting threats to animals.

Jenkin, M., Scherr, S. J., and Inbar, M. (2004). Markets for Biodiversity Services: Potential Roles and Challenges. *Environment* 46 (6): 32–42. An interesting look at ways the marketplace can help protect biodiversity.

LaBastille, A. (1999). *Jaguar Totem: New Wildlands and Wildlife*. Westport, NY: West of the Wind Publications. A delightful chronicle of an ecologist's studies of tropical rain forests. Insightful and delightful reading.

Laycock, G. (1966). *The Alien Animals*. Garden City, NJ: Natural History Press. A classic account of the troubles created by species introduction.

Lipske, M. (1994). Animal, Heal Thyself. *National Wildlife* 32 (1): 46–49. Excellent article on the value of naturally occurring medicinals.

McNeely, J. A. (2004). Strangers in Our Midst: The Problem of Invasive Alien Species. *Environment* 46 (6): 16–31. In-depth discussion of the impacts of accidental introduction of alien species on the ecosystems of the world.

Margoluis, R., and Salafsky, N. (1998). *Measures of Success: Designing, Managing, and Monitoring Conservation and Development Projects*. Washington, DC: Island Press. A guide for development projects that protects the environment

Monks, V. (1996). The Beauty of Wetlands. *National Wildlife* 34 (4): 20–27. An excellent article with new insights on wetlands and the value of protecting them.

Mooney, H. A., and Hobbs, R. J. (2000). *Invasive Species in a Changing World*. Washington, DC. Describes how changing patterns of global commerce are resulting in the spread of alien species and how climate change and other factors are influencing this spread.

Myers, N. (1983). *A Wealth of Wild Species: Storehouse for Human Welfare*. Boulder, CO: Westview Press. Excellent treatise on the many benefits of wild plants and animals.

Noss, R. E., and Cooperrider, A. Y. (1994). *Saving Nature's Legacy: Protecting and Restoring Biodiversity*. Washington, DC: Island Press. Important reading.

Owens, M., and Owens, D. (1985). *Cry of the Kalahari*. New York: Houghton Mifflin. Important information on behavior, ecology, and conservation.

Plotkin, M., and Famolare, L., eds. (1992). *Sustainable Harvest and Marketing of Rain Forest Products*. Washington, DC: Island Press. Outlines ways of sustainably harvesting tropical rain forest products while protecting native species.

Posey, D. A. (1996). Protecting Indigenous People's Rights to Biodiversity. *Environment* 38 (8): 6–9, 37–45. A look at the role of indigenous people in protecting biological resources.

Quammen, D. (1997). *Song of the Dodo: Island Biogeography in an Age of Extinctions*. New York: Touchstone. A massive book with much to offer on ecological science and extinction.

Regenstein, L. (1975). *Politics of Extinction*. New York: Macmillan. A classic work on wildlife extinction, well worth reading.

Reisner, M. (1991). *Game Wars: Undercover Pursuit of Game Poachers*. New York: Viking. Riveting account of poaching in the United States and efforts to stop it.

Sachs, J. S. (2004). Poisoning the Imperiled. *National Wildlife* 42 (1): 22–29. Looks at endangered species threatened by pesticides.

Sunquist, F. (1988). Zeroing in on Keystone Species. *International Wildlife* 18 (5): 18–23. Good reference on this new concept.

Temple, S. A. (1998). Easing the Travails of Migratory Birds. *Environment* 40 (1): 6–9, 28–32. Describes the need for coordinated efforts among countries to protect migratory bird species.

Timmer, V., and Juma, C. (2005). Taking Root: Biodiversity Conservation and Poverty Reduction Come Together in the Tropics. *Environment* 47 (4): 24–44. Detailed piece on the feasibility of sustainable ways to protect wild species and ecosystems while creating economic prosperity in Third World Nations.

Tolmé, P. (2005). Mercury Rising. *National Wildlife* 43 (5): 33–37. Describes the impacts of mercury pollution on numerous bird species.

Tudge, C. (1992). *Last Animals at the Zoo: How Mass Extinction Can Be Stopped*. Washington, DC: Island Press. Describes how captive breeding programs and restoration of natural habitat can be used to save endangered animals from extinction.

Tuxil, J. (1999). Appreciating the Benefits of Plant Biodiversity in *State of the World*, ed. L. Starke. New York: Norton. Great information on the value of plants.

Wilcove, D. (1990). Empty Skies. *The Nature Conservancy Magazine* 40 (1): 4–13. Excellent overview of factors causing the decline of songbirds in the United States.

Youth, H. (2003). *Winged Messengers: The Decline of Birds*. Worldwatch Paper 165. Washington, DC: Worldwatch Institute. Detailed account of the impacts of human society on bird populations.

Chapter 12 Grasslands, Forests, and Wilderness

Behan, R. W. (2001). *Plundered Promise: Capitalism, Politics, and the Fate of Federal Lands*. Washington, DC: Island Press. Offers an in-depth view of the history of public land management and what's wrong with it.

Brick, P., Snow, D., and Van de Wetering, S. (2000). *Across the Great Divide: Explorations in Collaborative Conservation and the American West*. Washington, DC: Island Press. Explores a new movement in conservation: collaborative conservation. A must-read for anyone interested in a career in environmental conservation or natural resource conservation.

Bright, C., and Sarin, R. (2003). *Venture Capitalism for a Tropical Forest: Cocoa in the Mata Atlantica*. Worldwatch Paper 168. Washington, DC: Worldwatch Institute. A great case study for those interested in learning ways to produce sustainable forest products.

Brown, B. (1982). *Mountain in the Clouds*. New York: Simon & Schuster. Classic account of deforestation practices in the Pacific Northwest.

Carvalho, G. O., et al. (2002). "Frontier Expansion in the Amazon: Balancing Development and Sustainability." *Environment* 44 (3): 34–45. A detailed look at Brazil's ongoing policy of tropical rainforest development that could lead to huge losses.

Chiras, D. D., and Reganold, J. P. (2005). *Natural Resource Conservation: Management for a Sustainable Future*. 9th ed. New York: Prentice-Hall. See chapters on rangeland and forest management.

Dagget, D. (2000). *Beyond the Rangeland Conflict: Toward a West That Works*. Reno: University of Nevada Press. A book of ten real world examples of environmentalists and ranchers working together to deal with issues of open space, endangered species, functional ecosystems, environmental restoration, holistic management, and conflict resolution.

Dobbs, D., and Ober, R. (1995). *The Northern Forest*. White River Junction, VT: Chelsea Green. Insightful and informative story of our view of land and the struggles between conservation and use.

Dunning, J. (1998). *From the Redwood Forest: Ancient Trees and the Bottom Line: A Headwaters Journey*. White River Junction, VT: Chelsea Green. A must read for anyone interested in learning more about the struggle to protect our old-growth forests.

Durning, A. T., and Brough, H. B. (1992). Reforming the Livestock Economy. In *State of the World 1992*, ed. L. Starke. New York: Norton. Outlines ways to create a sustainable livestock industry.

Grumbine, R. E. (1994). What Is Ecosystem Management? *Conservation Biology* 8 (1): 27–38. Overview of the history of ecosystem management and the basic underlying principles.

Hardin, G. (1968). The Tragedy of the Commons. *Science* 162: 1243–1248. A classic paper.

Heady, H. F., and Child, R. D. (2001). *Rangeland Ecology and Management*, 2d ed. Boulder, CO: Westview Press. Rangeland management textbook that focuses on the ecology of rangeland grazing, practical management of animals, and vegetational manipulation.

Holechek, J. L., Pieper, R. D., and Herbel, C. H. (2001). *Range Management: Principles and Practices*, 4th ed. Englewood Cliffs, NJ: Prentice Hall. Solid fundamental textbook covering almost all aspects of range science and management.

Hughes, J. D., and Thirgood, J. V. (1982). Deforestation in Ancient Greece and Rome: A Cause of Collapse. *Ecologist* 12 (5): 196–207. Detailed paper.

Jensen, D., and Draffan, G. (2003). *Strangely Like War: The Global Assault on Forests*. White River Junction, VT: Chelsea Green. A popular book with an important, somewhat alarming message.

Kemmis, D. (2001). *A New Vision for Governing the West*. Washington, DC: Island Press. Offers a radical new approach to managing resources regionally.

Kerasote, T. ed. (2002). *Return of the Wild: The Future of Our Natural Lands*. Washington, DC: Island Press. A collection of writings about the value of wilderness, current threats, and what can be done to truly protect it.

Lambin, E. F., and Geist, H. J. (2003). Regional Differences in Tropical Deforestation. *Environment* 45 (6): 22–36. Examines the many causes of tropical deforestation.

Myers, N. (1991). Trees by the Billions: A Blueprint for Cooling. *International Wildlife* 21 (5); 12–15. Excellent discussion of revegetation potential of tropical lands and benefits.

Nierenberg, D. (2005). *Happier Meals: Rethinking The Global Meat Industry*. Worldwatch Paper 171. Washington, DC: The Worldwatch Institute. Superb overview of ways to create a more sustainable meat-producing industry.

Pinchot, G. (1999). *Breaking New Ground*. Washington, DC: Island Press. The autobiography of Gifford Pinchot. Essential reading for those interested in learning the basis of our present national forest policy.

Shenk, T. M., and Franklin, A. B. (2001). *Modeling in Natural Resource Management*. Washington, DC: Island Press. For students interested in learning about one of the most useful tools in resource management: computer modeling.

Teitel, M. (1992). *Rain Forest in Your Kitchen*. Washington, DC: Island Press. Shows what you can do to help save the rain forest.

Watkins, T. H. (1988). Blueprints for Ruin. *Wilderness* 52 (182): 56–60. Extraordinary article on the fate of old-growth forests in North America.

Willers, B. (1999). *Unmanaged Landscapes: Voices for Untamed Nature*. Washington, DC: Island Press. An interesting book well worth reading. It examines the effects and implications of resource management and the need to keep some landscapes free from human interference.

Chapter 13 Water Resources

Abramovitz, J. N. (1996). Sustaining Freshwater Ecosystems. In *State of the World 1996*, ed. L. Starke. New York: Norton. Important reading.

Brown, L. R. (1991). The Aral Sea: Going, Going… *World-Watch* 4 (1): 20–27. Gripping story of what can happen when people abuse their water resources.

Brown, L. R. (2003). *Plan B: Rescuing a Planet Under Stress and a Civilization in Trouble*. New York: Norton. Lots of valuable information on water resources and sustainable solutions.

del Moral, L., and Sauri, D. (1999). Changing Course: Water Policy in Spain. *Environment* 41 (6): 12–15, 31–36. An in-depth look at alternative means of meeting demands for water in Spain.

Gardner, G. (1996). Preserving Agricultural Resources. In *State of the World 1996*, ed. L. Starke. New York: Norton. Excellent in-depth view of agricultural problems, including water supply.

Gleick, P. H. (2000). *The World's Water 2000-2001: The Biennial Report on Freshwater Resources*. Washington, DC: Island Press. Very valuable resources with tons of useful information.

Glennon, R. (2002). *Water Follies: Groundwater Pumping and the Fate of America's Fresh Waters*. Washington, DC: Island Press. An insightful look at groundwater overdraft and other problems associated with surface waters in the United States.

Hinrichsen, D. (2004). Water Pressure. *National Wildlife* 42 (4): 32–36. A personal look at water supply and water conservation. Be sure to check out other articles on water in this issue.

Khagram, S. (2003). Neither Temples Nor Tombs: A Global Analysis of Large Dams. *Environment* 45 (4): 28–37. An excellent look at the impacts and performance of large dams.

McGuane, T. (1993). Wild Rivers. *Audubon*, Nov.–Dec. Celebrates our wild western rivers. A tribute to the 25th anniversary of the National Wild and Scenic Rivers Act.

McPhee, J. (1989). *The Control of Nature*. New York: Farrar, Straus & Giroux. Excellent reading.

Okun, D. (1991). A Water and Sanitation Strategy for the Developing World. *Environment* 33 (8): 16–20, 38–43. Describes ways to provide clean drinking water in developing countries.

Olmstead, S. M. (2003). Water Supply and Poor Communities. *Environment* 45 (10): 22–35. This article looks at problems of supplying water to poor communities throughout the world—and potential solutions.

Platt, R. H., Barten, P. K., and Pfeffer, M. J. (2000). A Full, Clean Glass? Managing New York City's Watersheds. *Environment* 42 (5): 8–20. Useful case study for those interested in learning more about watershed management.

Postel, S. (1996). Forging a Sustainable Water Strategy. In *State of the World 1996*, ed. L. Starke. New York: Norton. Very important reading for students who want to learn more about this important topic.

Postel, S. (2000). Redesigning Irrigated Agriculture. In *State of the World 2000*, ed. L. Starke. New York: W.W. Norton. Excellent overview of what can be done to improve the efficiency of irrigation.

Postel, S. (2005). *Liquid Assets: The Critical Need to Safeguard Freshwater Ecosystems*. Worldwatch Paper 170. Washington, DC: Worldwatch Institute. Great overview.

Reisner, M. (1993). *Cadillac Desert: The American West and its Disappearing Water*. New York: Penguin. Fascinating book for anyone interested in learning about water and the political struggles that surround it in the arid American West.

Reissner, M., and Bates, S. F. (1989). *Overtapped Oasis: Reform or Revolution for Western Water*. Washington, DC: Island Press. Critique of western water policy with recommendations for change.

Rosengrant, M. W., Cai, X., and Cline, S. A. (2003). Will the World Run Dry? Global Water and Food Security. *Environment* 45 (7): 24–36. A look at global water demand and ways to meet rising demands sustainably.

Sheaffer, J. R., Mullan, J. D., and Hinch, N. B. (2002). Encouraging Wise Use of Floodplains with Market-Based Incentives. *Environment* 44 (1): 32–43. Insightful examination of ways that floodplains can be better managed.

Stauffer, J. (2003). *The Water You Drink*. Gabriola Island, BC: New Society. A popular book for those interested in learning more about drinking water.

Steinhart, P. (1993). Mud Wrestling. *Sierra* (Jan.-Feb.). Discusses the political barriers to saving our nation's valuable wetlands.

White, G. (2000). Water Science and Technology: Some Lessons from the 20th Century. *Environment* 42 (1): 30–38. Good overview of changes in water policy, many of which were discussed in this chapter.

Chapter 14   Nonrenewable Energy Resources

Brown, L. R. (2003). *Plan B: Rescuing a Planet Under Stress and a Civilization in Trouble*. New York: Norton. Lots of valuable information on energy resources and sustainable solutions.

Brown, L. R., Flavin, C., and Postel, S. (1991). *Saving the Planet*. New York: Norton. Contains an important overview of the need for an alternative, nonfossil-fuel-based economy.

Campbell, C. J. (1997). *The Coming Oil Crisis*. Brentwood, UK: Multi-Science Publishing, and Geneva: Petroconsultants SA. Detailed and authoritative analysis of world oil supplies. Paints a rather dismal future.

Campbell, C. J. (1997). *The Coming Oil Crisis*. Petroconsultants SA. Detailed and authoritative analysis of world oil supplies.

Chiras, D. D. (1992). *Lessons from Nature: Learning to Live Sustainably on the Planet*. Washington, DC: Island Press. Describes why fossil fuel combustion is not a sustainable activity.

Dawson, J. I., and Darst, R. G. (2005). Russia's Proposal for a Global Nuclear Waste Repository: Safe, Secure, and Environmentally Just? *Environment* 47 (4): 10–21. In-depth analysis of Russia's proposal to house huge quantities of high-level radioactive waste.

DeCarolis, J. F., Goble, R. L., and Hohenemser, C. (2000). Searching for Energy Efficiency on Campus: Clark University's 30-Year Quest. *Environment* 42 (4): 8–20. Interesting story about efforts that could be carried out on many college campuses.

Flavin, C. (1987). *Reassessing Nuclear Power: The Fallout from Chernobyl*. Worldwatch Paper 75. Washington, DC: Worldwatch Institute. Important look at nuclear power and its future.

Flavin, C. (1999). Rethinking the Energy System. In *State of the World 1999*, ed. L. Starke. New York: W. W. Norton. Part of an ongoing series on ways to restructure global energy systems that are good for people and the environment. Very important reference.

Flynn, J., Kasperson, R. E., Kunreuther, H., and Slovic, P. (1997). Redirecting the U.S. High-Level Nuclear Waste Program. *Environment* 39 (3): 6–11, 25–30. Very important reading.

Gofman, J. W., and Tamplin, A. R. (1979). *Poisoned Power: The Case against Nuclear Power Plants before and after Three Mile Island*. Emmaus, PA: Rodale Press. Well-written analysis.

Heinberg, R. (2003). *The Party's Over*. Gabriola Island, BC: New Society. A popular book for those interested in learning more about oil supplies and problems that may arise as reserves are depleted.

Hollander, J. M. (1992). *The Energy-Environment Connection*. Washington, DC: Island Press. Comprehensive survey of the problems created by energy use.

Jungk, R. (1979). *The New Tyranny*. New York: Warner Books. An important, well-written book.

Kammen, D. M. (1999). Bringing Power to the People: Promoting Appropriate Energy Technologies in the Developing World. *Environment* 41 (5): 10–15, 34–41. Looks at some of the key issues involved in creating more sustainable energy systems in less developed countries.

Kunreuther, H., Desvousges, W. H., and Slovic, P. (1988). Nevada's Predicament: Public Perceptions of Risk from the Proposed Nuclear Waste Repository. *Environment* 30 (8): 16–20, 30–33. Outlines problems with efforts to place a high-level radioactive waste disposal site in Nevada.

Lenssen, M. (1992). Confronting Nuclear Waste. In *State of the World 1992*, ed. L. Starke. New York: Norton. Detailed account of nuclear energy and its problems.

Roodman, D. M. (1997). Reforming Subsidies. In *State of the World 1997*, ed. L. Starke. New York: W. W. Norton. Looks at the ways subsidies give fossil fuels and other environmentally unsound activities an advantage and how changes in policy can make other, more environmentally acceptable strategies affordable.

Segerstahl, B., Akleyev, A., and Novikov, V. (1997). The Long Shadow of Soviet Plutonium Production. *Environment* 39 (1): 12–20. Looks at the impacts of another aspect of radiation pollution, notably weapons production.

Youngquist, W. (1997). *Geodestinies: The Inevitable Control of Earth Resources Over Nations and Individuals*. Portland, OR: National Book Co. Fun to read and thought-provoking look at renewable and nonrenewable resources.

Chapter 15   Conservation and Renewable Energy

Brown, L. R., Flavin, C., and Postel, S. (1991). *Saving the Planet: How to Shape an Environmentally Sustainable Global Economy*. New York: Norton. Contains important advice on creating a sustainable system of energy, including economic and employment opportunities.

Chiras, D. D. (1992). *Lessons from Nature: Learning to Live Sustainably on the Earth*. Washington, DC: Island Press. Outlines key tenets of sustainable energy and transportation systems, with numerous examples.

Chiras, D. D. (2000). *The Natural House: A Complete Guide to Healthy, Energy-Efficient, Environmental Homes*. White River Junction: VT: Chelsea Green. Contains overviews of renewable energy strategies for homes.

Chiras, D. D. (2002). *The Solar House: Passive Heating and Cooling*. White River Junction, VT: Chelsea Green. Describes ways to heat homes and cool homes naturally.

Chiras, D., and Wann, D. (2003). *Superbia! 31 Ways to Create Sustainable Neighborhoods*. Gabriola Island, BC: New Society. Describes actions people can take to build a more sustainable world right in their own neighborhood, including many steps to improve energy efficiency and reliance on renewable energy.

Chiras, D. D. (2004). *The New Ecological House*. White River Junction, VT: Chelsea Green. See the chapters on green power, passive solar heating and cooling to learn more about sustainable home energy systems.

Cole, N., and Skerret, P.J. (1995). *Renewables Are Ready: People Creating Renewable Energy Solutions*. White River Junction: VT: Chelsea Green. Full of interesting cases studies.

Davidson, J. (1987). *The New Solar Electric Home*. Ann Arbor, MI: Aatec Publications. Contains lots of practical information on photovoltaics.

Dunn, S. 2000. The Hydrogen Experiment. *World-Watch* 13 (6): 14–25. Tells how the country of Iceland is planning to convert to hydrogen power by the year 2030.

Dunn, S. (2001). Decarbonizing the Energy Economy. In *State of the World 2000*, ed. L. Starke. New York: W. W. Norton. Important reading on renewable energy.

Dunn, S., and Flavin, C. (2000). Sizing up Micropower. In *State of the World*, ed. L. Starke. New York: W. W. Norton. A critical examination of renewable energy.

Flavin, C. (1995). Harnessing the Sun and the Wind. In *State of the World 1995*, ed. L. Starke. New York: Norton. Important reading.

Flavin, C., and Lenssen, N. (1994). Reshaping the Power Industry. In *State of the World 1994*, ed. L. Starke. New York: Norton. Important insights on changes under way in the power industry, especially related to energy conservation.

Gipe, P. (1999). *Wind Energy Basics: A Guide to Small and Micro Wind Systems*. White River Junction, VT: Chelsea Green. A brief, but fairly technical overview of wind energy for homes.

Givoni, B. (1994). *Passive and Low Energy Cooling of Buildings*. New York: Wiley. Detailed account of ways to cool homes in hot climates using renewable energy or natural methods.

Goho, A. (2004). Solar Hydrogen: The Search for Water-Splitting Materials Brightens Up. *Science News* 166: 282–284. Excellent update on materials that can be used to split water into hydrogen using sunlight energy.

Hackleman, M., and C. Anderson. (2002). Harvest the Wind. *Mother Earth News* 192: 70–78. An easy-to-read and fact-filled article on wind power.

Jeffrey, K. (1995). *Independent Energy Guide: Electrical Power for Home, Boat, and RV*. Ashland, MA: Orwell Cove Press. A practical, fact-filled guide to renewable energy.

Johansson, T. B., Kelly, H., Reddy, A., and Williams, R. H. (1992). *Renewable Energy: Sources for Fuels and Electricity*. Washington, DC: Island Press. In-depth analysis of renewable energy options.

Katchadorian, J. (1997). *The Passive Solar House: Using Solar Design to Heat and Cool Your Home*. White River Junction: VT. Discusses principles of passive solar design and features the author's own design.

Kendall, H. W., and Nadis, S. J., eds. (1980). *Energy Strategies: Toward a Solar Future*. Cambridge, MA: Ballinger. Detailed survey of energy sources and their prospects for the future. Superb!

Pasqualetti, M. J. (2004). Wind Power: Obstacles and Opportunities. *Environment* 46 (7): 22–38. Describes the barriers that stand in the way of wind energy development and the potential of this vast resource.

Perlin, J. (1999). From Space to Earth. The Story of Solar Electricity. Ann Arbor: Aatec Publications. A wonderfully readable history of solar electricity.

Pollock, C. (1986). *Decommissioning: Nuclear Power's Missing Link*. Worldwatch Paper 69. Washington, DC: Worldwatch Institute. Authoritative coverage of the costs involved.

Potts, M. (1999). *The New Independent Home*. White River Junction: VT: Chelsea Green. Full of lots of interesting stories and ideas.

Renner, M. (1988). *Rethinking the Role of the Automobile*. Worldwatch Paper 84. Washington, DC: Worldwatch Institute. Detailed coverage of the growth in automobile use and alternative transportation systems.

Renner, M. (1992). Creating Sustainable Jobs in Industrial Countries. In *State of the World 1992*, ed. L. Starke. New York: Norton. Excellent data on the job potential and costs of renewable energy and energy efficiency.

Renner, M. (2000). Creating Jobs, Preserving the Environment. In *State of the World*, ed. L. Starke. New York: W. W. Norton. Shows the economic and employment benefits of a renewable energy strategy.

Renner, M. (2001). Employment in Wind Power. *World-Watch* 14 (1): 22–30. Fascinating look at the employment boom in the wind energy sector.

Reynolds, M. (1990). *Earthship: How to Build Your Own*. Taos, NM: Solar Survival Press. Delightful reading that shows how tire homes are built. Two subsequent volumes have been published to provide new details.

Rifkin, J. (2002). *The Hydrogen Economy*. New York: Tarcher/Putnam. A fascinating look at the potential for building an energy future based on hydrogen.

Romm, J. J. (2004). *The Hype about Hydrogen: Fact and Fiction in the Race to Save the Climate*. Washington, DC: Island Press. A critical look at the prospects of hydrogen as a future fuel.

Roodman, D. M. (2000). Reforming Subsidies. In *State of the World*, ed. L. Starke. New York: W. W. Norton. Extremely important reading.

Savin, J. (2003). Charting a New Energy Future. In State of the World 2003, ed. L. Starke. New York: Norton. A good overview of renewable energy potential and what other nations are doing to create a sustainable energy future.

Schaeffer, J. (2005). *Solar Living Source Book: The Complete Guide to Renewable Energy Technologies and Sustainable Living*. Ukiah, CA: Real Goods. Full of useful background information and product information. A must for anyone serious about living sustainably.

Short, W., and Blair, N. (2003). The Long-Term Potential of Wind Power in the U.S. *Solar Today* 17 (6): 28–29. Important study of wind power's potential.

Swain, J. 2003. Charting a New Energy Future. In *State of the World 2003*, ed. L. Starke. New York: Norton. Examines the potential of renewable energy resources and many other important subjects.

Wilson, A., and Morrill, J. (1998). *Consumer Guide to Home Energy Savings*. Washington, DC: American Council for an Energy-Efficient Economy. Superb reference.

Chapter 16   The Earth and Its Mineral Resources

Draper, D. (1998). *Our Environment: A Canadian Perspective*. Toronto: ITP Nelson. See the chapter on mining for an overview of Canadian issues and solutions.

Gardner, G., and Sampat, P. (1999). Forging a Sustainable Materials Economy. In *State of the World 1999*, ed. L. Starke. New York: W.W. Norton. Good overview of changes that need to be made to create a sustainable economy, especially changes related to recycling.

Kane, H. (1996). Shifting to Sustainable Industries. In *State of the World 1996*, ed. L. Starke. New York: Norton. Covers many important issues related to minerals and industry.

Mastny, L. (2003). *Purchasing Power: Harnessing Institutional Procurement for People and the Planet*. Worldwatch Paper 166. Washington, DC: Worldwatch Institute. Detailed report on the power of government procurement programs that help promote environmentally responsible products.

McDonough, W., and Braungart, M. (1999). *Cradle to Cradle: Remaking Ways We Make Things*. New York: North Point Press. Revolutionary new ideas on ways to improve the manufacturing and recycling of products in modern society.

Sampat, P. (2003). Scrapping Mining Dependence. In *State of the World 2003*, ed. L. Starke. An insightful look into mining and ways to create a more sustainable system of mineral production.

U.S. Bureau of Mines. (2005). *Mineral Commodity Summaries*. Washington, DC: U.S. Government Printing Office. Excellent source of information on minerals, published every year.

U.S. Bureau of Mines. (2005). *Mineral Facts and Problems*. Washington, DC: U.S. Government Printing Office. Superb reference.

Wilkinson, C. F. (1992) *Crossing the Next Meridian: Land, Water, and the Future of the West*. Washington, DC: Island Press. Discusses the outmoded laws of the past that govern today's policy.

Young, J. E. (1992). Free-Loading Off Uncle Sam. *World-Watch* 5(1): 34–35. Describes policies that give mining an unfair economic advantage.

Young, J. E. (1992). Mining the Earth. In *State of the World 1992*, ed. L. Starke. New York: Norton. Excellent source for more detailed information on minerals.

Young, J. E., and Sachs, A. (1995). Creating a Sustainable Materials Economy. In *State of the World 1995*, ed. L. Starke. New York: Norton. Excellent source of detailed information on recycling.

Chapter 17   Creating Sustainable Cities and Towns

Audirac, Ivonne, ed. (1997). *Rural Sustainable Development in America*. New York: Wiley. Excellent collection of essays on sustainable development.

Beatley, T., and Manning, K. (1997). *The Ecology of Place: Planning for Environment, Economy, and Community*. Washington, DC: Island Press. Examines current development patterns and outlines principles of sustainable development.

Bullard, R. D., Johnson, G. S., and Torres, A. O. (2000). *Sprawl City: Race, Politics, and Planning in Atlanta*. Washington, DC: Island Press. Examines a social aspect of sprawl that is often overlooked: economic and racial polarization.

Chiras, D., and Wann, D. (2003). *Superbia! 31 Ways to Create Sustainable Neighborhoods*. Gabriola Island, BC: New Society. Describes actions people can take to build a more sustainable world starting in their own neighborhood.

Christian, D. L. (2003). *Creating a Life Together: Practical Tools to Grow Ecovillages and Intentional Communities*. Gabriola Island, BC: New Society. A useful book that offers advice on creating ecovillages.

Corbett, J., and Corbett, M. (2000). *Designing Sustainable Communities: Learning from Village Homes*. Washington, DC: Island Press. A fabulous look into an innovative suburban development project that could be a model for all future development.

Daniels, T. (1999). *Where City and Country Collide: Managing Growth in the Metropolitan Fringe*. Washington, DC: Island Press. Describes ways of achieving more compact development.

Engwicht, D. (1999). *Street Reclaiming: Creating Livable Streets and Vibrant Communities*. Gabriola Island, BC: New Society Publishers. Very interesting ideas on ways to convert streets in urban neighborhoods into friendly peopleways rather than strictly highways for motorized traffic.

Fodor, E. (1999). *Better Not Bigger: How to Take Control of Urban Growth and Improve Your Community*. Gabriola Island, BC: New Society Publishers. Very important reading.

James, S., and Lahti, T. (2004). *The Natural Step for Communities: How Cities and Towns Can Change to Sustainable Practices*. Gabriola Island, BC: New Society. A fascinating look at what Sweden is doing to create more sustainable cities and towns.

Kelly, E. D., and Becker, B. (1999). *Community Planning: An Introduction to the Comprehensive Plan*. Washington, DC: Island Press. Good textbook on community planning.

Platt, R. H. (2004). Toward Ecological Cities: Adapting to the 21st Century Metropolis. *Environment* 46 (5): 10–27. Excellent reading.

Roseland, M., ed. (1997). *Eco-City Dimensions: Healthy Communities, Healthy Planet*. Gabriola Island, BC: New Society Publishers. Showcases ideas on creating more sustainable cities.

Roseland, M. (1998). *Toward Sustainable Communities: Resources for Citizens and Their Governments*. Gabriola Island, BC: New Society Publishers. Discusses many innovative ways to help create more sustainable communities.

Porter, D. R. (1997). *Managing Growth in America's Communities*. Washington, DC: Island Press. Describes proven approaches to managing growth and much more. See readings in Chapters 14–17 and 20–24.

Chapter 18 Toxicology and Risk Assessment

Amdur, M. O., Doull, J., and Klaassen, C. D. (1991). *Casarett and Doull's Toxicology: The Basic Science of Poisons*. 4th ed. New York: Pergamon Press. Superb reference.

Chiras, D. (1982). Risk and Risk Assessment in Environmental Education. *American Biology Teacher* 44 (4): 460–465. A more technical presentation of risk and risk assessment.

Chiras, D. D. (1992). *Lessons from Nature: Learning to Live Sustainably on the Earth*. Washington, DC: Island Press. Describes several market-based strategies to reduce pollution.

Chiras, D. D. (1993). *Biology: The Web of Life*. St. Paul: West. Contains an important Point/Counterpoint on the health effects of magnetic fields.

Eyles, J., and Consitt, N. (2004). What's at Risk? Environmental Influences on Human Health. *Environment* 47 (1): 24–39. Looks at the importance of environment on human health.

Goldbaum, E. (1987). Can Cell Cultures Predict Toxicity? *Industrial Chemist*, January: 34–37. Interesting look at an alternative way to test toxicity.

Kamarin, M. A. (1988). *Toxicology: A Primer on Toxicology Principles and Applications*. Chelsea, MI: Lewis Publishers. Good introduction to the subject.

Klaassen, C. D., and Eaton, D. L. (1991). Principles of Toxicology. In *Casarett and Doull's Toxicology: The Basic Science of Poisons*. 4th ed., eds. M. O. Amdur, J. Doull, and C. D. Klaassen. New York: Pergamon Press. Detailed coverage of many important concepts.

Manson, J. M., and Weisburger, L. D. (1991). Teratogens. In *Casarett and Doull's Toxicology: The Basic Science of Poisons*. 4th ed., eds. M. O. Amdur, J. Doull, and C. D. Klaassen. New York: Pergamon Press. Detailed coverage of teratology.

Mausner, J. S., and Kramer, S. (1985). *Epidemiology: An Introductory Text*. Philadelphia: Saunders. Excellent reference.

Mitchell, J. D. (1997). Nowhere to Hide: The Global Spread of High-Risk Synthetic Chemicals. *World-Watch* 10 (2): 26–36.

Mossman, B. T., Bignom, J., Corn, M., Seaton, A., and Gee, J. B. L. (1990). Asbestos: Scientific Developments and Implications for Public Policy. *Science* 247 (Jan 19): 294–301. Excellent review.

Postel, S. (1988). Controlling Toxic Chemicals. In *State of the World 1988*, ed. L. Starke. New York: W. W. Norton. Excellent overview of toxic chemicals and their control.

Stavins, R. N., and Whitehead, B. W. (1992). Dealing with Pollution. Market-Based Incentives for Environmental Protection. *Environment* 34 (7): 7–11, 29–42. Examines several different market-based policies.

Wilson, R., and Crouch, E. A. C. (1987). Risk Assessment and Comparisons: An Introduction. *Science* 236 (April 17): 267–270. Excellent introduction.

Chapter 19 Air Pollution

Amdur, M. O. (1991). Air Pollutants. In *Casarett and Doull's Toxicology: The Basic Science of Poisons*. New York: Pergamon Press. Detailed analysis of the toxic effects of air pollutants.

Beatly, T. (2000). *Green Urbanism: Learning from European Cities*. Washington, DC: Island Press. This book describes, among other things, how to revamp our cities to reduce air pollution and other environmental problems.

Bernstein, M., and Whitman, D. (2005). Smog Alert: The Challenges of Battling Ozone Pollution. *Environment* 47 (8): 28–41. Addresses a very important issue.

Bower, J. (1997). *The Healthy House: How to Buy One, How to Build One, and How to Cure a Sick One*. Bloomington, IN: The Healthy House Institute. Excellent, detailed reference on achieving clean air.

Bower, J., and Bower, L. M. (1997). *The Healthy House Answer Book: Answers to the 133 Most Commonly Asked Questions*. Bloomington, IN: The Healthy House Institute. Great overview to the subject of healthy homes. This is a good place to start.

Brown, L. R. (2003). *Plan B: Rescuing a Planet Under Stress and a Civilization in Trouble*. New York: Norton. Lots of valuable information on energy resources and sustainable solutions.

Bruemmer, F. (1991). In Praise of the Lowly Lichen. *International Wildlife* 21 (6): 30–33. Delightfully interesting account on lichens and their vulnerability to air pollution.

Cannon, J. (1990). *The Health Costs of Air Pollution: A Survey of Studies 1984 to 1989*. New York: American Lung Association. Excellent review.

Clayton, T., Spinardi, G. and Williams, R. (1999). *Policies for Cleaner Technology: A New Agenda for Government and Industry*. New York: Earthscan. A ground-breaking treatise on new ways to make industrial systems more sustainable by the businesses themselves.

Environmental Protection Agency. (2000). *National Air Pollutant Emission Trends, 1900–1997*. Washington, DC: U.S. EPA. Although this annual publication is always three years behind the time, it is a great resource.

Eyles, J., and Consitt, N. (2004). What's at Risk? Environmental Influences on Human Health. *Environment* 47 (1): 24–39. Looks at the importance of environment on human health.

Homes, D., Strain, L., Wilson, A., and Leibowitz, S. (1999). *GreenSpec: The Environmental Building News Product Directory and Guideline Specifications*. Brattleboro, VT: E Build, Inc. One of five books on green building materials that are good for the environment and easy on our health.

Lents, J. M., and Kelly, W. J. (1993). Clearing the Air in Los Angeles. *Scientific American* 269 (4): 32–39. Considerable progress has been made due to tough emission control standards.

Moore, C. (1992). Down Germany's Road to a Clean Tomorrow. *International Wildlife* 22 (5): 24–28. Extraordinary piece on the economic benefits of the sustainable strategy.

Newman, P., and Kenworthy, J. (1999). *Sustainability and Cities: Overcoming Automobile Dependence.* Washington, DC: Island Press, 1999.

Roome, N. J. (1998). *Sustainability Strategies for Industry: The Future of Corporate Practice.* Washington, DC: Island Press. Great overview of the subject.

Selin, N. E. (2005). Mercury Rising: Is Global Action Needed to Protect Human Health and the Environment? *Environment* 47 (1): 22–35. An important piece on an important topic.

Sheehan, M. O. (2001). Making Better Transportation Choices. In *State of the World 2001,* ed. L. Starke. New York: Norton. An excellent look at components of a more environmentally sustainable transportation system.

United Nations Environment Programme. (2000). *Global Environment Outlook.* New York: Earthscan. Great reference book on global environmental trends and policy changes needed to address them.

Chapter 20   Global Air Pollution

Ozone Depletion

EPA. (1995). *Stratospheric Ozone Depletion: A Focus on EPA's Research.* Washington, DC: Office of Research and Development, U.S. EPA. Very good summary of current understandings regarding ozone depletion.

Jestin, K. (1995). International Efforts to Abate the Depletion of the Ozone Layer. *Georgetown International Environmental Law Review* 7 (829): 82–845. Excellent survey of treaties aimed at reducing ozone depletion.

Meadows, D. H., Meadows, D. L., and Randers, J. (1992). *Beyond the Limits: Confronting Global Collapse, Envisioning a Sustainable Future.* Post Mills, VT: Chelsea Green. Chapter 5 contains an excellent historical view of the ozone issue.

Moran, J. M., and Morgan, M. D. (1994). *Meteorology: The Atmosphere and the Science of Weather,* 4th ed. New York: Macmillan. Excellent nontechnical coverage of global pollution problems.

Parson, E. A. (2003). *Protecting the Ozone Layer.* New York: Oxford University Press. A detailed look at efforts to protect the ozone layer from destruction.

Acid Deposition

Baker, L. A., Herlihy, A. T., Kaufmann, P. R., and Eilers, J. M. (1991). Acidic Lakes and Streams in the United States: The Role of Acidic Deposition. *Science* 252 (5007): 1151–1155. Excellent scientific paper on acid deposition.

Driscoll, C., et al. (2003). Nitrogen Pollution: Sources and Consequences in the U.S. Northeast. *Environment* 45 (7): 8–22. Detailed account of nitrogen pollution primarily from acid deposition.

Johnson, A. H. (1986). Acid Deposition: Trends, Relationships, and Effects. *Environment* 28 (4): 6–11, 34–43. Comprehensive summary of the National Academy of Sciences report. Well worth reading.

Luoma, J. R. (1987). Black Duck Decline: An Acid Rain Link. *Audubon* 89 (3): 19–24. Extraordinary piece on the links between acid deposition and the dramatic decline in black ducks.

McDonald, A. (1999). Combatting Acid Deposition and Climate Change: Priorities for Asia. *Environment* 41 (3): 4–11, 34–41. Gives a great perspective on what other countries are doing and have to do to address these problems.

Mello, R. A. (1987). *Last Stand of the Red Spruce.* Washington, DC: Natural Resources Defense Council and Island Press. Highly readable account of forest damage and the underlying causes.

National Acid Precipitation Assessment Program. (1990). *National Acid Precipitation Assessment Program: 1990 Integrated Assessment Report.* Washington, DC: National Acid Precipitation Assessment Program. Overview of a comprehensive analysis of acid deposition in the United States.

National Acid Precipitation Assessment Program. (1998). *NAPAP Biennial Report to Congress: An Integrated Assessment.* Washington,

DC: National Science and Technology Council. Full of useful information.

Nierenberg, D. (2001). "Nitrogen: The Other Cycle." *World-Watch* 14 (2): 30–38. A must read for those who want to learn more about the impacts of nitrogen pollution, especially nitrogen oxides.

Roberts, L. (1991). Acid Rain Program: Mixed Review. *Science* 252 (5004): 371. Penetrating analysis of the merits of the National Acid Precipitation Assessment program.

Ryan, J. C. (1992). When Nature Loses Its Cool. *World-Watch* 5 (5): 10–16. Interesting discussion of biological effects of global warming.

Wang, J., et al. (2004). Controlling Sulfur Dioxide in China: Will Emission Trading Work? *Environment* 46 (5): 28–39. An important case study.

Wilbanks, T. J., et al. (2003). Possible Responses to Global Climate Change: Integrating Mitigation and Adaptation. *Environment* 45 (5): 28–38. Examines options for addressing global warming.

Global Warming

Bird, E. (1993). *Submerging Coasts.* New York: Wiley. Examines the implications of rising sea level on coastlines and coastal communities.

Blair, T. (2003). Meeting the Sustainable Development Challenge. *Environment* 45 (4): 20–26. Read what Prime Minister Tony Blair has to say about global climate change.

Depledge, J. (1999). Coming of Age at Buenos Aires: The Climate Change Regime after Kyoto. *Environment* 41 (7): 15–20. Excellent discussion of climate change negotiations and treaties.

Dunn, S. (2002). *Reading the Weathervane: Climate Policy from Rio to Johannesburg.* Worldwatch Paper 60. Washington, DC: Worldwatch Institute. A great overview of efforts to create international agreements to control greenhouse gases.

Dunn, S., and Flavin, C. (2002). Moving the Climate Change Agenda Forward. In *State of the World 2002,* ed. L. Starke. New York: Norton. A must read for students interested in understanding more about the interplay between politics and science as it pertains to climate change.

Flavin, C. (1994). Storm Warnings: Climate Change Hits the Insurance Industry. *World-Watch* 7 (6): 10–20. A very telling tale of how seriously the insurance industry is taking warnings of global climate change.

Flavin, C. (1996). Facing Up to the Risks of Climate Change. In *State of the World 1996,* ed. L. Starke. New York: Norton. Excellent overview.

Flavin, C, and Tunali, O. (1995). Getting Warmer: Looking for a Way Out of the Climate Impasse. *World-Watch* 8 (2): 10–19. Excellent look at what nations aren't doing to address global climate change.

Gardiner, D., and L. Jacobson. (2002). Will Voluntary Programs Be Sufficient to Reduce U.S. Greenhouse Gas Emissions? *Environment* 44 (8): 24–33. Detailed examination of the Bush Administration's proposal to reduce greenhouse gas emissions voluntarily.

Jacobson, J. L. (1990). Holding Back the Sea. In *State of the World 1990,* ed. L. Starke. New York: Norton. Insightful look into the effects of rising sea level.

Kasemir, B., Schibli, D., Stoll, S. and Jaeger, C. C. (2000). Involving the Public in Climate and Energy Decisions, *Environment* 42 (3): 32–42. Explores an aspect of change we don't discuss often in this book.

Kruger, J., and Pizer, W. A. (2004). Greenhouse Gas Trading in Europe: The New Grand Policy Experiment. *Environment* 46 (8): 8–23. Looks at a brave new effort to help reduce global carbon dioxide emissions.

Moore, C. A. (1996). Warming Up to the Hot New Evidence. *National Wildlife* 27 (1): 21–25. A shocking article that looks at global warming predictions and whether they are coming true.

Moran, J. M., and Morgan, M. D. (1994). *Meteorology: The Atmosphere and the Science of Weather.* 4th ed. New York: Macmillan. Excellent nontechnical coverage of global pollution problems.

Moreira, N. (2005). The Wind and the Fury. *Science News* 168: 184–186. A look at the impacts of global climate change on hurricanes.

Myers, N. (1991). Trees by the Billions: A Blueprint for Cooling. *International Wildlife* 21 (5): 12–15. Superb article showing the potential to replant tropical forests to reduce global carbon dioxide levels.

Socolow, R., et al. (2004). Solving the Climate Problem: Technologies Available to Curb CO_2 emissions. *Environment* 46 (10): 8–19. Excellent reading.

Stavins, R. N. (2004). Forging a More Effective Global Climate Treaty. *Environment* 46 (10): 22–30. Excellent analysis of what's needed to address climate change.

Chapter 21   Water Pollution

Adler, R. W., Cameron, D. M., and Landman, J. (1993). *The Clean Water Act 20 Years Later.* Washington, DC: Island Press. Great summary of the Clean Water Act, including its strengths and weaknesses, by the Natural Resources Defense Council.

Borelli, P. (1989). Troubled Waters: Alaska's Rude Awakening to the Price of Oil Development. *Amicus Journal* 11 (3): 10–20. Excellent overview of the Alaskan oil spill.

Carey, J. (1996). Lessons from Loons. *National Wildlife* 34 (5): 12–19. An excellent look at the effects of water pollutants on wildlife.

Cheremisinoff, P. N. (1993). *Water Treatment and Waste Recovery: Advanced Technology and Applications.* Englewood Cliffs, NJ: Prentice-Hall. Authoritative survey of latest methods of water purification.

Doppelt, B., Scurlock, M., Frissell, C., and Karr, J. (1993). *Entering the Watershed: A New Approach to Save America's River Ecosystems.* Washington, DC: Island Press. Good overview of watershed protection.

Eder, T., and Jackson, J. (1988). *A Citizen's Guide to the Great Lakes Water Quality Agreement.* Buffalo: State University College. Excellent overview of the agreement between the U.S. and Canada to clean up the Great Lakes.

Garelik, G. (1996). Russia's Legacy of Death. *National Wildlife* 34 (4): 36–41. A telling story of the water pollution and other environmental atrocities caused by lax environmental enforcement.

Gleick, P. H. (2000). *The World's Water 2000-2001.* Washington, DC: Island Press. Contains a vast amount of information on the world's water resources, including waterborne diseases.

Hearne, S. A. (1996). Tracking Toxics: Chemical Use and the Public's Right to Know. *Environment* 38 (6): 4–9, 28–34. A description of an effective tool in controlling water pollution.

Maurits la Riviäre, J. W. (1989). Threats to the World's Water. *Scientific American* 261 (3): 80–94. Up-to-date survey of global pollution problems.

Platt, R. H., Barten, P. K., and Pfeffer, M. J. (2000). A Full, Clean Glass? Managing New York City's Watersheds. *Environment* 42 (5): 8–20. Useful case study for those interested in learning more about watershed management.

Riley, A. L. (1998). *Restoring Streams in Cities: A Guide for Planners and Citizens.* Washington, DC: Island Press. Examines a host of sustainable approaches to protecting watersheds and rivers.

Sampat, P. (2001). Uncovering Groundwater Pollution. In *State of the World 2001*, ed. L. Starke. New York: Norton. Contains lots of information on groundwater pollution in the United States.

Thompson, J. W., and Sorvig, K. (2000). *Sustainable Landscape Construction: A Guide to Green Building Outdoors.* Washington, DC: Island Press. Offers invaluable advice on landscape design in cities and towns that help protect watersheds.

UNEP. (1995). *The Pollution of Lakes and Reservoirs.* Nairobi, Kenya: United Nations Environment Programme. Documents pollution issues of the major lakes and reservoirs of the world and ways to address them.

UNEP. (1995). *Water Quality of the World River Basins.* Nairobi, Kenya: United Nations Environment Programme. Summarizes data on water quality from 82 major river basins around the world.

UNEP/GEMS. (1996). *Groundwater: A Threatened Resource.* Nairobi, Kenya: United Nations Environment Programme. Explains global groundwater pollution.

Weber, M. L. (2002). *From Abundance to Scarcity: A History of U.S. Marine Fisheries Policy.* Washington, DC: Island Press. An in-depth look at U.S. policy on marine fisheries.

Williams, T. (1994). Death in a Black Desert. *Audubon* 96 (1): 24. Discusses a California irrigation nightmare: Toxic runoff is poisoning the land and killing thousands of birds.

WRI and International Institute for Environment and Development. (1998). *World Resources 1998–99.* New York: Oxford University Press. See sections on water.

Chapter 22   Pests and Pesticides

Ames, B. N., and Gold, L. S. (1989). Pesticides, Risk, and Applesauce. *Science* 244 (4906): 755–767. Looks at natural carcinogens and important considerations for controlling human-produced pesticides.

Bosch, R. van der (1978). *The Pesticide Conspiracy.* New York: Doubleday. A classic study of the influence of pesticide manufacturers on integrated pest management.

Carson, R. (1962). *Silent Spring.* Boston: Houghton Mifflin. The book that raised worldwide alarm over the use of pesticides.

DiSilvestro, R. (1996). What's Killing the Swainson's Hawk? *International Wildlife* 26 (3): 38–43. Describes the impacts of pesticide use on a migratory species.

Dreistadt, S. H., and Dahlsten, D. L. (1986). California's Medfly Campaign: Lessons from the Field. *Environment* 28 (6): 18–20, 40–44. A sober look at pest control.

Gardner, G. (1996). Preserving Agricultural Resources. In *State of the World 1996*, ed. L. Starke. New York: Norton. Covers several important topics, including pest management.

Goodstein, C. (1996). Stood Up by the Birds and the Bees. *Amicus Journal* 18 (1): 26–30. A look at the ecological consequences of disappearing birds and insects, both pollinators of flowering plants.

Karlsson, S. I. (2004). Agricultural Pesticides in Developing Countries: A Multilevel Governance Challenge. *Environment* 46 (4): 22–39. An in-depth look at the effectiveness of policies designed to protect people and the environment in less developed nations from the effects of pesticides.

Kinley, D. H. (1998). Aerial Assault on the Tsetse Fly. *Environment* 40 (7), 14–18, 40–41. Tells of a successful effort to control disease-carrying tsetse flies using the sterile male technique.

Halweil, B. (1999). The Emperor's New Crops. *World-Watch* 12 (4): 21–29. Examines the controversy over genetically modified crops.

Lipske, M. (1990) Natural Farming Harvests New Support. *National Wildlife* 28 (3): 19–23. Highlights some alternative farming methods.

McGinn, A. P. (2000). POPs Culture. *World-Watch* 13(2): 26–36. Examines the problems created by persistent organic pollutants, among them pesticides.

McGinn, A. P. (2003). Combatting Malaria. In *State of the World 2003*, ed. L. Starke. New York: W. W. Norton. Offers important insight into pesticide use to combat insects and disease.

Olkowski, W., Daar, S., and Olkowski, H. (1991). *Common-Sense Pest Control.* Newtown, CT: Taunton Press. Over 700 pages of practical, least-toxic pest control solutions for home, garden, pets, and community.

Sachs, J. S. (2004). Poisoning the Imperiled. *National Wildlife* 42 (1): 22–29. Looks at endangered species whose future is threatened by pesticides.

Sewell, B. H., Whyatt, R. M., Hathaway, J., and Mott, L. (1989). *Intolerable Risk: Pesticides in Our Children's Food.* Washington, DC: Natural Resources Defense Council. Controversial paper, worth reading.

Tuxill, J. (2000). The Biodiversity that People Made. *World-Watch* 13 (3): 25–35. Superb article on the loss of genetic diversity in crops.

Weber, P. (1992). A Place for Pesticides? *World-Watch* 5 (3): 18–25. Superb overview of the pesticide issue, with a frank discussion of options.

Wilcox, F. A. (1983). *Waiting for an Army to Die: The Tragedy of Agent Orange*. New York: Vintage Books. A superb book on the troubles facing Vietnam veterans.

Youth, H. (1994). Flying into Trouble. *World-Watch* 7 (1): 10–19. Documents the startling decline in bird species and the role of pesticides in this tragic decline.

Chapter 23   Municipal Solid Waste and Hazardous Waste

Municipal Solid Waste

Abramovitz, J. N., and Mattoon, A. T. (2000). Recovering the Paper Landscape. In *State of the World 2000*. New York: Norton. Examines sustainable ways to produce paper, including recycling.

Andersen, M. S. (1998). Assessing the Effectiveness of Denmark's Waste Tax, *Environment* 40 (4): 10–15, 38–41. Examines experience in Denmark with waste taxes and explores additional options to promote recycling.

Carless, J. (1992). *Taking Out the Trash: A No-Nonsense Guide to Recycling*. Washington, DC: Island Press. A practical guide showing how individuals, businesses, and communities can help alleviate the solid waste crisis.

De Graaf, J. D. Wann, and Naylor, T. H. (2001). *Affluenza: The All-Consuming Epidemic*. San Francisco: Berrett-Koehler. A very popular treatment of the overconsumption and its impacts on our lives; based on a television documentary of the same name.

Durning, A. (1990). How Much Is Enough? *World-Watch* 3 (6): 12–19. Hard-hitting article on the impacts of and reasons for consumption.

EPA. (1995). *Municipal Solid Waste Factbook–Internet Version*. Washington, DC: EPA. An extremely valuable resource, though slightly out of date. Available on-line at www.epa.gov/epaoswer/non-hw/muncpl/factbook/internet/.

Gardner, G., and Sampat, P. (1999). Forging a Sustainable Materials Economy. In *State of the World 1999*. New York: Norton. Great look at consumption and ways to lessen it.

Kane, H. (1996). Shifting to Sustainable Industries. In *State of the World 1996*, ed. L. Starke. New York: W. W. Norton. Describes important aspects of recycling.

Kates, R. W. (2000). Population and Consumption: What We Know, What We Need to Know. *Environment* 42 (3): 10–19. An important report on the subject.

McDonough, W., and Braungart, M. (2002). *Cradle to Cradle: Remaking the Way We Make Things*. New York: North Point Press. Explores a new way of making products to eliminate toxicity and enhance their recyclability.

Renner, M. (1992). Creating Sustainable Jobs in Industrial Countries. In *State of the World 1992*, ed. L. Starke. New York: Norton. Discusses the job potential of a sustainable economy based in part on recycling.

Rosenblatt, R., ed. (1999). *Consuming Desires: Consumption, Culture, and the Pursuit of Happiness*. Washington, DC: Island Press. Collection of writings on one of the most challenging aspects of modern times, consumerism.

Taylor, B. (2003). *What Kids Really Want That Money Can't Buy. Tips for Parenting in a Commercial World*. New York: Warner Books. An interesting look at ways to raise children without endulging in modern consumerism.

Ueta, K., and Koizumi, H. (2001). Reducing Household Waste: Japan Learns from Germany. *Environment* 43 (9): 20–32. A very important reading on waste management in Japan and Germany.

Young, J. (1991). Reducing Waste, Saving Materials. In *State of the World 1992*, ed. L. Starke. New York: Norton. Describes several strategies to reduce our waste of material.

Young, J., and Sachs, A. (1995). Creating a Sustainable Materials Economy. In *State of the World 1995*, ed. L. Starke. New York: Norton. Addresses important issues regarding solid waste and recycling.

Hazardous Wastes

Carlin, A., Scodari, P. F., and Garner, D. H. (1992). Environmental Investments: The Cost of Cleaning Up. *Environment* 34 (2): 12–20, 38–45. Summary of U.S. EPA report to Congress.

Center for Neighborhood Technology. (1990). *Sustainable Manufacturing*. Chicago: Center for Neighborhood Technology. Excellent discussion of efforts needed to reduce hazardous waste.

Fischhoff, B. (1991). Report from Poland: Science and Politics in the Midst of Environmental Disaster. *Environment* 33 (2): 12–17, 37. Describes the dimensions of the hazardous waste problem in Poland.

French, H. (1990). A Most Deadly Trade. *World-Watch* 3 (4): 11–17. Documents the movement of hazardous materials to the developing countries and eastern Europe.

Frosch, R. A. (1995). Industrial Ecology: Adapting Technology for a Sustainable World. *Environment* 37 (10): 16–24, 34–37. Looks at ways to reduce waste production by factories.

Frosch, R. A., and Gallopoulos, N. F. (1989). Strategies for Manufacturing. *Scientific American* 261 (3): 144–152. Outlines the concept of the industrial ecosystem.

Gibbs, L. M. (1998). *Love Canal: The Story Continues*. Gabriola Island, BC: New Society Publishers. Fascinating account of the containment of wastes and resettlement of part of the contaminated area and more.

Krueger, J. (1999). What's to Become of Trade in Hazardous Wastes?: The Basel Convention One Decade Later, *Environment* 41 (9): 10–21. Examines the effects of an important international agreement on the export of hazardous wastes.

McGinn, A. P. (2000). Phasing Out Persistent Organic Pollutants. In *State of the World 2000*, ed. L. Starke. New York: W.W. Norton. Covers a variety of strategies for reducing and eliminating the production and release of persistent organic chemicals including those in hazardous waste.

McGinn, A. P. (2002). Reducing Our Toxic Burden. In *State of the World 2002*, ed. L. Starke. New York: Norton. Looks at toxic chemicals such as persistent organic pollutants and heavy metals.

Probst, K. N., and Bierle, T. C. (1999). Hazardous Waste Management: Lessons from Eight Countries. *Environment* 41 (9): 22–30. Valuable resource.

Renner, M. (1994). Cleaning Up After the Arms Race. In *State of the World 1994*, ed. L. Starke. New York: Norton. Discusses hazardous-waste problems at federal facilities.

Russell, M., Colglazier, E. W., and Tonn, B. E. (1992). The U.S. Hazardous Waste Legacy. *Environment* 34 (6): 12–15, 34–39. Discusses the cost of cleaning up America's hazardous wastes.

Sherlock, M. (2003). *Living Simply with Children*. New York: Three Rivers Press. A guide to voluntary simplicity for parents.

Stigliani, W. M., et al. (1991). Chemical Time Bombs: Predicting the Unpredictable. *Environment* 33 (4): 4–9, 26–30. Shows how chemical contamination builds and then surpasses critical threshold levels, suddenly creating poisonous conditions.

Chapter 24   Environmental Ethics

Atkisson, A. (1999). *Believing Cassandra: An Optimist Looks at a Pessimist's World*. White River Junction, VT: Chelsea Green. An entertaining and thoughtful book on sustainability.

Berry, W. (1987). *Home Economics*. Berkeley, CA: North Point Press. Thoughtful collection of essays on wise stewardship.

Chapman, A. R., Petersen, R. L., and Smith-Moran, B. (1999). *Consumption, Population, and Sustainability: Perspectives from Science and Religion*. Washington, DC. Important reading on ethics.

Chiras, D. D. (1990). *Beyond the Fray: Reshaping the American Environmental Response*. Boulder, CO: Johnson Books. Describes important changes needed in the environmental movement.

Chiras, D. D. (1992). An Inquiry into the Root Causes of the Environmental Crisis. *Environmental Carcinogenesis and Ecotoxicology Reviews* C10 (1): 73–119. Reviews the thinking on the root causes of the environmental crisis, including human attitudes.

Chiras, D. D. (1992). *Lessons from Nature: Learning to Live Sustainably on the Earth*. Washington, DC: Island Press. See Chapters 2 and 3 for a detailed discussion of sustainable ethics.

Chiras, D. D., ed. (1995.). *Voices for the Earth: Vital Ideas from America's Best Environmental Books*. Boulder, CO.: Johnson Books. A collection of essays on sustainability by a wide assortment of writers.

Derr, P. G., and McNamara, E. M. (2004). *Case Studies in Environmental Ethics.* Lanham, MD: Rowman and Littlefield. Forty-three brief case studies that encompass a wide range of issues.

DeVall, B. (1988). *Simple in Means, Rich in Ends: Practicing Deep Ecology.* Salt Lake City: Peregrine Smith. Important discussion of deep ecology and ways to put reverence for nature into action.

Dobbs, D., and Ober, R. (1995). *The Northern Forest.* White River Junction, VT: Chelsea Green. Insightful and informative story of our view of land and the struggles between conservation and use.

Durning, A. T. (1997). After the Deluge: The Changing World-view. *World-Watch* 10 (1): 25–31. Outlines key economic and ethical changes needed to build a sustainable society.

Gorke, M. (2003). *The Death of Our Planet's Species: A Challenge to Ecology and Ethics.* Washington, DC: Island Press. In this book, the author makes a case for a holistic environmental ethic to help save imperiled species of the world.

Hardin, G. (1985). *Filters against Folly.* New York: Viking. Thoughtful treatise on ethics. Important reading.

LaMay, C. L., and Dennis, E. E. (1991). *Media and the Environment.* Washington, DC: Island Press. Insightful and useful critique of environmental reporting by the media.

Leopold, A. (1966). *A Sand County Almanac.* New York: Ballantine. Collection of essays on nature and conservation.

Leopold, A. (1999). *For the Health of the Land: Previously Unpublished Essays and Other Writings.* Edited by J. Baird Callicott and Eric T. Freyfogle. Washington, DC: Island Press. A collection of Leopold's writings that focuses on the notion of land health and ways to protect private land.

Meadows, D. (1991). *The Global Citizen.* Washington, DC: Island Press. Collection of short, insightful essays on a variety of environmental topics, especially values.

Milbrath, L. W. (1989). *Envisioning a Sustainable Society: Learning Our Way Out.* Albany: State University of New York Press. Early chapters discuss values of the dominant paradigm (frontierism) and need for a new value system.

Nash, R. F. (1989). *The Rights of Nature: A History of Environmental Ethics.* Madison: University of Wisconsin Press. Important reading.

Regenstein, L. (1991). *Replenish the Earth.* New York: Crossroad. Discusses organized religion's treatment of animals and nature.

Rolston, H. (1987). *Environmental Ethics: Duties to and Values in the Natural World.* Philadelphia: Temple University Press. Explains the rights of other creatures.

Schumacher, E. F. (1973). *Small Is Beautiful: Economics as if People Mattered.* New York: Harper and Row. One of the best books ever written on the subject of a sustainable society and new ethical systems.

Van Matre, S., and Weiler, B. (1983). *The Earth Speaks.* Warrenville, IL: Institute for Earth Education. Superb collection of writings on nature.

Chapter 25 Sustainable Economics: More Developed Countries

Ayres, R. U. (2001). How Economists Have Misjudged Global Warming. *World-Watch* 14 (5): 12–25. Extremely important reading about the myths of economic strength and controls on emissions of carbon dioxide from U.S. industries.

Brown, L. R., Flavin, C., and Postel, S. (1991). *Saving the Planet: How to Shape an Environmentally Sustainable Global Economy.* New York: Norton. Basic account of ways to reshape the human economy.

Chiras, D. D. (1992). *Lessons from Nature: Learning to Live Sustainably on the Earth.* Washington, DC: Island Press. See Chapters 4 and 5 for a discussion of sustainable economics.

Chiras, D. D., ed. (1995). *Voices for the Earth: Vital Ideas from America's Best Environmental Books.* Boulder, CO: Johnson Books. A collection of essays on sustainability with several important contributions from ecological economists.

Cobb, C., Halstead, T. and Rowe, J. (1995). If the GDP is up, why is America down? *The Atlantic Monthly* 276 (4): 59–78. Takes a good hard look at alternative indicators of success.

Daly, H. E., and Cobb, J. B. (1989). *For the Common Good: Redirecting the Economy toward Community, the Environment, and a Sustainable Future.* Boston: Beacon Press. A challenging but worthwhile book for those interested in probing deeper.

Douthwaite, R. (1999). *The Growth Illusion: How Economic Growth Has Enriched the Few, Impoverished the Many, and Endangered the Planet.* Gabriola Island, BC: New Society Publishers. The subtitles says it all.

Durning, A. T. (1997). After the Deluge: The Changing World-view. *World-Watch* 10 (1): 25–31. Outlines key economic and ethical changes needed to build a sustainable society.

Frankel, C. (1999). *In Earth's Company: Business, Environment, and the Challenge of Sustainability.* Gabriola Island, BC: New Society Publishers. Shows how corporations can operate while practicing environmental protection.

Gardner, G. (2001). Accelerating the Shift to Sustainability. In *State of the World 2001,* ed. L. Starke. New York: Norton. Looks at the role of business, government, and citizens in shaping a sustainable future.

Goodstein, E. (1999). *The Trade-Off Myth: Fact and Fiction about Jobs and the Environment.* Washington, DC: Island Press. Examines a deeply held belief, notably that environmental protection threatens jobs.

Gowdy, J. (1998). *Limited Wants, Unlimited Means: A Reader on Hunter-Gatherer Economics and the Environment.* Washington, DC: Island Press. A detailed look at subsistence economics.

Hawken, P., Lovins, A., and Lovins, L. H. (1999). *Natural Capitalism: Creating the Next Industrial Revolution.* Boston: Little, Brown. An insightful look into a new way of doing business.

Johnston, C. M. (1992). The Wisdom of Limits. In *Context* 32: 48–51. Important reading on reaching cultural maturity by recognizing limits and making appropriate adjustments.

Krishnan, R., Harris, J. M., and Goodwin, N. R., eds. (1995). *A Survey of Ecological Economics.* Washington, DC: Island Press. A useful collection of writings on ecological economics.

Mastny, L. (2003). *Purchasing Power: Harnessing Institutional Procurement for People and the Planet.* Worldwatch Paper 166. Washington, DC: Worldwatch Institute. Detailed report on the power of government procurement programs that help promote environmentally responsible products.

Meadows, D. H., Meadows, D. L., and Randers, J. (1992). *Beyond the Limits: Confronting Global Collapse, Envisioning a Sustainable Future.* Post Mills, VT: Chelsea Green. Contains important information on economics.

Meeker-Lowry, S. (1988). *Economics as if the Earth Really Mattered.* Philadelphia: New Society Publishers. Contains ideas on ways individuals can contribute to creating a sustainable economic system.

Meyers, N., and Kent, J. (2001). *Perverse Subsidies: How Misued Tax Dollars Harm the Environment and the Economy.* Washington, DC: Island Press. Very important reading on the impacts of government subsidies on environmentally unfriendly activities.

Nattrass, B., and Altomare, M. (1999). *The Natural Step for Business: Wealth, Ecology, and the Evolutionary Corporation.* Gabriola Island, BC: New Society Publishers. Shows how four businesses are changing themselves to change the world, destroying the myth that companies must choose between profit and environmental protection.

Postel, S., and Flavin, C. (1991). Reshaping the Global Economy. In *State of the World 1991,* ed. L. Starke. New York: Norton. Important reading that outlines ways to help convert to a more sustainable economy.

Power, T. M., and Barrett, R. (2001). *Post-Cowboy Economics: Pay and Prosperity in the New American West.* Washington, DC: Island Press. Important reading for those who want to understand some of the myths propagated by anti-environmental factions in the United States.

Renner, M. G. (1992). Saving the Earth, Creating Jobs. *World-Watch* 5 (1): 10–17. Essential reading on the job potential of the sustainable strategy.

Renner, M. (2001). Employment in Wind Power. *World-Watch* 14 (1): 22–30. Fascinating look at the employment boom in the wind energy sector, showing that environmentally sound ideas make good economic sense.

Repetto, R. (1992). Earth in the Balance Sheet: Incorporating Natural Resources in National Income Accounts. *Environment* 34 (7): 12–20, 43–45. Very important reading.

Schumacher, E. F. (1977). *Small Is Beautiful: Economics as if People Mattered.* New York: Harper and Row. One of the best books ever written on the subject of a sustainable society and new ethical and economic systems.

Stavins, R. N., and Whitehead, B. W. (1992). Dealing with Pollution: Market-Based Incentives for Environmental Protection. *Environment* 34 (7): 6–11, 29–42. Detailed analysis of various market-based incentives, with an analysis of their effectiveness.

Chapter 26   Sustainable Economics: Less Developed Countries

Brown, L. R., Flavin, C., and Postel, S. (1991). *Saving the Planet: How to Shape an Environmentally Sustainable Global Economy.* New York: Norton. Contains many important ideas for sustainable economic development.

Chiras, D. D. (1992). *Lessons from Nature: Learning to Live Sustainably on the Earth.* Washington, DC: Island Press. See Chapters 6 and 7.

Dobb, E. (1992). Solar Cooker. *Audubon* 94 (6): 100–105. Interesting article that tells about an appropriate technology.

Holmberg, J. (1992). *Making Development Sustainable: Redefining Institutions, Policy, and Economics.* Washington, DC: Island Press. Important collection of essays on sustainable development.

Panayotou, T., and Ashton, P. S. (1992). *Not by Timber Alone: Economics and Ecology for Sustaining Tropical Forests.* Washington, DC: Island Press. Examines tropical rain forests as a source of many products that could be harvested sustainably.

Plotkin, M., and Famolare, L. (1992). *Sustainable Harvest and Marketing of Rain Forest Products.* Washington, DC: Island Press. Describes products that can be extracted with little if any damage to the forest ecosystem.

Reid, W. V. C., Barnes, J. N., and Blackwelder, B. (1988). *Bankrolling Success: A Portfolio of Sustainable Development Projects.* Washington, DC: Environmental Policy Institute and National Wildlife Federation. Study of successful development projects.

Sirolli, E. (1999). *Ripples from the Zambezi: Passion, Entrepreneurship, and the Rebirth of Local Economies.* Gabriola Island, BC: New Society Publishers. Critically examines old development approaches and alternative ways of bringing about development that is good for people and the environment of LDCs.

Timmer, V., and Juma, C. (2005). Taking Root: Biodiversity Conservation and Poverty Reduction Come Together in the Tropics. *Environment* 47 (4): 24–44. Detailed piece on the feasibility of sustainable ways to protect wild species and ecosystems while creating economic prosperity in Third World Nations.

World Commission on Environment and Development. (1987). *Our Common Future.* Oxford: Oxford University Press. See Chapter 2, on sustainable development.

Chapter 27   Law and Government

Adamson, J., Evans, M. M., and Stein, R. (2002). *The Environmental Justice Reader.* Tucson: University of Arizona Press. Important readings on this vital topic.

Agyeman, J. (2005). Alternatives for Community and Environment: Where Justice and Sustainability Meet. *Environment* 47 (6): 10–23. A good look at the issue of environmental justice.

Bennear, L. S., and Coglianese, C. (2005). Measuring Progress: Program Evaluation of Environmental Policies. *Environment* 46 (2): 22–43. Answers important questions about the effectiveness of different types of environmental policy.

Brick, P., Snow, D., and Van De Wetering, S., eds. (2001). *Across the Great Divide: Explorations on Collaborative Conservation and the American West.* Washington, DC: Island Press. A wonderful examination of collaboration in solving environmental problems.

Brown, L. R. (1992). Launching the Environmental Revolution. In *State of the World 1992*, ed. L. Starke. New York: Norton. Describes the pivotal role of governments in promoting a sustainable future.

Brown, L. R., Flavin, C., and Postel, S. (1991). *Saving the Planet: How to Shape an Environmentally Sustainable Global Economy.* New York: Norton. Highly readable account of ways to reshape the economy.

Buck, S. J. (1996). *Understanding Environmental Administration and Law.* Washington, DC: Island Press. A must read for anyone interested in understanding how environmental laws really work.

Cash, D. (2004). Innovative Natural Resource Management: Nebraska's Model for Linking Science and Decisionmaking. *Environment* 45 (10): 8–20. Looks at an innovative approach to manage natural resources sustainably using scientific information.

Chiras, D. D. (1990). *Beyond the Fray: Reshaping America's Environmental Response.* Boulder, CO: Johnson Books. Describes alternative governmental solutions to achieve sustainability.

Chiras, D. D. (1992). *Lessons from Nature: Learning to Live Sustainably on the Earth.* Washington, DC: Island Press. See Chapters 8 and 9 for a discussion of government.

Frankel, J. (2005). Climate and Trade: Links Between the Kyoto Protocol and WTO. *Environment* 47 (7): 8–19. Looks at ways that these important treaties could actually be shaped to improve environmental conditions.

French, H. F. (1992). Strengthening Global Environmental Governance. In *State of the World 1992*, ed. L. Starke. New York: Norton. Discusses ways to strengthen international cooperation.

Gillies, E. (1999). *A Guide to EC Environmental Law.* Washington, DC: Earthscan. Examines environmental law of the European Community.

Holmberg, J. (1992). *Making Development Sustainable: Redefining Institutions, Policy, and Economics.* Washington, DC: Island Press. Important collection of essays on sustainable development.

National Wildlife Federation. (2003). The Guide to Worldwide Environmental Organizations. Washington, DC: National Wildlife Federation. An important resource for students wanting to learn more about environmental groups and seeking employment.

Orr, D. W. (2004). *The Last Refuge: Patriotism, Politics, and the Environment in an Age of Terror.* Washington, DC: Island Press. Explains the political roots of the world's rapidly deteriorating environment.

Porter, G., and Islam, I. (1992). *The Road from Rio: An Agenda for U.S. Follow-Up to the Earth Summit.* Washington, DC: Environmental and Energy Study Institute. Excellent survey of the agreements reached at the Rio Conference.

Schreurs, M. A. (2003). Divergent Paths. Environmental Policy in Germany, the United States, and Japan. *Environment* 45 (8): 8–17. An important look at environmental policy differences of the three largest economies of the world and the impacts of such policies.

Vaughn, S. (2004). How Green Is NAFTA? Measuring the Impacts of Agricultural Trade. *Environment* 46 (2): 26–42. An in-depth look at the environmental impacts of free trade.

Many references and suggested readings in previous chapters are good resources for information on law and government.

About the Author: Skyler Chiras

Chapter 1

Chapter opener © Photodisc; 1-1 © Doug Scott/age fotostock; 1-3 Courtesy of Roger Taylor/National Renewable Energy Laboratory; 1-5 © Michael Redmer/Visuals Unlimited; 1-6 © Francois Gohier/Ardea London; 1-7 Courtesy of Roger Taylor/National Renewable Energy Laboratory; Spotlight 1-1 © AbleStock

Chapter 2

Chapter opener © Mark Lewis/Alamy Images; 2-2 Courtesy of Electrolux, used with permission; 2-7 © AbleStock; 2-9a Courtesy of Dan Chiras; 2-9b Courtesy of Dan Chiras; Spotlight 2-1 Courtesy of Dan Chiras; Spotlight 2-2 Courtesy of Dan Chiras; Spotlight 2-3a Courtesy of Dan Chiras; Spotlight 2-3b Courtesy of Dan Chiras; Spotlight 2-4 Courtesy of Dan Chiras

Chapter 3

Chapter opener © Steve Strickland/Visuals Unlimited; 3-1a © Geoffrey Morgan/Alamy Images; 3-1b © China Photos/Reuters/Landov; 3-2 © Jeff Greenberg/age fotostock; 3-3 © Anil Risal Singh/UNEP/Peter Arnold, Inc.; 3-4 © Dr. John D. Cunningham/Visuals Unlimited; 3-5a © Richard Ashley/Visuals Unlimited; 3-5b © G. Prance/Visuals Unlimited; 3-7a © Joerg Boethling/Peter Arnold, Inc.; 3-7b © Photodisc; 3-8 © P. Narayan/age fotostock

Chapter 4

Chapter opener © Photos.com; 4-3a Courtesy of U.S. Fish & Wildlife Services; 4-3b © Photos.com; 4-3c © Photos.com; 4-3d Courtesy of Tim McCabe/NRCS; 4-3e © Doug Sokell/Visuals Unlimited; 4-5a © Jim Wark/Peter Arnold, Inc.; 4-5b © Ken Lucas/Visuals Unlimited; 4-6 © Josh Meyer/ShutterStock, Inc.; Point 4-1 Courtesy of David M. Armstrong; Counterpoint 4-1 Courtesy of University of Illinois, Urbana-Champaign; Spotlight 4-1 Courtesy of John Todd, University of Vermont

Chapter 5

Chapter opener © AbleStock; 5-6 Courtesy of U.S. Fish & Wildlife Services; 5-7a © Photodisc; 5-7b © Photos.com; 5-8 © Steve McCutcheon/Visuals Unlimited; 5-9 © Kent Knudson/PhotoLink/Photodisc/Getty Images; 5-10 © Corbis; 5-11a © Photodisc; 5-11b © Photodisc; 5-12 © Steve Maslowski/Visuals Unlimited; 5-13 © Photodisc; 5-14 Courtesy of Erwin Cole/NRCS; 5-15 © SuperStock/age fotostock; 5-16a © Photos.com; 5-16b © Kim Pin Tan/ShutterStock, Inc.; 5-17a © Gerald and Buff Corsi/Visuals Unlimited; 5-17b © Beth Davidow/Visuals Unlimited; 5-18a © Photodisc; 5-18b © Photodisc; 5-20 © Tomas Kopecny/Alamy Images; 5-21a © Cabisco/Visuals Unlimited; 5-21b © T.E. Adams/Visuals Unlimited; 5-22 © John D. Cunningham/Visuals Unlimited; 5-25 Courtesy of National Estuarine Research Reserve Collection/NOAA; 5-26 Courtesy of NOAA National Estuarine Research Reserve Collection; 5-27 Courtesy of Mr. Mohammed Al Momany, Aqaba, Jordan/NOAA; 5-28a Courtesy of Mr. Mohammed Al Momany, Aqaba, Jordan/NOAA; 5-28b © Photos.com; Spotlight 5-1 © Joseph L. Fontenot/Visuals Unlimited

Chapter 6

Chapter opener © Steffen Foerster/ShutterStock, Inc.; 6-1 Courtesy of Nancy Mayberry/U.S. Army Corps of Engineers; 6-3 © Ryan Rehmeier, Konza Prairie LTER Program, Kansas State University; 6-7a Courtesy of Dan Chiras; 6-8a Courtesy of R. Ross Gilchrist/Copyright © 1999 Weyerhaeuser. All rights reserved; 6-8b Courtesy of R. Ross Gilchrist/Copyright © 1999 Weyerhaeuser. All rights reserved; 6-8c Courtesy of R. Ross Gilchrist/Copyright © 1999 Weyerhaeuser. All rights reserved; 6-8d Courtesy of Carl Thornber/USGS; 6-13 Courtesy of U.S. Fish & Wildlife Services; Spotlight 6-1 Courtesy of Gary Kramer/NRCS

Chapter 7

Chapter opener © Ana Vasileva/ShutterStock, Inc.; 7-4b © Khoo Si Lin/ShutterStock, Inc.; 7-4b © AbleStock; 7-5 © Jack Dagley/ShutterStock, Inc.; Spotlight 7-1 © Still Pictures/Peter Arnold, Inc.

Chapter 8

Chapter opener © Corbis; 8-3 © Karin Duthie/Alamy Images; 8-9 © Jeff Greenberg/Visuals Unlimited; Point 8-1 Courtesy of Garrett Hardin; Counterpoint 8-1 Courtesy of Julian L. Simon; Spotlight 8-1 © Barbara Walton/EPA/Landov

Chapter 9

Chapter opener © Ambient Image Inc./Alamy Images; 9-2 © Tim Graham/Alamy Images; 9-3 © Adrian Bradshaw/EPA/Landov; Spotlight 9-1 © Fary Braasch/Corbis

Chapter 10

Chapter opener © Cole/USDA; 10-1 Courtesy of Dr. Lyle Conrad/CDC; 10-2 © A. Vollan, WHOI/Visuals Unlimited; 10-4 Courtesy of Lynn Betts/NRCS; 10-6 © Bill Bowden, York Daily Record/AP Photos; 10-8 Courtesy of International Center for Maize and Wheat Improvement (CIMMYT); 10-9 © Ken Hammond/USDA; 10-11 © Inga Spence/Tom Stock & Associates, Inc.; 10-12 Courtesy of Tim McCabe/USDA ARS; 10-13a Courtesy of Tim McCabe/NRCS; 10-13b © Kim Pin Tan/ShutterStock, Inc.; 10-14 Courtesy of Erwin Cole/NRCS; 10-15 Courtesy of Lynn Betts/NRCS; 10-16a © Muriel Lasure/ShutterStock, Inc.; 10-16b Courtesy of Gene Alexander/NRCS; 10-17 Courtesy of Scott Bauer/USDA; Spotlight 10-1 © Forrest Anderson–Rouviere Media; Spotlight 10-2 Courtesy of The Food Project (www.thefoodproject.org)

Chapter 11

Chapter opener © Glenn Frank/ShutterStock, Inc.; 11-2 Courtesy of SFWMD; 11-3 © Paul Lovich Photography/Alamy Images; 11-5 Courtesy of Gary Zahm/U.S. Fish & Wildlife Services; 11-6 © Stephen J. Krasemann/Photo Researchers, Inc.; 11-7 © Richard L. Carton/Visuals Unlimited; 11-8 © Isifa Image Service S.P.O./Alamy Images; 11-9 Courtesy of NOAA; 11-10a Courtesy of U.S. Fish & Wildlife Services; 11-10b © Mark C. Burnett/Photo Researchers, Inc.; 11-11 Courtesy of Pedro Ramirez, Jr./U.S. Fish and Wildlife Services; 11-13 © Victor Englebert/Photo Researchers, Inc.; Counterpoint 11-1 Courtesy of John Echevarria; Point 11-2 Courtesy of L. Schenberger; Counterpoint 11-2 Courtesy of Craig Miller; Spotlight 11-1 Courtesy of Thirteen Mile Lamb & Wool Company (www.lambandwool.com); Spotlight 11-2 © Gregory G. Dimijiam/Photo Researchers, Inc.; Viewpoint 11-1 Courtesy of Norman Myers

Chapter 12

Chapter opener © WizData, Inc./ShutterStock, Inc.; 12-1 © Richard L. Carlton/Visuals Unlimited; 12-2 Courtesy of Lynn Betts/NRCS; 12-6 © Photodisc/Getty Images; 12-7a Courtesy of John and Karen Hollingsworth/U.S. Fish & Wildlife Services; 12-7b Courtesy of Dan Chiras; 12-8a Courtesy of Weyerhaeuser Company; 12-8b Courtesy of Weyerhaeuser Company; 12-9a © Charlie Ott/Photo Researchers, Inc.; 12-9b © Doug Sokell/Visuals Unlimited; 12-9c © G. Perkins/Visuals Unlimited; 12-10 © Photos.com; Point 12-1 Courtesy of Ralph Saperstein; Counterpoint 12-1 Courtesy of Victor Rozek; Spotlight 12-1 © Takeshi Takahara/Photo Researchers, Inc.

Chapter 13

Chapter opener © Theunis Jacobus Botha/ShutterStock, Inc.; 13-5 © Justin Wambold, The North Platte Telegraph/AP Photos; 13-6 © AP Photos; 13-7 Courtesy of John N. Rinne/USFS (www.rmrs.nau.edu/lab/4302); 13-8 © Sovfoto/Eastfoto; 13-13a Courtesy of Invisible Structures, Inc. (www.invisiblestructures.com); 13-13b Courtesy of Invisible Structures, Inc. (www.invisiblestructures.com); 13-14 Courtesy of GSFC/NASA; 13-19a Courtesy of NASA/GSFC/MITI/ERSDAC/JAROS, and U.S./Japan ASTER Science Team; 13-19b © Gregory Bull/AP Photos; 13-20b Courtesy of Hope Alexander, EPA Documeria/NOAA 13-22 Courtesy of Richard B. Mieremet, Senior Advisor, NOAA OSDIA; Spotlight 13-1 Courtesy South Florida Water Management District; Spotlight 13-2 © Alex S. MacLean/The Picture Cube/Index Stock Imagery, Inc.; Spotlight 13-2 Courtesy of Dan Chiras

Chapter 14

Chapter opener © LiquidLibrary; 14-8a Courtesy of H.E. Malde/USGS Photo Library, Denver, Colorado; 14-9a © Biofoto Associates/Photo Researchers, Inc.; 14-9b © Ken Eward/Science Source/Photo Researchers, Inc.; 14-10b Courtesy of D. Hardesty/USGS; 14-12 Reproduced from the BP Statistical Review of World Energy, 205; 14-13 Courtesy of Toyota Motor Sales; 14-20 © Novosti/SPL/Photo Researchers, Inc.; 14-21 Courtesy of U.S. Fish & Wildlife Service; Point 14-1 Courtesy of Dina Titus; Counterpoint 14-1 Courtesy of Frank H. Murkowski; Spotlight 14-1 © Steve McCutcheon/Visuals Unlimited